NEW AND FULL MOONS
1001 B.C. TO A.D. 1651

HERMAN H. GOLDSTINE

Fellow of the International Business Machines
Corporation
and Member of the Institute for Advanced Study

AMERICAN PHILOSOPHICAL SOCIETY
INDEPENDENCE SQUARE • PHILADELPHIA
1973

Copyright © 1973 by The American Philosophical Society

Library of Congress Catalog Card Number 72-89401
International Standard Book Number 0-87169-094-2

INTRODUCTION

In the following pages the reader will find tables giving the dates of all new and full moons during an historical era when these data were of considerable interest and importance. At present the scholar who needs such information must rather laboriously calculate the date and time using P. V. Neugebauer's tables or look them up in Ginzel's somewhat inconvenient ones.[1] As a result many scholars today have felt the need for a more convenient way to establish these chronological data. This has been frequently pointed out to me by Otto Neugebauer who has long urged the preparation of the present tables as a natural supplement to Bryant Tuckerman's very important numerical work on ancient chronology.[2]

It is therefore believed that these tables giving the dates and times of all lunar syzygies from 1001 B.C. through A.D. 1651 will be of real use to the scholars interested in this period. To make them more useful the longitudes of the moon at each of these times is also given, as is a consecutive enumeration of the conjunctions and a similar one of the oppositions.

All dates are reckoned in the Julian calendar and all times are given in hours and the nearest minute. These dates and times are calculated for an observer in Babylon, or equivalently Baghdad, since this location is fairly centrally located for the historians of the period. It is assumed by definition that the observer is exactly 3 hours east of Greenwich. Moreover, the time used is civil time and is based on a 24-hour clock with its origin at midnight; thus noon is 12 hours.

Since this volume may be considered as a supplement to Tuckerman's tables, I have taken all fundamental astronomical elements from them. So, for example, the formulas for the mean longitudes of the moon and sun, of the moon's ascending node and for the eccentricities of their orbits are taken from there.[3] However, for the completeness of this volume all formulas and elements utilized herein are summarized in §2 below.

In consonance with Tuckerman's approach the five most important perturbation terms affecting the moon's position have been applied. These include among others the evection, the variation, and the annual equation. However, since the perturbations applied by Tuckerman to the sun's position hardly influence the times of the moon's syzygies, they have not been included here. The lunar perturbations have been taken from P. V. Neugebauer's well-known astronomical chronologies and are those used by Tuckerman.[4]

While in a certain sense the present tables may be viewed as an extension and revision of both Ginzel's and P. V. Neugebauer's works on lunar syzygies, it is hoped that the data contained herein give somewhat better values for the times of the syzygies. The astronomical elements adopted here are improvements on those used by Ginzel. This matter is already discussed in Tuckerman's tables.

Inasmuch as Tuckerman in his tables has given a quite complete discussion of their accuracy, little more need be said here. Extensive spot checks have, however, been made to ensure the agreement between those and the present tables and the agreement is extremely close, the moon's positions in longitude agreeing exactly and the sun's differing occasionally by about .01 degree. This latter is explicable by the fact that the perturbations of the solar orbit mentioned above can in the worst possible case add up to about .016 degree. As still another check, we choose at random a half-dozen dates for new or full moons and used these data to interpolate for the sun's position in both volumes of Tuckerman's tables. Essentially complete agreement was reached in all cases. As a further and independent check the dates of the lunar syzygies for the year 1963 were calculated and compared to those in the *Nautical Almanac* for that year.[5] The greatest difference is about 10 minutes. As a further check the dates and times given in the

---

[1] P. V. Neugebauer, *Tafeln zur astronomischen Chronologie*, Vol. II; *Tafeln fur Sonne, Planeten und Mond*, Leipzig (1914) and F. K. Ginzel, *Handbuch der Mathematischen und Technischen Chronologie*, Vols. I and II (1906 and 1911, Leipzig). In the Ginzel tables new moons are given for the period 605 B.C. through 308 A.D. and full moons from 500 B.C. through 100 A.D.

[2] B. Tuckerman, *Planetary, Lunar, and Solar Positions 601 B.C. and A.D. 1*, Mem. Amer. Philos. Soc. **56** (1962) and *Planetary, Lunar and Solar Positions A.D. 2 to A.D. 1649*, Mem. Amer. Philos. Soc. **59** (1964). These will be referred to as Vols. I and II. In the Preface to Vol. I the reader will find an account by Neugebauer of some of the scholary usages for these tables.

[3] *Op. cit.*, Vol. II: p. 13.

[4] P. V. Neugebauer, *op. cit.* and *Astronomischen Chronologie* (Berlin and Leipzig, 1929).

[5] *The American Ephemeris and Nautical Almanac for the Year 1963* (Washington and London, 1963), p. 159.

Connaissance des Temps for 1953 for new and full moons were also used. The largest disagreement with those values was 19 minutes.[6] (This disagreement is in part explicable by the fact that the *Connaissance* includes the parallax of Paris.) The agreement of the present tables with those of Ginzel is not so close. In one sample of a year disagreements of slightly more than 20 minutes were occasionally noted. Comparable or even greater disagreements have also been noted in comparisons with P. V. Neugebauer's original—his 1914—chronology. This was before he modified his astronomical elements for the sun.

Since the moon moves in its orbit about 12 degrees per day, it moves nearly a hundredth of a degree each minute. For this reason lunar longitudes are here recorded to the nearest hundredth of a degree and the corresponding times to the nearest minute even though such precision is probably more nominal than real. It does, however, make possible the use of the results for further calculations prior to an ultimate rounding-off.

In closing this introduction I feel I would be unforgivably remiss were I not to acknowledge gratefully the constant encouragement and support given me by Otto Neugebauer; the excellent and essential programming help and enthusiastic cooperation of Roger C. Evans without which these tables would not have been possible; the many invaluable conversations on the subject I had with Bryant Tuckerman; and the noteworthy assistance given me by Marcus W. Hanlon, Robert K. McNeill, John H. Rooney, and Clifford C. van Fleet. I wish also to thank the International Business Machines Corporation and in particular T. Vincent Learson, Chairman of the Board, and Ralph E. Gomory, the Director of Research, for making it possible for me to undertake and carry to fruition this task.

Finally it should be said that the programming was first done in the well-known language, APL, and initial tests run on the IBM 360, Model 91, located in the Thomas J. Watson Research Center, Yorktown Heights, N. Y. For technical reasons connected with printing the final results, it was then reprogrammed in FORTRAN using double-precision arithmetic and then run on the same Model 91. The actual calculation of the entire table on that computer, exclusive of checking and printing, took, *mirabile dictu*, 132 seconds. This compares to about a quarter of an hour per syzygy for a human or about 17,500 hours for the entire table. In fairness it should be pointed out that various other clerical functions, such as checking and

manipulating Julian day numbers, took an additional 300 seconds.

1. UNDERLYING ASTRONOMICAL AND MATHEMATICAL CONCEPTS

The fundamental quantities that enter into the determination of the moon's syzygies are the longitudes of the sun and moon measured in the plane of the ecliptic by an observer at the earth's center. By definition a new moon occurs at the instant these are equal, and a full moon when they differ by 180°. Thus no account is taken of parallax.[7]

The calculation of these longitudes and the times at which they are equal modulo 180° are made possible by relatively simple formulas. There are several possible sets but we have used those of Tuckerman for consistency reasons. In these formulas quantities with "primes" on them refer to the sun and those without to the moon. The symbols used are L for mean longitude, P for the longitude of perihelion or perigee, e the eccentricity of the orbit, N the longitude of the moon's ascending node and i the inclination of the moon's orbit to the ecliptic. The symbol t is used to measure time in Julian years. In formulas (1)–(5) below this time is measured from the epoch Gregorian 1800 January 0, 12^h Greenwich civil time and in formula (7) it is Gregorian 1900, January 1, 12^h Greenwich civil time.

In terms of these the relevant formulas are the following:

(1) $L = 355°43'24''.37 + 17325643''.945t + 1''.22 \times 10^{-3}t^2 + 6''.6 \times 10^{-9}t^3,$

(2) $L' = 279°54'49''.60 + 1296027''.6437t + 2''.6 \times 10^{-4}t^2 + 0''.0 \times t^3,$

(3) $P = 225°23'50''.37 + 146485''.937t - 3''.7033 \times 10^{-3}t^2 - 4''.4 \times 10^{-8}t^3,$

(4) $P' = 279°30'12''.30 + 61''.8026t + 1''.71 \times 10^{-4}t^2 + 1''.0 \times 10^{-8}t^3,$

(5) $N = 33°16'22''.57 - 69629''.209t + 7''.489 \times 10^{-4}t^2 + 7''.5 \times 10^{-9}t^3,$

(6) $e = 0.05490807,$

(7) $e' = 0.0167498 - 4.258 \times 10^{-7}t - 1.37 \times 10^{-11}t^2,$

(8) $i = 5°8'39''.96.$[8]

To these formulas we must append the perturbations of the moon's orbit that have been taken into account. They are the same ones as used by Tucker-

---

[6] *Connaissance des temps, ou des movements celestes, pour l'an 1953* (Paris, Bureau des longitudes).

[7] *The American Ephemeris, op. cit.*, pp. 491–492.

[8] Tuckerman, *op. cit.*, Vol. II: p. 13. (The value of e in formula (6) above is that adopted by Hansen. P. V. Neugebauer, followed by Tuckerman, used the value 0.05490897, which is probably a typographical error. This introduces no measurable changes in times or longitudes.)

man and P. V. Neugebauer and are those of Hansen. They include, *inter alia*, the important and well-known effects such as the evection, the variation and the annual equation.

$$A_1 = 4467'' \sin(L - 2L' + P),$$
$$A_2 = 2145'' \sin2(L - L'),$$
$$(9) \qquad A_3 = 658'' \sin(L' - P' + 180°),$$
$$A_4 = 198'' \sin(L - 3L' + P + P'),$$
$$A_5 = 155'' \sin(2L - 3L' + P').$$

Then in the formulas (11) and (12) below L is replaced by

$$(10) \qquad L_M = L + A_1 + A_2 + A_3 + A_4 + A_5.$$

For notational simplicity we shall drop the subscript M on L and write simply L for the expression (10). This can cause no difficulties.

As mentioned above, in contradistinction to Tuckerman's work, we have introduced no perturbations in the sun's mean longitude.

Now given the corrected mean longitudes L and L' for the moon and sun we find the true longitudes λ and λ' of these bodies in their orbits. This is done with the help of a well-known approximation to the equation of center.[9] This formula is given below through terms of the order e^4.

$$(11) \qquad \lambda = L + \frac{180}{\pi} \times [(2e - \frac{1}{4} e^3) \sin (L - P) +$$
$$(\frac{5}{4} e^2 - \frac{11}{24} e^4) \sin 2 (L - P) +$$
$$\frac{13}{12} e^3 \sin 3 (L - P) +$$
$$\frac{103}{96} e^4 \sin 4 (L - P)],$$

with an exactly similar formula for the primed variables, i.e., for the sun. In this formula (11) above and its counterpart with primes it should be understood that λ and L are considered to be expressed in degrees.

In the case of the sun, of course, the longitude λ' is already measured in the plane of the ecliptic whereas that for the moon is not. In fact the longitude in orbit of the moon is really the sum of two angles: one measured in the plane of the ecliptic from vernal equinox to the ascending node and the other measured in the plane of its orbit from the ascending node to the moon's true position. The latter plane is inclined slightly more than 5°—see formula (8) above—to the former one, the ecliptic. The final step then consists of projecting the moon's position onto the plane of the ecliptic. This replaces λ by a new angle ℓ measured entirely in the plane of the ecliptic.

---

[9] Cf., e.g., D. Brouwer and G. M. Clemence, *Methods of Celestial Mechanics* (New York, 1961), p. 77.

This can be done by a simple approximate formula known as the reduction to the ecliptic. It is this:

$$(12) \qquad \ell = L - \frac{180}{\pi} \times [p^2 \sin2 (\lambda - N) -$$
$$\frac{1}{2} p^4 \sin4 (\lambda - N) + \frac{1}{3} p^6 \sin6 (\lambda - N)],$$
$$p = \tan i/2,$$

where i is given in (8) and N in (5) above and where ℓ, L, and N are expressed in degrees.[10] This formula is correct through sixth powers of p.

Now it remains only to describe how the times of the syzygies are found. This is done inductively. Suppose we have found the date and time T_n of the nth new or full moon. Then to a first approximation the next one occurs at the time

$$T_{n+1}^{(0)} = T_n + 29.53/365.25$$

where the quantities are reckoned in Julian years. Then the equations

$$\ell (T) - \lambda' (T) = \begin{cases} 0 \\ \pi \end{cases}$$

are solved in the neighborhood of $T_{n+1}^{(0)}$ for T_{n+1}. The procedure is a modified Newton one. The value $T_{n+1}^{(k+1)}$ of the $(k+1)$-st approximant to T_{n+1} is calculated with the help of the relations

$$T_{n+1}^{(k+1)} = T_{n+1}^{(k)} - \frac{\ell \left(T_{n+1}^{(k)}\right) - \lambda' \left(T_{n+1}^{(k)}\right)}{D \left(T_{n+1}^{(k)}\right)},$$
$$D(T) = d[L(T) - L'(T)]/dT.$$

In these formulas the denominator, $D(T)$, is very close to the value of the derivative, $d[\ell(T) - \lambda'(T)]/dT$. When $T_{n+1}^{(k+1)} - T_{n+1}^{(k)}$ is an absolute value less than 2×10^{-7}, $T_{n+1} = T_{n+1}^{(k+1)}$ by definition.

2. USE OF THE TABLES AND CONVERSIONS TO OTHER MERIDIA

The tables themselves are virtually self-explanatory. Each page contains two pairs of columns headed NEW MOONS and FULL MOONS. In each column there are four sub-columns headed NUMBER, DATE, TIME, AND LONGITUDE. These give, respectively, consecutive enumerations, starting at zero, of the lunations; the days; the months in a two-character

---

[10] *Ibid.*, p. 47.

form; the civil times of the syzygies for an observer in Babylon expressed by means of a 24-hour clock in hours and the nearest minute; and the longitudes of the moon at those times as calculated at the earth's center. The zero of the clock is midnight; times are recorded, for example, as 7;23 or 18;36 which mean, respectively, 7;23 A.M. and 6;36 P.M. The year is given in the blank rows over each pair of columns and is expressed as an astronomical date. (The conversion from astronomical to civil years is quite easy: for astronomical events bearing a minus sign add 1 to obtain the civil date B.C., and for those events without a sign simply append A.D. to the astronomical date. Thus −100 = 101 B.C., 0 = 1 B.C. and +1 = 1 A.D.) On each page a given pair of columns will be found to contain the syzygies for six consecutive years. Thus each page contains the events of 12 years.

For an observer at any other location on the earth than Babylon, the little table below can be used to calculate when the syzygy will occur according to his clock. It is essentially an abridgement of one of O. Neugebauer.[11] The + sign is to be taken to mean the syzygy occurs later in local civil time than at Babylon and the − earlier. Thus a syzygy occurring at 11:20 in Babylon civil time occurs at 8:20 Greenwich civil time and 9:50 Samarkand civil time.

Cities	Δt Babylon	Cities	Δt Babylon
Toledo	−3ʰ14ᵐ	Baghdad/Babylon	0ʰ 0ᵐ
Greenwich	−3ʰ 0ᵐ	Constantinople	+1ʰ 2ᵐ
Hveen	−2ʰ 7ᵐ	Samarkand	+1ʰ30ᵐ
Prague	−2ʰ 0ᵐ	Ujjain	+2ʰ 6ᵐ
Alexandria	−0ʰ58ᵐ	Peking	+7ʰ45ᵐ
Damascus	−0ʰ32ᵐ		

---

[11] Cf. Tuckerman, op. cit., Vol. II: p. v.

NEW AND FULL MOONS
FROM 1001 B.C. TO A.D. 1651

NEW MOONS / FULL MOONS — NEW MOONS / FULL MOONS

Left half — NEW MOONS

NUMBER	DATE	TIME	LONG.
			-1000
0	11JA	13;28	282.17
1	10FE	4;21	311.81
2	10MR	16;21	340.82
3	9AP	1;52	9.24
4	8MY	9;32	37.26
5	6JN	16;15	65.12
6	5JL	23;06	93.10
7	4AU	7;16	121.46
8	2SE	17;49	150.39
9	2OC	7;27	179.95
10	1NO	0;05	210.07
11	30NO	18;49	240.48
12	30DE	14;10	270.82
			-999
13	29JA	8;37	300.79
14	28FE	0;53	330.17
15	29MR	14;16	358.92
16	28AP	0;46	27.15
17	27MY	9;03	55.09
18	25JN	16;13	83.02
19	24JL	23;27	111.20
20	23AU	7;48	139.85
21	21SE	18;00	169.09
22	21OC	6;26	198.90
23	19NO	21;11	229.09
24	19DE	14;00	259.40
			-998
25	18JA	8;13	289.51
26	17FE	2;32	319.17
27	18MR	19;26	348.25
28	17AP	9;50	16.77
29	16MY	21;34	44.90
30	15JN	7;17	72.88
31	14JL	15;56	101.00
32	13AU	0;27	129.50
33	11SE	9;30	158.50
34	10OC	19;31	188.05
35	9NO	6;49	218.02
36	8DE	19;48	248.19
			-997
37	7JA	10;36	278.28
38	6FE	2;56	308.05
39	7MR	19;48	337.32
40	6AP	12;04	6.07
41	6MY	2;57	34.41
42	4JN	16;16	62.53
43	4JL	4;13	90.69
44	2AU	15;09	119.12
45	1SE	1;25	148.00
46	30SE	11;25	177.37
47	29OC	21;35	207.18
48	28NO	8;21	237.24
49	27DE	20;05	267.29
			-996
50	26JA	8;53	297.08
51	24FE	22;36	326.42
52	25MR	12;56	355.27
53	24AP	3;40	23.71
54	23MY	18;37	51.93
55	22JN	9;30	80.14
56	21JL	23;51	108.58
57	20AU	13;14	137.39
58	19SE	1;31	166.67
59	18OC	12;58	196.40
60	16NO	23;57	226.42
61	16DE	10;47	256.48
			-995
62	14JA	21;33	286.32
63	13FE	8;17	315.72
64	14MR	19;13	344.62
65	13AP	6;53	13.08
66	12MY	19;50	41.28
67	11JN	10;19	69.45
68	11JL	2;00	97.82
69	9AU	18;04	126.57
70	8SE	9;40	155.79
71	8OC	0;21	185.49
72	6NO	13;55	215.54
73	6DE	2;19	245.68

Left half — FULL MOONS

NUMBER	DATE	TIME	LONG.
0	25JA	15;11	116.31
1	24FE	3;17	145.60
2	24MR	16;28	174.41
3	23AP	6;41	202.85
4	22MY	21;31	231.06
5	21JN	12;31	259.28
6	21JL	3;21	287.72
7	19AU	17;47	316.56
8	18SE	7;39	345.90
9	17OC	20;37	15.68
10	16NO	8;29	45.73
11	15DE	19;20	75.80
12	14JA	5;29	105.62
13	12FE	15;23	135.00
14	14MR	1;32	163.87
15	12AP	12;19	192.31
16	12MY	0;08	220.47
17	10JN	13;18	248.59
18	10JL	4;04	276.91
19	8AU	20;18	305.65
20	7SE	13;14	334.92
21	7OC	5;45	4.68
22	5NO	20;47	34.78
23	5DE	9;56	64.96
24	3JA	21;21	94.93
25	2FE	7;23	124.49
26	3MR	16;26	153.51
27	2AP	0;55	182.01
28	1MY	9;29	210.13
29	30MY	19;02	238.11
30	29JN	6;33	266.23
31	28JL	20;44	294.73
32	27AU	13;30	323.79
33	26SE	7;51	353.44
34	26OC	2;14	23.55
35	24NO	19;14	53.87
36	24DE	10;09	84.07
37	22JA	22;44	113.89
38	21FE	9;03	143.15
39	22MR	17;30	171.81
40	21AP	0;46	200.00
41	20MY	7;51	227.93
42	18JN	15;58	255.86
43	18JL	2;15	284.08
44	16AU	15;25	312.81
45	15SE	7;32	342.17
46	15OC	1;55	12.12
47	13NO	21;18	42.47
48	13DE	16;07	72.88
49	12JA	8;56	103.01
50	10FE	22;47	132.60
51	11MR	9;33	161.54
52	9AP	17;51	189.91
53	9MY	0;44	217.90
54	7JN	7;24	245.76
55	6JL	14;58	273.77
56	5AU	0;20	302.18
57	3SE	12;10	331.17
58	3OC	2;52	0.80
59	1NO	20;24	30.96
60	1DE	15;58	61.40
61	31DE	11;47	91.77
62	30JA	5;46	121.70
63	28FE	20;36	151.01
64	30MR	8;09	179.67
65	28AP	17;10	207.84
66	28MY	0;44	235.75
67	26JN	7;49	263.67
68	25JL	15;21	291.87
69	24AU	0;04	320.55
70	22SE	10;46	349.82
71	22OC	0;08	19.67
72	20NO	16;28	49.94
73	20DE	11;04	80.32

Right half — NEW MOONS

NUMBER	DATE	TIME	LONG.
			-994
74	4JA	13;26	275.64
75	2FE	23;18	305.18
76	4MR	8;16	334.18
77	2AP	17;03	2.69
78	2MY	2;38	30.84
79	31MY	13;50	58.89
80	30JN	3;06	87.08
81	29JL	18;22	115.63
82	28AU	11;09	144.71
83	27SE	4;42	174.34
84	26OC	22;01	204.42
85	25NO	14;03	234.70
86	25DE	3;54	264.85
			-993
87	23JA	15;19	294.61
88	22FE	0;38	323.82
89	23MR	8;39	352.46
90	21AP	16;15	20.65
91	21MY	0;23	48.61
92	19JN	9;52	76.60
93	18JL	21;24	104.88
94	17AU	11;30	133.66
95	16SE	4;20	163.06
96	15OC	23;15	193.05
97	14NO	18;40	223.41
98	14DE	12;36	253.78
			-992
99	13JA	3;42	283.83
100	11FE	15;46	313.33
101	12MR	1;21	342.21
102	10AP	9;18	10.56
103	9MY	16;25	38.55
104	7JN	23;32	66.43
105	7JL	7;36	94.47
106	5AU	17;39	122.91
107	4SE	6;37	151.97
108	3OC	22;58	181.67
109	2NO	18;04	211.91
110	2DE	14;07	242.37
			-991
111	1JA	8;58	272.69
112	31JA	1;15	302.54
113	1MR	14;35	331.77
114	31MR	1;11	0.38
115	29AP	9;38	28.52
116	28MY	16;42	56.41
117	26JN	23;23	84.32
118	26JL	6;50	112.52
119	24AU	16;11	141.24
120	23SE	4;20	170.59
121	22OC	19;34	200.53
122	21NO	13;23	230.86
123	21DE	8;32	261.26
			-990
124	20JA	3;28	291.38
125	18FE	20;49	320.97
126	20MR	11;36	349.93
127	18AP	23;28	18.33
128	18MY	8;46	46.35
129	16JN	16;26	74.26
130	15JL	23;37	102.34
131	14AU	7;27	130.82
132	12SE	16;49	159.86
133	12OC	4;13	189.49
134	10NO	17;49	219.56
135	10DE	9;32	249.85
			-989
136	9JA	2;59	280.04
137	7FE	21;12	309.85
138	9MR	14;47	339.13
139	8AP	6;24	7.83
140	7MY	19;26	36.07
141	6JN	6;09	64.09
142	5JL	15;23	92.16
143	4AU	0;05	120.52
144	2SE	9;01	149.36
145	1OC	18;41	178.74
146	31OC	5;23	208.59
147	29NO	17;27	238.71
148	29DE	7;11	268.83

Right half — FULL MOONS

NUMBER	DATE	TIME	LONG.
74	19JA	6;18	110.46
75	18FE	0;16	140.10
76	19MR	15;49	169.10
77	18AP	4;43	197.55
78	17MY	15;20	225.63
79	16JN	0;17	253.59
80	15JL	8;19	281.69
81	13AU	16;14	310.17
82	12SE	0;56	339.17
83	11OC	11;12	8.74
84	9NO	23;39	38.76
85	9DE	14;22	69.00
86	8JA	6;49	99.16
87	7FE	0;02	128.95
88	8MR	17;02	158.21
89	7AP	9;02	186.94
90	6MY	23;26	215.25
91	5JN	11;57	243.34
92	4JL	22;39	271.46
93	3AU	8;07	299.84
94	1SE	17;11	328.68
95	1OC	2;42	358.04
96	30OC	13;12	27.87
97	29NO	0;52	57.97
98	28DE	13;36	88.06
99	27JA	3;20	117.87
100	25FE	18;01	147.24
101	26MR	9;29	176.12
102	25AP	1;15	204.60
103	24MY	16;29	232.83
104	23JN	6;32	261.01
105	22JL	19;14	289.39
106	21AU	6;53	318.14
107	19SE	18;02	347.39
108	19OC	5;02	17.11
109	17NO	15;57	47.13
110	17DE	2;48	77.19
111	15JA	13;43	107.02
112	14FE	1;06	136.44
113	15MR	13;24	165.39
114	14AP	2;51	193.91
115	13MY	17;16	222.16
116	12JN	8;13	250.35
117	11JL	23;18	278.70
118	10AU	14;14	307.42
119	9SE	4;44	336.61
120	8OC	18;28	6.29
121	7NO	7;06	36.29
122	6DE	18;29	66.39
123	5JA	4;51	96.32
124	3FE	14;38	125.84
125	5MR	0;21	154.87
126	3AP	10;30	183.41
127	2MY	21;31	211.62
128	1JN	9;50	239.71
129	30JN	23;46	267.93
130	30JL	15;26	296.52
131	29AU	8;24	325.62
132	28SE	1;35	355.25
133	27OC	17;43	25.28
134	26NO	7;58	55.49
135	25DE	20;13	85.57
136	24JA	6;48	115.29
137	22FE	16;07	144.48
138	24MR	0;37	173.13
139	22AP	8;51	201.34
140	21MY	17;38	229.33
141	20JN	3;57	257.35
142	19JL	16;45	285.69
143	18AU	8;24	314.54
144	17SE	2;20	344.01
145	16OC	21;08	14.01
146	15NO	15;08	44.31
147	15DE	7;15	74.60

NUMBER	DATE	TIME	LONG.	NUMBER	DATE	TIME	LONG.
		-988					
				148	13JA	20;59	104.58
149	27JA	22;34	298.70	149	12FE	8;18	134.04
150	26FE	15;00	328.13	150	12MR	17;29	162.90
151	27MR	7;27	357.04	151	11AP	1;05	191.22
152	25AP	22;58	25.49	152	10MY	7;59	219.20
153	25MY	13;05	53.67	153	8JN	15;22	247.09
154	24JN	1;50	81.81	154	8JL	0;26	275.17
155	23JL	13;28	110.15	155	6AU	12;09	303.70
156	22AU	0;19	138.89	156	5SE	2;56	332.83
157	20SE	10;39	168.11	157	4OC	20;26	2.60
158	19OC	20;52	197.81	158	3NO	15;37	32.85
159	18NO	7;20	227.81	159	3DE	10;59	63.29
160	17DE	18;29	257.88				
		-987					
				160	2JA	4;58	93.56
161	16JA	6;33	287.76	161	31JA	20;18	123.36
162	14FE	19;33	317.23	162	2MR	8;29	152.53
163	16MR	9;17	346.22	163	31MR	17;48	181.08
164	14AP	23;35	14.77	164	30AP	1;11	209.17
165	14MY	14;19	43.03	165	29MY	7;47	237.04
166	13JN	5;19	71.22	166	27JN	14;47	264.96
167	12JL	20;10	99.57	167	26JL	23;13	293.22
168	11AU	10;22	128.27	168	25AU	9;50	321.99
169	9SE	23;31	157.42	169	23SE	23;10	351.40
170	9OC	11;36	187.04	170	23OC	15;23	21.39
171	7NO	22;56	216.99	171	22NO	10;06	51.77
172	7DE	9;51	247.08	172	22DE	6;00	82.19
		-986					
173	5JA	20;33	277.01	173	21JA	1;02	112.31
174	4FE	7;05	306.55	174	19FE	17;26	141.84
175	5MR	17;35	335.60	175	21MR	6;29	170.72
176	4AP	4;31	4.17	176	19AP	16;36	199.04
177	3MY	16;32	32.41	177	19MY	0;46	227.02
178	2JN	6;07	60.55	178	17JN	8;02	254.91
179	1JL	21;18	88.82	179	16JL	15;22	283.00
180	31JL	13;28	117.44	180	14AU	23;31	311.50
181	30AU	5;41	146.52	181	13SE	9;16	340.57
182	28SE	21;10	176.10	182	12OC	21;18	10.23
183	28OC	11;33	206.07	183	11NO	12;10	40.37
184	27NO	0;41	236.23	184	11DE	5;41	70.73
185	26DE	12;27	266.29				
		-985					
				185	10JA	0;40	100.98
186	24JA	22;48	295.98	186	8FE	19;17	130.80
187	23FE	7;57	325.16	187	10MR	11;57	160.02
188	24MR	16;29	353.80	188	9AP	2;05	188.64
189	23AP	1;19	22.03	189	8MY	13;47	216.83
190	22MY	11;27	50.06	190	6JN	23;35	244.82
191	20JN	23;34	78.16	191	6JL	8;08	272.86
192	20JL	13;54	106.57	192	4AU	16;12	301.20
193	19AU	6;10	135.46	193	3SE	0;35	330.03
194	17SE	23;41	164.93	194	2OC	10;07	359.42
195	17OC	17;30	194.89	195	31OC	21;30	29.30
196	16NO	10;30	225.16	196	30NO	11;04	59.48
197	16DE	1;36	255.41	197	30DE	2;34	89.68
		-984					
198	14JA	14;12	285.33	198	28JA	19;15	119.59
199	13FE	0;22	314.73	199	27FE	12;12	149.03
200	13MR	8;45	343.55	200	28MR	4;35	177.92
201	11AP	16;16	11.86	201	26AP	19;46	206.35
202	10MY	23;53	39.86	202	26MY	9;18	234.50
203	9JN	8;30	67.80	203	24JN	21;02	262.60
204	8JL	18;54	95.94	204	24JL	7;15	290.90
205	7AU	7;44	124.52	205	22AU	16;40	319.59
206	5SE	23;20	153.71	206	21SE	2;06	348.79
207	5OC	17;30	183.52	207	20OC	12;13	18.48
208	4NO	13;05	213.79	208	18NO	23;20	48.52
209	4DE	8;05	244.21	209	18DE	11;28	78.63
		-983					
210	3JA	0;43	274.42	210	17JA	0;29	108.54
211	1FE	14;12	304.13	211	15FE	14;23	138.04
212	3MR	0;49	333.23	212	17MR	5;11	167.06
213	1AP	9;22	1.74	213	15AP	20;39	195.64
214	30AP	16;41	29.81	214	15MY	12;10	223.92
215	29MY	23;39	57.70	215	14JN	2;57	252.11
216	28JN	7;08	85.65	216	13JL	16;32	280.42
217	27JL	16;06	113.93	217	12AU	4;56	309.06
218	26AU	3;35	142.75	218	10SE	16;34	338.16
219	24SE	18;20	172.23	219	10OC	3;52	7.75
220	24OC	12;18	202.31	220	8NO	14;58	37.70
221	23NO	8;11	232.74	221	8DE	1;51	67.78
222	23DE	3;51	263.15				

NUMBER	DATE	TIME	LONG.	NUMBER	DATE	TIME	LONG.
		-982					
				222	6JA	12;34	97.71
223	21JA	21;27	293.19	223	4FE	23;25	127.26
224	20FE	12;10	322.64	224	6MR	10;55	156.33
225	22MR	0;00	351.46	225	4AP	23;29	184.96
226	20AP	9;23	19.74	226	4MY	13;14	213.26
227	19MY	17;01	47.69	227	3JN	3;53	241.44
228	17JN	23;48	75.57	228	2JL	19;02	269.72
229	17JL	6;49	103.65	229	1AU	10;17	298.31
230	15AU	15;12	132.17	230	31AU	1;21	327.36
231	14SE	1;58	161.30	231	29SE	15;50	356.91
232	13OC	15;44	191.04	232	29OC	5;17	26.85
233	12NO	8;23	221.26	233	27NO	17;23	56.96
234	12DE	2;58	251.66	234	27DE	4;11	86.98
		-981					
235	10JA	22;03	281.90	235	25JA	14;04	116.65
236	9FE	16;11	311.69	236	23FE	23;31	145.82
237	11MR	8;12	340.87	237	25MR	9;08	174.50
238	9AP	21;28	9.45	238	23AP	19;25	202.78
239	9MY	8;00	37.59	239	23MY	6;52	230.86
240	7JN	16;25	65.51	240	21JN	19;53	259.00
241	6JL	23;50	93.52	241	21JL	10;46	287.44
242	5AU	7;22	121.85	242	20AU	3;22	316.36
243	3SE	16;03	150.70	243	18SE	20;53	345.84
244	3OC	2;31	180.13	244	18OC	13;57	15.79
245	1NO	15;03	210.07	245	17NO	5;24	45.99
246	1DE	5;41	240.30	246	16DE	18;42	76.16
247	30DE	22;09	270.53				
		-980					
				247	15JA	6;01	106.02
248	29JA	15;53	300.48	248	13FE	15;46	135.40
249	28FE	9;44	329.94	249	14MR	0;26	164.21
250	29MR	2;19	358.82	250	12AP	8;30	192.54
251	27AP	16;38	27.21	251	11MY	16;44	220.56
252	27MY	4;30	55.30	252	10JN	2;03	248.52
253	25JN	14;31	83.33	253	9JL	13;29	276.71
254	24JL	23;35	111.59	254	8AU	3;47	305.37
255	23AU	8;33	140.27	255	6SE	20;52	334.63
256	21SE	18;00	169.48	256	6OC	15;39	4.48
257	21OC	4;17	199.19	257	5NO	10;26	34.73
258	19NO	15;38	229.24	258	5DE	3;42	65.08
259	19DE	4;24	259.38				
		-979					
				259	3JA	18;39	95.21
260	17JA	18;45	289.33	260	2FE	7;06	124.86
261	16FE	10;27	318.89	261	3MR	17;10	153.92
262	18MR	2;45	347.96	262	2AP	1;19	182.41
263	16AP	18;38	16.55	263	1MY	8;20	210.47
264	16MY	9;27	44.80	264	30MY	15;15	238.34
265	14JN	22;58	72.94	265	28JN	23;18	266.32
266	14JL	11;20	101.20	266	28JL	9;37	294.67
267	12AU	22;50	129.81	267	26AU	22;56	323.58
268	11SE	9;40	158.90	268	25SE	15;17	353.14
269	10OC	20;06	188.46	269	25OC	9;54	23.25
270	9NO	6;30	218.39	270	24NO	5;27	53.67
271	8DE	17;15	248.47	271	24DE	0;20	84.04
		-978					
272	7JA	4;41	278.42	272	22JA	17;03	114.04
273	5FE	16;58	308.02	273	21FE	6;45	143.44
274	7MR	6;03	337.14	274	22MR	17;19	172.19
275	5AP	19;47	5.80	275	21AP	1;28	200.41
276	5MY	10;07	34.12	276	20MY	8;16	228.33
277	4JN	0;59	62.31	277	18JN	14;57	256.20
278	3JL	16;05	90.59	278	17JL	22;36	284.31
279	2AU	6;56	119.18	279	16AU	8;08	312.90
280	31AU	20;58	148.20	280	14SE	20;10	342.08
281	30SE	9;51	177.69	281	14OC	11;01	11.88
282	29OC	21;43	207.58	282	13NO	4;36	42.15
283	28NO	8;54	237.65	283	13DE	0;04	72.59
284	27DE	19;39	267.66				
		-977					
				284	11JA	19;41	102.84
285	26JA	6;05	297.34	285	10FE	13;27	132.60
286	24FE	16;18	326.54	286	12MR	4;05	161.71
287	26MR	2;40	355.24	287	10AP	15;32	190.20
288	24AP	13;48	23.54	288	10MY	0;32	218.27
289	24MY	2;23	51.67	289	8JN	8;11	246.11
290	22JN	16;47	79.86	290	7JL	15;28	274.17
291	22JL	8;41	108.35	291	5AU	23;16	302.52
292	21AU	1;14	137.28	292	4SE	8;16	331.40
293	19SE	17;27	166.72	293	3OC	19;11	0.86
294	19OC	8;42	196.61	294	2NO	8;37	30.85
295	17NO	22;39	226.75	295	2DE	0;49	61.15
296	17DE	11;10	256.89	296	31DE	19;07	91.45

-976

NEW NUMBER	DATE	TIME	LONG.	FULL NUMBER	DATE	TIME	LONG.
297	15JA	22;08	286.73	297	30JA	13;56	121.43
298	14FE	7;40	316.08	298	29FE	7;31	150.85
299	14MR	16;11	344.88	299	29MR	22;50	179.68
300	13AP	0;32	13.21	300	28AP	11;41	208.00
301	12MY	9;42	41.26	301	27MY	22;26	236.04
302	10JN	20;39	69.29	302	26JN	7;40	264.05
303	10JL	9;52	97.56	303	25JL	16;04	292.28
304	9AU	1;19	126.27	304	24AU	0;24	320.95
305	7SE	18;31	155.55	305	22SE	9;26	350.15
306	7OC	12;34	185.38	306	21OC	19;55	19.88
307	6NO	6;20	215.60	307	20NO	8;22	49.98
308	5DE	22;38	245.91	308	19DE	22;50	80.19

-975

NEW NUMBER	DATE	TIME	LONG.	FULL NUMBER	DATE	TIME	LONG.
309	4JA	12;31	275.99	309	18JA	14;48	110.20
310	2FE	23;45	305.58	310	17FE	7;26	139.79
311	4MR	8;47	334.59	311	18MR	23;56	168.84
312	2AP	16;27	3.04	312	17AP	15;39	197.41
313	1MY	23;45	31.11	313	17MY	6;03	225.64
314	31MY	7;39	59.03	314	15JN	18;49	253.75
315	29JN	17;02	87.06	315	15JL	5;58	281.98
316	29JL	4;37	115.46	316	13AU	15;56	310.54
317	27AU	18;54	144.43	317	12SE	1;31	339.58
318	26SE	12;00	174.03	318	11OC	11;25	9.14
319	26OC	7;14	204.18	319	9NO	22;06	39.08
320	25NO	2;54	234.61	320	9DE	9;42	69.19
321	24DE	20;56	264.94				

-974

NEW NUMBER	DATE	TIME	LONG.	FULL NUMBER	DATE	TIME	LONG.
322	23JA	11;58	294.86	321	7JA	22;06	99.18
323	21FE	23;51	324.17	322	6FE	11;16	128.80
324	23MR	9;12	352.86	323	8MR	1;19	157.96
325	21AP	16;56	21.06	324	6AP	16;13	186.65
326	20MY	23;55	48.98	325	6MY	7;39	215.01
327	19JN	7;01	76.87	326	4JN	22;54	243.21
328	18JL	15;09	105.01	327	4JL	13;17	271.47
329	17AU	1;20	133.63	328	3AU	2;30	300.00
330	15SE	14;29	162.87	329	1SE	14;47	328.96
331	15OC	7;00	192.75	330	1OC	2;29	358.42
332	14NO	2;12	223.09	331	30OC	13;51	28.29
333	13DE	22;14	253.56	332	29NO	0;55	58.36
				333	28DE	11;37	88.36

-973

NEW NUMBER	DATE	TIME	LONG.	FULL NUMBER	DATE	TIME	LONG.
334	12JA	16;58	283.76	334	26JA	22;09	118.04
335	11FE	9;05	313.44	335	25FE	8;59	147.25
336	12MR	22;14	342.47	336	26MR	20;40	175.99
337	11AP	8;42	10.92	337	25AP	9;35	204.36
338	10MY	17;05	38.96	338	24MY	23;41	232.54
339	9JN	0;13	66.84	339	23JN	14;39	260.76
340	8JL	7;03	94.83	340	23JL	6;03	289.24
341	6AU	14;43	123.18	341	21AU	21;32	318.14
342	5SE	0;18	152.09	342	20SE	12;40	347.55
343	4OC	12;37	181.62	343	20OC	2;58	17.41
344	3NO	3;56	211.69	344	18NO	15;55	47.51
345	2DE	21;38	242.06	345	18DE	3;22	77.60

-972

NEW NUMBER	DATE	TIME	LONG.	FULL NUMBER	DATE	TIME	LONG.
346	1JA	16;31	272.38	346	16JA	13;33	107.40
347	31JA	11;07	302.34	347	14FE	22;57	136.74
348	1MR	4;10	331.73	348	15MR	8;10	165.56
349	30MR	18;46	0.51	349	13AP	17;49	193.93
350	29AP	6;36	28.78	350	13MY	4;25	222.03
351	28MY	16;02	56.76	351	11JN	16;29	250.11
352	26JN	23;56	84.73	352	11JL	6;25	278.41
353	26JL	7;26	112.93	353	9AU	22;22	307.16
354	24AU	15;37	141.60	354	8SE	15;49	336.46
355	23SE	1;17	170.84	355	8OC	9;34	6.30
356	22OC	12;51	200.63	356	7NO	2;10	36.47
357	21NO	2;25	230.78	357	6DE	16;41	66.70
358	20DE	17;52	261.02				

-971

NEW NUMBER	DATE	TIME	LONG.	FULL NUMBER	DATE	TIME	LONG.
359	19JA	10;51	291.07	358	5JA	4;55	96.71
360	18FE	4;32	320.68	359	3FE	15;17	126.26
361	19MR	21;42	349.75	360	5MR	0;15	155.25
362	18AP	13;08	18.30	361	3AP	8;22	183.71
363	18MY	2;14	46.48	362	2MY	16;14	211.80
364	16JN	13;13	74.52	363	1JN	0;45	239.73
365	15JL	22;51	102.69	364	30JN	10;58	267.80
366	14AU	8;01	131.23	365	29JL	23;48	296.27
367	12SE	17;24	160.27	366	28AU	15;40	325.31
368	12OC	3;24	189.84	367	27SE	9;59	354.98
369	10NO	14;14	219.80	368	27OC	5;11	25.14
370	10DE	2;11	249.92	369	25NO	23;29	55.51
				370	25DE	15;42	85.77

-970

NEW NUMBER	DATE	TIME	LONG.	FULL NUMBER	DATE	TIME	LONG.
371	8JA	15;32	279.94	371	24JA	5;22	115.61
372	7FE	6;20	309.62	372	22FE	16;28	144.88
373	8MR	22;09	338.83	373	24MR	1;23	173.54
374	7AP	14;07	7.56	374	22AP	8;43	201.72
375	7MY	5;25	35.90	375	21MY	15;28	229.62
376	5JN	19;38	64.06	376	19JN	22;45	257.53
377	5JL	8;44	92.28	377	19JL	7;50	285.71
378	3AU	20;55	120.78	378	17AU	19;40	314.41
379	2SE	8;20	149.72	379	16SE	10;39	343.74
380	1OC	19;12	179.15	380	16OC	4;22	13.69
381	31OC	5;43	208.99	381	14NO	23;43	44.04
382	29NO	16;17	239.04	382	14DE	19;09	74.47
383	29DE	3;14	269.05				

-969

NEW NUMBER	DATE	TIME	LONG.	FULL NUMBER	DATE	TIME	LONG.
384	27JA	14;52	298.77	383	13JA	13;04	104.63
385	26FE	3;16	328.03	384	12FE	4;16	134.26
386	27MR	16;23	356.81	385	13MR	16;16	163.23
387	26AP	6;11	25.21	386	12AP	1;26	191.62
388	25MY	20;41	53.40	387	11MY	8;43	219.61
389	24JN	11;46	81.63	388	9JN	15;19	247.47
390	24JL	3;02	110.11	389	8JL	22;25	275.47
391	22AU	17;50	139.01	390	7AU	7;00	303.87
392	21SE	7;37	168.37	391	5SE	17;50	332.84
393	20OC	20;13	198.17	392	5OC	7;20	2.44
394	19NO	7;49	228.22	393	3NO	23;37	32.56
395	18DE	18;46	258.28	394	3DE	18;16	62.97

-968

NEW NUMBER	DATE	TIME	LONG.	FULL NUMBER	DATE	TIME	LONG.
396	17JA	5;13	288.09	395	2JA	13;58	93.31
397	15FE	15;18	317.44	396	1FE	8;45	123.26
398	16MR	1;16	346.28	397	2MR	0;55	152.60
399	14AP	11;40	14.67	398	31MR	13;50	181.30
400	13MY	23;14	42.81	399	29AP	23;55	209.50
401	12JN	12;37	70.94	400	29MY	8;09	237.43
402	12JL	3;53	99.30	401	27JN	15;36	265.38
403	10AU	20;26	128.09	402	26JL	23;11	293.60
404	9SE	13;13	157.38	403	25AU	7;39	322.29
405	9OC	5;18	187.16	404	23SE	17;39	351.56
406	7NO	20;09	217.26	405	23OC	5;50	21.38
407	7DE	9;30	247.45	406	21NO	20;38	51.58
				407	21DE	13;53	81.90

-967

NEW NUMBER	DATE	TIME	LONG.	FULL NUMBER	DATE	TIME	LONG.
408	5JA	21;13	277.42	408	20JA	8;27	112.00
409	4FE	7;17	306.95	409	19FE	2;37	141.62
410	5MR	16;02	335.92	410	20MR	18;59	170.64
411	4AP	0;08	4.37	411	19AP	8;59	199.14
412	3MY	8;34	32.47	412	18MY	20;46	227.24
413	1JN	18;23	60.46	413	17JN	6;49	255.25
414	1JL	6;23	88.61	414	16JL	15;44	283.40
415	30JL	20;48	117.15	415	15AU	0;13	311.92
416	29AU	13;23	146.24	416	13SE	9;00	340.95
417	28SE	7;22	175.90	417	12OC	18;49	10.52
418	28OC	1;39	206.04	418	11NO	6;17	40.51
419	26NO	18;59	236.37	419	10DE	19;41	70.69
420	26DE	10;12	266.58				

-966

NEW NUMBER	DATE	TIME	LONG.	FULL NUMBER	DATE	TIME	LONG.
421	24JA	22;41	296.36	420	9JA	10;46	100.78
422	23FE	8;36	325.57	421	8FE	2;53	130.50
423	24MR	16;40	354.19	422	9MR	19;16	159.72
424	22AP	23;53	22.35	423	8AP	11;15	188.43
425	22MY	7;15	50.28	424	8MY	2;18	216.76
426	20JN	15;45	78.23	425	6JN	16;00	244.90
427	20JL	2;09	106.48	426	6JL	4;08	273.08
428	18AU	15;06	135.23	427	4AU	14;52	301.53
429	17SE	6;56	164.61	428	3SE	0;49	330.43
430	17OC	1;24	194.60	429	2OC	10;42	359.83
431	15NO	21;14	224.98	430	31OC	21;05	29.67
432	15DE	16;22	255.40	431	30NO	8;14	59.75
				432	29DE	20;07	89.79

-965

NEW NUMBER	DATE	TIME	LONG.	FULL NUMBER	DATE	TIME	LONG.
433	14JA	8;58	285.50	433	28JA	8;40	119.54
434	12FE	22;18	315.03	434	26FE	21;56	148.83
435	14MR	8;43	343.93	435	28MR	12;07	177.64
436	12AP	17;03	12.27	436	27AP	3;08	206.07
437	12MY	0;14	40.26	437	26MY	18;31	234.30
438	10JN	7;09	68.13	438	25JN	9;31	262.53
439	9JL	14;41	96.15	439	24JL	23;33	290.97
440	7AU	23;47	124.58	440	23AU	12;34	319.80
441	6SE	11;26	153.60	441	22SE	0;49	349.12
442	6OC	2;22	183.27	442	21OC	12;34	18.89
443	4NO	20;26	213.47	443	19NO	23;54	48.93
444	4DE	16;19	243.93	444	19DE	10;45	78.99

--------NEW MOONS-------- --------FULL MOONS------- --------NEW MOONS-------- --------FULL MOONS-------

NEW MOONS				FULL MOONS				NEW MOONS				FULL MOONS			
NUMBER	DATE	TIME	LONG.	NUMBER	DATE	TIME	LONG.	NUMBER	DATE	TIME	LONG.	NUMBER	DATE	TIME	LONG.
					-964								**-958**		
445	3JA	11;52	274.26	445	17JA	21;11	108.79	520	26JA	6;43	297.76	519	11JA	3;04	102.53
446	2FE	5;17	304.15	446	16FE	7;33	138.14	521	24FE	15;54	326.91	520	9FE	21;24	132.32
447	2MR	19;49	333.40	447	16MR	18;26	167.00	522	26MR	0;01	355.51	521	11MR	14;36	161.54
448	1AP	7;29	2.04	448	15AP	6;26	195.45	523	24AP	7;57	23.69	522	10AP	5;41	190.19
449	30AP	16;49	30.20	449	14MY	19;46	223.66	524	23MY	16;48	51.67	523	9MY	18;33	218.42
450	30MY	0;29	58.11	450	13JN	10;17	251.83	525	22JN	3;32	79.72	524	8JN	5;30	246.45
451	28JN	7;25	86.04	451	13JL	1;38	280.21	526	21JL	16;46	108.09	525	7JL	15;05	274.54
452	27JL	14;39	114.25	452	11AU	17;21	308.97	527	20AU	8;27	136.98	526	5AU	23;55	302.94
453	25AU	23;16	142.96	453	10SE	9;02	338.23	528	19SE	2;03	166.47	527	4SE	8;40	331.80
454	24SE	10;15	172.28	454	10OC	0;07	7.98	529	18OC	20;33	196.48	528	3OC	18;03	1.20
455	24OC	0;08	202.18	455	8NO	13;59	38.05	530	17NO	14;42	226.80	529	2NO	4;43	31.07
456	22NO	16;43	232.46	456	8DE	2;16	68.19	531	17DE	7;10	257.10	530	1DE	17;07	61.20
457	22DE	11;04	262.82									531	31DE	7;15	91.33
					-963								**-957**		
458	21JA	5;48	292.92	457	6JA	12;58	98.11	532	15JA	21;01	287.07	532	29JA	22;40	121.18
459	19FE	23;35	322.51	458	4FE	22;32	127.61	533	14FE	8;03	316.48	533	28FE	14;42	150.55
460	21MR	15;21	351.50	459	6MR	7;32	156.58	534	15MR	16;48	345.29	534	30MR	6;42	179.41
461	20AP	4;33	19.93	460	4AP	16;39	185.07	535	14AP	0;10	13.58	535	28AP	22;10	207.84
462	19MY	15;09	48.00	461	4MY	2;29	213.22	536	13MY	7;13	41.54	536	28MY	12;37	236.03
463	17JN	23;48	75.95	462	2JN	13;37	241.26	537	11JN	14;58	69.45	537	27JN	1;42	264.19
464	17JL	7;31	104.06	463	2JL	2;33	269.45	538	11JL	0;19	97.56	538	26JL	13;21	292.56
465	15AU	15;25	132.57	464	31JL	17;36	298.02	539	9AU	11;59	126.11	539	24AU	23;53	321.31
466	14SE	0;26	161.62	465	30AU	10;36	327.14	540	8SE	2;27	155.27	540	23SE	9;57	350.56
467	13OC	11;08	191.23	466	29SE	4;39	356.83	541	7OC	19;49	185.07	541	22OC	20;13	20.28
468	11NO	23;44	221.28	467	28OC	22;15	26.94	542	6NO	15;18	215.35	542	21NO	7;02	50.31
469	11DE	14;10	251.51	468	27NO	14;01	57.21	543	6DE	11;08	245.81	543	20DE	18;29	80.39
				469	27DE	3;24	87.33								
					-962								**-956**		
470	10JA	6;14	281.62	470	25JA	14;34	117.05	544	5JA	5;10	276.06	544	19JA	6;30	110.24
471	8FE	23;25	311.38	471	24FE	0;01	146.23	545	3FE	20;05	305.81	545	17FE	19;05	139.66
472	10MR	16;46	340.62	472	25MR	8;18	174.85	546	4MR	7;46	334.93	546	18MR	8;29	168.60
473	9AP	9;03	9.34	473	23AP	16;02	203.03	547	2AP	16;56	3.45	547	16AP	22;51	197.13
474	8MY	23;19	37.62	474	22MY	23;59	230.97	548	2MY	0;30	31.52	548	16MY	14;00	225.39
475	7JN	11;24	65.70	475	21JN	9;08	258.96	549	31MY	7;25	59.40	549	15JN	5;19	253.60
476	6JL	21;48	93.82	476	20JL	20;34	287.25	550	29JN	14;34	87.34	550	14JL	20;05	281.97
477	5AU	7;19	122.23	477	19AU	11;00	316.07	551	28JL	22;50	115.61	551	13AU	9;53	310.68
478	3SE	16;46	151.11	478	18SE	4;23	345.54	552	27AU	9;11	144.41	552	11SE	22;47	339.86
479	3OC	2;38	180.52	479	17OC	23;34	15.57	553	25SE	22;30	173.85	553	11OC	11;01	9.51
480	1NO	13;08	210.38	480	16NO	18;40	45.92	554	25OC	15;08	203.89	554	9NO	22;44	39.50
481	1DE	0;28	240.47	481	16DE	12;06	76.27	555	24NO	10;21	234.29	555	9DE	9;51	69.59
482	30DE	12;56	270.53					556	24DE	6;16	264.71				
					-961								**-955**		
483	29JA	2;46	300.32	482	15JA	3;02	106.28	557	23JA	0;49	294.77	556	7JA	20;22	99.49
484	27FE	17;50	329.67	483	13FE	15;18	135.76	558	21FE	16;43	324.26	557	6FE	6;29	128.99
485	29MR	9;33	358.53	484	15MR	1;08	164.62	559	23MR	5;42	353.10	558	7MR	16;45	157.99
486	28AP	1;07	26.98	485	13AP	9;02	192.94	560	21AP	16;05	21.41	559	6AP	3;51	186.54
487	27MY	15;53	55.18	486	12MY	15;52	220.91	561	21MY	0;30	49.38	560	5MY	16;17	214.78
488	26JN	5;40	83.36	487	10JN	22;41	248.77	562	19JN	7;45	77.28	561	4JN	6;08	242.92
489	25JL	18;32	111.77	488	10JL	6;44	276.83	563	18JL	14;49	105.38	562	3JL	21;09	271.21
490	24AU	6;37	140.57	489	8AU	17;09	305.32	564	16AU	22;45	133.91	563	2AU	12;55	299.83
491	22SE	18;01	169.87	490	7SE	6;38	334.42	565	15SE	8;34	163.01	564	1SE	4;58	328.94
492	22OC	4;52	199.61	491	6OC	23;11	4.17	566	14OC	21;03	192.72	565	30SE	20;44	358.56
493	20NO	15;27	229.62	492	5NO	17;59	34.41	567	13NO	12;21	222.89	566	30OC	11;31	28.58
494	20DE	2;07	259.67	493	5DE	13;36	64.86	568	13DE	5;52	253.25	567	29NO	0;45	58.74
												568	28DE	12;12	88.77
					-960								**-954**		
495	18JA	13;12	289.48	494	4JA	8;26	95.16	569	12JA	0;23	283.45	569	26JA	22;08	118.43
496	17FE	0;56	318.89	495	3FE	1;01	124.99	570	10FE	18;36	313.22	570	25FE	7;08	147.56
497	17MR	13;23	347.80	496	3MR	14;32	154.20	571	12MR	11;19	342.41	571	26MR	15;53	176.18
498	16AP	2;34	16.29	497	2AP	0;57	182.77	572	11AP	1;46	11.02	572	25AP	1;04	204.40
499	15MY	16;33	44.51	498	1MY	8;59	210.88	573	10MY	13;39	39.21	573	24MY	11;19	232.43
500	14JN	7;22	72.70	499	30MY	15;47	238.75	574	8JN	23;17	67.18	574	22JN	23;13	260.53
501	13JL	22;45	101.08	500	28JN	22;32	266.67	575	8JL	7;29	95.23	575	22JL	13;13	288.95
502	12AU	14;09	129.85	501	28JL	6;22	294.92	576	6AU	15;21	123.59	576	21AU	5;27	317.87
503	11SE	4;49	159.08	502	26AU	16;07	323.68	577	4SE	23;55	152.46	577	19SE	23;23	347.39
504	10OC	18;17	188.77	503	25SE	4;20	353.06	578	4OC	9;52	181.89	578	19OC	17;40	17.40
505	9NO	6;32	218.78	504	24OC	19;17	23.02	579	2NO	21;34	211.81	579	18NO	10;40	47.68
506	8DE	17;49	248.88	505	23NO	12;50	53.35	580	2DE	11;02	241.99	580	18DE	1;20	77.90
				506	23DE	8;07	83.74								
					-959								**-953**		
507	7JA	4;25	278.79	507	22JA	3;28	113.85	581	1JA	2;08	272.16	581	16JA	13;31	107.79
508	5FE	14;29	308.30	508	20FE	20;57	143.42	582	30JA	18;35	302.03	582	14FE	23;38	137.15
509	7MR	0;13	337.29	509	22MR	11;24	172.33	583	1MR	11;43	331.44	583	16MR	8;16	165.94
510	5AP	10;02	5.79	510	20AP	22;47	200.69	584	31MR	4;28	0.32	584	14AP	16;01	194.24
511	4MY	20;42	33.97	511	20MY	7;51	228.69	585	29AP	19;46	28.73	585	13MY	23;36	222.23
512	3JN	8;59	62.06	512	18JN	15;39	256.61	586	29MY	8;59	56.87	586	12JN	7;56	250.15
513	2JL	23;19	90.31	513	17JL	23;12	284.72	587	27JN	20;18	84.96	587	11JL	18;05	278.30
514	1AU	15;29	118.94	514	16AU	7;19	313.25	588	27JL	6;24	113.27	588	10AU	7;01	306.92
515	31AU	8;32	148.07	515	14SE	16;37	342.32	589	25AU	16;04	142.01	589	8SE	23;07	336.16
516	30SE	1;22	177.72	516	14OC	3;43	11.96	590	24SE	1;54	171.26	590	8OC	17;46	6.02
517	29OC	17;08	207.77	517	12NO	17;11	42.05	591	23OC	12;12	200.99	591	7NO	13;18	36.31
518	28NO	7;23	237.97	518	12DE	9;10	72.34	592	21NO	23;08	231.02	592	7DE	7;49	66.71
519	27DE	19;56	268.06					593	21DE	10;54	261.11				

	NEW MOONS				FULL MOONS		
NUMBER	DATE	TIME	LONG.	NUMBER	DATE	TIME	LONG.
			−952				
				593	6JA	0;04	96.89
594	19JA	23;48	290.99	594	4FE	13;36	126.57
595	18FE	13;59	320.47	595	5MR	0;29	155.64
596	19MR	5;10	349.47	596	3AP	9;09	184.13
597	17AP	20;40	18.03	597	2MY	16;19	212.18
598	17MY	11;49	46.29	598	31MY	22;56	240.04
599	16JN	2;11	74.46	599	30JN	6;13	268.00
600	15JL	15;42	102.79	600	29JL	15;23	296.31
601	14AU	4;28	131.47	601	28AU	3;22	325.19
602	12SE	16;29	160.62	602	26SE	18;32	354.72
603	12OC	3;50	190.24	603	26OC	12;25	24.82
604	10NO	14;38	220.20	604	25NO	7;51	55.24
605	10DE	1;12	250.27	605	25DE	3;15	85.62
			−951				
606	8JA	11;55	280.17	606	23JA	21;03	115.65
607	6FE	23;04	309.71	607	22FE	12;03	145.08
608	8MR	10;52	338.76	608	23MR	23;54	173.87
609	6AP	23;22	7.35	609	22AP	8;57	202.11
610	6MY	12;44	35.62	610	21MY	16;12	230.04
611	5JN	3;03	63.79	611	19JN	22;52	257.91
612	4JL	18;17	92.08	612	19JL	6;08	286.03
613	3AU	9;59	120.72	613	17AU	14;57	314.60
614	2SE	1;25	149.82	614	16SE	1;59	343.77
615	1OC	15;50	179.39	615	15OC	15;37	13.53
616	31OC	4;53	209.34	616	14NO	7;55	43.75
617	29NO	16;43	239.45	617	14DE	2;24	74.15
618	29DE	3;36	269.45				
			−950				
				618	12JA	21;49	104.38
619	27JA	13;47	299.11	619	11FE	16;17	134.15
620	25FE	23;25	328.26	620	13MR	8;13	163.29
621	27MR	8;51	356.89	621	11AP	21;02	191.83
622	25AP	18;45	25.14	622	11MY	7;08	219.93
623	25MY	5;58	53.20	623	9JN	15;31	247.86
624	23JN	19;11	81.36	624	8JL	23;14	275.89
625	23JL	10;35	109.84	625	7AU	7;08	304.26
626	22AU	3;32	138.81	626	5SE	15;56	333.15
627	20SE	20;54	168.31	627	5OC	2;11	2.61
628	20OC	13;32	198.27	628	3NO	14;27	32.56
629	19NO	4;46	228.48	629	3DE	5;07	62.79
630	18DE	18;15	258.65				
			−949				
				630	1JA	22;01	93.03
631	17JA	5;52	288.50	631	31JA	16;05	122.96
632	15FE	15;39	317.84	632	2MR	9;47	152.37
633	17MR	0;41	346.61	633	1AP	1;51	181.20
634	15AP	7;43	14.90	634	30AP	15;46	209.56
635	14MY	15;48	42.90	635	30MY	3;42	237.64
636	13JN	1;23	70.88	636	28JN	14;03	265.70
637	12JL	13;19	99.10	637	27JL	23;24	293.99
638	11AU	3;53	127.80	638	26AU	8;21	322.70
639	9SE	20;47	157.09	639	24SE	17;31	351.94
640	9OC	15;11	186.95	640	24OC	3;36	21.67
641	8NO	9;52	217.21	641	22NO	15;06	51.73
642	8DE	3;26	247.58	642	22DE	4;16	81.88
			−948				
643	6JA	18;41	277.70	643	20JA	18;53	111.82
644	5FE	7;02	307.32	644	19FE	10;23	141.34
645	5MR	16;41	336.33	645	20MR	2;11	170.35
646	4AP	0;28	4.77	646	18AP	17;48	198.90
647	3MY	7;26	32.81	647	18MY	8;46	227.15
648	1JN	14;38	60.70	648	16JN	22;42	255.31
649	30JN	23;05	88.70	649	16JL	11;17	283.60
650	30JL	9;32	117.08	650	14AU	22;36	312.24
651	28AU	22;39	146.01	651	13SE	9;05	341.34
652	27SE	14;42	175.59	652	12OC	19;23	10.93
653	27OC	9;24	205.74	653	11NO	6;01	40.88
654	26NO	5;24	236.18	654	10DE	17;08	70.97
655	26DE	0;34	266.55				
			−947				
				655	9JA	4;42	100.91
656	24JA	17;05	296.51	656	7FE	16;43	130.46
657	23FE	6;15	325.86	657	9MR	5;22	159.54
658	24MR	16;28	354.56	658	7AP	18;57	188.16
659	23AP	0;38	22.77	659	7MY	9;35	216.48
660	22MY	7;45	50.68	660	6JN	0;52	244.68
661	20JN	14;41	78.57	661	5JL	16;07	272.98
662	19JL	22;20	106.71	662	4AU	6;41	301.59
663	18AU	7;37	135.31	663	2SE	20;20	330.63
664	16SE	19;28	164.52	664	2OC	9;10	0.15
665	16OC	10;31	194.36	665	31OC	21;20	30.07
666	15NO	4;37	224.66	666	30NO	8;50	60.16
667	15DE	0;24	255.11	667	29DE	19;35	90.16

	NEW MOONS				FULL MOONS		
NUMBER	DATE	TIME	LONG.	NUMBER	DATE	TIME	LONG.
			−946				
668	13JA	19;45	285.33	668	28JA	5;41	119.80
669	12FE	12;57	315.03	669	26FE	15;32	148.95
670	14MR	3;17	344.09	670	28MR	1;52	177.61
671	12AP	14;51	12.57	671	26AP	13;20	205.91
672	12MY	0;10	40.63	672	26MY	2;18	234.04
673	10JN	7;57	68.54	673	24JN	16;45	262.25
674	9JL	15;05	96.55	674	24JL	8;20	290.74
675	7AU	22;35	124.92	675	23AU	0;33	319.70
676	6SE	7;30	153.82	676	21SE	16;50	349.01
677	5OC	18;41	183.33	677	21OC	8;29	19.10
678	4NO	8;37	213.36	678	19NO	22;44	49.27
679	4DE	1;04	243.67	679	19DE	11;06	79.39
			−945				
680	2JA	19;05	273.94	680	17JA	21;39	109.19
681	1FE	13;24	303.87	681	16FE	6;52	138.50
682	3MR	6;47	333.26	682	17MR	15;26	167.27
683	1AP	22;20	2.06	683	16AP	0;06	195.58
684	1MY	11;30	30.37	684	15MY	9;33	223.63
685	30MY	22;16	58.41	685	13JN	20;27	251.66
686	29JN	7;12	86.42	686	13JL	9;20	279.94
687	28JL	15;18	114.66	687	12AU	0;35	308.67
688	26AU	23;36	143.36	688	10SE	17;59	337.99
689	25SE	8;58	172.61	689	10OC	12;33	7.88
690	24OC	19;52	202.38	690	9NO	6;35	38.12
691	23NO	8;27	232.50	691	8DE	22;37	68.42
692	22DE	22;37	262.68				
			−944				
				692	7JA	12;00	98.46
693	21JA	14;11	292.65	693	5FE	22;59	128.01
694	20FE	6;47	322.21	694	6MR	8;08	156.99
695	20MR	23;38	351.24	695	4AP	16;05	185.43
696	19AP	15;39	19.80	696	3MY	23;31	213.48
697	19MY	5;57	48.02	697	2JN	7;15	241.39
698	17JN	18;18	76.12	698	1JL	16;20	269.42
699	17JL	5;08	104.35	699	31JL	3;47	297.85
700	15AU	15;11	132.94	700	29AU	18;24	326.86
701	14SE	1;07	162.03	701	28SE	12;05	356.52
702	13OC	11;21	191.63	702	28OC	7;34	26.70
703	11NO	22;02	221.59	703	27NO	2;54	57.12
704	11DE	9;18	251.68	704	26DE	20;25	87.42
			−943				
705	9JA	21;25	281.64	705	25JA	11;17	117.30
706	8FE	10;40	311.24	706	23FE	23;22	146.59
707	10MR	1;06	340.37	707	25MR	8;58	175.26
708	8AP	16;16	9.05	708	23AP	16;40	203.44
709	8MY	7;31	37.38	709	22MY	23;22	231.33
710	6JN	22;19	65.56	710	21JN	6;10	259.22
711	6JL	12;26	93.83	711	20JL	14;18	287.38
712	5AU	1;50	122.39	712	19AU	0;51	316.04
713	3SE	14;32	151.41	713	17SE	14;31	345.35
714	3OC	2;29	180.91	714	17OC	7;14	15.26
715	1NO	13;43	210.79	715	16NO	2;07	45.60
716	1DE	0;24	240.85	716	15DE	21;43	76.04
717	30DE	10;55	270.83				
			−942				
				717	14JA	16;25	106.22
718	28JA	21;37	300.49	718	13FE	8;49	135.88
719	27FE	8;47	329.68	719	14MR	22;09	164.89
720	28MR	20;38	358.39	720	13AP	8;27	193.30
721	27AP	9;18	26.73	721	12MY	16;27	221.31
722	26MY	22;59	54.89	722	10JN	23;18	249.18
723	25JN	13;48	83.11	723	10JL	6;13	277.19
724	25JL	5;31	111.63	724	8AU	14;16	305.59
725	23AU	21;28	140.58	725	7SE	0;14	334.55
726	22SE	12;47	170.04	726	6OC	12;38	4.11
727	22OC	2;48	199.90	727	5NO	3;39	34.19
728	20NO	15;22	230.00	728	4DE	21;04	64.55
729	20DE	2;42	260.08				
			−941				
				729	3JA	16;04	94.86
730	18JA	13;06	289.87	730	2FE	11;04	124.80
731	16FE	22;46	319.18	731	4MR	4;16	154.16
732	18MR	8;00	347.97	732	2AP	18;34	182.90
733	16AP	17;20	16.30	733	2MY	5;56	211.14
734	16MY	3;35	44.38	734	31MY	15;07	239.10
735	14JN	15;38	72.46	735	29JN	23;09	267.08
736	14JL	5;59	100.79	736	29JL	7;02	295.33
737	12AU	22;26	129.59	737	27AU	15;30	324.04
738	11SE	16;00	158.93	738	26SE	1;06	353.32
739	11OC	9;23	188.78	739	25OC	12;22	23.11
740	10NO	1;36	218.96	740	24NO	1;47	53.26
741	9DE	16;05	249.19	741	23DE	17;29	83.51

Left half

NEW NUMBER	DATE	TIME	LONG.	FULL NUMBER	DATE	TIME	LONG.
			-940				
742	8JA	4;37	279.19	742	22JA	10;54	113.55
743	6FE	15;11	308.71	743	21FE	4;42	143.13
744	7MR	0;01	337.66	744	21MR	21;30	172.16
745	5AP	7;44	6.08	745	20AP	12;25	200.66
746	4MY	15;20	34.14	746	20MY	1;21	228.82
747	2JN	23;55	62.08	747	18JN	12;34	256.87
748	2JL	10;33	90.18	748	17JL	22;34	285.08
749	31JL	23;49	118.69	749	16AU	7;53	313.66
750	30AU	15;45	147.77	750	14SE	17;05	342.73
751	29SE	9;43	177.45	751	14OC	2;47	12.31
752	29OC	4;37	207.62	752	12NO	13;35	42.28
753	27NO	23;03	238.01	753	12DE	1;50	72.42
754	27DE	15;37	268.26				
			-939				
				754	10JA	15;34	102.43
755	26JA	5;23	298.08	755	9FE	6;25	132.08
756	24FE	16;12	327.31	756	10MR	21;50	161.24
757	26MR	0;41	355.92	757	9AP	13;21	189.92
758	24AP	7;48	24.07	758	9MY	4;36	218.25
759	23MY	14;40	51.97	759	7JN	19;09	246.42
760	21JN	22;21	79.89	760	7JL	8;38	274.67
761	21JL	7;44	108.11	761	5AU	20;50	303.19
762	19AU	19;32	136.84	762	4SE	7;57	332.16
763	18SE	10;12	166.19	763	3OC	18;31	1.61
764	18OC	3;46	196.16	764	2NO	5;05	31.47
765	16NO	23;25	226.54	765	1DE	15;59	61.54
766	16DE	19;18	256.98	766	31DE	3;13	91.55
			-938				
767	15JA	13;16	287.13	767	29JA	14;47	121.23
768	14FE	4;02	316.70	768	28FE	2;47	150.45
769	15MR	15;32	345.62	769	29MR	15;33	179.19
770	14AP	0;32	13.97	770	28AP	5;26	207.56
771	13MY	8;01	41.96	771	27MY	20;21	235.77
772	11JN	14;57	69.83	772	26JN	11;47	264.02
773	10JL	22;12	97.86	773	26JL	2;58	292.52
774	9AU	6;39	126.28	774	24AU	17;23	321.43
775	7SE	17;12	155.28	775	23SE	6;55	350.82
776	7OC	6;41	184.90	776	22OC	19;39	20.64
777	5NO	23;23	215.06	777	21NO	7;38	50.72
778	5DE	18;30	245.49	778	20DE	18;45	80.79
			-937				
779	4JA	14;13	275.82	779	19JA	5;01	110.56
780	3FE	8;31	305.72	780	17FE	14;41	139.87
781	5MR	0;11	335.00	781	19MR	0;25	168.66
782	3AP	13;01	3.67	782	17AP	10;58	197.03
783	2MY	23;23	31.86	783	16MY	22;58	225.18
784	1JN	7;53	59.80	784	15JN	12;37	253.32
785	30JN	15;20	87.76	785	15JL	3;45	281.70
786	29JL	22;40	115.99	786	13AU	19;55	310.49
787	28AU	6;54	144.71	787	12SE	12;32	339.82
788	26SE	16;59	174.01	788	12OC	4;53	9.63
789	26OC	5;35	203.87	789	10NO	20;08	39.77
790	24NO	20;49	234.10	790	10DE	9;33	69.96
791	24DE	14;03	264.41				
			-936				
				791	8JA	20;56	99.90
792	23JA	8;08	294.46	792	7FE	6;36	129.38
793	22FE	1;54	324.03	793	7MR	15;13	158.31
794	22MR	18;19	353.03	794	5AP	23;31	186.75
795	21AP	8;39	21.50	795	5MY	8;17	214.84
796	20MY	20;38	49.61	796	3JN	18;15	242.84
797	19JN	6;32	77.62	797	3JL	6;03	270.99
798	18JL	15;06	105.78	798	1AU	20;09	299.54
799	16AU	23;23	134.32	799	31AU	12;42	328.66
800	15SE	8;20	163.39	800	30SE	7;06	358.38
801	14OC	18;34	193.00	801	30OC	1;50	28.55
802	13NO	6;21	223.02	802	28NO	19;09	58.89
803	12DE	19;38	253.20	803	28DE	9;55	89.06
			-935				
804	11JA	10;19	283.25	804	26JA	21;59	118.80
805	10FE	2;11	312.93	805	25FE	7;50	147.98
806	11MR	18;44	342.12	806	26MR	16;09	176.57
807	10AP	11;07	10.82	807	24AP	23;36	204.73
808	10MY	2;18	39.14	808	24MY	6;58	232.64
809	8JN	15;43	67.26	809	22JN	15;12	260.60
810	8JL	3;26	95.45	810	22JL	1;21	288.85
811	6AU	14;02	123.92	811	20AU	14;24	317.64
812	5SE	0;14	152.86	812	19SE	6;45	347.08
813	4OC	10;31	182.31	813	19OC	1;41	17.11
814	2NO	21;05	212.18	814	17NO	21;28	47.50
815	2DE	8;02	242.25	815	17DE	16;06	77.90
816	31DE	19;32	272.27				

Right half

NEW NUMBER	DATE	TIME	LONG.	FULL NUMBER	DATE	TIME	LONG.
			-934				
				816	16JA	8;18	107.95
817	30JA	7;57	301.98	817	14FE	21;41	137.46
818	28FE	21;30	331.24	818	16MR	8;21	166.33
819	30MR	12;04	0.04	819	14AP	16;50	194.66
820	29AP	3;09	28.46	820	13MY	23;51	222.62
821	28MY	18;11	56.66	821	12JN	6;27	250.48
822	27JN	8;45	84.89	822	11JL	13;47	278.51
823	26JL	22;45	113.35	823	9AU	23;05	306.98
824	25AU	12;07	142.23	824	8SE	11;14	336.05
825	24SE	0;45	171.60	825	8OC	2;34	5.77
826	23OC	12;33	201.39	826	6NO	20;33	35.99
827	21NO	23;34	231.43	827	6DE	15;59	66.43
828	21DE	10;06	261.46				
			-933				
				828	5JA	11;15	96.73
829	19JA	20;30	291.24	829	4FE	4;51	126.59
830	18FE	7;09	320.57	830	5MR	19;40	155.83
831	19MR	18;21	349.41	831	4AP	7;23	184.44
832	18AP	6;18	17.84	832	3MY	16;23	212.57
833	17MY	19;16	46.01	833	1JN	23;41	240.46
834	16JN	9;28	74.18	834	1JL	6;29	268.39
835	16JL	0;53	102.58	835	30JL	13;58	296.64
836	14AU	17;01	131.39	836	28AU	23;02	325.40
837	13SE	9;08	160.70	837	27SE	10;17	354.76
838	13OC	0;09	190.47	838	27OC	0;02	24.68
839	11NO	13;37	220.55	839	25NO	16;16	54.96
840	11DE	1;36	250.67	840	25DE	10;29	85.30
			-932				
841	9JA	12;22	280.58	841	24JA	5;32	115.39
842	7FE	22;13	310.06	842	22FE	23;39	144.96
843	8MR	7;24	339.00	843	23MR	15;21	173.90
844	6AP	16;21	7.45	844	22AP	4;06	202.30
845	6MY	1;49	35.57	845	21MY	14;18	230.34
846	4JN	12;44	63.60	846	19JN	22;54	258.30
847	4JL	1;51	91.81	847	19JL	6;55	286.44
848	2AU	17;26	120.43	848	17AU	15;12	314.99
849	1SE	10;48	149.60	849	16SE	0;21	344.08
850	1OC	4;42	179.31	850	15OC	10;50	13.72
851	30OC	21;51	209.43	851	13NO	23;08	43.77
852	29NO	13;23	239.69	852	13DE	13;36	73.99
853	29DE	2;56	269.81				
			-931				
				853	12JA	6;03	104.10
854	27JA	14;23	299.52	854	10FE	23;34	133.84
855	25FE	23;53	328.66	855	12MR	16;47	163.05
856	27MR	7;53	357.24	856	11AP	8;34	191.71
857	25AP	15;14	25.38	857	10MY	22;28	219.97
858	24MY	23;03	53.31	858	9JN	10;36	248.04
859	23JN	8;29	81.31	859	8JL	21;21	276.20
860	22JL	20;24	109.65	860	7AU	7;11	304.65
861	21AU	11;08	138.52	861	5SE	16;36	333.56
862	20SE	4;20	168.01	862	5OC	2;10	2.99
863	19OC	23;07	198.05	863	3NO	12;27	32.86
864	18NO	18;06	228.41	864	2DE	23;56	62.96
865	18DE	11;50	258.76				
			-930				
				865	1JA	12;47	93.02
866	17JA	3;03	288.77	866	31JA	2;52	122.79
867	15FE	13;13	318.21	867	1MR	17;45	152.10
868	17MR	0;38	347.02	868	31MR	8;59	180.91
869	15AP	8;11	15.30	869	30AP	0;15	209.33
870	14MY	14;57	43.25	870	29MY	15;12	237.53
871	12JN	22;04	71.13	871	28JN	5;25	265.74
872	12JL	6;31	99.21	872	27JL	18;31	294.18
873	10AU	17;05	127.74	873	26AU	6;26	323.01
874	9SE	6;22	156.87	874	24SE	17;28	352.32
875	8OC	22;33	186.64	875	24OC	4;10	22.08
876	7NO	17;29	216.90	876	22NO	14;58	52.11
877	7DE	13;32	247.37	877	22DE	1;59	82.17
			-929				
878	6JA	8;39	277.66	878	20JA	13;12	111.96
879	5FE	1;02	307.46	879	19FE	0;40	141.32
880	6MR	14;01	336.61	880	20MR	12;41	170.19
881	5AP	0;05	5.14	881	19AP	1;43	198.64
882	4MY	8;09	33.22	882	18MY	15;59	226.86
883	2JN	15;15	61.11	883	17JN	7;15	255.08
884	1JL	22;17	89.05	884	16JL	22;48	283.49
885	31JL	6;07	117.32	885	15AU	13;56	312.27
886	29AU	15;37	146.11	886	14SE	4;13	341.52
887	28SE	3;39	175.52	887	13OC	17;37	11.24
888	27OC	18;48	205.50	888	12NO	6;09	41.28
889	26NO	12;51	235.86	889	11DE	17;45	71.38
890	26DE	8;26	266.26				

NEW MOONS / **FULL MOONS**

-928

NUMBER	DATE	TIME	LONG.	NUMBER	DATE	TIME	LONG.
				890	10JA	4;21	101.28
891	25JA	3;30	296.33	891	8FE	14;04	130.74
892	23FE	20;26	325.84	892	8MR	23;25	159.68
893	24MR	10;34	354.71	893	7AP	9;12	188.15
894	22AP	22;05	23.05	894	6MY	20;12	216.33
895	22MY	7;27	51.05	895	5JN	8;53	244.43
896	20JN	15;25	78.99	896	4JL	23;19	272.70
897	19JL	22;50	107.11	897	3AU	15;10	301.34
898	18AU	6;40	135.66	898	2SE	7;54	330.50
899	16SE	15;52	164.76	899	2OC	0;47	0.18
900	16OC	3;14	194.44	900	31OC	16;56	30.27
901	14NO	17;11	224.56	901	30NO	7;28	60.49
902	14DE	9;24	254.86	902	29DE	19;52	90.55

-927

NUMBER	DATE	TIME	LONG.	NUMBER	DATE	TIME	LONG.
903	13JA	3;00	285.01	903	28JA	6;13	120.21
904	11FE	20;50	314.75	904	26FE	15;05	149.31
905	13MR	13;50	343.93	905	27MR	23;14	177.89
906	12AP	5;11	12.56	906	26AP	7;29	206.06
907	11MY	18;21	40.79	907	25MY	16;37	234.04
908	10JN	5;21	68.82	908	24JN	3;21	262.10
909	9JL	14;39	96.92	909	23JL	16;16	290.48
910	7AU	23;10	125.33	910	22AU	7;44	319.39
911	6SE	7;54	154.23	911	21SE	1;32	348.92
912	5OC	17;36	183.67	912	20OC	20;33	18.98
913	4NO	4;41	213.58	913	19NO	14;57	49.32
914	3DE	17;10	243.72	914	19DE	7;08	79.61

-926

NUMBER	DATE	TIME	LONG.	NUMBER	DATE	TIME	LONG.
915	2JA	7;00	273.82	915	17JA	20;29	109.53
916	31JA	22;02	303.62	916	16FE	7;16	138.90
917	2MR	14;01	332.96	917	17MR	16;07	167.68
918	1AP	6;22	1.80	918	15AP	23;46	195.95
919	30AP	22;09	30.23	919	15MY	6;58	223.91
920	30MY	12;31	58.40	920	13JN	14;35	251.81
921	29JN	1;13	86.56	921	12JL	23;38	279.93
922	28JL	12;33	114.94	922	11AU	11;11	308.51
923	26AU	23;09	143.73	923	10SE	1;59	337.72
924	25SE	9;35	173.03	924	9OC	19;55	7.56
925	24OC	20;10	202.79	925	8NO	15;39	37.87
926	23NO	6;58	232.82	926	8DE	11;08	68.32
927	22DE	18;05	262.88				

-925

NUMBER	DATE	TIME	LONG.	NUMBER	DATE	TIME	LONG.
				927	7JA	4;38	98.53
928	21JA	5;47	292.69	928	5FE	19;22	128.25
929	19FE	18;27	322.08	929	7MR	7;15	157.34
930	21MR	8;14	351.00	930	5AP	16;40	185.83
931	19AP	22;52	19.51	931	5MY	0;14	213.89
932	19MY	13;53	47.76	932	3JN	6;53	241.76
933	18JN	4;46	75.96	933	2JL	13;44	269.70
934	17JL	19;15	104.34	934	31JL	21;59	297.99
935	16AU	9;14	133.09	935	30AU	8;43	326.84
936	14SE	22;34	162.31	936	28SE	22;33	356.34
937	14OC	11;03	192.00	937	28OC	15;23	26.40
938	12NO	22;36	222.00	938	27NO	10;16	56.80
939	12DE	9;20	252.07	939	27DE	5;45	87.19

-924

NUMBER	DATE	TIME	LONG.	NUMBER	DATE	TIME	LONG.
940	10JA	19;38	281.95	940	26JA	0;14	117.22
941	9FE	5;54	311.42	941	24FE	16;26	146.69
942	9MR	16;31	340.41	942	25MR	5;37	175.51
943	8AP	3;47	8.93	943	23AP	15;51	203.78
944	7MY	15;59	37.15	944	22MY	23;53	231.73
945	6JN	5;26	65.28	945	21JN	6;51	259.63
946	5JL	20;19	93.57	946	20JL	13;59	287.75
947	4AU	12;23	122.23	947	18AU	22;18	316.32
948	3SE	4;56	151.40	948	17SE	8;32	345.48
949	2OC	20;52	181.06	949	16OC	21;05	15.22
950	1NO	11;23	211.08	950	15NO	12;04	45.39
951	1DE	0;11	241.23	951	15DE	5;17	75.73
952	30DE	11;31	271.24				

-923

NUMBER	DATE	TIME	LONG.	NUMBER	DATE	TIME	LONG.
				952	13JA	23;55	105.92
953	28JA	21;40	300.88	953	12FE	18;31	135.67
954	27FE	6;56	329.99	954	14MR	11;25	164.83
955	28MR	15;42	358.58	955	13AP	1;35	193.41
956	27AP	0;35	26.77	956	12MY	12;59	221.56
957	26MY	10;29	54.78	957	10JN	22;22	249.53
958	24JN	22;22	82.88	958	10JL	6;43	277.59
959	24JL	12;46	111.34	959	8AU	14;58	306.00
960	23AU	5;32	140.32	960	6SE	23;49	334.91
961	21SE	23;36	169.87	961	6OC	9;42	4.38
962	21OC	17;30	199.90	962	4NO	21;06	34.30
963	20NO	10;06	230.17	963	4DE	10;24	64.48
964	20DE	0;45	260.38				

-922

NUMBER	DATE	TIME	LONG.	NUMBER	DATE	TIME	LONG.
				964	3JA	1;44	94.64
965	18JA	13;12	290.26	965	1FE	18;36	124.50
966	16FE	23;30	319.60	966	3MR	11;51	153.88
967	18MR	8;00	348.35	967	2AP	4;16	182.71
968	16AP	15;23	16.60	968	1MY	19;03	211.09
969	15MY	22;41	44.57	969	31MY	8;06	239.21
970	14JN	7;06	72.50	970	29JN	19;40	267.33
971	13JL	17;40	100.68	971	29JL	6;08	295.68
972	12AU	7;03	129.35	972	27AU	15;58	324.45
973	10SE	23;14	158.63	973	26SE	1;37	353.72
974	10OC	17;32	188.50	974	25OC	11;37	23.47
975	9NO	12;45	218.80	975	23NO	22;28	53.51
976	9DE	7;23	249.21	976	23DE	10;31	83.60

-921

NUMBER	DATE	TIME	LONG.	NUMBER	DATE	TIME	LONG.
977	7JA	23;58	279.38	977	21JA	23;48	113.47
978	6FE	13;36	309.03	978	20FE	14;01	142.92
979	8MR	0;13	338.06	979	22MR	4;49	171.87
980	6AP	8;27	6.50	980	20AP	19;54	200.38
981	5MY	15;22	34.52	981	20MY	10;59	228.63
982	3JN	22;08	62.39	982	19JN	1;42	256.82
983	3JL	5;49	90.37	983	18JL	15;37	285.19
984	1AU	15;18	118.73	984	17AU	4;25	313.90
985	31AU	3;15	147.64	985	15SE	16;08	343.07
986	29SE	18;06	177.19	986	15OC	3;10	12.71
987	29OC	11;50	207.30	987	13NO	14;00	42.69
988	28NO	7;33	237.74	988	13DE	0;54	72.76
989	28DE	3;23	268.13				

-920

NUMBER	DATE	TIME	LONG.	NUMBER	DATE	TIME	LONG.
				989	11JA	11;53	102.66
990	26JA	21;13	298.13	990	9FE	22;57	132.16
991	25FE	11;48	327.51	991	10MR	10;21	161.16
992	25MR	23;09	356.24	992	8AP	22;32	189.71
993	24AP	8;02	24.46	993	8MY	11;58	217.97
994	23MY	15;30	52.38	994	7JN	2;43	246.15
995	21JN	22;31	80.28	995	6JL	18;19	274.48
996	21JL	5;56	108.42	996	5AU	9;58	303.14
997	19AU	14;37	137.02	997	4SE	1;01	332.26
998	18SE	1;23	166.21	998	3OC	15;09	1.85
999	17OC	14;59	196.01	999	2NO	4;21	31.83
1000	16NO	7;41	226.25	1000	1DE	16;32	61.95
1001	16DE	2;38	256.67	1001	31DE	3;35	91.95

-919

NUMBER	DATE	TIME	LONG.	NUMBER	DATE	TIME	LONG.
1002	14JA	22;03	286.87	1002	29JA	13;33	121.57
1003	13FE	16;02	316.59	1003	27FE	22;46	150.67
1004	15MR	7;28	345.68	1004	29MR	7;58	179.26
1005	13AP	20;12	14.18	1005	27AP	18;02	207.49
1006	13MY	6;35	42.29	1006	27MY	5;40	235.57
1007	11JN	15;15	70.23	1007	25JN	19;12	263.74
1008	10JL	22;58	98.27	1008	25JL	10;29	292.24
1009	9AU	6;38	126.67	1009	24AU	3;03	321.23
1010	7SE	15;12	155.58	1010	22SE	20;14	350.76
1011	7OC	1;32	185.07	1011	22OC	13;09	20.76
1012	5NO	14;12	215.06	1012	21NO	4;46	50.99
1013	5DE	5;17	245.31	1013	20DE	18;19	81.15

-918

NUMBER	DATE	TIME	LONG.	NUMBER	DATE	TIME	LONG.
1014	3JA	22;09	275.53	1014	19JA	5;34	110.97
1015	2FE	15;45	305.41	1015	17FE	14;57	140.26
1016	4MR	9;03	334.77	1016	18MR	23;09	168.99
1017	3AP	1;09	3.58	1017	17AP	7;04	197.26
1018	2MY	15;25	31.93	1018	16MY	15;30	225.26
1019	1JN	3;34	60.01	1019	15JN	1;16	253.25
1020	30JN	13;48	88.08	1020	14JL	13;00	281.49
1021	29JL	22;49	116.38	1021	13AU	3;15	310.20
1022	28AU	7;33	145.12	1022	11SE	20;07	339.52
1023	26SE	16;53	174.39	1023	11OC	14;56	9.43
1024	26OC	3;22	204.16	1024	10NO	10;04	39.73
1025	24NO	15;10	234.25	1025	10DE	3;37	70.09
1026	24DE	4;12	264.38				

-917

NUMBER	DATE	TIME	LONG.	NUMBER	DATE	TIME	LONG.
				1026	8JA	18;24	100.18
1027	22JA	18;24	294.27	1027	7FE	6;18	129.75
1028	21FE	9;39	323.75	1028	8MR	15;54	158.72
1029	23MR	1;38	352.74	1029	6AP	23;55	187.15
1030	21AP	17;38	21.28	1030	6MY	7;08	215.18
1031	21MY	8;46	49.52	1031	4JN	14;22	243.06
1032	19JN	22;26	77.68	1032	3JL	22;33	271.07
1033	19JL	10;37	105.98	1033	2AU	8;46	299.46
1034	17AU	21;47	134.64	1034	31AU	21;58	328.44
1035	16SE	8;31	163.78	1035	30SE	14;32	358.07
1036	15OC	19;14	193.42	1036	30OC	9;42	28.25
1037	14NO	6;01	223.40	1037	29NO	5;38	58.70
1038	13DE	16;54	253.47	1038	29DE	0;17	89.04

-916 / -910

NEW MOONS				FULL MOONS				NEW MOONS				FULL MOONS			
NUMBER	DATE	TIME	LONG.	NUMBER	DATE	TIME	LONG.	NUMBER	DATE	TIME	LONG.	NUMBER	DATE	TIME	LONG.
1039	12JA	4;06	283.37	1039	27JA	16;23	118.96	1113	5JA	16;22	277.36	1113	20JA	13;00	112.34
1040	10FE	15;59	312.89	1040	26FE	5;35	148.27	1114	4FE	11;06	307.26	1114	18FE	22;20	141.61
1041	11MR	4;53	341.94	1041	26MR	16;05	176.95	1115	6MR	3;44	336.57	1115	20MR	7;12	170.35
1042	9AP	18;52	10.55	1042	25AP	0;25	205.14	1116	4AP	17;43	5.26	1116	18AP	16;28	198.65
1043	9MY	9;35	38.86	1043	24MY	7;22	233.04	1117	4MY	5;12	33.49	1117	18MY	3;03	226.73
1044	8JN	0;33	67.05	1044	22JN	14;00	260.93	1118	2JN	14;43	61.47	1118	16JN	15;32	254.83
1045	7JL	15;24	95.35	1045	21JL	21;28	289.08	1119	1JL	22;57	89.46	1119	16JL	6;00	283.20
1046	6AU	5;54	123.98	1046	20AU	6;56	317.72	1120	31JL	6;42	117.73	1120	14AU	22;09	312.01
1047	4SE	19;55	153.07	1047	18SE	19;16	346.99	1121	29AU	14;52	146.47	1121	13SE	15;23	341.38
1048	4OC	9;08	182.64	1048	18OC	10;44	16.87	1122	28SE	0;22	175.77	1122	13OC	8;49	11.26
1049	2NO	21;20	212.58	1049	17NO	4;45	47.18	1123	27OC	11;54	205.60	1123	12NO	1;25	41.46
1050	2DE	8;31	242.66	1050	17DE	0;04	77.60	1124	26NO	1;46	235.78	1124	11DE	16;10	71.70
1051	31DE	18;55	272.63					1125	25DE	17;42	266.02				

-915 / -909

NEW MOONS				FULL MOONS				NEW MOONS				FULL MOONS			
1052	30JA	4;58	302.24	1051	15JA	19;08	107.79	1126	24JA	10;49	296.02	1125	10JA	4;32	101.67
1053	28FE	15;07	331.37	1052	14FE	12;29	137.46	1127	23FE	4;07	325.55	1126	8FE	14;39	131.15
1054	30MR	1;44	0.01	1053	16MR	3;07	166.50	1128	24MR	20;43	354.53	1127	9MR	23;10	160.05
1055	28AP	13;11	28.28	1054	14AP	14;43	194.95	1129	23AP	11;53	23.02	1128	8AP	6;56	188.45
1056	28MY	1;49	56.40	1055	13MY	23;44	223.00	1130	23MY	1;09	51.19	1129	7MY	14;51	216.50
1057	26JN	15;57	84.60	1056	12JN	7;09	250.89	1131	21JN	12;26	79.25	1130	5JN	23;44	244.45
1058	26JL	7;37	113.12	1057	11JL	14;11	278.91	1132	20JL	22;10	107.47	1131	5JL	10;22	272.56
1059	25AU	0;18	142.13	1058	9AU	21;56	307.32	1133	19AU	7;10	136.06	1132	3AU	23;22	301.09
1060	23SE	16;59	171.66	1059	8SE	7;17	336.27	1134	17SE	16;20	165.17	1133	2SE	15;04	330.19
1061	23OC	8;33	201.61	1060	7OC	18;44	5.82	1135	17OC	2;21	194.79	1134	2OC	9;14	359.92
1062	21NO	22;22	231.77	1061	6NO	8;32	35.86	1136	15NO	13;32	224.79	1135	1NO	4;38	30.13
1063	21DE	10;25	261.87	1062	6DE	0;36	66.16	1137	15DE	1;53	254.93	1136	30NO	23;19	60.53
												1137	30DE	15;35	90.76

-914 / -908

NEW MOONS				FULL MOONS				NEW MOONS				FULL MOONS			
1064	19JA	21;02	291.65	1063	4JA	18;30	96.41	1138	13JA	15;18	284.91	1138	29JA	4;50	120.53
1065	18FE	6;32	320.93	1064	3FE	13;06	126.32	1139	12FE	5;45	314.51	1139	27FE	15;23	149.71
1066	19MR	15;16	349.67	1065	5MR	6;50	155.69	1140	12MR	21;06	343.63	1140	27MR	23;58	178.30
1067	17AP	23;47	17.96	1066	3AP	22;20	184.46	1141	11AP	12;59	12.30	1141	26AP	7;23	206.43
1068	17MY	8;53	45.98	1067	3MY	11;03	212.74	1142	11MY	4;34	40.62	1142	25MY	14;25	234.33
1069	15JN	19;33	74.01	1068	1JN	21;25	240.75	1143	9JN	19;05	68.80	1143	23JN	21;59	262.26
1070	15JL	8;39	102.31	1069	1JL	6;19	268.77	1144	9JL	8;10	97.04	1144	23JL	7;04	290.49
1071	14AU	0;26	131.09	1070	30JL	14;43	297.05	1145	7AU	20;03	125.59	1145	21AU	18;45	319.25
1072	12SE	18;13	160.47	1071	28AU	23;24	325.80	1146	6SE	7;14	154.58	1146	20SE	9;45	348.65
1073	12OC	12;38	190.38	1072	27SE	8;53	355.09	1147	5OC	18;10	184.08	1147	20OC	3;53	18.67
1074	11NO	6;12	220.62	1073	26OC	19;34	24.88	1148	4NO	5;03	213.98	1148	18NO	23;46	49.07
1075	10DE	21;58	250.90	1074	25NO	7;51	54.99	1149	3DE	15;54	244.05	1149	18DE	19;17	79.49
				1075	24DE	22;02	85.16								

-913 / -907

NEW MOONS				FULL MOONS				NEW MOONS				FULL MOONS			
1076	9JA	11;31	280.93	1076	23JA	13;59	115.12	1150	2JA	2;48	274.03	1150	17JA	12;43	109.59
1077	7FE	22;47	310.46	1077	22FE	6;55	144.65	1151	31JA	14;03	303.67	1151	16FE	3;17	139.12
1078	9MR	7;58	339.41	1078	23MR	23;38	173.65	1152	2MR	2;06	332.85	1152	17MR	15;00	168.01
1079	7AP	15;39	7.80	1079	22AP	15;10	202.17	1153	31MR	15;15	1.58	1153	16AP	0;15	196.35
1080	6MY	22;43	35.83	1080	22MY	5;05	230.36	1154	30AP	5;26	29.94	1154	15MY	7;45	224.33
1081	5JN	6;20	63.73	1081	20JN	17;29	258.47	1155	29MY	20;14	58.14	1155	13JN	14;26	252.19
1082	4JL	15;40	91.79	1082	20JL	4;42	286.74	1156	28JN	11;15	86.39	1156	12JL	21;24	280.22
1083	3AU	3;38	120.26	1083	18AU	15;03	315.37	1157	28JL	2;10	114.90	1157	11AU	5;50	308.68
1084	1SE	18;33	149.32	1084	17SE	0;59	344.50	1158	26AU	16;46	143.85	1158	9SE	16;45	337.72
1085	1OC	12;03	179.00	1085	16OC	10;55	14.11	1159	25SE	6;44	173.29	1159	9OC	6;44	7.40
1086	31OC	7;08	209.19	1086	14NO	21;22	44.08	1160	24OC	19;42	203.15	1160	7NO	23;37	37.58
1087	30NO	2;21	239.61	1087	14DE	8;44	74.17	1161	23NO	7;30	233.23	1161	7DE	18;25	67.99
1088	29DE	20;09	269.91					1162	22DE	18;14	263.27				

-912 / -906

NEW MOONS				FULL MOONS				NEW MOONS				FULL MOONS			
1089	28JA	11;17	299.77	1088	12JA	21;13	104.12	1163	21JA	4;16	293.01	1162	6JA	13;41	98.29
1090	26FE	23;16	329.02	1089	11FE	10;44	133.70	1164	19FE	14;05	322.29	1163	5FE	7;54	128.16
1091	27MR	8;27	357.64	1090	12MR	0;58	162.79	1165	21MR	0;08	351.06	1164	6MR	23;52	157.42
1092	25AP	15;48	25.78	1091	10AP	15;40	191.42	1166	19AP	10;52	19.42	1165	5AP	12;55	186.06
1093	24MY	22;27	53.67	1092	10MY	6;39	219.72	1167	18MY	22;41	47.54	1166	4MY	23;08	214.23
1094	23JN	5;33	81.58	1093	8JN	21;38	247.91	1168	17JN	11;57	75.68	1167	3JN	7;16	242.15
1095	22JL	14;05	109.78	1094	8JL	12;11	276.21	1169	17JL	2;56	104.07	1168	2JL	14;27	270.11
1096	21AU	0;48	138.48	1095	7AU	1;50	304.82	1170	15AU	19;25	132.91	1169	31JL	21;51	298.37
1097	19SE	14;16	167.81	1096	5SE	14;23	333.86	1171	14SE	12;32	162.28	1170	30AU	6;28	327.14
1098	19OC	6;41	197.74	1097	5OC	1;58	3.37	1172	14OC	5;04	192.14	1171	28SE	16;57	356.49
1099	18NO	1;37	228.10	1098	3NO	13;01	33.27	1173	12NO	20;00	222.28	1172	28OC	5;37	26.38
1100	17DE	21;38	258.54	1099	2DE	23;55	63.34	1174	12DE	9;00	252.45	1173	26NO	20;32	56.60
												1174	26DE	13;27	86.88

-911 / -905

NEW MOONS				FULL MOONS				NEW MOONS				FULL MOONS			
1101	16JA	16;37	288.72	1100	1JA	10;46	93.32	1175	10JA	20;14	282.36	1175	25JA	7;39	116.92
1102	15FE	8;48	318.33	1101	30JA	21;34	122.96	1176	9FE	6;06	311.82	1176	24FE	1;48	146.47
1103	16MR	21;38	347.28	1102	1MR	8;29	152.10	1177	10MR	14;58	340.73	1177	25MR	18;23	175.44
1104	15AP	7;34	15.66	1103	30MR	19;54	180.77	1178	8AP	23;18	9.13	1178	24AP	8;27	203.87
1105	14MY	15;36	43.66	1104	29AP	8;25	209.08	1179	8MY	7;48	37.20	1179	23MY	19;58	231.96
1106	12JN	22;46	71.54	1105	28MY	22;24	237.24	1180	6JN	17;26	65.18	1180	22JN	5;37	259.96
1107	12JL	5;58	99.58	1106	27JN	13;41	265.50	1181	6JL	5;12	93.35	1181	21JL	14;21	288.15
1108	10AU	14;03	128.00	1107	27JL	5;36	294.04	1182	4AU	19;43	121.94	1182	19AU	23;02	316.74
1109	8SE	23;47	156.99	1108	25AU	21;18	323.02	1183	3SE	12;49	151.12	1183	18SE	8;15	345.86
1110	8OC	11;59	186.58	1109	24SE	12;13	352.49	1184	3OC	7;20	180.87	1184	17OC	18;25	15.50
1111	7NO	3;10	216.68	1110	24OC	2;09	22.38	1185	2NO	1;42	211.05	1185	16NO	5;53	45.52
1112	6DE	21;05	247.06	1111	22NO	14;59	52.50	1186	1DE	18;36	241.38	1186	15DE	18;59	75.68
				1112	22DE	2;38	82.58	1187	31DE	9;19	271.54				

-904

NEW: NUMBER	DATE	TIME	LONG.	FULL: NUMBER	DATE	TIME	LONG.
				1187	14JA	9;53	105.71
1188	29JA	21;38	301.26	1188	13FE	2;10	135.38
1189	28FE	7;41	330.41	1189	13MR	18;51	164.55
1190	28MR	15;52	358.97	1190	12AP	10;53	193.20
1191	26AP	22;58	27.08	1191	12MY	1;35	221.49
1192	26MY	6;04	54.99	1192	10JN	14;49	249.61
1193	24JN	14;21	82.95	1193	10JL	2;48	277.82
1194	24JL	0;56	111.24	1194	8AU	13;48	306.34
1195	22AU	10;27	140.08	1195	7SE	0;10	335.31
1196	21SE	6;53	169.56	1196	6OC	10;15	4.79
1197	21OC	1;28	199.61	1197	4NO	20;30	34.66
1198	19NO	20;55	229.99	1198	4DE	7;22	64.74
1199	19DE	15;38	260.38				

-903

NEW: NUMBER	DATE	TIME	LONG.	FULL: NUMBER	DATE	TIME	LONG.
				1199	2JA	19;08	94.75
1200	18JA	8;11	290.43	1200	1FE	7;55	124.44
1201	16FE	21;40	319.90	1201	2MR	21;30	153.67
1202	18MR	8;04	348.73	1202	1AP	11;42	182.43
1203	16AP	16;07	17.02	1203	1MY	2;22	210.81
1204	15MY	22;55	44.96	1204	30MY	17;20	239.01
1205	14JN	5;39	72.83	1205	29JN	8;17	267.26
1206	13JL	13;23	100.90	1206	28JL	22;42	295.76
1207	11AU	23;00	129.41	1207	27AU	12;07	324.67
1208	10SE	11;09	158.51	1208	26SE	0;26	354.06
1209	10OC	2;10	188.24	1209	25OC	11;55	23.87
1210	8NO	19;59	218.47	1210	23NO	22;57	53.92
1211	8DE	15;40	248.93	1211	23DE	9;46	83.95

-902

NEW: NUMBER	DATE	TIME	LONG.	FULL: NUMBER	DATE	TIME	LONG.
1212	7JA	11;22	279.23	1212	21JA	20;26	113.71
1213	6FE	5;00	309.06	1213	20FE	7;01	143.01
1214	7MR	19;24	338.24	1214	21MR	17;49	171.80
1215	6AP	6;37	6.80	1215	20AP	5;27	200.19
1216	5MY	15;28	34.91	1216	19MY	18;29	228.36
1217	3JN	22;58	62.81	1217	18JN	9;07	256.55
1218	3JL	6;08	90.77	1218	18JL	0;57	284.98
1219	1AU	13;48	119.05	1219	16AU	17;06	313.82
1220	30AU	22;44	147.84	1220	15SE	8;46	343.15
1221	29SE	9;42	177.22	1221	14OC	23;30	12.94
1222	28OC	23;25	207.16	1222	13NO	13;05	43.03
1223	27NO	16;01	237.46	1223	13DE	1;24	73.17
1224	27DE	10;42	267.81				

-901

NEW: NUMBER	DATE	TIME	LONG.	FULL: NUMBER	DATE	TIME	LONG.
				1224	11JA	12;20	103.07
1225	26JA	5;45	297.87	1225	9FE	21;58	132.50
1226	24FE	23;23	327.38	1226	11MR	6;44	161.40
1227	26MR	14;36	356.28	1227	9AP	15;27	189.81
1228	25AP	3;15	24.65	1228	9MY	1;05	217.92
1229	24MY	13;44	52.70	1229	7JN	12;25	245.97
1230	22JN	22;39	80.67	1230	7JL	1;53	274.20
1231	22JL	6;42	108.84	1231	5AU	17;21	302.85
1232	20AU	14;44	137.41	1232	4SE	10;21	332.04
1233	18SE	23;38	166.53	1233	4OC	4;04	1.77
1234	18OC	10;11	196.19	1234	2NO	21;28	31.92
1235	16NO	22;53	226.27	1235	2DE	13;23	62.21
1236	16DE	13;45	256.50				

-900

NEW: NUMBER	DATE	TIME	LONG.	FULL: NUMBER	DATE	TIME	LONG.
				1236	1JA	2;59	92.31
1237	15JA	6;10	286.59	1237	30JA	14;05	121.97
1238	13FE	23;13	316.28	1238	28FE	23;09	151.07
1239	14MR	16;02	345.43	1239	29MR	7;00	179.61
1240	13AP	7;51	14.07	1240	27AP	14;33	207.73
1241	12MY	22;06	42.33	1241	26MY	22;44	235.68
1242	11JN	10;29	70.42	1242	25JN	8;21	263.69
1243	10JL	21;07	98.58	1243	24JL	20;07	292.05
1244	9AU	6;37	127.05	1244	23AU	10;32	320.94
1245	7SE	15;50	155.99	1245	22SE	3;43	350.46
1246	7OC	1;33	185.46	1246	21OC	22;53	20.54
1247	5NO	12;14	215.36	1247	20NO	18;19	50.93
1248	4DE	23;59	245.47	1248	20DE	12;00	81.27

-899

NEW: NUMBER	DATE	TIME	LONG.	FULL: NUMBER	DATE	TIME	LONG.
1249	3JA	12;42	275.51	1249	19JA	2;45	111.24
1250	2FE	2;22	305.24	1250	17FE	14;29	140.63
1251	3MR	16;58	334.50	1251	18MR	23;49	169.40
1252	2AP	8;23	3.29	1252	17AP	7;36	197.67
1253	2MY	0;04	31.70	1253	16MY	14;39	225.61
1254	31MY	15;13	59.90	1254	14JN	21;48	253.50
1255	30JN	5;11	88.12	1255	14JL	6;01	281.59
1256	29JL	17;53	116.57	1256	12AU	16;20	310.14
1257	28AU	5;38	145.43	1257	11SE	5;43	339.31
1258	26SE	16;56	174.78	1258	10OC	22;28	9.13
1259	26OC	4;02	204.58	1259	9NO	17;47	39.43
1260	24NO	14;58	234.63	1260	9DE	13;46	69.89
1261	24DE	1;44	264.66				

-898

NEW: NUMBER	DATE	TIME	LONG.	FULL: NUMBER	DATE	TIME	LONG.
				1261	8JA	8;22	100.14
1262	22JA	12;34	294.42	1262	7FE	0;19	129.89
1263	20FE	23;53	323.73	1263	8MR	13;20	159.00
1264	22MR	12;10	352.58	1264	6AP	23;40	187.52
1265	21AP	1;35	21.02	1265	6MY	7;55	215.59
1266	20MY	15;59	49.24	1266	4JN	14;53	243.47
1267	19JN	6;57	77.45	1267	3JL	21;37	271.42
1268	18JL	22;07	105.86	1268	2AU	5;17	299.71
1269	17AU	13;11	134.67	1269	31AU	14;57	328.53
1270	16SE	3;49	163.97	1270	30SE	3;28	357.99
1271	15OC	17;36	193.74	1271	29OC	19;01	28.02
1272	14NO	6;10	223.79	1272	28NO	12;58	58.38
1273	13DE	17;26	253.88	1273	28DE	8;05	88.74

-897

NEW: NUMBER	DATE	TIME	LONG.	FULL: NUMBER	DATE	TIME	LONG.
1274	12JA	3;40	283.74	1274	27JA	2;51	118.78
1275	10FE	13;20	313.17	1275	25FE	19;56	148.25
1276	11MR	22;58	342.09	1276	27MR	10;23	177.11
1277	10AP	9;02	10.54	1277	25AP	21;57	205.43
1278	9MY	20;02	38.70	1278	25MY	7;03	233.41
1279	8JN	8;25	66.79	1279	23JN	14;39	261.34
1280	7JL	22;32	95.06	1280	22JL	21;57	289.48
1281	6AU	14;28	123.74	1281	21AU	6;01	318.07
1282	5SE	7;40	152.95	1282	19SE	15;40	347.22
1283	5OC	0;57	182.68	1283	19OC	3;18	16.94
1284	3NO	17;01	212.78	1284	17NO	17;05	47.07
1285	3DE	7;06	242.99	1285	17DE	8;56	77.35

-896

NEW: NUMBER	DATE	TIME	LONG.	FULL: NUMBER	DATE	TIME	LONG.
1286	1JA	19;11	273.02	1286	16JA	2;23	107.47
1287	31JA	5;34	302.65	1287	14FE	20;30	137.19
1288	29FE	14;43	331.73	1288	15MR	13;51	166.35
1289	29MR	23;02	0.28	1289	14AP	5;09	194.95
1290	28AP	7;10	28.43	1290	13MY	17;56	223.15
1291	27MY	15;58	56.39	1291	12JN	4;31	251.17
1292	26JN	2;29	84.45	1292	11JL	13;47	279.28
1293	25JL	15;36	112.86	1293	9AU	22;37	307.73
1294	24AU	7;36	141.83	1294	8SE	7;43	336.68
1295	23SE	1;48	171.41	1295	7OC	17;33	6.16
1296	22OC	20;39	201.49	1296	6NO	4;23	36.07
1297	21NO	14;35	231.82	1297	5DE	16;34	66.21
1298	21DE	6;30	262.08				

-895

NEW: NUMBER	DATE	TIME	LONG.	FULL: NUMBER	DATE	TIME	LONG.
				1298	4JA	6;24	96.29
1299	19JA	19;59	291.99	1299	2FE	21;47	126.07
1300	18FE	7;02	321.34	1300	4MR	14;06	155.39
1301	19MR	15;56	350.08	1301	3AP	6;21	184.20
1302	17AP	23;19	18.32	1302	2MY	21;40	212.59
1303	17MY	6;10	46.26	1303	1JN	11;40	240.75
1304	15JN	13;39	74.16	1304	1JL	0;25	268.92
1305	14JL	22;59	102.30	1305	30JL	12;08	297.33
1306	13AU	11;02	130.93	1306	28AU	23;03	326.17
1307	12SE	2;10	160.19	1307	27SE	9;29	355.50
1308	11OC	19;55	190.06	1308	26OC	19;45	25.27
1309	10NO	15;14	220.37	1309	25NO	6;18	55.31
1310	10DE	10;34	250.80	1310	24DE	17;30	85.35

-894

NEW: NUMBER	DATE	TIME	LONG.	FULL: NUMBER	DATE	TIME	LONG.
1311	9JA	4;21	281.01	1311	23JA	5;34	115.15
1312	7FE	19;21	310.71	1312	21FE	18;29	144.53
1313	9MR	7;08	339.76	1313	23MR	8;05	173.41
1314	7AP	16;09	8.21	1314	21AP	22;17	201.88
1315	6MY	23;22	36.24	1315	21MY	13;01	230.11
1316	5JN	5;58	64.10	1316	20JN	4;05	258.32
1317	4JL	13;07	92.07	1317	19JL	19;01	286.73
1318	2AU	21;47	120.41	1318	18AU	9;17	315.52
1319	1SE	8;42	149.29	1319	16SE	22;27	344.78
1320	30SE	22;20	178.81	1320	16OC	10;33	14.48
1321	30OC	14;51	208.89	1321	14NO	21;55	44.49
1322	29NO	9;46	239.29	1322	14DE	8;50	74.56
1323	29DE	5;39	269.68				

-893

NEW: NUMBER	DATE	TIME	LONG.	FULL: NUMBER	DATE	TIME	LONG.
				1323	12JA	19;28	104.43
1324	28JA	0;25	299.70	1324	11FE	5;50	133.88
1325	26FE	16;24	329.12	1325	12MR	16;11	162.81
1326	28MR	5;04	357.89	1326	11AP	3;03	191.29
1327	26AP	14;57	26.13	1327	10MY	15;06	219.49
1328	25MY	23;01	54.08	1328	9JN	4;51	247.63
1329	24JN	6;19	81.99	1329	8JL	20;12	275.96
1330	23JL	13;45	110.15	1330	7AU	12;30	304.66
1331	21AU	22;06	138.76	1331	6SE	4;48	333.85
1332	20SE	8;05	167.94	1332	5OC	20;20	3.52
1333	19OC	20;26	197.69	1333	4NO	10;44	33.56
1334	18NO	11;35	227.89	1334	3DE	23;49	63.73
1335	18DE	5;17	258.24				

Left half

NEW MOONS

NUMBER	DATE	TIME	LONG.
		-892	
1336	17JA	0;12	288.41
1337	15FE	18;31	318.12
1338	16MR	10;51	347.23
1339	15AP	0;42	15.77
1340	14MY	12;15	43.91
1341	12JN	21;58	71.89
1342	12JL	6;31	99.98
1343	10AU	14;40	128.41
1344	8SE	23;13	157.35
1345	8OC	8;59	186.84
1346	6NO	20;38	216.79
1347	6DE	10;22	246.99
		-891	
1348	5JA	1;55	277.14
1349	3FE	18;29	306.96
1350	5MR	11;15	336.28
1351	4AP	3;27	5.08
1352	3MY	18;29	33.45
1353	2JN	7;54	61.58
1354	1JL	19;33	89.71
1355	31JL	5;46	118.08
1356	29AU	15;16	146.87
1357	28SE	0;53	176.18
1358	27OC	11;11	205.95
1359	25NO	22;25	236.02
1360	25DE	10;33	266.10
		-890	
1361	23JA	23;30	295.93
1362	22FE	13;19	325.33
1363	24MR	4;04	354.25
1364	22AP	19;29	22.75
1365	22MY	10;56	51.00
1366	21JN	1;39	79.21
1367	20JL	15;12	107.57
1368	19AU	3;40	136.30
1369	17SE	15;26	165.51
1370	17OC	2;51	195.20
1371	15NO	13;59	225.20
1372	15DE	0;49	255.27
		-889	
1373	13JA	11;26	285.13
1374	11FE	22;12	314.58
1375	13MR	9;39	343.55
1376	11AP	22;12	12.09
1377	11MY	11;57	40.35
1378	10JN	2;37	68.52
1379	9JL	17;49	96.85
1380	8AU	9;12	125.53
1381	7SE	0;25	154.69
1382	6OC	15;00	184.34
1383	5NO	4;25	214.34
1384	4DE	16;24	244.46
		-888	
1385	3JA	3;03	274.42
1386	1FE	12;46	304.00
1387	1MR	22;08	333.07
1388	31MR	7;40	1.65
1389	29AP	17;55	29.87
1390	29MY	5;23	57.94
1391	27JN	18;33	86.10
1392	27JL	9;41	114.62
1393	26AU	2;34	143.65
1394	24SE	20;16	173.24
1395	24OC	13;21	203.27
1396	23NO	4;38	233.50
1397	22DE	17;44	263.63
		-887	
1398	21JA	4;51	293.42
1399	19FE	14;25	322.68
1400	20MR	22;53	351.39
1401	19AP	6;50	19.64
1402	18MY	15;01	47.62
1403	17JN	0;27	75.60
1404	16JL	12;11	103.86
1405	15AU	2;50	132.62
1406	13SE	20;15	161.99
1407	13OC	15;12	191.94
1408	12NO	9;57	222.24
1409	12DE	3;03	252.58

FULL MOONS

NUMBER	DATE	TIME	LONG.
1335	2JA	11;26	93.74
1336	31JA	21;33	123.34
1337	1MR	6;28	152.40
1338	30MR	14;52	180.95
1339	28AP	23;42	209.11
1340	28MY	9;57	237.13
1341	26JN	22;16	265.27
1342	26JL	12;49	293.75
1343	25AU	5;18	322.75
1344	23SE	23;01	352.32
1345	23OC	16;58	22.38
1346	22NO	9;56	52.67
1347	22DE	0;49	82.89
1348	20JA	13;06	112.73
1349	18FE	22;58	142.02
1350	20MR	7;08	170.73
1351	18AP	14;33	198.96
1352	17MY	22;11	226.92
1353	16JN	6;55	254.87
1354	15JL	17;31	283.08
1355	14AU	6;37	311.76
1356	12SE	22;34	341.06
1357	12OC	17;04	10.98
1358	11NO	12;47	41.31
1359	11DE	7;38	71.73
1360	9JA	23;55	101.86
1361	8FE	13;02	131.47
1362	9MR	23;22	160.45
1363	8AP	7;43	188.86
1364	7MY	14;57	216.89
1365	5JN	21;53	244.76
1366	5JL	5;28	272.75
1367	3AU	14;39	301.12
1368	2SE	2;30	330.06
1369	1OC	17;40	359.65
1370	31OC	11;57	29.81
1371	30NO	7;54	60.26
1372	30DE	3;22	90.63
1373	28JA	20;39	120.58
1374	27FE	11;02	149.91
1375	28MR	22;35	178.63
1376	27AP	7;44	206.83
1377	26MY	15;14	234.75
1378	24JN	22;01	262.65
1379	24JL	5;09	290.80
1380	22AU	13;49	319.43
1381	21SE	0;57	348.67
1382	20OC	15;04	18.51
1383	19NO	7;55	48.78
1384	19DE	2;32	79.17
1385	17JA	21;29	109.33
1386	16FE	15;24	139.01
1387	17MR	7;07	168.08
1388	15AP	20;05	196.57
1389	15MY	6;21	224.65
1390	13JN	14;40	252.58
1391	12JL	22;07	280.64
1392	11AU	5;51	309.06
1393	9SE	14;47	338.02
1394	9OC	1;31	7.56
1395	7NO	14;15	37.57
1396	7DE	5;00	67.81
1397	5JA	21;32	97.99
1398	4FE	15;13	127.85
1399	6MR	8;54	157.19
1400	5AP	1;12	185.98
1401	4MY	15;13	214.30
1402	3JN	2;55	242.36
1403	2JL	12;54	270.43
1404	31JL	22;04	298.77
1405	30AU	7;12	327.55
1406	28SE	16;50	356.87
1407	28OC	3;14	26.66
1408	26NO	14;42	56.74
1409	26DE	3;32	86.85

Right half

NEW MOONS

NUMBER	DATE	TIME	LONG.
		-886	
1410	10JA	17;46	282.64
1411	9FE	5;56	312.19
1412	10MR	15;43	341.14
1413	8AP	23;38	9.53
1414	8MY	6;30	37.53
1415	6JN	13;27	65.41
1416	5JL	21;43	93.43
1417	4AU	8;22	121.87
1418	2SE	22;02	150.89
1419	2OC	14;41	180.56
1420	1NO	9;30	210.75
1421	1DE	5;05	241.19
1422	30DE	23;49	271.52
		-885	
1423	29JA	16;15	301.42
1424	28FE	5;33	330.70
1425	29MR	15;47	359.35
1426	27AP	23;41	27.50
1427	27MY	6;26	55.38
1428	25JN	13;12	83.28
1429	24JL	21;04	111.47
1430	23AU	6;53	140.16
1431	21SE	19;12	169.46
1432	21OC	10;21	199.35
1433	20NO	4;11	229.67
1434	19DE	23;45	260.09
		-884	
1435	18JA	19;13	290.27
1436	17FE	12;36	319.92
1437	18MR	2;50	348.91
1438	16AP	13;57	17.31
1439	15MY	22;49	45.34
1440	14JN	6;26	73.24
1441	13JL	13;50	101.30
1442	11AU	21;46	129.74
1443	10SE	6;59	158.72
1444	9OC	18;10	188.29
1445	8NO	7;55	218.35
1446	8DE	0;21	248.66
		-883	
1447	6JA	18;40	278.91
1448	5FE	13;17	308.79
1449	7MR	6;33	338.10
1450	5AP	21;33	6.82
1451	5MY	10;12	35.08
1452	3JN	20;51	63.11
1453	3JL	6;04	91.15
1454	1AU	14;31	119.46
1455	30AU	22;58	148.23
1456	29SE	8;12	177.54
1457	28OC	18;56	207.36
1458	27NO	7;36	237.49
1459	26DE	22;09	267.67
		-882	
1460	25JA	14;03	297.60
1461	24FE	6;32	327.08
1462	25MR	22;51	356.03
1463	24AP	14;26	24.52
1464	24MY	4;43	52.72
1465	22JN	17;24	80.85
1466	22JL	4;30	109.13
1467	20AU	14;32	137.79
1468	19SE	0;14	166.94
1469	18OC	10;19	196.58
1470	16NO	21;09	226.58
1471	16DE	8;46	256.68
		-881	
1472	14JA	21;07	286.60
1473	13FE	10;12	316.13
1474	15MR	0;10	345.17
1475	13AP	15;03	13.78
1476	13MY	6;27	42.09
1477	11JN	21;39	70.29
1478	11JL	11;59	98.60
1479	10AU	1;15	127.22
1480	8SE	13;37	156.29
1481	8OC	1;27	185.84
1482	6NO	12;53	215.78
1483	5DE	23;55	245.86

FULL MOONS

NUMBER	DATE	TIME	LONG.
1410	24JA	17;56	116.73
1411	23FE	9;35	146.19
1412	25MR	1;43	175.15
1413	23AP	17;25	203.66
1414	23MY	8;04	231.87
1415	21JN	21;33	260.03
1416	21JL	10;00	288.36
1417	19AU	21;34	317.07
1418	18SE	8;29	346.25
1419	17OC	18;59	15.91
1420	16NO	5;26	45.88
1421	15DE	16;14	75.95
1422	14JA	3;41	105.84
1423	12FE	15;54	135.34
1424	14MR	4;52	164.36
1425	12AP	18;30	192.93
1426	12MY	8;48	221.20
1427	10JN	23;43	249.39
1428	10JL	14;56	277.73
1429	9AU	5;52	306.40
1430	7SE	19;56	335.53
1431	7OC	8;50	5.11
1432	5NO	20;43	35.06
1433	5DE	7;53	65.15
1434	3JA	18;35	95.11
1435	2FE	4;53	124.70
1436	2MR	14;57	153.79
1437	1AP	1;11	182.39
1438	30AP	12;19	210.63
1439	30MY	1;01	238.74
1440	28JN	15;36	266.98
1441	28JL	7;41	295.54
1442	27AU	0;21	324.58
1443	25SE	16;39	354.12
1444	25OC	7;55	24.08
1445	23NO	21;50	54.26
1446	23DE	10;13	84.36
1447	21JA	20;58	114.12
1448	20FE	6;15	143.36
1449	21MR	14;35	172.06
1450	19AP	22;52	200.31
1451	19MY	8;07	228.33
1452	17JN	19;14	256.38
1453	17JL	8;41	284.71
1454	16AU	0;23	313.53
1455	14SE	17;48	342.92
1456	14OC	12;01	12.85
1457	13NO	5;50	43.11
1458	12DE	21;58	73.41
1459	11JA	11;34	103.42
1460	9FE	22;28	132.91
1461	11MR	7;13	161.80
1462	9AP	14;44	190.16
1463	8MY	22;01	218.18
1464	7JN	6;00	246.09
1465	6JL	15;33	274.17
1466	5AU	3;22	302.67
1467	3SE	17;59	331.75
1468	3OC	11;27	1.46
1469	2NO	6;55	31.69
1470	2DE	2;34	62.13
1471	31DE	20;18	92.41
1472	30JA	10;57	122.23
1473	28FE	22;30	151.42
1474	30MR	7;37	180.01
1475	28AP	15;13	208.14
1476	27MY	22;09	236.04
1477	26JN	5;18	263.96
1478	25JL	13;36	292.17
1479	24AU	0;05	320.89
1480	22SE	13;38	350.26
1481	22OC	6;32	20.24
1482	21NO	1;55	50.62
1483	20DE	21;51	81.06

-880

NUMBER	DATE	TIME	LONG.	NUMBER	DATE	TIME	LONG.
1484	4JA	10;30	275.81	1484	19JA	16;18	111.18
1485	2FE	20;55	305.40	1485	18FE	8;04	140.75
1486	3MR	7;41	334.50	1486	18MR	20;55	169.67
1487	1AP	19;21	3.15	1487	17AP	7;08	198.03
1488	1MY	8;16	31.45	1488	16MY	15;22	226.02
1489	30MY	22;24	59.62	1489	14JN	22;25	253.91
1490	29JN	13;25	87.87	1490	14JL	5;20	281.95
1491	29JL	4;56	116.43	1491	12AU	13;13	310.40
1492	27AU	20;34	145.44	1492	10SE	23;08	339.42
1493	26SE	11;51	174.95	1493	10OC	11;49	9.06
1494	26OC	2;10	204.89	1494	9NO	3;23	39.20
1495	24NO	15;01	235.02	1495	8DE	21;12	69.58
1496	24DE	2;18	265.07				

-879

NUMBER	DATE	TIME	LONG.	NUMBER	DATE	TIME	LONG.
1497	22JA	12;18	294.79	1496	7JA	16;00	99.84
1498	20FE	21;34	324.02	1497	6FE	10;25	129.70
1499	22MR	6;42	352.74	1498	8MR	3;12	158.97
1500	20AP	16;17	21.03	1499	6AP	17;30	187.65
1501	20MY	2;54	49.11	1500	6MY	5;04	215.86
1502	18JN	15;04	77.20	1501	4JN	14;20	243.83
1503	18JL	5;14	105.57	1502	3JL	22;12	271.83
1504	16AU	21;29	134.42	1503	2AU	5;50	300.11
1505	15SE	15;11	163.84	1504	31AU	14;15	328.89
1506	15OC	9;01	193.76	1505	30SE	0;11	358.24
1507	14NO	1;31	223.98	1506	29OC	11;58	28.11
1508	13DE	15;48	254.20	1507	28NO	1;40	58.29
				1508	27DE	17;12	88.50

-878

NUMBER	DATE	TIME	LONG.	NUMBER	DATE	TIME	LONG.
1509	12JA	3;50	284.13	1509	26JA	10;10	118.47
1510	10FE	13;59	313.58	1510	25FE	3;45	147.98
1511	11MR	22;46	342.46	1511	26MR	20;42	176.94
1512	10AP	6;43	10.83	1512	25AP	11;52	205.41
1513	9MY	14;31	38.87	1513	25MY	0;44	233.55
1514	7JN	23;05	66.80	1514	23JN	11;38	261.60
1515	7JL	9;30	94.92	1515	22JL	21;19	289.84
1516	5AU	22;42	123.48	1516	21AU	6;37	318.48
1517	4SE	14;57	152.64	1517	19SE	16;10	347.63
1518	4OC	9;31	182.42	1518	19OC	2;18	17.29
1519	3NO	4;46	212.65	1519	17NO	13;15	47.29
1520	2DE	22;56	243.03	1520	17DE	1;16	77.41

-877

NUMBER	DATE	TIME	LONG.	NUMBER	DATE	TIME	LONG.
1521	1JA	14;56	273.23	1521	15JA	14;40	107.36
1522	31JA	4;19	302.98	1522	14FE	5;28	136.95
1523	1MR	15;08	332.13	1523	15MR	21;10	166.06
1524	30MR	23;46	0.69	1524	14AP	12;57	194.69
1525	29AP	6;56	28.80	1525	14MY	4;06	222.98
1526	28MY	13;38	56.68	1526	12JN	18;14	251.14
1527	26JN	21;04	84.61	1527	12JL	7;23	279.41
1528	26JL	6;25	112.88	1528	10AU	19;39	308.00
1529	24AU	18;37	141.68	1529	9SE	7;10	337.04
1530	23SE	9;56	171.13	1530	8OC	18;05	6.57
1531	23OC	3;54	201.17	1531	7NO	4;39	36.47
1532	21NO	23;21	231.56	1532	6DE	15;14	66.54
1533	21DE	18;43	261.97				

-876

NUMBER	DATE	TIME	LONG.	NUMBER	DATE	TIME	LONG.
1534	20JA	12;24	292.05	1533	5JA	2;12	96.50
1535	19FE	3;14	321.56	1534	3FE	13;48	126.13
1536	19MR	14;52	350.42	1535	4MR	2;06	155.28
1537	17AP	23;44	18.72	1536	2AP	15;05	183.97
1538	17MY	6;53	46.67	1537	2MY	4;50	212.30
1539	15JN	13;31	74.54	1538	31MY	19;22	240.48
1540	14JL	20;47	102.60	1539	30JN	10;34	268.75
1541	13AU	5;38	131.10	1540	30JL	1;58	297.31
1542	11SE	16;45	160.18	1541	28AU	16;50	326.30
1543	11OC	6;32	189.88	1542	27SE	8;55	355.77
1544	9NO	23;06	220.07	1543	26OC	19;13	25.64
1545	9DE	17;55	250.48	1544	25NO	6;49	55.72
				1545	24DE	17;43	85.75

-875

NUMBER	DATE	TIME	LONG.	NUMBER	DATE	TIME	LONG.
1546	8JA	13;34	280.78	1546	23JA	4;04	115.48
1547	7FE	8;04	310.62	1547	21FE	13;59	144.73
1548	8MR	23;49	339.84	1548	22MR	23;47	173.46
1549	7AP	12;22	8.44	1549	21AP	10;07	201.78
1550	6MY	22;14	36.57	1550	20MY	21;47	229.88
1551	5JN	6;24	64.50	1551	19JN	11;21	258.04
1552	4JL	13;55	92.48	1552	19JL	2;50	286.47
1553	2AU	21;39	120.78	1553	17AU	19;33	315.35
1554	1SE	6;18	149.58	1554	16SE	12;26	344.75
1555	30SE	16;32	178.96	1555	16OC	4;33	14.61
1556	30OC	4;59	208.86	1556	14NO	19;22	44.77
1557	28NO	20;03	239.10	1557	14DE	8;37	74.94
1558	28DE	13;25	269.38				

-874

NUMBER	DATE	TIME	LONG.	NUMBER	DATE	TIME	LONG.
				1558	12JA	20;08	104.84
1559	27JA	7;53	299.40	1559	11FE	5;58	134.27
1560	26FE	1;46	328.91	1560	12MR	14;29	163.13
1561	27MR	17;49	357.82	1561	10AP	22;27	191.49
1562	26AP	7;34	26.22	1562	10MY	6;55	219.55
1563	25MY	19;14	54.31	1563	8JN	16;53	247.54
1564	24JN	5;14	82.33	1564	8JL	5;06	275.74
1565	23JL	14;11	110.55	1565	6AU	19;47	304.37
1566	21AU	22;45	139.17	1566	5SE	12;37	333.57
1567	20SE	7;41	168.31	1567	5OC	6;47	3.34
1568	19OC	17;43	197.97	1568	4NO	1;10	33.54
1569	18NO	5;25	228.01	1569	3DE	18;25	63.88
1570	17DE	18;57	258.19				

-873

NUMBER	DATE	TIME	LONG.	NUMBER	DATE	TIME	LONG.
				1570	2JA	9;23	94.04
1571	16JA	10;02	288.21	1571	31JA	21;31	123.72
1572	15FE	2;01	317.83	1572	2MR	7;07	152.82
1573	16MR	18;13	346.94	1573	31MR	14;59	181.34
1574	15AP	10;03	15.56	1574	29AP	22;07	209.43
1575	15MY	1;01	43.84	1575	29MY	5;33	237.34
1576	13JN	14;38	71.98	1576	27JN	14;10	265.33
1577	13JL	2;42	100.21	1577	27JL	0;48	293.65
1578	11AU	13;28	128.75	1578	25AU	14;03	322.51
1579	9SE	23;30	157.75	1579	24SE	6;15	352.01
1580	9OC	9;32	187.25	1580	24OC	1;01	22.09
1581	7NO	20;05	217.16	1581	22NO	20;57	52.51
1582	7DE	7;18	247.25	1582	22DE	15;54	82.90

-872

NUMBER	DATE	TIME	LONG.	NUMBER	DATE	TIME	LONG.
1583	5JA	19;09	277.24	1583	21JA	8;08	112.91
1584	4FE	7;35	306.89	1584	19FE	21;05	142.33
1585	4MR	20;47	336.08	1585	20MR	7;12	171.11
1586	3AP	10;55	4.80	1586	18AP	15;22	199.38
1587	3MY	1;56	33.17	1587	17MY	22;28	227.32
1588	1JN	17;18	61.38	1588	16JN	5;24	255.20
1589	1JL	8;16	89.64	1589	15JL	13;03	283.28
1590	30JL	22;19	118.16	1590	13AU	22;24	311.81
1591	29AU	11;24	147.09	1591	12SE	10;25	340.95
1592	27SE	23;46	176.51	1592	12OC	1;45	10.72
1593	27OC	11;36	206.36	1593	10NO	20;06	40.99
1594	25NO	22;56	236.43	1594	10DE	16;01	71.45
1595	25DE	9;41	266.45				

-871

NUMBER	DATE	TIME	LONG.	NUMBER	DATE	TIME	LONG.
				1595	9JA	11;20	101.72
1596	23JA	19;58	296.17	1596	8FE	4;25	131.49
1597	22FE	6;13	325.42	1597	9MR	18;37	160.64
1598	23MR	17;05	354.18	1598	8AP	6;01	189.18
1599	22AP	5;05	22.56	1599	7MY	15;09	217.28
1600	21MY	18;27	50.73	1600	5JN	22;43	245.18
1601	20JN	9;02	78.93	1601	5JL	5;39	273.14
1602	20JL	0;28	107.37	1602	3AU	13;02	301.44
1603	18AU	16;22	136.23	1603	1SE	21;57	330.26
1604	17SE	8;12	165.60	1604	1OC	9;17	359.69
1605	16OC	23;21	195.44	1605	30OC	23;29	29.67
1606	15NO	13;10	225.55	1606	29NO	16;15	59.98
1607	15DE	1;16	255.67	1607	29DE	10;34	90.31

-870

NUMBER	DATE	TIME	LONG.	NUMBER	DATE	TIME	LONG.
1608	13JA	11;47	285.53	1608	28JA	5;09	120.32
1609	11FE	21;10	314.92	1609	26FE	22;42	149.79
1610	13MR	6;04	343.79	1610	28MR	14;13	178.67
1611	11AP	15;07	12.19	1611	27AP	3;07	207.03
1612	11MY	0;57	40.29	1612	26MY	13;31	235.06
1613	9JN	12;09	68.33	1613	24JN	22;04	263.04
1614	9JL	1;15	96.57	1614	24JL	5;51	291.21
1615	7AU	16;35	125.24	1615	22AU	13;58	319.82
1616	6SE	9;54	154.47	1616	20SE	23;14	348.98
1617	6OC	4;08	184.27	1617	20OC	10;11	18.69
1618	4NO	21;41	214.44	1618	18NO	22;55	48.78
1619	4DE	13;16	244.71	1619	18DE	13;26	79.00

-869

NUMBER	DATE	TIME	LONG.	NUMBER	DATE	TIME	LONG.
1620	3JA	2;24	274.78	1620	17JA	5;31	109.05
1621	1FE	13;20	304.41	1621	15FE	22;39	138.71
1622	2MR	22;35	333.48	1622	17MR	15;50	167.84
1623	1AP	6;42	2.00	1623	16AP	7;53	196.46
1624	30AP	14;19	30.11	1624	15MY	21;55	224.70
1625	29MY	22;15	58.04	1625	14JN	9;51	252.78
1626	28JN	7;34	86.05	1626	13JL	20;14	280.95
1627	27JL	19;18	114.43	1627	12AU	5;53	309.45
1628	26AU	10;08	143.36	1628	10SE	15;30	338.44
1629	25SE	3;52	172.94	1629	10OC	1;30	7.95
1630	24OC	23;10	203.06	1630	8NO	12;06	37.86
1631	23NO	18;12	233.44	1631	7DE	23;30	67.96
1632	23DE	11;26	263.75				

−868 / −862

New #	Date	Time	Long.	Full #	Date	Time	Long.	New #	Date	Time	Long.	Full #	Date	Time	Long.
				1632	6JA	12;01	97.98	1707	14JA	19;11	286.90	1707	30JA	0;05	122.16
1633	22JA	2;06	293.69	1633	5FE	1;51	127.68	1708	13FE	5;10	316.30	1708	28FE	15;39	151.53
1634	20FE	14;05	323.05	1634	5MR	16;52	156.92	1709	14MR	15;21	345.20	1709	30MR	4;20	180.27
1635	20MR	23;37	351.81	1635	4AP	8;27	185.69	1710	13AP	2;27	13.66	1710	28AP	14;30	208.50
1636	19AP	7;18	20.04	1636	3MY	23;51	214.07	1711	12MY	14;55	41.86	1711	27MY	22;46	236.44
1637	18MY	14;01	47.96	1637	2JN	14;31	242.26	1712	11JN	4;50	70.01	1712	26JN	5;59	264.37
1638	16JN	20;54	75.84	1638	2JL	4;19	270.48	1713	10JL	19;58	98.34	1713	25JL	13;08	292.53
1639	16JL	5;11	103.96	1639	31JL	17;17	298.96	1714	9AU	11;53	127.06	1714	23AU	21;18	321.16
1640	14AU	15;57	132.56	1640	30AU	5;27	327.87	1715	8SE	4;06	156.28	1715	22SE	7;27	350.39
1641	13SE	5;48	161.78	1641	28SE	16;55	357.26	1716	7OC	19;59	186.00	1716	21OC	20;16	20.19
1642	12OC	22;39	191.63	1642	28OC	3;48	27.07	1717	6NO	10;46	216.07	1717	20NO	11;49	50.41
1643	11NO	17;36	221.93	1643	26NO	14;23	57.12	1718	5DE	23;51	246.24	1718	20DE	5;23	80.75
1644	11DE	13;13	252.38	1644	26DE	1;03	87.13								

−867 / −861

New #	Date	Time	Long.	Full #	Date	Time	Long.	New #	Date	Time	Long.	Full #	Date	Time	Long.
1645	10JA	7;52	282.61	1645	24JA	12;07	116.87	1719	4JA	11;05	276.22	1719	18JA	23;48	110.88
1646	9FE	0;09	312.34	1646	22FE	23;47	146.17	1720	2FE	20;49	305.78	1720	17FE	17;48	140.54
1647	10MR	13;17	341.43	1647	24MR	12;07	174.98	1721	4MR	5;40	334.80	1721	19MR	10;17	169.62
1648	8AP	23;22	9.90	1648	23AP	1;13	203.39	1722	2AP	14;20	3.33	1722	18AP	0;28	198.15
1649	8MY	7;12	37.94	1649	22MY	15;12	231.59	1723	1MY	23;29	31.49	1723	17MY	12;07	226.28
1650	6JN	13;57	65.81	1650	21JN	6;07	259.80	1724	31MY	9;47	59.50	1724	15JN	21;36	254.26
1651	5JL	20;50	93.78	1651	20JL	21;40	288.25	1725	29JN	21;49	87.64	1725	15JL	5;48	282.35
1652	4AU	4;53	122.11	1652	19AU	13;10	317.11	1726	29JL	12;05	116.13	1726	13AU	13;49	310.81
1653	2SE	14;55	150.99	1653	18SE	3;53	346.45	1727	28AU	4;39	145.17	1727	11SE	22;37	339.79
1654	2OC	3;25	180.48	1654	17OC	17;20	16.22	1728	26SE	22;50	174.80	1728	11OC	8;49	9.32
1655	31OC	18;39	210.51	1655	16NO	5;33	46.28	1729	26OC	17;11	204.89	1729	9NO	20;42	39.31
1656	30NO	12;24	240.87	1656	15DE	16;47	76.36	1730	25NO	10;02	235.19	1730	9DE	10;16	69.50
1657	30DE	7;44	271.23					1731	25DE	0;27	265.38				

−866 / −860

New #	Date	Time	Long.	Full #	Date	Time	Long.	New #	Date	Time	Long.	Full #	Date	Time	Long.
				1657	14JA	3;18	106.21	1732	23JA	12;22	295.18	1731	8JA	1;24	99.61
1658	29JA	2;54	301.25	1658	12FE	13;13	135.62	1733	21FE	22;16	324.44	1732	6FE	17;48	129.39
1659	27FE	20;02	330.70	1659	13MR	22;46	164.50	1734	22MR	6;42	353.12	1733	7MR	10;50	158.69
1660	29MR	10;06	359.50	1660	12AP	8;29	192.92	1735	20AP	14;19	21.34	1734	6AP	3;25	187.47
1661	27AP	21;11	27.78	1661	11MY	19;10	221.05	1736	19MY	21;51	49.29	1735	5MY	18;28	215.82
1662	27MY	6;07	55.75	1662	10JN	7;36	249.14	1737	18JN	6;16	77.23	1736	4JN	7;30	243.94
1663	25JN	13;56	83.70	1663	9JL	22;11	277.44	1738	17JL	16;40	105.45	1737	3JL	18;46	272.07
1664	24JL	21;37	111.88	1664	8AU	14;34	306.16	1739	16AU	5;58	134.17	1738	2AU	4;56	300.46
1665	23AU	5;53	140.50	1665	7SE	7;46	335.41	1740	14SE	22;29	163.53	1739	31AU	14;45	329.30
1666	21SE	15;24	169.69	1666	7OC	0;39	5.16	1741	14OC	17;22	193.49	1740	30SE	0;45	358.65
1667	21OC	2;45	199.42	1667	5NO	16;24	35.26	1742	13NO	12;55	223.83	1741	29OC	11;09	28.46
1668	19NO	16;28	229.56	1668	5DE	6;34	65.48	1743	13DE	7;16	254.22	1742	27NO	22;08	58.52
1669	19DE	8;39	259.84									1743	27DE	9;55	88.58

−865 / −859

New #	Date	Time	Long.	Full #	Date	Time	Long.	New #	Date	Time	Long.	Full #	Date	Time	Long.
				1669	3JA	18;57	95.51	1744	11JA	23;15	284.32	1744	25JA	22;50	118.38
1670	18JA	2;32	289.96	1670	2FE	5;29	125.11	1745	10FE	12;29	313.90	1745	24FE	13;00	147.76
1671	16FE	20;40	319.64	1671	3MR	14;25	154.15	1746	11MR	23;05	342.86	1746	26MR	4;05	176.65
1672	18MR	13;33	348.75	1672	1AP	22;21	182.66	1747	10AP	7;30	11.25	1747	24AP	19;27	205.14
1673	17AP	4;22	17.31	1673	1MY	6;14	210.77	1748	9MY	14;29	39.25	1748	24MY	10;29	233.36
1674	16MY	17;04	45.49	1674	30MY	15;11	238.74	1749	7JN	21;06	67.11	1749	23JN	0;50	261.56
1675	15JN	3;57	73.53	1675	29JN	2;09	266.82	1750	7JL	4;33	95.11	1750	22JL	14;26	289.95
1676	14JL	13;33	101.67	1676	28JL	15;39	295.28	1751	5AU	14;01	123.51	1751	21AU	3;18	318.73
1677	12AU	22;26	130.15	1677	27AU	7;36	324.27	1752	4SE	2;23	152.51	1752	19SE	15;24	347.99
1678	11SE	7;19	159.13	1678	26SE	1;25	353.87	1753	3OC	17;53	182.15	1753	19OC	2;46	17.69
1679	10OC	16;53	188.62	1679	25OC	20;04	23.97	1754	2NO	11;59	212.32	1754	17NO	13;34	47.70
1680	9NO	3;45	218.56	1680	24NO	14;13	54.32	1755	2DE	7;29	242.76	1755	17DE	0;08	77.75
1681	8DE	16;19	248.71	1681	24DE	6;29	84.59								

−864 / −858

New #	Date	Time	Long.	Full #	Date	Time	Long.	New #	Date	Time	Long.	Full #	Date	Time	Long.
1682	7JA	6;29	278.79	1682	22JA	20;01	114.47	1756	1JA	2;47	273.10	1756	15JA	10;49	107.59
1683	5FE	21;50	308.54	1683	21FE	6;42	143.77	1757	30JA	20;19	303.03	1757	13FE	21;54	137.02
1684	6MR	13;42	337.80	1684	21MR	15;10	172.46	1758	1MR	10;58	332.35	1758	15MR	9;36	165.97
1685	5AP	5;33	6.57	1685	19AP	22;24	200.67	1759	30MR	22;26	1.02	1759	13AP	22;01	194.48
1686	4MY	20;54	34.94	1686	19MY	5;27	228.61	1760	29AP	7;13	29.19	1760	13MY	11;21	222.70
1687	3JN	11;18	63.11	1687	17JN	13;19	256.53	1761	28MY	14;22	57.09	1761	12JN	1;45	250.87
1688	3JL	0;20	91.30	1688	16JL	22;52	284.70	1762	26JN	21;06	85.00	1762	11JL	17;08	279.22
1689	1AU	11;58	119.75	1689	15AU	10;48	313.36	1763	26JL	4;33	113.19	1763	10AU	9;01	307.95
1690	30AU	22;33	148.60	1690	14SE	1;37	342.63	1764	24AU	13;38	141.86	1764	9SE	0;31	337.15
1691	29SE	8;45	177.95	1691	13OC	19;20	12.53	1765	23SE	0;58	171.15	1765	8OC	14;56	6.82
1692	28OC	19;09	207.75	1692	12NO	15;01	42.87	1766	22OC	14;52	201.00	1766	7NO	3;57	36.83
1693	27NO	6;05	237.81	1693	12DE	10;47	73.32	1767	21NO	7;24	231.27	1767	6DE	15;43	66.94
1694	26DE	17;31	267.86					1768	21DE	2;00	261.65				

−863 / −857

New #	Date	Time	Long.	Full #	Date	Time	Long.	New #	Date	Time	Long.	Full #	Date	Time	Long.
				1694	11JA	4;30	103.51					1768	5JA	2;31	96.90
1695	25JA	5;26	297.62	1695	9FE	19;00	133.15	1769	19JA	21;21	291.81	1769	3FE	12;33	126.46
1696	23FE	17;55	326.94	1696	11MR	6;21	162.15	1770	18FE	15;32	321.47	1770	4MR	22;00	155.50
1697	25MR	7;15	355.78	1697	9AP	15;17	190.57	1771	20MR	7;04	350.49	1771	3AP	7;18	184.04
1698	23AP	21;37	24.24	1698	8MY	22;45	218.59	1772	18AP	19;32	18.94	1772	2MY	17;10	212.22
1699	23MY	12;47	52.47	1699	7JN	5;39	246.46	1773	18MY	5;27	47.00	1773	1JN	4;29	240.28
1700	22JN	4;06	80.70	1700	6JL	12;53	274.45	1774	16JN	13;48	74.93	1774	30JN	17;57	268.47
1701	21JL	18;52	109.13	1701	4AU	21;20	302.81	1775	15JL	21;35	103.02	1775	30JL	9;36	297.03
1702	20AU	8;43	137.94	1702	3SE	8;00	331.72	1776	14AU	5;39	131.48	1776	29AU	2;45	326.11
1703	18SE	21;43	167.22	1703	2OC	21;43	1.27	1777	12SE	14;38	160.48	1777	27SE	20;12	355.72
1704	18OC	10;03	196.96	1704	1NO	14;43	31.39	1778	12OC	1;06	190.04	1778	27OC	12;51	25.75
1705	16NO	21;47	227.00	1705	1DE	10;04	61.82	1779	10NO	13;37	220.05	1779	26NO	4;01	55.99
1706	16DE	8;50	257.07	1706	31DE	5;51	92.19	1780	10DE	4;30	250.30	1780	25DE	17;21	86.12

--------NEW MOONS--------				--------FULL MOONS-------			
NUMBER	DATE	TIME	LONG.	NUMBER	DATE	TIME	LONG.
-856							
1781	8JA	21;29	280.48	1781	24JA	4;44	115.89
1782	7FE	15;26	310.32	1782	22FE	14;15	145.12
1783	8MR	8;51	339.62	1783	22MR	22;23	173.78
1784	7AP	0;37	8.35	1784	21AP	5;59	201.99
1785	6MY	14;20	36.64	1785	20MY	14;07	229.96
1786	5JN	2;10	64.71	1786	18JN	23;53	257.96
1787	4JL	12;32	92.81	1787	18JL	12;05	286.26
1788	2AU	21;55	121.18	1788	17AU	2;56	315.05
1789	1SE	6;57	150.00	1789	15SE	20;05	344.46
1790	30SE	16;17	179.33	1790	15OC	14;40	14.42
1791	30OC	2;32	209.14	1791	14NO	9;25	44.73
1792	28NO	14;13	239.24	1792	14DE	2;52	75.08
1793	28DE	3;28	269.35				
-855							
				1793	12JA	17;50	105.13
1794	26JA	18;03	299.21	1794	11FE	5;48	134.65
1795	25FE	9;25	328.62	1795	12MR	15;09	163.54
1796	27MR	1;04	357.53	1796	10AP	22;44	191.89
1797	25AP	16;34	26.01	1797	10MY	5;39	219.88
1798	25MY	7;29	54.23	1798	8JN	12;56	247.76
1799	23JN	21;22	82.40	1799	7JL	21;32	275.82
1800	23JL	9;56	110.76	1800	6AU	8;15	304.28
1801	21AU	21;16	139.49	1801	4SE	21;40	333.33
1802	20SE	7;51	168.70	1802	4OC	14;05	3.03
1803	19OC	18;17	198.38	1803	3NO	9;04	33.24
1804	18NO	5;01	228.38	1804	3DE	5;07	63.71
1805	17DE	16;10	258.45				
-854							
				1805	2JA	0;04	94.03
1806	16JA	3;40	288.32	1806	31JA	16;11	123.89
1807	14FE	15;33	317.78	1807	2MR	4;57	153.11
1808	16MR	4;07	346.75	1808	31MR	14;54	181.72
1809	14AP	17;41	15.29	1809	29AP	22;55	209.85
1810	14MY	8;20	43.56	1810	29MY	5;58	237.74
1811	12JN	23;40	71.77	1811	27JN	12;58	265.66
1812	12JL	14;56	100.12	1812	26JL	20;46	293.87
1813	11AU	5;32	128.81	1813	25AU	6;18	322.58
1814	9SE	19;15	157.96	1814	23SE	18;30	351.91
1815	9OC	8;11	187.58	1815	23OC	9;57	21.84
1816	7NO	20;25	217.56	1816	22NO	4;18	52.19
1817	7DE	7;52	247.66	1817	22DE	0;04	82.61
-853							
1818	5JA	18;28	277.60	1818	20JA	19;10	112.75
1819	4FE	4;23	307.14	1819	19FE	12;00	142.34
1820	5MR	14;08	336.19	1820	21MR	2;01	171.29
1821	4AP	0;25	4.76	1821	19AP	13;20	199.68
1822	3MY	11;35	33.00	1822	18MY	22;29	227.70
1823	2JN	0;59	61.12	1823	17JN	6;12	255.61
1824	1JL	15;32	89.36	1824	16JL	13;22	283.68
1825	31JL	7;15	117.94	1825	14AU	21;03	312.14
1826	29AU	23;39	147.00	1826	13SE	6;14	341.16
1827	28SE	16;06	176.58	1827	12OC	17;46	10.77
1828	28OC	7;48	206.59	1828	11NO	7;59	40.86
1829	26NO	21;55	236.77	1829	11DE	0;34	71.18
1830	26DE	10;05	266.86				
-852							
				1830	9JA	18;32	101.40
1831	24JA	20;24	296.57	1831	8FE	12;40	131.23
1832	23FE	5;26	325.77	1832	9MR	5;51	160.50
1833	23MR	13;53	354.44	1833	7AP	21;09	189.20
1834	21AP	22;30	22.68	1834	7MY	10;04	217.46
1835	21MY	7;58	50.70	1835	5JN	20;38	245.48
1836	19JN	18;59	78.75	1836	5JL	5;31	273.52
1837	19JL	8;05	107.09	1837	3AU	13;42	301.85
1838	17AU	23;39	135.93	1838	1SE	22;12	330.65
1839	16SE	17;23	165.37	1839	1OC	7;49	0.01
1840	16OC	12;06	195.35	1840	30OC	18;56	29.86
1841	15NO	6;04	225.64	1841	29NO	7;38	60.00
1842	14DE	21;51	255.92	1842	28DE	21;50	90.15
-851							
1843	13JA	10;58	285.88	1843	27JA	13;23	120.04
1844	11FE	21;41	315.33	1844	26FE	5;56	149.49
1845	13MR	6;38	344.20	1845	27MR	22;33	178.43
1846	11AP	14;25	12.55	1846	26AP	14;27	206.91
1847	10MY	21;46	40.55	1847	26MY	4;33	235.09
1848	9JN	5;32	68.46	1848	24JN	16;47	263.21
1849	8JL	14;47	96.54	1849	24JL	3;39	291.51
1850	7AU	2;35	125.06	1850	22AU	13;49	320.20
1851	5SE	17;36	154.19	1851	20SE	23;56	349.40
1852	5OC	11;37	183.96	1852	20OC	10;17	19.08
1853	4NO	7;14	214.21	1853	18NO	21;01	49.09
1854	4DE	2;27	244.64	1854	18DE	8;17	79.17

--------NEW MOONS--------				--------FULL MOONS-------			
NUMBER	DATE	TIME	LONG.	NUMBER	DATE	TIME	LONG.
-850							
1855	2JA	19;43	274.89	1855	16JA	20;24	109.05
1856	1FE	10;17	304.67	1856	15FE	9;40	138.56
1857	2MR	22;04	333.84	1857	17MR	0;02	167.59
1858	1AP	7;24	2.41	1858	15AP	15;05	196.18
1859	30AP	14;54	30.51	1859	15MY	6;14	224.46
1860	29MY	21;32	58.39	1860	13JN	20;59	252.65
1861	28JN	4;25	86.31	1861	13JL	11;08	280.97
1862	27JL	12;47	114.55	1862	12AU	0;40	309.62
1863	25AU	23;42	143.32	1863	10SE	13;28	338.74
1864	24SE	13;44	172.74	1864	10OC	1;27	8.33
1865	24OC	6;44	202.75	1865	8NO	12;40	38.28
1866	23NO	1;44	233.13	1866	7DE	23;20	68.34
1867	22DE	21;17	263.54				
-849							
				1867	6JA	9;48	98.27
1868	21JA	15;48	293.64	1868	4FE	20;27	127.84
1869	20FE	7;53	323.18	1869	6MR	7;32	156.93
1870	21MR	20;51	352.08	1870	4AP	19;16	185.54
1871	20AP	6;50	20.40	1871	4MY	7;53	213.82
1872	19MY	14;39	48.37	1872	2JN	21;37	241.97
1873	17JN	21;29	76.25	1873	2JL	12;35	270.23
1874	17JL	4;33	104.32	1874	1AU	4;30	298.83
1875	15AU	12;51	132.81	1875	30AU	20;36	327.89
1876	13SE	23;06	161.89	1876	29SE	11;57	357.44
1877	13OC	11;46	191.56	1877	29OC	1;54	27.38
1878	12NO	3;01	221.70	1878	27NO	14;24	57.50
1879	11DE	20;37	252.06	1879	27DE	1;38	87.54
-848							
1880	10JA	15;38	282.32	1880	25JA	11;55	117.25
1881	9FE	10;26	312.16	1881	23FE	21;25	146.46
1882	10MR	3;17	341.40	1882	24MR	6;29	175.14
1883	8AP	17;12	10.04	1883	22AP	15;43	203.40
1884	8MY	4;18	38.21	1884	22MY	2;01	231.45
1885	6JN	13;24	66.17	1885	20JN	14;16	259.55
1886	5JL	21;29	94.19	1886	20JL	4;54	287.96
1887	4AU	5;30	122.52	1887	18AU	21;36	316.86
1888	2SE	14;08	151.34	1888	17SE	15;19	346.31
1889	1OC	23;56	180.72	1889	17OC	8;44	16.25
1890	31OC	11;25	210.59	1890	16NO	0;54	46.47
1891	30NO	1;03	240.77	1891	15DE	15;16	76.68
1892	29DE	16;54	270.99				
-847							
				1892	14JA	3;36	106.61
1893	28JA	10;17	300.94	1893	12FE	13;53	136.03
1894	27FE	3;53	330.42	1894	13MR	22;27	164.87
1895	28MR	20;23	359.33	1895	12AP	6;01	193.20
1896	27AP	11;04	27.76	1896	11MY	13;35	221.21
1897	26MY	23;52	55.89	1897	9JN	22;18	249.15
1898	25JN	11;04	83.97	1898	9JL	9;11	277.30
1899	24JL	21;06	112.24	1899	7AU	22;46	305.91
1900	23AU	6;29	140.91	1900	6SE	14;59	335.10
1901	21SE	15;48	170.09	1901	6OC	9;10	4.89
1902	21OC	1;39	199.76	1902	5NO	4;11	35.13
1903	19NO	12;37	229.78	1903	4DE	22;34	65.52
1904	19DE	1;00	259.90				
-846							
				1904	3JA	14;55	95.73
1905	17JA	14;44	289.85	1905	2FE	4;19	125.45
1906	16FE	5;28	319.40	1906	3MR	14;54	154.55
1907	17MR	20;44	348.45	1907	1AP	23;00	183.07
1908	16AP	12;08	17.05	1908	1MY	6;00	211.14
1909	16MY	3;19	45.33	1909	30MY	12;54	239.03
1910	14JN	17;52	73.51	1910	28JN	20;41	266.99
1911	14JL	7;20	101.80	1911	28JL	6;20	295.28
1912	12AU	19;32	130.42	1912	26AU	18;25	324.12
1913	11SE	6;42	159.48	1913	25SE	9;26	353.59
1914	10OC	17;23	189.03	1914	25OC	3;20	23.65
1915	9NO	4;03	218.96	1915	23NO	23;09	54.07
1916	8DE	15;00	249.04	1916	23DE	18;55	84.48
-845							
1917	7JA	2;12	278.99	1917	22JA	12;32	114.54
1918	5FE	13;38	308.58	1918	21FE	7;53	143.99
1919	7MR	1;31	337.69	1919	22MR	14;04	172.80
1920	5AP	14;14	6.33	1920	20AP	22;51	201.07
1921	5MY	4;09	34.66	1921	20MY	6;15	229.02
1922	3JN	19;08	62.85	1922	18JN	13;12	256.90
1923	3JL	10;37	91.14	1923	17JL	20;33	284.99
1924	2AU	1;50	119.72	1924	16AU	5;12	313.51
1925	31AU	16;19	148.73	1925	14SE	16;04	342.62
1926	30SE	5;56	178.22	1926	14OC	5;56	12.35
1927	29OC	18;44	208.12	1927	12NO	22;58	42.58
1928	28NO	6;42	238.23	1928	12DE	18;12	73.01
1929	27DE	17;41	268.25				

−844 (left) / −838 (right)

NEW #	Date	Time	Long.	FULL #	Date	Time	Long.	NEW #	Date	Time	Long.	FULL #	Date	Time	Long.
				1929	11JA	13:45	103.27	2004	19JA	5:26	291.53	2003	5JA	2:00	97.26
1930	26JA	3:46	297.94	1930	10FE	7:43	133.07	2005	17FE	22:49	321.16	2004	3FE	13:12	126.87
1931	24FE	13:17	327.14	1931	10MR	23:03	162.23	2006	19MR	15:47	350.26	2005	4MR	22:24	155.90
1932	24MR	22:56	355.83	1932	9AP	11:36	190.80	2007	18AP	7:18	18.83	2006	3AP	6:11	184.38
1933	23AP	9:30	24.14	1933	8MY	21:46	218.94	2008	17MY	21:01	47.04	2007	2MY	13:27	212.46
1934	22MY	21:35	52.25	1934	7JN	6:10	246.87	2009	16JN	9:06	75.13	2008	31MY	21:21	240.38
1935	21JN	11:21	80.42	1935	6JL	13:36	274.86	2010	15JL	19:53	103.33	2009	30JN	7:00	268.42
1936	21JL	2:37	108.86	1936	4AU	21:03	303.18	2011	14AU	5:46	131.87	2010	29JL	19:13	296.84
1937	19AU	18:58	137.76	1937	3SE	5:31	332.00	2012	12SE	15:17	160.90	2011	28AU	10:15	325.82
1938	18SE	11:46	167.19	1938	2OC	15:55	1.42	2013	12OC	0:59	190.42	2012	27SE	3:43	355.42
1939	18OC	4:14	197.10	1939	1NO	4:50	31.36	2014	10NO	11:25	220.35	2013	26OC	22:40	25.54
1940	16NO	19:25	227.28	1940	30NO	20:15	61.62	2015	9DE	23:01	250.45	2014	25NO	17:41	55.93
1941	16DE	8:38	257.45	1941	30DE	13:30	91.89					2015	25DE	11:15	86.25

−843 (left) / −837 (right)

NEW #	Date	Time	Long.	FULL #	Date	Time	Long.	NEW #	Date	Time	Long.	FULL #	Date	Time	Long.
1942	14JA	19:46	287.32	1942	29JA	7:28	121.85	2016	8JA	11:56	280.47	2016	24JA	2:09	116.17
1943	13FE	5:13	316.69	1943	28FE	1:02	151.31	2017	7FE	1:57	310.14	2017	22FE	13:56	145.49
1944	14MR	13:40	345.52	1944	29MR	17:13	180.20	2018	8MR	16:41	339.34	2018	23MR	23:02	174.19
1945	12AP	21:54	13.86	1945	28AP	7:19	208.59	2019	7AP	7:47	8.06	2019	22AP	6:24	202.39
1946	12MY	6:41	41.92	1946	27MY	19:06	236.68	2020	6MY	22:59	36.42	2020	21MY	13:09	230.31
1947	10JN	16:43	69.91	1947	26JN	4:53	264.71	2021	5JN	13:56	64.61	2021	19JN	20:22	258.21
1948	10JL	4:41	98.12	1948	25JL	13:29	292.93	2022	5JL	4:09	92.86	2022	19JL	5:01	286.36
1949	8AU	19:05	126.76	1949	23AU	21:55	321.58	2023	3AU	17:15	121.37	2023	17AU	15:51	314.99
1950	7SE	11:59	156.00	1950	22SE	7:06	350.76	2024	2SE	5:12	150.31	2024	16SE	5:27	344.23
1951	7OC	6:38	185.82	1951	21OC	17:34	20.46	2025	1OC	16:19	179.72	2025	15OC	22:03	14.10
1952	6NO	1:24	216.06	1952	20NO	5:29	50.53	2026	31OC	3:07	209.55	2026	14NO	17:10	44.43
1953	5DE	18:32	246.40	1953	19DE	18:49	80.69	2027	29NO	13:58	239.61	2027	14DE	13:15	74.89
								2028	29DE	0:58	269.63				

−842 (left) / −836 (right)

NEW #	Date	Time	Long.	FULL #	Date	Time	Long.	NEW #	Date	Time	Long.	FULL #	Date	Time	Long.
1954	4JA	9:00	276.52	1954	18JA	9:29	110.67					2028	13JA	8:06	105.11
1955	2FE	20:46	306.16	1955	17FE	1:19	140.25	2029	27JA	12:04	299.35	2029	12FE	0:04	134.79
1956	4MR	6:24	335.22	1956	18MR	17:47	169.34	2030	25FE	23:24	328.59	2030	12MR	12:40	163.82
1957	2AP	14:32	3.72	1957	17AP	9:59	197.95	2031	26MR	11:21	357.36	2031	10AP	22:28	192.26
1958	1MY	21:52	31.81	1958	17MY	0:59	226.22	2032	25AP	0:23	25.75	2032	10MY	6:24	220.29
1959	31MY	5:13	59.71	1959	15JN	14:15	254.35	2033	24MY	14:44	53.95	2033	8JN	13:29	248.17
1960	29JN	13:33	87.69	1960	15JL	1:57	282.58	2034	23JN	6:04	82.18	2034	7JL	20:36	276.16
1961	28JL	23:59	116.03	1961	13AU	12:39	311.15	2035	22JL	21:42	110.66	2035	6AU	4:36	304.52
1962	27AU	13:26	144.93	1962	11SE	23:00	340.19	2036	21AU	12:52	139.53	2036	4SE	14:21	333.42
1963	26SE	6:10	174.49	1963	11OC	9:25	9.74	2037	20SE	3:14	168.89	2037	4OC	2:44	2.94
1964	26OC	1:20	204.61	1964	9NO	20:03	39.67	2038	19OC	16:42	198.70	2038	2NO	18:15	33.00
1965	24NO	21:06	235.03	1965	9DE	7:00	69.75	2039	18NO	5:15	228.78	2039	2DE	12:31	63.39
1966	24DE	15:31	265.39					2040	17DE	16:46	258.87				

−841 (left) / −835 (right)

NEW #	Date	Time	Long.	FULL #	Date	Time	Long.	NEW #	Date	Time	Long.	FULL #	Date	Time	Long.
				1966	7JA	18:30	99.71					2040	1JA	8:03	93.74
1967	23JA	7:26	295.36	1967	6FE	6:54	129.33	2041	16JA	3:10	288.69	2041	31JA	2:50	123.72
1968	21FE	20:30	324.75	1968	7MR	20:25	158.49	2042	14FE	12:42	318.05	2042	1MR	19:25	153.11
1969	23MR	6:54	353.51	1969	6AP	10:54	187.19	2043	15MR	21:55	346.88	2043	31MR	9:15	181.87
1970	21AP	15:08	21.75	1970	6MY	1:54	215.55	2044	14AP	7:41	15.27	2044	29AP	20:33	210.14
1971	20MY	22:01	49.68	1971	4JN	16:51	243.74	2045	13MY	18:45	43.41	2045	29MY	5:47	238.12
1972	19JN	4:38	77.55	1972	4JL	7:27	272.00	2046	12JN	7:34	71.51	2046	27JN	13:43	266.08
1973	18JL	12:10	105.65	1973	2AU	21:34	300.55	2047	11JL	22:09	99.84	2047	26JL	21:11	294.27
1974	16AU	21:47	134.22	1974	1SE	11:03	329.53	2048	10AU	14:10	128.57	2048	25AU	5:11	322.92
1975	15SE	10:19	163.41	1975	30SE	23:45	359.00	2049	9SE	7:05	157.84	2049	23SE	14:40	352.13
1976	15OC	1:58	193.23	1976	30OC	11:33	28.86	2050	9OC	0:07	187.63	2050	23OC	2:21	21.91
1977	13NO	20:09	223.50	1977	28NO	22:31	58.93	2051	7NO	16:17	217.77	2051	21NO	16:32	52.07
1978	13DE	15:35	253.94	1978	28DE	8:59	88.93	2052	7DE	6:40	247.99	2052	21DE	8:51	82.36

−840 (left) / −834 (right)

NEW #	Date	Time	Long.	FULL #	Date	Time	Long.	NEW #	Date	Time	Long.	FULL #	Date	Time	Long.
1979	12JA	10:44	284.18	1979	26JA	19:19	118.62	2053	5JA	18:48	278.00	2053	20JA	2:22	112.43
1980	11FE	4:03	313.94	1980	25FE	5:54	147.85	2054	4FE	4:54	307.55	2054	18FE	20:01	142.06
1981	11MR	18:31	343.06	1981	25MR	17:00	176.59	2055	5MR	13:34	336.55	2055	20MR	12:49	171.14
1982	10AP	5:52	11.57	1982	24AP	4:53	204.94	2056	3AP	21:37	5.03	2056	19AP	3:56	199.68
1983	9MY	14:38	39.64	1983	23MY	17:52	233.09	2057	3MY	5:50	33.14	2057	18MY	16:55	227.86
1984	7JN	21:51	67.52	1984	22JN	8:11	261.28	2058	1JN	15:02	61.11	2058	17JN	3:46	255.90
1985	7JL	4:45	95.50	1985	21JL	23:49	289.74	2059	1JL	1:54	89.20	2059	16JL	13:02	284.05
1986	5AU	12:27	123.83	1986	20AU	16:11	318.66	2060	30JL	15:00	117.67	2060	14AU	21:39	312.55
1987	3SE	21:47	152.70	1987	19SE	8:20	348.08	2061	29AU	6:53	146.69	2061	13SE	6:35	341.56
1988	3OC	9:19	182.18	1988	18OC	23:19	17.94	2062	28SE	1:00	176.34	2062	12OC	16:31	11.10
1989	1NO	23:18	212.17	1989	17NO	12:42	48.04	2063	27OC	20:10	206.48	2063	11NO	3:45	41.07
1990	1DE	15:43	242.47	1990	17DE	0:35	78.15	2064	26NO	14:27	236.84	2064	10DE	16:19	71.22
1991	31DE	10:02	272.78					2065	26DE	6:22	267.09				

−839 (left) / −833 (right)

NEW #	Date	Time	Long.	FULL #	Date	Time	Long.	NEW #	Date	Time	Long.	FULL #	Date	Time	Long.
				1991	15JA	11:14	107.99					2065	9JA	6:09	101.26
1992	30JA	4:59	302.78	1992	13FE	20:54	137.37	2066	24JA	19:24	296.92	2066	7FE	21:08	130.97
1993	28FE	22:49	332.23	1993	15MR	5:54	166.20	2067	23FE	5:54	326.18	2067	9MR	13:04	160.20
1994	30MR	14:09	1.07	1994	13AP	14:44	194.57	2068	24MR	14:33	354.85	2068	8AP	5:18	188.95
1995	29AP	2:35	29.39	1995	13MY	0:11	222.64	2069	22AP	22:04	23.05	2069	7MY	20:55	217.32
1996	28MY	12:37	57.41	1996	11JN	11:15	250.68	2070	22MY	5:12	50.97	2070	6JN	11:08	245.44
1997	26JN	21:13	85.39	1997	11JL	0:39	278.95	2071	20JN	12:52	78.89	2071	5JL	23:46	273.67
1998	26JL	5:21	113.60	1998	9AU	16:31	307.66	2072	19JL	22:08	107.07	2072	4AU	11:08	302.13
1999	24AU	13:48	142.26	1999	8SE	10:07	336.94	2073	18AU	10:02	135.76	2073	2SE	21:53	331.03
2000	22SE	23:07	171.45	2000	8OC	4:06	6.76	2074	17SE	1:16	165.09	2074	2OC	8:28	0.43
2001	22OC	9:47	201.18	2001	6NO	21:13	36.93	2075	16OC	19:31	195.04	2075	31OC	19:08	30.26
2002	20NO	22:18	231.27	2002	6DE	12:39	67.20	2076	15NO	15:20	225.40	2076	30NO	5:57	60.32
2003	20DE	12:56	261.48					2077	15DE	10:40	255.83	2077	29DE	17:01	90.34

-832

NEW MOONS				FULL MOONS			
2078	14JA	3;54	285.97	2078	28JA	4;42	120.07
2079	12FE	18;19	315.58	2079	26FE	17;20	149.36
2080	13MR	5;54	344.55	2080	27MR	7;04	178.18
2081	11AP	15;03	12.95	2081	25AP	21;39	206.62
2082	10MY	22;27	40.96	2082	25MY	12;35	234.84
2083	9JN	5;03	68.82	2083	24JN	3;28	263.06
2084	8JL	12;01	96.81	2084	23JL	18;02	291.51
2085	6AU	20;32	125.20	2085	22AU	8;09	320.35
2086	5SE	7;38	154.16	2086	20SE	21;35	349.69
2087	4OC	21;50	183.77	2087	20OC	10;05	19.46
2088	3NO	14;54	213.91	2088	18NO	21;34	49.50
2089	3DE	9;52	244.32	2089	18DE	8;14	79.55

-831

NEW MOONS				FULL MOONS			
2090	2JA	5;17	274.66	2090	16JA	18;28	109.36
2091	31JA	23;33	304.61	2091	15FE	4;39	138.73
2092	2MR	15;26	333.95	2092	16MR	15;10	167.61
2093	1AP	4;15	2.67	2093	15AP	2;21	196.05
2094	30AP	14;11	30.87	2094	14MY	14;33	224.22
2095	29MY	22;04	58.79	2095	13JN	4;05	252.36
2096	28JN	5;04	86.72	2096	12JL	19;09	280.71
2097	27JL	12;22	114.91	2097	11AU	11;28	309.47
2098	25AU	20;56	143.59	2098	10SE	4;09	338.74
2099	24SE	7;27	172.86	2099	9OC	20;07	8.50
2100	23OC	20;15	202.69	2100	8NO	10;32	38.57
2101	22NO	11;26	232.91	2101	7DE	23;14	68.73
2102	22DE	4;47	263.23				

-830

NEW MOONS				FULL MOONS			
				2102	6JA	10;25	98.68
2103	20JA	23;24	293.34	2103	4FE	20;24	128.22
2104	19FE	17;48	322.99	2104	6MR	5;30	157.23
2105	21MR	10;21	352.04	2105	4AP	14;06	185.72
2106	20AP	0;11	20.52	2106	3MY	22;55	213.85
2107	19MY	11;21	48.63	2107	2JN	8;55	241.85
2108	17JN	20;40	76.60	2108	1JL	21;01	269.99
2109	17JL	5;06	104.73	2109	31JL	11;45	298.53
2110	15AU	13;31	133.23	2110	30AU	4;48	327.62
2111	13SE	22;31	162.25	2111	28SE	23;00	357.29
2112	13OC	8;36	191.81	2112	28OC	16;55	27.39
2113	11NO	20;10	221.80	2113	27NO	9;26	57.68
2114	11DE	9;38	251.98	2114	26DE	23;54	87.86

-829

NEW MOONS				FULL MOONS			
2115	10JA	1;04	282.09	2115	25JA	12;07	117.64
2116	8FE	17;53	311.86	2116	23FE	22;08	146.87
2117	10MR	10;57	341.12	2117	25MR	6;22	175.52
2118	9AP	3;05	9.86	2118	23AP	13;36	203.70
2119	8MY	17;40	38.17	2119	22MY	20;55	231.63
2120	7JN	6;38	66.29	2120	21JN	5;29	259.58
2121	6JL	18;12	94.44	2121	20JL	16;21	287.84
2122	5AU	4;44	122.87	2122	19AU	6;03	316.61
2123	3SE	14;38	151.75	2123	17SE	22;32	346.00
2124	3OC	0;23	181.12	2124	17OC	17;03	15.97
2125	1NO	10;31	210.94	2125	16NO	12;21	46.32
2126	30NO	21;30	241.01	2126	16DE	6;54	76.71
2127	30DE	9;37	271.07				

-828

NEW MOONS				FULL MOONS			
				2127	14JA	23;13	106.81
2128	28JA	22;52	300.85	2128	13FE	12;28	136.35
2129	27FE	12;59	330.19	2129	13MR	22;43	165.26
2130	28MR	3;38	359.04	2130	12AP	6;43	193.61
2131	26AP	18;38	27.49	2131	11MY	13;33	221.59
2132	26MY	9;42	55.71	2132	9JN	20;22	249.46
2133	25JN	0;27	83.93	2133	9JL	4;13	277.49
2134	24JL	14;24	112.36	2134	7AU	13;57	305.93
2135	23AU	3;12	141.16	2135	6SE	2;12	334.96
2136	21SE	14;58	170.44	2136	5OC	17;23	4.62
2137	21OC	2;05	200.17	2137	4NO	11;25	34.81
2138	19NO	13;00	230.18	2138	4DE	7;17	65.26
2139	18DE	23;53	260.24				

-827

NEW MOONS				FULL MOONS			
				2139	3JA	2;58	95.61
2140	17JA	10;47	290.07	2140	1FE	20;26	125.50
2141	15FE	21;43	319.47	2141	3MR	10;36	154.76
2142	17MR	8;59	348.36	2142	1AP	21;37	183.39
2143	15AP	21;09	16.83	2143	1MY	6;20	211.54
2144	15MY	10;39	45.05	2144	30MY	13;44	239.45
2145	14JN	1;31	73.24	2145	28JN	20;47	267.37
2146	13JL	17;12	101.62	2146	28JL	4;21	295.59
2147	12AU	8;56	130.37	2147	26AU	13;14	324.29
2148	11SE	0;02	159.60	2148	25SE	0;19	353.60
2149	10OC	14;15	189.29	2149	24OC	14;17	23.48
2150	9NO	3;28	219.32	2150	23NO	7;15	53.78
2151	8DE	15;35	249.45	2151	23DE	2;17	84.17

-826

NEW MOONS				FULL MOONS			
2152	7JA	2;29	279.39	2152	21JA	21;3L	114.30
2153	5FE	12;14	308.91	2153	20FE	15;09	143.90
2154	6MR	21;17	337.90	2154	22MR	6;16	172.87
2155	5AP	6;25	6.40	2155	20AP	18;45	201.29
2156	4MY	16;31	34.58	2156	20MY	4;58	229.35
2157	3JN	4;16	62.65	2157	18JN	13;34	257.30
2158	2JL	17;58	90.86	2158	17JL	21;18	285.41
2159	1AU	9;25	119.44	2159	16AU	5;06	313.89
2160	31AU	2;12	148.53	2160	14SE	13;53	342.92
2161	29SE	19;34	178.18	2161	14OC	0;30	12.51
2162	29OC	12;33	208.25	2162	12NO	13;28	42.56
2163	28NO	4;04	238.50	2163	12DE	4;42	72.82
2164	27DE	17;22	268.62				

-825

NEW MOONS				FULL MOONS			
				2164	10JA	21;32	102.98
2165	26JA	4;21	298.35	2165	9FE	14;59	132.76
2166	24FE	13;29	327.53	2166	11MR	8;05	162.01
2167	25MR	21;33	356.16	2167	10AP	0;00	190.72
2168	24AP	5;24	24.36	2168	9MY	14;04	219.01
2169	23MY	13;52	52.33	2169	8JN	2;03	247.09
2170	21JN	23;44	80.34	2170	7JL	12;12	275.19
2171	21JL	11;42	108.65	2171	5AU	21;15	303.57
2172	20AU	2;15	137.46	2172	4SE	6;09	332.42
2173	18SE	19;29	166.90	2173	3OC	15;43	1.80
2174	18OC	14;32	196.91	2174	2NO	2;23	31.64
2175	17NO	9;40	227.25	2175	1DE	14;17	61.75
2176	17DE	2;59	257.59	2176	31DE	3;20	91.84

-824

NEW MOONS				FULL MOONS			
2177	15JA	17;26	287.60	2177	29JA	17;29	121.66
2178	14FE	5;02	317.07	2178	28FE	8;40	151.03
2179	14MR	14;23	345.93	2179	29MR	0;35	179.92
2180	12AP	22;15	14.26	2180	27AP	16;28	208.39
2181	12MY	5;23	42.25	2181	27MY	7;28	236.60
2182	10JN	12;36	70.13	2182	25JN	21;02	264.78
2183	9JL	20;56	98.19	2183	25JL	9;12	293.14
2184	8AU	7;27	126.67	2184	23AU	20;29	321.90
2185	6SE	21;04	155.77	2185	22SE	7;22	351.16
2186	6OC	14;01	185.51	2186	21OC	18;11	20.88
2187	5NO	9;24	215.77	2187	20NO	5;00	50.89
2188	5DE	5;16	246.22	2188	19DE	15;51	80.95

-823

NEW MOONS				FULL MOONS			
2189	3JA	23;40	276.51	2189	18JA	2;59	110.78
2190	2FE	15;28	306.32	2190	16FE	14;50	140.20
2191	4MR	4;21	335.52	2191	18MR	3;43	169.15
2192	2AP	14;34	4.10	2192	16AP	17;38	197.68
2193	1MY	22;41	32.22	2193	16MY	8;18	225.94
2194	31MY	5;32	60.10	2194	14JN	23;15	254.14
2195	29JN	12;12	88.02	2195	14JL	14;09	282.49
2196	28JL	19;53	116.25	2196	13AU	4;48	311.21
2197	27AU	5;41	145.00	2197	11SE	18;56	340.41
2198	25SE	18;24	174.38	2198	11OC	8;12	10.07
2199	25OC	10;22	204.35	2199	9NO	20;22	40.07
2200	24NO	4;20	234.70	2200	9DE	7;27	70.15
2201	23DE	23;38	265.10				

-822

NEW MOONS				FULL MOONS			
				2201	7JA	17;45	100.06
2202	22JA	18;32	295.20	2202	6FE	3;43	129.58
2203	21FE	11;36	324.77	2203	7MR	13;46	158.60
2204	23MR	1;54	353.69	2204	6AP	0;18	187.15
2205	21AP	13;11	22.06	2205	5MY	11;43	215.37
2206	20MY	21;59	50.06	2206	4JN	0;24	243.47
2207	19JN	5;21	77.96	2207	3JL	14;42	271.72
2208	18JL	12;29	106.05	2208	2AU	6;37	300.33
2209	16AU	20;28	134.55	2209	31AU	23;31	329.44
2210	15SE	6;06	163.62	2210	30SE	16;16	359.07
2211	14OC	17;48	193.27	2211	30OC	7;47	29.09
2212	13NO	7;48	223.37	2212	28NO	21;28	59.27
2213	13DE	0;02	253.67	2213	28DE	9;23	89.33

-821

NEW MOONS				FULL MOONS			
2214	11JA	17;58	283.87	2214	26JA	19;49	119.02
2215	10FE	12;28	313.68	2215	25FE	5;08	148.20
2216	12MR	5;55	342.93	2216	26MR	13;42	176.84
2217	10AP	21;04	11.60	2217	24AP	22;06	205.05
2218	10MY	9;31	39.81	2218	24MY	7;13	233.05
2219	8JN	19;45	67.82	2219	22JN	18;05	261.10
2220	8JL	4;40	95.88	2220	22JL	7;30	289.47
2221	6AU	13;12	124.25	2221	20AU	23;36	318.37
2222	4SE	22;04	153.10	2222	19SE	17;37	347.85
2223	4OC	7;43	182.50	2223	19OC	12;06	17.85
2224	2NO	18;33	212.36	2224	18NO	5;36	48.13
2225	2DE	7;00	242.49	2225	17DE	21;13	78.40
2226	31DE	21;18	272.63				

-820

NEW NUMBER	DATE	TIME	LONG.	FULL NUMBER	DATE	TIME	LONG.
				2226	16JA	10;33	108.35
2227	30JA	13;16	302.51	2227	14FE	21;31	137.78
2228	29FE	6;04	331.93	2228	15MR	6;26	166.61
2229	29MR	22;33	0.83	2229	13AP	13;54	194.92
2230	28AP	13;51	29.27	2230	12MY	20;54	222.89
2231	28MY	3;39	57.43	2231	11JN	4;37	250.80
2232	26JN	16;03	85.57	2232	10JL	14;13	278.91
2233	26JL	3;18	113.91	2233	9AU	2;30	307.48
2234	24AU	13;44	142.64	2234	7SE	17;45	336.65
2235	22SE	23;44	171.87	2235	7OC	11;31	6.45
2236	22OC	9;47	201.56	2236	6NO	6;44	36.70
2237	20NO	20;20	231.57	2237	6DE	1;56	67.13
2238	20DE	7;47	261.65				

-819

NEW NUMBER	DATE	TIME	LONG.	FULL NUMBER	DATE	TIME	LONG.
				2238	4JA	19;32	97.37
2239	18JA	20;17	291.53	2239	3FE	10;19	127.14
2240	17FE	9;43	321.01	2240	4MR	21;54	156.26
2241	18MR	23;49	349.99	2241	3AP	6;48	184.79
2242	17AP	14;24	18.54	2242	2MY	13;59	212.86
2243	17MY	5;21	46.81	2243	31MY	20;39	240.73
2244	15JN	20;22	75.00	2244	30JN	3;53	268.68
2245	15JL	10;59	103.36	2245	29JL	12;38	296.95
2246	14AU	0;40	132.05	2246	27AU	23;38	325.76
2247	12SE	13;14	161.20	2247	26SE	13;25	355.21
2248	12OC	0;52	190.80	2248	26OC	6;09	25.23
2249	10NO	12;00	220.76	2249	25NO	1;18	55.62
2250	9DE	22;54	250.84	2250	24DE	21;18	86.05

-818

NEW NUMBER	DATE	TIME	LONG.	FULL NUMBER	DATE	TIME	LONG.
2251	8JA	9;42	280.76	2251	23JA	16;00	116.13
2252	6FE	20;22	310.30	2252	22FE	7;46	145.62
2253	8MR	7;08	339.34	2253	23MR	20;13	174.46
2254	6AP	18;29	7.91	2254	22AP	5;55	202.76
2255	6MY	7;02	36.17	2255	21MY	13;51	230.72
2256	4JN	21;08	64.33	2256	19JN	21;01	258.62
2257	4JL	12;33	92.62	2257	19JL	4;20	286.71
2258	3AU	4;34	121.25	2258	17AU	12;35	315.24
2259	1SE	20;20	150.33	2259	15SE	22;34	344.33
2260	1OC	11;19	179.90	2260	15OC	11;06	14.03
2261	31OC	1;18	209.86	2261	14NO	2;37	44.20
2262	29NO	14;06	240.01	2262	13DE	20;43	74.57
2263	29DE	1;36	270.04				

-817

NEW NUMBER	DATE	TIME	LONG.	FULL NUMBER	DATE	TIME	LONG.
				2263	12JA	15;55	104.82
2264	27JA	11;46	299.72	2264	11FE	10;21	134.61
2265	25FE	20;52	328.88	2265	13MR	2;39	163.80
2266	27MR	5;37	357.51	2266	11AP	16;21	192.40
2267	25AP	14;54	25.75	2267	11MY	3;39	220.57
2268	25MY	1;35	53.81	2268	9JN	13;04	248.54
2269	23JN	14;13	81.93	2269	8JL	21;16	276.58
2270	23JL	4;53	110.36	2270	7AU	5;06	304.93
2271	21AU	21;15	139.28	2271	5SE	13;28	333.77
2272	20SE	14;40	168.76	2272	4OC	23;13	3.17
2273	20OC	8;15	198.73	2273	3NO	11;02	33.08
2274	19NO	0;49	228.98	2274	3DE	1;07	63.29
2275	18DE	15;22	259.19				

-816

NEW NUMBER	DATE	TIME	LONG.	FULL NUMBER	DATE	TIME	LONG.
				2275	1JA	17;04	93.49
2276	17JA	3;26	289.09	2276	31JA	10;05	123.41
2277	15FE	13;16	318.46	2277	1MR	3;13	152.83
2278	15MR	21;35	347.25	2278	30MR	19;37	181.71
2279	14AP	5;15	15.56	2279	29AP	10;37	210.12
2280	13MY	13;10	43.57	2280	28MY	23;47	238.26
2281	11JN	22;09	71.52	2281	27JN	10;54	266.35
2282	11JL	8;57	99.69	2282	26JL	20;37	294.63
2283	9AU	22;13	128.31	2283	25AU	5;43	323.33
2284	8SE	14;17	157.53	2284	23SE	15;05	352.54
2285	8OC	8;47	187.37	2285	23OC	1;18	22.25
2286	7NO	4;18	217.65	2286	21NO	12;37	52.29
2287	6DE	22;49	248.05	2287	21DE	0;59	82.41

-815

NEW NUMBER	DATE	TIME	LONG.	FULL NUMBER	DATE	TIME	LONG.
2288	5JA	14;47	278.21	2288	19JA	14;22	112.32
2289	4FE	3;41	307.88	2289	18FE	4;45	141.82
2290	5MR	13;57	336.95	2290	19MR	20;03	170.84
2291	3AP	22;21	5.44	2291	18AP	11;51	199.43
2292	3MY	5;38	33.51	2292	18MY	3;19	227.71
2293	1JN	12;39	61.40	2293	16JN	17;44	255.88
2294	30JN	20;17	89.36	2294	16JL	6;47	284.18
2295	30JL	5;37	117.67	2295	14AU	18;44	312.82
2296	28AU	17;40	146.54	2296	13SE	5;43	341.92
2297	27SE	9;05	176.06	2297	12OC	17;07	11.52
2298	27OC	3;32	206.16	2298	11NO	4;03	41.47
2299	25NO	23;28	236.59	2299	10DE	14;52	71.54
2300	25DE	18;48	266.98				

-814

NEW NUMBER	DATE	TIME	LONG.	FULL NUMBER	DATE	TIME	LONG.
				2300	9JA	1;41	101.46
2301	24JA	11;55	296.99	2301	7FE	12;52	131.01
2302	23FE	2;10	326.41	2302	9MR	0;54	160.09
2303	24MR	13;35	355.19	2303	7AP	14;01	188.72
2304	22AP	22;36	23.45	2304	7MY	4;09	217.04
2305	22MY	5;57	51.39	2305	5JN	18;56	245.22
2306	20JN	12;36	79.27	2306	5JL	10;00	273.51
2307	19JL	19;43	107.36	2307	4AU	1;02	302.11
2308	18AU	4;26	135.91	2308	2SE	15;47	331.16
2309	16SE	15;43	165.08	2309	2OC	5;50	0.70
2310	16OC	6;04	194.85	2310	31OC	18;47	30.63
2311	14NO	23;09	225.10	2311	30NO	6;29	60.73
2312	14DE	18;00	255.51	2312	29DE	17;05	90.73

-813

NEW NUMBER	DATE	TIME	LONG.	FULL NUMBER	DATE	TIME	LONG.
2313	13JA	13;09	285.74	2313	28JA	3;01	120.38
2314	12FE	7;09	315.50	2314	26FE	12;44	149.55
2315	13MR	22;48	344.65	2315	27MR	22;43	178.23
2316	12AP	11;31	13.19	2316	26AP	9;23	206.52
2317	11MY	21;28	41.30	2317	25MY	21;12	234.62
2318	10JN	5;29	69.22	2318	24JN	10;37	262.77
2319	9JL	12;42	97.22	2319	24JL	1;50	291.24
2320	7AU	20;18	125.57	2320	22AU	18;34	320.18
2321	6SE	5;11	154.45	2321	21SE	11;51	349.67
2322	5OC	15;56	183.91	2322	21OC	4;23	19.61
2323	4NO	4;49	213.87	2323	19NO	19;11	49.79
2324	3DE	19;53	244.12	2324	19DE	8;01	79.93

-812

NEW NUMBER	DATE	TIME	LONG.	FULL NUMBER	DATE	TIME	LONG.
2325	2JA	12;53	274.36	2325	17JA	19;05	109.77
2326	1FE	7;02	304.31	2326	16FE	4;46	139.12
2327	2MR	0;59	333.75	2327	16MR	13;27	167.92
2328	31MR	17;16	2.61	2328	14AP	21;38	196.25
2329	30AP	7;02	30.97	2329	14MY	6;06	224.27
2330	29MY	18;21	59.03	2330	12JN	15;52	252.26
2331	28JN	3;58	87.06	2331	12JL	3;54	280.48
2332	27JL	12;48	115.32	2332	10AU	18;46	309.18
2333	25AU	21;38	144.01	2333	9SE	12;09	338.47
2334	24SE	7;02	173.23	2334	9OC	6;50	8.32
2335	23OC	17;21	202.96	2335	8NO	1;10	38.56
2336	22NO	4;57	233.02	2336	7DE	17;55	68.89
2337	21DE	18;10	263.16				

-811

NEW NUMBER	DATE	TIME	LONG.	FULL NUMBER	DATE	TIME	LONG.
				2337	6JA	8;26	98.99
2338	20JA	9;08	293.13	2338	4FE	20;29	128.61
2339	19FE	1;21	322.70	2339	6MR	6;15	157.64
2340	20MR	17;51	351.75	2340	4AP	14;11	186.11
2341	19AP	9;40	20.33	2341	3MY	21;09	214.16
2342	19MY	0;12	48.57	2342	2JN	4;17	242.05
2343	17JN	13;23	76.70	2343	1JL	12;46	270.05
2344	17JL	1;24	104.96	2344	30JL	23;40	298.43
2345	15AU	12;29	133.57	2345	29AU	13;32	327.38
2346	13SE	22;55	162.65	2346	28SE	6;15	356.97
2347	13OC	9;05	192.22	2347	28OC	1;02	27.10
2348	11NO	19;25	222.15	2348	26NO	20;32	57.52
2349	11DE	6;22	252.23	2349	26DE	15;08	87.87

-810

NEW NUMBER	DATE	TIME	LONG.	FULL NUMBER	DATE	TIME	LONG.
2350	9JA	18;11	282.18	2350	25JA	7;23	117.83
2351	8FE	6;54	311.79	2351	23FE	20;28	147.19
2352	9MR	20;22	340.91	2352	25MR	6;31	175.91
2353	8AP	10;27	9.57	2353	23AP	14;20	204.11
2354	8MY	1;04	37.90	2354	22MY	21;04	232.02
2355	6JN	16;04	66.09	2355	21JN	3;54	259.90
2356	6JL	7;05	94.38	2356	20JL	11;50	288.04
2357	4AU	21;33	122.97	2357	18AU	21;43	316.65
2358	3SE	11;00	151.98	2358	17SE	10;09	345.87
2359	2OC	23;21	181.47	2359	17OC	1;30	15.70
2360	1NO	10;52	211.34	2360	15NO	19;35	45.99
2361	30NO	21;56	241.42	2361	15DE	15;22	76.44
2362	30DE	8;43	271.42				

-809

NEW NUMBER	DATE	TIME	LONG.	FULL NUMBER	DATE	TIME	LONG.
				2362	14JA	10;53	106.68
2363	28JA	19;16	301.09	2363	13FE	4;09	136.40
2364	27FE	5;41	330.28	2364	14MR	18;08	165.46
2365	28MR	16;22	358.97	2365	13AP	5;03	193.93
2366	27AP	4;00	27.29	2366	12MY	13;44	221.98
2367	26MY	17;09	55.44	2367	10JN	21;13	249.87
2368	25JN	7;57	83.65	2368	10JL	4;27	277.88
2369	24JL	23;54	112.16	2369	8AU	12;16	306.25
2370	23AU	16;09	141.10	2370	6SE	21;25	335.15
2371	22SE	7;53	170.53	2371	6OC	8;41	4.64
2372	21OC	22;39	200.41	2372	4NO	22;43	34.66
2373	20NO	12;14	230.54	2373	4DE	15;34	64.98
2374	20DE	0;27	260.65				

-808

NUMBER	DATE	TIME	LONG.	NUMBER	DATE	TIME	LONG.
				2374	3JA	10;17	95.29
2375	18JA	11;10	290.47	2375	2FE	5;07	125.25
2376	16FE	20;34	319.80	2376	2MR	22;25	154.65
2377	17MR	5;10	348.59	2377	1AP	13;20	183.44
2378	15AP	13;50	16.92	2378	1MY	1;46	211.74
2379	14MY	23;31	44.99	2379	30MY	12;08	239.76
2380	13JN	11;01	73.05	2380	28JN	20;59	267.77
2381	13JL	0;41	101.34	2381	28JL	5;05	296.00
2382	11AU	16;22	130.08	2382	26AU	13;16	324.68
2383	10SE	9;35	159.38	2383	24SE	22;23	353.90
2384	10OC	3;29	189.22	2384	24OC	9;12	23.66
2385	8NO	20;55	219.43	2385	22NO	22;08	53.78
2386	8DE	12;42	249.71	2386	22DE	13;06	84.00

-807

NUMBER	DATE	TIME	LONG.	NUMBER	DATE	TIME	LONG.
2387	7JA	2;01	279.75	2387	21JA	5;27	114.01
2388	5FE	12;48	309.32	2388	19FE	22;21	143.59
2389	6MR	21;37	338.30	2389	21MR	14;59	172.64
2390	5AP	5;19	6.74	2390	20AP	6;39	201.19
2391	4MY	12;51	34.81	2391	19MY	20;45	229.41
2392	2JN	21;05	62.74	2392	18JN	9;00	257.51
2393	2JL	6;51	90.80	2393	17JL	19;35	285.72
2394	31JL	18;51	119.24	2394	16AU	5;08	314.28
2395	30AU	9;36	148.24	2395	14SE	14;31	343.33
2396	29SE	3;09	177.87	2396	14OC	0;26	12.89
2397	28OC	22;33	208.04	2397	12NO	11;16	42.85
2398	27NO	17;56	238.46	2398	11DE	23;04	72.97
2399	27DE	11;21	268.76				

-806

NUMBER	DATE	TIME	LONG.	NUMBER	DATE	TIME	LONG.
				2399	10JA	11;45	102.95
2400	26JA	1;44	298.63	2400	9FE	1;21	132.58
2401	24FE	13;08	327.90	2401	10MR	15;54	161.74
2402	25MR	22;15	356.57	2402	9AP	7;16	190.44
2403	24AP	5;53	24.76	2403	8MY	22;52	218.79
2404	23MY	12;52	52.67	2404	7JN	13;55	246.99
2405	21JN	20;03	80.58	2405	7JL	3;50	275.24
2406	21JL	4;26	108.74	2406	5AU	16;33	303.77
2407	19AU	15;05	137.39	2407	4SE	4;26	332.73
2408	18SE	4;52	166.68	2408	3OC	15;51	2.19
2409	17OC	22;00	196.60	2409	2NO	3;02	32.06
2410	16NO	17;30	226.95	2410	1DE	13;57	62.12
2411	16DE	13;23	257.40	2411	31DE	0;38	92.12

-805

NUMBER	DATE	TIME	LONG.	NUMBER	DATE	TIME	LONG.
2412	15JA	7;42	287.58	2412	29JA	11;23	121.79
2413	13FE	23;19	317.22	2413	27FE	22;39	151.00
2414	15MR	12;02	346.22	2414	29MR	10;54	179.75
2415	13AP	22;07	14.64	2415	28AP	0;18	208.13
2416	13MY	6;10	42.66	2416	27MY	14;41	236.32
2417	11JN	13;04	70.53	2417	26JN	5;41	264.55
2418	10JL	19;52	98.53	2418	25JL	20;57	293.03
2419	9AU	3;45	126.91	2419	24AU	12;10	321.95
2420	7SE	13;46	155.85	2420	23SE	2;56	351.36
2421	7OC	2;39	185.42	2421	22OC	16;45	21.20
2422	5NO	18;29	215.52	2422	21NO	5;13	51.29
2423	5DE	12;33	245.90	2423	20DE	16;21	81.36

-804

NUMBER	DATE	TIME	LONG.	NUMBER	DATE	TIME	LONG.
2424	4JA	7;36	276.22	2424	19JA	2;26	111.14
2425	3FE	2;11	306.16	2425	17FE	11;59	140.47
2426	3MR	18;59	335.52	2426	17MR	21;32	169.28
2427	2AP	9;07	4.27	2427	16AP	7;33	197.66
2428	1MY	20;23	32.51	2428	15MY	18;32	225.78
2429	31MY	5;18	60.48	2429	14JN	7;00	253.87
2430	29JN	12;53	88.43	2430	13JL	21;02	282.20
2431	28JL	20;19	116.65	2431	12AU	13;33	310.98
2432	27AU	4;38	145.34	2432	11SE	6;59	340.30
2433	25SE	14;32	174.61	2433	11OC	0;19	10.13
2434	25OC	2;24	204.41	2434	9NO	16;18	40.28
2435	23NO	16;21	234.58	2435	9DE	6;13	70.49
2436	23DE	8;18	264.84				

-803

NUMBER	DATE	TIME	LONG.	NUMBER	DATE	TIME	LONG.
				2436	7JA	18;05	100.46
2437	22JA	1;47	294.89	2437	6FE	4;18	129.99
2438	20FE	19;46	324.50	2438	7MR	13;15	158.96
2439	22MR	12;52	353.55	2439	5AP	21;24	187.42
2440	21AP	3;52	22.07	2440	5MY	5;27	215.50
·2441	20MY	16;23	50.22	2441	3JN	14;18	243.46
2442	19JN	2;53	78.25	2442	3JL	1;02	271.56
2443	18JL	12;12	106.42	2443	1AU	14;30	300.06
2444	16AU	21;10	134.97	2444	31AU	6;51	329.14
2445	15SE	6;27	164.02	2445	30SE	1;17	358.83
2446	14OC	16;25	193.60	2446	29OC	20;11	28.99
2447	13NO	3;23	223.57	2447	28NO	14;00	59.34
2448	12DE	15;41	253.70	2448	28DE	5;43	89.56

-802

NUMBER	DATE	TIME	LONG.	NUMBER	DATE	TIME	LONG.
2449	11JA	5;35	283.73	2449	26JA	18;57	119.37
2450	9FE	20;58	313.42	2450	25FE	5;42	148.61
2451	11MR	13;10	342.63	2451	26MR	14;20	177.25
2452	10AP	5;12	11.34	2452	24AP	21;32	205.41
2453	9MY	20;20	39.67	2453	24MY	4;20	233.32
2454	8JN	10;15	67.83	2454	22JN	11;58	261.24
2455	7JL	23;01	96.04	2455	21JL	21;34	289.46
2456	6AU	10;49	124.54	2456	20AU	9;58	318.19
2457	4SE	21;49	153.48	2457	19SE	1;26	347.56
2458	4OC	8;18	182.91	2458	18OC	19;26	17.53
2459	2NO	18;39	212.75	2459	17NO	14;51	47.89
2460	2DE	5;16	242.80	2460	17DE	10;08	78.31
2461	31DE	16;30	272.81				

-801

NUMBER	DATE	TIME	LONG.	NUMBER	DATE	TIME	LONG.
				2461	16JA	3;41	108.45
2462	30JA	4;33	302.53	2462	14FE	18;19	138.03
2463	28FE	17;21	331.80	2463	16MR	5;43	166.96
2464	30MR	6;50	0.58	2464	14AP	14;27	195.32
2465	28AP	20;57	28.98	2465	13MY	21;32	223.30
2466	28MY	11;42	57.18	2466	12JN	4;10	251.16
2467	27JN	2;52	85.42	2467	11JL	11;29	279.19
2468	26JL	17;54	113.91	2468	9AU	20;37	307.62
2469	25AU	8;12	142.80	2469	8SE	7;35	336.62
2470	23SE	21;24	172.16	2470	7OC	21;32	6.24
2471	23OC	9;31	201.94	2471	6NO	14;20	36.39
2472	21NO	20;54	231.99	2472	6DE	9;26	66.82
2473	21DE	7;48	262.04				

-800

NUMBER	DATE	TIME	LONG.	NUMBER	DATE	TIME	LONG.
				2473	5JA	5;16	97.16
2474	19JA	18;20	291.84	2474	3FE	23;45	127.08
2475	18FE	4;33	321.18	2475	4MR	15;18	156.38
2476	18MR	14;45	350.01	2476	3AP	3;37	185.04
2477	17AP	1;33	18.41	2477	2MY	13;16	213.21
2478	16MY	13;41	46.57	2478	31MY	21;16	241.14
2479	15JN	3;35	74.72	2479	30JN	4;36	269.09
2480	14JL	19;05	103.11	2480	29JL	12;10	297.32
2481	13AU	11;34	131.90	2481	27AU	20;42	326.03
2482	12SE	3;56	161.20	2482	26SE	6;56	355.32
2483	11OC	19;31	190.97	2483	25OC	19;35	25.17
2484	10NO	9;56	221.06	2484	24NO	11;02	55.40
2485	9DE	22;56	251.22	2485	24DE	4;51	85.74

-799

NUMBER	DATE	TIME	LONG.	NUMBER	DATE	TIME	LONG.
2486	8JA	10;21	281.17	2486	22JA	23;40	115.83
2487	6FE	20;14	310.68	2487	21FE	17;41	145.43
2488	8MR	4;56	339.63	2488	23MR	9;42	174.42
2489	6AP	13;13	8.08	2489	21AP	23;18	202.88
2490	5MY	22;05	36.19	2490	21MY	10;42	230.98
2491	4JN	8;27	64.21	2491	19JN	20;21	258.97
2492	3JL	20;58	92.38	2492	19JL	4;55	287.12
2493	2AU	11;46	120.95	2493	17AU	13;08	315.65
2494	1SE	4;28	150.06	2494	15SE	21;52	344.69
2495	30SE	22;24	179.74	2495	15OC	7;53	14.28
2496	30OC	16;27	209.87	2496	13NO	19;46	44.29
2497	29NO	9;20	240.19	2497	13DE	9;40	74.49
2498	28DE	23;59	270.36				

-798

NUMBER	DATE	TIME	LONG.	NUMBER	DATE	TIME	LONG.
				2498	12JA	1;13	104.58
2499	27JA	11;56	300.11	2499	10FE	17;40	134.31
2500	25FE	21;30	329.29	2500	12MR	10;15	163.52
2501	27MR	5;29	357.89	2501	11AP	2;18	192.22
2502	25AP	12;50	26.05	2502	10MY	17;12	220.53
2503	24MY	20;29	53.99	2503	9JN	6;29	248.66
2504	23JN	5;20	81.96	2504	8JL	18;04	276.83
2505	22JL	16;09	110.24	2505	7AU	4;17	305.28
2506	21AU	5;33	139.02	2506	5SE	13;54	334.17
2507	19SE	21;52	168.44	2507	4OC	23;41	3.58
2508	19OC	16;40	198.46	2508	3NO	10;10	33.44
2509	18NO	12;28	228.84	2509	2DE	21;29	63.52
2510	18DE	7;08	259.23				

-797

NUMBER	DATE	TIME	LONG.	NUMBER	DATE	TIME	LONG.
				2510	1JA	9;36	93.56
2511	16JA	23;04	289.29	2511	30JA	22;28	123.31
2512	15FE	11;49	318.78	2512	1MR	12;14	152.60
2513	16MR	21;52	347.65	2513	31MR	2;56	181.42
2514	15AP	6;02	15.98	2514	29AP	18;19	209.86
2515	14MY	13;10	43.95	2515	29MY	9;42	238.09
2516	12JN	20;07	71.83	2516	28JN	0;21	266.31
2517	12JL	3;48	99.87	2517	27JL	13;53	294.75
2518	10AU	13;15	128.33	2518	26AU	2;26	323.58
2519	9SE	1;28	157.39	2519	24SE	14;20	352.89
2520	8OC	17;04	187.09	2520	24OC	1;51	22.66
2521	7NO	11;38	217.33	2521	22NO	12;59	52.70
2522	7DE	7;35	247.79	2522	21DE	23;44	82.74

-796 (left) / **-790** (right)

NUMBER	DATE	TIME	LONG.	NUMBER	DATE	TIME	LONG.	NUMBER	DATE	TIME	LONG.	NUMBER	DATE	TIME	LONG.
2523	6JA	2;50	278.10	2523	20JA	10;15	112.53								
2524	4FE	19;47	307.94	2524	18FE	20;56	141.88	2598	28JA	19;45	301.49	2597	13JA	18;11	106.36
2525	5MR	9;51	337.16	2525	19MR	8;20	170.75	2599	27FE	4;47	330.63	2598	12FE	12;34	136.14
2526	3AP	21;07	5.77	2526	17AP	20;54	199.21	2600	28MR	12;57	359.22	2599	14MR	5;31	165.33
2527	3MY	6;04	33.91	2527	17MY	10;38	227.43	2601	26AP	21;11	27.40	2600	12AP	20;14	193.96
2528	1JN	13;26	61.81	2528	16JN	1;20	255.62	2602	26MY	6;32	55.40	2601	12MY	8;42	222.16
2529	30JN	20;13	89.74	2529	15JL	16;37	284.00	2603	24JN	17;51	83.48	2602	10JN	19;16	250.18
2530	30JL	3;31	117.97	2530	14AU	8;10	312.77	2604	24JL	7;33	111.88	2603	10JL	4;28	278.27
2531	28AU	12;29	146.71	2531	12SE	23;32	342.04	2605	22AU	23;29	140.80	2604	8AU	12;58	306.66
2532	26SE	23;58	176.06	2532	12OC	14;10	11.78	2606	21SE	17;08	170.30	2605	6SE	21;33	335.54
2533	26OC	14;25	205.99	2533	11NO	3;32	41.84	2607	21OC	11;30	200.32	2606	6OC	7;00	4.95
2534	25NO	7;27	236.30	2534	10DE	15;23	71.95	2608	20NO	5;19	230.63	2607	4NO	17;58	34.84
2535	25DE	2;04	266.66					2609	19DE	21;16	260.91	2608	4DE	6;50	65.00

-795 (left) / **-789** (right)

NUMBER	DATE	TIME	LONG.	NUMBER	DATE	TIME	LONG.	NUMBER	DATE	TIME	LONG.	NUMBER	DATE	TIME	LONG.
				2535	9JA	1;52	101.86					2609	2JA	21;27	95.13
2536	23JA	20;53	296.75	2536	7FE	11;27	131.34	2610	18JA	10;32	290.83	2610	1FE	13;16	124.98
2537	22FE	14;33	326.32	2537	8MR	20;42	160.30	2611	16FE	21;06	320.21	2611	3MR	5;34	154.35
2538	24MR	5;59	355.27	2538	7AP	6;10	188.79	2612	18MR	5;37	348.99	2612	1AP	21;44	183.20
2539	22AP	18;38	23.68	2539	6MY	16;23	216.95	2613	16AP	13;00	17.27	2613	1MY	13;11	211.62
2540	22MY	4;41	51.72	2540	5JN	3;54	245.01	2614	15MY	20;16	45.45	2614	31MY	3;22	239.80
2541	20JN	12;54	79.66	2541	4JL	17;15	273.22	2615	14JN	4;21	73.17	2615	29JN	15;58	267.95
2542	19JL	20;25	107.78	2542	3AU	8;39	301.83	2616	13JL	14;05	101.30	2616	29JL	3;02	296.31
2543	18AU	4;21	136.30	2543	2SE	1;50	330.97	2617	12AU	2;10	129.89	2617	27AU	13;08	325.06
2544	16SE	13;34	165.37	2544	1OC	19;41	0.66	2618	10SE	17;08	159.09	2618	25SE	22;59	354.32
2545	16OC	0;32	195.01	2545	31OC	12;43	30.76	2619	10OC	10;57	188.91	2619	25OC	9;14	24.04
2546	14NO	13;27	225.07	2546	30NO	3;51	61.01	2620	9NO	6;38	219.21	2620	23NO	20;11	54.08
2547	14DE	4;18	255.31	2547	29DE	16;44	91.10	2621	9DE	2;11	249.65	2621	23DE	7;49	84.16

-794 (left) / **-788** (right)

NUMBER	DATE	TIME	LONG.	NUMBER	DATE	TIME	LONG.	NUMBER	DATE	TIME	LONG.	NUMBER	DATE	TIME	LONG.
2548	12JA	20;53	285.44	2548	28JA	3;38	120.79	2622	7JA	19;37	279.87	2622	21JA	20;06	114.00
2549	11FE	14;31	315.20	2549	26FE	13;00	149.95	2623	6FE	9;53	309.58	2623	20FE	9;05	143.43
2550	13MR	8;01	344.43	2550	27MR	21;18	178.56	2624	6MR	21;06	338.66	2624	20MR	23;00	172.37
2551	12AP	0;02	13.12	2551	26AP	5;07	206.73	2625	5AP	5;59	7.15	2625	19AP	13;51	200.91
2552	11MY	13;47	41.38	2552	25MY	13;18	234.69	2626	4MY	13;28	35.22	2626	19MY	5;14	229.18
2553	10JN	1;20	69.44	2553	23JN	22;53	262.69	2627	2JN	20;22	63.10	2627	17JN	20;23	257.39
2554	9JL	11;18	97.55	2554	23JL	10;55	291.02	2628	2JL	3;35	91.05	2628	17JL	10;42	285.75
2555	7AU	20;35	125.97	2555	22AU	1;57	319.89	2629	31JL	12;05	119.35	2629	16AU	0;00	314.46
2556	6SE	5;53	154.86	2556	20SE	19;41	349.38	2630	29AU	22;53	148.18	2630	14SE	12;30	343.63
2557	5OC	15;40	184.28	2557	20OC	14;46	19.42	2631	28SE	12;51	177.66	2631	14OC	0;26	13.28
2558	4NO	2;11	214.14	2558	19NO	9;27	49.76	2632	28OC	6;07	207.73	2632	12NO	11;55	43.27
2559	3DE	13;44	244.24	2559	19DE	2;22	80.08	2633	27NO	1;38	238.15	2633	11DE	22;52	73.35
								2634	26DE	21;26	268.55				

-793 (left) / **-787** (right)

NUMBER	DATE	TIME	LONG.	NUMBER	DATE	TIME	LONG.	NUMBER	DATE	TIME	LONG.	NUMBER	DATE	TIME	LONG.
2560	2JA	2;40	274.31	2560	17JA	16;51	110.06					2634	10JA	9;21	103.24
2561	31JA	17;05	304.11	2561	16FE	4;44	139.51	2635	25JA	15;35	298.59	2635	8FE	19;39	132.73
2562	2MR	8;40	333.46	2562	17MR	14;13	168.34	2636	24FE	7;01	328.03	2636	10MR	6;21	161.73
2563	1AP	0;38	2.32	2563	15AP	21;54	196.64	2637	25MR	19;33	356.85	2637	8AP	18;00	190.29
2564	30AP	16;08	30.76	2564	15MY	4;40	224.60	2638	24AP	5;33	25.13	2638	8MY	6;56	218.54
2565	30MY	6;41	58.95	2565	13JN	11;41	252.47	2639	23MY	13;36	53.09	2639	6JN	21;06	246.70
2566	28JN	20;10	87.13	2566	12JL	20;10	280.56	2640	21JN	20;37	80.98	2640	6JL	12;11	275.00
2567	28JL	8;40	115.53	2567	11AU	7;09	309.09	2641	21JL	3;37	109.00	2641	5AU	3;51	303.64
2568	26AU	20;20	144.34	2568	9SE	21;11	338.23	2642	19AU	11;45	137.64	2642	3SE	19;40	332.75
2569	25SE	7;19	173.63	2569	9OC	14;07	8.01	2643	17SE	22;00	166.78	2643	3OC	11;03	2.37
2570	24OC	17;52	203.37	2570	8NO	9;05	38.27	2644	17OC	11;02	196.52	2644	2NO	1;21	32.37
2571	23NO	4;22	233.38	2571	8DE	4;41	68.71	2645	16NO	2;51	226.72	2645	1DE	14;04	62.52
2572	22DE	15;12	263.43					2646	15DE	20;44	257.08	2646	31DE	1;10	92.53

-792 (left) / **-786** (right)

NUMBER	DATE	TIME	LONG.	NUMBER	DATE	TIME	LONG.	NUMBER	DATE	TIME	LONG.	NUMBER	DATE	TIME	LONG.
				2572	6JA	23;16	98.99					2647	29JA	11;00	122.16
2573	21JA	2;38	293.25	2573	5FE	15;23	128.78	2647	14JA	15;27	287.28	2648	27FE	20;09	151.28
2574	19FE	14;48	322.85	2574	6MR	4;18	157.95	2648	13FE	9;40	317.04	2649	29MR	5;11	179.90
2575	20MR	3;38	351.56	2575	4AP	14;11	186.49	2649	15MR	2;11	346.20	2650	27AP	14;44	208.13
2576	18AP	17;10	20.05	2576	3MY	21;53	214.57	2650	13AP	16;11	14.78	2651	27MY	1;22	236.18
2577	18MY	7;27	48.28	2577	2JN	4;35	242.44	2651	13MY	3;30	42.94	2652	25JN	13;40	264.30
2578	16JN	22;27	76.49	2578	1JL	11;29	270.38	2652	11JN	12;36	70.90	2653	25JL	4;05	292.74
2579	16JL	13;48	104.88	2579	30JL	19;34	298.65	2653	10JL	20;28	98.94	2654	23AU	20;38	321.70
2580	15AU	4;50	133.64	2580	29AU	5;39	327.44	2654	9AU	4;15	127.31	2655	22SE	14;35	351.23
2581	13SE	18;55	162.87	2581	27SE	18;16	356.86	2655	7SE	12;55	156.20	2656	22OC	8;28	21.24
2582	13OC	7;49	192.55	2582	27OC	9;43	26.84	2656	6OC	23;06	185.66	2657	21NO	0;50	51.49
2583	11NO	19;42	222.55	2583	26NO	3;47	57.19	2657	5NO	11;05	215.60	2658	20DE	14;54	81.68
2584	11DE	6;52	252.64	2584	25DE	23;24	87.59	2658	5DE	0;55	245.79				

-791 (left) / **-785** (right)

NUMBER	DATE	TIME	LONG.	NUMBER	DATE	TIME	LONG.	NUMBER	DATE	TIME	LONG.	NUMBER	DATE	TIME	LONG.
2585	9JA	17;29	282.54	2585	24JA	18;40	117.69	2659	3JA	16;30	275.97	2659	19JA	2;42	111.54
2586	8FE	3;38	312.04	2586	23FE	11;41	147.21	2660	2FE	9;28	305.85	2660	17FE	12;38	140.88
2587	9MR	13;32	341.02	2587	25MR	1;31	176.09	2661	4MR	2;56	335.25	2661	18MR	21;14	169.66
2588	7AP	23;40	9.52	2588	23AP	12;21	204.41	2662	2AP	19;39	4.11	2662	17AP	5;02	197.94
2589	7MY	10;49	37.71	2589	22MY	21;04	232.40	2663	2MY	10;32	32.50	2663	16MY	12;46	225.93
2590	5JN	23;04	65.82	2590	21JN	4;42	260.32	2664	31MY	23;13	60.62	2664	14JN	21;25	253.88
2591	5JL	14;28	94.10	2591	20JL	12;12	288.44	2665	30JN	10;02	88.70	2665	14JL	8;05	282.06
2592	4AU	6;44	122.75	2592	18AU	20;18	316.98	2666	29JL	19;48	117.01	2666	12AU	21;39	310.71
2593	2SE	23;31	151.89	2593	17SE	5;44	346.07	2667	28AU	5;16	145.75	2667	11SE	14;17	339.99
2594	2OC	15;51	181.54	2594	16OC	17;12	15.74	2668	26SE	14;58	175.01	2668	11OC	9;05	9.87
2595	1NO	7;08	211.57	2595	15NO	7;14	45.86	2669	26OC	1;13	204.75	2669	10NO	4;20	40.16
2596	30NO	21;00	241.76	2596	14DE	23;52	76.17	2670	24NO	12;15	234.79	2670	9DE	22;22	70.54
2597	30DE	9;14	271.82					2671	24DE	0;20	264.89				

NUMBER	DATE	TIME	LONG.	NUMBER	DATE	TIME	LONG.	NUMBER	DATE	TIME	LONG.	NUMBER	DATE	TIME	LONG.
		−784								**−778**					
				2671	8JA	14;07	100.68					2745	1JA	8;34	93.91
2672	22JA	13;47	294.77	2672	7FE	3;13	130.33	2746	16JA	10;44	289.16	2746	30JA	18;42	123.54
2673	21FE	4;33	324.26	2673	7MR	13;44	159.37	2747	15FE	3;29	318.82	2747	1MR	4;52	152.68
2674	21MR	20;08	353.26	2674	5AP	22;07	187.83	2748	16MR	17;22	347.85	2748	30MR	15;41	181.35
2675	20AP	11;45	21.81	2675	5MY	5;06	215.87	2749	15AP	4;31	16.30	2749	29AP	3;43	209.66
2676	20MY	2;45	50.06	2676	3JN	11;47	243.74	2750	14MY	13;27	44.35	2750	28MY	17;08	237.81
2677	18JN	16;52	78.23	2677	2JL	19;23	271.71	2751	12JN	20;55	72.24	2751	27JN	7;47	266.03
2678	18JL	6;04	106.56	2678	1AU	5;03	300.06	2752	12JL	3;54	100.26	2752	26JL	23;21	294.54
2679	16AU	18;26	135.24	2679	30AU	17;37	328.98	2753	10AU	11;28	128.64	2753	25AU	15;25	323.51
2680	15SE	6;01	164.39	2680	29SE	9;16	358.55	2754	8SE	20;40	157.57	2754	24SE	7;24	352.99
2681	14OC	16;58	194.01	2681	29OC	3;27	28.67	2755	8OC	8;22	187.12	2755	23OC	22;36	22.91
2682	13NO	3;34	223.96	2682	27NO	22;59	59.09	2756	6NO	22;51	217.17	2756	22NO	12;19	53.06
2683	12DE	14;10	254.02	2683	27DE	18;16	89.46	2757	6DE	15;45	247.50	2757	22DE	0;14	83.15
		−783								**−777**					
2684	11JA	1;08	283.93	2684	26JA	11;41	119.46	2758	5JA	10;03	277.78	2758	20JA	10;32	112.93
2685	9FE	12;41	313.46	2685	25FE	2;09	148.85	2759	4FE	4;28	307.69	2759	18FE	19;46	142.22
2686	11MR	0;53	342.51	2686	26MR	13;23	177.59	2760	5MR	21;48	337.06	2760	20MR	4;33	170.98
2687	9AP	13;46	11.11	2687	24AP	22;00	205.81	2761	4AP	13;01	5.83	2761	18AP	13;33	199.30
2688	9MY	3;28	39.39	2688	24MY	5;02	233.73	2762	4MY	1;40	34.12	2762	17MY	23;23	227.36
2689	7JN	18;04	67.57	2689	22JN	11;43	261.61	2763	2JN	11;51	62.13	2763	16JN	10;40	255.42
2690	7JL	9;24	95.88	2690	21JL	19;11	289.75	2764	1JL	20;21	90.13	2764	15JL	23;59	283.72
2691	6AU	0;55	124.52	2691	20AU	4;18	318.35	2765	31JL	4;13	118.38	2765	14AU	15;38	312.48
2692	4SE	15;52	153.62	2692	18SE	15;42	347.54	2766	29AU	12;32	147.10	2766	13SE	9;15	341.83
2693	4OC	5;40	183.18	2693	18OC	5;47	17.34	2767	27SE	22;04	176.37	2767	13OC	3;37	11.72
2694	2NO	18;14	213.12	2694	16NO	22;36	47.59	2768	27OC	9;14	206.16	2768	11NO	21;07	41.95
2695	2DE	5;49	243.22	2695	16DE	17;33	78.00	2769	25NO	22;07	236.29	2769	11DE	12;28	72.22
2696	31DE	16;38	273.21					2770	25DE	12;42	266.48				
		−782								**−776**					
				2696	15JA	13;07	108.22					2770	10JA	1;21	102.22
2697	30JA	2;51	302.85	2697	14FE	7;19	137.96	2771	24JA	4;47	296.46	2771	8FE	12;04	131.75
2698	28FE	12;36	331.99	2698	15MR	22;40	167.06	2772	22FE	21;51	326.01	2772	8MR	21;07	160.70
2699	29MR	22;16	0.62	2699	14AP	10;52	195.56	2773	23MR	14;52	355.04	2773	7AP	5;03	189.13
2700	28AP	8;34	28.87	2700	13MY	20;32	223.64	2774	22AP	6;40	23.58	2774	6MY	12;34	217.18
2701	27MY	20;20	56.96	2701	12JN	4;40	251.57	2775	21MY	20;29	51.78	2775	4JN	20;32	245.10
2702	26JN	10;07	85.14	2702	11JL	12;15	279.60	2776	20JN	8;18	79.86	2776	4JL	6;01	273.16
2703	26JL	1;50	113.65	2703	9AU	20;07	307.99	2777	19JL	18;42	108.09	2777	2AU	18;05	301.62
2704	24AU	18;42	142.63	2704	8SE	4;58	336.90	2778	18AU	4;29	136.69	2778	1SE	9;19	330.68
2705	23SE	11;40	172.14	2705	7OC	15;26	6.37	2779	16SE	14;16	165.79	2779	1OC	3;22	0.37
2706	23OC	3;49	202.09	2706	6NO	4;10	36.35	2780	16OC	0;24	195.39	2780	30OC	22;47	30.56
2707	21NO	18;36	232.28	2707	5DE	19;28	66.61	2781	14NO	11;04	225.36	2781	29NO	17;43	60.96
2708	21DE	7;43	262.42					2782	13DE	22;31	255.45	2782	29DE	10;44	91.23
		−781								**−775**					
				2708	4JA	12;56	96.86					2783	28JA	1;08	121.07
2709	19JA	19;01	292.25	2709	3FE	7;16	126.78	2783	12JA	11;04	285.41	2784	26FE	12;48	150.33
2710	18FE	4;35	321.57	2710	5MR	0;51	156.17	2784	11FE	0;55	315.02	2785	27MR	22;04	178.97
2711	19MR	12;53	350.32	2711	3AP	16;36	184.98	2785	12MR	15;52	344.16	2786	26AP	5;32	207.13
2712	17AP	20;45	18.60	2712	3MY	6;09	213.31	2786	11AP	7;18	12.83	2787	25MY	12;10	235.02
2713	17MY	5;15	46.62	2713	1JN	17;41	241.38	2787	10MY	22;33	41.16	2788	23JN	19;08	262.93
2714	15JN	15;24	74.62	2714	1JL	3;39	269.43	2788	9JN	13;10	69.34	2789	23JL	3;40	291.11
2715	15JL	3;51	102.88	2715	30JL	12;38	297.72	2789	9JL	2;59	97.60	2790	21AU	14;47	319.82
2716	13AU	18;48	131.61	2716	28AU	21;18	326.44	2790	7AU	16;02	126.17	2791	20SE	5;01	349.16
2717	12SE	11;52	160.92	2717	27SE	6;24	355.69	2791	6SE	4;19	155.18	2792	19OC	22;07	19.10
2718	12OC	6;14	190.79	2718	26OC	16;40	25.43	2792	5OC	15;49	184.68	2793	18NO	17;12	49.46
2719	11NO	0;42	221.05	2719	25NO	4;33	55.51	2793	4NO	2;43	214.55	2794	18DE	12;48	79.89
2720	10DE	17;50	251.40	2720	24DE	18;12	85.67	2794	3DE	13;19	244.61				
		−780								**−774**					
2721	9JA	8;30	281.48	2721	23JA	9;15	115.62	2795	1JA	23;58	274.59	2795	17JA	7;16	110.05
2722	7FE	20;18	311.07	2722	22FE	1;06	145.13	2796	31JA	11;00	304.24	2796	15FE	23;13	139.66
2723	8MR	5;36	340.04	2723	22MR	17;08	174.14	2797	1MR	22;34	333.43	2797	17MR	11;58	168.63
2724	6AP	13;16	8.47	2724	21AP	8;51	202.68	2798	31MR	10;48	2.15	2798	15AP	21;43	197.02
2725	5MY	20;21	36.51	2725	20MY	23;43	230.92	2799	29AP	23;50	30.49	2799	15MY	5;22	225.01
2726	4JN	3;50	64.41	2726	19JN	13;15	259.07	2800	29MY	13;51	58.67	2800	13JN	12;07	252.88
2727	3JL	12;37	92.43	2727	19JL	1;17	287.36	2801	28JN	4;54	86.91	2801	12JL	19;08	280.90
2728	1AU	23;29	120.84	2728	17AU	12;04	315.99	2802	27JL	20;36	115.43	2802	11AU	3;27	309.32
2729	31AU	13;03	149.81	2729	15SE	22;13	345.09	2803	26AU	12;13	144.39	2803	9SE	13;45	338.31
2730	30SE	5;37	179.43	2730	15OC	8;24	14.69	2804	25SE	2;57	173.84	2804	9OC	2;32	7.91
2731	30OC	0;40	209.59	2731	13NO	19;05	44.65	2805	24OC	16;23	203.69	2805	7NO	18;02	38.00
2732	28NO	20;39	240.04	2732	13DE	6;21	74.74	2806	23NO	4;34	233.78	2806	7DE	11;59	68.39
2733	28DE	15;22	270.38					2807	22DE	15;45	263.84				
		−779								**−773**					
				2733	11JA	18;08	104.67					2807	6JA	7;20	98.70
2734	27JA	7;13	300.30	2734	10FE	6;28	134.23	2808	21JA	2;08	293.61	2808	5FE	2;18	128.63
2735	25FE	19;48	329.60	2735	11MR	19;35	163.30	2809	19FE	11;53	322.91	2809	6MR	19;03	157.95
2736	27MR	5;39	358.28	2736	10AP	9;42	191.94	2810	20MR	21;16	351.69	2810	5AP	8;43	186.65
2737	25AP	13;38	26.47	2737	10MY	0;43	220.26	2811	19AP	6;54	20.02	2811	4MY	19;33	214.86
'2738	24MY	20;41	54.38	2738	8JN	16;04	248.47	2812	18MY	17;38	48.12	2812	3JN	4;23	242.82
2739	23JN	3;39	82.28	2739	8JL	7;01	276.77	2813	17JN	6;16	76.23	2813	2JL	12;15	270.80
2740	22JL	11;26	110.43	2740	6AU	21;05	305.37	2814	16JL	21;06	104.59	2814	31JL	20;02	299.05
2741	20AU	21;03	139.06	2741	5SE	10;16	334.41	2815	15AU	13;41	133.41	2815	30AU	4;29	327.78
2742	19SE	9;27	168.31	2742	4OC	22;44	3.93	2816	14SE	7;01	162.77	2816	28SE	14;12	357.07
2743	19OC	1;11	198.19	2743	3NO	10;38	33.84	2817	13OC	23;56	192.61	2817	28OC	1;48	26.89
2744	17NO	19;47	228.52	2744	2DE	21;56	63.93	2818	12NO	15;40	222.77	2818	26NO	15;46	57.07
2745	17DE	15;40	258.96					2819	12DE	5;44	252.98	2819	26DE	8;07	87.33

19

NUMBER	DATE	TIME	LONG.	NUMBER	DATE	TIME	LONG.
			−772				
2820	10JA	17;56	282.95	2820	25JA	1;58	117.37
2821	9FE	4;12	312.45	2821	23FE	19;52	146.95
2822	9MR	12;52	341.38	2822	24MR	12;27	175.94
2823	7AP	20;38	9.78	2823	23AP	3;01	204.42
2824	7MY	4;31	37.85	2824	22MY	15;34	232.56
2825	5JN	13;36	65.81	2825	21JN	2;24	260.61
2826	5JL	0;48	93.94	2826	20JL	12;00	288.82
2827	3AU	14;33	122.48	2827	18AU	20;58	317.39
2828	2SE	6;47	151.59	2828	17SE	5;59	346.47
2829	2OC	0;49	181.30	2829	16OC	15;44	16.07
2830	31OC	19;36	211.47	2830	15NO	2;48	46.05
2831	30NO	13;43	241.84	2831	14DE	15;30	76.20
2832	30DE	5;46	272.06				
			−771				
				2832	13JA	5;42	106.22
2833	28JA	18;56	301.84	2833	11FE	20;56	135.88
2834	27FE	5;16	331.03	2834	13MR	12;39	165.03
2835	28MR	13;30	359.62	2835	12AP	4;22	193.71
2836	26AP	20;37	27.76	2836	11MY	19;39	222.03
2837	26MY	3;42	55.67	2837	10JN	9;58	250.19
2838	24JN	11;41	83.62	2838	9JL	22;58	278.43
2839	23JL	21;27	111.86	2839	8AU	10;35	306.95
2840	22AU	9;40	140.62	2840	6SE	21;15	335.91
2841	21SE	0;50	170.01	2841	6OC	7;34	5.37
2842	20OC	18;53	200.01	2842	4NO	18;07	35.24
2843	19NO	14;45	230.40	2843	4DE	5;06	65.31
2844	19DE	10;23	260.83				
			−770				
				2844	2JA	16;31	95.31
2845	18JA	3;45	290.93	2845	1FE	4;19	124.99
2846	16FE	17;52	320.46	2846	2MR	16;42	154.21
2847	18MR	4;53	349.35	2847	1AP	6;00	182.95
2848	16AP	13;37	17.68	2848	30AP	20;23	211.34
2849	15MY	20;59	45.66	2849	30MY	11;34	239.56
2850	14JN	3;53	73.53	2850	29JN	2;53	267.81
2851	13JL	11;12	101.57	2851	28JL	17;40	296.31
2852	11AU	19;52	130.02	2852	27AU	7;35	325.22
2853	10SE	6;52	159.05	2853	25SE	20;41	354.61
2854	9OC	20;59	188.71	2854	25OC	9;06	24.43
2855	8NO	14;18	218.90	2855	23NO	20;50	54.50
2856	8DE	9;46	249.34	2856	23DE	7;46	84.54
			−769				
2857	7JA	5;23	279.66	2857	21JA	17;58	114.30
2858	5FE	23;18	309.53	2858	20FE	3;48	143.60
2859	7MR	14;31	338.78	2859	21MR	13;56	172.39
2860	6AP	2;55	7.41	2860	20AP	1;02	200.78
2861	5MY	12;53	35.58	2861	19MY	13;34	228.94
2862	3JN	21;01	63.51	2862	18JN	3;33	257.10
2863	3JL	4;13	91.46	2863	17JL	18;48	285.50
2864	1AU	11;29	119.71	2864	16AU	10;53	314.31
2865	30AU	19;54	148.44	2865	15SE	3;17	343.64
2866	29SE	6;23	177.78	2866	14OC	19;16	13.45
2867	28OC	19;31	207.67	2867	13NO	10;00	43.57
2868	27NO	11;16	237.93	2868	12DE	22;54	73.73
2869	27DE	4;51	268.24				
			−768				
				2869	11JA	9;55	103.65
2870	25JA	23;10	298.29	2870	9FE	19;27	133.11
2871	24FE	16;58	327.84	2871	10MR	4;10	162.03
2872	25MR	9;12	356.81	2872	8AP	12;45	190.46
2873	23AP	23;08	25.25	2873	7MY	21;54	218.57
2874	23MY	10;32	53.35	2874	6JN	8;15	246.58
2875	21JN	19;54	81.34	2875	5JL	20;27	274.75
2876	21JL	4;08	109.50	2876	4AU	11;00	303.34
2877	19AU	12;19	138.05	2877	3SE	3;54	332.49
2878	17SE	21;21	167.14	2878	2OC	22;20	2.23
2879	17OC	7;47	196.77	2879	1NO	16;41	32.39
2880	15NO	19;50	226.81	2880	1DE	9;22	62.70
2881	15DE	9;28	256.99	2881	30DE	23;31	92.84
			−767				
2882	14JA	0;37	287.05	2882	29JA	11;11	122.55
2883	12FE	17;00	316.73	2883	27FE	20;51	151.70
2884	14MR	9;56	345.93	2884	29MR	5;06	180.28
2885	13AP	2;18	14.61	2885	27AP	12;35	208.43
2886	12MY	17;07	42.91	2886	26MY	20;05	236.35
2887	11JN	6;00	71.02	2887	25JN	4;37	264.32
2888	10JL	17;14	99.19	2888	24JL	15;18	292.61
2889	9AU	3;29	127.67	2889	23AU	5;00	321.44
2890	7SE	13;28	156.61	2890	21SE	21;53	350.92
2891	6OC	23;36	186.07	2891	21OC	17;00	20.97
2892	5NO	10;06	215.94	2892	20NO	12;31	51.35
2893	4DE	21;07	246.02	2893	20DE	6;40	81.72

NUMBER	DATE	TIME	LONG.	NUMBER	DATE	TIME	LONG.
			−766				
2894	3JA	8;55	276.03	2894	18JA	22;24	111.74
2895	1FE	21;51	305.75	2895	17FE	11;20	141.21
2896	3MR	11;59	335.02	2896	18MR	21;38	170.06
2897	2AP	2;58	3.82	2897	17AP	5;48	198.36
2898	1MY	18;12	32.24	2898	16MY	12;39	226.31
2899	31MY	9;08	60.44	2899	14JN	19;17	254.17
2900	29JN	23;30	88.66	2900	14JL	2;55	282.23
2901	29JL	13;11	117.13	2901	12AU	12;42	310.74
2902	28AU	2;09	146.01	2902	11SE	1;26	339.85
2903	26SE	14;20	175.37	2903	10OC	17;16	9.60
2904	26OC	1;43	205.16	2904	9NO	11;34	39.84
2905	24NO	12;30	235.19	2905	9DE	7;06	70.28
2906	23DE	23;02	265.22				
			−765				
				2906	8JA	2;16	100.57
2907	22JA	9;41	294.99	2907	6FE	19;32	130.39
2908	20FE	20;42	324.32	2908	8MR	9;48	159.59
2909	22MR	8;17	353.16	2909	6AP	20;55	188.16
2910	20AP	20;37	21.59	2910	6MY	5;27	216.27
2911	20MY	9;57	49.78	2911	4JN	12;31	244.15
2912	19JN	0;28	77.96	2912	3JL	19;20	272.10
2913	18JL	16;02	106.38	2913	2AU	3;01	300.36
2914	17AU	8;04	135.20	2914	31AU	12;22	329.15
2915	15SE	23;39	164.51	2915	29SE	23;16	358.54
2916	15OC	14;02	194.27	2916	29OC	14;08	28.49
2917	14NO	3;00	224.33	2917	28NO	6;53	58.79
2918	13DE	14;42	254.43	2918	28DE	1;36	89.14
			−764				
2919	12JA	1;24	284.33	2919	26JA	20;50	119.22
2920	10FE	11;16	313.79	2920	25FE	14;42	148.76
2921	10MR	20;32	342.72	2921	26MR	5;51	177.67
2922	9AP	5;42	11.17	2922	24AP	18;00	206.04
2923	8MY	15;34	39.30	2923	24MY	3;45	234.06
2924	7JN	3;02	67.35	2924	22JN	12;05	262.01
2925	6JL	16;44	95.60	2925	21JL	19;58	290.16
2926	5AU	8;40	124.25	2926	20AU	4;12	318.73
2927	4SE	2;00	153.43	2927	18SE	13;22	347.83
2928	3OC	19;31	183.15	2928	18OC	0;03	17.49
2929	2NO	12;10	213.25	2929	16NO	12;48	47.56
2930	2DE	3;15	243.49	2930	16DE	3;53	77.80
2931	31DE	16;25	273.58				
			−763				
				2931	14JA	20;55	107.93
2932	30JA	3;33	303.26	2932	13FE	14;43	137.66
2933	28FE	12;47	332.38	2933	15MR	7;52	166.84
2934	29MR	20;43	0.94	2934	13AP	23;21	195.48
2935	28AP	4;13	29.08	2935	13MY	12;54	223.72
2936	27MY	12;26	57.03	2936	12JN	0;39	251.79
2937	25JN	22;25	85.06	2937	11JL	11;00	279.94
2938	25JL	10;53	113.43	2938	9AU	20;27	308.39
2939	24AU	2;01	142.33	2939	8SE	5;35	337.31
2940	22SE	19;26	171.85	2940	7OC	15;03	6.75
2941	22OC	14;12	201.90	2941	6NO	1;30	36.62
2942	21NO	8;59	232.25	2942	5DE	13;20	66.74
2943	21DE	2;17	262.58				
			−762				
				2943	4JA	2;40	96.81
2944	19JA	16;54	292.55	2944	2FE	17;10	126.58
2945	18FE	4;31	321.95	2945	4MR	8;24	155.89
2946	19MR	13;33	350.73	2946	2AP	23;54	184.69
2947	17AP	20;58	18.99	2947	2MY	15;19	213.11
2948	17MY	3;51	46.94	2948	1JN	6;11	241.31
2949	15JN	11;14	74.84	2949	30JN	20;03	269.51
2950	14JL	20;01	102.95	2950	30JL	8;36	297.94
2951	13AU	6;59	131.51	2951	28AU	19;58	326.77
2952	11SE	20;44	160.68	2952	27SE	6;38	356.08
2953	11OC	13;30	190.48	2953	26OC	17;12	25.84
2954	10NO	8;45	220.76	2954	25NO	4;01	55.88
2955	10DE	4;49	251.23	2955	24DE	15;10	85.93
			−761				
2956	8JA	23;30	281.49	2956	23JA	2;34	115.72
2957	7FE	15;12	311.24	2957	21FE	14;20	145.07
2958	9MR	3;37	340.35	2958	23MR	2;50	173.94
2959	7AP	13;17	8.85	2959	21AP	16;24	202.41
2960	6MY	21;10	36.92	2960	21MY	7;06	230.65
2961	5JN	4;12	64.80	2961	19JN	22;28	258.87
2962	4JL	11;14	92.76	2962	19JL	13;45	287.28
2963	2AU	19;12	121.05	2963	18AU	4;24	316.06
2964	1SE	5;01	149.87	2964	16SE	18;13	345.31
2965	30SE	17;35	179.31	2965	16OC	7;14	15.02
2966	30OC	9;24	209.33	2966	14NO	19;29	45.05
2967	29NO	3;59	239.72	2967	14DE	6;51	75.15
2968	28DE	23;41	270.11				

-760

NUMBER	DATE	TIME	LONG.	NUMBER	DATE	TIME	LONG.
				2968	12JA	17;18	105.02
2969	27JA	18;30	300.15	2969	11FE	3;03	134.47
2970	26FE	11;00	329.62	2970	11MR	12;41	163.41
2971	27MR	0;43	358.46	2971	9AP	22;57	191.89
2972	25AP	11;48	26.78	2972	9MY	10;30	220.08
2973	24MY	20;48	54.77	2973	7JN	23;39	248.20
2974	23JN	4;26	82.69	2974	7JL	14;20	276.49
2975	22JL	11;40	110.82	2975	6AU	6;13	305.15
2976	20AU	19;32	139.38	2976	4SE	22;48	334.32
2977	19SE	5;01	168.52	2977	4OC	15;24	4.01
2978	18OC	16;53	198.23	2978	3NO	7;06	34.08
2979	17NO	7;22	228.37	2979	2DE	21;05	64.28
2980	17DE	0;02	258.69				

-759

NUMBER	DATE	TIME	LONG.	NUMBER	DATE	TIME	LONG.
				2980	1JA	9;00	94.31
2981	15JA	17;55	288.84	2981	30JA	19;06	123.94
2982	14FE	11;54	318.56	2982	1MR	3;57	153.03
2983	16MR	4;51	347.72	2983	30MR	12;18	181.60
2984	14AP	19;55	16.34	2984	28AP	20;53	209.77
2985	14MY	8;35	44.54	2985	28MY	6;23	237.77
2986	12JN	19;00	72.55	2986	26JN	17;31	265.85
2987	12JL	3;51	100.64	2987	26JL	6;53	294.27
2988	10AU	12;08	129.06	2988	24AU	22;46	323.21
2989	8SE	20;51	157.97	2989	23SE	16;49	352.77
2990	8OC	6;42	187.43	2990	23OC	11;40	22.83
2991	6NO	18;00	217.35	2991	22NO	5;31	53.15
2992	6DE	6;48	247.51	2992	21DE	21;03	83.40

-758

NUMBER	DATE	TIME	LONG.	NUMBER	DATE	TIME	LONG.
2993	4JA	21;01	277.61	2993	20JA	9;52	113.29
2994	3FE	12;34	307.42	2994	18FE	20;21	142.63
2995	5MR	5;02	336.76	2995	20MR	5;05	171.39
2996	3AP	21;36	5.60	2996	18AP	12;43	199.65
2997	3MY	13;12	34.01	2997	17MY	19;59	227.61
2998	2JN	3;07	62.17	2998	16JN	3;48	255.53
2999	1JL	15;17	90.31	2999	15JL	13;16	283.67
3000	31JL	2;11	118.69	3000	14AU	1;26	312.29
3001	29AU	12;30	147.48	3001	12SE	16;52	341.54
3002	27SE	22;46	176.79	3002	12OC	11;11	11.42
3003	27OC	9;13	206.55	3003	11NO	6;53	41.73
3004	25NO	19;59	236.59	3004	11DE	1;58	72.16
3005	25DE	7;16	266.64				

-757

NUMBER	DATE	TIME	LONG.	NUMBER	DATE	TIME	LONG.
				3005	9JA	18;59	102.34
3006	23JA	19;23	296.45	3006	8FE	9;15	132.02
3007	22FE	8;37	325.85	3007	9MR	20;44	161.07
3008	23MR	22;55	354.78	3008	8AP	5;47	189.54
3009	22AP	13;52	23.30	3009	7MY	13;06	217.59
3010	22MY	4;55	51.54	3010	5JN	19;40	245.45
3011	20JN	19;39	79.74	3011	5JL	2;41	273.41
3012	20JL	9;53	108.12	3012	3AU	11;20	301.73
3013	18AU	23;31	136.87	3013	1SE	22;37	330.62
3014	17SE	12;25	166.10	3014	1OC	13;00	0.15
3015	17OC	0;26	195.78	3015	31OC	6;14	30.25
3016	15NO	11;37	225.77	3016	30NO	1;21	60.65
3017	14DE	22;14	255.83	3017	29DE	20;51	91.03

-756

NUMBER	DATE	TIME	LONG.	NUMBER	DATE	TIME	LONG.
3018	13JA	8;40	285.70	3018	28JA	15;08	121.04
3019	11FE	19;14	315.17	3019	27FE	6;53	150.47
3020	12MR	6;14	344.15	3020	27MR	19;28	179.25
3021	10AP	17;53	12.68	3021	26AP	5;09	207.50
3022	10MY	6;28	40.90	3022	25MY	12;49	235.43
3023	8JN	20;16	69.05	3023	23JN	19;40	263.33
3024	8JL	11;25	97.36	3024	23JL	2;54	291.47
3025	7AU	3;32	126.05	3025	21AU	11;28	320.07
3026	5SE	19;45	155.21	3026	19SE	22;00	349.25
3027	5OC	11;06	184.86	3027	19OC	10;56	19.02
3028	4NO	1;01	214.87	3028	18NO	2;24	49.21
3029	3DE	13;25	245.00	3029	17DE	20;09	79.57

-755

NUMBER	DATE	TIME	LONG.	NUMBER	DATE	TIME	LONG.
3030	2JA	0;33	275.00	3030	16JA	15;10	109.76
3031	31JA	10;41	304.61	3031	15FE	9;45	139.49
3032	1MR	20;00	333.71	3032	17MR	2;13	168.62
3033	31MR	4;54	2.29	3033	15AP	15;47	197.16
3034	29AP	14;05	30.49	3034	15MY	2;40	225.28
3035	29MY	0;28	58.52	3035	13JN	11;41	253.24
3036	27JN	12;56	86.65	3036	12JL	19;50	281.31
3037	27JL	3;52	115.14	3037	11AU	4;00	309.73
3038	25AU	20;49	144.15	3038	9SE	12;48	338.66
3039	24SE	14;39	173.71	3039	8OC	22;48	8.14
3040	24OC	8;06	203.73	3040	7NO	10;30	38.08
3041	23NO	0;12	233.98	3041	7DE	0;20	68.28
3042	22DE	14;25	264.17				

-754

NUMBER	DATE	TIME	LONG.	NUMBER	DATE	TIME	LONG.
				3042	5JA	16;17	98.45
3043	21JA	2;31	294.01	3043	4FE	9;37	128.32
3044	19FE	12;31	323.32	3044	6MR	3;00	157.68
3045	20MR	20;50	352.05	3045	4AP	19;13	186.49
3046	19AP	4;15	20.30	3046	4MY	9;41	214.85
3047	18MY	11;49	48.27	3047	2JN	22;23	242.96
3048	16JN	20;43	76.23	3048	2JL	9;33	271.07
3049	16JL	7;52	104.45	3049	31JL	19;38	299.42
3050	14AU	21;45	133.15	3050	30AU	5;05	328.19
3051	13SE	14;14	162.46	3051	28SE	14;31	357.47
3052	13OC	8;39	192.35	3052	28OC	0;33	27.23
3053	12NO	3;46	222.65	3053	26NO	11;40	57.28
3054	11DE	22;05	253.04	3054	26DE	0;08	87.38

-753

NUMBER	DATE	TIME	LONG.	NUMBER	DATE	TIME	LONG.
3055	10JA	14;09	283.17	3055	24JA	13;52	117.25
3056	9FE	3;11	312.79	3056	23FE	4;29	146.70
3057	10MR	13;18	341.78	3057	24MR	19;35	175.65
3058	8AP	21;16	10.19	3058	23AP	10;54	204.16
3059	8MY	4;11	38.22	3059	23MY	2;03	232.41
3060	6JN	11;08	66.09	3060	21JN	16;35	260.60
3061	5JL	19;07	94.10	3061	21JL	6;02	288.96
3062	4AU	4;58	122.48	3062	19AU	18;15	317.67
3063	2SE	17;20	151.42	3063	18SE	5;29	346.84
3064	2OC	8;42	181.01	3064	17OC	16;15	16.48
3065	1NO	2;55	211.15	3065	16NO	3;02	46.45
3066	30NO	22;53	241.60	3066	15DE	14;00	76.53
3067	30DE	18;30	271.97				

-752

NUMBER	DATE	TIME	LONG.	NUMBER	DATE	TIME	LONG.
				3067	14JA	1;08	106.42
3068	29JA	11;45	301.93	3068	12FE	12;26	135.91
3069	28FE	1;41	331.27	3069	13MR	0;12	164.91
3070	28MR	12;32	359.97	3070	11AP	12;54	193.47
3071	26AP	21;09	28.16	3071	11MY	2;52	221.74
3072	26MY	4;29	56.08	3072	9JN	17;55	249.94
3073	24JN	11;27	83.99	3073	9JL	9;27	278.28
3074	23JL	18;55	112.14	3074	8AU	0;43	306.94
3075	22AU	3;48	140.77	3075	6SE	15;16	336.05
3076	20SE	15;00	169.99	3076	6OC	4;59	5.64
3077	20OC	5;14	199.82	3077	4NO	17;50	35.61
3078	18NO	22;34	230.10	3078	4DE	5;44	65.72
3079	18DE	17;52	260.52				

-751

NUMBER	DATE	TIME	LONG.	NUMBER	DATE	TIME	LONG.
				3079	2JA	16;35	95.70
3080	17JA	13;13	290.71	3080	1FE	2;28	125.30
3081	16FE	6;51	320.39	3081	2MR	11;50	154.39
3082	17MR	21;51	349.45	3082	31MR	21;25	182.99
3083	16AP	10;09	17.93	3083	30AP	8;02	211.23
3084	15MY	20;08	46.01	3084	29MY	20;12	239.33
3085	14JN	4;26	73.94	3085	28JN	10;06	267.53
3086	13JL	11;53	101.98	3086	28JL	1;32	296.05
3087	11AU	19;28	130.39	3087	26AU	18;04	325.05
3088	10SE	4;11	159.32	3088	25SE	11;03	354.59
3089	9OC	14;54	188.84	3089	25OC	3;35	24.58
3090	8NO	4;06	218.86	3090	23NO	18;41	54.79
3091	7DE	19;41	249.13	3091	23DE	7;41	84.93

-750

NUMBER	DATE	TIME	LONG.	NUMBER	DATE	TIME	LONG.
3092	6JA	12;55	279.35	3092	21JA	18;33	114.71
3093	5FE	6;45	309.23	3093	20FE	3;47	143.98
3094	7MR	0;07	338.57	3094	21MR	12;05	172.70
3095	5AP	16;05	7.36	3095	19AP	20;15	200.97
3096	5MY	5;57	35.69	3096	19MY	5;03	228.99
3097	3JN	17;33	63.75	3097	17JN	15;11	256.99
3098	3JL	3;14	91.81	3098	17JL	3;22	285.26
3099	1AU	11;53	120.11	3099	15AU	18;04	314.01
3100	30AU	20;29	148.86	3100	14SE	11;19	343.36
3101	29SE	5;54	178.15	3101	14OC	6;12	13.29
3102	28OC	16;34	207.93	3102	13NO	0;58	43.58
3103	27NO	4;36	238.03	3103	12DE	17;52	73.91
3104	26DE	17;59	268.16				

-749

NUMBER	DATE	TIME	LONG.	NUMBER	DATE	TIME	LONG.
				3104	11JA	8;01	103.95
3105	25JA	8;38	298.07	3105	9FE	19;31	133.49
3106	24FE	0;24	327.55	3106	11MR	4;55	162.44
3107	25MR	16;47	356.53	3107	9AP	12;52	190.85
3108	24AP	8;50	25.07	3108	8MY	20;06	218.88
3109	23MY	23;39	53.30	3109	7JN	3;27	246.77
3110	22JN	12;48	81.44	3110	6JL	11;55	274.80
3111	22JL	0;29	109.73	3111	4AU	22;39	303.22
3112	20AU	11;18	138.39	3112	3SE	12;31	332.24
3113	18SE	21;48	167.55	3113	3OC	5;38	1.92
3114	18OC	8;20	197.18	3114	2NO	1;00	32.12
3115	16NO	19;01	227.16	3115	1DE	20;42	62.55
3116	16DE	5;58	257.23	3116	31DE	14;54	92.86

Left half

	NEW MOONS				FULL MOONS		
NUMBER	DATE	TIME	LONG.	NUMBER	DATE	TIME	LONG.
-748							
3117	14JA	17;26	287.13	3117	30JA	6;31	122.74
3118	13FE	5;49	316.66	3118	28FE	19;17	152.02
3119	13MR	19;18	345.72	3119	29MR	5;23	180.68
3120	12AP	9;43	14.33	3120	27AP	13;23	208.84
3121	12MY	0;37	42.64	3121	26MY	20;10	236.74
3122	10JN	15;32	70.83	3122	25JN	2;49	264.63
3123	10JL	6;11	99.13	3123	24JL	10;34	292.80
3124	8AU	20;24	127.77	3124	22AU	20;31	321.48
3125	7SE	10;01	156.86	3125	21SE	9;26	350.79
3126	6OC	22;46	186.42	3126	21OC	1;23	20.70
3127	5NO	10;32	216.35	3127	19NO	19;44	51.03
3128	4DE	21;27	246.42	3128	19DE	15;11	81.45
-747							
3129	3JA	7;50	276.37	3129	18JA	10;10	111.62
3130	1FE	18;07	305.98	3130	17FE	3;12	141.26
3131	3MR	4;36	335.11	3131	18MR	17;18	170.26
3132	1AP	15;36	3.75	3132	17AP	4;19	198.68
3133	1MY	3;26	32.03	3133	16MY	12;51	226.70
3134	30MY	16;27	60.16	3134	14JN	20;01	254.59
3135	29JN	6;56	88.39	3135	14JL	3;02	282.62
3136	28JL	22;47	116.93	3136	12AU	10;58	311.05
3137	27AU	15;21	145.95	3137	10SE	20;35	340.03
3138	26SE	7;33	175.48	3138	10OC	8;22	9.61
3139	25OC	22;29	205.41	3139	8NO	22;35	39.67
3140	24NO	11;47	235.55	3140	8DE	15;11	69.99
3141	23DE	23;32	265.63				
-746							
				3141	7JA	9;34	100.25
3142	22JA	10;03	295.39	3142	6FE	4;23	130.15
3143	20FE	19;33	324.66	3143	7MR	21;54	159.49
3144	22MR	4;22	353.39	3144	6AP	12;52	188.23
3145	20AP	13;05	21.67	3145	6MY	1;01	216.48
3146	19MY	22;33	49.71	3146	4JN	10;56	244.47
3147	18JN	9;48	77.76	3147	3JL	19;33	272.49
3148	17JL	23;29	106.10	3148	2AU	3;48	300.78
3149	16AU	15;40	134.92	3149	31AU	12;24	329.54
3150	15SE	9;27	164.31	3150	29SE	21;54	358.84
3151	15OC	3;30	194.21	3151	29OC	8;46	28.65
3152	13NO	20;34	224.44	3152	27NO	21;28	58.78
3153	13DE	11;53	254.70	3153	27DE	12;15	88.97
-745							
3154	12JA	1;02	284.70	3154	26JA	4;47	118.94
3155	10FE	11;57	314.20	3155	24FE	22;01	148.46
3156	11MR	20;52	343.12	3156	26MR	14;42	177.45
3157	10AP	4;27	11.50	3157	25AP	5;59	205.94
3158	9MY	11;39	39.53	3158	24MY	19;35	234.12
3159	7JN	19;39	67.44	3159	23JN	7;37	262.22
3160	7JL	5;32	95.53	3160	22JL	18;25	290.48
3161	5AU	18;04	124.04	3161	21AU	4;23	319.12
3162	4SE	9;25	153.13	3162	19SE	13;59	348.25
3163	4OC	3;08	182.85	3163	18OC	23;49	17.87
3164	2NO	22;14	213.04	3164	17NO	10;23	47.84
3165	2DE	17;16	243.45	3165	16DE	22;06	77.95
-744							
3166	1JA	10;38	273.73	3166	15JA	11;02	107.90
3167	31JA	1;10	303.55	3167	14FE	0;58	137.48
3168	29FE	12;35	332.75	3168	14MR	15;33	166.57
3169	29MR	21;23	1.35	3169	13AP	6;33	195.20
3170	28AP	4;36	29.48	3170	12MY	21;42	223.51
3171	27MY	11;20	57.37	3171	11JN	12;39	251.70
3172	25JN	18;41	85.29	3172	11JL	2;53	279.99
3173	25JL	3;32	113.52	3173	9AU	16;00	308.59
3174	23AU	14;39	142.25	3174	8SE	3;58	337.63
3175	22SE	4;35	171.62	3175	7OC	15;10	7.14
3176	21OC	21;31	201.58	3176	6NO	2;04	37.04
3177	20NO	16;52	231.95	3177	5DE	12;58	67.11
3178	20DE	12;55	262.40				
-743							
				3178	3JA	23;55	97.08
3179	19JA	7;29	292.54	3179	2FE	10;54	126.71
3180	17FE	23;02	322.11	3180	3MR	22;06	155.85
3181	19MR	11;16	351.02	3181	2AP	9;58	184.52
3182	17AP	20;49	19.37	3182	1MY	23;02	212.84
3183	17MY	4;39	47.36	3183	31MY	13;29	241.03
3184	15JN	11;43	75.24	3184	30JN	4;55	269.30
3185	14JL	18;55	103.29	3185	29JL	20;36	297.84
3186	13AU	3;06	131.74	3186	28AU	11;50	326.82
3187	11SE	13;08	160.75	3187	27SE	2;16	356.29
3188	11OC	1;52	190.37	3188	26OC	15;49	26.17
3189	9NO	17;43	220.51	3189	25NO	4;21	56.28
3190	9DE	12;10	250.91	3190	24DE	15;44	86.34

Right half

	NEW MOONS				FULL MOONS		
NUMBER	DATE	TIME	LONG.	NUMBER	DATE	TIME	LONG.
-742							
3191	8JA	7;36	281.21	3191	23JA	1;56	116.08
3192	7FE	2;06	311.08	3192	21FE	11;16	145.33
3193	8MR	18;20	340.35	3193	22MR	20;24	174.07
3194	7AP	7;54	9.02	3194	21AP	6;09	202.38
3195	6MY	18;59	37.22	3195	20MY	17;18	230.48
3196	5JN	4;06	65.18	3196	19JN	6;15	258.61
3197	4JL	11;59	93.18	3197	18JL	20;59	286.99
3198	2AU	19;33	121.45	3198	17AU	13;13	315.83
3199	1SE	3;44	150.20	3199	16SE	6;20	345.20
3200	30SE	13;30	179.53	3200	15OC	23;30	15.09
3201	30OC	1;30	209.39	3201	14NO	15;38	45.28
3202	28NO	15;54	239.59	3202	14DE	5;49	75.49
3203	28DE	8;16	269.84				
-741							
				3203	12JA	17;41	105.43
3204	27JA	1;41	299.84	3204	11FE	3;32	134.88
3205	25FE	19;10	329.36	3205	12MR	12;01	163.77
3206	27MR	11;45	358.33	3206	10AP	19;58	192.15
3207	26AP	2;40	26.80	3207	10MY	4;11	220.21
3208	25MY	15;27	54.94	3208	8JN	13;26	248.18
3209	24JN	2;10	82.99	3209	8JL	0;29	276.32
3210	23JL	11;25	111.20	3210	6AU	13;55	304.88
3211	21AU	20;09	139.80	3211	5SE	6;05	334.02
3212	20SE	5;18	168.91	3212	5OC	0;31	3.77
3213	19OC	15;27	198.55	3213	3NO	19;47	33.99
3214	18NO	2;50	228.57	3214	3DE	13;56	64.36
3215	17DE	15;27	258.71				
-740							
				3215	2JA	5;32	94.55
3216	16JA	5;15	288.69	3216	31JA	18;15	124.29
3217	14FE	20;12	318.30	3217	1MR	4;29	153.43
3218	15MR	12;04	347.43	3218	30MR	12;56	182.01
3219	14AP	4;12	16.09	3219	28AP	20;19	210.13
3220	13MY	19;39	44.41	3220	28MY	3;25	238.04
3221	12JN	9;44	72.56	3221	26JN	11;10	265.98
3222	11JL	22;19	100.80	3222	25JL	20;39	294.24
3223	10AU	9;45	129.34	3223	24AU	8;57	323.03
3224	8SE	20;38	158.35	3224	23SE	0;35	352.48
3225	8OC	7;22	187.85	3225	22OC	19;08	22.52
3226	6NO	18;06	217.75	3226	21NO	15;00	52.92
3227	6DE	4;54	247.81	3227	21DE	10;10	83.33
-739							
3228	4JA	15;56	277.78	3228	20JA	3;06	113.39
3229	3FE	3;35	307.43	3229	18FE	17;13	142.88
3230	4MR	16;12	336.62	3230	20MR	4;30	171.75
3231	3AP	5;53	5.35	3231	18AP	13;24	200.06
3232	2MY	20;23	33.72	3232	17MY	20;37	228.02
3233	1JN	11;16	61.92	3233	16JN	3;12	255.89
3234	1JL	2;10	90.17	3234	15JL	10;19	283.94
3235	30JL	16;51	118.69	3235	13AU	19;08	312.42
3236	29AU	7;07	147.65	3236	12SE	6;36	341.50
3237	27SE	20;38	177.08	3237	11OC	21;09	11.21
3238	27OC	9;07	206.93	3238	10NO	14;26	41.42
3239	25NO	20;32	237.00	3239	10DE	9;28	71.84
3240	25DE	7;07	267.02				
-738							
				3240	9JA	4;47	102.13
3241	23JA	17;15	296.75	3241	7FE	22;49	131.97
3242	22FE	3;21	326.02	3242	9MR	14;22	161.20
3243	23MR	13;47	354.80	3243	8AP	2;50	189.81
3244	22AP	0;54	23.16	3244	7MY	12;29	217.94
3245	21MY	13;05	51.30	3245	5JN	20;15	245.86
3246	20JN	2;45	79.45	3246	5JL	3;17	273.82
3247	19JL	18;02	107.87	3247	3AU	10;47	302.10
3248	18AU	10;35	136.73	3248	1SE	19;37	330.88
3249	17SE	3;24	166.11	3249	1OC	6;24	0.26
3250	16OC	19;21	195.95	3250	30OC	19;26	30.17
3251	15NO	9;41	226.07	3251	29NO	10;49	60.42
3252	14DE	22;14	256.22	3252	29DE	4;16	90.72
-737							
3253	13JA	9;17	286.11	3253	27JA	22;51	120.75
3254	11FE	19;06	315.55	3254	26FE	17;01	150.28
3255	13MR	4;00	344.44	3255	28MR	9;14	179.22
3256	11AP	12;27	12.84	3256	26AP	22;44	207.62
3257	10MY	21;14	40.92	3257	26MY	9;43	235.69
3258	9JN	7;21	68.92	3258	24JN	19;10	263.69
3259	8JL	19;43	97.12	3259	24JL	3;31	291.88
3260	7AU	10;47	125.75	3260	22AU	12;04	320.48
3261	6SE	4;06	154.95	3261	20SE	21;15	349.61
3262	5OC	22;26	184.72	3262	20OC	7;30	19.26
3263	4NO	16;21	214.89	3263	18NO	19;14	49.30
3264	4DE	8;44	245.19	3264	18DE	8;52	79.48

-736 (New & Full Moons, left) / -730 (New & Full Moons, right)

NEW No.	Date	Time	Long.	FULL No.	Date	Time	Long.	NEW No.	Date	Time	Long.	FULL No.	Date	Time	Long.
3265	2JA	23;01	275.32	3265	17JA	0;23	109.52	3340	25JA	1;51	298.17	3339	10JA	23;00	103.96
3266	1FE	10;58	305.01	3266	15FE	17;07	139.19	3341	23FE	13;39	327.49	3340	9FE	14;29	133.67
3267	1MR	20;42	334.13	3267	16MR	9;58	168.35	3342	25MR	2;30	356.34	3341	11MR	3;04	162.75
3268	31MR	4;41	2.67	3268	15AP	1;52	196.99	3343	23AP	16;23	24.80	3342	9AP	13;00	191.24
3269	29AP	11;48	30.78	3269	14MY	16;17	225.25	3344	23MY	6;59	53.02	3343	8MY	20;55	219.29
3270	28MY	19;09	58.69	3270	13JN	5;10	253.37	3345	21JN	21;57	81.23	3344	7JN	3;41	247.16
3271	27JN	3;55	86.68	3271	12JL	16;46	281.57	3346	21JL	12;57	109.65	3345	6JL	10;26	275.12
3272	26JL	15;04	115.01	3272	11AU	3;21	310.09	3347	20AU	3;44	138.47	3346	4AU	18;21	303.43
3273	25AU	5;06	143.89	3273	9SE	13;20	339.07	3348	18SE	17;59	167.77	3347	3SE	4;30	332.30
3274	23SE	21;52	173.40	3274	8OC	23;10	8.54	3349	18OC	7;17	197.53	3348	2OC	17;35	1.80
3275	23OC	16;35	203.46	3275	7NO	9;25	38.42	3350	16NO	19;23	227.56	3349	1NO	9;38	31.85
3276	22NO	11;57	233.84	3276	6DE	20;31	68.51	3351	16DE	6;21	257.64	3350	1DE	3;55	62.23
3277	22DE	6;23	264.21									3351	30DE	23;11	92.59

-735 / -729

NEW No.	Date	Time	Long.	FULL No.	Date	Time	Long.	NEW No.	Date	Time	Long.	FULL No.	Date	Time	Long.
3278	20JA	22;24	294.23	3277	5JA	8;42	98.52	3352	14JA	16;33	287.48	3352	29JA	17;54	122.60
3279	19FE	11;16	323.66	3278	3FE	21;54	128.22	3353	13FE	2;25	316.90	3353	28FE	10;41	152.05
3280	20MR	21;10	352.45	3279	5MR	11;53	157.45	3354	14MR	12;23	345.82	3354	30MR	0;37	180.87
3281	19AP	4;56	20.71	3280	4AP	2;25	186.20	3355	12AP	22;50	14.28	3355	28AP	11;36	209.15
3282	18MY	11;43	48.65	3281	3MY	17;21	214.59	3356	12MY	10;13	42.45	3356	27MY	20;12	237.12
3283	16JN	18;37	76.53	3282	2JN	8;26	242.79	3357	10JN	22;59	70.56	3357	26JN	3;33	265.04
3284	16JL	2;39	104.62	3283	1JL	23;13	271.04	3358	10JL	13;29	98.85	3358	25JL	10;49	293.20
3285	14AU	12;38	133.16	3284	31JL	13;11	299.54	3359	9AU	5;39	127.55	3359	23AU	19;03	321.80
3286	13SE	1;11	162.30	3285	30AU	2;01	328.45	3360	7SE	22;46	156.78	3360	22SE	4;56	350.99
3287	12OC	16;43	192.07	3286	28SE	13;49	357.83	3361	7OC	15;35	186.50	3361	21OC	16;54	20.73
3288	11NO	11;02	222.33	3287	28OC	1;01	27.63	3362	6NO	7;00	216.59	3362	20NO	7;05	50.88
3289	11DE	7;00	252.78	3288	26NO	11;59	57.68	3363	5DE	20;33	246.77	3363	19DE	23;27	81.17
				3289	25DE	22;52	87.71								

-734 / -728

NEW No.	Date	Time	Long.	FULL No.	Date	Time	Long.	NEW No.	Date	Time	Long.	FULL No.	Date	Time	Long.
3290	10JA	2;30	283.07	3290	24JA	9;39	117.46	3364	4JA	8;18	276.78	3364	18JA	17;25	111.30
3291	8FE	19;34	312.86	3291	22FE	20;25	146.75	3365	2FE	18;35	306.38	3365	17FE	11;46	141.01
3292	10MR	9;20	342.00	3292	24MR	7;36	175.55	3366	3MR	3;42	335.45	3366	18MR	4;57	170.15
3293	8AP	20;03	10.53	3293	22AP	19;45	203.94	3367	1AP	12;06	3.99	3367	16AP	19;45	198.72
3294	8MY	4;36	38.62	3294	22MY	9;20	232.13	3368	30AP	20;24	32.14	3368	16MY	7;57	226.89
3295	6JN	11;58	66.51	3295	21JN	0;19	260.34	3369	30MY	5;34	60.12	3369	14JN	18;05	254.89
3296	5JL	19;04	94.48	3296	20JL	16;07	288.79	3370	28JN	16;39	88.20	3370	14JL	3;03	283.01
3297	4AU	2;46	122.77	3297	19AU	7;54	317.63	3371	28JL	6;23	116.66	3371	12AU	11;43	311.47
3298	2SE	11;53	151.58	3298	17SE	23;05	346.96	3372	26AU	22;49	145.66	3372	10SE	20;44	340.43
3299	1OC	23;17	181.00	3299	17OC	13;22	16.74	3373	25SE	17;03	175.26	3373	10OC	6;33	9.92
3300	31OC	13;36	210.97	3300	16NO	2;36	46.82	3374	25OC	11;34	205.33	3374	8NO	17;33	39.84
3301	30NO	6;51	241.30	3301	15DE	14;37	76.94	3375	24NO	4;59	235.88	3375	8DE	6;08	69.99
3302	30DE	1;54	271.66					3376	23DE	20;26	265.88				

-733 / -727

NEW No.	Date	Time	Long.	FULL No.	Date	Time	Long.	NEW No.	Date	Time	Long.	FULL No.	Date	Time	Long.
3303	28JA	20;54	301.70	3302	14JA	1;19	106.81	3377	22JA	9;31	295.75	3376	6JA	20;33	100.09
3304	27FE	14;13	331.18	3303	12FE	10;51	136.23	3378	20FE	20;12	325.07	3377	5FE	12;31	129.88
3305	29MR	5;01	0.05	3304	13MR	19;45	165.11	3379	22MR	4;50	353.79	3378	7MR	5;10	159.19
3306	27AP	17;16	28.39	3305	12AP	4;50	193.53	3380	20AP	12;07	22.02	3379	5AP	21;25	187.99
3307	27MY	3;21	56.42	3306	11MY	15;00	221.66	3381	19MY	19;05	49.96	3380	5MY	12;31	216.36
3308	25JN	11;52	84.39	3307	10JN	2;53	249.73	3382	18JN	2;56	77.88	3381	4JN	2;13	244.51
3309	24JL	19;38	112.56	3308	9JL	16;44	277.99	3383	17JL	12;48	106.05	3382	3JL	14;37	272.68
3310	23AU	3;34	141.14	3309	8AU	8;24	306.66	3384	16AU	1;25	134.72	3383	2AU	1;55	301.09
3311	21SE	12;36	170.28	3310	7SE	1;23	335.86	3385	14SE	16;59	164.01	3384	31AU	12;25	329.93
3312	20OC	23;31	199.97	3311	6OC	18;56	5.61	3386	14OC	10;59	193.91	3385	29SE	22;30	359.26
3313	19NO	12;44	230.07	3312	5NO	11;57	35.75	3387	13NO	6;20	224.22	3386	29OC	8;39	29.03
3314	19DE	4;05	260.32	3313	5DE	3;20	66.01	3388	13DE	1;30	254.65	3387	27NO	19;18	59.07
												3388	27DE	6;50	89.12

-732 / -726

NEW No.	Date	Time	Long.	FULL No.	Date	Time	Long.	NEW No.	Date	Time	Long.	FULL No.	Date	Time	Long.
3315	17JA	20;52	290.41	3314	3JA	16;23	96.08	3389	11JA	18;52	284.82	3389	25JA	19;19	118.93
3316	16FE	14;10	320.09	3315	2FE	3;04	125.71	3390	10FE	9;16	314.48	3390	24FE	8;38	148.30
3317	17MR	7;05	349.23	3316	2MR	11;58	154.78	3391	11MR	20;29	343.49	3391	25MR	22;36	177.18
3318	15AP	22;48	17.85	3317	31MR	19;54	183.31	3392	10AP	5;06	11.91	3392	24AP	13;07	205.65
3319	15MY	12;42	46.10	3318	30AP	3;42	211.44	3393	9MY	12;09	39.93	3393	24MY	4;04	233.89
3320	14JN	0;32	74.17	3319	29MY	12;13	239.40	3394	7JN	18;50	67.79	3394	22JN	19;08	262.11
3321	13JL	10;37	102.32	3320	27JN	22;13	267.44	3395	7JL	2;13	95.78	3395	22JL	9;48	290.52
3322	11AU	19;44	130.79	3321	27JL	10;25	295.82	3396	5AU	11;12	124.15	3396	20AU	23;31	319.31
3323	10SE	4;48	159.74	3322	26AU	1;19	324.75	3397	3SE	22;30	153.07	3397	19SE	12;06	348.56
3324	9OC	14;34	189.22	3323	24SE	18;54	354.30	3398	3OC	12;36	182.62	3398	18OC	23;48	18.25
3325	8NO	1;25	219.13	3324	24OC	14;10	24.40	3399	2NO	5;39	212.73	3399	17NO	10;58	48.26
3326	7DE	13;23	249.25	3325	23NO	9;15	54.78	3400	2DE	1;00	243.15	3400	16DE	21;53	78.32
				3326	23DE	2;18	85.08	3401	31DE	20;57	273.53				

-731 / -725

NEW No.	Date	Time	Long.	FULL No.	Date	Time	Long.	NEW No.	Date	Time	Long.	FULL No.	Date	Time	Long.
3327	6JA	2;25	279.29	3327	21JA	16;25	115.01	3402	30JA	15;19	303.52	3401	15JA	8;35	108.18
3328	4FE	16;31	309.02	3328	20FE	3;43	144.36	3403	1MR	6;40	332.90	3402	13FE	19;06	137.62
3329	6MR	7;40	338.29	3329	21MR	12;50	173.12	3404	30MR	18;45	1.63	3403	15MR	5;44	166.55
3330	4AP	23;30	7.08	3330	19AP	20;33	201.37	3405	29AP	4;14	29.84	3404	13AP	17;02	195.04
3331	4MY	15;16	35.49	3331	19MY	3;35	229.31	3406	28MY	12;05	57.78	3405	13MY	5;39	223.25
3332	3JN	6;08	63.68	3332	17JN	10;51	257.21	3407	26JN	19;17	85.70	3406	11JN	19;53	251.42
3333	2JL	19;37	91.89	3333	16JL	19;21	285.32	3408	26JL	2;42	113.87	3407	11JL	11;27	279.76
3334	1AU	7;49	120.33	3334	15AU	6;11	313.91	3409	24AU	11;08	142.49	3408	10AU	3;33	308.47
3335	30AU	19;13	149.19	3335	13SE	20;13	343.12	3410	22SE	21;24	171.70	3409	8SE	19;24	337.66
3336	29SE	6;15	178.55	3336	13OC	13;33	12.97	3411	22OC	10;16	201.49	3410	8OC	10;27	7.33
3337	28OC	17;09	208.35	3337	12NO	9;05	43.29	3412	21NO	2;06	231.71	3411	7NO	0;27	37.35
3338	27NO	3;58	238.39	3338	12DE	4;52	73.74	3413	20DE	20;19	262.08	3412	6DE	13;12	67.51
3339	26DE	14;46	268.42												

-724

NUMBER	DATE	TIME	LONG.	NUMBER	DATE	TIME	LONG.
				3413	5JA	0;32	97.49
3414	19JA	15;24	292.25	3414	3FE	10;27	127.07
3415	18FE	9;32	321.93	3415	3MR	19;22	156.12
3416	19MR	1;30	351.01	3416	2AP	4;00	184.66
3417	17AP	14;57	19.52	3417	1MY	13;18	212.84
3418	17MY	2;05	47.64	3418	31MY	0;07	240.88
3419	15JN	11;24	75.61	3419	29JN	12;55	269.04
3420	14JL	19;36	103.70	3420	29JL	3;47	297.55
3421	13AU	3;32	132.14	3421	27AU	20;22	326.57
3422	11SE	12;05	161.09	3422	26SE	14;00	356.16
3423	10OC	22;07	190.60	3423	26OC	7;41	26.21
3424	9NO	10;11	220.58	3424	25NO	0;11	56.49
3425	9DE	0;26	250.80	3425	24DE	14;30	86.67

-723

NUMBER	DATE	TIME	LONG.	NUMBER	DATE	TIME	LONG.
3426	7JA	16;25	280.95	3426	23JA	2;15	116.48
3427	6FE	9;18	310.77	3427	21FE	11;49	145.74
3428	8MR	2;16	340.08	3428	22MR	19;57	174.43
3429	6AP	18;30	8.86	3429	21AP	3;33	202.66
3430	6MY	9;19	37.22	3430	20MY	11;28	230.64
3431	4JN	22;16	65.34	3431	18JN	20;34	258.61
3432	4JL	9;21	93.46	3432	18JL	7;34	286.84
3433	2AU	19;04	121.82	3433	16AU	21;08	315.56
3434	1SE	4;18	150.61	3434	15SE	13;34	344.90
3435	30SE	13;51	179.93	3435	15OC	8;22	14.83
3436	30OC	0;16	209.72	3436	14NO	3;57	45.17
3437	28NO	11;41	239.80	3437	13DE	22;17	75.56
3438	28DE	0;04	269.88				

-722

NUMBER	DATE	TIME	LONG.	NUMBER	DATE	TIME	LONG.
				3438	12JA	13;54	105.65
3439	26JA	13;23	299.71	3439	11FE	2;28	135.22
3440	25FE	3;43	329.11	3440	12MR	12;29	164.17
3441	26MR	18;59	358.03	3441	10AP	20;41	192.57
3442	25AP	10;41	26.54	3442	10MY	3;52	220.58
3443	25MY	2;04	54.79	3443	8JN	10;51	248.46
3444	23JN	16;22	82.98	3444	7JL	18;36	276.47
3445	23JL	5;25	111.34	3445	6AU	4;12	304.87
3446	21AU	17;26	140.07	3446	4SE	16;38	333.85
3447	20SE	4;54	169.28	3447	4OC	8;28	3.48
3448	19OC	16;04	198.97	3448	3NO	3;11	33.67
3449	18NO	3;02	228.97	3449	2DE	23;08	64.12
3450	17DE	13;47	259.03				

-721

NUMBER	DATE	TIME	LONG.	NUMBER	DATE	TIME	LONG.
				3450	1JA	18;16	94.46
3451	16JA	0;32	288.88	3451	31JA	11;04	124.37
3452	14FE	11;40	318.33	3452	2MR	1;00	153.68
3453	15MR	23;39	347.31	3453	31MR	12;08	182.36
3454	14AP	12;45	15.86	3454	29AP	20;55	210.54
3455	14MY	2;51	44.12	3455	29MY	4;07	238.45
3456	12JN	17;38	72.31	3456	27JN	10;47	266.35
3457	12JL	8;46	100.64	3457	26JL	18;03	294.52
3458	10AU	23;57	129.33	3458	25AU	3;05	323.18
3459	9SE	14;50	158.49	3459	23SE	14;05	352.45
3460	9OC	4;57	188.13	3460	23OC	5;25	22.33
3461	7NO	17;52	218.12	3461	21NO	22;42	52.62
3462	7DE	5;27	248.23	3462	21DE	17;33	83.01

-720

NUMBER	DATE	TIME	LONG.	NUMBER	DATE	TIME	LONG.
3463	5JA	15;55	278.17	3463	20JA	12;35	113.17
3464	4FE	1;44	307.74	3464	19FE	6;20	142.82
3465	4MR	11;21	336.80	3465	19MR	21;40	171.85
3466	2AP	21;15	5.38	3466	18AP	10;03	200.31
3467	2MY	7;52	33.61	3467	17MY	19;45	228.37
3468	31MY	19;44	61.69	3468	16JN	3;40	256.29
3469	30JN	9;18	89.89	3469	15JL	10;58	284.35
3470	30JL	0;47	118.43	3470	13AU	18;46	312.79
3471	28AU	17;47	147.48	3471	12SE	3;55	341.78
3472	27SE	11;12	177.08	3472	11OC	14;55	11.34
3473	27OC	3;41	207.09	3473	10NO	4;00	41.37
3474	25NO	18;21	237.29	3474	9DE	19;13	71.62
3475	25DE	7;00	267.40				

-719

NUMBER	DATE	TIME	LONG.	NUMBER	DATE	TIME	LONG.
				3475	8JA	12;17	101.82
3476	23JA	17;54	297.16	3476	7FE	6;24	131.67
3477	22FE	3;24	326.41	3477	9MR	0;08	161.00
3478	23MR	11;53	355.10	3478	7AP	16;06	189.76
3479	21AP	19;56	23.35	3479	7MY	5;34	218.05
3480	21MY	4;24	51.34	3480	5JN	16;54	246.10
3481	19JN	14;18	79.34	3481	5JL	2;20	274.16
3482	19JL	2;38	107.63	3482	3AU	11;16	302.50
3483	17AU	17;52	136.43	3483	1SE	20;16	331.30
3484	16SE	11;32	165.84	3484	1OC	5;50	0.63
3485	16OC	6;20	195.79	3485	30OC	16;17	30.43
3486	15NO	0;38	226.08	3486	29NO	4;01	60.52
3487	14DE	17;14	256.39	3487	28DE	17;21	90.64

-718

NUMBER	DATE	TIME	LONG.	NUMBER	DATE	TIME	LONG.
3488	13JA	7;31	286.42	3488	27JA	8;21	120.53
3489	11FE	19;17	315.94	3489	26FE	0;29	149.99
3490	13MR	4;45	344.86	3490	27MR	16;48	178.94
3491	11AP	12;27	13.23	3491	26AP	8;24	207.44
3492	10MY	19;19	41.23	3492	25MY	22;48	235.64
3493	9JN	2;30	69.11	3493	24JN	11;58	263.79
3494	8JL	11;13	97.16	3494	24JL	0;02	292.12
3495	6AU	22;26	125.64	3495	22AU	11;11	320.83
3496	5SE	12;39	154.70	3496	20SE	21;41	350.01
3497	5OC	5;39	184.40	3497	20OC	7;55	19.67
3498	4NO	0;36	214.61	3498	18NO	18;21	49.65
3499	3DE	20;08	245.04	3499	18DE	5;22	79.71

-717

NUMBER	DATE	TIME	LONG.	NUMBER	DATE	TIME	LONG.
3500	2JA	14;35	275.35	3500	16JA	17;11	109.61
3501	1FE	6;31	305.21	3501	15FE	5;50	139.11
3502	2MR	19;12	334.45	3502	16MR	19;11	168.13
3503	1AP	4;55	3.06	3503	15AP	9;09	196.70
3504	30AP	12;32	31.19	3504	14MY	23;45	224.98
3505	29MY	19;14	59.08	3505	13JN	14;49	253.18
3506	28JN	2;10	86.99	3506	13JL	5;55	281.52
3507	27JL	10;19	115.20	3507	11AU	20;26	310.19
3508	25AU	20;28	143.92	3508	10SE	9;54	339.31
3509	24SE	9;12	173.25	3509	9OC	22;17	8.89
3510	24OC	0;52	203.18	3510	8NO	9;50	38.83
3511	22NO	19;12	233.52	3511	7DE	20;55	68.91
3512	22DE	15;03	263.95				

-716

NUMBER	DATE	TIME	LONG.	NUMBER	DATE	TIME	LONG.
				3512	6JA	7;39	98.86
3513	21JA	10;22	294.11	3513	4FE	18;03	128.44
3514	20FE	3;13	323.71	3514	5MR	4;18	157.53
3515	20MR	16;49	352.66	3515	3AP	14;54	186.12
3516	19AP	3;26	21.04	3516	3MY	2;32	214.38
3517	18MY	11;59	49.04	3517	1JN	15;49	242.52
3518	16JN	19;27	76.95	3518	1JL	6;47	270.77
3519	16JL	2;47	105.01	3519	30JL	22;53	299.35
3520	14AU	10;45	133.47	3520	29AU	15;13	328.39
3521	12SE	20;08	162.48	3521	28SE	7;02	357.93
3522	12OC	7;42	192.07	3522	27OC	21;50	27.89
3523	10NO	22;04	222.16	3523	26NO	11;23	58.04
3524	10DE	15;08	252.50	3524	25DE	23;27	88.13

-715

NUMBER	DATE	TIME	LONG.	NUMBER	DATE	TIME	LONG.
3525	9JA	9;50	282.75	3525	24JA	9;57	117.86
3526	8FE	4;26	312.61	3526	22FE	19;07	147.08
3527	9MR	21;24	341.90	3527	24MR	3;33	175.77
3528	8AP	12;01	10.59	3528	22AP	12;11	204.03
3529	8MY	0;16	38.83	3529	21MY	21;58	232.06
3530	6JN	10;31	66.84	3530	20JN	9;39	260.14
3531	5JL	19;20	94.87	3531	19JL	23;31	288.50
3532	4AU	3;29	123.19	3532	18AU	15;26	317.34
3533	2SE	11;48	151.97	3533	17SE	8;52	346.75
3534	1OC	21;10	181.30	3534	17OC	2;55	16.69
3535	31OC	8;15	211.13	3535	15NO	20;22	46.94
3536	29NO	21;24	241.29	3536	15DE	11;58	77.21
3537	29DE	12;25	271.48				

-714

NUMBER	DATE	TIME	LONG.	NUMBER	DATE	TIME	LONG.
				3537	14JA	0;58	107.18
3538	28JA	4;42	301.41	3538	12FE	11;27	136.64
3539	26FE	21;27	330.88	3539	13MR	20;02	165.51
3540	28MR	13;55	359.82	3540	12AP	3;37	193.86
3541	27AP	5;25	28.30	3541	11MY	11;07	221.88
3542	26MY	19;22	56.49	3542	9JN	19;26	249.82
3543	25JN	7;31	84.60	3543	9JL	5;22	277.92
3544	24JL	18;04	112.88	3544	7AU	17;38	306.45
3545	23AU	3;41	141.53	3545	6SE	8;44	335.56
3546	21SE	13;13	170.69	3546	6OC	2;38	5.31
3547	20OC	23;20	200.35	3547	4NO	22;14	35.55
3548	19NO	10;18	230.35	3548	4DE	17;32	65.98
3549	18DE	22;08	260.45				

-713

NUMBER	DATE	TIME	LONG.	NUMBER	DATE	TIME	LONG.
				3549	3JA	10;38	96.22
3550	17JA	10;46	290.37	3550	2FE	0;39	126.00
3551	16FE	0;18	319.90	3551	3MR	11;45	155.16
3552	17MR	14;48	348.96	3552	1AP	20;39	183.72
3553	16AP	6;07	17.57	3553	1MY	4;09	211.84
3554	15MY	21;39	45.88	3554	30MY	11;04	239.73
3555	14JN	12;37	74.08	3555	28JN	18;19	267.67
3556	14JL	2;29	102.38	3556	28JL	2;54	295.90
3557	12AU	15;15	130.99	3557	26AU	13;52	324.66
3558	11SE	3;15	160.06	3558	25SE	4;05	354.07
3559	10OC	14;48	189.62	3559	24OC	21;35	24.09
3560	9NO	2;01	219.55	3560	23NO	17;12	54.48
3561	8DE	12;54	249.62	3561	23DE	12;57	84.91

```
        --------NEW MOONS--------   --------FULL MOONS--------   --------NEW MOONS--------   --------FULL MOONS--------
        NUMBER  DATE   TIME   LONG.  NUMBER  DATE   TIME   LONG.  NUMBER  DATE   TIME   LONG.  NUMBER  DATE   TIME   LONG.
```

-712 / -706

NUMBER	DATE	TIME	LONG.	NUMBER	DATE	TIME	LONG.	NUMBER	DATE	TIME	LONG.	NUMBER	DATE	TIME	LONG.
3562	6JA	23;30	279.56	3562	22JA	6;59	115.00					3636	15JA	9;14	108.60
3563	5FE	10;09	309.14	3563	20FE	22;17	144.52	3637	29JA	23;04	303.23	3637	13FE	18;51	137.99
3564	5MR	21;22	338.25	3564	21MR	10;41	173.41	3638	28FE	16;48	332.71	3638	15MR	3;21	166.84
3565	4AP	9;36	6.90	3565	19AP	20;30	201.75	3639	30MR	8;30	1.60	3639	13AP	11;33	195.20
3566	3MY	23;00	35.22	3566	19MY	4;23	229.73	3640	28AP	21;52	29.98	3640	12MY	20;27	223.27
3567	2JN	13;23	63.40	3567	17JN	11;13	257.60	3641	28MY	9;08	58.05	3641	11JN	6;59	251.29
3568	2JL	4;25	91.67	3568	16JL	18;08	285.66	3642	26JN	18;44	86.06	3642	10JL	19;43	279.52
3569	31JL	19;49	120.23	3569	15AU	2;16	314.13	3643	26JL	3;18	114.28	3643	9AU	10;44	308.17
3570	30AU	11;12	149.25	3570	13SE	12;38	343.19	3644	24AU	11;38	142.90	3644	8SE	3;41	337.40
3571	29SE	2;05	178.76	3571	13OC	1;52	12.87	3645	22SE	20;33	172.05	3645	7OC	21;48	7.18
3572	28OC	15;53	208.68	3572	11NO	17;57	43.03	3646	22OC	6;49	201.73	3646	6NO	15;56	37.38
3573	27NO	4;15	238.79	3573	11DE	12;06	73.42	3647	20NO	18;56	231.80	3647	6DE	8;44	67.70
3574	26DE	15;13	268.82					3648	20DE	8;58	261.99				

-711 / -705

NUMBER	DATE	TIME	LONG.	NUMBER	DATE	TIME	LONG.	NUMBER	DATE	TIME	LONG.	NUMBER	DATE	TIME	LONG.
				3574	10JA	7;04	103.68					3648	4JA	23;05	97.82
3575	25JA	1;10	298.53	3575	9FE	1;28	133.52	3649	19JA	0;28	292.01	3649	3FE	10;42	127.47
3576	23FE	10;37	327.75	3576	10MR	17;59	162.76	3650	17FE	16;47	321.63	3650	4MR	19;59	156.53
3577	24MR	20;04	356.46	3577	9AP	7;48	191.41	3651	19MR	9;13	350.73	3651	3AP	3;47	185.04
3578	23AP	6;01	24.77	3578	8MY	18;47	219.59	3652	18AP	1;07	19.35	3652	2MY	11;05	213.13
3579	22MY	17;01	52.85	3579	7JN	3;32	247.54	3653	17MY	15;53	47.62	3653	31MY	18;47	241.05
3580	21JN	5;36	80.97	3580	6JL	11;07	275.53	3654	16JN	5;04	75.74	3654	30JN	3;46	269.06
3581	20JL	20;10	109.36	3581	4AU	18;43	303.84	3655	15JL	16;35	103.96	3655	29JL	14;48	297.41
3582	19AU	12;41	138.24	3582	3SE	3;16	332.63	3656	14AU	2;50	132.50	3656	28AU	4;32	326.31
3583	18SE	6;19	167.67	3583	2OC	13;26	2.01	3657	12SE	12;34	161.50	3657	26SE	21;13	355.85
3584	17OC	23;42	197.59	3584	1NO	1;31	31.89	3658	11OC	22;32	191.01	3658	26OC	16;19	25.95
3585	16NO	15;33	227.79	3585	30NO	15;37	62.09	3659	10NO	9;10	220.93	3659	25NO	12;09	56.37
3586	16DE	5;17	257.98	3586	30DE	7;39	92.32	3660	9DE	20;33	251.02	3660	25DE	6;35	86.73

-710 / -704

NUMBER	DATE	TIME	LONG.	NUMBER	DATE	TIME	LONG.	NUMBER	DATE	TIME	LONG.	NUMBER	DATE	TIME	LONG.
3587	14JA	16;58	287.88	3587	29JA	1;08	122.29	3661	8JA	8;37	281.01	3661	23JA	22;09	116.69
3588	13FE	2;59	317.31	3588	27FE	19;00	151.79	3662	6FE	21;24	310.66	3662	22FE	10;33	146.07
3589	14MR	11;44	346.18	3589	29MR	11;49	180.74	3663	7MR	11;06	339.85	3663	22MR	20;20	174.83
3590	12AP	19;44	14.54	3590	28AP	2;31	209.17	3664	6AP	1;46	8.58	3664	21AP	4;20	203.08
3591	12MY	3;42	42.57	3591	27MY	14;50	237.29	3665	5MY	17;07	36.96	3665	20MY	11;23	231.02
3592	10JN	12;38	70.53	3592	26JN	1;15	265.34	3666	4JN	8;27	65.17	3666	18JN	18;21	258.90
3593	9JL	23;37	98.68	3593	25JL	10;38	293.58	3667	3JL	23;02	93.43	3667	18JL	2;10	287.00
3594	8AU	13;27	127.28	3594	23AU	19;46	322.22	3668	2AU	12;36	121.94	3668	16AU	11;53	315.56
3595	7SE	6;09	156.48	3595	22SE	5;12	351.38	3669	1SE	1;14	150.87	3669	15SE	0;30	344.74
3596	7OC	0;47	186.27	3596	21OC	15;18	21.05	3670	30SE	13;16	180.29	3670	14OC	16;30	14.55
3597	5NO	19;42	216.49	3597	20NO	2;23	51.06	3671	30OC	0;51	210.13	3671	13NO	11;19	44.85
3598	5DE	13;24	246.85	3598	19DE	14;47	81.19	3672	28NO	11;59	240.20	3672	13DE	7;15	75.31
								3673	27DE	22;38	270.21				

-709 / -703

NUMBER	DATE	TIME	LONG.	NUMBER	DATE	TIME	LONG.	NUMBER	DATE	TIME	LONG.	NUMBER	DATE	TIME	LONG.
3599	4JA	4;54	277.02	3599	18JA	4;45	111.16					3673	12JA	2;15	105.55
3600	2FE	17;52	306.74	3600	16FE	20;07	140.75	3674	26JA	9;01	299.91	3674	10FE	18;52	135.29
3601	4MR	4;19	335.86	3601	18MR	12;10	169.85	3675	24FE	19;37	329.16	3675	12MR	8;37	164.39
3602	2AP	12;40	4.40	3602	17AP	4;00	198.48	3676	26MR	7;01	357.93	3676	10AP	19;37	192.91
3603	1MY	19;42	32.49	3603	16MY	18;59	226.76	3677	24AP	19;34	26.32	3677	10MY	4;21	220.99
3604	31MY	2;30	60.38	3604	15JN	8;51	254.91	3678	24MY	9;19	54.51	3678	8JN	11;37	248.88
3605	29JN	10;17	88.33	3605	14JL	21;39	283.18	3679	23JN	0;03	82.72	3679	7JL	18;26	276.85
3606	28JL	20;11	116.63	3606	13AU	9;31	311.76	3680	22JL	15;27	111.16	3680	6AU	1;55	305.16
3607	27AU	8;57	145.48	3607	11SE	20;35	340.81	3681	21AU	7;09	140.04	3681	4SE	11;11	334.01
3608	26SE	0;43	174.96	3608	11OC	7;08	10.33	3682	19SE	22;40	169.41	3682	3OC	23;03	3.48
3609	25OC	18;57	205.02	3609	9NO	17;32	40.23	3683	19OC	13;22	199.24	3683	2NO	13;47	33.49
3610	24NO	14;28	235.42	3610	9DE	4;13	70.30	3684	18NO	2;39	229.34	3684	2DE	6;58	63.82
3611	24DE	9;41	265.81					3685	17DE	14;19	259.44				

-708 / -702

NUMBER	DATE	TIME	LONG.	NUMBER	DATE	TIME	LONG.	NUMBER	DATE	TIME	LONG.	NUMBER	DATE	TIME	LONG.
				3611	7JA	15;29	100.26					3685	1JA	1;34	94.14
3612	23JA	2;58	295.86	3612	6FE	3;29	129.89	3686	16JA	0;38	289.27	3686	30JA	20;15	124.14
3613	21FE	17;13	325.33	3613	6MR	16;11	159.05	3687	14FE	10;05	318.65	3687	1MR	13;40	153.60
3614	22MR	4;15	354.15	3614	5AP	5;33	187.73	3688	15MR	19;14	347.51	3688	31MR	4;48	182.45
3615	20AP	12;42	22.42	3615	4MY	19;37	216.07	3689	14AP	4;37	15.92	3689	29AP	17;09	210.77
3616	19MY	19;40	50.36	3616	3JN	10;25	244.27	3690	13MY	14;50	44.03	3690	29MY	2;59	238.78
3617	18JN	2;22	78.24	3617	3JL	1;40	272.54	3691	12JN	2;26	72.09	3691	27JN	11;08	266.75
3618	17JL	9;52	106.32	3618	1AU	16;48	301.11	3692	11JL	15;58	100.36	3692	26JL	18;45	294.93
3619	15AU	19;01	134.85	3619	31AU	7;08	330.10	3693	10AU	7;41	129.06	3693	25AU	2;54	323.56
3620	14SE	6;30	163.96	3620	29SE	20;21	359.55	3694	9SE	1;08	158.31	3694	23SE	12;22	352.74
3621	13OC	20;45	193.69	3621	29OC	8;30	29.41	3695	8OC	19;06	188.11	3695	22OC	23;34	22.47
3622	12NO	13;51	223.91	3622	27NO	19;53	59.48	3696	7NO	12;05	218.26	3696	21NO	12;38	52.58
3623	12DE	9;07	254.34	3623	27DE	6;45	89.51	3697	7DE	3;01	248.51	3697	21DE	3;36	82.81

-707 / -701

NUMBER	DATE	TIME	LONG.	NUMBER	DATE	TIME	LONG.	NUMBER	DATE	TIME	LONG.	NUMBER	DATE	TIME	LONG.
3624	11JA	4;51	284.62	3624	25JA	17;09	119.22	3698	5JA	15;41	278.55	3698	19JA	20;13	112.87
3625	9FE	23;00	314.44	3625	24FE	3;12	148.46	3699	4FE	2;23	308.15	3699	18FE	13;47	142.52
3626	11MR	14;08	343.61	3626	25MR	13;16	177.19	3700	5MR	11;33	337.20	3700	20MR	7;04	171.65
3627	10AP	2;06	12.18	3627	24AP	0;02	205.52	3701	3AP	19;40	5.70	3701	18AP	22;49	200.25
3628	9MY	11;33	40.29	3628	23MY	12;16	233.64	3702	3MY	3;23	33.81	3702	18MY	12;19	228.46
3629	7JN	19;30	68.21	3629	22JN	2;21	261.82	3703	1JN	11;34	61.75	3703	16JN	23;44	256.52
3630	7JL	2;54	96.20	3630	21JL	18;05	290.28	3704	30JN	21;21	89.79	3704	16JL	9;44	284.68
3631	5AU	10;36	124.51	3631	20AU	10;39	319.17	3705	30JL	9;42	118.20	3705	14AU	19;08	313.19
3632	3SE	19;19	153.33	3632	19SE	3;06	348.57	3706	29AU	1;08	147.19	3706	13SE	4;36	342.19
3633	3OC	5;49	182.73	3633	18OC	18;43	18.42	3707	27SE	19;09	176.79	3707	12OC	14;31	11.71
3634	1NO	18;47	212.66	3634	17NO	9;07	48.56	3708	27OC	14;19	206.91	3708	11NO	1;08	41.63
3635	1DE	10;30	242.92	3635	16DE	22;01	78.72	3709	26NO	8;56	237.28	3709	10DE	12;47	71.74
3636	31DE	4;25	273.22					3710	26DE	1;40	267.56				

−700 (left) / −694 (right)

NUMBER	DATE	TIME	LONG.	NUMBER	DATE	TIME	LONG.	NUMBER	DATE	TIME	LONG.	NUMBER	DATE	TIME	LONG.
				3710	9JA	1;46	101.76					3784	2JA	20;58	96.03
3711	24JA	15;53	297.47	3711	7FE	16;13	131.47	3785	17JA	8;09	290.65	3785	1FE	14;48	125.97
3712	23FE	3;27	326.80	3712	8MR	7;43	160.72	3786	15FE	18;20	320.04	3786	3MR	5;54	155.30
3713	23MR	12;39	355.52	3713	6AP	23;30	189.48	3787	17MR	4;59	348.94	3787	1AP	18;09	184.01
3714	21AP	20;07	23.74	3714	6MY	14;50	217.85	3788	15AP	16;38	17.42	3788	1MY	3;55	212.22
3715	21MY	2;48	51.66	3715	5JN	5;18	246.03	3789	15MY	5;36	45.63	3789	30MY	11;49	240.15
3716	19JN	9;55	79.55	3716	4JL	18;48	274.25	3790	13JN	19;48	73.79	3790	28JN	18;48	268.07
3717	18JL	18;39	107.70	3717	3AU	7;23	302.73	3791	13JL	10;59	102.14	3791	28JL	1;56	296.25
3718	17AU	5;58	136.33	3718	1SE	19;08	331.64	3792	12AU	2;48	130.87	3792	26AU	10;19	324.90
3719	15SE	20;22	165.59	3719	1OC	6;10	1.03	3793	10SE	18;47	160.10	3793	24SE	20;55	354.16
3720	15OC	13;34	195.47	3720	30OC	16;45	30.84	3794	10OC	10;17	189.81	3794	24OC	10;17	24.00
3721	14NO	8;42	225.79	3721	29NO	3;17	60.88	3795	9NO	0;32	219.87	3795	23NO	2;19	54.24
3722	14DE	4;18	256.23	3722	28DE	14;09	90.89	3796	8DE	13;06	250.01	3796	22DE	20;14	84.59

−699 (left) / −693 (right)

NUMBER	DATE	TIME	LONG.	NUMBER	DATE	TIME	LONG.	NUMBER	DATE	TIME	LONG.	NUMBER	DATE	TIME	LONG.
3723	12JA	22;41	286.44	3723	27JA	1;34	120.63	3797	7JA	0;00	279.97	3797	21JA	14;51	114.71
3724	11FE	14;27	316.13	3724	25FE	13;38	149.93	3798	5FE	9;40	309.51	3798	20FE	8;52	144.36
3725	13MR	2;58	345.17	3725	27MR	2;21	178.74	3799	6MR	18;41	338.53	3799	22MR	1;07	173.41
3726	11AP	12;31	13.61	3726	25AP	15;49	207.16	3800	5AP	3;39	7.05	3800	20AP	14;50	201.90
3727	10MY	20;03	41.64	3727	25MY	6;07	235.37	3801	4MY	13;09	35.22	3801	20MY	1;53	230.01
3728	9JN	2;45	69.51	3728	23JN	21;14	263.59	3802	2JN	23;50	63.25	3802	18JN	10;51	257.97
3729	8JL	9;46	97.49	3729	23JL	12;42	292.05	3803	2JL	12;18	91.41	3803	17JL	18;45	286.07
3730	6AU	18;06	125.85	3730	22AU	3;48	320.91	3804	1AU	2;59	119.93	3804	16AU	2;43	314.54
3731	5SE	4;27	154.75	3731	20SE	17;55	350.24	3805	30AU	19;52	149.00	3805	14SE	11;38	343.54
3732	4OC	17;22	184.27	3732	20OC	6;49	20.01	3806	29SE	14;01	178.64	3806	13OC	22;03	13.10
3733	3NO	9;06	214.34	3733	18NO	18;42	50.05	3807	29OC	7;55	208.73	3807	12NO	10;13	43.09
3734	3DE	3;23	244.72	3734	18DE	5;49	80.12	3808	28NO	0;07	239.00	3808	12DE	0;09	73.30
								3809	27DE	13;57	269.15				

−698 (left) / −692 (right)

NUMBER	DATE	TIME	LONG.	NUMBER	DATE	TIME	LONG.	NUMBER	DATE	TIME	LONG.	NUMBER	DATE	TIME	LONG.
3735	1JA	23;02	275.08	3735	16JA	16;20	109.96					3809	10JA	15;47	103.42
3736	31JA	18;04	305.08	3736	15FE	2;20	139.35	3810	26JA	1;31	298.93	3810	9FE	8;44	133.21
3737	2MR	10;41	334.49	3737	16MR	12;03	168.23	3811	24FE	11;15	328.16	3811	10MR	2;04	162.50
3738	1AP	0;08	3.25	3738	14AP	22;07	196.65	3812	24MR	19;39	356.83	3812	8AP	18;33	191.26
3739	30AP	10;42	31.50	3739	14MY	9;19	224.79	3813	23AP	3;18	25.04	3813	8MY	9;10	219.59
3740	29MY	19;19	59.46	3740	12JN	22;21	252.91	3814	22MY	11;00	53.00	3814	6JN	21;40	247.69
3741	28JN	2;59	87.41	3741	12JL	13;21	281.24	3815	20JN	19;46	80.96	3815	6JL	8;28	275.81
3742	27JL	10;35	115.60	3742	11AU	5;47	309.98	3816	20JL	6;43	109.21	3816	4AU	18;19	304.20
3743	25AU	18;51	144.24	3743	9SE	22;41	339.23	3817	18AU	20;40	137.97	3817	3SE	3;56	333.05
3744	24SE	4;31	173.44	3744	9OC	15;05	8.98	3818	17SE	13;39	167.37	3818	2OC	13;47	2.41
3745	23OC	16;15	203.20	3745	8NO	6;22	39.07	3819	17OC	8;39	197.35	3819	1NO	0;09	32.22
3746	22NO	6;34	233.37	3746	7DE	20;09	69.27	3820	16NO	3;53	227.68	3820	30NO	11;14	62.29
3747	21DE	23;22	263.67					3821	15DE	21;45	258.04	3821	29DE	23;23	92.35

−697 (left) / −691 (right)

NUMBER	DATE	TIME	LONG.	NUMBER	DATE	TIME	LONG.	NUMBER	DATE	TIME	LONG.	NUMBER	DATE	TIME	LONG.
				3747	6JA	8;12	99.27	3822	14JA	13;16	288.11	3822	28JA	12;52	122.16
3748	20JA	17;39	293.79	3748	4FE	18;28	128.85	3823	13FE	2;04	317.66	3823	27FE	3;35	151.55
3749	19FE	11;48	323.46	3749	6MR	3;15	157.87	3824	14MR	12;17	346.58	3824	28MR	19;03	180.44
3750	21MR	4;25	352.54	3750	4AP	11;16	186.36	3825	12AP	20;24	14.95	3825	27AP	10;30	208.92
3751	19AP	18;53	21.08	3751	3MY	19;30	214.48	3826	12MY	3;15	42.94	3826	27MY	1;24	237.14
3752	19MY	7;12	49.24	3752	2JN	4;58	242.47	3827	10JN	9;57	70.80	3827	25JN	15;30	265.33
3753	17JN	17;41	77.26	3753	1JL	16;30	270.59	3828	9JL	17;45	98.83	3828	25JL	4;46	293.72
3754	17JL	2;52	105.40	3754	31JL	6;26	299.07	3829	8AU	3;44	127.27	3829	23AU	17;13	322.50
3755	15AU	11;26	133.89	3755	29AU	22;37	328.10	3830	6SE	16;40	156.31	3830	22SE	4;52	351.76
3756	13SE	20;10	162.87	3756	28SE	16;30	357.71	3831	6OC	8;38	185.98	3831	21OC	15;51	21.46
3757	13OC	5;50	192.38	3757	28OC	11;01	27.82	3832	5NO	3;01	216.17	3832	20NO	2;28	51.46
3758	11NO	17;02	222.34	3758	27NO	4;48	58.15	3833	4DE	22;36	246.61	3833	19DE	13;06	81.50
3759	11DE	6;03	252.50	3759	26DE	20;31	88.39								

−696 (left) / −690 (right)

NUMBER	DATE	TIME	LONG.	NUMBER	DATE	TIME	LONG.	NUMBER	DATE	TIME	LONG.	NUMBER	DATE	TIME	LONG.
3760	9JA	20;42	282.59	3760	25JA	9;27	118.23	3834	3JA	17;46	276.94	3834	18JA	0;03	111.34
3761	8FE	12;25	312.34	3761	23FE	19;41	147.49	3835	2FE	10;54	306.84	3835	16FE	11;32	140.78
3762	9MR	4;34	341.60	3762	24MR	3;58	176.17	3836	4MR	0;59	336.11	3836	17MR	23;37	169.73
3763	7AP	20;36	10.35	3763	22AP	11;15	204.37	3837	2AP	11;52	4.75	3837	16AP	12;24	198.24
3764	7MY	11;56	38.72	3764	21MY	18;31	232.31	3838	1MY	20;13	32.90	3838	16MY	2;05	226.47
3765	6JN	2;01	66.88	3765	20JN	2;43	260.25	3839	31MY	3;10	60.79	3839	14JN	16;47	254.66
3766	5JL	14;32	95.06	3766	19JL	12;38	288.45	3840	29JN	9;57	88.70	3840	14JL	8;16	283.02
3767	4AU	1;36	123.50	3767	18AU	1;01	317.14	3841	28JL	17;37	116.91	3841	12AU	23;55	311.76
3768	2SE	11;46	152.35	3768	16SE	16;20	346.45	3842	27AU	3;00	145.62	3842	11SE	14;54	340.96
3769	1OC	21;46	181.71	3769	16OC	10;30	16.38	3843	25SE	14;40	174.93	3843	11OC	4;42	10.62
3770	31OC	8;11	211.52	3770	15NO	6;20	46.74	3844	25OC	5;02	204.82	3844	9NO	17;15	40.61
3771	29NO	19;13	241.58	3771	15DE	1;47	77.17	3845	23NO	22;07	235.11	3845	9DE	4;48	70.71
3772	29DE	6;50	271.63					3846	23DE	17;11	265.50				

−695 (left) / −689 (right)

NUMBER	DATE	TIME	LONG.	NUMBER	DATE	TIME	LONG.	NUMBER	DATE	TIME	LONG.	NUMBER	DATE	TIME	LONG.
				3772	13JA	18;52	107.31					3846	7JA	15;32	100.65
3773	27JA	19;01	301.39	3773	12FE	8;44	136.91	3847	22JA	12;38	295.65	3847	6FE	1;36	130.20
3774	26FE	7;56	330.71	3774	13MR	19;38	165.88	3848	21FE	6;29	325.27	3848	7MR	11;10	159.23
3775	27MR	21;49	359.56	3775	12AP	4;20	194.28	3849	22MR	21;26	354.26	3849	5AP	20;42	187.76
3776	26AP	12;39	28.02	3776	11MY	11;41	222.29	3850	21AP	9;19	22.67	3850	5MY	7;00	215.96
3777	26MY	4;00	56.26	3777	9JN	18;34	250.16	3851	20MY	18;49	50.71	3851	3JN	18;54	244.04
3778	24JN	19;07	84.49	3778	9JL	1;52	278.16	3852	19JN	2;56	78.64	3852	3JL	8;54	272.26
3779	24JL	9;26	112.92	3779	7AU	10;35	306.55	3853	18JL	10;36	106.73	3853	2AU	0;51	300.85
3780	22AU	22;48	141.72	3780	5SE	21;44	335.49	3854	16AU	18;37	135.22	3854	31AU	17;53	329.94
3781	21SE	11;24	171.00	3781	5OC	12;07	5.09	3855	15SE	3;39	164.23	3855	30SE	10;55	359.55
3782	20OC	23;27	200.74	3782	4NO	5;43	35.24	3856	14OC	14;22	193.81	3856	30OC	3;05	29.57
3783	19NO	10;56	230.77	3783	4DE	1;20	65.68	3857	13NO	3;22	223.85	3857	28NO	17;48	59.78
3784	18DE	21;48	260.83					3858	12DE	18;54	254.12	3858	28DE	6;46	89.90

-688

NUMBER	DATE	TIME	LONG.	NUMBER	DATE	TIME	LONG.
3859	11JA	12;25	284.32	3859	26JA	17;50	119.64
3860	10FE	6;35	314.14	3860	25FE	3;08	148.84
3861	10MR	23;53	343.42	3861	25MR	11;13	177.49
3862	9AP	15;21	12.13	3862	23AP	19;01	205.70
3863	9MY	4;42	40.40	3863	23MY	3;35	233.68
3864	7JN	16;08	68.45	3864	21JN	13;55	261.71
3865	7JL	2;05	96.55	3865	21JL	2;38	290.04
3866	5AU	11;06	124.91	3866	19AU	17;51	318.87
3867	3SE	19;52	153.74	3867	18SE	11;10	348.29
3868	3OC	5;09	183.08	3868	18OC	5;43	18.26
3869	1NO	15;37	212.91	3869	17NO	0;14	48.57
3870	1DE	3;42	243.02	3870	16DE	17;13	78.90
3871	30DE	17;25	273.14				

-682

NUMBER	DATE	TIME	LONG.	NUMBER	DATE	TIME	LONG.
3933	5JA	9;59	278.69	3933	19JA	10;06	112.83
3934	4FE	0;06	308.44	3934	17FE	23;56	142.34
3935	5MR	11;28	337.58	3935	19MR	14;48	171.37
3936	3AP	20;27	6.12	3936	18AP	6;06	199.96
3937	3MY	3;43	34.21	3937	17MY	21;14	228.24
3938	1JN	10;18	62.08	3938	16JN	11;48	256.43
3939	30JN	17;23	90.02	3939	16JL	1;41	284.74
3940	30JL	2;12	118.29	3940	14AU	14;50	313.40
3941	28AU	13;41	147.10	3941	13SE	3;12	342.52
3942	27SE	4;15	176.56	3942	12OC	14;44	12.11
3943	26OC	21;37	206.59	3943	11NO	1;38	42.04
3944	25NO	16;49	236.98	3944	10DE	12;13	72.10
3945	25DE	12;23	267.39				

-687

NUMBER	DATE	TIME	LONG.	NUMBER	DATE	TIME	LONG.
				3871	15JA	7;34	108.91
3872	29JA	8;25	303.01	3872	13FE	19;00	138.39
3873	28FE	0;07	332.42	3873	15MR	4;01	167.25
3874	29MR	16;01	1.32	3874	13AP	11;32	195.58
3875	28AP	7;38	29.79	3875	12MY	18;34	223.58
3876	27MY	22;24	58.00	3876	11JN	2;08	251.48
3877	26JN	11;53	86.17	3877	10JL	11;05	279.55
3878	25JL	23;53	114.52	3878	8AU	22;12	308.05
3879	24AU	10;41	143.25	3879	7SE	12;06	337.14
3880	22SE	20;57	172.46	3880	7OC	5;01	6.87
3881	22OC	7;18	202.14	3881	6NO	0;21	37.11
3882	20NO	18;05	232.15	3882	5DE	20;20	67.56
3883	20DE	5;22	262.22				

-681

NUMBER	DATE	TIME	LONG.	NUMBER	DATE	TIME	LONG.
				3945	8JA	22;52	102.03
3946	24JA	6;38	297.46	3946	7FE	9;50	131.59
3947	22FE	22;13	326.97	3947	8MR	21;19	160.68
3948	24MR	10;35	355.82	3948	7AP	9;26	189.30
3949	22AP	20;02	24.12	3949	6MY	22;26	217.58
3950	22MY	3;32	52.07	3950	5JN	12;31	245.75
3951	20JN	10;17	79.95	3951	5JL	3;43	274.03
3952	19JL	17;29	108.03	3952	3AU	19;35	302.64
3953	18AU	2;02	136.56	3953	2SE	11;18	331.70
3954	16SE	12;37	165.66	3954	2OC	2;02	1.24
3955	16OC	1;41	195.36	3955	31OC	15;26	31.17
3956	14NO	17;26	225.90	3956	30NO	3;34	61.28
3957	14DE	11;33	255.90	3957	29DE	14;40	91.30

-686

NUMBER	DATE	TIME	LONG.	NUMBER	DATE	TIME	LONG.
				3883	4JA	14;47	97.86
3884	18JA	17;04	292.09	3884	3FE	6;14	127.67
3885	17FE	5;18	321.54	3885	4MR	18;27	156.85
3886	18MR	18;22	350.52	3886	3AP	4;03	185.43
3887	17AP	8;28	19.07	3887	2MY	11;54	213.55
3888	16MY	23;30	47.35	3888	31MY	18;53	241.44
3889	15JN	14;50	75.56	3889	30JN	1;54	269.37
3890	15JL	5;46	103.92	3890	29JL	9;51	297.60
3891	13AU	19;53	132.60	3891	27AU	19;45	326.33
3892	12SE	9;09	161.74	3892	26SE	8;32	355.70
3893	11OC	21;44	191.36	3893	26OC	0;39	25.68
3894	10NO	9;41	221.33	3894	24NO	19;29	56.05
3895	9DE	20;55	251.42	3895	24DE	15;18	86.47

-680

NUMBER	DATE	TIME	LONG.	NUMBER	DATE	TIME	LONG.
3958	13JA	6;55	286.16	3958	28JA	0;56	120.99
3959	12FE	1;37	315.98	3959	26FE	10;30	150.19
3960	12MR	17;59	345.19	3960	26MR	19;43	178.86
3961	11AP	7;18	13.79	3961	25AP	5;17	207.12
3962	10MY	17;53	41.94	3962	24MY	16;06	235.19
3963	9JN	2;39	69.89	3963	23JN	4;57	263.32
3964	8JL	10;34	97.91	3964	22JL	20;02	291.76
3965	6AU	18;29	126.25	3965	21AU	12;51	320.68
3966	5SE	3;06	155.08	3966	20SE	6;17	350.14
3967	4OC	13;02	184.48	3967	19OC	23;14	20.07
3968	3NO	0;53	214.38	3968	18NO	14;56	50.27
3969	2DE	15;05	244.58	3969	18DE	4;53	80.47

-685

NUMBER	DATE	TIME	LONG.	NUMBER	DATE	TIME	LONG.
3896	8JA	7;24	281.35	3896	23JA	10;05	116.58
3897	6FE	17;24	310.88	3897	22FE	2;30	146.13
3898	8MR	3;28	339.92	3898	23MR	16;04	175.04
3899	6AP	14;16	8.50	3899	22AP	2;59	203.41
3900	6MY	2;20	36.75	3900	21MY	11;44	231.41
3901	4JN	15;48	64.90	3901	19JN	19;08	259.32
3902	4JL	6;33	93.15	3902	19JL	2;10	287.39
3903	2AU	22;16	121.74	3903	17AU	9;55	315.87
3904	1SE	14;30	150.81	3904	15SE	19;27	344.92
3905	1OC	6;38	180.40	3905	15OC	7;29	14.56
3906	30OC	21;51	210.39	3906	13NO	22;14	44.68
3907	29NO	11;26	240.56	3907	13DE	15;14	75.01
3908	28DE	23;08	270.62				

-679

NUMBER	DATE	TIME	LONG.	NUMBER	DATE	TIME	LONG.
3970	1JA	7;34	274.81	3970	16JA	16;51	110.37
3971	31JA	1;21	304.77	3971	15FE	2;51	139.76
3972	1MR	19;00	334.23	3972	16MR	11;16	168.58
3973	31MR	11;17	3.12	3973	14AP	18;54	196.90
3974	30AP	1;38	31.52	3974	14MY	2;47	224.92
3975	29MY	14;04	59.64	3975	12JN	12;01	252.88
3976	28JN	0;51	87.71	3976	11JL	23;28	281.07
3977	27JL	10;29	115.98	3977	10AU	13;30	309.71
3978	25AU	19;31	144.65	3978	9SE	5;59	338.93
3979	24SE	4;40	173.84	3979	9OC	0;16	8.74
3980	23OC	14;36	203.52	3980	7NO	19;10	38.98
3981	22NO	1;52	233.56	3981	7DE	13;12	69.36
3982	21DE	14;41	263.69				

-684

NUMBER	DATE	TIME	LONG.	NUMBER	DATE	TIME	LONG.
				3908	12JA	9;29	105.23
3909	27JA	9;15	300.31	3909	11FE	3;45	135.05
3910	25FE	18;19	329.49	3910	11MR	20;50	164.30
3911	26MR	3;00	358.16	3911	10AP	11;47	192.98
3912	24AP	11;57	26.40	3912	10MY	0;09	221.20
3913	23MY	21;48	54.44	3913	8JN	10;10	249.20
3914	22JN	9;12	82.51	3914	7JL	18;37	277.24
3915	21JL	22;46	110.88	3915	6AU	2;37	305.57
3916	20AU	14;44	139.75	3916	4SE	11;09	334.39
3917	19SE	8;38	169.21	3917	3OC	20;56	3.78
3918	19OC	3;08	199.20	3918	2NO	8;17	33.64
3919	17NO	20;31	229.46	3919	1DE	21;17	63.80
3920	17DE	11;38	259.71	3920	31DE	11;56	93.95

-678

NUMBER	DATE	TIME	LONG.	NUMBER	DATE	TIME	LONG.
				3982	6JA	4;59	99.52
3983	20JA	4;52	293.64	3983	4FE	17;47	129.20
3984	18FE	19;59	323.19	3984	6MR	3;47	158.27
3985	20MR	11;33	352.24	3985	4AP	11;47	186.77
3986	19AP	3;10	20.83	3986	3MY	18;50	214.84
3987	18MY	18;22	49.11	3987	2JN	1;56	242.73
3988	17JN	8;39	77.28	3988	1JL	10;04	270.71
3989	16JL	21;36	105.57	3989	30JL	20;03	299.04
3990	15AU	9;14	134.18	3990	29AU	8;34	327.91
3991	13SE	19;57	163.24	3991	28SE	0;06	357.42
3992	13OC	6;25	192.80	3992	27OC	18;29	27.50
3993	11NO	17;05	222.73	3993	26NO	14;28	57.93
3994	11DE	4;07	252.80	3994	26DE	9;57	88.33

-683

NUMBER	DATE	TIME	LONG.	NUMBER	DATE	TIME	LONG.
3921	16JA	0;16	289.64	3921	30JA	4;02	123.85
3922	14FE	10;45	319.06	3922	28FE	21;01	153.30
3923	15MR	19;35	347.91	3923	30MR	13;52	182.23
3924	14AP	3;22	16.25	3924	29AP	5;25	210.69
3925	13MY	10;47	44.25	3925	28MY	19;01	238.85
3926	11JN	18;48	72.17	3926	27JN	6;44	266.96
3927	11JL	4;30	100.28	3927	26JL	17;12	295.25
3928	9AU	16;56	128.84	3928	25AU	3;07	323.95
3929	8SE	8;34	158.01	3929	23SE	13;03	353.16
3930	8OC	2;54	187.81	3930	22OC	23;18	22.85
3931	6NO	22;24	218.07	3931	21NO	10;02	52.85
3932	6DE	17;13	248.48	3932	20DE	21;31	82.94

-677

NUMBER	DATE	TIME	LONG.	NUMBER	DATE	TIME	LONG.
3995	9JA	15;28	282.75	3995	25JA	2;57	118.34
3996	8FE	3;10	312.34	3996	23FE	16;40	147.75
3997	9MR	15;27	341.45	3997	25MR	3;23	176.53
3998	8AP	4;44	10.10	3998	23AP	11;55	204.78
3999	7MY	19;08	38.44	3999	22MY	19;12	232.72
4000	6JN	10;21	66.64	4000	21JN	2;07	260.61
4001	6JL	1;40	94.94	4001	20JL	9;32	288.71
4002	4AU	16;28	123.51	4002	18AU	18;26	317.26
4003	3SE	6;28	152.52	4003	17SE	5;46	346.40
4004	2OC	19;40	182.01	4004	16OC	20;17	16.17
4005	1NO	8;10	211.91	4005	15NO	13;55	46.43
4006	30NO	19;51	242.00	4006	15DE	9;27	76.86
4007	30DE	6;39	272.01				

-676

NUMBER	DATE	TIME	LONG.	NUMBER	DATE	TIME	LONG.
				4007	14JA	4;52	107.11
4008	28JA	16;41	301.68	4008	12FE	22;27	136.87
4009	27FE	2;24	330.87	4009	13MR	13;20	166.00
4010	27MR	12;29	359.57	4010	12AP	1;29	194.55
4011	25AP	23;36	27.88	4011	11MY	11;14	222.66
4012	25MY	12;11	56.02	4012	9JN	19;15	250.58
4013	24JN	2;17	84.20	4013	9JL	2;27	278.57
4014	23JL	17;39	112.67	4014	7AU	9;51	306.90
4015	22AU	9;56	141.58	4015	5SE	18;32	335.75
4016	21SE	2;31	171.02	4016	5OC	5;21	5.19
4017	20OC	18;34	200.92	4017	3NO	18;48	35.17
4018	19NO	9;13	231.08	4018	3DE	10;42	65.44
4019	18DE	21;55	261.22				

-675

NUMBER	DATE	TIME	LONG.	NUMBER	DATE	TIME	LONG.
				4019	2JA	4;18	95.72
4020	17JA	8;42	291.06	4020	31JA	22;29	125.68
4021	15FE	18;03	320.41	4021	2MR	16;06	155.12
4022	17MR	2;38	349.23	4022	1AP	8;05	183.99
4023	15AP	11;09	17.58	4023	30AP	21;44	212.35
4024	14MY	20;17	45.64	4024	30MY	8;56	240.42
4025	13JN	6;43	73.66	4025	28JN	18;12	268.43
4026	12JL	19;07	101.89	4026	28JL	2;29	296.66
4027	11AU	9;58	130.57	4027	26AU	10;50	325.31
4028	10SE	3;12	159.84	4028	24SE	20;07	354.51
4029	9OC	21;51	189.68	4029	24OC	6;47	24.23
4030	8NO	16;11	219.90	4030	22NO	18;57	54.31
4031	8DE	8;40	250.21	4031	22DE	8;39	84.48

-674

NUMBER	DATE	TIME	LONG.	NUMBER	DATE	TIME	LONG.
4032	6JA	22;32	280.29	4032	20JA	23;49	114.47
4033	5FE	9;56	309.90	4033	19FE	16;10	144.05
4034	6MR	19;23	338.94	4034	21MR	8;58	173.14
4035	5AP	3;28	7.43	4035	20AP	1;09	201.74
4036	4MY	10;49	35.51	4036	19MY	15;45	229.99
4037	2JN	18;19	63.41	4037	18JN	4;29	258.10
4038	2JL	2;59	91.42	4038	17JL	15;43	286.33
4039	31JL	13;58	119.79	4039	16AU	2;05	314.90
4040	30AU	4;05	148.74	4040	14SE	12;13	343.95
4041	28SE	21;20	178.33	4041	13OC	22;29	13.50
4042	28OC	16;37	208.47	4042	12NO	9;02	43.43
4043	27NO	12;06	238.88	4043	11DE	20;05	73.51
4044	27DE	6;03	269.21				

-673

NUMBER	DATE	TIME	LONG.	NUMBER	DATE	TIME	LONG.
				4044	10JA	7;55	103.47
4045	25JA	21;29	299.14	4045	8FE	20;51	133.10
4046	24FE	10;07	328.50	4046	10MR	10;56	162.27
4047	25MR	20;07	357.23	4047	9AP	1;49	190.98
4048	24AP	4;03	25.45	4048	8MY	16;55	219.33
4049	23MY	10;47	53.37	4049	7JN	7;48	247.52
4050	21JN	17;27	81.25	4050	6JL	22;11	275.79
4051	21JL	1;19	109.37	4051	5AU	11;58	304.33
4052	19AU	11;26	137.98	4052	4SE	1;03	333.32
4053	18SE	0;33	167.21	4053	3OC	13;16	2.78
4054	17OC	16;40	197.06	4054	2NO	0;39	32.63
4055	16NO	11;09	227.36	4055	1DE	11;25	62.69
4056	16DE	6;42	257.79	4056	30DE	21;55	92.68

-672

NUMBER	DATE	TIME	LONG.	NUMBER	DATE	TIME	LONG.
4057	15JA	1;44	288.02	4057	29JA	8;31	122.37
4058	13FE	18;41	317.74	4058	27FE	19;27	151.60
4059	14MR	8;35	346.82	4059	28MR	6;56	180.34
4060	12AP	19;20	15.29	4060	26AP	19;12	208.69
4061	12MY	3;39	43.34	4061	26MY	8;33	236.86
4062	10JN	10;40	71.22	4062	24JN	23;12	265.07
4063	9JL	17;36	99.21	4063	24JL	14;58	293.55
4064	8AU	1;30	127.56	4064	23AU	7;09	322.48
4065	6SE	11;07	156.46	4065	21SE	22;47	351.89
4066	5OC	23;00	185.96	4066	21OC	13;08	21.73
4067	4NO	13;25	215.98	4067	20NO	2;03	51.83
4068	4DE	6;23	246.31	4068	19DE	13;40	81.92

-671

NUMBER	DATE	TIME	LONG.	NUMBER	DATE	TIME	LONG.
4069	3JA	1;10	276.63	4069	18JA	0;14	111.74
4070	1FE	20;16	306.61	4070	16FE	9;56	141.10
4071	3MR	13;48	336.04	4071	17MR	19;01	169.92
4072	2AP	4;34	4.84	4072	16AP	4;05	198.29
4073	1MY	16;25	33.13	4073	15MY	13;57	226.37
4074	31MY	2;02	61.13	4074	14JN	1;36	254.44
4075	29JN	10;23	89.11	4075	13JL	15;34	282.74
4076	28JL	18;23	117.33	4076	12AU	7;46	311.49
4077	27AU	2;45	145.99	4077	11SE	1;16	340.78
4078	25SE	12;07	175.21	4078	10OC	18;52	10.59
4079	24OC	23;01	204.95	4079	9NO	11;29	40.75
4080	23NO	12;00	235.06	4080	9DE	2;28	71.00
4081	23DE	3;16	265.30				

-670

NUMBER	DATE	TIME	LONG.	NUMBER	DATE	TIME	LONG.
				4081	7JA	15;27	101.03
4082	21JA	20;19	295.35	4082	6FE	2;18	130.61
4083	20FE	13;56	324.98	4083	7MR	11;16	159.62
4084	22MR	6;49	354.05	4084	5AP	19;00	188.08
4085	20AP	22;03	22.61	4085	5MY	2;27	216.16
4086	20MY	11;26	50.80	4086	3JN	10;45	244.10
4087	18JN	23;08	78.88	4087	2JL	20;58	272.17
4088	18JL	9;30	107.07	4088	1AU	9;43	300.62
4089	16AU	18;59	135.62	4089	31AU	1;08	329.63
4090	15SE	4;13	164.64	4090	29SE	18;48	359.26
4091	14OC	13;51	194.18	4091	29OC	13;44	29.39
4092	13NO	0;29	224.12	4092	28NO	8;32	59.78
4093	12DE	12;28	254.24	4093	28DE	1;38	90.06

-669

NUMBER	DATE	TIME	LONG.	NUMBER	DATE	TIME	LONG.
4094	11JA	1;49	284.26	4094	26JA	15;55	119.94
4095	9FE	16;14	313.93	4095	25FE	3;09	149.23
4096	11MR	7;19	343.13	4096	26MR	11;54	177.90
4097	9AP	22;42	11.84	4097	24AP	19;11	206.09
4098	9MY	14;03	40.20	4098	24MY	2;03	234.01
4099	8JN	4;54	68.39	4099	22JN	9;32	261.92
4100	7JL	18;44	96.64	4100	21JL	18;31	290.10
4101	6AU	7;16	125.14	4101	20AU	5;45	318.76
4102	4SE	18;40	154.07	4102	18SE	19;50	348.04
4103	4OC	5;27	183.48	4103	18OC	12;58	17.94
4104	2NO	16;08	213.32	4104	17NO	8;27	48.29
4105	2DE	3;01	243.38	4105	17DE	4;29	78.75
4106	31DE	14;08	273.39				

-668

NUMBER	DATE	TIME	LONG.	NUMBER	DATE	TIME	LONG.
				4106	15JA	22;52	108.93
4107	30JA	1;26	303.10	4107	14FE	14;10	138.57
4108	28FE	13;05	332.35	4108	15MR	2;12	167.56
4109	29MR	1;31	1.12	4109	13AP	11;39	195.97
4110	27AP	15;06	29.52	4110	12MY	19;24	223.99
4111	27MY	5;51	57.73	4111	11JN	2;24	251.87
4112	25JN	21;16	85.98	4112	10JL	9;32	279.88
4113	25JL	12;35	114.45	4113	8AU	17;40	308.25
4114	24AU	3;17	143.33	4114	7SE	3;46	337.18
4115	22SE	17;12	172.69	4115	6OC	16;43	6.74
4116	22OC	6;18	202.49	4116	5NO	8;54	36.84
4117	20NO	18;33	232.56	4117	5DE	3;39	67.24
4118	20DE	5;49	262.63				

-667

NUMBER	DATE	TIME	LONG.	NUMBER	DATE	TIME	LONG.
				4118	3JA	23;16	97.59
4119	18JA	16;04	292.43	4119	2FE	17;47	127.54
4120	17FE	1;40	321.78	4120	4MR	9;57	156.89
4121	18MR	11;12	350.61	4121	2AP	23;22	185.62
4122	16AP	21;28	19.01	4122	2MY	10;14	213.87
4123	16MY	9;05	47.16	4123	31MY	19;05	241.83
4124	14JN	22;20	75.29	4124	30JN	2;40	269.79
4125	14JL	13;08	103.63	4125	29JL	9;59	297.99
4126	13AU	5;12	132.39	4126	27AU	18;03	326.65
4127	11SE	21;59	161.67	4127	26SE	3;51	355.90
4128	11OC	14;43	191.45	4128	25OC	16;02	25.70
4129	10NO	6;24	221.58	4129	24NO	6;44	55.89
4130	9DE	20;12	251.78	4130	23DE	23;28	86.18

-666

NUMBER	DATE	TIME	LONG.	NUMBER	DATE	TIME	LONG.
4131	8JA	7;53	281.75	4131	22JA	17;17	116.26
4132	6FE	17;44	311.28	4132	21FE	11;05	145.88
4133	8MR	2;26	340.26	4133	23MR	3;49	174.93
4134	6AP	10;41	8.74	4134	21AP	18;38	203.46
4135	5MY	19;14	36.86	4135	21MY	7;05	231.61
4136	4JN	4;48	64.84	4136	19JN	17;21	259.63
4137	3JL	16;05	92.96	4137	19JL	2;11	287.78
4138	2AU	5;42	121.46	4138	17AU	10;36	316.28
4139	31AU	21;57	150.52	4139	15SE	19;32	345.31
4140	30SE	16;17	180.19	4140	15OC	5;37	14.87
4141	30OC	11;14	210.33	4141	13NO	17;05	44.85
4142	29NO	4;57	240.67	4142	13DE	5;56	75.01
4143	28DE	20;11	270.88				

-665

NUMBER	DATE	TIME	LONG.	NUMBER	DATE	TIME	LONG.
				4143	11JA	20;11	105.06
4144	27JA	8;44	300.67	4144	10FE	11;42	134.77
4145	25FE	18;57	329.91	4145	12MR	4;06	164.00
4146	27MR	3;30	358.56	4146	10AP	20;31	192.75
4147	25AP	10;59	26.75	4147	10MY	11;55	221.10
4148	24MY	18;12	54.68	4148	9JN	1;40	249.25
4149	23JN	2;05	82.61	4149	8JL	13;47	277.43
4150	22JL	11;47	110.82	4150	7AU	0;45	305.88
4151	21AU	0;20	139.55	4151	5SE	11;13	334.78
4152	19SE	16;11	168.92	4152	4OC	21;37	4.19
4153	19OC	10;47	198.89	4153	3NO	8;09	34.03
4154	18NO	6;31	229.26	4154	2DE	18;56	64.08
4155	18DE	1;27	259.66				

Years −664 (left) / −658 (right)

NUMBER	DATE	TIME	LONG.	NUMBER	DATE	TIME	LONG.	NUMBER	DATE	TIME	LONG.	NUMBER	DATE	TIME	LONG.
				4155	1JA	6;12	94.10	4230	9JA	15;26	283.14	4230	24JA	12;37	118.13
4156	16JA	18;12	289.77	4156	30JA	18;19	123.83	4231	8FE	1;07	312.64	4231	23FE	5;55	147.70
4157	15FE	8;10	319.34	4157	29FE	7;32	153.13	4232	9MR	10;21	341.62	4232	24MR	20;36	176.64
4158	15MR	19;20	348.28	4158	29MR	21;47	181.96	4233	7AP	19;54	10.13	4233	23AP	8;40	205.03
4159	14AP	4;07	16.66	4159	28AP	12;37	210.40	4234	7MY	6;33	38.32	4234	22MY	18;29	233.08
4160	13MY	11;15	44.65	4160	28MY	3;35	238.62	4235	5JN	18;49	66.41	4235	21JN	2;42	261.02
4161	11JN	17;48	72.51	4161	26JN	18;19	266.85	4236	5JL	8;51	94.65	4236	20JL	10;10	289.12
4162	11JL	0;58	100.52	4162	26JL	8;39	295.29	4237	4AU	0;28	123.25	4237	18AU	17;54	317.62
4163	9AU	9;54	128.94	4163	24AU	22;24	324.14	4238	2SE	17;12	152.36	4238	17SE	2;53	346.67
4164	7SE	21;34	157.95	4164	23SE	11;23	353.48	4239	2OC	10;22	182.01	4239	16OC	13;55	16.29
4165	7OC	12;18	187.59	4165	22OC	23;24	23.24	4240	1NO	2;57	212.07	4240	15NO	3;23	46.36
4166	6NO	5;46	217.75	4166	21NO	10;33	53.27	4241	30NO	17;55	242.30	4241	14DE	19;05	76.64
4167	6DE	0;57	248.18	4167	20DE	21;07	83.31	4242	30DE	6;40	272.39				

Years −663 (left) / −657 (right)

NUMBER	DATE	TIME	LONG.	NUMBER	DATE	TIME	LONG.	NUMBER	DATE	TIME	LONG.	NUMBER	DATE	TIME	LONG.
4168	4JA	20;22	278.51	4168	19JA	7;29	113.11					4242	13JA	12;17	106.80
4169	3FE	14;25	308.42	4169	17FE	17;59	142.48	4243	28JA	17;16	302.09	4243	12FE	5;58	136.58
4170	5MR	5;49	337.73	4170	19MR	4;53	171.35	4244	27FE	2;18	331.25	4244	13MR	23;09	165.81
4171	3AP	18;02	6.41	4171	17AP	16;27	199.80	4245	28MR	10;29	359.87	4245	12AP	14;54	194.51
4172	3MY	3;26	34.58	4172	17MY	5;01	227.98	4246	26AP	18;36	28.07	4246	12MY	4;33	222.77
4173	1JN	10;58	62.49	4173	15JN	18;56	256.14	4247	26MY	3;25	56.05	4247	10JN	15;58	250.83
4174	30JN	17;53	90.42	4174	15JL	10;17	284.51	4248	24JN	13;40	84.09	4248	10JL	1;35	278.92
4175	30JL	1;18	118.64	4175	14AU	2;35	313.29	4249	24JL	2;04	112.42	4249	8AU	10;18	307.30
4176	28AU	10;07	147.34	4176	12SE	18;55	342.56	4250	22AU	17;06	141.28	4250	6SE	19;06	336.16
4177	26SE	20;55	176.64	4177	12OC	10;16	12.31	4251	21SE	10;42	170.75	4251	6OC	4;44	5.56
4178	26OC	10;06	206.49	4178	11NO	0;06	42.36	4252	21OC	5;47	200.77	4252	4NO	15;35	35.42
4179	25NO	1;48	236.73	4179	10DE	12;25	72.50	4253	20NO	0;30	231.10	4253	4DE	3;43	65.53
4180	24DE	19;41	267.07					4254	19DE	17;09	261.40				

Years −662 (left) / −656 (right)

NUMBER	DATE	TIME	LONG.	NUMBER	DATE	TIME	LONG.	NUMBER	DATE	TIME	LONG.	NUMBER	DATE	TIME	LONG.
				4180	8JA	23;27	102.44					4254	2JA	17;06	95.63
4181	23JA	14;39	297.18	4181	7FE	9;24	131.96	4255	18JA	7;00	291.37	4255	1FE	7;44	125.45
4182	22FE	8;59	326.81	4182	8MR	18;32	160.95	4256	16FE	18;13	320.81	4256	1MR	23;28	154.83
4183	24MR	1;06	355.82	4183	7AP	3;17	189.43	4257	17MR	3;23	349.64	4257	31MR	15;45	183.71
4184	22AP	14;20	24.27	4184	6MY	12;25	217.57	4258	15AP	11;10	17.96	4258	30AP	7;38	212.18
4185	22MY	1;00	52.35	4185	4JN	22;55	245.59	4259	14MY	18;19	45.95	4259	29MY	22;17	240.38
4186	20JN	9;58	80.32	4186	4JL	11;39	273.77	4260	13JN	1;40	73.84	4260	28JN	11;20	268.54
4187	19JL	18;13	108.45	4187	3AU	2;52	302.35	4261	12JL	10;19	101.92	4261	27JL	23;02	296.90
4188	18AU	2;31	136.96	4188	1SE	20;03	331.46	4262	10AU	21;23	130.44	4262	26AU	9;58	325.66
4189	16SE	11;29	166.00	4189	1OC	14;01	1.13	4263	9SE	11;40	159.58	4263	24SE	20;38	354.92
4190	15OC	21;40	195.58	4190	31OC	7;28	31.22	4264	9OC	5;09	189.36	4264	24OC	7;16	24.64
4191	14NO	9;35	225.58	4191	29NO	23;29	61.49	4265	8NO	0;40	219.63	4265	22NO	17;58	54.66
4192	13DE	23;37	255.78	4192	29DE	13;33	91.64	4266	7DE	20;17	250.07	4266	22DE	4;53	84.71

Years −661 (left) / −655 (right)

NUMBER	DATE	TIME	LONG.	NUMBER	DATE	TIME	LONG.	NUMBER	DATE	TIME	LONG.	NUMBER	DATE	TIME	LONG.
4193	12JA	15;40	285.91	4193	28JA	1;23	121.40	4267	6JA	14;14	280.33	4267	20JA	16;21	114.54
4194	11FE	8;54	315.67	4194	26FE	11;06	150.60	4268	5FE	5;33	310.11	4268	19FE	4;42	143.97
4195	13MR	2;02	344.93	4195	27MR	19;10	179.22	4269	6MR	18;00	339.27	4269	20MR	18;09	172.92
4196	11AP	18;00	13.64	4196	26AP	2;27	207.39	4270	5AP	3;49	7.82	4270	19AP	8;29	201.46
4197	11MY	8;17	41.94	4197	25MY	10;04	235.34	4271	4MY	11;36	35.92	4271	18MY	23;18	229.72
4198	9JN	20;54	70.04	4198	23JN	19;09	263.32	4272	2JN	18;17	63.80	4272	17JN	14;12	257.92
4199	9JL	8;04	98.19	4199	23JL	6;34	291.61	4273	2JL	1;01	91.72	4273	17JL	4;56	286.28
4200	7AU	18;11	126.62	4200	21AU	20;45	320.41	4274	31JL	9;00	119.98	4274	15AU	19;17	315.01
4201	6SE	3;43	155.49	4201	20SE	13;32	349.84	4275	29AU	19;19	148.76	4275	14SE	9;00	344.20
4202	5OC	13;16	184.88	4202	20OC	8;09	19.82	4276	28SE	8;35	178.19	4276	13OC	21;46	13.86
4203	3NO	23;27	214.70	4203	19NO	3;21	50.17	4277	28OC	0;50	208.19	4277	12NO	9;30	43.84
4204	3DE	10;43	244.78	4204	18DE	21;33	80.54	4278	26NO	19;19	238.55	4278	11DE	20;21	73.91
								4279	26DE	14;44	268.95				

Years −660 (left) / −654 (right)

NUMBER	DATE	TIME	LONG.	NUMBER	DATE	TIME	LONG.	NUMBER	DATE	TIME	LONG.	NUMBER	DATE	TIME	LONG.
4205	1JA	23;16	274.85	4205	17JA	13;19	110.60					4279	10JA	6;40	103.81
4206	31JA	12;56	304.64	4206	16FE	1;58	140.11	4280	25JA	9;33	299.03	4280	8FE	16;52	133.32
4207	1MR	3;26	333.98	4207	16MR	11;45	168.98	4281	24FE	2;17	328.56	4281	10MR	3;16	162.34
4208	30MR	18;24	2.82	4208	14AP	19;30	197.31	4282	25MR	16;01	357.45	4282	8AP	14;10	190.89
4209	29AP	9;38	31.27	4209	14MY	2;22	225.28	4283	24AP	2;42	25.78	4283	8MY	1;58	219.12
4210	29MY	0;46	59.49	4210	12JN	9;23	253.16	4284	23MY	11;03	53.76	4284	6JN	15;03	247.24
4211	27JN	15;18	87.71	4211	11JL	17;32	281.22	4285	21JN	18;12	81.66	4285	6JL	5;43	275.51
4212	27JL	4;45	116.13	4212	10AU	3;37	309.69	4286	21JL	1;21	109.76	4286	4AU	21;48	304.14
4213	25AU	16;58	144.94	4213	8SE	16;18	338.75	4287	19AU	9;30	138.29	4287	3SE	14;31	333.27
4214	24SE	4;16	174.21	4214	8OC	8;01	8.45	4288	17SE	19;24	167.38	4288	3OC	6;47	2.89
4215	23OC	15;09	203.93	4215	7NO	2;33	38.66	4289	17OC	7;26	197.06	4289	1NO	21;39	32.89
4216	22NO	2;01	233.95	4216	6DE	22;36	69.12	4290	15NO	21;53	227.18	4290	1DE	10;50	63.05
4217	21DE	12;59	264.01					4291	15DE	14;38	257.50	4291	30DE	22;28	93.09

Years −659 (left) / −653 (right)

NUMBER	DATE	TIME	LONG.	NUMBER	DATE	TIME	LONG.	NUMBER	DATE	TIME	LONG.	NUMBER	DATE	TIME	LONG.
				4217	5JA	18;01	99.45	4292	14JA	9;04	287.70	4292	29JA	8;49	122.77
4218	20JA	0;01	293.83	4218	4FE	10;53	129.30	4293	13FE	3;44	317.50	4293	27FE	18;08	151.93
4219	18FE	11;11	323.22	4219	6MR	0;25	158.52	4294	14MR	20;56	346.73	4294	29MR	2;47	180.56
4220	19MR	22;52	352.11	4220	4AP	10;58	187.12	4295	13AP	11;33	15.36	4295	27AP	11;23	208.77
4221	18AP	11;34	20.59	4221	3MY	19;25	215.25	4296	12MY	23;36	43.55	4296	26MY	20;55	236.78
4222	18MY	1;35	48.83	4222	2JN	2;41	243.15	4297	11JN	9;14	71.54	4297	25JN	8;22	264.86
4223	16JN	16;43	77.04	4223	1JL	9;42	271.08	4298	10JL	17;53	99.60	4298	24JL	22;22	293.27
4224	16JL	8;18	105.42	4224	30JL	17;19	299.31	4299	9AU	2;16	127.98	4299	23AU	14;50	322.19
4225	14AU	23;38	134.18	4225	29AU	2;26	328.04	4300	7SE	11;02	156.85	4300	22SE	8;48	351.70
4226	13SE	14;15	163.39	4226	27SE	13;58	357.38	4301	6OC	20;42	186.25	4301	22OC	2;54	21.69
4227	13OC	4;03	193.08	4227	27OC	4;35	27.30	4302	5NO	7;45	216.13	4302	20NO	19;55	51.95
4228	11NO	16;56	223.11	4228	25NO	22;10	57.62	4303	4DE	20;39	246.28	4303	20DE	11;05	82.19
4229	11DE	4;45	253.22	4229	25DE	17;30	88.02								

-652 / -646

NUMBER	DATE	TIME	LONG.	NUMBER	DATE	TIME	LONG.	NUMBER	DATE	TIME	LONG.	NUMBER	DATE	TIME	LONG.
4304	3JA	11;34	276.44	4304	19JA	0;00	112.11					4378	11JA	14;49	105.22
4305	2FE	4;05	306.32	4305	17FE	10;38	141.51	4379	27JA	2;16	300.79	4379	10FE	2;26	134.77
4306	2MR	21;09	335.74	4306	17MR	19;17	170.32	4380	25FE	16;03	330.17	4380	11MR	15;01	163.86
4307	1AP	13;35	4.62	4307	16AP	2;41	198.61	4381	27MR	3;03	358.92	4381	10AP	4;39	192.50
4308	1MY	4;39	33.04	4308	15MY	9;51	226.59	4382	25AP	11;41	27.16	4382	9MY	19;06	220.82
4309	30MY	18;08	61.19	4309	13JN	17;59	254.52	4383	24MY	18;46	55.08	4383	8JN	9;57	249.01
4310	29JN	6;09	89.32	4310	13JL	4;07	282.66	4384	23JN	1;21	82.96	4384	8JL	0;54	277.30
4311	28JL	16;59	117.66	4311	11AU	16;58	311.27	4385	22JL	8;38	111.08	4385	6AU	15;43	305.90
4312	27AU	3;00	146.39	4312	10SE	8;36	340.48	4386	20AU	17;46	139.67	4386	5SE	6;06	334.96
4313	25SE	12;42	175.62	4313	10OC	2;35	10.29	4387	19SE	5;36	168.86	4387	4OC	19;40	4.50
4314	24OC	22;39	205.32	4314	8NO	21;50	40.56	4388	18OC	20;29	198.68	4388	3NO	8;08	34.41
4315	23NO	9;22	235.34	4315	8DE	16;49	70.97	4389	17NO	13;58	228.94	4389	2DE	19;28	64.49
4316	22DE	21;11	265.43					4390	17DE	9;03	259.35				

-651 / -645

NUMBER	DATE	TIME	LONG.	NUMBER	DATE	TIME	LONG.	NUMBER	DATE	TIME	LONG.	NUMBER	DATE	TIME	LONG.
				4316	7JA	9;57	101.19					4390	1JA	5;58	94.48
4317	21JA	10;07	295.31	4317	6FE	0;07	130.91	4391	16JA	4;15	289.57	4391	30JA	16;00	124.12
4318	19FE	23;56	324.79	4318	7MR	11;09	160.00	4392	14FE	22;02	319.31	4392	1MR	2;01	153.29
4319	21MR	14;23	353.77	4319	5AP	19;41	188.49	4393	16MR	13;14	348.42	4393	30MR	12;21	181.97
4320	20AP	5;18	22.32	4320	5MY	2;05	216.55	4394	15AP	1;21	16.94	4394	28AP	23;24	210.26
4321	19MY	20;26	50.59	4321	3JN	9;32	244.43	4395	14MY	10;45	45.02	4395	28MY	11;37	238.38
4322	18JN	11;24	78.79	4322	2JL	17;01	272.39	4396	12JN	18;25	72.92	4396	27JN	1;26	266.56
4323	18JL	1;38	107.14	4323	1AU	2;05	300.70	4397	12JL	1;32	100.93	4397	26JL	16;58	295.05
4324	16AU	14;46	135.83	4324	30AU	13;29	329.54	4398	10AU	9;14	129.30	4398	25AU	9;44	324.01
4325	15SE	2;46	164.97	4325	29SE	3;45	359.02	4399	8SE	18;20	158.20	4399	24SE	2;40	353.50
4326	14OC	14;03	194.57	4326	28OC	21;02	29.07	4400	8OC	5;22	187.68	4400	23OC	18;36	23.42
4327	13NO	1;01	224.53	4327	27NO	16;34	59.49	4401	6NO	18;37	217.67	4401	22NO	8;48	53.58
4328	12DE	11;57	254.60	4328	27DE	12;34	89.90	4402	6DE	10;11	247.94	4402	21DE	21;14	83.70

-650 / -644

NUMBER	DATE	TIME	LONG.	NUMBER	DATE	TIME	LONG.	NUMBER	DATE	TIME	LONG.	NUMBER	DATE	TIME	LONG.
4329	10JA	22;50	284.52	4329	26JA	6;48	119.95	4403	5JA	3;44	278.19	4403	20JA	8;07	113.52
4330	9FE	9;40	314.05	4330	24FE	21;55	149.40	4404	3FE	22;16	308.14	4404	18FE	17;45	142.85
4331	10MR	20;44	343.08	4331	26MR	9;48	178.20	4405	4MR	16;11	337.56	4405	19MR	2;27	171.64
4332	9AP	8;34	11.66	4332	24AP	19;08	206.47	4406	3AP	8;03	6.39	4406	17AP	10;46	199.96
4333	8MY	21;42	39.94	4333	24MY	2;52	234.42	4407	2MY	21;15	34.72	4407	16MY	19;32	227.99
4334	7JN	12;14	68.12	4334	22JN	9;57	262.32	4408	1JN	8;04	62.76	4408	15JN	5;48	256.00
4335	7JL	3;46	96.42	4335	21JL	17;15	290.43	4409	30JN	17;19	90.78	4409	14JL	18;28	284.26
4336	5AU	19;31	125.05	4336	20AU	1;37	318.98	4410	30JL	1;57	119.05	4410	13AU	9;51	312.99
4337	4SE	10;49	154.13	4337	18SE	11;57	348.10	4411	28AU	10;39	147.75	4411	12SE	3;25	342.31
4338	4OC	1;21	183.70	4338	18OC	1;03	17.83	4412	26SE	20;00	176.99	4412	11OC	21;52	12.17
4339	2NO	14;56	213.66	4339	16NO	17;14	48.03	4413	26OC	6;24	206.72	4413	10NO	15;45	42.40
4340	2DE	3;25	243.78	4340	16DE	11;48	78.43	4414	24NO	18;19	236.80	4414	10DE	8;02	72.70
4341	31DE	14;39	273.80					4415	24DE	8;06	266.96				

-649 / -643

NUMBER	DATE	TIME	LONG.	NUMBER	DATE	TIME	LONG.	NUMBER	DATE	TIME	LONG.	NUMBER	DATE	TIME	LONG.
				4341	15JA	7;07	108.66					4415	8JA	22;06	102.77
4342	30JA	0;39	303.45	4342	14FE	1;18	138.42	4416	22JA	23;40	296.94	4416	7FE	9;47	132.36
4343	28FE	9;48	332.60	4343	15MR	17;13	167.58	4417	21FE	16;18	326.51	4417	8MR	19;12	161.36
4344	29MR	18;50	1.23	4344	14AP	6;30	196.15	4418	23MR	8;56	355.55	4418	7AP	2;58	189.81
4345	28AP	4;36	29.48	4345	13MY	17;24	224.30	4419	22AP	0;36	24.11	4419	6MY	9;58	217.85
4346	27MY	15;51	57.56	4346	12JN	2;24	252.26	4420	21MY	14;52	52.33	4420	4JN	17;23	245.76
4347	26JN	4;56	85.71	4347	11JL	10;16	280.29	4421	20JN	3;43	80.46	4421	4JL	2;22	273.79
4348	25JL	19;51	114.17	4348	9AU	17;55	308.65	4422	19JL	15;20	108.72	4422	2AU	13;50	302.20
4349	24AU	12;17	143.10	4349	8SE	2;20	337.51	4423	18AU	1;59	137.33	4423	1SE	4;11	331.19
4350	23SE	5;37	172.59	4350	7OC	12;23	6.94	4424	16SE	12;02	166.41	4424	30SE	21;13	0.81
4351	22OC	22;54	202.56	4351	6NO	0;40	36.88	4425	15OC	21;59	195.98	4425	30OC	16;08	30.96
4352	21NO	14;58	232.79	4352	5DE	15;15	67.10	4426	14NO	8;21	225.91	4426	29NO	11;34	61.37
4353	21DE	4;56	262.98					4427	13DE	19;33	256.00	4427	29DE	5;50	91.70

-648 / -642

NUMBER	DATE	TIME	LONG.	NUMBER	DATE	TIME	LONG.	NUMBER	DATE	TIME	LONG.	NUMBER	DATE	TIME	LONG.
				4353	4JA	7;38	97.31	4428	12JA	7;45	285.96	4428	27JA	21;32	121.62
4354	19JA	16;31	292.84	4354	3FE	0;57	127.22	4429	10FE	20;53	315.56	4429	26FE	10;00	150.94
4355	18FE	2;06	322.18	4355	3MR	18;15	156.63	4430	12MR	10;44	344.69	4430	27MR	19;34	179.62
4356	18MR	10;25	350.96	4356	2AP	10;39	185.50	4431	11AP	1;10	13.35	4431	26AP	3;08	207.81
4357	16AP	18;17	19.27	4357	2MY	1;22	213.90	4432	10MY	16;03	41.68	4432	25MY	9;52	235.71
4358	16MY	2;30	47.29	4358	31MY	13;57	242.01	4433	9JN	7;10	69.88	4433	23JN	16;53	263.61
4359	14JN	11;51	75.26	4359	30JN	0;33	270.08	4434	8JL	21;59	98.17	4434	23JL	1;06	291.77
4360	13JL	23;04	103.45	4360	29JL	9;49	298.37	4435	7AU	11;59	126.75	4435	21AU	11;20	320.41
4361	12AU	12;49	132.11	4361	27AU	18;42	327.06	4436	6SE	0;50	155.76	4436	20SE	0;13	349.67
4362	11SE	5;21	161.36	4362	26SE	4;03	356.29	4437	5OC	12;42	185.24	4437	19OC	16;05	19.53
4363	11OC	0;04	191.23	4363	25OC	14;24	26.02	4438	3NO	23;58	215.11	4438	18NO	10;41	49.85
4364	9NO	19;24	221.50	4364	24NO	1;54	56.07	4439	3DE	10;58	245.18	4439	18DE	6;42	80.30
4365	9DE	13;21	251.88	4365	23DE	14;33	86.19								

-647 / -641

NUMBER	DATE	TIME	LONG.	NUMBER	DATE	TIME	LONG.	NUMBER	DATE	TIME	LONG.	NUMBER	DATE	TIME	LONG.
4366	8JA	4;38	282.00	4366	22JA	4;20	116.10	4440	1JA	21;48	275.17	4440	17JA	1;58	110.51
4367	6FE	17;02	311.64	4367	20FE	19;15	145.61	4441	31JA	8;27	304.84	4441	15FE	18;38	140.19
4368	8MR	3;02	340.67	4368	22MR	11;03	174.64	4442	1MR	19;05	334.02	4442	17MR	8;00	169.21
4369	6AP	11;17	9.15	4369	21AP	3;03	203.22	4443	31MR	6;10	2.71	4443	15AP	18;26	197.65
4370	5MY	18;33	37.21	4370	20MY	18;22	231.49	4444	29AP	18;10	31.05	4444	15MY	2;51	225.69
4371	4JN	1;37	65.10	4371	19JN	8;20	259.66	4445	29MY	8;02	59.21	4445	13JN	10;11	253.58
4372	3JL	9;28	93.08	4372	18JL	20;53	287.95	4446	27JN	23;08	87.45	4446	12JL	17;22	281.60
4373	1AU	19;13	121.42	4373	17AU	8;25	316.58	4447	27JL	15;02	115.97	4447	11AU	1;13	309.98
4374	31AU	7;54	150.33	4374	15SE	19;26	345.69	4448	26AU	6;54	144.91	4448	9SE	10;35	338.90
4375	29SE	23;58	179.89	4375	15OC	6;17	15.29	4449	24SE	22;10	174.34	4449	8OC	22;18	8.42
4376	29OC	18;47	210.02	4376	13NO	17;03	45.24	4450	24OC	12;30	204.21	4450	7NO	12;58	38.48
4377	28NO	14;39	240.45	4377	13DE	3;50	75.30	4451	23NO	1;43	234.33	4451	7DE	6;26	68.82
4378	28DE	9;37	270.82					4452	22DE	13;37	264.42				

-640

NUMBER	DATE	TIME	LONG.	NUMBER	DATE	TIME	LONG.
				4452	6JA	1;28	99.14
4453	21JA	0;07	294.22	4453	4FE	20;14	129.08
4454	19FE	9;26	323.53	4454	5MR	13;13	158.45
4455	19MR	18;11	352.31	4455	4AP	3;43	187.21
4456	18AP	3;14	20.65	4456	3MY	15;45	215.48
4457	17MY	13;28	48.73	4457	2JN	1;42	243.49
4458	16JN	1;30	76.82	4458	1JL	10;10	271.49
4459	15JL	15;32	105.14	4459	30JL	17;59	299.73
4460	14AU	7;25	133.90	4460	29AU	2;04	328.41
4461	13SE	0;37	163.22	4461	27SE	11;22	357.66
4462	12OC	18;20	193.06	4462	26OC	22;34	27.44
4463	11NO	11;22	223.26	4463	25NO	12;00	57.58
4464	11DE	2;35	253.51	4464	25DE	3;26	87.81

-634

NUMBER	DATE	TIME	LONG.	NUMBER	DATE	TIME	LONG.
4527	13JA	19;47	287.53	4527	29JA	8;26	123.13
4528	12FE	11;43	317.23	4528	27FE	18;49	152.34
4529	14MR	4;12	346.43	4529	29MR	3;11	180.96
4530	12AP	20;14	15.13	4530	27AP	10;17	209.11
4531	12MY	11;10	43.45	4531	26MY	17;15	237.02
4532	11JN	0;48	71.59	4532	25JN	1;16	264.96
4533	10JL	13;12	99.80	4533	24JL	11;25	293.21
4534	9AU	0;34	128.30	4534	23AU	0;23	321.99
4535	7SE	11;08	157.24	4535	21SE	16;15	351.39
4536	6OC	21;17	186.67	4536	21OC	10;29	21.38
4537	5NO	7;31	216.51	4537	20NO	5;57	51.75
4538	4DE	18;17	246.57	4538	20DE	1;03	82.15

-639

NUMBER	DATE	TIME	LONG.	NUMBER	DATE	TIME	LONG.
4465	9JA	15;19	283.52	4465	23JA	20;09	117.83
4466	8FE	1;43	313.05	4466	22FE	13;18	147.40
4467	9MR	10;25	342.01	4467	24MR	6;03	176.43
4468	7AP	18;13	10.45	4468	22AP	21;35	204.98
4469	7MY	2;00	38.52	4469	22MY	11;18	233.17
4470	5JN	10;34	66.47	4470	20JN	22;59	261.25
4471	4JL	20;43	94.55	4471	20JL	9;01	289.46
4472	3AU	9;10	123.02	4472	18AU	18;13	318.02
4473	2SE	0;25	152.06	4473	17SE	3;28	347.08
4474	1OC	18;22	181.73	4474	16OC	13;27	16.66
4475	31OC	13;49	211.90	4475	15NO	0;27	46.63
4476	30NO	8;49	242.30	4476	14DE	12;28	76.75
4477	30DE	1;34	272.56				

-633

NUMBER	DATE	TIME	LONG.	NUMBER	DATE	TIME	LONG.
4539	3JA	5;51	276.58	4539	18JA	18;09	112.26
4540	1FE	18;18	306.30	4540	17FE	8;10	141.80
4541	3MR	7;30	335.57	4541	18MR	19;00	170.69
4542	1AP	21;21	4.35	4542	17AP	3;21	199.02
4543	1MY	11;48	32.76	4543	16MY	10;18	227.00
4544	31MY	2;47	60.97	4544	14JN	17;02	254.86
4545	29JN	17;55	89.22	4545	14JL	0;35	282.91
4546	29JL	8;37	117.70	4546	12AU	9;49	311.36
4547	27AU	22;22	146.59	4547	10SE	21;23	340.40
4548	26SE	10;59	175.94	4548	10OC	11;49	10.06
4549	25OC	22;44	205.71	4549	9NO	5;11	40.24
4550	24NO	9;57	235.75	4550	9DE	0;42	70.68
4551	23DE	20;50	265.80				

-638

NUMBER	DATE	TIME	LONG.	NUMBER	DATE	TIME	LONG.
				4477	13JA	1;28	106.73
4478	28JA	15;20	302.39	4478	11FE	15;32	136.36
4479	27FE	2;20	331.64	4479	13MR	6;38	165.53
4480	28MR	11;15	0.28	4480	11AP	22;24	194.23
4481	26AP	18;48	28.46	4481	11MY	14;03	222.58
4482	26MY	1;47	56.37	4482	10JN	4;47	250.77
4483	24JN	9;05	84.29	4483	9JL	18;12	279.01
4484	23JL	17;47	112.47	4484	8AU	6;27	307.53
4485	22AU	4;58	141.16	4485	6SE	17;59	336.50
4486	20SE	19;26	170.50	4486	6OC	5;09	5.96
4487	20OC	13;07	200.45	4487	4NO	16;07	35.83
4488	19NO	8;46	230.82	4488	4DE	2;54	65.89
4489	19DE	4;25	261.25				

-632

NUMBER	DATE	TIME	LONG.	NUMBER	DATE	TIME	LONG.
				4551	7JA	20;32	101.01
4552	22JA	7;26	295.59	4552	6FE	14;34	130.89
4553	20FE	17;47	324.92	4553	7MR	5;30	160.15
4554	21MR	4;17	353.75	4554	5AP	17;14	188.78
4555	19AP	15;34	22.16	4555	5MY	2;31	216.93
4556	19MY	4;16	50.33	4556	3JN	10;18	244.85
4557	17JN	18;39	78.51	4557	2JL	17;33	272.80
4558	17JL	10;21	106.92	4558	1AU	1;05	301.04
4559	16AU	2;34	135.72	4559	30AU	9;43	329.77
4560	14SE	18;29	165.01	4560	28SE	20;17	359.09
4561	14OC	9;36	194.77	4561	28OC	9;29	28.97
4562	12NO	23;37	224.85	4562	27NO	1;36	59.24
4563	12DE	12;16	255.00	4563	26DE	19;55	89.58

-637

NUMBER	DATE	TIME	LONG.	NUMBER	DATE	TIME	LONG.
				4489	2JA	13;39	95.87
4490	17JA	22;17	291.39	4490	1FE	0;40	125.54
4491	16FE	13;27	320.99	4491	2MR	12;27	154.76
4492	18MR	1;43	349.96	4492	1AP	1;16	183.51
4493	16AP	11;24	18.36	4493	30AP	15;06	211.90
4494	15MY	19;07	46.36	4494	30MY	5;40	240.10
4495	14JN	1;49	74.23	4495	28JN	20;40	268.34
4496	13JL	8;40	102.24	4496	28JL	11;46	296.83
4497	11AU	16;50	130.64	4497	27AU	2;43	325.75
4498	10SE	3;21	159.62	4498	25SE	17;03	355.16
4499	9OC	16;47	189.23	4499	25OC	6;21	24.99
4500	8NO	9;05	219.36	4500	23NO	18;22	55.06
4501	8DE	3;29	249.75	4501	23DE	5;14	85.11

-631

NUMBER	DATE	TIME	LONG.	NUMBER	DATE	TIME	LONG.
4564	10JA	23;24	284.93	4564	25JA	14;50	119.67
4565	9FE	9;06	314.41	4565	24FE	8;40	149.24
4566	10MR	17;49	343.35	4566	26MR	0;19	178.20
4567	9AP	2;23	11.80	4567	24AP	13;31	206.63
4568	8MY	11;43	39.92	4568	24MY	0;30	234.71
4569	6JN	22;39	67.96	4569	22JN	9;44	262.70
4570	6JL	11;39	96.16	4570	21JL	17;57	290.84
4571	5AU	2;43	124.76	4571	20AU	1;58	319.38
4572	3SE	19;32	153.89	4572	18SE	10;44	348.44
4573	3OC	13;22	183.59	4573	17OC	21;02	18.05
4574	2NO	7;09	213.71	4574	16NO	9;22	48.08
4575	1DE	23;32	244.01	4575	15DE	23;45	78.30
4576	31DE	13;35	274.14				

-636

NUMBER	DATE	TIME	LONG.	NUMBER	DATE	TIME	LONG.
4502	6JA	22;41	280.06	4502	21JA	15;20	114.88
4503	5FE	17;13	309.98	4503	20FE	1;06	144.20
4504	6MR	9;42	339.31	4504	20MR	10;57	173.02
4505	4AP	23;17	8.02	4505	18AP	21;20	201.40
4506	4MY	9;57	36.24	4506	18MY	8;43	229.53
4507	2JN	18;24	64.18	4507	16JN	21;36	257.64
4508	2JL	1;45	92.14	4508	16JL	12;19	286.00
4509	31JL	9;11	120.37	4509	15AU	4;45	314.80
4510	29AU	17;40	149.08	4510	13SE	22;02	344.13
4511	28SE	3;49	178.37	4511	13OC	14;53	13.96
4512	27OC	16;00	208.20	4512	12NO	6;13	44.09
4513	26NO	6;22	238.39	4513	11DE	19;36	74.27
4514	25DE	22;51	268.66				

-630

NUMBER	DATE	TIME	LONG.	NUMBER	DATE	TIME	LONG.
				4576	14JA	15;43	108.40
4577	30JA	1;02	303.86	4577	13FE	8;28	138.11
4578	28FE	10;20	333.01	4578	15MR	1;16	167.32
4579	29MR	18;18	1.60	4579	13AP	17;21	196.01
4580	28AP	1;49	29.75	4580	13MY	8;00	224.31
4581	27MY	9;47	57.70	4581	11JN	20;48	252.42
4582	25JN	18;59	85.70	4582	11JL	7;48	280.58
4583	25JL	6;12	114.00	4583	9AU	17;33	309.02
4584	23AU	20;06	142.83	4584	8SE	2;55	337.92
4585	22SE	12;54	172.28	4585	7OC	12;40	7.34
4586	22OC	7;59	202.32	4586	5NO	23;15	37.20
4587	21NO	3;36	232.70	4587	5DE	10;45	67.30
4588	20DE	21;43	263.06				

-635

NUMBER	DATE	TIME	LONG.	NUMBER	DATE	TIME	LONG.
				4514	10JA	7;11	104.22
4515	24JA	16;50	298.72	4515	8FE	17;17	133.72
4516	23FE	11;02	328.32	4516	10MR	2;13	162.68
4517	25MR	3;54	357.35	4517	8AP	10;26	191.13
4518	23AP	18;23	25.84	4518	7MY	18;40	219.22
4519	23MY	6;22	53.96	4519	6JN	3;56	247.19
4520	21JN	16;26	81.98	4520	5JL	15;15	275.32
4521	21JL	1;27	110.15	4521	4AU	5;20	303.86
4522	19AU	10;16	138.71	4522	2SE	22;05	332.98
4523	17SE	19;27	167.77	4523	2OC	16;29	2.68
4524	17OC	5;24	197.36	4524	1NO	11;02	32.83
4525	15NO	16;33	227.34	4525	1DE	4;22	63.16
4526	15DE	5;17	257.49	4526	30DE	19;37	93.36

-629

NUMBER	DATE	TIME	LONG.	NUMBER	DATE	TIME	LONG.
				4588	3JA	23;06	97.34
4589	19JA	12;58	293.07	4589	2FE	12;22	127.08
4590	18FE	1;12	322.53	4590	4MR	2;39	156.38
4591	19MR	10;57	351.37	4591	2AP	17;52	185.21
4592	17AP	18;59	19.68	4592	2MY	9;31	213.65
4593	17MY	2;04	47.65	4593	1JN	0;46	241.87
4594	15JN	9;04	75.53	4594	30JN	15;01	270.09
4595	14JL	16;57	103.59	4595	30JL	4;03	298.52
4596	13AU	2;49	132.08	4596	28AU	16;11	327.35
4597	11SE	15;40	161.18	4597	27SE	3;46	356.67
4598	11OC	7;54	190.93	4598	26OC	15;02	26.43
4599	10NO	2;52	221.19	4599	25NO	2;00	56.46
4600	9DE	22;47	251.64	4600	24DE	12;41	86.50

-628

NUMBER	DATE	TIME	LONG.	NUMBER	DATE	TIME	LONG.
4601	8JA	17;40	281.93	4601	22JA	23;20	116.28
4602	7FE	10;10	311.74	4602	21FE	10;25	145.63
4603	7MR	23;47	340.92	4603	21MR	22;23	174.51
4604	6AP	10;37	9.50	4604	20AP	11;28	202.98
4605	5MY	19;11	37.62	4605	20MY	1;32	231.21
4606	4JN	2;16	65.51	4606	18JN	16;20	259.40
4607	3JL	8;58	93.44	4607	18JL	7;33	287.80
4608	1AU	16;26	121.69	4608	16AU	22;53	316.58
4609	31AU	1;47	150.46	4609	15SE	13;55	345.85
4610	29SE	13;49	179.85	4610	15OC	4;04	15.58
4611	29OC	4;47	209.82	4611	13NO	16;55	45.62
4612	27NO	22;13	240.14	4612	13DE	4;22	75.71
4613	27DE	17;05	270.51				

-627

NUMBER	DATE	TIME	LONG.	NUMBER	DATE	TIME	LONG.
4614	26JA	11;59	300.58	4613	11JA	14;43	105.60
4615	25FE	5;29	330.12	4614	10FE	0;25	135.07
4616	26MR	20;29	359.04	4615	11MR	9;56	164.03
4617	25AP	8;32	27.42	4616	9AP	19;44	192.52
4618	24MY	18;01	55.44	4617	9MY	6;20	220.69
4619	23JN	1;52	83.37	4618	7JN	18;16	248.77
4620	22JL	9;16	111.49	4619	7JL	8;02	277.01
4621	20AU	17;17	140.03	4620	5AU	23;47	305.64
4622	19SE	2;42	169.13	4621	4SE	17;02	334.81
4623	18OC	13;56	198.79	4622	4OC	10;33	4.50
4624	17NO	3;12	228.87	4623	3NO	2;59	34.58
4625	16DE	18;33	259.13	4624	2DE	17;30	64.80

-626

NUMBER	DATE	TIME	LONG.	NUMBER	DATE	TIME	LONG.
4626	15JA	11;41	289.26	4625	1JA	5;58	94.86
4627	14FE	5;43	319.02	4626	30JA	16;40	124.54
4628	15MR	23;13	348.24	4627	1MR	1;58	153.67
4629	14AP	14;52	16.90	4628	30MR	10;16	182.27
4630	14MY	4;04	45.13	4629	28AP	18;12	210.44
4631	12JN	15;06	73.17	4630	28MY	2;40	238.40
4632	12JL	0;43	101.28	4631	26JN	12;46	266.44
4633	10AU	9;47	129.71	4632	26JL	1;25	294.81
4634	8SE	18;56	158.61	4633	24AU	17;01	323.71
4635	8OC	4;38	188.04	4634	23SE	10;57	353.23
4636	6NO	15;13	217.91	4635	23OC	5;50	23.27
4637	6DE	3;04	248.02	4636	22NO	0;04	53.59
				4637	21DE	16;31	83.88

-625

NUMBER	DATE	TIME	LONG.	NUMBER	DATE	TIME	LONG.
4638	4JA	16;31	278.10	4638	20JA	6;33	113.84
4639	3FE	7;33	307.91	4639	18FE	18;02	143.25
4640	4MR	23;34	337.26	4640	20MR	3;11	172.06
4641	3AP	15;41	6.11	4641	18AP	10;40	200.34
4642	3MY	7;06	34.54	4642	17MY	17;27	228.29
4643	1JN	21;24	62.72	4643	16JN	0;44	256.18
4644	1JL	10;34	90.90	4644	15JL	9;42	284.30
4645	30JL	22;41	119.30	4645	13AU	21;15	312.87
4646	29AU	9;54	148.10	4646	12SE	11;48	342.05
4647	27SE	20;28	177.39	4647	12OC	5;04	11.85
4648	27OC	6;46	207.13	4648	11NO	0;12	42.12
4649	25NO	17;16	237.14	4649	10DE	19;45	72.56
4650	25DE	4;21	267.19				

-624

NUMBER	DATE	TIME	LONG.	NUMBER	DATE	TIME	LONG.
4651	23JA	16;10	297.01	4650	9JA	14;00	102.81
4652	22FE	4;43	326.42	4651	8FE	5;34	132.57
4653	22MR	17;56	355.33	4652	8MR	17;52	161.69
4654	21AP	7;50	23.82	4653	7AP	3;15	190.20
4655	20MY	22;26	52.07	4654	6MY	10;42	218.27
4656	19JN	13;34	80.28	4655	4JN	17;23	246.14
4657	19JL	4;45	108.68	4656	4JL	0;27	274.09
4658	17AU	19;19	137.44	4657	2AU	8;49	302.39
4659	16SE	8;49	166.66	4658	31AU	19;14	331.21
4660	15OC	21;13	196.33	4659	30SE	8;17	0.66
4661	14NO	8;49	226.32	4660	30OC	0;14	30.67
4662	13DE	19;53	256.40	4661	28NO	18;51	61.05
				4662	28DE	14;43	91.45

-623

NUMBER	DATE	TIME	LONG.	NUMBER	DATE	TIME	LONG.
4663	12JA	6;31	286.29	4663	27JA	9;46	121.52
4664	10FE	16;46	315.78	4664	26FE	2;13	151.00
4665	12MR	2;52	344.75	4665	27MR	15;25	179.84
4666	10AP	13;23	13.26	4666	26AP	1;47	208.13
4667	10MY	1;05	41.47	4667	25MY	10;13	236.11
4668	8JN	14;30	69.60	4668	23JN	17;42	264.03
4669	8JL	5;39	97.90	4669	23JL	1;07	292.16
4670	6AU	21;53	126.57	4670	21AU	9;15	320.71
4671	5SE	14;19	155.71	4671	19SE	18;53	349.83
4672	5OC	6;12	185.36	4672	19OC	6;46	19.53
4673	3NO	21;02	215.38	4673	17NO	21;26	49.67
4674	3DE	10;31	245.55	4674	17DE	14;41	80.01

-622

NUMBER	DATE	TIME	LONG.	NUMBER	DATE	TIME	LONG.
4675	1JA	22;25	275.59	4675	16JA	9;20	110.20
4676	31JA	8;41	305.23	4676	15FE	3;41	139.96
4677	1MR	17;37	334.34	4677	16MR	20;20	169.12
4678	31MR	1;55	2.93	4678	15AP	10;41	197.72
4679	29AP	10;32	31.12	4679	14MY	22;45	225.91
4680	28MY	20;25	59.14	4680	13JN	8;54	253.91
4681	27JN	8;17	87.24	4681	12JL	17;41	281.99
4682	26JL	22;22	115.68	4682	11AU	1;54	310.39
4683	25AU	14;31	144.62	4683	9SE	10;23	339.28
4684	24SE	8;11	174.15	4684	8OC	19;58	8.71
4685	24OC	2;24	204.17	4685	7NO	7;19	38.62
4686	22NO	19;49	234.46	4686	6DE	20;39	68.80
4687	22DE	11;11	264.71				

-621

NUMBER	DATE	TIME	LONG.	NUMBER	DATE	TIME	LONG.
4688	20JA	23;52	294.59	4687	5JA	11;43	98.94
4689	19FE	10;02	323.94	4688	4FE	3;54	128.79
4690	20MR	18;24	352.70	4689	5MR	20;29	158.15
4691	19AP	1;53	20.97	4690	4AP	12;48	186.99
4692	18MY	9;23	48.95	4691	4MY	4;10	215.41
4693	16JN	17;47	76.89	4692	2JN	17;59	243.57
4694	16JL	3;54	105.05	4693	2JL	6;01	271.71
4695	14AU	16;26	133.68	4694	31JL	16;33	300.05
4696	13SE	7;55	162.91	4695	30AU	2;16	328.81
4697	13OC	2;10	192.77	4696	28SE	11;58	358.07
4698	11NO	21;55	223.07	4697	27OC	22;16	27.81
4699	11DE	17;06	253.50	4698	26NO	9;20	57.85
				4699	25DE	21;10	87.93

-620

NUMBER	DATE	TIME	LONG.	NUMBER	DATE	TIME	LONG.
4700	10JA	9;52	283.68	4700	24JA	9;45	117.78
4701	8FE	23;31	313.34	4701	22FE	23;12	147.20
4702	9MR	10;19	342.39	4702	23MR	13;40	176.16
4703	7AP	19;00	10.86	4703	22AP	4;57	204.70
4704	7MY	2;22	38.92	4704	21MY	20;25	232.97
4705	5JN	9;16	66.80	4705	20JN	11;18	261.17
4706	4JL	16;35	94.77	4706	20JL	1;09	289.53
4707	3AU	1;23	123.09	4707	18AU	13;59	318.23
4708	1SE	12;43	151.96	4708	17SE	2;06	347.41
4709	1OC	3;21	181.48	4709	16OC	13;46	17.06
4710	30OC	21;11	211.59	4710	15NO	1;01	47.04
4711	29NO	16;54	242.01	4711	14DE	11;49	77.11
4712	29DE	12;29	272.40				

-619

NUMBER	DATE	TIME	LONG.	NUMBER	DATE	TIME	LONG.
4713	28JA	6;12	302.40	4712	12JA	22;19	106.99
4714	26FE	21;11	331.81	4713	11FE	8;53	136.47
4715	28MR	9;17	0.59	4714	12MR	20;04	165.48
4716	26AP	18;52	28.85	4715	11AP	8;17	194.05
4717	26MY	2;35	56.79	4716	10MY	21;40	222.32
4718	24JN	9;23	84.68	4717	9JN	12;04	250.49
4719	23JL	16;25	112.80	4718	9JL	3;11	278.80
4720	22AU	0;49	141.38	4719	7AU	18;43	307.44
4721	20SE	11;33	170.56	4720	6SE	10;16	336.57
4722	20OC	1;07	200.33	4721	6OC	1;14	6.18
4723	18NO	17;25	230.55	4722	4NO	15;00	36.17
4724	18DE	11;37	260.93	4723	4DE	3;15	66.29

-618

NUMBER	DATE	TIME	LONG.	NUMBER	DATE	TIME	LONG.
4725	17JA	6;31	291.12	4724	2JA	14;04	96.28
4726	16FE	0;43	320.85	4725	31JA	23;52	125.89
4727	17MR	16;56	349.99	4726	2MR	9;12	155.01
4728	16AP	6;25	18.54	4727	31MR	18;33	183.63
4729	15MY	17;08	46.67	4728	30AP	4;28	211.86
4730	14JN	1;45	74.61	4729	29MY	15;29	239.92
4731	13JL	9;22	102.65	4730	28JN	4;13	268.07
4732	11AU	17;08	131.04	4731	27JL	19;03	296.54
4733	10SE	1;57	159.95	4732	26AU	11;52	325.53
4734	9OC	12;22	189.43	4733	25SE	5;42	355.07
4735	8NO	0;38	219.39	4734	24OC	23;07	25.07
4736	7DE	14;53	249.60	4735	23NO	14;48	55.30
				4736	23DE	4;19	85.46

-617

NUMBER	DATE	TIME	LONG.	NUMBER	DATE	TIME	LONG.
4737	6JA	7;00	279.78	4737	21JA	15;48	115.29
4738	5FE	0;28	309.67	4738	20FE	1;37	144.61
4739	6MR	18;10	339.06	4739	21MR	10;11	173.37
4740	5AP	10;43	7.90	4740	19AP	18;01	201.65
4741	5MY	1;07	36.27	4741	19MY	1;56	229.64
4742	3JN	13;15	64.36	4742	17JN	10;59	257.61
4743	2JL	23;38	92.44	4743	16JL	22;15	285.82
4744	1AU	9;06	120.75	4744	15AU	12;27	314.52
4745	30AU	18;23	149.50	4745	14SE	5;30	343.84
4746	29SE	3;58	178.77	4746	14OC	0;18	13.73
4747	28OC	14;12	208.51	4747	12NO	19;13	44.01
4748	27NO	1;23	238.56	4748	12DE	12;46	74.36
4749	26DE	13;53	268.67				

-616 / -610

NEW	MOONS			FULL	MOONS			NEW	MOONS			FULL	MOONS		
				4749	11JA	4;03	104.47					4823	3JA	21;29	97.66
4750	25JA	3;54	298.56	4750	9FE	16;44	134.09	4824	19JA	1;36	292.98	4824	2FE	7;45	127.28
4751	23FE	19;13	328.06	4751	10MR	2;52	163.09	4825	17FE	17;54	322.61	4825	3MR	18;17	156.42
4752	24MR	11;07	357.05	4752	8AP	10;58	191.53	4826	19MR	7;20	351.60	4826	2AP	5;39	185.10
4753	23AP	2;46	25.60	4753	7MY	17;51	219.56	4827	17AP	18;04	20.03	4827	1MY	18;13	213.42
4754	22MY	17;37	53.84	4754	6JN	0;40	247.44	4828	17MY	2;36	48.06	4828	31MY	7;59	241.59
4755	21JN	7;27	82.00	4755	5JL	8;39	275.44	4829	15JN	9;47	75.94	4829	29JN	22;47	269.83
4756	20JL	20;19	110.33	4756	3AU	18;51	303.82	4830	14JL	16;39	103.97	4830	29JL	14;19	298.35
4757	19AU	8;15	139.01	4757	2SE	7;58	332.78	4831	13AU	0;22	132.37	4831	28AU	6;12	327.32
4758	17SE	19;23	168.16	4758	2OC	0;03	2.38	4832	11SE	9;57	161.34	4832	26SE	21;51	356.81
4759	17OC	5;59	197.77	4759	31OC	18;30	32.52	4833	10OC	22;10	190.91	4833	26OC	12;33	26.72
4760	15NO	16;26	227.72	4760	30NO	14;06	62.94	4834	9NO	13;10	221.00	4834	25NO	1;43	56.84
4761	15DE	3;10	257.79	4761	30DE	9;12	93.30	4835	9DE	6;28	251.34	4835	24DE	13;14	86.91

-615 / -609

NEW	MOONS			FULL	MOONS			NEW	MOONS			FULL	MOONS		
4762	13JA	14;26	287.69	4762	29JA	2;11	123.26	4836	8JA	1;02	281.61	4836	22JA	23;23	116.67
4763	12FE	2;22	317.22	4763	27FE	16;03	152.61	4837	6FE	19;34	311.52	4837	21FE	8;41	145.95
4764	13MR	14;57	346.28	4764	29MR	2;42	181.32	4838	8MR	12;45	340.86	4838	22MR	17;44	174.70
4765	12AP	4;13	14.88	4765	27AP	10;55	209.52	4839	7AP	3;34	9.60	4839	21AP	3;03	203.03
4766	11MY	18;16	43.16	4766	26MY	17;49	237.42	4840	6MY	15;37	37.86	4840	20MY	13;15	231.10
4767	10JN	9;08	71.35	4767	25JN	0;35	265.32	4841	5JN	1;15	65.85	4841	19JN	0;58	259.18
4768	10JL	0;30	99.68	4768	24JL	8;16	293.47	4842	4JL	9;22	93.85	4842	18JL	14;44	287.51
4769	8AU	15;43	128.32	4769	22AU	17;41	322.10	4843	2AU	17;06	122.11	4843	17AU	6;45	316.31
4770	7SE	6;06	157.42	4770	21SE	5;27	351.33	4844	1SE	1;29	150.84	4844	16SE	0;29	345.68
4771	6OC	19;19	186.97	4771	20OC	20;01	21.16	4845	30SE	11;12	180.13	4845	15OC	18;33	15.57
4772	5NO	7;29	216.90	4772	19NO	13;24	51.43	4846	29OC	22;36	209.94	4846	14NO	11;25	45.77
4773	4DE	18;52	246.98	4773	19DE	8;47	81.85	4847	28NO	11;50	240.09	4847	14DE	2;10	76.01
								4848	28DE	2;53	270.29				

-614 / -608

NEW	MOONS			FULL	MOONS			NEW	MOONS			FULL	MOONS		
4774	3JA	5;40	276.96	4774	18JA	4;23	112.06					4848	12JA	14;36	105.98
4775	1FE	15;55	306.59	4775	16FE	22;10	141.77	4849	26JA	19;31	300.28	4849	11FE	1;05	135.49
4776	3MR	1;48	335.72	4776	18MR	12;54	170.83	4850	25FE	13;00	329.83	4850	11MR	10;03	164.42
4777	1AP	11;44	4.35	4777	17AP	0;33	199.30	4851	26MR	6;05	358.85	4851	9AP	18;00	192.84
4778	30AP	22;31	32.61	4778	16MY	9;50	227.36	4852	24AP	21;32	27.36	4852	9MY	1;36	220.89
4779	30MY	10;52	60.72	4779	14JN	17;45	255.28	4853	24MY	10;49	55.53	4853	7JN	9;51	248.82
4780	29JN	1;09	88.93	4780	14JL	1;13	283.32	4854	22JN	22;08	83.60	4854	6JL	19;50	276.91
4781	28JL	17;03	117.46	4781	12AU	9;02	311.72	4855	22JL	8;11	111.83	4855	5AU	8;33	305.41
4782	27AU	9;45	146.45	4782	10SE	17;58	340.64	4856	20AU	17;43	140.44	4856	4SE	0;21	334.51
4783	26SE	2;16	175.96	4783	10OC	4;45	10.15	4857	19SE	3;20	169.54	4857	3OC	18;38	4.22
4784	25OC	17;57	205.90	4784	8NO	18;00	40.15	4858	18OC	13;22	199.15	4858	2NO	13;53	34.41
4785	24NO	8;19	236.07	4785	8DE	9;58	70.44	4859	17NO	0;05	229.13	4859	2DE	8;23	64.80
4786	23DE	21;04	266.20					4860	16DE	11;48	259.23				

-613 / -607

NEW	MOONS			FULL	MOONS			NEW	MOONS			FULL	MOONS		
				4786	7JA	3;56	100.70					4860	1JA	0;55	95.04
4787	22JA	8;03	296.00	4787	5FE	22;25	130.61	4861	15JA	0;52	289.20	4861	30JA	14;52	124.85
4788	20FE	17;25	325.29	4788	7MR	15;50	159.98	4862	13FE	15;18	318.81	4862	1MR	2;08	154.07
4789	22MR	1;43	354.03	4789	6AP	7;15	188.76	4863	15MR	6;42	347.95	4863	30MR	11;02	182.68
4790	20AP	9;51	22.31	4790	5MY	20;25	217.07	4864	13AP	22;19	16.62	4864	28AP	18;18	210.82
4791	19MY	18;49	50.34	4791	4JN	7;33	245.12	4865	13MY	13;31	44.94	4865	28MY	0;56	238.72
4792	18JN	5;31	78.37	4792	3JL	17;06	273.16	4866	12JN	3;55	73.11	4866	26JN	8;10	266.64
4793	17JL	18;28	106.66	4793	2AU	1;42	301.45	4867	11JL	17;27	101.38	4867	25JL	17;10	294.86
4794	16AU	9;44	135.42	4794	31AU	10;09	330.18	4868	10AU	6;07	129.94	4868	24AU	4;51	323.60
4795	15SE	2;56	164.75	4795	29SE	19;16	359.44	4869	8SE	17;56	158.96	4869	22SE	19;35	352.98
4796	14OC	21;15	194.64	4796	29OC	5;47	29.20	4870	8OC	5;00	188.44	4870	22OC	13;03	22.95
4797	13NO	15;27	224.90	4797	27NO	18;06	59.30	4871	6NO	15;38	218.32	4871	21NO	8;19	53.31
4798	13DE	8;05	255.21	4798	27DE	8;13	89.47	4872	6DE	2;12	248.37	4872	21DE	3;53	83.74

-612 / -606

NEW	MOONS			FULL	MOONS			NEW	MOONS			FULL	MOONS		
4799	11JA	22;08	285.26	4799	25JA	23;41	119.42	4873	4JA	13;05	278.35	4873	19JA	22;03	113.87
4800	10FE	9;24	314.81	4800	24FE	15;51	148.93	4874	3FE	0;28	308.00	4874	18FE	13;27	143.45
4801	10MR	18;25	343.76	4801	25MR	8;08	177.93	4875	4MR	12;25	337.19	4875	20MR	1;35	172.37
4802	9AP	2;04	12.17	4802	23AP	23;55	206.47	4876	3AP	1;02	5.91	4876	18AP	10;49	200.73
4803	8MY	9;19	40.21	4803	23MY	14;34	234.70	4877	2MY	14;26	34.26	4877	17MY	18;12	228.71
4804	6JN	17;05	68.12	4804	22JN	3;38	262.84	4878	1JN	4;48	62.45	4878	16JN	0;55	256.58
4805	6JL	2;12	96.17	4805	21JL	15;06	291.11	4879	30JN	20;02	90.70	4879	15JL	8;06	284.62
4806	4AU	13;29	124.61	4806	20AU	1;23	319.74	4880	30JL	11;37	119.24	4880	13AU	16;40	313.06
4807	3SE	3;33	153.62	4807	18SE	11;15	348.85	4881	29AU	2;47	148.20	4881	12SE	3;18	342.08
4808	2OC	20;37	183.27	4808	17OC	21;24	18.45	4882	27SE	16;55	177.63	4882	11OC	16;30	11.71
4809	1NO	15;59	213.46	4809	16NO	8;09	48.42	4883	27OC	5;49	207.47	4883	10NO	8;31	41.85
4810	1DE	11;49	243.89	4810	15DE	19;34	78.51	4884	25NO	17;41	237.55	4884	10DE	3;00	72.24
4811	31DE	5;59	274.21					4885	25DE	4;46	267.60				

-611 / -605

NEW	MOONS			FULL	MOONS			NEW	MOONS			FULL	MOONS		
				4811	14JA	7;35	108.44					4885	8JA	22;38	102.55
4812	29JA	21;09	304.08	4812	12FE	20;18	138.00	4886	23JA	15;09	297.36	4886	7FE	17;25	132.45
4813	28FE	9;13	333.34	4813	14MR	9;56	167.08	4887	22FE	0;58	326.65	4887	9MR	9;37	161.74
4814	29MR	18;45	2.00	4814	13AP	0;35	195.72	4888	23MR	10;32	355.41	4888	7AP	22;41	190.40
4815	28AP	2;36	30.17	4815	12MY	15;54	224.05	4889	21AP	20;32	23.76	4889	7MY	9;02	218.58
4816	27MY	9;35	58.08	4816	11JN	7;10	252.26	4890	21MY	7;49	51.87	4890	5JN	17;34	246.53
4817	25JN	16;34	85.99	4817	10JL	21;44	280.56	4891	19JN	21;02	80.01	4891	5JL	1;16	274.51
4818	25JL	0;33	114.16	4818	9AU	11;19	309.15	4892	19JL	12;16	108.40	4892	3AU	8;59	302.78
4819	23AU	10;34	142.82	4819	8SE	0;04	338.19	4893	18AU	4;53	137.24	4893	1SE	17;26	331.52
4820	21SE	23;35	172.11	4820	7OC	12;13	7.71	4894	16SE	21;53	166.59	4894	1OC	3;19	0.84
4821	21OC	15;59	202.03	4821	5NO	23;51	37.62	4895	16OC	14;20	196.43	4895	30OC	15;20	30.68
4822	20NO	11;00	232.38	4822	5DE	10;56	67.69	4896	15NO	5;36	226.57	4896	29NO	5;55	60.88
4823	20DE	6;52	262.82					4897	14DE	19;17	256.76	4897	28DE	22;52	91.17

33

---------NEW MOONS--------- ---------FULL MOONS-------- ---------NEW MOONS--------- ---------FULL MOONS--------
NUMBER DATE TIME LONG. | NUMBER DATE TIME LONG. | NUMBER DATE TIME LONG. | NUMBER DATE TIME LONG.

-604

NUMBER	DATE	TIME	LONG.	NUMBER	DATE	TIME	LONG.
4898	13JA	7;07	286.71	4898	27JA	17;04	121.21
4899	11FE	17;07	316.18	4899	26FE	10;57	150.76
4900	12MR	1;40	345.09	4900	27MR	3;17	179.73
4901	10AP	9;33	13.49	4901	25AP	17;31	208.19
4902	9MY	17;48	41.56	4902	25MY	5;40	236.31
4903	8JN	3;25	69.55	4903	23JN	16;05	264.35
4904	7JL	15;09	97.71	4904	23JL	1;17	292.55
4905	6AU	5;20	126.28	4905	21AU	9;56	321.13
4906	4SE	21;48	155.42	4906	19SE	18;49	350.22
4907	4OC	15;54	185.14	4907	19OC	4;41	19.83
4908	3NO	10;32	215.32	4908	17NO	16;07	49.84
4909	3DE	4;15	245.67	4909	17DE	5;16	80.00

-603

NUMBER	DATE	TIME	LONG.	NUMBER	DATE	TIME	LONG.
4910	1JA	19;43	275.86	4910	15JA	19;54	110.02
4911	31JA	8;18	305.60	4911	14FE	11;30	139.67
4912	1MR	18;13	334.75	4912	16MR	3;31	168.83
4913	31MR	2;17	3.33	4913	14AP	19;25	197.50
4914	29AP	9;28	31.46	4914	14MY	10;39	225.81
4915	28MY	16;46	59.38	4915	13JN	0;39	253.97
4916	27JN	1;05	87.34	4916	12JL	13;06	282.19
4917	26JL	11;12	115.61	4917	11AU	0;10	310.71
4918	24AU	23;53	144.41	4918	9SE	10;25	339.67
4919	23SE	15;35	173.84	4919	8OC	20;35	9.13
4920	23OC	10;05	203.87	4920	7NO	7;08	39.00
4921	22NO	6;03	234.26	4921	6DE	18;14	69.08
4922	22DE	1;19	264.67				

-602

NUMBER	DATE	TIME	LONG.	NUMBER	DATE	TIME	LONG.
				4922	5JA	5;49	99.08
4923	20JA	18;03	294.74	4923	3FE	17;54	128.75
4924	19FE	7;32	324.22	4924	5MR	6;45	157.97
4925	20MR	18;08	353.07	4925	3AP	20;36	186.73
4926	19AP	2;38	21.39	4926	3MY	11;26	215.13
4927	18MY	9;54	49.36	4927	2JN	2;46	243.35
4928	16JN	16;46	77.24	4928	1JL	17;51	271.60
4929	16JL	0;11	105.29	4929	31JL	8;10	300.10
4930	14AU	9;08	133.77	4930	29AU	21;37	329.00
4931	12SE	20;39	162.83	4931	28SE	10;20	358.39
4932	12OC	11;26	192.54	4932	27OC	22;28	28.21
4933	11NO	5;20	222.76	4933	26NO	9;56	58.27
4934	11DE	1;01	253.20	4934	25DE	20;42	88.30

-601

NUMBER	DATE	TIME	LONG.	NUMBER	DATE	TIME	LONG.
4935	9JA	20;27	283.50	4935	24JA	6;54	118.04
4936	8FE	13;57	313.33	4936	22FE	16;59	147.33
4937	10MR	4;44	342.55	4937	24MR	3;35	176.13
4938	8AP	16;42	11.15	4938	22AP	15;15	204.53
4939	8MY	2;15	39.30	4939	22MY	4;14	232.71
4940	6JN	10;01	67.21	4940	20JN	18;30	260.89
4941	5JL	17;00	95.17	4941	20JL	9;48	289.30
4942	4AU	0;17	123.43	4942	19AU	1;48	318.12
4943	2SE	8;57	152.19	4943	17SE	17;57	347.46
4944	1OC	19;53	181.56	4944	17OC	9;31	17.27
4945	31OC	9;34	211.48	4945	15NO	23;42	47.37
4946	30NO	1;46	241.76	4946	15DE	12;05	77.51
4947	29DE	19;43	272.08				

-600

NUMBER	DATE	TIME	LONG.	NUMBER	DATE	TIME	LONG.
				4947	13JA	22;48	107.39
4948	28JA	14;13	302.12	4948	12FE	8;18	136.83
4949	27FE	8;02	331.65	4949	12MR	17;11	165.75
4950	28MR	0;00	0.60	4950	11AP	2;04	194.18
4951	26AP	13;25	29.01	4951	10MY	11;33	222.30
4952	26MY	0;14	57.08	4952	8JN	22;18	250.33
4953	24JN	9;06	85.06	4953	8JL	10;57	278.53
4954	23JL	17;04	113.22	4954	7AU	1;56	307.15
4955	22AU	1;13	141.78	4955	5SE	19;08	336.33
4956	20SE	10;22	170.89	4956	5OC	13;29	6.08
4957	19OC	21;01	200.55	4957	4NO	7;21	36.23
4958	18NO	9;20	230.60	4958	3DE	23;22	66.51
4959	17DE	23;22	260.79				

-599

NUMBER	DATE	TIME	LONG.	NUMBER	DATE	TIME	LONG.
				4959	2JA	12;58	96.62
4960	16JA	15;03	290.86	4960	1FE	0;18	126.30
4961	15FE	7;57	320.55	4961	2MR	9;49	155.42
4962	17MR	1;09	349.73	4962	31MR	18;02	183.99
4963	15AP	17;23	18.40	4963	30AP	1;33	212.13
4964	15MY	7;46	46.68	4964	29MY	9;13	240.06
4965	13JN	20;07	74.77	4965	27JN	18;08	268.05
4966	13JL	6;54	102.94	4966	27JL	5;24	296.38
4967	11AU	16;51	131.42	4967	25AU	19;45	325.25
4968	10SE	2;38	160.37	4968	24SE	13;04	354.77
4969	9OC	12;37	189.83	4969	24OC	8;14	24.83
4970	7NO	23;04	219.71	4970	23NO	3;26	55.20
4971	7DE	10;13	249.79	4971	22DE	21;07	85.54

-598

NUMBER	DATE	TIME	LONG.	NUMBER	DATE	TIME	LONG.
4972	5JA	22;25	279.81	4972	21JA	12;22	115.53
4973	4FE	11;55	309.53	4973	20FE	0;52	144.96
4974	6MR	2;35	338.81	4974	21MR	10;47	173.78
4975	4AP	17;55	7.61	4975	19AP	18;39	202.06
4976	4MY	9;13	36.02	4976	19MY	1;23	230.00
4977	3JN	0;02	64.22	4977	17JN	8;08	257.88
4978	2JL	14;09	92.44	4978	16JL	16;08	285.96
4979	1AU	3;30	120.91	4979	15AU	2;27	314.50
4980	30AU	16;02	149.79	4980	13SE	15;45	343.65
4981	29SE	3;45	179.15	4981	13OC	8;01	13.43
4982	28OC	14;45	208.93	4982	12NO	2;35	43.69
4983	27NO	1;23	238.95	4983	11DE	22;13	74.13
4984	26DE	12;00	268.98				

-597

NUMBER	DATE	TIME	LONG.	NUMBER	DATE	TIME	LONG.
				4984	10JA	17;14	104.40
4985	24JA	22;56	298.74	4985	9FE	10;04	134.19
4986	23FE	10;20	328.08	4986	10MR	23;45	163.35
4987	24MR	22;18	356.92	4987	9AP	10;16	191.89
4988	23AP	11;00	25.35	4988	8MY	18;24	219.97
4989	23MY	0;43	53.55	4989	7JN	1;19	247.85
4990	21JN	15;32	81.76	4990	6JL	8;12	275.81
4991	21JL	7;10	110.18	4991	4AU	16;05	304.10
4992	19AU	22;55	139.01	4992	3SE	1;43	332.91
4993	18SE	13;57	168.32	4993	2OC	13;41	2.33
4994	18OC	3;44	198.06	4994	1NO	4;20	32.31
4995	16NO	16;16	228.10	4995	30NO	21;38	62.63
4996	16DE	3;45	258.20	4996	30DE	16;48	93.00

-596

NUMBER	DATE	TIME	LONG.	NUMBER	DATE	TIME	LONG.
4997	14JA	14;24	288.08	4997	29JA	12;06	123.06
4998	13FE	0;18	317.52	4998	28FE	5;35	152.57
4999	13MR	9;41	346.45	4999	28MR	20;08	181.44
5000	11AP	19;06	14.89	5000	27AP	7;44	209.77
5001	11MY	5;26	43.04	5001	26MY	17;06	237.78
5002	9JN	17;29	71.12	5002	25JN	1;12	265.73
5003	9JL	7;44	99.39	5003	24JL	8;58	293.89
5004	7AU	23;54	128.07	5004	22AU	17;07	322.46
5005	6SE	17;05	157.26	5005	21SE	2;22	351.59
5006	6OC	10;11	186.98	5006	20OC	13;19	21.26
5007	5NO	2;21	217.07	5007	19NO	2;36	51.36
5008	4DE	17;00	247.29	5008	18DE	18;19	81.63

-595

NUMBER	DATE	TIME	LONG.	NUMBER	DATE	TIME	LONG.
5009	3JA	5;48	277.36	5009	17JA	11;51	111.76
5010	1FE	16;35	307.00	5010	16FE	5;50	141.48
5011	3MR	1;37	336.10	5011	17MR	22;51	170.64
5012	1AP	9;32	4.64	5012	16AP	14;03	199.26
5013	30AP	17;16	32.79	5013	16MY	3;14	227.48
5014	30MY	1;56	60.75	5014	14JN	14;35	255.54
5015	28JN	12;29	88.81	5015	14JL	0;31	283.67
5016	28JL	1;27	117.22	5016	12AU	9;34	312.13
5017	26AU	16;56	146.15	5017	10SE	18;27	341.05
5018	25SE	10;30	175.69	5018	10OC	3;55	10.50
5019	25OC	5;14	205.75	5019	8NO	14;36	40.40
5020	23NO	23;46	236.10	5020	8DE	2;51	70.52
5021	23DE	16;34	266.39				

-594

NUMBER	DATE	TIME	LONG.	NUMBER	DATE	TIME	LONG.
				5021	6JA	16;37	100.60
5022	22JA	6;34	296.32	5022	5FE	7;32	130.38
5023	20FE	17;38	325.68	5023	6MR	23;06	159.68
5024	22MR	2;23	354.44	5024	5AP	14;52	188.48
5025	20AP	9;45	22.69	5025	5MY	6;23	216.89
5026	19MY	16;47	50.64	5026	3JN	21;06	245.09
5027	18JN	0;26	78.55	5027	3JL	10;30	273.29
5028	17JL	9;33	106.69	5028	1AU	22;28	301.71
5029	15AU	20;56	135.28	5029	31AU	9;20	330.53
5030	14SE	11;11	164.49	5030	29SE	19;43	359.84
5031	14OC	4;29	194.32	5031	29OC	6;12	29.61
5032	13NO	0;03	224.63	5032	27NO	17;05	59.65
5033	12DE	20;00	255.09	5033	27DE	4;21	89.70

-593

NUMBER	DATE	TIME	LONG.	NUMBER	DATE	TIME	LONG.
5034	11JA	14;08	285.31	5034	25JA	15;58	119.48
5035	10FE	5;11	315.02	5035	24FE	4;06	148.83
5036	11MR	17;04	344.09	5036	25MR	17;06	177.71
5037	10AP	2;25	12.57	5037	24AP	7;13	206.19
5038	9MY	10;08	40.63	5038	23MY	22;16	234.44
5039	7JN	17;05	68.51	5039	22JN	13;36	262.66
5040	7JL	0;09	96.47	5040	22JL	4;32	291.07
5041	5AU	8;17	124.78	5041	20AU	18;42	319.85
5042	3SE	18;29	153.63	5042	19SE	8;05	349.10
5043	3OC	7;41	183.12	5043	18OC	20;45	18.81
5044	2NO	0;10	213.18	5044	17NO	8;43	48.83
5045	1DE	19;09	243.57	5045	16DE	19;51	78.91
5046	31DE	14;52	273.96				

34

-592

NEW MOONS				FULL MOONS			
5047	30JA	9;23	303.97	5046	15JA	6;11	108.77
5048	29FE	1;27	333.41	5047	13FE	16;03	138.21
5049	29MR	14;44	2.22	5048	14MR	2;03	167.14
5050	28AP	1;24	30.50	5049	12AP	12;50	195.63
5051	27MY	9;59	58.48	5050	12MY	0;56	223.84
5052	25JN	17;19	86.40	5051	10JN	14;28	251.98
5053	25JL	0;26	114.54	5052	10JL	5;19	280.29
5054	23AU	8;25	143.12	5053	8AU	21;12	308.96
5055	21SE	18;16	172.28	5054	7SE	13;38	338.14
5056	21OC	6;38	202.02	5055	7OC	5;54	7.83
5057	19NO	21;37	232.20	5056	5NO	21;06	37.89
5058	19DE	14;42	262.52	5057	5DE	10;31	68.06

-591

NEW MOONS				FULL MOONS			
5059	18JA	8;53	292.67	5058	3JA	22;01	98.07
5060	17FE	2;59	322.38	5059	2FE	7;55	127.67
5061	18MR	19;50	351.52	5060	3MR	16;51	156.75
5062	17AP	10;30	20.11	5061	2AP	1;26	185.31
5063	16MY	22;37	48.28	5062	1MY	10;20	213.49
5064	15JN	8;28	76.28	5063	30MY	20;12	241.51
5065	14JL	16;55	104.36	5064	29JN	7;45	269.61
5066	13AU	1;02	132.79	5065	28JL	21;35	298.06
5067	11SE	9;48	161.72	5066	27AU	13;53	327.04
5068	10OC	19;49	191.20	5067	26SE	8;04	356.62
5069	9NO	7;21	221.14	5068	26OC	2;38	26.68
5070	8DE	20;27	251.30	5069	24NO	19;53	56.98
				5070	24DE	10;46	87.19

-590

NEW MOONS				FULL MOONS			
5071	7JA	11;08	281.41	5071	22JA	23;08	117.04
5072	6FE	3;14	311.22	5072	21FE	9;23	146.36
5073	7MR	20;09	340.57	5073	22MR	18;01	175.10
5074	6AP	12;48	9.39	5074	21AP	1;38	203.35
5075	6MY	4;07	37.79	5075	20MY	8;59	231.32
5076	4JN	17;31	65.93	5076	18JN	17;04	259.25
5077	4JL	5;11	94.06	5077	18JL	3;01	287.42
5078	2AU	15;43	122.43	5078	16AU	15;50	316.08
5079	1SE	1;47	151.24	5079	15SE	7;52	345.37
5080	30SE	11;51	180.55	5080	15OC	2;27	15.28
5081	29OC	22;11	210.31	5081	13NO	22;00	45.59
5082	28NO	8;58	240.35	5082	13DE	16;40	75.99
5083	27DE	20;30	270.41				

-589

NEW MOONS				FULL MOONS			
5084	26JA	9;07	300.23	5083	12JA	9;12	106.14
5085	24FE	22;56	329.64	5084	10FE	23;01	135.79
5086	26MR	13;42	358.57	5085	12MR	10;05	164.81
5087	25AP	4;52	27.08	5086	10AP	18;46	193.25
5088	24MY	19;53	55.33	5087	10MY	1;52	221.28
5089	23JN	10;27	83.52	5088	8JN	8;25	249.14
5090	23JL	0;25	111.90	5089	7JL	15;40	277.12
5091	21AU	13;40	140.66	5090	6AU	0;47	305.48
5092	20SE	2;05	169.88	5091	4SE	12;37	334.41
5093	19OC	13;39	199.55	5092	4OC	3;31	3.98
5094	18NO	0;32	229.53	5093	2NO	21;08	34.09
5095	17DE	11;07	259.59	5094	2DE	16;26	64.51

-588

NEW MOONS				FULL MOONS			
5096	15JA	21;44	289.45	5095	1JA	11;56	94.87
5097	14FE	8;38	318.92	5096	31JA	5;56	124.86
5098	14MR	20;01	347.90	5097	29FE	21;09	154.25
5099	13AP	8;02	16.43	5098	30MR	9;08	182.99
5100	12MY	21;00	44.67	5099	28AP	18;17	211.21
5101	11JN	11;11	72.83	5100	28MY	1;40	239.13
5102	11JL	2;33	101.17	5101	26JN	8;29	267.03
5103	9AU	18;35	129.86	5102	25JL	15;51	295.19
5104	8SE	10;23	159.03	5103	24AU	0;39	323.81
5105	8OC	1;08	188.67	5104	22SE	11;30	353.03
5106	6NO	14;29	218.66	5105	22OC	0;51	22.82
5107	6DE	2;34	248.77	5106	20NO	16;51	53.04
				5107	20DE	11;08	83.41

-587

NEW MOONS				FULL MOONS			
5108	4JA	13;34	278.75	5108	19JA	6;27	113.60
5109	2FE	23;41	308.36	5109	18FE	0;53	143.31
5110	4MR	9;03	337.44	5110	19MR	16;51	172.40
5111	2AP	18;07	6.02	5111	18AP	5;49	200.91
5112	2MY	3;39	34.21	5112	17MY	16;12	229.01
5113	31MY	14;35	62.27	5113	16JN	0;55	256.96
5114	30JN	3;40	90.44	5114	15JL	8;54	285.03
5115	29JL	19;01	118.95	5115	13AU	16;57	313.46
5116	28AU	12;01	147.98	5116	12SE	1;45	342.40
5117	27SE	5;34	177.55	5117	11OC	11;53	11.90
5118	26OC	22;33	207.55	5118	9NO	23;59	41.87
5119	25NO	14;11	237.79	5119	9DE	14;25	72.09
5120	25DE	4;00	267.95				

-586

NEW MOONS				FULL MOONS			
5121	23JA	15;43	297.77	5120	8JA	6;59	102.28
5122	22FE	1;26	327.05	5121	7FE	0;40	132.14
5123	23MR	9;37	355.76	5122	8MR	18;04	161.49
5124	21AP	17;08	24.00	5123	7AP	10;05	190.28
5125	21MY	1;03	51.99	5124	7MY	0;14	218.62
5126	19JN	10;28	79.97	5125	5JN	12;33	246.71
5127	18JL	22;10	108.22	5126	4JL	23;19	274.82
5128	17AU	12;29	136.96	5127	3AU	8;58	303.16
5129	16SE	5;15	166.29	5128	1SE	18;05	331.94
5130	15OC	23;44	196.20	5129	1OC	3;22	1.23
5131	14NO	18;44	226.50	5130	30OC	13;29	30.99
5132	14DE	12;40	256.87	5131	29NO	0;56	61.06
				5132	28DE	13;51	91.17

-585

NEW MOONS				FULL MOONS			
5133	13JA	4;09	286.97	5133	27JA	4;00	121.05
5134	11FE	16;34	316.54	5134	25FE	18;59	150.49
5135	13MR	2;15	345.50	5135	27MR	10;25	179.44
5136	11AP	10;03	13.89	5136	26AP	1;56	207.95
5137	10MY	17;01	41.91	5137	25MY	17;05	236.20
5138	9JN	0;11	69.80	5138	24JN	7;19	264.38
5139	8JL	8;28	97.83	5139	23JL	20;14	292.73
5140	6AU	18;40	126.24	5140	22AU	7;53	321.43
5141	5SE	7;31	155.22	5141	20SE	18;42	350.60
5142	4OC	23;25	184.84	5142	20OC	5;17	20.24
5143	3NO	18;06	215.01	5143	18NO	16;03	50.22
5144	3DE	14;12	245.46	5144	18DE	3;06	80.29

-584

NEW MOONS				FULL MOONS			
5145	2JA	9;28	275.81	5145	16JA	14;23	110.18
5146	1FE	2;05	305.73	5146	15FE	1;57	139.66
5147	1MR	15;24	335.02	5147	15MR	14;10	168.67
5148	31MR	1;49	3.69	5148	14AP	3;26	197.24
5149	29AP	10;11	31.87	5149	13MY	17;52	225.53
5150	28MY	17;24	59.79	5150	12JN	9;07	253.74
5151	27JN	0;20	87.70	5151	12JL	0;28	282.07
5152	26JL	7;53	115.87	5152	10AU	15;18	310.73
5153	24AU	17;02	144.52	5153	9SE	5;23	339.84
5154	23SE	4;44	173.78	5154	8OC	18;41	9.44
5155	22OC	19;39	203.65	5155	7NO	7;13	39.40
5156	21NO	13;33	233.95	5156	6DE	18;51	69.50
5157	21DE	9;05	264.37				

-583

NEW MOONS				FULL MOONS			
5158	20JA	4;17	294.54	5157	5JA	5;30	99.45
5159	18FE	21;32	324.19	5158	3FE	15;22	129.04
5160	20MR	12;07	353.21	5159	5MR	0;58	158.12
5161	18AP	23;59	21.67	5160	3AP	11;00	186.72
5162	18MY	9;33	49.73	5161	2MY	22;10	214.98
5163	16JN	17;28	77.65	5162	1JN	10;49	243.10
5164	16JL	0;41	105.70	5163	1JL	1;01	271.32
5165	14AU	8;15	134.11	5164	30JL	16;33	299.85
5166	12SE	17;13	163.08	5165	29AU	9;01	328.87
5167	12OC	4;22	192.63	5166	28SE	1;46	358.42
5168	10NO	18;05	222.67	5167	27OC	17;53	28.40
5169	10DE	10;08	252.96	5168	26NO	8;25	58.59
				5169	25DE	20;53	88.69

-582

NEW MOONS				FULL MOONS			
5170	9JA	3;43	283.18	5170	24JA	7;26	118.45
5171	7FE	21;46	313.05	5171	22FE	16;36	147.70
5172	9MR	15;11	342.38	5172	24MR	1;04	176.41
5173	8AP	6;55	11.14	5173	22AP	9;31	204.69
5174	7MY	20;19	39.45	5174	21MY	18;39	232.71
5175	6JN	7;19	67.49	5175	20JN	5;11	260.74
5176	5JL	16;30	95.53	5176	19JL	17;48	289.04
5177	4AU	0;51	123.84	5177	18AU	8;59	317.82
5178	2SE	9;25	152.60	5178	17SE	2;33	347.21
5179	1OC	18;55	181.91	5179	16OC	21;23	17.14
5180	31OC	5;47	211.71	5180	15NO	15;40	47.42
5181	29NO	18;04	241.82	5181	15DE	7;55	77.72
5182	29DE	7;49	271.96				

-581

NEW MOONS				FULL MOONS			
5183	27JA	23;00	301.87	5182	13JA	21;30	107.72
5184	26FE	15;18	331.35	5183	12FE	8;39	137.23
5185	28MR	7;59	0.34	5184	13MR	17;53	166.16
5186	26AP	23;57	28.86	5185	12AP	1;47	194.55
5187	26MY	14;21	57.07	5186	11MY	9;02	222.58
5188	25JN	2;59	85.20	5187	9JN	16;32	250.48
5189	24JL	14;13	113.49	5188	9JL	1;23	278.53
5190	23AU	0;43	142.15	5189	7AU	12;42	307.00
5191	21SE	11;00	171.31	5190	6SE	3;13	336.06
5192	20OC	21;22	200.95	5191	5OC	20;49	5.77
5193	19NO	7;58	230.93	5192	4NO	16;15	35.98
5194	18DE	19;01	261.00	5193	4DE	11;39	66.40

-580

NEW MOONS				FULL MOONS			
NUMBER	DATE	TIME	LONG.	NUMBER	DATE	TIME	LONG.
				5194	3JA	5;22	96.68
5195	17JA	6;52	290.90	5195	1FE	20;32	126.53
5196	15FE	19;48	320.43	5196	2MR	8;51	155.77
5197	16MR	9;50	349.50	5197	31MR	18;33	184.40
5198	15AP	0;36	18.12	5198	30AP	2;15	212.54
5199	14MY	15;37	46.42	5199	29MY	8;53	240.43
5200	13JN	6;27	74.61	5200	27JN	15;39	268.34
5201	12JL	20;53	102.92	5201	26JL	23;45	296.54
5202	11AU	10;48	131.56	5202	25AU	10;13	325.25
5203	9SE	23;57	160.64	5203	23SE	23;41	354.59
5204	9OC	12;13	190.20	5204	23OC	16;06	24.54
5205	7NO	23;35	220.12	5205	22NO	10;44	54.88
5206	7DE	10;19	250.18	5206	22DE	6;17	85.30

-579

NEW MOONS				FULL MOONS			
5207	5JA	20;47	280.13	5207	21JA	1;09	115.45
5208	4FE	7;19	309.73	5208	19FE	17;47	145.06
5209	5MR	18;10	338.85	5209	21MR	7;17	174.02
5210	4AP	5;32	7.50	5210	19AP	17;43	202.40
5211	3MY	17;45	35.79	5211	19MY	1;50	230.40
5212	2JN	7;10	63.94	5212	17JN	8;50	258.29
5213	1JL	21;59	92.18	5213	16JL	15;54	286.34
5214	31JL	13;57	120.75	5214	15AU	0;01	314.79
5215	30AU	6;16	149.77	5215	13SE	9;54	343.80
5216	28SE	21;55	179.29	5216	12OC	22;03	13.40
5217	28OC	12;14	209.21	5217	11NO	12;44	43.49
5218	27NO	1;04	239.33	5218	11DE	5;53	73.83
5219	26DE	12;36	269.39				

-578

NEW MOONS				FULL MOONS			
				5219	10JA	0;44	104.09
5220	24JA	23;02	299.13	5220	8FE	19;39	133.98
5221	23FE	8;33	328.39	5221	10MR	12;50	163.29
5222	24MR	17;27	357.11	5222	9AP	3;13	191.99
5223	23AP	2;25	25.39	5223	8MY	14;49	220.21
5224	22MY	12;21	53.45	5224	7JN	0;19	248.19
5225	21JN	0;11	81.53	5225	6JL	8;42	276.21
5226	20JL	14;27	109.90	5226	4AU	16;48	304.51
5227	19AU	6;54	138.75	5227	3SE	1;20	333.28
5228	18SE	0;34	168.15	5228	2OC	10;53	2.60
5229	17OC	18;12	198.05	5229	31OC	22;00	32.43
5230	16NO	10;48	228.26	5230	30NO	11;13	62.57
5231	16DE	1;40	258.50	5231	30DE	2;38	92.78

-577

NEW MOONS				FULL MOONS			
5232	14JA	14;25	288.46	5232	28JA	19;40	122.76
5233	13FE	0;59	317.94	5233	27FE	13;06	152.28
5234	14MR	9;41	346.83	5234	29MR	5;42	181.24
5235	12AP	17;14	15.20	5235	27AP	20;43	209.72
5236	12MY	0;40	43.23	5236	27MY	9;58	237.88
5237	10JN	9;06	71.17	5237	25JN	21;37	265.97
5238	9JL	19;32	99.29	5238	25JL	7;59	294.23
5239	8AU	8;35	127.83	5239	23AU	17;33	322.87
5240	7SE	0;18	156.96	5240	22SE	2;53	352.00
5241	6OC	18;13	186.69	5241	21OC	12;40	21.62
5242	5NO	13;18	216.90	5242	19NO	23;28	51.61
5243	5DE	8;05	247.30	5243	19DE	11;34	81.73

-576

NEW MOONS				FULL MOONS			
5244	4JA	0;57	277.54	5244	18JA	0;56	111.69
5245	2FE	14;52	307.32	5245	16FE	15;15	141.26
5246	3MR	1;44	336.49	5246	17MR	6;12	170.35
5247	1AP	10;13	5.06	5247	15AP	21;29	198.98
5248	30AP	17;22	33.17	5248	15MY	12;47	227.29
5249	30MY	0;15	61.07	5249	14JN	3;37	255.48
5250	28JN	7;51	89.01	5250	13JL	17;25	283.77
5251	27JL	17;02	117.26	5251	12AU	5;57	312.36
5252	26AU	4;33	146.03	5252	10SE	17;24	341.39
5253	24SE	19;00	175.43	5253	10OC	4;17	10.91
5254	24OC	12;29	205.43	5254	8NO	15;05	40.81
5255	23NO	8;10	235.82	5255	8DE	2;01	70.88
5256	23DE	4;07	266.25				

-575

NEW MOONS				FULL MOONS			
				5256	6JA	13;04	100.84
5257	21JA	22;10	296.36	5257	5FE	0;14	130.46
5258	20FE	13;03	325.88	5258	6MR	11;47	159.60
5259	22MR	0;45	354.76	5259	5AP	0;10	188.28
5260	20AP	9;58	23.08	5260	4MY	13;48	216.62
5261	19MY	17;37	51.06	5261	3JN	4;37	244.82
5262	18JN	0;37	78.95	5262	2JL	20;04	273.09
5263	17JL	7;49	107.00	5263	1AU	11;26	301.65
5264	15AU	16;09	135.47	5264	31AU	2;12	330.62
5265	14SE	2;34	164.52	5265	29SE	16;12	0.08
5266	13OC	15;55	194.18	5266	29OC	5;23	29.96
5267	12NO	8;26	224.35	5267	27NO	17;36	60.06
5268	12DE	3;18	254.76	5268	27DE	4;43	90.10

-574

NEW MOONS				FULL MOONS			
5269	10JA	22;47	285.05	5269	25JA	14;48	119.82
5270	9FE	17;00	314.90	5270	24FE	0;14	149.06
5271	11MR	8;50	344.13	5271	25MR	9;41	177.79
5272	9AP	21;59	12.77	5272	23AP	19;58	206.12
5273	9MY	8;38	40.95	5273	23MY	7;39	234.24
5274	7JN	17;20	68.90	5274	21JN	21;01	262.39
5275	7JL	0;54	96.89	5275	21JL	11;58	290.79
5276	5AU	8;19	125.17	5276	20AU	4;14	319.65
5277	3SE	16;37	153.94	5277	18SE	21;13	349.04
5278	3OC	2;43	183.30	5278	18OC	14;04	18.92
5279	1NO	15;12	213.18	5279	17NO	5;41	49.09
5280	1DE	6;06	243.41	5280	16DE	19;17	79.27
5281	30DE	22;52	273.67				

-573

NEW MOONS				FULL MOONS			
				5281	15JA	6;42	109.18
5282	29JA	16;35	303.66	5282	13FE	16;20	138.60
5283	28FE	10;14	333.17	5283	15MR	0;53	167.48
5284	30MR	2;45	2.12	5284	13AP	9;03	195.87
5285	28AP	17;19	30.57	5285	12MY	17;34	223.93
5286	28MY	5;33	58.69	5286	11JN	3;12	251.92
5287	26JN	15;41	86.72	5287	10JL	14;39	280.08
5288	26JL	0;32	114.93	5288	9AU	4;35	308.67
5289	24AU	9;05	143.53	5289	7SE	21;11	337.85
5290	22SE	18;15	172.67	5290	7OC	15;48	7.63
5291	22OC	4;33	202.32	5291	6NO	10;48	37.84
5292	20NO	16;09	232.35	5292	6DE	4;20	68.19
5293	20DE	5;04	262.50				

-572

NEW MOONS				FULL MOONS			
				5293	4JA	19;17	98.34
5294	18JA	19;19	292.49	5294	3FE	7;32	128.04
5295	17FE	10;49	322.10	5295	3MR	17;31	157.16
5296	18MR	3;08	351.23	5296	2AP	1;52	185.72
5297	16AP	19;24	19.89	5297	1MY	9;13	213.83
5298	16MY	10;37	48.19	5298	30MY	16;23	241.74
5299	15JN	0;13	76.33	5299	29JN	0;23	269.70
5300	14JL	12;17	104.56	5300	28JL	10;21	297.99
5301	12AU	23;21	133.10	5301	26AU	23;17	326.83
5302	11SE	9;58	162.11	5302	25SE	15;33	356.32
5303	10OC	20;29	191.62	5303	25OC	10;24	26.38
5304	9NO	7;04	221.51	5304	24NO	6;08	56.78
5305	8DE	17;52	251.58	5305	24DE	0;53	87.16

-571

NEW MOONS				FULL MOONS			
5306	7JA	5;08	281.55	5306	22JA	17;22	117.19
5307	5FE	17;14	311.19	5307	21FE	7;01	146.65
5308	7MR	6;26	340.39	5308	22MR	17;53	175.48
5309	5AP	20;36	9.13	5309	21AP	2;24	203.77
5310	5MY	11;21	37.51	5310	20MY	9;23	231.72
5311	4JN	2;15	65.71	5311	18JN	15;57	259.59
5312	3JL	17;01	93.96	5312	17JL	23;17	287.66
5313	2AU	7;27	122.49	5313	16AU	8;32	316.17
5314	31AU	21;20	151.44	5314	14SE	20;33	345.29
5315	30SE	10;21	180.87	5315	14OC	11;37	15.04
5316	29OC	22;22	210.71	5316	13NO	5;18	45.27
5317	28NO	9;28	240.76	5317	13DE	0;32	75.69
5318	27DE	19;59	270.77				

-570

NEW MOONS				FULL MOONS			
				5318	11JA	19;52	105.97
5319	26JA	6;17	300.50	5319	10FE	13;40	135.78
5320	24FE	16;42	329.77	5320	12MR	4;41	164.97
5321	26MR	3;30	358.54	5321	10AP	16;32	193.55
5322	24AP	14;59	26.91	5322	10MY	1;40	221.66
5323	24MY	3;34	55.06	5323	8JN	9;07	249.56
5324	22JN	17;38	83.24	5324	7JL	16;06	277.53
5325	22JL	9;11	111.68	5325	5AU	23;43	305.82
5326	21AU	1;42	140.55	5326	4SE	8;47	334.64
5327	19SE	18;06	169.93	5327	3OC	19;52	4.04
5328	19OC	9;26	199.76	5328	2NO	9;18	33.98
5329	17NO	23;11	229.86	5329	2DE	1;12	64.25
5330	17DE	11;24	259.99	5330	31DE	19;12	94.56

-569

NEW MOONS				FULL MOONS			
5331	15JA	22;17	289.86	5331	30JA	14;07	124.59
5332	14FE	8;04	319.28	5332	1MR	8;10	154.10
5333	15MR	17;01	348.16	5333	30MR	23;54	183.00
5334	14AP	1;37	16.56	5334	29AP	12;49	211.37
5335	13MY	10;45	44.64	5335	28MY	23;19	239.42
5336	11JN	21;24	72.67	5336	27JN	8;17	267.41
5337	11JL	10;23	100.90	5337	26JL	16;36	295.61
5338	10AU	1;55	129.57	5338	25AU	1;03	324.22
5339	8SE	19;19	158.80	5339	23SE	10;11	353.36
5340	8OC	13;23	188.56	5340	22OC	20;34	23.03
5341	7NO	6;50	218.72	5341	21NO	8;41	53.08
5342	6DE	22;46	249.00	5342	20DE	22;54	83.28

Left half

NEW MOONS NUMBER	DATE	TIME	LONG.	FULL MOONS NUMBER	DATE	TIME	LONG.
		-568					
5343	5JA	12;37	279.10	5343	19JA	15;00	113.34
5344	4FE	0;11	308.76	5344	18FE	8;08	143.01
5345	4MR	9;36	337.85	5345	19MR	1;02	172.15
5346	2AP	17;27	6.37	5346	17AP	16;45	200.77
5347	2MY	0;39	34.48	5347	17MY	6;52	229.02
5348	31MY	8;19	62.40	5348	15JN	19;25	257.12
5349	29JN	17;36	90.42	5349	15JL	6;35	285.33
5350	29JL	5;19	118.79	5350	13AU	16;44	313.84
5351	27AU	19;48	147.70	5351	12SE	2;21	342.82
5352	26SE	12;51	177.24	5352	11OC	12;02	12.30
5353	26OC	7;41	207.31	5353	9NO	22;22	42.19
5354	25NO	2;57	237.70	5354	9DE	9;46	72.28
5355	24DE	21;00	268.04				
		-567					
				5355	7JA	22;21	102.30
5356	23JA	12;26	298.02	5356	6FE	11;58	132.00
5357	22FE	0;41	327.40	5357	8MR	2;19	161.23
5358	23MR	10;08	356.17	5358	6AP	17;12	189.98
5359	21AP	17;43	24.41	5359	6MY	8;22	218.37
5360	21MY	0;32	52.35	5360	4JN	23;30	246.58
5361	19JN	7;38	80.24	5361	4JL	14;01	274.83
5362	18JL	15;58	108.36	5362	3AU	3;28	303.33
5363	17AU	2;18	136.92	5363	1SE	15;43	332.23
5364	15SE	15;19	166.10	5364	1OC	3;05	1.60
5365	15OC	7;23	195.90	5365	30OC	14;04	31.40
5366	14NO	2;12	226.19	5366	29NO	1;00	61.45
5367	13DE	22;18	256.65	5367	28DE	11;56	91.48
		-566					
5368	12JA	17;29	286.90	5368	26JA	22;51	121.22
5369	11FE	9;57	316.65	5369	25FE	9;53	150.50
5370	12MR	23;05	345.75	5370	26MR	21;29	179.30
5371	11AP	9;22	14.25	5371	25AP	10;12	207.71
5372	10MY	17;40	42.32	5372	25MY	0;18	235.92
5373	9JN	0;54	70.21	5373	23JN	15;31	264.14
5374	8JL	7;57	98.19	5374	23JL	7;10	292.59
5375	6AU	15;43	126.50	5375	21AU	22;33	321.44
5376	5SE	1;05	155.34	5376	20SE	13;16	350.76
5377	4OC	12;59	184.79	5377	20OC	3;08	20.54
5378	3NO	3;58	214.80	5378	18NO	16;01	50.61
5379	2DE	21;47	245.15	5379	18DE	3;44	80.71
		-565					
5380	1JA	17;05	275.51	5380	16JA	14;14	110.56
5381	31JA	11;58	305.53	5381	14FE	23;43	139.96
5382	2MR	4;56	334.98	5382	16MR	8;49	168.84
5383	31MR	19;19	3.81	5383	14AP	18;21	197.26
5384	30AP	7;09	32.13	5384	14MY	5;03	225.40
5385	29MY	16;48	60.14	5385	12JN	17;26	253.50
5386	28JN	0;57	88.11	5386	12JL	7;37	281.78
5387	27JL	8;28	116.28	5387	10AU	23;25	310.47
5388	25AU	16;22	144.87	5388	9SE	16;23	339.69
5389	24SE	1;37	174.03	5389	9OC	9;43	9.45
5390	23OC	12;58	203.75	5390	8NO	2;18	39.57
5391	22NO	2;40	233.88	5391	7DE	17;07	69.81
5392	21DE	18;28	264.14				
		-564					
				5392	6JA	5;36	99.85
5393	20JA	11;37	294.23	5393	4FE	15;57	129.45
5394	19FE	5;10	323.90	5394	5MR	0;46	158.50
5395	19MR	22;09	353.03	5395	3AP	8;50	187.02
5396	18AP	13;41	21.64	5396	2MY	16;55	215.16
5397	18MY	3;07	49.86	5397	1JN	1;46	243.12
5398	16JN	14;22	77.91	5398	30JN	12;09	271.19
5399	15JL	23;56	106.05	5399	30JL	0;48	299.60
5400	14AU	8;44	134.52	5400	28AU	16;12	328.57
5401	12SE	17;44	163.48	5401	27SE	10;09	358.15
5402	12OC	3;35	192.98	5402	27OC	5;23	28.26
5403	10NO	14;36	222.91	5403	26NO	0;01	58.62
5404	10DE	2;48	253.03	5404	25DE	16;23	88.89
		-563					
5405	8JA	16;12	283.08	5405	24JA	5;55	118.78
5406	7FE	6;49	312.81	5406	22FE	16;52	148.10
5407	8MR	22;30	342.08	5407	24MR	1;49	176.83
5408	7AP	14;40	10.88	5408	22AP	9;26	205.07
5409	7MY	6;24	39.28	5409	21MY	16;30	233.01
5410	5JN	20;54	67.46	5410	19JN	23;54	260.92
5411	5JL	9;52	95.65	5411	19JL	8;45	289.06
5412	3AU	21;37	124.09	5412	17AU	20;10	317.68
5413	2SE	8;41	152.95	5413	16SE	10;52	346.94
5414	1OC	19;28	182.32	5414	16OC	4;42	16.83
5415	31OC	6;11	212.12	5415	15NO	0;20	47.15
5416	29NO	16;54	242.16	5416	14DE	19;49	77.59
5417	29DE	3;47	272.18				

Right half

NEW MOONS NUMBER	DATE	TIME	LONG.	FULL MOONS NUMBER	DATE	TIME	LONG.
		-562					
				5417	13JA	13;31	107.78
5418	27JA	15;13	301.93	5418	12FE	4;32	137.45
5419	26FE	3;34	331.26	5419	13MR	16;40	166.50
5420	27MR	16;58	0.11	5420	12AP	2;12	194.95
5421	26AP	7;13	28.58	5421	11MY	9;47	222.99
5422	25MY	21;59	56.81	5422	9JN	16;25	250.86
5423	24JN	12;53	85.02	5423	8JL	23;15	278.83
5424	24JL	3;43	113.45	5424	7AU	7;30	307.17
5425	22AU	18;12	142.27	5425	5SE	18;09	336.07
5426	21SE	8;00	171.57	5426	5OC	7;47	5.61
5427	20OC	20;47	201.31	5427	4NO	0;17	35.69
5428	19NO	8;27	231.34	5428	3DE	18;53	66.08
5429	18DE	19;13	261.39				
		-561					
				5429	2JA	14;17	96.43
5430	17JA	5;28	291.23	5430	1FE	8;54	126.43
5431	15FE	15;35	320.64	5431	3MR	1;18	155.84
5432	17MR	1;53	349.56	5432	1AP	14;40	184.62
5433	15AP	12;42	18.02	5433	1MY	1;03	212.88
5434	15MY	0;28	46.20	5434	30MY	9;13	240.83
5435	13JN	13;39	74.33	5435	28JN	16;23	268.75
5436	13JL	4;32	102.65	5436	27JL	23;41	296.92
5437	11AU	20;51	131.37	5437	26AU	8;05	325.55
5438	10SE	13;44	160.61	5438	24SE	18;14	354.75
5439	10OC	6;00	190.33	5439	24OC	6;32	24.52
5440	8NO	20;48	220.39	5440	22NO	21;11	54.69
5441	8DE	9;52	250.55	5441	22DE	14;06	85.00
		-560					
5442	6JA	21;22	280.54	5442	21JA	8;32	115.14
5443	5FE	7;33	310.12	5443	20FE	3;02	144.83
5444	5MR	16;40	339.18	5444	20MR	19;54	173.94
5445	4AP	1;08	7.71	5445	19AP	10;10	202.48
5446	3MY	9;41	35.85	5446	18MY	21;49	230.63
5447	1JN	19;18	63.85	5447	17JN	7;32	258.62
5448	1JL	6;59	91.97	5448	16JL	16;15	286.74
5449	30JL	21;18	120.46	5449	15AU	0;46	315.20
5450	29AU	14;03	149.49	5450	13SE	9;41	344.17
5451	28SE	8;11	179.10	5451	12OC	19;31	13.68
5452	28OC	2;19	209.18	5452	11NO	6;45	43.62
5453	26NO	19;16	239.47	5453	10DE	19;50	73.79
5454	26DE	10;16	269.68				
		-559					
				5454	9JA	10;52	103.90
5455	24JA	22;56	299.52	5455	8FE	3;20	133.69
5456	23FE	9;15	328.80	5456	9MR	20;13	163.00
5457	24MR	17;38	357.50	5457	8AP	12;25	191.78
5458	23AP	0;52	25.71	5458	8MY	3;17	220.14
5459	22MY	8;02	53.66	5459	6JN	16;41	248.27
5460	20JN	16;19	81.60	5460	6JL	4;41	276.44
5461	20JL	2;44	109.81	5461	4AU	15;33	304.85
5462	18AU	15;53	138.52	5462	3SE	1;38	333.68
5463	17SE	7;50	167.84	5463	2OC	11;26	3.01
5464	17OC	2;03	197.76	5464	31OC	21;30	32.79
5465	15NO	21;26	228.08	5465	30NO	8;21	62.84
5466	15DE	16;22	258.49	5466	29DE	20;15	92.90
		-558					
5467	14JA	9;14	288.63	5467	28JA	9;09	122.71
5468	12FE	23;00	318.24	5468	26FE	22;51	152.08
5469	14MR	9;40	347.22	5469	28MR	13;11	180.96
5470	12AP	17;56	15.62	5470	27AP	4;01	209.43
5471	12MY	0;55	43.63	5471	26MY	19;09	237.68
5472	10JN	7;44	71.50	5472	25JN	10;08	265.90
5473	9JL	15;22	99.51	5473	25JL	0;24	294.31
5474	8AU	0;40	127.90	5474	23AU	13;32	323.09
5475	6SE	12;21	156.85	5475	22SE	1;35	352.33
5476	6OC	2;58	186.44	5476	21OC	12;56	22.02
5477	4NO	20;34	216.58	5477	19NO	23;59	52.02
5478	4DE	16;17	247.02	5478	19DE	10;55	82.09
		-557					
5479	3JA	12;09	277.38	5479	17JA	21;42	111.94
5480	2FE	6;02	307.34	5480	16FE	8;23	141.37
5481	3MR	20;44	336.67	5481	17MR	19;21	170.30
5482	2AP	8;17	5.36	5482	16AP	7;09	198.79
5483	1MY	17;25	33.56	5483	15MY	20;21	227.02
5484	31MY	1;05	61.49	5484	14JN	11;00	255.21
5485	29JN	8;11	89.41	5485	14JL	2;37	283.57
5486	28JL	15;36	117.59	5486	12AU	18;27	312.28
5487	27AU	0;10	146.24	5487	11SE	9;50	341.47
5488	25SE	10;48	175.48	5488	11OC	0;26	11.13
5489	25OC	0;16	205.30	5489	9NO	14;03	41.15
5490	23NO	16;45	235.55	5490	9DE	2;28	71.28
5491	23DE	11;25	265.93				

-556

NEW NUMBER	DATE	TIME	LONG.	FULL NUMBER	DATE	TIME	LONG.
				5491	7JA	13;31	101.25
5492	22JA	6;34	296.09	5492	5FE	23;18	130.81
5493	21FE	0;27	325.75	5493	6MR	8;17	159.84
5494	21MR	16;02	354.79	5494	4AP	17;15	188.38
5495	20AP	5;05	23.27	5495	4MY	3;03	216.57
5496	19MY	15;47	51.37	5496	2JN	14;24	244.64
5497	18JN	0;42	79.33	5497	2JL	3;38	272.82
5498	17JL	8;33	107.42	5498	31JL	18;45	301.36
5499	15AU	16;19	135.87	5499	30AU	11;24	330.40
5500	14SE	0;57	164.84	5500	29SE	4;56	0.00
5501	13OC	11;18	194.37	5501	28OC	22;18	30.05
5502	11NO	23;51	224.38	5502	27NO	14;17	60.31
5503	11DE	14;35	254.61	5503	27DE	3;59	90.46

-555

NEW NUMBER	DATE	TIME	LONG.	FULL NUMBER	DATE	TIME	LONG.
5504	10JA	6;58	284.77	5504	25JA	15;17	120.23
5505	9FE	0;10	314.59	5505	24FE	0;38	149.46
5506	10MR	17;19	343.89	5506	25MR	8;48	178.14
5507	9AP	9;31	12.66	5507	23AP	16;35	206.37
5508	9MY	0;02	40.99	5508	23MY	0;49	234.35
5509	7JN	12;26	69.09	5509	21JN	10;15	262.34
5510	6JL	22;56	97.19	5510	20JL	21;41	290.60
5511	5AU	8;14	125.55	5511	19AU	11;45	319.36
5512	3SE	17;16	154.36	5512	18SE	4;40	348.74
5513	3OC	2;50	183.69	5513	17OC	23;40	18.70
5514	1NO	13;22	213.49	5514	16NO	19;00	49.03
5515	1DE	0;58	243.58	5515	16DE	12;44	79.38
5516	30DE	13;37	273.66				

-554

NEW NUMBER	DATE	TIME	LONG.	FULL NUMBER	DATE	TIME	LONG.
				5516	15JA	3;42	109.44
5517	29JA	3;22	303.50	5517	13FE	15;47	138.96
5518	27FE	18;15	332.90	5518	15MR	1;32	167.89
5519	29MR	10;00	1.83	5519	13AP	9;36	196.27
5520	28AP	1;53	30.34	5520	12MY	16;45	224.28
5521	27MY	17;03	58.57	5521	10JN	23;47	252.17
5522	26JN	6;55	86.75	5522	10JL	7;47	280.19
5523	25JL	19;28	115.11	5523	8AU	17;50	308.62
5524	24AU	7;06	143.84	5524	7SE	6;56	337.65
5525	22SE	18;16	173.06	5525	6OC	23;23	7.33
5526	22OC	5;12	202.74	5526	5NO	18;26	37.53
5527	20NO	16;00	232.73	5527	5DE	14;17	67.98
5528	20DE	2;44	262.79				

-553

NEW NUMBER	DATE	TIME	LONG.	FULL NUMBER	DATE	TIME	LONG.
				5528	4JA	9;01	98.29
5529	18JA	13;40	292.64	5529	3FE	1;22	128.17
5530	17FE	1;15	322.09	5530	4MR	14;51	157.44
5531	18MR	13;49	351.07	5531	3AP	1;33	186.09
5532	17AP	3;24	19.63	5532	2MY	9;57	214.25
5533	16MY	17;47	47.90	5533	31MY	16;54	242.14
5534	15JN	8;37	76.10	5534	29JN	23;31	270.05
5535	14JL	23;39	104.44	5535	29JL	7;00	298.24
5536	13AU	14;36	133.13	5536	27AU	16;27	326.93
5537	12SE	5;07	162.30	5537	26SE	4;40	356.25
5538	11OC	18;44	191.93	5538	25OC	19;50	26.16
5539	10NO	7;08	221.90	5539	24NO	13;30	56.46
5540	9DE	18;23	251.99	5540	24DE	8;36	86.86

-552

NEW NUMBER	DATE	TIME	LONG.	FULL NUMBER	DATE	TIME	LONG.
5541	8JA	4;46	281.92	5541	23JA	3;41	117.00
5542	6FE	14;44	311.48	5542	21FE	21;12	146.63
5543	7MR	0;38	340.54	5543	22MR	12;03	175.63
5544	5AP	10;53	9.12	5544	20AP	23;49	204.05
5545	4MY	21;53	37.35	5545	20MY	8;59	232.08
5546	3JN	10;10	65.45	5546	18JN	16;35	260.00
5547	3JL	0;09	93.67	5547	17JL	23;48	288.06
5548	1AU	15;57	122.24	5548	16AU	7;43	316.52
5549	31AU	8;56	151.32	5549	14SE	17;05	345.53
5550	30SE	1;57	180.91	5550	14OC	4;21	15.12
5551	29OC	17;49	210.91	5551	12NO	17;50	45.17
5552	28NO	7;54	241.08	5552	12DE	9;33	75.45
5553	27DE	20;11	271.17				

-551

NEW NUMBER	DATE	TIME	LONG.	FULL NUMBER	DATE	TIME	LONG.
				5553	11JA	3;11	105.65
5554	26JA	6;54	300.91	5554	9FE	21;38	135.50
5555	24FE	16;21	330.14	5555	11MR	15;17	164.81
5556	26MR	0;52	358.82	5556	10AP	9;49	193.54
5557	24AP	9;03	27.06	5557	9MY	19;43	221.80
5558	23MY	17;50	55.06	5558	8JN	6;24	249.83
5559	22JN	4;17	83.09	5559	7JL	15;10	277.89
5560	21JL	17;15	111.42	5560	6AU	0;24	306.24
5561	20AU	8;59	140.26	5561	4SE	9;16	335.04
5562	19SE	2;47	169.68	5562	3OC	18;46	4.38
5563	18OC	21;19	199.64	5563	2NO	5;20	34.20
5564	17NO	15;10	229.91	5564	1DE	17;24	64.30
5565	17DE	7;18	260.20	5565	31DE	7;19	94.44

-550

NEW NUMBER	DATE	TIME	LONG.	FULL NUMBER	DATE	TIME	LONG.
5566	15JA	21;09	290.20	5566	29JA	22;55	124.34
5567	14FE	8;31	319.69	5567	28FE	15;26	153.80
5568	15MR	17;39	348.57	5568	30MR	7;50	182.74
5569	14AP	1;11	16.93	5569	28AP	23;18	211.22
5570	13MY	8;07	44.92	5570	28MY	13;27	239.41
5571	11JN	15;38	72.83	5571	27JN	2;17	267.56
5572	11JL	0;51	100.91	5572	26JL	13;56	295.88
5573	9AU	12;38	129.42	5573	25AU	0;37	324.58
5574	8SE	3;18	158.52	5574	23SE	10;45	353.77
5575	7OC	20;37	188.25	5575	22OC	20;47	23.43
5576	6NO	15;43	218.47	5576	21NO	7;16	53.41
5577	6DE	11;10	248.89	5577	20DE	18;34	83.49

-549

NEW NUMBER	DATE	TIME	LONG.	FULL NUMBER	DATE	TIME	LONG.
5578	5JA	5;15	279.18	5578	19JA	6;47	113.38
5579	3FE	20;35	309.00	5579	17FE	19;49	142.89
5580	5MR	8;39	338.19	5580	19MR	9;33	171.90
5581	3AP	17;54	6.77	5581	17AP	23;52	200.48
5582	3MY	1;19	34.89	5582	17MY	14;45	228.77
5583	1JN	8;02	62.78	5583	16JN	5;54	256.97
5584	30JN	15;09	90.70	5584	15JL	20;47	285.32
5585	29JL	23;35	118.94	5585	14AU	10;47	313.98
5586	28AU	10;05	147.69	5586	12SE	23;40	343.09
5587	26SE	23;17	177.06	5587	12OC	11;35	12.67
5588	26OC	15;29	207.02	5588	11NO	22;55	42.60
5589	25NO	10;21	237.38	5589	10DE	9;56	72.68
5590	25DE	6;22	267.81				

-548

NEW NUMBER	DATE	TIME	LONG.	FULL NUMBER	DATE	TIME	LONG.
				5590	8JA	20;42	102.62
5591	24JA	1;23	297.94	5591	7FE	7;12	132.19
5592	22FE	17;38	327.50	5592	7MR	17;41	161.26
5593	23MR	6;37	356.41	5593	6AP	4;42	189.87
5594	21AP	16;48	24.76	5594	5MY	16;55	218.14
5595	21MY	1;05	52.75	5595	4JN	6;44	246.30
5596	19JN	8;25	80.66	5596	3JL	21;59	274.57
5597	18JL	15;40	108.73	5597	2AU	13;59	303.17
5598	16AU	23;41	137.20	5598	1SE	5;56	332.21
5599	15SE	9;18	166.24	5599	30SE	21;16	1.75
5600	14OC	21;21	195.87	5600	30OC	11;39	31.69
5601	13NO	12;22	225.99	5601	29NO	0;50	61.83
5602	13DE	6;01	256.34	5602	28DE	12;35	91.89

-547

NEW NUMBER	DATE	TIME	LONG.	FULL NUMBER	DATE	TIME	LONG.
5603	12JA	0;59	286.60	5603	26JA	22;50	121.60
5604	10FE	19;30	316.44	5604	25FE	7;57	150.81
5605	12MR	12;09	345.69	5605	26MR	16;35	179.48
5606	11AP	2;23	14.35	5606	25AP	1;37	207.75
5607	10MY	14;13	42.57	5607	24MY	11;57	235.81
5608	9JN	0;02	70.56	5608	23JN	0;08	263.91
5609	8JL	8;28	98.60	5609	22JL	14;22	292.30
5610	6AU	16;20	126.91	5610	21AU	6;27	321.16
5611	5SE	0;37	155.71	5611	19SE	23;54	350.59
5612	4OC	10;10	185.06	5612	19OC	17;45	20.53
5613	2NO	21;39	214.92	5613	18NO	10;46	50.77
5614	2DE	11;17	245.09	5614	18DE	1;46	81.01

-546

NEW NUMBER	DATE	TIME	LONG.	FULL NUMBER	DATE	TIME	LONG.
5615	1JA	2;45	275.29	5615	16JA	14;13	110.94
5616	30JA	19;24	305.22	5616	15FE	0;20	140.37
5617	1MR	12;24	334.69	5617	16MR	8;49	169.22
5618	31MR	4;59	3.62	5618	14AP	16;31	197.57
5619	29AP	20;20	32.09	5619	14MY	0;17	225.60
5620	29MY	9;52	60.25	5620	12JN	8;55	253.54
5621	27JN	21;26	88.35	5621	11JL	19;14	281.67
5622	27JL	7;27	116.62	5622	10AU	7;58	310.23
5623	25AU	16;44	145.28	5623	8SE	23;36	339.39
5624	24SE	2;11	174.44	5624	8OC	17;52	9.17
5625	23OC	12;21	204.11	5625	7NO	13;28	39.42
5626	21NO	23;28	234.13	5626	7DE	8;20	69.82
5627	21DE	11;31	264.23				

-545

NEW NUMBER	DATE	TIME	LONG.	FULL NUMBER	DATE	TIME	LONG.
				5627	6JA	0;46	100.03
5628	20JA	0;29	294.15	5628	4FE	14;12	129.76
5629	18FE	14;31	323.69	5629	6MR	0;55	158.89
5630	20MR	5;34	352.75	5630	4AP	9;37	187.44
5631	18AP	21;16	21.37	5631	3MY	17;02	215.54
5632	18MY	12;48	49.67	5632	1JN	23;57	243.44
5633	17JN	3;26	77.86	5633	1JL	7;20	271.38
5634	16JL	16;49	106.15	5634	30JL	16;15	299.64
5635	15AU	5;08	134.76	5635	29AU	3;48	328.44
5636	13SE	16;47	163.83	5636	27SE	18;42	357.90
5637	13OC	4;03	193.39	5637	27OC	12;42	27.94
5638	11NO	15;04	223.32	5638	26NO	8;27	58.35
5639	11DE	1;49	253.38	5639	26DE	3;56	88.75

NEW MOONS / FULL MOONS

Left column group

−544

NEW MOONS

NUMBER	DATE	TIME	LONG.
5640	9JA	12;29	283.31
5641	7FE	23;28	312.89
5642	8MR	11;12	342.01
5643	6AP	23;59	10.67
5644	6MY	13;47	39.00
5645	5JN	4;21	67.19
5646	4JL	19;23	95.46
5647	3AU	10;38	124.03
5648	2SE	1;43	153.05
5649	1OC	16;09	182.57
5650	31OC	5;25	212.47
5651	29NO	17;20	242.56
5652	29DE	4;05	272.58

FULL MOONS

NUMBER	DATE	TIME	LONG.
5640	24JA	21;31	118.81
5641	23FE	12;22	148.30
5642	24MR	0;20	177.15
5643	22AP	9;44	205.46
5644	21MY	17;17	233.43
5645	19JN	23;57	261.30
5646	19JL	6;56	289.37
5647	17AU	15;23	317.87
5648	16SE	2;15	346.97
5649	15OC	16;02	16.68
5650	14NO	8;34	46.87
5651	14DE	3;02	77.26

−543

NEW MOONS

NUMBER	DATE	TIME	LONG.
5653	27JA	14;04	302.27
5654	25FE	23;43	331.48
5655	27MR	9;29	0.20
5656	25AP	19;48	28.50
5657	25MY	7;12	56.60
5658	23JN	20;13	84.74
5659	23JL	11;12	113.17
5660	22AU	3;54	142.07
5661	20SE	21;21	171.51
5662	20OC	14;12	201.42
5663	19NO	5;24	231.59
5664	18DE	18;38	261.75

FULL MOONS

NUMBER	DATE	TIME	LONG.
5652	12JA	22;10	107.52
5653	11FE	16;30	137.34
5654	13MR	8;39	166.55
5655	11AP	21;54	195.17
5656	11MY	8;17	223.32
5657	9JN	16;36	251.25
5658	8JL	23;59	279.25
5659	7AU	7;35	307.56
5660	5SE	16;18	336.38
5661	5OC	2;43	5.78
5662	3NO	15;07	35.69
5663	3DE	5;40	65.90

−542

NEW MOONS

NUMBER	DATE	TIME	LONG.
5665	17JA	6;03	291.64
5666	15FE	15;57	321.04
5667	17MR	0;41	349.89
5668	15AP	8;44	18.25
5669	14MY	16;55	46.29
5670	13JN	2;18	74.26
5671	12JL	13;53	102.45
5672	11AU	4;19	131.09
5673	9SE	21;23	160.32
5674	9OC	15;57	190.12
5675	8NO	10;30	220.34
5676	8DE	3;43	250.68

FULL MOONS

NUMBER	DATE	TIME	LONG.
5664	1JA	22;15	96.14
5665	31JA	16;13	126.12
5666	2MR	10;15	155.61
5667	1AP	2;48	184.53
5668	30AP	16;59	212.94
5669	30MY	4;45	241.03
5670	28JN	14;46	269.07
5671	27JL	23;53	297.31
5672	26AU	8;50	325.97
5673	24SE	18;10	355.14
5674	24OC	4;16	24.81
5675	22NO	15;33	54.84
5676	22DE	4;25	84.98

−541

NEW MOONS

NUMBER	DATE	TIME	LONG.
5677	6JA	18;46	280.82
5678	5FE	7;19	310.50
5679	6MR	17;23	339.59
5680	5AP	1;28	8.10
5681	4MY	8;26	36.19
5682	2JN	15;26	64.08
5683	1JL	23;38	92.06
5684	31JL	10;04	120.39
5685	29AU	23;22	149.28
5686	28SE	15;33	178.80
5687	28OC	10;01	208.87
5688	27NO	5;34	239.27
5689	27DE	0;34	269.65

FULL MOONS

NUMBER	DATE	TIME	LONG.
5677	20JA	19;00	114.96
5678	19FE	10;53	144.55
5679	21MR	3;11	173.65
5680	19AP	19;00	202.26
5681	19MY	9;47	230.53
5682	17JN	23;22	258.68
5683	17JL	11;48	286.94
5684	15AU	23;13	315.53
5685	14SE	9;51	344.57
5686	13OC	20;05	14.09
5687	12NO	6;24	44.00
5688	11DE	17;15	74.06

−540

NEW MOONS

NUMBER	DATE	TIME	LONG.
5690	25JA	17;22	299.67
5691	24FE	6;59	329.09
5692	24MR	17;27	357.88
5693	23AP	1;34	26.13
5694	22MY	8;27	54.06
5695	20JN	15;15	81.94
5696	19JL	22;58	110.05
5697	18AU	8;26	138.60
5698	16SE	20;19	167.75
5699	16OC	11;05	197.51
5700	15NO	4;45	227.76
5701	15DE	0;23	258.20

FULL MOONS

NUMBER	DATE	TIME	LONG.
5689	10JA	4;51	104.03
5690	8FE	17;15	133.66
5691	9MR	6;20	162.82
5692	7AP	20;04	191.51
5693	7MY	10;29	219.85
5694	6JN	1;30	248.06
5695	5JL	16;43	276.34
5696	4AU	7;28	304.90
5697	2SE	21;14	333.89
5698	2OC	9;53	3.34
5699	31OC	21;40	33.19
5700	30NO	8;55	63.25
5701	29DE	19;46	93.27

−539

NEW MOONS

NUMBER	DATE	TIME	LONG.
5702	13JA	20;05	288.47
5703	12FE	13;45	318.24
5704	14MR	4;15	347.38
5705	12AP	15;41	15.91
5706	12MY	0;47	44.00
5707	10JN	8;32	71.91
5708	9JL	15;49	99.91
5709	7AU	23;29	128.24
5710	6SE	8;20	157.07
5711	5OC	19;11	186.50
5712	4NO	8;44	216.46
5713	4DE	1;06	246.76

FULL MOONS

NUMBER	DATE	TIME	LONG.
5702	28JA	6;13	122.97
5703	26FE	16;25	152.20
5704	28MR	2;48	180.92
5705	26AP	14;05	209.26
5706	26MY	2;54	237.41
5707	24JN	17;26	265.62
5708	24JL	9;17	294.09
5709	23AU	1;35	322.99
5710	21SE	17;36	352.38
5711	21OC	8;46	22.24
5712	19NO	22;46	52.36
5713	19DE	11;18	82.49

Right column group

−538

NEW MOONS

NUMBER	DATE	TIME	LONG.
5714	2JA	19;28	277.07
5715	1FE	14;12	307.06
5716	3MR	7;43	336.52
5717	1AP	23;05	5.37
5718	1MY	12;04	33.73
5719	30MY	22;54	61.78
5720	29JN	8;04	89.79
5721	28JL	16;17	118.00
5722	27AU	0;27	146.64
5723	25SE	9;26	175.80
5724	24OC	20;00	205.51
5725	23NO	8;33	235.59
5726	22DE	23;03	265.80

FULL MOONS

NUMBER	DATE	TIME	LONG.
5714	17JA	22;13	112.35
5715	16FE	7;41	141.72
5716	17MR	16;13	170.55
5717	16AP	0;43	198.92
5718	15MY	10;07	227.00
5719	13JN	21;12	255.04
5720	13JL	10;23	283.30
5721	12AU	1;41	311.99
5722	10SE	18;44	341.23
5723	10OC	12;47	11.03
5724	9NO	6;36	41.22
5725	8DE	22;52	71.52

−537

NEW MOONS

NUMBER	DATE	TIME	LONG.
5727	21JA	14;58	295.82
5728	20FE	7;36	325.44
5729	22MR	0;14	354.53
5730	20AP	16;09	23.14
5731	20MY	6;39	51.90
5732	18JN	19;19	79.50
5733	18JL	6;15	107.71
5734	16AU	16;03	136.24
5735	15SE	1;34	165.25
5736	14OC	11;31	194.77
5737	12NO	22;14	224.70
5738	12DE	9;47	254.79

FULL MOONS

NUMBER	DATE	TIME	LONG.
5726	7JA	12;37	101.60
5727	5FE	23;44	131.21
5728	7MR	8;47	160.25
5729	5AP	16;36	188.74
5730	5MY	0;05	216.84
5731	3JN	8;04	244.77
5732	2JL	17;24	272.80
5733	1AU	4;52	301.18
5734	30AU	19;06	330.12
5735	29SE	12;18	359.69
5736	29OC	7;37	29.82
5737	28NO	3;14	60.22
5738	27DE	21;04	90.55

−536

NEW MOONS

NUMBER	DATE	TIME	LONG.
5739	10JA	22;07	284.79
5740	9FE	11;19	314.44
5741	10MR	1;34	343.63
5742	8AP	16;44	12.37
5743	8MY	8;18	40.75
5744	6JN	23;28	68.96
5745	6JL	13;39	97.21
5746	5AU	2;43	125.71
5747	3SE	14;57	154.65
5748	3OC	2;40	184.07
5749	1NO	14;00	213.91
5750	1DE	0;56	243.96
5751	30DE	11;33	273.96

FULL MOONS

NUMBER	DATE	TIME	LONG.
5739	26JA	11;58	120.48
5740	24FE	23;53	149.82
5741	25MR	9;24	178.55
5742	23AP	17;15	206.78
5743	23MY	0;15	234.72
5744	21JN	7;16	262.61
5745	20JL	15;18	290.73
5746	19AU	1;29	319.33
5747	17SE	14;45	348.54
5748	17OC	7;23	18.40
5749	16NO	2;32	48.71
5750	15DE	22;24	79.16

−535

NEW MOONS

NUMBER	DATE	TIME	LONG.
5752	28JA	22;07	303.66
5753	27FE	9;09	332.91
5754	28MR	21;05	1.69
5755	27AP	10;08	30.09
5756	27MY	0;12	58.29
5757	25JN	15;03	86.51
5758	25JL	6;23	114.97
5759	23AU	21;53	143.85
5760	22SE	13;01	173.22
5761	22OC	3;12	203.04
5762	20NO	15;57	233.12
5763	20DE	3;16	263.20

FULL MOONS

NUMBER	DATE	TIME	LONG.
5751	14JA	17;02	109.38
5752	13FE	9;13	139.08
5753	14MR	22;31	168.15
5754	13AP	9;04	196.63
5755	12MY	17;25	224.69
5756	11JN	0;25	252.58
5757	10JL	7;10	280.55
5758	8AU	14;51	308.89
5759	7SE	0;32	337.77
5760	6OC	12;55	7.27
5761	5NO	4;10	37.32
5762	4DE	21;44	67.66

−534

NEW MOONS

NUMBER	DATE	TIME	LONG.
5764	18JA	13;29	293.01
5765	16FE	23;03	322.38
5766	18MR	8;28	351.24
5767	16AP	18;12	19.65
5768	16MY	4;46	47.77
5769	14JN	16;48	75.85
5770	14JL	6;47	104.15
5771	12AU	22;50	132.88
5772	11SE	16;20	162.15
5773	11OC	9;55	191.95
5774	10NO	2;16	222.08
5775	9DE	16;36	252.30

FULL MOONS

NUMBER	DATE	TIME	LONG.
5763	3JA	16;35	97.99
5764	2FE	11;20	127.97
5765	4MR	4;34	157.40
5766	2AP	19;14	186.22
5767	2MY	6;59	214.51
5768	31MY	16;16	242.50
5769	30JN	0;04	270.46
5770	29JL	7;36	298.65
5771	27AU	15;51	327.30
5772	26SE	1;30	356.50
5773	25OC	12;58	26.25
5774	24NO	2;24	56.38
5775	23DE	17;53	86.62

−533

NEW MOONS

NUMBER	DATE	TIME	LONG.
5776	8JA	4;53	282.31
5777	6FE	15;24	311.89
5778	8MR	0;29	340.92
5779	6AP	8;37	9.41
5780	5MY	16;27	37.52
5781	4JN	0;57	65.47
5782	3JL	11;16	93.54
5783	2AU	0;16	122.00
5784	31AU	16;13	151.02
5785	30SE	10;24	180.65
5786	30OC	5;21	210.76
5787	28NO	23;30	241.11
5788	28DE	15;46	271.37

FULL MOONS

NUMBER	DATE	TIME	LONG.
5776	22JA	11;03	116.69
5777	21FE	4;59	146.35
5778	22MR	22;14	175.45
5779	21AP	13;34	204.02
5780	21MY	2;32	232.21
5781	19JN	13;28	260.26
5782	18JL	23;07	288.42
5783	17AU	8;18	316.93
5784	15SE	17;37	345.95
5785	15OC	3;27	15.47
5786	13NO	14;09	45.40
5787	13DE	2;08	75.52

-532 / -526

NEW MOONS				FULL MOONS				NEW MOONS				FULL MOONS			
5789	27JA	5;33	301.24	5788	11JA	15;40	105.55	5863	19JA	21;06	294.40	5862	5JA	11;58	99.87
5790	25FE	16;42	330.54	5789	10FE	6;42	135.27	5864	18FE	7;35	323.79	5863	4FE	5;23	129.78
5791	26MR	1;34	359.23	5790	10MR	22;36	164.52	5865	19MR	18;44	352.69	5864	5MR	20;02	159.07
5792	24AP	8;51	27.43	5791	9AP	14;31	193.27	5866	18AP	6;57	21.18	5865	4AP	7;51	187.75
5793	23MY	15;35	55.35	5792	9MY	5;46	221.63	5867	17MY	20;19	49.40	5866	3MY	17;11	215.93
5794	21JN	23;00	83.27	5793	7JN	20;00	249.80	5868	16JN	10;45	77.58	5867	2JN	0;45	243.85
5795	21JL	8;13	111.44	5794	7JL	9;11	278.02	5869	16JL	1;58	105.94	5868	1JL	7;33	271.77
5796	19AU	20;07	140.12	5795	5AU	21;21	306.50	5870	14AU	17;40	134.68	5869	30JL	14;44	299.97
5797	18SE	10;59	169.41	5796	4SE	8;38	335.40	5871	13SE	9;23	163.91	5870	28AU	23;25	328.65
5798	18OC	4;31	199.32	5797	3OC	19;15	4.79	5872	13OC	0;25	193.62	5871	27SE	10;30	357.94
5799	16NO	23;48	229.65	5798	2NO	5;38	34.60	5873	11NO	14;07	223.66	5872	27OC	0;24	27.81
5800	16DE	19;20	260.08	5799	1DE	16;12	64.64	5874	11DE	2;12	253.79	5873	25NO	16;53	58.07
				5800	31DE	3;19	94.66					5874	25DE	11;08	88.43

-531 / -525

NEW MOONS				FULL MOONS				NEW MOONS				FULL MOONS			
5801	15JA	13;23	290.26	5801	29JA	15;06	124.40	5875	9JA	12;52	283.72	5875	24JA	5;56	118.55
5802	14FE	4;34	319.91	5802	28FE	3;33	153.70	5876	7FE	22;32	313.24	5876	22FE	23;55	148.17
5803	15MR	16;28	348.91	5803	29MR	16;39	182.51	5877	9MR	7;45	342.25	5877	24MR	15;50	177.19
5804	14AP	1;32	17.32	5804	28AP	6;29	210.93	5878	7AP	17;01	10.78	5878	23AP	5;00	205.66
5805	13MY	8;51	45.34	5805	27MY	21;07	239.15	5879	7MY	2;52	38.95	5879	22MY	15;27	233.74
5806	11JN	15;33	73.21	5806	26JN	12;21	267.39	5880	5JN	13;57	67.00	5880	20JN	23;58	261.69
5807	10JL	22;45	101.21	5807	26JL	3;37	295.85	5881	5JL	2;52	95.18	5881	20JL	7;39	289.79
5808	9AU	7;20	129.59	5808	24AU	18;14	324.71	5882	3AU	18;00	123.74	5882	18AU	15;36	318.27
5809	7SE	18;02	158.52	5809	23SE	7;45	354.03	5883	2SE	11;06	152.84	5883	17SE	0;40	347.29
5810	7OC	7;24	188.08	5810	22OC	20;10	23.79	5884	2OC	5;06	182.49	5884	16OC	11;18	16.87
5811	5NO	23;42	218.17	5811	21NO	7;47	53.82	5885	31OC	22;27	212.56	5885	14NO	23;46	46.89
5812	5DE	18;30	248.57	5812	20DE	18;50	83.89	5886	30NO	14;00	242.81	5886	14DE	14;08	77.10
								5887	30DE	3;20	272.93				

-530 / -524

NEW MOONS				FULL MOONS				NEW MOONS				FULL MOONS			
5813	4JA	14;20	278.93	5813	19JA	5;22	113.71					5887	13JA	6;19	107.24
5814	3FE	9;07	308.91	5814	17FE	15;26	143.09	5888	28JA	14;36	302.68	5888	11FE	23;46	137.03
5815	5MR	1;09	338.27	5815	19MR	1;23	171.96	5889	27FE	0;12	331.89	5889	12MR	17;18	166.31
5816	3AP	13;59	7.00	5816	17AP	11;51	200.38	5890	27MR	8;34	0.54	5890	11AP	9;33	195.05
5817	3MY	0;07	35.22	5817	16MY	23;37	228.55	5891	25AP	16;16	28.75	5891	10MY	23;41	223.36
5818	1JN	8;27	63.18	5818	15JN	13;13	256.69	5892	25MY	0;10	56.71	5892	9JN	11;40	251.44
5819	30JN	15;58	91.12	5819	15JL	4;32	285.05	5893	23JN	9;22	84.69	5893	8JL	22;02	279.55
5820	29JL	23;29	119.32	5820	13AU	20;55	313.80	5894	22JL	20;56	112.98	5894	7AU	7;36	307.95
5821	28AU	7;47	147.98	5821	12SE	13;27	343.06	5895	21AU	11;30	141.79	5895	5SE	17;02	336.80
5822	26SE	17;39	177.21	5822	12OC	5;23	12.79	5896	20SE	4;52	171.22	5896	5OC	2;45	6.17
5823	26OC	5;51	207.00	5823	10NO	20;14	42.88	5897	19OC	23;50	201.20	5897	3NO	13;05	35.99
5824	24NO	20;49	237.19	5824	10DE	9;38	73.05	5898	18NO	18;43	231.53	5898	3DE	0;22	66.06
5825	24DE	14;13	267.51					5899	18DE	12;07	261.87				

-529 / -523

NEW MOONS				FULL MOONS				NEW MOONS				FULL MOONS			
5826	23JA	8;47	297.63	5825	8JA	21;20	103.03	5900	17JA	3;10	291.90	5899	1JA	12;58	96.14
5827	22FE	2;51	327.28	5826	7FE	7;20	132.58	5901	15FE	15;32	321.41	5900	31JA	3;02	125.96
5828	23MR	19;12	356.33	5827	8MR	16;03	161.58	5902	17MR	1;22	350.31	5901	1MR	18;17	155.34
5829	22AP	9;18	24.84	5828	7AP	0;14	190.07	5903	15AP	9;12	18.65	5902	31MR	10;01	184.24
5830	21MY	21;12	52.98	5829	6MY	8;52	218.20	5904	14MY	15;58	46.63	5903	30AP	1;29	212.71
5831	20JN	7;16	80.99	5830	4JN	18;53	246.21	5905	12JN	22;51	74.51	5904	29MY	16;14	240.92
5832	19JL	16;03	109.13	5831	4JL	6;56	274.36	5906	12JL	7;02	102.56	5905	28JN	6;05	269.11
5833	18AU	0;20	137.62	5832	2AU	21;15	302.87	5907	10AU	17;34	131.04	5906	27JL	18;59	297.50
5834	16SE	8;59	166.61	5833	1SE	13;39	331.93	5908	9SE	7;02	160.11	5907	26AU	6;59	326.28
5835	15OC	18;49	196.15	5834	1OC	7;33	1.56	5909	8OC	23;25	189.82	5908	24SE	18;11	355.53
5836	14NO	6;24	226.12	5835	31OC	1;53	31.66	5910	7NO	18;04	220.03	5909	24OC	4;49	25.23
5837	13DE	19;53	256.30	5836	29NO	19;14	61.98	5911	7DE	13;43	250.47	5910	22NO	15;20	55.22
				5837	29DE	10;22	92.19					5911	22DE	2;06	85.27

-528 / -522

NEW MOONS				FULL MOONS				NEW MOONS				FULL MOONS			
5838	12JA	10;58	286.39	5838	27JA	22;43	121.98	5912	6JA	8;41	280.78	5912	20JA	13;22	115.11
5839	11FE	3;02	316.14	5839	26FE	8;35	151.22	5913	5FE	1;21	310.64	5913	19FE	1;13	144.54
5840	11MR	19;29	345.40	5840	26MR	16;44	179.87	5914	6MR	14;48	339.87	5914	20MR	13;41	173.49
5841	10AP	11;40	14.15	5841	25AP	0;08	208.07	5915	5AP	1;06	8.47	5915	19AP	2;51	202.00
5842	10MY	2;53	42.50	5842	24MY	7;38	236.02	5916	4MY	9;06	36.60	5916	18MY	16;55	230.25
5843	8JN	16;34	70.65	5843	22JN	16;09	263.98	5917	2JN	15;57	64.49	5917	17JN	7;52	258.45
5844	8JL	4;31	98.82	5844	22JL	2;27	292.21	5918	1JL	22;49	92.41	5918	16JL	23;22	286.83
5845	6AU	15;03	127.25	5845	20AU	15;18	320.93	5919	31JL	6;42	120.64	5919	15AU	14;39	315.56
5846	5SE	0;52	156.10	5846	19SE	7;10	350.28	5920	29AU	16;23	149.38	5920	14SE	5;04	344.76
5847	4OC	10;45	185.47	5847	19OC	1;44	20.24	5921	28SE	4;26	178.72	5921	13OC	18;17	14.41
5848	2NO	21;12	215.29	5848	17NO	21;36	50.60	5922	27OC	19;19	208.64	5922	12NO	6;27	44.39
5849	2DE	8;21	245.36	5849	17DE	16;37	81.01	5923	26NO	12;57	238.96	5923	11DE	17;50	74.48
5850	31DE	20;10	275.40					5924	26DE	8;26	269.36				

-527 / -521

NEW MOONS				FULL MOONS				NEW MOONS				FULL MOONS			
				5850	16JA	9;02	111.11					5924	10JA	4;32	104.40
5851	30JA	8;40	305.16	5851	14FE	22;19	140.67	5925	25JA	3;52	299.49	5925	8FE	14;38	133.94
5852	28FE	22;05	334.49	5852	16MR	8;50	169.60	5926	23FE	21;16	329.08	5926	10MR	0;20	162.96
5853	30MR	12;31	3.34	5853	14AP	17;19	197.99	5927	25MR	11;36	358.02	5927	8AP	10;11	191.49
5854	29AP	3;46	31.81	5854	14MY	0;35	225.99	5928	23AP	22;57	26.41	5928	7MY	20;58	219.70
5855	28MY	19;10	60.05	5855	12JN	7;27	253.87	5929	23MY	8;06	54.43	5929	6JN	9;28	247.81
5856	27JN	9;59	88.28	5856	11JL	14;52	281.88	5930	21JN	15;59	82.36	5930	5JL	23;58	276.06
5857	26JL	23;50	116.70	5857	9AU	23;54	310.29	5931	20JL	23;32	110.45	5931	4AU	16;03	304.67
5858	25AU	12;44	145.50	5858	8SE	11;37	339.28	5932	19AU	7;30	138.95	5932	3SE	8;52	333.76
5859	24SE	0;59	174.78	5859	8OC	2;41	8.92	5933	17SE	16;39	167.98	5933	3OC	1;29	3.37
5860	23OC	12;44	204.52	5860	6NO	20;48	39.10	5934	17OC	3;42	197.59	5934	1NO	17;10	33.39
5861	21NO	23;59	234.54	5861	6DE	16;34	69.54	5935	15NO	17;16	227.66	5935	1DE	7;29	63.58
5862	21DE	10;43	264.59					5936	15DE	9;26	257.95	5936	30DE	20;05	93.67

| ---------NEW MOONS--------- | | | | ---------FULL MOONS-------- | | | | ---------NEW MOONS--------- | | | | ---------FULL MOONS-------- | | | |
NUMBER	DATE	TIME	LONG.	NUMBER	DATE	TIME	LONG.	NUMBER	DATE	TIME	LONG.	NUMBER	DATE	TIME	LONG.
			-520								-514				
5937	14JA	3;25	288.15	5937	29JA	6;49	123.38	6011	8JA	0;08	282.50	6011	21JA	23;56	116.61
5938	12FE	21;42	317.97	5938	27FE	15;56	152.56	6012	6FE	13;48	312.21	6012	20FE	14;21	146.13
5939	13MR	14;49	347.22	5939	28MR	0;03	181.20	6013	8MR	0;45	341.31	6013	22MR	5;38	175.17
5940	12AP	5;58	15.90	5940	26AP	8;08	209.41	6014	6AP	9;22	9.83	6014	20AP	21;06	203.75
5941	11MY	18;57	44.15	5941	25MY	17;12	237.41	6015	5MY	16;26	37.90	6015	20MY	12;10	232.03
5942	10JN	5;58	72.20	5942	24JN	4;05	265.47	6016	3JN	23;03	65.78	6016	19JN	2;33	260.21
5943	9JL	15;29	100.28	5943	23JL	17;16	293.83	6017	3JL	6;26	93.74	6017	18JL	16;08	288.52
5944	8AU	0;07	128.65	5944	22AU	8;47	322.69	6018	1AU	15;44	122.04	6018	17AU	4;53	317.18
5945	6SE	8;42	157.48	5945	21SE	2;14	352.13	6019	31AU	3;46	150.89	6019	15SE	16;45	346.30
5946	5OC	18;01	186.84	5946	20OC	20;44	22.12	6020	29SE	18;50	180.38	6020	15OC	3;52	15.88
5947	4NO	4;46	216.68	5947	19NO	14;57	52.42	6021	29OC	12;33	210.44	6021	13NO	14;31	45.81
5948	3DE	17;16	246.81	5948	19DE	7;24	82.71	6022	28NO	7;55	240.84	6022	13DE	1;07	75.86
								6023	28DE	3;26	271.23				
			-519								-513				
5949	2JA	7;27	276.95	5949	17JA	21;07	112.69					6023	11JA	11;59	105.79
5950	31JA	22;51	306.81	5950	16FE	8;03	142.12	6024	26JA	21;22	301.28	6024	9FE	23;19	135.35
5951	2MR	14;53	336.22	5951	17MR	16;49	170.96	6025	25FE	12;23	330.74	6025	11MR	11;10	164.44
5952	1AP	7;02	5.11	5952	16AP	0;19	199.29	6026	27MR	0;07	359.56	6026	9AP	23;40	193.06
5953	30AP	22;41	33.58	5953	15MY	7;32	227.28	6027	25AP	9;04	27.83	6027	9MY	13;02	221.35
5954	30MY	13;14	61.78	5954	13JN	15;22	255.20	6028	24MY	16;20	55.77	6028	8JN	3;29	249.53
5955	29JN	2;12	89.94	5955	13JL	0;40	283.30	6029	22JN	23;06	83.65	6029	7JL	18;51	277.83
5956	28JL	13;37	118.28	5956	11AU	12;12	311.83	6030	22JL	6;27	111.76	6030	6AU	10;34	306.45
5957	26AU	23;58	147.00	5957	10SE	2;38	340.95	6031	20AU	15;15	140.30	6031	5SE	1;48	335.51
5958	25SE	9;59	176.22	5958	9OC	20;05	10.72	6032	19SE	2;09	169.44	6032	4OC	15;56	5.04
5959	24OC	20;17	205.91	5959	8NO	15;40	40.97	6033	18OC	15;40	199.16	6033	3NO	4;49	34.95
5960	23NO	7;09	235.92	5960	8DE	11;27	71.42	6034	17NO	7;58	229.36	6034	2DE	16;40	65.05
5961	22DE	18;34	265.99					6035	17DE	2;38	259.76				
			-518								-512				
				5961	7JA	5;19	101.68					6035	1JA	3;40	95.06
5962	21JA	6;31	295.85	5962	5FE	20;06	131.45	6036	15JA	22;12	290.01	6036	30JA	13;56	124.74
5963	19FE	19;09	325.31	5963	7MR	7;50	160.59	6037	14FE	16;41	319.80	6037	28FE	23;33	153.92
5964	21MR	8;45	354.29	5964	5AP	17;08	189.15	6038	15MR	8;29	348.97	6038	29MR	8;59	182.59
5965	19AP	23;23	22.85	5965	5MY	0;50	217.25	6039	13AP	21;12	17.53	6039	27AP	18;57	210.86
5966	19MY	14;40	51.14	5966	3JN	7;45	245.14	6040	13MY	7;21	45.66	6040	27MY	6;20	238.95
5967	18JN	5;54	79.35	5967	2JL	14;48	273.08	6041	11JN	15;49	73.60	6041	25JN	19;46	267.11
5968	17JL	20;26	107.70	5968	31JL	22;57	301.33	6042	10JL	23;34	101.63	6042	25JL	11;13	295.58
5969	16AU	10;05	136.38	5969	30AU	9;18	330.10	6043	9AU	7;24	129.98	6043	24AU	4;00	324.51
5970	14SE	22;56	165.53	5970	28SE	22;44	359.51	6044	7SE	16;01	158.83	6044	22SE	21;06	353.98
5971	14OC	11;11	195.14	5971	28OC	15;29	29.52	6045	7OC	2;09	188.25	6045	22OC	13;35	23.90
5972	12NO	22;51	225.11	5972	27NO	10;40	59.91	6046	5NO	14;27	218.17	6046	21NO	4;50	54.08
5973	12DE	9;52	255.19	5973	27DE	6;27	90.32	6047	5DE	5;17	248.40	6047	20DE	18;23	84.25
			-517								-511				
5974	10JA	20;17	285.10	5974	26JA	0;54	120.40	6048	3JA	22;20	278.65	6048	19JA	5;59	114.12
5975	9FE	6;27	314.62	5975	24FE	16;53	149.91	6049	2FE	16;26	308.60	6049	17FE	15;43	143.49
5976	10MR	16;55	343.66	5976	26MR	6;01	178.80	6050	4MR	10;03	338.04	6050	19MR	0;02	172.29
5977	9AP	4;16	12.25	5977	24AP	16;29	207.13	6051	3AP	2;05	6.91	6051	17AP	7;49	200.61
5978	8MY	16;50	40.52	5978	24MY	0;50	235.12	6052	2MY	16;06	35.29	6052	16MY	16;06	228.64
5979	7JN	6;39	68.68	5979	22JN	7;56	263.02	6053	1JN	4;08	63.39	6053	15JN	1;52	256.63
5980	6JL	21;32	96.95	5980	21JL	14;54	291.10	6054	30JN	14;30	91.45	6054	14JL	13;50	284.85
5981	5AU	13;13	125.55	5981	19AU	22;50	319.60	6055	29JL	23;42	119.72	6055	13AU	4;17	313.51
5982	4SE	5;17	154.63	5982	18SE	8;45	348.68	6056	28AU	8;26	148.39	6056	11SE	21;00	342.76
5983	3OC	21;03	184.22	5983	17OC	21;18	18.36	6057	26SE	17;29	177.59	6057	11OC	15;20	12.59
5984	2NO	11;44	214.20	5984	16NO	12;34	48.51	6058	26OC	3;34	207.29	6058	10NO	10;05	42.83
5985	2DE	0;46	244.34	5985	16DE	5;58	78.85	6059	24NO	15;12	237.34	6059	10DE	3;41	73.19
5986	31DE	12;05	274.37					6060	24DE	4;28	267.49				
			-516								-510				
				5986	15JA	0;28	109.07					6060	8JA	18;52	103.32
5987	29JA	22;04	304.05	5987	13FE	18;50	138.87	6061	22JA	19;04	297.44	6061	7FE	7;04	132.96
5988	28FE	7;15	333.22	5988	14MR	11;46	168.10	6062	21FE	10;33	326.99	6062	8MR	16;41	161.99
5989	28MR	16;11	1.88	5989	13AP	2;16	196.75	6063	23MR	2;26	356.04	6063	7AP	0;33	190.47
5990	27AP	1;27	30.13	5990	12MY	14;03	224.94	6064	21AP	18;14	24.62	6064	6MY	7;41	218.54
5991	26MY	11;40	58.18	5991	10JN	23;30	252.92	6065	21MY	9;22	52.90	6065	4JN	15;01	246.44
5992	24JN	23;31	86.27	5992	10JL	7;37	280.96	6066	19JN	23;17	81.06	6066	3JL	23;27	274.45
5993	24JL	13;33	114.67	5993	8AU	15;29	309.30	6067	19JL	11;40	109.34	6067	2AU	9;49	302.80
5994	23AU	5;53	143.58	5994	7SE	0;06	338.14	6068	17AU	22;44	137.94	6068	31AU	22;49	331.70
5995	21SE	23;52	173.06	5995	6OC	10;04	7.54	6069	16SE	9;06	167.00	6069	30SE	14;53	1.25
5996	21OC	17;59	203.04	5996	4NO	21;39	37.43	6070	15OC	19;25	196.56	6070	30OC	9;42	31.37
5997	20NO	10;45	233.29	5997	4DE	11;01	67.59	6071	14NO	6;06	226.50	6071	29NO	5;45	61.80
5998	20DE	1;16	263.50					6072	13DE	17;14	256.58	6072	29DE	0;49	92.17
			-515								-509				
				5998	3JA	2;09	97.76	6073	12JA	4;45	286.52	6073	27JA	17;10	122.14
5999	18JA	13;29	293.40	5999	1FE	18;48	127.67	6074	10FE	16;45	316.10	6074	26FE	6;17	151.51
6000	16FE	23;45	322.80	6000	3MR	12;11	157.12	6075	12MR	5;31	345.21	6075	27MR	16;36	180.25
6001	18MR	8;30	351.63	6001	2AP	5;02	186.03	6076	10AP	19;22	13.88	6076	26AP	0;55	208.49
6002	16AP	16;17	19.95	6002	1MY	20;13	214.47	6077	10MY	10;12	42.22	6077	25MY	8;05	236.42
6003	15MY	23;49	47.96	6003	31MY	9;18	242.61	6078	9JN	1;31	70.44	6078	23JN	14;58	264.32
6004	14JN	8;07	75.89	6004	29JN	20;32	270.70	6079	8JL	16;35	98.73	6079	22JL	22;30	292.43
6005	13JL	18;21	104.04	6005	29JL	6;39	299.00	6080	7AU	6;56	127.30	6080	21AU	7;43	321.01
6006	12AU	7;26	132.64	6006	27AU	16;20	327.70	6081	5SE	20;28	156.31	6081	19SE	19;36	350.19
6007	10SE	23;37	161.85	6007	26SE	2;05	356.92	6082	5OC	9;18	185.80	6082	19OC	10;48	20.00
6008	10OC	18;09	191.67	6008	25OC	12;14	26.61	6083	3NO	21;29	215.69	6083	18NO	4;58	50.28
6009	9NO	13;27	221.93	6009	23NO	23;02	56.62	6084	3DE	8;54	245.77	6084	18DE	0;39	80.72
6010	9DE	7;50	252.31	6010	23DE	10;50	86.71								

41

```
--------NEW MOONS--------      --------FULL MOONS--------      --------NEW MOONS--------      --------FULL MOONS--------
NUMBER  DATE  TIME  LONG.      NUMBER  DATE  TIME  LONG.      NUMBER  DATE  TIME  LONG.      NUMBER  DATE  TIME  LONG.
```

-508 / -502

NEW MOONS				FULL MOONS				NEW MOONS				FULL MOONS			
6085	1JA	19:33	275.76	6085	16JA	19:52	110.95					6159	10JA	4:46	104.80
6086	31JA	5:36	305.42	6086	15FE	13:03	140.67	6160	24JA	11:15	299.18	6160	8FE	15:17	134.35
6087	29FE	15:36	334.61	6087	16MR	3:31	169.77	6161	23FE	5:02	328.80	6161	10MR	0:03	163.33
6088	30MR	2:10	3.31	6088	14AP	15:13	198.29	6162	24MR	21:45	357.85	6162	8AP	7:48	191.78
6089	28AP	13:50	31.64	6089	14MY	0:32	226.37	6163	23AP	12:43	26.38	6163	7MY	15:31	219.87
6090	28MY	2:52	59.79	6090	12JN	8:12	254.28	6164	23MY	1:45	54.56	6164	6JN	0:18	247.83
6091	26JN	17:12	87.99	6091	11JL	15:12	282.28	6165	21JN	13:02	82.62	6165	5JL	11:03	275.93
6092	26JL	8:39	116.47	6092	9AU	22:39	310.63	6166	20JL	22:57	110.82	6166	4AU	0:17	304.41
6093	25AU	0:51	145.40	6093	8SE	7:36	339.50	6167	19AU	8:03	139.36	6167	2SE	16:03	333.46
6094	23SE	17:10	174.84	6094	7OC	18:54	8.98	6168	17SE	17:04	168.39	6168	2OC	9:52	3.11
6095	23OC	8:46	204.74	6095	6NO	8:52	38.98	6169	17OC	2:43	197.94	6169	1NO	4:47	33.25
6096	21NO	22:50	234.88	6096	6DE	1:13	69.28	6170	15NO	13:36	227.89	6170	30NO	23:18	63.62
6097	21DE	11:02	264.99					6171	15DE	1:59	258.02	6171	30DE	15:51	93.88

-507 / -501

NEW MOONS				FULL MOONS				NEW MOONS				FULL MOONS			
				6097	4JA	19:09	99.55	6172	13JA	15:47	288.05	6172	29JA	5:30	123.71
6098	19JA	21:33	294.80	6098	3FE	13:33	129.51	6173	12FE	6:36	317.73	6173	27FE	16:13	152.96
6099	18FE	6:53	324.14	6099	5MR	7:09	158.93	6174	13MR	22:01	346.92	6174	29MR	0:42	181.61
6100	19MR	15:39	352.95	6100	3AP	22:51	187.77	6175	12AP	13:41	15.63	6175	27AP	7:57	209.79
6101	18AP	0:28	21.30	6101	3MY	11:58	216.11	6176	12MY	5:08	43.99	6176	26MY	14:59	237.71
6102	17MY	9:56	49.37	6102	1JN	22:34	244.15	6177	10JN	19:46	72.18	6177	24JN	22:44	265.64
6103	15JN	20:46	77.41	6103	1JL	7:22	272.15	6178	10JL	9:08	100.41	6178	24JL	8:03	293.84
6104	15JL	9:38	105.67	6104	30JL	15:24	300.38	6179	8AU	21:05	128.91	6179	22AU	19:43	322.54
6105	14AU	0:57	134.39	6105	28AU	23:45	329.05	6180	7SE	8:00	157.83	6180	21SE	10:20	351.86
6106	12SE	18:27	163.68	6106	27SE	9:09	358.27	6181	6OC	18:30	187.25	6181	21OC	4:00	21.80
6107	12OC	12:58	193.53	6107	26OC	20:00	28.01	6182	5NO	5:08	217.09	6182	19NO	23:46	52.16
6108	11NO	6:47	223.74	6108	25NO	8:28	58.10	6183	4DE	16:04	247.15	6183	19DE	19:37	82.60
6109	10DE	22:35	254.02	6109	24DE	22:35	88.28								

-506 / -500

NEW MOONS				FULL MOONS				NEW MOONS				FULL MOONS			
6110	9JA	11:56	284.06	6110	23JA	14:17	118.28	6184	3JA	3:18	277.16	6184	18JA	13:26	112.75
6111	7FE	23:02	313.65	6111	22FE	7:09	147.87	6185	1FE	14:48	306.86	6185	17FE	4:04	142.34
6112	9MR	8:20	342.66	6112	24MR	0:11	176.95	6186	2MR	2:51	336.11	6186	17MR	15:37	171.30
6113	7AP	16:21	11.13	6113	22AP	16:11	205.53	6187	31MR	15:50	4.89	6187	16AP	0:45	199.69
6114	6MY	23:45	39.21	6114	22MY	6:20	233.76	6188	30AP	5:58	33.30	6188	15MY	8:21	227.70
6115	5JN	7:26	67.13	6115	20JN	18:33	261.86	6189	29MY	21:01	61.53	6189	13JN	15:16	255.58
6116	4JL	16:32	95.16	6116	20JL	5:21	290.08	6190	28JN	12:22	89.77	6190	12JL	22:25	283.59
6117	3AU	4:07	123.57	6117	18AU	15:26	318.65	6191	28JL	3:19	118.25	6191	11AU	6:44	311.99
6118	1SE	18:52	152.56	6118	17SE	1:21	347.71	6192	26AU	17:33	147.13	6192	9SE	17:17	340.95
6119	1OC	12:32	182.18	6119	16OC	11:27	17.27	6193	25SE	7:07	176.48	6193	9OC	6:52	10.55
6120	31OC	7:49	212.33	6120	14NO	21:58	47.20	6194	24OC	19:48	206.27	6194	7NO	23:42	40.69
6121	30NO	2:57	242.73	6121	14DE	9:11	77.28	6195	23NO	7:44	236.33	6195	7DE	18:49	71.10
6122	29DE	20:26	273.03					6196	22DE	18:46	266.39				

-505 / -499

NEW MOONS				FULL MOONS				NEW MOONS				FULL MOONS			
				6122	12JA	21:26	107.25					6196	6JA	14:24	101.43
6123	28JA	11:25	302.93	6123	11FE	10:56	136.88	6197	21JA	4:56	296.17	6197	5FE	8:37	131.35
6124	26FE	23:37	332.25	6124	13MR	1:33	166.06	6198	19FE	14:40	325.51	6198	7MR	0:22	160.67
6125	28MR	9:13	0.95	6125	11AP	16:44	194.77	6199	21MR	0:35	354.34	6199	5AP	13:21	189.38
6126	26AP	16:51	29.15	6126	11MY	7:55	223.12	6200	19AP	11:23	22.76	6200	4MY	23:47	217.59
6127	25MY	23:29	57.06	6127	9JN	22:40	251.31	6201	18MY	23:31	50.93	6201	3JN	8:13	245.55
6128	24JN	6:19	84.96	6128	9JL	12:49	279.57	6202	17JN	13:08	79.07	6202	2JL	15:31	273.49
6129	23JL	14:34	113.11	6129	8AU	2:16	308.12	6203	17JL	4:07	107.43	6203	31JL	22:44	301.71
6130	22AU	1:13	141.75	6130	6SE	14:53	337.10	6204	15AU	20:11	136.21	6204	30AU	6:57	330.39
6131	20SE	14:52	171.02	6131	6OC	2:37	6.55	6205	14SE	12:49	165.50	6205	28SE	17:08	359.67
6132	20OC	7:25	200.90	6132	4NO	13:38	36.40	6206	14OC	5:11	195.28	6206	28OC	5:49	29.50
6133	19NO	2:10	231.21	6133	4DE	0:16	66.45	6207	12NO	20:20	225.39	6207	26NO	21:00	59.71
6134	18DE	21:49	261.65					6208	12DE	9:34	255.56	6208	26DE	14:08	90.02

-504 / -498

NEW MOONS				FULL MOONS				NEW MOONS				FULL MOONS			
				6134	2JA	10:54	96.44	6209	10JA	20:50	285.51	6209	25JA	8:14	120.09
6135	17JA	16:41	291.85	6135	31JA	21:47	126.12	6210	9FE	6:33	315.01	6210	24FE	2:10	149.69
6136	16FE	9:10	321.53	6136	1MR	9:05	155.35	6211	10MR	15:20	343.98	6211	25MR	18:47	178.73
6137	16MR	22:27	350.57	6137	30MR	20:56	184.09	6212	8AP	23:49	12.46	6212	24AP	9:11	207.23
6138	15AP	8:38	19.01	6138	29AP	9:35	212.46	6213	8MY	8:41	40.58	6213	23MY	21:03	235.36
6139	14MY	16:34	47.04	6139	28MY	23:20	240.63	6214	6JN	18:36	68.58	6214	22JN	6:45	263.36
6140	12JN	23:28	74.92	6140	27JN	14:18	268.86	6215	6JL	6:20	96.73	6215	21JL	15:13	291.50
6141	12JL	6:28	102.92	6141	27JL	6:06	297.37	6216	4AU	20:27	125.26	6216	19AU	23:30	320.02
6142	10AU	14:34	131.30	6142	25AU	21:57	326.29	6217	3SE	13:06	154.36	6217	18SE	8:30	349.06
6143	9SE	0:28	160.23	6143	24SE	13:00	355.70	6218	3OC	7:33	184.04	6218	17OC	18:44	18.64
6144	8OC	12:43	189.76	6144	24OC	2:47	25.53	6219	2NO	2:09	214.18	6219	16NO	6:25	48.63
6145	7NO	3:39	219.81	6145	22NO	15:16	55.61	6220	1DE	19:13	244.50	6220	15DE	19:36	78.80
6146	6DE	21:11	250.15	6146	22DE	2:42	85.68	6221	31DE	9:51	274.67				

-503 / -497

NEW MOONS				FULL MOONS				NEW MOONS				FULL MOONS			
6147	5JA	16:24	280.48	6147	20JA	13:13	115.49					6221	14JA	10:20	108.86
6148	4FE	11:30	310.45	6148	18FE	22:56	144.83	6222	29JA	21:57	304.43	6222	13FE	2:25	138.58
6149	6MR	4:37	339.84	6149	20MR	8:08	173.65	6223	28FE	7:58	333.64	6223	14MR	19:14	167.81
6150	4AP	18:47	8.60	6150	18AP	17:29	202.01	6224	29MR	16:24	2.27	6224	13AP	11:42	196.55
6151	4MY	6:07	36.86	6151	18MY	3:51	230.12	6225	27AP	23:52	30.45	6225	13MY	2:47	224.88
6152	2JN	15:22	64.85	6152	16JN	16:06	258.21	6226	27MY	7:11	58.38	6226	11JN	15:01	253.01
6153	1JL	23:29	92.82	6153	16JL	6:36	286.54	6227	25JN	15:22	86.33	6227	11JL	3:39	281.18
6154	31JL	7:20	121.05	6154	14AU	22:59	315.31	6228	25JL	1:35	114.58	6228	9AU	14:16	309.64
6155	29AU	15:39	149.73	6155	13SE	16:18	344.62	6229	23AU	14:47	143.35	6229	8SE	0:28	338.54
6156	28SE	1:06	178.97	6156	13OC	9:28	14.42	6230	22SE	7:12	172.76	6230	7OC	10:40	7.96
6157	27OC	12:19	208.73	6157	12NO	1:38	44.57	6231	22OC	2:02	202.76	6231	5NO	21:05	37.79
6158	26NO	1:50	238.87	6158	11DE	16:11	74.79	6232	20NO	21:35	233.11	6232	5DE	7:54	67.85
6159	25DE	17:44	269.12					6233	20DE	16:06	263.50				

```
        --------NEW MOONS---------    --------FULL MOONS--------    --------NEW MOONS---------    --------FULL MOONS--------
        NUMBER DATE   TIME   LONG.    NUMBER DATE   TIME   LONG.    NUMBER DATE   TIME   LONG.    NUMBER DATE   TIME   LONG.
```

-496 / -490

NUMBER	DATE	TIME	LONG.	NUMBER	DATE	TIME	LONG.	NUMBER	DATE	TIME	LONG.	NUMBER	DATE	TIME	LONG.
				6233	3JA	19;28	97.87	6308	12JA	4;18	286.89	6308	27JA	3;38	121.96
6234	19JA	8;23	293.57	6234	2FE	8;05	127.61	6309	10FE	14;00	316.37	6309	25FE	20;34	151.49
6235	17FE	21;54	323.11	6235	2MR	21;53	156.92	6310	11MR	23;29	345.35	6310	27MR	10;50	180.40
6236	18MR	8;38	352.02	6236	1AP	12;33	185.75	6311	10AP	9;30	13.87	6311	25AP	22;28	208.77
6237	16AP	17;03	20.37	6237	1MY	3;36	214.19	6312	9MY	20;42	42.07	6312	25MY	7;51	236.80
6238	15MY	23;59	48.35	6238	30MY	18;32	242.41	6313	8JN	9;26	70.18	6313	23JN	15;41	264.73
6239	14JN	6;33	76.21	6239	29JN	9;07	270.63	6314	7JL	23;45	98.44	6314	22JL	22;57	292.84
6240	13JL	13;59	104.24	6240	28JL	23;10	299.08	6315	6AU	15;28	127.06	6315	21AU	6;42	321.35
6241	11AU	23;24	132.70	6241	27AU	12;31	327.93	6316	5SE	8;09	156.19	6316	19SE	15;56	350.42
6242	10SE	11;36	161.74	6242	26SE	0;59	357.26	6317	5OC	1;04	185.84	6317	19OC	3;25	20.08
6243	10OC	2;50	191.42	6243	25OC	12;33	27.01	6318	3NO	17;11	215.89	6318	17NO	17;23	50.18
6244	8NO	20;39	221.60	6244	23NO	23;26	57.03	6319	3DE	7;34	246.10	6319	17DE	9;33	80.47
6245	8DE	16;02	252.03	6245	23DE	10;00	87.06								

-495 / -489

NUMBER	DATE	TIME	LONG.	NUMBER	DATE	TIME	LONG.	NUMBER	DATE	TIME	LONG.	NUMBER	DATE	TIME	LONG.
6246	7JA	11;27	282.35	6246	21JA	20;34	116.86	6320	1JA	19;48	276.16	6320	16JA	3;05	110.63
6247	6FE	5;11	312.24	6247	20FE	7;24	146.23	6321	31JA	6;07	305.83	6321	14FE	21;00	140.40
6248	7MR	20;01	341.51	6248	21MR	18;39	175.11	6322	1MR	15;07	334.97	6322	16MR	14;13	169.62
6249	6AP	7;37	10.14	6249	20AP	6;36	203.56	6323	30MR	23;27	3.58	6323	15AP	5;42	198.29
6250	5MY	16;31	38.29	6250	19MY	19;35	231.75	6324	29AP	7;51	31.79	6324	14MY	18;51	226.53
6251	3JN	23;48	66.19	6251	18JN	9;53	259.93	6325	28MY	17;00	59.78	6325	13JN	5;40	254.57
6252	3JL	6;41	94.13	6252	18JL	1;26	288.32	6326	27JN	3;39	87.84	6326	12JL	14;48	282.65
6253	1AU	14;15	122.36	6253	16AU	17;38	317.11	6327	26JL	16;32	116.21	6327	10AU	23;15	311.04
6254	30AU	23;18	151.09	6254	15SE	9;29	346.38	6328	25AU	8;04	145.10	6328	9SE	8;01	339.91
6255	29SE	10;26	180.42	6255	15OC	0;13	16.11	6329	24SE	1;58	174.60	6329	8OC	17;45	9.32
6256	29OC	0;02	210.30	6256	13NO	13;32	46.15	6330	23OC	20;56	204.63	6330	7NO	4;47	39.19
6257	27NO	16;17	240.56	6257	13DE	1;32	76.27	6331	22NO	15;08	234.94	6331	6DE	17;10	69.32
6258	27DE	10;42	270.91					6332	22DE	7;07	265.21				

-494 / -488

NUMBER	DATE	TIME	LONG.	NUMBER	DATE	TIME	LONG.	NUMBER	DATE	TIME	LONG.	NUMBER	DATE	TIME	LONG.
				6258	11JA	12;26	106.19	6333	20JA	20;26	295.14	6332	5JA	6;58	99.43
6259	26JA	5;56	301.02	6259	9FE	22;22	135.70	6334	19FE	7;19	324.55	6333	3FE	22;08	129.25
6260	25FE	0;04	330.62	6260	11MR	7;33	164.67	6335	19MR	16;20	353.36	6334	4MR	14;24	158.63
6261	26MR	15;39	359.60	6261	9AP	16;29	193.15	6336	18AP	0;03	21.67	6335	3AP	6;56	187.52
6262	25AP	4;18	28.02	6262	9MY	2;00	221.30	6337	17MY	7;13	49.65	6336	2MY	22;42	215.97
6263	24MY	14;31	56.08	6263	7JN	13;05	249.35	6338	15JN	14;45	77.55	6337	1JN	12;55	244.15
6264	22JN	23;12	84.04	6264	7JL	2;25	277.56	6339	14JL	23;48	105.66	6338	1JL	1;28	272.30
6265	22JL	7;15	112.18	6265	5AU	18;02	306.16	6340	13AU	11;29	134.22	6339	30JL	12;45	300.66
6266	20AU	15;26	140.70	6266	4SE	11;14	335.30	6341	12SE	2;25	163.40	6340	28AU	23;22	329.42
6267	19SE	0;24	169.75	6267	4OC	4;52	4.97	6342	11OC	20;20	193.22	6341	27SE	9;47	358.69
6268	18OC	10;46	199.35	6268	2NO	21;52	35.05	6343	10NO	15;52	223.49	6342	26OC	20;15	28.41
6269	16NO	23;06	229.37	6269	2DE	13;26	65.30	6344	10DE	11;10	253.92	6343	25NO	6;53	58.42
6270	16DE	13;45	259.60									6344	24DE	17;57	88.47

-493 / -487

NUMBER	DATE	TIME	LONG.	NUMBER	DATE	TIME	LONG.	NUMBER	DATE	TIME	LONG.	NUMBER	DATE	TIME	LONG.
				6270	1JA	3;04	95.42	6345	9JA	4;40	284.14	6345	23JA	5;48	118.31
6271	15JA	6;23	289.73	6271	30JA	14;32	125.15	6346	7FE	19;32	313.89	6346	21FE	18;44	147.74
6272	13FE	23;56	319.50	6272	28FE	23;57	154.32	6347	9MR	7;32	343.01	6347	23MR	8;42	176.70
6273	15MR	17;05	348.73	6273	30MR	7;55	182.93	6348	7AP	16;56	11.54	6348	21AP	23;22	205.24
6274	14AP	8;50	17.42	6274	28AP	15;20	211.10	6349	7MY	0;25	39.62	6349	21MY	14;17	233.51
6275	13MY	22;49	45.71	6275	27MY	23;20	239.05	6350	5JN	7;00	67.49	6350	20JN	5;07	261.71
6276	12JN	11;02	73.80	6276	26JN	8;56	267.06	6351	4JL	13;52	95.43	6351	19JL	19;38	290.07
6277	11JL	21;47	101.94	6277	25JL	20;53	295.39	6352	2AU	22;13	123.72	6352	18AU	9;39	318.80
6278	10AU	7;27	130.36	6278	24AU	11;30	324.23	6353	1SE	9;03	152.54	6353	16SE	22;53	347.99
6279	8SE	16;39	159.24	6279	23SE	4;32	353.68	6354	30SE	22;52	182.00	6354	16OC	11;10	17.64
6280	8OC	2;06	188.63	6280	22OC	23;14	23.68	6355	30OC	15;33	212.03	6355	14NO	22;30	47.61
6281	6NO	12;24	218.47	6281	21NO	18;18	54.02	6356	29NO	10;19	242.41	6356	14DE	9;12	77.67
6282	6DE	0;00	248.56	6282	21DE	12;05	84.37	6357	29DE	5;51	272.80				

-492 / -486

NUMBER	DATE	TIME	LONG.	NUMBER	DATE	TIME	LONG.	NUMBER	DATE	TIME	LONG.	NUMBER	DATE	TIME	LONG.
6283	4JA	12;59	278.64	6283	20JA	3;14	114.39					6357	12JA	19;37	107.56
6284	3FE	3;04	308.43	6284	18FE	15;17	143.85	6358	28JA	0;31	302.86	6358	11FE	6;05	137.07
6285	3MR	17;56	337.77	6285	19MR	0;39	172.70	6359	26FE	16;49	332.35	6359	12MR	16;49	166.09
6286	2AP	9;15	6.61	6286	17AP	8;16	201.01	6360	28MR	5;56	1.21	6360	11AP	4;06	194.64
6287	2MY	0;42	35.06	6287	16MY	15;12	228.98	6361	26AP	16;02	29.50	6361	10MY	16;17	222.88
6288	31MY	15;48	63.28	6288	14JN	22;26	256.87	6362	25MY	23;59	57.47	6362	9JN	5;47	251.02
6289	30JN	6;00	91.49	6289	14JL	6;52	284.95	6363	24JN	7;00	85.37	6363	8JL	20;47	279.31
6290	29JL	18;53	119.91	6290	12AU	17;20	313.45	6364	23JL	14;13	113.48	6364	7AU	12;57	307.96
6291	28AU	6;33	148.70	6291	11SE	6;30	342.55	6365	21AU	22;34	142.03	6365	6SE	5;24	337.09
6292	26SE	17;27	177.98	6292	10OC	22;47	12.28	6366	20SE	8;43	171.15	6366	5OC	21;04	6.71
6293	26OC	4;10	207.70	6293	9NO	17;46	42.53	6367	19OC	21;07	200.85	6367	4NO	11;20	36.69
6294	24NO	15;02	237.72	6294	9DE	13;54	72.99	6368	18NO	12;03	231.00	6368	4DE	0;06	66.83
6295	24DE	2;04	267.77					6369	18DE	5;23	261.34				

-491 / -485

NUMBER	DATE	TIME	LONG.	NUMBER	DATE	TIME	LONG.	NUMBER	DATE	TIME	LONG.	NUMBER	DATE	TIME	LONG.
				6295	8JA	8;55	103.28					6369	2JA	11;31	96.85
6296	22JA	13;15	297.59	6296	7FE	1;08	133.09	6370	17JA	0;15	291.55	6370	31JA	21;47	126.51
6297	21FE	0;42	326.97	6297	8MR	14;04	162.27	6371	15FE	18;58	321.33	6371	2MR	7;06	155.65
6298	22MR	12;51	355.87	6298	7AP	0;14	190.84	6372	17MR	11;47	350.52	6372	31MR	15;51	184.27
6299	21AP	2;07	24.37	6299	6MY	8;26	218.95	6373	16AP	1;49	19.12	6373	30AP	0;44	212.49
6300	20MY	16;37	52.62	6300	4JN	15;35	246.85	6374	15MY	13;11	47.29	6374	29MY	10;45	240.52
6301	19JN	7;54	80.83	6301	3JL	22;33	274.79	6375	13JN	22;36	75.27	6375	27JN	22;49	268.63
6302	18JL	23;16	109.23	6302	2AU	6;15	303.04	6376	13JL	7;01	103.33	6376	27JL	13;22	297.08
6303	17AU	14;09	137.98	6303	31AU	15;40	331.80	6377	11AU	15;15	131.72	6377	26AU	6;03	326.03
6304	16SE	4;20	167.19	6304	30SE	3;45	1.17	6378	9SE	23;56	160.59	6378	24SE	23;52	355.54
6305	15OC	17;44	196.88	6305	29OC	19;03	31.13	6379	9OC	9;40	190.02	6379	24OC	17;34	25.53
6306	14NO	6;17	226.89	6306	28NO	13;11	61.48	6380	7NO	21;01	219.91	6380	23NO	10;07	55.77
6307	13DE	17;49	256.99	6307	28DE	8;41	91.87	6381	7DE	10;26	250.08	6381	23DE	0;50	85.99

--------NEW MOONS--------				--------FULL MOONS--------			
NUMBER	DATE	TIME	LONG.	NUMBER	DATE	TIME	LONG.

-484

NUMBER	DATE	TIME	LONG.	NUMBER	DATE	TIME	LONG.
6382	6JA	1;59	280.26	6382	21JA	13;21	115.88
6383	4FE	18;58	310.15	6383	19FE	23;37	145.25
6384	5MR	12;12	339.55	6384	20MR	8;03	174.03
6385	4AP	4;33	8.41	6385	18AP	15;27	202.31
6386	3MY	19;21	36.82	6386	17MY	22;52	230.30
6387	2JN	8;30	64.96	6387	16JN	7;27	258.25
6388	1JL	20;07	93.07	6388	15JL	18;09	286.43
6389	31JL	6;30	121.41	6389	14AU	7;29	315.07
6390	29AU	16;07	150.14	6390	12SE	23;29	344.31
6391	28SE	1;34	179.38	6391	12OC	17;39	14.14
6392	27OC	11;31	209.09	6392	11NO	12;54	44.42
6393	25NO	22;28	239.11	6393	11DE	7;37	74.82
6394	25DE	10;40	269.21				

-483

NUMBER	DATE	TIME	LONG.	NUMBER	DATE	TIME	LONG.
				6394	10JA	0;13	105.00
6395	24JA	0;00	299.10	6395	8FE	13;44	134.67
6396	22FE	14;13	328.57	6396	10MR	0;14	163.72
6397	24MR	5;02	357.56	6397	8AP	8;29	192.19
6398	22AP	20;15	26.11	6398	7MY	15;32	220.25
6399	22MY	11;31	54.38	6399	5JN	22;27	248.14
6400	21JN	2;19	82.58	6400	5JL	6;11	276.12
6401	20JL	16;06	110.93	6401	3AU	15;35	304.45
6402	19AU	4;39	139.61	6402	2SE	3;24	333.32
6403	17SE	16;09	168.74	6403	1OC	18;12	2.84
6404	17OC	3;08	198.34	6404	31OC	12;02	32.93
6405	15NO	14;02	228.30	6405	30NO	7;53	63.35
6406	15DE	0;59	258.37	6406	30DE	3;43	93.75

-482

NUMBER	DATE	TIME	LONG.	NUMBER	DATE	TIME	LONG.
6407	13JA	11;57	288.28	6407	28JA	21;24	123.76
6408	11FE	22;59	317.80	6408	27FE	11;52	153.17
6409	13MR	10;26	346.83	6409	28MR	23;15	181.93
6410	11AP	22;49	15.42	6410	27AP	8;16	210.18
6411	11MY	12;30	43.71	6411	26MY	15;49	238.13
6412	10JN	3;23	71.91	6412	24JN	22;49	266.03
6413	9JL	18;53	100.23	6413	24JL	6;08	294.15
6414	8AU	10;18	128.86	6414	22AU	14;40	322.72
6415	7SE	1;09	157.94	6415	21SE	1;25	351.88
6416	6OC	15;15	187.50	6416	20OC	15;09	21.64
6417	5NO	4;28	217.45	6417	19NO	7;58	51.88
6418	4DE	16;38	247.56	6418	19DE	2;56	82.28

-481

NUMBER	DATE	TIME	LONG.	NUMBER	DATE	TIME	LONG.
6419	3JA	3;35	277.56	6419	17JA	22;15	112.50
6420	1FE	13;29	307.19	6420	16FE	16;10	142.24
6421	2MR	22;45	336.32	6421	18MR	7;41	171.36
6422	1AP	8;09	4.96	6422	16AP	20;33	199.90
6423	30AP	18;27	33.22	6423	16MY	7;00	228.03
6424	30MY	6;13	61.32	6424	14JN	15;35	255.97
6425	28JN	19;42	89.49	6425	13JL	23;09	284.01
6426	28JL	10;50	117.97	6426	12AU	6;41	312.37
6427	27AU	3;18	146.93	6427	10SE	15;13	341.25
6428	25SE	20;30	176.43	6428	10OC	1;38	10.72
6429	25OC	13;25	206.39	6429	8NO	14;24	40.68
6430	24NO	4;57	236.60	6430	8DE	5;28	70.92
6431	23DE	18;19	266.76				

-480

NUMBER	DATE	TIME	LONG.	NUMBER	DATE	TIME	LONG.
				6431	6JA	22;15	101.14
6432	22JA	5;28	296.59	6432	5FE	15;52	131.04
6433	20FE	14;54	325.91	6433	6MR	9;20	160.44
6434	20MR	23;17	354.68	6434	5AP	1;38	189.29
6435	19AP	7;22	22.98	6435	4MY	15;58	217.67
6436	18MY	15;53	51.01	6436	3JN	3;59	245.76
6437	17JN	1;36	79.00	6437	2JL	14;01	273.82
6438	16JL	13;16	107.22	6438	31JL	22;55	302.10
6439	15AU	3;31	135.91	6439	30AU	7;37	330.81
6440	13SE	20;29	165.21	6440	28SE	17;00	0.05
6441	13OC	15;21	195.09	6441	28OC	3;30	29.79
6442	12NO	10;21	225.36	6442	26NO	15;12	59.86
6443	12DE	3;40	255.70	6443	26DE	4;10	89.98

-479

NUMBER	DATE	TIME	LONG.	NUMBER	DATE	TIME	LONG.
6444	10JA	18;20	285.79	6444	24JA	18;25	119.90
6445	9FE	6;18	315.39	6445	23FE	9;54	149.41
6446	10MR	16;03	344.40	6446	25MR	22;08	178.44
6447	9AP	0;11	12.85	6447	23AP	18;14	207.01
6448	8MY	7;25	40.91	6448	23MY	9;16	235.27
6449	6JN	14;34	68.81	6449	21JN	22;45	263.42
6450	5JL	22;42	96.81	6450	21JL	10;49	291.71
6451	4AU	8;58	125.18	6451	19AU	21;59	320.34
6452	2SE	22;19	154.13	6452	18SE	8;44	349.46
6453	2OC	14;57	183.74	6453	17OC	19;21	19.06
6454	1NO	10;01	213.89	6454	16NO	5;59	49.00
6455	1DE	5;44	244.31	6455	15DE	16;46	79.07
6456	31DE	0;18	274.65				

--------NEW MOONS--------				--------FULL MOONS--------			
NUMBER	DATE	TIME	LONG.	NUMBER	DATE	TIME	LONG.

-478

NUMBER	DATE	TIME	LONG.	NUMBER	DATE	TIME	LONG.
				6456	14JA	4;02	108.98
6457	29JA	16;29	304.59	6457	12FE	16;08	138.54
6458	28FE	5;50	333.93	6458	14MR	5;17	167.63
6459	29MR	16;23	2.65	6459	12AP	19;22	196.28
6460	28AP	0;38	30.87	6460	12MY	10;02	224.60
6461	27MY	7;30	58.78	6461	11JN	0;55	252.79
6462	25JN	14;05	86.67	6462	10JL	15;44	281.09
6463	24JL	21;38	114.80	6463	9AU	6;18	309.70
6464	23AU	7;13	143.42	6464	7SE	20;17	338.76
6465	21SE	19;36	172.66	6465	7OC	9;20	8.29
6466	21OC	10;58	202.51	6466	5NO	21;19	38.19
6467	20NO	4;50	232.79	6467	5DE	8;22	68.26
6468	20DE	0;07	263.21				

-477

NUMBER	DATE	TIME	LONG.	NUMBER	DATE	TIME	LONG.
				6468	3JA	18;49	98.23
6469	18JA	19;20	293.41	6469	2FE	5;03	127.87
6470	17FE	12;50	323.12	6470	3MR	15;22	157.03
6471	19MR	3;30	352.19	6471	2AP	2;03	185.71
6472	17AP	14;59	20.67	6472	1MY	13;29	214.01
6473	16MY	23;53	48.73	6473	31MY	2;07	242.14
6474	15JN	7;16	76.63	6474	29JN	16;21	270.35
6475	14JL	14;21	104.65	6475	29JL	8;08	298.86
6476	12AU	22;11	133.03	6476	28AU	0;50	327.84
6477	11SE	7;13	161.96	6477	26SE	17;18	357.33
6478	10OC	18;50	191.47	6478	26OC	8;36	27.23
6479	9NO	8;30	221.48	6479	24NO	22;16	57.37
6480	9DE	0;37	251.76	6480	24DE	10;22	87.47

-476

NUMBER	DATE	TIME	LONG.	NUMBER	DATE	TIME	LONG.
6481	7JA	18;43	282.03	6481	22JA	21;06	117.27
6482	6FE	13;31	311.97	6482	21FE	6;42	146.59
6483	7MR	7;17	341.37	6483	21MR	15;26	175.36
6484	5AP	22;39	10.17	6484	19AP	23;55	203.67
6485	5MY	11;16	38.47	6485	19MY	9;04	231.72
6486	3JN	21;38	66.49	6486	17JN	19;54	259.76
6487	3JL	6;36	94.51	6487	17JL	9;11	288.05
6488	1AU	15;01	122.78	6488	16AU	1;00	316.82
6489	30AU	23;36	151.49	6489	14SE	18;37	346.15
6490	29SE	8;55	180.74	6490	14OC	12;46	16.02
6491	28OC	19;28	210.50	6491	13NO	6;12	46.23
6492	27NO	7;48	240.59	6492	12DE	22;01	76.51
6493	26DE	22;10	270.77				

-475

NUMBER	DATE	TIME	LONG.	NUMBER	DATE	TIME	LONG.
				6493	11JA	11;40	106.55
6494	25JA	14;19	300.76	6494	9FE	22;56	136.11
6495	24FE	7;18	330.32	6495	11MR	8;04	165.08
6496	25MR	23;57	359.35	6496	9AP	15;41	193.51
6497	24AP	15;27	27.89	6497	8MY	22;49	221.55
6498	24MY	5;27	56.10	6498	7JN	6;35	249.47
6499	22JN	17;56	84.22	6499	6JL	16;05	277.53
6500	22JL	5;07	112.48	6500	5AU	4;06	305.99
6501	20AU	15;19	141.08	6501	3SE	18;53	335.01
6502	19SE	1;00	170.16	6502	3OC	12;13	4.66
6503	18OC	10;49	199.74	6503	2NO	7;14	34.82
6504	16NO	21;17	229.68	6504	2DE	2;33	65.22
6505	16DE	8;48	259.78	6505	31DE	20;24	95.53

-474

NUMBER	DATE	TIME	LONG.	NUMBER	DATE	TIME	LONG.
6506	14JA	21;25	289.74	6506	30JA	11;29	125.41
6507	13FE	10;57	319.34	6507	28FE	23;20	154.68
6508	15MR	1;10	348.47	6508	30MR	8;29	183.33
6509	13AP	15;57	17.13	6509	28AP	15;54	211.50
6510	13MY	7;06	45.47	6510	27MY	22;42	239.41
6511	11JN	22;14	73.67	6511	26JN	5;54	267.33
6512	11JL	12;45	101.96	6512	25JL	14;25	295.51
6513	10AU	2;12	130.54	6513	24AU	1;01	324.18
6514	8SE	14;28	159.54	6514	22SE	14;22	353.47
6515	8OC	1;55	189.01	6515	22OC	6;48	23.37
6516	6NO	13;00	218.89	6516	21NO	1;53	53.71
6517	5DE	23;58	248.95	6517	20DE	21;59	84.16

-473

NUMBER	DATE	TIME	LONG.	NUMBER	DATE	TIME	LONG.
6518	4JA	10;51	278.93	6518	19JA	16;53	114.35
6519	2FE	21;38	308.59	6519	18FE	8;56	143.98
6520	4MR	8;32	337.77	6520	19MR	21;42	172.96
6521	2AP	20;05	6.47	6521	18AP	7;44	201.37
6522	2MY	8;50	34.81	6522	17MY	15;53	229.49
6523	31MY	23;01	63.00	6523	15JN	23;06	257.29
6524	30JN	14;19	91.25	6524	15JL	6;13	285.31
6525	30JL	6;03	119.77	6525	13AU	14;09	313.71
6526	28AU	21;30	148.72	6526	11SE	23;47	342.66
6527	27SE	12;18	178.15	6527	11OC	12;02	12.22
6528	27OC	2;14	208.01	6528	10NO	3;23	42.31
6529	25NO	15;06	238.11	6529	9DE	21;24	72.68
6530	25DE	2;41	268.19				

NEW MOONS / FULL MOONS tables

	NEW MOONS				FULL MOONS		
NUMBER	DATE	TIME	LONG.	NUMBER	DATE	TIME	LONG.

-472

NEW MOONS				FULL MOONS			
				6530	8JA	16;38	102.98
6531	23JA	12;58	297.96	6531	7FE	11;15	132.90
6532	21FE	22;16	327.26	6532	8MR	3;53	162.24
6533	22MR	7;16	356.03	6533	6AP	18;00	190.97
6534	20AP	16;47	24.38	6534	6MY	5;37	219.23
6535	20MY	3;33	52.48	6535	4JN	15;07	247.22
6536	18JN	16;04	80.59	6536	3JL	23;12	275.20
6537	18JL	6;25	108.94	6537	2AU	6;46	303.45
6538	16AU	22;26	137.72	6538	31AU	14;52	332.15
6539	15SE	15;37	167.05	6539	30SE	0;24	1.42
6540	15OC	9;05	196.90	6540	29OC	12;02	31.23
6541	14NO	1;40	227.08	6541	28NO	1;58	61.39
6542	13DE	16;16	257.31	6542	27DE	17;50	91.63

-471

NEW MOONS				FULL MOONS			
6543	12JA	4;29	287.28	6543	26JA	10;54	121.65
6544	10FE	14;34	316.78	6544	25FE	4;18	151.21
6545	11MR	23;12	345.72	6545	26MR	21;07	180.24
6546	10AP	7;10	14.16	6546	25AP	12;26	208.76
6547	9MY	15;12	42.24	6547	25MY	1;39	236.94
6548	8JN	0;06	70.20	6548	23JN	12;45	265.00
6549	7JL	10;39	98.30	6549	22JL	22;18	293.19
6550	5AU	23;35	126.81	6550	21AU	7;13	321.76
6551	4SE	15;21	155.88	6551	19SE	16;25	350.83
6552	4OC	9;38	185.58	6552	19OC	2;28	20.43
6553	3NO	5;00	215.77	6553	17NO	13;37	50.41
6554	2DE	23;29	246.14	6554	17DE	1;52	80.53

-470

NEW MOONS				FULL MOONS			
6555	1JA	15;34	276.37	6555	15JA	15;16	110.52
6556	31JA	4;47	306.16	6556	14FE	5;52	140.16
6557	1MR	15;28	335.37	6557	15MR	21;30	169.32
6558	31MR	0;12	4.00	6558	14AP	13;34	198.03
6559	29AP	7;40	32.16	6559	14MY	5;09	226.37
6560	28MY	14;39	60.08	6560	12JN	19;29	254.55
6561	26JN	22;08	88.00	6561	12JL	8;25	282.78
6562	26JL	7;13	116.22	6562	10AU	20;14	311.30
6563	24AU	19;00	144.95	6563	9SE	7;26	340.27
6564	23SE	10;08	174.32	6564	8OC	18;20	9.73
6565	23OC	4;16	204.31	6565	7NO	5;06	39.60
6566	21NO	23;59	234.68	6566	6DE	15;48	69.65
6567	21DE	19;20	265.10				

-469

NEW MOONS				FULL MOONS			
				6567	5JA	2;40	99.63
6568	20JA	12;46	295.21	6568	3FE	14;04	129.30
6569	19FE	3;28	324.77	6569	5MR	2;23	158.53
6570	20MR	15;17	353.70	6570	3AP	15;44	187.29
6571	19AP	0;32	22.07	6571	3MY	5;56	215.68
6572	18MY	7;56	50.06	6572	1JN	20;39	243.89
6573	16JN	14;31	77.93	6573	1JL	11;35	272.13
6574	15JL	21;30	105.95	6574	31JL	2;32	300.63
6575	14AU	6;01	134.39	6575	29AU	17;09	329.55
6576	12SE	17;03	163.40	6576	28SE	7;01	358.96
6577	12OC	7;01	193.05	6577	27OC	19;47	28.78
6578	10NO	23;45	223.20	6578	26NO	7;23	58.83
6579	10DE	18;27	253.60	6579	25DE	18;05	88.87

-468

NEW MOONS				FULL MOONS			
6580	9JA	13;48	283.90	6580	24JA	4;15	118.63
6581	8FE	8;12	313.80	6581	22FE	14;15	147.94
6582	9MR	0;16	343.10	6582	23MR	0;27	176.76
6583	7AP	13;16	11.78	6583	21AP	11;11	205.14
6584	6MY	23;20	39.96	6584	20MY	22;58	233.28
6585	5JN	7;23	67.89	6585	19JN	12;17	261.42
6586	4JL	14;35	95.85	6586	19JL	3;23	289.81
6587	2AU	22;03	124.09	6587	17AU	19;57	318.63
6588	1SE	6;42	152.83	6588	16SE	12;58	347.97
6589	30SE	17;06	182.15	6589	16OC	5;14	17.78
6590	30OC	5;38	212.00	6590	14NO	19;57	47.89
6591	28NO	20;30	242.21	6591	14DE	8;53	78.04
6592	28DE	13;33	272.49				

-467

NEW MOONS				FULL MOONS			
				6592	12JA	20;14	107.97
6593	27JA	7;59	302.56	6593	11FE	6;14	137.46
6594	26FE	2;16	332.14	6594	12MR	15;09	166.40
6595	27MR	18;48	1.14	6595	10AP	23;27	194.84
6596	26AP	8;44	29.60	6596	10MY	7;57	222.93
6597	25MY	20;11	57.70	6597	8JN	17;41	250.93
6598	24JN	5;51	85.71	6598	8JL	5;38	279.09
6599	23JL	14;38	113.88	6599	6AU	20;17	307.67
6600	21AU	23;17	142.45	6600	5SE	13;18	336.82
6601	20SE	8;21	171.52	6601	5OC	7;35	6.53
6602	19OC	18;21	201.12	6602	4NO	1;44	36.67
6603	18NO	5;46	231.12	6603	3DE	18;36	66.98
6604	17DE	19;01	261.28				

-466

NEW MOONS				FULL MOONS			
				6604	2JA	9;25	97.15
6605	16JA	10;08	291.34	6605	31JA	21;48	126.90
6606	15FE	2;33	321.04	6606	2MR	7;49	156.07
6607	16MR	19;13	350.24	6607	31MR	15;56	184.66
6608	15AP	11;12	18.92	6608	29AP	23;02	212.80
6609	15MY	1;54	47.23	6609	29MY	6;14	240.72
6610	13JN	15;14	75.36	6610	27JN	14;41	268.69
6611	13JL	3;15	103.56	6611	27JL	1;23	296.98
6612	11AU	14;09	132.06	6612	25AU	14;51	325.79
6613	10SE	0;17	160.99	6613	24SE	7;06	355.23
6614	9OC	10;10	190.43	6614	24OC	1;34	25.24
6615	7NO	20;23	220.27	6615	22NO	21;02	55.61
6616	7DE	7;21	250.34	6616	22DE	15;53	86.00

-465

NEW MOONS				FULL MOONS			
6617	5JA	19;17	280.36	6617	21JA	8;27	116.06
6618	4FE	8;08	310.08	6618	19FE	21;49	145.56
6619	5MR	21;43	339.35	6619	21MR	8;07	174.42
6620	4AP	11;56	8.13	6620	19AP	16;10	202.73
6621	4MY	2;43	36.54	6621	18MY	23;04	230.69
6622	2JN	17;53	64.76	6622	17JN	5;56	258.57
6623	2JL	8;54	93.01	6623	16JL	13;43	286.63
6624	31JL	23;10	121.49	6624	14AU	23;16	315.12
6625	30AU	12;19	150.37	6625	13SE	11;15	344.19
6626	29SE	0;25	179.71	6626	13OC	2;13	13.88
6627	28OC	11;51	209.49	6627	11NO	20;09	44.09
6628	26NO	22;57	239.52	6628	11DE	16;00	74.54
6629	26DE	9;51	269.56				

-464

NEW MOONS				FULL MOONS			
				6629	10JA	11;42	104.86
6630	24JA	20;31	299.34	6630	9FE	5;12	134.70
6631	23FE	7;03	328.66	6631	9MR	19;30	163.91
6632	23MR	17;55	357.49	6632	8AP	6;44	192.51
6633	22AP	5;44	25.91	6633	7MY	15;41	220.64
6634	21MY	19;01	54.11	6634	5JN	23;17	248.55
6635	20JN	9;47	82.31	6635	5JL	6;26	276.51
6636	20JL	1;30	110.72	6636	3AU	13;57	304.77
6637	18AU	17;24	139.53	6637	1SE	22;45	333.52
6638	17SE	8;53	168.82	6638	1OC	9;42	2.87
6639	16OC	23;34	198.58	6639	30OC	23;32	32.79
6640	15NO	13;11	228.65	6640	29NO	16;17	63.08
6641	15DE	1;29	258.78	6641	29DE	11;00	93.43

-463

NEW MOONS				FULL MOONS			
6642	13JA	12;20	288.68	6642	28JA	5;57	123.50
6643	11FE	21;54	318.14	6643	26FE	23;32	153.05
6644	13MR	6;44	347.07	6644	28MR	14;49	181.98
6645	11AP	15;38	15.52	6645	27AP	3;37	210.38
6646	11MY	1;29	43.66	6646	26MY	14;10	238.44
6647	9JN	12;57	71.72	6647	24JN	22;58	266.42
6648	9JL	2;22	99.95	6648	24JL	6;51	294.57
6649	7AU	17;41	128.57	6649	22AU	14;45	323.11
6650	6SE	10;34	157.72	6650	20SE	23;37	352.19
6651	6OC	4;18	187.43	6651	20OC	10;16	21.82
6652	4NO	21;43	217.55	6652	18NO	23;03	51.88
6653	4DE	13;34	247.82	6653	18DE	13;55	82.12

-462

NEW MOONS				FULL MOONS			
6654	3JA	3;00	277.92	6654	17JA	6;16	112.21
6655	1FE	14;00	307.60	6655	15FE	23;21	141.93
6656	2MR	23;07	336.72	6656	17MR	16;20	171.12
6657	1AP	7;08	5.30	6657	16AP	8;21	199.80
6658	30AP	14;52	33.46	6658	15MY	22;40	228.08
6659	29MY	23;07	61.42	6659	14JN	10;54	256.17
6660	28JN	8;40	89.44	6660	13JL	21;20	284.32
6661	27JL	20;20	117.77	6661	12AU	6;41	312.76
6662	26AU	10;45	146.64	6662	10SE	15;52	341.67
6663	25SE	4;02	176.13	6663	10OC	1;38	11.10
6664	24OC	23;16	206.18	6664	8NO	12;20	40.98
6665	23NO	18;36	236.55	6665	8DE	0;00	71.08
6666	23DE	12;04	266.88				

-461

NEW MOONS				FULL MOONS			
				6666	6JA	12;40	101.12
6667	22JA	2;42	296.86	6667	5FE	2;23	130.87
6668	20FE	14;29	326.27	6668	6MR	17;14	160.17
6669	21MR	23;59	355.09	6669	5AP	8;55	189.00
6670	20AP	7;53	23.39	6670	5MY	0;41	217.44
6671	19MY	14;55	51.35	6671	3JN	15;43	245.66
6672	17JN	21;59	79.24	6672	3JL	5;30	273.86
6673	17JL	6;08	107.32	6673	1AU	18;04	302.29
6674	15AU	16;30	135.85	6674	31AU	5;49	331.12
6675	14SE	6;01	164.99	6675	29SE	17;07	0.44
6676	13OC	22;51	194.78	6676	29OC	4;08	30.21
6677	12NO	18;05	225.05	6677	27NO	14;55	60.23
6678	12DE	13;53	255.50	6678	27DE	1;36	90.26

```
        --------NEW MOONS---------      ---------FULL MOONS--------      --------NEW MOONS---------      ---------FULL MOONS--------
        NUMBER  DATE   TIME   LONG.     NUMBER   DATE   TIME   LONG.     NUMBER   DATE   TIME   LONG.    NUMBER   DATE   TIME   LONG.
```

-460 / -454

NUMBER	DATE	TIME	LONG.	NUMBER	DATE	TIME	LONG.	NUMBER	DATE	TIME	LONG.	NUMBER	DATE	TIME	LONG.
6679	11JA	8:23	285.76	6679	25JA	12:30	120.04	6753	4JA	11:29	279.35	6753	19JA	0:28	114.04
6680	10FE	0:26	315.53	6680	24FE	0:03	149.39	6754	2FE	21:31	308.97	6754	17FE	18:42	143.78
6681	10MR	13:36	344.68	6681	24MR	12:34	178.28	6755	4MR	6:25	338.06	6755	19MR	11:02	172.91
6682	8AP	23:59	13.23	6682	23AP	2:06	206.76	6756	2AP	14:57	6.64	6756	18AP	1:01	201.49
6683	8MY	8:09	41.33	6683	22MY	16:27	234.99	6757	2MY	0:00	34.84	6757	17MY	12:40	229.65
6684	6JN	15:01	69.21	6684	21JN	7:18	263.19	6758	31MY	10:26	62.89	6758	15JN	22:22	257.64
6685	5JL	21:42	97.15	6685	20JL	22:26	291.60	6759	29JN	22:47	91.02	6759	15JL	6:46	285.72
6686	4AU	5:24	125.42	6686	19AU	13:32	320.38	6760	29JL	13:12	119.48	6760	13AU	14:43	314.12
6687	2SE	15:11	154.23	6687	18SE	4:10	349.65	6761	28AU	5:32	148.45	6761	11SE	23:11	343.02
6688	2OC	3:45	183.66	6688	17OC	17:47	19.38	6762	26SE	23:13	177.99	6762	11OC	9:00	12.48
6689	31OC	19:13	213.65	6689	16NO	6:08	49.40	6763	26OC	17:12	208.01	6763	9NO	20:45	42.41
6690	30NO	13:02	243.99	6690	15DE	17:16	79.48	6764	25NO	10:09	238.29	6764	9DE	10:33	72.60
6691	30DE	8:08	274.35					6765	25DE	0:55	268.50				

-459 / -453

NUMBER	DATE	TIME	LONG.	NUMBER	DATE	TIME	LONG.	NUMBER	DATE	TIME	LONG.	NUMBER	DATE	TIME	LONG.
6692	29JA	3:04	304.41	6691	14JA	3:34	109.35	6766	23JA	13:03	298.35	6765	8JA	2:03	102.76
6693	27FE	20:19	333.93	6692	12FE	13:25	138.81	6767	21FE	22:53	327.67	6766	6FE	18:35	132.60
6694	29MR	10:48	2.81	6693	13MR	23:13	167.77	6768	23MR	7:11	356.41	6767	8MR	11:27	161.96
6695	27AP	22:14	31.15	6694	12AP	9:22	196.26	6769	21AP	14:47	24.68	6768	7AP	3:52	190.79
6696	27MY	7:12	59.15	6695	11MY	20:21	224.44	6770	20MY	22:33	52.67	6769	6MY	19:03	219.19
6697	25JN	14:46	87.08	6696	10JN	8:42	252.54	6771	19JN	7:16	80.62	6770	5JN	8:25	247.33
6698	24JL	22:06	115.21	6697	9JL	22:55	280.80	6772	18JL	17:46	108.81	6771	4JL	19:52	275.46
6699	23AU	6:14	143.77	6698	8AU	14:58	309.46	6773	17AU	6:14	137.47	6772	3AU	5:53	303.80
6700	21SE	15:51	172.89	6699	7SE	8:10	338.64	6774	15SE	22:49	166.74	6773	1SE	15:18	332.55
6701	21OC	3:22	202.58	6700	7OC	1:15	8.34	6775	15OC	17:26	196.63	6774	1OC	0:56	1.83
6702	19NO	17:02	232.68	6701	5NO	17:02	38.40	6776	14NO	13:08	226.94	6775	30OC	11:16	31.58
6703	19DE	8:56	262.95	6702	5DE	6:59	68.59	6777	14DE	7:49	257.34	6776	28NO	22:29	61.63
												6777	28DE	10:32	91.71

-458 / -452

NUMBER	DATE	TIME	LONG.	NUMBER	DATE	TIME	LONG.	NUMBER	DATE	TIME	LONG.	NUMBER	DATE	TIME	LONG.
6704	18JA	2:37	293.10	6703	3JA	19:07	98.63	6778	12JA	23:55	287.48	6778	26JA	23:28	121.56
6705	16FE	20:57	322.85	6704	2FE	5:39	128.28	6779	11FE	13:00	317.10	6779	25FE	13:27	150.99
6706	18MR	14:19	352.04	6705	3MR	14:53	157.40	6780	11MR	23:27	346.12	6780	26MR	4:29	179.95
6707	17AP	5:31	20.68	6706	1AP	23:13	185.98	6781	10AP	7:57	14.57	6781	24AP	20:06	208.49
6708	16MY	18:10	48.89	6707	1MY	7:18	214.15	6782	9MY	15:14	42.62	6782	24MY	11:32	236.76
6709	15JN	4:44	76.91	6708	30MY	16:08	242.13	6783	7JN	22:07	70.50	6783	23JN	2:04	264.96
6710	14JL	14:02	105.02	6709	29JN	2:47	270.19	6784	7JL	5:36	98.49	6784	22JL	15:26	293.30
6711	12AU	22:53	133.45	6710	28JL	16:05	298.60	6785	5AU	14:46	126.83	6785	21AU	3:50	322.01
6712	11SE	7:54	162.36	6711	27AU	8:00	327.54	6786	4SE	2:42	155.75	6786	19SE	15:36	351.19
6713	10OC	17:32	191.80	6712	26SE	2:10	357.08	6787	3OC	18:01	185.32	6787	19OC	2:59	20.84
6714	9NO	4:16	221.68	6713	25OC	20:46	27.12	6788	2NO	12:19	215.45	6788	17NO	14:00	50.81
6715	8DE	16:30	251.81	6714	24NO	14:34	57.43	6789	2DE	8:06	245.88	6789	17DE	0:42	80.87
				6715	24DE	6:33	87.69								

-457 / -451

NUMBER	DATE	TIME	LONG.	NUMBER	DATE	TIME	LONG.	NUMBER	DATE	TIME	LONG.	NUMBER	DATE	TIME	LONG.
6716	7JA	6:32	281.90	6716	22JA	20:08	117.61	6790	1JA	3:25	276.24	6790	15JA	11:18	110.74
6717	5FE	22:08	311.72	6717	21FE	7:12	146.99	6791	30JA	20:43	306.21	6791	13FE	22:14	140.23
6718	7MR	14:31	341.07	6718	22MR	16:02	175.77	6792	1MR	11:14	335.58	6792	15MR	9:56	169.24
6719	6AP	6:42	9.91	6719	20AP	23:23	204.03	6793	30MR	22:54	4.33	6793	13AP	22:41	197.81
6720	5MY	21:58	38.32	6720	20MY	6:16	231.99	6794	29AP	8:02	.32.56	6794	13MY	12:27	226.09
6721	4JN	12:02	66.49	6721	18JN	13:54	259.90	6795	28MY	15:25	60.49	6795	12JN	3:01	254.28
6722	4JL	0:51	94.67	6722	17JL	23:22	288.04	6796	26JN	22:05	88.39	6796	11JL	18:08	282.59
6723	2AU	12:32	123.07	6723	16AU	11:28	316.65	6797	26JL	5:14	116.52	6797	10AU	9:32	311.25
6724	31AU	23:17	151.87	6724	15SE	2:28	345.87	6798	24AU	13:58	145.13	6798	9SE	0:46	340.38
6725	30SE	9:28	181.15	6725	14OC	20:03	15.70	6799	23SE	1:12	174.34	6799	8OC	15:15	9.99
6726	29OC	19:37	210.89	6726	13NO	15:19	45.98	6800	22OC	15:18	204.15	6800	7NO	4:28	39.96
6727	28NO	6:12	240.91	6727	13DE	10:45	76.42	6801	21NO	8:02	234.39	6801	6DE	16:16	70.06
6728	27DE	17:34	270.96					6802	21DE	2:34	264.77				

-456 / -450

NUMBER	DATE	TIME	LONG.	NUMBER	DATE	TIME	LONG.	NUMBER	DATE	TIME	LONG.	NUMBER	DATE	TIME	LONG.
6729	26JA	5:46	300.79	6728	12JA	4:37	106.64	6803	19JA	21:37	294.96	6802	5JA	2:54	100.03
6730	24FE	18:42	330.19	6729	10FE	19:34	136.35	6804	18FE	15:43	324.67	6803	3FE	12:46	129.63
6731	25MR	8:18	359.10	6730	11MR	7:15	165.43	6805	20MR	7:34	353.78	6804	4MR	22:19	158.74
6732	23AP	22:34	27.60	6731	9AP	16:12	193.91	6806	18AP	20:27	22.29	6805	3AP	7:59	187.36
6733	23MY	13:27	55.85	6732	8MY	23:28	221.97	6807	18MY	6:34	50.39	6806	2MY	18:14	215.60
6734	22JN	4:40	84.07	6733	7JN	6:12	249.84	6808	16JN	14:46	78.32	6807	1JN	5:40	243.68
6735	21JL	19:35	112.48	6734	6JL	13:18	277.81	6809	15JL	22:13	106.37	6808	30JN	18:52	271.85
6736	20AU	9:37	141.23	6735	4AU	22:05	306.13	6810	14AU	6:01	134.77	6809	30JL	10:07	300.36
6737	18SE	22:31	170.45	6736	3SE	8:52	334.98	6811	12SE	14:59	163.71	6810	29AU	3:05	329.36
6738	18OC	10:29	200.11	6737	2OC	22:23	4.46	6812	12OC	1:38	193.21	6811	27SE	20:40	358.91
6739	16NO	21:52	230.10	6738	1NO	14:56	34.51	6813	10NO	14:14	223.18	6812	27OC	13:30	28.90
6740	16DE	8:53	260.17	6739	1DE	10:01	64.91	6814	10DE	4:57	253.41	6813	26NO	4:34	59.10
				6740	31DE	6:00	95.31					6814	25DE	17:38	89.24

-455 / -449

NUMBER	DATE	TIME	LONG.	NUMBER	DATE	TIME	LONG.	NUMBER	DATE	TIME	LONG.	NUMBER	DATE	TIME	LONG.
6741	14JA	19:33	290.05	6741	30JA	0:43	125.34	6815	8JA	21:38	283.61	6815	24JA	4:51	119.04
6742	13FE	5:54	319.52	6742	28FE	16:33	154.79	6816	7FE	15:34	313.50	6816	22FE	14:33	148.34
6743	14MR	16:15	348.49	6743	30MR	5:10	183.58	6817	9MR	9:24	342.88	6817	23MR	23:05	177.08
6744	13AP	3:13	17.00	6744	28AP	15:07	211.86	6818	8AP	1:38	11.69	6818	22AP	7:00	205.36
6745	12MY	15:30	45.23	6745	27MY	23:18	239.82	6819	7MY	15:31	40.03	6819	21MY	15:10	233.36
6746	11JN	5:27	73.39	6746	26JN	6:18	267.74	6820	6JN	3:08	68.11	6820	20JN	0:41	261.35
6747	10JL	20:50	101.71	6747	25JL	13:59	295.88	6821	5JL	13:08	96.18	6821	19JL	12:33	289.60
6748	9AU	12:56	130.39	6748	23AU	22:10	324.45	6822	3AU	22:20	124.49	6822	18AU	3:22	318.34
6749	8SE	4:59	159.53	6749	22SE	8:04	353.60	6823	2SE	7:26	153.25	6823	16SE	20:43	347.68
6750	7OC	20:24	189.17	6750	21OC	20:28	23.33	6824	1OC	16:54	182.53	6824	16OC	15:25	17.59
6751	6NO	10:49	219.18	6751	20NO	11:47	53.50	6825	31OC	3:08	212.28	6825	15NO	9:58	47.85
6752	5DE	23:55	249.34	6752	20DE	5:35	83.86	6826	29NO	14:34	242.34	6826	15DE	3:03	78.18
								6827	29DE	3:33	272.46				

Band 1: **−448** (left) / **−442** (right)

NUMBER	DATE	TIME	LONG.	NUMBER	DATE	TIME	LONG.	NUMBER	DATE	TIME	LONG.	NUMBER	DATE	TIME	LONG.
				6827	13JA	17;53	108.26					6901	6JA	10;22	101.41
6828	27JA	18;11	302.37	6828	12FE	6;07	137.84	6902	21JA	16;20	296.81	6902	4FE	20;52	131.03
6829	26FE	9;59	331.86	6829	12MR	15;52	166.82	6903	20FE	8;12	326.40	6903	6MR	7;51	160.17
6830	27MR	2;06	0.85	6830	10AP	23;43	195.23	6904	21MR	21;12	355.36	6904	4AP	19;46	188.86
6831	25AP	17;44	29.38	6831	10MY	6;34	223.26	6905	20AP	7;28	23.75	6905	4MY	8;47	217.20
6832	25MY	8;24	57.62	6832	8JN	13;37	251.15	6906	19MY	15;37	51.76	6906	2JN	22;51	245.37
6833	23JN	21;58	85.78	6833	7JL	22;01	279.17	6907	17JN	22;32	79.65	6907	2JL	13;46	273.62
6834	23JL	10;26	114.10	6834	6AU	8;46	307.59	6908	17JL	5;23	107.68	6908	1AU	5;14	302.16
6835	21AU	21;53	142.78	6835	4SE	22;24	336.69	6909	15AU	13;18	136.11	6909	30AU	20;54	331.14
6836	20SE	8;35	171.92	6836	4OC	14;53	6.22	6910	13SE	23;19	165.10	6910	29SE	12;10	0.62
6837	19OC	18;53	201.54	6837	3NO	9;34	36.38	6911	13OC	12;03	194.71	6911	29OC	2;19	30.52
6838	18NO	5;17	231.49	6838	3DE	5;12	66.80	6912	12NO	3;33	224.83	6912	27NO	14;58	60.62
6839	17DE	16;12	261.55					6913	11DE	21;15	255.18	6913	27DE	2;08	90.67

Band 2: **−447** (left) / **−441** (right)

NUMBER	DATE	TIME	LONG.	NUMBER	DATE	TIME	LONG.	NUMBER	DATE	TIME	LONG.	NUMBER	DATE	TIME	LONG.
				6839	2JA	0;04	97.14	6914	10JA	16;04	285.46				
6840	16JA	3;49	291.46	6840	31JA	16;32	127.06	6915	9FE	10;39	315.34	6914	25JA	12;12	120.41
6841	14FE	16;08	320.99	6841	2MR	5;44	156.37	6916	11MR	3;37	344.66	6915	23FE	21;39	149.68
6842	16MR	5;06	350.05	6842	31MR	15;51	185.04	6917	9AP	17;56	13.37	6916	25MR	6;58	178.43
6843	14AP	18;44	18.65	6843	29AP	23;45	213.22	6918	9MY	5;23	41.60	6917	23AP	16;37	206.76
6844	14MY	9;09	46.95	6844	29MY	6;35	241.12	6919	7JN	14;29	69.57	6918	23MY	3;12	234.85
6845	13JN	0;14	75.15	6845	27JN	13;28	269.03	6920	6JL	22;17	97.56	6919	21JN	15;21	262.94
6846	12JL	15;32	103.48	6846	26JL	21;23	297.20	6921	5AU	5;57	125.83	6920	21JL	5;35	291.31
6847	11AU	6;20	132.13	6847	25AU	7;06	325.86	6922	3SE	14;26	154.58	6921	19AU	21;57	320.13
6848	9SE	20;07	161.21	6848	23SE	19;17	355.12	6923	3OC	0;20	183.90	6922	18SE	15;39	349.52
6849	9OC	8;48	190.76	6849	23OC	10;23	24.98	6924	1NO	12;00	213.73	6923	18OC	9;18	19.41
6850	7NO	20;37	220.67	6850	22NO	4;20	55.28	6925	1DE	1;36	243.89	6924	17NO	1;32	49.59
6851	7DE	7;53	250.75	6851	22DE	0;04	85.71	6926	30DE	17;12	274.11	6925	16DE	15;41	79.80

Band 3: **−446** (left) / **−440** (right)

NUMBER	DATE	TIME	LONG.	NUMBER	DATE	TIME	LONG.	NUMBER	DATE	TIME	LONG.	NUMBER	DATE	TIME	LONG.
6852	5JA	18;39	280.72	6852	20JA	19;34	115.91					6926	15JA	3;46	109.74
6853	4FE	4;58	310.34	6853	19FE	12;50	145.57	6927	29JA	10;24	304.10	6927	13FE	14;04	139.22
6854	5MR	15;00	339.46	6854	21MR	2;57	174.59	6928	28FE	4;13	333.65	6928	13MR	22;57	168.14
6855	4AP	1;17	8.09	6855	19AP	14;05	203.03	6929	28MR	21;12	2.65	6929	12AP	6;54	196.54
6856	3MY	12;36	36.36	6856	18MY	23;02	231.08	6930	27AP	12;14	31.14	6930	11MY	14;39	224.60
6857	2JN	1;32	64.50	6857	17JN	6;45	258.99	6931	27MY	0;59	59.29	6931	9JN	23;15	252.54
6858	1JL	16;15	92.73	6858	16JL	14;06	287.04	6932	25JN	11;50	87.35	6932	9JL	9;48	280.66
6859	31JL	8;13	121.28	6859	14AU	21;55	315.45	6933	24JL	21;33	115.57	6933	7AU	21;09	309.21
6860	30AU	0;39	150.28	6860	13SE	6;59	344.40	6934	23AU	6;52	144.18	6934	6SE	15;27	338.34
6861	28SE	16;44	179.78	6861	12OC	18;08	13.93	6935	21SE	16;19	173.30	6935	6OC	9;51	8.08
6862	28OC	7;57	209.71	6862	11NO	8;00	43.97	6936	21OC	2;16	202.92	6936	5NO	4;51	38.27
6863	26NO	21;55	239.87	6863	11DE	0;36	74.28	6937	19NO	13;06	232.90	6937	4DE	22;55	68.63
6864	26DE	10;18	269.97					6938	19DE	1;11	263.01				

Band 4: **−445** (left) / **−439** (right)

NUMBER	DATE	TIME	LONG.	NUMBER	DATE	TIME	LONG.	NUMBER	DATE	TIME	LONG.	NUMBER	DATE	TIME	LONG.
				6864	9JA	18;59	104.54					6938	3JA	14;59	98.84
6865	24JA	20;59	299.75	6865	8FE	13;31	134.44	6939	17JA	14;48	292.99	6939	2FE	4;29	128.62
6866	23FE	6;12	329.01	6866	10MR	6;43	163.78	6940	16FE	5;49	322.61	6940	3MR	15;19	157.80
6867	24MR	14;35	357.74	6867	8AP	21;48	192.53	6941	17MR	21;35	351.75	6941	1AP	23;54	186.39
6868	22AP	23;03	26.03	6868	8MY	10;35	220.82	6942	16AP	13;19	20.41	6942	1MY	7;00	214.52
6869	22MY	8;31	54.08	6869	6JN	21;17	248.86	6943	16MY	4;25	48.72	6943	30MY	13;43	242.42
6870	20JN	19;45	82.14	6870	6JL	6;23	276.89	6944	14JN	18;36	76.89	6944	28JN	21;16	270.36
6871	20JL	9;09	110.45	6871	4AU	14;39	305.18	6945	14JL	7;49	105.15	6945	28JL	6;47	298.61
6872	19AU	0;41	139.24	6872	2SE	22;56	333.92	6946	12AU	20;03	133.72	6946	26AU	19;00	327.39
6873	17SE	18;00	168.59	6873	2OC	8;09	3.19	6947	11SE	7;22	162.73	6947	25SE	10;12	356.80
6874	17OC	12;13	198.49	6874	31OC	18;59	32.98	6948	10OC	18;03	192.21	6948	25OC	4;00	26.80
6875	16NO	6;04	228.74	6875	30NO	7;45	63.10	6949	9NO	4;29	222.08	6949	23NO	23;25	57.17
6876	15DE	22;08	259.02	6876	29DE	22;19	93.28	6950	8DE	15;07	252.14	6950	23DE	18;54	87.59

Band 5: **−444** (left) / **−438** (right)

NUMBER	DATE	TIME	LONG.	NUMBER	DATE	TIME	LONG.	NUMBER	DATE	TIME	LONG.	NUMBER	DATE	TIME	LONG.
6877	14JA	11;35	289.03	6877	28JA	14;10	123.23	6951	7JA	2;15	282.11	6951	22JA	12;42	117.69
6878	12FE	22;23	318.55	6878	27FE	6;40	152.74	6952	5FE	14;00	311.77	6952	21FE	3;29	147.22
6879	13MR	7;12	347.47	6879	27MR	21;31	181.73	6953	7MR	2;20	340.96	6953	22MR	14;59	176.11
6880	11AP	14;53	15.87	6880	26AP	14;57	210.26	6954	5AP	15;19	9.68	6954	20AP	23;47	204.44
6881	10MY	22;19	43.92	6881	26MY	5;17	238.48	6955	5MY	5;07	38.04	6955	20MY	6;58	232.40
6882	9JN	6;22	71.84	6882	24JN	17;49	266.60	6956	3JN	19;43	66.24	6956	18JN	13;43	260.28
6883	8JL	15;51	99.92	6883	24JL	4;42	294.86	6957	3JL	11;09	94.50	6957	17JL	21;05	288.34
6884	7AU	3;34	128.39	6884	22AU	14;34	323.49	6958	2AU	2;31	123.05	6958	16AU	5;54	316.81
6885	5SE	18;10	157.43	6885	21SE	0;15	352.60	6959	31AU	17;09	152.00	6959	14SE	16;52	345.86
6886	5OC	11;44	187.12	6886	20OC	10;22	22.22	6960	30SE	6;41	181.43	6960	14OC	6;33	15.52
6887	4NO	7;17	217.33	6887	18NO	21;13	52.19	6961	29OC	19;08	211.26	6961	12NO	23;09	45.69
6888	4DE	2;50	247.75	6888	18DE	8;47	82.29	6962	28NO	6;45	241.32	6962	12DE	18;09	76.10
								6963	27DE	17;45	271.36				

Band 6: **−443** (left) / **−437** (right)

NUMBER	DATE	TIME	LONG.	NUMBER	DATE	TIME	LONG.	NUMBER	DATE	TIME	LONG.	NUMBER	DATE	TIME	LONG.
6889	2JA	20;23	278.03	6889	16JA	21;05	112.22					6963	11JA	13;56	106.41
6890	1FE	10;55	307.86	6890	15FE	10;14	141.77	6964	26JA	4;09	301.11	6964	10FE	8;23	136.27
6891	2MR	22;31	337.08	6891	17MR	0;27	170.86	6965	24FE	14;03	330.38	6965	12MR	0;00	165.52
6892	1AP	7;47	5.71	6892	15AP	15;35	199.51	6966	25MR	23;51	359.15	6966	10AP	12;29	194.15
6893	30AP	15;29	33.87	6893	15MY	7;05	227.85	6967	24AP	10;18	27.50	6967	9MY	22;25	222.31
6894	29MY	22;25	61.78	6894	13JN	22;09	256.05	6968	23MY	22;10	55.63	6968	8JN	6;41	250.24
6895	28JN	5;28	89.70	6895	13JL	12;17	284.34	6969	22JN	11;56	83.79	6969	7JL	14;13	278.23
6896	27JL	13;42	117.89	6896	12AU	1;25	312.93	6970	22JL	3;27	112.21	6970	5AU	21;51	306.50
6897	26AU	0;12	146.59	6897	10SE	13;46	341.97	6971	20AU	19;58	141.06	6971	4SE	6;20	335.27
6898	24SE	13;54	175.93	6898	10OC	1;36	11.49	6972	19SE	12;35	170.42	6972	3OC	16;28	4.60
6899	24OC	6;53	205.88	6899	8NO	12;58	41.40	6973	19OC	4;35	200.25	6973	2NO	4;59	34.48
6900	23NO	2;12	236.24	6900	7DE	23;51	71.46	6974	17NO	19;26	230.38	6974	1DE	20;13	64.71
6901	22DE	21;58	266.67					6975	17DE	8;43	260.55	6975	31DE	13;44	95.01

-436 / -430

NUMBER	DATE	TIME	LONG.	NUMBER	DATE	TIME	LONG.	NUMBER	DATE	TIME	LONG.	NUMBER	DATE	TIME	LONG.
6976	15JA	20:12	290.46	6976	30JA	8:10	125.04	7050	8JA	12:02	283.59	7050	24JA	2:13	119.32
6977	14FE	5:57	319.91	6977	29FE	1:58	154.57	7051	7FE	2:07	313.33	7051	22FE	14:17	148.71
6978	14MR	14:27	348.80	6978	29MR	18:01	183.52	7052	8MR	17:17	342.61	7052	23MR	23:48	177.50
6979	12AP	22:32	17.20	6979	28AP	7:54	211.95	7053	7AP	8:52	11.40	7053	22AP	7:24	205.76
6980	12MY	7:12	45.28	6980	27MY	19:39	240.06	7054	7MY	0:11	39.81	7054	21MY	14:05	233.70
6981	10JN	17:21	73.29	6981	26JN	5:37	268.08	7055	5JN	14:51	68.00	7055	19JN	21:02	261.59
6982	10JL	5:36	101.49	6982	25JL	14:24	296.28	7056	5JL	4:43	96.22	7056	19JL	5:28	289.70
6983	8AU	20:09	130.09	6983	23AU	22:46	324.86	7057	3AU	17:42	124.69	7057	17AU	16:19	318.27
6984	7SE	12:49	159.25	6984	22SE	7:37	353.97	7058	2SE	5:45	153.57	7058	16SE	6:08	347.46
6985	7OC	6:57	188.99	6985	21OC	17:42	23.60	7059	1OC	16:59	182.92	7059	15OC	22:48	17.27
6986	6NO	1:23	219.17	6986	20NO	5:30	53.62	7060	31OC	3:40	212.69	7060	14NO	17:38	47.55
6987	5DE	18:39	249.50	6987	19DE	19:07	83.80	7061	29NO	14:13	242.72	7061	14DE	13:20	77.99
								7062	29DE	1:01	272.74				

-435 / -429

NUMBER	DATE	TIME	LONG.	NUMBER	DATE	TIME	LONG.	NUMBER	DATE	TIME	LONG.	NUMBER	DATE	TIME	LONG.
6988	4JA	9:29	279.65	6988	18JA	10:10	113.83					7062	13JA	8:08	108.24
6989	2FE	21:29	309.35	6989	17FE	2:08	143.48	7063	27JA	12:15	302.51	7063	12FE	0:27	137.99
6990	4MR	7:04	338.48	6990	18MR	18:27	172.63	7064	26FE	0:01	331.84	7064	13MR	13:29	167.11
6991	2AP	15:02	7.03	6991	17AP	10:30	201.29	7065	27MR	12:21	0.68	7065	11AP	23:27	195.61
6992	1MY	22:21	35.16	6992	17MY	1:35	229.59	7066	26AP	1:28	29.12	7066	11MY	7:15	223.67
6993	31MY	5:53	63.09	6993	15JN	15:09	257.74	7067	25MY	15:34	57.33	7067	9JN	14:05	251.55
6994	29JN	14:30	91.07	6994	15JL	3:01	285.95	7068	24JN	6:38	85.55	7068	8JL	21:05	279.52
6995	29JL	1:02	119.38	6995	13AU	13:33	314.46	7069	23JL	22:15	113.99	7069	7AU	5:10	307.83
6996	27AU	14:12	148.21	6996	11SE	23:30	343.42	7070	22AU	13:37	142.82	7070	5SE	15:06	336.68
6997	26SE	6:27	177.68	6997	11OC	9:33	12.89	7071	21SE	4:02	172.12	7071	5OC	3:27	6.13
6998	26OC	1:21	207.73	6998	9NO	20:09	42.78	7072	20OC	17:16	201.86	7072	3NO	18:39	36.13
6999	24NO	21:17	238.13	6999	9DE	7:21	72.85	7073	19NO	5:26	231.88	7073	3DE	12:32	66.48
7000	24DE	16:04	268.51					7074	18DE	16:46	261.97				

-434 / -428

NUMBER	DATE	TIME	LONG.	NUMBER	DATE	TIME	LONG.	NUMBER	DATE	TIME	LONG.	NUMBER	DATE	TIME	LONG.
				7000	7JA	19:08	102.85					7074	2JA	8:04	96.85
7001	23JA	8:08	298.53	7001	6FE	7:34	132.53	7075	17JA	3:23	291.83	7075	1FE	3:17	126.90
7002	21FE	21:04	327.98	7002	7MR	20:55	161.75	7076	15FE	13:18	321.26	7076	1MR	20:18	156.37
7003	23MR	7:18	356.80	7003	6AP	11:20	190.51	7077	15MR	22:50	350.18	7077	31MR	10:15	185.20
7004	21AP	15:36	25.10	7004	6MY	2:33	218.92	7078	14AP	8:35	18.62	7078	29AP	21:19	213.50
7005	20MY	22:45	53.06	7005	4JN	17:54	247.14	7079	13MY	19:26	46.78	7079	29MY	6:21	241.50
7006	19JN	5:37	80.94	7006	4JL	8:40	275.39	7080	12JN	8:06	74.89	7080	27JN	14:14	269.45
7007	18JL	13:09	109.01	7007	2AU	22:32	303.88	7081	11JL	22:48	103.20	7081	26JL	21:52	297.61
7008	16AU	22:28	137.52	7008	1SE	11:32	332.79	7082	10AU	15:05	131.89	7082	25AU	6:00	326.20
7009	15SE	10:35	166.62	7009	30SE	23:54	2.17	7083	9SE	8:01	161.10	7083	23SE	15:21	355.35
7010	15OC	2:03	196.37	7010	30OC	11:43	31.99	7084	9OC	0:42	190.80	7084	23OC	2:41	25.05
7011	13NO	20:26	226.62	7011	28NO	22:56	62.04	7085	7NO	16:25	220.88	7085	21NO	16:32	55.17
7012	13DE	16:12	257.06	7012	28DE	9:34	92.06	7086	7DE	6:39	251.09	7086	21DE	8:54	85.46

-433 / -427

NUMBER	DATE	TIME	LONG.	NUMBER	DATE	TIME	LONG.	NUMBER	DATE	TIME	LONG.	NUMBER	DATE	TIME	LONG.
7013	12JA	11:23	287.34	7013	26JA	19:51	121.79	7087	5JA	19:03	281.13	7087	20JA	2:51	115.59
7014	11FE	4:30	317.14	7014	25FE	6:16	151.08	7088	4FE	5:31	310.75	7088	18FE	20:55	145.30
7015	12MR	18:50	346.32	7015	26MR	17:22	179.88	7089	5MR	14:23	339.82	7089	20MR	13:45	174.44
7016	11AP	6:21	14.89	7016	25AP	5:34	208.30	7090	3AP	22:21	8.35	7090	19AP	4:38	203.03
7017	10MY	15:27	43.01	7017	24MY	18:58	236.49	7091	3MY	6:25	36.50	7091	18MY	17:27	231.24
7018	8JN	22:54	70.92	7018	23JN	9:26	264.68	7092	1JN	15:34	64.49	7092	17JN	4:23	259.28
7019	8JL	5:43	98.88	7019	23JL	0:47	293.10	7093	1JL	2:39	92.58	7093	16JL	13:51	287.41
7020	6AU	13:05	127.15	7020	21AU	16:04	321.94	7094	30JL	16:05	121.01	7094	14AU	22:33	315.86
7021	4SE	22:04	155.94	7021	20SE	8:31	351.28	7095	29AU	7:51	149.98	7095	13SE	7:16	344.80
7022	4OC	9:30	185.35	7022	19OC	23:36	21.08	7096	28SE	1:34	179.54	7096	12OC	16:48	14.26
7023	2NO	23:41	215.30	7023	18NO	13:12	51.16	7097	27OC	20:15	209.60	7097	11NO	3:46	44.17
7024	2DE	16:20	245.59	7024	18DE	1:08	81.27	7098	26NO	14:26	239.94	7098	10DE	16:26	74.32
								7099	26DE	6:40	270.20				

-432 / -426

NUMBER	DATE	TIME	LONG.	NUMBER	DATE	TIME	LONG.	NUMBER	DATE	TIME	LONG.	NUMBER	DATE	TIME	LONG.
7025	1JA	10:36	275.92	7025	16JA	11:37	111.14					7099	9JA	6:39	104.41
7026	31JA	5:18	305.95	7026	14FE	21:10	140.57	7100	24JA	20:02	300.09	7100	7FE	21:57	134.18
7027	29FE	23:03	335.47	7027	15MR	6:15	169.47	7101	23FE	6:38	329.42	7101	9MR	13:52	163.48
7028	30MR	14:40	4.38	7028	13AP	15:26	197.91	7102	24MR	15:09	358.15	7102	8AP	5:53	192.28
7029	29AP	3:31	32.76	7029	13MY	1:16	226.03	7103	22AP	22:33	26.39	7103	7MY	21:27	220.68
7030	28MY	13:45	60.81	7030	11JN	12:25	254.08	7104	22MY	5:45	54.35	7104	6JN	11:53	248.87
7031	26JN	22:11	88.77	7031	11JL	1:33	282.31	7105	20JN	13:40	82.28	7105	6JL	0:46	277.05
7032	26JL	5:57	116.94	7032	9AU	16:59	310.97	7106	19JL	23:09	110.44	7106	4AU	12:09	305.47
7033	24AU	14:06	145.52	7033	8SE	10:22	340.17	7107	18AU	10:58	139.06	7107	2SE	22:34	334.29
7034	22SE	23:24	174.65	7034	8OC	4:31	9.93	7108	17SE	1:46	168.31	7108	2OC	8:44	3.61
7035	22OC	10:16	204.33	7035	6NO	21:49	40.06	7109	16OC	19:35	198.18	7109	31OC	19:11	33.38
7036	20NO	22:53	234.39	7036	6DE	13:11	70.32	7110	15NO	15:22	228.50	7110	30NO	6:08	63.42
7037	20DE	13:23	264.60					7111	15DE	11:03	258.94	7111	29DE	17:32	93.47

-431 / -425

NUMBER	DATE	TIME	LONG.	NUMBER	DATE	TIME	LONG.	NUMBER	DATE	TIME	LONG.	NUMBER	DATE	TIME	LONG.
				7037	5JA	2:18	100.39					7112	28JA	5:24	123.25
7038	19JA	5:38	294.68	7038	3FE	13:21	130.04	7112	14JA	4:35	289.13	7113	26FE	17:58	152.60
7039	17FE	23:01	324.37	7039	4MR	22:45	159.15	7113	12FE	18:59	318.79	7114	28MR	7:32	181.48
7040	19MR	16:22	353.54	7040	3AP	6:54	187.70	7114	14MR	6:24	347.82	7115	26AP	22:10	209.98
7041	18AP	8:21	22.19	7041	2MY	14:29	215.84	7115	12AP	15:28	16.28	7116	26MY	13:26	238.23
7042	17MY	22:14	50.44	7042	31MY	22:23	243.78	7116	11MY	23:02	44.33	7117	25JN	4:37	266.46
7043	16JN	10:04	78.52	7043	30JN	7:46	271.79	7117	10JN	5:55	72.21	7118	24JL	19:09	294.87
7044	15JL	20:27	106.68	7044	29JL	19:39	300.16	7118	9JL	13:02	100.19	7119	23AU	8:52	323.64
7045	14AU	6:08	135.16	7045	28AU	10:37	329.07	7119	7AU	21:23	128.52	7120	21SE	21:50	352.89
7046	12SE	15:42	164.12	7046	27SE	4:17	358.62	7120	6SE	8:05	157.40	7121	21OC	10:11	22.60
7047	12OC	1:33	193.59	7047	26OC	23:22	28.69	7121	5OC	16:56	186.93	7122	19NO	21:51	52.61
7048	10NO	11:59	223.48	7048	25NO	18:13	59.05	7122	4NO	15:02	217.03	7123	19DE	8:46	82.67
7049	9DE	23:22	253.56	7049	25DE	11:26	89.36	7123	4DE	10:20	247.44				

-424

NEW NUMBER	DATE	TIME	LONG.	FULL NUMBER	DATE	TIME	LONG.
7124	3JA	5:58	277.81	7124	17JA	19:03	112.52
7125	2FE	0:09	307.79	7125	16FE	5:06	141.95
7126	2MR	15:48	337.20	7126	16MR	15:32	170.88
7127	1AP	4:38	5.97	7127	15AP	2:52	199.39
7128	30AP	14:51	34.23	7128	14MY	15:27	227.61
7129	29MY	23:02	62.19	7129	13JN	5:18	255.77
7130	28JN	6:06	90.11	7130	12JL	20:18	284.09
7131	27JL	13:10	118.26	7131	11AU	12:10	312.78
7132	25AU	21:21	146.86	7132	10SE	4:25	341.97
7133	24SE	7:36	176.05	7133	9OC	20:17	11.66
7134	23OC	20:29	205.83	7134	8NO	10:55	41.70
7135	22NO	11:57	236.03	7135	7DE	23:47	71.84
7136	22DE	5:25	266.36				

-423

NEW NUMBER	DATE	TIME	LONG.	FULL NUMBER	DATE	TIME	LONG.
				7136	6JA	10:55	101.82
7137	20JA	23:52	296.50	7137	4FE	20:44	131.41
7138	19FE	18:04	326.20	7138	6MR	5:46	160.47
7139	21MR	10:44	355.32	7139	4AP	14:37	189.04
7140	20AP	0:56	23.88	7140	3MY	23:50	217.22
7141	19MY	12:26	52.02	7141	2JN	10:05	245.25
7142	17JN	21:45	80.00	7142	1JL	22:05	273.38
7143	17JL	5:52	108.08	7143	31JL	12:24	301.86
7144	15AU	13:55	136.52	7144	30AU	5:05	330.88
7145	13SE	22:45	165.47	7145	28SE	23:17	0.47
7146	13OC	8:56	194.97	7146	28OC	17:26	30.53
7147	11NO	20:42	224.92	7147	27NO	10:02	60.80
7148	11DE	10:11	255.10	7148	27DE	0:20	90.98

-422

NEW NUMBER	DATE	TIME	LONG.	FULL NUMBER	DATE	TIME	LONG.
7149	10JA	1:23	285.23	7149	25JA	12:19	120.80
7150	8FE	18:03	315.04	7150	23FE	22:22	150.09
7151	10MR	11:19	344.38	7151	25MR	6:54	178.81
7152	9AP	3:56	13.19	7152	23AP	14:31	207.06
7153	8MY	18:52	41.57	7153	22MY	22:00	235.03
7154	7JN	7:45	69.69	7154	21JN	6:25	262.97
7155	6JL	18:58	97.81	7155	20JL	16:55	291.18
7156	5AU	5:09	126.18	7156	19AU	6:23	319.89
7157	3SE	14:58	154.99	7157	17SE	22:56	349.22
7158	3OC	0:51	184.31	7158	17OC	17:41	19.14
7159	1NO	11:05	214.08	7159	16NO	13:00	49.45
7160	30NO	21:58	244.12	7160	16DE	7:15	79.83
7161	30DE	9:50	274.19				

-421

NEW NUMBER	DATE	TIME	LONG.	FULL NUMBER	DATE	TIME	LONG.
				7161	14JA	23:19	109.94
7162	28JA	22:59	304.01	7162	13FE	12:40	139.55
7163	27FE	13:22	333.43	7163	14MR	23:18	168.54
7164	29MR	4:32	2.36	7164	13AP	7:38	196.96
7165	27AP	19:50	30.87	7165	12MY	14:33	224.98
7166	27MY	10:48	59.11	7166	10JN	21:11	252.85
7167	26JN	1:11	87.31	7167	10JL	4:45	280.85
7168	25JL	14:50	115.69	7168	8AU	14:20	309.24
7169	24AU	3:40	144.44	7169	7SE	2:44	338.20
7170	22SE	15:35	173.66	7170	6OC	18:06	7.81
7171	22OC	2:43	203.32	7171	5NO	12:03	37.95
7172	20NO	13:24	233.30	7172	5DE	7:32	68.37
7173	20DE	0:01	263.35				

-420

NEW NUMBER	DATE	TIME	LONG.	FULL NUMBER	DATE	TIME	LONG.
				7173	4JA	2:58	98.72
7174	18JA	10:52	293.21	7174	2FE	20:37	128.68
7175	16FE	22:07	322.68	7175	3MR	11:15	158.02
7176	17MR	9:51	351.66	7176	1AP	22:35	186.72
7177	15AP	22:16	20.19	7177	1MY	7:17	214.92
7178	15MY	11:38	48.44	7178	30MY	14:27	242.83
7179	14JN	2:11	76.62	7179	28JN	21:17	270.74
7180	13JL	17:42	104.97	7180	28JL	4:49	298.92
7181	12AU	9:32	133.68	7181	26AU	13:52	327.57
7182	11SE	0:49	162.84	7182	25SE	1:03	356.81
7183	10OC	14:57	192.47	7183	24OC	14:51	26.63
7184	9NO	3:50	222.44	7184	23NO	7:26	56.88
7185	8DE	15:39	252.55	7185	23DE	2:14	87.27

-419

NEW NUMBER	DATE	TIME	LONG.	FULL NUMBER	DATE	TIME	LONG.
7186	7JA	2:33	282.51	7186	21JA	21:43	117.45
7187	5FE	12:39	312.10	7187	20FE	15:52	147.13
7188	6MR	22:05	341.17	7188	22MR	7:17	176.18
7189	5AP	7:22	9.74	7189	20AP	19:40	204.65
7190	4MY	17:20	37.95	7190	20MY	5:38	232.73
7191	3JN	4:52	66.03	7191	18JN	14:04	260.68
7192	2JL	18:31	94.23	7192	17JL	21:52	288.76
7193	1AU	10:11	122.77	7193	16AU	5:50	317.19
7194	31AU	3:08	151.81	7194	14SE	14:39	346.15
7195	29SE	20:20	181.38	7195	14OC	1:01	15.68
7196	29OC	12:52	211.38	7196	12NO	13:35	45.67
7197	28NO	4:04	241.60	7197	12DE	4:40	75.91
7198	27DE	17:27	271.73				

-418

NEW NUMBER	DATE	TIME	LONG.	FULL NUMBER	DATE	TIME	LONG.
				7198	10JA	21:47	106.11
7199	26JA	4:48	301.52	7199	9FE	15:44	135.97
7200	24FE	14:15	330.78	7200	11MR	9:05	165.30
7201	25MR	22:22	359.47	7201	10AP	0:51	194.06
7202	24AP	6:04	27.71	7202	9MY	14:41	222.38
7203	23MY	14:24	55.71	7203	8JN	2:36	250.47
7204	22JN	0:20	83.72	7204	7JL	12:55	278.56
7205	21JL	12:33	112.00	7205	5AU	22:08	306.90
7206	20AU	3:16	140.77	7206	4SE	6:57	335.68
7207	18SE	20:16	170.13	7207	3OC	16:10	4.98
7208	18OC	14:49	200.06	7208	2NO	2:29	34.76
7209	17NO	9:38	230.35	7209	1DE	14:17	64.85
7210	17DE	3:06	260.70	7210	31DE	3:38	94.97

-417

NEW NUMBER	DATE	TIME	LONG.	FULL NUMBER	DATE	TIME	LONG.
7211	15JA	17:56	290.76	7211	29JA	18:11	124.84
7212	14FE	5:47	320.29	7212	28FE	9:33	154.29
7213	15MR	15:06	349.21	7213	30MR	1:18	183.23
7214	13AP	22:48	17.60	7214	28AP	17:01	211.74
7215	13MY	5:52	45.62	7215	28MY	8:04	239.98
7216	11JN	13:16	73.51	7216	26JN	21:54	268.16
7217	10JL	21:51	101.56	7217	26JL	10:14	296.50
7218	9AU	8:26	130.00	7218	24AU	21:20	325.19
7219	7SE	21:47	159.02	7219	23SE	7:49	354.36
7220	7OC	14:15	188.68	7220	22OC	18:16	24.01
7221	6NO	9:22	218.88	7221	21NO	5:04	53.99
7222	6DE	5:26	249.33	7222	20DE	16:12	84.06

-416

NEW NUMBER	DATE	TIME	LONG.	FULL NUMBER	DATE	TIME	LONG.
7223	5JA	0:15	279.65	7223	19JA	3:38	113.94
7224	3FE	16:12	309.52	7224	17FE	15:33	143.43
7225	4MR	4:58	338.77	7225	18MR	4:16	172.44
7226	2AP	15:01	7.42	7226	16AP	18:06	201.02
7227	1MY	23:10	35.58	7227	16MY	8:57	229.32
7228	31MY	6:16	63.49	7228	15JN	0:16	257.53
7229	29JN	13:10	91.41	7229	14JL	15:20	285.87
7230	28JL	20:50	119.59	7230	13AU	5:43	314.53
7231	27AU	6:20	148.27	7231	11SE	19:22	343.64
7232	25SE	18:37	177.57	7232	11OC	8:18	13.23
7233	25OC	10:13	207.48	7233	9NO	20:30	43.18
7234	24NO	4:37	237.81	7234	9DE	7:51	73.27
7235	24DE	0:16	268.23				

-415

NEW NUMBER	DATE	TIME	LONG.	FULL NUMBER	DATE	TIME	LONG.
				7235	7JA	18:21	103.21
7236	22JA	19:14	298.38	7236	6FE	4:16	132.78
7237	21FE	12:06	327.99	7237	7MR	14:11	161.86
7238	23MR	2:16	356.98	7238	6AP	0:43	190.47
7239	21AP	13:41	25.41	7239	5MY	12:25	218.74
7240	20MY	22:48	53.45	7240	4JN	1:29	246.88
7241	19JN	6:23	81.36	7241	3JL	15:56	275.11
7242	18JL	13:25	109.41	7242	2AU	7:32	303.67
7243	16AU	21:03	137.85	7243	31AU	23:56	332.70
7244	15SE	6:19	166.83	7244	30SE	16:24	2.25
7245	14OC	17:56	196.42	7245	30OC	8:01	32.22
7246	13NO	8:10	226.49	7246	28NO	21:57	62.38
7247	13DE	0:39	256.79	7247	28DE	9:56	92.46

-414

NEW NUMBER	DATE	TIME	LONG.	FULL NUMBER	DATE	TIME	LONG.
7248	11JA	18:34	287.02	7248	26JA	20:15	122.19
7249	10FE	12:50	316.87	7249	25FE	5:26	151.43
7250	12MR	6:13	346.19	7250	26MR	14:05	180.13
7251	10AP	21:38	14.93	7251	24AP	22:49	208.41
7252	10MY	10:28	43.20	7252	24MY	8:18	236.45
7253	8JN	20:53	71.22	7253	22JN	19:14	264.50
7254	8JL	5:37	99.26	7254	22JL	8:21	292.83
7255	6AU	13:46	127.57	7255	21AU	0:00	321.64
7256	4SE	22:19	156.34	7256	19SE	17:49	351.06
7257	4OC	7:57	185.67	7257	19OC	12:27	21.00
7258	2NO	18:59	215.49	7258	18NO	6:11	51.25
7259	2DE	7:34	245.61	7259	17DE	21:46	81.52
7260	31DE	21:46	275.76				

-413

NEW NUMBER	DATE	TIME	LONG.	FULL NUMBER	DATE	TIME	LONG.
				7260	16JA	10:52	111.49
7261	30JA	13:30	305.68	7261	14FE	21:43	140.98
7262	1MR	6:19	335.17	7262	16MR	6:48	169.88
7263	30MR	23:11	4.14	7263	14AP	14:38	198.26
7264	29AP	14:56	32.65	7264	13MY	21:56	226.28
7265	29MY	4:52	60.84	7265	12JN	5:39	254.20
7266	27JN	17:00	88.95	7266	11JL	14:57	282.28
7267	27JL	3:50	117.24	7267	10AU	2:54	310.78
7268	25AU	14:02	145.91	7268	8SE	18:03	339.89
7269	24SE	0:06	175.07	7269	8OC	12:00	9.62
7270	23OC	10:18	204.72	7270	7NO	7:24	39.84
7271	21NO	20:52	234.69	7271	7DE	2:27	70.25
7272	21DE	8:08	264.77				

-412

NUMBER	DATE	TIME	LONG.		NUMBER	DATE	TIME	LONG.
7273	19JA	20;25	294.68		7272	5JA	19;44	100.50
7274	18FE	9;56	324.22		7273	4FE	10;26	130.31
7275	19MR	0;28	353.28		7274	4MR	22;18	159.51
7276	17AP	15;31	21.90		7275	3AP	7;35	188.11
7277	17MY	6;35	50.21		7276	2MY	15;00	216.24
7278	15JN	21;18	78.40		7277	31MY	21;35	244.13
7279	15JL	11;31	106.71		7278	30JN	4;32	272.05
7280	14AU	1;03	135.34		7279	29JL	13;02	300.28
7281	12SE	13;44	164.43		7280	28AU	0;02	329.03
7282	12OC	1;30	193.98		7281	26SE	14;01	358.41
7283	10NO	12;31	223.89		7282	26OC	6;51	28.38
7284	9DE	23;09	253.94		7283	25NO	1;46	58.74
					7284	24DE	21;24	89.15

-411

NUMBER	DATE	TIME	LONG.		NUMBER	DATE	TIME	LONG.
7285	8JA	9;46	283.88		7285	23JA	16;04	119.28
7286	6FE	20;35	313.48		7286	22FE	8;12	148.85
7287	8MR	7;47	342.60		7287	23MR	21;05	177.77
7288	6AP	19;32	11.25		7288	22AP	6;56	206.12
7289	6MY	8;09	39.55		7289	21MY	14;43	234.11
7290	4JN	21;58	67.72		7290	19JN	21;37	261.99
7291	4JL	13;06	95.99		7291	19JL	4;46	290.05
7292	3AU	5;04	124.57		7292	17AU	13;05	318.53
7293	1SE	21;01	153.59		7293	15SE	23;15	347.57
7294	1OC	12;05	183.10		7294	15OC	11;47	17.20
7295	31OC	1;49	213.00		7295	14NO	2;59	47.31
7296	29NO	14;16	243.11		7296	13DE	20;44	77.67
7297	29DE	1;37	273.15					

-410

NUMBER	DATE	TIME	LONG.		NUMBER	DATE	TIME	LONG.
7298	27JA	12;00	302.88		7297	12JA	15;58	107.95
7299	25FE	21;31	332.12		7298	11FE	10;50	137.81
7300	27MR	6;33	0.83		7299	13MR	3;35	167.09
7301	25AP	15;50	29.12		7300	11AP	17;23	195.75
7302	25MY	2;17	57.19		7301	11MY	4;28	223.95
7303	23JN	14;45	85.31		7302	9JN	13;37	251.92
7304	23JL	5;29	113.71		7303	8JL	21;46	279.93
7305	21AU	22;05	142.58		7304	7AU	5;44	308.24
7306	20SE	15;33	171.99		7305	5SE	14;13	337.03
7307	20OC	8;46	201.89		7306	4OC	23;51	6.37
7308	19NO	0;55	232.08		7307	3NO	11;19	36.21
7309	18DE	15;21	262.29		7308	3DE	1;06	66.38

-409

NUMBER	DATE	TIME	LONG.		NUMBER	DATE	TIME	LONG.
7310	17JA	3;41	292.23		7309	1JA	17;09	96.61
7311	15FE	13;55	321.68		7310	31JA	10;36	126.59
7312	16MR	22;26	350.54		7311	2MR	4;09	156.09
7313	15AP	6;01	18.91		7312	31MR	20;37	185.04
7314	14MY	13;45	46.94		7313	30AP	11;21	213.49
7315	12JN	22;40	74.90		7314	30MY	0;16	241.64
7316	12JL	9;39	103.05		7315	28JN	11;29	269.72
7317	10AU	23;10	131.63		7316	27JL	21;24	297.97
7318	9SE	15;12	160.79		7317	26AU	6;34	326.61
7319	9OC	9;17	190.55		7318	24SE	15;42	355.75
7320	8NO	4;21	220.76		7319	24OC	1;33	25.39
7321	7DE	22;48	251.14		7320	22NO	12;37	55.39
					7321	22DE	1;07	85.52

-408

NUMBER	DATE	TIME	LONG.		NUMBER	DATE	TIME	LONG.
7322	6JA	15;06	281.35		7322	20JA	14;54	115.48
7323	5FE	4;22	311.08		7323	19FE	5;36	145.06
7324	5MR	14;44	340.22		7324	19MR	20;55	174.14
7325	3AP	22;59	8.76		7325	18AP	12;29	202.77
7326	3MY	6;08	36.87		7326	18MY	3;52	231.08
7327	1JN	13;11	64.78		7327	16JN	18;27	259.27
7328	30JN	21;03	92.74		7328	16JL	7;45	287.55
7329	30JL	6;34	121.01		7329	14AU	19;42	316.13
7330	28AU	18;32	149.82		7330	13SE	6;41	345.16
7331	27SE	9;32	179.25		7331	12OC	17;19	14.67
7332	27OC	3;33	209.29		7332	11NO	4;04	44.58
7333	25NO	23;29	239.69		7333	10DE	15;03	74.65
7334	25DE	19;12	270.11					

-407

NUMBER	DATE	TIME	LONG.		NUMBER	DATE	TIME	LONG.
7335	24JA	12;39	300.17		7334	9JA	2;12	104.61
7336	23FE	2;54	329.65		7335	7FE	13;36	134.22
7337	24MR	14;07	358.49		7336	9MR	1;34	163.36
7338	22AP	23;03	26.80		7337	7AP	14;32	192.05
7339	22MY	6;32	54.77		7338	7MY	4;41	220.40
7340	20JN	13;27	82.65		7339	5JN	19;46	248.61
7341	19JL	20;41	110.72		7340	5JL	11;08	276.90
7342	18AU	5;14	139.22		7341	4AU	2;07	305.45
7343	16SE	16;07	168.29		7342	2SE	16;26	334.42
7344	16OC	6;07	198.00		7343	2OC	6;01	3.88
7345	14NO	23;15	228.20		7344	31OC	18;50	33.75
7346	14DE	18;28	258.62		7345	30NO	6;44	63.84
					7346	29DE	17;37	93.86

-406

NUMBER	DATE	TIME	LONG.		NUMBER	DATE	TIME	LONG.
7347	13JA	13;53	288.90		7347	28JA	3;38	123.56
7348	12FE	7;47	318.71		7348	26FE	13;14	152.79
7349	13MR	23;13	347.92		7349	27MR	23;06	181.53
7350	12AP	11;56	16.52		7350	26AP	9;55	209.87
7351	11MY	22;08	44.68		7351	25MY	22;06	238.01
7352	10JN	6;25	72.61		7352	24JN	11;48	266.17
7353	9JL	13;42	100.60		7353	24JL	2;57	294.60
7354	7AU	21;03	128.89		7354	22AU	19;13	323.47
7355	6SE	5;32	157.69		7355	21SE	12;03	352.87
7356	5OC	16;02	187.07		7356	21OC	4;30	22.75
7357	4NO	5;01	216.99		7357	19NO	19;32	52.90
7358	3DE	20;23	247.23		7358	19DE	8;34	83.05

-405

NUMBER	DATE	TIME	LONG.		NUMBER	DATE	TIME	LONG.
7359	2JA	13;32	277.50		7359	17JA	19;37	112.93
7360	1FE	7;33	307.49		7360	16FE	5;08	142.33
7361	3MR	1;19	336.99		7361	17MR	13;46	171.20
7362	1AP	17;41	5.91		7362	15AP	22;11	199.58
7363	1MY	7;48	34.33		7363	15MY	7;01	227.66
7364	30MY	19;26	62.43		7364	13JN	17;00	255.66
7365	29JN	5;02	90.45		7365	13JL	4;55	283.86
7366	28JL	13;32	118.66		7366	11AU	19;22	312.48
7367	26AU	21;59	147.27		7367	10SE	12;23	341.69
7368	25SE	7;13	176.42		7368	10OC	7;03	11.49
7369	24OC	17;39	206.10		7369	9NO	1;38	41.69
7370	23NO	5;28	236.14		7370	8DE	18;32	72.01
7371	22DE	18;44	266.29					

-404

NUMBER	DATE	TIME	LONG.		NUMBER	DATE	TIME	LONG.
7372	21JA	9;30	296.29		7371	7JA	8;53	102.13
7373	20FE	1;34	325.91		7372	5FE	20;44	131.79
7374	20MR	18;16	355.04		7373	6MR	6;30	160.89
7375	19AP	10;32	23.68		7374	4AP	14;44	189.43
7376	19MY	1;24	51.97		7375	3MY	22;04	217.54
7377	17JN	14;31	80.10		7376	2JN	5;21	245.54
7378	17JL	2;08	108.32		7377	1JL	13;40	273.43
7379	15AU	12;51	136.86		7378	31JL	0;12	301.76
7380	13SE	23;11	165.87		7379	29AU	13;48	330.64
7381	13OC	9;30	195.38		7380	28SE	6;36	0.16
7382	11NO	19;58	225.28		7381	28OC	1;37	30.25
7383	11DE	6;50	255.35		7382	26NO	21;09	60.64
					7383	26DE	15;30	90.99

-403

NUMBER	DATE	TIME	LONG.		NUMBER	DATE	TIME	LONG.
7384	9JA	18;25	285.32		7384	25JA	7;31	120.99
7385	8FE	7;03	314.97		7385	23FE	20;42	150.41
7386	9MR	20;48	344.17		7386	25MR	7;07	179.21
7387	8AP	11;22	12.91		7387	23AP	15;17	207.48
7388	8MY	2;17	41.29		7388	22MY	22;05	235.41
7389	6JN	17;11	69.49		7389	21JN	4;41	263.29
7390	6JL	7;48	97.75		7390	20JL	12;19	291.38
7391	4AU	21;57	126.28		7391	18AU	22;03	319.93
7392	3SE	11;24	155.23		7392	17SE	10;37	349.09
7393	2OC	23;54	184.66		7393	17OC	2;09	18.87
7394	1NO	11;28	214.49		7394	15NO	20;11	49.12
7395	30NO	22;19	244.53		7395	15DE	15;38	79.55
7396	30DE	8;51	274.53					

-402

NUMBER	DATE	TIME	LONG.		NUMBER	DATE	TIME	LONG.
7397	28JA	19;23	304.25		7396	14JA	10;56	109.81
7398	27FE	6;07	333.52		7397	13FE	4;23	139.59
7399	28MR	17;15	2.29		7398	14MR	18;49	168.74
7400	27AP	5;08	30.67		7399	13AP	6;03	197.28
7401	26MY	18;09	58.83		7400	12MY	14;43	225.37
7402	25JN	8;36	87.03		7401	10JN	21;56	253.26
7403	25JL	0;22	115.49		7402	10JL	4;55	281.24
7404	23AU	16;42	144.38		7403	8AU	12;41	309.56
7405	22SE	8;36	173.75		7404	6SE	21;59	338.40
7406	21OC	23;19	203.57		7405	6OC	9;22	7.83
7407	20NO	12;34	233.65		7406	4NO	23;15	37.79
7408	20DE	0;30	263.76		7407	4DE	15;44	68.08

-401

NUMBER	DATE	TIME	LONG.		NUMBER	DATE	TIME	LONG.
7409	18JA	11;16	293.61		7408	3JA	10;16	98.40
7410	16FE	21;01	323.02		7409	2FE	5;23	128.43
7411	18MR	6;00	351.89		7410	3MR	23;11	157.91
7412	16AP	14;49	20.28		7411	2AP	14;23	186.78
7413	16MY	0;21	48.38		7412	2MY	2;44	215.12
7414	14JN	11;36	76.43		7413	31MY	12;48	243.15
7415	14JL	1;12	104.70		7414	29JN	21;29	271.13
7416	12AU	17;05	133.40		7415	29JL	5;37	299.33
7417	11SE	10;28	162.64		7416	27AU	13;56	327.96
7418	11OC	4;12	192.40		7417	25SE	23;05	357.12
7419	9NO	21;12	222.55		7418	25OC	9;40	26.80
7420	9DE	12;41	252.81		7419	23NO	22;14	56.88
					7420	23DE	13;05	87.10

-400

NUMBER	DATE	TIME	LONG.		NUMBER	DATE	TIME	LONG.
7421	8JA	2:06	282.88		7421	22JA	5:45	117.17
7422	6FE	13:17	312.51		7422	20FE	23:09	146.83
7423	6MR	22:25	341.57		7423	21MR	16:02	175.95
7424	5AP	6:10	10.08		7424	20AP	7:33	204.55
7425	4MY	13:32	38.18		7425	19MY	21:22	232.79
7426	2JN	21:36	66.12		7426	18JN	9:32	260.88
7427	2JL	7:25	94.17		7427	17JL	20:15	289.07
7428	31JL	19:39	122.57		7428	16AU	5:57	317.59
7429	30AU	10:33	151.52		7429	14SE	15:15	346.57
7430	29SE	3:52	181.08		7430	14OC	0:51	16.06
7431	28OC	22:47	211.17		7431	12NO	11:20	45.96
7432	27NO	17:53	241.55		7432	11DE	23:05	76.07
7433	27DE	11:28	271.87					

-399

NUMBER	DATE	TIME	LONG.		NUMBER	DATE	TIME	LONG.
					7433	10JA	12:05	106.09
7434	26JA	2:16	301.80		7434	9FE	2:06	135.79
7435	24FE	13:56	331.15		7435	10MR	16:50	165.02
7436	25MR	23:00	359.88		7436	9AP	8:03	193.77
7437	24AP	6:28	28.11		7437	8MY	23:27	222.16
7438	23MY	13:22	56.05		7438	7JN	14:31	250.37
7439	21JN	20:41	83.96		7439	7JL	4:40	278.62
7440	21JL	5:18	112.10		7440	5AU	17:32	307.10
7441	19AU	16:01	140.70		7441	4SE	5:14	336.00
7442	18SE	5:32	169.91		7442	3OC	16:15	5.37
7443	17OC	22:12	199.75		7443	2NO	3:04	35.18
7444	16NO	17:27	230.05		7444	1DE	14:00	65.22
7445	16DE	13:34	260.51		7445	31DE	1:00	95.24

-398

NUMBER	DATE	TIME	LONG.		NUMBER	DATE	TIME	LONG.
7446	15JA	8:18	290.74		7446	29JA	12:03	124.97
7447	14FE	0:07	320.44		7447	27FE	23:24	154.25
7448	15MR	12:42	349.50		7448	29MR	11:30	183.06
7449	13AP	22:36	17.97		7449	28AP	0:48	211.48
7450	13MY	6:40	46.03		7450	27MY	15:21	239.71
7451	11JN	13:46	73.92		7451	26JN	6:40	267.94
7452	10JL	20:47	101.90		7452	25JL	22:05	296.40
7453	9AU	4:39	130.23		7453	24AU	13:02	325.24
7454	7SE	14:21	159.10		7454	23SE	3:18	354.56
7455	7OC	2:49	188.59		7455	22OC	16:47	24.34
7456	5NO	18:29	218.64		7456	21NO	5:20	54.39
7457	5DE	12:48	249.01		7457	20DE	16:45	84.48

-397

NUMBER	DATE	TIME	LONG.		NUMBER	DATE	TIME	LONG.
7458	4JA	8:15	279.36		7458	19JA	3:03	114.31
7459	3FE	2:56	309.35		7459	17FE	12:35	143.69
7460	4MR	19:32	338.77		7460	18MR	21:59	172.57
7461	3AP	9:32	7.58		7461	17AP	7:59	201.00
7462	2MY	20:55	35.87		7462	16MY	19:14	229.16
7463	1JN	6:07	63.87		7463	15JN	8:03	257.27
7464	30JN	13:53	91.82		7464	14JL	22:31	285.58
7465	29JL	21:13	119.99		7465	13AU	14:26	314.29
7466	28AU	5:10	148.61		7466	12SE	7:20	343.53
7467	26SE	14:42	177.80		7467	12OC	0:24	13.28
7468	26OC	2:30	207.54		7468	10NO	16:29	43.40
7469	24NO	16:42	237.69		7469	10DE	6:41	73.60
7470	24DE	8:56	267.97					

-396

NUMBER	DATE	TIME	LONG.		NUMBER	DATE	TIME	LONG.
					7470	8JA	18:40	103.61
7471	23JA	2:25	298.06		7471	7FE	4:46	133.18
7472	21FE	20:12	327.72		7472	7MR	13:35	162.22
7473	22MR	13:13	356.84		7473	5AP	21:49	190.73
7474	21AP	4:27	25.42		7474	5MY	6:10	218.88
7475	20MY	17:21	53.61		7475	3JN	15:21	246.86
7476	19JN	4:00	81.65		7476	3JL	2:09	274.95
7477	18JL	13:06	109.78		7477	1AU	15:18	303.40
7478	16AU	21:41	138.26		7478	31AU	7:12	332.40
7479	15SE	6:39	167.24		7479	30SE	1:25	2.01
7480	14OC	16:37	196.75		7480	29OC	20:30	32.12
7481	13NO	3:47	226.69		7481	28NO	14:34	62.46
7482	12DE	16:15	256.82		7482	28DE	6:17	92.69

-395

NUMBER	DATE	TIME	LONG.		NUMBER	DATE	TIME	LONG.
7483	11JA	6:04	286.88		7483	26JA	19:18	122.54
7484	9FE	21:15	316.61		7484	25FE	5:56	151.84
7485	11MR	13:28	345.89		7485	26MR	14:44	180.55
7486	10AP	5:52	14.68		7486	24AP	22:17	208.77
7487	9MY	21:25	43.07		7487	24MY	5:22	236.71
7488	8JN	11:28	71.23		7488	22JN	12:58	264.63
7489	7JL	23:58	99.42		7489	21JL	22:16	292.81
7490	6AU	11:19	127.85		7490	20AU	10:19	321.47
7491	4SE	22:04	156.72		7491	19SE	1:40	350.77
7492	4OC	8:36	186.09		7492	18OC	19:52	20.69
7493	2NO	19:07	215.88		7493	17NO	15:29	51.02
7494	2DE	5:47	245.92		7494	17DE	10:40	81.44
7495	31DE	16:52	275.94					

-394

NUMBER	DATE	TIME	LONG.		NUMBER	DATE	TIME	LONG.
					7495	16JA	3:25	111.59
7496	30JA	4:43	305.70		7496	14FE	18:28	141.23
7497	28FE	17:37	335.03		7497	16MR	6:09	170.24
7498	30MR	7:31	3.89		7498	14AP	15:15	198.67
7499	28AP	22:05	32.36		7499	13MY	22:33	226.69
7500	28MY	12:56	60.59		7500	12JN	5:06	254.56
7501	27JN	3:47	88.81		7501	11JL	12:06	282.55
7502	26JL	18:24	117.24		7502	9AU	20:44	310.92
7503	25AU	8:32	146.07		7503	8SE	7:55	339.85
7504	23SE	21:50	175.36		7504	7OC	22:04	9.43
7505	23OC	10:06	205.10		7505	6NO	15:00	39.53
7506	21NO	21:24	235.11		7506	6DE	9:53	69.93
7507	21DE	8:03	265.15					

-393

NUMBER	DATE	TIME	LONG.		NUMBER	DATE	TIME	LONG.
					7507	5JA	5:24	100.28
7508	19JA	18:25	294.98		7508	3FE	23:51	130.25
7509	18FE	4:48	324.39		7509	5MR	15:47	159.63
7510	19MR	15:26	353.30		7510	4AP	4:30	188.37
7511	18AP	2:37	21.77		7511	3MY	14:19	216.60
7512	17MY	14:48	49.96		7512	1JN	22:08	244.53
7513	16JN	4:25	78.11		7513	1JL	5:10	272.46
7514	15JL	19:38	106.46		7514	30JL	12:33	300.64
7515	14AU	12:00	135.20		7515	28AU	21:09	329.30
7516	13SE	4:33	164.44		7516	27SE	7:33	358.53
7517	12OC	20:13	194.15		7517	26OC	20:13	28.32
7518	11NO	10:26	224.19		7518	25NO	11:23	58.52
7519	10DE	23:06	254.33		7519	25DE	4:54	88.84

-392

NUMBER	DATE	TIME	LONG.		NUMBER	DATE	TIME	LONG.
7520	9JA	10:24	284.30		7520	23JA	23:45	118.98
7521	7FE	20:30	313.87		7521	22FE	18:13	148.66
7522	8MR	5:36	342.90		7522	23MR	10:41	177.74
7523	6AP	14:11	11.42		7523	22AP	0:23	206.25
7524	5MY	23:02	39.58		7524	21MY	11:31	234.37
7525	4JN	9:10	67.60		7525	19JN	20:54	262.35
7526	3JL	21:28	95.75		7526	19JL	5:22	290.46
7527	2AU	11:19	124.27		7527	17AU	13:43	318.94
7528	1SE	5:15	153.33		7528	15SE	22:34	347.93
7529	30SE	23:12	182.95		7529	15OC	8:29	17.45
7530	30OC	16:56	213.02		7530	13NO	20:02	47.41
7531	29NO	9:25	243.29		7531	13DE	9:40	77.59
7532	28DE	23:59	273.47					

-391

NUMBER	DATE	TIME	LONG.		NUMBER	DATE	TIME	LONG.
					7532	12JA	1:19	107.72
7533	27JA	12:13	303.28		7533	10FE	18:14	137.51
7534	25FE	22:11	332.53		7534	12MR	11:14	166.81
7535	27MR	6:22	1.21		7535	11AP	3:20	195.57
7536	25AP	13:37	29.41		7536	10MY	17:58	223.91
7537	24MY	21:05	57.37		7537	9JN	7:02	252.04
7538	23JN	5:50	85.34		7538	8JL	18:37	280.19
7539	22JL	16:47	113.58		7539	7AU	5:01	308.60
7540	21AU	6:25	142.32		7540	5SE	14:42	337.43
7541	19SE	22:43	171.68		7541	5OC	0:16	6.77
7542	19OC	17:08	201.61		7542	3NO	10:22	36.56
7543	18NO	12:29	231.94		7543	2DE	21:28	66.62
7544	18DE	7:07	262.33					

-390

NUMBER	DATE	TIME	LONG.		NUMBER	DATE	TIME	LONG.
					7544	1JA	9:44	96.68
7545	16JA	23:25	292.44		7545	30JA	23:02	126.49
7546	15FE	12:32	322.00		7546	1MR	13:08	155.86
7547	16MR	22:32	350.94		7547	31MR	3:50	184.74
7548	15AP	6:43	19.32		7548	29AP	18:59	213.22
7549	14MY	13:41	47.32		7549	29MY	10:15	241.47
7550	12JN	20:39	75.21		7550	28JN	1:02	269.69
7551	12JL	4:32	103.23		7551	27JL	14:48	298.10
7552	10AU	14:09	131.65		7552	26AU	3:21	326.87
7553	9SE	2:17	160.64		7553	24SE	14:55	356.10
7554	8OC	17:28	190.27		7554	24OC	2:01	25.80
7555	7NO	11:37	220.44		7555	22NO	12:59	55.80
7556	7DE	7:36	250.88		7556	21DE	23:56	85.85

-389

NUMBER	DATE	TIME	LONG.		NUMBER	DATE	TIME	LONG.
7557	6JA	3:15	281.23		7557	20JA	10:48	115.69
7558	4FE	20:33	311.14		7558	18FE	21:42	145.12
7559	6MR	10:38	340.43		7559	20MR	9:03	174.05
7560	4AP	21:42	9.09		7560	18AP	21:26	202.56
7561	4MY	6:32	37.27		7561	18MY	11:11	230.80
7562	2JN	14:01	65.19		7562	17JN	2:09	259.01
7563	1JL	21:02	93.12		7563	16JL	17:43	287.37
7564	31JL	4:27	121.31		7564	15AU	9:11	316.09
7565	29AU	13:13	149.98		7565	14SE	0:08	345.28
7566	28SE	0:19	179.26		7566	13OC	14:19	14.94
7567	27OC	14:25	209.12		7567	12NO	3:33	44.94
7568	26NO	7:32	239.40		7568	11DE	15:38	75.06
7569	26DE	2:32	269.79					

Left half

NEW MOONS

NUMBER	DATE	TIME	LONG.
-388			
7570	24JA	21;39	299.93
7571	23FE	15;16	329.56
7572	24MR	6;28	358.57
7573	22AP	19;05	27.02
7574	22MY	5;21	55.10
7575	20JN	13;49	83.05
7576	19JL	21;23	111.14
7577	18AU	5;04	139.60
7578	16SE	13;52	168.59
7579	16OC	0;35	198.15
7580	14NO	13;37	228.18
7581	14DE	4;49	258.43
-387			
7582	12JA	21;35	288.60
7583	11FE	15;05	318.41
7584	13MR	8;23	347.70
7585	12AP	0;29	16.45
7586	11MY	14;35	44.76
7587	10JN	2;24	72.84
7588	9JL	12;21	100.93
7589	7AU	21;17	129.29
7590	6SE	6;11	158.10
7591	5OC	15;47	187.45
7592	4NO	2;27	217.27
7593	3DE	14;15	247.36
-386			
7594	2JA	3;14	277.45
7595	31JA	17;29	307.29
7596	2MR	8;56	336.70
7597	1AP	1;06	5.63
7598	30AP	17;02	34.13
7599	30MY	7;54	62.36
7600	28JN	21;17	90.52
7601	28JL	9;22	118.87
7602	26AU	20;39	147.61
7603	25SE	7;32	176.83
7604	24OC	18;14	206.51
7605	23NO	4;53	236.50
7606	22DE	15;40	266.55
-385			
7607	21JA	2;54	296.40
7608	19FE	14;59	325.86
7609	21MR	4;06	354.85
7610	19AP	18;06	23.41
7611	19MY	8;42	51.69
7612	17JN	23;34	79.89
7613	17JL	14;29	108.24
7614	16AU	5;10	136.93
7615	14SE	19;15	166.10
7616	14OC	8;20	195.72
7617	12NO	20;16	225.68
7618	12DE	7;15	255.75
-384			
7619	10JA	17;38	285.67
7620	9FE	3;47	315.22
7621	9MR	14;00	344.28
7622	8AP	0;35	12.88
7623	7MY	11;58	41.11
7624	6JN	0;41	69.22
7625	5JL	15;06	97.47
7626	4AU	7;08	126.07
7627	3SE	0;00	155.15
7628	2OC	16;31	184.74
7629	1NO	7;45	214.72
7630	30NO	21;19	244.87
7631	30DE	9;18	274.94
-383			
7632	28JA	19;52	304.66
7633	27FE	5;16	333.87
7634	28MR	13;49	2.54
7635	26AP	22;11	30.78
7636	26MY	7;23	58.79
7637	24JN	18;25	86.85
7638	24JL	8;01	115.22
7639	23AU	0;07	144.09
7640	21SE	17;57	173.53
7641	21OC	12;10	203.49
7642	20NO	5;34	233.74
7643	19DE	21;15	264.01

FULL MOONS

NUMBER	DATE	TIME	LONG.
-388			
7569	10JA	2;25	105.00
7570	8FE	12;06	134.54
7571	8MR	21;14	163.57
7572	7AP	6;36	192.11
7573	6MY	16;55	220.32
7574	5JN	4;47	248.40
7575	4JL	18;24	276.61
7576	3AU	9;43	305.17
7577	2SE	2;25	334.23
7578	1OC	19;49	3.84
7579	31OC	12;47	33.88
7580	30NO	4;11	64.12
7581	29DE	17;18	94.23
-387			
7582	28JA	4;12	123.97
7583	26FE	13;24	153.19
7584	27MR	21;39	181.85
7585	26AP	5;40	210.09
7586	25MY	14;12	238.08
7587	24JN	0;00	266.09
7588	23JL	11;54	294.38
7589	22AU	2;30	323.17
7590	20SE	19;51	352.58
7591	20OC	14;55	22.56
7592	19NO	9;53	52.88
7593	19DE	2;58	83.20
-386			
7594	17JA	17;20	113.22
7595	16FE	5;00	142.71
7596	17MR	14;31	171.61
7597	15AP	22;28	199.98
7598	15MY	5;35	227.99
7599	13JN	12;44	255.88
7600	12JL	21;02	283.93
7601	11AU	7;38	312.39
7602	9SE	21;23	341.46
7603	9OC	14;24	11.18
7604	8NO	9;39	41.40
7605	8DE	5;19	71.84
-385			
7606	6JA	23;39	102.12
7607	5FE	15;33	131.96
7608	7MR	4;34	161.20
7609	5AP	14;49	189.81
7610	4MY	22;50	217.95
7611	3JN	5;36	245.84
7612	2JL	12;15	273.76
7613	31JL	20;01	301.97
7614	30AU	5;56	330.70
7615	28SE	18;40	0.05
7616	28OC	10;20	29.99
7617	27NO	4;22	60.32
7618	26DE	23;41	90.71
-384			
7619	25JA	18;45	120.84
7620	24FE	11;58	150.44
7621	25MR	2;14	179.40
7622	23AP	13;23	207.78
7623	22MY	22;04	235.80
7624	21JN	5;25	263.70
7625	20JL	12;38	291.78
7626	18AU	20;40	320.26
7627	17SE	6;15	349.30
7628	16OC	17;49	18.91
7629	15NO	7;44	48.98
7630	15DE	0;02	79.28
-383			
7631	13JA	18;12	109.49
7632	12FE	12;53	139.34
7633	14MR	6;20	168.62
7634	12AP	21;20	197.31
7635	12MY	9;42	225.55
7636	10JN	19;57	253.57
7637	10JL	4;55	281.62
7638	8AU	13;27	309.97
7639	6SE	22;10	338.79
7640	6OC	7;39	8.14
7641	4NO	18;24	37.97
7642	4DE	6;55	68.10

Right half

NEW MOONS

NUMBER	DATE	TIME	LONG.
-382			
7644	18JA	10;39	293.98
7645	16FE	21;37	323.43
7646	18MR	6;27	352.29
7647	16AP	13;53	20.63
7648	15MY	20;58	48.63
7649	14JN	4;52	76.55
7650	13JL	14;36	104.66
7651	12AU	2;55	133.21
7652	10SE	18;01	162.34
7653	10OC	11;37	192.10
7654	9NO	6;49	222.33
7655	9DE	2;07	252.75
-381			
7656	7JA	19;46	283.00
7657	6FE	10;27	312.78
7658	7MR	21;55	341.93
7659	6AP	6;47	10.49
7660	5MY	14;03	38.59
7661	3JN	20;51	66.48
7662	3JL	4;10	94.42
7663	1AU	12;53	122.68
7664	30AU	23;45	151.46
7665	29SE	13;27	180.87
7666	29OC	6;15	210.86
7667	28NO	1;34	241.25
7668	27DE	21;38	271.67
-380			
7669	26JA	16;13	301.77
7670	25FE	7;51	331.28
7671	25MR	20;16	0.15
7672	24AP	6;03	28.48
7673	23MY	14;06	56.46
7674	21JN	21;53	84.37
7675	21JL	4;30	112.45
7676	19AU	12;36	140.94
7677	17SE	22;32	170.00
7678	17OC	11;09	199.67
7679	16NO	2;50	229.82
7680	15DE	20;59	260.20
-379			
7681	14JA	16;07	290.44
7682	13FE	10;28	320.26
7683	15MR	2;48	349.48
7684	13AP	16;38	18.12
7685	13MY	4;02	46.31
7686	11JN	13;24	74.29
7687	10JL	21;26	102.32
7688	9AU	5;06	130.64
7689	7SE	13;24	159.45
7690	6OC	23;14	188.83
7691	5NO	11;08	218.71
7692	5DE	1;15	248.90
-378			
7693	3JA	17;09	279.11
7694	2FE	10;09	309.04
7695	4MR	3;25	338.50
7696	2AP	20;03	7.42
7697	2MY	11;09	35.87
7698	1JN	0;10	64.02
7699	30JN	11;08	92.09
7700	29JL	20;40	120.36
7701	28AU	5;43	149.02
7702	26SE	15;07	178.20
7703	26OC	1;22	207.88
7704	24NO	12;38	237.91
7705	24DE	0;55	268.02
-377			
7706	22JA	14;18	297.94
7707	21FE	4;53	327.48
7708	22MR	22;32	356.54
7709	21AP	12;26	25.16
7710	21MY	3;51	53.46
7711	19JN	18;04	81.64
7712	19JL	6;59	109.92
7713	17AU	18;53	138.53
7714	16SE	6;12	167.60
7715	15OC	17;13	197.16
7716	14NO	4;01	227.27
7717	13DE	14;41	257.14

FULL MOONS

NUMBER	DATE	TIME	LONG.
-382			
7643	2JA	21;26	98.25
7644	1FE	13;35	128.16
7645	3MR	6;25	157.61
7646	1AP	22;50	186.54
7647	1MY	14;08	215.00
7648	31MY	4;01	243.19
7649	29JN	16;28	271.32
7650	29JL	3;40	299.64
7651	27AU	13;54	328.34
7652	25SE	23;41	357.53
7653	25OC	9;37	27.19
7654	23NO	20;14	57.18
7655	23DE	7;50	87.26
-381			
7656	21JA	20;26	117.16
7657	20FE	9;52	146.67
7658	21MR	23;59	175.68
7659	20AP	14;41	204.26
7660	20MY	5;49	232.56
7661	18JN	20;58	260.77
7662	18JL	11;30	289.11
7663	17AU	0;56	317.77
7664	15SE	13;15	346.87
7665	15OC	0;47	16.44
7666	13NO	11;56	46.38
7667	12DE	22;55	76.45
-380			
7668	11JA	9;43	106.38
7669	9FE	20;22	135.94
7670	10MR	7;08	165.01
7671	8AP	18;39	193.62
7672	8MY	7;27	221.91
7673	6JN	21;45	250.09
7674	6JL	13;08	278.38
7675	5AU	4;56	306.98
7676	3SE	20;29	336.02
7677	3OC	11;22	5.56
7678	2NO	1;22	35.49
7679	1DE	14;10	65.62
7680	31DE	1;35	95.66
-379			
7681	29JA	11;39	125.34
7682	27FE	20;47	154.53
7683	29MR	5;40	183.21
7684	27AP	15;12	211.48
7685	27MY	2;03	239.56
7686	25JN	14;42	267.69
7687	25JL	5;14	296.10
7688	23AU	21;28	324.99
7689	22SE	14;53	354.43
7690	22OC	8;29	24.37
7691	21NO	1;00	54.60
7692	20DE	15;23	84.80
-378			
7693	19JA	3;18	114.70
7694	17FE	13;08	144.10
7695	18MR	21;36	172.93
7696	17AP	5;28	201.28
7697	16MY	13;29	229.31
7698	14JN	22;27	257.28
7699	14JL	9;10	285.43
7700	12AU	22;25	314.03
7701	11SE	14;34	343.22
7702	11OC	9;09	13.03
7703	10NO	4;37	43.28
7704	9DE	22;55	73.66
-377			
7705	8JA	14;42	103.83
7706	7FE	3;36	133.52
7707	8MR	14;01	162.62
7708	6AP	22;32	191.15
7709	6MY	5;52	219.25
7710	4JN	12;48	247.14
7711	3JL	20;22	275.10
7712	2AU	5;43	303.39
7713	31AU	17;54	332.24
7714	30SE	9;26	1.73
7715	30OC	3;51	31.81
7716	28NO	23;37	62.21
7717	28DE	18;48	92.59

-376 | **-370**

7718	12JA	1;31	287.07	7718	27JA	11;57	122.62	7792	5JA	10;33	280.92	7792	20JA	11;06	116.10
7719	10FE	12;54	316.65	7719	26FE	2;21	152.07	7793	4FE	5;17	310.90	7793	18FE	20;27	145.45
7720	11MR	1;11	345.78	7720	26MR	13;51	180.89	7794	5MR	22;33	340.32	7794	20MR	5;08	174.27
7721	9AP	14;29	14.45	7721	24AP	22;49	209.18	7795	4AP	13;33	9.15	7795	18AP	14;00	202.64
7722	9MY	4;36	42.78	7722	24MY	6;03	237.13	7796	4MY	2;08	37.48	7796	17MY	23;55	230.74
7723	7JN	19;18	70.98	7723	22JN	12;38	265.01	7797	2JN	12;31	65.52	7797	16JN	11;31	258.81
7724	7JL	10;19	99.26	7724	21JL	19;46	293.09	7798	1JL	21;15	93.51	7798	16JL	1;06	287.09
7725	6AU	1;23	127.84	7725	20AU	4;36	321.63	7799	31JL	5;09	121.73	7799	14AU	16;38	315.81
7726	4SE	16;08	156.86	7726	18SE	15;58	350.76	7800	29AU	13;12	150.37	7800	13SE	9;47	345.07
7727	4OC	6;03	186.37	7727	18OC	6;16	20.50	7801	27SE	22;20	179.56	7801	13OC	3;42	14.88
7728	2NO	18;47	216.26	7728	16NO	23;14	50.72	7802	27OC	9;15	209.29	7802	11NO	21;09	45.06
7729	2DE	6;18	246.33	7729	16DE	18;00	81.12	7803	25NO	22;16	239.39	7803	11DE	12;48	75.33
7730	31DE	16;54	276.33					7804	25DE	13;12	269.61				

-375 | **-369**

7731	30JA	2;59	306.01	7730	15JA	13;17	111.36	7805	24JA	5;31	299.64	7804	10JA	1;57	105.37
7732	28FE	12;54	335.22	7731	14FE	7;28	141.16	7806	22FE	22;28	329.25	7805	8FE	12;39	134.95
7733	29MR	22;58	3.93	7732	15MR	23;10	170.34	7807	24MR	15;18	358.34	7806	9MR	21;33	163.97
7734	28AP	9;39	32.25	7733	14AP	11;48	198.92	7808	23AP	7;09	26.93	7807	8AP	5;26	192.45
7735	27MY	21;27	60.36	7734	13MY	21;36	227.04	7809	22MY	21;16	55.17	7808	7MY	13;07	220.55
7736	26JN	10;56	88.53	7735	12JN	5;32	254.96	7810	21JN	9;21	83.26	7809	5JN	21;24	248.50
7737	26JL	2;18	116.98	7736	11JL	12;48	282.96	7811	20JL	19;43	111.46	7810	5JL	7;06	276.55
7738	24AU	19;05	145.90	7737	9AU	20;28	311.29	7812	19AU	5;09	139.99	7811	3AU	19;02	304.96
7739	23SE	12;13	175.35	7738	8SE	5;21	340.14	7813	17SE	14;31	169.00	7812	2SE	9;49	333.94
7740	23OC	4;28	205.25	7739	7OC	16;00	9.56	7814	17OC	0;29	198.54	7813	2OC	3;28	3.55
7741	21NO	19;04	235.39	7740	6NO	4;45	39.49	7815	15NO	11;18	228.47	7814	31OC	22;54	33.69
7742	21DE	7;53	265.53	7741	5DE	19;49	69.72	7816	14DE	23;01	258.57	7815	30NO	18;09	64.08
												7816	30DE	11;21	94.37

-374 | **-368**

7743	19JA	19;04	295.39	7742	4JA	12;59	99.98	7817	13JA	11;40	288.57	7817	29JA	1;38	124.25
7744	18FE	4;52	324.78	7743	3FE	7;23	129.96	7818	12FE	1;22	318.22	7818	27FE	13;08	153.56
7745	19MR	13;35	353.61	7744	5MR	1;26	159.43	7819	12MR	16;11	347.42	7819	27MR	22;24	182.27
7746	17AP	21;44	21.96	7745	3AP	17;37	188.32	7820	11AP	7;48	16.16	7820	26AP	6;07	210.49
7747	17MY	6;12	50.01	7746	3MY	7;15	216.70	7821	10MY	23;28	44.55	7821	25MY	13;04	238.42
7748	15JN	16;05	78.01	7747	1JN	18;32	244.77	7822	9JN	14;22	72.75	7822	23JN	20;10	266.32
7749	15JL	4;19	106.23	7748	1JL	4;11	272.80	7823	9JL	4;05	100.99	7823	23JL	4;30	294.47
7750	13AU	19;18	134.91	7749	30JL	13;03	301.05	7824	7AU	16;42	129.49	7824	21AU	15;13	323.10
7751	12SE	12;35	164.16	7750	28AU	21;49	329.71	7825	6SE	4;34	158.42	7825	20SE	5;09	352.36
7752	12OC	7;00	193.98	7751	27SE	7;02	358.90	7826	5OC	15;59	187.85	7826	19OC	22;21	22.25
7753	11NO	1;09	224.18	7752	26OC	17;12	28.58	7827	4NO	3;03	217.68	7827	18NO	17;44	52.58
7754	10DE	17;55	254.50	7753	25NO	4;48	58.62	7828	3DE	13;49	247.73	7828	18DE	13;26	83.02
				7754	24DE	18;12	88.78								

-373 | **-367**

7755	9JA	8;31	284.61	7755	23JA	9;23	118.77	7829	2JA	0;27	277.72	7829	17JA	7;42	113.20
7756	7FE	20;37	314.26	7756	22FE	1;42	148.37	7830	31JA	11;17	307.42	7830	15FE	23;26	142.87
7757	9MR	6;19	343.32	7757	23MR	18;11	177.45	7831	1MR	22;48	336.67	7831	17MR	12;17	171.91
7758	7AP	14;11	11.81	7758	22AP	9;56	206.05	7832	31MR	11;18	5.46	7832	15AP	22;22	200.36
7759	6MY	21;10	39.89	7759	22MY	0;31	234.31	7833	30AP	0;47	33.87	7833	15MY	6;20	228.40
7760	5JN	4;25	67.79	7760	20JN	13;47	262.45	7834	29MY	15;06	62.08	7834	13JN	13;07	256.28
7761	4JL	13;05	95.80	7761	20JL	1;19	290.71	7835	28JN	6;00	90.30	7835	12JL	19;53	284.26
7762	3AU	0;04	124.17	7762	18AU	12;45	319.29	7836	27JL	21;16	118.77	7836	11AU	3;50	312.62
7763	1SE	13;51	153.09	7763	16SE	22;57	348.33	7837	26AU	12;31	147.66	7837	9SE	13;58	341.54
7764	1OC	6;24	182.64	7764	16OC	8;56	17.86	7838	25SE	3;13	177.03	7838	9OC	2;53	11.08
7765	31OC	1;05	212.73	7765	14NO	19;16	47.76	7839	24OC	16;51	206.84	7839	7NO	18;36	41.15
7766	29NO	20;39	243.13	7766	14DE	6;20	77.84	7840	23NO	5;06	236.90	7840	7DE	12;33	71.51
7767	29DE	15;22	273.49					7841	22DE	16;08	266.96				

-372 | **-366**

7768	28JA	7;35	303.47	7767	12JA	18;17	107.81	7842	21JA	2;19	296.76	7841	6JA	7;39	101.84
7769	26FE	20;33	332.85	7768	11FE	7;04	137.44	7843	19FE	12;04	326.12	7842	5FE	2;25	131.80
7770	27MR	6;31	1.60	7769	11MR	22;06	166.59	7844	20MR	21;46	354.98	7843	6MR	19;22	161.20
7771	25AP	14;21	29.83	7770	10AP	10;39	195.29	7845	19AP	7;49	23.38	7844	5AP	9;29	189.98
7772	24MY	21;12	57.76	7771	10MY	1;25	223.64	7846	18MY	18;47	51.52	7845	4MY	20;36	218.25
7773	23JN	4;09	85.65	7772	8JN	16;37	251.85	7847	17JN	7;16	79.62	7846	3JN	5;23	246.22
7774	22JL	12;07	113.78	7773	8JL	7;40	280.14	7848	16JL	21;42	107.95	7847	2JL	12;57	274.17
7775	20AU	21;54	142.36	7774	6AU	21;57	308.70	7849	15AU	14;02	136.71	7848	31JL	20;26	302.38
7776	19SE	10;12	171.54	7775	5SE	11;07	337.67	7850	14SE	7;26	166.00	7849	30AU	4;48	331.04
7777	19OC	1;32	201.34	7776	4OC	23;16	7.12	7851	14OC	0;33	195.78	7850	28SE	14;39	0.27
7778	17NO	19;45	231.62	7777	3NO	10;46	36.97	7852	12NO	16;15	225.90	7851	28OC	2;23	30.04
7779	17DE	15;41	262.07	7778	2DE	21;55	67.03	7853	12DE	6;03	256.09	7852	26NO	16;16	60.19
												7853	26DE	8;18	90.45

-371 | **-365**

7780	16JA	11;11	292.31	7779	1JA	8;45	97.03	7854	10JA	18;00	286.08	7854	25JA	2;01	120.53
7781	15FE	4;18	322.05	7780	30JA	19;17	126.72	7855	9FE	4;21	315.63	7855	23FE	20;13	150.17
7782	16MR	18;12	351.14	7781	1MR	5;40	155.94	7856	10MR	13;23	344.64	7856	25MR	13;18	179.26
7783	15AP	5;08	19.64	7782	30MR	16;27	184.67	7857	8AP	21;31	13.13	7857	24AP	4;09	207.80
7784	14MY	13;56	47.72	7783	29AP	4;17	213.02	7858	8MY	5;32	41.23	7858	23MY	16;35	235.96
7785	12JN	21;29	75.62	7784	28MY	17;40	241.20	7859	6JN	14;26	69.20	7859	22JN	3;04	264.00
7786	12JL	4;40	103.63	7785	27JN	8;34	269.42	7860	6JL	1;20	97.31	7860	21JL	12;26	292.16
7787	10AU	12;20	131.97	7786	27JL	0;23	297.90	7861	4AU	14;58	125.80	7861	19AU	21;23	320.68
7788	8SE	21;22	160.83	7787	25AU	16;23	326.80	7862	3SE	7;21	154.85	7862	18SE	6;32	349.70
7789	8OC	8;39	190.29	7788	24SE	7;57	356.20	7863	3OC	1;34	184.50	7863	17OC	16;20	19.23
7790	6NO	22;50	220.28	7789	23OC	22;42	26.05	7864	1NO	20;14	214.62	7864	16NO	3;12	49.18
7791	6DE	15;49	250.60	7790	22NO	12;18	56.16	7865	1DE	13;57	244.95	7865	15DE	15;36	79.31
				7791	22DE	0;28	86.27	7866	31DE	5;45	275.17				

-364

NEW MOONS NUMBER	DATE	TIME	LONG.	FULL MOONS NUMBER	DATE	TIME	LONG.
				7866	14JA	5;44	109.36
7867	29JA	19;05	305.01	7867	12FE	21;18	139.08
7868	28FE	5;49	334.28	7868	13MR	13;32	168.32
7869	28MR	14;22	2.94	7869	12AP	5;30	197.06
7870	26AP	21;32	31.13	7870	11MY	20;37	225.42
7871	26MY	4;24	59.06	7871	10JN	10;37	253.58
7872	24JN	12;11	86.99	7872	9JL	23;26	281.79
7873	23JL	21;55	115.20	7873	8AU	11;09	310.27
7874	22AU	10;20	143.91	7874	6SE	21;57	339.17
7875	21SE	1;39	173.24	7875	6OC	8;13	8.56
7876	20OC	19;30	203.17	7876	4NO	18;27	38.37
7877	19NO	14;55	233.51	7877	4DE	5;08	68.41
7878	19DE	10;19	263.93				

-363

NEW MOONS NUMBER	DATE	TIME	LONG.	FULL MOONS NUMBER	DATE	TIME	LONG.
				7878	2JA	16;33	98.43
7879	18JA	3;56	294.08	7879	1FE	4;42	128.17
7880	16FE	18;29	323.69	7880	2MR	17;32	157.47
7881	18MR	5;46	352.65	7881	1AP	7;01	186.28
7882	16AP	14;26	21.03	7882	30AP	21;14	214.71
7883	15MY	21;36	49.04	7883	30MY	12;10	242.94
7884	14JN	4;22	76.91	7884	29JN	3;27	271.19
7885	13JL	11;45	104.93	7885	28JL	18;24	299.65
7886	11AU	20;36	133.34	7886	27AU	8;27	328.51
7887	10SE	7;40	162.30	7887	25SE	21;23	357.82
7888	9OC	21;32	191.89	7888	25OC	9;24	27.57
7889	8NO	14;25	222.02	7889	23NO	20;49	57.60
7890	8DE	9;42	252.44	7890	23DE	7;49	87.65

-362

NEW MOONS NUMBER	DATE	TIME	LONG.	FULL MOONS NUMBER	DATE	TIME	LONG.
7891	7JA	5;36	282.79	7891	21JA	18;22	117.46
7892	5FE	23;59	312.73	7892	20FE	4;33	146.83
7893	7MR	15;24	342.05	7893	21MR	14;46	175.70
7894	6AP	3;41	10.74	7894	20AP	1;43	204.13
7895	5MY	13;25	38.95	7895	19MY	14;05	232.32
7896	3JN	21;31	66.89	7896	18JN	4;11	260.49
7897	3JL	4;52	94.84	7897	17JL	19;42	288.86
7898	1AU	12;18	123.05	7898	16AU	11;55	317.63
7899	30AU	20;41	151.72	7899	15SE	4;03	346.88
7900	29SE	6;51	180.98	7900	14OC	19;33	16.61
7901	28OC	19;36	210.80	7901	13NO	9;59	46.68
7902	27NO	11;13	241.02	7902	12DE	22;59	76.84
7903	27DE	5;08	271.36				

-361

NEW MOONS NUMBER	DATE	TIME	LONG.	FULL MOONS NUMBER	DATE	TIME	LONG.
				7903	11JA	10;21	106.79
7904	25JA	23;53	301.47	7904	9FE	20;08	136.32
7905	24FE	17;50	331.09	7905	11MR	4;51	165.30
7906	26MR	9;53	0.12	7906	9AP	13;17	193.79
7907	24AP	23;37	28.61	7907	8MY	22;22	221.93
7908	24MY	11;05	56.73	7908	7JN	8;55	249.97
7909	22JN	20;41	84.72	7909	6JL	21;26	278.14
7910	22JL	5;04	112.86	7910	5AU	12;05	306.68
7911	20AU	13;07	141.35	7911	4SE	4;40	335.76
7912	18SE	21;47	170.36	7912	3OC	22;35	5.41
7913	18OC	7;52	199.92	7913	2NO	16;40	35.51
7914	16NO	19;51	229.91	7914	2DE	9;31	65.81
7915	16DE	9;48	260.11				

-360

NEW MOONS NUMBER	DATE	TIME	LONG.	FULL MOONS NUMBER	DATE	TIME	LONG.
				7915	1JA	0;00	95.98
7916	15JA	1;17	290.21	7916	30JA	11;49	125.74
7917	13FE	17;44	319.95	7917	28FE	21;23	154.94
7918	14MR	10;28	349.20	7918	29MR	5;30	183.58
7919	13AP	2;44	17.94	7919	27AP	13;02	211.78
7920	12MY	17;45	46.29	7920	26MY	20;48	239.74
7921	11JN	6;57	74.42	7921	25JN	5;37	267.71
7922	10JL	18;18	102.58	7922	24JL	16;20	295.97
7923	9AU	4;20	130.99	7923	23AU	5;42	324.73
7924	7SE	13;53	159.86	7924	21SE	22;07	354.12
7925	6OC	23;42	189.24	7925	21OC	17;01	24.11
7926	5NO	10;12	219.06	7926	20NO	12;46	54.47
7927	4DE	21;29	249.13	7927	20DE	7;14	84.85

-359

NEW MOONS NUMBER	DATE	TIME	LONG.	FULL MOONS NUMBER	DATE	TIME	LONG.
7928	3JA	9;31	279.18	7928	18JA	23;00	114.91
7929	1FE	22;25	308.94	7929	17FE	11;46	144.43
7930	3MR	12;22	338.27	7930	18MR	21;57	173.33
7931	2AP	3;21	7.13	7931	17AP	6;15	201.70
7932	1MY	18;54	35.61	7932	16MY	13;24	229.69
7933	31MY	10;14	63.85	7933	14JN	20;16	257.57
7934	30JN	0;41	92.06	7934	14JL	3;52	285.61
7935	29JL	14;04	120.48	7935	12AU	13;19	314.05
7936	28AU	2;34	149.28	7936	11SE	1;40	343.08
7937	26SE	14;28	178.56	7937	10OC	17;23	12.76
7938	26OC	1;55	208.30	7938	9NO	11;56	42.96
7939	24NO	12;56	238.31	7939	9DE	7;44	73.40
7940	23DE	23;34	268.35				

-358

NEW MOONS NUMBER	DATE	TIME	LONG.	FULL MOONS NUMBER	DATE	TIME	LONG.
				7940	8JA	2;51	103.71
7941	22JA	10;05	298.15	7941	6FE	19;51	133.58
7942	20FE	20;57	327.54	7942	8MR	10;03	162.84
7943	22MR	8;38	356.45	7943	6AP	21;24	191.49
7944	20AP	21;21	24.95	7944	6MY	6;18	219.65
7945	20MY	11;06	53.18	7945	4JN	13;33	247.56
7946	19JN	1;41	81.37	7946	3JL	20;14	275.48
7947	18JL	16;55	109.74	7947	2AU	3;33	303.69
7948	17AU	8;29	138.49	7948	31AU	12;36	332.41
7949	15SE	23;51	167.73	7949	30SE	0;11	1.73
7950	15OC	14;22	197.43	7950	29OC	14;35	31.63
7951	14NO	3;31	227.45	7951	28NO	7;30	61.91
7952	13DE	15;11	257.55	7952	28DE	2;04	92.27

-357

NEW MOONS NUMBER	DATE	TIME	LONG.	FULL MOONS NUMBER	DATE	TIME	LONG.
7953	12JA	1;40	287.47	7953	26JA	21;02	122.38
7954	10FE	11;25	316.98	7954	25FE	14;54	151.99
7955	11MR	20;52	345.98	7955	27MR	6;24	180.98
7956	10AP	6;26	14.51	7956	25AP	18;57	209.41
7957	9MY	16;39	42.69	7957	25MY	4;50	237.46
7958	8JN	4;09	70.76	7958	23JN	12;57	265.40
7959	7JL	17;32	98.97	7959	22JL	20;29	293.51
7960	6AU	9;05	127.56	7960	21AU	4;29	322.01
7961	5SE	2;19	156.68	7961	19SE	13;42	351.05
7962	4OC	20;01	186.33	7962	19OC	0;34	20.65
7963	3NO	12;47	216.39	7963	17NO	13;21	50.68
7964	3DE	3;43	246.61	7964	17DE	4;14	80.92

-356

NEW MOONS NUMBER	DATE	TIME	LONG.	FULL MOONS NUMBER	DATE	TIME	LONG.
7965	1JA	16;36	276.71	7965	15JA	21;00	111.07
7966	31JA	3;38	306.43	7966	14FE	14;53	140.86
7967	29FE	13;07	335.62	7967	15MR	8;29	170.13
7968	29MR	21;26	4.26	7968	14AP	0;25	198.84
7969	28AP	5;13	32.46	7969	13MY	14;02	227.12
7970	27MY	13;24	60.43	7970	12JN	1;30	255.18
7971	25JN	23;05	88.44	7971	11JL	11;31	283.29
7972	25JL	11;18	116.76	7972	9AU	20;49	311.69
7973	24AU	2;27	145.61	7973	8SE	6;02	340.55
7974	22SE	20;04	175.07	7974	7OC	15;38	9.93
7975	22OC	14;54	205.06	7975	6NO	2;01	39.76
7976	21NO	9;25	235.37	7976	5DE	13;34	69.85
7977	21DE	2;21	265.69				

-355

NEW MOONS NUMBER	DATE	TIME	LONG.	FULL MOONS NUMBER	DATE	TIME	LONG.
				7977	4JA	2;41	99.93
7978	19JA	16;57	295.69	7978	2FE	17;20	129.76
7979	18FE	4;52	325.16	7979	4MR	9;03	159.15
7980	19MR	14;18	354.03	7980	3AP	0;59	188.03
7981	17AP	21;54	22.35	7981	2MY	16;26	216.49
7982	17MY	4;41	50.33	7982	1JN	7;00	244.70
7983	15JN	11;48	78.22	7983	30JN	20;34	272.88
7984	14JL	20;27	106.30	7984	30JL	9;04	301.27
7985	13AU	7;31	134.81	7985	28AU	20;35	330.05
7986	11SE	21;28	163.93	7986	27SE	7;19	359.29
7987	11OC	14;14	193.66	7987	26OC	17;41	28.99
7988	10NO	9;08	223.89	7988	25NO	4;11	58.99
7989	10DE	4;49	254.33	7989	24DE	15;09	89.04

-354

NEW MOONS NUMBER	DATE	TIME	LONG.	FULL MOONS NUMBER	DATE	TIME	LONG.
7990	8JA	23;31	284.61	7990	23JA	2;45	118.88
7991	7FE	15;37	314.43	7991	21FE	14;58	148.31
7992	9MR	4;25	343.62	7992	23MR	3;49	177.26
7993	7AP	14;11	12.19	7993	21AP	17;23	205.78
7994	6MY	21;54	40.30	7994	21MY	7;49	234.03
7995	5JN	4;43	68.19	7995	19JN	23;00	262.25
7996	4JL	11;43	96.13	7996	19JL	14;22	290.63
7997	2AU	19;49	124.38	7997	18AU	5;13	319.37
7998	1SE	5;47	153.14	7998	16SE	19;01	348.55
7999	30SE	18;17	182.52	7999	16OC	7;43	18.19
8000	30OC	9;43	212.47	8000	14NO	19;35	48.17
8001	29NO	3;56	242.81	8001	14DE	6;49	78.25
8002	28DE	23;43	273.22				

-353

NEW MOONS NUMBER	DATE	TIME	LONG.	FULL MOONS NUMBER	DATE	TIME	LONG.
				8002	12JA	17;30	108.16
8003	27JA	18;59	303.33	8003	11FE	3;39	137.68
8004	26FE	11;52	332.88	8004	12MR	13;32	166.69
8005	28MR	1;35	1.78	8005	10AP	23;45	195.23
8006	26AP	12;27	30.14	8006	10MY	11;06	223.45
8007	25MY	21;17	58.15	8007	9JN	0;11	251.59
8008	24JN	4;58	86.07	8008	8JL	15;04	279.86
8009	23JL	12;24	114.18	8009	7AU	7;12	308.49
8010	21AU	20;21	142.68	8010	5SE	23;43	337.59
8011	20SE	5;39	171.74	8011	5OC	15;53	7.20
8012	19OC	17;07	201.38	8012	4NO	7;09	37.20
8013	18NO	7;19	231.48	8013	3DE	21;03	67.38
8014	18DE	0;06	261.79				

−352 / −346

NUMBER	DATE	TIME	LONG.	NUMBER	DATE	TIME	LONG.	NUMBER	DATE	TIME	LONG.	NUMBER	DATE	TIME	LONG.
8015	16JA	18;27	292.00	8014	2JA	9;16	97.44	8089	10JA	14;10	286.30	8089	24JA	13;55	120.41
8016	15FE	12;46	321.79	8015	31JA	19;41	127.13	8090	9FE	3;22	315.98	8090	23FE	4;53	149.93
8017	16MR	5;40	351.02	8016	1MR	4;41	156.28	8091	10MR	13;52	345.05	8091	24MR	20;31	178.96
8018	14AP	20;29	19.68	8017	30MR	12;55	184.91	8092	8AP	22;10	13.54	8092	23AP	12;04	207.54
8019	14MY	9;05	47.91	8018	28AP	21;21	213.13	8093	8MY	5;07	41.60	8093	23MY	3;03	235.81
8020	12JN	19;39	75.94	8019	28MY	6;55	241.15	8094	6JN	11;51	69.48	8094	21JN	17;13	263.99
8021	12JL	4;43	104.01	8020	26JN	18;20	269.24	8095	5JL	19;35	97.46	8095	21JL	6;28	292.30
8022	10AU	13;01	132.38	8021	26JL	7;56	297.62	8096	4AU	5;23	125.80	8096	19AU	18;45	320.96
8023	8SE	21;28	161.22	8022	24AU	23;43	326.51	8097	2SE	17;57	154.69	8097	18SE	6;08	350.07
8024	8OC	6;55	190.60	8023	23SE	17;17	355.98	8098	2OC	9;27	184.21	8098	17OC	16;51	19.65
8025	6NO	17;59	220.47	8024	23OC	11;42	25.97	8099	1NO	3;30	214.30	8099	16NO	3;21	49.57
8026	6DE	6;56	250.61	8025	22NO	5;32	56.26	8100	30NO	23;02	244.70	8100	15DE	14;02	79.63
				8026	21DE	21;23	86.52	8101	30DE	18;27	275.08				

−351 / −345

NUMBER	DATE	TIME	LONG.	NUMBER	DATE	TIME	LONG.	NUMBER	DATE	TIME	LONG.	NUMBER	DATE	TIME	LONG.
8027	4JA	21;33	280.75	8027	20JA	10;29	116.45					8101	14JA	1;11	109.55
8028	3FE	13;20	310.62	8028	18FE	20;59	145.86	8102	29JA	11;57	305.10	8102	12FE	12;51	139.11
8029	5MR	5;43	340.02	8029	20MR	5;34	174.68	8103	28FE	2;20	334.52	8103	14MR	1;04	168.20
8030	3AP	22;05	8.91	8030	18AP	13;08	202.99	8104	29MR	13;27	3.29	8104	12AP	13;57	196.83
8031	3MY	13;43	37.37	8031	17MY	20;33	230.99	8105	27AP	22;00	31.54	8105	12MY	3;45	225.13
8032	2JN	3;54	65.56	8032	16JN	4;39	258.92	8106	27MY	5;06	59.47	8106	10JN	18;31	253.33
8033	1JL	16;19	93.70	8033	15JL	14;18	287.05	8107	25JN	11;54	87.36	8107	10JL	9;59	281.64
8034	31JL	3;10	122.03	8034	14AU	2;19	315.61	8108	24JL	19;26	115.48	8108	9AU	1;25	310.26
8035	29AU	13;07	150.75	8035	12SE	17;18	344.78	8109	23AU	4;29	144.06	8109	7SE	16;05	339.31
8036	27SE	22;57	179.98	8036	12OC	11;14	14.57	8110	21SE	15;45	173.22	8110	7OC	5;38	8.84
8037	27OC	9;15	209.68	8037	11NO	6;58	44.85	8111	21OC	5;44	202.98	8111	5NO	18;06	38.74
8038	25NO	20;12	239.70	8038	11DE	2;24	75.27	8112	19NO	22;39	233.21	8112	5DE	5;43	68.82
8039	25DE	7;46	269.77					8113	19DE	17;48	263.62				

−350 / −344

NUMBER	DATE	TIME	LONG.	NUMBER	DATE	TIME	LONG.	NUMBER	DATE	TIME	LONG.	NUMBER	DATE	TIME	LONG.
				8039	9JA	19;38	105.49					8113	3JA	16;39	98.82
8040	23JA	20;00	299.63	8040	8FE	9;48	135.22	8114	18JA	13;28	293.86	8114	2FE	2;54	128.48
8041	22FE	9;07	329.09	8041	9MR	21;06	164.33	8115	17FE	7;35	323.62	8115	2MR	12;37	157.65
8042	23MR	23;18	358.07	8042	8AP	6;09	192.86	8116	17MR	22;47	352.75	8116	31MR	22;18	186.32
8043	22AP	14;24	26.65	8043	7MY	13;42	220.96	8117	16AP	10;57	21.28	8117	30AP	8;44	214.60
8044	22MY	5;50	54.94	8044	5JN	20;34	248.85	8118	15MY	20;42	49.38	8118	29MY	20;44	242.71
8045	20JN	20;50	83.15	8045	5JL	3;40	276.79	8119	14JN	4;55	77.32	8119	28JN	10;42	270.90
8046	20JL	10;57	111.49	8046	3AU	12;07	305.07	8120	13JL	12;29	105.35	8120	28JL	2;23	299.39
8047	19AU	0;08	140.17	8047	1SE	22;59	333.88	8121	11AU	20;14	133.71	8121	26AU	19;02	328.34
8048	17SE	12;37	169.31	8048	1OC	13;06	3.33	8122	10SE	4;55	162.58	8122	25SE	11;45	357.81
8049	17OC	0;32	198.93	8049	31OC	6;26	33.38	8123	9OC	15;19	192.02	8123	25OC	3;48	27.72
8050	15NO	11;55	228.89	8050	30NO	1;52	63.77	8124	8NO	4;09	221.97	8124	23NO	18;38	57.89
8051	14DE	22;44	258.95	8051	29DE	21;30	94.17	8125	7DE	19;38	252.23	8125	23DE	7;46	88.04

−349 / −343

NUMBER	DATE	TIME	LONG.	NUMBER	DATE	TIME	LONG.	NUMBER	DATE	TIME	LONG.	NUMBER	DATE	TIME	LONG.
8052	13JA	9;09	288.85	8052	28JA	15;36	124.22	8126	6JA	13;13	282.49	8126	21JA	19;00	117.88
8053	11FE	19;34	318.37	8053	27FE	7;09	153.70	8127	5FE	7;30	312.43	8127	20FE	4;29	147.22
8054	13MR	6;31	347.41	8054	28MR	19;49	182.55	8128	7MR	1;02	341.85	8128	21MR	12;48	176.00
8055	11AP	18;24	16.01	8055	27AP	5;49	210.86	8129	5AP	16;48	10.69	8129	19AP	20;49	204.32
8056	11MY	7;25	44.29	8056	26MY	13;47	238.83	8130	5MY	6;28	39.05	8130	19MY	5;32	232.36
8057	9JN	21;30	72.46	8057	24JN	20;40	266.73	8131	3JN	18;05	67.14	8131	17JN	15;50	260.38
8058	9JL	12;30	100.75	8058	24JL	3;37	294.82	8132	3JL	3;59	95.19	8132	17JL	4;18	288.63
8059	8AU	4;08	129.37	8059	22AU	11;48	323.35	8133	1AU	12;46	123.45	8133	15AU	19;06	317.63
8060	6SE	19;59	158.46	8060	20SE	22;10	352.46	8134	30AU	21;14	152.13	8134	14SE	12;02	346.60
8061	6OC	11;19	188.04	8061	20OC	11;13	22.17	8135	29SE	6;17	181.34	8135	14OC	6;24	16.45
8062	5NO	1;26	218.00	8062	19NO	2;57	52.34	8136	28OC	16;36	211.06	8136	13NO	0;54	46.68
8063	4DE	13;56	248.12	8063	18DE	20;44	82.70	8137	27NO	4;37	241.13	8137	12DE	18;01	77.01
								8138	26DE	18;19	271.29				

−348 / −342

NUMBER	DATE	TIME	LONG.	NUMBER	DATE	TIME	LONG.	NUMBER	DATE	TIME	LONG.	NUMBER	DATE	TIME	LONG.
8064	3JA	0;57	278.13	8064	17JA	15;30	112.91					8138	11JA	8;32	107.10
8065	1FE	10;53	307.79	8065	16FE	9;55	142.70	8139	25JA	9;20	301.25	8139	9FE	20;12	136.70
8066	1MR	20;13	336.95	8066	17MR	2;36	171.90	8140	24FE	1;11	330.79	8140	11MR	5;30	165.71
8067	31MR	5;25	5.61	8067	15AP	16;35	200.51	8141	25MR	17;23	359.84	8141	9AP	13;18	194.17
8068	29AP	15;01	33.86	8068	15MY	3;44	228.68	8142	24AP	9;18	28.42	8142	8MY	20;33	222.25
8069	29MY	1;37	61.92	8069	13JN	12;41	256.64	8143	24MY	0;17	56.68	8143	7JN	4;08	250.16
8070	27JN	13;55	90.05	8070	12JL	20;31	284.68	8144	22JN	13;43	84.83	8144	6JL	12;53	278.18
8071	27JL	4;26	118.48	8071	11AU	4;21	313.03	8145	22JL	1;31	113.10	8145	4AU	23;38	306.57
8072	25AU	21;06	147.42	8072	9SE	13;03	341.89	8146	20AU	12;05	141.69	8146	3SE	13;10	335.51
8073	24SE	15;00	176.91	8073	8OC	23;11	11.31	8147	18SE	22;09	170.76	8147	3OC	5;49	5.09
8074	24OC	8;40	206.88	8074	7NO	11;03	41.22	8148	18OC	8;23	200.33	8148	2NO	1;00	35.24
8075	23NO	0;46	237.10	8075	7DE	0;48	71.40	8149	16NO	19;06	230.27	8149	1DE	20;57	65.66
8076	22DE	14;44	267.29					8150	16DE	6;20	260.35	8150	31DE	15;29	96.00

−347 / −341

NUMBER	DATE	TIME	LONG.	NUMBER	DATE	TIME	LONG.	NUMBER	DATE	TIME	LONG.	NUMBER	DATE	TIME	LONG.
				8076	5JA	16;30	101.58								
8077	21JA	2;37	297.16	8077	4FE	9;43	131.50	8151	14JA	18;03	290.29	8151	30JA	7;10	125.93
8078	19FE	12;42	326.54	8078	6MR	3;23	160.94	8152	13FE	6;25	319.87	8152	28FE	19;45	155.26
8079	20MR	21;23	355.35	8079	4AP	20;07	189.83	8153	14MR	19;44	348.99	8153	30MR	5;44	183.98
8080	19AP	5;09	23.66	8080	4MY	10;51	218.24	8154	13AP	10;08	17.66	8154	28AP	13;51	212.20
8081	18MY	12;51	51.67	8081	2JN	23;24	246.36	8155	13MY	1;19	46.02	8155	27MY	20;55	240.13
8082	16JN	21;33	79.62	8082	2JL	10;13	274.45	8156	11JN	16;37	74.23	8156	26JN	3;47	268.02
8083	16JL	8;22	107.80	8083	31JL	20;00	302.74	8157	11JL	7;21	102.52	8157	25JL	11;28	296.16
8084	14AU	22;06	136.44	8084	30AU	5;27	331.46	8158	9AU	21;15	131.09	8158	23AU	21;05	324.77
8085	13SE	14;44	165.69	8085	28SE	15;02	0.68	8159	8SE	10;22	160.10	8159	22SE	9;36	353.99
8086	13OC	9;20	195.53	8086	28OC	1;06	30.38	8160	7OC	22;50	189.59	8160	22OC	1;28	23.84
8087	12NO	4;22	225.78	8087	26NO	12;02	60.40	8161	6NO	10;42	219.47	8161	20NO	20;04	54.15
8088	11DE	22;19	256.15	8088	26DE	0;14	90.50	8162	5DE	21;52	249.54	8162	20DE	15;48	84.58

-340

NEW MOONS Number	Date	Time	Long.	FULL MOONS Number	Date	Time	Long.
8163	4JA	8:23	279.52	8163	19JA	10:46	114.78
8164	2FE	18:33	309.16	8164	18FE	3:34	144.48
8165	3MR	4:54	338.35	8165	18MR	17:35	173.54
8166	1AP	15:59	7.06	8166	17AP	4:49	202.02
8167	1MY	4:11	35.41	8167	16MY	13:42	230.09
8168	30MY	17:35	63.57	8168	14JN	21:02	257.99
8169	29JN	8:08	91.79	8169	14JL	3:54	285.99
8170	28JL	23:38	120.28	8170	12AU	11:28	314.36
8171	27AU	15:42	149.22	8171	10SE	20:46	343.26
8172	26SE	7:42	178.67	8172	10OC	8:32	12.77
8173	25OC	22:47	208.55	8173	8NO	23:00	42.80
8174	24NO	12:16	238.67	8174	8DE	15:47	73.11
8175	24DE	0:02	268.76				

-339

NEW MOONS Number	Date	Time	Long.	FULL MOONS Number	Date	Time	Long.
				8175	7JA	10:03	103.39
8176	22JA	10:21	298.55	8176	6FE	4:38	133.34
8177	20FE	19:45	327.87	8177	7MR	22:09	162.74
8178	22MR	4:43	356.68	8178	6AP	13:28	191.55
8179	20AP	13:49	25.03	8179	6MY	1:59	219.86
8180	19MY	23:38	53.11	8180	4JN	12:00	247.88
8181	18JN	10:54	81.17	8181	3JL	20:23	275.87
8182	18JL	0:16	109.46	8182	2AU	4:16	304.11
8183	16AU	16:01	138.21	8183	31AU	12:38	332.80
8184	15SE	9:42	167.53	8184	29SE	22:10	2.04
8185	15OC	3:56	197.38	8185	29OC	9:14	31.79
8186	13NO	21:10	227.57	8186	27NO	22:00	61.90
8187	13DE	12:20	257.82	8187	27DE	12:36	92.09

-338

NEW MOONS Number	Date	Time	Long.	FULL MOONS Number	Date	Time	Long.
8188	12JA	1:13	287.84	8188	26JA	4:54	122.10
8189	10FE	12:04	317.39	8189	24FE	22:14	151.69
8190	11MR	21:14	346.38	8190	26MR	15:22	180.76
8191	10AP	5:12	14.84	8191	25AP	7:05	209.32
8192	9MY	12:40	42.92	8192	24MY	20:44	237.52
8193	7JN	20:37	70.85	8193	23JN	8:28	265.61
8194	7JL	6:12	98.90	8194	22JL	18:54	293.83
8195	5AU	18:26	127.36	8195	21AU	4:41	322.40
8196	4SE	9:46	156.39	8196	19SE	14:23	351.47
8197	4OC	3:43	186.04	8197	19OC	0:21	21.03
8198	2NO	22:55	216.19	8198	17NO	10:53	50.97
8199	2DE	17:41	246.57	8199	16DE	22:21	81.06

-337

NEW MOONS Number	Date	Time	Long.	FULL MOONS Number	Date	Time	Long.
8200	1JA	10:43	276.85	8200	15JA	11:05	111.04
8201	31JA	1:14	306.71	8201	14FE	1:11	140.68
8202	1MR	12:58	336.00	8202	15MR	16:15	169.85
8203	30MR	22:10	4.67	8203	14AP	7:40	198.56
8204	29AP	5:33	32.85	8204	13MY	22:51	226.90
8205	28MY	12:10	60.76	8205	12JN	13:29	255.09
8206	26JN	19:15	88.67	8206	12JL	3:22	283.35
8207	26JL	3:55	116.85	8207	10AU	16:25	311.90
8208	24AU	15:07	145.53	8208	9SE	4:32	340.88
8209	23SE	5:15	174.84	8209	8OC	15:48	10.33
8210	22OC	22:13	204.75	8210	7NO	2:31	40.17
8211	21NO	17:14	235.07	8211	6DE	13:07	70.21
8212	21DE	12:56	265.50				

-336

NEW MOONS Number	Date	Time	Long.	FULL MOONS Number	Date	Time	Long.
				8212	4JA	23:55	100.20
8213	20JA	7:32	295.68	8213	3FE	11:07	129.88
8214	18FE	23:29	325.33	8214	3MR	22:46	159.11
8215	19MR	12:06	354.32	8215	2AP	10:59	187.86
8216	17AP	21:45	22.73	8216	2MY	0:03	216.23
8217	17MY	5:24	50.74	8217	31MY	14:13	244.42
8218	15JN	12:14	78.62	8218	30JN	5:25	272.67
8219	14JL	19:21	106.64	8219	29JL	21:09	301.18
8220	13AU	3:39	135.04	8220	28AU	12:35	330.11
8221	11SE	13:51	164.00	8221	27SE	3:02	359.50
8222	11OC	2:30	193.56	8222	26OC	16:16	29.32
8223	9NO	18:00	223.64	8223	25NO	4:25	59.39
8224	9DE	12:07	254.01	8224	24DE	15:42	89.45

-335

NEW MOONS Number	Date	Time	Long.	FULL MOONS Number	Date	Time	Long.
8225	8JA	7:40	284.33	8225	23JA	2:10	119.24
8226	7FE	2:38	314.28	8226	21FE	11:54	148.57
8227	8MR	19:15	343.63	8227	22MR	21:16	177.38
8228	7AP	8:50	12.36	8228	21AP	6:58	205.74
8229	6MY	19:40	40.59	8229	20MY	17:54	233.87
8230	5JN	4:36	68.57	8230	19JN	6:46	261.99
8231	4JL	12:30	96.54	8231	18JL	21:41	290.35
8232	2AU	20:13	124.78	8232	17AU	14:08	319.14
8233	1SE	4:30	153.48	8233	16SE	7:12	348.45
8234	30SE	14:05	182.73	8234	15OC	23:57	18.25
8235	30OC	1:42	212.52	8235	14NO	15:40	48.39
8236	28NO	15:50	242.69	8236	14DE	5:47	78.59
8237	28DE	8:21	272.96				

-334

NEW MOONS Number	Date	Time	Long.	FULL MOONS Number	Date	Time	Long.
				8237	12JA	17:58	108.57
8238	27JA	2:15	303.02	8238	11FE	4:09	138.09
8239	25FE	20:04	332.62	8239	12MR	12:47	167.05
8240	27MR	12:38	1.64	8240	10AP	20:37	195.49
8241	26AP	3:17	30.16	8241	10MY	4:41	223.58
8242	25MY	15:57	58.32	8242	8JN	13:58	251.56
8243	24JN	2:47	86.37	8243	8JL	1:15	279.70
8244	23JL	12:14	114.56	8244	6AU	14:55	308.22
8245	21AU	20:59	143.10	8245	5SE	6:59	337.29
8246	20SE	5:51	172.14	8246	5OC	0:57	6.96
8247	19OC	15:36	201.70	8247	3NO	19:47	37.11
8248	18NO	2:48	231.67	8248	3DE	13:55	67.46
8249	17DE	15:36	261.82				

-333

NEW MOONS Number	Date	Time	Long.	FULL MOONS Number	Date	Time	Long.
				8249	2JA	5:52	97.68
8250	16JA	5:48	291.85	8250	31JA	18:53	127.48
8251	14FE	21:01	321.53	8251	2MR	5:10	156.69
8252	16MR	12:48	350.72	8252	31MR	13:28	185.32
8253	15AP	4:43	19.43	8253	29AP	20:45	213.49
8254	14MY	20:11	47.78	8254	29MY	3:58	241.42
8255	13JN	10:31	75.96	8255	27JN	11:58	269.37
8256	12JL	23:19	104.18	8256	26JL	21:38	297.60
8257	11AU	10:41	132.67	8257	25AU	9:46	326.33
8258	9SE	21:12	161.60	8258	24SE	0:58	355.69
8259	9OC	7:31	191.02	8259	23OC	19:08	25.66
8260	7NO	18:06	220.86	8260	22NO	15:04	56.03
8261	7DE	5:06	250.92	8261	22DE	10:36	86.45

-332

NEW MOONS Number	Date	Time	Long.	FULL MOONS Number	Date	Time	Long.
8262	5JA	16:27	280.93	8262	21JA	3:47	116.56
8263	4FE	4:14	310.63	8263	19FE	17:49	146.11
8264	4MR	16:44	339.88	8264	20MR	4:55	175.03
8265	3AP	6:18	8.66	8265	18AP	13:47	203.40
8266	2MY	20:56	37.09	8266	17MY	21:13	231.40
8267	1JN	12:10	65.32	8267	16JN	4:04	259.28
8268	1JL	3:20	93.57	8268	15JL	11:16	287.31
8269	30JL	17:53	122.05	8269	13AU	19:52	315.73
8270	29AU	7:41	150.92	8270	12SE	6:55	344.73
8271	27SE	20:46	180.27	8271	11OC	21:11	14.37
8272	27OC	9:11	210.06	8272	10NO	14:36	44.54
8273	25NO	20:49	240.11	8273	10DE	9:59	74.96
8274	25DE	7:37	270.15				

-331

NEW MOONS Number	Date	Time	Long.	FULL MOONS Number	Date	Time	Long.
				8274	9JA	5:27	105.28
8275	23JA	17:46	299.92	8275	7FE	23:20	135.17
8276	22FE	3:44	329.25	8276	9MR	14:41	164.46
8277	23MR	14:06	358.09	8277	8AP	3:13	193.13
8278	22AP	1:26	26.51	8278	7MY	13:11	221.32
8279	21MY	14:02	54.70	8279	5JN	21:12	249.26
8280	20JN	3:58	82.86	8280	5JL	4:15	277.20
8281	19JL	19:05	111.24	8281	3AU	11:27	305.43
8282	18AU	11:09	140.03	8282	1SE	19:55	334.14
8283	17SE	3:35	169.33	8283	1OC	6:30	3.45
8284	16OC	19:31	199.11	8284	30OC	19:41	33.31
8285	15NO	10:04	229.20	8285	29NO	11:21	63.54
8286	14DE	22:46	259.34	8286	29DE	4:52	93.85

-330

NEW MOONS Number	Date	Time	Long.	FULL MOONS Number	Date	Time	Long.
8287	13JA	9:42	289.26	8287	27JA	23:14	123.92
8288	11FE	19:21	318.75	8288	26FE	17:15	153.52
8289	13MR	4:15	347.71	8289	28MR	9:39	182.52
8290	11AP	13:00	16.18	8290	26AP	23:33	210.99
8291	10MY	22:10	44.31	8291	26MY	10:48	239.10
8292	9JN	8:29	72.33	8292	24JN	19:59	267.08
8293	8JL	20:41	100.50	8293	24JL	4:09	295.23
8294	7AU	11:18	129.07	8294	22AU	12:22	323.76
8295	6SE	4:19	158.20	8295	20SE	21:27	352.82
8296	5OC	22:43	187.90	8296	20OC	7:50	22.42
8297	4NO	16:52	218.03	8297	18NO	19:46	52.42
8298	4DE	9:18	248.31	8298	18DE	9:21	82.60

-329

NEW MOONS Number	Date	Time	Long.	FULL MOONS Number	Date	Time	Long.
8299	2JA	23:21	278.45	8299	17JA	0:37	112.67
8300	1FE	11:06	308.18	8300	15FE	17:16	142.39
8301	2MR	20:55	337.37	8301	17MR	10:24	171.63
8302	1AP	5:15	5.98	8302	16AP	2:47	200.34
8303	30AP	12:43	34.16	8303	15MY	17:28	228.66
8304	29MY	20:10	62.09	8304	14JN	6:12	256.77
8305	28JN	4:44	90.06	8305	13JL	17:24	284.94
8306	27JL	15:32	118.35	8306	12AU	3:41	313.39
8307	26AU	5:24	147.16	8307	10SE	13:38	342.30
8308	24SE	22:18	176.61	8308	9OC	23:38	11.72
8309	24OC	17:13	206.62	8309	8NO	9:57	41.56
8310	23NO	12:32	236.97	8310	7DE	20:54	71.62
8311	23DE	6:38	267.33				

-328

NEW NUMBER	DATE	TIME	LONG.	FULL NUMBER	DATE	TIME	LONG.
				8311	6JA	8;49	101.65
8312	21JA	22;27	297.37	8312	4FE	22;00	131.39
8313	20FE	11;29	326.87	8313	5MR	12;20	160.71
8314	20MR	21;47	355.75	8314	4AP	3;22	189.54
8315	19AP	5;52	24.08	8315	3MY	18;33	217.98
8316	18MY	12;39	52.05	8316	2JN	9;27	246.20
8317	16JN	19;19	79.92	8317	1JL	23;50	274.42
8318	16JL	3;05	107.97	8318	31JL	13;34	302.87
8319	14AU	13;00	136.46	8319	30AU	2;28	331.72
8320	13SE	1;43	165.54	8320	28SE	14;25	1.04
8321	12OC	17;25	195.25	8321	28OC	1;34	30.79
8322	11NO	11;35	225.46	8322	26NO	12;16	60.80
8323	11DE	7;09	255.89	8323	25DE	22;54	90.83

-327

NEW NUMBER	DATE	TIME	LONG.	FULL NUMBER	DATE	TIME	LONG.
8324	10JA	2;29	286.20	8324	24JA	9;43	120.62
8325	8FE	19;49	316.05	8325	22FE	20;52	149.99
8326	10MR	10;02	345.28	8326	24MR	8;29	178.86
8327	8AP	21;00	13.87	8327	22AP	20;49	207.32
8328	8MY	5;28	42.00	8328	22MY	10;14	235.53
8329	6JN	12;35	69.90	8329	21JN	0;54	263.72
8330	5JL	19;30	97.84	8330	20JL	16;36	292.13
8331	4AU	3;13	126.10	8331	19AU	8;32	320.93
8332	2SE	12;31	154.85	8332	17SE	23;51	350.20
8333	1OC	23;58	184.21	8333	17OC	13;58	19.91
8334	31OC	14;04	214.12	8334	16NO	2;50	49.94
8335	30NO	6;55	244.41	8335	15DE	14;36	80.04
8336	30DE	1;51	274.77				

-326

NEW NUMBER	DATE	TIME	LONG.	FULL NUMBER	DATE	TIME	LONG.
				8336	14JA	1;24	109.95
8337	28JA	21;11	304.87	8337	12FE	11;19	139.44
8338	27FE	15;00	334.44	8338	13MR	20;34	168.40
8339	29MR	6;00	3.37	8339	12AP	5;44	196.88
8340	27AP	18;06	31.76	8340	11MY	15;43	225.04
8341	27MY	3;55	59.81	8341	10JN	3;25	253.11
8342	25JN	12;20	87.76	8342	9JL	17;18	281.36
8343	24JL	20;12	115.90	8343	8AU	9;12	309.99
8344	23AU	4;17	144.44	8344	7SE	2;18	339.13
8345	21SE	13;17	173.50	8345	6OC	19;35	8.80
8346	20OC	23;54	203.12	8346	5NO	12;09	38.87
8347	19NO	12;45	233.17	8347	5DE	3;16	69.11
8348	19DE	4;02	263.42				

-325

NEW NUMBER	DATE	TIME	LONG.	FULL NUMBER	DATE	TIME	LONG.
				8348	3JA	16;29	99.20
8349	17JA	21;11	293.56	8349	2FE	3;33	128.90
8350	16FE	14;58	323.32	8350	3MR	12;43	158.04
8351	18MR	8;03	352.53	8351	1AP	20;39	186.63
8352	16AP	23;35	21.21	8352	1MY	4;17	214.81
8353	16MY	13;14	49.47	8353	30MY	12;42	242.78
8354	15JN	1;03	77.55	8354	28JN	22;50	270.82
8355	14JL	11;19	105.68	8355	28JL	11;17	299.17
8356	12AU	20;34	134.11	8356	27AU	2;17	328.04
8357	11SE	5;30	162.99	8357	25SE	19;33	357.52
8358	10OC	14;54	192.39	8358	25OC	14;19	27.54
8359	9NO	1;25	222.25	8359	24NO	9;11	57.88
8360	8DE	13;23	252.35	8360	24DE	2;27	88.20

-324

NEW NUMBER	DATE	TIME	LONG.	FULL NUMBER	DATE	TIME	LONG.
8361	7JA	2;46	282.43	8361	22JA	16;57	118.18
8362	5FE	17;16	312.23	8362	21FE	4;26	147.60
8363	6MR	8;30	341.56	8363	21MR	13;28	176.41
8364	5AP	0;09	10.41	8364	19AP	21;01	204.71
8365	4MY	15;47	38.85	8365	19MY	4;03	232.69
8366	3JN	6;46	67.07	8366	17JN	11;30	260.59
8367	2JL	20;31	95.27	8367	16JL	20;15	288.69
8368	1AU	8;49	123.68	8368	15AU	7;06	317.23
8369	30AU	19;57	152.47	8369	13SE	20;48	346.36
8370	29SE	6;33	181.74	8370	13OC	10;13	16.13
8371	28OC	17;09	211.47	8371	12NO	9;03	46.40
8372	27NO	4;01	241.49	8372	12DE	5;06	76.85
8373	26DE	15;08	271.54				

-323

NEW NUMBER	DATE	TIME	LONG.	FULL NUMBER	DATE	TIME	LONG.
				8373	10JA	23;37	107.12
8374	25JA	2;29	301.35	8374	9FE	15;11	136.88
8375	23FE	14;18	330.73	8375	11MR	3;23	166.02
8376	25MR	2;59	359.64	8376	9AP	13;23	194.56
8377	23AP	16;50	28.15	8377	8MY	21;23	222.66
8378	23MY	7;42	56.41	8378	7JN	4;25	250.56
8379	21JN	23;00	84.63	8379	6JL	11;22	278.51
8380	21JL	14;05	113.03	8380	4AU	19;12	306.77
8381	20AU	4;32	141.77	8381	3SE	5;00	335.56
8382	18SE	18;17	170.99	8382	2OC	17;41	4.98
8383	18OC	7;19	200.67	8383	1NO	9;39	34.98
8384	16NO	19;31	230.68	8384	1DE	4;14	65.34
8385	16DE	6;46	260.76	8385	30DE	23;49	95.73

-322

NEW NUMBER	DATE	TIME	LONG.	FULL NUMBER	DATE	TIME	LONG.
8386	14JA	17;07	290.64	8386	29JA	18;33	125.79
8387	13FE	2;54	320.11	8387	28FE	11;06	155.29
8388	14MR	12;43	349.09	8388	30MR	0;57	184.17
8389	12AP	23;15	17.61	8389	28AP	12;07	212.51
8390	12MY	10;58	45.83	8390	27MY	21;03	240.52
8391	11JN	0;06	73.96	8391	26JN	4;32	268.44
8392	10JL	14;49	102.24	8392	25JL	11;39	296.55
8393	9AU	6;28	130.88	8393	23AU	19;30	325.09
8394	7SE	23;04	160.02	8394	22SE	5;04	354.19
8395	7OC	15;40	189.67	8395	21OC	17;01	23.87
8396	6NO	7;15	219.71	8396	20NO	7;29	54.00
8397	5DE	21;02	249.89	8397	20DE	0;03	84.30

-321

NEW NUMBER	DATE	TIME	LONG.	FULL NUMBER	DATE	TIME	LONG.
8398	4JA	8;48	279.92	8398	18JA	17;57	114.46
8399	2FE	18;55	309.57	8399	17FE	12;05	144.22
8400	4MR	3;56	338.70	8400	19MR	5;15	173.42
8401	2AP	12;29	7.30	8401	17AP	20;22	202.07
8402	1MY	21;09	35.51	8402	17MY	8;56	230.28
8403	31MY	6;39	63.52	8403	15JN	19;10	258.30
8404	29JN	17;44	91.60	8404	15JL	3;53	286.38
8405	29JL	7;07	120.00	8405	13AU	12;09	314.77
8406	27AU	23;08	148.93	8406	11SE	20;55	343.66
8407	26SE	17;14	178.45	8407	11OC	6;46	13.09
8408	26OC	11;57	208.48	8408	9NO	17;59	42.98
8409	25NO	5;34	238.77	8409	9DE	6;40	73.12
8410	24DE	20;54	269.01				

-320

NEW NUMBER	DATE	TIME	LONG.	FULL NUMBER	DATE	TIME	LONG.
				8410	7JA	20;56	103.23
8411	23JA	9;45	298.91	8411	6FE	12;41	133.06
8412	21FE	20;22	328.29	8412	7MR	5;26	162.45
8413	22MR	5;13	357.08	8413	5AP	21;07	191.32
8414	20AP	12;52	25.38	8414	5MY	13;38	219.75
8415	19MY	20;05	53.35	8415	4JN	3;23	247.92
8416	18JN	3;52	81.27	8416	3JL	15;27	276.06
8417	17JL	13;25	109.41	8417	2AU	2;21	304.42
8418	16AU	1;44	138.01	8418	31AU	12;41	333.18
8419	14SE	17;17	167.24	8419	29SE	22;51	2.45
8420	14OC	11;31	197.08	8420	29OC	9;08	32.18
8421	13NO	6;59	227.36	8421	27NO	19;46	62.19
8422	13DE	1;56	257.77	8422	27DE	7;04	92.25

-319

NEW NUMBER	DATE	TIME	LONG.	FULL NUMBER	DATE	TIME	LONG.
8423	11JA	18;59	287.96	8423	25JA	19;24	122.08
8424	10FE	9;23	317.67	8424	24FE	8;53	151.53
8425	11MR	20;55	346.76	8425	25MR	23;20	180.49
8426	10AP	5;54	15.26	8426	24AP	14;15	209.03
8427	9MY	13;08	43.32	8427	24MY	5;14	237.29
8428	7JN	19;40	71.19	8428	22JN	19;58	265.50
8429	7JL	2;45	99.15	8429	22JL	10;15	293.87
8430	5AU	11;32	127.46	8430	20AU	23;52	322.59
8431	3SE	22;54	156.33	8431	19SE	12;36	351.78
8432	3OC	13;13	185.82	8432	19OC	0;23	21.42
8433	2NO	6;18	215.88	8433	17NO	11;24	51.38
8434	2DE	1;21	246.27	8434	16DE	22;01	81.43
8435	31DE	20;58	276.65				

-318

NEW NUMBER	DATE	TIME	LONG.	FULL NUMBER	DATE	TIME	LONG.
				8435	15JA	8;36	111.32
8436	30JA	15;24	306.69	8436	13FE	19;21	140.82
8437	1MR	7;10	336.14	8437	15MR	6;26	169.84
8438	30MR	19;38	4.95	8438	13AP	18;05	198.40
8439	29AP	5;12	33.22	8439	13MY	6;41	226.65
8440	28MY	14;37	61.17	8440	11JN	20;37	254.81
8441	26JN	19;46	89.08	8441	11JL	11;55	283.12
8442	26JL	3;05	117.20	8442	10AU	4;03	311.79
8443	24AU	11;38	145.78	8443	8SE	20;05	340.92
8444	22SE	22;04	174.93	8444	8OC	11;09	10.52
8445	22OC	10;52	204.66	8445	7NO	0;52	40.49
8446	21NO	2;21	234.83	8446	6DE	13;16	70.61
8447	20DE	20;17	265.19				

-317

NEW NUMBER	DATE	TIME	LONG.	FULL NUMBER	DATE	TIME	LONG.
				8447	5JA	0;31	100.61
8448	19JA	15;30	295.40	8448	3FE	10;43	130.25
8449	18FE	10;06	325.16	8449	4MR	20;02	159.38
8450	20MR	2;28	354.32	8450	3AP	4;55	188.00
8451	18AP	15;55	22.88	8451	2MY	14;09	216.22
8452	18MY	2;48	51.03	8452	1JN	0;43	244.27
8453	16JN	11;53	78.99	8453	30JN	13;25	272.41
8454	15JL	20;05	107.06	8454	30JL	4;25	300.89
8455	14AU	4;09	135.45	8455	28AU	21;14	329.86
8456	12SE	12;47	164.34	8456	27SE	14;49	359.38
8457	11OC	22;38	193.78	8457	27OC	8;05	29.36
8458	10NO	10;22	223.70	8458	26NO	0;11	59.59
8459	10DE	0;22	253.90	8459	25DE	14;28	89.78

-316

NEW NUMBER	DATE	TIME	LONG.	FULL NUMBER	DATE	TIME	LONG.
8460	8JA	16;32	284.08	8460	24JA	2;33	119.64
8461	7FE	9;54	313.97	8461	22FE	12;29	148.98
8462	8MR	3;13	343.37	8462	22MR	20;45	177.74
8463	6AP	19;26	12.21	8463	21AP	4;14	206.02
8464	6MY	9;59	40.59	8464	20MY	11;59	234.02
8465	4JN	22;46	68.72	8465	18JN	21;03	261.99
8466	4JL	9;56	96.83	8466	18JL	8;17	290.20
8467	2AU	19;51	125.15	8467	16AU	22;04	318.87
8468	1SE	5;05	153.89	8468	15SE	14;24	348.14
8469	30SE	14;22	183.13	8469	15OC	8;44	18.00
8470	30OC	0;23	212.85	8470	14NO	3;55	48.28
8471	28NO	11;38	242.90	8471	13DE	22;17	78.66
8472	28DE	0;13	273.00				

-315

NEW NUMBER	DATE	TIME	LONG.	FULL NUMBER	DATE	TIME	LONG.
8473	26JA	13;58	302.89	8472	12JA	14;16	108.80
8474	25FE	4;34	332.36	8473	11FE	3;09	138.43
8475	26MR	19;46	1.35	8474	12MR	13;12	167.45
8476	25AP	11;16	29.90	8475	10AP	21;14	195.90
8477	25MY	2;36	58.17	8476	10MY	4;18	223.95
8478	23JN	17;07	86.37	8477	8JN	11;23	251.85
8479	23JL	6;23	114.71	8478	7JL	19;23	279.85
8480	21AU	18;20	143.38	8479	6AU	5;07	308.20
8481	20SE	5;24	172.51	8480	4SE	17;24	337.12
8482	19OC	16;10	202.12	8481	4OC	8;47	6.67
8483	18NO	3;00	232.07	8482	3NO	3;09	36.79
8484	17DE	14;00	262.14	8483	2DE	23;11	67.22

-314

NEW NUMBER	DATE	TIME	LONG.	FULL NUMBER	DATE	TIME	LONG.
8485	16JA	1;04	292.04	8484	1JA	18;43	97.60
8486	14FE	12;21	321.56	8485	31JA	11;47	127.57
8487	16MR	0;15	350.59	8486	2MR	1;39	156.93
8488	14AP	13;13	19.19	8487	31MR	12;35	185.67
8489	14MY	3;24	47.50	8488	29AP	21;19	213.90
8490	12JN	18;31	75.71	8489	29MY	4;43	241.83
8491	12JL	9;54	104.03	8490	27JN	11;37	269.74
8492	11AU	0;56	132.66	8491	26JL	18;58	297.87
8493	9SE	15;21	161.74	8492	25AU	3;46	326.47
8494	9OC	5;02	191.30	8493	23SE	15;01	355.66
8495	7NO	17;53	221.24	8494	23OC	5;24	25.47
8496	7DE	5;43	251.34	8495	21NO	22;50	55.73
				8496	21DE	18;04	86.14

-313

NEW NUMBER	DATE	TIME	LONG.	FULL NUMBER	DATE	TIME	LONG.
8497	5JA	16;26	281.32	8497	20JA	13;18	116.34
8498	4FE	2;17	310.93	8498	19FE	6;54	146.05
8499	5MR	11;46	340.06	8499	20MR	22;02	175.14
8500	3AP	21;36	8.70	8500	19AP	10;28	203.65
8501	3MY	8;26	36.97	8501	18MY	20;27	231.76
8502	1JN	20;40	65.09	8502	17JN	4;37	259.69
8503	1JL	10;29	93.28	8503	16JL	11;54	287.72
8504	31JL	1;48	121.78	8504	14AU	19;24	316.10
8505	29AU	18;18	150.76	8505	13SE	4;10	345.01
8506	28SE	11;18	180.27	8506	12OC	14;59	14.50
8507	28OC	3;48	210.23	8507	11NO	4;14	44.49
8508	26NO	18;44	240.41	8508	10DE	19;45	74.75
8509	26DE	7;32	270.54				

-312

NEW NUMBER	DATE	TIME	LONG.	FULL NUMBER	DATE	TIME	LONG.
8510	24JA	18;20	300.33	8509	9JA	12;55	104.97
8511	23FE	3;40	329.63	8510	8FE	6;50	134.87
8512	23MR	12;11	358.39	8511	9MR	0;25	164.26
8513	21AP	20;30	26.70	8512	7AP	16;28	193.09
8514	21MY	5;19	54.73	8513	7MY	6;24	221.44
8515	19JN	15;25	82.75	8514	5JN	17;48	249.51
8516	19JL	3;33	111.00	8515	5JL	2;56	277.55
8517	17AU	18;20	139.73	8516	3AU	11;53	305.83
8518	16SE	11;42	169.06	8517	1SE	20;31	334.55
8519	16OC	6;34	198.95	8518	1OC	5;58	3.81
8520	15NO	1;07	229.21	8519	30OC	16;35	33.57
8521	14DE	17;47	259.52	8520	29NO	4;31	63.64
				8521	28DE	17;50	93.77

-311

NEW NUMBER	DATE	TIME	LONG.	FULL NUMBER	DATE	TIME	LONG.
8522	13JA	7;52	289.57	8522	27JA	8;38	123.70
8523	11FE	19;28	319.14	8523	26FE	0;41	153.22
8524	13MR	5;00	348.12	8524	27MR	17;17	182.25
8525	11AP	13;02	16.56	8525	26AP	9;21	210.81
8526	10MY	20;14	44.62	8526	26MY	0;00	239.05
8527	9JN	3;31	72.52	8527	24JN	13;00	267.19
8528	8JL	12;00	100.54	8528	24JL	0;38	295.47
8529	6AU	22;51	128.95	8529	22AU	11;27	324.11
8530	5SE	12;52	157.95	8530	20SE	21;56	353.22
8531	5OC	6;01	187.59	8531	20OC	8;20	22.83
8532	4NO	1;13	217.75	8532	18NO	18;51	52.77
8533	3DE	20;43	248.17	8533	18DE	5;44	82.84

-310

NEW NUMBER	DATE	TIME	LONG.	FULL NUMBER	DATE	TIME	LONG.
8534	2JA	14;52	278.48	8534	16JA	17;20	112.75
8535	1FE	6;36	308.38	8535	15FE	5;58	142.31
8536	2MR	19;28	337.69	8536	16MR	19;40	171.41
8537	1AP	5;33	6.38	8537	15AP	10;08	200.06
8538	30AP	13;28	34.57	8538	15MY	0;58	228.39
8539	29MY	20;10	62.48	8539	13JN	15;50	256.58
8540	28JN	2;50	90.37	8540	13JL	6;31	284.89
8541	27JL	10;42	118.54	8541	11AU	20;46	313.50
8542	25AU	20;46	147.19	8542	10SE	10;17	342.55
8543	24SE	9;41	176.47	8543	9OC	22;49	12.08
8544	24OC	1;30	206.34	8544	8NO	10;22	41.97
8545	22NO	19;44	236.65	8545	7DE	21;12	72.02
8546	22DE	15;13	267.07				

-309

NEW NUMBER	DATE	TIME	LONG.	FULL NUMBER	DATE	TIME	LONG.
8547	21JA	10;23	297.25	8546	6JA	7;42	101.99
8548	20FE	3;30	326.93	8547	4FE	18;09	131.62
8549	21MR	17;33	355.96	8548	6MR	4;47	160.78
8550	20AP	4;25	24.40	8549	4AP	15;49	189.46
8551	19MY	12;53	52.44	8550	4MY	3;38	217.77
8552	17JN	20;04	80.33	8551	2JN	16;43	245.92
8553	17JL	3;10	108.36	8552	2JL	7;21	274.15
8554	15AU	11;08	136.77	8553	31JL	23;19	302.68
8555	13SE	20;41	165.72	8554	30AU	15;48	331.67
8556	13OC	8;20	195.26	8555	29SE	7;44	1.15
8557	11NO	22;29	225.29	8556	28OC	22;24	31.04
8558	11DE	15;12	255.60	8557	27NO	11;36	61.16
				8558	26DE	23;26	91.24

-308

NEW NUMBER	DATE	TIME	LONG.	FULL NUMBER	DATE	TIME	LONG.
8559	10JA	9;48	285.88	8559	25JA	10;03	121.02
8560	9FE	4;46	315.81	8560	23FE	19;36	150.32
8561	9MR	22;14	345.18	8561	24MR	4;24	179.08
8562	8AP	13;04	13.94	8562	22AP	13;07	207.39
8563	8MY	1;09	42.21	8563	21MY	22;42	235.45
8564	6JN	11;06	70.22	8564	20JN	10;09	263.52
8565	5JL	19;46	98.24	8565	20JL	0;01	291.85
8566	4AU	4;00	126.51	8566	18AU	16;09	320.65
8567	2SE	12;28	155.24	8567	17SE	9;43	350.00
8568	1OC	21;47	184.50	8568	17OC	3;32	19.86
8569	31OC	8;35	214.27	8569	15NO	20;32	50.06
8570	29NO	21;24	244.39	8570	15DE	11;54	80.32
8571	29DE	12;24	274.59				

-307

NEW NUMBER	DATE	TIME	LONG.	FULL NUMBER	DATE	TIME	LONG.
8572	28JA	5;04	304.58	8571	14JA	1;05	110.32
8573	26FE	22;17	334.14	8572	12FE	11;58	139.85
8574	28MR	14;56	3.15	8573	13MR	20;49	168.80
8575	27AP	6;14	31.67	8574	12AP	4;23	197.21
8576	26MY	19;56	59.87	8575	11MY	11;43	225.26
8577	25JN	8;01	87.98	8576	9JN	19;54	253.20
8578	24JL	18;44	116.23	8577	9JL	5;56	281.29
8579	23AU	4;28	144.83	8578	7AU	18;26	309.78
8580	21SE	13;52	173.92	8579	6SE	9;38	338.83
8581	20OC	23;37	203.50	8580	6OC	3;14	8.51
8582	19NO	10;17	233.46	8581	4NO	22;20	38.68
8583	18DE	22;08	263.56	8582	4DE	17;27	69.08

-306

NEW NUMBER	DATE	TIME	LONG.	FULL NUMBER	DATE	TIME	LONG.
8584	17JA	11;09	293.53	8583	3JA	10;49	99.35
8585	16FE	1;04	323.13	8584	2FE	1;14	129.19
8586	17MR	15;41	352.26	8585	3MR	12;31	158.42
8587	16AP	6;49	20.92	8586	1AP	21;19	187.05
8588	15MY	22;11	49.26	8587	1MY	4;38	215.21
8589	14JN	13;14	77.46	8588	30MY	11;32	243.12
8590	14JL	3;21	105.75	8589	28JN	18;57	271.05
8591	12AU	16;12	134.32	8590	28JL	3;45	299.25
8592	11SE	3;57	163.31	8591	26AU	14;44	327.96
8593	10OC	15;03	192.79	8592	25SE	4;37	357.29
8594	9NO	2;00	222.66	8593	24OC	21;39	27.22
8595	8DE	12;57	252.72	8594	23NO	17;09	57.58
				8595	23DE	13;12	88.03

-305

NEW NUMBER	DATE	TIME	LONG.	FULL NUMBER	DATE	TIME	LONG.
8596	6JA	23;53	282.70	8596	22JA	7;37	118.17
8597	5FE	10;49	312.34	8597	20FE	23;02	147.76
8598	6MR	22;04	341.52	8598	22MR	11;16	176.71
8599	5AP	10;08	10.23	8599	20AP	20;56	205.09
8600	4MY	23;29	38.59	8600	20MY	4;52	233.10
8601	3JN	14;05	66.80	8601	18JN	11;56	260.99
8602	3JL	5;27	95.06	8602	17JL	19;01	289.03
8603	1AU	20;55	123.58	8603	16AU	3;04	317.45
8604	31AU	11;57	152.52	8604	14SE	13;05	346.43
8605	30SE	2;19	181.95	8605	14OC	1;56	16.03
8606	29OC	15;52	211.81	8606	12NO	17;56	46.15
8607	28NO	4;22	241.90	8607	12DE	12;25	76.53
8608	27DE	15;38	271.95				

```
--------NEW MOONS--------        --------FULL MOONS-------        --------NEW MOONS--------        --------FULL MOONS-------
NUMBER  DATE   TIME   LONG.      NUMBER  DATE   TIME   LONG.      NUMBER  DATE   TIME   LONG.      NUMBER  DATE   TIME   LONG.
```

-304 / -298

NEW MOONS				FULL MOONS				NEW MOONS				FULL MOONS			
8609	26JA	1;45	301.71	8608	11JA	7;45	106.83	8683	19JA	0;37	295.16	8682	4JA	23;04	100.94
8610	24FE	11;08	330.99	8609	10FE	2;10	136.73	8684	17FE	17;26	324.86	8683	3FE	11;02	130.65
8611	24MR	20;26	359.76	8610	10MR	18;28	166.03	8685	19MR	10;13	354.04	8684	4MR	20;41	159.80
8612	23AP	6;27	28.12	8611	9AP	8;10	194.74	8686	18AP	2;05	22.71	8685	3AP	4;37	188.37
8613	22MY	17;45	56.24	8612	8MY	19;19	222.97	8687	17MY	16;35	51.00	8686	2MY	11;47	216.51
8614	21JN	6;41	84.37	8613	7JN	4;22	250.94	8688	16JN	5;34	79.13	8687	31MY	19;17	244.44
8615	20JL	21;18	112.74	8614	6JL	12;05	278.92	8689	15JL	17;08	107.32	8688	30JN	4;14	272.43
8616	19AU	13;26	141.55	8615	4AU	19;30	307.17	8690	14AU	3;33	135.81	8689	29JL	15;28	300.75
8617	18SE	6;34	170.89	8616	3SE	3;40	335.89	8691	12SE	13;17	164.75	8690	28AU	5;24	329.60
8618	17OC	23;44	200.74	8617	2OC	13;31	5.19	8692	11OC	22;59	194.19	8691	26SE	21;59	359.07
8619	16NO	15;47	230.91	8618	1NO	1;36	35.02	8693	10NO	9;15	224.04	8692	26OC	16;39	29.10
8620	16DE	5;46	261.10	8619	30NO	16;00	65.21	8694	9DE	20;29	254.12	8693	25NO	12;06	59.47
				8620	30DE	8;17	95.46					8694	25DE	6;35	89.84

-303 / -297

NEW MOONS				FULL MOONS				NEW MOONS				FULL MOONS			
8621	14JA	17;29	291.04	8621	29JA	1;43	125.47	8695	8JA	8;47	284.14	8695	23JA	22;33	119.86
8622	13FE	3;21	320.52	8622	27FE	19;22	155.03	8696	6FE	22;01	313.87	8696	22FE	11;16	149.32
8623	14MR	12;01	349.45	8623	29MR	12;10	184.04	8697	8MR	12;00	343.13	8697	23MR	21;05	178.14
8624	12AP	20;09	17.87	8624	28AP	3;09	212.54	8698	7AP	2;36	11.92	8698	22AP	4;55	206.43
8625	12MY	4;27	45.96	8625	27MY	15;49	240.69	8699	6MY	17;43	40.33	8699	21MY	11;50	234.39
8626	10JN	13;41	73.93	8626	26JN	2;19	268.74	8700	5JN	8;59	68.56	8700	19JN	18;51	262.28
8627	10JL	0;40	102.07	8627	25JL	11;26	296.93	8701	4JL	23;46	96.81	8701	19JL	2;53	290.37
8628	8AU	14;08	130.60	8628	23AU	20;09	325.50	8702	3AU	13;31	125.28	8702	17AU	12;45	318.88
8629	7SE	6;24	159.72	8629	22SE	5;20	354.59	8703	2SE	2;04	154.15	8703	16SE	1;13	347.98
8630	7OC	0;55	189.44	8630	21OC	15;29	24.20	8704	1OC	13;43	183.49	8704	15OC	16;46	17.72
8631	5NO	20;03	219.63	8631	20NO	2;47	54.19	8705	31OC	0;54	213.26	8705	14NO	11;14	47.96
8632	5DE	13;57	249.98	8632	19DE	15;19	84.32	8706	29NO	11;56	243.30	8706	14DE	7;18	78.41
								8707	28DE	22;50	273.33				

-302 / -296

NEW MOONS				FULL MOONS				NEW MOONS				FULL MOONS			
8633	4JA	5;24	280.17	8633	18JA	5;10	114.31	8708	27JA	9;35	303.09	8707	13JA	2;44	108.70
8634	2FE	18;08	309.92	8634	16FE	20;21	143.96	8709	25FE	20;21	332.41	8708	11FE	19;38	138.51
8635	4MR	4;31	339.10	8635	18MR	12;29	173.13	8710	26MR	7;39	1.24	8709	12MR	9;20	167.68
8636	2AP	13;05	7.71	8636	17AP	4;44	201.83	8711	24AP	20;03	29.68	8710	10AP	20;07	196.24
8637	1MY	20;29	35.87	8637	16MY	20;07	230.16	8712	24MY	9;52	57.89	8711	10MY	4;46	224.36
8638	31MY	3;30	63.78	8638	15JN	10;01	258.32	8713	23JN	0;55	86.11	8712	8JN	12;12	252.26
8639	29JN	11;12	91.72	8639	14JL	22;28	286.55	8714	22JL	16;33	114.54	8713	7JL	19;14	280.23
8640	28JL	20;46	119.97	8640	13AU	9;54	315.07	8715	21AU	8;06	143.35	8714	6AU	2;47	308.50
8641	27AU	9;12	148.74	8641	11SE	20;47	344.04	8716	19SE	23;08	172.64	8715	4SE	11;49	337.27
8642	26SE	0;57	178.16	8642	11OC	7;26	13.50	8717	19OC	13;24	202.39	8716	3OC	23;15	6.66
8643	25OC	19;26	208.17	8643	9NO	18;00	43.37	8718	18NO	2;38	232.44	8717	2NO	13;44	36.61
8644	24NO	15;06	238.55	8644	9DE	4;40	73.42	8719	17DE	14;35	262.55	8718	2DE	7;05	66.93
8645	24DE	10;07	268.94												

-301 / -295

NEW MOONS				FULL MOONS				NEW MOONS				FULL MOONS			
8646	23JA	3;07	299.02	8645	7JA	15;45	103.40	8720	16JA	1;11	292.43	8719	1JA	2;06	97.29
8647	21FE	17;22	328.55	8646	6FE	3;36	133.07	8721	14FE	10;41	321.87	8720	30JA	21;00	127.34
8648	23MR	4;42	357.44	8647	7MR	16;28	162.30	8722	15MR	19;41	350.79	8721	1MR	14;18	156.85
8649	21AP	13;32	25.79	8648	6AP	6;18	191.07	8723	14AP	5;01	19.25	8722	31MR	5;13	185.75
8650	20MY	20;39	53.76	8649	5MY	20;47	219.47	8724	13MY	15;23	47.41	8723	29AP	17;36	214.13
8651	19JN	3;11	81.63	8650	4JN	11;35	247.68	8725	12JN	3;21	75.49	8724	29MY	3;40	242.18
8652	18JL	10;22	109.67	8651	4JL	2;29	275.93	8726	11JL	17;07	103.74	8725	27JN	12;03	270.14
8653	16AU	19;18	138.14	8652	2AU	17;13	304.43	8727	10AU	8;39	132.39	8726	26JL	19;38	298.29
8654	15SE	6;50	167.19	8653	1SE	7;27	333.36	8728	9SE	1;35	161.56	8727	25AU	3;29	326.84
8655	14OC	21;19	196.87	8654	30SE	20;48	2.75	8729	8OC	19;10	191.27	8728	23SE	12;33	355.95
8656	13NO	14;28	227.05	8655	30OC	9;02	32.56	8730	7NO	12;09	221.39	8729	22OC	23;34	25.61
8657	13DE	9;28	257.46	8656	28NO	20;18	62.60	8731	7DE	3;23	251.63	8730	21NO	12;50	55.69
				8657	28DE	6;54	92.63					8731	21DE	4;08	85.94

-300 / -294

NEW MOONS				FULL MOONS				NEW MOONS				FULL MOONS			
8658	12JA	4;55	287.75	8658	26JA	17;12	122.38	8732	5JA	16;14	281.69	8732	19JA	20;52	116.04
8659	10FE	23;07	317.63	8659	25FE	3;29	151.69	8733	4FE	2;52	311.34	8733	18FE	14;16	145.75
8660	11MR	14;41	346.89	8660	25MR	14;00	180.50	8734	5MR	11;53	340.45	8734	20MR	7;25	174.93
8661	10AP	3;01	15.52	8661	24AP	1;06	208.89	8735	3AP	20;00	9.02	8735	18AP	23;18	203.59
8662	9MY	12;33	43.68	8662	23MY	13;19	237.05	8736	3MY	3;57	37.18	8736	18MY	13;09	231.85
8663	7JN	20;16	71.60	8663	22JN	3;05	265.21	8737	1JN	12;29	65.15	8737	17JN	0;48	259.92
8664	7JL	3;22	99.56	8664	21JL	18;31	293.62	8738	30JN	22;25	93.19	8738	16JL	10;41	288.06
8665	5AU	10;56	127.83	8665	20AU	11;05	322.46	8739	30JL	10;35	121.56	8739	14AU	19;42	316.50
8666	3SE	19;46	156.58	8666	19SE	3;43	351.80	8740	29AU	1;33	150.46	8740	13SE	4;47	345.42
8667	3OC	6;25	185.93	8667	18OC	19;23	21.60	8741	27SE	19;15	179.98	8741	12OC	14;36	14.87
8668	1NO	19;20	215.80	8668	17NO	9;31	51.69	8742	27OC	14;31	210.05	8742	11NO	1;24	44.76
8669	1DE	10;44	246.03	8669	16DE	22;05	81.82	8743	26NO	9;24	240.40	8743	10DE	13;16	74.86
8670	31DE	4;23	276.34					8744	26DE	2;14	270.70				

-299 / -293

NEW MOONS				FULL MOONS				NEW MOONS				FULL MOONS			
8671	29JA	23;12	306.40	8670	15JA	9;14	111.73	8745	24JA	16;16	300.63	8744	9JA	2;17	104.91
8672	28FE	17;25	335.97	8671	13FE	19;09	141.20	8746	23FE	3;40	330.02	8745	7FE	16;32	134.66
8673	30MR	9;30	4.93	8672	15MR	4;03	170.13	8747	24MR	12;57	358.81	8746	9MR	7;58	163.97
8674	28AP	22;53	33.36	8673	13AP	12;29	198.56	8748	22AP	20;43	27.09	8747	8AP	0;01	192.81
8675	28MY	9;51	61.44	8674	12MY	21;19	226.66	8749	22MY	3;43	55.05	8748	7MY	15;48	221.24
8676	26JN	19;19	89.44	8675	11JN	7;35	254.67	8750	20JN	10;54	82.95	8749	6JN	6;30	249.44
8677	26JL	3;44	117.62	8676	10JL	20;10	282.88	8751	19JL	19;24	111.06	8750	5JL	19;49	277.64
8678	24AU	12;12	146.19	8677	9AU	11;18	311.49	8752	18AU	6;20	139.63	8751	4AU	7;57	306.06
8679	22SE	21;17	175.28	8678	8SE	4;29	340.66	8753	16SE	20;32	168.81	8752	2SE	19;21	334.89
8680	22OC	7;19	204.89	8679	7OC	22;34	10.38	8754	16OC	13;53	198.64	8753	2OC	6;21	4.22
8681	20NO	19;05	234.91	8680	6NO	16;19	40.52	8755	15NO	9;16	228.92	8754	31OC	17;07	33.98
8682	20DE	8;54	265.09	8681	6DE	8;43	70.80	8756	15DE	4;52	259.36	8755	30NO	3;46	64.00
												8756	29DE	14;31	94.02

-292

NUMBER	DATE	TIME	LONG.	NUMBER	DATE	TIME	LONG.
8757	13JA	22:59	289.58	8757	28JA	1:45	123.80
8758	12FE	14:35	319.32	8758	26FE	13:49	153.16
8759	13MR	3:16	348.44	8759	27MR	2:52	182.05
8760	11AP	13:12	16.96	8760	25AP	16:48	210.54
8761	10MY	21:00	45.03	8761	25MY	7:21	238.78
8762	9JN	3:41	72.91	8762	23JN	22:15	266.99
8763	8JL	10:26	100.86	8763	23JL	13:16	295.40
8764	6AU	18:26	129.16	8764	22AU	4:05	324.19
8765	5SE	4:42	158.00	8765	20SE	18:14	353.46
8766	4OC	17:47	187.46	8766	20OC	7:19	23.17
8767	3NO	9:43	217.49	8767	18NO	19:12	53.18
8768	3DE	3:54	247.84	8768	18DE	6:06	83.24

-286

NUMBER	DATE	TIME	LONG.	NUMBER	DATE	TIME	LONG.
8831	7JA	0:26	283.11	8831	21JA	15:34	117.89
8832	5FE	10:17	312.71	8832	20FE	9:37	147.60
8833	6MR	19:14	341.79	8833	22MR	1:39	176.70
8834	5AP	4:04	10.37	8834	20AP	15:14	205.25
8835	4MY	13:36	38.58	8835	20MY	2:26	233.39
8836	3JN	0:33	66.64	8836	18JN	11:40	261.37
8837	2JL	13:21	94.80	8837	17JL	19:41	289.44
8838	1AU	4:04	123.29	8838	16AU	3:27	317.85
8839	30AU	20:34	152.28	8839	14SE	11:59	346.77
8840	29SE	14:12	181.84	8840	13OC	22:05	16.25
8841	29OC	7:54	211.86	8841	12NO	10:16	46.21
8842	28NO	0:19	242.12	8842	12DE	0:31	76.42
8843	27DE	14:26	272.29				

-291

NUMBER	DATE	TIME	LONG.	NUMBER	DATE	TIME	LONG.
8769	1JA	23:14	278.21	8769	16JA	16:25	113.10
8770	31JA	18:08	308.25	8770	15FE	2:28	142.55
8771	2MR	11:01	337.73	8771	16MR	12:34	171.51
8772	1AP	0:54	6.58	8772	14AP	23:03	200.00
8773	30AP	11:43	34.88	8773	14MY	10:26	228.19
8774	29MY	20:14	62.86	8774	12JN	23:15	256.31
8775	28JN	3:35	90.79	8775	12JL	13:54	284.61
8776	27JL	10:56	118.93	8776	11AU	6:10	313.29
8777	25AU	19:11	147.52	8777	9SE	23:11	342.48
8778	24SE	5:01	176.66	8778	9OC	15:44	12.17
8779	23OC	16:50	206.37	8779	8NO	6:54	42.21
8780	22NO	6:58	236.49	8780	7DE	20:22	72.38
8781	21DE	23:27	266.78				

-285

NUMBER	DATE	TIME	LONG.	NUMBER	DATE	TIME	LONG.
				8843	10JA	16:26	106.58
8844	26JA	2:04	302.11	8844	9FE	9:21	136.42
8845	24FE	11:40	331.40	8845	11MR	2:29	165.77
8846	25MR	19:58	0.13	8846	9AP	18:56	194.59
8847	24AP	3:44	28.39	8847	9MY	9:50	222.97
8848	23MY	11:44	56.39	8848	7JN	22:39	251.10
8849	21JN	20:47	84.36	8849	7JL	9:30	279.20
8850	21JL	7:43	112.58	8850	5AU	19:04	307.54
8851	19AU	21:18	141.27	8851	4SE	4:16	336.30
8852	18SE	13:51	170.59	8852	3OC	13:51	5.59
8853	18OC	8:43	200.50	8853	2NO	0:17	35.36
8854	17NO	4:13	230.80	8854	1DE	11:37	65.41
8855	16DE	22:19	261.17	8855	30DE	23:55	95.49

-290

NUMBER	DATE	TIME	LONG.	NUMBER	DATE	TIME	LONG.
				8781	6JA	8:12	102.40
8782	20JA	17:40	296.94	8782	4FE	18:36	132.03
8783	19FE	12:11	326.68	8783	6MR	3:46	161.13
8784	21MR	5:18	355.85	8784	4AP	12:08	189.70
8785	19AP	19:58	24.45	8785	3MY	20:26	217.87
8786	19MY	8:06	52.63	8786	2JN	5:43	245.86
8787	17JN	18:15	80.65	8787	1JL	16:59	273.96
8788	17JL	3:16	108.75	8788	31JL	6:53	302.40
8789	15AU	11:54	137.19	8789	29AU	23:17	331.38
8790	13SE	20:46	166.11	8790	28SE	17:17	0.93
8791	13OC	6:24	195.56	8791	28OC	11:35	30.97
8792	11NO	17:21	225.46	8792	27NO	4:56	61.26
8793	11DE	6:03	255.60	8793	26DE	20:27	91.50

-284

NUMBER	DATE	TIME	LONG.	NUMBER	DATE	TIME	LONG.
8856	15JA	13:47	291.27	8856	29JA	13:18	125.34
8857	14FE	2:23	320.86	8857	28FE	3:52	154.78
8858	14MR	12:31	349.85	8858	28MR	19:24	183.75
8859	12AP	20:50	18.29	8859	27AP	11:15	212.29
8860	12MY	4:02	46.33	8860	27MY	2:32	240.55
8861	10JN	10:57	74.21	8861	25JN	16:39	268.74
8862	9JL	18:38	102.21	8862	25JL	5:33	297.08
8863	8AU	4:16	130.59	8863	23AU	17:34	325.78
8864	6SE	16:51	159.55	8864	22SE	5:01	354.96
8865	6OC	8:48	189.16	8865	21OC	16:06	24.61
8866	5NO	3:27	219.31	8866	20NO	2:54	54.58
8867	4DE	23:13	249.74	8867	19DE	13:33	84.63

-289

NUMBER	DATE	TIME	LONG.	NUMBER	DATE	TIME	LONG.
8794	9JA	20:41	285.72	8794	25JA	9:35	121.39
8795	8FE	12:49	315.54	8795	23FE	20:14	150.73
8796	10MR	5:28	344.88	8796	25MR	4:47	179.48
8797	8AP	21:40	13.71	8797	23AP	12:03	207.74
8798	8MY	12:47	42.10	8798	22MY	19:08	235.70
8799	7JN	2:34	70.27	8799	21JN	3:09	263.63
8800	6JL	15:01	98.43	8800	20JL	13:09	291.80
8801	5AU	2:13	126.83	8801	19AU	1:46	320.45
8802	3SE	12:30	155.63	8802	17SE	17:10	349.70
8803	2OC	22:22	184.92	8803	17OC	11:03	19.56
8804	1NO	8:26	214.66	8804	16NO	6:25	49.85
8805	30NO	19:11	244.68	8805	16DE	1:41	80.27
8806	30DE	6:51	274.74				

-283

NUMBER	DATE	TIME	LONG.	NUMBER	DATE	TIME	LONG.
8868	3JA	18:14	280.08	8868	18JA	0:20	114.50
8869	2FE	11:06	310.01	8869	16FE	11:41	143.99
8870	4MR	1:10	339.35	8870	17MR	23:56	173.01
8871	2AP	12:21	8.07	8871	16AP	13:11	201.59
8872	1MY	21:04	36.27	8872	16MY	3:15	229.88
8873	31MY	4:09	64.19	8873	14JN	17:57	258.07
8874	29JN	10:45	92.09	8874	14JL	9:03	286.39
8875	28JL	18:05	120.25	8875	13AU	0:17	315.06
8876	27AU	3:13	148.89	8876	11SE	15:08	344.19
8877	25SE	14:57	178.13	8877	11OC	5:05	13.79
8878	25OC	5:33	207.98	8878	9NO	17:46	43.75
8879	23NO	22:42	238.24	8879	9DE	5:12	73.83
8880	23DE	17:33	268.63				

-288

NUMBER	DATE	TIME	LONG.	NUMBER	DATE	TIME	LONG.
				8806	14JA	19:04	110.46
8807	28JA	19:25	304.57	8807	13FE	9:21	140.13
8808	27FE	8:45	333.97	8808	13MR	20:27	169.17
8809	27MR	22:44	2.88	8809	12AP	5:02	197.62
8810	26AP	13:23	31.39	8810	11MY	12:12	225.67
8811	26MY	4:32	59.65	8811	9JN	19:01	253.55
8812	24JN	19:43	87.87	8812	9JL	2:28	281.53
8813	24JL	10:15	116.27	8813	7AU	11:23	309.88
8814	22AU	23:42	145.02	8814	5SE	22:33	338.76
8815	21SE	12:03	174.23	8815	5OC	11:23	8.28
8816	20OC	23:39	203.89	8816	4NO	5:45	38.36
8817	19NO	10:52	233.87	8817	4DE	1:16	68.78
8818	18DE	21:51	263.94				

-282

NUMBER	DATE	TIME	LONG.	NUMBER	DATE	TIME	LONG.
				8880	7JA	15:42	103.78
8881	22JA	12:44	298.80	8881	6FE	1:41	133.38
8882	21FE	6:40	328.49	8882	7MR	11:29	162.48
8883	22MR	22:01	357.56	8883	5AP	21:28	191.10
8884	21AP	10:16	26.04	8884	5MY	8:05	219.35
8885	20MY	19:50	54.11	8885	3JN	19:57	247.44
8886	19JN	3:41	82.04	8886	3JL	9:37	275.64
8887	18JL	11:02	110.09	8887	2AU	1:15	304.17
8888	16AU	18:54	138.51	8888	31AU	18:16	333.20
8889	15SE	4:02	167.46	8889	30SE	11:29	2.75
8890	14OC	14:55	196.99	8890	30OC	3:41	32.73
8891	13NO	3:53	226.99	8891	28NO	18:11	62.90
8892	12DE	19:08	257.24	8892	28DE	6:51	93.01

-287

NUMBER	DATE	TIME	LONG.	NUMBER	DATE	TIME	LONG.
				8818	2JA	21:14	99.17
8819	17JA	8:33	293.81	8819	1FE	15:29	129.17
8820	15FE	19:02	323.27	8820	3MR	6:42	158.57
8821	17MR	5:42	352.24	8821	1AP	18:47	187.33
8822	15AP	17:12	20.76	8822	1MY	4:21	215.58
8823	15MY	6:05	49.00	8823	30MY	12:18	243.53
8824	13JN	20:29	77.18	8824	28JN	19:29	271.45
8825	13JL	11:58	105.52	8825	28JL	2:47	299.60
8826	12AU	3:51	134.20	8826	26AU	11:05	328.20
8827	10SE	19:29	163.35	8827	24SE	21:19	357.37
8828	10OC	10:28	192.98	8828	24OC	10:18	27.13
8829	9NO	0:29	222.98	8829	23NO	2:17	57.34
8830	8DE	13:12	253.12	8830	22DE	20:34	87.71

-281

NUMBER	DATE	TIME	LONG.	NUMBER	DATE	TIME	LONG.
8893	11JA	12:25	287.45	8893	26JA	17:52	122.80
8894	10FE	6:45	317.34	8894	25FE	3:27	152.07
8895	12MR	0:33	346.70	8895	26MR	11:57	180.80
8896	10AP	16:24	15.48	8896	24AP	19:58	209.07
8897	10MY	5:44	43.79	8897	24MY	4:28	237.08
8898	8JN	16:52	71.85	8898	22JN	14:31	265.10
8899	8JL	2:31	99.91	8899	22JL	3:03	293.38
8900	6AU	11:28	128.23	8900	20AU	18:21	322.16
8901	4SE	20:21	157.00	8901	19SE	11:54	351.53
8902	4OC	5:45	186.29	8902	19OC	6:26	21.44
8903	2NO	16:04	216.05	8903	18NO	0:35	51.70
8904	2DE	3:50	246.13	8904	17DE	17:13	82.01
8905	31DE	17:23	276.26				

NUMBER	DATE	TIME	LONG.	NUMBER	DATE	TIME	LONG.	NUMBER	DATE	TIME	LONG.	NUMBER	DATE	TIME	LONG.
						−280								**−274**	
				8905	16JA	7;35	112.05					8979	8JA	23;15	105.17
8906	30JA	8;36	306.18	8906	14FE	19;22	141.60	8980	24JA	6;58	300.63	8980	7FE	10;03	134.78
8907	29FE	0;48	335.67	8907	15MR	4;45	170.54	8981	22FE	22;24	330.19	8981	8MR	21;32	163.93
8908	29MR	17;05	4.65	8908	13AP	12;23	198.94	8982	24MR	10;55	359.11	8982	7AP	9;59	192.63
8909	28AP	8;39	33.17	8909	12MY	19;18	226.96	8983	22AP	20;43	27.48	8983	6MY	23;26	220.97
8910	27MY	23;07	61.40	8910	11JN	2;38	254.86	8984	22MY	4;29	55.47	8984	5JN	13;44	249.16
8911	26JN	12;22	89.55	8911	10JL	11;30	282.92	8985	20JN	11;13	83.35	8985	5JL	4;43	277.42
8912	26JL	0;23	117.87	8912	8AU	22;47	311.37	8986	19JL	18;06	111.39	8986	3AU	20;07	305.97
8913	24AU	11;22	146.54	8913	7SE	12;54	340.40	8987	18AU	2;19	139.85	8987	2SE	11;31	334.96
8914	22SE	21;37	175.68	8914	7OC	5;44	10.06	8988	16SE	12;48	168.88	8988	2OC	2;18	4.44
8915	22OC	7;42	205.30	8915	6NO	0;39	40.24	8989	16OC	2;02	198.53	8989	31OC	15;53	34.32
8916	20NO	18;09	235.26	8916	5DE	20;16	70.66	8990	14NO	18;00	228.66	8990	30NO	4;03	64.40
8917	20DE	5;18	265.33					8991	14DE	12;04	259.03	8991	29DE	14;57	94.43
						−279								**−273**	
				8917	4JA	14;48	100.98	8992	13JA	7;08	289.30	8992	28JA	1;02	124.16
8918	18JA	17;16	295.24	8918	3FE	6;40	130.86	8993	12FE	1;43	319.17	8993	26FE	10;40	153.42
8919	17FE	5;57	324.77	8919	4MR	19;14	160.12	8994	13MR	18;21	348.46	8994	27MR	20;15	182.17
8920	18MR	19;18	353.82	8920	3AP	4;51	188.76	8995	12AP	8;06	17.14	8995	26AP	6;14	210.50
8921	17AP	9;21	22.43	8921	2MY	12;31	216.92	8996	11MY	18;55	45.33	8996	25MY	17;13	238.60
8922	17MY	0;07	50.74	8922	31MY	19;20	244.82	8997	10JN	3;34	73.29	8997	24JN	5;50	266.72
8923	15JN	15;21	78.95	8923	30JN	2;22	272.74	8998	9JL	11;08	101.28	8998	23JL	20;33	295.11
8924	15JL	6;27	107.28	8924	29JL	10;31	300.94	8999	7AU	18;47	129.56	8999	22AU	13;09	323.97
8925	13AU	20;45	135.92	8925	27AU	20;33	329.63	9000	6SE	3;23	158.33	9000	21SE	6;43	353.37
8926	12SE	9;56	165.00	8926	26SE	9;11	358.92	9001	5OC	13;28	187.67	9001	20OC	23;50	23.24
8927	11OC	22;08	194.54	8927	26OC	0;52	28.82	9002	4NO	1;25	217.52	9002	19NO	15;26	53.41
8928	10NO	9;42	224.45	8928	24NO	19;23	59.15	9003	3DE	15;29	247.70	9003	19DE	5;05	83.58
8929	9DE	20;51	254.52	8929	24DE	15;21	89.58								
						−278								**−272**	
8930	8JA	7;37	284.48	8930	23JA	10;37	119.75	9004	2JA	7;40	277.94	9004	17JA	16;52	113.51
8931	6FE	18;00	314.08	8931	22FE	3;19	149.37	9005	1FE	1;24	307.94	9005	16FE	3;00	142.96
8932	8MR	4;14	343.20	8932	23MR	16;50	178.35	9006	1MR	19;25	337.48	9006	16MR	11;49	171.86
8933	6AP	14;57	11.83	8933	22AP	3;31	206.76	9007	31MR	12;12	6.45	9007	14AP	19;47	200.26
8934	6MY	2;50	40.12	8934	21MY	12;10	234.79	9008	30AP	2;45	34.91	9008	14MY	3;44	228.31
8935	4JN	16;21	68.28	8935	19JN	19;41	262.70	9009	29MY	14;59	63.04	9009	12JN	12;45	256.28
8936	4JL	7;22	96.54	8936	19JL	2;55	290.75	9010	28JN	1;25	91.09	9010	11JL	23;55	284.43
8937	2AU	23;18	125.10	8937	17AU	10;44	319.18	9011	27JL	10;50	119.32	9011	10AU	13;54	313.02
8938	1SE	15;24	154.10	8938	15SE	20;01	348.16	9012	25AU	19;55	147.93	9012	9SE	6;35	342.18
8939	1OC	7;03	183.60	8939	15OC	7;39	17.72	9013	24SE	5;12	177.06	9013	9OC	1;00	11.94
8940	30OC	21;51	213.52	8940	13NO	22;10	47.79	9014	23OC	15;08	206.68	9014	7NO	19;42	42.12
8941	29NO	11;25	243.66	8941	13DE	15;21	78.12	9015	22NO	2;09	236.68	9015	7DE	13;20	72.47
8942	28DE	23;25	273.74					9016	21DE	14;41	266.80				
						−277								**−271**	
				8942	12JA	10;02	108.39					9016	6JA	4;56	102.65
8943	27JA	9;49	303.49	8943	11FE	4;33	138.27	9017	20JA	4;54	296.79	9017	4FE	17;58	132.39
8944	25FE	18;57	332.74	8944	12MR	21;31	167.58	9018	18FE	20;26	326.42	9018	6MR	4;22	161.54
8945	27MR	3;30	1.46	8945	11AP	12;15	196.31	9019	20MR	12;29	355.55	9019	4AP	12;38	190.10
8946	25AP	12;21	29.76	8946	11MY	0;37	224.57	9020	19AP	4;16	24.21	9020	3MY	19;39	218.22
8947	24MY	22;21	57.82	8947	9JN	10;51	252.59	9021	18MY	19;15	52.51	9021	2JN	2;32	246.12
8948	23JN	10;05	85.91	8948	8JL	19;31	280.62	9022	17JN	9;12	80.67	9022	1JL	10;29	274.09
8949	22JL	23;52	114.25	8949	7AU	3;28	308.91	9023	16JL	22;03	108.93	9023	30JL	20;31	302.37
8950	21AU	15;39	143.06	8950	5SE	11;41	337.66	9024	15AU	9;48	137.49	9024	29AU	9;15	331.19
8951	20SE	9;02	172.44	8951	4OC	21;04	6.96	9025	13SE	20;38	166.49	9025	28SE	0;53	0.64
8952	20OC	3;08	202.34	8952	3NO	8;15	36.77	9026	13OC	6;58	195.98	9026	27OC	18;59	30.66
8953	18NO	20;33	232.57	8953	2DE	21;28	66.91	9027	11NO	17;18	225.85	9027	26NO	14;32	61.04
8954	18DE	12;00	262.83					9028	11DE	4;04	255.91	9028	26DE	9;52	91.44
						−276								**−270**	
				8954	1JA	12;28	97.10	9029	9JA	15;30	285.88	9029	25JA	3;11	121.51
8955	17JA	0;50	292.80	8955	31JA	4;43	127.05	9030	8FE	3;36	315.54	9030	23FE	17;19	151.00
8956	15FE	11;16	322.28	8956	29FE	21;34	156.55	9031	9MR	16;18	344.73	9031	25MR	4;13	179.84
8957	15MR	19;57	351.19	8957	30MR	14;15	185.54	9032	8AP	5;42	13.45	9032	23AP	12;39	208.14
8958	14AP	3;43	19.58	8958	29AP	5;55	214.05	9033	7MY	19;54	41.82	9033	22MY	19;43	236.11
8959	13MY	11;21	47.63	8959	28MY	19;51	242.25	9034	6JN	10;53	70.03	9034	21JN	2;32	263.99
8960	11JN	19;41	75.57	8960	27JN	7;47	270.36	9035	6JL	2;14	98.31	9035	20JL	10;04	292.06
8961	11JL	5;32	103.67	8961	26JL	18;07	298.61	9036	4AU	17;14	126.85	9036	18AU	19;10	320.57
8962	9AU	17;12	132.17	8962	25AU	3;38	327.24	9037	3SE	7;18	155.80	9037	17SE	6;31	349.64
8963	8SE	8;56	161.26	8963	23SE	13;11	356.36	9038	2OC	20;16	185.22	9038	16OC	20;43	19.34
8964	8OC	2;57	190.98	8964	22OC	23;20	25.99	9039	1NO	8;20	215.04	9039	15NO	13;56	49.54
8965	6NO	22;33	221.20	8965	21NO	10;16	55.97	9040	30NO	19;47	245.10	9040	15DE	9;22	79.97
8966	6DE	17;41	251.60	8966	20DE	22;01	86.07	9041	30DE	6;43	275.12				
						−275								**−269**	
8967	5JA	10;35	281.84	8967	19JA	10;38	116.00					9041	14JA	5;09	110.26
8968	4FE	0;32	311.63	8968	18FE	0;19	145.56	9042	28JA	17;07	304.86	9042	12FE	23;11	140.09
8969	5MR	11;44	340.83	8969	19MR	15;06	174.66	9043	27FE	3;08	334.12	9043	14MR	14;12	169.30
8970	3AP	20;46	9.43	8970	18AP	6;39	203.31	9044	28MR	13;15	2.88	9044	13AP	2;09	197.89
8971	3MY	4;19	37.58	8971	17MY	22;12	231.64	9045	27AP	0;12	31.25	9045	12MY	11;42	226.03
8972	1JN	11;12	65.48	8972	16JN	13;00	259.84	9046	26MY	12;41	59.40	9046	10JN	19;43	253.96
8973	30JN	18;21	93.41	8973	16JL	2;41	288.12	9047	25JN	2;56	87.59	9047	10JL	3;06	281.94
8974	30JL	2;55	121.63	8974	14AU	15;22	316.71	9048	24JL	18;36	116.03	9048	8AU	10;39	310.23
8975	28AU	14;00	150.37	8975	13SE	3;21	345.75	9049	23AU	10;55	144.89	9049	6SE	19;14	339.02
8976	27SE	4;21	179.75	8976	12OC	14;52	15.27	9050	22SE	3;09	174.25	9050	6OC	5;42	8.38
8977	26OC	21;53	209.74	8977	11NO	1;58	45.17	9051	21OC	18;43	204.07	9051	4NO	18;47	38.29
8978	25NO	17;23	240.11	8978	10DE	12;42	75.22	9052	20NO	9;09	234.19	9052	4DE	10;40	68.55
8979	25DE	12;58	270.52					9053	19DE	22;01	264.33				

NEW MOONS (left)

NUMBER	DATE	TIME	LONG.
		−268	
9054	18JA	9;09	294.22
9055	16FE	18;42	323.64
9056	17MR	3;14	352.52
9057	15AP	11;36	20.92
9058	14MY	20;44	49.02
9059	13JN	7;24	77.05
9060	12JL	20;07	105.27
9061	11AU	11;00	133.90
9062	10SE	3;51	163.09
9063	9OC	21;59	192.85
9064	8NO	16;09	223.02
9065	8DE	8;51	253.32
		−267	
9066	6JA	23;02	283.44
9067	5FE	10;31	313.10
9068	6MR	19;50	342.20
9069	5AP	3;48	10.74
9070	4MY	11;16	38.87
9071	2JN	19;02	66.81
9072	2JL	3;58	94.81
9073	31JL	14;55	123.15
9074	30AU	4;39	152.02
9075	28SE	21;27	181.53
9076	28OC	16;39	211.60
9077	27NO	12;24	242.00
9078	27DE	6;37	272.35
		−266	
9079	25JA	22;02	302.32
9080	24FE	10;28	331.73
9081	25MR	20;23	0.53
9082	24AP	4;30	28.81
9083	23MY	11;33	56.76
9084	21JN	18;25	84.65
9085	21JL	2;10	112.74
9086	19AU	11;55	141.28
9087	18SE	0;41	170.43
9088	17OC	16;47	200.22
9089	16NO	11;33	230.49
9090	16DE	7;19	260.93
		−265	
9091	15JA	2;13	291.17
9092	13FE	18;56	320.94
9093	15MR	8;49	350.09
9094	13AP	19;51	18.63
9095	13MY	4;31	46.73
9096	11JN	11;39	74.62
9097	10JL	18;23	102.59
9098	9AU	1;55	130.88
9099	7SE	11;17	159.71
9100	6OC	23;12	189.14
9101	5NO	13;53	219.13
9102	5DE	6;58	249.44
		−264	
9103	4JA	1;33	279.76
9104	2FE	20;24	309.79
9105	3MR	14;01	339.28
9106	2AP	5;11	8.16
9107	1MY	17;24	36.51
9108	31MY	3;03	64.53
9109	29JN	11;08	92.49
9110	28JL	18;47	120.66
9111	27AU	2;59	149.26
9112	25SE	12;26	178.42
9113	24OC	23;31	208.11
9114	23NO	12;29	238.19
9115	23DE	3;30	268.42
		−263	
9116	21JA	20;21	298.50
9117	20FE	14;10	328.20
9118	22MR	7;31	357.36
9119	20AP	23;09	25.98
9120	20MY	12;31	54.21
9121	18JN	23;52	82.27
9122	18JL	9;54	110.43
9123	16AU	19;19	138.91
9124	15SE	4;40	167.88
9125	14OC	14;24	197.36
9126	13NO	0;54	227.25
9127	12DE	12;35	257.35

FULL MOONS (left)

NUMBER	DATE	TIME	LONG.
9053	3JA	4;39	98.85
9054	1FE	23;14	128.88
9055	2MR	16;55	158.39
9056	1AP	8;41	187.31
9057	30AP	22;11	215.72
9058	30MY	9;29	243.81
9059	28JN	18;59	271.82
9060	28JL	3;22	300.01
9061	26AU	11;32	328.60
9062	24SE	20;25	357.72
9063	24OC	6;46	27.37
9064	22NO	18;59	57.42
9065	22DE	9;01	87.61
9066	21JA	0;30	117.64
9067	19FE	16;50	147.29
9068	21MR	9;27	176.43
9069	20AP	1;34	205.09
9070	19MY	16;25	233.38
9071	18JN	5;28	261.51
9072	17JL	16;44	289.71
9073	16AU	2;48	318.21
9074	14SE	12;30	347.18
9075	13OC	22;30	16.66
9076	12NO	9;08	46.55
9077	11DE	20;28	76.63
9078	10JA	8;28	106.63
9079	8FE	21;20	136.31
9080	10MR	11;15	165.53
9081	9AP	2;13	194.30
9082	8MY	17;41	222.71
9083	7JN	8;55	250.94
9084	6JL	23;19	279.18
9085	5AU	12;44	307.67
9086	4SE	1;20	336.57
9087	3OC	13;21	5.96
9088	2NO	0;51	35.77
9089	1DE	11;50	65.81
9090	30DE	22;23	95.82
9091	29JA	8;50	125.54
9092	27FE	19;39	154.83
9093	29MR	7;18	183.64
9094	27AP	19;59	212.07
9095	27MY	9;43	240.27
9096	26JN	0;22	268.47
9097	25JL	15;44	296.91
9098	24AU	7;28	325.76
9099	22SE	22;57	355.10
9100	22OC	13;29	24.89
9101	21NO	2;32	54.96
9102	20DE	14;04	85.04
9103	19JA	0;25	114.89
9104	17FE	10;03	144.30
9105	17MR	19;22	173.21
9106	16AP	4;51	201.64
9107	15MY	15;02	229.77
9108	14JN	2;38	257.84
9109	13JL	16;15	286.11
9110	12AU	8;06	314.79
9111	11SE	1;35	344.02
9112	10OC	19;22	13.77
9113	9NO	12;04	43.89
9114	9DE	2;50	74.12
9115	7JA	15;32	104.16
9116	6FE	2;22	133.79
9117	7MR	11;37	162.87
9118	5AP	19;45	191.41
9119	5MY	3;25	219.54
9120	3JN	11;38	247.50
9121	2JL	21;32	275.54
9122	1AU	10;05	303.95
9123	31AU	1;35	332.91
9124	29SE	19;28	2.47
9125	29OC	14;24	32.55
9126	28NO	8;52	62.90
9127	28DE	1;38	93.18

NEW MOONS (right)

NUMBER	DATE	TIME	LONG.
		−262	
9128	11JA	1;47	287.39
9129	9FE	16;27	317.13
9130	11MR	8;03	346.41
9131	9AP	23;48	15.20
9132	9MY	15;06	43.60
9133	8JN	5;37	71.79
9134	7JL	19;11	100.00
9135	6AU	7;44	128.47
9136	4SE	19;17	157.34
9137	4OC	6;04	186.69
9138	2NO	16;30	216.46
9139	2DE	3;04	246.48
9140	31DE	14;05	276.51
		−261	
9141	30JA	1;39	306.27
9142	28FE	13;46	335.60
9143	30MR	2;29	4.45
9144	28AP	16;01	32.89
9145	28MY	6;30	61.12
9146	26JN	21;46	89.36
9147	26JL	13;13	117.80
9148	25AU	4;06	146.63
9149	23SE	17;55	175.92
9150	23OC	6;40	205.64
9151	21NO	18;33	235.66
9152	21DE	5;45	265.74
		−260	
9153	19JA	16;19	295.59
9154	18FE	2;17	325.00
9155	18MR	12;00	353.91
9156	16AP	22;10	22.36
9157	16MY	9;36	50.54
9158	14JN	22;52	78.68
9159	14JL	13;55	107.01
9160	13AU	6;11	135.72
9161	11SE	22;49	164.93
9162	11OC	15;05	194.63
9163	10NO	6;22	224.70
9164	9DE	20;10	254.88
		−259	
9165	8JA	8;10	284.89
9166	6FE	18;20	314.49
9167	8MR	3;06	343.54
9168	6AP	11;13	12.07
9169	5MY	19;39	40.22
9170	4JN	5;20	68.23
9171	3JL	16;55	96.35
9172	2AU	6;45	124.81
9173	31AU	22;48	153.81
9174	30SE	16;38	183.39
9175	30OC	11;12	213.46
9176	29NO	4;58	243.78
9177	28DE	20;34	274.01
		−258	
9178	27JA	9;19	303.86
9179	25FE	19;31	333.15
9180	27MR	3;54	1.86
9181	25AP	11;21	30.10
9182	24MY	18;45	58.06
9183	23JN	2;57	86.01
9184	22JL	12;47	114.19
9185	21AU	1;06	142.85
9186	19SE	16;29	172.14
9187	19OC	10;47	202.04
9188	18NO	6;39	232.37
9189	18DE	1;55	262.79
		−257	
9190	16JA	18;49	292.94
9191	15FE	8;38	322.56
9192	16MR	19;38	351.56
9193	15AP	4;28	20.00
9194	14MY	11;52	48.04
9195	12JN	18;42	75.92
9196	12JL	1;53	103.91
9197	10AU	10;34	132.27
9198	8SE	21;49	161.19
9199	8OC	12;21	190.76
9200	7NO	5;59	220.88
9201	7DE	1;30	251.30

FULL MOONS (right)

NUMBER	DATE	TIME	LONG.
9128	26JA	15;57	123.10
9129	25FE	3;33	152.46
9130	26MR	12;40	181.22
9131	24AP	20;04	209.46
9132	24MY	2;47	237.40
9133	22JN	10;01	265.30
9134	21JL	18;54	293.44
9135	20AU	6;17	322.06
9136	18SE	20;35	351.28
9137	18OC	13;38	21.12
9138	17NO	8;43	51.41
9139	17DE	4;25	81.85
9140	15JA	22;55	112.07
9141	14FE	14;38	141.79
9142	16MR	3;01	170.86
9143	14AP	12;29	199.33
9144	13MY	20;02	227.38
9145	12JN	2;51	255.26
9146	11JL	9;58	283.24
9147	9AU	18;16	311.58
9148	8SE	4;31	340.45
9149	7OC	17;19	9.93
9150	6NO	9;05	39.97
9151	6DE	3;33	70.34
9152	4JA	23;21	100.71
9153	3FE	18;21	130.73
9154	4MR	10;49	160.16
9155	3AP	0;10	188.96
9156	2MY	10;48	217.24
9157	31MY	19;31	245.22
9158	30JN	3;12	273.16
9159	29JL	10;42	301.33
9160	27AU	18;49	329.94
9161	26SE	4;22	359.11
9162	25OC	16;09	28.84
9163	24NO	6;38	58.99
9164	23DE	23;36	89.30
9165	22JA	17;52	119.44
9166	21FE	11;56	149.13
9167	23MR	4;34	178.24
9168	21AP	19;09	206.81
9169	21MY	7;33	234.99
9170	19JN	18;01	263.02
9171	19JL	3;02	291.14
9172	17AU	11;24	319.60
9173	15SE	20;01	348.54
9174	15OC	5;42	18.03
9175	13NO	17;01	47.96
9176	13DE	6;07	78.12
9177	11JA	20;44	108.21
9178	10FE	12;26	137.99
9179	12MR	4;42	167.28
9180	10AP	20;57	196.08
9181	10MY	12;27	224.48
9182	9JN	2;30	252.65
9183	8JL	14;48	280.82
9184	7AU	1;38	309.22
9185	5SE	11;41	338.04
9186	4OC	21;42	7.37
9187	3NO	8;09	37.15
9188	2DE	19;10	67.20
9189	1JA	6;43	97.24
9190	30JA	18;53	127.02
9191	1MR	7;58	156.38
9192	30MR	22;07	185.27
9193	29AP	13;11	213.77
9194	29MY	4;33	242.03
9195	27JN	19;30	270.25
9196	27JL	9;37	298.66
9197	25AU	22;53	327.43
9198	24SE	11;29	356.68
9199	23OC	23;29	26.38
9200	22NO	10;51	56.39
9201	21DE	21;35	86.44

Page 63

NEW MOONS / FULL MOONS — left half

NUMBER	DATE	TIME	LONG.		NUMBER	DATE	TIME	LONG.
				-256				
9202	5JA	20;59	281.66		9202	20JA	7;54	116.27
9203	4FE	14;48	311.61		9203	18FE	18;15	145.69
9204	5MR	6;02	340.98		9204	19MR	5;09	174.64
9205	3AP	18;24	9.72		9205	17AP	17;01	203.15
9206	3MY	4;08	37.96		9206	17MY	6;02	231.38
9207	1JN	11;55	65.90		9207	15JN	20;09	259.55
9208	30JN	18;47	93.81		9208	15JL	11;15	287.89
9209	30JL	1;53	121.98		9209	14AU	3;05	316.60
9210	28AU	10;21	150.61		9210	12SE	19;05	345.79
9211	26SE	21;03	179.84		9211	12OC	10;29	15.48
9212	26OC	10;25	209.64		9212	11NO	0;32	45.50
9213	25NO	2;21	239.86		9213	10DE	12;53	75.62
9214	24DE	20;12	270.20					
				-255				
9215	23JA	14;55	300.35		9214	8JA	23;45	105.58
9216	22FE	9;09	330.03		9215	7FE	9;32	135.15
9217	24MR	1;31	359.12		9216	8MR	18;45	164.20
9218	22AP	15;10	27.64		9217	7AP	3;50	192.77
9219	22MY	2;03	55.76		9218	6MY	13;23	220.96
9220	20JN	10;53	83.72		9219	5JN	0;02	249.00
9221	19JL	18;46	111.81		9220	4JL	12;31	277.16
9222	18AU	2;46	140.25		9221	3AU	3;20	305.68
9223	16SE	11;42	169.22		9222	1SE	20;18	334.72
9224	15OC	22;04	198.75		9223	1OC	14;23	4.32
9225	14NO	10;06	228.72		9224	31OC	8;02	34.37
9226	14DE	0;00	258.91		9225	29NO	23;59	64.62
					9226	29DE	13;45	94.77
				-254				
9227	12JA	15;47	289.05		9227	28JA	1;25	124.56
9228	11FE	9;00	318.87		9228	26FE	11;17	153.83
9229	13MR	2;31	348.20		9229	27MR	19;44	182.53
9230	11AP	18;58	16.99		9230	26AP	3;22	210.77
9231	11MY	9;26	45.34		9231	25MY	11;02	238.74
9232	9JN	21;49	73.44		9232	23JN	19;52	266.71
9233	9JL	8;37	101.56		9233	23JL	6;59	294.95
9234	7AU	18;29	129.93		9234	21AU	21;05	323.70
9235	6SE	4;03	158.75		9235	20SE	14;03	353.06
9236	5OC	13;45	188.07		9236	20OC	8;50	23.00
9237	3NO	23;56	217.85		9237	19NO	3;52	53.30
9238	3DE	10;59	247.90		9238	18DE	21;42	83.66
				-253				
9239	1JA	23;17	277.97		9239	17JA	13;18	113.74
9240	31JA	13;00	307.81		9240	16FE	2;11	143.31
9241	2MR	3;54	337.23		9241	17MR	12;22	172.27
9242	31MR	19;23	6.16		9242	15AP	20;23	200.67
9243	30AP	10;47	34.66		9243	15MY	3;12	228.67
9244	30MY	1;41	62.90		9244	13JN	9;59	256.55
9245	28JN	15;51	91.09		9245	12JL	17;55	284.58
9246	28JL	5;09	119.47		9246	11AU	4;01	313.00
9247	26AU	17;29	148.22		9247	9SE	16;55	342.01
9248	25SE	4;53	177.43		9248	9OC	8;44	11.64
9249	24OC	15;39	207.10		9249	8NO	3;01	41.80
9250	23NO	2;13	237.07		9250	7DE	22;40	72.23
9251	22DE	12;56	267.12					
				-252				
9252	21JA	0;04	296.98		9251	6JA	17;58	102.57
9253	19FE	11;39	326.44		9252	5FE	11;09	132.49
9254	19MR	23;45	355.42		9253	6MR	1;07	161.79
9255	18AP	12;34	23.96		9254	4AP	11;52	190.45
9256	18MY	2;22	52.22		9255	3MY	20;11	218.63
9257	16JN	17;15	80.42		9256	2JN	3;13	246.53
9258	16JL	8;50	108.78		9257	1JL	10;06	274.46
9259	15AU	0;20	137.49		9258	30JL	17;48	302.65
9260	13SE	15;02	166.65		9259	29AU	3;06	331.33
9261	13OC	4;35	196.27		9260	27SE	14;39	0.60
9262	11NO	17;04	226.23		9261	27OC	4;58	30.45
9263	11DE	4;40	256.32		9262	25NO	22;10	60.73
					9263	25DE	17;26	91.13
				-251				
9264	9JA	15;30	286.27		9264	24JA	12;57	121.30
9265	8FE	1;35	315.84		9265	23FE	6;42	150.94
9266	9MR	11;07	344.90		9266	24MR	21;31	179.96
9267	7AP	20;42	13.47		9267	23AP	9;23	208.40
9268	7MY	7;10	41.70		9268	22MY	18;58	236.46
9269	5JN	19;18	69.80		9269	21JN	3;08	264.40
9270	5JL	9;28	98.03		9270	20JL	10;46	292.48
9271	4AU	1;20	126.59		9271	18AU	18;38	320.93
9272	2SE	18;08	155.65		9272	17SE	3;31	349.91
9273	2OC	10;57	185.21		9273	16OC	14;12	19.46
9274	1NO	3;03	215.21		9274	15NO	3;20	49.48
9275	30NO	17;49	245.40		9275	14DE	19;03	79.75
9276	30DE	6;47	275.52					

NEW MOONS / FULL MOONS — right half

NUMBER	DATE	TIME	LONG.		NUMBER	DATE	TIME	LONG.
				-250				
					9276	13JA	12;39	109.95
9277	28JA	17;45	305.27		9277	12FE	6;46	139.80
9278	27FE	2;59	334.50		9278	14MR	0;01	169.11
9279	28MR	11;06	3.18		9279	12AP	15;33	197.85
9280	26AP	19;04	31.43		9280	12MY	5;01	226.15
9281	26MY	3;51	59.44		9281	10JN	16;30	254.22
9282	24JN	14;20	87.47		9282	10JL	2;20	282.30
9283	24JL	3;01	115.79		9283	8AU	11;08	310.64
9284	22AU	18;05	144.59		9284	6SE	19;44	339.42
9285	21SE	11;17	173.98		9285	6OC	4;59	8.74
9286	21OC	5;52	203.92		9286	4NO	15;32	38.54
9287	20NO	0;25	234.21		9287	4DE	3;44	68.64
9288	19DE	17;20	264.52					
				-249				
9289	18JA	7;32	294.54		9288	2JA	17;29	98.77
9290	16FE	18;50	324.03		9289	1FE	8;27	128.65
9291	18MR	3;53	352.93		9290	3MR	0;12	158.09
9292	16AP	11;33	21.30		9291	1AP	16;16	187.03
9293	15MY	18;45	49.32		9292	1MY	8;06	215.54
9294	14JN	2;22	77.23		9293	30MY	22;57	243.77
9295	13JL	11;16	105.30		9294	29JN	12;17	271.94
9296	11AU	22;17	133.77		9295	29JL	0;01	300.26
9297	10SE	12;11	162.83		9296	27AU	10;39	328.95
9298	10OC	5;14	192.53		9297	25SE	20;51	358.13
9299	9NO	0;39	222.75		9298	25OC	7;14	27.78
9300	8DE	20;35	253.19		9299	23NO	18;03	57.77
					9300	23DE	5;16	87.84
				-248				
9301	7JA	14;50	283.48		9301	21JA	16;56	117.71
9302	6FE	6;09	313.31		9302	20FE	5;14	147.20
9303	6MR	18;24	342.53		9303	20MR	18;31	176.21
9304	5AP	4;07	11.14		9304	19AP	8;55	204.81
9305	4MY	12;04	39.29		9305	19MY	0;04	233.11
9306	2JN	19;03	67.19		9306	17JN	15;19	261.33
9307	2JL	1;57	95.12		9307	17JL	6;03	289.67
9308	31JL	9;49	123.33		9308	15AU	20;00	318.32
9309	29AU	19;44	152.04		9309	14SE	9;13	347.43
9310	28SE	8;40	181.38		9310	13OC	21;48	17.02
9311	28OC	0;54	211.33		9311	12NO	9;40	46.96
9312	26NO	19;42	241.68		9312	11DE	20;45	77.03
9313	26DE	15;22	272.09					
				-247				
					9313	10JA	7;09	106.96
9314	25JA	10;05	302.21		9314	8FE	17;13	136.52
9315	24FE	2;35	331.79		9315	10MR	3;30	165.60
9316	25MR	16;17	0.75		9316	8AP	14;34	194.22
9317	24AP	3;14	29.14		9317	8MY	2;45	222.51
9318	23MY	11;54	57.16		9318	6JN	16;12	250.66
9319	21JN	19;09	85.07		9319	6JL	6;51	278.91
9320	21JL	2;06	113.12		9320	4AU	22;31	307.48
9321	19AU	9;53	141.58		9321	3SE	14;47	336.52
9322	17SE	19;30	170.60		9322	3OC	6;54	6.08
9323	17OC	7;36	200.22		9323	1NO	21;57	36.04
9324	15NO	22;19	230.31		9324	1DE	11;18	66.17
9325	15DE	15;13	260.63		9325	30DE	22;53	96.23
				-246				
9326	14JA	9;29	290.86		9326	29JA	9;02	125.94
9327	13FE	3;56	320.71		9327	27FE	18;17	155.16
9328	14MR	21;12	350.00		9328	29MR	3;09	183.86
9329	13AP	12;11	18.71		9329	27AP	12;10	212.14
9330	13MY	0;25	46.95		9330	26MY	22;00	240.19
9331	11JN	10;15	74.95		9331	25JN	9;23	268.26
9332	10JL	18;37	102.98		9332	24JL	23;01	296.62
9333	9AU	2;22	131.30		9333	23AU	15;07	325.48
9334	7SE	11;12	160.09		9334	22SE	9;03	354.91
9335	6OC	20;58	189.44		9335	22OC	3;22	24.85
9336	5NO	8;13	219.27		9336	20NO	20;29	55.09
9337	4DE	21;07	249.41		9337	20DE	11;27	85.32
				-245				
9338	3JA	11;50	279.57		9338	19JA	0;07	115.26
9339	2FE	4;10	309.50		9339	17FE	10;44	144.72
9340	3MR	21;25	338.99		9340	18MR	19;40	173.60
9341	2AP	14;20	7.95		9341	17AP	3;27	201.97
9342	2MY	5;46	36.43		9342	16MY	10;49	229.99
9343	31MY	19;13	64.61		9343	14JN	18;50	257.92
9344	30JN	6;53	92.71		9344	14JL	4;40	286.03
9345	29JL	17;21	120.99		9345	12AU	17;16	314.58
9346	28AU	3;16	149.66		9346	11SE	8;59	343.72
9347	26SE	13;05	178.83		9347	11OC	3;11	13.48
9348	25OC	23;09	208.48		9348	9NO	22;27	43.70
9349	24NO	9;46	238.47		9349	9DE	17;09	74.09
9350	23DE	21;19	268.55					

Left half

NEW#	Date	Time	Long.	FULL#	Date	Time	Long.
−244							
				9350	8JA	9;59	104.32
9351	22JA	10;07	298.46	9351	7FE	0;12	134.09
9352	21FE	0;12	328.01	9352	7MR	11;35	163.26
9353	21MR	15;09	357.08	9353	5AP	20;29	191.83
9354	20AP	6;25	25.70	9354	5MY	3;40	219.94
9355	19MY	21;30	54.00	9355	3JN	10;16	247.83
9356	18JN	12;07	82.19	9356	2JL	17;28	275.77
9357	18JL	2;03	110.50	9357	1AU	2;25	304.03
9358	16AU	15;10	139.13	9358	30AU	13;57	332.81
9359	15SE	3;19	168.21	9359	29SE	4;25	2.23
9360	14OC	14;37	197.75	9360	28OC	21;39	32.23
9361	13NO	1;22	227.66	9361	27NO	16;49	62.60
9362	12DE	11;59	257.71	9362	27DE	12;31	93.01
−243							
9363	10JA	22;48	287.65	9363	26JA	6;53	123.11
9364	9FE	9;55	317.24	9364	24FE	22;26	152.64
9365	10MR	21;27	346.36	9365	26MR	10;39	181.52
9366	9AP	9;35	15.01	9366	24AP	20;00	209.84
9367	8MY	22;38	43.33	9367	24MY	3;31	237.81
9368	7JN	12;53	71.51	9368	22JN	10;23	265.70
9369	7JL	4;14	99.79	9369	21JL	17;39	293.78
9370	5AU	20;05	128.38	9370	20AU	2;10	322.28
9371	4SE	11;34	157.41	9371	18SE	12;38	351.34
9372	4OC	2;01	186.91	9372	18OC	1;35	21.00
9373	2NO	15;15	216.80	9373	16NO	17;23	51.15
9374	2DE	3;24	246.89	9374	16DE	11;43	81.53
9375	31DE	14;36	276.92				
−242							
				9375	15JA	7;13	111.80
9376	30JA	0;55	306.63	9376	14FE	1;55	141.65
9377	28FE	10;27	335.85	9377	15MR	18;08	170.88
9378	29MR	19;40	4.56	9378	14AP	7;22	199.51
9379	28AP	5;20	32.85	9379	13MY	17;59	227.68
9380	27MY	16;23	60.95	9380	12JN	2;50	255.64
9381	26JN	5;26	89.09	9381	11JL	10;46	283.66
9382	25JL	20;35	117.52	9382	9AU	18;35	311.97
9383	24AU	13;13	146.41	9383	8SE	3;02	340.77
9384	23SE	6;24	175.83	9384	7OC	12;51	10.13
9385	22OC	23;12	205.72	9385	6NO	0;45	40.00
9386	21NO	14;54	235.90	9386	5DE	15;09	70.20
9387	21DE	4;54	266.09				
−241							
				9387	4JA	7;47	100.44
9388	19JA	16;49	296.00	9388	3FE	1;35	130.42
9389	18FE	2;44	325.41	9389	4MR	19;10	159.91
9390	19MR	11;07	354.26	9390	3AP	11;27	188.83
9391	17AP	18;51	22.62	9391	3MY	1;55	217.27
9392	17MY	2;56	50.66	9392	1JN	14;26	245.40
9393	15JN	12;22	78.65	9393	1JL	1;11	273.47
9394	14JL	23;52	106.83	9394	30JL	10;37	301.72
9395	13AU	13;48	135.44	9395	28AU	19;27	330.36
9396	12SE	6;08	164.63	9396	27SE	4;28	359.50
9397	12OC	0;22	194.40	9397	26OC	14;26	29.16
9398	10NO	19;20	224.62	9398	25NO	1;50	59.18
9399	10DE	13;22	254.98	9399	24DE	14;44	89.31
−240							
9400	9JA	5;02	285.15	9400	23JA	4;55	119.28
9401	7FE	17;41	314.84	9401	21FE	20;02	148.86
9402	8MR	3;38	343.94	9402	22MR	11;42	177.94
9403	6AP	11;44	12.47	9403	21AP	3;32	206.57
9404	5MY	18;56	40.58	9404	20MY	18;54	234.88
9405	4JN	2;09	68.49	9405	19JN	9;08	263.05
9406	3JL	10;17	96.47	9406	18JL	21;52	291.32
9407	1AU	20;10	124.77	9407	17AU	9;15	319.90
9408	31AU	8;37	153.61	9408	15SE	19;51	348.93
9409	30SE	0;13	183.09	9409	15OC	6;19	18.44
9410	29OC	18;44	213.15	9410	13NO	17;02	48.35
9411	28NO	14;46	243.56	9411	13DE	4;03	78.42
9412	28DE	10;06	273.95				
−239							
				9412	11JA	15;21	108.37
9413	27JA	2;55	303.98	9413	10FE	3;02	137.98
9414	25FE	16;35	333.42	9414	11MR	15;29	167.13
9415	27MR	3;23	2.23	9415	10AP	5;02	195.83
9416	25AP	12;03	30.51	9416	9MY	19;40	224.19
9417	24MY	19;23	58.47	9417	8JN	10;54	252.41
9418	23JN	2;13	86.36	9418	8JL	2;03	280.70
9419	22JL	11;45	114.45	9419	6AU	16;38	309.25
9420	20AU	18;22	142.97	9420	5SE	6;32	338.22
9421	19SE	5;48	172.08	9421	4OC	19;44	7.68
9422	18OC	20;29	201.83	9422	3NO	8;11	37.54
9423	17NO	14;10	232.06	9423	2DE	19;46	67.61
9424	17DE	9;36	262.49				

Right half

NEW#	Date	Time	Long.	FULL#	Date	Time	Long.
−238							
				9424	1JA	6;26	97.62
9425	16JA	4;53	292.74	9425	30JA	16;27	127.31
9426	14FE	22;29	322.53	9426	1MR	2;19	156.53
9427	16MR	13;30	351.70	9427	30MR	12;39	185.27
9428	15AP	1;45	20.27	9428	28AP	23;59	213.63
9429	14MY	11;28	48.40	9429	28MY	12;37	241.78
9430	12JN	19;21	76.33	9430	27JN	2;37	269.97
9431	12JL	2;25	104.31	9431	26JL	17;55	298.41
9432	10AU	9;46	132.62	9432	25AU	10;11	327.30
9433	8SE	18;31	161.44	9433	24SE	2;47	356.70
9434	8OC	5;26	190.86	9434	23OC	18;45	26.57
9435	6NO	18;54	220.80	9435	22NO	9;12	56.70
9436	6DE	10;43	251.06	9436	21DE	21;42	86.83
−237							
9437	5JA	4;16	281.34	9437	20JA	8;26	116.68
9438	3FE	22;34	311.32	9438	18FE	17;55	146.07
9439	5MR	16;24	340.81	9439	20MR	2;42	174.92
9440	4AP	8;31	9.71	9440	18AP	11;21	203.31
9441	3MY	22;06	38.10	9441	17MY	20;30	231.39
9442	2JN	9;07	66.17	9442	16JN	6;53	259.41
9443	1JL	18;13	94.18	9443	15JL	19;18	287.64
9444	31JL	2;28	122.39	9444	14AU	10;16	316.30
9445	29AU	10;52	151.02	9445	13SE	3;37	345.54
9446	27SE	20;10	180.19	9446	12OC	22;11	15.35
9447	27OC	6;45	209.88	9447	11NO	16;17	45.54
9448	25NO	18;49	239.93	9448	11DE	8;31	75.83
9449	25DE	8;29	270.09				
−236							
				9449	9JA	22;20	105.91
9450	23JA	23;50	300.10	9450	8FE	9;51	135.55
9451	22FE	16;27	329.73	9451	8MR	19;26	164.62
9452	23MR	9;27	358.85	9452	7AP	3;34	193.14
9453	22AP	1;35	27.48	9453	6MY	10;53	221.24
9454	21MY	16;02	55.74	9454	4JN	18;20	249.16
9455	20JN	4;39	83.86	9455	4JL	3;04	277.17
9456	19JL	15;51	112.08	9456	2AU	14;12	305.53
9457	18AU	2;14	140.62	9457	1SE	4;27	334.46
9458	16SE	12;19	169.64	9458	30SE	21;41	4.02
9459	15OC	22;25	199.15	9459	30OC	16;47	34.11
9460	14NO	8;48	229.05	9460	29NO	12;03	64.50
9461	13DE	19;49	259.12	9461	29DE	5;59	94.83
−235							
9462	12JA	7;48	289.10	9462	27JA	21;32	124.78
9463	10FE	20;59	318.76	9463	26FE	10;15	154.17
9464	12MR	11;16	347.96	9464	27MR	20;13	182.94
9465	11AP	2;10	16.70	9465	26AP	4;02	211.18
9466	10MY	17;13	45.08	9466	25MY	10;43	239.11
9467	9JN	8;05	73.28	9467	23JN	17;28	267.00
9468	8JL	22;31	101.54	9468	23JL	1;27	295.12
9469	7AU	12;20	130.07	9469	21AU	11;41	323.70
9470	6SE	1;17	159.02	9470	20SE	0;46	352.90
9471	5OC	13;16	188.44	9471	19OC	16;45	22.71
9472	4NO	0;26	218.26	9472	18NO	11;08	52.98
9473	3DE	11;08	248.30	9473	18DE	6;45	83.41
−234							
9474	1JA	21;46	278.29	9474	17JA	1;56	113.65
9475	31JA	8;32	308.01	9475	15FE	18;57	143.40
9476	1MR	19;35	337.27	9476	17MR	8;45	172.51
9477	31MR	7;05	6.05	9477	15AP	19;22	201.01
9478	29AP	19;22	34.43	9478	15MY	3;38	229.08
9479	29MY	8;50	62.61	9479	13JN	10;42	256.97
9480	27JN	23;39	90.83	9480	12JL	17;44	284.96
9481	27JL	15;31	119.31	9481	11AU	1;39	313.29
9482	26AU	7;33	148.20	9482	9SE	11;11	342.16
9483	24SE	22;54	177.57	9483	8OC	22;55	11.62
9484	24OC	13;00	207.37	9484	7NO	13;19	41.61
9485	23NO	1;50	237.44	9485	7DE	6;25	71.93
9486	22DE	13;32	267.53				
−233							
				9486	6JA	1;25	102.26
9487	21JA	0;12	297.37	9487	4FE	20;36	132.27
9488	19FE	9;55	326.76	9488	6MR	14;03	161.72
9489	20MR	18;59	355.61	9489	5AP	4;40	190.55
9490	19AP	4;04	24.01	9490	4MY	16;30	218.86
9491	18MY	14;07	52.12	9491	3JN	2;11	246.88
9492	17JN	1;59	80.20	9492	2JL	10;35	274.86
9493	16JL	16;06	108.50	9493	31JL	18;32	303.07
9494	15AU	8;14	137.22	9494	30AU	2;45	331.70
9495	14SE	1;30	166.48	9495	28SE	11;57	0.87
9496	13OC	18;52	196.25	9496	27OC	22;49	30.58
9497	12NO	11;26	226.38	9497	26NO	11;56	60.68
9498	12DE	2;28	256.62	9498	26DE	3;24	90.92

------- NEW MOONS -------				------- FULL MOONS -------			
NUMBER	DATE	TIME	LONG.	NUMBER	DATE	TIME	LONG.
-232							
9499	10JA	15;27	286.66	9499	24JA	20;33	121.00
9500	9FE	2;14	316.26	9500	23FE	14;09	150.65
9501	9MR	11;08	345.29	9501	24MR	6;58	179.75
9502	7AP	18;53	13.78	9502	22AP	22;17	208.34
9503	7MY	2;29	41.89	9503	22MY	11;47	236.56
9504	5JN	11;01	69.85	9504	20JN	23;30	264.64
9505	4JL	21;20	97.93	9505	20JL	9;44	292.82
9506	3AU	10;03	126.36	9506	18AU	19;01	321.33
9507	2SE	1;20	155.34	9507	17SE	4;04	350.32
9508	1OC	18;54	184.93	9508	16OC	13;39	19.82
9509	31OC	13;52	215.04	9509	15NO	0;22	49.74
9510	30NO	8;43	245.40	9510	14DE	12;28	79.85
9511	30DE	1;46	275.69				
-231							
				9511	13JA	1;52	109.89
9512	28JA	15;54	305.58	9512	11FE	16;17	139.59
9513	27FE	3;01	334.89	9513	13MR	7;25	168.81
9514	28MR	11;47	3.60	9514	11AP	22;58	197.57
9515	26AP	19;12	31.82	9515	11MY	14;32	225.96
9516	26MY	2;13	59.76	9516	10JN	5;27	254.16
9517	24JN	9;45	87.68	9517	9JL	19;08	282.40
9518	23JL	18;40	115.84	9518	8AU	7;24	310.88
9519	22AU	5;48	144.47	9519	6SE	18;36	339.76
9520	20SE	19;53	173.72	9520	6OC	5;19	9.14
9521	20OC	13;08	203.60	9521	4NO	16;03	38.95
9522	19NO	8;44	233.93	9522	4DE	2;58	69.00
9523	19DE	4;43	264.37				
-230							
				9523	2JA	14;02	99.01
9524	17JA	22;55	294.57	9524	1FE	1;17	128.74
9525	16FE	14;06	324.22	9525	2MR	13;01	158.02
9526	18MR	2;10	353.25	9526	1AP	1;41	186.83
9527	16AP	11;44	21.69	9527	30AP	15;33	215.27
9528	15MY	19;35	49.74	9528	30MY	6;26	243.50
9529	14JN	2;34	77.63	9529	28JN	21;45	271.74
9530	13JL	9;34	105.62	9530	28JL	12;51	300.20
9531	11AU	17;36	133.97	9531	27AU	3;23	329.04
9532	10SE	3;43	162.87	9532	25SE	17;14	358.36
9533	9OC	16;49	192.40	9533	25OC	6;20	28.13
9534	8NO	9;08	222.48	9534	23NO	18;31	58.18
9535	8DE	3;52	252.87	9535	23DE	5;39	88.24
-229							
9536	6JA	23;21	283.21	9536	21JA	15;50	118.05
9537	5FE	17;48	313.18	9537	20FE	1;29	147.43
9538	7MR	10;03	342.57	9538	21MR	11;14	176.30
9539	5AP	23;36	11.34	9539	19AP	21;45	204.74
9540	5MY	10;31	39.61	9540	19MY	9;30	232.92
9541	3JN	19;15	67.59	9541	17JN	22;43	261.06
9542	3JL	2;41	95.53	9542	17JL	13;25	289.39
9543	1AU	9;54	123.72	9543	16AU	5;26	318.11
9544	30AU	17;59	152.35	9544	14SE	22;14	347.36
9545	29SE	3;52	181.57	9545	14OC	14;57	17.12
9546	28OC	16;07	211.34	9546	13NO	6;29	47.22
9547	27NO	6;47	241.51	9547	12DE	20;04	77.39
9548	26DE	23;26	271.80				
-228							
				9548	11JA	7;37	107.37
9549	25JA	17;17	301.90	9549	9FE	17;32	136.92
9550	24FE	11;17	331.55	9550	10MR	2;24	165.94
9551	25MR	4;13	0.64	9551	8AP	10;50	194.45
9552	23AP	19;04	29.20	9552	7MY	19;27	222.60
9553	23MY	7;22	57.37	9553	6JN	4;59	250.60
9554	21JN	17;26	85.38	9554	5JL	16;14	278.71
9555	21JL	2;10	113.51	9555	4AU	5;56	307.20
9556	19AU	10;34	142.00	9556	2SE	22;18	336.24
9557	17SE	19;33	170.99	9557	2OC	16;41	5.87
9558	17OC	5;37	200.52	9558	1NO	11;27	35.98
9559	15NO	16;59	230.47	9559	1DE	4;54	66.29
9560	15DE	5;45	260.62	9560	30DE	20;00	96.49
-227							
9561	13JA	20;04	290.68	9561	29JA	8;35	126.31
9562	12FE	11;51	320.43	9562	27FE	18;57	155.57
9563	14MR	4;31	349.70	9563	29MR	3;35	184.26
9564	12AP	21;01	18.48	9564	27AP	11;04	212.48
9565	12MY	12;18	46.85	9565	26MY	18;13	240.42
9566	11JN	1;53	75.00	9566	25JN	2;06	268.36
9567	10JL	13;55	103.18	9567	24JL	11;55	296.56
9568	9AU	0;53	131.61	9568	23AU	0;37	325.27
9569	7SE	11;20	160.48	9569	21SE	16;33	354.61
9570	6OC	21;37	189.86	9570	21OC	11;03	24.55
9571	5NO	7;59	219.65	9571	20NO	6;33	54.88
9572	4DE	18;39	249.69	9572	20DE	1;23	85.28

------- NEW MOONS -------				------- FULL MOONS -------			
NUMBER	DATE	TIME	LONG.	NUMBER	DATE	TIME	LONG.
-226							
9573	3JA	6;00	279.71	9573	18JA	18;12	115.40
9574	1FE	18;21	309.48	9574	17FE	8;17	145.00
9575	3MR	7;48	338.82	9575	18MR	19;28	173.98
9576	1AP	22;09	7.68	9576	17AP	4;10	202.38
9577	1MY	12;57	36.15	9577	16MY	11;13	230.39
9578	31MY	3;52	64.38	9578	14JN	17;46	258.26
9579	29JN	18;38	92.60	9579	14JL	1;01	286.27
9580	29JL	9;00	121.04	9580	12AU	10;05	314.67
9581	27AU	22;43	149.86	9581	10SE	21;47	343.65
9582	26SE	11;29	179.15	9582	10OC	12;26	13.25
9583	25OC	23;15	208.88	9583	9NO	5;46	43.39
9584	24NO	10;16	238.88	9584	9DE	0;56	73.79
9585	23DE	20;53	268.91				
-225							
				9585	7JA	20;31	104.14
9586	22JA	7;25	298.74	9586	6FE	14;41	134.08
9587	20FE	18;05	328.14	9587	8MR	6;03	163.42
9588	22MR	5;02	357.05	9588	6AP	18;08	192.12
9589	20AP	16;36	25.53	9589	6MY	3;25	220.31
9590	20MY	5;13	53.73	9590	4JN	10;58	248.24
9591	18JN	19;18	81.90	9591	3JL	17;57	276.17
9592	18JL	10;47	110.27	9592	2AU	1;26	304.37
9593	17AU	3;05	139.03	9593	31AU	10;13	333.05
9594	15SE	19;11	168.26	9594	29SE	20;54	2.31
9595	15OC	10;14	197.96	9595	29OC	9;59	32.13
9596	13NO	23;54	227.98	9596	28NO	1;44	62.35
9597	13DE	12;14	258.11	9597	27DE	19;50	92.69
-224							
9598	11JA	23;22	288.06	9598	26JA	14;59	122.83
9599	10FE	9;24	317.61	9599	25FE	9;19	152.48
9600	10MR	18;31	346.63	9600	26MR	1;17	181.53
9601	9AP	3;15	15.14	9601	24AP	14;25	210.00
9602	8MY	12;28	43.31	9602	24MY	1;07	238.10
9603	6JN	23;11	71.35	9603	22JN	10;09	266.08
9604	6JL	12;07	99.54	9604	21JL	18;24	294.20
9605	5AU	3;23	128.09	9605	20AU	2;35	322.68
9606	3SE	20;24	157.17	9606	18SE	11;23	351.68
9607	3OC	14;06	186.80	9607	17OC	21;27	21.22
9608	2NO	7;25	216.85	9608	16NO	9;25	51.20
9609	1DE	23;28	247.11	9609	15DE	23;39	81.41
9610	31DE	13;34	277.26				
-223							
				9610	14JA	15;53	111.54
9611	30JA	1;22	307.04	9611	13FE	9;09	141.34
9612	28FE	11;00	336.26	9612	15MR	2;14	170.62
9613	29MR	19;02	4.92	9613	13AP	18;12	199.36
9614	28AP	2;25	33.12	9614	13MY	8;35	227.69
9615	27MY	10;13	61.09	9615	11JN	21;16	255.80
9616	25JN	19;28	89.08	9616	11JL	8;23	283.95
9617	25JL	6;56	117.36	9617	9AU	18;18	312.35
9618	23AU	21;01	146.13	9618	8SE	3;37	341.19
9619	22SE	13;38	175.52	9619	7OC	13;03	10.53
9620	22OC	8;13	205.47	9620	5NO	23;15	40.33
9621	21NO	3;30	235.80	9621	5DE	10;39	70.40
9622	20DE	21;44	266.17				
-222							
				9622	3JA	23;18	100.47
9623	19JA	13;23	296.24	9623	2FE	13;00	130.28
9624	18FE	1;53	325.76	9624	4MR	3;29	159.65
9625	19MR	11;36	354.66	9625	2AP	18;35	188.54
9626	17AP	19;27	23.03	9626	2MY	10;01	217.02
9627	17MY	2;28	51.03	9627	1JN	1;19	245.26
9628	15JN	9;35	78.92	9628	30JN	15;48	273.48
9629	14JL	17;43	106.97	9629	30JL	5;00	301.88
9630	13AU	3;42	135.42	9630	28AU	16;58	330.64
9631	11SE	16;19	164.44	9631	27SE	4;08	359.88
9632	11OC	8;06	194.11	9632	26OC	15;02	29.57
9633	10NO	2;46	224.31	9633	25NO	1;57	59.57
9634	9DE	22;53	254.76	9634	24DE	12;55	89.62
-221							
9635	8JA	18;11	285.08	9635	22JA	23;53	119.46
9636	7FE	10;52	314.95	9636	21FE	11;04	148.87
9637	9MR	0;21	344.20	9637	22MR	22;54	177.81
9638	7AP	11;00	12.83	9638	21AP	11;53	206.33
9639	6MY	19;34	40.99	9639	21MY	2;07	.234.59
9640	5JN	2;52	68.90	9640	19JN	17;16	262.81
9641	4JL	9;48	96.83	9641	19JL	8;41	291.18
9642	2AU	17;17	125.04	9642	17AU	23;47	319.90
9643	1SE	2;20	153.74	9643	16SE	14;17	349.08
9644	30SE	13;57	183.05	9644	16OC	4;04	18.74
9645	30OC	4;44	212.95	9645	14NO	16;56	48.74
9646	28NO	22;24	243.26	9646	14DE	4;39	78.84
9647	28DE	17;39	273.65				

```
---------NEW MOONS---------        ---------FULL MOONS--------       ---------NEW MOONS---------        ---------FULL MOONS-------
NUMBER  DATE   TIME   LONG.    NUMBER  DATE   TIME   LONG.     NUMBER  DATE   TIME   LONG.    NUMBER  DATE   TIME   LONG.

                     -220                                                              -214
                          9647  12JA  15;13  108.76    9722  21JA   0;01  297.75    9721   5JA  11;42  102.07
9648  27JA  12;40  303.77   9648  11FE   0;53  138.28    9723  19FE  10;35  327.17    9722   4FE   4;20  131.98
9649  26FE   5;58  333.36   9649  11MR  10;16  167.30    9724  20MR  19;09  356.01    9723   5MR  21;23  161.43
9650  26MR  20;48    2.34   9650   9AP  20;04  195.85    9725  19AP   2;34   24.33    9724   4AP  13;47  190.34
9651  25AP   8;58   30.77   9651   9MY   6;55  224.07    9726  18MY   9;53   52.34    9725   4MY   4;54  218.79
9652  24MY  18;44   58.83   9652   7JN  19;14  252.18    9727  16JN  18;12   80.28    9726   2JN  18;28  246.96
9653  23JN   2;48   86.77   9653   7JL   9;11  280.41    9728  16JL   4;28  108.42    9727   2JL   6;31  275.08
9654  22JL  10;07  114.86   9654   6AU   0;41  308.99    9729  14AU  17;16  137.00    9728  31JL  17;13  303.40
9655  20AU  17;47  143.33   9655   4SE  17;25  338.07    9730  13SE   8;46  166.17    9729  30AU   3;00  332.10
9656  19SE   2;50  172.35   9656   4OC  10;36    7.68    9731  13OC   2;38  195.95    9730  28SE  12;30    1.29
9657  18OC  13;57  201.94   9657   3NO   3;07   37.72    9732  11NO  21;55  226.19    9731  27OC  22;25   30.95
9658  17NO   3;27  231.99   9658   2DE  17;53   67.92    9733  11DE  17;00  256.60    9732  26NO   9;14   60.96
9659  16DE  19;05  262.26                                                            9733  25DE  21;11   91.05

                     -219                                                              -213
                          9659   1JA   6;27   98.01    9734  10JA  10;06  286.82    9734  24JA  10;10  120.95
9660  15JA  12;15  292.43   9660  30JA  17;01  127.72    9735   9FE   0;07  316.55    9735  22FE  23;59  150.45
9661  14FE   6;04  322.23   9661   1MR   2;11  156.91    9736  10MR  11;02  345.67    9736  24MR  14;30  179.47
9662  15MR  23;29  351.51   9662  30MR  10;33  185.57    9737   8AP  19;34   14.20    9737  23AP   5;34  208.06
9663  14AP  15;21   20.24   9663  28AP  18;48  213.80    9738   8MY   2;47   42.29    9738  22MY  20;55  236.35
9664  14MY   4;56   48.53   9664  28MY   3;37  241.81    9739   6JN   9;41   70.19    9739  21JN  11;57  264.56
9665  12JN  16;09   76.58   9665  26JN  13;49  269.84    9740   5JL  17;13   98.15    9740  21JL   2;02  292.90
9666  12JL   1;36  104.67   9666  26JL   2;14  298.16    9741   4AU   2;13  126.43    9741  19AU  14;53  321.55
9667  10AU  10;15  133.03   9667  24AU  17;22  327.00    9742   2SE  13;30  155.24    9742  18SE   2;40  350.65
9668   8SE  19;05  161.85   9668  23SE  11;05  356.44    9743   2OC   3;45  184.69    9743  17OC  13;53   20.22
9669   8OC   4;44  191.22   9669  23OC   6;06   26.43    9744  31OC  21;10  214.72    9744  16NO   0;55   50.15
9670   6NO  15;31  221.05   9670  22NO   0;35   56.73    9745  30NO  16;51  245.12    9745  15DE  11;53   80.22
9671   6DE   3;33  251.15   9671  21DE  17;01   87.02    9746  30DE  12;48  275.53

                     -218                                                              -212
9672   4JA  16;56  281.25   9672  20JA   6;49  117.00                                 9746  13JA  22;44  110.14
9673   3FE   7;45  311.09   9673  18FE  18;09  146.46    9747  29JA   6;52  305.59    9747  12FE   9;32  139.69
9674   4MR  23;46  340.51   9674  20MR   3;27  175.34    9748  27FE  21;52  335.06    9748  12MR  20;41  168.76
9675   3AP  16;14    9.44   9675  18AP  11;17  203.69    9749  28MR   9;47    3.90    9749  11AP   8;45  197.38
9676   3MY   8;06   37.93   9676  17MY  18;22  231.69    9750  26AP  19;14   32.20    9750  10MY  22;08  225.69
9677   1JN  22;35   66.14   9677  16JN   1;40  259.59    9751  26MY   3;03   60.17    9751   9JN  12;49  253.89
9678   1JL  11;29   94.29   9678  15JL  10;22  287.67    9752  24JN  10;06   88.08    9752   9JL   4;14  282.19
9679  30JL  23;10  122.64   9679  13AU  21;34  316.17    9753  23JL  17;16  116.17    9753   7AU  19;45  310.79
9680  29AU  10;06  151.37   9680  12SE  12;00  345.28    9754  22AU   1;31  144.69    9754   6SE  10;53  339.83
9681  27SE  20;41  180.60   9681  12OC   5;28   15.03    9755  20SE  11;52  173.78    9755   6OC   1;22    9.36
9682  27OC   7;10  210.28   9682  11NO   0;48   45.27    9756  20OC   1;06  203.48    9756   4NO  14;57   39.29
9683  25NO  17;43  240.27   9683  10DE  20;14   75.69    9757  18NO  17;25  233.67    9757   4DE   3;22   69.40
9684  25DE   4;37  270.32                                9758  18DE  12;00  264.05

                     -217                                                              -211
                          9684   9JA  14;11  105.95    9759  17JA   7;12  294.29    9758   2JA  14;29   99.42
9685  23JA  16;14  300.17   9685   8FE   5;37  135.76    9760  16FE   1;21  324.08    9759   1FE   0;24  129.09
9686  22FE   4;52  329.63   9686   9MR  17;09  164.95    9761  17MR  17;21  353.27    9760   2MR   9;38  158.26
9687  23MR  18;29  358.63   9687   8AP   3;56  193.54    9762  16AP   6;46   21.88    9761  31MR  18;52  186.94
9688  22AP   8;51   27.20   9688   7MY  11;37  221.66    9763  15MY  17;42   50.05    9762  30AP   4;53  215.22
9689  21MY  23;37   55.48   9689   5JN  18;14  249.54    9764  14JN   2;35   78.01    9763  29MY  16;16  243.32
9690  20JN  14;29   83.68   9690   5JL   1;00  277.47    9765  13JL  10;17  106.04    9764  28JN   5;19  271.47
9691  20JL   5;15  112.04   9691   3AU   9;07  305.71    9766  11AU  17;48  134.37    9765  27JL  20;07  299.91
9692  18AU  19;37  140.74   9692   1SE  19;31  334.48    9767  10SE   2;13  163.20    9766  26AU  12;29  328.82
9693  17SE   9;12  169.90   9693   1OC   8;46    3.86    9768   9OC  12;22  192.60    9767  25SE   5;50  358.28
9694  16OC  21;44  199.51   9694  31OC   0;53   33.83    9769   8NO   0;43  222.51    9768  24OC  23;06   28.22
9695  15NO   9;15  229.46   9695  29NO  19;16   64.17    9770   7DE  15;17  252.72    9769  23NO  15;02   58.42
9696  14DE  20;03  259.52   9696  29DE  14;48   94.57                                 9770  23DE   4;48   88.59

                     -216                                                              -210
9697  13JA   6;30  289.43   9697  28JA   9;47  124.68    9771   6JA   7;36  282.94    9771  21JA  16;16  118.46
9698  11FE  16;53  318.97   9698  27FE   2;34  154.24    9772   5FE   0;58  312.87    9772  20FE   1;54  147.83
9699  12MR   3;24  348.03   9699  27MR  16;12  183.16    9773   6MR  18;29  342.32    9773  21MR  10;24  176.65
9700  10AP  14;20   16.61   9700  26AP   2;45  211.51    9774   5AP  11;05   11.22    9774  19AP  18;26  205.00
9701  10MY   2;08   44.87   9701  25MY  11;01  239.51    9775   5MY   1;49   39.65    9775  19MY   2;43  233.04
9702   8JN  15;19   73.00   9702  23JN  18;13  267.41    9776   3JN  14;15   67.77    9776  17JN  12;01  261.01
9703   8JL   6;08  101.28   9703  23JL   1;27  295.50    9777   3JL   0;38   95.84    9777  16JL  23;12  289.20
9704   6AU  22;19  129.89   9704  21AU   9;38  324.01    9778   1AU   9;47  124.10    9778  15AU  13;00  317.83
9705   5SE  14;55  158.98   9705  19SE  19;25  353.07    9779  30AU  18;39  152.77    9779  14SE   5;39  347.07
9706   5OC   6;52  188.56   9706  19OC   7;20   22.70    9780  29SE   4;02  181.96    9780  14OC   0;26   16.90
9707   3NO  21;30  218.52   9707  17NO  21;45   52.80    9781  28OC  14;22  211.66    9781  12NO  19;36   47.14
9708   3DE  10;37  248.66   9708  17DE  14;40   83.12    9782  27NO   1;47  241.69    9782  12DE  13;19   77.49
                                                         9783  26DE  14;22  271.81

                     -215                                                              -209
9709   1JA  22;20  278.71   9709  16JA   9;19  113.34                                 9783  11JA   4;28  107.62
9710  31JA   8;48  308.41   9710  15FE   4;06  143.17    9784  25JA   4;14  301.74    9784   9FE  16;55  137.28
9711   1MR  18;08  337.60   9711  16MR  21;13  172.43    9785  23FE  19;24  331.28    9785  11MR   3;02  166.35
9712  31MR   2;45    6.26   9712  15AP  11;41  201.09    9786  25MR  11;28    0.35    9786   9AP  11;24  194.86
9713  29AP  11;23   34.50   9713  14MY  23;31  229.30    9787  24AP   3;34   28.97    9787   8MY  18;38  222.95
9714  28MY  21;04   62.53   9714  13JN   9;23  257.30    9788  23MY  18;46   57.25    9788   7JN   1;38  250.85
9715  27JN   8;44   90.62   9715  12JL  18;05  285.36    9789  22JN   8;33   85.41    9789   6JL   9;28  278.83
9716  26JL  22;53  119.03   9716  11AU   2;24  313.71    9790  21JL  21;00  113.69    9790   4AU  19;19  307.15
9717  25AU  15;16  147.92   9717   9SE  10;55  342.54    9791  20AU   8;31  142.30    9791   3SE   8;09  336.04
9718  24SE   9;00  177.38   9718   8OC  20;30   11.91    9792  18SE  19;32  171.38    9792   3OC   0;18    5.58
9719  24OC   2;53  207.33   9719   7NO   7;32   41.76    9793  18OC   6;16  200.94    9793   1NO  19;01   35.67
9720  22NO  19;52  237.58   9720   6DE  20;34   71.90    9794  16NO  16;52  230.86    9794   1DE  14;41   66.08
9721  22DE  11;05  267.81                                9795  16DE   3;32  260.91    9795  31DE   9;33   96.44
```

66

-208

NUMBER	DATE	TIME	LONG.		NUMBER	DATE	TIME	LONG.
9796	14JA	14;36	290.84		9796	30JA	2;17	126.43
9797	13FE	2;27	320.42		9797	28FE	16;12	155.85
9798	13MR	15;17	349.55		9798	29MR	3;12	184.63
9799	12AP	5;02	18.23		9799	27AP	11;45	212.89
9800	11MY	19;26	46.57		9800	26MY	18;44	240.83
9801	10JN	10;14	74.77		9801	25JN	1;18	268.71
9802	10JL	1;12	103.06		9802	24JL	8;40	296.82
9803	8AU	16;03	131.64		9803	22AU	17;55	325.38
9804	7SE	6;23	160.67		9804	21SE	5;48	354.55
9805	6OC	19;45	190.17		9805	20OC	20;35	24.33
9806	5NO	7;58	220.04		9806	19NO	13;57	54.57
9807	4DE	19;10	250.10		9807	19DE	9;02	84.97

-207

NUMBER	DATE	TIME	LONG.		NUMBER	DATE	TIME	LONG.
9808	3JA	5;43	280.09		9808	18JA	4;24	115.21
9809	1FE	15;56	309.76		9809	16FE	22;20	144.98
9810	3MR	2;07	338.97		9810	18MR	13;30	174.12
9811	1AP	12;31	7.68		9811	17AP	1;29	202.66
9812	30AP	23;34	36.00		9812	16MY	10;45	230.76
9813	30MY	11;49	64.13		9813	14JN	18;24	258.67
9814	29JN	1;46	92.32		9814	14JL	1;35	286.68
9815	28JL	17;27	120.80		9815	12AU	9;20	315.03
9816	27AU	10;12	149.74		9816	10SE	18;24	343.89
9817	26SE	2;54	179.19		9817	10OC	5;18	13.34
9818	25OC	18;32	209.07		9818	8NO	18;28	43.29
9819	24NO	8;35	239.19		9819	8DE	10;06	73.55
9820	23DE	21;02	269.31					

-206

NUMBER	DATE	TIME	LONG.		NUMBER	DATE	TIME	LONG.
					9820	7JA	3;52	103.82
9821	22JA	8;03	299.15		9821	5FE	22;36	133.80
9822	20FE	17;45	328.51		9822	7MR	16;32	163.25
9823	22MR	2;27	357.33		9823	6AP	8;16	192.11
9824	20AP	10;45	25.68		9824	5MY	21;21	220.46
9825	19MY	19;35	53.74		9825	4JN	8;11	248.51
9826	18JN	6;02	81.76		9826	3JL	17;30	276.54
9827	17JL	18;54	110.02		9827	2AU	2;07	304.79
9828	16AU	10;20	138.73		9828	31AU	10;42	333.46
9829	15SE	3;44	168.01		9829	29SE	19;52	2.65
9830	14OC	21;56	197.83		9830	29OC	6;09	32.36
9831	13NO	15;41	228.02		9831	27NO	18;08	62.42
9832	13DE	8;00	258.32		9832	27DE	8;08	92.59

-205

NUMBER	DATE	TIME	LONG.		NUMBER	DATE	TIME	LONG.
9833	11JA	22;08	288.40		9833	25JA	23;53	122.59
9834	10FE	9;46	318.01		9834	24FE	16;34	152.18
9835	11MR	19;07	347.04		9835	26MR	9;09	181.26
9836	10AP	2;50	15.51		9836	25AP	0;49	209.84
9837	9MY	9;55	43.59		9837	24MY	15;10	238.09
9838	7JN	17;30	71.51		9838	23JN	4;06	266.22
9839	7JL	2;39	99.54		9839	22JL	15;39	294.47
9840	5AU	14;10	127.95		9840	21AU	2;05	323.05
9841	4SE	4;24	156.90		9841	19SE	11;54	352.09
9842	3OC	21;18	186.48		9842	18OC	21;43	21.62
9843	2NO	16;11	216.60		9843	17NO	8;08	51.53
9844	2DE	11;42	247.00		9844	16DE	19;29	81.62

-204

NUMBER	DATE	TIME	LONG.		NUMBER	DATE	TIME	LONG.
9845	1JA	6;01	277.33		9845	15JA	7;48	111.59
9846	30JA	21;36	307.27		9846	13FE	20;57	141.22
9847	29FE	9;56	336.60		9847	14MR	10;49	170.38
9848	29MR	19;26	5.32		9848	13AP	1;21	199.07
9849	28AP	3;05	33.53		9849	12MY	16;26	227.43
9850	27MY	9;58	61.46		9850	11JN	7;42	255.65
9851	25JN	17;04	89.37		9851	10JL	22;29	283.94
9852	25JL	1;16	117.52		9852	9AU	12;14	312.49
9853	23AU	11;23	146.13		9853	8SE	0;49	341.46
9854	22SE	0;11	175.35		9854	7OC	12;32	10.90
9855	21OC	16;07	205.18		9855	5NO	23;49	40.74
9856	20NO	10;53	235.49		9856	5DE	10;52	70.80
9857	20DE	6;58	265.93					

-203

NUMBER	DATE	TIME	LONG.		NUMBER	DATE	TIME	LONG.
					9857	3JA	21;43	100.79
9858	19JA	2;09	296.15		9858	2FE	8;20	130.48
9859	17FE	18;39	325.85		9859	3MR	18;58	159.69
9860	19MR	7;58	354.90		9860	2AP	6;13	188.42
9861	17AP	18;29	23.37		9861	1MY	18;39	216.79
9862	17MY	3;00	51.43		9862	31MY	8;34	244.98
9863	15JN	10;22	79.34		9863	29JN	23;41	273.22
9864	14JL	17;27	107.34		9864	29JL	15;24	301.72
9865	13AU	1;09	135.70		9865	28AU	7;02	330.62
9866	11SE	10;27	164.59		9866	26SE	22;11	0.02
9867	10OC	22;15	194.08		9867	26OC	12;31	29.85
9868	9NO	13;05	224.11		9868	25NO	1;43	59.95
9869	9DE	6;39	254.45		9869	24DE	13;31	90.04

-202

NUMBER	DATE	TIME	LONG.		NUMBER	DATE	TIME	LONG.
9870	8JA	1;37	284.77		9870	22JA	23;54	119.84
9871	6FE	20;17	314.73		9871	21FE	9;12	149.18
9872	8MR	13;18	344.13		9872	22MR	18;06	178.00
9873	7AP	3;56	12.93		9873	21AP	3;24	206.38
9874	6MY	16;04	41.23		9874	20MY	13;51	234.49
9875	5JN	1;58	69.25		9875	19JN	1;55	262.58
9876	4JL	10;16	97.24		9876	18JL	15;51	290.89
9877	2AU	17;54	125.45		9877	17AU	7;36	319.63
9878	1SE	1;56	154.11		9878	16SE	0;48	348.91
9879	30SE	11;16	183.33		9879	15OC	18;33	18.73
9880	29OC	22;35	213.08		9880	14NO	11;31	48.90
9881	28NO	12;03	243.21		9881	14DE	2;32	79.13
9882	28DE	3;25	273.43					

-201

NUMBER	DATE	TIME	LONG.		NUMBER	DATE	TIME	LONG.
					9882	12JA	15;07	109.14
9883	26JA	20;08	303.46		9883	11FE	1;29	138.69
9884	25FE	13;25	333.07		9884	12MR	10;18	167.69
9885	27MR	6;24	2.15		9885	10AP	18;18	196.16
9886	25AP	22;04	30.72		9886	10MY	2;12	224.27
9887	25MY	11;42	58.94		9887	8JN	10;46	252.23
9888	23JN	23;11	87.01		9888	7JL	20;51	280.30
9889	23JL	9;02	115.20		9889	6AU	9;18	308.75
9890	21AU	18;09	143.73		9890	5SE	0;39	337.77
9891	20SE	3;26	172.76		9891	4OC	18;42	7.41
9892	19OC	13;26	202.31		9892	3NO	14;07	37.55
9893	18NO	0;22	232.25		9893	3DE	8;53	67.93
9894	17DE	12;17	262.36					

-200

NUMBER	DATE	TIME	LONG.		NUMBER	DATE	TIME	LONG.
					9894	2JA	1;26	98.18
9895	16JA	1;18	292.36		9895	31JA	15;10	128.03
9896	14FE	15;33	322.02		9896	1MR	2;17	157.30
9897	15MR	6;57	351.23		9897	30MR	11;20	185.99
9898	13AP	22;54	19.97		9898	28AP	18;55	214.19
9899	13MY	14;32	48.34		9899	28MY	1;50	242.12
9900	12JN	5;06	76.53		9900	26JN	9;04	270.04
9901	11JL	18;22	104.76		9901	25JL	17;47	298.21
9902	10AU	6;34	133.26		9902	24AU	5;07	326.88
9903	8SE	18;05	162.20		9903	22SE	19;43	356.19
9904	8OC	5;11	191.62		9904	22OC	13;24	26.11
9905	6NO	16;00	221.46		9905	21NO	8;53	56.45
9906	6DE	2;38	251.50		9906	21DE	4;23	86.87

-199

NUMBER	DATE	TIME	LONG.		NUMBER	DATE	TIME	LONG.
9907	4JA	13;22	281.48		9907	19JA	22;15	117.03
9908	3FE	0;34	311.18		9908	18FE	13;33	146.66
9909	4MR	1;54	340.44		9909	20MR	1;54	175.66
9910	3AP	1;37	9.23		9910	18AP	11;31	204.08
9911	2MY	15;29	37.65		9911	17MY	19;07	232.11
9912	1JN	5;59	65.86		9912	16JN	1;45	259.98
9913	30JN	20;56	94.10		9913	15JL	8;38	287.98
9914	30JL	12;05	122.58		9914	13AU	16;54	316.37
9915	29AU	3;02	151.47		9915	12SE	3;31	345.32
9916	27SE	17;15	180.84		9916	11OC	16;56	14.90
9917	27OC	6;18	210.63		9917	10NO	9;06	44.99
9918	25NO	18;06	240.68		9918	10DE	3;26	75.37
9919	25DE	4;56	270.72					

-198

NUMBER	DATE	TIME	LONG.		NUMBER	DATE	TIME	LONG.
					9919	8JA	22;45	105.69
9920	23JA	15;10	300.51		9920	7FE	17;28	135.64
9921	22FE	1;07	329.87		9921	9MR	10;01	165.00
9922	23MR	11;06	358.72		9922	7AP	23;31	193.74
9923	21AP	21;30	27.13		9923	7MY	10;01	221.98
9924	21MY	8;53	55.28		9924	5JN	18;23	249.93
9925	19JN	21;50	83.40		9925	5JL	1;45	277.89
9926	19JL	12;43	111.76		9926	3AU	9;15	306.10
9927	18AU	5;14	140.54		9927	1SE	17;45	334.79
9928	16SE	22;24	169.84		9928	1OC	3;48	4.05
9929	16OC	14;57	199.61		9929	30OC	15;52	33.84
9930	15NO	6;02	229.71		9930	29NO	6;13	64.01
9931	14DE	19;23	259.88		9931	28DE	22;52	94.28

-197

NUMBER	DATE	TIME	LONG.		NUMBER	DATE	TIME	LONG.
9932	13JA	7;04	289.84		9932	27JA	17;06	124.37
9933	11FE	17;16	319.38		9933	26FE	11;25	154.00
9934	13MR	2;13	348.37		9934	28MR	4;13	183.06
9935	11AP	10;25	16.84		9935	26AP	18;33	211.57
9936	10MY	18;40	44.95		9936	26MY	6;28	239.71
9937	9JN	4;03	72.94		9937	24JN	16;34	267.73
9938	8JL	15;35	101.08		9938	24JL	1;38	295.90
9939	7AU	5;48	129.61		9939	22AU	10;22	324.42
9940	5SE	22;29	158.69		9940	20SE	19;23	353.45
9941	5OC	16;39	188.35		9941	20OC	5;11	23.00
9942	4NO	11;00	218.47		9942	18NO	16;18	52.96
9943	4DE	4;17	248.78		9943	18DE	5;11	83.10

NUMBER	DATE	TIME	LONG.	NUMBER	DATE	TIME	LONG.	NUMBER	DATE	TIME	LONG.	NUMBER	DATE	TIME	LONG.
						−196								**−190**	
9944	2JA	19;38	278.98	9944	16JA	19;55	113.17	10019	24JA	23;08	301.91	10018	10JA	17;37	107.55
9945	1FE	8;28	308.78	9945	15FE	11;59	142.89	10020	23FE	10;27	331.30	10019	9FE	10;12	137.39
9946	1MR	18;47	338.01	9946	16MR	4;27	172.13	10021	24MR	22;40	0.22	10020	10MR	23;57	166.61
9947	31MR	3;04	6.66	9947	14AP	20;27	200.86	10022	23AP	11;51	28.72	10021	9AP	10;48	195.22
9948	29AP	10;11	34.83	9948	14MY	11;25	229.20	10023	23MY	1;53	56.97	10022	8MY	19;15	223.36
9949	28MY	17;17	62.77	9949	13JN	1;08	257.36	10024	21JN	16;37	85.17	10023	7JN	2;14	251.26
9950	27JN	1;28	90.72	9950	12JL	13;34	285.56	10025	21JL	7;50	113.55	10024	6JL	8;53	279.19
9951	26JL	11;44	118.96	9951	11AU	0;47	314.03	10026	19AU	23;12	142.31	10025	4AU	16;25	307.42
9952	25AU	0;39	147.71	9952	9SE	11;06	342.93	10027	18SE	14;10	171.54	10026	3SE	1;53	336.17
9953	23SE	16;22	177.08	9953	8OC	21;04	12.32	10028	18OC	4;08	201.24	10027	2OC	13;57	5.53
9954	23OC	10;31	207.03	9954	7NO	7;15	42.13	10029	16NO	16;44	231.24	10028	1NO	4;51	35.46
9955	22NO	6;02	237.38	9955	6DE	18;07	72.18	10030	16DE	4;04	261.32	10029	30NO	22;11	65.76
9956	22DE	1;14	267.78									10030	30DE	17;04	96.13
						−195								**−189**	
9957	20JA	18;18	297.90	9956	5JA	5;51	102.21	10031	14JA	14;28	291.22	10031	29JA	12;09	126.23
9958	19FE	8;11	327.46	9957	3FE	18;22	131.95	10032	13FE	0;21	320.72	10032	28FE	5;48	155.80
9959	20MR	18;54	356.38	9958	5MR	7;34	161.25	10033	14MR	10;02	349.72	10033	29MR	20;47	184.76
9960	19AP	3;15	24.74	9959	3AP	21;28	190.06	10034	12AP	19;54	18.25	10034	28AP	8;42	213.15
9961	18MY	10;19	52.74	9960	3MY	12;05	218.50	10035	12MY	6;29	46.44	10035	27MY	18;02	241.18
9962	16JN	17;11	80.62	9961	2JN	3;16	246.74	10036	10JN	18;27	74.53	10036	26JN	1;51	269.12
9963	16JL	0;46	108.66	9962	1JL	18;28	274.99	10037	10JL	8;20	102.77	10037	25JL	9;18	297.23
9964	14AU	9;54	137.09	9963	31JL	9;00	303.45	10038	9AU	0;14	131.39	10038	23AU	17;22	325.75
9965	12SE	21;22	166.09	9964	29AU	22;28	332.30	10039	7SE	17;28	160.52	10039	22SE	2;44	354.81
9966	12OC	11;47	195.72	9965	28SE	10;52	1.61	10040	7OC	10;46	190.18	10040	21OC	13;50	24.44
9967	11NO	5;17	225.88	9966	27OC	22;33	31.35	10041	6NO	2;54	220.22	10041	20NO	3;01	54.49
9968	11DE	0;57	256.31	9967	26NO	9;49	61.37	10042	5DE	17;16	250.41	10042	19DE	18;27	84.75
				9968	25DE	20;46	91.42								
						−194								**−188**	
9969	9JA	20;47	286.65	9969	24JA	7;20	121.22	10043	4JA	5;47	280.48	10043	18JA	11;50	114.90
9970	8FE	14;40	316.55	9970	22FE	17;40	150.58	10044	2FE	16;36	310.18	10044	17FE	6;04	144.69
9971	10MR	5;29	345.83	9971	24MR	4;14	179.44	10045	3MR	1;59	339.35	10045	17MR	23;36	173.94
9972	8AP	17;15	14.49	9972	22AP	15;44	207.89	10046	1AP	10;16	7.97	10046	16AP	15;07	202.63
9973	8MY	2;38	42.67	9973	22MY	4;43	236.10	10047	30AP	18;11	36.17	10047	16MY	4;12	230.88
9974	6JN	10;29	70.60	9974	20JN	19;13	264.28	10048	30MY	2;42	64.15	10048	14JN	15;13	258.93
9975	5JL	17;41	98.55	9975	20JL	10;49	292.68	10049	28JN	12;58	92.20	10049	14JL	0;53	287.03
9976	4AU	1;05	126.77	9976	19AU	2;47	321.45	10050	28JL	1;50	120.56	10050	12AU	9;55	315.44
9977	2SE	9;35	155.47	9977	17SE	18;31	350.70	10051	26AU	17;28	149.44	10051	10SE	18;57	344.31
9978	1OC	20;09	184.76	9978	17OC	9;35	20.43	10052	25SE	11;14	178.92	10052	10OC	4;27	13.70
9979	31OC	9;30	214.61	9979	15NO	23;37	50.49	10053	25OC	5;52	208.92	10053	8NO	14;56	43.53
9980	30NO	1;46	244.87	9980	15DE	12;13	80.62	10054	24NO	0;00	239.22	10054	8DE	2;52	73.63
9981	29DE	20;06	275.21					10055	23DE	16;29	269.51				
						−193								**−187**	
				9981	13JA	23;14	110.55					10055	6JA	16;33	103.73
9982	28JA	14;57	305.31	9982	12FE	8;52	140.05	10056	22JA	6;35	299.48	10056	5FE	7;46	133.57
9983	27FE	8;44	334.91	9983	13MR	17;39	169.02	10057	20FE	18;03	328.91	10057	6MR	23;51	162.95
9984	29MR	0;28	3.91	9984	12AP	2;25	197.51	10058	22MR	3;07	357.75	10058	5AP	15;55	191.83
9985	27AP	13;48	32.37	9985	11MY	12;00	225.67	10059	20AP	10;33	26.06	10059	5MY	7;19	220.28
9986	27MY	0;49	60.47	9986	9JN	23;03	253.73	10060	19MY	17;24	54.03	10060	3JN	21;42	248.48
9987	25JN	9;56	88.45	9987	9JL	12;00	281.93	10061	18JN	0;51	81.94	10061	3JL	10;56	276.66
9988	24JL	17;56	116.58	9988	8AU	2;57	310.50	10062	17JL	9;58	110.05	10062	1AU	22;59	305.05
9989	23AU	1;50	145.09	9989	6SE	19;42	339.60	10063	15AU	21;33	138.60	10063	31AU	9;59	333.81
9990	21SE	10;35	174.11	9990	6OC	13;34	9.26	10064	14SE	11;58	167.75	10064	29SE	20;19	3.06
9991	20OC	20;59	203.70	9991	5NO	7;20	39.36	10065	14OC	5;07	197.51	10065	29OC	6;30	32.76
9992	19NO	9;24	233.71	9992	4DE	23;36	69.63	10066	13NO	0;14	227.76	10066	27NO	17;03	62.76
9993	18DE	23;46	263.92					10067	12DE	19;53	258.19	10067	27DE	4;16	92.81
						−192								**−186**	
				9993	3JA	13;27	99.76					10068	25JA	16;12	122.65
9994	17JA	15;41	294.03	9994	2FE	0;47	129.49	10068	11JA	14;12	288.45	10069	24FE	4;47	152.08
9995	16FE	8;30	323.77	9995	2MR	10;09	158.67	10069	10FE	5;41	318.23	10070	25MR	18;01	181.03
9996	17MR	1;31	353.02	9996	31MR	18;17	187.30	10070	11MR	17;49	347.37	10071	24AP	8;01	209.56
9997	15AP	17;47	21.75	9997	30AP	1;58	215.49	10071	10AP	3;08	15.91	10072	23MY	22;49	237.83
9998	15MY	8;29	50.07	9998	29MY	9;59	243.46	10072	9MY	10;39	44.00	10073	22JN	14;07	266.05
9999	13JN	21;07	78.18	9999	27JN	19;08	271.45	10073	7JN	17;28	71.89	10074	22JL	5;15	294.44
10000	13JL	7;53	106.32	10000	27JL	6;18	299.74	10074	7JL	0;37	99.85	10075	20AU	19;34	323.17
10001	11AU	17;29	134.74	10001	25AU	20;14	328.54	10075	5AU	8;57	128.12	10076	19SE	8;46	352.35
10002	10SE	2;50	163.61	10002	24SE	13;11	357.97	10076	3SE	19;15	156.91	10077	18OC	21;02	21.98
10003	9OC	12;38	193.00	10003	24OC	8;19	27.98	10077	3OC	8;13	186.32	10078	17NO	8;39	51.95
10004	7NO	23;12	222.84	10004	23NO	3;48	58.33	10078	2NO	0;16	216.31	10079	16DE	19;47	82.02
10005	7DE	10;36	252.91	10005	22DE	21;40	88.68	10079	1DE	19;02	246.68				
								10080	31DE	15;00	277.08				
						−191								**−185**	
10006	5JA	22;55	282.96	10006	21JA	12;48	118.70	10081	30JA	9;58	307.17	10080	15JA	6;27	111.92
10007	4FE	12;17	312.73	10007	20FE	1;06	148.18	10082	1MR	2;16	336.67	10081	13FE	16;39	141.43
10008	6MR	2;49	342.06	10008	21MR	10;59	177.06	10083	30MR	15;25	5.54	10082	15MR	2;46	170.43
10009	4AP	18;18	10.93	10009	19AP	19;06	205.41	10084	29AP	1;51	33.87	10083	13AP	13;26	198.98
10010	4MY	10;02	39.41	10010	19MY	2;33	233.40	10085	28MY	10;23	61.86	10084	13MY	1;23	227.22
10011	3JN	1;11	67.64	10011	17JN	9;04	261.28	10086	26JN	17;52	89.79	10085	11JN	15;02	255.35
10012	2JL	15;14	95.85	10012	16JL	16;55	289.34	10087	26JL	1;11	117.90	10086	11JL	6;11	283.67
10013	1AU	4;10	124.26	10013	15AU	2;52	317.81	10088	24AU	9;09	146.42	10087	9AU	22;14	312.31
10014	30AU	16;16	153.06	10014	13SE	15;52	346.88	10089	22SE	18;43	175.51	10088	8SE	14;26	341.42
10015	29SE	3;50	182.34	10015	13OC	8;12	16.60	10090	22OC	6;41	205.17	10089	8OC	6;11	11.02
10016	28OC	14;59	212.07	10016	12NO	3;04	46.83	10091	20NO	21;31	235.31	10090	6NO	21;01	41.01
10017	27NO	1;47	242.08	10017	11DE	22;48	77.27	10092	20DE	14;53	265.64	10091	6DE	10;30	71.17
10018	26DE	12;23	272.11												

-184 / -178

NEW NUMBER	DATE	TIME	LONG.	FULL NUMBER	DATE	TIME	LONG.	NEW NUMBER	DATE	TIME	LONG.	FULL NUMBER	DATE	TIME	LONG.
				10092	4JA	22;19	101.21	10167	13JA	4;04	290.10	10167	27JA	4;03	124.21
10093	19JA	9;29	295.84	10093	3FE	8;28	130.87	10168	11FE	16;47	319.74	10168	25FE	19;30	153.74
10094	18FE	3;45	325.62	10094	3MR	17;24	160.01	10169	13MR	2;51	348.78	10169	27MR	11;23	182.76
10095	18MR	20;26	354.82	10095	2AP	1;51	188.63	10170	11AP	10;52	17.24	10170	26AP	3;00	211.33
10096	17AP	10;55	23.45	10096	1MY	10;42	216.86	10171	10MY	17;45	45.30	10171	25MY	17;53	239.60
10097	16MY	23;04	51.66	10097	30MY	20;47	244.90	10172	9JN	0;41	73.19	10172	24JN	7;48	267.77
10098	15JN	9;10	79.67	10098	29JN	8;40	273.01	10173	8JL	8;49	101.20	10173	23JL	20;40	296.08
10099	14JL	17;46	107.74	10099	28JL	22;39	301.43	10174	6AU	19;08	129.56	10174	22AU	8;27	324.73
10100	13AU	1;48	136.11	10100	27AU	14;41	330.34	10175	5SE	8;12	158.49	10175	20SE	19;20	353.84
10101	11SE	10;12	164.96	10101	26SE	8;21	359.83	10176	5OC	0;08	188.05	10176	20OC	5;43	23.41
10102	10OC	19;51	194.37	10102	26OC	2;35	29.82	10177	3NO	18;30	218.16	10177	18NO	16;09	53.34
10103	9NO	7;18	224.26	10103	24NO	19;57	60.09	10178	3DE	14;10	248.57	10178	18DE	3;00	83.40
10104	8DE	20;40	254.42	10104	24DE	11;09	90.33								

-183 / -177

NEW NUMBER	DATE	TIME	LONG.	FULL NUMBER	DATE	TIME	LONG.	NEW NUMBER	DATE	TIME	LONG.	FULL NUMBER	DATE	TIME	LONG.
10105	7JA	11;42	284.57	10105	22JA	23;40	120.22	10179	2JA	9;23	278.93	10179	16JA	14;26	113.33
10106	6FE	3;53	314.43	10106	21FE	9;49	149.59	10180	1FE	2;22	308.91	10180	15FE	2;26	142.88
10107	7MR	20;37	343.83	10107	22MR	18;18	178.39	10181	2MR	16;05	338.29	10181	16MR	15;02	171.97
10108	6AP	13;10	12.72	10108	21AP	1;58	206.70	10182	1AP	2;37	7.02	10182	15AP	4;20	200.60
10109	6MY	4;40	41.16	10109	20MY	9;35	234.71	10183	30AP	10;50	35.24	10183	14MY	18;33	228.92
10110	4JN	18;24	69.33	10110	18JN	17;58	262.65	10184	29MY	17;50	63.17	10184	13JN	9;37	257.13
10111	4JL	6;13	97.46	10111	18JL	4;00	290.80	10185	28JN	0;43	91.08	10185	13JL	1;02	285.45
10112	2AU	16;32	125.78	10112	16AU	16;32	319.40	10186	27JL	8;25	119.22	10186	11AU	16;05	314.07
10113	1SE	2;10	154.51	10113	15SE	8;06	348.61	10187	25AU	17;44	147.82	10187	10SE	6;10	343.12
10114	30SE	11;54	183.74	10114	15OC	2;28	18.44	10188	24SE	5;24	177.01	10188	9OC	19;10	12.63
10115	29OC	22;12	213.45	10115	13NO	22;11	48.72	10189	23OC	19;57	206.81	10189	8NO	7;16	42.52
10116	28NO	9;14	243.47	10116	13DE	17;10	79.12	10190	22NO	13;29	237.06	10190	7DE	18;43	72.60
10117	27DE	20;59	273.55					10191	22DE	9;02	267.49				

-182 / -176

NEW NUMBER	DATE	TIME	LONG.	FULL NUMBER	DATE	TIME	LONG.	NEW NUMBER	DATE	TIME	LONG.	FULL NUMBER	DATE	TIME	LONG.
				10117	12JA	9;45	109.30					10191	6JA	5;35	102.59
10118	26JA	9;34	303.41	10118	10FE	23;22	138.99	10192	21JA	4;39	297.71	10192	4FE	15;50	132.23
10119	24FE	23;14	332.87	10119	12MR	10;17	168.07	10193	19FE	22;18	327.44	10193	5MR	1;41	161.39
10120	26MR	13;59	1.87	10120	10AP	19;05	196.58	10194	20MR	12;55	356.52	10194	3AP	11;42	190.06
10121	25AP	5;28	30.45	10121	10MY	2;30	224.66	10195	19AP	0;34	25.02	10195	2MY	22;41	218.35
10122	24MY	20;55	58.73	10122	8JN	9;19	252.55	10196	18MY	9;57	53.11	10196	1JN	11;17	246.49
10123	23JN	11;37	86.94	10123	7JL	16;33	280.51	10197	16JN	17;55	81.04	10197	1JL	1;42	274.71
10124	23JL	1;18	115.27	10124	6AU	1;21	308.82	10198	16JL	1;20	109.07	10198	30JL	17;30	303.21
10125	21AU	14;04	143.95	10125	4SE	12;49	337.67	10199	14AU	9;00	137.44	10199	29AU	9;57	332.17
10126	20SE	2;10	173.10	10126	4OC	3;36	7.16	10200	12SE	17;48	166.33	10200	28SE	2;17	1.64
10127	19OC	13;46	202.71	10127	2NO	21;26	37.24	10201	12OC	4;35	195.80	10201	27OC	17;55	31.54
10128	18NO	0;52	232.66	10128	2DE	17;00	67.64	10202	10NO	18;00	225.79	10202	26NO	8;18	61.70
10129	17DE	11;32	262.72					10203	10DE	10;07	256.07	10203	25DE	21;01	91.82

-181 / -175

NEW NUMBER	DATE	TIME	LONG.	FULL NUMBER	DATE	TIME	LONG.	NEW NUMBER	DATE	TIME	LONG.	FULL NUMBER	DATE	TIME	LONG.
				10129	1JA	12;27	98.02	10204	9JA	4;08	286.33	10204	24JA	7;54	121.63
10130	15JA	22;02	292.61	10130	31JA	6;11	128.04	10205	7FE	22;32	316.27	10205	22FE	17;12	150.95
10131	14FE	8;47	322.13	10131	1MR	21;18	157.49	10206	9MR	15;56	345.66	10206	24MR	1;34	179.72
10132	15MR	20;14	351.17	10132	31MR	9;29	186.30	10207	8AP	7;26	14.48	10207	22AP	9;53	208.04
10133	14AP	8;39	19.78	10133	29AP	19;01	214.58	10208	7MY	20;43	42.82	10208	21MY	19;05	236.10
10134	13MY	22;03	48.07	10134	29MY	2;36	242.54	10209	6JN	7;52	70.88	10209	20JN	5;54	264.14
10135	12JN	12;22	76.25	10135	27JN	9;18	270.43	10210	5JL	17;17	98.92	10210	19JL	18;48	292.42
10136	12JL	3;27	104.55	10136	26JL	16;21	298.54	10211	4AU	1;41	127.18	10211	18AU	9;56	321.15
10137	10AU	19;01	133.18	10137	25AU	0;51	327.10	10212	2SE	9;59	155.88	10212	17SE	3;04	350.45
10138	9SE	10;34	162.28	10138	23SE	11;40	356.24	10213	1OC	19;05	185.10	10213	16OC	21;25	20.30
10139	9OC	1;24	191.85	10139	23OC	1;13	25.99	10214	31OC	5;42	214.84	10214	15NO	15;37	50.54
10140	7NO	14;56	221.80	10140	21NO	17;24	56.18	10215	29NO	18;06	244.93	10215	15DE	8;08	80.84
10141	7DE	2;58	251.90	10141	21DE	11;34	86.55	10216	29DE	8;14	275.10				

-180 / -174

NEW NUMBER	DATE	TIME	LONG.	FULL NUMBER	DATE	TIME	LONG.	NEW NUMBER	DATE	TIME	LONG.	FULL NUMBER	DATE	TIME	LONG.
10142	5JA	13;45	281.89	10142	20JA	6;35	116.76					10216	13JA	22;01	110.89
10143	3FE	23;43	311.53	10143	19FE	0;59	146.53	10217	27JA	23;39	305.06	10217	12FE	9;11	140.45
10144	4MR	9;14	340.69	10144	19MR	17;17	175.70	10218	26FE	15;54	334.60	10218	13MR	18;15	169.44
10145	2AP	18;42	9.34	10145	18AP	6;40	204.27	10219	28MR	8;24	3.64	10219	12AP	2;37	197.88
10146	2MY	4;37	37.60	10146	17MY	17;13	232.41	10220	27AP	0;23	32.22	10220	11MY	9;28	225.96
10147	31MY	15;38	65.68	10147	16JN	1;44	260.36	10221	26MY	15;04	60.47	10221	9JN	17;16	253.88
10148	30JN	4;27	93.83	10148	15JL	9;21	288.40	10222	25JN	3;58	88.60	10222	9JL	2;21	281.92
10149	29JL	19;26	122.30	10149	13AU	17;11	316.77	10223	24JL	15;10	116.86	10223	7AU	13;30	310.34
10150	28AU	12;19	151.26	10150	12SE	2;00	345.65	10224	23AU	1;18	145.45	10224	6SE	3;39	339.32
10151	27SE	6;02	180.76	10151	11OC	12;19	15.09	10225	21SE	11;09	174.53	10225	5OC	20;52	8.95
10152	26OC	23;08	210.72	10152	10NO	0;28	45.01	10226	20OC	21;19	204.10	10226	4NO	16;17	39.11
10153	25NO	14;37	240.92	10153	9DE	14;42	75.21	10227	19NO	8;04	234.05	10227	4DE	12;00	69.53
10154	25DE	4;06	271.07					10228	18DE	19;25	264.13				

-179 / -173

NEW NUMBER	DATE	TIME	LONG.	FULL NUMBER	DATE	TIME	LONG.	NEW NUMBER	DATE	TIME	LONG.	FULL NUMBER	DATE	TIME	LONG.
				10154	8JA	7;01	105.41					10228	3JA	5;57	99.83
10155	23JA	15;41	300.92	10155	7FE	0;45	135.33	10229	17JA	7;23	294.07	10229	1FE	21;00	129.72
10156	22FE	1;37	330.27	10156	8MR	18;35	164.76	10230	15FE	20;12	323.65	10230	3MR	9;07	159.01
10157	23MR	10;12	359.07	10157	7AP	11;03	193.63	10231	17MR	10;07	352.78	10231	1AP	18;48	187.70
10158	21AP	18;00	27.37	10158	7MY	1;19	222.02	10232	16AP	1;02	21.46	10232	1MY	2;43	215.91
10159	21MY	1;56	55.39	10159	5JN	13;22	250.12	10233	15MY	16;27	49.82	10233	30MY	9;40	243.83
10160	19JN	11;05	83.36	10160	4JL	23;47	278.19	10234	14JN	7;36	78.03	10234	28JN	16;34	271.74
10161	18JL	22;33	111.58	10161	3AU	9;16	306.49	10235	13JL	21;57	106.31	10235	28JL	0;29	299.89
10162	17AU	12;53	140.26	10162	1SE	18;28	335.21	10236	12AU	11;25	134.88	10236	26AU	10;34	328.53
10163	16SE	5;52	169.54	10163	1OC	3;53	4.44	10237	11SE	0;08	163.89	10237	24SE	23;45	357.80
10164	16OC	0;26	199.39	10164	30OC	13;56	34.15	10238	10OC	12;15	193.37	10238	24OC	16;14	27.69
10165	14NO	19;10	229.64	10165	29NO	1;06	64.18	10239	8NO	23;47	223.25	10239	23NO	11;11	58.01
10166	14DE	12;42	259.98	10166	28DE	13;47	94.29	10240	8DE	10;43	253.31	10240	23DE	6;53	88.44

Left column — NEW MOONS | FULL MOONS

-172

NUMBER	DATE	TIME	LONG.	NUMBER	DATE	TIME	LONG.
10241	6JA	21;10	283.27	10241	22JA	1;34	118.62
10242	5FE	7;33	312.92	10242	20FE	17;58	148.28
10243	5MR	18;19	342.10	10243	21MR	7;32	177.31
10244	4AP	5;56	10.82	10244	19AP	18;16	205.76
10245	3MY	18;36	39.18	10245	19MY	2;42	233.80
10246	2JN	8;20	67.36	10246	17JN	9;44	261.70
10247	1JL	23;03	95.59	10247	16JL	16;33	289.71
10248	31JL	14;35	124.09	10248	15AU	0;19	318.09
10249	30AU	6;29	153.04	10249	13SE	10;01	347.03
10250	28SE	22;05	182.49	10250	12OC	22;16	16.57
10251	28OC	12;35	212.36	10251	11NO	13;12	46.63
10252	27NO	1;31	242.46	10252	11DE	6;25	76.96
10253	26DE	12;55	272.52				

-171

NUMBER	DATE	TIME	LONG.	NUMBER	DATE	TIME	LONG.
				10253	10JA	1;01	107.24
10254	24JA	23;08	302.30	10254	8FE	19;45	137.18
10255	23FE	8;38	331.61	10255	10MR	13;05	166.55
10256	24MR	17;50	0.41	10256	9AP	3;54	195.33
10257	23AP	3;13	28.76	10257	8MY	15;48	223.61
10258	22MY	13;24	56.86	10258	7JN	1;16	251.60
10259	21JN	1;08	84.94	10259	6JL	9;19	279.59
10260	20JL	15;01	113.26	10260	4AU	17;06	307.84
10261	19AU	7;11	142.04	10261	3SE	1;31	336.54
10262	18SE	0;53	171.38	10262	2OC	11;12	5.80
10263	17OC	18;44	201.23	10263	31OC	22;29	35.58
10264	16NO	11;20	231.40	10264	30NO	11;37	65.70
10265	16DE	1;56	261.62	10265	30DE	2;47	95.91

-170

NUMBER	DATE	TIME	LONG.	NUMBER	DATE	TIME	LONG.
10266	14JA	14;26	291.61	10266	28JA	19;41	125.93
10267	13FE	1;02	321.14	10267	27FE	13;23	155.52
10268	14MR	10;05	350.11	10268	29MR	6;29	184.57
10269	12AP	18;00	18.55	10269	27AP	21;49	213.10
10270	12MY	1;35	46.63	10270	27MY	10;58	241.29
10271	10JN	9;52	74.57	10271	25JN	22;15	269.36
10272	9JL	20;01	102.66	10272	25JL	8;19	297.58
10273	8AU	8;54	131.15	10273	23AU	17;50	326.16
10274	7SE	0;46	160.22	10274	22SE	3;19	355.23
10275	6OC	18;53	189.90	10275	21OC	13;10	24.80
10276	5NO	13;54	220.05	10276	19NO	23;47	54.74
10277	5DE	8;18	250.42	10277	19DE	11;36	84.84

-169

NUMBER	DATE	TIME	LONG.	NUMBER	DATE	TIME	LONG.
10278	4JA	0;53	280.66	10278	18JA	0;53	114.83
10279	2FE	14;55	310.49	10279	16FE	15;32	144.48
10280	4MR	2;10	339.74	10280	18MR	7;00	173.65
10281	2AP	10;59	8.39	10281	16AP	22;35	202.35
10282	1MY	18;11	36.55	10282	16MY	13;45	230.69
10283	31MY	0;52	64.46	10283	15JN	4;14	258.88
10284	29JN	8;15	92.39	10284	14JL	17;49	287.14
10285	28JL	17;24	120.60	10285	13AU	6;24	315.68
10286	27AU	5;06	149.32	10286	11SE	18;00	344.65
10287	25SE	19;43	178.66	10287	11OC	4;49	14.10
10288	25OC	13;03	208.60	10288	9NO	15;20	43.94
10289	24NO	8;19	238.94	10289	9DE	1;58	73.98
10290	24DE	4;01	269.37				

-168

NUMBER	DATE	TIME	LONG.	NUMBER	DATE	TIME	LONG.
				10290	7JA	13;00	103.97
10291	22JA	22;15	299.51	10291	6FE	0;30	133.65
10292	21FE	13;35	329.11	10292	6MR	12;30	162.88
10293	22MR	1;33	358.07	10293	5AP	1;07	191.62
10294	20AP	10;44	26.45	10294	4MY	14;38	220.00
10295	19MY	18;09	54.45	10295	3JN	5;11	248.21
10296	18JN	0;59	82.33	10296	2JL	20;33	276.47
10297	17JL	8;14	110.37	10297	1AU	12;26	304.99
10298	15AU	16;45	138.79	10298	31AU	3;00	333.92
10299	14SE	3;16	167.77	10299	29SE	16;51	3.30
10300	13OC	16;24	197.37	10300	29OC	5;36	33.11
10301	12NO	8;30	227.48	10301	27NO	17;30	63.17
10302	12DE	3;11	257.87	10302	27DE	4;39	93.22

-167

NUMBER	DATE	TIME	LONG.	NUMBER	DATE	TIME	LONG.
10303	10JA	22;56	288.19	10303	25JA	15;05	122.99
10304	9FE	17;38	318.11	10304	24FE	0;52	152.31
10305	11MR	9;42	347.43	10305	25MR	10;27	181.11
10306	9AP	22;42	16.12	10306	23AP	20;35	209.49
10307	9MY	9;07	44.33	10307	23MY	8;07	237.63
10308	7JN	17;44	72.29	10308	21JN	21;33	265.78
10309	7JL	1;25	100.27	10309	21JL	11;42	294.16
10310	5AU	9;01	128.51	10310	20AU	5;12	322.97
10311	3SE	17;18	157.22	10311	18SE	21;57	352.28
10312	3OC	3;07	186.50	10312	18OC	14;17	22.08
10313	1NO	15;13	216.32	10313	17NO	5;34	52.20
10314	1DE	5;59	246.51	10314	16DE	19;16	82.38
10315	30DE	23;04	276.80				

Right column — NEW MOONS | FULL MOONS

-166

NUMBER	DATE	TIME	LONG.	NUMBER	DATE	TIME	LONG.
				10315	15JA	7;01	112.33
10316	29JA	17;14	306.86	10316	13FE	16;55	141.83
10317	28FE	11;04	336.44	10317	15MR	1;28	170.77
10318	30MR	3;25	5.44	10318	13AP	9;29	199.21
10319	28AP	17;46	33.93	10319	12MY	17;57	227.31
10320	28MY	6;01	62.07	10320	11JN	3;45	255.31
10321	26JN	16;22	90.11	10321	10JL	15;31	283.47
10322	26JL	1;21	118.29	10322	9AU	5;36	312.02
10323	24AU	9;48	146.84	10323	7SE	21;56	341.13
10324	22SE	18;36	175.89	10324	7OC	16;01	10.81
10325	22OC	4;32	205.47	10325	6NO	10;42	40.97
10326	20NO	16;04	235.46	10326	6DE	4;23	71.30
10327	20DE	5;16	265.62				

-165

NUMBER	DATE	TIME	LONG.	NUMBER	DATE	TIME	LONG.
				10327	4JA	19;41	101.49
10328	18JA	19;54	295.66	10328	3FE	8;06	131.24
10329	17FE	11;30	325.33	10329	4MR	18;00	160.42
10330	19MR	3;40	354.52	10330	3AP	2;11	189.03
10331	17AP	19;48	23.23	10331	2MY	9;33	217.20
10332	17MY	11;10	51.57	10332	31MY	16;57	245.13
10333	16JN	1;05	79.74	10333	30JN	1;14	273.10
10334	15JL	13;17	107.95	10334	29JL	11;17	301.36
10335	14AU	0;08	136.43	10335	27AU	23;56	330.12
10336	12SE	10;17	165.36	10336	26SE	15;44	359.53
10337	11OC	20;28	194.79	10337	26OC	10;22	29.53
10338	10NO	7;03	224.64	10338	25NO	6;19	59.90
10339	9DE	18;07	254.70	10339	25DE	1;24	90.30

-164

NUMBER	DATE	TIME	LONG.	NUMBER	DATE	TIME	LONG.
10340	8JA	5;37	284.70	10340	23JA	17;56	120.37
10341	6FE	17;44	314.40	10341	22FE	7;25	149.88
10342	7MR	6;47	343.65	10342	22MR	18;07	178.77
10343	5AP	20;55	12.45	10343	21AP	2;45	207.12
10344	5MY	11;58	40.88	10344	20MY	10;01	235.11
10345	4JN	3;16	69.12	10345	18JN	16;49	262.99
10346	3JL	18;10	97.37	10346	18JL	0;07	291.03
10347	2AU	8;18	125.84	10347	16AU	9;04	319.49
10348	31AU	21;41	154.71	10348	14SE	20;42	348.53
10349	30SE	10;23	184.07	10349	14OC	11;39	18.20
10350	29OC	22;27	213.85	10350	13NO	5;34	48.40
10351	28NO	9;47	243.89	10351	13DE	1;06	78.83
10352	27DE	20;25	273.91				

-163

NUMBER	DATE	TIME	LONG.	NUMBER	DATE	TIME	LONG.
				10352	11JA	20;25	109.13
10353	26JA	6;37	303.67	10353	10FE	13;58	138.99
10354	24FE	16;53	333.00	10354	12MR	4;53	168.24
10355	26MR	3;45	1.84	10355	10AP	16;56	196.88
10356	24AP	15;36	30.28	10356	10MY	3;24	225.04
10357	24MY	4;37	58.47	10357	8JN	10;03	252.97
10358	22JN	18;48	86.65	10358	7JL	16;55	280.91
10359	22JL	10;03	115.05	10359	6AU	0;11	309.15
10360	21AU	2;04	143.85	10360	4SE	8;56	337.89
10361	19SE	18;14	173.16	10361	3OC	19;58	7.23
10362	19OC	9;40	202.93	10362	2NO	9;38	37.13
10363	17NO	23;36	233.00	10363	2DE	1;44	67.38
10364	17DE	11;48	263.12	10364	31DE	19;39	97.70

-162

NUMBER	DATE	TIME	LONG.	NUMBER	DATE	TIME	LONG.
10365	15JA	22;30	293.01	10365	30JA	14;18	127.76
10366	14FE	8;09	322.49	10366	1MR	8;20	157.34
10367	15MR	17;14	351.44	10367	31MR	0;23	186.32
10368	14AP	2;13	19.91	10368	29AP	13;42	214.76
10369	13MY	11;43	48.04	10369	29MY	0;20	242.83
10370	11JN	22;26	76.08	10370	27JN	9;05	270.81
10371	11JL	11;09	104.29	10371	26JL	17;02	298.95
10372	10AU	2;17	132.89	10372	25AU	1;13	327.50
10373	8SE	19;33	162.05	10373	23SE	10;23	356.58
10374	8OC	13;47	191.76	10374	22OC	20;57	26.19
10375	7NO	7;23	221.87	10375	21NO	9;09	56.22
10376	6DE	23;11	252.13	10376	20DE	23;11	86.41

-161

NUMBER	DATE	TIME	LONG.	NUMBER	DATE	TIME	LONG.
10377	5JA	12;44	282.23	10377	19JA	15;04	116.49
10378	4FE	0;11	311.94	10378	18FE	8;15	146.22
10379	5MR	9;49	341.10	10379	20MR	1;35	175.44
10380	3AP	18;03	9.70	10380	18AP	17;45	204.14
10381	3MY	1;32	37.86	10381	18MY	7;59	232.43
10382	1JN	9;12	65.81	10382	16JN	20;14	260.53
10383	30JN	18;17	93.81	10383	16JL	7;01	288.69
10384	30JL	5;39	122.13	10384	14AU	16;59	317.15
10385	28AU	20;08	150.98	10385	13SE	2;41	346.06
10386	27SE	13;24	180.46	10386	12OC	12;29	15.49
10387	27OC	8;20	210.48	10387	10NO	22;46	45.33
10388	26NO	3;22	240.83	10388	10DE	9;55	75.40
10389	25DE	21;03	271.16				

-160

NEW MOONS				FULL MOONS			
NUMBER	DATE	TIME	LONG.	NUMBER	DATE	TIME	LONG.
				10389	8JA	22;18	105.43
10390	24JA	12;24	301.17	10390	7FE	12;03	135.19
10391	23FE	0;56	330.63	10391	8MR	2;53	164.50
10392	23MR	10;47	359.48	10392	6AP	18;12	193.34
10393	21AP	18;34	27.78	10393	6MY	9;28	221.77
10394	21MY	1;17	55.74	10394	5JN	0;18	249.98
10395	19JN	8;07	83.63	10395	4JL	14;29	278.21
10396	18JL	16;17	111.71	10396	3AU	3;51	306.66
10397	17AU	2;42	140.23	10397	1SE	16;13	335.50
10398	15SE	15;56	169.35	10398	1OC	3;40	4.82
10399	15OC	8;03	199.09	10399	30OC	14;28	34.56
10400	14NO	2;34	229.32	10400	29NO	1;04	64.57
10401	13DE	22;17	259.76	10401	28DE	11;50	94.59

-159

NEW MOONS				FULL MOONS			
10402	12JA	17;26	290.04	10402	26JA	22;55	124.38
10403	11FE	10;16	319.86	10403	25FE	10;24	153.75
10404	12MR	23;49	349.04	10404	26MR	22;23	182.62
10405	11AP	10;13	17.60	10405	25AP	11;08	211.09
10406	10MY	18;20	45.71	10406	25MY	1;00	239.31
10407	9JN	1;19	73.60	10407	23JN	16;00	267.53
10408	8JL	8;18	101.56	10408	23JL	7;42	295.95
10409	6AU	16;12	129.83	10409	21AU	23;17	324.75
10410	5SE	1;44	158.62	10410	20SE	14;00	354.01
10411	4OC	13;35	188.00	10411	20OC	3;34	23.71
10412	3NO	4;14	217.94	10412	18NO	16;02	53.73
10413	2DE	21;42	248.26	10413	18DE	3;37	83.82

-158

NEW MOONS				FULL MOONS			
10414	1JA	17;03	278.63	10414	16JA	14;19	113.71
10415	31JA	12;23	308.72	10415	15FE	0;12	143.18
10416	2MR	5;45	338.25	10416	16MR	9;34	172.13
10417	31MR	20;11	7.15	10417	14AP	19;04	200.61
10418	30AP	7;46	35.51	10418	14MY	5;35	228.79
10419	29MY	17;13	63.53	10419	12JN	17;54	256.89
10420	28JN	1;22	91.49	10420	12JL	8;16	285.16
10421	27JL	9;03	119.63	10421	11AU	0;19	313.81
10422	25AU	17;04	148.18	10422	9SE	17;15	342.97
10423	24SE	2;10	177.26	10423	9OC	10;10	12.64
10424	23OC	13;07	206.91	10424	8NO	2;18	42.70
10425	22NO	2;33	236.99	10425	7DE	16;59	72.91
10426	21DE	18;28	267.26				

-157

NEW MOONS				FULL MOONS			
				10426	6JA	5;44	102.98
10427	20JA	12;03	297.40	10427	4FE	16;26	132.65
10428	19FE	5;59	327.15	10428	6MR	1;24	161.77
10429	20MR	22;58	356.34	10429	4AP	9;23	190.35
10430	19AP	14;15	24.99	10430	3MY	17;18	218.52
10431	19MY	3;33	53.24	10431	2JN	2;11	246.51
10432	17JN	14;55	81.30	10432	1JL	12;50	274.58
10433	17JL	0;41	109.43	10433	31JL	1;45	302.97
10434	15AU	9;31	137.85	10434	29AU	17;06	331.87
10435	13SE	18;15	166.74	10435	28SE	10;36	1.37
10436	13OC	3;42	196.16	10436	28OC	5;23	31.40
10437	11NO	14;29	226.03	10437	26NO	23;56	61.73
10438	11DE	2;50	256.15	10438	26DE	16;37	92.02

-156

NEW MOONS				FULL MOONS			
10439	9JA	16;37	286.23	10439	25JA	6;27	121.96
10440	8FE	7;31	316.03	10440	23FE	17;26	151.34
10441	8MR	23;10	345.36	10441	24MR	2;13	180.13
10442	7AP	15;08	14.21	10442	22AP	9;45	208.42
10443	7MY	6;52	42.65	10443	21MY	16;56	236.40
10444	5JN	21;36	70.86	10444	20JN	0;37	264.32
10445	5JL	10;50	99.05	10445	19JL	9;40	292.44
10446	3AU	22;32	127.44	10446	17AU	20;58	321.01
10447	2SE	9;13	156.23	10447	16SE	11;15	350.18
10448	1OC	19;34	185.51	10448	16OC	4;41	19.99
10449	31OC	6;06	215.25	10449	15NO	0;20	50.28
10450	29NO	16;59	245.27	10450	14DE	20;11	80.72
10451	29DE	4;11	275.31				

-155

NEW MOONS				FULL MOONS			
				10451	13JA	14;07	110.94
10452	27JA	15;46	305.12	10452	12FE	5;03	140.67
10453	26FE	4;01	334.50	10453	13MR	16;59	169.77
10454	27MR	17;17	3.42	10454	12AP	2;28	198.29
10455	26AP	7;41	31.94	10455	11MY	10;16	226.37
10456	25MY	22;48	60.21	10456	9JN	17;10	254.26
10457	24JN	14;01	88.43	10457	9JL	0;08	282.22
10458	24JL	4;45	116.82	10458	7AU	8;11	310.51
10459	22AU	18;46	145.57	10459	5SE	18;27	339.33
10460	21SE	8;07	174.79	10460	5OC	7;48	8.79
10461	20OC	20;46	204.47	10461	4NO	0;23	38.83
10462	19NO	8;38	234.46	10462	3DE	19;19	69.21
10463	18DE	19;37	264.52				

-154

NEW MOONS				FULL MOONS			
				10463	2JA	14;54	99.58
10464	17JA	5;53	294.39	10464	1FE	9;22	129.62
10465	15FE	15;51	323.86	10465	3MR	1;33	159.09
10466	17MR	2;05	352.83	10466	1AP	14;57	187.93
10467	15AP	13;07	21.37	10467	1MY	1;37	216.25
10468	15MY	1;19	49.60	10468	30MY	10;05	244.23
10469	13JN	14;48	77.75	10469	28JN	17;16	272.15
10470	13JL	5;35	106.04	10470	28JL	0;19	300.27
10471	11AU	21;26	134.70	10471	26AU	8;20	328.83
10472	10SE	13;54	163.85	10472	24SE	18;17	357.96
10473	10OC	6;07	193.51	10473	24OC	6;42	27.68
10474	8NO	21;07	223.53	10474	22NO	21;38	57.83
10475	8DE	10;19	253.68	10475	22DE	14;38	88.14

-153

NEW MOONS				FULL MOONS			
10476	6JA	21;42	283.68	10476	21JA	8;52	118.31
10477	5FE	7;41	313.31	10477	20FE	3;11	148.05
10478	6MR	16;48	342.43	10478	21MR	20;12	177.23
10479	5AP	1;33	11.03	10479	20AP	10;52	205.85
10480	4MY	10;19	39.24	10480	19MY	22;49	234.03
10481	2JN	20;21	67.27	10481	18JN	8;29	262.03
10482	2JL	7;54	95.37	10482	17JL	16;51	290.11
10483	31JL	21;50	123.81	10483	16AU	1;01	318.51
10484	30AU	14;16	152.77	10484	14SE	9;49	347.41
10485	29SE	8;27	182.31	10485	13OC	19;47	16.86
10486	29OC	2;48	212.34	10486	12NO	7;12	46.77
10487	27NO	19;47	242.60	10487	11DE	20;14	76.91
10488	27DE	10;32	272.81				

-152

NEW MOONS				FULL MOONS			
				10488	10JA	11;02	107.04
10489	25JA	22;58	302.68	10489	9FE	3;24	136.88
10490	24FE	9;20	332.03	10490	9MR	20;32	166.27
10491	24MR	18;03	0.80	10491	8AP	13;14	195.13
10492	23AP	1;39	29.08	10492	8MY	4;25	223.54
10493	22MY	8;58	57.06	10493	6JN	17;41	251.69
10494	20JN	17;04	85.00	10494	6JL	5;18	279.82
10495	20JL	3;10	113.17	10495	4AU	15;51	308.17
10496	18AU	16;09	141.82	10496	3SE	1;52	336.95
10497	17SE	8;14	171.08	10497	2OC	11;48	6.22
10498	17OC	2;40	200.94	10498	31OC	21;57	35.95
10499	15NO	21;59	231.22	10499	30NO	8;39	65.97
10500	15DE	16;35	261.61	10500	29DE	20;17	96.03

-151

NEW MOONS				FULL MOONS			
10501	14JA	9;11	291.77	10501	28JA	9;08	125.87
10502	12FE	23;06	321.44	10502	26FE	23;10	155.32
10503	14MR	10;08	350.50	10503	28MR	14;01	184.29
10504	12AP	18;44	18.97	10504	27AP	5;08	212.82
10505	12MY	1;45	47.02	10505	26MY	20;08	241.09
10506	10JN	8;21	74.90	10506	25JN	10;45	269.29
10507	9JL	15;43	102.88	10507	25JL	0;46	297.66
10508	8AU	0;58	131.22	10508	23AU	13;56	326.38
10509	6SE	12;50	160.12	10509	22SE	2;08	355.57
10510	6OC	3;37	189.65	10510	21OC	13;26	25.20
10511	4NO	21;06	219.73	10511	20NO	0;13	55.15
10512	4DE	16;26	250.13	10512	19DE	10;52	85.20

-150

NEW MOONS				FULL MOONS			
10513	3JA	12;04	280.51	10513	17JA	21;39	115.08
10514	2FE	6;10	310.52	10514	16FE	8;41	144.58
10515	3MR	21;19	339.93	10515	17MR	20;06	173.60
10516	2AP	9;18	8.70	10516	16AP	8;08	202.16
10517	1MY	18;13	36.94	10517	15MY	21;12	230.42
10518	31MY	1;37	64.88	10518	14JN	11;34	258.60
10519	29JN	8;32	92.79	10519	14JL	3;04	286.93
10520	28JL	15;58	120.93	10520	12AU	19;03	315.61
10521	27AU	0;43	149.53	10521	11SE	10;35	344.73
10522	25SE	11;26	178.71	10522	11OC	1;02	14.33
10523	25OC	0;42	208.47	10523	9NO	14;15	44.28
10524	23NO	16;48	238.67	10524	9DE	2;22	74.39
10525	23DE	11;18	269.05				

-149

NEW MOONS				FULL MOONS			
				10525	7JA	13;28	104.38
10526	22JA	6;45	299.25	10526	5FE	23;36	134.00
10527	21FE	1;07	328.99	10527	7MR	8;57	163.12
10528	22MR	16;57	358.10	10528	5AP	18;02	191.72
10529	21AP	5;51	26.64	10529	5MY	4;13	219.95
10530	20MY	16;18	54.76	10530	3JN	14;52	248.03
10531	19JN	1;04	82.72	10531	3JL	4;09	276.21
10532	18JL	9;02	110.78	10532	1AU	19;30	304.71
10533	16AU	16;58	139.19	10533	31AU	12;19	333.70
10534	15SE	1;35	168.09	10534	30SE	5;37	3.22
10535	14OC	11;39	197.56	10535	29OC	22;29	33.20
10536	12NO	23;50	227.50	10536	28NO	14;09	63.41
10537	12DE	14;28	257.72	10537	28DE	3;58	93.58

-148

NUMBER	DATE	TIME	LONG.	NUMBER	DATE	TIME	LONG.
10538	11JA	7;11	287.92	10538	26JA	15;37	123.40
10539	10FE	0;51	317.80	10539	25FE	1;15	152.71
10540	10MR	18;12	347.18	10540	25MR	9;25	181.45
10541	9AP	10;15	16.00	10541	23AP	17;04	209.73
10542	9MY	0;31	44.37	10542	23MY	1;12	237.74
10543	7JN	12;54	72.48	10543	21JN	10;47	265.73
10544	6JL	23;35	100.58	10544	20JL	22;30	293.98
10545	5AU	9;01	128.90	10545	19AU	12;42	322.68
10546	3SE	17;55	157.64	10546	18SE	5;20	351.98
10547	3OC	3;07	186.89	10547	17OC	23;50	21.86
10548	1NO	13;18	216.63	10548	16NO	18;53	52.14
10549	1DE	0;52	246.68	10549	16DE	12;48	82.50
10550	30DE	13;51	276.80				

-147

NUMBER	DATE	TIME	LONG.	NUMBER	DATE	TIME	LONG.
10551	29JA	3;59	306.69	10550	15JA	4;07	112.60
10552	27FE	19;00	336.16	10551	13FE	16;24	142.19
10553	29MR	10;35	5.15	10552	15MR	2;03	171.18
10554	28AP	2;20	33.70	10553	13AP	9;58	199.61
10555	27MY	17;37	61.97	10554	12MY	17;06	227.66
10556	26JN	7;45	90.15	10555	11JN	0;21	255.56
10557	25JL	20;26	118.48	10556	10JL	8;36	283.58
10558	24AU	7;50	147.14	10557	8AU	18;43	311.97
10559	22SE	18;32	176.28	10558	7SE	7;31	340.92
10560	22OC	5;09	205.89	10559	6OC	23;31	10.51
10561	20NO	15;57	235.85	10560	5NO	18;21	40.66
10562	20DE	2;58	265.91	10561	5DE	14;27	71.10

-146

NUMBER	DATE	TIME	LONG.	NUMBER	DATE	TIME	LONG.
10563	18JA	14;11	295.81	10562	4JA	9;33	101.45
10564	17FE	1;48	325.32	10563	3FE	1;59	131.37
10565	18MR	14;12	354.36	10564	4MR	15;17	160.70
10566	17AP	3;46	22.98	10565	3AP	1;49	189.40
10567	16MY	18;24	51.29	10566	2MY	10;18	217.61
10568	15JN	9;38	79.51	10567	31MY	17;31	245.54
10569	15JL	0;47	107.83	10568	30JN	0;22	273.45
10570	13AU	15;25	136.46	10569	29JL	7;48	301.60
10571	12SE	5;25	165.54	10570	27AU	16;56	330.22
10572	11OC	18;43	195.10	10571	26SE	4;45	359.46
10573	10NO	7;11	225.03	10572	25OC	19;49	29.30
10574	9DE	18;41	255.11	10573	24NO	13;45	59.59
				10574	24DE	9;11	90.00

-145

NUMBER	DATE	TIME	LONG.	NUMBER	DATE	TIME	LONG.
10575	8JA	5;13	285.07	10575	23JA	4;16	120.18
10576	6FE	15;06	314.67	10576	21FE	21;34	149.86
10577	8MR	0;52	343.80	10577	23MR	12;17	178.92
10578	6AP	11;10	12.44	10578	22AP	0;14	207.40
10579	5MY	22;31	40.73	10579	21MY	9;44	235.48
10580	4JN	11;11	68.87	10580	19JN	17;30	263.40
10581	4JL	1;18	97.08	10581	19JL	0;35	291.44
10582	2AU	16;46	125.60	10582	17AU	8;08	319.83
10583	1SE	9;15	154.59	10583	15SE	17;10	348.77
10584	1OC	2;01	184.11	10584	15OC	4;24	18.29
10585	30OC	18;00	214.05	10585	13NO	18;07	48.31
10586	29NO	8;18	244.21	10586	13DE	10;05	78.58
10587	28DE	20;36	274.31				

-144

NUMBER	DATE	TIME	LONG.	NUMBER	DATE	TIME	LONG.
10588	27JA	7;08	304.08	10587	12JA	3;39	108.81
10589	25FE	16;27	333.37	10588	10FE	21;52	138.71
10590	26MR	1;07	2.12	10589	11MR	15;30	168.08
10591	24AP	9;40	30.42	10590	10AP	7;20	196.88
10592	23MY	18;48	58.47	10591	9MY	20;37	225.20
10593	22JN	5;18	86.50	10592	8JN	7;25	253.25
10594	21JL	17;58	114.79	10593	7JL	16;28	281.28
10595	20AU	9;17	143.55	10594	6AU	0;47	309.57
10596	19SE	2;57	172.91	10595	4SE	9;23	338.30
10597	18OC	21;40	202.81	10596	3OC	18;54	7.58
10598	17NO	15;41	233.05	10597	2NO	5;40	37.35
10599	17DE	7;43	263.33	10598	1DE	17;51	67.44
				10599	31DE	7;37	97.58

-143

NUMBER	DATE	TIME	LONG.	NUMBER	DATE	TIME	LONG.
10600	15JA	21;17	293.35	10600	29JA	23;01	127.51
10601	14FE	8;33	322.89	10601	28FE	15;36	157.04
10602	15MR	17;54	351.85	10602	30MR	8;26	186.06
10603	14AP	1;49	20.28	10603	29AP	0;20	214.61
10604	13MY	9;01	48.32	10604	28MY	14;35	242.83
10605	11JN	16;30	76.23	10605	27JN	3;06	270.95
10606	11JL	1;26	104.28	10606	26JL	14;20	299.23
10607	9AU	12;55	132.74	10607	25AU	0;49	327.87
10608	8SE	3;34	161.78	10608	23SE	11;01	356.99
10609	7OC	21;07	191.45	10609	22OC	21;12	26.60
10610	6NO	16;20	221.62	10610	21NO	7;40	56.54
10611	6DE	11;34	252.02	10611	20DE	18;43	86.61

-142

NUMBER	DATE	TIME	LONG.	NUMBER	DATE	TIME	LONG.
10612	5JA	5;19	282.31	10612	19JA	6;45	116.53
10613	3FE	20;35	312.18	10613	17FE	19;57	146.10
10614	5MR	8;57	341.45	10614	19MR	10;08	175.20
10615	3AP	18;35	10.11	10615	18AP	0;55	203.85
10616	3MY	2;10	38.27	10616	17MY	15;52	232.18
10617	1JN	8;47	66.18	10617	16JN	6;43	260.38
10618	30JN	15;37	94.09	10618	15JL	21;13	288.68
10619	29JL	23;51	122.28	10619	14AU	11;07	317.29
10620	28AU	10;25	150.97	10620	13SE	0;07	346.35
10621	26SE	23;50	180.28	10621	12OC	12;07	15.86
10622	26OC	16;06	210.19	10622	11NO	23;17	45.74
10623	25NO	10;41	240.51	10623	10DE	10;00	75.79
10624	25DE	6;21	270.93				

-141

NUMBER	DATE	TIME	LONG.	NUMBER	DATE	TIME	LONG.
10625	24JA	1;22	301.10	10624	8JA	20;36	105.75
10626	22FE	18;00	330.73	10625	7FE	7;18	135.38
10627	24MR	7;23	359.72	10626	8MR	18;14	164.54
10628	22AP	17;40	28.13	10627	7AP	5;38	193.22
10629	22MY	1;46	56.15	10628	6MY	17;53	221.53
10630	20JN	8;50	84.04	10629	5JN	7;26	249.70
10631	19JL	15;59	112.09	10630	4JL	22;26	277.95
10632	18AU	0;06	140.51	10631	3AU	14;27	306.50
10633	16SE	9;53	169.49	10632	2SE	6;36	335.50
10634	15OC	21;54	199.05	10633	1OC	21;57	4.97
10635	14NO	12;36	229.12	10634	31OC	12;02	34.84
10636	14DE	5;56	259.45	10635	30NO	0;50	64.95
				10636	29DE	12;27	95.01

-140

NUMBER	DATE	TIME	LONG.	NUMBER	DATE	TIME	LONG.
10637	13JA	0;59	289.74	10637	27JA	22;57	124.77
10638	11FE	19;57	319.65	10638	26FE	8;28	154.05
10639	12MR	13;01	348.99	10639	26MR	17;22	182.81
10640	11AP	3;17	17.71	10640	25AP	2;23	211.12
10641	10MY	14;52	45.96	10641	24MY	12;30	239.20
10642	9JN	0;27	73.95	10642	23JN	0;34	267.30
10643	8JL	8;51	101.97	10643	22JL	14;57	295.66
10644	6AU	16;53	130.25	10644	21AU	7;17	324.48
10645	5SE	1;15	158.98	10645	20SE	0;43	353.84
10646	4OC	10;39	188.26	10646	19OC	18;10	23.70
10647	2NO	21;46	218.06	10647	18NO	10;44	53.89
10648	2DE	11;09	248.20	10648	18DE	1;38	84.12

-139

NUMBER	DATE	TIME	LONG.	NUMBER	DATE	TIME	LONG.
10649	1JA	2;46	278.42	10649	16JA	14;23	114.10
10650	30JA	19;52	308.41	10650	15FE	0;51	143.59
10651	1MR	13;16	337.96	10651	16MR	9;29	172.51
10652	31MR	5;51	6.95	10652	14AP	17;06	200.92
10653	29AP	20;57	35.46	10653	14MY	0;41	228.98
10654	29MY	10;18	63.64	10654	12JN	9;20	256.93
10655	27JN	21;57	91.74	10655	11JL	19;52	285.05
10656	27JL	8;10	119.98	10656	10AU	8;51	313.57
10657	25AU	17;28	148.58	10657	9SE	0;26	342.66
10658	24SE	2;39	177.67	10658	8OC	18;16	12.36
10659	23OC	12;25	207.27	10659	7NO	13;25	42.54
10660	21NO	23;20	237.24	10660	7DE	8;15	72.93
10661	21DE	11;32	267.35				

-138

NUMBER	DATE	TIME	LONG.	NUMBER	DATE	TIME	LONG.
10662	20JA	0;56	297.32	10661	6JA	1;01	103.17
10663	18FE	15;15	326.93	10662	4FE	14;46	132.96
10664	20MR	6;17	356.05	10663	6MR	1;32	162.16
10665	18AP	21;46	24.72	10664	4AP	10;04	190.76
10666	18MY	13;17	53.06	10665	3MY	17;22	218.91
10667	17JN	4;07	81.26	10666	2JN	0;23	246.83
10668	16JL	17;44	109.53	10667	1JL	8;01	274.77
10669	15AU	6;00	138.09	10668	30JL	17;07	303.00
10670	13SE	17;16	167.08	10669	29AU	4;33	331.74
10671	13OC	4;07	196.56	10670	27SE	19;01	1.11
10672	11NO	14;57	226.44	10671	27OC	12;39	31.09
10673	11DE	1;54	256.50	10672	26NO	8;26	61.46
				10673	26DE	4;18	91.89

-137

NUMBER	DATE	TIME	LONG.	NUMBER	DATE	TIME	LONG.
10674	9JA	12;54	286.47	10674	24JA	22;10	122.00
10675	8FE	0;02	316.10	10675	23FE	12;57	151.54
10676	9MR	11;42	345.28	10676	25MR	0;42	180.45
10677	8AP	0;21	14.00	10677	23AP	10;02	208.81
10678	7MY	14;15	42.38	10678	22MY	17;45	236.81
10679	6JN	5;10	70.59	10679	21JN	0;42	264.70
10680	5JL	20;29	98.86	10680	20JL	7;47	292.75
10681	4AU	11;38	127.39	10681	18AU	16;01	321.19
10682	3SE	2;15	156.33	10682	17SE	2;29	350.21
10683	2OC	16;13	185.76	10683	16OC	15;59	19.84
10684	1NO	5;22	215.61	10684	15NO	8;37	50.00
10685	30NO	17;30	245.68	10685	15DE	3;28	80.40
10686	30DE	4;29	275.72				

−136

NUMBER	DATE	TIME	LONG.	NUMBER	DATE	TIME	LONG.
				10686	13JA	22;49	110.69
10687	28JA	14;30	305.45	10687	12FE	17;00	140.55
10688	27FE	0;02	334.72	10688	13MR	8;57	169.83
10689	27MR	9;44	3.50	10689	11AP	22;13	198.50
10690	25AP	20;15	31.87	10690	11MY	8;52	226.70
10691	25MY	8;02	60.01	10691	9JN	17;27	254.66
10692	23JN	21;20	88.16	10692	9JL	0;51	282.64
10693	23JL	12;12	116.55	10693	7AU	8;10	310.90
10694	22AU	4;26	145.37	10694	5SE	16;30	339.64
10695	20SE	21;27	174.73	10695	5OC	2;42	8.97
10696	20OC	14;15	204.58	10696	3NO	15;14	38.83
10697	19NO	5;41	234.72	10697	3DE	6;05	69.03
10698	18DE	19;04	264.89				

−135

NUMBER	DATE	TIME	LONG.	NUMBER	DATE	TIME	LONG.
				10698	1JA	22;48	99.29
10699	17JA	6;23	294.80	10699	31JA	16;36	129.31
10700	15FE	16;07	324.25	10700	2MR	10;27	158.86
10701	17MR	0;50	353.17	10701	1AP	3;09	187.84
10702	15AP	9;10	21.59	10702	30AP	17;43	216.32
10703	14MY	17;44	49.68	10703	30MY	5;46	244.45
10704	13JN	3;19	77.68	10704	28JN	15;42	272.47
10705	12JL	14;46	105.84	10705	28JL	0;27	300.67
10706	11AU	4;48	134.41	10706	26AU	9;02	329.25
10707	9SE	21;32	163.57	10707	24SE	18;14	358.35
10708	9OC	16;09	193.31	10708	24OC	4;29	27.97
10709	8NO	10;57	223.49	10709	22NO	15;58	57.97
10710	8DE	4;13	253.81	10710	22DE	4;50	88.11

−134

NUMBER	DATE	TIME	LONG.	NUMBER	DATE	TIME	LONG.
10711	6JA	19;03	283.96	10711	20JA	19;12	118.12
10712	5FE	7;22	313.68	10712	19FE	10;59	147.77
10713	6MR	17;30	342.84	10713	21MR	3;33	176.94
10714	5AP	1;55	11.43	10714	19AP	19;51	205.63
10715	4MY	9;14	39.57	10715	19MY	10;56	233.95
10716	2JN	16;21	67.49	10716	18JN	0;23	262.09
10717	2JL	0;21	95.45	10717	17JL	12;24	290.31
10718	31JL	10;28	123.73	10718	15AU	23;27	318.83
10719	29AU	23;34	152.55	10719	14SE	10;02	347.81
10720	28SE	15;53	182.01	10720	13OC	20;24	17.28
10721	28OC	10;35	212.04	10721	12NO	6;49	47.14
10722	27NO	6;07	242.41	10722	11DE	17;32	77.19
10723	27DE	0;48	272.78				

−133

NUMBER	DATE	TIME	LONG.	NUMBER	DATE	TIME	LONG.
				10723	10JA	4;54	107.17
10724	25JA	17;22	302.83	10724	8FE	17;16	136.85
10725	24FE	7;07	332.32	10725	10MR	6;41	166.09
10726	25MR	17;58	1.18	10726	8AP	20;55	194.86
10727	24AP	2;23	29.50	10727	8MY	11;38	223.26
10728	23MY	9;17	57.46	10728	7JN	2;30	251.47
10729	21JN	15;51	85.34	10729	6JL	17;19	279.72
10730	20JL	23;18	113.40	10730	5AU	7;48	308.23
10731	19AU	8;41	141.90	10731	3SE	21;35	337.16
10732	17SE	20;44	170.99	10732	3OC	10;22	6.55
10733	17OC	11;41	200.70	10733	1NO	22;08	36.35
10734	16NO	5;15	230.90	10734	1DE	9;07	66.38
10735	16DE	0;32	261.32	10735	30DE	19;43	96.39

−132

NUMBER	DATE	TIME	LONG.	NUMBER	DATE	TIME	LONG.
10736	14JA	20;01	291.61	10736	29JA	6;12	126.14
10737	13FE	13;55	321.45	10737	27FE	16;45	155.44
10738	14MR	4;53	350.67	10738	28MR	3;35	184.25
10739	12AP	16;34	19.27	10739	26AP	15;05	212.65
10740	12MY	1;36	47.40	10740	26MY	3;46	240.82
10741	10JN	9;04	75.31	10741	24JN	17;59	269.01
10742	9JL	16;08	103.28	10742	24JL	9;41	297.44
10743	7AU	23;48	131.56	10743	23AU	2;08	326.29
10744	6SE	8;49	160.34	10744	21SE	18;17	355.63
10745	5OC	19;46	189.71	10745	21OC	9;19	25.41
10746	4NO	9;07	219.61	10746	19NO	22;56	55.49
10747	4DE	1;08	249.87	10747	19DE	11;12	85.60

−131

NUMBER	DATE	TIME	LONG.	NUMBER	DATE	TIME	LONG.
10748	2JA	19;22	280.19	10748	17JA	22;11	115.49
10749	1FE	14;26	310.25	10749	16FE	8;01	144.94
10750	3MR	8;27	339.79	10750	17MR	16;55	173.85
10751	2AP	0;02	8.72	10751	16AP	1;32	202.28
10752	1MY	12;53	37.11	10752	15MY	10;47	230.39
10753	30MY	23;25	65.18	10753	13JN	21;39	258.44
10754	29JN	8;26	93.17	10754	13JL	10;51	286.67
10755	28JL	16;43	121.35	10755	12AU	2;23	315.32
10756	27AU	1;02	149.93	10756	10SE	19;35	344.50
10757	25SE	10;00	179.03	10757	10OC	13;24	14.23
10758	24OC	20;17	208.67	10758	9NO	6;45	44.35
10759	23NO	8;30	238.71	10759	8DE	22;43	74.62
10760	22DE	22;56	268.91				

−130

NUMBER	DATE	TIME	LONG.	NUMBER	DATE	TIME	LONG.
				10760	7JA	12;36	104.73
10761	21JA	15;12	298.99	10761	6FE	0;06	134.41
10762	20FE	8;20	328.68	10762	7MR	9;26	163.52
10763	22MR	1;11	357.85	10763	5AP	17;16	192.08
10764	20AP	16;56	26.50	10764	5MY	0;34	220.21
10765	20MY	7;10	54.78	10765	3JN	8;27	248.16
10766	18JN	19;46	82.89	10766	2JL	17;53	276.18
10767	18JL	6;51	111.08	10767	1AU	5;37	304.54
10768	16AU	16;47	139.57	10768	30AU	19;59	333.42
10769	15SE	2;10	168.51	10769	29SE	12;55	2.92
10770	14OC	11;45	197.95	10770	29OC	7;45	32.96
10771	12NO	22;09	227.82	10771	28NO	3;06	63.33
10772	12DE	9;41	257.90	10772	27DE	21;08	93.67

−129

NUMBER	DATE	TIME	LONG.	NUMBER	DATE	TIME	LONG.
10773	10JA	22;21	287.93	10773	26JA	12;26	123.66
10774	9FE	11;58	317.66	10774	25FE	0;32	153.07
10775	11MR	2;21	346.92	10775	26MR	9;57	181.86
10776	9AP	17;22	15.71	10776	24AP	17;38	210.14
10777	9MY	8;46	44.13	10777	24MY	0;36	238.10
10778	8JN	0;02	72.36	10778	22JN	7;47	266.00
10779	7JL	14;28	100.60	10779	21JL	16;04	294.11
10780	6AU	3;39	129.07	10780	20AU	2;18	322.65
10781	4SE	15;38	157.93	10781	18SE	15;17	351.79
10782	4OC	2;54	187.27	10782	18OC	7;27	21.56
10783	2NO	13;54	217.04	10783	17NO	2;26	51.83
10784	2DE	0;53	247.07	10784	16DE	22;33	82.28
10785	31DE	11;48	277.10				

−128

NUMBER	DATE	TIME	LONG.	NUMBER	DATE	TIME	LONG.
				10785	15JA	17;35	112.54
10786	29JA	22;39	306.85	10786	14FE	9;53	142.31
10787	28FE	9;44	336.17	10787	14MR	23;00	171.44
10788	28MR	21;32	5.00	10788	13AP	9;23	199.97
10789	27AP	10;32	33.45	10789	12MY	17;47	228.07
10790	27MY	0;49	61.69	10790	11JN	1;01	255.97
10791	25JN	16;01	89.91	10791	10JL	7;58	283.94
10792	25JL	7;28	118.35	10792	8AU	15;36	312.23
10793	23AU	22;39	147.16	10793	7SE	0;57	341.04
10794	22SE	13;16	176.45	10794	6OC	12;57	10.46
10795	22OC	3;08	206.19	10795	5NO	4;06	40.45
10796	20NO	15;58	236.24	10796	4DE	21;58	70.79
10797	20DE	3;34	266.33				

−127

NUMBER	DATE	TIME	LONG.	NUMBER	DATE	TIME	LONG.
				10797	3JA	17;11	101.14
10798	18JA	13;56	296.18	10798	2FE	11;58	131.18
10799	16FE	23;27	325.61	10799	4MR	4;59	160.66
10800	18MR	8;44	354.53	10800	2AP	19;31	189.53
10801	16AP	18;31	22.99	10801	2MY	7;25	217.88
10802	16MY	5;24	51.16	10802	31MY	17;00	245.90
10803	14JN	17;48	79.27	10803	30JN	0;58	273.86
10804	14JL	7;54	107.54	10804	29JL	8;20	302.01
10805	12AU	23;37	136.21	10805	27AU	16;12	330.58
10806	11SE	16;35	165.40	10806	26SE	1;32	359.71
10807	11OC	9;56	195.12	10807	25OC	12;58	29.40
10808	10NO	2;25	225.22	10808	24NO	2;41	59.51
10809	9DE	17;00	255.43	10809	23DE	18;25	89.76

−126

NUMBER	DATE	TIME	LONG.	NUMBER	DATE	TIME	LONG.
10810	8JA	5;19	285.46	10810	22JA	11;34	119.87
10811	6FE	15;40	315.09	10811	21FE	5;17	149.58
10812	8MR	0;38	344.17	10812	22MR	22;30	178.74
10813	6AP	8;53	12.74	10813	21AP	14;07	207.38
10814	5MY	17;04	40.90	10814	21MY	3;27	235.62
10815	4JN	1;54	68.88	10815	19JN	14;29	263.67
10816	3JL	12;15	96.95	10816	18JL	23;53	291.79
10817	2AU	0;56	125.35	10817	17AU	8;39	320.24
10818	31AU	16;28	154.29	10818	15SE	17;41	349.18
10819	30SE	10;30	183.85	10819	15OC	3;32	18.64
10820	30OC	5;39	213.92	10820	13NO	14;28	48.54
10821	29NO	0;01	244.25	10821	13DE	2;34	78.65
10822	28DE	16;11	274.51				

−125

NUMBER	DATE	TIME	LONG.	NUMBER	DATE	TIME	LONG.
				10822	11JA	16;00	108.71
10823	27JA	5;43	304.41	10823	10FE	6;50	138.47
10824	25FE	16;47	333.77	10824	11MR	22;49	167.78
10825	27MR	1;51	2.53	10825	10AP	15;09	196.61
10826	25AP	9;29	30.80	10826	10MY	6;49	225.04
10827	24MY	16;29	58.76	10827	8JN	21;08	253.22
10828	22JN	23;51	86.67	10828	8JL	9;59	281.41
10829	22JL	8;46	114.81	10829	6AU	21;43	309.83
10830	20AU	20;20	143.41	10830	5SE	8;46	338.66
10831	19SE	11;11	172.64	10831	4OC	19;28	7.99
10832	19OC	4;57	202.50	10832	3NO	6;00	37.75
10833	18NO	0;24	232.79	10833	2DE	16;34	67.77
10834	17DE	19;45	263.21				

-124 / -118

NUMBER	DATE	TIME	LONG.	NUMBER	DATE	TIME	LONG.	NUMBER	DATE	TIME	LONG.	NUMBER	DATE	TIME	LONG.
				10834	1JA	3:29	97.79	10909	9JA	13:17	286.87	10909	24JA	6:37	121.73
10835	16JA	13:28	293.41	10835	30JA	15:07	127.57	10910	7FE	23:00	316.45	10910	23FE	0:29	151.41
10836	15FE	4:36	323.11	10836	29FE	3:43	156.94	10911	9MR	8:06	345.52	10911	24MR	16:11	180.49
10837	15MR	16:47	352.19	10837	29MR	17:16	185.83	10912	7AP	17:17	14.10	10912	23AP	5:20	209.01
10838	14AP	2:15	20.68	10838	28AP	7:33	214.32	10913	7MY	3:19	42.33	10913	22MY	16:02	237.13
10839	13MY	9:43	48.74	10839	27MY	22:15	242.57	10914	5JN	14:46	70.41	10914	21JN	0:48	265.09
10840	11JN	16:18	76.61	10840	26JN	13:09	270.79	10915	5JL	3:57	98.59	10915	20JL	8:29	293.16
10841	10JL	23:12	104.58	10841	26JL	4:02	299.20	10916	3AU	18:58	127.10	10916	18AU	16:08	321.58
10842	9AU	7:34	132.91	10842	24AU	18:30	328.00	10917	2SE	11:35	156.12	10917	17SE	0:48	350.52
10843	7SE	18:19	161.78	10843	23SE	8:08	357.26	10918	2OC	5:09	185.69	10918	16OC	11:15	20.03
10844	7OC	7:54	191.28	10844	22OC	20:39	26.96	10919	31OC	22:28	215.70	10919	14NO	23:52	50.01
10845	6NO	0:16	221.33	10845	21NO	8:08	56.95	10920	30NO	14:17	245.93	10920	14DE	14:34	80.24
10846	5DE	18:49	251.70	10846	20DE	18:54	87.01	10921	30DE	3:46	276.07				

-123 / -117

NUMBER	DATE	TIME	LONG.	NUMBER	DATE	TIME	LONG.	NUMBER	DATE	TIME	LONG.	NUMBER	DATE	TIME	LONG.
10847	4JA	14:21	282.06	10847	19JA	5:18	116.86					10921	13JA	6:54	110.40
10848	3FE	9:08	312.09	10848	17FE	15:35	146.30	10922	28JA	14:59	305.86	10922	12FE	0:11	140.24
10849	5MR	1:34	341.53	10849	19MR	1:58	175.26	10923	27FE	0:25	335.13	10923	13MR	17:33	169.59
10850	3AP	14:48	10.34	10850	17AP	12:48	203.75	10924	28MR	8:46	3.84	10924	12AP	9:57	198.39
10851	3MY	1:01	38.61	10851	17MY	0:36	231.96	10925	26AP	16:42	32.11	10925	12MY	0:27	226.75
10852	1JN	9:09	66.58	10852	15JN	13:55	260.10	10926	26MY	0:58	60.11	10926	10JN	12:40	254.85
10853	30JN	16:22	94.50	10853	15JL	4:58	288.42	10927	24JN	10:22	88.10	10927	9JL	22:57	282.95
10854	29JL	23:45	122.66	10854	13AU	21:20	317.12	10928	23JL	21:47	116.35	10928	8AU	8:09	311.28
10855	28AU	8:09	151.26	10855	12SE	14:03	346.32	10929	22AU	11:56	145.09	10929	6SE	17:11	340.06
10856	26SE	18:11	180.43	10856	12OC	6:01	15.99	10930	21SE	4:58	174.44	10930	6OC	2:46	9.36
10857	26OC	6:21	210.16	10857	10NO	20:35	46.02	10931	20OC	23:59	204.37	10931	4NO	13:16	39.13
10858	24NO	21:01	240.32	10858	10DE	9:37	76.17	10932	19NO	19:08	234.66	10932	4DE	0:46	69.20
10859	24DE	14:08	270.63					10933	19DE	12:37	265.00				

-122 / -116

NUMBER	DATE	TIME	LONG.	NUMBER	DATE	TIME	LONG.	NUMBER	DATE	TIME	LONG.	NUMBER	DATE	TIME	LONG.
				10859	8JA	21:13	106.16					10933	2JA	13:23	99.28
10860	23JA	8:48	300.79	10860	7FE	7:29	135.77	10934	18JA	3:29	295.07	10934	1FE	3:16	129.14
10861	22FE	3:21	330.51	10861	8MR	16:36	164.86	10935	16FE	15:39	324.62	10935	1MR	18:26	158.59
10862	23MR	20:07	359.65	10862	7AP	1:03	193.42	10936	17MR	1:31	353.58	10936	31MR	10:25	187.55
10863	22AP	10:15	28.22	10863	6MY	9:38	221.59	10937	15AP	9:40	22.00	10937	30AP	2:22	216.09
10864	21MY	21:53	56.38	10864	4JN	19:26	249.61	10938	14MY	16:46	50.03	10938	29MY	17:23	244.34
10865	20JN	7:40	84.38	10865	4JL	7:20	277.74	10939	12JN	23:45	77.92	10939	28JN	7:05	272.52
10866	19JL	16:24	112.49	10866	2AU	21:47	306.22	10940	12JL	7:44	105.94	10940	27JL	19:33	300.85
10867	18AU	0:49	140.93	10867	1SE	14:25	335.22	10941	10AU	17:54	134.36	10941	26AU	7:11	329.56
10868	16SE	9:34	169.86	10868	1OC	8:18	4.79	10942	9SE	7:10	163.36	10942	24SE	18:18	358.74
10869	15OC	19:15	199.33	10869	31OC	2:15	34.82	10943	8OC	23:41	193.01	10943	24OC	5:06	28.39
10870	14NO	6:30	229.25	10870	29NO	19:11	65.09	10944	7NO	18:36	223.18	10944	22NO	15:43	58.35
10871	13DE	19:45	259.41	10871	29DE	10:14	95.30	10945	7DE	14:15	253.60	10945	22DE	2:23	88.40

-121 / -115

NUMBER	DATE	TIME	LONG.	NUMBER	DATE	TIME	LONG.	NUMBER	DATE	TIME	LONG.	NUMBER	DATE	TIME	LONG.
10872	12JA	10:59	289.54	10872	27JA	22:54	125.16	10946	6JA	8:56	283.92	10946	20JA	13:27	118.26
10873	11FE	3:34	319.36	10873	26FE	9:08	154.47	10947	5FE	1:23	313.82	10947	19FE	1:17	147.76
10874	12MR	20:24	348.69	10874	27MR	17:27	183.19	10948	6MR	14:58	343.12	10948	20MR	14:04	176.78
10875	11AP	12:35	17.50	10875	26AP	0:44	211.44	10949	5AP	1:39	11.80	10949	19AP	3:44	205.37
10876	11MY	3:32	45.89	10876	25MY	8:03	239.41	10950	4MY	9:56	39.99	10950	18MY	18:04	233.66
10877	9JN	17:01	74.04	10877	23JN	16:32	267.36	10951	2JN	16:48	67.89	10951	17JN	8:53	261.86
10878	9JL	5:01	102.20	10878	23JL	3:02	295.57	10952	1JL	23:25	95.80	10952	16JL	23:56	290.20
10879	7AU	15:43	130.59	10879	21AU	14:07	324.25	10953	31JL	6:59	123.98	10953	15AU	14:55	318.87
10880	6SE	1:32	159.38	10880	20SE	7:56	353.53	10954	29AU	16:34	152.65	10954	14SE	5:21	348.00
10881	5OC	11:10	188.68	10881	20OC	2:05	23.42	10955	28SE	4:47	181.94	10955	13OC	18:43	17.59
10882	3NO	21:14	218.42	10882	18NO	21:32	53.72	10956	27OC	19:52	211.81	10956	12NO	6:53	47.53
10883	3DE	8:12	248.46	10883	18DE	16:31	84.12	10957	26NO	13:26	242.09	10957	11DE	18:02	77.60
								10958	26DE	8:35	272.49				

-120 / -114

NUMBER	DATE	TIME	LONG.	NUMBER	DATE	TIME	LONG.	NUMBER	DATE	TIME	LONG.	NUMBER	DATE	TIME	LONG.
10884	1JA	20:12	278.53	10884	17JA	9:19	114.27					10958	10JA	4:30	107.54
10885	31JA	9:09	308.35	10885	15FE	22:56	143.90	10959	25JA	3:50	302.65	10959	8FE	14:39	137.13
10886	29FE	22:52	337.75	10886	16MR	9:29	172.90	10960	23FE	21:29	332.31	10960	10MR	0:42	166.23
10887	30MR	13:16	6.67	10887	14AP	17:48	201.33	10961	25MR	12:16	1.34	10961	8AP	10:59	194.84
10888	29AP	4:19	35.18	10888	14MY	0:55	229.37	10962	23AP	23:53	29.78	10962	7MY	22:00	223.10
10889	28MY	19:39	63.45	10889	12JN	7:51	257.26	10963	23MY	8:56	57.83	10963	6JN	10:20	251.21
10890	27JN	10:39	91.67	10890	11JL	15:15	285.26	10964	21JN	16:32	85.75	10964	6JL	0:30	279.44
10891	27JL	0:43	120.06	10891	10AU	0:42	313.63	10965	20JL	23:49	113.81	10965	4AU	16:25	308.00
10892	25AU	13:34	148.81	10892	8SE	12:18	342.55	10966	19AU	7:46	142.25	10966	3SE	9:21	337.04
10893	24SE	1:25	178.01	10893	8OC	2:57	12.11	10967	17SE	17:04	171.23	10967	3OC	2:06	6.59
10894	23OC	12:44	207.67	10894	6NO	20:43	42.23	10968	17OC	4:13	200.78	10968	1NO	17:41	36.55
10895	21NO	23:51	237.65	10895	6DE	16:32	72.65	10969	15NO	17:37	230.80	10969	1DE	7:39	66.70
10896	21DE	10:47	267.71					10970	15DE	9:28	261.07	10970	30DE	19:59	96.79

-119 / -113

NUMBER	DATE	TIME	LONG.	NUMBER	DATE	TIME	LONG.	NUMBER	DATE	TIME	LONG.	NUMBER	DATE	TIME	LONG.
				10896	5JA	12:21	103.02	10971	14JA	3:21	291.29	10971	29JA	6:48	126.55
10897	19JA	21:32	297.57	10897	4FE	6:03	132.99	10972	12FE	21:58	321.18	10972	27FE	16:17	155.81
10898	18FE	8:12	327.02	10898	5MR	20:40	162.34	10973	14MR	15:36	350.51	10973	29MR	0:47	184.52
10899	19MR	19:16	355.99	10899	4AP	8:15	191.07	10974	13AP	6:59	19.27	10974	27AP	8:58	212.79
10900	18AP	7:21	24.53	10900	3MY	17:30	219.30	10975	12MY	19:48	47.55	10975	26MY	17:52	240.81
10901	17MY	20:48	52.79	10901	2JN	1:13	247.25	10976	11JN	6:30	75.59	10976	25JN	4:31	268.86
10902	16JN	11:32	80.98	10902	1JL	8:15	275.17	10977	10JL	15:49	103.65	10977	24JL	17:40	297.18
10903	16JL	3:02	109.33	10903	30JL	15:32	303.33	10978	9AU	0:30	131.98	10978	23AU	9:24	325.99
10904	14AU	18:38	138.02	10904	29AU	0:00	331.94	10979	7SE	9:13	160.75	10979	22SE	3:01	355.38
10905	13SE	9:51	167.17	10905	27SE	10:41	1.15	10980	6OC	18:32	190.05	10980	21OC	21:19	25.30
10906	13OC	0:26	196.80	10906	27OC	0:18	30.95	10981	5NO	5:01	219.83	10981	20NO	15:04	55.54
10907	11NO	14:02	226.79	10907	25NO	16:55	61.19	10982	4DE	17:13	249.92	10982	20DE	7:15	85.82
10908	11DE	2:21	256.91	10908	25DE	11:34	91.57								

−112

| NEW MOONS | | | | FULL MOONS | | | |
NUMBER	DATE	TIME	LONG.	NUMBER	DATE	TIME	LONG.
10983	3JA	7;21	280.07	10983	18JA	21;08	115.84
10984	1FE	23;07	310.00	10984	17FE	8;27	145.35
10985	2MR	15;40	339.49	10985	17MR	17;30	174.26
10986	1AP	8;01	8.46	10986	16AP	1;01	202.64
10987	30AP	23;30	36.96	10987	15MY	8;03	230.67
10988	30MY	13;45	65.18	10988	13JN	15;44	258.59
10989	29JN	2;38	93.32	10989	13JL	1;06	286.67
10990	28JL	14;11	121.63	10990	11AU	12;54	315.16
10991	27AU	0;39	150.30	10991	10SE	3;27	344.23
10992	25SE	10;32	179.45	10992	9OC	20;39	13.92
10993	24OC	20;29	209.07	10993	8NO	15;45	44.11
10994	23NO	7;02	239.03	10994	8DE	11;18	74.53
10995	22DE	18;28	269.11				

−111

| NEW MOONS | | | | FULL MOONS | | | |
NUMBER	DATE	TIME	LONG.	NUMBER	DATE	TIME	LONG.
				10995	7JA	5;24	104.81
10996	21JA	6;46	299.02	10996	5FE	20;36	134.65
10997	19FE	19;50	328.55	10997	7MR	8;31	163.87
10998	21MR	9;35	357.60	10998	5AP	17;44	192.48
10999	20AP	0;04	26.22	10999	5MY	1;14	220.63
11000	19MY	15;10	54.53	11000	3JN	8;06	248.53
11001	18JN	6;26	82.75	11001	2JL	15;17	276.46
11002	17JL	21;13	111.08	11002	31JL	23;40	304.68
11003	16AU	10;57	139.72	11003	30AU	10;03	333.40
11004	14SE	23;34	168.79	11004	28SE	23;12	2.73
11005	14OC	11;22	198.32	11005	28OC	15;31	32.67
11006	12NO	22;44	228.23	11006	27NO	10;33	63.02
11007	12DE	9;48	258.30	11007	27DE	6;37	93.45

−110

| NEW MOONS | | | | FULL MOONS | | | |
NUMBER	DATE	TIME	LONG.	NUMBER	DATE	TIME	LONG.
11008	10JA	20;33	288.25	11008	26JA	1;30	123.59
11009	9FE	7;01	317.84	11009	24FE	17;36	153.17
11010	10MR	17;32	346.94	11010	26MR	6;33	182.11
11011	9AP	4;45	15.59	11011	24AP	16;50	210.49
11012	8MY	17;15	43.90	11012	24MY	1;12	238.51
11013	7JN	7;15	72.08	11013	22JN	8;31	266.42
11014	6JL	22;28	100.35	11014	21JL	15;40	294.48
11015	5AU	14;16	128.91	11015	19AU	23;32	322.92
11016	4SE	6;00	157.92	11016	18SE	9;07	351.92
11017	3OC	21;15	187.42	11017	17OC	21;17	21.52
11018	2NO	11;38	217.34	11018	16NO	12;28	51.63
11019	2DE	0;46	247.46	11019	16DE	6;11	81.98
11020	31DE	12;24	277.51				

−109

| NEW MOONS | | | | FULL MOONS | | | |
NUMBER	DATE	TIME	LONG.	NUMBER	DATE	TIME	LONG.
				11020	15JA	1;05	112.24
11021	29JA	22;34	307.24	11021	13FE	19;31	142.10
11022	28FE	7;41	336.47	11022	15MR	12;14	171.38
11023	29MR	16;30	5.19	11023	14AP	2;36	200.08
11024	28AP	1;47	33.49	11024	13MY	14;30	228.33
11025	27MY	12;17	61.57	11025	12JN	0;14	256.33
11026	26JN	0;29	89.68	11026	11JL	8;28	284.35
11027	25JL	14;36	118.05	11027	9AU	16;11	312.64
11028	24AU	6;37	146.89	11028	8SE	0;25	341.40
11029	23SE	0;04	176.29	11029	7OC	10;02	10.73
11030	22OC	17;57	206.19	11030	5NO	21;37	40.56
11031	21NO	10;52	236.41	11031	5DE	11;16	70.72
11032	21DE	1;39	266.63				

−108

| NEW MOONS | | | | FULL MOONS | | | |
NUMBER	DATE	TIME	LONG.	NUMBER	DATE	TIME	LONG.
				11032	4JA	2;41	100.92
11033	19JA	13;56	296.57	11033	2FE	19;21	130.87
11034	18FE	0;03	326.02	11034	3MR	12;32	160.37
11035	18MR	8;41	354.91	11035	2AP	5;21	189.35
11036	16AP	16;34	23.30	11036	1MY	20;48	217.84
11037	16MY	0;26	51.35	11037	31MY	10;13	246.02
11038	14JN	9;03	79.30	11038	29JN	21;33	274.11
11039	13JL	19;19	107.43	11039	29JL	7;23	302.36
11040	12AU	8;04	135.97	11040	27AU	16;38	330.99
11041	10SE	23;49	165.10	11041	26SE	2;06	0.12
11042	10OC	18;12	194.85	11042	25OC	12;16	29.76
11043	9NO	13;43	225.07	11043	23NO	23;19	59.75
11044	9DE	8;20	255.45	11044	23DE	11;16	89.85

−107

| NEW MOONS | | | | FULL MOONS | | | |
NUMBER	DATE	TIME	LONG.	NUMBER	DATE	TIME	LONG.
11045	8JA	0;35	285.65	11045	22JA	0;17	119.78
11046	6FE	14;00	315.40	11046	20FE	14;32	149.35
11047	8MR	0;51	344.57	11047	22MR	5;54	178.46
11048	6AP	9;40	13.15	11048	20AP	21;45	207.11
11049	5MY	17;05	41.29	11049	20MY	13;14	235.44
11050	3JN	23;56	69.19	11050	19JN	3;41	263.62
11051	3JL	7;16	97.13	11051	18JL	16;55	291.90
11052	1AU	16;14	125.38	11052	17AU	5;12	320.49
11053	31AU	3;56	154.16	11053	15SE	16;50	349.53
11054	29SE	18;57	183.59	11054	15OC	4;01	19.05
11055	29OC	12;56	213.60	11055	13NO	14;52	48.95
11056	28NO	8;30	243.98	11056	13DE	1;29	78.99
11057	28DE	3;52	274.38				

−106

| NEW MOONS | | | | FULL MOONS | | | |
NUMBER	DATE	TIME	LONG.	NUMBER	DATE	TIME	LONG.
				11057	11JA	12;11	108.93
11058	26JA	21;29	304.45	11058	9FE	23;21	138.54
11059	25FE	12;28	333.97	11059	11MR	11;22	167.70
11060	27MR	0;28	2.86	11060	10AP	0;19	196.40
11061	25AP	9;48	31.20	11061	9MY	14;07	224.76
11062	24MY	17;13	59.18	11062	8JN	4;37	252.95
11063	22JN	23;50	87.05	11063	7JL	19;39	281.22
11064	22JL	6;51	115.11	11064	6AU	10;56	309.78
11065	20AU	15;25	143.60	11065	5SE	2;00	338.77
11066	19SE	2;22	172.67	11066	4OC	16;16	8.24
11067	18OC	16;06	202.34	11067	3NO	5;16	38.11
11068	17NO	8;31	232.51	11068	2DE	17;00	68.18
11069	17DE	2;57	262.89				

−105

| NEW MOONS | | | | FULL MOONS | | | |
NUMBER	DATE	TIME	LONG.	NUMBER	DATE	TIME	LONG.
				11069	1JA	3;45	98.19
11070	15JA	22;15	293.16	11070	30JA	13;53	127.91
11071	14FE	16;45	323.01	11071	28FE	23;44	157.16
11072	16MR	8;57	352.26	11072	30MR	9;35	185.91
11073	14AP	22;03	20.89	11073	28AP	19;54	214.24
11074	14MY	8;17	49.06	11074	28MY	7;19	242.36
11075	12JN	16;31	77.00	11075	26JN	20;27	270.51
11076	11JL	23;57	105.00	11076	26JL	11;36	298.93
11077	10AU	7;37	133.29	11077	25AU	4;21	327.81
11078	8SE	16;19	162.09	11078	23SE	21;38	357.21
11079	8OC	2;37	191.45	11079	23OC	14;11	27.08
11080	6NO	14;54	221.32	11080	22NO	5;10	57.22
11081	6DE	5;29	251.52	11081	21DE	18;23	87.37

−104

| NEW MOONS | | | | FULL MOONS | | | |
NUMBER	DATE	TIME	LONG.	NUMBER	DATE	TIME	LONG.
11082	4JA	22;17	281.77	11082	20JA	5;54	117.27
11083	3FE	16;30	311.78	11083	18FE	15;53	146.70
11084	4MR	10;36	341.30	11084	19MR	0;37	175.59
11085	3AP	3;03	10.25	11085	17AP	8;39	203.97
11086	2MY	17;06	38.68	11086	16MY	16;53	232.03
11087	1JN	4;50	66.79	11087	15JN	2;24	260.02
11088	30JN	14;53	94.83	11088	14JL	14;13	288.21
11089	30JL	0;01	123.06	11089	13AU	4;45	316.84
11090	28AU	8;52	151.68	11090	11SE	21;43	346.03
11091	26SE	18;00	180.82	11091	11OC	16;02	15.80
11092	26OC	3;58	210.45	11092	10NO	10;25	45.98
11093	24NO	15;16	240.46	11093	10DE	3;38	76.30
11094	24DE	4;19	270.60				

−103

| NEW MOONS | | | | FULL MOONS | | | |
NUMBER	DATE	TIME	LONG.	NUMBER	DATE	TIME	LONG.
				11094	8JA	18;45	106.45
11095	22JA	19;08	300.60	11095	7FE	7;17	136.15
11096	21FE	11;07	330.23	11096	8MR	17;17	165.27
11097	23MR	3;23	359.36	11097	7AP	1;18	193.81
11098	21AP	19;11	28.00	11098	6MY	8;18	221.92
11099	21MY	10;02	56.29	11099	4JN	15;26	249.84
11100	19JN	23;43	84.45	11100	3JL	23;49	277.82
11101	19JL	12;07	112.70	11101	2AU	10;20	306.14
11102	17AU	23;21	141.26	11102	31AU	23;34	335.00
11103	16SE	9;44	170.26	11103	30SE	15;36	4.48
11104	15OC	19;47	199.74	11104	30OC	10;01	34.52
11105	14NO	6;06	229.62	11105	29NO	5;40	64.91
11106	13DE	17;04	259.68	11106	29DE	0;44	95.29

−102

| NEW MOONS | | | | FULL MOONS | | | |
NUMBER	DATE	TIME	LONG.	NUMBER	DATE	TIME	LONG.
11107	12JA	4;48	289.66	11107	27JA	17;28	125.32
11108	10FE	17;15	319.32	11108	26FE	6;56	154.76
11109	12MR	6;21	348.51	11109	27MR	17;18	183.57
11110	10AP	20;57	17.23	11110	26AP	1;26	211.85
11111	10MY	10;47	45.61	11111	25MY	8;26	239.81
11112	9JN	2;00	73.83	11112	23JN	15;21	267.70
11113	8JL	17;14	102.11	11113	22JL	23;05	295.80
11114	7AU	7;47	130.65	11114	21AU	8;27	324.32
11115	5SE	21;15	159.59	11115	19SE	20;14	353.44
11116	5OC	9;42	189.00	11116	19OC	11;01	23.17
11117	3NO	21;27	218.82	11117	18NO	4;51	53.40
11118	3DE	8;45	248.88	11118	18DE	0;38	83.84

−101

| NEW MOONS | | | | FULL MOONS | | | |
NUMBER	DATE	TIME	LONG.	NUMBER	DATE	TIME	LONG.
11119	1JA	19;38	278.89	11119	16JA	20;17	114.11
11120	31JA	6;03	308.61	11120	15FE	13;47	143.91
11121	1MR	16;15	337.87	11121	17MR	4;12	173.07
11122	31MR	2;44	6.63	11122	15AP	15;40	201.63
11123	29AP	14;16	35.01	11123	15MY	0;53	229.75
11124	29MY	3;20	63.19	11124	13JN	8;40	257.68
11125	27JN	17;58	91.39	11125	12JL	15;53	285.66
11126	27JL	9;40	119.85	11126	10AU	23;24	313.96
11127	26AU	1;45	148.71	11127	9SE	8;08	342.77
11128	24SE	17;35	178.07	11128	8OC	19;02	12.17
11129	24OC	8;44	207.89	11129	7NO	8;44	42.11
11130	22NO	22;44	237.99	11130	7DE	1;15	72.39
11131	22DE	11;11	268.11				

-100

New Moons

NUMBER	DATE	TIME	LONG.
11132	20JA	21;59	297.98
11133	19FE	7;23	327.38
11134	19MR	16;02	356.24
11135	18AP	0;45	24.65
11136	17MY	10;23	52.76
11137	15JN	21;33	80.81
11138	15JL	10;41	109.06
11139	14AU	1;52	137.72
11140	12SE	18;53	166.94
11141	12OC	12;58	196.70
11142	11NO	6;46	226.87
11143	10DE	22;51	257.14

Full Moons

NUMBER	DATE	TIME	LONG.
11131	5JA	19;37	102.70
11132	4FE	14;16	132.72
11133	5MR	7;46	162.20
11134	3AP	23;15	191.09
11135	3MY	12;20	219.48
11136	1JN	23;10	247.55
11137	1JL	8;10	275.55
11138	30JL	16;12	303.74
11139	29AU	0;14	332.34
11140	27SE	9;14	1.48
11141	26OC	19;54	31.16
11142	25NO	8;32	61.22
11143	24DE	23;00	91.42

-99

New Moons

NUMBER	DATE	TIME	LONG.
11144	9JA	12;24	287.22
11145	7FE	23;27	316.85
11146	9MR	8;35	345.92
11147	7AP	16;34	14.46
11148	7MY	0;12	42.59
11149	5JN	8;13	70.53
11150	4JL	17;30	98.56
11151	3AU	4;55	126.92
11152	1SE	19;14	155.83
11153	1OC	12;34	185.38
11154	31OC	7;55	215.48
11155	30NO	3;21	245.86
11156	29DE	20;58	276.17

Full Moons

NUMBER	DATE	TIME	LONG.
11144	23JA	14;54	121.46
11145	22FE	7;38	151.11
11146	24MR	0;30	180.24
11147	22AP	16;36	208.89
11148	22MY	7;05	237.16
11149	20JN	19;33	265.27
11150	20JL	6;15	293.46
11151	18AU	15;56	321.96
11152	17SE	1;27	350.94
11153	16OC	11;25	20.43
11154	14NO	22;07	50.33
11155	14DE	9;34	80.41

-98

New Moons

NUMBER	DATE	TIME	LONG.
11157	28JA	11;46	306.11
11158	26FE	23;46	335.48
11159	28MR	9;24	4.26
11160	26AP	17;19	32.52
11161	26MY	0;16	60.47
11162	24JN	7;12	88.36
11163	23JL	15;13	116.48
11164	22AU	1;31	145.05
11165	20SE	14;56	174.24
11166	20OC	7;38	204.07
11167	19NO	2;40	234.35
11168	18DE	22;22	264.79

Full Moons

NUMBER	DATE	TIME	LONG.
11156	12JA	21;52	110.41
11157	11FE	11;13	140.09
11158	13MR	1;45	169.33
11159	11AP	17;11	198.10
11160	11MY	8;48	226.51
11161	9JN	23;49	254.73
11162	9JL	13;49	282.97
11163	8AU	2;48	311.45
11164	6SE	15;01	340.36
11165	6OC	2;41	9.74
11166	4NO	13;53	39.55
11167	4DE	0;39	69.58

-97

New Moons

NUMBER	DATE	TIME	LONG.
11169	17JA	16;58	295.01
11170	16FE	9;15	324.74
11171	17MR	22;39	353.85
11172	16AP	9;12	22.36
11173	15MY	17;25	50.44
11174	14JN	0;18	78.33
11175	13JL	7;02	106.30
11176	11AU	14;48	134.62
11177	10SE	0;35	163.48
11178	9OC	13;00	192.95
11179	8NO	4;09	222.96
11180	7DE	21;39	253.29

Full Moons

NUMBER	DATE	TIME	LONG.
11168	2JA	11;11	99.58
11169	31JA	21;53	129.30
11170	2MR	9;11	158.59
11171	31MR	21;21	187.41
11172	30AP	10;29	215.84
11173	30MY	0;30	244.06
11174	28JN	15;17	272.27
11175	28JL	6;49	300.72
11176	26AU	22;10	329.58
11177	25SE	13;14	358.92
11178	25OC	3;10	28.70
11179	23NO	15;41	58.74
11180	23DE	2;54	88.81

-96

New Moons

NUMBER	DATE	TIME	LONG.
11181	6JA	16;34	283.62
11182	5FE	11;31	313.63
11183	6MR	4;53	343.10
11184	4AP	19;29	11.94
11185	4MY	7;04	40.26
11186	2JN	16;13	68.25
11187	2JL	0;00	96.21
11188	31JL	7;35	124.39
11189	29AU	15;52	153.01
11190	28SE	1;27	182.18
11191	27OC	12;47	211.90
11192	26NO	2;11	242.00
11193	25DE	17;47	272.25

Full Moons

NUMBER	DATE	TIME	LONG.
11181	21JA	13;13	118.64
11182	19FE	22;58	148.05
11183	20MR	8;32	176.94
11184	18AP	18;18	205.38
11185	18MY	4;53	233.52
11186	16JN	16;58	261.62
11187	16JL	6;49	289.91
11188	14AU	23;17	318.62
11189	13SE	16;43	347.87
11190	13OC	10;03	17.62
11191	12NO	2;07	47.72
11192	11DE	16;20	77.92

-95

New Moons

NUMBER	DATE	TIME	LONG.
11194	24JA	11;13	302.34
11195	23FE	5;20	332.03
11196	24MR	22;34	1.17
11197	23AP	13;46	29.76
11198	23MY	2;38	57.97
11199	21JN	13;33	86.02
11200	20JL	23;15	114.17
11201	19AU	8;23	142.66
11202	17SE	17;32	171.64
11203	17OC	3;11	201.12
11204	15NO	13;50	231.02
11205	15DE	1;55	261.14

Full Moons

NUMBER	DATE	TIME	LONG.
11193	10JA	4;41	107.94
11194	8FE	15;18	137.54
11195	10MR	0;27	166.60
11196	8AP	8;33	195.13
11197	7MY	16;22	223.26
11198	6JN	0;58	251.23
11199	5JL	11;28	279.31
11200	4AU	0;39	307.75
11201	2SE	16;36	336.74
11202	2OC	10;36	6.33
11203	1NO	5;19	36.41
11204	30NO	23;24	66.74
11205	30DE	15;43	97.00

-94

New Moons

NUMBER	DATE	TIME	LONG.
11206	13JA	15;42	291.19
11207	12FE	6;54	320.94
11208	13MR	22;51	350.22
11209	12AP	14;43	19.00
11210	12MY	5;59	47.39
11211	10JN	20;18	75.57
11212	10JL	9;32	103.79
11213	8AU	21;35	132.24
11214	7SE	8;37	161.11
11215	6OC	19;01	190.46
11216	5NO	5;18	220.23
11217	4DE	15;57	250.26

Full Moons

NUMBER	DATE	TIME	LONG.
11206	29JA	5;33	126.88
11207	27FE	16;40	156.21
11208	29MR	1;26	184.93
11209	27AP	8;41	213.16
11210	26MY	15;30	241.10
11211	24JN	23;05	269.03
11212	24JL	8;27	297.20
11213	22AU	20;20	325.85
11214	21SE	11;05	355.11
11215	21OC	4;31	24.98
11216	19NO	23;50	55.28
11217	19DE	19;28	85.71

-93

New Moons

NUMBER	DATE	TIME	LONG.
11218	3JA	3;12	280.28
11219	1FE	15;06	310.05
11220	3MR	3;34	339.38
11221	1AP	16;42	8.22
11222	1MY	6;40	36.68
11223	30MY	21;31	64.92
11224	29JN	12;53	93.16
11225	29JL	4;03	121.61
11226	27AU	18;23	150.43
11227	26SE	7;37	179.71
11228	25OC	19;56	209.43
11229	24NO	7;35	239.45
11230	23DE	18;41	269.51

Full Moons

NUMBER	DATE	TIME	LONG.
11218	18JA	13;32	115.91
11219	17FE	4;37	145.57
11220	18MR	16;21	174.60
11221	17AP	1;23	203.04
11222	16MY	8;46	231.08
11223	14JN	15;36	258.96
11224	13JL	22;52	286.96
11225	12AU	7;24	315.32
11226	10SE	17;58	344.22
11227	10OC	7;17	13.74
11228	8NO	23;41	43.82
11229	8DE	18;41	74.21

-92

New Moons

NUMBER	DATE	TIME	LONG.
11231	22JA	5;13	299.34
11232	20FE	15;16	328.75
11233	21MR	1;15	357.65
11234	19AP	11;54	26.11
11235	18MY	23;56	54.31
11236	17JN	13;43	82.47
11237	17JL	5;01	110.82
11238	15AU	21;11	139.54
11239	14SE	13;29	168.76
11240	14OC	5;20	198.46
11241	12NO	20;11	228.51
11242	12DE	9;34	258.68

Full Moons

NUMBER	DATE	TIME	LONG.
11230	7JA	14;36	104.58
11231	6FE	9;15	134.57
11232	7MR	1;08	163.95
11233	5AP	13;57	192.71
11234	5MY	0;10	220.97
11235	3JN	8;35	248.94
11236	2JL	16;04	276.88
11237	31JL	23;27	305.06
11238	30AU	7;36	333.69
11239	28SE	17;26	2.88
11240	28OC	5;45	32.64
11241	26NO	20;54	62.83
11242	26DE	14;22	93.15

-91

New Moons

NUMBER	DATE	TIME	LONG.
11243	10JA	21;09	288.66
11244	9FE	7;04	318.23
11245	10MR	15;48	347.26
11246	9AP	0;09	15.79
11247	8MY	9;01	43.95
11248	6JN	19;12	71.98
11249	6JL	7;15	100.13
11250	4AU	21;28	128.62
11251	3SE	13;47	157.64
11252	3OC	7;42	187.24
11253	2NO	2;03	217.32
11254	1DE	19;19	247.62
11255	31DE	10;15	277.81

Full Moons

NUMBER	DATE	TIME	LONG.
11243	25JA	8;53	123.28
11244	24FE	2;54	152.95
11245	25MR	19;19	182.04
11246	24AP	9;33	210.58
11247	23MY	21;30	238.75
11248	22JN	7;28	266.76
11249	21JL	16;02	294.88
11250	20AU	0;09	323.34
11251	18SE	8;45	352.30
11252	17OC	18;39	21.80
11253	16NO	6;21	51.76
11254	15DE	19;51	81.93

-90

New Moons

NUMBER	DATE	TIME	LONG.
11256	29JA	22;26	307.62
11257	28FE	8;19	336.88
11258	29MR	16;37	5.58
11259	28AP	0;11	33.81
11260	27MY	7;47	61.78
11261	25JN	16;15	89.74
11262	25JL	2;30	117.95
11263	23AU	15;22	146.65
11264	22SE	7;20	175.98
11265	22OC	2;02	205.91
11266	20NO	21;50	236.25
11267	20DE	16;37	266.64

Full Moons

NUMBER	DATE	TIME	LONG.
11255	14JA	10;54	112.03
11256	13FE	3;01	141.80
11257	14MR	19;38	171.09
11258	13AP	12;03	199.88
11259	13MY	3;22	228.27
11260	11JN	16;56	256.42
11261	11JL	4;38	284.58
11262	9AU	14;58	312.98
11263	8SE	0;43	341.80
11264	7OC	10;38	11.14
11265	5NO	21;05	40.93
11266	5DE	8;10	70.98

-89

New Moons

NUMBER	DATE	TIME	LONG.
11268	19JA	8;52	296.75
11269	17FE	22;09	326.33
11270	19MR	8;47	355.30
11271	17AP	17;22	23.71
11272	17MY	0;38	51.74
11273	15JN	7;25	79.62
11274	14JL	14;46	107.63
11275	12AU	23;51	136.02
11276	11SE	11;42	164.99
11277	11OC	2;54	194.60
11278	9NO	21;00	224.75
11279	9DE	16;36	255.17

Full Moons

NUMBER	DATE	TIME	LONG.
11267	3JA	19;55	101.02
11268	2FE	8;29	130.80
11269	3MR	22;07	160.17
11270	2AP	12;51	189.07
11271	2MY	4;16	217.57
11272	31MY	19;36	245.83
11273	30JN	10;14	274.04
11274	29JL	23;56	302.44
11275	28AU	12;47	331.22
11276	27SE	1;01	0.47
11277	26OC	12;40	30.17
11278	24NO	23;45	60.16
11279	24DE	10;22	90.20

Left half

NUMBER	DATE	TIME	LONG.		NUMBER	DATE	TIME	LONG.
				-88				
11280	8JA	11;54	285.51		11280	22JA	20;47	120.03
11281	7FE	5;22	315.43		11281	21FE	7;29	149.45
11282	7MR	20;09	344.76		11282	21MR	18;54	178.40
11283	6AP	8;01	13.47		11283	20AP	7;16	206.92
11284	5MY	17;15	41.68		11284	19MY	20;39	235.17
11285	4JN	0;41	69.61		11285	18JN	11;01	263.35
11286	3JL	7;24	97.52		11286	18JL	2;12	291.70
11287	1AU	14;37	125.70		11287	16AU	17;57	320.42
11288	30AU	23;25	154.36		11288	15SE	9;38	349.62
11289	29SE	10;34	183.63		11289	15OC	0;30	19.29
11290	29OC	0;26	213.46		11290	13NO	13;57	49.30
11291	27NO	16;48	243.70		11291	13DE	1;52	79.40
11292	27DE	11;03	274.05					
				-87				
					11292	11JA	12;32	109.34
11293	26JA	6;01	304.19		11293	9FE	22;22	138.89
11294	25FE	0;11	333.85		11294	11MR	7;45	167.94
11295	26MR	16;09	2.91		11295	9AP	17;07	196.50
11296	25AP	5;11	31.40		11296	9MY	2;59	224.70
11297	24MY	15;28	59.49		11297	7JN	14;04	252.76
11298	22JN	23;53	87.44		11298	7JL	3;05	280.95
11299	22JL	7;35	115.53		11299	5AU	18;22	309.50
11300	20AU	15;36	144.00		11300	4SE	11;31	338.57
11301	19SE	0;39	172.98		11301	4OC	5;21	8.18
11302	18OC	11;11	202.53		11302	2NO	22;25	38.21
11303	16NO	23;32	232.51		11303	2DE	13;46	68.43
11304	16DE	13;56	262.72					
				-86				
					11304	1JA	3;04	98.55
11305	15JA	6;21	292.88		11305	30JA	14;28	128.32
11306	14FE	0;03	322.70		11306	1MR	0;09	157.56
11307	15MR	17;40	352.02		11307	30MR	8;31	186.25
11308	14AP	9;50	20.79		11308	28AP	16;11	214.48
11309	13MY	23;50	49.11		11309	28MY	0;07	242.46
11310	12JN	11;45	77.20		11310	26JN	9;27	270.45
11311	11JL	22;09	105.31		11311	25JL	21;13	298.74
11312	10AU	7;43	133.69		11312	24AU	11;54	327.52
11313	8SE	17;02	162.51		11313	23SE	5;10	356.92
11314	8OC	2;34	191.84		11314	22OC	23;54	26.86
11315	6NO	12;45	221.62		11315	21NO	18;38	57.16
11316	6DE	0;04	251.68		11316	21DE	12;02	87.49
				-85				
11317	4JA	12;51	281.76		11317	20JA	3;10	117.54
11318	3FE	3;09	311.61		11318	18FE	15;32	147.08
11319	4MR	18;31	341.03		11319	20MR	1;17	176.00
11320	3AP	10;15	9.96		11320	18AP	9;02	204.37
11321	3MY	1;42	38.45		11321	17MY	15;50	232.38
11322	1JN	16;30	66.69		11322	15JN	22;50	260.27
11323	1JL	6;25	94.88		11323	15JL	7;11	288.32
11324	30JL	19;18	123.25		11324	13AU	17;48	316.78
11325	29AU	7;06	152.00		11325	12SE	7;11	345.82
11326	27SE	18;02	181.21		11326	11OC	23;26	15.49
11327	27OC	4;30	210.87		11327	10NO	18;02	45.67
11328	25NO	15;00	240.84		11328	10DE	13;47	76.10
11329	25DE	1;54	270.89					
				-84				
					11329	9JA	8;52	106.42
11330	23JA	13;19	300.75		11330	8FE	1;29	136.30
11331	22FE	1;14	330.21		11331	8MR	14;47	165.55
11332	22MR	13;43	359.19		11332	7AP	0;58	194.18
11333	21AP	2;58	27.74		11333	6MY	8;59	222.33
11334	20MY	17;13	56.01		11334	4JN	15;56	250.24
11335	19JN	8;22	84.23		11335	3JL	22;54	278.17
11336	18JL	23;52	112.60		11336	2AU	6;47	306.39
11337	17AU	14;57	141.30		11337	31AU	16;21	335.09
11338	16SE	5;03	170.45		11338	30SE	4;19	4.39
11339	15OC	18;04	200.06		11339	29OC	19;13	34.29
11340	14NO	6;13	230.02		11340	28NO	13;02	64.59
11341	13DE	17;39	260.10		11341	28DE	8;41	95.00
				-83				
11342	12JA	4;24	290.03		11342	27JA	4;05	125.15
11343	10FE	14;29	319.59		11343	25FE	21;21	154.76
11344	12MR	0;10	348.64		11344	27MR	11;35	183.73
11345	10AP	10;07	17.21		11345	25AP	22;58	212.14
11346	9MY	21;09	45.45		11346	25MY	8;12	240.18
11347	8JN	9;54	73.58		11347	23JN	16;07	268.12
11348	8JL	0;29	101.83		11348	22JL	23;34	296.20
11349	6AU	16;26	130.42		11349	21AU	7;23	324.67
11350	5SE	9;00	159.48		11350	19SE	16;25	353.67
11351	5OC	1;27	189.05		11351	19OC	3;30	23.24
11352	3NO	17;07	219.03		11352	17NO	17;14	53.29
11353	3DE	7;26	249.21		11353	17DE	9;34	83.59

Right half

NUMBER	DATE	TIME	LONG.		NUMBER	DATE	TIME	LONG.
				-82				
11354	1JA	19;58	279.29		11354	16JA	3;34	113.80
11355	31JA	6;35	309.02		11355	14FE	21;47	143.63
11356	1MR	15;39	338.23		11356	16MR	14;54	172.92
11357	30MR	23;53	6.90		11357	15AP	6;09	201.63
11358	29AP	8;10	35.15		11358	14MY	19;14	229.91
11359	28MY	17;26	63.18		11359	13JN	6;15	257.96
11360	27JN	4;25	91.24		11360	12JL	15;35	286.04
11361	26JL	17;32	119.58		11361	11AU	0;00	314.38
11362	25AU	8;56	148.41		11362	9SE	8;27	343.17
11363	24SE	2;20	177.83		11363	8OC	17;48	12.50
11364	23OC	20;53	207.78		11364	7NO	4;39	42.32
11365	22NO	15;05	238.06		11365	6DE	17;14	72.44
11366	22DE	7;23	268.34					
				-81				
					11366	5JA	7;24	102.58
11367	20JA	20;55	298.32		11367	3FE	22;47	132.46
11368	19FE	7;46	327.78		11368	5MR	14;56	161.90
11369	20MR	16;37	356.65		11369	4AP	7;18	190.84
11370	19AP	0;18	25.01		11370	3MY	23;09	219.34
11371	18MY	7;39	53.03		11371	2JN	13;41	247.56
11372	16JN	15;30	80.96		11372	2JL	2;27	275.71
11373	16JL	0;44	109.05		11373	31JL	13;37	304.02
11374	14AU	12;14	137.56		11374	29AU	23;49	332.71
11375	13SE	2;43	166.66		11375	28SE	9;50	1.89
11376	12OC	20;19	196.39		11376	27OC	20;10	31.56
11377	11NO	15;57	226.63		11377	26NO	7;00	61.55
11378	11DE	11;34	257.05		11378	25DE	18;20	91.61
				-80				
11379	10JA	5;13	287.31		11379	24JA	6;16	121.49
11380	8FE	19;55	317.09		11380	22FE	19;03	150.97
11381	9MR	7;43	346.27		11381	23MR	8;57	180.00
11382	7AP	17;09	14.86		11382	21AP	23;50	208.60
11383	7MY	0;54	43.00		11383	21MY	15;10	236.91
11384	5JN	7;46	70.90		11384	20JN	6;15	265.13
11385	4JL	14;43	98.83		11385	19JL	20;36	293.46
11386	2AU	22;50	127.06		11386	18AU	10;08	322.11
11387	1SE	9;17	155.81		11387	16SE	22;58	351.23
11388	30SE	22;53	185.20		11388	16OC	11;11	20.81
11389	30OC	15;42	215.18		11389	14NO	22;42	50.74
11390	29NO	10;48	245.54		11390	14DE	9;34	80.80
11391	29DE	6;25	275.95					
				-79				
					11391	12JA	19;56	110.72
11392	28JA	0;51	306.04		11392	11FE	6;13	140.27
11393	26FE	16;57	335.59		11393	12MR	16;57	169.35
11394	28MR	6;11	4.51		11394	11AP	4;32	197.98
11395	26AP	16;37	32.87		11395	10MY	17;11	226.28
11396	26MY	0;51	60.88		11396	9JN	6;56	254.44
11397	24JN	7;50	88.77		11397	8JL	21;46	282.71
11398	23JL	14;45	116.84		11398	7AU	13;28	311.30
11399	21AU	22;45	145.32		11399	6SE	5;33	340.35
11400	20SE	8;47	174.38		11400	5OC	21;14	9.90
11401	19OC	21;21	204.02		11401	4NO	11;41	39.85
11402	18NO	12;32	234.14		11402	4DE	0;29	69.96
11403	18DE	5;51	264.48					
				-78				
					11403	2JA	11;44	99.99
11404	17JA	0;28	294.71		11404	31JA	21;49	129.69
11405	15FE	19;02	324.54		11405	2MR	7;11	158.90
11406	17MR	12;06	353.81		11406	31MR	16;16	187.59
11407	16AP	2;34	22.48		11407	30AP	1;34	215.87
11408	15MY	14;09	50.70		11408	29MY	11;46	243.93
11409	13JN	23;28	78.68		11409	27JN	23;39	272.04
11410	13JL	7;32	106.71		11410	27JL	13;49	300.43
11411	11AU	15;27	135.03		11411	26AU	6;18	329.32
11412	10SE	0;05	163.84		11412	25SE	0;13	358.77
11413	9OC	9;59	193.21		11413	24OC	18;06	28.70
11414	7NO	21;27	223.06		11414	23NO	10;34	58.91
11415	7DE	10;45	253.21		11415	23DE	0;59	89.12
				-77				
11416	6JA	2;02	283.39		11416	21JA	13;17	119.03
11417	4FE	18;58	313.33		11417	19FE	23;40	148.46
11418	6MR	12;33	342.81		11418	21MR	8;28	177.32
11419	5AP	5;24	11.76		11419	19AP	16;13	205.68
11420	4MY	20;27	40.22		11420	18MY	23;44	233.70
11421	3JN	9;24	68.37		11421	17JN	8;07	261.65
11422	2JL	20;38	96.46		11422	16JL	18;31	289.79
11423	1AU	6;46	124.74		11423	15AU	7;47	318.38
11424	30AU	16;23	153.42		11424	13SE	23;59	347.56
11425	29SE	1;59	182.60		11425	13OC	18;19	17.35
11426	28OC	11;56	212.25		11426	12NO	13;24	47.57
11427	26NO	22;40	242.24		11427	12DE	7;43	77.94
11428	26DE	10;36	272.33					

Left Panel

NEW MOONS NUMBER	DATE	TIME	LONG.	FULL MOONS NUMBER	DATE	TIME	LONG.
−76							
				11428	11JA	0;06	108.13
11429	24JA	23;57	302.26	11429	9FE	13;49	137.87
11430	23FE	14;34	331.81	11430	10MR	0;43	167.00
11431	24MR	5;54	0.88	11431	8AP	9;14	195.54
11432	22AP	21;19	29.49	11432	7MY	16;16	223.64
11433	22MY	12;24	57.79	11433	5JN	22;58	251.53
11434	21JN	2;51	85.98	11434	5JL	6;30	279.49
11435	20JL	16;28	114.29	11435	3AU	15;55	307.78
11436	19AU	5;06	142.92	11436	2SE	3;57	336.61
11437	17SE	16;43	171.99	11437	1OC	18;53	6.06
11438	17OC	3;36	201.53	11438	31OC	12;30	36.09
11439	15NO	14;10	231.43	11439	30NO	7;56	66.47
11440	15DE	0;51	261.48	11440	30DE	3;35	96.87
−75							
11441	13JA	11;53	291.42	11441	28JA	21;33	126.94
11442	11FE	23;18	321.01	11442	27FE	12;27	156.42
11443	13MR	11;11	350.13	11443	29MR	0;02	185.26
11444	11AP	23;43	18.78	11444	27AP	8;56	213.56
11445	11MY	13;14	47.11	11445	26MY	16;21	241.52
11446	10JN	3;53	75.31	11446	24JN	23;08	269.41
11447	9JL	19;22	103.61	11447	24JL	6;32	297.51
11448	8AU	10;59	132.20	11448	22AU	15;16	326.03
11449	7SE	1;55	161.22	11449	21SE	2;03	355.12
11450	6OC	15;47	190.71	11450	20OC	15;30	24.82
11451	5NO	4;34	220.59	11451	19NO	7;56	55.00
11452	4DE	16;27	250.67	11452	19DE	2;48	85.39
−74							
11453	3JA	3;32	280.69	11453	17JA	22;28	115.66
11454	1FE	13;47	310.38	11454	16FE	16;51	145.47
11455	2MR	23;23	339.59	11455	18MR	8;31	174.67
11456	1AP	8;51	8.29	11456	16AP	21;12	203.26
11457	30AP	18;59	36.60	11457	16MY	7;24	231.41
11458	30MY	6;38	64.71	11458	14JN	15;57	259.36
11459	28JN	20;15	92.89	11459	13JL	23;39	287.38
11460	28JL	11;40	121.34	11460	12AU	7;20	315.71
11461	27AU	4;14	150.24	11461	10SE	15;49	344.52
11462	25SE	21;07	179.66	11462	10OC	1;54	13.91
11463	25OC	13;31	209.55	11463	8NO	14;18	43.81
11464	24NO	4;47	239.72	11464	8DE	5;20	74.03
11465	23DE	18;19	269.88				
−73							
				11465	6JA	22;30	104.29
11466	22JA	5;49	299.76	11466	5FE	16;33	134.26
11467	20FE	15;27	329.14	11467	7MR	10;08	163.73
11468	21MR	23;48	357.98	11468	6AP	2;14	192.63
11469	20AP	7;44	26.33	11469	5MY	16;22	221.04
11470	19MY	16;14	54.39	11470	4JN	4;27	249.16
11471	18JN	2;10	82.40	11471	3JL	14;42	277.21
11472	17JL	14;09	110.61	11472	1AU	23;41	305.46
11473	16AU	4;28	139.25	11473	31AU	8;13	334.10
11474	14SE	21;06	168.47	11474	29SE	17;13	3.26
11475	14OC	15;27	198.26	11475	29OC	3;23	32.93
11476	13NO	10;14	228.48	11476	27NO	15;07	62.97
11477	13DE	3;46	258.82	11477	27DE	4;25	93.11
−72							
11478	11JA	18;45	288.95	11478	25JA	19;00	123.09
11479	10FE	6;49	318.60	11479	24FE	10;32	152.66
11480	10MR	16;26	347.67	11480	25MR	2;36	181.75
11481	9AP	0;27	16.18	11481	23AP	18;38	210.37
11482	8MY	7;44	44.29	11482	23MY	9;52	238.67
11483	6JN	15;09	72.21	11483	21JN	23;39	266.83
11484	5JL	23;33	100.20	11484	21JL	11;47	295.09
11485	4AU	9;49	128.54	11485	19AU	22;38	323.66
11486	2SE	22;50	157.42	11486	18SE	8;56	352.69
11487	2OC	15;02	186.94	11487	17OC	19;16	22.22
11488	1NO	9;58	217.03	11488	16NO	5;58	52.13
11489	1DE	5;58	247.44	11489	15DE	17;02	82.20
11490	31DE	0;49	277.80				
−71							
				11490	14JA	4;30	112.15
11491	29JA	17;00	307.78	11491	12FE	16;33	141.75
11492	28FE	6;08	337.18	11492	14MR	5;34	170.91
11493	29MR	16;34	5.96	11493	12AP	19;42	199.61
11494	28AP	0;59	34.23	11494	12MY	10;43	227.99
11495	27MY	8;09	62.18	11495	11JN	1;58	256.21
11496	25JN	14;56	90.07	11496	10JL	16;50	284.49
11497	24JL	22;22	118.17	11497	9AU	7;01	313.04
11498	23AU	7;36	146.73	11498	7SE	20;30	342.02
11499	21SE	19;39	175.88	11499	7OC	9;19	11.47
11500	21OC	10;59	205.67	11500	5NO	21;24	41.33
11501	20NO	5;09	235.93	11501	5DE	8;40	71.39
11502	20DE	0;41	266.35				

Right Panel

NEW MOONS NUMBER	DATE	TIME	LONG.	FULL MOONS NUMBER	DATE	TIME	LONG.
−70							
				11502	3JA	19;11	101.38
11503	18JA	19;49	296.59	11503	2FE	5;18	131.06
11504	17FE	13;04	326.34	11504	3MR	15;30	160.28
11505	19MR	3;40	355.48	11505	2AP	2;19	189.03
11506	17AP	15;24	24.02	11506	1MY	14;10	217.39
11507	17MY	0;38	52.13	11507	31MY	3;12	245.56
11508	15JN	8;09	80.04	11508	29JN	17;27	273.76
11509	14JL	15;03	108.03	11509	29JL	8;52	302.22
11510	12AU	22;30	136.35	11510	28AU	1;06	331.13
11511	11SE	7;34	165.20	11511	26SE	17;24	0.54
11512	10OC	18;56	194.65	11512	26OC	8;50	30.40
11513	9NO	8;51	224.62	11513	24NO	22;39	60.50
11514	9DE	1;08	254.90	11514	24DE	10;41	90.60
−69							
11515	7JA	19;05	285.18	11515	22JA	21;13	120.43
11516	6FE	13;39	315.16	11516	21FE	6;43	149.80
11517	8MR	7;27	344.63	11517	22MR	15;40	178.65
11518	6AP	23;12	13.50	11518	21AP	0;34	207.04
11519	6MY	12;11	41.86	11519	20MY	10;02	235.13
11520	4JN	22;36	69.91	11520	18JN	20;52	263.17
11521	4JL	7;17	97.90	11521	18JL	9;49	291.43
11522	2AU	15;19	126.12	11522	17AU	1;17	320.13
11523	31AU	23;43	154.76	11523	15SE	18;51	349.40
11524	30SE	9;06	183.95	11524	15OC	13;12	19.21
11525	29OC	19;51	213.66	11525	14NO	6;44	49.38
11526	28NO	8;12	243.72	11526	13DE	22;20	79.64
11527	27DE	22;22	273.90				
−68							
				11527	12JA	11;41	109.69
11528	26JA	14;19	303.92	11528	10FE	22;54	139.30
11529	25FE	7;27	333.55	11529	11MR	8;18	168.35
11530	26MR	0;36	2.67	11530	9AP	16;19	196.85
11531	24AP	16;30	31.28	11531	8MY	23;40	224.95
11532	24MY	6;30	59.52	11532	7JN	7;22	252.88
11533	22JN	18;39	87.62	11533	6JL	16;35	280.91
11534	22JL	5;27	115.83	11534	5AU	4;22	309.32
11535	20AU	15;31	144.38	11535	3SE	19;14	338.28
11536	19SE	1;19	173.40	11536	3OC	12;47	7.87
11537	18OC	11;14	202.92	11537	2NO	7;51	37.98
11538	16NO	21;37	232.82	11538	2DE	2;51	68.35
11539	16DE	8;52	262.90	11539	31DE	20;22	98.66
−67							
11540	14JA	21;19	292.88	11540	30JA	11;26	128.58
11541	13FE	11;04	322.55	11541	28FE	23;38	157.93
11542	15MR	1;48	351.76	11542	30MR	9;09	186.66
11543	13AP	16;59	20.50	11543	28AP	16;42	214.89
11544	13MY	8;08	48.87	11544	27MY	23;20	242.81
11545	11JN	22;56	77.08	11545	26JN	6;17	270.71
11546	11JL	13;09	105.34	11546	25JL	14;41	298.86
11547	10AU	2;34	133.86	11547	24AU	1;25	327.48
11548	8SE	14;59	162.81	11548	22SE	14;59	356.72
11549	8OC	2;27	192.22	11549	22OC	7;24	26.56
11550	6NO	13;17	222.03	11550	21NO	2;07	56.85
11551	5DE	23;56	252.07	11551	20DE	21;53	87.28
−66							
11552	4JA	10;42	282.06	11552	19JA	16;52	117.50
11553	2FE	21;44	311.77	11553	18FE	9;19	147.21
11554	4MR	9;06	341.03	11554	19MR	22;27	176.27
11555	2AP	20;58	9.81	11555	18AP	8;31	204.73
11556	2MY	9;42	38.20	11556	17MY	16;27	232.79
11557	31MY	23;38	66.40	11557	15JN	23;26	260.68
11558	30JN	14;46	94.64	11558	15JL	6;32	288.68
11559	30JL	6;35	123.13	11559	13AU	14;37	317.04
11560	28AU	22;14	152.02	11560	12SE	0;24	345.93
11561	27SE	12;58	181.38	11561	11OC	12;33	15.42
11562	27OC	2;32	211.17	11562	10NO	3;31	45.44
11563	25NO	15;01	241.23	11563	9DE	21;15	75.79
11564	25DE	2;31	271.30				
−65							
				11564	8JA	16;38	106.12
11565	23JA	13;05	301.13	11565	7FE	11;44	136.11
11566	21FE	22;47	330.50	11566	9MR	4;43	165.53
11567	23MR	7;59	359.34	11567	7AP	18;47	194.32
11568	21AP	17;25	27.74	11568	7MY	6;08	222.61
11569	21MY	4;01	55.87	11569	5JN	15;28	250.61
11570	19JN	16;30	83.98	11570	4JL	23;36	278.59
11571	19JL	7;05	112.31	11571	3AU	7;21	306.80
11572	17AU	23;20	141.06	11572	1SE	15;30	335.44
11573	16SE	16;24	170.32	11573	1OC	0;50	4.64
11574	16OC	9;25	200.09	11574	30OC	12;05	34.37
11575	15NO	1;33	230.21	11575	29NO	1;47	64.50
11576	14DE	16;08	260.42	11576	28DE	17;52	94.76

-64 (left) / -58 (right)

NUMBER	DATE	TIME	LONG.	NUMBER	DATE	TIME	LONG.	NUMBER	DATE	TIME	LONG.	NUMBER	DATE	TIME	LONG.
11577	13JA	4;39	290.43	11577	27JA	11;25	124.84	11651	5JA	19;14	283.49	11651	21JA	8;22	119.22
11578	11FE	15;04	320.00	11578	26FE	5;08	154.47	11652	4FE	8;07	313.26	11652	19FE	21;56	148.78
11579	11MR	23;47	349.01	11579	26MR	21;52	183.56	11653	5MR	22;06	342.61	11653	21MR	8;38	177.72
11580	10AP	7;37	17.50	11580	25AP	12;56	212.12	11654	4AP	12;49	11.48	11654	19AP	16;57	206.10
11581	9MY	15;32	45.61	11581	25MY	2;04	240.33	11655	4MY	3;50	39.94	11655	18MY	23;49	234.09
11582	8JN	0;32	73.59	11582	23JN	13;19	268.39	11656	2JN	18;46	68.17	11656	17JN	6;27	261.97
11583	7JL	11;22	101.69	11583	22JL	23;03	296.57	11657	2JL	9;25	96.40	11657	16JL	14;00	290.00
11584	6AU	0;32	130.16	11584	21AU	7;55	325.08	11658	31JL	23;30	124.84	11658	14AU	23;32	318.43
11585	4SE	16;09	159.17	11585	19SE	16;48	354.07	11659	30AU	12;43	153.66	11659	13SE	11;45	347.45
11586	4OC	9;57	188.79	11586	19OC	2;28	23.59	11660	29SE	0;56	182.94	11660	13OC	2;51	17.09
11587	3NO	4;55	218.90	11587	17NO	13;27	53.52	11661	28OC	12;16	212.66	11661	11NO	20;35	47.24
11588	2DE	23;26	249.26	11588	17DE	1;56	83.65	11662	26NO	23;04	242.65	11662	11DE	16;02	77.66
								11663	26DE	9;43	272.68				

-63 (left) / -57 (right)

NUMBER	DATE	TIME	LONG.	NUMBER	DATE	TIME	LONG.	NUMBER	DATE	TIME	LONG.	NUMBER	DATE	TIME	LONG.
11589	1JA	15;51	279.51	11589	15JA	15;43	113.69					11663	10JA	11;36	107.99
11590	31JA	5;19	309.35	11590	14FE	6;33	143.39	11664	24JA	20;28	302.50	11664	9FE	5;24	137.90
11591	1MR	15;57	338.63	11591	15MR	22;05	172.62	11665	23FE	7;24	331.90	11665	10MR	20;07	167.20
11592	31MR	0;31	7.31	11592	14AP	13;59	201.37	11666	24MR	18;41	0.81	11666	9AP	7;34	195.86
11593	29AP	7;56	35.52	11593	14MY	5;37	229.76	11667	23AP	6;40	29.29	11667	8MY	16;23	224.03
11594	28MY	15;05	63.47	11594	12JN	20;14	257.95	11668	22MY	19;46	57.51	11668	6JN	23;44	251.95
11595	26JN	22;51	91.40	11595	12JL	9;22	286.18	11669	21JN	10;16	85.70	11669	6JL	6;43	279.89
11596	26JL	8;05	119.59	11596	10AU	21;04	314.64	11670	21JL	1;57	114.09	11670	4AU	14;18	308.11
11597	24AU	19;42	148.26	11597	9SE	7;50	343.53	11671	19AU	18;02	142.85	11671	2SE	23;17	336.81
11598	23SE	10;23	177.55	11598	8OC	18;20	12.92	11672	18SE	9;36	172.08	11672	2OC	10;17	6.09
11599	23OC	4;12	207.46	11599	7NO	4;59	42.73	11673	18OC	0;03	201.77	11673	31OC	23;51	35.94
11600	22NO	0;01	237.81	11600	6DE	15;54	72.77	11674	16NO	13;15	231.78	11674	30NO	16;14	66.19
11601	21DE	19;44	268.24					11675	16DE	1;19	261.89	11675	30DE	10;53	96.55

-62 (left) / -56 (right)

NUMBER	DATE	TIME	LONG.	NUMBER	DATE	TIME	LONG.	NUMBER	DATE	TIME	LONG.	NUMBER	DATE	TIME	LONG.
				11601	5JA	3;04	102.78	11676	14JA	12;18	291.82	11676	29JA	6;13	126.68
11602	20JA	13;21	298.39	11602	3FE	14;34	132.50	11677	12FE	22;14	321.35	11677	28FE	0;16	156.31
11603	19FE	3;55	328.00	11603	5MR	2;46	161.79	11678	13MR	7;24	350.36	11678	28MR	15;42	185.31
11604	20MR	15;32	356.99	11604	3AP	16;02	190.61	11679	11AP	16;21	18.87	11679	27AP	4;18	213.76
11605	19AP	0;47	25.42	11605	3MY	6;25	219.06	11680	11MY	2;03	47.05	11680	26MY	14;35	241.84
11606	18MY	8;26	53.45	11606	1JN	21;32	247.30	11681	9JN	13;22	75.12	11681	24JN	23;19	269.81
11607	16JN	15;19	81.34	11607	1JL	12;43	275.55	11682	9JL	2;52	103.33	11682	24JL	7;19	297.93
11608	15JL	22;19	109.34	11608	31JL	3;28	304.00	11683	7AU	18;28	131.92	11683	22AU	15;21	326.42
11609	14AU	6;35	137.72	11609	29AU	17;35	332.84	11684	6SE	11;27	161.01	11684	21SE	0;09	355.43
11610	12SE	17;13	166.65	11610	28SE	7;03	2.17	11685	6OC	4;52	190.64	11685	20OC	10;28	24.99
11611	12OC	6;58	196.22	11611	27OC	19;45	31.93	11686	4NO	21;47	220.69	11686	18NO	22;56	55.00
11612	11NO	23;53	226.34	11612	26NO	7;34	61.96	11687	4DE	13;23	250.93	11687	18DE	13;47	85.23
11613	10DE	18;55	256.74	11613	25DE	18;27	92.01								

-61 (left) / -55 (right)

NUMBER	DATE	TIME	LONG.	NUMBER	DATE	TIME	LONG.	NUMBER	DATE	TIME	LONG.	NUMBER	DATE	TIME	LONG.
11614	9JA	14;23	287.07	11614	24JA	4;35	121.80	11688	3JA	3;00	281.05	11688	17JA	6;32	115.38
11615	8FE	8;35	317.01	11615	22FE	14;27	151.17	11689	1FE	14;22	310.79	11689	16FE	0;05	145.17
11616	10MR	0;27	346.36	11616	24MR	0;37	180.05	11690	2MR	23;42	339.99	11690	17MR	17;11	174.43
11617	8AP	13;33	15.10	11617	22AP	11;39	208.50	11691	1AP	7;41	8.63	11691	16AP	9;00	203.16
11618	7MY	23;57	43.35	11618	21MY	23;52	236.69	11692	30AP	15;15	36.83	11692	15MY	23;05	231.47
11619	6JN	8;14	71.30	11619	20JN	13;25	264.84	11693	29MY	23;27	64.81	11693	14JN	11;22	259.57
11620	5JL	15;23	99.25	11620	20JL	4;20	293.19	11694	28JN	9;12	92.83	11694	13JL	21;59	287.70
11621	3AU	22;33	127.44	11621	18AU	20;24	321.94	11695	27JL	21;10	121.14	11695	12AU	7;25	316.10
11622	2SE	6;50	156.10	11622	17SE	13;04	351.20	11696	26AU	11;39	149.95	11696	10SE	16;25	344.94
11623	1OC	17;07	185.35	11623	17OC	5;21	20.95	11697	25SE	4;35	179.37	11697	10OC	1;48	14.29
11624	31OC	5;49	215.15	11624	15NO	20;16	51.03	11698	24OC	23;19	209.34	11698	8NO	12;11	44.11
11625	29NO	20;57	245.34	11625	15DE	9;17	81.18	11699	23NO	18;27	239.67	11699	7DE	23;54	74.19
11626	29DE	14;02	275.64					11700	23DE	12;10	270.01				

-60 (left) / -54 (right)

NUMBER	DATE	TIME	LONG.	NUMBER	DATE	TIME	LONG.	NUMBER	DATE	TIME	LONG.	NUMBER	DATE	TIME	LONG.
				11626	13JA	20;28	111.13					11700	6JA	12;55	104.27
11627	28JA	8;14	305.74	11627	12FE	6;18	140.66	11701	22JA	3;09	300.03	11701	5FE	3;01	134.08
11628	27FE	2;23	335.38	11628	12MR	15;16	169.67	11702	20FE	15;03	329.51	11702	6MR	17;55	163.45
11629	27MR	19;09	4.45	11629	10AP	23;54	198.18	11703	22MR	0;25	358.39	11703	5AP	9;25	192.34
11630	26AP	9;30	32.97	11630	10MY	8;47	226.33	11704	20AP	8;10	26.73	11704	5MY	1;06	220.82
11631	25MY	21;10	61.12	11631	8JN	18;41	254.34	11705	19MY	15;15	54.74	11705	3JN	16;19	249.06
11632	24JN	6;43	89.11	11632	8JL	6;26	282.49	11706	17JN	22;33	82.64	11706	3JL	6;22	277.27
11633	23JL	15;07	117.24	11633	6AU	20;42	311.01	11707	17JL	6;56	110.71	11707	1AU	18;59	305.66
11634	21AU	23;26	145.74	11634	5SE	13;29	340.08	11708	15AU	17;18	139.19	11708	31AU	6;26	334.41
11635	20SE	8;27	174.75	11635	5OC	7;52	9.73	11709	14SE	6;28	168.25	11709	29SE	17;15	3.65
11636	19OC	18;37	204.30	11636	4NO	2;14	39.83	11710	13OC	22;53	197.96	11710	29OC	4;00	33.35
11637	18NO	6;11	234.26	11637	3DE	19;03	70.12	11711	12NO	18;00	228.18	11711	27NO	14;52	63.35
11638	17DE	19;20	264.42					11712	12DE	14;06	258.63	11712	27DE	1;52	93.40

-59 (left) / -53 (right)

NUMBER	DATE	TIME	LONG.	NUMBER	DATE	TIME	LONG.	NUMBER	DATE	TIME	LONG.	NUMBER	DATE	TIME	LONG.
				11638	2JA	9;34	100.29	11713	11JA	8;56	288.92	11713	25JA	12;59	123.22
11639	16JA	10;13	294.49	11639	31JA	21;46	130.07	11714	10FE	1;00	318.75	11714	24FE	0;31	152.63
11640	15FE	2;36	324.25	11640	2MR	7;54	159.31	11715	11MR	13;57	347.96	11715	25MR	12;54	181.58
11641	16MR	19;37	353.52	11641	31MR	16;23	187.98	11716	10AP	0;13	16.56	11716	24AP	2;28	210.11
11642	15AP	12;05	22.29	11642	29AP	23;49	216.19	11717	9MY	8;31	44.70	11717	23MY	17;07	238.39
11643	15MY	3;01	50.64	11643	29MY	7;05	244.13	11718	7JN	15;39	72.61	11718	22JN	8;21	266.61
11644	13JN	16;08	78.77	11644	27JN	15;20	272.09	11719	6JL	22;30	100.55	11719	21JL	23;31	294.99
11645	13JL	3;44	106.94	11645	27JL	1;43	300.33	11720	5AU	6;06	128.77	11720	20AU	14;13	323.71
11646	11AU	14;21	135.38	11646	25AU	15;05	329.08	11721	3SE	15;32	157.50	11721	19SE	4;20	352.89
11647	10SE	0;30	164.25	11647	24SE	7;31	358.46	11722	3OC	3;45	186.85	11722	18OC	17;43	22.54
11648	9OC	10;32	193.62	11648	24OC	2;10	28.42	11723	1NO	19;11	216.79	11723	17NO	6;11	52.53
11649	7NO	20;46	223.43	11649	22NO	21;31	58.75	11724	1DE	13;20	247.12	11724	16DE	17;34	82.61
11650	7DE	7;32	253.47	11650	22DE	16;00	89.12	11725	31DE	8;43	277.51				

−52

NEW MOONS NUMBER	DATE	TIME	LONG.	FULL MOONS NUMBER	DATE	TIME	LONG.
				11725	15JA	3;57	112.51
11726	30JA	3;36	307.61	11726	13FE	13;42	142.02
11727	28FE	20;36	337.17	11727	13MR	23;23	171.04
11728	29MR	11;01	6.12	11728	12AP	9;40	199.60
11729	27AP	22;41	34.52	11729	11MY	21;01	227.83
11730	27MY	7;57	62.56	11730	10JN	9;45	255.96
11731	25JN	15;37	90.49	11731	9JL	23;59	284.21
11732	24JL	22;46	118.58	11732	8AU	15;40	312.81
11733	23AU	6;31	147.07	11733	7SE	8;23	341.91
11734	21SE	15;51	176.11	11734	7OC	1;17	11.53
11735	21OC	3;24	205.74	11735	5NO	17;14	41.54
11736	19NO	17;22	235.81	11736	5DE	7;22	71.72
11737	19DE	9;26	266.09				

−51

NEW MOONS NUMBER	DATE	TIME	LONG.	FULL MOONS NUMBER	DATE	TIME	LONG.
				11737	3JA	19;27	101.78
11738	18JA	3;01	296.27	11738	2FE	5;48	131.47
11739	16FE	21;07	326.07	11739	3MR	14;57	160.64
11740	18MR	14;33	355.33	11740	1AP	23;28	189.30
11741	17AP	6;05	24.03	11741	1MY	7;57	217.53
11742	16MY	19;05	52.29	11742	30MY	17;05	245.54
11743	15JN	5;42	80.33	11743	29JN	3;44	273.60
11744	14JL	14;42	108.40	11744	28JL	16;41	301.96
11745	12AU	23;09	136.77	11745	27AU	8;22	330.83
11746	11SE	7;57	165.61	11746	26SE	2;20	0.30
11747	10OC	17;40	194.99	11747	25OC	21;09	30.29
11748	9NO	4;35	224.83	11748	24NO	15;04	60.57
11749	8DE	16;53	254.94	11749	24DE	6;52	90.83

−50

NEW MOONS NUMBER	DATE	TIME	LONG.	FULL MOONS NUMBER	DATE	TIME	LONG.
11750	7JA	6;45	285.05	11750	22JA	20;11	120.78
11751	5FE	22;10	314.91	11751	21FE	7;12	150.21
11752	7MR	14;43	344.33	11752	22MR	16;18	179.06
11753	6AP	7;22	13.25	11753	21AP	0;01	207.40
11754	5MY	23;02	41.72	11754	20MY	7;08	235.40
11755	4JN	13;06	69.92	11755	18JN	14;40	263.31
11756	4JL	1;33	98.06	11756	17JL	23;50	291.41
11757	2AU	12;50	126.41	11757	16AU	11;41	319.96
11758	31AU	23;26	155.14	11758	15SE	2;44	349.12
11759	30SE	9;44	184.37	11759	14OC	20;34	18.90
11760	29OC	20;00	214.05	11760	13NO	15;54	49.14
11761	28NO	6;30	244.04	11761	13DE	11;03	79.55
11762	27DE	17;37	274.09				

−49

NEW MOONS NUMBER	DATE	TIME	LONG.	FULL MOONS NUMBER	DATE	TIME	LONG.
				11762	12JA	4;36	109.78
11763	26JA	5;42	303.95	11763	10FE	19;34	139.55
11764	24FE	18;52	333.42	11764	12MR	7;34	168.70
11765	26MR	8;58	2.42	11765	10AP	16;53	197.26
11766	24AP	23;38	30.99	11766	10MY	0;16	225.36
11767	24MY	14;31	59.27	11767	8JN	6;50	253.24
11768	23JN	5;22	87.48	11768	7JL	13;48	281.19
11769	22JL	19;57	115.84	11769	5AU	22;18	309.46
11770	21AU	9;56	144.54	11770	4SE	9;12	338.26
11771	19SE	22;58	173.69	11771	3OC	22;56	7.68
11772	19OC	10;58	203.30	11772	2NO	15;29	37.68
11773	17NO	22;08	233.23	11773	2DE	10;15	68.04
11774	17DE	8;50	263.29				

−48

NEW MOONS NUMBER	DATE	TIME	LONG.	FULL MOONS NUMBER	DATE	TIME	LONG.
				11774	1JA	5;56	98.43
11775	15JA	19;25	293.19	11775	31JA	0;44	128.52
11776	14FE	6;02	322.73	11776	29FE	17;00	158.04
11777	14MR	16;51	351.78	11777	30MR	5;58	186.92
11778	13AP	4;08	20.37	11778	28AP	15;56	215.24
11779	12MY	16;23	48.63	11779	27MY	23;53	243.22
11780	11JN	6;04	76.79	11780	26JN	6;58	271.13
11781	10JL	21;15	105.09	11781	25JL	14;15	299.23
11782	9AU	13;26	133.72	11782	23AU	22;35	327.75
11783	8SE	5;39	162.81	11783	22SE	8;37	356.84
11784	7OC	21;01	192.38	11784	21OC	20;55	26.50
11785	6NO	11;05	222.33	11785	20NO	11;54	56.63
11786	5DE	23;49	252.45	11786	20DE	5;27	86.97

−47

NEW MOONS NUMBER	DATE	TIME	LONG.	FULL MOONS NUMBER	DATE	TIME	LONG.
11787	4JA	11;20	282.48	11787	19JA	0;30	117.20
11788	2FE	21;40	312.16	11788	17FE	19;14	147.01
11789	4MR	6;58	341.33	11789	19MR	11;56	176.23
11790	2AP	15;41	9.98	11790	18AP	1;51	204.85
11791	2MY	0;40	38.23	11791	17MY	13;13	233.04
11792	31MY	10;54	66.28	11792	15JN	22;43	261.03
11793	29JN	23;12	94.41	11793	15JL	7;08	289.09
11794	29JL	13;50	122.84	11794	13AU	15;15	317.45
11795	28AU	6;23	151.76	11795	11SE	23;45	346.29
11796	26SE	23;57	181.23	11796	11OC	9;22	15.67
11797	26OC	17;29	211.18	11797	9NO	20;45	45.55
11798	25NO	10;02	241.41	11798	9DE	10;22	75.71
11799	25DE	0;47	271.62				

−46

NEW MOONS NUMBER	DATE	TIME	LONG.	FULL MOONS NUMBER	DATE	TIME	LONG.
				11799	8JA	2;06	105.90
11800	23JA	13;15	301.52	11800	6FE	19;08	135.81
11801	21FE	23;25	330.91	11801	8MR	12;20	165.24
11802	23MR	7;48	359.72	11802	7AP	4;40	194.14
11803	21AP	15;16	28.04	11803	6MY	19;35	222.57
11804	20MY	22;53	56.05	11804	5JN	8;49	250.73
11805	19JN	7;40	84.02	11805	4JL	20;24	278.84
11806	18JL	18;26	112.19	11806	3AU	6;35	307.15
11807	17AU	7;41	140.81	11807	1SE	15;57	335.85
11808	15SE	23;34	170.01	11808	1OC	1;16	5.05
11809	15OC	17;42	199.81	11809	30OC	11;14	34.73
11810	14NO	13;00	230.06	11810	28NO	22;18	64.74
11811	14DE	7;45	260.46	11811	28DE	10;35	94.84

−45

NEW MOONS NUMBER	DATE	TIME	LONG.	FULL MOONS NUMBER	DATE	TIME	LONG.
11812	13JA	0;13	290.63	11812	26JA	23;57	124.74
11813	11FE	13;34	320.33	11813	25FE	14;11	154.25
11814	13MR	0;00	349.40	11814	27MR	5;07	183.27
11815	11AP	8;19	17.91	11815	25AP	20;33	211.86
11816	10MY	15;31	46.00	11816	25MY	12;00	240.15
11817	8JN	22;32	73.90	11817	24JN	2;47	268.36
11818	8JL	6;16	101.88	11818	23JL	16;21	296.69
11819	6AU	15;35	130.19	11819	22AU	4;37	325.33
11820	5SE	3;20	159.04	11820	20SE	15;57	354.43
11821	4OC	18;12	188.52	11821	20OC	2;56	24.00
11822	3NO	12;12	218.58	11822	18NO	13;52	53.93
11823	3DE	8;08	249.00	11823	18DE	0;48	83.99

−44

NEW MOONS NUMBER	DATE	TIME	LONG.	FULL MOONS NUMBER	DATE	TIME	LONG.
11824	2JA	3;50	279.39	11824	16JA	11;43	113.91
11825	31JA	21;20	309.41	11825	14FE	22;45	143.45
11826	1MR	11;44	338.84	11826	15MR	10;21	172.52
11827	30MR	23;01	7.64	11827	13AP	23;01	201.16
11828	29AP	8;19	35.92	11828	13MY	12;57	229.48
11829	28MY	15;54	63.89	11829	12JN	3;53	257.69
11830	26JN	22;49	91.79	11830	11JL	19;14	286.00
11831	26JL	6;00	119.89	11831	10AU	10;26	314.60
11832	24AU	14;29	148.43	11832	9SE	1;10	343.65
11833	23SE	1;19	177.56	11833	8OC	15;14	13.17
11834	22OC	15;12	207.31	11834	7NO	4;24	43.09
11835	21NO	8;08	237.52	11835	6DE	16;27	73.19
11836	21DE	3;02	267.92				

−43

NEW MOONS NUMBER	DATE	TIME	LONG.	FULL MOONS NUMBER	DATE	TIME	LONG.
				11836	5JA	3;16	103.18
11837	19JA	22;14	298.14	11837	3FE	13;08	132.83
11838	18FE	16;09	327.90	11838	4MR	22;32	162.00
11839	20MR	7;48	357.06	11839	3AP	8;11	190.68
11840	18AP	20;46	25.64	11840	2MY	18;43	218.98
11841	18MY	7;11	53.79	11841	1JN	6;33	247.09
11842	16JN	15;37	81.73	11842	30JN	19;58	275.26
11843	15JL	23;00	109.75	11843	30JL	11;01	303.73
11844	14AU	6;28	138.10	11844	29AU	3;30	332.65
11845	12SE	15;04	166.95	11845	27SE	20;43	2.12
11846	12OC	1;35	196.39	11846	27OC	13;34	32.06
11847	10NO	14;23	226.32	11847	26NO	4;52	62.24
11848	10DE	5;23	256.55	11848	25DE	18;02	92.38

−42

NEW MOONS NUMBER	DATE	TIME	LONG.	FULL MOONS NUMBER	DATE	TIME	LONG.
11849	8JA	22;08	286.77	11849	24JA	5;07	122.22
11850	7FE	15;52	316.70	11850	22FE	14;39	151.57
11851	9MR	9;34	346.14	11851	23MR	23;13	180.38
11852	8AP	2;02	15.03	11852	22AP	7;27	208.72
11853	7MY	16;18	43.43	11853	21MY	15;59	236.76
11854	6JN	4;08	71.53	11854	20JN	1;40	264.76
11855	5JL	13;58	99.57	11855	19JL	13;21	292.98
11856	3AU	22;47	127.84	11856	18AU	3;43	321.65
11857	2SE	7;31	156.52	11857	16SE	20;50	350.92
11858	1OC	16;56	185.73	11858	16OC	15;39	20.77
11859	31OC	3;21	215.44	11859	15NO	10;25	51.00
11860	29NO	14;57	245.48	11860	15DE	3;30	81.32
11861	29DE	3;53	275.60				

−41

NEW MOONS NUMBER	DATE	TIME	LONG.	FULL MOONS NUMBER	DATE	TIME	LONG.
				11861	13JA	18;04	111.42
11862	27JA	18;18	305.54	11862	12FE	6;07	141.04
11863	26FE	10;05	335.09	11863	13MR	16;00	170.09
11864	28MR	2;33	4.16	11864	12AP	0;11	198.58
11865	26AP	18;39	32.77	11865	11MY	7;22	226.66
11866	26MY	9;32	61.04	11866	9JN	14;27	254.56
11867	24JN	22;52	89.19	11867	8JL	22;38	282.56
11868	24JL	10;54	117.46	11868	7AU	9;03	310.93
11869	22AU	22;03	146.07	11869	5SE	22;34	339.86
11870	21SE	8;44	175.15	11870	5OC	15;14	9.43
11871	20OC	19;12	204.71	11871	4NO	10;08	39.54
11872	19NO	5;39	234.63	11872	4DE	5;40	69.94
11873	18DE	16;23	264.68				

--------NEW MOONS-------- / --------FULL MOONS------- (left half, years −40 to −35)

NUMBER	DATE	TIME	LONG.	NUMBER	DATE	TIME	LONG.
−40							
				11873	3JA	0;12	100.28
11874	17JA	3;47	294.61	11874	1FE	16;29	130.24
11875	15FE	16;09	324.20	11875	2MR	5;53	159.62
11876	16MR	5;31	353.34	11876	31MR	16;24	188.37
11877	14AP	19;39	22.01	11877	30AP	0;33	216.60
11878	14MY	10;17	50.36	11878	29MY	7;20	244.53
11879	13JN	1;09	78.56	11879	27JN	13;58	272.42
11880	12JL	16;02	106.86	11880	26JL	21;37	300.55
11881	11AU	6;37	135.45	11881	25AU	7;19	329.15
11882	9SE	20;28	164.47	11882	23SE	19;42	358.36
11883	9OC	9;16	193.96	11883	23OC	10;57	28.17
11884	7NO	21;00	223.83	11884	22NO	4;44	58.42
11885	7DE	7;58	253.88	11885	22DE	0;07	88.84
−39							
11886	5JA	18;32	283.85	11886	20JA	19;29	119.06
11887	4FE	4;57	313.52	11887	19FE	13;04	148.80
11888	5MR	15;23	342.72	11888	21MR	3;37	177.90
11889	4AP	2;06	11.43	11889	19AP	14;57	206.40
11890	3MY	13;33	39.76	11890	18MY	23;46	234.48
11891	2JN	2;18	67.91	11891	17JN	7;11	262.38
11892	1JL	16;43	96.12	11892	16JL	14;21	290.40
11893	31JL	8;37	124.63	11893	14AU	22;12	318.77
11894	30AU	1;12	153.57	11894	13SE	7;27	347.66
11895	28SE	17;23	183.01	11895	12OC	18;39	17.13
11896	28OC	8;24	212.88	11896	11NO	8;17	47.11
11897	26NO	21;58	242.99	11897	11DE	0;33	77.39
11898	26DE	10;08	273.09				
−38							
				11898	9JA	18;53	107.68
11899	24JA	20;58	302.91	11899	8FE	13;49	137.64
11900	23FE	6;34	332.25	11900	10MR	7;30	167.07
11901	24MR	15;17	1.06	11901	8AP	22;44	195.89
11902	22AP	23;48	29.40	11902	8MY	11;18	224.21
11903	22MY	9;05	57.47	11903	6JN	21;42	252.25
11904	20JN	20;09	85.53	11904	6JL	6;42	280.27
11905	20JL	9;37	113.82	11905	4AU	15;04	308.52
11906	19AU	1;24	142.56	11906	2SE	23;29	337.20
11907	17SE	18;48	171.86	11907	2OC	8;38	6.41
11908	17OC	12;44	201.68	11908	31OC	19;09	36.13
11909	16NO	6;06	231.87	11909	30NO	7;36	66.22
11910	15DE	21;57	262.14	11910	29DE	22;12	96.41
−37							
11911	14JA	11;36	292.18	11911	28JA	14;29	126.41
11912	12FE	22;47	321.76	11912	27FE	7;27	156.00
11913	14MR	7;49	350.76	11913	29MR	0;05	185.07
11914	12AP	15;28	19.22	11914	27AP	15;38	213.64
11915	11MY	22;43	47.30	11915	27MY	5;44	241.87
11916	10JN	6;42	75.24	11916	25JN	18;16	269.99
11917	9JL	16;20	103.30	11917	25JL	5;18	298.23
11918	8AU	4;20	131.74	11918	23AU	15;15	326.80
11919	6SE	19;00	160.73	11919	22SE	0;44	355.85
11920	6OC	12;14	190.34	11920	21OC	10;29	25.39
11921	5NO	7;18	220.47	11921	19NO	21;03	55.31
11922	5DE	2;41	250.86	11922	19DE	8;41	85.40
−36							
11923	3JA	20;30	281.17	11923	17JA	21;22	115.38
11924	2FE	11;24	311.06	11924	16FE	10;54	145.01
11925	2MR	23;08	340.35	11925	17MR	1;11	174.17
11926	1AP	8;16	9.04	11926	15AP	16;08	202.87
11927	30AP	15;48	37.24	11927	15MY	7;31	231.23
11928	29MY	22;44	65.17	11928	13JN	22;44	259.45
11929	28JN	6;00	93.09	11929	13JL	13;08	287.73
11930	27JL	14;27	121.26	11930	12AU	2;17	316.28
11931	26AU	0;57	149.90	11931	10SE	14;20	345.24
11932	24SE	14;17	179.16	11932	10OC	1;42	14.68
11933	24OC	6;52	209.04	11933	8NO	12;48	44.52
11934	23NO	2;06	239.36	11934	7DE	23;48	74.58
11935	22DE	22;12	269.80				
−35							
				11935	6JA	10;38	104.56
11936	21JA	16;55	299.99	11936	4FE	21;23	134.23
11937	20FE	8;49	329.64	11937	6MR	8;21	163.44
11938	21MR	21;36	358.66	11938	4AP	20;08	192.19
11939	20AP	7;44	27.10	11939	4MY	9;10	220.57
11940	19MY	15;58	55.15	11940	2JN	23;32	248.78
11941	17JN	23;09	83.05	11941	2JL	14;46	277.03
11942	17JL	6;09	111.06	11942	1AU	6;16	305.53
11943	15AU	13;58	139.44	11943	30AU	21;33	334.44
11944	13SE	23;36	168.36	11944	29SE	12;17	3.83
11945	13OC	12;00	197.89	11945	29OC	2;12	33.66
11946	12NO	3;30	227.96	11946	27NO	14;59	63.74
11947	11DE	21;32	258.31	11947	27DE	2;26	93.81

--------NEW MOONS-------- / --------FULL MOONS------- (right half, years −34 to −29)

NUMBER	DATE	TIME	LONG.	NUMBER	DATE	TIME	LONG.
−34							
11948	10JA	16;40	288.62	11948	25JA	12;37	123.59
11949	9FE	11;14	318.56	11949	23FE	21;58	152.92
11950	11MR	3;57	347.93	11950	25MR	7;10	181.73
11951	9AP	18;12	16.70	11951	23AP	16;56	210.12
11952	9MY	5;50	44.98	11952	23MY	3;52	238.25
11953	7JN	15;14	72.98	11953	21JN	16;22	266.36
11954	6JL	23;08	100.96	11954	21JL	6;38	294.70
11955	5AU	6;34	129.18	11955	19AU	22;35	323.46
11956	3SE	14;39	157.86	11956	18SE	15;48	352.76
11957	3OC	0;17	187.10	11957	18OC	9;17	22.58
11958	1NO	12;00	216.88	11958	17NO	1;42	52.73
11959	1DE	1;54	247.02	11959	16DE	16;04	82.93
11960	30DE	17;43	277.26				
−33							
				11960	15JA	4;08	112.91
11961	29JA	10;50	307.29	11961	13FE	14;15	142.43
11962	28FE	4;27	336.90	11962	14MR	23;03	171.41
11963	29MR	21;28	5.96	11963	13AP	7;11	199.88
11964	28AP	12;51	34.51	11964	12MY	15;18	227.99
11965	28MY	1;55	62.70	11965	11JN	0;11	255.96
11966	26JN	12;48	90.76	11966	10JL	10;42	284.06
11967	25JL	22;12	118.94	11967	8AU	23;42	312.55
11968	24AU	7;05	147.48	11968	7SE	15;37	341.60
11969	22SE	16;19	176.52	11969	7OC	9;58	11.27
11970	22OC	2;21	206.08	11970	6NO	5;12	41.42
11971	20NO	13;24	236.04	11971	5DE	23;25	71.77
11972	20DE	1;34	266.15				
−32							
				11972	4JA	15;20	101.99
11973	18JA	15;03	296.15	11973	3FE	4;34	131.80
11974	17FE	5;54	325.82	11974	3MR	15;22	161.05
11975	17MR	21;50	355.04	11975	2AP	0;11	189.71
11976	16AP	14;01	23.77	11976	1MY	7;39	217.90
11977	16MY	5;30	52.14	11977	30MY	14;34	245.83
11978	14JN	19;40	80.31	11978	28JN	22;01	273.76
11979	14JL	8;30	108.54	11979	28JL	7;12	301.96
11980	12AU	20;18	137.04	11980	26AU	19;10	330.68
11981	11SE	7;28	165.97	11981	25SE	10;24	0.03
11982	10OC	18;15	195.40	11982	25OC	4;28	29.98
11983	9NO	4;50	225.23	11983	23NO	23;59	60.32
11984	8DE	15;25	255.27	11984	23DE	19;13	90.72
−31							
11985	7JA	2;20	285.25	11985	22JA	12;43	120.85
11986	5FE	13;58	314.95	11986	21FE	3;31	150.45
11987	7MR	2;32	344.22	11987	22MR	15;21	179.41
11988	5AP	16;01	13.02	11988	21AP	0;31	207.80
11989	5MY	6;12	41.44	11989	20MY	7;48	235.81
11990	3JN	20;52	69.66	11990	18JN	14;21	263.68
11991	3JL	11;51	97.90	11991	17JL	21;24	291.70
11992	2AU	2;50	126.39	11992	16AU	6;04	320.12
11993	31AU	17;24	155.28	11993	14SE	17;09	349.11
11994	30SE	7;04	184.65	11994	14OC	7;03	18.72
11995	29OC	19;34	214.43	11995	12NO	23;40	48.84
11996	28NO	7;00	244.45	11996	12DE	18;22	79.23
11997	27DE	17;43	274.49				
−30							
				11997	11JA	13;53	109.55
11998	26JA	4;03	304.27	11998	10FE	8;26	139.47
11999	24FE	14;13	333.62	11999	12MR	0;29	168.80
12000	26MR	0;29	2.46	12000	10AP	13;19	197.50
12001	24AP	11;14	30.88	12001	9MY	23;16	225.70
12002	23MY	23;04	59.04	12002	8JN	7;16	253.65
12003	22JN	12;33	87.19	12003	7JL	14;32	281.60
12004	22JL	3;49	115.58	12004	5AU	22;04	309.84
12005	20AU	20;24	144.37	12005	4SE	6;41	338.54
12006	19SE	13;12	173.68	12006	3OC	16;58	7.82
12007	19OC	5;09	203.44	12007	2NO	5;24	37.64
12008	17NO	19;40	233.52	12008	1DE	20;19	67.83
12009	17DE	8;37	263.67	12009	31DE	13;36	98.13
−29							
12010	15JA	20;04	293.61	12010	30JA	8;15	128.22
12011	14FE	6;07	323.12	12011	1MR	2;33	157.83
12012	15MR	15;01	352.09	12012	30MR	18;58	186.86
12013	13AP	23;18	20.56	12013	29AP	8;47	215.34
12014	13MY	7;53	48.68	12014	28MY	20;14	243.46
12015	11JN	17;48	76.69	12015	27JN	5;57	271.47
12016	11JL	5;59	104.86	12016	26JL	14;44	299.64
12017	9AU	20;43	133.43	12017	24AU	23;15	328.17
12018	8SE	13;36	162.54	12018	23SE	8;08	357.21
12019	8OC	7;38	192.21	12019	22OC	18;01	26.77
12020	7NO	1;38	222.32	12020	21NO	5;29	56.75
12021	6DE	18;30	252.61	12021	20DE	18;56	86.91

```
--------NEW MOONS--------    --------FULL MOONS-------    --------NEW MOONS--------    --------FULL MOONS-------
NUMBER DATE  TIME   LONG.    NUMBER DATE  TIME   LONG.    NUMBER DATE  TIME   LONG.    NUMBER DATE  TIME   LONG.
                          -28                                                      -22
12022   5JA   9;21  282.78  12022  19JA  10;15  116.99                             12096  13JA   8;17  111.39
12023   3FE  21;43  312.55  12023  18FE   2;44  146.72  12097  27JA  12;16  305.68 12097  12FE   0;27  141.19
12024   4MR   7;38  341.74  12024  18MR  19;22  175.94  12098  26FE   0;04  335.07 12098  13MR  13;41  170.38
12025   2AP  15;41   10.37  12025  17AP  11;21  204.66  12099  27MR  12;48    3.99 12099  12AP   0;02  198.96
12026   1MY  22;52   38.54  12026  17MY   2;09  232.99  12100  26AP   2;24   32.50 12100  11MY   8;04  227.07
12027  31MY   6;14   66.48  12027  15JN  15;34  261.13  12101  25MY  16;42   60.76 12101   9JN  14;51  254.96
12028  29JN  14;52   94.46  12028  15JL   3;31  289.33  12102  24JN   7;32   88.97 12102   8JL  21;33  282.90
12029  29JL   1;38  122.74  12029  13AU  14;13  317.79  12103  23JL  22;43  117.36 12103   7AU   5;22  311.16
12030  27AU  15;01  151.52  12030  12SE   0;06  346.69  12104  22AU  13;50  146.12 12104   5SE  15;15  339.94
12031  26SE   7;08  180.92  12031  11OC   9;50   16.09  12105  21SE   4;19  175.36 12105   5OC   3;49    9.34
12032  26OC   1;34  210.90  12032   9NO  20;04   45.91  12106  20OC  17;41  205.04 12106   3NO  19;10   39.30
12033  24NO  21;08  241.25  12033   9DE   7;09   75.97  12107  19NO   5;48  235.03 12107   3DE  12;55   69.62
12034  24DE  16;01  271.64                              12108  18DE  16;52  265.09
                          -27                                                      -21
12035  23JA   8;28  301.71  12034   7JA  19;12  105.99  12109  17JA   3;16  294.98 12108   2JA   8;08   99.99
12036  21FE  21;40  331.22  12035   6FE   8;05  135.74  12110  15FE  13;19  324.47 12109   1FE   3;15  130.07
12037  23MR   7;53    0.11  12036   7MR  21;41  165.03  12111  16MR  23;14  353.47 12110   2MR  20;35  159.62
12038  21AP  16;00   28.45  12037   6AP  12;01  193.85  12112  15AP   9;25   21.99 12111   1AP  10;57  188.53
12039  20MY  23;03   56.44  12038   6MY   3;02  222.30  12113  14MY  20;24   50.19 12112  30AP  22;14  216.89
12040  19JN   6;01   84.34  12039   4JN  18;22  250.54  12114  13JN   8;52   78.30 12113  30MY   7;05  244.90
12041  18JL  13;47  112.39  12040   4JL   9;22  278.79  12115  12JL  23;15  106.57 12114  28JN  14;40  272.84
12042  16AU  23;14  140.85  12041   2AU  23;24  307.25  12116  11AU  15;25  135.22 12115  27JL  22;05  300.96
12043  15SE  11;09  169.88  12042   1SE  12;16  336.09  12117  10SE   8;31  164.37 12116  26AU   6;14  329.50
12044  15OC   2;11  199.55  12043   1OC   0;12    5.39  12118  10OC   1;18  194.02 12117  24SE  15;46  358.58
12045  13NO  20;17  229.74  12044  30OC  11;37   35.13  12119   8NO  16;50  224.04 12118  24OC   3;09   28.23
12046  13DE  16;14  260.19  12045  28NO  22;46   65.15  12120   8DE   6;41  254.21 12119  22NO  16;47   58.31
                           12046  28DE   9;40   95.19                             12120  22DE   8;51   88.58
                          -26                                                      -20
12047  12JA  11;50  290.50  12047  26JA  20;17  124.98  12121   6JA  18;53  284.26 12121  21JA   2;47  118.75
12048  11FE   5;10  320.36  12048  25FE   6;49  154.33  12122   5FE   5;31  313.93 12122  19FE  21;16  148.53
12049  12MR  19;23  349.60  12049  26MR  17;50  183.20  12123   5MR  14;47  343.60 12123  20MR  14;35  177.76
12050  11AP   6;41   18.23  12050  25AP   5;55  211.66  12124   3AP  23;04   11.69 12124  19AP   5;38  206.41
12051  10MY  15;45   46.39  12051  24MY  19;27  239.88  12125   3MY   7;11   39.89 12125  18MY  18;12  234.64
12052   8JN  23;22   74.32  12052  23JN  10;17  268.09  12126   1JN  16;08   67.96 12126  17JN   4;48  262.67
12053   8JL   6;25  102.27  12053  23JL   1;50  296.49  12127   1JL   3;01   95.96 12127  16JL  14;08  290.78
12054   6AU  13;48  130.50  12054  21AU  17;31  325.27  12128  30JL  16;29  124.36 12128  14AU  22;55  319.18
12055   4SE  22;32  159.23  12055  20SE   8;52  354.52  12129  29AU   8;31  153.28 12129  13SE   7;45  348.06
12056   4OC   9;34  188.55  12056  19OC  23;33   24.24  12130  28SE   2;19  182.78 12130  12OC  17;14   17.46
12057   2NO  23;33  218.44  12057  18NO  13;07   54.29  12131  27OC  20;43  212.78 12131  11NO   3;54   47.31
12058   2DE  16;25  248.72  12058  18DE   1;18   84.40  12132  26NO  14;27  243.06 12132  10DE  16;17   77.43
                                                        12133  26DE   6;29  273.32
                          -25                                                      -19
12059   1JA  11;06  279.07  12059  16JA  12;01  114.31                             12133   9JA   6;33  107.54
12060  31JA   5;57  309.15  12060  14FE  21;33  143.79  12134  24JA  20;05  303.26 12134   7FE  22;18  137.39
12061   1MR  23;33  338.73  12061  16MR   6;31  172.75  12135  23FE   7;04  332.66 12135   9MR  14;41  166.77
12062  31MR  14;58    7.70  12062  14AP  15;40  201.25  12136  24MR  15;49    1.47 12136   8AP   6;51  195.64
12063  30AP   3;52   36.13  12063  14MY   1;44  229.42  12137  22AP  23;09   29.76 12137   7MY  22;10  224.08
12064  29MY  14;21   64.21  12064  12JN  13;17  257.50  12138  22MY   6;10   57.74 12138   6JN  12;20  252.22
12065  27JN  23;00   92.18  12065  12JL   2;37  285.72  12139  20JN  13;59   85.67 12139   6JL   1;11  280.44
12066  27JL   6;42  120.31  12066  10AU  17;51  314.32  12140  19JL  23;35  113.81 12140   4AU  12;43  308.82
12067  25AU  14;30  148.82  12067   9SE  10;44  343.44  12141  18AU  11;40  142.39 12141   2SE  23;12  337.58
12068  23SE  23;26  177.87  12068   9OC   4;30   13.11  12142  17SE   2;32  171.57 12142   2OC   9;11    6.83
12069  23OC  10;10  207.48  12069   7NO  21;51   43.20  12143  16OC  20;02  201.37 12143  31OC  19;15   36.53
12070  21NO  23;00  237.52  12070   7DE  13;29   73.45  12144  15NO  15;21  231.63 12144  30NO   5;56   66.53
12071  21DE  13;49  267.74                              12145  15DE  10;53  262.05 12145  29DE  17;26   96.59
                          -24                                                      -18
12072  20JA   6;09  297.85  12071   6JA   2;42  103.54  12146  14JA   4;43  292.28 12146  28JA   5;42  126.43
12073  18FE  23;22  327.60  12072   4FE  13;39  133.24  12147  12FE  19;31  322.01 12147  26FE  18;40  155.86
12074  19MR  16;36  356.83  12073   4MR  22;53  162.40  12148  14MR   7;03  351.12 12148  28MR   8;19  184.81
12075  18AP   8;47   25.55  12074   3AP   7;05  191.02  12149  12AP  15;59   19.63 12149  26AP  22;46  213.35
12076  17MY  23;02   53.85  12075   2MY  14;56  219.21  12150  11MY  23;22   47.71 12150  26MY  13;53  241.63
12077  16JN  11;03   81.94  12076  31MY  23;12  247.19  12151  10JN   6;14   75.60 12151  25JN   5;11  269.86
12078  15JL  21;17  110.07  12077  30JN   8;43  275.20  12152   9JL  13;31  103.57 12152  24JL  19;57  298.24
12079  14AU   6;32  138.49  12078  29JL  20;23  303.52  12153   7AU  22;09  131.87 12153  23AU   9;42  326.96
12080  12SE  15;45  167.37  12079  28AU  10;55  332.36  12154   6SE   8;45  160.69 12154  21SE  22;21  356.14
12081  12OC   1;32  196.77  12080  27SE   4;20    1.83  12155   5OC  22;16  190.14 12155  21OC  10;14   25.76
12082  10NO  12;09  226.62  12081  26OC  23;33   31.85  12156   4NO  14;58  220.17 12156  19NO  21;39   55.73
12083   9DE  23;44  256.70  12082  25NO  18;39   62.19  12157   4DE  10;13  250.55 12157  19DE   8;42   85.79
                           12083  25DE  11;54   92.51
                          -23                                                      -17
12084   8JA  12;22  286.75  12084  24JA   2;27  122.49  12158   3JA   6;13  280.95 12158  17JA  19;20  115.68
12085   7FE   2;17  316.52  12085  22FE  14;20  151.94  12159   2FE   0;46  311.00 12159  16FE   5;39  145.18
12086   8MR  17;26  345.87  12086  23MR  23;57  180.79  12160   3MR  16;29  340.47 12160  17MR  16;04  174.18
12087   7AP   9;20   14.74  12087  22AP   7;53  209.12  12161   2AP   5;06    9.30 12161  16AP   3;16  202.74
12088   7MY   1;07   43.21  12088  21MY  14;52  237.10  12162   1MY  15;08   37.60 12162  15MY  15;50  231.00
12089   5JN  16;00   71.43  12089  19JN  21;52  265.00  12163  30MY  23;23   65.58 12163  14JN   5;57  259.17
12090   5JL   5;37   99.63  12090  19JL   6;02  293.07  12164  29JN   6;41   93.50 12164  13JL  21;16  287.48
12091   3AU  18;08  128.04  12091  17AU  16;33  321.58  12165  28JL  13;53  121.62 12165  12AU  13;09  316.13
12092   2SE   5;52  156.84  12092  16SE   6;14  350.70  12166  26AU  21;57  150.17 12166  11SE   5;00  345.24
12093   1OC  17;05  186.12  12093  15OC  23;06   20.46  12167  25SE   7;50  179.28 12167  10OC  20;21   14.84
12094  31OC   3;56  215.85  12094  14NO  18;11   50.70  12168  24OC  20;22  208.98 12168   9NO  10;45   44.83
12095  29NO  14;34  245.85  12095  14DE  13;48   81.13  12169  23NO  11;52  239.15 12169   8DE  23;48   74.97
12096  29DE   1;12  275.88                              12170  23DE   5;43  269.49
```

Left half — NEW MOONS

NUMBER	DATE	TIME	LONG.
−16			
12171	22JA	0;31	299.69
12172	20FE	18;42	329.45
12173	21MR	11;08	358.62
12174	20AP	1;14	27.23
12175	19MY	12;55	55.41
12176	17JN	22;29	83.41
12177	17JL	6;41	111.47
12178	15AU	14;29	139.85
12179	13SE	22;56	168.72
12180	13OC	8;50	198.14
12181	11NO	20;40	228.05
12182	11DE	10;28	258.23
−15			
12183	10JA	1;55	288.39
12184	8FE	18;32	318.26
12185	10MR	11;36	347.65
12186	9AP	4;15	16.53
12187	8MY	19;29	44.95
12188	7JN	8;42	73.11
12189	6JL	19;54	101.22
12190	5AU	5;45	129.53
12191	3SE	15;08	158.26
12192	3OC	0;48	187.51
12193	1NO	11;08	217.23
12194	30NO	22;15	247.26
12195	30DE	10;13	277.33
−14			
12196	28JA	23;15	307.20
12197	27FE	13;30	336.67
12198	29MR	4;49	5.67
12199	27AP	20;34	34.25
12200	27MY	11;54	62.53
12201	26JN	2;15	90.72
12202	25JL	15;30	119.06
12203	24AU	3;52	147.74
12204	22SE	15;37	176.88
12205	22OC	2;52	206.49
12206	20NO	13;43	236.44
12207	20DE	0;18	266.48
−13			
12208	18JA	10;58	296.37
12209	16FE	22;07	325.89
12210	18MR	10;05	354.94
12211	16AP	22;58	23.56
12212	16MY	12;44	51.86
12213	15JN	3;15	80.05
12214	14JL	18;23	108.36
12215	13AU	9;49	137.00
12216	12SE	1;01	166.10
12217	11OC	15;17	195.67
12218	10NO	4;14	225.60
12219	9DE	15;52	255.68
−12			
12220	8JA	2;32	285.65
12221	6FE	12;34	315.28
12222	6MR	22;17	344.43
12223	5AP	8;01	13.08
12224	4MY	18;17	41.35
12225	3JN	5;46	69.44
12226	2JL	19;06	97.62
12227	1AU	10;31	126.12
12228	31AU	3;30	155.10
12229	29SE	20;53	184.61
12230	29OC	13;24	214.55
12231	28NO	4;17	244.73
12232	27DE	17;21	274.86
−11			
12233	26JA	4;41	304.68
12234	24FE	14;27	334.01
12235	25MR	22;58	2.79
12236	24AP	6;51	31.09
12237	23MY	15;05	59.11
12238	22JN	0;47	87.11
12239	21JL	12;53	115.36
12240	20AU	3;45	144.08
12241	18SE	20;58	173.39
12242	18OC	15;27	203.25
12243	17NO	9;51	233.49
12244	17DE	2;57	263.81

Left half — FULL MOONS

NUMBER	DATE	TIME	LONG.
−16			
12170	7JA	11;14	104.98
12171	5FE	21;10	134.61
12172	6MR	6;08	163.74
12173	4AP	14;51	192.36
12174	4MY	0;09	220.60
12175	2JN	10;44	248.66
12176	1JL	23;04	276.79
12177	31JL	13;24	305.24
12178	30AU	5;40	334.18
12179	28SE	23;23	3.68
12180	28OC	17;22	33.68
12181	27NO	10;11	63.93
12182	27DE	0;43	94.12
−15			
12183	25JA	12;42	123.98
12184	23FE	22;35	153.33
12185	25MR	7;02	182.11
12186	23AP	14;49	210.42
12187	22MY	22;38	238.43
12188	21JN	7;20	266.39
12189	20JL	17;47	294.57
12190	19AU	6;53	323.20
12191	17SE	23;03	352.45
12192	17OC	17;44	22.31
12193	16NO	13;18	52.59
12194	16DE	7;44	82.97
−14			
12195	14JA	23;41	113.11
12196	13FE	12;48	142.76
12197	14MR	23;22	171.81
12198	13AP	7;57	200.30
12199	12MY	15;13	228.37
12200	10JN	22;02	256.26
12201	10JL	5;28	284.24
12202	8AU	14;43	312.57
12203	7SE	2;49	341.47
12204	6OC	18;15	11.00
12205	5NO	12;28	41.11
12206	5DE	8;05	71.51
−13			
12207	4JA	3;19	101.87
12208	2FE	20;41	131.86
12209	4MR	11;19	161.27
12210	2AP	22;59	190.05
12211	2MY	8;02	218.30
12212	31MY	15;17	246.24
12213	29JN	21;55	274.14
12214	29JL	5;07	302.27
12215	27AU	13;58	330.85
12216	26SE	1;15	0.03
12217	25OC	15;18	29.81
12218	24NO	7;55	60.03
12219	24DE	2;28	90.40
−12			
12220	22JA	21;42	120.61
12221	21FE	15;59	150.36
12222	22MR	7;48	179.49
12223	20AP	20;23	208.03
12224	20MY	6;30	236.14
12225	18JN	14;39	264.08
12226	17JL	22;09	292.12
12227	16AU	6;00	320.51
12228	14SE	14;56	349.41
12229	14OC	1;27	18.87
12230	12NO	13;58	48.82
12231	12DE	4;45	79.04
−11			
12232	10JA	21;41	109.25
12233	9FE	15;51	139.17
12234	11MR	9;43	168.58
12235	10AP	1;50	197.43
12236	9MY	15;36	225.78
12237	8JN	3;11	253.87
12238	7JL	13;13	281.94
12239	5AU	22;25	310.24
12240	4SE	7;22	338.96
12241	3OC	16;39	8.20
12242	2NO	2;46	37.92
12243	1DE	14;15	67.97
12244	31DE	3;27	98.09

Right half — NEW MOONS

NUMBER	DATE	TIME	LONG.
−10			
12245	15JA	17;50	293.90
12246	14FE	6;03	323.50
12247	15MR	15;42	352.51
12248	13AP	23;29	20.95
12249	13MY	6;24	49.00
12250	11JN	13;36	76.91
12251	10JL	22;10	104.94
12252	9AU	8;59	133.34
12253	7SE	22;31	162.31
12254	7OC	14;52	191.90
12255	6NO	9;33	222.03
12256	6DE	5;17	252.44
−9			
12257	5JA	0;12	282.78
12258	3FE	16;34	312.72
12259	5MR	5;37	342.05
12260	3AP	15;39	10.75
12261	2MY	23;35	38.95
12262	1JN	6;33	66.88
12263	30JN	13;32	94.79
12264	29JL	21;25	122.95
12265	28AU	7;02	151.58
12266	26SE	19;00	180.81
12267	26OC	10;18	210.64
12268	25NO	4;26	240.93
12269	25DE	0;17	271.36
−8			
12270	23JA	19;43	301.56
12271	22FE	12;50	331.25
12272	23MR	2;52	0.29
12273	21AP	14;04	28.76
12274	20MY	23;07	56.83
12275	19JN	6;50	84.75
12276	18JL	14;05	112.79
12277	16AU	21;44	141.18
12278	15SE	6;43	170.09
12279	14OC	17;57	199.59
12280	13NO	8;00	229.61
12281	13DE	0;43	259.92
−7			
12282	11JA	19;05	290.18
12283	10FE	13;32	320.10
12284	12MR	6;46	349.47
12285	10AP	21;58	18.26
12286	10MY	10;51	46.58
12287	8JN	21;29	74.63
12288	8JL	6;25	102.65
12289	6AU	14;28	130.92
12290	4SE	22;40	159.62
12291	4OC	7;55	188.87
12292	2NO	18;51	218.63
12293	2DE	7;40	248.74
12294	31DE	22;12	278.91
−6			
12295	30JA	14;04	308.88
12296	1MR	6;43	338.42
12297	30MR	23;28	7.46
12298	29AP	15;23	36.02
12299	29MY	5;40	64.25
12300	27JN	17;59	92.37
12301	27JL	4;38	120.61
12302	25AU	14;24	149.21
12303	24SE	0;05	178.29
12304	23OC	10;14	207.87
12305	21NO	21;01	237.83
12306	21DE	8;30	267.91
−5			
12307	19JA	20;47	297.85
12308	18FE	10;08	327.44
12309	20MR	0;39	356.57
12310	18AP	16;01	25.26
12311	18MY	7;31	53.62
12312	16JN	22;27	81.82
12313	16JL	12;24	110.10
12314	15AU	1;27	138.67
12315	13SE	13;47	167.68
12316	13OC	1;33	197.16
12317	11NO	12;45	227.03
12318	10DE	23;29	257.08

Right half — FULL MOONS

NUMBER	DATE	TIME	LONG.
−10			
12245	29JA	18;18	128.02
12246	28FE	10;11	157.55
12247	30MR	2;17	186.57
12248	28AP	17;54	215.13
12249	28MY	8;39	243.39
12250	26JN	22;17	271.55
12251	26JL	10;42	299.86
12252	24AU	21;56	328.50
12253	23SE	8;22	357.61
12254	22OC	18;31	27.18
12255	21NO	4;57	57.11
12256	20DE	16;00	87.18
−9			
12257	19JA	3;44	117.10
12258	17FE	16;05	146.66
12259	19MR	5;04	175.75
12260	17AP	18;50	204.38
12261	17MY	9;28	232.71
12262	16JN	0;44	260.93
12263	15JL	16;00	289.25
12264	14AU	6;33	317.87
12265	12SE	20;03	346.91
12266	12OC	8;33	16.42
12267	10NO	20;22	46.31
12268	10DE	7;41	76.38
−8			
12269	8JA	18;28	106.35
12270	7FE	4;44	135.98
12271	7MR	14;46	165.14
12272	6AP	1;13	193.80
12273	5MY	12;47	222.12
12274	4JN	1;58	250.28
12275	3JL	16;44	278.51
12276	2AU	8;33	307.04
12277	1SE	0;44	336.00
12278	30SE	16;41	5.46
12279	30OC	7;53	35.36
12280	28NO	21;50	65.50
12281	28DE	10;07	95.60
−7			
12282	26JA	20;40	125.38
12283	25FE	5;52	154.68
12284	26MR	14;23	183.44
12285	24AP	23;05	211.77
12286	24MY	8;46	239.84
12287	22JN	20;04	267.91
12288	22JL	9;22	296.22
12289	21AU	0;49	324.97
12290	19SE	18;07	354.30
12291	19OC	12;23	24.16
12292	18NO	6;11	54.38
12293	17DE	22;03	84.65
−6			
12294	16JA	11;18	114.66
12295	14FE	22;03	144.20
12296	16MR	8;05	173.16
12297	14AP	14;50	201.60
12298	13MY	22;23	229.67
12299	12JN	6;27	257.61
12300	11JL	15;52	285.67
12301	10AU	3;35	314.13
12302	8SE	18;18	343.15
12303	8OC	12;00	12.81
12304	7NO	7;33	42.99
12305	7DE	2;53	73.39
−5			
12306	5JA	20;13	103.66
12307	4FE	10;41	133.50
12308	5MR	22;23	162.76
12309	4AP	7;46	191.43
12310	3MY	15;30	219.62
12311	1JN	22;22	247.54
12312	1JL	5;20	275.45
12313	30JL	13;34	303.64
12314	29AU	0;13	332.31
12315	27SE	14;04	1.63
12316	27OC	7;06	31.55
12317	26NO	2;16	61.88
12318	25DE	21;53	92.30

NEW MOONS / FULL MOONS / NEW MOONS / FULL MOONS

NUMBER	DATE	TIME	LONG.		NUMBER	DATE	TIME	LONG.
				-4				
12319	9JA	9:58	287.03		12319	24JA	16:16	122.45
12320	7FE	20:37	316.68		12320	23FE	8:15	152.08
12321	8MR	7:52	345.86		12321	23MR	21:19	181.07
12322	6AP	20:00	14.59		12322	22AP	7:32	209.49
12323	6MY	9:05	42.95		12323	21MY	15:32	237.52
12324	4JN	23:06	71.15		12324	19JN	22:22	265.40
12325	4JL	13:59	99.39		12325	19JL	5:12	293.43
12326	3AU	5:30	127.91		12326	17AU	13:14	321.84
12327	1SE	21:11	156.87		12327	15SE	23:21	350.81
12328	1OC	12:18	186.31		12328	15OC	12:05	20.38
12329	31OC	2:12	216.17		12329	14NO	3:29	50.47
12330	29NO	14:37	246.24		12330	13DE	21:07	80.81
12331	29DE	1:43	276.29					
				-3				
12332	27JA	11:55	306.05		12331	12JA	16:03	111.10
12333	25FE	21:33	335.35		12332	11FE	10:51	141.02
12334	27MR	6:59	4.14		12333	13MR	3:54	170.37
12335	25AP	16:40	32.50		12334	11AP	18:08	199.10
12336	25MY	3:15	60.61		12335	11MY	5:24	227.35
12337	23JN	15:29	88.71		12336	9JN	14:23	255.33
12338	23JL	5:54	117.07		12337	8JL	22:11	283.32
12339	21AU	22:22	145.88		12338	7AU	5:54	311.57
12340	20SE	15:59	175.23		12339	5SE	14:23	340.30
12341	20OC	9:20	205.08		12340	5OC	0:12	9.58
12342	19NO	1:18	235.22		12341	3NO	11:45	39.37
12343	18DE	15:23	265.42		12342	3DE	1:20	69.51
				-2				
12344	17JA	3:33	295.38		12343	1JA	17:06	99.74
12345	15FE	13:57	324.89		12344	31JA	10:35	129.76
12346	16MR	22:51	353.84		12345	2MR	4:33	159.35
12347	15AP	6:46	22.27		12346	31MR	21:29	188.38
12348	14MY	14:31	50.34		12347	30AP	12:23	216.89
12349	12JN	23:14	78.31		12348	30MY	1:02	245.05
12350	12JL	9:59	106.43		12349	28JN	11:54	273.11
12351	10AU	23:31	134.96		12350	27JL	21:38	301.33
12352	9SE	15:47	164.07		12351	26AU	6:53	329.91
12353	9OC	9:59	193.76		12352	24SE	16:09	358.99
12354	8NO	4:47	223.92		12353	24OC	1:56	28.56
12355	7DE	22:48	254.26		12354	22NO	12:43	58.52
					12355	22DE	0:58	88.63
				-1				
12356	6JA	14:56	284.48		12356	20JA	14:49	118.63
12357	5FE	4:27	314.27		12357	19FE	5:59	148.29
12358	6MR	15:13	343.48		12358	20MR	21:47	177.46
12359	4AP	23:41	12.10		12359	19AP	13:29	206.15
12360	4MY	6:47	40.26		12360	19MY	4:37	234.49
12361	2JN	13:36	68.17		12361	17JN	18:54	262.67
12362	1JL	21:20	96.12		12362	17JL	8:08	290.92
12363	31JL	6:58	124.36		12363	15AU	20:13	319.46
12364	29AU	19:10	153.12		12364	14SE	7:16	348.43
12365	28SE	10:14	182.50		12365	13OC	17:43	17.87
12366	28OC	3:57	212.46		12366	12NO	4:06	47.71
12367	26NO	23:26	242.81		12367	11DE	14:50	77.76
12368	26DE	19:02	273.23					
				0				
12369	25JA	12:49	303.34		12368	10JA	2:07	107.75
12370	24FE	3:28	332.90		12369	8FE	13:56	137.43
12371	24MR	14:49	1.81		12370	9MR	2:18	166.65
12372	22AP	23:36	30.16		12371	7AP	15:21	195.40
12373	22MY	6:53	58.16		12372	7MY	5:19	223.79
12374	20JN	13:44	86.04		12373	5JN	20:14	252.01
12375	19JL	21:08	114.09		12374	5JL	11:40	280.29
12376	18AU	5:52	142.54		12375	4AU	2:52	308.81
12377	16SE	16:44	171.56		12376	2SE	17:13	337.72
12378	16OC	6:24	201.18		12377	2OC	6:29	7.10
12379	14NO	23:09	231.33		12378	31OC	18:51	36.90
12380	14DE	18:20	261.74		12379	30NO	6:31	66.95
					12380	29DE	17:34	96.99
				1				
12381	13JA	14:09	292.06		12381	28JA	3:57	126.75
12382	12FE	8:28	321.94		12382	26FE	13:49	156.05
12383	13MR	23:57	351.21		12383	27MR	23:41	184.85
12384	12AP	12:26	19.87		12384	26AP	10:21	213.24
12385	11MY	22:27	48.06		12385	25MY	22:30	241.40
12386	10JN	6:46	76.01		12386	24JN	12:26	269.58
12387	9JL	14:15	103.99		12387	24JL	3:52	297.98
12388	7AU	21:44	132.24		12388	22AU	20:09	326.80
12389	6SE	6:05	160.97		12389	21SE	12:35	356.12
12390	5OC	16:13	190.28		12390	21OC	4:31	25.91
12391	4NO	4:52	220.13		12391	19NO	19:21	56.02
12392	3DE	20:17	250.35		12392	19DE	8:35	86.18

NUMBER	DATE	TIME	LONG.		NUMBER	DATE	TIME	LONG.
12393	2JA	13:51	280.65	**2**	12393	17JA	19:57	116.10
12394	1FE	8:14	310.70		12394	16FE	5:36	145.56
12395	3MR	2:00	340.26		12395	17MR	14:10	174.49
12396	1AP	18:08	9.24		12396	15AP	22:27	202.93
12397	1MY	8:08	37.70		12397	15MY	7:20	231.04
12398	30MY	19:55	65.83		12398	13JN	17:38	259.07
12399	29JN	5:45	93.85		12399	13JL	5:52	287.25
12400	28JL	14:19	122.03		12400	11AU	20:19	315.84
12401	26AU	22:31	150.58		12401	10SE	12:55	344.97
12402	25SE	7:20	179.65		12402	10OC	7:05	14.68
12403	24OC	17:30	209.25		12403	9NO	1:32	44.82
12404	23NO	5:24	239.26		12404	8DE	18:40	75.14
12405	22DE	19:01	269.43					
				3				
12406	21JA	10:03	299.47		12405	7JA	9:17	105.28
12407	20FE	2:06	329.15		12406	5FE	21:09	135.00
12408	21MR	18:37	358.34		12407	7MR	6:46	164.15
12409	20AP	10:54	27.04		12408	5AP	14:54	192.75
12410	20MY	2:03	55.37		12409	4MY	22:23	220.91
12411	18JN	15:27	83.51		12410	3JN	5:58	248.85
12412	18JL	3:03	111.71		12411	2JL	14:33	276.84
12413	16AU	13:25	140.19		12412	1AU	1:01	305.12
12414	14SE	23:18	169.12		12413	30AU	14:15	333.93
12415	14OC	9:23	198.56		12414	29SE	6:38	3.37
12416	12NO	19:58	228.41		12415	29OC	1:37	33.40
12417	12DE	7:06	258.48		12416	27NO	21:27	63.78
					12417	27DE	16:00	94.14
				4				
12418	10JA	18:49	288.48		12418	26JA	7:55	124.17
12419	9FE	7:21	318.18		12419	24FE	20:53	153.65
12420	9MR	20:59	347.44		12420	25MR	7:14	182.51
12421	8AP	11:41	16.25		12421	23AP	15:37	210.83
12422	8MY	3:02	44.68		12422	22MY	22:45	238.82
12423	6JN	18:17	72.92		12423	21JN	5:31	266.70
12424	6JL	8:52	101.16		12424	20JL	13:00	294.76
12425	4AU	22:35	129.63		12425	18AU	22:23	323.25
12426	3SE	11:33	158.50		12426	17SE	10:39	352.33
12427	2OC	23:53	187.86		12427	17OC	2:14	22.05
12428	1NO	11:35	217.64		12428	15NO	20:34	52.27
12429	30NO	22:37	247.66		12429	15DE	16:11	82.70
12430	30DE	9:09	277.68					
				5				
12431	28JA	19:30	307.43		12430	14JA	11:18	112.98
12432	27FE	6:09	336.75		12431	13FE	4:29	142.80
12433	28MR	17:31	5.60		12432	14MR	18:57	172.02
12434	27AP	5:52	34.05		12433	13AP	6:29	200.62
12435	26MY	19:14	62.26		12434	12MY	15:28	228.77
12436	25JN	9:39	90.45		12435	10JN	22:46	256.68
12437	25JL	1:00	118.86		12436	10JL	5:31	284.63
12438	23AU	16:56	147.68		12437	8AU	12:56	312.89
12439	22SE	8:44	176.98		12438	6SE	22:02	341.66
12440	21OC	23:36	206.75		12439	6OC	9:31	11.03
12441	20NO	12:57	236.80		12440	4NO	23:39	40.95
12442	20DE	0:44	266.89		12441	4DE	16:13	71.23
				6				
12443	18JA	11:16	296.77		12442	3JA	10:31	101.55
12444	16FE	20:58	326.23		12443	2FE	5:24	131.61
12445	18MR	6:14	355.17		12444	3MR	23:20	161.16
12446	16AP	15:29	23.64		12445	2AP	14:57	190.11
12447	16MY	1:19	51.79		12446	2MY	3:38	218.51
12448	14JN	12:30	79.85		12447	31MY	13:41	246.57
12449	14JL	1:45	108.08		12448	29JN	22:03	274.53
12450	12AU	17:21	136.72		12449	29JL	5:51	302.68
12451	11SE	10:46	165.90		12450	27AU	14:03	331.24
12452	11OC	4:42	195.61		12451	25SE	23:19	0.34
12453	9NO	21:42	225.71		12452	25OC	10:03	29.98
12454	9DE	12:53	255.94		12453	23NO	22:35	60.02
					12454	23DE	13:10	90.23
				7				
12455	8JA	2:01	286.02		12455	22JA	5:40	120.32
12456	6FE	13:12	315.70		12456	20FE	23:19	150.05
12457	7MR	22:39	344.83		12457	22MR	16:43	179.26
12458	6AP	6:48	13.42		12458	21AP	8:35	207.94
12459	5MY	14:20	41.58		12459	20MY	22:20	236.20
12460	3JN	22:18	69.53		12460	19JN	10:07	264.29
12461	3JL	7:50	97.56		12461	18JL	20:32	292.44
12462	1AU	19:56	125.92		12462	17AU	6:11	320.90
12463	31AU	10:58	154.81		12463	15SE	15:37	349.82
12464	30SE	4:31	184.32		12464	15OC	1:16	19.25
12465	29OC	23:22	214.35		12465	13NO	11:35	49.11
12466	28NO	18:05	244.68		12466	12DE	23:02	79.18
12467	28DE	11:21	274.99					

```
        --------NEW MOONS--------        --------FULL MOONS-------        --------NEW MOONS--------        --------FULL MOONS-------
        NUMBER  DATE   TIME   LONG.      NUMBER  DATE   TIME   LONG.      NUMBER  DATE   TIME   LONG.      NUMBER  DATE   TIME   LONG.
```

8

NEW NUMBER	DATE	TIME	LONG.	FULL NUMBER	DATE	TIME	LONG.
				12467	11JA	11;55	109.23
12468	27JA	2;12	304.97	12468	10FE	2;14	138.99
12469	25FE	14;14	334.39	12469	10MR	17;30	168.31
12470	25MR	23;38	3.20	12470	9AP	9;03	197.14
12471	24AP	7;10	31.49	12471	9MY	0;22	225.57
12472	23MY	13;54	59.45	12472	7JN	15;07	253.78
12473	21JN	21;00	87.35	12473	7JL	5;02	282.00
12474	21JL	5;35	115.46	12474	5AU	17;57	310.44
12475	19AU	16;29	144.01	12475	4SE	5;47	339.28
12476	18SE	6;12	173.17	12476	3OC	16;45	8.59
12477	17OC	22;46	202.94	12477	2NO	3;17	38.33
12478	16NO	17;36	233.19	12478	1DE	13;52	68.34
12479	16DE	13;25	263.63	12479	31DE	0;48	98.37

9

NEW NUMBER	DATE	TIME	LONG.	FULL NUMBER	DATE	TIME	LONG.
12480	15JA	8;17	293.89	12480	29JA	12;10	128.16
12481	14FE	0;31	323.66	12481	27FE	23;59	157.51
12482	15MR	13;24	352.80	12482	29MR	12;21	186.39
12483	13AP	23;16	21.33	12483	28AP	1;34	214.86
12484	13MY	7;06	49.42	12484	27MY	15;52	243.11
12485	11JN	14;03	77.31	12485	26JN	7;07	271.34
12486	10JL	21;07	105.28	12486	25JL	22;42	299.76
12487	9AU	5;10	133.57	12487	24AU	13;49	328.56
12488	7SE	14;59	162.38	12488	23SE	3;57	357.81
12489	7OC	3;16	191.80	12489	22OC	17;00	27.51
12490	5NO	18;32	221.78	12490	21NO	5;10	57.51
12491	5DE	12;37	252.12	12491	20DE	16;34	87.59

10

NEW NUMBER	DATE	TIME	LONG.	FULL NUMBER	DATE	TIME	LONG.
12492	4JA	8;17	282.50	12492	19JA	3;11	117.47
12493	3FE	3;27	312.56	12493	17FE	13;05	146.92
12494	4MR	20;19	342.05	12494	18MR	22;37	175.87
12495	3AP	10;11	10.92	12495	17AP	8;31	204.35
12496	2MY	21;19	39.25	12496	16MY	19;37	232.55
12497	1JN	6;25	67.26	12497	15JN	8;32	260.67
12498	30JN	14;19	95.21	12498	14JL	23;17	288.97
12499	29JL	21;49	123.36	12499	13AU	15;23	317.64
12500	28AU	5;47	151.92	12500	12SE	8;05	346.81
12501	26SE	15;03	181.03	12501	12OC	0;38	16.48
12502	26OC	2;27	210.70	12502	10NO	16;20	46.53
12503	24NO	16;30	240.81	12503	10DE	6;33	76.72
12504	24DE	9;00	271.10				

11

NEW NUMBER	DATE	TIME	LONG.	FULL NUMBER	DATE	TIME	LONG.
				12504	8JA	18;52	106.76
12505	23JA	2;57	301.25	12505	7FE	5;13	136.39
12506	21FE	20;58	330.98	12506	8MR	14;03	165.49
12507	23MR	13;49	0.15	12507	6AP	22;09	194.06
12508	22AP	4;50	28.78	12508	6MY	6;26	222.25
12509	21MY	17;44	57.00	12509	4JN	15;48	250.26
12510	20JN	4;36	85.05	12510	4JL	2;56	278.35
12511	19JL	13;52	113.17	12511	2AU	16;16	306.77
12512	17AU	22;21	141.59	12512	1SE	7;57	335.70
12513	16SE	6;57	170.49	12513	1OC	1;40	5.23
12514	15OC	16;32	199.92	12514	30OC	20;23	35.27
12515	14NO	3;37	229.81	12515	29NO	14;32	65.58
12516	13DE	16;21	259.95	12516	29DE	6;35	95.84

12

NEW NUMBER	DATE	TIME	LONG.	FULL NUMBER	DATE	TIME	LONG.
12517	12JA	6;32	290.04	12517	27JA	19;46	125.73
12518	10FE	21;51	319.84	12518	26FE	6;19	155.08
12519	11MR	13;55	349.17	12519	26MR	14;56	183.85
12520	10AP	6;11	18.02	12520	24AP	22;30	212.13
12521	9MY	21;54	46.45	12521	24MY	5;49	240.11
12522	8JN	12;17	74.65	12522	22JN	13;49	268.04
12523	8JL	0;56	102.82	12523	21JL	23;08	296.19
12524	6AU	12;04	131.21	12524	20AU	10;57	324.80
12525	4SE	22;22	160.00	12525	19SE	1;51	354.01
12526	4OC	8;32	189.28	12526	18OC	19;49	23.86
12527	2NO	19;01	219.03	12527	17NO	15;36	54.15
12528	2DE	5;55	249.05	12528	17DE	11;06	84.58
12529	31DE	17;14	279.09				

13

NEW NUMBER	DATE	TIME	LONG.	FULL NUMBER	DATE	TIME	LONG.
				12529	16JA	4;25	114.76
12530	30JA	5;07	308.89	12530	14FE	18;47	144.45
12531	28FE	17;52	338.28	12531	16MR	6;17	173.52
12532	30MR	7;45	7.21	12532	14AP	15;28	202.01
12533	28AP	22;16	35.73	12533	13MY	23;03	230.08
12534	28MY	13;53	64.01	12534	12JN	5;52	257.97
12535	27JN	4;55	92.23	12535	11JL	12;52	285.94
12536	26JL	19;16	120.62	12536	9AU	21;14	314.26
12537	25AU	8;53	149.37	12537	8SE	8;02	343.12
12538	23SE	21;50	178.59	12538	7OC	22;03	12.62
12539	23OC	10;06	208.26	12539	6NO	15;12	42.68
12540	21NO	21;37	238.24	12540	6DE	10;24	73.08
12541	21DE	8;23	268.29				

14

NEW NUMBER	DATE	TIME	LONG.	FULL NUMBER	DATE	TIME	LONG.
				12541	5JA	5;54	103.44
12542	19JA	18;39	298.15	12542	4FE	0;05	133.45
12543	18FE	4;52	327.61	12543	5MR	15;52	162.88
12544	19MR	15;33	356.59	12544	4AP	4;47	191.70
12545	18AP	3;07	25.13	12545	3MY	14;56	219.98
12546	17MY	15;45	53.38	12546	1JN	22;58	247.95
12547	16JN	5;33	81.54	12547	1JL	5;55	275.86
12548	15JL	20;30	109.85	12548	30JL	12;58	304.00
12549	14AU	12;24	138.53	12549	28AU	21;14	332.58
12550	13SE	4;39	167.69	12550	27SE	7;35	1.75
12551	12OC	20;24	197.34	12551	26OC	20;28	31.49
12552	11NO	10;46	227.34	12552	25NO	11;51	61.66
12553	10DE	23;25	257.46	12553	25DE	5;17	91.99

15

NEW NUMBER	DATE	TIME	LONG.	FULL NUMBER	DATE	TIME	LONG.
12554	9JA	10;30	287.45	12554	23JA	23;52	122.15
12555	7FE	20;27	317.06	12555	22FE	18;17	151.89
12556	9MR	5;41	346.16	12556	24MR	11;03	181.04
12557	7AP	14;38	14.76	12557	23AP	1;10	209.62
12558	6MY	23;52	42.97	12558	22MY	12;28	237.78
12559	5JN	10;07	71.02	12559	20JN	21;39	265.76
12560	4JL	22;12	99.15	12560	20JL	5;45	293.83
12561	3AU	12;41	127.62	12561	18AU	13;50	322.25
12562	2SE	5;28	156.61	12562	16SE	22;41	351.17
12563	1OC	23;35	186.17	12563	16OC	8;47	20.64
12564	31OC	17;27	216.19	12564	14NO	20;25	50.56
12565	30NO	9;48	246.43	12565	14DE	9;53	80.72
12566	30DE	0;01	276.60				

16

NEW NUMBER	DATE	TIME	LONG.	FULL NUMBER	DATE	TIME	LONG.
				12566	13JA	1;18	110.86
12567	28JA	12;06	306.44	12567	11FE	18;14	140.71
12568	26FE	22;15	335.77	12568	12MR	11;41	170.09
12569	27MR	6;49	4.52	12569	11AP	4;15	198.93
12570	25AP	14;23	32.79	12570	10MY	19;02	227.32
12571	24MY	21;51	60.78	12571	9JN	7;50	255.45
12572	23JN	6;23	88.74	12572	8JL	19;02	283.58
12573	22JL	17;05	116.94	12573	7AU	5;13	311.93
12574	21AU	6;43	145.63	12574	5SE	14;57	340.71
12575	19SE	23;14	174.93	12575	5OC	0;39	9.99
12576	19OC	17;47	204.81	12576	3NO	10;43	39.72
12577	18NO	12;54	235.09	12577	2DE	21;34	69.74
12578	18DE	7;07	265.46				

17

NEW NUMBER	DATE	TIME	LONG.	FULL NUMBER	DATE	TIME	LONG.
				12578	1JA	9;36	99.81
12579	16JA	23;16	295.59	12579	30JA	23;00	129.67
12580	15FE	12;39	325.22	12580	1MR	13;33	159.12
12581	16MR	23;12	354.24	12581	31MR	4;44	188.09
12582	15AP	7;27	22.68	12582	29AP	20;01	216.62
12583	14MY	14;20	50.72	12583	29MY	11;02	244.88
12584	12JN	21;04	78.60	12584	28JN	1;29	273.08
12585	12JL	4;47	106.61	12585	27JL	15;09	301.45
12586	10AU	14;29	134.98	12586	26AU	3;49	330.17
12587	9SE	2;51	163.92	12587	24SE	15;27	359.35
12588	8OC	18;06	193.49	12588	24OC	2;22	28.97
12589	7NO	11;59	223.59	12589	22NO	13;00	58.92
12590	7DE	7;33	254.00	12590	21DE	23;43	88.97

18

NEW NUMBER	DATE	TIME	LONG.	FULL NUMBER	DATE	TIME	LONG.
12591	6JA	3;07	284.37	12591	20JA	10;44	118.85
12592	4FE	20;46	314.34	12592	18FE	22;04	148.35
12593	6MR	11;15	343.71	12593	20MR	9;49	177.36
12594	4AP	22;27	12.44	12594	18AP	22;17	205.93
12595	4MY	7;07	40.66	12595	18MY	11;50	234.20
12596	2JN	14;22	68.59	12596	17JN	2;35	262.40
12597	1JL	21;17	96.50	12597	16JL	18;12	290.75
12598	31JL	4;51	124.67	12598	15AU	9;53	319.43
12599	29AU	13;48	153.29	12599	14SE	0;51	348.55
12600	28SE	0;52	182.49	12600	13OC	14;44	18.13
12601	27OC	14;40	212.29	12601	12NO	3;32	48.08
12602	26NO	7;24	242.52	12602	11DE	15;24	78.17
12603	26DE	2;25	272.91				

19

NEW NUMBER	DATE	TIME	LONG.	FULL NUMBER	DATE	TIME	LONG.
				12603	10JA	2;22	108.14
12604	24JA	21;57	303.11	12604	8FE	12;26	137.75
12605	23FE	15;59	332.81	12605	9MR	21;51	166.85
12606	25MR	7;16	1.89	12606	8AP	7;13	195.46
12607	23AP	19;38	30.39	12607	7MY	17;23	223.70
12608	23MY	5;41	58.49	12608	6JN	5;10	251.80
12609	21JN	14;09	86.45	12609	5JL	18;59	280.01
12610	20JL	21;54	114.51	12610	4AU	10;35	308.54
12611	19AU	5;41	142.92	12611	3SE	3;18	337.54
12612	17SE	14;21	171.85	12612	2OC	20;18	7.07
12613	17OC	0;43	201.33	12613	1NO	12;46	37.03
12614	15NO	13;26	231.31	12614	1DE	3;59	67.23
12615	15DE	4;42	261.55	12615	30DE	17;19	97.37

20 / 26

NEW MOONS				FULL MOONS				NEW MOONS				FULL MOONS			
12616	13JA	21;54	291.76	12616	29JA	4;33	127.16	12690	7JA	19;39	286.14	12690	21JA	20;19	120.32
12617	12FE	15;49	321.64	12617	27FE	13;55	156.44	12691	6FE	10;25	315.97	12691	20FE	10;03	149.89
12618	13MR	9;08	350.99	12618	27MR	22;05	185.17	12692	7MR	22;16	345.20	12692	22MR	0;41	179.00
12619	12AP	1;00	19.80	12619	26AP	5;58	213.45	12693	6AP	7;27	13.83	12693	20AP	15;43	207.65
12620	11MY	14;56	48.14	12620	25MY	14;31	241.47	12694	5MY	14;47	41.98	12694	20MY	6;46	235.97
12621	10JN	2;52	76.24	12621	24JN	0;36	269.49	12695	3JN	21;24	69.88	12695	18JN	21;34	264.17
12622	9JL	13;02	104.32	12622	23JL	12;47	297.76	12696	3JL	4;28	97.81	12696	18JL	11;50	292.48
12623	7AU	22;01	132.64	12623	22AU	3;23	326.50	12697	1AU	13;07	126.03	12697	17AU	1;18	321.09
12624	6SE	6;40	161.38	12624	20SE	20;19	355.83	12698	31AU	0;10	154.76	12698	15SE	13;45	350.14
12625	5OC	15;52	190.65	12625	20OC	14;55	25.73	12699	29SE	14;04	184.10	12699	15OC	1;14	19.64
12626	4NO	2;16	220.41	12626	19NO	9;46	56.00	12700	29OC	6;46	214.04	12700	13NO	12;06	49.52
12627	3DE	14;10	250.48	12627	19DE	3;07	86.33	12701	28NO	1;42	244.38	12701	12DE	22;47	79.56
								12702	27DE	21;29	274.79				

21 / 27

NEW MOONS				FULL MOONS				NEW MOONS				FULL MOONS			
12628	2JA	3;32	280.60	12628	17JA	17;45	116.39					12702	11JA	9;33	109.52
12629	31JA	18;05	310.49	12629	16FE	5;28	145.94	12703	26JA	16;14	304.94	12703	9FE	20;30	139.14
12630	2MR	9;31	339.97	12630	17MR	14;49	174.90	12704	25FE	8;18	334.53	12704	11MR	7;45	168.30
12631	1AP	1;29	8.95	12631	15AP	22;40	203.32	12705	26MR	21;01	3.48	12705	9AP	19;31	196.98
12632	30AP	17;25	37.50	12632	15MY	5;53	231.37	12706	25AP	6;46	31.86	12706	9MY	8;14	225.31
12633	30MY	8;32	65.76	12633	13JN	13;20	259.28	12707	24MY	14;34	59.86	12707	7JN	22;16	253.49
12634	28JN	22;12	93.94	12634	12JL	21;52	287.32	12708	22JN	21;34	87.76	12708	7JL	13;34	281.77
12635	28JL	10;16	122.25	12635	11AU	8;24	315.74	12709	22JL	4;47	115.81	12709	6AU	5;30	310.33
12636	26AU	21;11	150.91	12636	9SE	21;46	344.73	12710	20AU	13;04	144.26	12710	4SE	21;12	339.32
12637	25SE	7;36	180.05	12637	9OC	14;24	14.37	12711	18SE	23;06	173.26	12711	4OC	11;57	8.78
12638	24OC	18;05	209.67	12638	8NO	9;36	44.54	12712	18OC	11;33	202.86	12712	3NO	1;32	38.65
12639	23NO	4;52	239.63	12639	8DE	5;36	74.97	12713	17NO	2;51	232.95	12713	2DE	13;59	68.73
12640	22DE	15;56	269.69					12714	16DE	20;48	263.31				

22 / 28

NEW MOONS				FULL MOONS				NEW MOONS				FULL MOONS			
				12640	7JA	0;11	105.29					12714	1JA	1;24	98.78
12641	21JA	3;20	299.58	12641	5FE	16;00	135.17	12715	15JA	16;12	293.60	12715	30JA	11;48	128.53
12642	19FE	15;20	329.09	12642	7MR	4;48	164.46	12716	14FE	11;02	323.49	12716	28FE	21;18	157.79
12643	21MR	4;20	358.14	12643	5AP	14;58	193.13	12717	15MR	3;38	352.79	12717	29MR	6;20	186.54
12644	19AP	18;28	26.77	12644	4MY	23;12	221.33	12718	13AP	17;21	21.48	12718	27AP	15;45	214.86
12645	19MY	9;26	55.09	12645	3JN	6;15	249.25	12719	13MY	4;28	49.70	12719	27MY	2;27	242.96
12646	18JN	0;39	83.31	12646	2JL	13;03	277.16	12720	11JN	13;42	77.68	12720	25JN	15;08	271.09
12647	17JL	15;31	111.63	12647	31JL	20;39	305.33	12721	10JL	21;50	105.70	12721	25JL	5;56	299.48
12648	16AU	5;46	140.27	12648	30AU	6;12	333.99	12722	9AU	5;40	133.98	12722	23AU	22;22	328.32
12649	14SE	19;21	169.34	12649	28SE	18;39	3.26	12723	7SE	13;58	162.73	12723	22SE	15;35	357.69
12650	14OC	8;16	198.90	12650	28OC	10;22	33.15	12724	6OC	23;31	192.03	12724	22OC	8;41	27.55
12651	12NO	20;21	228.82	12651	27NO	4;44	63.46	12725	5NO	11;04	221.85	12725	21NO	0;49	57.72
12652	12DE	7;32	258.89	12652	27DE	0;14	93.87	12726	5DE	1;02	252.02	12726	20DE	15;15	87.93

23 / 29

NEW MOONS				FULL MOONS				NEW MOONS				FULL MOONS			
12653	10JA	17;56	288.83	12653	25JA	19;09	124.02	12727	3JA	17;14	282.25	12727	19JA	3;31	117.87
12654	9FE	3;56	318.42	12654	24FE	12;07	153.67	12728	2FE	10;43	312.25	12728	17FE	13;38	147.33
12655	10MR	14;04	347.54	12655	26MR	2;24	182.70	12729	4MR	4;13	341.78	12729	18MR	22;07	176.23
12656	9AP	0;52	16.20	12656	24AP	13;50	211.15	12730	2AP	20;43	10.76	12730	17AP	5;50	204.63
12657	8MY	12;42	44.50	12657	23MY	22;50	239.20	12731	2MY	11;34	39.24	12731	16MY	13;46	232.70
12658	7JN	1;46	72.65	12658	22JN	6;14	267.12	12732	1JN	0;33	67.41	12732	14JN	22;52	260.67
12659	6JL	16;08	100.88	12659	21JL	13;12	295.16	12733	30JN	11;42	95.49	12733	14JL	9;54	288.82
12660	5AU	7;44	129.42	12660	19AU	20;52	323.57	12734	29JL	21;24	123.73	12734	12AU	23;20	317.38
12661	4SE	0;10	158.42	12661	18SE	6;14	352.53	12735	28AU	6;20	152.33	12735	11SE	15;16	346.50
12662	3OC	16;36	187.94	12662	17OC	17;55	22.09	12736	26SE	15;22	181.43	12736	11OC	9;21	16.22
12663	2NO	8;00	217.88	12663	16NO	8;06	52.13	12737	26OC	1;15	211.04	12737	10NO	4;28	46.41
12664	1DE	21;41	248.01	12664	16DE	0;30	82.42	12738	24NO	12;26	241.03	12738	9DE	22;54	76.78
12665	31DE	9;32	278.08					12739	24DE	1;00	271.15				

24 / 30

NEW MOONS				FULL MOONS				NEW MOONS				FULL MOONS			
				12665	14JA	18;29	112.66					12739	8JA	15;01	106.98
12666	29JA	19;54	307.83	12666	13FE	12;57	142.55	12740	22JA	14;47	301.12	12740	7FE	4;07	136.73
12667	28FE	5;15	337.10	12667	14MR	6;32	171.90	12741	21FE	5;31	330.73	12741	8MR	14;26	165.89
12668	28MR	14;04	5.85	12668	12AP	21;57	200.67	12742	22MR	20;59	359.85	12742	6AP	22;47	194.48
12669	26AP	22;52	34.15	12669	12MY	10;37	228.96	12743	21AP	12;48	28.52	12743	6MY	6;06	222.62
12670	26MY	8;20	62.21	12670	10JN	20;50	256.98	12744	21MY	4;21	56.86	12744	4JN	13;15	250.54
12671	24JN	19;18	90.27	12671	10JL	5;29	285.01	12745	19JN	18;52	85.05	12745	3JL	21;06	278.50
12672	24JL	8;32	118.59	12672	8AU	13;39	313.30	12746	19JL	7;55	113.31	12746	2AU	6;32	306.76
12673	23AU	0;21	147.39	12673	6SE	22;14	342.05	12747	17AU	19;36	141.87	12747	31AU	18;28	335.54
12674	21SE	18;11	176.77	12674	6OC	7;49	11.35	12748	16SE	6;28	170.86	12748	30SE	9;34	4.95
12675	21OC	12;37	206.67	12675	4NO	18;45	41.13	12749	15OC	17;06	200.34	12749	30OC	3;45	34.95
12676	20NO	6;03	236.89	12676	4DE	7;15	71.23	12750	14NO	3;52	230.22	12750	28NO	23;42	65.34
12677	19DE	21;28	267.14					12751	13DE	14;49	260.27	12751	28DE	19;15	95.75

25 / 31

NEW MOONS				FULL MOONS				NEW MOONS				FULL MOONS			
				12677	2JA	21;32	101.39					12752	27JA	12;30	125.81
12678	18JA	10;36	297.13	12678	1FE	13;33	131.34	12752	12JA	1;54	290.24	12753	26FE	2;42	155.32
12679	16FE	21;34	326.64	12679	3MR	6;37	160.86	12753	10FE	13;19	319.87	12754	27MR	14;01	184.20
12680	18MR	6;43	355.58	12680	1AP	23;33	189.87	12754	12MR	1;29	349.05	12755	25AP	23;04	212.54
12681	16AP	14;32	23.99	12681	1MY	15;11	218.40	12755	10AP	14;45	17.78	12756	25MY	6;34	240.53
12682	15MY	21;47	52.03	12682	31MY	4;59	246.61	12756	10MY	5;08	46.17	12757	23JN	13;23	268.42
12683	14JN	5;32	79.96	12683	29JN	17;04	274.72	12757	8JN	20;14	74.40	12758	22JL	20;30	296.47
12684	13JL	14;59	108.04	12684	29JL	3;55	302.99	12758	8JL	11;25	102.67	12759	21AU	5;02	324.94
12685	12AU	3;08	136.53	12685	27AU	14;04	331.63	12759	7AU	2;13	131.20	12760	19SE	16;02	353.99
12686	10SE	18;22	165.61	12686	25SE	23;59	0.76	12760	5SE	16;26	160.14	12761	19OC	6;12	23.67
12687	10OC	12;12	195.31	12687	25OC	9;59	30.37	12761	5OC	6;00	189.56	12762	17NO	23;24	53.86
12688	9NO	7;23	225.49	12688	23NO	20;27	60.32	12762	3NO	18;45	219.40	12763	17DE	18;30	84.26
12689	9DE	2;19	255.88	12689	23DE	7;47	90.39	12763	3DE	6;30	249.46				

```
--------NEW MOONS--------   --------FULL MOONS-------        --------NEW MOONS--------   --------FULL MOONS-------
NUMBER  DATE   TIME  LONG.   NUMBER  DATE   TIME  LONG.      NUMBER  DATE   TIME  LONG.   NUMBER  DATE   TIME  LONG.

                          32                                                          38
12764   1JA  17;14  279.48   12764  16JA  13;49  114.54                                  12838  10JA   1;59  108.52
12765  31JA   3;14  309.20   12765  15FE   7;45  144.38      12839  24JA   5;52  302.82   12839   8FE  13;03  138.16
12766  29FE  13;00  338.47   12766  15MR  23;19  173.62      12840  22FE  23;14  332.51   12840   9MR  22;06  167.25
12767  29MR  23;08    7.24   12767  14AP  12;06  202.26      12841  24MR  16;07    1.66   12841   8AP   5;54  195.79
12768  28AP  10;09   35.62   12768  13MY  22;14  230.44      12842  23AP   7;43   30.30   12842   7MY  13;26  223.93
12769  27MY  22;24   63.78   12769  12JN   6;22  258.38      12843  22MY  21;39   58.56   12843   5JN  21;43  251.89
12770  26JN  12;02   91.95   12770  11JL  13;30  286.35      12844  21JN   9;49   86.66   12844   5JL   7;39  279.94
12771  26JL   3;07  120.36   12771   9AU  20;49  314.63      12845  20JL  20;22  114.84   12845   3AU  19;51  308.33
12772  24AU  19;26  149.21   12772   8SE   5;23  343.40      12846  19AU   5;50  143.32   12846   2SE  10;38  337.24
12773  23SE  12;16  178.58   12773   7OC  15;58   12.75      12847  17SE  14;57  172.26   12847   2OC   3;53    6.77
12774  23OC   4;35  208.42   12774   6NO   4;57   42.64      12848  17OC   0;30  201.72   12848  31OC  22;51   36.83
12775  21NO  19;23  238.54   12775   5DE  20;15   72.86      12849  15NO  11;05  231.60   12849  30NO  18;00   67.20
12776  21DE   8;12  268.67                                   12850  14DE  22;56  261.70   12850  30DE  11;30   97.51

                          33                                                          39
                            12776   4JA  13;24  103.13       12851  13JA  11;58  291.73   12851  29JA   2;06  127.45
12777  19JA  19;12  298.56   12777   3FE   7;34  133.15      12852  12FE   1;59  321.45   12852  27FE  13;38  156.82
12778  18FE   4;52  327.99   12778   5MR   1;33  162.68      12853  13MR  16;48  350.72   12853  28MR  22;44  185.58
12779  19MR  13;41  356.90   12779   3AP  18;02  191.65      12854  12AP   8;14   19.51   12854  27AP   6;20  213.85
12780  17AP  22;12   25.32   12780   3MY   8;04  220.09      12855  11MY  23;52   47.93   12855  26MY  13;23  241.81
12781  17MY   7;03   53.42   12781   1JN  19;30  248.19      12856  10JN  15;00   76.15   12856  24JN  20;44  269.72
12782  15JN  17;02   81.43   12782   1JL   4;56  276.21      12857  10JL   4;58  104.39   12857  24JL   5;17  297.85
12783  15JL   5;01  109.62   12783  30JL  13;24  304.40      12858   8AU  17;34  132.85   12858  22AU  15;56  326.42
12784  13AU  19;37  138.23   12784  28AU  21;53  332.99      12859   7SE   5;03  161.70   12859  21SE   5;29  355.61
12785  12SE  12;44  167.42   12785  27SE   7;06    2.11      12860   6OC  16;00  191.05   12860  20OC  22;18   25.42
12786  12OC   7;19  197.18   12786  26OC  17;27   31.75      12861   5NO   2;52  220.82   12861  19NO  17;40   55.71
12787  11NO   1;39  227.34   12787  25NO   5;09   61.77      12862   4DE  13;47  250.85   12862  19DE  13;43   86.15
12788  10DE  18;17  257.64   12788  24DE  18;26   91.92

                          34                                                          40
12789   9JA   8;35  287.75   12789  23JA   9;24  121.93      12863   3JA   0;43  280.87   12863  18JA   8;15  116.38
12790   7FE  20;32  317.45   12790  22FE   1;46  151.59      12864   1FE  11;45  310.62   12864  16FE  23;56  146.10
12791   9MR   6;25  346.58   12791  23MR  18;39  180.76      12865   1MR  23;12  339.93   12865  17MR  12;33  175.20
12792   7AP  14;40   15.14   12792  22AP  10;53  209.44      12866  31MR  11;34    8.78   12866  15AP  22;34  203.70
12793   6MY  21;56   43.28   12793  22MY   1;36  237.73      12867  30AP   1;09   37.24   12867  15MY   6;42  231.79
12794   5JN   5;12   71.20   12794  20JN  14;36  265.86      12868  29MY  15;50   65.49   12868  13JN  13;45  259.68
12795   4JL  13;36   99.19   12795  20JL   2;12  294.08      12869  28JN   7;04   93.72   12869  12JL  20;39  287.66
12796   3AU   0;19  127.51   12796  18AU  12;54  322.60      12870  27JL  22;16  122.16   12870  11AU   4;26  315.97
12797   1SE  14;05  156.37   12797  16SE  23;09  351.58      12871  26AU  13;04  150.97   12871   9SE  14;11  344.81
12798   1OC   6;51  185.86   12798  16OC   9;16   21.05      12872  25SE   3;17  180.26   12872   9OC   2;48   14.27
12799  31OC   1;41  215.91   12799  14NO  19;34   50.91      12873  24OC  16;44  210.00   12873   7NO  18;36   44.29
12800  29NO  21;03  246.28   12800  14DE   6;25   80.96      12874  23NO   5;09  240.03   12874   7DE  12;54   74.65
12801  29DE  15;23  276.63                                   12875  22DE  16;25  270.10

                          35                                                          41
12802  28JA   7;29  306.64   12801  12JA  18;10  110.95                                  12875   6JA   8;13  105.00
12803  26FE  20;43  336.09   12802  11FE   7;04  140.64      12876  21JA   2;38  299.94   12876   5FE   2;52  135.01
12804  28MR   7;03    4.92   12803  12MR  20;59  169.88      12877  19FE  12;15  329.35   12877   6MR  19;35  164.47
12805  26AP  15;06   33.21   12804  11AP  11;35  198.65      12878  20MR  21;53  358.27   12878   5AP   9;41  193.31
12806  25MY  21;52   61.16   12805  11MY   2;29  227.05      12879  19AP   8;08   26.74   12879   4MY  21;05  221.63
12807  24JN   4;33   89.05   12806   9JN  17;24  255.27      12880  18MY  19;31   54.93   12880   3JN   6;09  249.63
12808  23JL  12;19  117.14   12807   9JL   8;06  283.52      12881  17JN   8;19   83.05   12881   2JL  13;45  277.58
12809  21AU  22;10  145.67   12808   7AU  22;15  312.04      12882  16JL  22;42  111.35   12882  31JL  20;58  305.73
12810  20SE  10;42  174.80   12809   6SE  11;32  340.95      12883  15AU  14;36  140.04   12883  30AU   4;57  334.33
12811  20OC   2;07  204.54   12810   5OC  23;45   10.34      12884  14SE   7;33  169.25   12884  28SE  14;35    3.48
12812  18NO  20;05  234.76   12811   4NO  11;05   40.13      12885  14OC   0;34  198.97   12885  28OC   2;26   33.20
12813  18DE  15;38  265.19   12812   3DE  21;54   70.15      12886  12NO  16;28  229.05   12886  26NO  16;36   63.33
                                                             12887  12DE   6;25  259.23   12887  26DE   8;46   93.60

                          36                                                          42
12814  17JA  11;05  295.46   12813   2JA   8;33  100.16      12888  10JA  18;16  289.23   12888  25JA   2;20  123.71
12815  16FE   4;33  325.27   12814  31JA  19;14  129.90      12889   9FE   4;25  318.83   12889  23FE  20;21  153.41
12816  16MR  18;52  354.45   12815   1MR   6;04  159.20      12890  10MR  13;24  347.90   12890  25MR  13;33  182.56
12817  15AP   5;56   23.00   12816  30MR  17;14  188.01      12891   8AP  21;48   16.46   12891  24AP   4;48  211.17
12818  14MY  14;32   51.11   12817  29AP   5;10  216.41      12892   8MY   6;12   44.63   12892  23MY  17;31  239.38
12819  12JN  21;50   79.02   12818  28MY  18;20  244.60      12893   6JN  15;22   72.62   12893  22JN   3;58  267.41
12820  12JL   4;54  107.00   12819  27JN   9;00  272.81      12894   6JL   2;11  100.71   12894  21JL  12;58  295.53
12821  10AU  12;40  135.30   12820  27JL   0;50  301.26      12895   4AU  15;27  129.15   12895  19AU  21;33  323.99
12822   8SE  21;53  164.10   12821  25AU  17;02  330.12      12896   3SE   7;30  158.13   12896  18SE   6;32  352.94
12823   8OC   9;09  193.50   12822  24SE   8;37  359.45      12897   3OC   1;45  187.71   12897  17OC  16;27   22.42
12824   6NO  23;02  223.43   12823  23OC  23;04   29.22      12898   1NO  20;38  217.78   12898  16NO   3;31   52.32
12825   6DE  15;41  253.72   12824  22NO  12;15   59.28      12899   1DE  14;25  248.09   12899  15DE  15;55   82.45
                            12825  22DE   0;15   89.38       12900  31DE   5;59  278.32

                          37                                                          43
12826   5JA  10;27  284.05   12826  20JA  11;05  119.25                                  12900  14JA   5;51  112.51
12827   4FE   5;38  314.10   12827  18FE  20;49  148.68      12901  29JA  19;03  308.18   12901  12FE  21;19  142.28
12828   5MR  23;20  343.61   12828  20MR   5;47  177.58      12902  28FE   5;48  337.51   12902  14MR  13;47  171.60
12829   4AP  14;24   12.50   12829  18AP  14;39  206.01      12903  29MR  14;40    6.26   12903  13AP   6;15  200.42
12830   4MY   2;43   40.87   12830  18MY   0;23  234.13      12904  27AP  22;11   34.51   12904  12MY  21;42  228.83
12831   2JN  12;51   68.91   12831  16JN  11;53  262.21      12905  27MY   5;13   62.47   12905  11JN  11;36  257.00
12832   1JL  21;33   96.90   12832  16JL   1;38  290.47      12906  25JN  12;50   90.39   12906  11JL   0;01  285.18
12833  31JL   5;37  125.09   12833  14AU  17;27  319.15      12907  24JL  22;16  118.56   12907   9AU  11;22  313.60
12834  29AU  13;46  153.67   12834  13SE  10;36  348.35      12908  23AU  10;30  147.21   12908   7SE  22;04  342.44
12835  27SE  22;46  182.79   12835  13OC   4;08   18.08      12909  22SE   1;56  176.48   12909   7OC   8;28   11.76
12836  27OC   9;20  212.45   12836  11NO  21;05   48.19      12910  21OC  20;02  206.36   12910   5NO  18;47   41.53
12837  25NO  22;04  242.51   12837  11DE  12;35   78.44      12911  20NO  15;27  236.66   12911   5DE   5;21   71.54
12838  25DE  13;06  272.74                                   12912  20DE  10;32  267.07
```

```
                                44                                                                50
                                   12912   3JA  16;30  101.57     12987  12JA   2;01  290.63        12987  26JA  21;36  125.58
 12913  19JA   3;51  297.23        12913   2FE   4;36  131.35     12988  10FE  11;43  320.19        12988  25FE  15;15  155.23
 12914  17FE  18;29  326.90        12914   2MR  17;45  160.72     12989  11MR  21;01  349.25        12989  27MR   6;36  184.28
 12915  18MR   6;08  355.94        12915   1AP   7;45  189.62     12990  10AP   6;37   17.84        12990  25AP  19;17  212.77
 12916  16AP  15;08   24.40        12916   3UAP 22;18  218.11     12991   9MY  17;09   46.08        12991  25MY   5;28  240.87
 12917  15MY  22;21   52.44        12917  3UMY  13;08  246.36     12992   8JN   5;04   74.18        12992  23JN  13;46  268.82
 12918  14JN   4;54   80.32        12918  29JN   4;02  274.59     12993   7JL  18;37  102.38        12993  22JL  21;10  296.89
 12919  13JL  12;01  108.30        12919  28JL  18;43  303.01     12994   6AU   9;52  130.92        12994  21AU   4;48  325.32
 12920  11AU  20;47  136.66        12920  27AU   8;46  331.80     12995   5SE   2;36  159.96        12995  19SE  13;41  354.28
 12921  10SE   8;01  165.57        12921  25SE  21;49    1.06     12996   4OC  20;00  189.54        12996  19OC   0;29   23.82
 12922   9OC  22;04  195.11        12922  25OC   9;49   30.75     12997   3NO  12;52  219.55        12997  17NO  13;32   53.83
 12923   8NO  14;54  225.18        12923  23NO  20;58   60.73     12998   3DE   4;01  249.75        12998  17DE   4;40   84.06
 12924   8DE   9;49  255.57        12924  23DE   7;40   90.77

                                45                                                                51
 12925   7JA   5;29  285.93        12925  21JA  18;12  120.62     12999   1JA  16;56  279.86        12999  15JA  21;26  114.24
 12926   6FE   0;03  315.92        12926  20FE   4;43  150.06     13000  31JA   3;48  309.61        13000  14FE  15;07  144.08
 12927   7MR  15;55  345.33        12927  21MR  15;24  179.01     13001   1MR  13;09  338.86        13001  16MR   8;39  173.41
 12928   6AP   4;29   14.09        12928  20AP   2;37  207.51     13002  30MR  21;34    7.57        13002  15AP   0;52  202.19
 12929   5MY  14;10   42.34        12929  19MY  14;54  235.73     13003  29AP   5;42   35.83        13003  14MY  14;52  230.53
 12930   3JN  21;59   70.29        12930  18JN   4;42  263.89     13004  28MY  14;14   63.84        13004  13JN   2;28  258.61
 12931   3JL   5;06   98.22        12931  17JL  20;05  292.24     13005  27JN   0;01   91.86        13005  12JL  12;14  286.69
 12932   1AU  12;33  126.39        12932  16AU  12;26  320.95     13006  26JL  11;57  120.14        13006  10AU  21;07  315.03
 12933  30AU  21;05  155.02        12933  15SE   4;43  350.15     13007  25AU   2;42  148.91        13007   9SE   6;03  343.81
 12934  29SE   7;22  184.21        12934  14OC  20;05   19.81     13008  23SE  20;10  178.30        13008   8OC  15;39   13.13
 12935  28OC  19;57  213.97        12935  13NO  10;07   49.82     13009  23OC  15;10  208.24        13009   7NO   2;13   42.91
 12936  27NO  11;13  244.15        12936  12DE  22;48   79.95     13010  22NO   9;53  238.52        13010   6DE  13;55   72.99
 12937  27DE   4;58  274.48                                       13011  22DE   2;44  268.83

                                46                                                                52
                                   12937  11JA  10;11  109.93     13011   5JA   2;55  103.08
 12938  25JA  23;59  304.64        12938   9FE  20;19  139.52     13012  20JA  17;02  298.86        13012   3FE  17;23  132.94
 12939  24FE  18;27  334.35        12939  11MR   5;24  168.59     13013  19FE   4;49  328.38        13013   4MR   9;09  162.40
 12940  26MR  10;46    3.45        12940   9AP  13;59  197.14     13014  19MR  14;26  357.31        13014   3AP   1;29  191.37
 12941  25AP   0;22   31.98        12941   8MY  22;56  225.32     13015  17AP  22;24   25.71        13015   2MY  17;24  219.89
 12942  24MY  11;33   60.13        12942   7JN   9;18  253.36     13016  17MY   5;27   53.74        13016   1JN   8;06  248.13
 12943  22JN  20;58   88.12        12943   6JL  21;50  281.53     13017  15JN  12;34   81.63        13017  30JN  21;22  276.29
 12944  22JL   5;25  116.22        12944   5AU  12;44  310.04     13018  14JL  20;56  109.68        13018  30JL   9;25  304.63
 12945  20AU  13;37  144.67        12945   4SE   5;30  339.06     13019  13AU   7;42  138.14        13019  28AU  20;41  333.33
 12946  18SE  22;17  173.62        12946   3OC  23;13    8.64     13020  11SE  21;37  167.18        13020  27SE   7;28    2.52
 12947  18OC   8;07  203.10        12947   2NO  16;49   38.67     13021  11OC  14;38  196.87        13021  26OC  17;59   32.16
 12948  16NO  19;45  233.04        12948   2DE   9;19   68.92     13022  10NO   9;41  227.05        13022  25NO   4;28   62.13
 12949  16DE   9;35  263.23        12949  31DE  23;53   99.11     13023  10DE   5;12  257.47        13023  24DE  15;14   92.17

                                47                                                                53
 12950  15JA   1;24  293.36        12950  30JA  12;03  128.93     13024   8JA  23;33  287.76        13024  23JA   2;39  122.04
 12951  13FE  18;21  323.19        12951  28FE  21;54  158.20     13025   7FE  15;33  317.63        13025  21FE  15;00  151.53
 12952  15MR  11;20  352.51        12952  30MR   6;03  186.91     13026   9MR   4;36  346.89        13026  23MR   4;18  180.57
 12953  14AP   3;27   21.30        12953  28AP  13;26  215.15     13027   7AP  14;46   15.54        13027  21AP  18;21  209.16
 12954  13MY  18;12   49.68        12954  27MY  21;05  243.13     13028   6MY  22;40   43.70        13028  21MY   8;54  237.46
 12955  12JN   7;20   77.82        12955  26JN   6;01  271.11     13029   5JN   5;23   71.60        13029  19JN  23;47  265.66
 12956  11JL  18;51  105.96        12956  25JL  17;01  299.35     13030   4JL  12;05   99.52        13030  19JL  14;46  294.01
 12957  10AU   5;01  134.34        12957  24AU   6;33  328.06     13031   2AU  19;59  127.72        13031  18AU   5;27  322.68
 12958   8SE  14;27  163.14        12958  22SE  22;45  357.38     13032   1SE   6;00  156.43        13032  16SE  19;22  351.81
 12959   7OC  23;54  192.44        12959  22OC  17;10   27.28     13033  30SE  18;42  185.74        13033  16OC   8;09   21.39
 12960   6NO  10;03  222.20        12960  21NO  12;36   57.59     13034  30OC  10;15  215.65        13034  14NO  19;52   51.31
 12961   5DE  21;16  252.25        12961  21DE   7;13   87.97     13035  29NO   4;15  245.95        13035  14DE   6;49   81.37
                                                                  13036  28DE  23;41  276.35

                                48                                                                54
 12962   4JA   9;37  282.32        12962  19JA  23;21  118.08     13037  27JA  18;55  306.50        13036  12JA  17;19  111.30
 12963   2FE  22;56  312.15        12963  18FE  12;18  147.67     13038  26FE  12;10  336.13        13037  11FE   3;39  140.88
 12964   3MR  13;03  341.54        12964  18MR  22;24  176.63     13039  28MR   2;18    5.11        13038  12MR  13;57  169.98
 12965   2AP   3;55   10.47        12965  17AP   6;31  205.05     13040  26AP  13;17   33.52        13039  11AP   0;34  198.59
 12966   1MY  19;19   38.98        12966  16MY  13;38  233.08     13041  25MY  21;54   61.55        13040  10MY  12;00  226.86
 12967  31MY  10;44   67.25        12967  14JN  20;41  260.97     13042  24JN   5;18   89.47        13041   9JN   0;51  255.00
 12968  30JN   1;27   95.47        12968  14JL   4;33  289.00     13043  23JL  12;35  117.54        13042   8JL  15;28  283.25
 12969  29JL  14;58  123.85        12969  12AU  14;05  317.39     13044  21AU  20;38  146.00        13043   7AU   7;35  311.83
 12970  28AU   3;14  152.59        12970  11SE   2;10  346.35     13045  20SE   6;06  175.00        13044   6SE   0;18  340.88
 12971  26SE  14;41  181.79        12971  10OC  17;27   15.95     13046  19OC  17;34  204.57        13045   5OC  16;31   10.42
 12972  26OC   1;46  211.45        12972   9NO  11;47   46.10     13047  18NO   7;29  234.62        13046   4NO   7;30   40.36
 12973  24NO  12;46  241.43        12973   9DE   7;48   76.53     13048  17DE  23;58  264.91        13047   3DE  21;00   70.50
 12974  23DE  23;41  271.48

                                49                                                                55
                                   12974   8JA   3;19  106.88                                         13048   2JA   9;02  100.57
 12975  22JA  10;29  301.33        12975   6FE  20;26  136.80     13049  16JA  18;23  295.15        13049  31JA  19;41  130.31
 12976  20FE  21;25  330.78        12976   8MR  10;28  166.12     13050  15FE  13;09  325.02        13050   2MR   5;04  159.54
 12977  22MR   8;58  359.75        12977   6AP  21;37  194.81     13051  17MR   6;30  354.33        13051  31MR  13;36  188.24
 12978  20AP  21;39   28.30        12978   6MY   6;33  223.02     13052  15AP  21;23   23.05        13052  29AP  22;02  216.51
 12979  20MY  11;38   56.58        12979   4JN  14;02  250.96     13053  15MY   9;42   51.31        13053  29MY   7;24  244.55
 12980  19JN   2;36   84.79        12980   3JL  20;57  278.88     13054  13JN  20;00   79.33        13054  27JN  18;41  272.63
 12981  18JL  17;59  113.14        12981   2AU   4;15  307.06     13055  13JL   5;00  107.39        13055  27JL   8;25  300.99
 12982  17AU   9;16  141.83        12982  31JL  12;59  335.70     13056  11AU  13;26  135.72        13056  26AU   0;18  329.83
 12983  16SE   0;06  170.98        12983  30SE   0;11    4.94     13057   9SE  21;58  164.50        13057  24SE  18;03  359.24
 12984  15OC  14;16  200.61        12984  29OC  14;28   34.78     13058   9OC   7;18  193.81        13058  24OC  12;05   29.15
 12985  14NO   3;26  230.59        12985  28NO   7;39   65.05     13059   7NO  18;02  223.61        13059  23NO   5;27   59.38
 12986  13DE  15;22  260.69        12986  28DE   2;35   95.43     13060   7DE   6;43  253.73        13060  22DE  21;10   89.64
```

New Moons

Year	Number	Date	Time	Long.
56	13061	5JA	21;27	283.89
	13062	4FE	13;43	313.82
	13063	5MR	6;31	343.30
	13064	3AP	22;56	12.26
	13065	3MY	14;19	40.76
	13066	2JN	4;18	68.96
	13067	1JL	16;45	97.10
	13068	31JL	3;46	125.40
	13069	29AU	13;45	154.06
	13070	27SE	23;20	183.21
	13071	27OC	9;14	212.84
	13072	25NO	19;57	242.81
	13073	25DE	7;41	272.90
57	13074	23JA	20;20	302.81
	13075	22FE	9;46	332.34
	13076	23MR	23;58	1.39
	13077	22AP	14;53	30.01
	13078	22MY	6;15	58.33
	13079	20JN	21;27	86.55
	13080	20JL	11;48	114.88
	13081	19AU	0;57	143.51
	13082	17SE	13;03	172.57
	13083	17OC	0;30	202.11
	13084	15NO	11;42	232.01
	13085	14DE	22;41	262.07
58	13086	13JA	9;27	292.01
	13087	11FE	20;03	321.59
	13088	13MR	6;56	350.70
	13089	11AP	18;43	19.35
	13090	11MY	7;48	47.68
	13091	9JN	22;13	75.87
	13092	9JL	13;31	104.16
	13093	8AU	5;06	132.73
	13094	6SE	20;29	161.74
	13095	6OC	11;19	191.24
	13096	5NO	1;17	221.14
	13097	4DE	13;59	251.25
59	13098	3JA	1;15	281.28
	13099	1FE	11;14	310.98
	13100	2MR	20;27	340.20
	13101	1AP	5;35	8.92
	13102	30AP	15;20	37.23
	13103	30MY	2;20	65.34
	13104	28JN	14;57	93.47
	13105	28JL	5;24	121.86
	13106	26AU	21;37	150.72
	13107	25SE	15;04	180.14
	13108	25OC	8;38	210.05
	13109	24NO	0;58	240.24
	13110	23DE	15;06	270.43
60	13111	22JA	2;54	300.34
	13112	20FE	12;48	329.76
	13113	20MR	21;26	358.63
	13114	19AP	5;27	27.02
	13115	18MY	13;31	55.07
	13116	16JN	22;27	83.04
	13117	16JL	9;11	111.19
	13118	14AU	22;31	139.77
	13119	13SE	14;50	168.94
	13120	13OC	9;27	198.72
	13121	12NO	4;44	228.94
	13122	11DE	22;47	259.29
61	13123	10JA	14;25	289.46
	13124	9FE	3;22	319.17
	13125	10MR	13;54	348.31
	13126	8AP	22;29	16.87
	13127	8MY	5;47	45.00
	13128	6JN	12;39	72.90
	13129	5JL	20;13	100.86
	13130	4AU	5;41	129.14
	13131	2SE	18;03	157.97
	13132	2OC	9;40	187.43
	13133	1NO	3;59	217.47
	13134	30NO	23;33	247.85
	13135	30DE	18;40	278.23
62	13136	29JA	11;54	308.27
	13137	28FE	2;23	337.76
	13138	29MR	13;52	6.61
	13139	27AP	22;43	34.92
	13140	27MY	5;51	62.88
	13141	25JN	12;25	90.76
	13142	24JL	19;39	118.84
	13143	23AU	4;36	147.36
	13144	21SE	16;01	176.46
	13145	21OC	6;14	206.17
	13146	19NO	23;06	236.36
	13147	19DE	17;55	266.75
63	13148	18JA	13;23	297.01
	13149	17FE	7;41	326.84
	13150	18MR	23;21	356.05
	13151	17AP	11;47	24.65
	13152	16MY	21;28	52.79
	13153	15JN	5;23	80.72
	13154	14JL	12;43	108.72
	13155	12AU	20;25	137.03
	13156	11SE	5;15	165.85
	13157	10OC	15;46	195.23
	13158	9NO	4;27	225.13
	13159	8DE	19;37	255.35
64	13160	7JA	13;03	285.62
	13161	6FE	7;39	315.63
	13162	7MR	1;41	345.13
	13163	5AP	17;45	14.05
	13164	5MY	7;16	42.45
	13165	3JN	18;34	70.54
	13166	3JL	4;15	98.57
	13167	1AU	13;05	126.80
	13168	30AU	21;42	155.43
	13169	29SE	6;44	184.58
	13170	28OC	16;48	214.23
	13171	27NO	4;29	244.25
	13172	26DE	18;06	274.41
65	13173	25JA	9;28	304.42
	13174	24FE	1;51	334.05
	13175	25MR	18;17	3.17
	13176	24AP	10;05	31.80
	13177	24MY	0;46	60.08
	13178	22JN	14;06	88.23
	13179	22JL	2;02	116.47
	13180	20AU	12;43	145.02
	13181	18SE	22;40	174.02
	13182	18OC	8;32	203.51
	13183	16NO	18;55	233.40
	13184	16DE	6;07	263.47
66	13185	14JA	18;10	293.45
	13186	13FE	6;58	323.10
	13187	14MR	20;28	352.29
	13188	13AP	10;45	21.02
	13189	13MY	1;45	49.41
	13190	11JN	17;06	77.64
	13191	11JL	8;05	105.92
	13192	9AU	22;06	134.45
	13193	8SE	11;00	163.39
	13194	7OC	23;00	192.79
	13195	6NO	10;30	222.61
	13196	5DE	21;41	252.66
67	13197	4JA	8;30	282.66
	13198	2FE	18;59	312.37
	13199	4MR	5;24	341.62
	13200	2AP	16;22	10.38
	13201	2MY	4;30	38.78
	13202	31MY	18;07	66.98
	13203	30JN	9;01	95.20
	13204	30JL	0;40	123.66
	13205	28AU	16;27	152.54
	13206	27SE	7;54	181.90
	13207	26OC	22;37	211.71
	13208	25NO	12;10	241.79
	13209	25DE	0;13	271.89

Full Moons

Year	Number	Date	Time	Long.
56	13061	21JA	10;32	119.62
	13062	19FE	21;24	149.09
	13063	20MR	6;09	177.98
	13064	18AP	13;37	206.35
	13065	17MY	20;52	234.38
	13066	16JN	4;57	262.32
	13067	15JL	14;49	290.43
	13068	14AU	3;05	318.95
	13069	12SE	18;04	348.06
	13070	12OC	11;36	17.77
	13071	11NO	6;53	47.98
	13072	11DE	2;15	78.39
57	13073	9JA	19;48	108.64
	13074	8FE	10;19	138.44
	13075	9MR	21;39	167.62
	13076	8AP	6;32	196.20
	13077	7MY	13;56	224.33
	13078	5JN	20;52	252.24
	13079	5JL	4;12	280.19
	13080	3AU	12;50	308.43
	13081	1SE	23;38	337.18
	13082	1OC	13;21	6.55
	13083	31OC	6;19	36.53
	13084	30NO	1;46	66.90
	13085	29DE	21;47	97.32
58	13086	28JA	16;12	127.42
	13087	27FE	7;42	156.96
	13088	28MR	20;09	185.86
	13089	27AP	6;03	214.22
	13090	26MY	14;09	242.23
	13091	24JN	21;16	270.13
	13092	24JL	4;20	298.20
	13093	22AU	12;21	326.67
	13094	20SE	22;19	355.70
	13095	20OC	11;05	25.34
	13096	19NO	2;55	55.47
	13097	18DE	21;05	85.84
59	13098	17JA	16;07	116.09
	13099	16FE	10;26	145.93
	13100	18MR	2;52	175.19
	13101	16AP	16;50	203.86
	13102	16MY	4;13	232.07
	13103	14JN	13;26	260.06
	13104	13JL	21;17	288.07
	13105	12AU	4;50	316.37
	13106	10SE	13;09	345.15
	13107	9OC	23;04	14.50
	13108	8NO	11;03	44.36
	13109	8DE	1;07	74.54
60	13110	6JA	16;59	104.75
	13111	5FE	10;04	134.70
	13112	6MR	3;34	164.20
	13113	4AP	20;28	193.16
	13114	4MY	11;31	221.63
	13115	3JN	0;21	249.79
	13116	2JL	11;06	277.86
	13117	31JL	20;31	306.10
	13118	30AU	5;33	334.74
	13119	28SE	14;58	3.89
	13120	28OC	1;11	33.54
	13121	26NO	12;20	63.54
	13122	26DE	0;34	93.64
61	13123	24JA	14;04	123.58
	13124	23FE	4;56	153.16
	13125	24MR	20;48	182.27
	13126	23AP	12;51	210.92
	13127	23MY	4;09	239.23
	13128	21JN	18;12	267.41
	13129	21JL	7;02	295.68
	13130	19AU	18;55	324.27
	13131	18SE	6;11	353.31
	13132	17OC	17;03	22.83
	13133	16NO	3;39	52.72
	13134	15DE	14;14	82.76
62	13135	14JA	1;09	112.70
	13136	12FE	12;47	142.32
	13137	14MR	1;19	171.48
	13138	12AP	14;43	200.18
	13139	12MY	4;50	228.54
	13140	10JN	19;30	256.75
	13141	10JL	10;33	285.03
	13142	9AU	1;40	313.59
	13143	7SE	16;20	342.59
	13144	7OC	6;01	12.05
	13145	5NO	18;28	41.90
	13146	5DE	5;51	71.95
63	13147	3JA	16;31	101.96
	13148	2FE	2;46	131.66
	13149	3MR	12;49	160.91
	13150	1AP	22;57	189.65
	13151	1MY	9;39	217.99
	13152	30MY	21;33	246.13
	13153	29JN	11;12	274.30
	13154	29JL	2;43	302.75
	13155	27AU	19;29	331.65
	13156	26SE	12;22	1.05
	13157	26OC	4;18	30.91
	13158	24NO	18;45	61.03
	13159	24DE	7;35	91.16
64	13160	22JA	18;52	121.04
	13161	21FE	4;42	150.45
	13162	21MR	13;23	179.31
	13163	19AP	21;32	207.69
	13164	19MY	6;07	235.76
	13165	17JN	16;13	263.78
	13166	17JL	4;39	292.00
	13167	15AU	19;41	320.66
	13168	14SE	12;48	349.88
	13169	14OC	6;59	19.65
	13170	13NO	1;02	49.83
	13171	12DE	17;48	80.13
65	13172	11JA	8;26	110.25
	13173	9FE	20;28	139.92
	13174	11MR	6;03	169.00
	13175	9AP	13;53	197.52
	13176	8MY	20;58	225.63
	13177	7JN	4;24	253.56
	13178	6JL	13;14	281.57
	13179	5AU	0;15	309.92
	13180	3SE	13;57	338.81
	13181	3OC	6;23	8.32
	13182	2NO	1;05	38.39
	13183	1DE	20;45	68.78
	13184	31DE	15;28	99.14
66	13185	30JA	7;33	129.12
	13186	28FE	20;21	158.53
	13187	30MR	6;14	187.30
	13188	28AP	14;10	215.56
	13189	27MY	21;09	243.52
	13190	26JN	4;10	271.42
	13191	25JL	12;06	299.53
	13192	23AU	21;47	328.09
	13193	22SE	10;03	357.24
	13194	22OC	1;29	27.01
	13195	20NO	19;54	57.27
	13196	20DE	15;53	87.71
67	13197	19JA	11;16	117.96
	13198	18FE	4;12	147.72
	13199	19MR	18;03	176.84
	13200	18AP	5;05	205.37
	13201	17MY	13;58	233.48
	13202	15JN	21;31	261.39
	13203	15JL	4;34	289.38
	13204	13AU	12;06	317.70
	13205	11SE	21;06	346.53
	13206	11OC	8;29	15.96
	13207	9NO	22;50	45.94
	13208	9DE	15;55	76.25

68

NEW MOONS				FULL MOONS			
NUMBER	DATE	TIME	LONG.	NUMBER	DATE	TIME	LONG.
				13209	8JA	10;35	106.56
13210	23JA	10;43	301.73	13210	7FE	5;15	136.56
13211	21FE	20;04	331.11	13211	7MR	22;34	166.02
13212	22MR	4;55	359.97	13212	6AP	13;43	194.88
13213	20AP	14;02	28.38	13213	6MY	2;21	223.24
13214	20MY	0;09	56.51	13214	4JN	12;39	251.29
13215	18JN	11;47	84.59	13215	3JL	21;11	279.28
13216	18JL	1;17	112.86	13216	2AU	4;55	307.47
13217	16AU	16;46	141.55	13217	31AU	12;54	336.08
13218	15SE	9;56	170.78	13218	29SE	22;06	5.25
13219	15OC	3;52	200.56	13219	29OC	9;07	34.95
13220	13NO	21;13	230.71	13220	27NO	22;09	65.03
13221	13DE	12;38	260.96	13221	27DE	13;03	95.24

69

NEW MOONS				FULL MOONS			
13222	12JA	1;35	291.00	13222	26JA	5;23	125.29
13223	10FE	12;16	320.60	13223	24FE	22;31	154.93
13224	11MR	21;18	349.65	13224	26MR	15;36	184.06
13225	10AP	5;22	18.17	13225	25AP	7;34	212.69
13226	9MY	13;09	46.30	13226	24MY	21;35	240.94
13227	7JN	21;26	74.26	13227	23JN	9;26	269.03
13228	7JL	7;05	102.31	13228	22JL	19;36	297.21
13229	5AU	19;03	130.71	13229	21AU	4;57	325.71
13230	4SE	9;58	159.66	13230	19SE	14;21	354.70
13231	4OC	3;45	189.25	13231	19OC	0;18	24.20
13232	2NO	23;08	219.35	13232	17NO	11;03	54.11
13233	2DE	18;08	249.72	13233	16DE	22;41	84.20

70

NEW MOONS				FULL MOONS			
13234	1JA	11;07	280.00	13234	15JA	11;21	114.20
13235	31JA	1;22	309.90	13235	14FE	1;16	143.89
13236	1MR	12;58	339.24	13236	15MR	16;24	173.13
13237	30MR	22;20	7.98	13237	14AP	8;12	201.91
13238	29AP	6;04	36.23	13238	13MY	23;51	230.32
13239	28MY	12;57	64.18	13239	12JN	14;35	258.52
13240	26JN	19;59	92.08	13240	12JL	4;10	286.75
13241	26JL	4;23	120.22	13241	10AU	16;44	315.24
13242	24AU	15;15	148.83	13242	9SE	4;35	344.14
13243	23SE	5;21	178.07	13243	8OC	15;53	13.53
13244	22OC	22;33	207.93	13244	7NO	2;46	43.33
13245	21NO	17;46	238.23	13245	6DE	13;23	73.35
13246	21DE	13;19	268.65				

71

NEW MOONS				FULL MOONS			
				13246	5JA	0;01	103.34
13247	20JA	7;36	298.85	13247	3FE	11;03	133.07
13248	18FE	23;27	328.54	13248	4MR	22;50	162.36
13249	20MR	12;20	357.62	13249	3AP	11;30	191.19
13250	18AP	22;21	26.09	13250	3MY	1;02	219.63
13251	18MY	6;11	54.15	13251	1JN	15;18	247.85
13252	16JN	12;53	82.03	13252	1JL	6;13	276.08
13253	15JL	19;42	110.02	13253	30JL	21;32	304.54
13254	14AU	3;46	138.36	13254	29AU	12;46	333.40
13255	12SE	13;59	167.26	13255	28SE	3;19	2.73
13256	12OC	2;52	196.76	13256	27OC	16;39	32.50
13257	10NO	18;29	226.80	13257	26NO	4;41	62.53
13258	10DE	12;25	257.15	13258	25DE	15;41	92.58

72

NEW MOONS				FULL MOONS			
13259	9JA	7;39	287.48	13259	24JA	2;00	122.40
13260	8FE	2;36	317.48	13260	22FE	11;56	151.79
13261	8MR	19;36	346.91	13261	22MR	21;44	180.68
13262	7AP	9;35	15.71	13262	21AP	7;49	209.12
13263	6MY	20;32	43.99	13263	20MY	18;49	237.28
13264	5JN	5;14	71.98	13264	19JN	7;25	265.40
13265	4JL	12;49	99.93	13265	18JL	22;03	293.72
13266	2AU	20;22	128.12	13266	17AU	14;28	322.46
13267	1SE	4;43	156.76	13267	16SE	7;43	351.71
13268	30SE	14;28	185.96	13268	16OC	0;31	21.46
13269	30OC	2;06	215.69	13269	14NO	15;58	51.54
13270	28NO	15;59	245.82	13270	14DE	5;43	81.71
13271	28DE	8;13	276.09				

73

NEW MOONS				FULL MOONS			
				13271	12JA	17;45	111.71
13272	27JA	2;12	306.19	13272	11FE	4;11	141.29
13273	25FE	20;30	335.87	13273	12MR	13;12	170.33
13274	27MR	13;30	4.98	13274	10AP	21;19	198.85
13275	26AP	4;14	33.54	13275	10MY	5;22	226.98
13276	25MY	16;36	61.73	13276	8JN	14;26	254.97
13277	24JN	3;07	89.76	13277	8JL	1;34	283.08
13278	23JL	12;29	117.92	13278	6AU	15;20	311.56
13279	21AU	21;21	146.41	13279	5SE	7;39	340.59
13280	20SE	6;18	175.39	13280	5OC	1;39	10.19
13281	19OC	15;56	204.89	13281	3NO	20;08	40.27
13282	18NO	2;49	234.81	13282	3DE	13;50	70.58
13283	17DE	15;23	264.94				

74

NEW MOONS				FULL MOONS			
				13283	2JA	5;40	100.81
13284	16JA	5;44	295.00	13284	31JA	18;59	130.66
13285	14FE	21;26	324.75	13285	2MR	5;37	159.95
13286	16MR	13;40	354.03	13286	31MR	14;05	188.65
13287	15AP	5;38	22.80	13287	29AP	21;16	216.87
13288	14MY	20;49	51.18	13288	29MY	4;17	244.81
13289	13JN	10;55	79.36	13289	27JN	12;15	272.76
13290	12JL	23;44	107.56	13290	26JL	22;05	300.96
13291	11AU	11;15	136.02	13291	25AU	10;28	329.65
13292	9SE	21;47	164.88	13292	24SE	1;39	358.94
13293	9OC	7;51	194.22	13293	23OC	19;27	28.84
13294	7NO	18;03	224.00	13294	22NO	14;57	59.16
13295	7DE	4;51	254.04	13295	22DE	10;27	89.58

75

NEW MOONS				FULL MOONS			
13296	5JA	16;23	284.07	13296	21JA	3;59	119.73
13297	4FE	4;35	313.83	13297	19FE	18;22	149.36
13298	5MR	17;26	343.16	13298	21MR	5;31	178.34
13299	4AP	7;01	12.01	13299	19AP	14;12	206.76
13300	3MY	21;27	40.47	13300	18MY	21;28	234.79
13301	2JN	12;36	68.72	13301	17JN	4;21	262.68
13302	2JL	3;56	96.97	13302	16JL	11;46	290.69
13303	31JL	18;42	125.42	13303	14AU	20;32	319.08
13304	30AU	8;27	154.23	13304	13SE	7;30	348.01
13305	28SE	21;09	183.51	13305	12OC	21;23	17.57
13306	28OC	9;06	213.22	13306	11NO	14;27	47.67
13307	26NO	20;34	243.23	13307	11DE	9;53	78.09
13308	26DE	7;34	273.28				

76

NEW MOONS				FULL MOONS			
				13308	10JA	5;46	108.44
13309	24JA	18;05	303.10	13309	8FE	23;59	138.39
13310	23FE	4;15	332.50	13310	9MR	15;17	167.74
13311	23MR	14;34	1.40	13311	8AP	3;35	196.47
13312	22AP	1;47	29.87	13312	7MY	13;25	224.70
13313	21MY	14;26	58.09	13313	5JN	21;33	252.66
13314	20JN	4;39	86.27	13314	5JL	4;50	280.60
13315	19JL	20;04	114.64	13315	3AU	12;08	308.79
13316	18AU	12;04	143.37	13316	1SE	20;24	337.43
13317	17SE	4;02	172.59	13317	1OC	6;36	6.66
13318	16OC	19;28	202.28	13318	30OC	19;30	36.46
13319	15NO	9;53	232.33	13319	29NO	11;17	66.67
13320	14DE	22;48	262.47	13320	29DE	5;13	97.00

77

NEW MOONS				FULL MOONS			
13321	13JA	10;01	292.43	13321	27JA	23;53	127.13
13322	11FE	19;44	321.97	13322	26FE	17;49	156.77
13323	13MR	4;32	350.99	13323	28MR	9;59	185.84
13324	11AP	13;11	19.51	13324	26AP	23;50	214.36
13325	10MY	22;30	47.69	13325	26MY	11;17	242.50
13326	9JN	9;10	75.74	13326	24JN	20;44	270.49
13327	8JL	21;40	103.91	13327	24JL	4;54	298.61
13328	7AU	12;13	132.44	13328	22AU	12;49	327.07
13329	6SE	4;47	161.48	13329	20SE	21;29	356.05
13330	5OC	22;44	191.10	13330	20OC	7;40	25.58
13331	4NO	16;48	221.18	13331	18NO	19;44	55.56
13332	4DE	9;28	251.45	13332	18DE	9;39	85.74

78

NEW MOONS				FULL MOONS			
13333	2JA	23;44	281.61	13333	17JA	1;08	115.85
13334	1FE	11;25	311.38	13334	15FE	17;41	145.62
13335	2MR	21;03	340.62	13335	17MR	10;39	174.91
13336	1AP	5;21	9.30	13336	16AP	3;07	203.69
13337	30AP	13;01	37.53	13337	15MY	18;09	232.06
13338	29MY	20;50	65.50	13338	14JN	7;09	260.19
13339	28JN	5;36	93.48	13339	13JL	18;16	288.34
13340	27JL	16;18	121.73	13340	12AU	4;09	316.73
13341	26AU	5;46	150.47	13341	10SE	13;41	345.56
13342	24SE	22;20	179.83	13342	9OC	23;31	14.91
13343	24OC	17;17	209.79	13343	8NO	9;59	44.71
13344	23NO	12;53	240.12	13344	7DE	21;10	74.76
13345	23DE	7;06	270.48				

79

NEW MOONS				FULL MOONS			
				13345	6JA	9;09	104.80
13346	21JA	22;44	300.55	13346	4FE	22;11	134.59
13347	20FE	11;32	330.09	13347	6MR	12;26	163.96
13348	21MR	21;51	359.04	13348	5AP	3;42	192.87
13349	20AP	6;12	27.43	13349	4MY	19;20	221.37
13350	19MY	13;19	55.45	13350	3JN	10;33	249.62
13351	17JN	20;06	83.33	13351	3JL	0;50	277.83
13352	17JL	3;41	111.36	13352	1AU	14;06	306.23
13353	15AU	13;15	139.78	13353	31AU	2;34	335.01
13354	14SE	1;46	168.79	13354	29SE	14;25	4.25
13355	13OC	17;34	198.44	13355	29OC	1;44	33.95
13356	12NO	12;02	228.62	13356	27NO	12;33	63.94
13357	12DE	7;39	259.04	13357	26DE	23;06	93.97

Left half (years 80–85)

NEW MOONS NUMBER	DATE	TIME	LONG.	FULL MOONS NUMBER	DATE	TIME	LONG.
80							
13358	11JA	2:43	289.35	13358	25JA	9:43	123.79
13359	9FE	19:49	319.25	13359	23FE	20:51	153.21
13360	10MR	10:07	348.54	13360	24MR	8:46	182.17
13361	8AP	21:27	17.22	13361	22AP	21:36	210.70
13362	8MY	6:13	45.40	13362	22MY	11:19	238.95
13363	6JN	13:20	73.31	13363	21JN	1:53	267.15
13364	5JL	20:00	101.24	13364	20JL	17:09	295.51
13365	4AU	3:24	129.44	13365	19AU	8:45	324.24
13366	2SE	12:34	158.13	13366	18SE	0:02	353.44
13367	2OC	0:11	187.42	13367	17OC	14:19	23.11
13368	31OC	14:30	217.29	13368	16NO	3:11	53.09
13369	30NO	7:21	247.55	13369	15DE	14:44	83.17
13370	30DE	1:59	277.91				
81							
				13370	14JA	1:17	113.10
13371	28JA	21:08	308.04	13371	12FE	11:13	142.64
13372	27FE	15:09	337.68	13372	13MR	20:48	171.68
13373	29MR	6:36	6.70	13373	12AP	6:25	200.24
13374	27AP	18:58	35.15	13374	11MY	16:39	228.45
13375	27MY	4:42	63.22	13375	10JN	4:13	256.53
13376	25JN	12:48	91.16	13376	9JL	17:47	284.75
13377	24JL	20:23	119.26	13377	8AU	9:29	313.33
13378	23AU	4:25	147.74	13378	7SE	2:41	342.41
13379	21SE	13:34	176.75	13379	6OC	20:09	12.03
13380	21OC	0:18	206.31	13380	5NO	12:36	42.04
13381	19NO	13:02	236.32	13381	5DE	3:23	72.24
13382	19DE	4:01	266.55				
82							
				13382	3JA	16:18	102.34
13383	17JA	21:04	296.72	13383	2FE	3:26	132.08
13384	16FE	15:10	326.54	13384	3MR	12:57	161.30
13385	18MR	8:46	355.84	13385	1AP	21:15	189.97
13386	17AP	0:34	24.58	13386	1MY	5:01	218.19
13387	16MY	14:04	52.88	13387	30MY	13:17	246.19
13388	15JN	1:32	80.95	13388	28JN	23:11	274.21
13389	14JL	11:34	109.06	13389	28JL	11:36	302.53
13390	12AU	20:50	137.44	13390	27AU	2:48	331.36
13391	11SE	5:53	166.26	13391	25SE	20:16	0.77
13392	10OC	15:18	195.60	13392	25OC	14:52	30.73
13393	9NO	1:35	225.40	13393	24NO	9:17	61.01
13394	8DE	13:14	255.47	13394	24DE	2:15	91.32
83							
13395	7JA	2:34	285.57	13395	22JA	16:52	121.34
13396	5FE	17:26	315.42	13396	21FE	4:44	150.83
13397	7MR	9:12	344.84	13397	22MR	14:04	179.73
13398	6AP	1:07	13.76	13398	20AP	21:37	208.08
13399	5MY	16:35	42.25	13399	20MY	4:28	236.08
13400	4JN	7:16	70.48	13400	18JN	11:46	263.99
13401	3JL	20:52	98.66	13401	17JL	20:34	292.07
13402	2AU	9:17	127.03	13402	16AU	7:39	320.56
13403	31AU	20:32	155.77	13403	14SE	21:31	349.64
13404	30SE	7:01	184.98	13404	14OC	14:11	19.34
13405	29OC	17:16	214.64	13405	13NO	9:06	49.54
13406	28NO	3:49	244.61	13406	13DE	4:54	79.97
13407	27DE	14:55	274.67				
84							
				13407	11JA	23:37	110.26
13408	26JA	2:37	304.53	13408	10FE	15:36	140.10
13409	24FE	14:53	333.99	13409	11MR	4:14	169.31
13410	25MR	3:45	2.97	13410	9AP	13:56	197.91
13411	23AP	17:29	31.52	13411	8MY	21:43	226.05
13412	23MY	8:09	59.81	13412	7JN	4:39	253.95
13413	21JN	23:29	88.03	13413	6JL	11:43	281.89
13414	21JL	14:47	116.41	13414	4AU	19:47	310.13
13415	20AU	5:21	145.11	13415	3SE	5:38	338.86
13416	18SE	18:52	174.25	13416	2OC	18:05	8.20
13417	18OC	7:26	203.85	13417	1NO	9:38	38.13
13418	16NO	19:18	233.80	13418	1DE	4:03	68.46
13419	16DE	6:35	263.88	13419	30DE	23:55	98.87
85							
13420	14JA	17:15	293.80	13420	29JA	19:05	128.99
13421	13FE	3:21	323.34	13421	28FE	11:48	158.56
13422	14MR	13:16	352.38	13422	30MR	1:29	187.50
13423	12AP	23:40	20.96	13423	28AP	12:25	215.88
13424	12MY	11:18	49.22	13424	27MY	21:19	243.91
13425	11JN	0:37	77.37	13425	26JN	4:59	271.84
13426	10JL	15:30	105.64	13426	25JL	12:17	299.93
13427	9AU	7:27	134.24	13427	23AU	20:05	328.41
13428	7SE	23:45	163.31	13428	22SE	5:21	357.43
13429	7OC	15:49	192.88	13429	21OC	16:55	27.04
13430	6NO	7:04	222.85	13430	20NO	7:17	57.12
13431	5DE	20:55	253.01	13431	20DE	0:11	87.43

Right half (years 86–91)

NEW MOONS NUMBER	DATE	TIME	LONG.	FULL MOONS NUMBER	DATE	TIME	LONG.
86							
13432	4JA	9:00	283.07	13432	18JA	18:30	117.65
13433	2FE	19:19	312.77	13433	17FE	12:45	147.47
13434	4MR	4:18	341.96	13434	19MR	5:43	176.73
13435	2AP	12:42	10.62	13435	17AP	20:40	205.42
13436	1MY	21:23	38.88	13436	17MY	9:19	233.68
13437	31MY	7:08	66.93	13437	15JN	19:48	261.71
13438	29JN	18:35	95.01	13438	15JL	4:39	289.77
13439	29JL	8:06	123.39	13439	13AU	12:46	318.11
13440	27AU	23:49	152.24	13440	11SE	21:08	346.92
13441	26SE	17:24	181.68	13441	11OC	6:39	16.27
13442	26OC	11:51	211.64	13442	9NO	17:49	46.11
13443	25NO	5:35	241.90	13443	9DE	6:48	76.25
13444	24DE	21:13	272.16				
87							
				13444	7JA	21:23	106.39
13445	23JA	10:08	302.09	13445	6FE	13:12	136.28
13446	21FE	20:36	331.52	13446	8MR	5:46	165.72
13447	23MR	5:19	0.38	13447	6AP	22:23	194.65
13448	21AP	13:03	28.73	13448	6MY	14:08	223.14
13449	20MY	20:34	56.75	13449	5JN	4:14	251.34
13450	19JN	4:40	84.69	13450	4JL	16:24	279.47
13451	18JL	14:16	112.81	13451	3AU	3:02	307.78
13452	17AU	2:18	141.35	13452	1SE	12:54	336.47
13453	15SE	17:25	170.49	13453	30SE	22:45	5.66
13454	15OC	11:29	200.26	13454	30OC	9:03	35.33
13455	14NO	7:10	230.51	13455	28NO	19:55	65.33
13456	14DE	2:23	260.91	13456	28DE	7:25	95.39
88							
13457	12JA	19:24	291.13	13457	26JA	19:41	125.27
13458	11FE	9:33	320.88	13458	25FE	9:01	154.76
13459	11MR	20:57	350.02	13459	25MR	23:31	183.80
13460	10AP	6:06	18.59	13460	24AP	14:49	212.41
13461	9MY	13:39	46.71	13461	24MY	6:13	240.72
13462	7JN	20:26	74.61	13462	22JN	21:04	268.92
13463	7JL	3:28	102.55	13463	22JL	11:01	297.25
13464	5AU	11:57	130.81	13464	21AU	0:09	325.91
13465	3SE	22:58	159.60	13465	19SE	12:36	355.02
13466	3OC	13:14	189.03	13466	19OC	0:25	24.60
13467	2NO	6:35	219.04	13467	17NO	11:37	54.53
13468	2DE	1:52	249.42	13468	16DE	22:17	84.57
13469	31DE	21:23	279.81				
89							
				13469	15JA	8:43	114.48
13470	30JA	15:31	309.87	13470	13FE	19:19	144.03
13471	1MR	7:11	339.39	13471	15MR	6:32	173.12
13472	30MR	19:55	8.27	13472	13AP	18:37	201.75
13473	29AP	5:49	36.60	13473	13MY	7:40	230.06
13474	28MY	13:38	64.59	13474	11JN	21:42	258.24
13475	26JN	20:25	92.48	13475	11JL	12:42	286.52
13476	26JL	3:24	120.57	13476	10AU	4:23	315.12
13477	24AU	11:42	149.08	13477	8SE	20:13	344.18
13478	22SE	22:08	178.16	13478	8OC	11:23	13.73
13479	22OC	11:10	207.84	13479	7NO	1:13	43.65
13480	21NO	2:49	237.98	13480	6DE	13:31	73.75
13481	20DE	20:35	268.33				
90							
				13481	5JA	0:31	103.76
13482	19JA	15:31	298.56	13482	3FE	10:35	133.44
13483	18FE	10:08	328.38	13483	4MR	20:06	162.64
13484	20MR	2:52	357.62	13484	3AP	5:24	191.33
13485	18AP	16:43	26.25	13485	2MY	15:00	219.61
13486	18MY	3:41	54.44	13486	1JN	1:38	247.69
13487	16JN	12:32	82.40	13487	30JN	14:03	275.82
13488	15JL	20:22	110.43	13488	30JL	4:45	304.25
13489	14AU	4:15	138.77	13489	28AU	21:30	333.16
13490	12SE	12:57	167.60	13490	27SE	15:16	2.62
13491	11OC	22:58	196.99	13491	27OC	8:37	32.55
13492	10NO	10:43	226.86	13492	26NO	0:29	62.74
13493	10DE	0:30	257.03	13493	25DE	14:24	92.91
91							
13494	8JA	16:25	287.22	13494	24JA	2:22	122.80
13495	7FE	9:54	317.17	13495	22FE	12:33	152.21
13496	9MR	3:42	346.64	13496	23MR	21:13	181.05
13497	7AP	20:21	15.56	13497	22AP	4:57	209.39
13498	7MY	10:57	44.00	13498	21MY	12:40	237.42
13499	5JN	23:23	72.13	13499	19JN	21:31	265.39
13500	5JL	10:16	100.22	13500	19JL	8:33	293.57
13501	3AU	20:03	128.50	13501	17AU	22:26	322.20
13502	2SE	5:23	157.18	13502	16SE	15:00	351.41
13503	1OC	14:46	186.36	13503	16OC	9:23	21.21
13504	31OC	0:41	216.02	13504	15NO	4:15	51.43
13505	29NO	11:38	246.02	13505	14DE	22:11	81.78
13506	29DE	0:00	276.13				

92

New Moons

NUMBER	DATE	TIME	LONG.
13507	27JA	13:55	306.06
13508	25FE	5:02	335.62
13509	26MR	20:40	4.68
13510	25AP	12:13	33.29
13511	25MY	3:15	61.58
13512	23JN	17:31	89.77
13513	23JL	6:45	118.07
13514	21AU	18:50	146.70
13515	20SE	5:56	175.77
13516	19OC	16:27	205.30
13517	18NO	2:55	235.20
13518	17DE	13:44	265.26

Full Moons

NUMBER	DATE	TIME	LONG.
13506	13JA	14:05	111.95
13507	12FE	3:17	141.64
13508	12MR	13:42	170.74
13509	10AP	21:54	199.26
13510	10MY	4:51	227.34
13511	8JN	11:43	255.24
13512	7JL	19:37	283.23
13513	6AU	5:31	311.55
13514	4SE	18:02	340.41
13515	4OC	9:25	9.90
13516	3NO	3:26	39.95
13517	2DE	23:04	70.35

93

New Moons

NUMBER	DATE	TIME	LONG.
13519	16JA	1:00	295.19
13520	14FE	12:44	324.78
13521	16MR	0:59	353.90
13522	14AP	13:58	22.56
13523	14MY	3:57	50.90
13524	12JN	18:57	79.11
13525	12JL	10:27	107.42
13526	11AU	1:42	136.02
13527	9SE	16:04	165.03
13528	9OC	5:22	194.51
13529	7NO	17:47	224.38
13530	7DE	5:27	254.45

Full Moons

NUMBER	DATE	TIME	LONG.
13518	1JA	18:35	100.73
13519	31JA	12:02	130.76
13520	2MR	2:15	160.20
13521	31MR	13:14	189.00
13522	29AP	21:47	217.27
13523	29MY	4:58	245.23
13524	27JN	11:53	273.13
13525	26JL	19:25	301.24
13526	25AU	4:22	329.79
13527	23SE	15:32	358.91
13528	23OC	5:34	28.64
13529	21NO	22:39	58.85
13530	21DE	17:58	89.27

94

New Moons

NUMBER	DATE	TIME	LONG.
13531	5JA	16:24	284.46
13532	4FE	2:37	314.13
13533	5MR	12:19	343.33
13534	3AP	22:06	12.03
13535	3MY	8:47	40.35
13536	1JN	21:04	68.49
13537	1JL	11:09	96.69
13538	31JL	2:44	125.16
13539	29AU	19:10	154.08
13540	28SE	11:42	183.50
13541	28OC	3:43	213.38
13542	26NO	18:31	243.53
13543	26DE	7:34	273.67

Full Moons

NUMBER	DATE	TIME	LONG.
13531	20JA	13:39	119.52
13532	19FE	7:36	149.30
13533	20MR	22:42	178.45
13534	19AP	10:53	207.01
13535	18MY	20:43	235.14
13536	17JN	4:57	263.09
13537	16JL	12:27	291.11
13538	14AU	20:02	319.44
13539	13SE	4:35	348.28
13540	12OC	15:01	17.69
13541	11NO	4:01	47.62
13542	10DE	19:40	77.87

95

New Moons

NUMBER	DATE	TIME	LONG.
13544	24JA	18:40	303.52
13545	23FE	4:06	332.88
13546	24MR	12:29	1.70
13547	22AP	20:43	30.05
13548	22MY	5:39	58.13
13549	20JN	16:04	86.16
13550	20JL	4:30	114.40
13551	18AU	19:12	143.07
13552	17SE	12:06	172.32
13553	17OC	6:31	202.13
13554	16NO	1:01	232.34
13555	15DE	17:58	262.65

Full Moons

NUMBER	DATE	TIME	LONG.
13543	9JA	13:17	108.13
13544	8FE	7:31	138.10
13545	10MR	1:02	167.55
13546	8AP	16:56	196.42
13547	8MY	6:42	224.82
13548	6JN	18:18	252.91
13549	6JL	4:02	280.95
13550	4AU	12:35	309.20
13551	2SE	20:55	337.85
13552	2OC	5:07	7.02
13553	31OC	16:22	36.71
13554	30NO	4:28	66.77
13555	29DE	18:09	96.92

96

New Moons

NUMBER	DATE	TIME	LONG.
13556	14JA	8:16	292.74
13557	12FE	19:49	322.36
13558	13MR	5:11	351.40
13559	11AP	13:09	19.90
13560	10MY	20:33	48.00
13561	9JN	4:09	75.93
13562	8JL	12:51	103.95
13563	6AU	23:34	132.31
13564	5SE	13:11	161.23
13565	5OC	5:59	190.79
13566	4NO	1:14	220.91
13567	3DE	21:03	251.31

Full Moons

NUMBER	DATE	TIME	LONG.
13556	28JA	9:10	126.90
13557	27FE	1:08	156.48
13558	27MR	17:34	185.56
13559	26AP	9:43	214.18
13560	26MY	0:42	242.46
13561	24JN	13:56	270.61
13562	24JL	1:29	298.86
13563	22AU	11:53	327.42
13564	20SE	21:56	356.46
13565	20OC	8:10	25.99
13566	18NO	18:51	55.91
13567	18DE	6:00	85.98

97

New Moons

NUMBER	DATE	TIME	LONG.
13568	2JA	15:21	281.64
13569	1FE	6:55	311.58
13570	2MR	19:33	340.94
13571	1AP	5:39	9.70
13572	30AP	13:49	37.94
13573	29MY	20:50	65.89
13574	28JN	3:36	93.79
13575	27JL	11:16	121.91
13576	25AU	20:58	150.49
13577	24SE	9:39	179.69
13578	24OC	1:37	209.52
13579	22NO	20:09	239.80
13580	22DE	15:44	270.22

Full Moons

NUMBER	DATE	TIME	LONG.
13568	16JA	17:41	115.92
13569	15FE	6:12	145.53
13570	16MR	19:48	174.70
13571	15AP	10:29	203.41
13572	15MY	1:46	231.79
13573	13JN	16:56	260.01
13574	13JL	7:29	288.29
13575	11AU	21:16	316.84
13576	10SE	10:21	345.81
13577	9OC	22:46	15.27
13578	8NO	10:29	45.12
13579	7DE	21:28	75.16

98

New Moons

NUMBER	DATE	TIME	LONG.
13581	21JA	10:39	300.43
13582	20FE	3:33	330.15
13583	21MR	17:41	359.25
13584	20AP	4:53	27.76
13585	19MY	13:38	55.85
13586	17JN	20:49	83.75
13587	17JL	3:39	111.74
13588	15AU	11:17	140.09
13589	13SE	20:41	168.97
13590	13OC	8:30	198.45
13591	11NO	22:53	228.45
13592	11DE	15:37	258.75

Full Moons

NUMBER	DATE	TIME	LONG.
13580	6JA	7:54	105.14
13581	4FE	18:11	134.81
13582	6MR	4:47	164.04
13583	4AP	16:07	192.79
13584	4MY	4:25	221.16
13585	2JN	17:49	249.35
13586	2JL	8:19	277.56
13587	31JL	23:50	306.04
13588	30AU	15:57	334.96
13589	29SE	7:52	4.37
13590	28OC	22:42	34.21
13591	27NO	11:56	64.30
13592	26DE	23:34	94.38

99

New Moons

NUMBER	DATE	TIME	LONG.
13593	10JA	9:58	289.04
13594	9FE	4:46	319.01
13595	10MR	22:26	348.45
13596	9AP	13:42	17.29
13597	9MY	2:03	45.62
13598	7JN	11:54	73.64
13599	6JL	20:14	101.63
13600	5AU	4:08	129.85
13601	3SE	12:32	158.51
13602	2OC	22:00	187.72
13603	1NO	8:56	217.44
13604	30NO	21:39	247.53
13605	30DE	12:23	277.73

Full Moons

NUMBER	DATE	TIME	LONG.
13593	25JA	9:58	124.19
13594	23FE	19:32	153.55
13595	25MR	4:40	182.39
13596	23AP	13:48	210.77
13597	22MY	23:38	238.87
13598	21JN	10:57	266.94
13599	21JL	0:28	295.23
13600	19AU	16:23	323.96
13601	18SE	10:03	353.25
13602	18OC	4:02	23.06
13603	16NO	20:58	53.22
13604	16DE	12:00	83.45

100

New Moons

NUMBER	DATE	TIME	LONG.
13606	29JA	4:58	307.76
13607	27FE	22:32	337.39
13608	28MR	15:41	6.48
13609	27AP	7:16	35.07
13610	26MY	20:48	63.29
13611	25JN	8:30	91.38
13612	24JL	18:57	119.59
13613	23AU	4:41	148.14
13614	21SE	14:13	177.16
13615	20OC	23:58	206.69
13616	19NO	10:25	236.60
13617	18DE	22:00	266.68

Full Moons

NUMBER	DATE	TIME	LONG.
13605	15JA	0:56	113.47
13606	13FE	11:52	143.05
13607	13MR	21:05	172.08
13608	12AP	5:01	200.57
13609	11MY	12:28	228.66
13610	9JN	20:29	256.61
13611	9JL	6:15	284.67
13612	7AU	18:42	313.12
13613	6SE	10:05	342.12
13614	6OC	3:53	11.74
13615	4NO	22:51	41.85
13616	4DE	17:32	72.21

101

New Moons

NUMBER	DATE	TIME	LONG.
13618	17JA	10:58	296.68
13619	16FE	1:16	326.35
13620	17MR	16:25	355.56
13621	16AP	7:49	24.30
13622	15MY	23:01	52.67
13623	14JN	13:44	80.87
13624	14JL	3:41	109.13
13625	12AU	16:37	137.65
13626	11SE	4:28	166.59
13627	10OC	15:28	196.00
13628	9NO	2:04	225.81
13629	8DE	12:44	255.84

Full Moons

NUMBER	DATE	TIME	LONG.
13617	3JA	10:37	102.48
13618	2FE	1:11	132.37
13619	3MR	12:51	161.68
13620	1AP	21:56	190.38
13621	1MY	5:16	218.59
13622	30MY	11:58	246.52
13623	28JN	19:11	274.44
13624	28JL	4:00	302.61
13625	26AU	15:13	331.27
13626	25SE	5:16	0.54
13627	24OC	22:07	30.41
13628	23NO	17:11	60.72
13629	23DE	13:01	91.15

102

New Moons

NUMBER	DATE	TIME	LONG.
13630	6JA	23:41	285.83
13631	5FE	10:59	315.54
13632	6MR	22:40	344.80
13633	5AP	10:56	13.58
13634	5MY	0:09	41.98
13635	3JN	14:32	70.20
13636	3JL	5:54	98.45
13637	1AU	21:33	126.95
13638	31AU	12:42	155.84
13639	30SE	2:51	185.19
13640	29OC	15:56	214.97
13641	28NO	4:07	245.02
13642	27DE	15:27	275.08

Full Moons

NUMBER	DATE	TIME	LONG.
13630	22JA	7:40	121.34
13631	20FE	23:30	151.01
13632	22MR	11:57	180.03
13633	20AP	21:31	208.46
13634	20MY	5:13	236.50
13635	18JN	12:09	264.39
13636	17JL	19:20	292.40
13637	16AU	3:35	320.78
13638	14SE	13:40	349.70
13639	14OC	2:16	19.23
13640	12NO	17:53	49.29
13641	12DE	12:13	79.65

103

New Moons

NUMBER	DATE	TIME	LONG.
13643	26JA	1:55	304.88
13644	24FE	11:37	334.24
13645	25MR	21:01	3.08
13646	24AP	6:53	31.48
13647	23MY	18:06	59.64
13648	22JN	7:10	87.77
13649	21JL	22:06	116.13
13650	20AU	14:22	144.89
13651	19SE	7:12	174.16
13652	18OC	23:51	203.92
13653	17NO	15:33	234.03
13654	17DE	5:39	264.23

Full Moons

NUMBER	DATE	TIME	LONG.
13642	11JA	7:52	109.99
13643	10FE	2:45	139.95
13644	11MR	19:13	169.33
13645	10AP	8:45	198.09
13646	9MY	19:39	226.35
13647	8JN	4:38	254.34
13648	7JL	12:30	282.31
13649	5AU	20:05	310.53
13650	4SE	4:12	339.19
13651	3OC	13:44	8.41
13652	2NO	1:27	38.17
13653	1DE	15:47	68.33
13654	31DE	8:25	98.60

```
--------NEW MOONS---------        ---------FULL MOONS--------        ---------NEW MOONS--------        ---------FULL  MOONS--------
NUMBER  DATE   TIME   LONG.       NUMBER  DATE   TIME   LONG.        NUMBER  DATE   TIME   LONG.       NUMBER  DATE   TIME   LONG.
```

104 / 110

NEW MOONS				FULL MOONS				NEW MOONS				FULL MOONS			
13655	15JA	17;42	294.21	13655	30JA	2;18	128.68	13729	8JA	8;35	287.28	13729	23JA	22;23	123.02
13656	14FE	3;47	323.74	13656	28FE	20;05	158.30	13730	6FE	22;01	317.06	13730	22FE	11;26	152.55
13657	14MR	12;24	352.73	13657	29MR	12;42	187.36	13731	8MR	12;29	346.41	13731	23MR	21;37	181.45
13658	12AP	20;24	21.21	13658	28AP	3;29	215.91	13732	7AP	3;32	15.27	13732	22AP	5;36	209.81
13659	12MY	4;41	49.34	13659	27MY	16;12	244.09	13733	6MY	18;42	43.74	13733	21MY	12;23	237.80
13660	10JN	14;09	77.34	13660	26JN	2;56	272.14	13734	5JN	9;39	71.97	13734	19JN	19;10	265.68
13661	10JL	1;28	105.47	13661	25JL	12;10	300.31	13735	5JL	0;08	100.20	13735	19JL	3;05	293.74
13662	8AU	15;03	133.97	13662	23AU	20;43	328.82	13736	3AU	13;51	128.63	13736	17AU	13;05	322.20
13663	7SE	7;02	163.01	13663	22SE	5;30	357.83	13737	2SE	2;32	157.45	13737	16SE	1;46	351.25
13664	7OC	1;02	192.65	13664	21OC	15;19	27.36	13738	1OC	14;12	186.72	13738	15OC	17;20	20.93
13665	5NO	19;54	222.77	13665	20NO	2;36	57.31	13739	31OC	1;09	216.43	13739	14NO	11;29	51.11
13666	5DE	13;58	253.11	13666	19DE	15;27	87.45	13740	29NO	11;50	246.42	13740	14DE	7;10	81.53
								13741	28DE	22;35	276.45				

105 / 111

NEW MOONS				FULL MOONS				NEW MOONS				FULL MOONS			
13667	4JA	5;43	283.32	13667	18JA	5;38	117.49					13741	13JA	2;37	111.85
13668	2FE	18;33	313.13	13668	16FE	20;54	147.20	13742	27JA	9;33	306.27	13742	11FF	19;55	141.72
13669	4MR	4;48	342.37	13669	18MR	12;52	176.43	13743	25FE	20;46	335.66	13743	13MR	9;58	170.97
13670	2AP	13;13	11.03	13670	17AP	5;03	205.18	13744	27MR	8;25	4.57	13744	11AP	20;48	199.60
13671	1MY	20;41	39.23	13671	16MY	20;38	233.56	13745	25AP	20;50	33.06	13745	11MY	5;15	227.75
13672	31MY	3;58	67.19	13672	15JN	10;51	261.74	13746	25MY	10;26	61.30	13746	9JN	12;27	255.66
13673	29JN	11;58	95.14	13673	14JL	23;24	289.95	13747	24JN	1;19	89.51	13747	8JL	19;28	283.61
13674	28JL	21;34	123.35	13674	13AU	10;33	318.41	13748	23JL	17;03	117.91	13748	7AU	3;10	311.84
13675	27AU	9;43	152.06	13675	11SE	20;57	347.30	13749	22AU	8;48	146.68	13749	5SE	12;22	340.57
13676	26SE	1;02	181.39	13676	11OC	7;17	16.69	13750	20SE	23;48	175.90	13750	4OC	23;43	9.89
13677	25OC	19;21	211.34	13677	9NO	17;53	46.51	13751	20OC	13;42	205.18	13751	3NO	13;51	39.77
13678	24NO	15;15	241.69	13678	9DE	4;49	76.55	13752	19NO	2;30	235.58	13752	3DE	6;54	70.05
13679	24DE	10;35	272.09					13753	18DE	14;19	265.67				

106 / 112

NEW MOONS				FULL MOONS				NEW MOONS				FULL MOONS			
				13679	7JA	16;05	106.56					13753	2JA	2;01	100.42
13680	23JA	3;34	302.21	13680	6FE	3;55	136.28	13754	17JA	1;09	295.59	13754	31JA	21;23	130.54
13681	21FE	17;35	331.78	13681	7MR	16;39	165.57	13755	15FE	11;02	325.10	13755	1MR	15;03	160.13
13682	23MR	4;47	0.74	13682	6AP	6;32	194.40	13756	15MR	20;16	354.09	13756	31MR	5;57	189.09
13683	21AP	13;45	29.14	13683	5MY	21;21	222.86	13757	14AP	5;33	22.61	13757	29AP	18;03	217.51
13684	20MY	21;11	57.16	13684	4JN	12;35	251.10	13758	13MY	15;46	50.80	13758	29MY	3;56	245.57
13685	19JN	3;56	85.05	13685	4JL	3;34	279.35	13759	12JN	3;43	78.89	13759	27JN	12;22	273.53
13686	18JL	11;03	113.06	13686	2AU	17;57	307.80	13760	11JL	17;44	107.14	13760	26JL	20;08	301.66
13687	16AU	19;40	141.47	13687	1SE	7;40	336.65	13761	10AU	9;31	135.75	13761	25AU	4;03	330.16
13688	15SE	6;51	170.44	13688	30SE	20;44	5.97	13762	9SE	2;24	164.86	13762	23SE	12;55	359.19
13689	14OC	21;17	200.06	13689	30OC	9;02	35.72	13763	8OC	19;31	194.49	13763	22OC	23;34	28.78
13690	13NO	14;43	230.20	13690	28NO	20;30	65.74	13764	7NO	12;02	224.53	13764	21NO	12;35	58.82
13691	13DE	9;58	260.61	13691	28DE	7;10	95.77	13765	7DE	3;09	254.75	13765	21DE	4;03	89.06

107 / 113

NEW MOONS				FULL MOONS				NEW MOONS				FULL MOONS			
13692	12JA	5;21	290.92	13692	26JA	17;20	125.56	13766	5JA	16;17	284.84	13766	19JA	21;16	119.22
13693	10FE	23;17	320.84	13693	25FE	3;29	154.92	13767	4FE	3;14	314.55	13767	18FE	15;00	149.00
13694	12MR	14;45	350.16	13694	26MR	14;08	183.80	13768	5MR	12;20	343.72	13768	20MR	8;06	178.24
13695	11AP	3;20	18.86	13695	25AP	1;39	212.26	13769	3AP	20;20	12.35	13769	18AP	23;44	206.95
13696	10MY	13;12	47.08	13696	24MY	14;17	240.47	13770	3MY	4;11	40.55	13770	18MY	13;29	235.24
13697	8JN	21;04	75.02	13697	23JN	4;09	268.64	13771	1JN	12;48	68.55	13771	17JN	1;17	263.33
13698	8JL	4;00	102.96	13698	22JL	19;16	297.01	13772	30JN	23;03	96.59	13772	16JL	11;22	291.45
13699	6AU	11;13	131.17	13699	21AU	11;22	325.77	13773	30JL	11;28	124.94	13773	14AU	20;22	319.84
13700	4SE	19;46	159.86	13700	20SE	3;48	355.04	13774	29AU	2;21	153.78	13774	13SE	5;09	348.69
13701	4OC	6;26	189.13	13701	19OC	19;34	24.78	13775	27SE	19;36	183.22	13775	12OC	14;33	18.06
13702	2NO	19;36	218.97	13702	18NO	9;50	54.84	13776	27OC	14;25	213.21	13776	11NO	1;09	47.89
13703	2DE	11;11	249.18	13703	17DE	22;20	84.96	13777	26NO	9;17	243.53	13777	10DE	13;13	77.99
								13778	26DE	2;25	273.84				

108 / 114

NEW MOONS				FULL MOONS				NEW MOONS				FULL MOONS			
13704	1JA	4;42	279.49	13704	16JA	9;15	114.89					13778	9JA	2;36	108.07
13705	30JA	23;15	309.58	13705	14FE	19;03	144.41	13779	24JA	16;42	303.82	13779	7FE	17;06	137.88
13706	29FE	17;29	339.21	13706	15MR	4;08	173.40	13780	23FE	4;04	333.26	13780	9MR	8;28	167.25
13707	30MR	9;57	8.25	13707	13AP	12;59	201.91	13781	24MR	13;10	2.11	13781	8AP	0;22	196.14
13708	28AP	23;42	36.75	13708	12MY	22;10	230.07	13782	22AP	20;52	30.45	13782	7MY	16;12	224.62
13709	28MY	10;46	64.86	13709	11JN	8;29	258.10	13783	22MY	4;02	58.45	13783	6JN	7;11	252.86
13710	26JN	19;50	92.85	13710	10JL	20;47	286.27	13784	20JN	11;31	86.36	13784	5JL	20;44	281.05
13711	26JL	4;00	120.98	13711	9AU	11;35	314.83	13785	19JL	20;11	114.45	13785	4AU	8;46	309.43
13712	24AU	12;15	149.49	13712	8SE	4;41	343.93	13786	18AU	7;00	142.96	13786	2SE	19;44	338.18
13713	22SE	21;18	178.51	13713	7OC	22;57	13.60	13787	16SE	20;47	172.07	13787	2OC	6;18	7.43
13714	22OC	7;35	208.08	13714	6NO	16;48	43.68	13788	16OC	13;48	201.82	13788	31OC	16;55	37.13
13715	20NO	19;24	238.06	13715	6DE	9;00	73.94	13789	15NO	9;16	232.06	13789	30NO	3;45	67.13
13716	20DE	9;01	268.23					13790	15DE	5;12	262.50	13790	29DE	14;48	97.17

109 / 115

NEW MOONS				FULL MOONS				NEW MOONS				FULL MOONS			
				13716	4JA	23;01	104.08	13791	13JA	23;30	292.76	13791	28JA	2;07	126.99
13717	19JA	0;32	298.32	13717	3FE	10;52	133.84	13792	12FE	14;57	322.54	13792	26FE	14;05	156.41
13718	17FE	17;28	328.08	13718	4MR	20;46	163.05	13793	14MR	3;25	351.71	13793	28MR	3;04	185.36
13719	19MR	10;44	357.35	13719	3AP	5;06	191.70	13794	12AP	13;20	20.29	13794	26AP	17;12	213.91
13720	18AP	3;03	26.09	13720	2MY	12;31	219.90	13795	11MY	21;22	48.42	13795	26MY	8;09	242.19
13721	17MY	17;35	54.42	13721	31MY	19;59	247.85	13796	10JN	4;20	76.32	13796	24JN	23;20	270.42
13722	16JN	6;15	82.54	13722	30JN	4;40	275.83	13797	9JL	11;10	104.26	13797	24JL	14;13	298.79
13723	15JL	17;26	110.70	13723	29JL	15;41	304.11	13798	7AU	18;58	132.51	13798	23AU	4;32	327.51
13724	14AU	3;42	139.14	13724	28AU	5;41	332.90	13799	6SE	4;50	161.28	13799	21SE	18;15	356.70
13725	12SE	13;32	168.02	13725	26SE	22;32	2.31	13800	5OC	17;42	190.67	13800	21OC	7;13	26.34
13726	11OC	23;20	197.39	13726	26OC	17;15	32.29	13801	4NO	9;46	220.64	13801	19NO	19;17	56.32
13727	10NO	9;30	227.20	13727	25NO	12;24	62.61	13802	4DE	4;18	250.99	13802	19DE	6;22	86.38
13728	9DE	20;27	257.25	13728	25DE	6;30	92.97								

116 / 122

NUMBER	DATE	TIME	LONG.	NUMBER	DATE	TIME	LONG.	NUMBER	DATE	TIME	LONG.	NUMBER	DATE	TIME	LONG.
13803	2JA	23:45	281.37	13803	17JA	16:38	116.27								
13804	1FE	18:27	311.45	13804	16FE	2:33	145.77	13878	26JA	2:18	305.29	13877	10JA	16:35	109.73
13805	2MR	11:08	340.98	13805	16MR	12:37	174.79	13879	24FE	12:08	334.65	13878	9FE	9:58	139.64
13806	1AP	1:05	9.90	13806	14AP	23:23	203.35	13880	25MR	20:24	3.44	13879	11MR	3:16	169.06
13807	30AP	12:13	38.26	13807	14MY	11:13	231.60	13881	24AP	4:01	31.75	13880	9AP	19:31	197.94
13808	29MY	20:59	66.28	13808	13JN	0:19	259.74	13882	23MY	11:59	59.78	13881	9MY	10:12	226.36
13809	28JN	4:19	94.20	13809	12JL	14:50	288.01	13883	21JN	21:14	87.76	13882	7JN	23:02	254.50
13810	27JL	11:22	122.30	13810	11AU	6:39	316.63	13884	21JL	8:28	115.97	13883	7JL	10:05	282.60
13811	25AU	19:17	150.81	13811	9SE	23:17	345.75	13885	19AU	22:10	144.62	13884	5AU	19:46	310.90
13812	24SE	4:57	179.89	13812	9OC	15:48	15.37	13886	18SE	14:25	173.85	13885	4SE	4:46	339.60
13813	23OC	16:56	209.54	13813	8NO	7:10	45.37	13887	18OC	8:47	203.68	13886	3OC	13:58	8.81
13814	22NO	7:20	239.64	13814	7DE	20:41	75.52	13888	17NO	4:02	233.93	13887	2NO	0:04	38.50
13815	21DE	23:52	269.93					13889	16DE	22:20	264.30	13888	1DE	11:25	68.53
												13889	31DE	0:03	98.64

117 / 123

NUMBER	DATE	TIME	LONG.	NUMBER	DATE	TIME	LONG.	NUMBER	DATE	TIME	LONG.	NUMBER	DATE	TIME	LONG.
				13815	6JA	8:20	105.55	13890	15JA	14:08	294.44	13890	29JA	13:48	128.54
13816	20JA	17:52	300.11	13816	4FE	18:32	135.22	13891	14FE	2:50	324.09	13891	28FE	4:28	158.05
13817	19FE	12:13	329.91	13817	6MR	3:44	164.38	13892	15MR	12:51	353.14	13892	29MR	19:51	187.07
13818	21MR	5:33	359.15	13818	4AP	12:25	193.02	13893	13AP	21:01	21.63	13893	28AP	11:36	215.66
13819	19AP	20:38	27.82	13819	3MY	21:08	221.26	13894	13MY	4:14	49.71	13894	28MY	3:03	243.96
13820	19MY	9:01	56.05	13820	2JN	6:37	249.29	13895	11JN	11:23	77.62	13895	26JN	17:28	272.15
13821	17JN	19:04	84.06	13821	1JL	17:45	277.37	13896	10JL	19:21	105.61	13896	26JL	6:27	300.47
13822	17JL	3:42	112.13	13822	31JL	7:17	305.76	13897	9AU	5:01	133.95	13897	24AU	18:10	329.11
13823	15AU	12:00	140.51	13823	29AU	23:27	334.67	13898	7SE	17:18	162.84	13898	23SE	5:08	358.20
13824	13SE	20:47	169.36	13824	28SE	17:33	4.16	13899	7OC	8:49	192.36	13899	22OC	15:54	27.78
13825	13OC	6:34	198.76	13825	28OC	12:02	34.15	13900	6NO	3:20	222.46	13900	21NO	2:45	57.71
13826	11NO	17:39	228.62	13826	27NO	5:21	64.41	13901	5DE	23:22	252.88	13901	20DE	13:41	87.77
13827	11DE	6:18	258.74	13827	26DE	20:34	94.64								

118 / 124

NUMBER	DATE	TIME	LONG.	NUMBER	DATE	TIME	LONG.	NUMBER	DATE	TIME	LONG.	NUMBER	DATE	TIME	LONG.
13828	9JA	20:42	288.87	13828	25JA	9:27	124.55	13902	4JA	18:42	283.24	13902	19JA	0:42	117.68
13829	8FE	12:46	318.73	13829	23FE	20:11	153.96	13903	3FE	11:36	313.22	13903	17FE	12:03	147.22
13830	10MR	5:44	348.15	13830	25MR	5:05	182.79	13904	4MR	1:27	342.61	13904	18MR	0:10	176.30
13831	8AP	22:27	17.06	13831	23AP	12:42	211.11	13905	2AP	12:28	11.39	13905	16AP	13:26	204.94
13832	8MY	13:50	45.52	13832	22MY	19:53	239.11	13906	1MY	21:18	39.65	13906	16MY	3:50	233.28
13833	7JN	3:28	73.69	13833	21JN	3:44	267.04	13907	31MY	4:40	67.60	13907	14JN	18:56	261.49
13834	6JL	15:29	101.83	13834	20JL	13:26	295.17	13908	29JN	11:28	95.50	13908	14JL	10:07	289.80
13835	5AU	2:23	130.17	13835	19AU	1:57	323.77	13909	28JL	18:43	123.62	13909	13AU	0:59	318.41
13836	3SE	12:39	158.91	13836	17SE	17:33	352.96	13910	27AU	3:31	152.19	13910	11SE	15:19	347.46
13837	2OC	22:39	188.14	13837	17OC	11:38	22.76	13911	25SE	14:54	181.36	13911	11OC	4:58	16.98
13838	1NO	8:44	217.83	13838	16NO	6:54	53.01	13912	25OC	5:28	211.14	13912	9NO	17:44	46.89
13839	30NO	19:18	247.82	13839	16DE	1:47	83.40	13913	23NO	22:55	241.38	13913	9DE	5:23	76.97
13840	30DE	6:43	277.87					13914	23DE	18:03	271.78				

119 / 125

NUMBER	DATE	TIME	LONG.	NUMBER	DATE	TIME	LONG.	NUMBER	DATE	TIME	LONG.	NUMBER	DATE	TIME	LONG.
				13840	14JA	18:54	113.60					13914	7JA	15:59	106.94
13841	28JA	19:17	307.74	13841	13FE	9:21	143.34	13915	22JA	13:12	301.99	13915	6FE	1:51	136.58
13842	27FE	8:59	337.21	13842	14MR	20:49	172.45	13916	21FE	6:53	331.73	13916	7MR	11:32	165.74
13843	28MR	23:30	6.22	13843	13AP	5:42	200.98	13917	22MR	22:08	0.85	13917	5AP	21:37	194.42
13844	27AP	14:25	34.78	13844	12MY	12:51	229.07	13918	21AP	10:36	29.40	13918	5MY	8:38	222.73
13845	27MY	5:24	63.07	13845	10JN	19:27	256.95	13919	20MY	20:29	57.52	13919	3JN	20:55	250.87
13846	25JN	20:13	91.28	13846	10JL	2:40	284.91	13920	19JN	4:29	85.45	13920	3JL	10:40	279.06
13847	25JL	10:33	119.64	13847	8AU	11:35	313.22	13921	18JL	11:38	113.47	13921	2AU	1:57	307.54
13848	24AU	0:03	148.34	13848	6SE	22:57	342.05	13922	16AU	19:08	141.83	13922	31AU	18:29	336.49
13849	22SE	12:31	177.48	13849	6OC	13:11	11.51	13923	15SE	3:58	170.71	13923	30SE	11:30	5.97
13850	22OC	0:02	207.08	13850	5NO	6:10	41.53	13924	14OC	14:52	200.18	13924	30OC	3:50	35.89
13851	20NO	10:56	237.01	13851	5DE	1:17	71.91	13925	13NO	4:06	230.14	13925	28NO	18:29	66.05
13852	19DE	21:38	267.06					13926	12DE	19:33	260.38	13926	28DE	7:06	96.16

120 / 126

NUMBER	DATE	TIME	LONG.	NUMBER	DATE	TIME	LONG.	NUMBER	DATE	TIME	LONG.	NUMBER	DATE	TIME	LONG.
				13852	3JA	21:04	102.30								
13853	18JA	8:22	296.96	13853	2FE	15:34	132.36	13927	11JA	12:45	290.61	13927	26JA	17:54	125.97
13854	16FE	19:14	326.49	13854	3MR	7:13	161.84	13928	10FE	6:52	320.54	13928	25FE	3:23	155.31
13855	17MR	6:21	355.54	13855	1AP	19:31	190.67	13929	12MR	0:41	349.97	13929	26MR	12:04	184.10
13856	15AP	18:02	24.13	13856	1MY	4:59	218.96	13930	10AP	16:52	18.83	13930	24AP	20:29	212.44
13857	15MY	6:47	52.41	13857	30MY	12:40	246.93	13931	10MY	6:35	47.20	13931	24MY	5:18	240.50
13858	13JN	20:56	80.59	13858	28JN	19:41	274.84	13932	8JN	17:47	75.27	13932	22JN	15:23	268.52
13859	13JL	12:23	108.91	13859	28JL	3:03	302.96	13933	8JL	3:09	103.31	13933	22JL	3:38	296.77
13860	12AU	4:27	137.55	13860	26AU	11:32	331.51	13934	6AU	11:42	131.58	13934	20AU	18:35	325.48
13861	10SE	20:11	166.64	13861	24SE	21:50	0.62	13935	4SE	20:22	160.27	13935	19SE	12:02	354.78
13862	10OC	10:57	196.20	13862	24OC	10:35	30.31	13936	4OC	5:47	189.50	13936	19OC	6:46	24.63
13863	9NO	0:31	226.13	13863	23NO	2:12	60.47	13937	2NO	16:18	219.22	13937	18NO	1:03	54.86
13864	8DE	12:57	256.24	13864	22DE	20:22	90.83	13938	2DE	4:08	249.27	13938	17DE	17:29	85.15
								13939	31DE	17:31	279.40				

121 / 127

NUMBER	DATE	TIME	LONG.	NUMBER	DATE	TIME	LONG.	NUMBER	DATE	TIME	LONG.	NUMBER	DATE	TIME	LONG.
13865	7JA	0:16	286.25	13865	21JA	15:43	121.06					13939	16JA	7:33	115.21
13866	5FE	10:28	315.91	13866	20FE	10:15	150.85	13940	30JA	8:33	309.36	13940	14FE	19:14	144.80
13867	6MR	19:46	345.07	13867	22MR	2:28	180.03	13941	1MR	0:54	338.92	13941	16MR	4:52	173.82
13868	5AP	4:40	13.71	13868	20AP	15:52	208.62	13942	30MR	17:38	7.98	13942	14AP	12:54	202.29
13869	4MY	14:04	41.96	13869	20MY	2:47	236.79	13943	29AP	9:38	36.57	13943	13MY	20:02	230.37
13870	3JN	0:54	70.04	13870	18JN	11:56	264.76	13944	29MY	0:09	64.82	13944	12JN	3:19	258.28
13871	2JL	13:48	98.20	13871	17JL	20:05	292.82	13945	27JN	13:03	92.96	13945	11JL	11:55	286.30
13872	1AU	4:49	126.66	13872	16AU	3:59	321.19	13946	27JL	0:40	121.23	13946	9AU	22:58	314.71
13873	30AU	21:26	155.60	13873	14SE	12:27	350.04	13947	25AU	11:28	149.84	13947	8SE	13:07	343.68
13874	29SE	14:47	185.08	13874	13OC	22:15	19.45	13948	23SE	21:49	178.92	13948	8OC	6:13	13.29
13875	29OC	7:58	215.02	13875	12NO	10:05	49.35	13949	23OC	8:01	208.49	13949	7NO	1:13	43.41
13876	28NO	0:04	245.24	13876	12DE	0:18	79.54	13950	21NO	18:22	238.40	13950	6DE	20:34	73.81
13877	27DE	14:20	275.42					13951	21DE	5:17	268.46				

	NEW MOONS				FULL MOONS		
NUMBER	DATE	TIME	LONG.	NUMBER	DATE	TIME	LONG.

128

NEW MOONS NUMBER	DATE	TIME	LONG.	FULL MOONS NUMBER	DATE	TIME	LONG.
				13951	5JA	14;44	104.12
13952	19JA	17;05	298.39	13952	4FE	6;33	134.04
13953	18FE	5;59	327.99	13953	4MR	19;25	163.38
13954	18MR	19;49	357.13	13954	3AP	5;25	192.10
13955	17AP	10;19	25.81	13955	2MY	13;13	220.31
13956	17MY	1;08	54.15	13956	31MY	19;54	248.23
13957	15JN	16;03	82.36	13957	30JN	2;40	276.14
13958	15JL	6;48	110.66	13958	29JL	10;40	304.29
13959	13AU	21;02	139.25	13959	27AU	20;49	332.93
13960	12SE	10;20	168.27	13960	26SE	9;41	2.17
13961	11OC	22;35	197.75	13961	26OC	1;24	32.01
13962	10NO	9;55	227.61	13962	24NO	19;36	62.29
13963	9DE	20;44	257.65	13963	24DE	15;14	92.71

129

NEW MOONS NUMBER	DATE	TIME	LONG.	FULL MOONS NUMBER	DATE	TIME	LONG.
13964	8JA	7;22	287.62	13964	23JA	10;32	122.91
13965	6FE	17;59	317.28	13965	22FE	3;38	152.61
13966	8MR	4;41	346.47	13966	23MR	17;31	181.67
13967	6AP	15;45	15.18	13967	22AP	4;15	210.14
13968	6MY	3;39	43.52	13968	21MY	12;40	238.19
13969	4JN	16;55	71.69	13969	19JN	19;56	266.10
13970	4JL	7;45	99.93	13970	19JL	3;06	294.12
13971	2AU	23;46	128.45	13971	17AU	11;04	322.51
13972	1SE	16;03	157.41	13972	15SE	20;30	351.43
13973	1OC	7;40	186.84	13973	15OC	8;03	20.93
13974	30OC	22;06	216.69	13974	13NO	22;14	50.94
13975	29NO	11;15	246.79	13975	13DE	15;09	81.24
13976	28DE	23;09	276.87				

130

NEW MOONS NUMBER	DATE	TIME	LONG.	FULL MOONS NUMBER	DATE	TIME	LONG.
				13976	12JA	9;59	111.54
13977	27JA	9;49	306.67	13977	11FE	4;59	141.49
13978	25FE	19;20	335.99	13978	12MR	22;19	170.88
13979	27MR	4;07	4.78	13979	11AP	13;02	199.67
13980	25AP	12;55	33.13	13980	11MY	1;07	227.97
13981	24MY	22;44	61.22	13981	9JN	11;07	255.99
13982	23JN	10;26	89.30	13982	8JL	19;48	284.01
13983	23JL	0;25	117.63	13983	7AU	3;55	312.26
13984	21AU	16;28	146.40	13984	5SE	12;12	340.95
13985	20SE	9;47	175.71	13985	4OC	21;23	10.18
13986	20OC	3;26	205.53	13986	3NO	8;13	39.92
13987	18NO	20;24	235.71	13987	2DE	21;13	70.03
13988	18DE	11;46	265.95				

131

NEW MOONS NUMBER	DATE	TIME	LONG.	FULL MOONS NUMBER	DATE	TIME	LONG.
				13988	1JA	12;24	100.23
13989	17JA	0;54	295.97	13989	31JA	5;09	130.25
13990	15FE	11;39	325.51	13990	1MR	22;21	159.83
13991	16MR	20;27	354.49	13991	31MR	13;00	188.87
13992	15AP	4;05	22.93	13992	30AP	6;24	217.43
13993	14MY	11;36	51.02	13993	29MY	20;12	245.65
13994	12JN	19;59	78.97	13994	28JN	8;16	273.76
13995	12JL	6;07	107.06	13995	27JL	18;46	301.99
13996	10AU	18;35	135.53	13996	26AU	4;15	330.56
13997	9SE	9;40	164.56	13997	24SE	13;29	359.61
13998	9OC	3;14	194.20	13998	23OC	23;14	29.16
13999	7NO	22;25	224.34	13999	22NO	10;00	59.10
14000	7DE	17;33	254.73	14000	21DE	21;57	89.20

132

NEW MOONS NUMBER	DATE	TIME	LONG.	FULL MOONS NUMBER	DATE	TIME	LONG.
14001	6JA	10;47	285.00	14001	20JA	10;59	119.17
14002	5FE	1;00	314.84	14002	19FE	0;55	148.80
14003	5MR	12;10	344.10	14003	19MR	15;39	177.96
14004	3AP	21;01	12.76	14004	18AP	7;02	206.67
14005	3MY	4;29	40.95	14005	17MY	22;37	235.04
14006	1JN	11;30	68.88	14006	16JN	13;40	263.25
14007	30JN	18;56	96.82	14007	16JL	3;34	291.52
14008	30JL	3;39	125.01	14008	14AU	16;47	320.06
14009	28AU	14;36	153.69	14009	13SE	3;42	349.02
14010	27SE	4;33	182.98	14010	12OC	14;46	18.46
14011	26OC	21;45	212.90	14011	11NO	1;44	48.31
14012	25NO	17;20	243.24	14012	10DE	12;40	78.35
14013	25DE	13;18	273.67				

133

NEW MOONS NUMBER	DATE	TIME	LONG.	FULL MOONS NUMBER	DATE	TIME	LONG.
				14013	8JA	23;32	108.33
14014	24JA	7;31	303.82	14014	7FE	10;27	138.00
14015	22FE	22;49	333.44	14015	8MR	21;51	167.21
14016	24MR	11;06	2.42	14016	7AP	10;13	195.96
14017	22AP	20;53	30.83	14017	6MY	23;50	224.36
14018	22MY	4;51	58.87	14018	5JN	14;32	252.58
14019	20JN	11;51	86.76	14019	5JL	5;46	280.84
14020	19JL	18;48	114.78	14020	3AU	21;02	309.34
14021	18AU	2;48	143.18	14021	2SE	11;56	338.25
14022	16SE	12;53	172.13	14022	2OC	2;16	7.65
14023	16OC	1;54	201.70	14023	31OC	15;45	37.47
14024	14NO	18;01	231.80	14024	30NO	4;07	67.53
14025	14DE	12;27	262.18	14025	29DE	15;13	97.58

134

NEW MOONS NUMBER	DATE	TIME	LONG.	FULL MOONS NUMBER	DATE	TIME	LONG.
14026	13JA	7;41	292.48	14026	28JA	1;17	127.34
14027	12FE	2;06	322.40	14027	26FE	10;47	156.66
14028	13MR	18;31	351.74	14028	27MR	20;20	185.47
14029	12AP	8;19	20.48	14029	26AP	6;35	213.87
14030	11MY	19;26	48.73	14030	25MY	17;59	242.02
14031	10JN	4;19	76.71	14031	24JN	6;53	270.15
14032	9JL	11;52	104.68	14032	23JL	21;27	298.51
14033	7AU	19;12	132.91	14033	22AU	13;35	327.29
14034	6SE	3;25	161.61	14034	21SE	6;45	356.61
14035	5OC	13;21	190.87	14035	20OC	23;52	26.42
14036	4NO	1;29	220.68	14036	19NO	15;40	56.55
14037	3DE	15;50	250.85	14037	19DE	5;24	86.73

135

NEW MOONS NUMBER	DATE	TIME	LONG.	FULL MOONS NUMBER	DATE	TIME	LONG.
14038	2JA	8;05	281.10	14038	17JA	17;02	116.67
14039	1FE	1;38	311.13	14039	16FE	2;59	146.18
14040	2MR	19;31	340.73	14040	17MR	11;49	175.14
14041	1AP	12;30	9.77	14041	15AP	20;05	203.61
14042	1MY	3;27	38.30	14042	15MY	4;26	231.72
14043	30MY	15;55	66.46	14043	13JN	13;39	259.70
14044	29JN	2;13	94.50	14044	13JL	0;40	287.83
14045	28JL	11;15	122.68	14045	11AU	14;16	316.36
14046	26AU	19;58	151.23	14046	10SE	6;41	345.45
14047	25SE	5;10	180.29	14047	10OC	1;12	15.14
14048	24OC	15;15	209.86	14048	8NO	20;07	45.29
14049	23NO	2;26	239.82	14049	8DE	13;45	75.61
14050	22DE	14;56	269.94				

136

NEW MOONS NUMBER	DATE	TIME	LONG.	FULL MOONS NUMBER	DATE	TIME	LONG.
				14050	7JA	5;04	105.80
14051	21JA	4;56	299.96	14051	5FE	17;52	135.57
14052	19FE	20;25	329.64	14052	6MR	4;21	164.79
14053	20MR	12;48	358.85	14053	4AP	12;58	193.43
14054	19AP	5;06	27.58	14054	3MY	20;19	221.61
14055	18MY	20;20	55.93	14055	2JN	3;18	249.54
14056	17JN	10;06	84.09	14056	1JL	11;02	277.49
14057	16JL	22;30	112.31	14057	30JL	20;45	305.73
14058	15AU	9;55	140.81	14058	29AU	9;23	334.48
14059	13SE	20;44	169.75	14059	28SE	1;11	3.87
14060	13OC	7;12	199.18	14060	27OC	19;31	33.85
14061	11NO	17;35	229.01	14061	26NO	15;00	64.19
14062	11DE	4;11	259.04	14062	26DE	9;58	94.58

137

NEW MOONS NUMBER	DATE	TIME	LONG.	FULL MOONS NUMBER	DATE	TIME	LONG.
14063	9JA	15;23	289.03	14063	25JA	3;03	124.67
14064	8FE	3;29	318.73	14064	23FE	17;21	154.23
14065	9MR	16;35	348.00	14065	25MR	4;38	183.16
14066	8AP	6;29	16.81	14066	23AP	13;20	211.52
14067	7MY	20;57	45.23	14067	22MY	20;23	239.51
14068	6JN	11;46	73.45	14068	21JN	2;57	267.39
14069	6JL	2;44	101.71	14069	20JL	10;15	295.43
14070	4AU	17;30	130.20	14070	18AU	19;18	323.88
14071	3SE	7;37	159.09	14071	17SE	6;52	352.91
14072	2OC	20;41	188.45	14072	16OC	21;14	22.55
14073	1NO	8;40	218.21	14073	15NO	14;19	52.70
14074	30NO	19;49	248.23	14074	15DE	9;23	83.10
14075	30DE	6;29	278.25				

138

NEW MOONS NUMBER	DATE	TIME	LONG.	FULL MOONS NUMBER	DATE	TIME	LONG.
				14075	14JA	5;01	113.41
14076	28JA	16;57	308.03	14076	12FE	23;18	143.31
14077	27FE	3;22	337.37	14077	14MR	14;45	172.59
14078	28MR	13;55	6.21	14078	13AP	2;55	201.25
14079	27AP	1;04	34.64	14079	12MY	12;21	229.43
14080	26MY	13;24	62.82	14080	10JN	20;05	257.36
14081	25JN	3;16	90.99	14081	10JL	3;16	285.33
14082	24JL	18;58	119.40	14082	8AU	10;52	313.57
14083	23AU	11;27	148.21	14083	6SE	19;37	342.30
14084	22SE	3;48	177.51	14084	6OC	6;09	11.61
14085	21OC	19;10	207.26	14085	4NO	19;01	41.45
14086	20NO	9;09	237.32	14086	4DE	10;34	71.67
14087	19DE	21;45	267.45				

139

NEW MOONS NUMBER	DATE	TIME	LONG.	FULL MOONS NUMBER	DATE	TIME	LONG.
				14087	3JA	4;28	101.99
14088	18JA	8;59	297.37	14088	1FE	23;25	132.07
14089	16FE	18;55	326.87	14089	3MR	17;35	161.66
14090	18MR	3;47	355.82	14090	2AP	9;33	190.65
14091	16AP	12;14	24.28	14091	1MY	22;51	219.10
14092	15MY	21;12	52.41	14092	31MY	9;51	247.21
14093	14JN	7;44	80.45	14093	29JN	19;14	275.21
14094	13JL	20;31	108.66	14094	29JL	3;43	303.38
14095	12AU	11;40	137.26	14095	27AU	12;01	331.92
14096	11SE	4;39	166.39	14096	25SE	20;50	0.97
14097	10OC	22;31	196.07	14097	25OC	6;53	30.55
14098	9NO	16;10	226.17	14098	23NO	18;47	60.55
14099	9DE	8;36	256.44	14099	23DE	8;48	90.73

140

NEW MOONS NUMBER	DATE	TIME	LONG.	FULL MOONS NUMBER	DATE	TIME	LONG.
14100	7JA	22;56	286.58	14100	22JA	0;40	120.82
14101	6FE	10;47	316.31	14101	20FE	17;30	150.54
14102	6MR	20;20	345.48	14102	21MR	10;17	179.76
14103	5AP	4;16	14.08	14103	20AP	2;13	208.46
14104	4MY	11;34	42.25	14104	19MY	16;49	236.78
14105	2JN	19;16	70.21	14105	18JN	5;51	264.91
14106	2JL	4;23	98.21	14106	17JL	17;17	293.10
14107	31JL	15;37	126.52	14107	16AU	3;27	321.55
14108	30AU	5;27	155.34	14108	14SE	12;58	350.45
14109	28SE	21;58	184.77	14109	13OC	22;34	19.85
14110	28OC	16;40	214.77	14110	12NO	8;54	49.68
14111	27NO	12;12	245.12	14111	11DE	20;15	79.75
14112	27DE	6;38	275.48				

141

NEW MOONS NUMBER	DATE	TIME	LONG.	FULL MOONS NUMBER	DATE	TIME	LONG.
				14112	10JA	8;37	109.79
14113	25JA	22;25	305.51	14113	8FE	21;51	139.53
14114	24FE	10;59	334.99	14114	10MR	11;54	168.82
14115	25MR	20;46	3.84	14115	9AP	2;42	197.65
14116	24AP	4;42	32.17	14116	8MY	18;04	226.10
14117	23MY	11;45	60.16	14117	7JN	9;27	254.35
14118	21JN	18;50	88.05	14118	7JL	0;07	282.59
14119	21JL	2;50	116.13	14119	5AU	13;35	311.04
14120	19AU	12;37	144.62	14120	4SE	1;54	339.87
14121	18SE	1;04	173.69	14121	3OC	13;26	9.17
14122	17OC	16;45	203.40	14122	2NO	0;37	38.92
14123	16NO	11;24	233.62	14123	1DE	11;40	68.93
14124	16DE	7;28	264.06	14124	30DE	22;31	98.96

142

NEW MOONS NUMBER	DATE	TIME	LONG.	FULL MOONS NUMBER	DATE	TIME	LONG.
14125	15JA	2;44	294.35	14125	29JA	9;13	128.74
14126	13FE	19;29	324.17	14126	27FE	20;03	158.09
14127	15MR	9;09	353.37	14127	29MR	7;34	186.96
14128	13AP	20;01	21.97	14128	27AP	20;16	215.43
14129	13MY	4;46	50.11	14129	27MY	10;18	243.68
14130	11JN	12;09	78.03	14130	26JN	1;19	271.90
14131	10JL	19;04	105.99	14131	25JL	16;45	300.31
14132	9AU	2;31	134.24	14132	24AU	8;07	329.08
14133	7SE	11;32	162.99	14133	22SE	23;05	358.34
14134	6OC	23;07	192.34	14134	22OC	13;19	28.06
14135	5NO	13;46	222.27	14135	21NO	2;28	58.09
14136	5DE	7;09	252.58	14136	20DE	14;15	88.18

143

NEW MOONS NUMBER	DATE	TIME	LONG.	FULL MOONS NUMBER	DATE	TIME	LONG.
14137	4JA	2;05	282.93	14137	19JA	0;43	118.07
14138	2FE	20;55	313.00	14138	17FE	10;15	147.53
14139	4MR	14;18	342.54	14139	18MR	19;27	176.49
14140	3AP	5;21	11.48	14140	17AP	5;02	204.99
14141	2MY	17;46	39.89	14141	16MY	15;35	233.17
14142	1JN	3;43	67.95	14142	15JN	3;34	261.27
14143	30JN	11;54	95.91	14143	14JL	17;16	289.52
14144	29JL	19;21	124.04	14144	13AU	8;46	318.14
14145	28AU	3;10	152.56	14145	12SE	1;45	347.29
14146	26SE	12;19	181.64	14146	11OC	19;20	16.97
14147	25OC	23;26	211.28	14147	10NO	12;10	47.04
14148	24NO	12;41	241.34	14148	10DE	3;08	77.26
14149	24DE	3;56	271.57				

144

NEW MOONS NUMBER	DATE	TIME	LONG.	FULL MOONS NUMBER	DATE	TIME	LONG.
				14149	8JA	15;48	107.32
14150	22JA	20;43	301.69	14150	7FE	2;26	136.99
14151	21FE	14;19	331.43	14151	7MR	11;36	166.13
14152	22MR	7;42	0.65	14152	5AP	19;53	194.74
14153	20AP	23;39	29.35	14153	5MY	3;56	222.93
14154	20MY	13;22	57.62	14154	3JN	12;27	250.92
14155	19JN	0;47	85.69	14155	2JL	22;23	278.96
14156	18JL	10;31	113.82	14156	1AU	10;37	307.31
14157	16AU	19;30	142.24	14157	31AU	1;45	336.20
14158	15SE	4;36	171.13	14158	29SE	19;33	5.70
14159	14OC	14;23	200.54	14159	29OC	14;42	35.73
14160	13NO	1;06	230.40	14160	28NO	9;19	66.05
14161	12DE	12;53	260.49	14161	28DE	1;56	96.33

145

NEW MOONS NUMBER	DATE	TIME	LONG.	FULL MOONS NUMBER	DATE	TIME	LONG.
14162	11JA	1;57	290.55	14162	26JA	15;57	126.28
14163	9FE	16;27	320.33	14163	25FE	3;28	155.70
14164	11MR	8;11	349.68	14164	26MR	12;50	184.52
14165	10AP	0;24	18.55	14165	24AP	20;35	212.83
14166	9MY	16;07	47.01	14166	24MY	3;31	240.81
14167	8JN	6;40	75.22	14167	22JN	10;41	268.71
14168	7JL	19;52	103.41	14168	21JL	19;16	296.82
14169	6AU	7;58	131.81	14169	20AU	6;24	325.37
14170	4SE	19;20	160.62	14170	18SE	20;44	354.53
14171	4OC	6;12	189.90	14171	18OC	14;03	24.32
14172	2NO	16;46	219.63	14172	17NO	9;15	54.58
14173	2DE	3;17	249.63	14173	17DE	4;43	84.99
14174	31DE	14;04	279.65				

146

NEW MOONS NUMBER	DATE	TIME	LONG.	FULL MOONS NUMBER	DATE	TIME	LONG.
				14174	15JA	22;52	115.23
14175	30JA	1;30	309.45	14175	14FE	14;34	144.99
14176	28FE	13;49	338.85	14176	16MR	3;15	174.14
14177	30MR	3;03	7.78	14177	14AP	13;05	202.69
14178	28AP	17;00	36.29	14178	13MY	20;46	230.78
14179	28MY	7;31	64.55	14179	12JN	3;24	258.67
14180	26JN	22;27	92.77	14180	11JL	10;14	286.62
14181	26JL	13;32	121.17	14181	9AU	18;22	314.91
14182	25AU	4;20	149.94	14182	8SE	4;42	343.72
14183	23SE	18;16	179.17	14183	7OC	17;45	13.15
14184	23OC	7;03	208.83	14184	6NO	9;34	43.14
14185	21NO	18;44	238.81	14185	6DE	3;45	73.48
14186	21DE	5;38	268.86				

147

NEW MOONS NUMBER	DATE	TIME	LONG.	FULL MOONS NUMBER	DATE	TIME	LONG.
				14186	4JA	23;14	103.85
14187	19JA	16;05	298.74	14187	3FE	18;18	133.92
14188	18FE	2;18	328.22	14188	5MR	11;11	163.43
14189	19MR	12;29	357.21	14189	4AP	0;54	192.30
14190	17AP	22;59	25.74	14190	3MY	11;34	220.63
14191	17MY	10;26	53.96	14191	1JN	20;02	248.62
14192	15JN	23;25	82.09	14192	1JL	3;26	276.55
14193	15JL	14;16	110.39	14193	30JL	10;51	304.69
14194	14AU	6;35	139.06	14194	28AU	19;05	333.25
14195	12SE	23;25	168.21	14195	27SE	4;47	2.35
14196	12OC	15;39	197.85	14196	26OC	16;30	32.02
14197	11NO	6;35	227.85	14197	25NO	6;41	62.13
14198	10DE	20;00	258.01	14198	24DE	23;24	92.43

148

NEW MOONS NUMBER	DATE	TIME	LONG.	FULL MOONS NUMBER	DATE	TIME	LONG.
14199	9JA	7;55	288.03	14199	23JA	17;51	122.60
14200	7FE	18;22	317.68	14200	22FE	12;25	152.37
14201	8MR	3;31	346.81	14201	23MR	5;25	181.57
14202	6AP	11;52	15.41	14202	21AP	19;58	210.19
14203	5MY	20;14	43.61	14203	21MY	8;05	238.40
14204	4JN	5;43	71.63	14204	19JN	18;17	266.42
14205	3JL	17;14	99.74	14205	19JL	3;17	294.52
14206	2AU	7;15	128.18	14206	17AU	11;48	322.93
14207	31AU	23;33	157.12	14207	15SE	20;28	351.81
14208	30SE	17;20	186.63	14208	15OC	5;58	21.23
14209	30OC	11;27	216.63	14209	13NO	16;57	51.10
14210	29NO	4;48	246.90	14210	13DE	5;51	81.24
14211	28DE	20;20	277.14				

149

NEW MOONS NUMBER	DATE	TIME	LONG.	FULL MOONS NUMBER	DATE	TIME	LONG.
				14211	11JA	20;41	111.37
14212	27JA	9;25	307.04	14212	10FE	12;54	141.21
14213	25FE	19;57	336.40	14213	12MR	5;32	170.58
14214	27MR	4;26	5.18	14214	10AP	21;45	199.44
14215	25AP	11;45	33.47	14215	10MY	12;58	227.87
14216	24MY	19;00	61.46	14216	9JN	2;51	256.05
14217	23JN	3;13	89.40	14217	8JL	15;15	284.21
14218	22JL	13;18	117.57	14218	7AU	2;15	312.58
14219	21AU	1;51	146.19	14219	5SE	12;15	341.34
14220	19SE	17;10	175.41	14220	4OC	21;57	10.59
14221	19OC	11;01	205.23	14221	3NO	8;01	40.30
14222	18NO	6;29	235.50	14222	2DE	18;52	70.32
14223	18DE	1;47	265.92				

150

NEW MOONS NUMBER	DATE	TIME	LONG.	FULL MOONS NUMBER	DATE	TIME	LONG.
				14223	1JA	6;39	100.38
14224	16JA	19;02	296.11	14224	30JA	19;16	130.22
14225	15FE	9;09	325.80	14225	1MR	8;36	159.65
14226	16MR	20;07	354.86	14226	30MR	22;43	188.60
14227	15AP	4;45	23.35	14227	29AP	13;37	217.15
14228	14MY	12;03	51.42	14228	29MY	4;59	245.43
14229	12JN	18;59	79.32	14229	27JN	20;09	273.66
14230	12JL	2;26	107.30	14230	27JL	10;28	302.04
14231	10AU	11;15	135.62	14231	25AU	23;37	330.76
14232	8SE	22;21	164.48	14232	24SE	11;47	359.92
14233	8OC	12;29	193.97	14233	23OC	23;21	29.55
14234	7NO	5;49	224.03	14234	22NO	10;36	59.51
14235	7DE	1;27	254.43	14235	21DE	21;34	89.57

151

NEW MOONS NUMBER	DATE	TIME	LONG.	FULL MOONS NUMBER	DATE	TIME	LONG.
14236	5JA	21;20	284.82	14236	20JA	8;12	119.45
14237	4FE	15;24	314.83	14237	18FE	18;41	148.93
14238	6MR	6;31	344.25	14238	20MR	5;30	177.93
14239	4AP	18;39	13.05	14239	18AP	17;16	206.50
14240	4MY	4;19	41.33	14240	18MY	6;26	234.78
14241	2JN	12;17	69.30	14241	16JN	20;55	262.97
14242	1JL	19;23	97.22	14242	16JL	12;17	291.30
14243	31JL	2;32	125.35	14243	15AU	3;57	319.95
14244	29AU	10;46	153.92	14244	13SE	19;27	349.07
14245	27SE	21;04	183.06	14245	13OC	10;23	18.66
14246	27OC	10;14	212.80	14246	12NO	0;21	48.64
14247	26NO	2;20	242.99	14247	11DE	12;57	78.76
14248	25DE	20;36	273.35				

152

New Moons

NUMBER	DATE	TIME	LONG.
14249	24JA	15;30	303.54
14250	23FE	9;35	333.28
14251	24MR	1;44	2.42
14252	22AP	15;25	31.00
14253	22MY	2;34	59.16
14254	20JN	11;38	87.14
14255	19JL	19;27	115.20
14256	18AU	3;08	143.58
14257	16SE	11;41	172.47
14258	15OC	21;53	201.93
14259	14NO	10;07	231.86
14260	14DE	0;20	262.05

Full Moons

NUMBER	DATE	TIME	LONG.
14248	10JA	0;01	108.74
14249	8FE	9;49	138.36
14250	8MR	18;54	167.47
14251	7AP	3;57	196.09
14252	6MY	13;43	224.34
14253	5JN	0;47	252.42
14254	4JL	13;32	280.58
14255	3AU	4;12	309.05
14256	1SE	20;41	338.02
14257	1OC	14;22	7.54
14258	31OC	8;01	37.53
14259	30NO	0;12	67.76
14260	29DE	14;05	97.92

153

New Moons

NUMBER	DATE	TIME	LONG.
14261	12JA	16;14	292.22
14262	11FE	9;17	322.09
14263	13MR	2;40	351.48
14264	11AP	19;18	20.34
14265	11MY	10;09	48.74
14266	9JN	22;46	76.87
14267	9JL	9;24	104.97
14268	7AU	18;52	133.28
14269	6SE	4;03	162.02
14270	5OC	13;39	191.28
14271	4NO	0;01	221.01
14272	3DE	11;15	251.04

Full Moons

NUMBER	DATE	TIME	LONG.
14261	28JA	1;37	127.75
14262	26FE	11;18	157.07
14263	27MR	19;47	185.84
14264	26AP	3;41	214.14
14265	25MY	11;43	242.15
14266	23JN	20;44	270.13
14267	23JL	7;41	298.34
14268	21AU	21;24	327.02
14269	20SE	14;06	356.31
14270	20OC	8;59	26.19
14271	19NO	4;16	56.46
14272	18DE	22;06	86.81

154

New Moons

NUMBER	DATE	TIME	LONG.
14273	1JA	23;32	281.12
14274	31JA	13;04	310.99
14275	2MR	3;57	340.47
14276	31MR	19;44	9.48
14277	30AP	11;38	38.05
14278	30MY	2;46	66.33
14279	28JN	16;44	94.51
14280	28JL	5;35	122.84
14281	26AU	17;34	151.52
14282	25SE	4;56	180.67
14283	24OC	15;51	210.28
14284	23NO	2;28	240.21
14285	22DE	13;03	270.25

Full Moons

NUMBER	DATE	TIME	LONG.
14273	17JA	13;27	116.91
14274	16FE	2;07	146.53
14275	17MR	12;23	175.55
14276	15AP	20;44	204.02
14277	15MY	3;52	232.08
14278	13JN	10;43	259.97
14279	12JL	18;26	287.97
14280	11AU	4;13	316.34
14281	9SE	16;59	345.28
14282	9OC	8;59	14.85
14283	8NO	3;31	44.98
14284	7DE	23;07	75.38

155

New Moons

NUMBER	DATE	TIME	LONG.
14286	20JA	23;58	300.14
14287	19FE	11;34	329.66
14288	21MR	0;03	358.72
14289	19AP	13;22	27.34
14290	19MY	3;26	55.65
14291	17JN	18;08	83.85
14292	17JL	9;18	112.17
14293	16AU	0;33	140.82
14294	14SE	15;17	169.92
14295	14OC	4;58	199.47
14296	12NO	17;22	229.39
14297	12DE	4;42	259.45

Full Moons

NUMBER	DATE	TIME	LONG.
14285	6JA	18;05	105.72
14286	5FE	11;03	135.68
14287	7MR	1;11	165.05
14288	5AP	12;19	193.79
14289	4MY	20;53	222.02
14290	3JN	3;53	249.95
14291	2JL	10;30	277.85
14292	31JL	17;56	306.00
14293	30AU	3;11	334.62
14294	28SE	14;55	3.83
14295	28OC	5;26	33.64
14296	26NO	22;31	63.88
14297	26DE	17;28	94.27

156

New Moons

NUMBER	DATE	TIME	LONG.
14298	10JA	15;18	289.41
14299	9FE	1;27	319.04
14300	9MR	11;22	348.18
14301	7AP	21;23	16.82
14302	7MY	8;02	45.10
14303	5JN	20;01	73.22
14304	5JL	9;53	101.42
14305	4AU	1;40	129.95
14306	2SE	18;36	158.95
14307	2OC	11;33	188.45
14308	1NO	3;27	218.38
14309	30NO	17;49	248.54
14310	30DE	6;31	278.64

Full Moons

NUMBER	DATE	TIME	LONG.
14298	25JA	12;50	124.47
14299	24FE	6;52	154.18
14300	24MR	22;07	183.28
14301	23AP	10;12	211.78
14302	22MY	19;39	239.87
14303	21JN	3;30	267.80
14304	20JL	10;55	295.85
14305	18AU	18;48	324.25
14306	17SE	3;51	353.17
14307	16OC	14;36	22.66
14308	15NO	3;32	52.63
14309	14DE	18;57	82.87

157

New Moons

NUMBER	DATE	TIME	LONG.
14311	28JA	17;37	308.45
14312	27FE	3;13	337.75
14313	28MR	11;41	6.51
14314	26AP	19;43	34.81
14315	26MY	4;21	62.84
14316	24JN	14;38	90.87
14317	24JL	3;22	119.16
14318	22AU	18;41	147.92
14319	21SE	12;02	177.25
14320	21OC	6;21	207.11
14321	20NO	0;25	237.34
14322	19DE	17;05	267.64

Full Moons

NUMBER	DATE	TIME	LONG.
14310	13JA	12;30	113.10
14311	12FE	7;00	143.02
14312	14MR	0;45	172.41
14313	12AP	16;28	201.22
14314	12MY	5;44	229.55
14315	10JN	16;53	257.62
14316	10JL	2;34	285.68
14317	8AU	11;26	313.98
14318	6SE	20;10	342.71
14319	6OC	5;21	11.96
14320	4NO	15;37	41.70
14321	4DE	3;30	71.76

158

New Moons

NUMBER	DATE	TIME	LONG.
14323	18JA	7;27	297.70
14324	16FE	19;08	327.26
14325	18MR	4;25	356.24
14326	16AP	12;03	24.66
14327	15MY	19;04	52.71
14328	14JN	2;36	80.63
14329	13JL	11;37	108.69
14330	11AU	22;55	137.13
14331	10SE	12;55	166.13
14332	10OC	5;41	195.75
14333	9NO	0;38	225.90
14334	8DE	20;22	256.31

Full Moons

NUMBER	DATE	TIME	LONG.
14322	2JA	17;16	101.90
14323	1FE	8;39	131.84
14324	3MR	0;54	161.37
14325	1AP	17;09	190.38
14326	1MY	8;47	218.93
14327	30MY	23;22	247.18
14328	29JN	12;39	275.33
14329	29JL	0;32	303.63
14330	27AU	11;15	332.27
14331	25SE	21;16	1.37
14332	25OC	7;16	30.95
14333	23NO	17;46	60.89
14334	23DE	5;03	90.96

159

New Moons

NUMBER	DATE	TIME	LONG.
14335	7JA	14;52	286.63
14336	6FE	6;34	316.52
14337	7MR	18;57	345.81
14338	6AP	4;48	14.48
14339	5MY	12;18	42.67
14340	3JN	19;15	70.59
14341	3JL	2;21	98.51
14342	1AU	10;26	126.70
14343	30AU	20;22	155.35
14344	29SE	8;59	184.62
14345	29OC	0;49	214.49
14346	27NO	19;31	244.80
14347	27DE	15;31	275.23

Full Moons

NUMBER	DATE	TIME	LONG.
14335	21JA	17;05	120.89
14336	20FE	5;48	150.44
14337	21MR	19;12	179.53
14338	20AP	9;27	208.17
14339	20MY	0;28	236.51
14340	18JN	15;50	264.74
14341	18JL	6;48	293.06
14342	16AU	20;49	321.67
14343	15SE	9;44	350.71
14344	14OC	21;49	20.21
14345	13NO	9;24	50.10
14346	12DE	20;34	80.16

160

New Moons

NUMBER	DATE	TIME	LONG.
14348	26JA	10;38	305.41
14349	25FE	3;11	335.05
14350	25MR	16;40	4.06
14351	24AP	3;26	32.50
14352	23MY	12;10	60.56
14353	21JN	19;38	88.47
14354	21JL	2;44	116.51
14355	19AU	10;26	144.91
14356	17SE	19;42	173.86
14357	17OC	7;27	203.39
14358	15NO	22;09	233.45
14359	15DE	15;24	263.77

Full Moons

NUMBER	DATE	TIME	LONG.
14347	11JA	7;18	110.12
14348	9FE	17;38	139.74
14349	10MR	3;56	168.88
14350	8AP	14;52	197.56
14351	8MY	3;04	225.89
14352	6JN	16;46	254.07
14353	6JL	7;46	282.32
14354	4AU	23;30	310.86
14355	3SE	15;24	339.83
14356	3OC	6;58	9.29
14357	1NO	21;45	39.19
14358	1DE	11;13	69.30
14359	30DE	23;04	99.37

161

New Moons

NUMBER	DATE	TIME	LONG.
14360	14JA	10;02	294.04
14361	13FE	4;30	323.94
14362	14MR	21;33	353.29
14363	13AP	12;25	22.05
14364	13MY	0;48	50.34
14365	11JN	10;54	78.37
14366	10JL	19;22	106.39
14367	9AU	3;09	134.65
14368	7SE	11;21	163.37
14369	6OC	20;48	192.64
14370	5NO	8;05	222.42
14371	4DE	21;18	252.55

Full Moons

NUMBER	DATE	TIME	LONG.
14360	29JA	9;22	129.13
14361	27FE	18;32	158.41
14362	29MR	3;16	187.17
14363	27AP	12;23	215.50
14364	26MY	22;32	243.60
14365	25JN	10;18	271.69
14366	25JL	0;00	300.02
14367	23AU	15;44	328.81
14368	22SE	9;10	358.15
14369	22OC	3;17	28.02
14370	20NO	20;33	58.23
14371	20DE	11;45	88.47

162

New Moons

NUMBER	DATE	TIME	LONG.
14372	3JA	12;15	282.73
14373	2FE	4;34	312.70
14374	3MR	21;38	342.25
14375	2AP	14;34	11.27
14376	2MY	6;18	39.82
14377	31MY	20;05	68.03
14378	30JN	7;47	96.13
14379	29JL	17;57	124.36
14380	28AU	3;25	152.96
14381	26SE	12;58	182.06
14382	25OC	23;06	211.65
14383	24NO	9;56	241.61
14384	23DE	21;36	271.69

Full Moons

NUMBER	DATE	TIME	LONG.
14372	19JA	0;25	118.44
14373	17FE	10;50	147.94
14374	18MR	19;41	176.89
14375	17AP	3;37	205.32
14376	16MY	11;20	233.39
14377	14JN	19;39	261.34
14378	14JL	5;28	289.43
14379	12AU	17;45	317.92
14380	11SE	9;05	346.99
14381	11OC	3;13	16.68
14382	9NO	22;43	46.86
14383	9DE	17;35	77.24

163

New Moons

NUMBER	DATE	TIME	LONG.
14385	22JA	10;18	301.64
14386	21FE	0;14	331.24
14387	22MR	15;20	0.38
14388	21AP	7;02	29.07
14389	20MY	22;32	57.42
14390	19JN	13;10	85.62
14391	19JL	2;43	113.89
14392	17AU	15;22	142.46
14393	16SE	3;19	171.46
14394	15OC	14;42	200.95
14395	14NO	1;36	230.81
14396	13DE	12;11	260.85

Full Moons

NUMBER	DATE	TIME	LONG.
14384	8JA	10;17	107.48
14385	7FE	0;14	137.29
14386	8MR	11;33	166.52
14387	6AP	20;40	195.16
14388	6MY	4;12	223.33
14389	4JN	11;00	251.25
14390	3JL	18;07	279.17
14391	2AU	2;45	307.38
14392	31AU	14;01	336.10
14393	30SE	4;31	5.46
14394	29OC	22;01	35.41
14395	28NO	17;20	65.76
14396	28DE	12;49	96.16

	NEW MOONS				FULL MOONS		
NUMBER	DATE	TIME	LONG.	NUMBER	DATE	TIME	LONG.

164

NEW NUMBER	DATE	TIME	LONG.	FULL NUMBER	DATE	TIME	LONG.
14397	11JA	22;47	290.80	14397	27JA	6;52	126.29
14398	10FE	9;48	320.44	14398	25FE	22;24	155.87
14399	10MR	21;33	349.63	14399	26MR	10;56	184.83
14400	9AP	10;10	18.37	14400	24AP	20;37	213.22
14401	8MY	23;38	46.74	14401	24MY	4;15	241.23
14402	7JN	13;54	74.94	14402	22JN	10;56	269.11
14403	7JL	4;55	103.20	14403	21JL	17;53	297.15
14404	5AU	20;22	131.73	14404	20AU	2;13	325.59
14405	4SE	11;44	160.69	14405	18SE	12;46	354.59
14406	4OC	2;19	190.13	14406	18OC	1;57	24.20
14407	2NO	15;36	219.97	14407	16NO	17;50	54.31
14408	2DE	3;34	250.03	14408	16DE	11;55	84.67
14409	31DE	14;29	280.06				

165

NEW NUMBER	DATE	TIME	LONG.	FULL NUMBER	DATE	TIME	LONG.
14410	30JA	0;42	309.80	14409	15JA	7;09	114.96
14411	28FE	10;30	339.10	14410	14FE	1;55	144.86
14412	29MR	20;10	7.89	14411	15MR	18;33	174.17
14413	28AP	6;10	36.24	14412	14AP	8;08	202.87
14414	27MY	17;12	64.37	14413	13MY	18;48	231.09
14415	26JN	6;00	92.50	14414	12JN	3;22	259.05
14416	25JL	20;53	120.89	14415	11JL	10;59	287.04
14417	24AU	13;33	149.72	14416	9AU	18;41	315.31
14418	23SE	6;56	179.08	14417	8SE	3;14	344.05
14419	22OC	23;44	208.92	14418	7OC	13;13	13.35
14420	21NO	15;06	239.04	14419	6NO	1;04	43.17
14421	21DE	4;44	269.21	14420	5DE	15;11	73.34

166

NEW NUMBER	DATE	TIME	LONG.	FULL NUMBER	DATE	TIME	LONG.
14422	19JA	16;35	299.15	14421	4JA	7;36	103.58
14423	18FE	2;47	328.63	14422	3FE	1;35	133.61
14424	19MR	11;35	357.56	14423	4MR	19;41	163.18
14425	17AP	19;31	25.98	14424	3AP	12;22	192.18
14426	17MY	3;32	54.07	14425	3MY	2;47	220.67
14427	15JN	12;44	82.05	14426	1JN	14;59	248.81
14428	15JL	0;08	110.21	14427	1JL	1;27	276.86
14429	13AU	14;14	138.78	14428	30JL	10;51	305.08
14430	12SE	6;49	167.91	14429	28AU	19;47	333.66
14431	12OC	1;00	197.62	14430	27SE	4;52	2.75
14432	10NO	19;33	227.78	14431	26OC	14;40	32.34
14433	10DE	13;11	258.11	14432	25NO	1;44	62.31
				14433	24DE	14;28	92.44

167

NEW NUMBER	DATE	TIME	LONG.	FULL NUMBER	DATE	TIME	LONG.
14434	9JA	4;49	288.29	14434	23JA	4;54	122.45
14435	7FE	17;48	318.05	14435	21FE	20;31	152.10
14436	9MR	4;06	347.22	14436	23MR	12;35	181.27
14437	7AP	12;18	15.82	14437	22AP	4;23	209.95
14438	6MY	19;21	43.96	14438	21MY	19;27	238.28
14439	5JN	2;24	71.89	14439	20JN	9;30	266.46
14440	4JL	10;32	99.86	14440	19JL	22;17	294.70
14441	2AU	20;38	128.13	14441	18AU	9;49	323.24
14442	1SE	9;18	156.93	14442	16SE	20;22	352.20
14443	1OC	0;49	186.33	14443	16OC	6;31	21.64
14444	30OC	18;55	216.33	14444	14NO	16;51	51.49
14445	29NO	14;35	246.69	14445	14DE	3;45	81.54
14446	29DE	9;58	277.09				

168

NEW NUMBER	DATE	TIME	LONG.	FULL NUMBER	DATE	TIME	LONG.
14447	28JA	3;11	307.17	14446	12JA	15;18	111.53
14448	26FE	17;08	336.68	14447	11FE	3;26	141.20
14449	27MR	3;56	5.55	14448	11MR	16;10	170.42
14450	25AP	12;23	33.88	14449	10AP	5;41	199.18
14451	24MY	19;34	61.87	14450	9MY	20;07	227.59
14452	23JN	2;29	89.76	14451	8JN	11;19	255.82
14453	22JL	10;01	117.83	14452	8JL	2;41	284.10
14454	20AU	19;00	146.30	14453	6AU	17;27	312.61
14455	19SE	6;17	175.35	14454	5SE	7;13	341.52
14456	18OC	20;33	205.01	14455	4OC	19;58	10.90
14457	17NO	13;58	235.19	14456	3NO	8;00	40.69
14458	17DE	9;32	265.62	14457	2DE	19;29	70.73

169

NEW NUMBER	DATE	TIME	LONG.	FULL NUMBER	DATE	TIME	LONG.
14459	16JA	5;16	295.92	14458	1JA	6;25	100.76
14460	14FE	23;07	325.77	14459	30JA	16;46	130.50
14461	16MR	14;03	355.00	14460	1MR	2;47	159.80
14462	15AP	2;02	23.62	14461	30MR	13;02	188.60
14463	14MY	11;40	51.79	14462	29AP	0;16	217.00
14464	12JN	19;43	79.73	14463	28MY	13;01	245.18
14465	12JL	2;59	107.71	14464	27JN	3;22	273.38
14466	10AU	10;23	135.97	14465	26JL	18;54	301.80
14467	8SE	18;53	164.72	14466	25AU	11;00	330.63
14468	8OC	5;25	194.06	14467	24SE	3;05	359.95
14469	6NO	18;40	223.95	14468	23OC	18;37	29.74
14470	6DE	10;42	254.19	14469	22NO	9;00	59.83
				14470	21DE	21;45	89.97

170

NEW NUMBER	DATE	TIME	LONG.	FULL NUMBER	DATE	TIME	LONG.
14471	5JA	4;41	284.50	14471	20JA	8;44	119.86
14472	3FE	23;12	314.54	14472	18FE	18;14	149.30
14473	5MR	16;53	344.08	14473	20MR	2;53	178.22
14474	4AP	8;47	13.04	14474	18AP	11;29	206.66
14475	3MY	22;23	41.48	14475	17MY	20;51	234.78
14476	2JN	9;38	69.58	14476	16JN	7;37	262.83
14477	1JL	18;57	97.59	14477	15JL	20;16	291.05
14478	31JL	3;07	125.76	14478	14AU	11;05	319.66
14479	29AU	11;11	154.32	14479	13SE	3;56	348.82
14480	27SE	20;05	183.41	14480	12OC	22;07	18.54
14481	27OC	6;32	213.04	14481	11NO	16;14	48.68
14482	25NO	18;48	243.06	14482	11DE	8;44	78.97
14483	25DE	8;49	273.24				

171

NEW NUMBER	DATE	TIME	LONG.	FULL NUMBER	DATE	TIME	LONG.
14484	24JA	0;18	303.29	14483	9JA	22;41	109.07
14485	22FE	16;47	332.97	14484	8FE	10;05	138.76
14486	24MR	9;39	2.16	14485	9MR	19;29	167.88
14487	23AP	1;57	30.85	14486	8AP	3;37	196.47
14488	22MY	16;46	59.16	14487	7MY	11;13	224.63
14489	21JN	5;35	87.28	14488	5JN	19;00	252.58
14490	20JL	16;38	115.47	14489	5JL	3;54	280.58
14491	19AU	2;35	143.95	14490	3AU	14;51	308.90
14492	17SE	12;16	172.89	14491	2SE	4;42	337.75
14493	16OC	22;16	202.33	14492	1OC	21;40	7.23
14494	15NO	8;51	232.20	14493	31OC	16;52	37.28
14495	14DE	20;04	262.26	14494	30NO	12;26	67.65
				14495	30DE	6;25	97.98

172

NEW NUMBER	DATE	TIME	LONG.	FULL NUMBER	DATE	TIME	LONG.
14496	13JA	8;04	292.26	14496	28JA	21;44	127.97
14497	11FE	21;06	321.97	14497	27FE	10;14	157.41
14498	12MR	11;20	351.24	14498	27MR	20;16	186.24
14499	11AP	2;33	20.05	14499	26AP	4;24	214.55
14500	10MY	18;05	48.49	14500	25MY	11;23	242.52
14501	9JN	9;11	76.72	14501	23JN	18;11	270.41
14502	8JL	23;24	104.96	14502	23JL	1;56	298.50
14503	7AU	12;44	133.42	14503	21AU	11;50	327.02
14504	6SE	1;19	162.30	14504	20SE	0;45	356.14
14505	5OC	13;15	191.65	14505	19OC	16;56	25.90
14506	4NO	0;35	221.42	14506	18NO	11;35	56.14
14507	3DE	11;22	251.44	14507	18DE	7;12	86.57

173

NEW NUMBER	DATE	TIME	LONG.	FULL NUMBER	DATE	TIME	LONG.
14508	1JA	21;52	281.44	14508	17JA	2;06	116.82
14509	31JA	8;27	311.19	14509	15FE	18;54	146.62
14510	1MR	19;33	340.52	14510	17MR	8;52	175.79
14511	31MR	7;25	9.37	14511	15AP	19;51	204.37
14512	29AP	20;11	37.82	14512	15MY	4;21	232.48
14513	29MY	9;54	66.04	14513	13JN	11;23	260.38
14514	28JN	0;32	94.25	14514	12JL	18;07	288.35
14515	27JL	15;57	122.68	14515	11AU	1;44	316.63
14516	26AU	7;43	151.51	14516	9SE	11;12	345.43
14517	24SE	23;05	180.81	14517	8OC	23;09	14.83
14518	24OC	13;20	210.56	14518	7NO	13;44	44.78
14519	23NO	2;07	240.59	14519	7DE	6;46	75.08
14520	22DE	13;33	270.67				

174

NEW NUMBER	DATE	TIME	LONG.	FULL NUMBER	DATE	TIME	LONG.
14521	21JA	0;01	300.53	14520	6JA	1;28	105.41
14522	19FE	9;49	329.97	14521	4FE	20;32	135.46
14523	20MR	19;16	358.91	14522	6MR	14;16	164.99
14524	19AP	4;47	27.38	14523	5AP	5;20	193.90
14525	18MY	14;59	55.54	14524	4MY	17;21	222.26
14526	17JN	2;41	83.62	14525	3JN	2;53	250.30
14527	16JL	16;30	111.89	14526	2JL	10;56	278.26
14528	15AU	8;29	140.56	14527	31JL	18;38	306.42
14529	14SE	1;52	169.75	14528	30AU	2;51	334.99
14530	13OC	19;25	199.46	14529	28SE	12;13	4.11
14531	12NO	11;49	229.54	14530	27OC	23;10	33.77
14532	12DE	2;28	259.75	14531	26NO	12;07	63.83
				14532	26DE	3;18	94.06

175

NEW NUMBER	DATE	TIME	LONG.	FULL NUMBER	DATE	TIME	LONG.
14533	10JA	15;12	289.80	14533	24JA	20;26	124.17
14534	9FE	2;07	319.45	14534	23FE	14;25	153.89
14535	10MR	11;24	348.56	14535	25MR	7;45	183.08
14536	8AP	19;29	17.13	14536	23AP	23;14	211.73
14537	8MY	3;10	45.29	14537	23MY	12;31	239.97
14538	6JN	11;30	73.26	14538	21JN	23;53	268.04
14539	5JL	21;37	101.32	14539	21JL	9;56	296.19
14540	4AU	10;21	129.72	14540	19AU	19;16	324.66
14541	3SE	1;52	158.65	14541	18SE	4;26	353.58
14542	2OC	19;35	188.18	14542	17OC	13;58	23.02
14543	1NO	14;18	218.22	14543	16NO	0;25	52.88
14544	1DE	8;42	248.54	14544	15DE	12;14	82.98
14545	31DE	1;31	278.82				

176

NUMBER	DATE	TIME	LONG.	NUMBER	DATE	TIME	LONG.
				14545	14JA	1;41	113.03
14546	29JA	15;51	308.76	14546	12FE	16;31	142.80
14547	28FE	3;21	338.14	14547	13MR	8;09	172.12
14548	28MR	12;22	6.92	14548	11AP	23;54	200.94
14549	26AP	19;44	35.19	14549	11MY	15;15	229.36
14550	26MY	2;33	63.16	14550	10JN	5;52	257.57
14551	24JN	9;57	91.07	14551	9JL	19;28	285.79
14552	23JL	18;59	119.21	14552	8AU	7;52	314.23
14553	22AU	6;22	147.80	14553	6SE	19;09	343.06
14554	20SE	20;33	176.99	14554	6OC	5;41	12.37
14555	20OC	13;32	206.80	14555	4NO	16;02	42.11
14556	19NO	8;41	237.07	14556	4DE	2;40	72.12
14557	19DE	4;30	267.50				

177

NUMBER	DATE	TIME	LONG.	NUMBER	DATE	TIME	LONG.
				14557	2JA	13;49	102.15
14558	17JA	22;59	297.73	14558	1FE	1;28	131.93
14559	16FE	14;33	327.46	14559	2MR	13;37	161.29
14560	18MR	2;46	356.55	14560	1AP	2;24	190.17
14561	16AP	12;12	25.05	14561	30AP	16;07	218.65
14562	15MY	19;50	53.13	14562	30MY	6;50	246.91
14563	14JN	2;45	81.02	14563	28JN	22;14	275.15
14564	13JL	9;55	109.01	14564	28JL	13;34	303.58
14565	11AU	18;09	137.32	14565	27AU	4;09	332.37
14566	10SE	4;17	166.16	14566	25SE	17;42	1.62
14567	9OC	17;05	195.61	14567	25OC	6;19	31.30
14568	8NO	9;00	225.63	14568	23NO	18;13	61.30
14569	8DE	3;40	255.99	14569	23DE	5;28	91.37

178

NUMBER	DATE	TIME	LONG.	NUMBER	DATE	TIME	LONG.
14570	6JA	23;31	286.37	14570	21JA	16;00	121.23
14571	5FE	18;23	316.40	14571	20FE	1;56	150.67
14572	7MR	10;43	345.86	14572	21MR	11;43	179.61
14573	6AP	0;03	14.68	14573	19AP	22;06	208.10
14574	5MY	10;45	42.99	14574	19MY	9;49	236.32
14575	3JN	19;31	70.99	14575	17JN	23;16	264.47
14576	3JL	3;09	98.93	14576	17JL	14;18	292.79
14577	1AU	10;30	127.09	14577	16AU	6;22	321.47
14578	30AU	18;28	155.66	14578	14SE	22;48	350.64
14579	29SE	4;00	184.80	14579	14OC	14;58	20.31
14580	28OC	15;55	214.50	14580	13NO	6;14	50.35
14581	27NO	6;36	244.64	14581	12DE	19;58	80.52
14582	26DE	23;38	274.95				

179

NUMBER	DATE	TIME	LONG.	NUMBER	DATE	TIME	LONG.
				14582	11JA	7;50	110.53
14583	25JA	17;52	305.09	14583	9FE	17;53	140.14
14584	24FE	11;54	334.81	14584	11MR	2;41	169.22
14585	26MR	4;37	3.96	14585	9AP	10;59	197.79
14586	24AP	19;19	32.56	14586	8MY	19;40	225.98
14587	24MY	7;46	60.77	14587	7JN	7;46	254.01
14588	22JN	18;05	88.80	14588	6JL	17;06	282.13
14589	22JL	2;53	116.90	14589	5AU	6;51	310.57
14590	20AU	11;04	145.33	14590	3SE	22;52	339.54
14591	18SE	19;38	174.24	14591	3OC	16;44	9.09
14592	18OC	5;24	203.70	14592	2NO	11;19	39.13
14593	16NO	16;49	233.61	14593	2DE	4;58	69.43
14594	16DE	5;55	263.76	14594	31DE	20;19	99.65

180

NUMBER	DATE	TIME	LONG.	NUMBER	DATE	TIME	LONG.
14595	14JA	20;31	293.86	14595	30JA	8;55	129.50
14596	13FE	12;18	323.65	14596	28FE	19;07	158.82
14597	14MR	4;47	352.99	14597	29MR	3;38	187.57
14598	12AP	21;17	21.83	14598	27AP	11;15	215.84
14599	12MY	12;51	50.25	14599	26MY	18;43	243.83
14600	11JN	2;46	78.43	14600	25JN	2;53	271.78
14601	10JL	14;49	106.59	14601	24JL	12;41	299.95
14602	9AU	1;27	134.97	14602	23AU	1;03	328.59
14603	7SE	11;26	163.76	14603	21SE	16;36	357.85
14604	6OC	21;27	193.06	14604	21OC	11;00	27.73
14605	5NO	7;53	222.81	14605	20NO	6;47	58.03
14606	4DE	18;49	252.83	14606	20DE	1;50	88.43

181

NUMBER	DATE	TIME	LONG.	NUMBER	DATE	TIME	LONG.
14607	3JA	6;17	282.87	14607	18JA	18;32	118.58
14608	1FE	18;33	312.67	14608	17FE	8;22	148.23
14609	3MR	7;53	342.07	14609	18MR	19;28	177.26
14610	1AP	22;22	11.00	14610	17AP	4;22	205.73
14611	1MY	13;36	39.54	14611	16MY	11;45	233.79
14612	31MY	4;54	67.81	14612	14JN	18;29	261.68
14613	29JN	19;40	96.03	14613	14JL	1;38	289.67
14614	29JL	9;39	124.42	14614	12AU	10;22	318.01
14615	27AU	22;52	153.17	14615	10SE	21;47	346.91
14616	26SE	11;25	182.38	14616	10OC	12;27	16.45
14617	25OC	23;18	212.05	14617	9NO	6;05	46.55
14618	24NO	10;28	242.02	14618	9DE	1;26	76.95
14619	23DE	21;04	272.06				

182

NUMBER	DATE	TIME	LONG.	NUMBER	DATE	TIME	LONG.
				14619	7JA	20;50	107.30
14620	22JA	7;26	301.91	14620	6FE	14;44	137.28
14621	20FE	18;00	331.37	14621	8MR	6;04	166.68
14622	22MR	5;10	0.35	14622	6AP	18;27	195.45
14623	20AP	17;12	28.90	14623	6MY	4;04	223.71
14624	20MY	6;13	57.16	14624	4JN	11;42	251.66
14625	18JN	20;19	85.33	14625	3JL	18;29	279.58
14626	18JL	11;26	113.67	14626	2AU	1;38	307.73
14627	17AU	3;19	142.35	14627	31AU	10;12	336.34
14628	15SE	19;17	171.52	14628	29SE	20;58	5.53
14629	15OC	10;28	201.16	14629	29OC	10;18	35.31
14630	14NO	0;14	231.14	14630	28NO	2;09	65.50
14631	13DE	12;24	261.25	14631	27DE	20;02	95.84

183

NUMBER	DATE	TIME	LONG.	NUMBER	DATE	TIME	LONG.
14632	11JA	23;16	291.21	14632	26JA	14;57	126.01
14633	10FE	9;13	320.80	14633	25FE	9;22	155.72
14634	11MR	18;35	349.90	14634	27MR	1;45	184.84
14635	10AP	3;46	18.49	14635	25AP	15;14	213.39
14636	9MY	13;19	46.71	14636	25MY	1;56	241.52
14637	8JN	0;00	74.77	14637	23JN	10;41	269.49
14638	7JL	12;39	102.94	14638	22JL	18;35	297.56
14639	6AU	3;39	131.44	14639	21AU	2;37	325.99
14640	4SE	20;40	160.46	14640	19SE	11;32	354.93
14641	4OC	14;34	190.03	14641	18OC	21;46	24.42
14642	3NO	7;54	220.03	14642	17NO	9;42	54.36
14643	2DE	23;39	250.25	14643	16DE	23;41	84.54

184

NUMBER	DATE	TIME	LONG.	NUMBER	DATE	TIME	LONG.
14644	1JA	13;24	280.40	14644	15JA	15;43	114.70
14645	31JA	1;09	310.21	14645	14FE	9;12	144.55
14646	29FE	11;05	339.51	14646	15MR	2;48	173.92
14647	29MR	19;31	8.25	14647	13AP	19;09	202.74
14648	28AP	3;06	36.51	14648	13MY	9;29	231.10
14649	27MY	10;49	64.50	14649	11JN	21;50	259.22
14650	25JN	19;49	92.48	14650	11JL	8;38	287.33
14651	25JL	7;10	120.73	14651	9AU	18;29	315.69
14652	23AU	21;23	149.45	14652	8SE	3;54	344.47
14653	22SE	14;15	178.78	14653	7OC	13;24	13.75
14654	22OC	8;49	208.68	14654	5NO	23;27	43.49
14655	21NO	3;43	238.95	14655	5DE	10;32	73.53
14656	20DE	21;33	269.30				

185

NUMBER	DATE	TIME	LONG.	NUMBER	DATE	TIME	LONG.
				14656	3JA	23;02	103.61
14657	19JA	13;12	299.40	14657	2FE	13;00	133.47
14658	18FE	2;03	328.99	14658	4MR	4;01	162.93
14659	19MR	12;07	357.97	14659	2AP	19;30	191.89
14660	17AP	20;03	26.39	14660	2MY	10;54	220.42
14661	17MY	2;54	54.43	14661	1JN	1;53	248.68
14662	15JN	9;49	82.32	14662	30JN	16;08	276.88
14663	14JL	17;55	110.35	14663	30JL	5;23	305.24
14664	13AU	4;06	138.76	14664	28AU	17;29	333.96
14665	11SE	16;56	167.73	14665	27SE	4;36	3.13
14666	11OC	8;39	197.33	14666	26OC	15;11	32.75
14667	10NO	2;56	227.46	14667	25NO	1;45	62.70
14668	9DE	22;41	257.88	14668	24DE	12;36	92.75

186

NUMBER	DATE	TIME	LONG.	NUMBER	DATE	TIME	LONG.
14669	8JA	18;04	288.23	14669	22JA	23;51	122.63
14670	7FE	11;10	318.16	14670	21FE	11;29	152.12
14671	9MR	0;58	347.48	14671	22MR	23;37	181.13
14672	7AP	11;35	16.17	14672	21AP	12;33	209.71
14673	6MY	19;55	44.37	14673	21MY	2;35	238.00
14674	5JN	3;04	72.30	14674	19JN	17;41	266.21
14675	4JL	10;02	100.22	14675	19JL	9;16	294.57
14676	2AU	17;43	128.40	14676	18AU	0;32	323.25
14677	1SE	2;54	157.05	14677	16SE	14;56	352.36
14678	30SE	14;22	186.29	14678	16OC	4;16	21.94
14679	30OC	4;46	216.12	14679	14NO	16;43	51.87
14680	28NO	22;10	246.38	14680	14DE	4;22	81.96
14681	28DE	17;36	276.78				

187

NUMBER	DATE	TIME	LONG.	NUMBER	DATE	TIME	LONG.
				14681	12JA	15;12	111.91
14682	27JA	13;05	306.96	14682	11FE	1;14	141.50
14683	26FE	6;40	336.63	14683	12MR	10;46	170.59
14684	27MR	21;24	5.67	14684	10AP	20;29	199.20
14685	26AP	9;18	34.14	14685	10MY	7;13	227.46
14686	25MY	18;57	62.23	14686	8JN	19;38	255.58
14687	24JN	3;08	90.17	14687	8JL	9;53	283.81
14688	23JL	10;39	118.24	14688	7AU	1;37	312.36
14689	21AU	18;21	146.66	14689	5SE	18;11	341.38
14690	20SE	3;08	175.61	14690	5OC	10;52	10.91
14691	19OC	13;53	205.12	14691	4NO	2;55	40.87
14692	18NO	3;11	235.13	14692	3DE	17;40	71.05
14693	17DE	19;04	265.39				

188 / 194

NUMBER	DATE	TIME	LONG.	NUMBER	DATE	TIME	LONG.	NUMBER	DATE	TIME	LONG.	NUMBER	DATE	TIME	LONG.
14694	16JA	12;41	295.61	14693	2JA	6;31	101.15	14768	10JA	9;52	289.96	14768	24JA	10;00	124.12
14695	15FE	6;45	325.48	14694	31JA	17;21	130.92	14769	9FE	0;06	319.76	14769	23FE	0;16	153.70
14696	16MR	0;02	354.81	14695	1MR	2;32	160.17	14770	10MR	11;24	348.95	14770	24MR	15;17	182.80
14697	14AP	15;41	23.59	14696	30MR	10;46	188.89	14771	8AP	20;11	17.55	14771	23AP	6;32	211.45
14698	14MY	5;15	51.92	14697	28AP	18;57	217.17	14772	8MY	3;20	45.68	14772	22MY	21;40	239.77
14699	12JN	16;40	80.00	14698	28MY	3;57	245.21	14773	6JN	10;01	73.59	14773	21JN	12;22	267.97
14700	12JL	2;18	108.07	14699	26JN	14;31	273.26	14774	5JL	17;23	101.54	14774	21JL	2;20	296.28
14701	10AU	10;52	136.38	14700	26JL	3;09	301.56	14775	4AU	2;28	129.79	14775	19AU	15;18	324.88
14702	8SE	19;21	165.13	14701	24AU	18;08	330.33	14776	2SE	14;00	158.55	14776	18SE	3;10	353.92
14703	8OC	4;37	194.42	14702	23SE	11;21	359.69	14777	2OC	4;22	187.93	14777	17OC	14;12	23.42
14704	6NO	15;16	224.19	14703	23OC	5;59	29.60	14778	31OC	21;31	217.90	14778	16NO	0;52	53.30
14705	6DE	3;31	254.28	14704	22NO	0;30	59.86	14779	30NO	16;47	248.25	14779	15DE	11;35	83.34
				14705	21DE	17;13	90.16	14780	30DE	12;36	278.66				

189 / 195

NUMBER	DATE	TIME	LONG.	NUMBER	DATE	TIME	LONG.	NUMBER	DATE	TIME	LONG.	NUMBER	DATE	TIME	LONG.
14706	4JA	17;16	284.41	14706	20JA	7;11	120.18					14780	13JA	22;32	113.29
14707	3FE	8;15	314.31	14707	18FE	18;25	149.70	14781	29JA	6;59	308.78	14781	12FE	9;45	142.91
14708	5MR	0;09	343.78	14708	20MR	3;33	178.63	14782	27FE	22;23	338.33	14782	13MR	21;19	172.06
14709	3AP	16;29	12.77	14709	18AP	11;22	207.04	14783	29MR	10;26	7.23	14783	12AP	9;30	200.75
14710	3MY	8;30	41.31	14710	17MY	18;42	235.08	14784	27AP	19;44	35.58	14784	11MY	22;44	229.09
14711	1JN	23;19	69.56	14711	16JN	2;19	263.01	14785	27MY	3;18	63.57	14785	10JN	13;13	257.30
14712	1JL	12;25	97.72	14712	15JL	11;10	291.07	14786	25JN	10;16	91.47	14786	10JL	4;42	285.58
14713	30JL	23;55	126.02	14713	13AU	22;10	319.52	14787	24JL	17;34	119.54	14787	8AU	20;26	314.15
14714	29AU	10;24	154.68	14714	12SE	12;11	348.55	14788	23AU	2;01	148.01	14788	7SE	11;36	343.14
14715	27SE	20;35	183.82	14715	12OC	5;24	18.23	14789	21SE	12;22	177.04	14789	7OC	1;47	12.59
14716	27OC	6;58	213.44	14716	11NO	0;52	48.42	14790	21OC	1;19	206.67	14790	5NO	14;54	42.45
14717	25NO	17;44	243.41	14717	10DE	20;37	78.84	14791	19NO	17;16	236.80	14791	5DE	3;04	72.52
14718	25DE	4;52	273.47					14792	19DE	11;49	267.18				

190 / 196

NUMBER	DATE	TIME	LONG.	NUMBER	DATE	TIME	LONG.	NUMBER	DATE	TIME	LONG.	NUMBER	DATE	TIME	LONG.
14719	23JA	16;31	303.35	14718	9JA	14;38	109.12					14792	3JA	14;18	102.56
14720	22FE	5;00	332.87	14719	8FE	5;52	138.97	14793	18JA	7;24	297.46	14793	2FE	0;36	132.28
14721	23MR	18;37	1.93	14720	9MR	18;11	168.22	14794	17FE	1;59	327.33	14794	2MR	10;06	161.53
14722	22AP	9;16	30.57	14721	8AP	4;01	196.87	14795	17MR	18;04	356.59	14795	31MR	19;23	190.27
14723	22MY	0;29	58.90	14722	7MY	11;59	225.04	14796	16AP	7;16	25.24	14796	30AP	5;15	218.60
14724	20JN	15;35	87.11	14723	5JN	18;53	252.96	14797	15MY	17;57	53.44	14797	29MY	16;34	246.73
14725	20JL	6;08	115.44	14724	5JL	1;42	280.88	14798	14JN	2;50	81.41	14798	28JN	5;49	274.88
14726	18AU	19;59	144.07	14725	3AU	9;33	309.07	14799	13JL	10;42	109.43	14799	27JL	20;56	303.30
14727	17SE	9;11	173.15	14726	1SE	19;36	337.77	14800	11AU	18;21	137.72	14800	26AU	13;22	332.15
14728	16OC	21;40	202.70	14727	1OC	8;42	7.08	14801	10SE	2;39	166.48	14801	25SE	6;21	1.53
14729	15NO	9;22	232.61	14728	31OC	1;01	37.00	14802	9OC	12;27	195.81	14802	24OC	23;05	31.39
14730	14DE	20;17	262.66	14729	29NO	19;43	67.33	14803	8NO	0;29	225.66	14803	23NO	14;46	61.55
				14730	29DE	15;15	97.73	14804	7DE	15;05	255.85	14804	23DE	4;42	91.73

191 / 197

NUMBER	DATE	TIME	LONG.	NUMBER	DATE	TIME	LONG.	NUMBER	DATE	TIME	LONG.	NUMBER	DATE	TIME	LONG.
14731	13JA	6;38	292.59	14731	28JA	9;58	127.87	14805	6JA	7;48	286.10	14805	21JA	16;29	121.64
14732	11FE	16;51	322.18	14732	27FE	2;35	157.48	14806	5FE	1;35	316.09	14806	20FE	2;18	151.07
14733	13MR	3;24	351.30	14733	28MR	16;22	186.47	14807	6MR	19;09	345.61	14807	21MR	10;43	179.96
14734	11AP	14;41	19.96	14734	27AP	3;15	214.89	14808	5AP	11;32	14.56	14808	19AP	18;37	208.35
14735	11MY	2;57	48.27	14735	26MY	11;46	242.93	14809	5MY	2;07	43.03	14809	19MY	2;56	236.43
14736	9JN	16;22	76.44	14736	24JN	18;53	270.83	14810	3JN	14;39	71.18	14810	17JN	12;31	264.43
14737	9JL	7;00	104.69	14737	24JL	1;48	298.88	14811	3JL	1;16	99.24	14811	17JL	0;01	292.60
14738	7AU	22;43	133.24	14738	22AU	9;40	327.32	14812	1AU	10;28	127.47	14812	15AU	13;52	321.19
14739	6SE	15;00	162.26	14739	20SE	19;23	356.31	14813	30AU	19;05	156.08	14813	14SE	6;10	350.35
14740	6OC	7;00	191.78	14740	20OC	7;30	25.89	14814	29SE	4;04	185.19	14814	14OC	0;26	20.09
14741	4NO	21;47	221.70	14741	18NO	22;08	55.96	14815	28OC	14;07	214.82	14815	12NO	19;26	50.28
14742	4DE	10;53	251.81	14742	18DE	15;00	86.27	14816	27NO	1;36	244.82	14816	12DE	13;22	80.63
								14817	26DE	14;31	274.95				

192 / 198

NUMBER	DATE	TIME	LONG.	NUMBER	DATE	TIME	LONG.	NUMBER	DATE	TIME	LONG.	NUMBER	DATE	TIME	LONG.
14743	2JA	22;22	281.85	14743	17JA	9;24	116.51					14817	11JA	4;48	110.79
14744	1FE	8;38	311.59	14744	16FE	4;04	146.39	14818	25JA	4;42	304.93	14818	9FE	17;17	140.50
14745	1MR	18;04	340.84	14745	16MR	21;28	175.72	14819	23FE	19;53	334.54	14819	11MR	3;14	169.63
14746	31MR	3;03	9.58	14746	15AP	12;22	204.45	14820	25MR	11;48	3.66	14820	9AP	11;28	198.19
14747	29AP	12;06	37.89	14747	15MY	0;24	232.71	14821	24AP	3;53	32.33	14821	8MY	18;50	226.33
14748	28MY	21;56	65.96	14748	13JN	10;05	260.72	14822	23MY	19;20	60.66	14822	7JN	2;07	254.26
14749	27JN	9;25	94.04	14749	12JL	18;25	288.74	14823	22JN	9;55	88.84	14823	6JL	10;12	282.24
14750	26JL	23;14	122.40	14750	11AU	2;27	317.04	14824	21JL	21;53	117.09	14824	4AU	20;02	310.53
14751	25AU	15;29	151.23	14751	9SE	11;03	345.81	14825	20AU	9;03	145.64	14825	3SE	8;32	339.34
14752	24SE	9;21	180.63	14752	8OC	20;43	15.12	14826	18SE	19;34	174.63	14826	3OC	0;17	8.79
14753	24OC	3;23	210.53	14753	7NO	7;50	44.92	14827	18OC	6;03	204.12	14827	1NO	18;56	38.83
14754	22NO	20;13	240.73	14754	6DE	20;44	75.04	14828	16NO	16;44	234.00	14828	1DE	14;54	69.22
14755	22DE	11;05	270.95					14829	16DE	3;41	264.05	14829	31DE	10;01	99.60

193 / 199

NUMBER	DATE	TIME	LONG.	NUMBER	DATE	TIME	LONG.	NUMBER	DATE	TIME	LONG.	NUMBER	DATE	TIME	LONG.
14756	20JA	23;48	300.91	14755	5JA	11;37	105.21	14830	14JA	14;55	294.01	14830	30JA	2;40	129.63
14757	19FE	10;30	330.39	14756	4FE	4;15	135.17	14831	13FE	2;42	323.64	14831	28FE	16;20	159.10
14758	20MR	19;28	359.31	14757	5MR	21;41	164.70	14832	14MR	15;25	352.83	14832	30MR	3;15	187.94
14759	19AP	3;12	27.70	14758	4AP	14;36	193.69	14833	13AP	5;17	21.57	14833	28AP	11;59	216.26
14760	18MY	10;34	55.75	14759	4MY	5;54	222.19	14834	12MY	20;05	49.97	14834	27MY	19;16	244.24
14761	16JN	18;41	83.69	14760	2JN	19;15	250.38	14835	11JN	11;15	78.20	14835	26JN	2;00	272.13
14762	16JL	4;43	111.80	14761	2JL	6;54	278.48	14836	11JL	2;13	106.47	14836	25JL	9;14	300.20
14763	14AU	17;30	140.34	14762	31JL	17;22	306.75	14837	9AU	16;40	135.00	14837	23AU	18;08	328.70
14764	13SE	9;14	169.45	14763	30AU	3;12	335.40	14838	8SE	6;29	163.94	14838	22SE	5;44	357.79
14765	13OC	3;16	199.17	14764	28SE	12;49	4.53	14839	7OC	19;38	193.37	14839	21OC	20;33	27.51
14766	11NO	22;20	229.36	14765	27OC	22;42	34.14	14840	6NO	7;58	223.25	14840	20NO	14;14	57.72
14767	11DE	16;58	259.73	14766	26NO	9;16	64.10	14841	5DE	19;22	253.25	14841	20DE	9;32	88.13
				14767	25DE	20;57	94.18								

200

NEW MOONS				FULL MOONS			
NUMBER	DATE	TIME	LONG.	NUMBER	DATE	TIME	LONG.
14842	4JA	5;55	283.24	14842	19JA	4;45	118.39
14843	2FE	15;59	312.95	14843	17FE	22;26	148.20
14844	3MR	2;04	342.22	14844	18MR	13;34	177.41
14845	1AP	12;40	11.00	14845	17AP	1;49	206.02
14846	1MY	0;10	39.39	14846	16MY	11;25	234.16
14847	30MY	12;49	67.56	14847	14JN	19;08	262.09
14848	29JN	2;46	95.75	14848	14JL	2;06	290.07
14849	28JL	18;04	124.18	14849	12AU	9;29	318.36
14850	27AU	10;23	153.04	14850	10SE	18;20	347.16
14851	26SE	2;57	182.42	14851	10OC	5;19	16.54
14852	25OC	18;44	212.25	14852	8NO	18;44	46.46
14853	24NO	8;53	242.34	14853	8DE	10;30	76.70
14854	23DE	21;12	272.46				

201

NEW MOONS				FULL MOONS			
NUMBER	DATE	TIME	LONG.	NUMBER	DATE	TIME	LONG.
				14854	7JA	4;06	106.98
14855	22JA	7;58	302.32	14855	5FE	22;37	137.00
14856	20FE	17;36	331.74	14856	7MR	16;39	166.52
14857	22MR	2;33	0.63	14857	6AP	8;47	195.45
14858	20AP	11;16	29.05	14858	5MY	22;12	223.86
14859	19MY	20;25	57.15	14859	4JN	9;01	251.94
14860	18JN	6;51	85.19	14860	3JL	18;01	279.94
14861	17JL	19;24	113.41	14861	2AU	2;16	308.14
14862	16AU	10;33	142.06	14862	31AU	10;41	336.75
14863	15SE	3;57	171.28	14863	29SE	19;57	5.88
14864	14OC	22;21	201.04	14864	29OC	6;25	35.53
14865	13NO	16;09	231.19	14865	27NO	18;23	65.56
14866	13DE	8;10	261.46	14866	27DE	8;09	95.73

202

NEW MOONS				FULL MOONS			
NUMBER	DATE	TIME	LONG.	NUMBER	DATE	TIME	LONG.
14867	11JA	21;59	291.55	14867	25JA	23;45	125.76
14868	10FE	9;36	321.21	14868	24FE	16;40	155.42
14869	11MR	19;14	350.31	14869	26MR	9;45	184.58
14870	10AP	3;21	18.86	14870	25AP	1;48	213.24
14871	9MY	10;37	46.99	14871	24MY	16;07	241.52
14872	7JN	18;06	74.93	14872	23JN	4;40	269.63
14873	7JL	2;59	102.94	14873	22JL	15;52	297.84
14874	5AU	14;20	131.29	14874	21AU	2;13	326.36
14875	4SE	4;42	160.20	14875	19SE	12;08	355.35
14876	3OC	21;51	189.72	14876	18OC	22;02	24.82
14877	2NO	16;44	219.78	14877	17NO	8;17	54.69
14878	2DE	11;53	250.14	14878	16DE	19;21	84.75

203

NEW MOONS				FULL MOONS			
NUMBER	DATE	TIME	LONG.	NUMBER	DATE	TIME	LONG.
14879	1JA	5;51	280.47	14879	15JA	7;33	114.74
14880	30JA	21;28	310.45	14880	13FE	20;59	144.43
14881	1MR	10;08	339.86	14881	15MR	11;23	173.68
14882	30MR	19;59	8.65	14882	14AP	2;17	202.45
14883	29AP	3;43	36.92	14883	13MY	17;21	230.85
14884	28MY	10;25	64.87	14884	12JN	8;17	259.07
14885	26JN	17;17	92.77	14885	11JL	22;48	287.33
14886	26JL	1;25	120.88	14886	10AU	12;33	315.84
14887	24AU	11;43	149.45	14887	9SE	1;16	344.75
14888	23SE	0;44	178.61	14888	8OC	12;57	14.12
14889	22OC	16;37	208.39	14889	6NO	23;56	43.91
14890	21NO	11;01	238.63	14890	6DE	10;39	73.92
14891	21DE	6;47	269.06				

204

NEW MOONS				FULL MOONS			
NUMBER	DATE	TIME	LONG.	NUMBER	DATE	TIME	LONG.
				14891	4JA	21;25	103.93
14892	20JA	2;04	299.32	14892	3FE	8;19	133.67
14893	18FE	19;00	329.08	14893	3MR	19;26	162.96
14894	19MR	8;37	358.22	14894	2AP	6;58	191.77
14895	17AP	19;07	26.74	14895	1MY	19;27	220.18
14896	17MY	3;23	54.83	14896	31MY	9;03	248.39
14897	15JN	10;33	82.74	14897	30JN	0;04	276.62
14898	14JL	17;38	110.72	14898	29JL	15;56	305.09
14899	13AU	1;32	139.04	14899	28AU	7;45	333.95
14900	11SE	10;57	167.87	14900	26SE	22;46	3.27
14901	10OC	22;37	197.30	14901	26OC	12;40	33.03
14902	9NO	13;05	227.27	14902	25NO	1;28	63.08
14903	9DE	6;24	257.58	14903	24DE	13;13	93.16

205

NEW MOONS				FULL MOONS			
NUMBER	DATE	TIME	LONG.	NUMBER	DATE	TIME	LONG.
14904	8JA	1;35	287.92	14904	22JA	23;54	123.02
14905	6FE	20;45	317.95	14905	21FE	9;35	152.43
14906	8MR	14;03	347.42	14906	22MR	18;39	181.31
14907	7AP	4;35	16.28	14907	21AP	3;52	209.74
14908	6MY	16;26	44.62	14908	20MY	14;09	237.89
14909	5JN	2;11	72.65	14909	19JN	2;17	265.99
14910	4JL	10;34	100.63	14910	18JL	16;30	294.29
14911	2AU	18;23	128.82	14911	17AU	8;29	322.98
14912	1SE	2;26	157.42	14912	16SE	1;31	352.20
14913	30SE	11;32	186.56	14913	15OC	18;45	21.95
14914	29OC	22;51	216.24	14914	14NO	11;18	52.03
14915	28NO	11;46	246.33	14915	14DE	2;19	82.26
14916	28DE	3;24	276.57				

206

NEW MOONS				FULL MOONS			
NUMBER	DATE	TIME	LONG.	NUMBER	DATE	TIME	LONG.
				14916	12JA	15;12	112.30
14917	26JA	20;35	306.66	14917	11FE	1;50	141.91
14918	25FE	14;09	336.34	14918	12MR	10;42	170.97
14919	27MR	7;00	5.47	14919	10AP	18;33	199.50
14920	25AP	22;26	34.09	14920	10MY	2;22	227.65
14921	25MY	12;01	62.34	14921	8JN	11;06	255.63
14922	23JN	23;41	90.42	14922	7JL	21;30	283.71
14923	23JL	9;42	118.59	14923	6AU	10;10	312.13
14924	21AU	18;44	147.07	14924	5SE	1;21	341.07
14925	20SE	3;39	176.02	14925	4OC	18;55	10.63
14926	19OC	13;15	205.49	14926	3NO	13;57	40.71
14927	18NO	0;04	235.38	14927	3DE	8;47	71.06
14928	17DE	12;15	265.49				

207

NEW MOONS				FULL MOONS			
NUMBER	DATE	TIME	LONG.	NUMBER	DATE	TIME	LONG.
				14928	2JA	1;39	101.34
14929	16JA	1;39	295.53	14929	31JA	15;34	131.23
14930	14FE	16;05	325.26	14930	2MR	2;36	160.56
14931	16MR	7;23	354.52	14931	31MR	11;28	189.31
14932	14AP	23;12	23.32	14932	29AP	19;02	217.56
14933	14MY	14;57	51.74	14933	29MY	2;09	245.52
14934	13JN	5;50	79.95	14934	27JN	9;42	273.45
14935	12JL	19;16	108.17	14935	26JL	18;32	301.60
14936	11AU	7;17	136.62	14936	25AU	5;39	330.21
14937	9SE	18;20	165.48	14937	23SE	19;51	359.43
14938	9OC	5;01	194.82	14938	23OC	13;16	29.28
14939	7NO	15;46	224.60	14939	22NO	8;55	59.59
14940	7DE	2;38	254.63	14940	22DE	4;45	90.02

208

NEW MOONS				FULL MOONS			
NUMBER	DATE	TIME	LONG.	NUMBER	DATE	TIME	LONG.
14941	5JA	13;37	284.64	14941	20JA	22;44	120.21
14942	4FE	0;53	314.39	14942	19FE	13;50	149.89
14943	4MR	12;48	343.70	14943	20MR	1;59	178.95
14944	3AP	1;46	12.56	14944	18AP	11;38	207.43
14945	2MY	15;54	41.03	14945	17MY	19;30	235.50
14946	1JN	6;51	69.29	14946	16JN	2;24	263.39
14947	30JN	22;01	97.53	14947	15JL	9;17	291.38
14948	30JL	12;55	125.96	14948	13AU	17;18	319.71
14949	29AU	3;21	154.78	14949	12SE	3;32	348.59
14950	27SE	17;10	184.07	14950	11OC	16;48	18.09
14951	27OC	6;11	213.80	14951	10NO	9;12	48.15
14952	25NO	18;12	243.82	14952	10DE	3;51	78.52
14953	25DE	5;10	273.87				

209

NEW MOONS				FULL MOONS			
NUMBER	DATE	TIME	LONG.	NUMBER	DATE	TIME	LONG.
				14953	8JA	23;14	108.86
14954	23JA	15;19	303.69	14954	7FE	17;42	138.85
14955	22FE	1;07	333.10	14955	9MR	10;05	168.27
14956	23MR	11;08	2.01	14956	7AP	23;43	197.08
14957	21AP	21;52	30.49	14957	7MY	10;33	225.37
14958	21MY	9;42	58.70	14958	5JN	19;07	253.35
14959	19JN	22;53	86.84	14959	5JL	2;24	281.30
14960	19JL	13;33	115.16	14960	3AU	9;34	309.46
14961	18AU	5;36	143.87	14961	1SE	17;44	338.08
14962	16SE	22;26	173.09	14962	1OC	3;42	7.26
14963	16OC	15;02	202.81	14963	30OC	15;58	37.01
14964	15NO	6;18	232.87	14964	29NO	6;35	67.16
14965	14DE	19;38	263.02	14965	28DE	23;13	97.44

210

NEW MOONS				FULL MOONS			
NUMBER	DATE	TIME	LONG.	NUMBER	DATE	TIME	LONG.
14966	13JA	7;07	293.00	14966	27JA	17;13	127.55
14967	11FE	17;08	322.58	14967	26FE	11;27	157.25
14968	13MR	2;11	351.64	14968	28MR	4;31	186.38
14969	11AP	10;44	20.19	14969	26AP	19;17	214.96
14970	10MY	19;23	48.36	14970	26MY	7;22	243.14
14971	9JN	4;55	76.37	14971	24JN	17;17	271.15
14972	8JL	16;14	104.49	14972	24JL	1;57	299.27
14973	7AU	6;06	132.96	14973	22AU	10;23	327.74
14974	5SE	22;37	161.98	14974	20SE	19;22	356.70
14975	5OC	16;56	191.58	14975	20OC	5;20	26.19
14976	4NO	11;27	221.65	14976	18NO	16;34	56.11
14977	4DE	4;38	251.93	14977	18DE	5;21	86.25

211

NEW MOONS				FULL MOONS			
NUMBER	DATE	TIME	LONG.	NUMBER	DATE	TIME	LONG.
14978	2JA	19;38	282.13	14978	16JA	19;51	116.33
14979	1FE	8;17	311.96	14979	15FE	11;56	146.11
14980	2MR	18;45	341.26	14980	17MR	4;49	175.43
14981	1AP	3;24	9.98	14981	15AP	21;18	204.23
14982	30AP	10;50	38.22	14982	15MY	12;27	232.62
14983	29MY	17;58	66.18	14983	14JN	1;55	260.78
14984	28JN	1;56	94.13	14984	13JL	13;55	288.95
14985	27JL	11;56	122.33	14985	12AU	0;53	317.37
14986	26AU	0;49	151.02	14986	10SE	11;14	346.21
14987	24SE	16;46	180.33	14987	9OC	21;20	15.54
14988	24OC	11;05	210.23	14988	8NO	7;30	45.30
14989	23NO	6;25	240.53	14989	7DE	18;08	75.32
14990	23DE	1;13	270.92				

NEW MOONS / FULL MOONS / NEW MOONS / FULL MOONS

212

NEW #	DATE	TIME	LONG.	FULL #	DATE	TIME	LONG.
				14990	6JA	5:37	105.35
14991	21JA	18:06	301.06	14991	4FE	18:13	135.14
14992	20FE	8:12	330.68	14992	5MR	7:53	164.52
14993	20MR	19:19	359.69	14993	3AP	22:17	193.42
14994	19AP	3:53	28.12	14994	3MY	13:05	221.91
14995	18MY	10:54	56.15	14995	2JN	4:02	250.16
14996	16JN	17:30	84.03	14996	1JL	18:53	278.39
14997	16JL	0:54	112.03	14997	31JL	9:17	306.81
14998	14AU	10:06	140.42	14998	29AU	22:50	335.61
14999	12SE	21:47	169.37	14999	28SE	11:18	4.85
15000	12OC	12:19	198.94	15000	27OC	22:49	34.53
15001	11NO	5:36	229.04	15001	26NO	9:45	64.51
15002	11DE	0:53	259.44	15002	25DE	20:27	94.55

213

NEW #	DATE	TIME	LONG.	FULL #	DATE	TIME	LONG.
15003	9JA	20:36	289.79	15003	24JA	7:09	124.39
15004	8FE	14:49	319.76	15004	22FE	17:54	153.82
15005	10MR	6:02	349.12	15005	24MR	4:54	182.76
15006	8AP	17:57	17.84	15006	22AP	16:31	211.27
15007	8MY	3:10	46.06	15007	22MY	5:19	239.51
15008	6JN	10:45	74.01	15008	20JN	19:37	267.69
15009	5JL	17:50	101.94	15009	20JL	11:14	296.06
15010	4AU	1:20	130.13	15010	19AU	3:24	324.79
15011	2SE	10:01	158.77	15011	17SE	19:11	353.98
15012	1OC	20:35	188.00	15012	17OC	9:58	23.63
15013	31OC	9:40	217.79	15013	15NO	23:31	53.63
15014	30NO	1:35	248.00	15014	15DE	11:54	83.74
15015	29DE	19:55	278.35				

214

NEW #	DATE	TIME	LONG.	FULL #	DATE	TIME	LONG.
				15015	13JA	23:04	113.70
15016	28JA	15:10	308.50	15016	12FE	9:05	143.27
15017	27FE	9:24	338.18	15017	13MR	18:10	172.32
15018	29MR	1:15	7.25	15018	12AP	2:57	200.87
15019	27AP	14:20	35.75	15019	11MY	12:22	229.06
15020	27MY	1:05	63.87	15020	9JN	23:21	257.13
15021	25JN	10:54	91.85	15021	9JL	12:28	285.32
15022	24JL	18:19	119.96	15022	8AU	3:43	313.87
15023	23AU	2:20	148.41	15023	6SE	20:31	342.91
15024	21SE	10:58	177.37	15024	6OC	14:02	12.49
15025	20OC	21:01	206.88	15025	5NO	7:16	42.52
15026	19NO	9:08	236.85	15026	4DE	23:19	72.75
15027	18DE	23:34	267.05				

215

NEW #	DATE	TIME	LONG.	FULL #	DATE	TIME	LONG.
				15027	3JA	13:22	102.90
15028	17JA	15:54	297.20	15028	2FE	1:02	132.69
15029	16FE	9:10	327.02	15029	3MR	10:34	161.94
15030	18MR	2:15	356.33	15030	1AP	18:38	190.63
15031	16AP	18:18	25.11	15031	1MY	2:11	218.86
15032	16MY	8:48	53.46	15032	30MY	10:12	246.86
15033	14JN	21:30	81.59	15033	28JN	19:36	274.86
15034	14JL	8:28	109.72	15034	28JL	7:04	303.13
15035	12AU	18:08	138.10	15035	26AU	21:03	331.88
15036	11SE	3:14	166.90	15036	25SE	13:37	1.23
15037	10OC	12:36	196.21	15037	25OC	8:16	31.15
15038	8NO	22:54	225.98	15038	24NO	3:36	61.46
15039	8DE	10:24	256.04	15039	23DE	21:44	91.82

216

NEW #	DATE	TIME	LONG.	FULL #	DATE	TIME	LONG.
15040	6JA	23:05	286.12	15040	22JA	13:10	121.89
15041	5FE	12:47	315.94	15041	21FE	1:30	151.43
15042	6MR	3:21	345.34	15042	21MR	11:14	180.36
15043	4AP	18:41	14.27	15043	19AP	19:13	208.76
15044	4MY	10:23	42.79	15044	19MY	2:21	236.79
15045	3JN	1:45	71.05	15045	17JN	9:32	264.69
15046	2JL	16:05	99.26	15046	16JL	17:37	292.74
15047	1AU	5:00	127.64	15047	15AU	3:21	321.16
15048	30AU	16:45	156.37	15048	13SE	16:11	350.16
15049	29SE	3:50	185.57	15049	13OC	8:07	19.80
15050	28OC	14:44	215.23	15050	12NO	2:57	49.98
15051	27NO	1:38	245.21	15051	11DE	23:01	80.41
15052	26DE	12:32	275.26				

217

NEW #	DATE	TIME	LONG.	FULL #	DATE	TIME	LONG.
				15052	10JA	18:06	110.73
15053	24JA	23:27	305.10	15053	9FE	10:38	140.61
15054	23FE	10:44	334.54	15054	11MR	0:08	169.89
15055	24MR	22:50	3.52	15055	9AP	10:53	198.56
15056	23AP	12:07	32.09	15056	8MY	19:30	226.74
15057	23MY	2:32	60.38	15057	7JN	2:46	254.67
15058	21JN	17:37	88.60	15058	6JL	9:34	282.60
15059	21JL	8:50	116.95	15059	4AU	16:58	310.79
15060	19AU	23:47	145.65	15060	3SE	2:03	339.46
15061	18SE	14:14	174.80	15061	2OC	13:50	8.75
15062	18OC	3:58	204.42	15062	1NO	4:46	38.62
15063	16NO	16:42	234.39	15063	30NO	22:26	68.91
15064	16DE	4:15	264.47	15064	30DE	17:35	99.29

218

NEW #	DATE	TIME	LONG.	FULL #	DATE	TIME	LONG.
15065	14JA	14:41	294.39	15065	29JA	12:33	129.43
15066	13FE	0:26	323.94	15066	28FE	5:57	159.05
15067	14MR	10:01	353.00	15067	29MR	20:54	188.07
15068	12AP	20:05	21.59	15068	28AP	9:04	216.53
15069	12MY	7:06	49.84	15069	27MY	18:42	244.60
15070	10JN	19:25	77.96	15070	26JN	2:35	272.54
15071	10JL	9:19	106.19	15071	25JL	9:47	300.61
15072	9AU	0:49	134.75	15072	23AU	17:28	329.06
15073	7SE	17:36	163.80	15073	22SE	2:36	358.05
15074	7OC	10:45	193.39	15074	21OC	13:48	27.62
15075	6NO	3:04	223.38	15075	20NO	3:16	57.64
15076	5DE	17:33	253.56	15076	19DE	18:51	87.90

219

NEW #	DATE	TIME	LONG.	FULL #	DATE	TIME	LONG.
15077	4JA	5:57	283.64	15077	18JA	12:05	118.08
15078	2FE	16:33	313.37	15078	17FE	6:08	147.92
15079	4MR	1:52	342.60	15079	18MR	23:45	177.23
15080	2AP	10:24	11.29	15080	17AP	15:39	205.99
15081	1MY	18:43	39.56	15081	17MY	5:04	234.30
15082	31MY	3:32	67.58	15082	15JN	16:04	262.36
15083	29JN	13:46	95.61	15083	15JL	1:23	290.43
15084	29JL	2:17	123.93	15084	13AU	10:02	318.77
15085	27AU	17:37	152.75	15085	11SE	18:53	347.57
15086	26SE	11:23	182.16	15086	11OC	4:30	16.90
15087	26OC	6:14	212.11	15087	9NO	15:10	46.70
15088	25NO	0:25	242.38	15088	9DE	3:06	76.78
15089	24DE	16:40	272.65				

220

NEW #	DATE	TIME	LONG.	FULL #	DATE	TIME	LONG.
				15089	7JA	16:35	106.88
15090	23JA	6:28	302.64	15090	6FE	7:41	136.76
15091	21FE	17:54	332.14	15091	6MR	23:59	166.22
15092	22MR	3:16	1.05	15092	5AP	16:34	195.18
15093	20AP	11:04	29.43	15093	5MY	8:20	223.69
15094	19MY	18:06	57.45	15094	3JN	22:40	251.91
15095	18JN	1:26	85.36	15095	3JL	11:30	280.07
15096	17JL	10:16	113.43	15096	1AU	23:10	308.40
15097	15AU	21:40	141.93	15097	31AU	10:03	337.11
15098	14SE	12:12	171.01	15098	29SE	20:29	6.29
15099	14OC	5:36	200.73	15099	29OC	6:45	35.94
15100	13NO	0:44	230.93	15100	27NO	17:11	65.90
15101	12DE	20:04	261.33	15101	27DE	4:08	95.95

221

NEW #	DATE	TIME	LONG.	FULL #	DATE	TIME	LONG.
15102	11JA	14:03	291.60	15102	25JA	15:59	125.82
15103	10FE	5:34	321.43	15103	24FE	4:51	155.32
15104	11MR	18:04	350.65	15104	25MR	18:37	184.36
15105	10AP	3:43	19.26	15105	24AP	8:59	212.95
15106	9MY	11:18	47.40	15106	23MY	23:45	241.25
15107	7JN	17:55	75.30	15107	22JN	14:42	269.47
15108	7JL	0:48	103.24	15108	22JL	5:32	297.82
15109	5AU	9:03	131.47	15109	20AU	19:50	326.49
15110	3SE	19:31	160.21	15110	19SE	9:10	355.61
15111	3OC	8:42	189.56	15111	18OC	21:24	25.18
15112	2NO	0:43	219.50	15112	17NO	8:44	55.10
15113	1DE	19:08	249.82	15113	16DE	19:34	85.15
15114	31DE	14:49	280.22				

222

NEW #	DATE	TIME	LONG.	FULL #	DATE	TIME	LONG.
				15114	15JA	6:10	115.07
15115	30JA	9:55	310.35	15115	13FE	16:41	144.64
15116	1MR	2:39	339.94	15116	15MR	3:15	173.73
15117	30MR	16:07	8.88	15117	13AP	14:13	202.35
15118	29AP	2:31	37.26	15118	13MY	2:07	230.63
15119	28MY	10:47	65.27	15119	11JN	15:31	258.79
15120	26JN	18:03	93.18	15120	11JL	6:32	287.06
15121	26JL	1:20	121.26	15121	9AU	22:43	315.66
15122	24AU	9:29	149.74	15122	8SE	15:05	344.72
15123	22SE	19:09	178.77	15123	8OC	6:43	14.25
15124	22OC	6:59	208.37	15124	6NO	21:08	44.18
15125	20NO	21:29	238.45	15125	6DE	10:14	74.30
15126	20DE	14:38	268.77				

223

NEW #	DATE	TIME	LONG.	FULL #	DATE	TIME	LONG.
				15126	4JA	22:01	104.34
15127	19JA	9:29	299.01	15127	3FE	8:30	134.06
15128	18FE	4:16	328.86	15128	4MR	17:48	163.28
15129	19MR	21:15	358.14	15129	3AP	2:25	191.97
15130	18AP	11:37	26.83	15130	2MY	11:10	220.24
15131	17MY	23:29	55.06	15131	31MY	21:05	248.30
15132	16JN	9:23	83.07	15132	30JN	9:00	276.41
15133	15JL	18:03	111.13	15133	29JL	23:14	304.81
15134	14AU	2:14	139.46	15134	28AU	15:30	333.67
15135	12SE	10:39	168.25	15135	27SE	9:00	3.09
15136	11OC	20:03	197.58	15136	27OC	2:45	33.00
15137	10NO	7:08	227.40	15137	25NO	19:43	63.22
15138	9DE	20:22	257.54	15138	25DE	10:56	93.46

NEW MOONS / FULL MOONS / NEW MOONS / FULL MOONS

NUMBER	DATE	TIME	LONG.		NUMBER	DATE	TIME	LONG.
				224				
15139	8JA	11;41	287.72		15139	23JA	23;46	123.40
15140	7FE	4;22	317.65		15140	22FE	10;12	152.84
15141	7MR	21;24	347.13		15141	22MR	18;44	181.70
15142	6AP	13;50	16.07		15142	21AP	2;15	210.06
15143	6MY	5;04	44.55		15143	20MY	9;46	238.10
15144	4JN	18;44	72.74		15144	18JN	18;16	266.06
15145	4JL	6;42	100.87		15145	18JL	4;36	294.20
15146	2AU	17;10	129.16		15146	16AU	17;20	322.76
15147	1SE	2;42	157.82		15147	15SE	8;44	351.89
15148	30SE	12;04	186.97		15148	15OC	2;38	21.64
15149	29OC	21;59	216.61		15149	13NO	22;00	51.86
15150	28NO	8;55	246.60		15150	13DE	17;04	82.25
15151	27DE	20;57	276.69					
				225				
					15151	12JA	10;00	112.47
15152	26JA	9;57	306.60		15152	10FE	23;49	142.22
15153	24FE	23;48	336.13		15153	12MR	10;39	171.36
15154	26MR	14;28	5.19		15154	10AP	19;16	199.92
15155	25AP	5;49	33.82		15155	10MY	2;37	228.05
15156	24MY	21;21	62.14		15156	8JN	9;37	255.96
15157	23JN	12;20	90.36		15157	7JL	17;08	283.92
15158	23JL	2;10	118.67		15158	6AU	2;03	312.19
15159	21AU	14;44	147.29		15159	4SE	13;18	340.97
15160	20SE	2;23	176.35		15160	4OC	3;40	10.38
15161	19OC	13;34	205.89		15161	2NO	21;16	40.39
15162	18NO	0;36	235.80		15162	2DE	17;01	70.78
15163	17DE	11;32	265.85					
				226				
					15163	1JA	12;50	101.18
15164	15JA	22;19	295.78		15164	31JA	6;43	131.25
15165	14FE	9;08	325.35		15165	1MR	21;38	160.75
15166	15MR	20;28	354.46		15166	31MR	9;37	189.62
15167	14AP	8;50	23.13		15167	29AP	19;09	217.95
15168	13MY	22;29	51.47		15168	29MY	2;59	245.94
15169	12JN	13;13	79.68		15169	27JN	9;56	273.85
15170	12JL	4;30	107.97		15170	26JL	16;58	301.92
15171	10AU	19;49	136.55		15171	25AU	1;11	330.42
15172	9SE	10;50	165.56		15172	23SE	11;38	359.48
15173	9OC	1;16	195.06		15173	23OC	1;03	29.16
15174	7NO	14;47	224.95		15174	21NO	17;28	59.32
15175	7DE	3;03	255.04		15175	21DE	11;59	89.70
				227				
15176	5JA	13;59	285.05		15176	20JA	7;06	119.95
15177	3FE	23;53	314.73		15177	19FE	1;17	149.76
15178	5MR	9;16	343.94		15178	20MR	17;24	178.99
15179	3AP	18;46	12.66		15179	19AP	6;54	207.63
15180	3MY	5;00	40.98		15180	18MY	17;45	235.82
15181	1JN	16;27	69.10		15181	17JN	2;28	263.78
15182	1JL	5;28	97.26		15182	16JL	9;59	291.80
15183	30JL	20;14	125.68		15183	14AU	17;27	320.11
15184	29AU	12;38	154.56		15184	13SE	1;56	348.91
15185	28SE	6;01	183.99		15185	12OC	12;10	18.29
15186	27OC	23;10	213.89		15186	11NO	0;33	48.17
15187	26NO	14;51	244.07		15187	10DE	15;03	78.36
15188	26DE	4;22	274.22					
				228				
					15188	9JA	7;22	108.58
15189	24JA	15;45	304.10		15189	8FE	0;54	138.54
15190	23FE	1;31	333.50		15190	8MR	18;40	168.03
15191	23MR	10;11	2.36		15191	7AP	11;24	196.97
15192	21AP	18;21	30.74		15192	7MY	2;04	225.42
15193	21MY	2;38	58.80		15193	5JN	14;17	253.55
15194	19JN	11;56	86.79		15194	5JL	0;29	281.60
15195	18JL	23;10	114.97		15195	3AU	9;33	309.85
15196	17AU	13;08	143.59		15196	1SE	18;25	338.49
15197	16SE	5;56	172.80		15197	1OC	3;48	7.66
15198	16OC	0;40	202.59		15198	30OC	14;03	37.32
15199	14NO	19;35	232.80		15199	29NO	1;21	67.33
15200	14DE	13;02	263.13		15200	28DE	13;56	97.44
				229				
15201	13JA	4;06	293.26		15201	27JA	4;00	127.39
15202	11FE	16;37	322.94		15202	25FE	19;30	156.98
15203	13MR	2;51	352.05		15203	27MR	11;47	186.09
15204	11AP	11;13	20.59		15204	26AP	3;53	214.73
15205	10MY	18;24	48.70		15205	25MY	18;56	243.03
15206	9JN	1;21	76.61		15206	24JN	8;36	271.19
15207	8JL	9;15	104.59		15207	23JL	21;00	299.46
15208	6AU	19;17	132.91		15208	22AU	8;30	328.05
15209	5SE	8;18	161.78		15209	20SE	19;24	357.09
15210	5OC	0;28	191.28		15210	20OC	5;56	26.61
15211	3NO	19;02	221.34		15211	18NO	16;22	56.50
15212	3DE	14;32	251.72		15212	18DE	3;00	86.54

NUMBER	DATE	TIME	LONG.		NUMBER	DATE	TIME	LONG.
				230				
15213	2JA	9;23	282.08		15213	16JA	14;14	116.48
15214	1FE	2;12	312.09		15214	15FE	2;20	146.10
15215	2MR	16;09	341.54		15215	16MR	15;22	175.26
15216	1AP	3;04	10.35		15216	15AP	5;11	203.97
15217	30AP	11;30	38.63		15217	14MY	19;34	232.34
15218	29MY	18;25	66.59		15218	13JN	10;24	260.55
15219	28JN	1;01	94.48		15219	13JL	1;26	288.84
15220	27JL	8;31	122.58		15220	11AU	16;18	317.41
15221	25AU	17;52	151.13		15221	10SE	6;29	346.40
15222	24SE	5;45	180.26		15222	9OC	19;33	15.85
15223	23OC	20;26	210.01		15223	8NO	7;31	45.69
15224	22NO	13;46	240.22		15224	7DE	18;38	75.73
15225	22DE	8;58	270.62					
				231				
					15225	6JA	5;17	105.72
15226	21JA	4;30	300.87		15226	4FE	15;40	135.42
15227	19FE	22;29	330.67		15227	6MR	1;57	164.66
15228	21MR	13;31	359.84		15228	4AP	12;23	193.40
15229	20AP	1;19	28.40		15229	3MY	23;29	221.75
15230	19MY	10;31	56.52		15230	2JN	11;54	249.91
15231	17JN	18;11	84.44		15231	2JL	2;04	278.11
15232	17JL	1;27	112.44		15232	31JL	17;52	306.58
15233	15AU	9;12	140.77		15233	30AU	10;30	335.49
15234	13SE	18;10	169.61		15234	29SE	2;54	4.89
15235	13OC	4;58	199.02		15235	28OC	18;15	34.73
15236	11NO	18;07	228.94		15236	27NO	8;11	64.83
15237	11DE	9;56	259.20		15237	26DE	20;41	94.95
				232				
15238	10JA	3;58	289.48		15238	25JA	7;45	124.80
15239	8FE	22;48	319.48		15239	23FE	17;27	154.19
15240	9MR	16;40	348.96		15240	24MR	2;07	183.04
15241	8AP	8;16	17.84		15241	22AP	10;27	211.41
15242	7MY	21;18	46.22		15242	21MY	19;28	239.50
15243	6JN	8;10	74.29		15243	20JN	6;11	267.54
15244	5JL	17;30	102.31		15244	19JL	19;14	295.80
15245	4AU	2;01	130.54		15245	18AU	10;38	324.49
15246	2SE	10;25	159.18		15246	17SE	3;50	353.73
15247	1OC	19;25	188.34		15247	16OC	21;50	23.51
15248	31OC	5;41	218.01		15248	15NO	15;32	53.68
15249	29NO	17;49	248.06		15249	15DE	7;51	83.96
15250	29DE	8;01	278.23					
				233				
					15250	13JA	21;56	114.04
15251	27JA	23;55	308.25		15251	12FE	9;28	143.67
15252	26FE	16;36	337.87		15252	13MR	18;43	172.73
15253	28MR	9;11	6.98		15253	12AP	2;27	201.24
15254	27AP	0;56	35.60		15254	11MY	9;41	229.34
15255	26MY	15;24	63.87		15255	9JN	17;28	257.28
15256	25JN	4;21	92.01		15256	9JL	2;45	285.32
15257	24JL	15;44	120.24		15257	7AU	14;15	313.71
15258	23AU	1;54	148.79		15258	6SE	4;24	342.63
15259	21SE	11;30	177.79		15259	5OC	21;15	12.18
15260	20OC	21;15	207.28		15260	4NO	16;12	42.27
15261	19NO	7;45	237.18		15261	4DE	11;47	72.65
15262	18DE	19;12	267.25					
				234				
					15262	3JA	6;01	102.98
15263	17JA	7;34	297.24		15263	1FE	21;25	132.92
15264	15FE	20;44	326.89		15264	3MR	9;35	162.28
15265	17MR	10;41	356.08		15265	1AP	19;05	191.03
15266	16AP	1;27	24.82		15266	1MY	2;51	219.28
15267	15MY	16;48	53.21		15267	30MY	9;51	247.23
15268	14JN	8;09	81.44		15268	28JN	17;00	275.14
15269	13JL	22;46	109.72		15269	28JL	1;08	303.28
15270	12AU	12;13	138.25		15270	26AU	11;11	331.86
15271	11SE	0;34	167.17		15271	25SE	0;00	1.05
15272	10OC	12;12	196.58		15272	24OC	16;06	30.86
15273	8NO	23;29	226.40		15273	23NO	11;02	61.15
15274	8DE	10;33	256.44		15274	23DE	7;06	91.59
				235				
15275	6JA	21;20	286.43		15275	22JA	2;05	121.81
15276	5FE	7;54	316.13		15276	20FE	18;27	151.52
15277	6MR	18;38	345.37		15277	22MR	7;46	180.61
15278	5AP	6;08	14.15		15278	20AP	18;23	209.11
15279	4MY	18;53	42.56		15279	20MY	2;57	237.20
15280	3JN	8;58	70.77		15280	18JN	10;15	265.11
15281	3JL	0;02	99.01		15281	17JL	17;12	293.11
15282	1AU	15;33	127.48		15282	16AU	0;48	321.44
15283	31AU	7;01	156.36		15283	14SE	10;08	350.30
15284	29SE	22;05	185.72		15284	13OC	22;05	19.77
15285	29OC	12;23	215.52		15285	12NO	13;05	49.78
15286	28NO	1;28	245.60		15286	12DE	6;40	80.11
15287	27DE	13;06	275.67					

236 / 242

NUMBER	DATE	TIME	LONG.	NUMBER	DATE	TIME	LONG.	NUMBER	DATE	TIME	LONG.	NUMBER	DATE	TIME	LONG.
				15287	11JA	1;33	110.42	15362	18JA	19;54	298.83	15361	4JA	19;28	104.63
15288	25JA	23;22	305.48	15288	9FE	20;12	140.40	15363	17FE	12;02	328.58	15362	3FE	8;14	134.44
15289	24FE	8;45	334.85	15289	10MR	13;18	169.83	15364	19MR	4;29	357.84	15363	4MR	18;26	163.69
15290	24MR	17;52	3.71	15290	9AP	4;04	198.67	15365	17AP	20;31	26.61	15364	3AP	2;39	192.37
15291	23AP	3;25	32.12	15291	8MY	16;11	227.00	15366	17MY	11;37	54.98	15365	2MY	9;52	220.58
15292	22MY	14;00	60.27	15292	7JN	1;56	255.02	15367	16JN	1;25	83.14	15366	31MY	17;08	248.53
15293	21JN	2;05	88.37	15293	6JL	10;02	283.00	15368	15JL	13;45	111.34	15367	30JN	1;30	276.50
15294	20JL	15;57	116.67	15294	4AU	17;33	311.20	15369	14AU	0;43	139.78	15368	29JL	11;49	304.73
15295	19AU	7;43	145.38	15295	3SE	1;35	339.83	15370	12SE	10;46	168.64	15369	28AU	0;40	333.45
15296	18SE	0;58	174.64	15296	2OC	11;01	9.02	15371	11OC	20;35	197.99	15370	26SE	16;18	2.78
15297	17OC	18;40	204.42	15297	31OC	22;23	38.74	15372	10NO	6;48	227.78	15371	26OC	10;28	32.71
15298	16NO	11;27	234.55	15298	30NO	11;50	68.85	15373	9DE	17;47	257.82	15372	25NO	6;05	63.04
15299	16DE	2;13	264.77	15299	30DE	3;10	99.07					15373	25DE	1;18	93.44

237 / 243

NUMBER	DATE	TIME	LONG.	NUMBER	DATE	TIME	LONG.	NUMBER	DATE	TIME	LONG.	NUMBER	DATE	TIME	LONG.
15300	14JA	14;37	294.78	15300	28JA	19;58	129.12	15374	8JA	5;36	287.85	15374	23JA	18;13	123.55
15301	13FE	1;02	324.35	15301	27FE	13;30	158.76	15375	6FE	18;08	317.61	15375	22FE	7;54	153.13
15302	14MR	10;00	353.38	15302	29MR	6;41	187.88	15376	8MR	7;24	346.94	15376	23MR	18;32	182.08
15303	12AP	18;09	21.90	15303	27AP	22;23	216.49	15377	6AP	21;27	15.80	15377	22AP	2;57	210.48
15304	12MY	2;07	50.03	15304	27MY	11;51	244.71	15378	6MY	12;00	44.27	15378	21MY	10;09	238.50
15305	10JN	10;41	78.00	15305	25JN	23;06	272.78	15379	5JN	3;42	72.53	15379	19JN	17;06	266.40
15306	9JL	20;46	106.07	15306	25JL	8;49	300.96	15380	4JL	18;52	100.78	15380	19JL	0;39	294.42
15307	8AU	9;19	134.51	15307	23AU	17;54	329.47	15381	3AU	9;09	129.22	15381	17AU	9;42	322.83
15308	7SE	0;51	163.50	15308	22SE	3;12	358.47	15382	1SE	22;19	158.03	15382	15SE	21;08	351.80
15309	6OC	18;58	193.12	15309	21OC	13;09	27.98	15383	1OC	10;32	187.30	15383	15OC	11;39	21.40
15310	5NO	14;14	223.23	15310	19NO	23;58	57.89	15384	30OC	22;12	217.01	15384	14NO	5;21	51.54
15311	5DE	8;43	253.57	15311	19DE	11;49	87.99	15385	29NO	9;29	247.01	15385	14DE	1;06	81.97
								15386	28DE	20;24	277.05				

238 / 244

NUMBER	DATE	TIME	LONG.	NUMBER	DATE	TIME	LONG.	NUMBER	DATE	TIME	LONG.	NUMBER	DATE	TIME	LONG.
15312	4JA	1;05	283.82	15312	18JA	0;57	118.00					15386	12JA	20;50	112.30
15313	2FE	14;50	313.68	15313	16FE	15;29	147.70	15387	27JA	6;55	306.86	15387	11FE	14;32	142.22
15314	4MR	2;04	342.99	15314	18MR	7;10	176.95	15388	25FE	17;16	336.25	15388	12MR	5;16	171.53
15315	2AP	11;10	11.71	15315	16AP	23;15	205.72	15389	26MR	4;01	5.15	15389	10AP	17;06	200.22
15316	1MY	18;43	39.94	15316	16MY	14;48	234.12	15390	24AP	15;49	33.64	15390	10MY	2;34	228.43
15317	31MY	1;34	67.88	15317	15JN	5;12	262.31	15391	24MY	5;03	61.88	15391	8JN	10;25	256.38
15318	29JN	8;49	95.80	15318	14JL	18;23	290.54	15392	22JN	19;37	90.08	15392	7JL	17;30	284.32
15319	28JL	17;39	123.97	15319	13AU	6;33	319.02	15393	22JL	11;04	118.46	15393	6AU	0;45	312.52
15320	27AU	5;10	152.63	15320	11SE	18;01	347.92	15394	21AU	2;50	147.20	15394	4SE	9;13	341.19
15321	25SE	19;53	181.91	15321	11OC	4;57	17.31	15395	19SE	18;27	176.41	15395	3OC	19;53	10.45
15322	25OC	13;29	211.79	15322	9NO	15;33	47.11	15396	19OC	9;29	206.11	15396	2NO	9;24	40.28
15323	24NO	8;48	242.10	15323	9DE	2;05	77.13	15397	17NO	23;25	236.14	15397	2DE	1;47	70.52
15324	24DE	4;12	272.51					15398	17DE	11;53	266.26	15398	31DE	20;05	100.86

239 / 245

NUMBER	DATE	TIME	LONG.	NUMBER	DATE	TIME	LONG.	NUMBER	DATE	TIME	LONG.	NUMBER	DATE	TIME	LONG.
				15324	7JA	12;53	107.12	15399	15JA	22;45	296.19	15399	30JA	14;51	130.97
15325	22JA	22;09	302.68	15325	6FE	0;19	136.84	15400	14FE	8;21	325.71	15400	1MR	8;41	160.60
15326	21FE	13;31	332.34	15326	7MR	12;36	166.14	15401	15MR	17;18	354.72	15401	31MR	0;33	189.64
15327	23MR	1;50	1.37	15327	6AP	1;45	194.97	15402	14AP	2;19	23.25	15402	29AP	13;58	218.13
15328	21AP	11;20	29.82	15328	5MY	15;37	223.41	15403	13MY	12;06	51.44	15403	29MY	0;53	246.25
15329	20MY	18;49	57.86	15329	4JN	6;08	251.64	15404	11JN	23;14	79.51	15404	27JN	9;48	274.23
15330	19JN	1;26	85.74	15330	3JL	21;08	279.88	15405	11JL	12;08	107.70	15405	26JL	17;38	302.34
15331	18JL	8;24	113.74	15331	2AU	12;21	308.35	15406	10AU	3;02	136.26	15406	25AU	1;27	330.81
15332	16AU	16;48	142.12	15332	1SE	3;13	337.22	15407	8SE	19;49	165.33	15407	23SE	10;16	359.81
15333	15SE	3;28	171.04	15333	30SE	17;11	6.55	15408	8OC	13;43	194.96	15408	22OC	20;45	29.37
15334	14OC	16;49	200.58	15334	30OC	5;56	36.29	15409	7NO	7;23	225.03	15409	21NO	9;11	59.36
15335	13NO	8;54	230.65	15335	28NO	17;35	66.31	15410	6DE	23;24	255.27	15410	20DE	23;31	89.56
15336	13DE	3;17	261.01	15336	28DE	4;26	96.35								

240 / 246

NUMBER	DATE	TIME	LONG.	NUMBER	DATE	TIME	LONG.	NUMBER	DATE	TIME	LONG.	NUMBER	DATE	TIME	LONG.
15337	11JA	22;46	291.34	15337	26JA	14;49	126.16	15411	5JA	13;01	285.40	15411	19JA	15;27	119.68
15338	10FE	17;38	321.32	15338	25FE	0;55	155.55	15412	4FE	0;17	315.14	15412	18FE	8;27	149.45
15339	11MR	10;08	350.71	15339	25MR	10;57	184.43	15413	5MR	9;46	344.35	15413	20MR	1;43	178.74
15340	9AP	23;27	19.48	15340	23AP	21;23	212.87	15414	3AP	18;05	13.02	15414	18AP	18;09	207.50
15341	9MY	9;49	47.73	15341	23MY	8;51	241.05	15415	3MY	1;53	41.25	15415	18MY	8;45	235.85
15342	7JN	18;09	75.70	15342	21JN	22;01	269.19	15416	1JN	9;54	69.23	15416	16JN	21;09	263.96
15343	7JL	1;35	103.65	15343	21JL	13;06	297.54	15417	30JN	19;01	97.23	15417	16JL	7;43	292.09
15344	5AU	9;07	131.86	15344	20AU	5;37	326.30	15418	30JL	6;14	125.50	15418	14AU	17;14	320.48
15345	3SE	17;33	160.52	15345	18SE	22;33	355.56	15419	28AU	20;20	154.29	15419	13SE	2;35	349.33
15346	3OC	3;30	189.73	15346	18OC	14;47	25.29	15420	27SE	13;25	183.69	15420	12OC	12;22	18.68
15347	1NO	15;28	219.49	15347	17NO	5;40	55.36	15421	27OC	8;31	213.67	15421	10NO	22;51	48.49
15348	1DE	5;56	249.65	15348	16DE	19;00	85.51	15422	26NO	3;46	243.99	15422	10DE	10;09	78.55
15349	30DE	22;50	279.93					15423	25DE	21;23	274.32				

241 / 247

NUMBER	DATE	TIME	LONG.	NUMBER	DATE	TIME	LONG.	NUMBER	DATE	TIME	LONG.	NUMBER	DATE	TIME	LONG.
				15349	15JA	6;45	115.48					15423	8JA	22;28	108.59
15350	29JA	17;16	310.04	15350	13FE	16;59	145.04	15424	24JA	12;27	304.35	15424	7FE	12;03	138.39
15351	28FE	11;37	339.70	15351	15MR	1;54	174.06	15425	23FE	0;49	333.86	15425	9MR	2;55	167.77
15352	30MR	4;17	8.79	15352	13AP	10;05	202.57	15426	24MR	10;49	2.77	15426	7AP	18;38	196.68
15353	28AP	18;31	37.32	15353	12MY	18;27	230.71	15427	22AP	18;56	31.15	15427	7MY	10;22	225.18
15354	28MY	6;26	65.48	15354	11JN	4;03	258.72	15428	22MY	1;56	59.16	15428	6JN	1;22	253.42
15355	26JN	16;34	93.51	15355	10JL	15;49	286.86	15429	20JN	8;47	87.05	15429	5JL	15;17	281.62
15356	26JL	1;36	121.66	15356	9AU	6;08	315.38	15430	19JL	16;41	115.10	15430	4AU	4;09	310.02
15357	24AU	10;11	150.16	15357	7SE	22;41	344.43	15431	18AU	2;48	143.56	15431	2SE	16;14	338.79
15358	22SE	18;59	179.15	15358	7OC	16;38	14.05	15432	16SE	15;58	172.61	15432	2OC	3;41	8.04
15359	22OC	4;41	208.66	15359	6NO	10;50	44.13	15433	16OC	8;20	202.30	15433	31OC	14;39	37.73
15360	20NO	15;52	238.60	15360	6DE	4;08	74.43	15434	15NO	3;03	232.49	15434	30NO	1;16	67.71
15361	20DE	4;58	268.75					15435	14DE	22;39	262.91	15435	29DE	11;50	97.74

NEW MOONS / FULL MOONS / NEW MOONS / FULL MOONS

248

NEW MOONS NUMBER	DATE	TIME	LONG.	FULL MOONS NUMBER	DATE	TIME	LONG.
15436	13JA	17;28	293.20	15436	27JA	22;45	127.56
15437	12FE	10;09	323.06	15437	26FE	10;20	156.98
15438	12MR	23;56	352.32	15438	26MR	22;45	185.94
15439	11AP	10;42	20.96	15439	25AP	12;00	214.48
15440	10MY	19;01	49.11	15440	25MY	2;01	242.74
15441	9JN	1;54	77.02	15441	23JN	16;46	270.95
15442	8JL	8;36	104.95	15442	23JL	8;04	299.33
15443	6AU	16;15	133.18	15443	21AU	23;27	328.07
15444	5SE	1;48	161.90	15444	20SE	14;15	357.26
15445	4OC	13;52	191.23	15445	20OC	3;54	26.91
15446	3NO	4;40	221.13	15446	18NO	16;15	56.88
15447	2DE	21;58	251.41	15447	18DE	3;31	86.95

249

NEW MOONS NUMBER	DATE	TIME	LONG.	FULL MOONS NUMBER	DATE	TIME	LONG.
15448	1JA	17;00	281.78	15448	16JA	14;02	116.86
15449	31JA	12;16	311.90	15449	15FE	0;05	146.39
15450	2MR	5;59	341.51	15450	16MR	9;52	175.43
15451	31MR	20;50	10.49	15451	14AP	19;46	203.98
15452	30AP	8;34	38.90	15452	14MY	6;24	232.20
15453	29MY	17;48	66.95	15453	12JN	18;31	260.31
15454	28JN	1;38	94.89	15454	12JL	8;37	288.55
15455	27JL	9;08	122.99	15455	11AU	0;38	317.16
15456	25AU	17;12	151.49	15456	9SE	17;45	346.26
15457	24SE	2;28	180.52	15457	9OC	10;44	15.87
15458	23OC	13;27	210.10	15458	8NO	2;36	45.87
15459	22NO	2;39	240.13	15459	7DE	16;52	76.05
15460	21DE	18;16	270.39				

250

NEW MOONS NUMBER	DATE	TIME	LONG.	FULL MOONS NUMBER	DATE	TIME	LONG.
				15460	6JA	5;26	106.12
15461	20JA	11;55	300.57	15461	4FE	16;19	135.84
15462	19FE	6;18	330.39	15462	6MR	1;41	165.03
15463	20MR	23;45	359.66	15463	4AP	9;57	193.69
15464	19AP	15;08	28.38	15464	3MY	17;53	221.92
15465	19MY	4;10	56.65	15465	2JN	2;35	249.92
15466	17JN	15;12	84.71	15466	1JL	13;05	277.98
15467	17JL	0;52	112.81	15467	31JL	2;07	306.34
15468	15AU	9;49	141.19	15468	29AU	17;44	335.19
15469	13SE	18;39	170.02	15469	28SE	11;18	4.63
15470	13OC	3;59	199.37	15470	28OC	5;45	34.59
15471	11NO	14;27	229.18	15471	26NO	23;49	64.87
15472	11DE	2;32	259.27	15472	26DE	16;20	95.15

251

NEW MOONS NUMBER	DATE	TIME	LONG.	FULL MOONS NUMBER	DATE	TIME	LONG.
15473	9JA	16;26	289.38	15473	25JA	6;25	125.13
15474	8FE	7;48	319.24	15474	23FE	17;45	154.59
15475	9MR	23;54	348.66	15475	25MR	2;43	183.45
15476	8AP	15;59	17.57	15476	23AP	10;11	211.79
15477	8MY	7;28	46.05	15477	22MY	17;09	239.79
15478	6JN	21;57	74.27	15478	21JN	0;47	267.72
15479	6JL	11;11	102.45	15479	20JL	10;02	295.82
15480	4AU	23;03	130.81	15480	18AU	21;36	324.35
15481	3SE	9;46	159.54	15481	17SE	11;55	353.47
15482	2OC	19;52	188.75	15482	17OC	5;01	23.20
15483	1NO	6;00	218.41	15483	16NO	0;13	53.42
15484	30NO	16;38	248.40	15484	15DE	19;58	83.84
15485	30DE	3;59	278.45				

252

NEW MOONS NUMBER	DATE	TIME	LONG.	FULL MOONS NUMBER	DATE	TIME	LONG.
				15485	14JA	14;13	114.11
15486	28JA	15;58	308.31	15486	13FE	5;30	143.90
15487	27FE	4;35	337.77	15487	13MR	17;30	173.07
15488	27MR	17;54	6.75	15488	12AP	2;48	201.64
15489	26AP	8;08	35.31	15489	11MY	10;25	229.76
15490	25MY	23;11	63.61	15490	9JN	17;21	257.66
15491	24JN	14;34	91.85	15491	9JL	1;03	285.62
15492	24JL	5;32	120.22	15492	7AU	8;46	313.88
15493	22AU	19;32	148.91	15493	5SE	18;59	342.64
15494	21SE	8;30	178.05	15494	5OC	7;59	12.01
15495	20OC	20;40	207.65	15495	4NO	0;13	41.98
15496	19NO	8;18	237.59	15496	3DE	19;09	72.34
15497	18DE	19;27	267.66				

253

NEW MOONS NUMBER	DATE	TIME	LONG.	FULL MOONS NUMBER	DATE	TIME	LONG.
				15497	2JA	15;07	102.74
15498	17JA	6;03	297.56	15498	1FE	9;56	132.83
15499	15FE	16;14	327.09	15499	3MR	2;05	162.36
15500	17MR	2;26	356.13	15500	1AP	15;14	191.26
15501	15AP	13;22	24.72	15501	1MY	1;46	219.62
15502	15MY	1;37	52.99	15502	30MY	10;20	247.64
15503	13JN	15;25	81.17	15503	28JN	17;45	275.56
15504	13JL	6;31	109.46	15504	28JL	0;55	303.66
15505	11AU	22;22	138.07	15505	26AU	8;47	332.15
15506	10SE	14;23	167.14	15506	24SE	18;21	1.21
15507	10OC	6;04	196.71	15507	24OC	6;28	30.85
15508	8NO	20;52	226.68	15508	22NO	21;29	60.96
15509	8DE	10;14	256.81	15509	22DE	14;53	91.29

254

NEW MOONS NUMBER	DATE	TIME	LONG.	FULL MOONS NUMBER	DATE	TIME	LONG.
15510	6JA	21;53	286.84	15510	21JA	9;26	121.50
15511	5FE	7;57	316.52	15511	20FE	3;41	151.30
15512	6MR	16;57	345.69	15512	21MR	20;28	180.54
15513	5AP	1;36	14.35	15513	20AP	11;04	209.20
15514	4MY	10;42	42.62	15514	19MY	23;13	237.44
15515	2JN	20;55	70.68	15515	18JN	9;09	265.45
15516	2JL	8;49	98.79	15516	17JL	17;33	293.51
15517	31JL	22;44	127.20	15517	16AU	1;25	321.85
15518	30AU	14;45	156.08	15518	14SE	9;49	350.67
15519	29SE	8;28	185.54	15519	13OC	19;33	20.05
15520	29OC	2;42	215.51	15520	12NO	7;04	49.91
15521	27NO	19;53	245.75	15521	11DE	20;26	80.06
15522	27DE	10;49	275.96				

255

NEW MOONS NUMBER	DATE	TIME	LONG.	FULL MOONS NUMBER	DATE	TIME	LONG.
				15522	10JA	11;26	110.21
15523	25JA	23;11	305.86	15523	9FE	3;43	140.10
15524	24FE	9;22	335.26	15524	10MR	20;42	169.54
15525	25MR	18;01	4.10	15525	9AP	13;29	198.47
15526	24AP	1;49	32.45	15526	9MY	5;01	226.94
15527	23MY	9;30	60.47	15527	7JN	18;35	255.12
15528	21JN	17;52	88.43	15528	7JL	6;09	283.24
15529	21JL	3;53	116.57	15529	5AU	16;18	311.54
15530	19AU	15;11	145.15	15530	4SE	1;54	340.23
15531	18SE	8;15	174.34	15531	3OC	11;38	9.43
15532	18OC	2;42	204.14	15532	1NO	21;54	39.11
15533	16NO	22;17	234.38	15533	1DE	8;49	69.11
15534	16DE	17;00	264.77	15534	30DE	20;31	99.18

256

NEW MOONS NUMBER	DATE	TIME	LONG.	FULL MOONS NUMBER	DATE	TIME	LONG.
15535	15JA	9;25	294.95	15535	29JA	9;13	129.06
15536	13FE	23;03	324.66	15536	27FE	23;10	158.56
15537	14MR	10;05	353.77	15537	28MR	14;14	187.60
15538	12AP	18;56	22.31	15538	27AP	5;50	216.21
15539	12MY	2;18	50.42	15539	26MY	21;11	244.52
15540	10JN	9;02	78.32	15540	25JN	11;44	272.72
15541	9JL	16;16	106.28	15541	25JL	1;18	301.04
15542	8AU	1;10	134.57	15542	23AU	14;02	329.70
15543	6SE	12;49	163.40	15543	22SE	2;05	358.81
15544	6OC	3;43	192.87	15544	21OC	13;31	28.39
15545	4NO	21;30	222.91	15545	20NO	0;25	58.31
15546	4DE	16;54	253.29	15546	19DE	10;58	88.34

257

NEW MOONS NUMBER	DATE	TIME	LONG.	FULL MOONS NUMBER	DATE	TIME	LONG.
15547	3JA	12;17	283.67	15547	17JA	21;33	118.24
15548	2FE	6;06	313.71	15548	16FE	8;32	147.80
15549	3MR	21;18	343.18	15549	17MR	20;14	176.89
15550	2AP	9;27	12.03	15550	16AP	8;47	205.53
15551	1MY	18;51	40.33	15551	15MY	22;12	233.84
15552	31MY	2;18	68.30	15552	14JN	12;31	262.04
15553	29JN	8;58	96.20	15553	14JL	3;38	290.33
15554	28JL	16;06	124.30	15554	12AU	19;15	318.95
15555	27AU	0;42	152.83	15555	11SE	10;44	348.01
15556	25SE	11;34	181.95	15556	11OC	1;19	17.55
15557	25OC	1;04	211.66	15557	9NO	14;33	47.45
15558	23NO	17;10	241.83	15558	9DE	2;26	77.53
15559	23DE	11;24	272.19				

258

NEW MOONS NUMBER	DATE	TIME	LONG.	FULL MOONS NUMBER	DATE	TIME	LONG.
				15559	7JA	13;15	107.52
15560	22JA	6;37	302.42	15560	5FE	23;22	137.19
15561	21FE	1;10	332.22	15561	7MR	9;02	166.38
15562	22MR	17;26	1.42	15562	5AP	18;34	195.07
15563	21AP	6;38	30.02	15563	5MY	4;30	223.36
15564	20MY	17;01	58.17	15564	3JN	15;36	251.45
15565	19JN	1;29	86.13	15565	3JL	4;36	279.61
15566	18JL	9;10	114.16	15566	1AU	19;47	308.07
15567	16AU	17;01	142.51	15567	31AU	12;41	337.01
15568	15SE	1;47	171.36	15568	30SE	6;09	6.48
15569	14OC	11;58	200.77	15569	29OC	22;56	36.39
15570	13NO	0;03	230.66	15570	28NO	14;13	66.56
15571	12DE	14;24	260.85	15571	28DE	3;43	96.71

259

NEW MOONS NUMBER	DATE	TIME	LONG.	FULL MOONS NUMBER	DATE	TIME	LONG.
15572	11JA	6;58	291.07	15572	26JA	15;23	126.57
15573	10FE	0;56	321.01	15573	25FE	1;20	155.95
15574	11MR	18;48	350.47	15574	26MR	9;53	184.77
15575	10AP	11;09	19.37	15575	24AP	17;41	213.11
15576	10MY	1;18	47.78	15576	24MY	1;42	241.15
15577	8JN	13;21	75.89	15577	22JN	11;04	269.14
15578	7JL	23;47	103.97	15578	21JL	22;45	297.36
15579	6AU	9;13	132.25	15579	20AU	13;10	326.02
15580	4SE	18;15	160.94	15580	19SE	6;01	355.27
15581	4OC	3;28	190.12	15581	19OC	0;23	25.08
15582	2NO	13;25	219.80	15582	17NO	18;59	55.29
15583	2DE	0;39	249.81	15583	17DE	12;32	85.63
15584	31DE	13;33	279.93				

260

NEW MOONS				FULL MOONS				NEW MOONS				FULL MOONS			
				15584	16JA	3;56	115.75					15658	8JA	20;37	108.90
15585	30JA	4;01	309.88	15585	14FE	16;34	145.41	15659	24JA	1;25	304.27	15659	7FE	7;10	138.57
15586	28FE	19;33	339.43	15586	15MR	2;31	174.47	15660	22FE	17;56	333.96	15660	8MR	18;12	167.80
15587	29MR	11;27	8.49	15587	13AP	10;28	202.97	15661	24MR	7;32	3.03	15661	7AP	6;01	196.56
15588	28AP	3;06	37.09	15588	12MY	17;25	231.05	15662	22AP	18;11	31.50	15662	6MY	18;45	224.94
15589	27MY	18;05	65.38	15589	11JN	0;31	258.96	15663	22MY	2;28	59.57	15663	5JN	8;28	253.14
15590	26JN	8;05	93.56	15590	10JL	8;50	286.97	15664	20JN	9;24	87.46	15664	4JL	23;12	281.37
15591	25JL	20;51	121.86	15591	8AU	19;11	315.33	15665	19JL	16;14	115.47	15665	3AU	14;47	309.87
15592	24AU	8;22	150.48	15592	7SE	8;11	344.23	15666	18AU	0;06	143.84	15666	2SE	6;43	338.79
15593	22SE	18;58	179.54	15593	7OC	0;02	13.75	15667	16SE	9;53	172.75	15667	1OC	22;09	8.20
15594	22OC	5;13	209.08	15594	5NO	18;25	43.83	15668	15OC	22;08	202.26	15668	31OC	12;21	38.03
15595	20NO	15;40	238.98	15595	5DE	14;12	74.22	15669	14NO	12;59	232.29	15669	30NO	1;02	68.09
15596	20DE	2;39	269.04					15670	14DE	6;11	262.60	15670	29DE	12;22	98.15

261 (left) — **267** (right)

NEW MOONS				FULL MOONS				NEW MOONS				FULL MOONS			
				15596	4JA	9;28	104.59	15671	13JA	0;57	292.90	15671	27JA	22;42	127.94
15597	18JA	14;11	298.97	15597	3FE	2;18	134.58	15672	11FE	19;53	322.86	15672	26FE	8;22	157.29
15598	17FE	2;13	328.56	15598	4MR	15;50	163.97	15673	13MR	13;18	352.27	15673	27MR	17;41	186.12
15599	18MR	14;52	357.68	15599	3AP	2;17	192.74	15674	12AP	3;59	21.07	15674	26AP	3;06	214.51
15600	17AP	4;20	26.34	15600	2MY	10;32	220.99	15675	11MY	15;41	49.37	15675	25MY	13;19	242.62
15601	16MY	18;48	54.69	15601	31MY	17;39	248.94	15676	10JN	1;03	77.37	15676	24JN	1;10	270.72
15602	15JN	10;03	82.92	15602	30JN	0;37	276.85	15677	9JL	9;06	105.36	15677	23JL	15;16	299.04
15603	15JL	1;26	111.24	15603	29JL	8;17	304.98	15678	7AU	16;55	133.59	15678	22AU	7;33	327.81
15604	13AU	16;13	139.83	15604	27AU	17;30	333.55	15679	6SE	1;20	162.27	15679	21SE	1;09	357.11
15605	12SE	6;00	168.84	15605	26SE	5;07	2.71	15680	5OC	10;54	191.49	15680	20OC	18;41	26.91
15606	11OC	18;49	198.31	15606	25OC	19;47	32.48	15681	3NO	22;03	221.24	15681	19NO	11;00	57.05
15607	10NO	6;54	228.18	15607	24NO	13;31	62.72	15682	3DE	11;13	251.34	15682	19DE	1;31	87.25
15608	9DE	18;23	258.24	15608	24DE	9;11	93.14								

262 (left) — **268** (right)

NEW MOONS				FULL MOONS				NEW MOONS				FULL MOONS			
15609	8JA	5;13	288.22	15609	23JA	4;43	123.38	15683	2JA	2;35	281.56	15683	17JA	14;05	117.25
15610	6FE	15;25	317.89	15610	21FE	22;11	153.12	15684	31JA	19;46	311.60	15684	16FE	0;46	146.81
15611	8MR	1;17	347.08	15611	23MR	12;44	182.23	15685	1MR	13;37	341.22	15685	16MR	9;48	175.81
15612	6AP	11;29	15.78	15612	22AP	0;27	210.76	15686	31MR	6;41	10.30	15686	14AP	17;41	204.28
15613	5MY	22;45	44.12	15613	21MY	9;54	238.88	15687	29AP	21;52	38.86	15687	14MY	1;17	232.38
15614	4JN	11;37	72.28	15614	19JN	17;51	266.81	15688	29MY	10;57	67.06	15688	12JN	9;43	260.34
15615	4JL	2;05	100.50	15615	19JL	1;08	294.83	15689	27JN	22;14	95.14	15689	11JL	20;05	288.44
15616	2AU	17;44	128.99	15616	17AU	8;39	323.18	15690	27JL	8;18	123.34	15690	10AU	9;10	316.92
15617	1SE	9;58	157.91	15617	15SE	17;24	352.04	15691	25AU	17;43	151.90	15691	9SE	0;59	345.96
15618	1OC	2;11	187.34	15618	15OC	4;16	21.48	15692	24SE	2;59	180.93	15692	8OC	18;55	15.60
15619	30OC	17;47	217.21	15619	13NO	17;52	51.45	15693	23OC	12;39	210.46	15693	7NO	13;45	45.72
15620	29NO	8;06	247.34	15620	13DE	10;07	81.72	15694	21NO	23;15	240.38	15694	7DE	8;07	76.06
15621	28DE	20;40	277.46					15695	21DE	11;14	270.48				

263 (left) — **269** (right)

NEW MOONS				FULL MOONS				NEW MOONS				FULL MOONS			
				15621	12JA	4;06	111.99					15695	6JA	0;45	106.31
15622	27JA	7;25	307.27	15622	10FE	22;28	141.94	15696	20JA	0;46	300.49	15696	4FE	14;45	136.16
15623	25FE	16;42	336.61	15623	12MR	15;55	171.37	15697	18FE	15;34	330.17	15697	6MR	1;54	165.43
15624	27MR	1;13	5.43	15624	11AP	7;33	200.22	15698	20MR	7;04	359.37	15698	4AP	10;36	194.10
15625	25AP	9;47	33.79	15625	10MY	20;55	228.59	15699	18AP	22;39	28.10	15699	3MY	17;48	222.30
15626	24MY	19;10	61.87	15626	9JN	7;58	256.66	15700	18MY	13;54	56.47	15700	2JN	0;37	250.23
15627	23JN	6;03	89.93	15627	8JL	17;10	284.69	15701	17JN	4;29	84.67	15701	1JL	8;10	278.17
15628	22JL	18;55	118.19	15628	7AU	1;21	312.93	15702	16JL	18;04	112.92	15702	30JL	17;25	306.37
15629	21AU	9;59	146.90	15629	5SE	9;34	341.59	15703	15AU	6;28	141.44	15703	29AU	5;07	335.07
15630	20SE	3;09	176.17	15630	4OC	18;43	10.79	15704	13SE	17;46	170.37	15704	27SE	19;37	4.37
15631	19OC	21;32	206.00	15631	3NO	5;25	40.50	15705	13OC	4;22	199.77	15705	27OC	12;56	34.27
15632	18NO	15;39	236.20	15632	2DE	17;52	70.57	15706	11NO	14;49	229.59	15706	26NO	8;17	64.60
15633	18DE	7;56	266.48					15707	11DE	1;32	259.62	15707	26DE	4;05	95.02

264 (left) — **270** (right)

NEW MOONS				FULL MOONS				NEW MOONS				FULL MOONS			
				15633	1JA	7;58	100.74	15708	9JA	12;42	289.61	15708	24JA	22;17	125.18
15634	16JA	21;35	296.53	15634	30JA	23;26	130.72	15709	8FE	0;16	319.32	15709	23FE	13;26	154.80
15635	15FE	8;42	326.11	15635	29FE	15;52	160.30	15710	9MR	12;18	348.57	15710	25MR	1;15	183.78
15636	15MR	17;53	355.13	15636	30MR	8;37	189.38	15711	8AP	1;01	17.35	15711	23AP	10;24	212.18
15637	14AP	1;52	23.62	15637	29AP	0;45	217.99	15712	7MY	14;44	45.77	15712	22MY	17;55	240.21
15638	13MY	9;22	51.72	15638	28MY	15;21	246.25	15713	6JN	5;32	74.00	15713	21JN	0;51	268.10
15639	11JN	17;10	79.66	15639	27JN	4;01	274.38	15714	5JL	21;00	102.27	15714	20JL	8;07	296.13
15640	11JL	2;12	107.70	15640	26JL	15;00	302.62	15715	4AU	12;22	130.77	15715	18AU	16;33	324.54
15641	9AU	13;27	136.10	15641	25AU	1;02	331.19	15716	3SE	2;58	159.65	15716	17SE	2;58	353.48
15642	8SE	3;42	165.06	15642	23SE	10;52	0.23	15717	2OC	16;33	189.00	15717	16OC	16;07	23.05
15643	7OC	21;03	194.66	15643	22OC	21;02	29.77	15718	1NO	5;14	218.77	15718	15NO	8;25	53.14
15644	6NO	16;28	224.79	15644	21NO	7;42	59.69	15719	30NO	17;09	248.81	15719	15DE	3;17	83.53
15645	6DE	11;58	255.18	15645	20DE	18;57	89.76	15720	30DE	4;19	278.85				

265 (left) — **271** (right)

NEW MOONS				FULL MOONS				NEW MOONS				FULL MOONS			
15646	5JA	5;41	285.48	15646	19JA	6;57	119.71					15720	13JA	23;03	113.86
15647	3FE	20;40	315.38	15647	17FE	19;59	149.32	15721	28JA	14;41	308.64	15721	12FE	17;37	143.79
15648	5MR	8;52	344.70	15648	19MR	10;13	178.49	15722	27FE	0;27	337.98	15722	14MR	9;33	173.13
15649	3AP	18;38	13.43	15649	18AP	1;22	207.22	15723	28MR	10;07	6.82	15723	12AP	22;34	201.85
15650	3MY	2;34	41.66	15650	17MY	16;47	235.60	15724	26AP	20;30	35.24	15724	12MY	9;03	230.09
15651	1JN	9;26	69.60	15651	16JN	7;47	263.81	15725	26MY	8;20	63.41	15725	10JN	17;42	258.06
15652	30JN	16;16	97.50	15652	15JL	22;00	292.09	15726	24JN	21;55	91.57	15726	10JL	1;18	286.04
15653	30JL	0;13	125.65	15653	14AU	11;24	320.64	15727	24JL	13;06	119.95	15727	8AU	8;43	314.26
15654	28AU	10;28	154.27	15654	13SE	0;04	349.61	15728	23AU	5;19	148.72	15728	6SE	16;53	342.94
15655	26SE	23;48	183.52	15655	12OC	12;05	19.06	15729	21SE	21;53	178.00	15729	6OC	2;43	12.18
15656	26OC	16;19	213.37	15656	10NO	23;25	48.90	15730	21OC	14;09	207.76	15730	4NO	14;58	41.98
15657	25NO	11;09	243.67	15657	10DE	10;10	78.94	15731	20NO	5;25	237.86	15731	4DE	5;55	72.16
15658	25DE	6;44	274.09					15732	19DE	19;00	268.03				

272

NEW MOONS				FULL MOONS			
NUMBER	DATE	TIME	LONG.	NUMBER	DATE	TIME	LONG.
				15732	2JA	23;03	102.45
15733	18JA	6;36	297.98	15733	1FE	17;11	132.52
15734	16FE	16;25	327.49	15734	2MR	11;01	162.13
15735	17MR	1;02	356.46	15735	1AP	3;29	191.17
15736	15AP	9;15	24.94	15736	30AP	17;57	219.69
15737	14MY	17;57	53.08	15737	30MY	6;11	247.86
15738	13JN	3;53	81.10	15738	28JN	16;21	275.89
15739	12JL	15;39	109.25	15739	28JL	1;07	304.05
15740	11AU	5;39	137.78	15740	26AU	9;24	332.57
15741	9SE	21;58	166.86	15741	24SE	18;11	1.59
15742	9OC	16;07	196.52	15742	24OC	4;12	31.14
15743	8NO	10;48	226.64	15743	22NO	15;48	61.11
15744	8DE	4;19	256.95	15744	22DE	5;01	91.26

273

NEW MOONS				FULL MOONS			
15745	6JA	19;22	287.13	15745	20JA	19;38	121.31
15746	5FE	7;38	316.89	15746	19FE	11;21	151.01
15747	6MR	17;34	346.10	15747	21MR	3;46	180.24
15748	5AP	1;54	14.75	15748	19AP	20;08	208.99
15749	4MY	9;25	42.95	15749	19MY	11;33	237.36
15750	2JN	16;52	70.91	15750	18JN	1;16	265.52
15751	2JL	1;07	98.87	15751	17JL	13;13	293.72
15752	31JL	11;08	127.12	15752	15AU	23;53	322.18
15753	29AU	23;52	155.86	15753	14SE	10;00	351.07
15754	28SE	15;50	185.24	15754	13OC	20;11	20.47
15755	28OC	10;34	215.22	15755	12NO	6;43	50.29
15756	27NO	6;24	245.56	15756	11DE	17;41	80.33
15757	27DE	1;14	275.95				

274

NEW MOONS				FULL MOONS			
				15757	10JA	5;09	110.34
15758	25JA	17;38	306.02	15758	8FE	17;23	140.06
15759	24FE	7;07	335.56	15759	10MR	6;43	169.36
15760	25MR	17;56	4.48	15760	8AP	21;10	198.20
15761	24AP	2;36	32.86	15761	8MY	12;20	226.66
15762	23MY	9;50	60.87	15762	7JN	3;33	254.91
15763	21JN	16;32	88.76	15763	6JL	18;17	283.15
15764	20JL	23;48	116.79	15764	5AU	8;18	311.60
15765	19AU	8;51	145.23	15765	3SE	21;38	340.45
15766	17SE	20;39	174.25	15766	3OC	10;16	9.77
15767	17OC	11;43	203.90	15767	1NO	22;10	39.52
15768	16NO	5;36	234.07	15768	1DE	9;18	69.52
15769	16DE	1;00	264.48	15769	30DE	19;50	99.54

275

NEW MOONS				FULL MOONS			
15770	14JA	20;15	294.78	15770	29JA	6;08	129.32
15771	13FE	13;53	324.66	15771	27FE	16;38	158.68
15772	15MR	4;54	353.95	15772	29MR	3;45	187.56
15773	13AP	16;56	22.62	15773	27AP	15;44	216.04
15774	13MY	2;16	50.80	15774	27MY	4;46	244.25
15775	11JN	9;45	78.73	15775	25JN	18;56	272.44
15776	10JL	16;34	106.68	15776	25JL	10;13	300.83
15777	8AU	23;53	134.90	15777	24AU	2;17	329.61
15778	7SE	8;45	163.62	15778	22SE	18;23	358.88
15779	6OC	19;50	192.93	15779	22OC	9;34	28.61
15780	5NO	9;26	222.79	15780	20NO	23;13	58.64
15781	5DE	1;29	253.03	15781	20DE	11;15	88.75

276

NEW MOONS				FULL MOONS			
15782	3JA	19;29	283.35	15782	18JA	22;00	118.65
15783	2FE	14;21	313.44	15783	17FE	7;49	148.15
15784	3MR	8;33	343.05	15784	17MR	17;02	177.14
15785	2AP	0;34	12.06	15785	16AP	2;05	205.65
15786	1MY	13;42	40.51	15786	15MY	11;36	233.81
15787	31MY	0;10	68.60	15787	13JN	22;23	261.86
15788	29JN	8;50	96.58	15788	13JL	11;16	290.07
15789	28JL	16;49	124.71	15789	12AU	2;36	318.66
15790	27AU	1;02	153.24	15790	10SE	19;53	347.79
15791	25SE	10;08	182.28	15791	10OC	13;53	17.46
15792	24OC	20;34	211.86	15792	9NO	7;11	47.53
15793	23NO	8;42	241.86	15793	8DE	22;47	77.77
15794	22DE	22;51	272.05				

277

NEW MOONS				FULL MOONS			
				15794	7JA	12;21	107.87
15795	21JA	15;01	302.15	15795	5FE	23;53	137.60
15796	20FE	8;27	331.92	15796	7MR	9;34	166.79
15797	22MR	1;50	1.17	15797	5AP	17;46	195.42
15798	20AP	17;53	29.90	15798	5MY	1;13	223.61
15799	20MY	8;00	58.20	15799	3JN	8;57	251.58
15800	18JN	20;13	86.31	15800	2JL	18;09	279.58
15801	18JL	7;01	114.46	15801	1AU	5;49	307.90
15802	16AU	16;55	142.90	15802	30AU	20;23	336.73
15803	15SE	2;26	171.78	15803	29SE	13;32	6.17
15804	14OC	12;02	201.16	15804	29OC	8;16	36.16
15805	12NO	22;13	230.97	15805	28NO	3;11	66.48
15806	12DE	9;27	261.03	15806	27DE	20;53	96.81

278

NEW MOONS				FULL MOONS			
15807	10JA	22;04	291.08	15807	26JA	12;16	126.83
15808	9FE	12;02	320.86	15808	25FE	0;45	156.31
15809	11MR	2;57	350.21	15809	26MR	10;28	185.18
15810	9AP	18;17	19.08	15810	24AP	18;10	213.52
15811	9MY	9;35	47.54	15811	24MY	0;57	241.50
15812	8JN	0;31	75.77	15812	22JN	7;57	269.40
15813	7JL	14;46	104.00	15813	21JL	16;15	297.48
15814	6AU	4;01	132.43	15814	20AU	2;43	325.98
15815	4SE	16;08	161.24	15815	18SE	15;52	355.07
15816	4OC	3;17	190.51	15816	18OC	7;55	24.78
15817	2NO	13;57	220.21	15817	17NO	2;28	54.98
15818	2DE	0;35	250.20	15818	16DE	22;19	85.41
15819	31DE	11;28	280.23				

279

NEW MOONS				FULL MOONS			
				15819	15JA	17;32	115.71
15820	29JA	22;40	310.04	15820	14FE	10;14	145.54
15821	28FE	10;11	339.43	15821	15MR	23;36	174.74
15822	29MR	22;13	8.34	15822	14AP	9;53	203.33
15823	28AP	11;08	36.84	15823	13MY	18;03	231.46
15824	28MY	1;14	65.10	15824	12JN	1;09	259.38
15825	26JN	16;25	93.32	15825	11JL	8;11	287.33
15826	26JL	8;05	121.73	15826	9AU	16;02	315.59
15827	24AU	23;24	150.50	15827	8SE	1;28	344.34
15828	23SE	13;48	179.71	15828	7OC	13;15	13.69
15829	23OC	3;12	209.38	15829	6NO	4;01	43.61
15830	21NO	15;39	239.37	15830	5DE	21;43	73.91
15831	21DE	3;16	269.45				

280

NEW MOONS				FULL MOONS			
				15831	4JA	17;12	104.29
15832	19JA	13;57	299.35	15832	3FE	12;27	134.39
15833	17FE	23;48	328.85	15833	4MR	5;40	163.94
15834	18MR	9;11	357.83	15834	2AP	20;02	192.87
15835	16AP	18;51	26.35	15835	2MY	7;40	221.26
15836	16MY	5;39	54.56	15836	31MY	17;11	249.31
15837	14JN	18;13	82.68	15837	30JN	1;18	277.26
15838	14JL	8;38	110.95	15838	29JL	8;51	305.39
15839	13AU	0;32	139.58	15839	27AU	16;41	333.90
15840	11SE	17;15	168.70	15840	26SE	1;43	2.96
15841	11OC	10;03	198.33	15841	25OC	12;46	32.57
15842	10NO	2;09	228.36	15842	24NO	2;23	62.64
15843	9DE	16;47	258.56	15843	23DE	18;26	92.91

281

NEW MOONS				FULL MOONS			
15844	8JA	5;24	288.62	15844	22JA	12;02	123.06
15845	6FE	15;58	318.30	15845	21FE	5;56	152.83
15846	8MR	0;55	347.45	15846	22MR	22;58	182.06
15847	6AP	9;02	16.07	15847	21AP	14;23	210.75
15848	5MY	17;11	44.28	15848	21MY	3;45	239.02
15849	4JN	2;16	72.29	15849	19JN	15;01	267.09
15850	3JL	12;58	100.36	15850	19JL	0;34	295.19
15851	2AU	1;49	128.74	15851	17AU	9;11	323.59
15852	31AU	17;06	157.61	15852	15SE	17;49	352.44
15853	30SE	10;38	187.08	15853	15OC	3;18	21.83
15854	30OC	5;29	217.08	15854	13NO	14;11	51.68
15855	28NO	23;58	247.39	15855	13DE	2;34	81.79
15856	28DE	16;26	277.67				

282

NEW MOONS				FULL MOONS			
				15856	11JA	16;21	111.88
15857	27JA	6;03	307.61	15857	10FE	7;17	141.70
15858	25FE	16;58	337.01	15858	11MR	23;08	171.07
15859	27MR	1;52	5.84	15859	10AP	15;23	199.96
15860	25AP	9;33	34.16	15860	10MY	7;16	228.43
15861	24MY	16;49	62.17	15861	8JN	21;55	256.65
15862	23JN	0;30	90.09	15862	8JL	10;53	284.83
15863	22JL	9;30	118.20	15863	6AU	22;21	313.20
15864	20AU	20;49	146.75	15864	5SE	8;56	341.95
15865	19SE	11;15	175.90	15865	4OC	19;16	11.20
15866	19OC	4;51	205.69	15866	3NO	5;47	40.91
15867	18NO	0;30	235.95	15867	2DE	16;36	70.91
15868	17DE	20;08	266.37				

283

NEW MOONS				FULL MOONS			
				15868	1JA	3;43	100.95
15869	16JA	13;52	296.59	15869	30JA	15;19	130.77
15870	15FE	4;45	326.34	15870	1MR	3;48	160.19
15871	16MR	16;46	355.47	15871	30MR	17;23	189.15
15872	15AP	2;20	24.02	15872	29AP	8;01	217.71
15873	14MY	10;07	52.13	15873	28MY	23;10	246.00
15874	12JN	16;56	80.03	15874	27JN	14;13	274.22
15875	11JL	23;48	107.99	15875	27JL	4;47	302.60
15876	10AU	7;53	136.26	15876	25AU	18;44	331.32
15877	8SE	18;17	165.06	15877	24SE	8;02	0.50
15878	8OC	7;48	194.49	15878	23OC	20;35	30.14
15879	7NO	0;26	224.50	15879	22NO	8;14	60.10
15880	6DE	19;16	254.86	15880	21DE	19;04	90.15

284

NEW MOONS				FULL MOONS			
15881	5JA	14;45	285.23	15881	20JA	5;19	120.03
15882	4FE	9;15	315.29	15882	18FE	15;28	149.52
15883	5MR	1;33	344.79	15883	19MR	1;58	178.54
15884	3AP	14;59	13.67	15884	17AP	13;13	207.12
15885	3MY	1;34	42.00	15885	17MY	1;28	235.38
15886	1JN	9;52	70.00	15886	15JN	14;56	263.53
15887	30JN	16;56	97.92	15887	15JL	5;42	291.82
15888	29JL	23;58	126.03	15888	13AU	21;38	320.47
15889	28AU	8;05	154.56	15889	12SE	14;06	349.59
15890	26SE	18;07	183.67	15890	12OC	6;09	19.20
15891	26OC	6;32	213.35	15891	10NO	20;52	49.19
15892	24NO	21;23	243.47	15892	10DE	9;49	79.31
15893	24DE	14;23	273.78				

285

NEW MOONS				FULL MOONS			
				15893	8JA	21;09	109.32
15894	23JA	8;48	303.97	15894	7FE	7;16	138.97
15895	22FE	3;21	333.75	15895	8MR	16;33	168.12
15896	23MR	20;27	2.96	15896	7AP	1;24	196.76
15897	22AP	10;59	31.60	15897	6MY	10;22	224.99
15898	21MY	22;44	59.80	15898	4JN	20;14	253.04
15899	20JN	8;16	87.80	15899	4JL	7;55	281.15
15900	19JL	16;37	115.87	15900	2AU	22;02	309.58
15901	18AU	0;48	144.25	15901	1SE	14;37	338.52
15902	16SE	9;35	173.12	15902	1OC	8;41	8.03
15903	15OC	19;27	202.54	15903	31OC	2;44	38.01
15904	14NO	6;45	232.41	15904	29NO	19;27	68.25
15905	13DE	19;48	262.55	15905	29DE	10;08	98.44

286

NEW MOONS				FULL MOONS			
15906	12JA	10;49	292.69	15906	27JA	22;38	128.33
15907	11FE	3;29	322.57	15907	26FE	9;05	157.71
15908	12MR	20;48	351.98	15908	27MR	17;47	186.51
15909	11AP	13;27	20.87	15909	26AP	1;20	214.82
15910	11MY	4;30	49.31	15910	25MY	8;39	242.82
15911	9JN	17;41	77.47	15911	23JN	16;54	270.77
15912	9JL	5;18	105.59	15912	23JL	3;13	298.95
15913	7AU	15;49	133.93	15913	21AU	16;22	327.58
15914	6SE	1;43	162.68	15914	20SE	8;25	356.81
15915	5OC	11;27	191.91	15915	20OC	2;41	26.63
15916	3NO	21;25	221.60	15916	18NO	21;50	56.88
15917	3DE	8;06	251.60	15917	18DE	16;24	87.26

287

NEW MOONS				FULL MOONS			
15918	1JA	19;54	281.66	15918	17JA	9;04	117.43
15919	31JA	9;00	311.54	15919	15FE	22;57	147.12
15920	1MR	23;13	341.02	15920	17MR	9;53	176.20
15921	31MR	14;06	10.02	15921	15AP	18;22	204.70
15922	30AP	5;14	38.58	15922	15MY	1;23	232.77
15923	29MY	20;18	66.87	15923	13JN	8;05	260.66
15924	28JN	11;01	95.08	15924	12JL	15;37	288.65
15925	28JL	1;00	123.44	15925	11AU	0;57	316.98
15926	26AU	13;59	152.13	15926	9SE	12;48	345.85
15927	25SE	1;52	181.27	15927	9OC	3;29	15.34
15928	24OC	12;57	210.86	15928	7NO	20;57	45.40
15929	22NO	23;41	240.79	15929	7DE	16;23	75.79
15930	22DE	10;25	270.83				

288

NEW MOONS				FULL MOONS			
				15930	6JA	12;09	106.17
15931	20JA	21;21	300.73	15931	5FE	6;13	136.19
15932	19FE	8;28	330.26	15932	5MR	21;12	165.63
15933	19MR	19;54	359.30	15933	4AP	8;52	194.42
15934	18AP	8;02	27.90	15934	3MY	17;54	222.69
15935	17MY	21;18	56.19	15935	2JN	1;23	250.65
15936	16JN	11;54	84.40	15936	1JL	8;24	278.56
15937	16JL	3;30	112.73	15937	30JL	15;49	306.70
15938	14AU	19;19	141.38	15938	29AU	0;29	335.26
15939	13SE	10;31	170.46	15939	27SE	11;06	4.40
15940	13OC	0;43	200.01	15940	27OC	0;24	34.13
15941	11NO	13;51	229.93	15941	25NO	16;41	64.32
15942	11DE	1;59	260.03	15942	25DE	11;24	94.70

289

NEW MOONS				FULL MOONS			
15943	9JA	13;07	290.02	15943	24JA	6;53	124.92
15944	7FE	23;13	319.66	15944	23FE	1;08	154.68
15945	9MR	8;32	348.80	15945	24MR	16;50	183.82
15946	7AP	17;43	17.45	15946	23AP	5;44	212.38
15947	7MY	3;36	45.71	15947	22MY	16;14	240.53
15948	5JN	15;04	73.82	15948	21JN	1;02	268.50
15949	5JL	4;30	102.00	15949	20JL	8;54	296.55
15950	3AU	19;48	130.49	15950	18AU	16;39	324.93
15951	2SE	12;24	159.44	15951	17SE	1;08	353.80
15952	2OC	5;31	188.93	15952	16OC	11;12	23.23
15953	31OC	22;20	218.87	15953	14NO	23;33	53.15
15954	30NO	13;59	249.06	15954	14DE	14;23	83.37
15955	30DE	3;42	279.22				

290

NEW MOONS				FULL MOONS			
				15955	13JA	7;10	113.58
15956	28JA	15;13	309.05	15956	12FE	0;49	143.48
15957	27FE	0;45	338.38	15957	13MR	18;11	172.89
15958	28MR	8;59	7.16	15958	12AP	10;20	201.74
15959	26AP	16;49	35.47	15959	12MY	0;43	230.14
15960	26MY	1;11	63.51	15960	10JN	13;05	258.27
15961	24JN	10;53	91.52	15961	9JL	23;35	286.36
15962	23JL	22;36	119.75	15962	8AU	8;46	314.65
15963	22AU	12;43	148.44	15963	6SE	17;30	343.35
15964	21SE	5;20	177.70	15964	6OC	2;40	12.57
15965	20OC	23;54	207.55	15965	4NO	12;56	42.28
15966	19NO	18;58	237.81	15966	4DE	0;35	72.33
15967	19DE	12;43	268.15				

291

NEW MOONS				FULL MOONS			
				15967	2JA	13;34	102.44
15968	18JA	3;49	298.25	15968	1FE	3;43	132.35
15969	16FE	15;57	327.85	15969	2MR	18;51	161.85
15970	18MR	1;38	356.87	15970	1AP	10;41	190.88
15971	16AP	9;41	25.34	15971	1MY	2;41	219.47
15972	15MY	16;57	53.42	15972	30MY	18;00	247.76
15973	14JN	0;15	81.34	15973	29JN	7;58	275.94
15974	13JL	8;27	109.35	15974	28JL	20;21	304.25
15975	11AU	18;31	137.72	15975	27AU	7;34	332.88
15976	10SE	7;24	166.65	15976	25SE	18;13	1.98
15977	9OC	23;35	196.22	15977	25OC	4;50	31.56
15978	8NO	18;32	226.34	15978	23NO	15;36	61.49
15979	8DE	14;31	256.75	15979	23DE	2;32	91.55

292

NEW MOONS				FULL MOONS			
15980	7JA	9;23	287.09	15980	21JA	13;42	121.45
15981	6FE	1;42	317.03	15981	20FE	1;26	150.99
15982	6MR	15;02	346.39	15982	20MR	14;09	180.08
15983	5AP	1;40	15.13	15983	19AP	4;01	208.73
15984	4MY	10;11	43.37	15984	18MY	18;47	237.08
15985	2JN	17;20	71.31	15985	17JN	9;55	265.30
15986	2JL	0;04	99.22	15986	17JL	0;53	293.61
15987	31JL	7;27	127.36	15987	15AU	15;24	322.22
15988	29AU	16;40	155.96	15988	14SE	5;21	351.27
15989	28SE	4;39	185.17	15989	13OC	18;34	20.79
15990	27OC	19;51	214.98	15990	12NO	6;53	50.69
15991	26NO	13;45	245.25	15991	11DE	18;12	80.75
15992	26DE	9;04	275.65				

293

NEW MOONS				FULL MOONS			
				15992	10JA	4;38	110.70
15993	25JA	4;07	305.84	15993	8FE	14;36	140.33
15994	23FE	21;31	335.55	15994	10MR	0;37	169.49
15995	25MR	12;20	4.64	15995	8AP	11;10	198.18
15996	24AP	0;16	33.16	15996	7MY	22;39	226.50
15997	23MY	9;36	61.25	15997	6JN	11;20	254.65
15998	21JN	17;12	89.17	15998	6JL	1;25	282.86
15999	21JL	0;13	117.20	15999	4AU	16;55	311.37
16000	19AU	7;48	145.57	16000	3SE	9;27	340.34
16001	17SE	16;56	174.48	16001	3OC	2;09	9.82
16002	17OC	4;14	203.97	16002	1NO	17;53	39.72
16003	15NO	17;54	233.96	16003	1DE	7;54	69.85
16004	15DE	9;49	264.22	16004	30DE	20;03	99.94

294

NEW MOONS				FULL MOONS			
16005	14JA	3;29	294.46	16005	29JA	6;38	129.73
16006	12FE	21;56	324.39	16006	27FE	16;07	159.05
16007	14MR	15;45	353.80	16007	29MR	0;55	187.84
16008	13AP	7;33	22.63	16008	27AP	9;32	216.17
16009	12MY	20;39	50.97	16009	26MY	18;40	244.24
16010	11JN	7;15	79.02	16010	25JN	5;14	272.28
16011	10JL	16;13	107.05	16011	24JL	18;04	300.57
16012	9AU	0;33	135.32	16012	23AU	9;34	329.32
16013	7SE	9;10	164.03	16013	22SE	3;15	358.64
16014	6OC	18;37	193.27	16014	21OC	21;45	28.50
16015	5NO	5;15	223.00	16015	20NO	15;28	58.71
16016	4DE	17;23	253.07	16016	20DE	7;19	88.97

295

NEW MOONS				FULL MOONS			
16017	3JA	7;17	283.22	16017	18JA	20;55	119.00
16018	1FE	22;58	313.18	16018	17FE	8;16	148.56
16019	3MR	15;49	342.75	16019	18MR	17;39	177.56
16020	2AP	8;42	11.80	16020	17AP	1;32	206.01
16021	2MY	0;30	40.37	16021	16MY	8;42	234.08
16022	31MY	14;37	68.61	16022	14JN	16;14	262.00
16023	30JN	3;05	96.73	16023	14JL	1;20	290.06
16024	29JL	14;19	125.00	16024	12AU	13;02	318.50
16025	28AU	0;44	153.61	16025	11SE	3;46	347.52
16026	26SE	10;45	182.70	16026	10OC	21;12	17.15
16027	25OC	20;44	212.26	16027	9NO	16;14	47.29
16028	24NO	7;05	242.18	16028	9DE	11;22	77.67
16029	23DE	18;14	272.24				

--------NEW MOONS--------				--------FULL MOONS-------				--------NEW MOONS--------				--------FULL MOONS-------			
NUMBER	DATE	TIME	LONG.	NUMBER	DATE	TIME	LONG.	NUMBER	DATE	TIME	LONG.	NUMBER	DATE	TIME	LONG.
			296								302				
				16029	8JA	5;10	107.96					16103	1JA	3;55	101.35
16030	22JA	6;30	302.18	16030	6FE	20;28	137.85	16104	15JA	22;40	296.34	16104	30JA	13;56	131.10
16031	20FE	19;55	331.79	16031	7MR	8;46	167.14	16105	14FE	16;55	326.23	16105	28FE	23;39	160.40
16032	21MR	10;13	0.92	16032	5AP	18;17	195.83	16106	16MR	8;59	355.54	16106	30MR	9;37	189.22
16033	20AP	1;00	29.61	16033	5MY	1;48	224.02	16107	14AP	22;17	24.25	16107	28AP	20;20	217.62
16034	19MY	15;59	57.95	16034	3JN	8;27	251.94	16108	14MY	8;50	52.47	16108	28MY	8;11	245.79
16035	18JN	6;56	86.17	16035	2JL	15;25	279.86	16109	12JN	17;14	80.43	16109	26JN	21;27	273.94
16036	17JL	21;30	114.47	16036	31JL	23;48	308.04	16110	12JL	0;30	108.40	16110	26JL	12;19	302.32
16037	16AU	11;17	143.06	16037	30AU	10;24	336.71	16111	10AU	7;48	136.64	16111	25AU	4;36	331.13
16038	15SE	0;00	172.07	16038	28SE	23;44	5.99	16112	8SE	16;13	165.36	16112	23SE	21;38	0.46
16039	14OC	11;43	201.54	16039	28OC	15;55	35.86	16113	8OC	2;30	194.66	16113	23OC	14;16	30.27
16040	12NO	22;44	231.39	16040	27NO	10;33	66.16	16114	6NO	15;02	224.49	16114	22NO	5;25	60.38
16041	12DE	9;29	261.42	16041	27DE	6;23	96.58	16115	6DE	5;49	254.68	16115	21DE	18;34	90.52
			297								303				
16042	10JA	20;14	291.39	16042	26JA	1;28	126.77	16116	4JA	22;33	284.94	16116	20JA	5;51	120.44
16043	9FE	7;03	321.05	16043	24FE	18;00	156.42	16117	3FE	16;32	314.98	16117	18FE	15;42	149.92
16044	10MR	18;02	350.24	16044	26MR	7;12	185.44	16118	5MR	10;38	344.56	16118	20MR	0;35	178.88
16045	9AP	5;28	18.95	16045	24AP	17;22	213.87	16119	4AP	3;26	13.59	16119	18AP	9;01	207.33
16046	8MY	17;52	47.30	16046	24MY	1;29	241.91	16120	3MY	17;52	42.08	16120	17MY	17;36	235.45
16047	7JN	7;40	75.49	16047	22JN	8;38	269.82	16121	2JN	5;42	70.22	16121	16JN	3;12	263.45
16048	6JL	22;51	103.75	16048	21JL	15;50	297.85	16122	1JL	15;29	98.25	16122	15JL	14;45	291.61
16049	5AU	14;50	132.28	16049	19AU	23;54	326.26	16123	31JL	0;12	126.42	16123	14AU	4;58	320.18
16050	4SE	6;42	161.24	16050	18SE	9;34	355.20	16124	29AU	8;48	154.98	16124	12SE	21;50	349.31
16051	3OC	21;44	190.66	16051	17OC	21;32	24.73	16125	27SE	17;58	184.05	16125	12OC	16;21	19.02
16052	2NO	11;39	220.50	16052	16NO	12;21	54.77	16126	27OC	4;07	213.64	16126	11NO	10;52	49.15
16053	2DE	0;26	250.59	16053	16DE	5;55	85.11	16127	25NO	15;29	243.61	16127	11DE	3;53	79.45
16054	31DE	12;06	280.65					16128	25DE	4;23	273.75				
			298								304				
				16054	15JA	1;08	115.40					16128	9JA	18;39	109.60
16055	29JA	22;36	310.43	16055	13FE	20;03	145.34	16129	23JA	18;59	303.77	16129	8FE	7;03	139.35
16056	28FE	8;04	339.73	16056	15MR	12;59	174.70	16130	22FE	11;05	333.46	16130	8MR	17;15	168.54
16057	29MR	16;59	8.52	16057	14AP	3;10	203.45	16131	23MR	3;50	2.68	16131	7AP	1;39	197.15
16058	28AP	2;09	36.87	16058	13MY	14;47	231.72	16132	21AP	20;06	31.39	16132	6MY	8;55	225.32
16059	27MY	12;33	64.98	16059	12JN	0;25	259.73	16133	21MY	11;02	59.72	16133	4JN	16;02	253.26
16060	26JN	0;52	93.09	16060	11JL	8;47	287.74	16134	20JN	0;23	87.87	16134	4JL	0;09	281.23
16061	25JL	15;18	121.45	16061	9AU	16;39	316.00	16135	19JL	12;23	116.09	16135	2AU	10;28	309.50
16062	24AU	7;29	150.24	16062	8SE	0;50	344.70	16136	17AU	23;25	144.59	16136	31AU	23;44	338.30
16063	23SE	0;41	179.56	16063	7OC	10;10	13.95	16137	16SE	9;51	173.53	16137	30SE	16;02	7.73
16064	22OC	18;01	209.38	16064	5NO	21;23	43.72	16138	15OC	20;01	202.95	16138	30OC	10;34	37.72
16065	21NO	10;34	239.55	16065	5DE	10;57	73.85	16139	14NO	6;15	232.78	16139	29NO	5;57	68.06
16066	21DE	1;26	269.77					16140	13DE	16;58	262.82	16140	29DE	0;38	98.43
			299								305				
				16066	4JA	2;43	104.07					16141	27JA	17;15	128.49
16067	19JA	14;03	299.75	16067	2FE	19;51	134.08	16141	12JA	4;31	292.81	16142	26FE	7;00	158.01
16068	18FE	0;24	329.26	16068	4MR	13;14	163.66	16142	10FE	17;08	322.52	16143	27MR	17;44	186.90
16069	19MR	9;01	358.21	16069	3AP	5;53	192.69	16143	12MR	6;43	351.80	16144	26AP	2;02	215.24
16070	17AP	16;45	26.65	16070	2MY	21;06	221.22	16144	10AP	21;01	20.60	16145	25MY	8;55	243.22
16071	17MY	0;34	54.74	16071	1JN	10;32	249.43	16145	10MY	11;44	49.03	16146	23JN	15;34	271.11
16072	15JN	9;23	82.71	16072	30JN	22;04	277.52	16146	9JN	2;40	77.26	16147	22JL	23;10	299.17
16073	14JL	19;58	110.83	16073	30JL	8;02	305.74	16147	8JL	17;34	105.51	16148	21AU	8;38	327.65
16074	13AU	8;53	139.34	16074	28AU	17;06	334.31	16148	7AU	8;02	134.01	16149	19SE	20;39	356.71
16075	12SE	0;24	168.39	16075	27SE	2;11	3.37	16149	5SE	21;37	162.90	16150	19OC	11;30	26.38
16076	11OC	18;17	198.06	16076	26OC	12;00	32.93	16150	5OC	10;06	192.24	16151	18NO	5;03	56.56
16077	10NO	13;30	228.22	16077	24NO	23;00	62.89	16151	3NO	21;37	222.00	16152	18DE	0;28	86.97
16078	10DE	8;16	258.59	16078	24DE	11;16	92.99	16152	3DE	8;34	252.01				
			300								306				
16079	9JA	0;50	288.82	16079	23JA	0;39	122.98	16153	1JA	19;16	282.03	16153	16JA	20;07	117.27
16080	7FE	14;23	318.62	16080	21FE	15;02	152.61	16154	31JA	5;54	311.79	16154	15FE	14;00	147.14
16081	8MR	1;05	347.84	16081	22MR	6;15	181.77	16155	1MR	16;32	341.13	16155	17MR	4;48	176.38
16082	6AP	9;44	16.48	16082	20AP	22;01	210.48	16156	31MR	3;24	9.97	16156	15AP	16;20	205.00
16083	5MY	17;10	44.66	16083	20MY	13;41	238.85	16157	29AP	14;59	38.40	16157	15MY	1;19	233.15
16084	4JN	0;16	72.60	16084	19JN	4;27	267.05	16158	29MY	3;51	66.60	16158	13JN	8;50	261.08
16085	3JL	7;52	100.55	16085	18JL	17;48	295.31	16159	27JN	18;19	94.80	16159	12JL	15;59	289.05
16086	1AU	16;55	128.76	16086	17AU	5;48	323.84	16160	27JL	10;06	123.23	16160	10AU	23;38	317.31
16087	31AU	4;21	157.48	16087	15SE	16;57	352.80	16161	26AU	2;23	152.05	16161	9SE	8;33	346.06
16088	29SE	18;58	186.82	16088	15OC	3;47	22.24	16162	24SE	18;12	181.34	16162	8OC	19;23	15.40
16089	29OC	12;46	216.77	16089	13NO	14;37	52.09	16163	24OC	8;59	211.08	16163	7NO	8;46	45.27
16090	28NO	8;34	247.13	16090	13DE	1;29	82.13	16164	22NO	22;31	241.13	16164	7DE	1;00	75.52
16091	28DE	4;16	277.54					16165	22DE	10;49	271.24				
			301								307				
				16091	11JA	12;25	112.10					16165	5JA	19;27	105.85
16092	26JA	21;56	307.65	16092	9FE	23;36	141.76	16166	20JA	21;50	301.14	16166	4FE	14;35	135.93
16093	25FE	12;40	337.22	16093	11MR	11;29	170.98	16167	19FE	7;38	330.61	16167	6MR	8;29	165.49
16094	27MR	0;30	6.17	16094	10AP	0;28	199.74	16168	20MR	16;31	359.55	16168	4AP	23;58	194.45
16095	25AP	9;55	34.56	16095	9MY	14;36	228.15	16169	19AP	1;13	28.01	16169	4MY	12;47	222.87
16096	24MY	17;37	62.58	16096	8JN	5;31	256.39	16170	18MY	10;41	56.16	16170	2JN	23;22	250.95
16097	23JN	0;27	90.47	16097	7JL	20;41	284.65	16171	16JN	21;50	84.22	16171	2JL	8;23	278.95
16098	22JL	7;26	118.51	16098	6AU	11;39	313.16	16172	16JL	11;11	112.46	16172	31JL	16;34	307.11
16099	20AU	15;41	146.93	16099	5SE	2;12	342.07	16173	15AU	2;39	141.09	16173	30AU	0;41	335.66
16100	19SE	2;17	175.92	16100	4OC	16;07	11.46	16174	13SE	19;39	170.24	16174	28SE	9;31	4.73
16101	18OC	15;57	205.53	16101	3NO	5;09	41.27	16175	13OC	13;17	199.92	16175	27OC	19;49	34.33
16102	17NO	8;39	235.66	16102	2DE	17;06	71.32	16176	12NO	6;35	230.02	16176	26NO	8;12	64.35
16103	17DE	3;24	266.05					16177	11DE	22;33	260.27	16177	25DE	22;49	94.56

308

NEW MOONS				FULL MOONS			
16178	10JA	12;21	290.38	16178	24JA	15;11	124.65
16179	8FE	23;43	320.07	16179	23FE	8;18	154.37
16180	9MR	8;57	349.21	16180	24MR	1;11	183.57
16181	7AP	16;50	17.80	16181	22AP	17;02	212.26
16182	7MY	0;20	45.97	16182	22MY	7;23	240.56
16183	5JN	8;25	73.94	16183	20JN	19;58	268.68
16184	4JL	17;58	101.97	16184	20JL	6;51	296.86
16185	3AU	5;41	130.31	16185	18AU	16;30	325.31
16186	1SE	19;58	159.16	16186	17SE	1;43	354.21
16187	1OC	12;53	188.62	16187	16OC	11;16	23.62
16188	31OC	7;47	218.64	16188	14NO	21;45	53.47
16189	30NO	3;10	248.99	16189	14DE	9;22	83.55
16190	29DE	21;04	279.33				

309

NEW MOONS				FULL MOONS			
16191	28JA	12;08	309.31	16190	12JA	22;04	113.58
16192	27FE	0;06	338.74	16191	11FE	11;42	143.33
16193	28MR	9;33	7.57	16192	13MR	2;13	172.62
16194	26AP	17;22	35.88	16193	11AP	17;29	201.45
16195	26MY	0;28	63.87	16194	11MY	9;08	229.91
16196	24JN	7;40	91.78	16195	10JN	0;26	258.15
16197	23JL	15;53	119.87	16196	9JL	14;40	286.39
16198	22AU	2;04	148.39	16197	8AU	3;34	314.83
16199	20SE	15;07	177.50	16198	6SE	15;21	343.65
16200	20OC	7;29	207.25	16199	6OC	2;33	12.96
16201	19NO	2;34	237.50	16200	4NO	13;34	42.70
16202	18DE	22;37	267.94	16201	4DE	0;30	72.72

310

NEW MOONS				FULL MOONS			
16203	17JA	17;27	298.20	16202	2JA	11;20	102.73
16204	16FE	9;37	327.98	16203	31JA	22;10	132.50
16205	17MR	22;46	357.14	16204	2MR	9;23	161.85
16206	16AP	9;15	25.71	16205	31MR	21;28	190.73
16207	15MY	17;40	53.84	16206	30AP	10;46	219.22
16208	14JN	0;50	81.75	16207	30MY	1;12	247.48
16209	13JL	7;39	109.71	16208	28JN	16;19	275.71
16210	11AU	15;14	137.98	16209	28JL	7;34	304.12
16211	10SE	0;38	166.76	16210	26AU	22;37	332.90
16212	9OC	12;49	196.15	16211	25SE	13;11	2.16
16213	8NO	4;06	226.12	16212	25OC	2;58	31.87
16214	7DE	21;57	256.44	16213	23NO	15;39	61.89
				16214	23DE	3;04	91.96

311

NEW MOONS				FULL MOONS			
16215	6JA	17;04	286.79	16215	21JA	13;21	121.82
16216	5FE	11;51	316.84	16216	19FE	22;58	151.28
16217	7MR	4;58	346.36	16217	21MR	8;29	180.23
16218	5AP	19;36	15.27	16218	19AP	18;31	208.74
16219	5MY	7;28	43.65	16219	19MY	5;32	236.94
16220	3JN	16;53	71.68	16220	17JN	17;56	265.05
16221	3JL	0;40	99.63	16221	17JL	7;59	293.32
16222	1AU	7;57	127.76	16222	15AU	23;44	321.97
16223	30AU	15;51	156.31	16223	14SE	16;46	351.14
16224	29SE	1;16	185.41	16224	14OC	10;02	20.82
16225	28OC	12;45	215.08	16225	13NO	2;17	50.88
16226	27NO	2;25	245.16	16226	12DE	16;35	81.07
16227	26DE	18;08	275.41				

312

NEW MOONS				FULL MOONS			
16228	25JA	11;23	305.53	16227	11JA	4;46	111.10
16229	24FE	5;21	335.27	16228	9FE	15;10	140.74
16230	24MR	22;46	4.48	16229	10MR	0;19	169.87
16231	23AP	14;23	33.15	16230	8AP	8;42	198.47
16232	23MY	3;30	61.39	16231	7MY	16;56	226.66
16233	21JN	14;20	89.44	16232	6JN	1;46	254.66
16234	20JL	23;38	117.56	16233	5JL	12;09	282.72
16235	19AU	8;24	145.99	16234	4AU	1;00	311.12
16236	17SE	17;26	174.89	16235	2SE	16;42	340.04
16237	17OC	3;13	204.32	16236	2OC	10;46	9.57
16238	15NO	14;01	234.19	16237	1NO	5;43	39.60
16239	15DE	2;04	264.28	16238	30NO	23;47	69.90
				16239	30DE	15;48	100.15

313

NEW MOONS				FULL MOONS			
16240	13JA	15;39	294.35	16240	29JA	5;21	130.06
16241	12FE	6;47	324.15	16241	27FE	16;31	159.45
16242	13MR	23;02	353.50	16242	29MR	2;19	188.25
16243	12AP	15;27	22.37	16243	27AP	9;13	216.55
16244	12MY	7;00	50.81	16244	26MY	16;09	244.52
16245	10JN	21;11	79.01	16245	24JN	23;33	272.44
16246	10JL	9;59	107.19	16246	24JL	8;38	300.57
16247	8AU	21;41	135.59	16247	22AU	20;25	329.17
16248	7SE	8;40	164.39	16248	21SE	11;20	358.37
16249	6OC	19;10	193.68	16249	21OC	5;01	28.19
16250	5NO	5;30	223.41	16250	20NO	0;16	58.45
16251	4DE	15;58	253.40	16251	19DE	19;33	88.86

314

NEW MOONS				FULL MOONS			
16252	3JA	2;59	283.43	16252	18JA	13;20	119.07
16253	1FE	14;51	313.23	16253	17FE	4;31	148.79
16254	3MR	3;42	342.63	16254	18MR	16;39	177.90
16255	1AP	17;22	11.57	16255	17AP	1;58	206.41
16256	1MY	7;38	40.09	16256	16MY	9;21	234.49
16257	30MY	22;22	68.35	16257	14JN	15;57	262.38
16258	29JN	13;22	96.58	16258	13JL	22;58	290.35
16259	29JL	4;17	124.98	16259	12AU	7;28	318.66
16260	27AU	18;39	153.75	16260	10SE	18;14	347.51
16261	26SE	8;00	182.97	16261	10OC	7;45	16.98
16262	25OC	20;14	212.63	16262	9NO	0;03	46.99
16263	24NO	7;34	242.59	16263	8DE	18;41	77.35
16264	23DE	18;22	272.64				

315

NEW MOONS				FULL MOONS			
16265	22JA	4;55	302.51	16264	7JA	14;23	107.72
16266	20FE	15;20	331.98	16265	6FE	9;16	137.77
16267	22MR	1;46	0.97	16266	8MR	1;36	167.24
16268	20AP	12;38	29.50	16267	6AP	14;38	196.07
16269	20MY	0;35	57.73	16268	6MY	0;45	224.36
16270	18JN	14;07	85.89	16269	4JN	8;53	252.35
16271	18JL	5;21	114.21	16270	3JL	16;10	280.28
16272	16AU	21;41	142.89	16271	1AU	23;35	308.42
16273	15SE	14;08	172.05	16272	31AU	7;55	337.00
16274	15OC	5;47	201.68	16273	29SE	17;50	6.13
16275	13NO	20;11	231.67	16274	29OC	5;57	35.83
16276	13DE	9;13	261.80	16275	27NO	20;45	65.96
				16276	27DE	14;07	96.28

316

NEW MOONS				FULL MOONS			
16277	11JA	20;52	291.81	16277	26JA	8;57	126.46
16278	10FE	7;08	321.44	16278	25FE	3;29	156.21
16279	10MR	16;13	350.55	16279	25MR	20;07	185.38
16280	9AP	0;40	19.14	16280	24AP	10;10	213.97
16281	8MY	9;24	47.35	16281	23MY	21;49	242.15
16282	6JN	19;27	75.39	16282	22JN	7;38	270.16
16283	6JL	7;36	103.53	16283	21JL	16;19	298.26
16284	4AU	22;05	132.00	16284	20AU	0;34	326.67
16285	3SE	14;34	160.97	16285	18SE	9;07	355.57
16286	3OC	8;15	190.49	16286	17OC	18;44	25.00
16287	2NO	2;06	220.49	16287	16NO	6;05	54.90
16288	1DE	19;01	250.75	16288	15DE	19;32	85.05
16289	31DE	10;02	280.95				

317

NEW MOONS				FULL MOONS			
16290	29JA	22;35	310.81	16289	14JA	10;56	115.20
16291	28FE	8;41	340.15	16290	13FE	3;33	145.04
16292	29MR	16;59	8.90	16291	14MR	20;23	174.41
16293	28AP	0;23	37.18	16292	13AP	12;38	203.25
16294	27MY	7;55	65.18	16293	13MY	3;43	231.66
16295	25JN	16;33	93.15	16294	11JN	17;15	259.83
16296	25JL	3;06	121.35	16295	11JL	5;08	287.98
16297	23AU	16;07	150.00	16296	9AU	15;34	316.34
16298	22SE	7;52	179.25	16297	8SE	1;09	345.10
16299	22OC	2;04	209.10	16298	7OC	10;40	14.36
16300	20NO	21;35	239.38	16299	5NO	20;46	44.08
16301	20DE	16;33	269.78	16300	5DE	7;51	74.10

318

NEW MOONS				FULL MOONS			
16302	19JA	9;08	299.93	16301	3JA	19;55	104.17
16303	17FE	22;34	329.57	16302	2FE	8;52	134.01
16304	19MR	9;04	358.60	16303	3MR	22;39	163.44
16305	17AP	17;44	27.07	16304	2AP	13;16	192.40
16306	17MY	0;44	55.13	16305	2MY	4;34	220.95
16307	15JN	7;44	83.03	16306	31MY	20;04	249.24
16308	14JL	15;20	111.03	16307	30JN	10;59	277.47
16309	13AU	0;28	139.38	16308	30JL	0;47	305.84
16310	11SE	12;04	168.28	16309	28AU	13;21	334.55
16311	11OC	2;51	197.81	16310	27SE	1;05	3.71
16312	9NO	20;48	227.90	16311	26OC	12;23	33.34
16313	9DE	16;40	258.32	16312	24NO	23;29	63.30
				16313	24DE	10;22	93.35

319

NEW MOONS				FULL MOONS			
16314	8JA	12;19	288.68	16314	22JA	21;02	123.21
16315	7FE	5;51	318.65	16315	21FE	7;46	152.69
16316	8MR	20;24	348.06	16316	22MR	19;03	181.70
16317	7AP	8;05	16.80	16317	21AP	7;27	210.28
16318	6MY	17;24	45.06	16318	20MY	21;08	238.57
16319	5JN	1;05	73.02	16319	19JN	11;54	266.78
16320	4JL	8;00	100.94	16320	19JL	3;13	295.12
16321	2AU	15;09	129.08	16321	17AU	18;38	323.77
16322	31AU	23;38	157.67	16322	16SE	9;46	352.89
16323	30SE	10;26	186.85	16323	16OC	0;18	22.49
16324	30OC	0;14	216.63	16324	14NO	13;48	52.44
16325	28NO	16;55	246.85	16325	14DE	1;57	82.54
16326	28DE	11;30	277.22				

320

NUMBER	DATE	TIME	LONG.	NUMBER	DATE	TIME	LONG.
				16326	12JA	12;43	112.51
16327	27JA	6;28	307.40	16327	10FE	22;27	142.10
16328	26FE	0;24	337.11	16328	11MR	7;43	171.21
16329	26MR	16;14	6.22	16329	9AP	17;10	199.84
16330	25AP	5;27	34.77	16330	9MY	3;24	228.09
16331	24MY	16;02	62.91	16331	7JN	14;54	256.19
16332	23JN	0;35	90.87	16332	7JL	4;03	284.37
16333	22JL	8;07	118.92	16333	5AU	19;02	312.87
16334	20AU	15;45	147.32	16334	4SE	11;43	341.86
16335	19SE	0;29	176.23	16335	4OC	5;18	11.40
16336	18OC	11;00	205.71	16336	2NO	22;28	41.38
16337	16NO	23;37	235.67	16337	2DE	14;00	71.58
16338	16DE	14;16	265.88				

321

NUMBER	DATE	TIME	LONG.	NUMBER	DATE	TIME	LONG.
				16338	1JA	3;16	101.71
16339	15JA	6;38	296.05	16339	30JA	14;26	131.51
16340	14FE	0;08	325.92	16340	1MR	0;00	160.81
16341	15MR	17;46	355.31	16341	30MR	8;31	189.56
16342	14AP	10;15	24.15	16342	28AP	16;33	217.86
16343	14MY	0;38	52.53	16343	28MY	0;50	245.88
16344	12JN	12;37	80.63	16344	26JN	10;13	273.88
16345	11JL	22;44	108.72	16345	25JL	21;43	302.12
16346	10AU	7;52	137.03	16346	24AU	12;04	330.84
16347	8SE	16;55	165.78	16347	23SE	5;14	0.17
16348	8OC	2;28	195.05	16348	23OC	0;09	30.06
16349	6NO	12;52	224.79	16349	21NO	19;03	60.32
16350	6DE	0;16	254.83	16350	21DE	12;17	90.65

322

NUMBER	DATE	TIME	LONG.	NUMBER	DATE	TIME	LONG.
16351	4JA	12;55	284.92	16351	20JA	3;05	120.71
16352	3FE	3;03	314.80	16352	18FE	15;20	150.29
16353	4MR	18;32	344.29	16353	20MR	1;18	179.29
16354	3AP	10;43	13.30	16354	18AP	9;25	207.74
16355	3MY	2;38	41.86	16355	17MY	16;28	235.79
16356	1JN	17;31	70.12	16356	15JN	23;25	263.69
16357	1JL	7;05	98.30	16357	15JL	7;30	291.71
16358	30JL	19;32	126.62	16358	13AU	17;52	320.12
16359	29AU	7;07	155.30	16359	12SE	7;17	349.10
16360	27SE	18;06	184.45	16360	11OC	23;47	18.71
16361	27OC	4;41	214.06	16361	10NO	18;32	48.85
16362	25NO	15;08	243.99	16362	10DE	14;04	79.25
16363	25DE	1;48	274.03				

323

NUMBER	DATE	TIME	LONG.	NUMBER	DATE	TIME	LONG.
				16363	9JA	8;46	109.57
16364	23JA	13;03	303.92	16364	8FE	1;18	139.50
16365	22FE	1;09	333.44	16365	9MR	14;53	168.83
16366	23MR	14;07	2.51	16366	8AP	1;27	197.53
16367	22AP	3;50	31.13	16367	7MY	9;36	225.74
16368	21MY	18;11	59.44	16368	5JN	16;25	253.66
16369	20JN	9;02	87.65	16369	4JL	23;06	281.57
16370	20JL	0;11	115.99	16370	3AU	6;49	309.74
16371	18AU	15;09	144.64	16371	1SE	16;28	338.39
16372	17SE	5;22	173.73	16372	1OC	4;40	7.64
16373	16OC	18;26	203.28	16373	30OC	19;39	37.48
16374	15NO	6;21	233.18	16374	29NO	13;13	67.74
16375	14DE	17;27	263.23	16375	29DE	8;32	98.14

324

NUMBER	DATE	TIME	LONG.	NUMBER	DATE	TIME	LONG.
16376	13JA	4;03	293.18	16376	28JA	3;57	128.32
16377	11FE	14;21	322.79	16377	26FE	21;36	158.01
16378	12MR	0;29	351.93	16378	27MR	12;13	187.06
16379	10AP	10;48	20.57	16379	25AP	23;39	215.53
16380	9MY	21;53	48.86	16380	25MY	8;39	243.60
16381	8JN	10;25	77.00	16381	23JN	16;17	271.52
16382	8JL	0;48	105.23	16382	22JL	23;38	299.58
16383	6AU	16;49	133.78	16383	21AU	7;34	328.00
16384	5SE	9;35	162.79	16384	19SE	16;45	356.94
16385	5OC	2;01	192.29	16385	19OC	3;48	26.45
16386	3NO	17;20	222.21	16386	17NO	17;14	56.45
16387	3DE	7;13	252.34	16387	17DE	9;19	86.72

325

NUMBER	DATE	TIME	LONG.	NUMBER	DATE	TIME	LONG.
16388	1JA	19;36	282.43	16388	16JA	3;26	116.96
16389	31JA	6;28	312.21	16389	14FE	22;08	146.87
16390	1MR	15;55	341.49	16390	16MR	15;40	176.23
16391	31MR	0;23	10.24	16391	15AP	6;56	205.01
16392	29AP	8;39	38.53	16392	14MY	19;43	233.32
16393	28MY	17;45	66.58	16393	13JN	6;27	261.37
16394	27JN	4;39	94.64	16394	12JL	15;46	289.43
16395	26JL	17;58	122.97	16395	11AU	0;20	317.73
16396	25AU	9;39	151.75	16396	9SE	8;51	346.47
16397	24SE	3;03	181.10	16397	8OC	18;01	15.73
16398	23OC	21;10	210.98	16398	7NO	4;31	45.48
16399	22NO	14;53	241.20	16399	6DE	16;53	75.57
16400	22DE	7;05	271.47				

326

NUMBER	DATE	TIME	LONG.	NUMBER	DATE	TIME	LONG.
				16400	5JA	7;13	105.72
16401	20JA	20;53	301.49	16401	3FE	23;06	135.67
16402	19FE	8;04	331.02	16402	5MR	15;39	165.19
16403	20MR	17;02	359.96	16403	4AP	8;02	194.19
16404	19AP	0;35	28.37	16404	3MY	23;38	222.73
16405	18MY	7;48	56.43	16405	2JN	13;59	250.97
16406	16JN	15;41	84.37	16406	2JL	2;50	279.11
16407	16JL	1;09	112.45	16407	31JL	14;10	307.41
16408	14AU	12;55	140.92	16408	30AU	0;21	336.04
16409	13SE	3;23	169.96	16409	28SE	10;03	5.14
16410	12OC	20;34	199.61	16410	27OC	19;58	34.73
16411	11NO	15;46	229.78	16411	26NO	6;37	64.68
16412	11DE	11;22	260.19	16412	25DE	18;09	94.75

327

NUMBER	DATE	TIME	LONG.	NUMBER	DATE	TIME	LONG.
16413	10JA	5;20	290.47	16413	24JA	6;29	124.68
16414	8FE	20;20	320.32	16414	22FE	19;34	154.23
16415	10MR	8;07	349.56	16415	24MR	9;27	183.32
16416	8AP	17;21	18.20	16416	23AP	0;11	211.97
16417	8MY	0;59	46.38	16417	22MY	15;32	240.32
16418	6JN	7;57	74.31	16418	21JN	6;51	268.55
16419	5JL	15;09	102.24	16419	20JL	21;26	296.87
16420	3AU	23;27	130.44	16420	19AU	10;52	325.47
16421	2SE	9;47	159.13	16421	17SE	23;16	354.50
16422	1OC	23;00	188.44	16422	17OC	11;00	24.00
16423	31OC	15;30	218.35	16423	15NO	22;22	53.88
16424	30NO	10;40	248.68	16424	15DE	9;24	83.94
16425	30DE	6;41	279.11				

328

NUMBER	DATE	TIME	LONG.	NUMBER	DATE	TIME	LONG.
				16425	13JA	20;05	113.89
16426	29JA	1;22	309.25	16426	12FE	6;32	143.50
16427	27FE	17;21	338.85	16427	12MR	17;12	172.64
16428	28MR	6;21	7.83	16428	11AP	4;41	201.33
16429	26AP	16;42	36.24	16429	10MY	17;29	229.67
16430	26MY	1;06	64.28	16430	9JN	7;38	257.87
16431	24JN	8;20	92.19	16431	8JL	22;45	286.14
16432	23JL	15;20	120.24	16432	7AU	14;21	314.68
16433	21AU	23;07	148.66	16433	6SE	5;57	343.66
16434	20SE	8;46	177.63	16434	5OC	21;08	13.12
16435	19OC	21;06	207.21	16435	4NO	11;27	43.00
16436	18NO	12;26	237.29	16436	4DE	0;27	73.10
16437	18DE	6;09	267.63				

329

NUMBER	DATE	TIME	LONG.	NUMBER	DATE	TIME	LONG.
				16437	2JA	11;54	103.15
16438	17JA	0;59	297.90	16438	31JA	21;59	132.89
16439	15FE	19;25	327.78	16439	2MR	7;12	162.15
16440	17MR	12;15	357.10	16440	31MR	16;15	190.91
16441	16AP	2;43	25.83	16441	30AP	1;47	219.25
16442	15MY	14;35	54.10	16442	29MY	12;24	247.36
16443	14JN	0;08	82.10	16443	28JN	0;36	275.47
16444	13JL	8;10	110.11	16444	27JL	14;40	303.83
16445	11AU	15;46	138.39	16445	26AU	6;42	332.64
16446	10SE	0;01	167.12	16446	25SE	0;13	2.01
16447	9OC	9;44	196.42	16447	24OC	18;02	31.89
16448	7NO	21;22	226.23	16448	23NO	10;43	62.07
16449	7DE	10;59	256.37	16449	23DE	1;14	92.27

330

NUMBER	DATE	TIME	LONG.	NUMBER	DATE	TIME	LONG.
16450	6JA	2;24	286.56	16450	21JA	13;24	122.21
16451	4FE	19;11	316.54	16451	19FE	23;34	151.69
16452	6MR	12;38	346.08	16452	21MR	8;22	180.62
16453	5AP	5;38	15.09	16453	19AP	16;24	209.04
16454	4MY	21;05	43.62	16454	19MY	0;17	237.12
16455	3JN	10;17	71.80	16455	17JN	8;53	265.08
16456	2JL	21;24	99.89	16456	16JL	19;11	293.20
16457	1AU	7;07	128.12	16457	15AU	8;05	321.73
16458	30AU	16;21	156.72	16458	14SE	0;00	350.84
16459	29SE	1;49	185.83	16459	13OC	18;26	20.56
16460	28OC	11;55	215.43	16460	12NO	13;45	50.74
16461	26NO	22;50	245.39	16461	12DE	8;06	81.10
16462	26DE	10;45	275.48				

331

NUMBER	DATE	TIME	LONG.	NUMBER	DATE	TIME	LONG.
				16462	11JA	0;12	111.30
16463	24JA	23;56	305.44	16463	9FE	13;40	141.07
16464	23FE	14;30	335.04	16464	11MR	0;36	170.27
16465	25MR	6;08	4.19	16465	9AP	9;27	198.88
16466	23AP	22;04	32.88	16466	8MY	16;49	227.04
16467	23MY	13;26	61.22	16467	6JN	23;36	254.96
16468	22JN	3;44	89.41	16468	6JL	6;57	282.90
16469	21JL	16;54	117.68	16469	4AU	16;04	311.14
16470	20AU	5;10	146.24	16470	3SE	3;58	339.91
16471	18SE	16;42	175.25	16471	2OC	19;05	9.30
16472	18OC	3;42	204.73	16472	1NO	12;57	39.28
16473	16NO	14;20	234.59	16473	1DE	8;21	69.63
16474	16DE	0;52	264.62	16474	31DE	3;40	100.03

NUMBER	DATE	TIME	LONG.	NUMBER	DATE	TIME	LONG.	NUMBER	DATE	TIME	LONG.	NUMBER	DATE	TIME	LONG.
332								**338**							
16475	14JA	11;40	294.57	16475	29JA	21;23	130.12	16549	7JA	19;32	288.36	16549	22JA	21;25	123.62
16476	12FE	23;06	324.22	16476	28FE	12;25	159.67	16550	6FE	14;09	318.39	16550	21FE	6;51	153.04
16477	13MR	11;21	353.41	16477	29MR	0;21	188.59	16551	8MR	7;44	347.91	16551	22MR	15;40	181.95
16478	12AP	0;24	22.15	16478	27AP	9;32	216.95	16552	6AP	23;20	16.84	16552	21AP	0;38	210.39
16479	11MY	14;13	50.53	16479	26MY	16;51	244.94	16553	6MY	12;28	45.25	16553	20MY	10;27	238.54
16480	10JN	4;44	78.74	16480	24JN	23;28	272.82	16554	4JN	23;10	73.33	16554	18JN	21;40	266.60
16481	9JL	19;50	107.01	16481	24JL	6;36	300.88	16555	4JL	7;58	101.32	16555	18JL	10;45	294.84
16482	8AU	11;11	135.55	16482	22AU	15;17	329.35	16556	2AU	15;49	129.50	16556	17AU	1;55	323.48
16483	7SE	2;08	164.52	16483	21SE	2;15	358.38	16557	31AU	23;48	158.07	16557	15SE	18;59	352.67
16484	6OC	16;07	193.94	16484	20OC	15;55	28.03	16558	30SE	8;52	187.18	16558	15OC	13;05	22.40
16485	5NO	4;50	223.77	16485	19NO	8;16	58.16	16559	29OC	19;37	216.83	16559	14NO	6;45	52.54
16486	4DE	16;25	253.81	16486	19DE	2;48	88.54	16560	28NO	8;16	246.87	16560	13DE	22;34	82.79
								16561	27DE	22;42	277.07				
333								**339**							
16487	3JA	3;13	283.83	16487	17JA	22;17	118.82					16561	12JA	11;54	112.86
16488	1FE	13;31	313.56	16488	16FE	16;54	148.70	16562	26JA	14;38	307.12	16562	10FE	22;55	142.51
16489	2MR	23;29	342.85	16489	18MR	9;01	177.98	16563	25FE	7;36	336.80	16563	12MR	8;11	171.62
16490	1AP	9;23	11.63	16490	16AP	21;56	206.64	16564	27MR	0;44	5.98	16564	10AP	16;20	200.19
16491	30AP	19;45	40.00	16491	16MY	8;01	234.82	16565	25AP	16;57	34.66	16565	10MY	0;03	228.35
16492	30MY	7;17	68.14	16492	14JN	16;15	262.78	16566	25MY	7;18	62.95	16566	8JN	8;04	256.31
16493	28JN	20;38	96.30	16493	13JL	23;44	290.77	16567	23JN	19;31	91.05	16567	7JL	17;19	284.33
16494	28JL	11;58	124.72	16494	12AU	7;25	319.05	16568	23JL	6;02	119.23	16568	6AU	4;50	312.69
16495	27AU	4;41	153.57	16495	10SE	16;03	347.81	16569	21AU	15;38	147.71	16569	4SE	19;19	341.58
16496	25SE	21;42	182.93	16496	10OC	2;13	17.13	16570	20SE	1;09	176.65	16570	4OC	12;47	11.10
16497	25OC	13;55	212.75	16497	8NO	14;28	46.98	16571	19OC	11;06	206.11	16571	3NO	8;04	41.16
16498	24NO	4;45	242.86	16498	8DE	5;11	77.17	16572	17NO	21;41	235.98	16572	3DE	3;15	71.51
16499	23DE	17;58	273.01					16573	17DE	9;03	266.05				
334								**340**							
				16499	6JA	22;15	107.43					16573	1JA	20;38	101.82
16500	22JA	5;33	302.92	16500	5FE	16;40	137.46	16574	15JA	21;24	296.05	16574	31JA	11;24	131.77
16501	20FE	15;33	332.38	16501	7MR	10;45	167.02	16575	14FE	11;00	325.77	16575	29FE	23;29	161.17
16502	22MR	0;14	1.29	16502	6AP	3;05	195.99	16576	15MR	1;52	355.05	16576	30MR	9;12	189.98
16503	20AP	8;17	29.71	16503	5MY	17;02	224.44	16577	13AP	17;29	23.86	16577	28AP	17;06	218.27
16504	19MY	16;38	57.80	16504	4JN	4;47	252.57	16578	13MY	9;05	52.30	16578	27MY	23;58	246.23
16505	18JN	2;24	85.81	16505	3JL	14;52	280.61	16579	11JN	23;57	80.52	16579	26JN	6;51	274.13
16506	17JL	14;26	114.00	16506	1AU	23;55	308.82	16580	11JL	13;49	108.75	16580	25JL	14;57	302.24
16507	16AU	5;02	142.61	16507	31AU	8;35	337.42	16581	10AU	2;45	137.22	16581	24AU	1;25	330.80
16508	14SE	21;50	171.77	16508	29SE	17;32	6.51	16582	8SE	14;56	166.09	16582	22SE	15;01	359.97
16509	14OC	15;57	201.49	16509	29OC	3;25	36.11	16583	8OC	2;28	195.44	16583	22OC	7;42	29.76
16510	13NO	10;14	231.64	16510	27NO	14;49	66.10	16584	6NO	13;26	225.21	16584	21NO	2;36	60.02
16511	13DE	3;27	261.95	16511	27DE	4;06	96.25	16585	6DE	0;03	255.21	16585	20DE	22;10	90.44
335								**341**							
16512	11JA	18;34	292.10	16512	25JA	19;04	126.27	16586	4JA	10;36	285.21	16586	19JA	16;48	120.67
16513	10FE	6;59	321.82	16513	24FE	11;07	155.92	16587	2FE	21;30	314.96	16587	18FE	9;12	150.43
16514	11MR	16;51	350.96	16514	26MR	3;24	185.08	16588	4MR	9;03	344.29	16588	19MR	22;36	179.57
16515	10AP	0;50	19.53	16515	24AP	19;16	213.76	16589	2AP	21;24	13.15	16589	18AP	9;01	208.11
16516	9MY	7;57	47.67	16516	24MY	10;14	242.08	16590	2MY	10;35	41.61	16590	17MY	17;06	236.20
16517	7JN	15;16	75.61	16517	22JN	23;58	270.24	16591	1JN	0;36	69.84	16591	15JN	23;55	264.10
16518	6JL	23;48	103.60	16518	22JL	12;15	298.48	16592	30JN	15;26	98.06	16592	15JL	6;42	292.07
16519	5AU	10;22	131.91	16519	20AU	23;12	327.01	16593	30JL	6;52	126.50	16593	13AU	14;36	320.37
16520	3SE	23;31	160.74	16520	19SE	9;19	355.97	16594	28AU	22;23	155.33	16594	12SE	0;28	349.21
16521	3OC	15;30	190.18	16521	18OC	19;15	25.41	16595	27SE	13;14	184.64	16595	11OC	12;50	18.64
16522	2NO	9;58	220.20	16522	17NO	5;37	55.27	16596	27OC	2;51	214.37	16596	10NO	3;54	48.62
16523	2DE	5;42	250.57	16523	16DE	16;42	85.32	16597	25NO	15;08	244.38	16597	9DE	21;25	78.94
								16598	25DE	2;19	274.44				
336								**342**							
16524	1JA	0;46	280.95	16524	15JA	4;31	115.31					16598	8JA	16;31	109.28
16525	30JA	17;19	310.98	16525	13FE	16;58	144.99	16599	23JA	12;46	304.29	16599	7FE	11;39	139.31
16526	29FE	6;36	340.45	16526	14MR	6;08	174.21	16600	21FE	22;41	333.73	16600	9MR	5;02	168.81
16527	29MR	16;54	9.29	16527	12AP	20;09	202.97	16601	23MR	8;19	2.66	16601	7AP	19;29	197.68
16528	28AP	1;07	37.60	16528	12MY	11;03	231.39	16602	21AP	18;07	31.12	16602	7MY	6;52	226.01
16529	27MY	8;15	65.58	16529	11JN	2;25	259.63	16603	21MY	4;46	59.29	16603	5JN	15;56	254.02
16530	25JN	15;34	93.48	16530	10JL	17;34	287.91	16604	19JN	17;01	87.40	16604	4JL	23;45	281.98
16531	24JL	22;54	121.56	16531	9AU	7;50	316.42	16605	19JL	7;23	115.70	16605	3AU	7;23	310.15
16532	23AU	8;11	150.07	16532	7SE	21;01	345.32	16606	17AU	23;39	144.40	16606	1SE	15;38	338.75
16533	21SE	19;57	179.15	16533	7OC	9;20	14.69	16607	16SE	16;55	173.60	16607	1OC	1;06	7.88
16534	21OC	10;53	208.85	16534	5NO	21;04	44.48	16608	16OC	9;56	203.31	16608	30OC	12;20	37.56
16535	20NO	4;55	239.07	16535	5DE	8;22	74.52	16609	15NO	1;44	233.37	16609	29NO	1;46	67.65
16536	20DE	0;45	269.50					16610	14DE	15;54	263.56	16610	28DE	17;36	97.90
337								**343**							
				16536	3JA	19;12	104.53	16611	13JA	4;18	293.58	16611	27JA	11;19	128.02
16537	18JA	20;16	299.78	16537	2FE	5;34	134.26	16612	11FE	14;58	323.21	16612	26FE	5;32	157.74
16538	17FE	13;36	329.59	16538	3MR	15;48	163.55	16613	13MR	0;05	352.30	16613	27MR	22;41	186.90
16539	19MR	3;59	358.78	16539	2AP	2;31	192.35	16614	11AP	8;09	20.85	16614	26AP	13;45	215.52
16540	17AP	15;31	27.37	16540	1MY	14;22	220.77	16615	10MY	16;02	49.01	16615	26MY	2;34	243.74
16541	17MY	0;47	55.52	16541	31MY	3;40	248.98	16616	9JN	0;49	77.00	16616	24JN	13;32	271.80
16542	15JN	8;32	83.45	16542	29JN	18;19	277.19	16617	8JL	11;34	105.09	16617	23JL	23;12	299.95
16543	14JL	15;37	111.43	16543	29JL	9;51	305.63	16618	7AU	0;55	133.53	16618	22AU	8;12	328.41
16544	12AU	23;00	139.71	16544	28AU	1;44	334.46	16619	5SE	16;48	162.49	16619	20SE	17;08	357.34
16545	11SE	7;43	168.49	16545	26SE	17;29	3.78	16620	5OC	10;36	192.04	16620	20OC	2;38	26.79
16546	10OC	18;44	197.86	16546	26OC	8;35	33.57	16621	4NO	5;09	222.08	16621	18NO	13;17	56.67
16547	9NO	8;37	227.78	16547	24NO	22;29	63.64	16622	3DE	23;13	252.39	16622	18DE	1;34	86.78
16548	9DE	1;13	258.05	16548	24DE	10;46	93.75								

--------NEW MOONS-------- --------FULL MOONS-------- --------NEW MOONS-------- --------FULL MOONS--------
NUMBER DATE TIME LONG. NUMBER DATE TIME LONG. NUMBER DATE TIME LONG. NUMBER DATE TIME LONG.

344 / 350

NEW MOONS				FULL MOONS				NEW MOONS				FULL MOONS			
NUMBER	DATE	TIME	LONG.	NUMBER	DATE	TIME	LONG.	NUMBER	DATE	TIME	LONG.	NUMBER	DATE	TIME	LONG.
16623	2JA	15;33	282.65	16623	16JA	15;34	116.85	16698	24JA	20;17	305.67	16697	10JA	11;43	111.16
16624	1FE	5;19	312.55	16624	15FE	6;54	146.62	16699	23FE	7;14	335.13	16698	9FE	5;16	141.11
16625	1MR	16;18	341.89	16625	15MR	22;51	175.93	16700	24MR	18;53	4.12	16699	10MR	20;08	170.47
16626	31MR	0;58	10.65	16626	14AP	14;46	204.75	16701	23AP	7;22	32.68	16700	9AP	7;55	199.21
16627	29AP	8;15	38.90	16627	14MY	6;07	233.16	16702	22MY	20;45	60.94	16701	8MY	17;01	227.44
16628	28MY	15;14	66.87	16628	12JN	20;33	261.37	16703	21JN	11;07	89.14	16702	7JN	0;19	255.37
16629	26JN	23;00	94.81	16629	12JL	9;44	289.57	16704	21JL	2;23	117.48	16703	6JL	7;02	283.29
16630	26JL	8;28	122.97	16630	10AU	21;35	318.01	16705	19AU	18;11	146.19	16704	4AU	14;20	311.46
16631	24AU	20;19	151.60	16631	9SE	8;19	346.83	16706	18SE	9;45	175.35	16705	2SE	23;14	340.10
16632	23SE	10;58	180.83	16632	8OC	18;30	16.14	16707	18OC	0;20	204.98	16706	2OC	10;25	9.33
16633	23OC	4;24	210.66	16633	7NO	4;45	45.88	16708	16NO	13;29	234.94	16707	1NO	0;12	39.13
16634	21NO	23;49	240.95	16634	6DE	15;30	75.90	16709	16DE	1;16	265.03	16708	30NO	16;32	69.35
16635	21DE	19;32	271.37									16709	30DE	10;53	99.70

345 / 351

NEW MOONS				FULL MOONS				NEW MOONS				FULL MOONS			
16636	20JA	13;30	301.57	16635	5JA	2;53	105.93	16710	14JA	12;00	294.98	16710	29JA	6;04	129.87
16637	19FE	4;22	331.25	16636	3FE	14;49	135.71	16711	12FE	22;00	324.56	16711	28FE	0;22	159.56
16638	20MR	15;58	0.30	16637	5MR	3;19	165.07	16712	14MR	7;31	353.64	16712	29MR	16;15	188.65
16639	19AP	1;01	28.78	16638	3AP	16;35	193.95	16713	12AP	16;55	22.23	16713	28AP	5;05	217.15
16640	18MY	8;31	56.84	16639	3MY	6;48	222.45	16714	12MY	2;49	50.46	16714	27MY	15;13	245.26
16641	16JN	15;26	84.74	16640	1JN	21;53	250.71	16715	10JN	14;01	78.54	16715	25JN	23;37	273.22
16642	15JL	22;42	112.74	16641	1JL	13;18	278.96	16716	10JL	3;14	106.74	16716	25JL	7;22	301.30
16643	14AU	7;09	141.07	16642	31JL	4;16	307.39	16717	8AU	18;42	135.28	16717	23AU	15;23	329.74
16644	12SE	17;39	169.94	16643	29AU	18;17	336.18	16718	7SE	11;50	164.32	16718	22SE	0;20	358.70
16645	12OC	7;02	199.43	16644	28SE	7;17	5.42	16719	7OC	5;24	193.89	16719	21OC	10;45	28.20
16646	10NO	23;38	229.49	16645	27OC	19;32	35.10	16720	5NO	22;09	223.88	16720	19NO	23;03	58.16
16647	10DE	18;48	259.87	16646	26NO	7;12	65.09	16721	5DE	13;20	254.07	16721	19DE	13;37	88.37
				16647	25DE	18;17	95.15								

346 / 352

NEW MOONS				FULL MOONS				NEW MOONS				FULL MOONS			
16648	9JA	14;40	290.24	16648	24JA	4;45	124.99	16722	4JA	2;40	284.19	16722	18JA	6;19	118.54
16649	8FE	9;09	320.24	16649	22FE	14;47	154.42	16723	2FE	14;07	313.98	16723	17FE	0;14	148.40
16650	10MR	0;56	349.65	16650	24MR	0;54	183.36	16724	2MR	23;50	343.25	16724	17MR	17;51	177.75
16651	8AP	13;46	18.45	16651	22AP	11;50	211.86	16725	1AP	8;09	11.97	16725	16AP	9;54	206.54
16652	8MY	0;03	46.73	16652	22MY	0;10	240.10	16726	30AP	15;49	40.23	16726	15MY	23;48	234.88
16653	6JN	8;29	74.71	16653	20JN	14;05	268.27	16727	29MY	23;51	68.23	16727	14JN	11;42	262.98
16654	5JL	15;52	102.66	16654	20JL	5;17	296.61	16728	28JN	9;25	96.24	16728	13JL	22;07	291.09
16655	3AU	23;06	130.82	16655	18AU	21;15	325.30	16729	27JL	21;24	124.52	16729	12AU	7;36	319.45
16656	2SE	7;10	159.41	16656	17SE	13;24	354.48	16730	26AU	12;09	153.28	16730	10SE	16;44	348.23
16657	1OC	17;03	188.58	16657	17OC	5;12	24.15	16731	25SE	5;16	182.64	16731	10OC	2;04	17.52
16658	31OC	5;32	218.32	16658	15NO	20;00	54.18	16732	24OC	23;47	212.55	16732	8NO	12;10	47.27
16659	29NO	20;49	248.49	16659	15DE	9;14	84.32	16733	23NO	18;26	242.81	16733	7DE	23;36	77.32
16660	29DE	14;20	278.80					16734	23DE	11;52	273.14				

347 / 353

NEW MOONS				FULL MOONS				NEW MOONS				FULL MOONS			
16661	28JA	8;47	308.95	16660	13JA	20;39	114.30	16735	22JA	2;59	303.20	16734	6JA	12;37	107.41
16662	27FE	2;49	338.64	16661	12FE	6;30	143.88	16736	20FE	15;15	332.75	16735	5FE	3;06	137.29
16663	28MR	19;21	7.77	16662	13MR	15;20	172.95	16737	22MR	0;52	1.70	16736	6MR	18;32	166.74
16664	27AP	9;41	36.35	16663	11AP	23;55	201.52	16738	20AP	8;35	30.11	16737	5AP	10;16	195.70
16665	26MY	21;36	64.53	16664	11MY	9;00	229.72	16739	19MY	15;29	58.14	16738	5MY	1;47	224.22
16666	25JN	7;22	92.54	16665	9JN	19;18	257.77	16740	17JN	22;39	86.04	16739	3JN	16;42	252.47
16667	24JL	15;44	120.64	16666	9JL	7;21	285.91	16741	17JL	7;09	114.10	16740	3JL	6;40	280.67
16668	22AU	23;43	149.07	16667	7AU	21;30	314.39	16742	15AU	17;47	142.54	16741	1AU	19;25	309.03
16669	21SE	8;20	178.00	16668	6SE	13;50	343.38	16743	14SE	7;06	171.55	16742	31AU	6;56	337.74
16670	20OC	18;19	207.48	16669	6OC	7;48	12.95	16744	13OC	23;17	201.18	16743	29SE	17;35	6.90
16671	19NO	6;04	237.41	16670	5NO	2;08	43.00	16745	12NO	17;57	231.34	16744	29OC	3;56	36.53
16672	18DE	19;33	267.57	16671	4DE	19;11	73.27	16746	12DE	13;50	261.76	16745	27NO	14;30	66.48
												16746	27DE	1;32	96.53

348 / 354

NEW MOONS				FULL MOONS				NEW MOONS				FULL MOONS			
16673	17JA	10;35	297.68	16672	3JA	9;50	103.45	16747	11JA	8;54	292.08	16747	25JA	13;01	126.41
16674	16FE	2;51	327.48	16673	1FE	21;55	133.27	16748	10FE	1;21	321.97	16748	24FE	0;58	155.89
16675	16MR	19;45	356.82	16674	2MR	7;51	162.56	16749	11MR	14;28	351.25	16749	25MR	13;30	184.91
16676	15AP	12;22	25.65	16675	31MR	16;19	191.30	16750	10AP	0;35	19.91	16750	24AP	2;57	213.49
16677	15MY	3;41	54.05	16676	30AP	0;00	219.56	16751	9MY	8;40	48.09	16751	23MY	17;28	241.80
16678	13JN	17;02	82.21	16677	29MY	7;38	247.55	16752	7JN	15;44	76.02	16752	22JN	8;47	270.03
16679	13JL	4;30	110.35	16678	27JN	16;05	275.52	16753	6JL	22;45	103.95	16753	22JL	0;12	298.39
16680	11AU	14;41	138.73	16679	27JL	2;20	303.72	16754	5AU	6;34	132.15	16754	20AU	15;00	327.06
16681	10SE	0;25	167.53	16680	25AU	15;19	332.40	16755	3SE	16;02	160.82	16755	19SE	4;48	356.17
16682	9OC	10;19	196.83	16681	24SE	7;29	1.71	16756	3OC	3;59	190.09	16756	18OC	17;41	25.74
16683	7NO	20;42	226.59	16682	24OC	2;14	31.61	16757	1NO	19;02	219.96	16757	17NO	5;49	55.67
16684	7DE	7;41	256.62	16683	22NO	21;51	61.91	16758	1DE	13;05	250.25	16758	16DE	17;16	85.74
				16684	22DE	16;22	92.29	16759	31DE	8;48	280.66				

349 / 355

NEW MOONS				FULL MOONS				NEW MOONS				FULL MOONS			
16685	5JA	19;24	286.66	16685	21JA	8;30	122.40	16760	30JA	4;05	310.82	16759	15JA	3;58	115.68
16686	4FE	8;07	316.46	16686	19FE	21;49	152.00	16761	28FE	21;11	340.45	16760	13FE	14;00	145.25
16687	5MR	22;04	345.87	16687	21MR	8;33	181.01	16762	30MR	11;23	9.45	16761	14MR	23;44	174.34
16688	4AP	13;06	14.81	16688	19AP	17;10	209.46	16763	28AP	22;50	37.89	16762	13AP	9;54	202.95
16689	4MY	4;35	43.35	16689	19MY	0;22	237.50	16764	28MY	8;07	65.96	16763	12MY	21;14	231.22
16690	2JN	19;50	71.62	16690	17JN	7;04	265.39	16765	26JN	15;59	93.90	16764	11JN	10;13	259.37
16691	2JL	10;18	99.83	16691	16JL	14;26	293.40	16766	25JL	23;17	121.97	16765	11JL	0;49	287.63
16692	31JL	23;54	128.21	16692	14AU	23;38	321.77	16767	24AU	6;57	150.40	16766	9AU	16;36	316.19
16693	30AU	12;44	156.96	16693	13SE	11;41	350.72	16768	22SE	15;57	179.37	16767	8SE	8;58	345.22
16694	29SE	0;52	186.18	16694	13OC	2;58	20.30	16769	22OC	3;10	208.92	16768	8OC	1;19	14.75
16695	28OC	12;20	215.84	16695	11NO	21;00	50.42	16770	20NO	17;05	238.96	16769	6NO	16;57	44.70
16696	26NO	23;13	245.80	16696	11DE	16;27	80.82	16771	20DE	9;31	269.24	16770	6DE	7;11	74.86
16697	26DE	9;44	275.83												

356 / 362

NUMBER	DATE	TIME	LONG.	NUMBER	DATE	TIME	LONG.	NUMBER	DATE	TIME	LONG.	NUMBER	DATE	TIME	LONG.
				16771	4JA	19;32	104.94	16846	12JA	23;56	293.79	16846	26JA	23;49	127.92
16772	19JA	3;29	299.46	16772	3FE	6;02	134.67	16847	11FE	13;36	323.54	16847	25FE	14;34	157.50
16773	17FE	21;41	329.32	16773	3MR	15;07	163.91	16848	13MR	0;23	352.70	16848	27MR	5;55	186.61
16774	18MR	14;53	358.63	16774	1AP	23;31	192.62	16849	11AP	8;48	21.27	16849	25AP	21;22	215.25
16775	17AP	6;16	27.39	16775	1MY	8;03	220.90	16850	10MY	15;51	49.39	16850	25MY	12;33	243.57
16776	16MY	19;24	55.69	16776	30MY	17;30	248.96	16851	8JN	22;41	77.30	16851	24JN	3;06	271.77
16777	15JN	6;16	83.75	16777	29JN	4;30	277.03	16852	8JL	6;23	105.27	16852	23JL	16;41	300.07
16778	14JL	15;22	111.81	16778	28JL	17;34	305.36	16853	6AU	15;54	133.55	16853	22AU	5;05	328.67
16779	12AU	23;36	140.12	16779	27AU	8;56	334.16	16854	5SE	3;54	162.35	16854	20SE	16;24	357.70
16780	11SE	8;00	168.89	16780	26SE	2;25	3.55	16855	4OC	18;44	191.77	16855	20OC	3;03	27.20
16781	10OC	17;24	198.19	16781	25OC	20;59	33.47	16856	3NO	12;21	221.76	16856	18NO	13;36	57.07
16782	9NO	4;20	227.98	16782	24NO	15;04	63.72	16857	3DE	7;54	252.13	16857	18DE	0;24	87.12
16783	8DE	16;56	258.09	16783	24DE	7;06	93.99								

357 / 363

NUMBER	DATE	TIME	LONG.	NUMBER	DATE	TIME	LONG.	NUMBER	DATE	TIME	LONG.	NUMBER	DATE	TIME	LONG.
16784	7JA	7;05	288.22	16784	22JA	20;26	123.96	16858	2JA	3;39	282.53	16858	16JA	11;32	117.07
16785	5FE	22;32	318.13	16785	21FE	7;15	153.45	16859	31JA	21;32	312.61	16859	14FE	23;01	146.68
16786	7MR	14;54	347.60	16786	22MR	16;14	182.36	16860	2MR	12;14	342.12	16860	16MR	10;56	175.83
16787	6AP	7;33	16.59	16787	21AP	0;04	210.76	16861	31MR	23;41	10.98	16861	14AP	23;36	204.52
16788	5MY	23;31	45.12	16788	20MY	7;30	238.80	16862	30AP	8;35	39.30	16862	14MY	13;21	232.88
16789	4JN	13;55	73.35	16789	18JN	15;20	266.74	16863	29MY	16;00	67.29	16863	13JN	4;14	261.10
16790	4JL	2;26	101.49	16790	18JL	0;32	294.82	16864	27JN	22;58	95.19	16864	12JL	19;47	289.40
16791	2AU	13;23	129.78	16791	16AU	12;05	323.31	16865	27JL	6;21	123.28	16865	11AU	11;11	317.97
16792	31AU	23;30	158.45	16792	15SE	2;46	352.39	16866	25AU	14;59	151.77	16866	10SE	1;48	346.96
16793	30SE	9;30	187.59	16793	14OC	20;30	22.10	16867	24SE	1;41	180.83	16867	9OC	15;26	16.40
16794	29OC	19;49	217.23	16794	13NO	16;04	52.30	16868	23OC	15;13	210.50	16868	8NO	4;09	46.25
16795	28NO	6;33	247.19	16795	13DE	11;27	82.71	16869	22NO	7;51	240.66	16869	7DE	16;03	76.31
16796	27DE	17;48	277.25					16870	22DE	2;54	271.06				

358 / 364

NUMBER	DATE	TIME	LONG.	NUMBER	DATE	TIME	LONG.	NUMBER	DATE	TIME	LONG.	NUMBER	DATE	TIME	LONG.
				16796	12JA	4;54	112.95					16870	6JA	3;07	106.33
16797	26JA	5;48	307.14	16797	10FE	19;34	142.76	16871	20JA	22;33	301.33	16871	4FE	13;20	136.04
16798	24FE	18;50	336.66	16798	12MR	7;28	171.98	16872	19FE	16;46	331.16	16872	4MR	22;55	165.27
16799	26MR	9;04	5.73	16799	10AP	16;58	200.60	16873	20MR	8;20	0.38	16873	3AP	8;30	194.01
16800	25AP	0;09	34.37	16800	10MY	0;41	228.76	16874	18AP	21;02	29.00	16874	2MY	18;55	222.36
16801	24MY	15;28	62.70	16801	8JN	7;28	256.67	16875	18MY	7;19	57.18	16875	1JN	6;51	250.50
16802	23JN	6;24	90.92	16802	7JL	14;21	284.60	16876	16JN	15;51	85.15	16876	30JN	20;36	278.69
16803	22JL	20;36	119.24	16803	5AU	22;32	312.82	16877	15JL	23;27	113.15	16877	30JL	11;55	307.12
16804	21AU	10;05	147.87	16804	4SE	9;09	341.55	16878	14AU	6;58	141.45	16878	29AU	4;17	335.99
16805	19SE	22;52	176.95	16805	3OC	22;22	10.91	16879	12SE	15;20	170.24	16879	27SE	21;00	5.38
16806	19OC	10;56	206.49	16806	2NO	15;44	40.86	16880	12OC	1;28	199.59	16880	27OC	13;42	35.23
16807	17NO	22;15	236.39	16807	2DE	10;42	71.20	16881	10NO	14;03	229.47	16881	26NO	4;34	65.37
16808	17DE	8;57	266.43					16882	10DE	5;15	259.69	16882	25DE	17;59	95.52

359 / 365

NUMBER	DATE	TIME	LONG.	NUMBER	DATE	TIME	LONG.	NUMBER	DATE	TIME	LONG.	NUMBER	DATE	TIME	LONG.
				16808	1JA	6;13	101.60	16883	8JA	22;27	289.95	16883	24JA	5;19	125.41
16809	15JA	19;20	296.36	16809	31JA	0;43	131.71	16884	7FE	16;28	319.93	16884	22FE	14;53	154.81
16810	14FE	5;50	325.94	16810	1MR	16;55	161.29	16885	9MR	10;04	349.43	16885	23MR	23;20	183.68
16811	15MR	16;50	355.06	16811	31MR	6;10	190.24	16886	8AP	2;18	18.37	16886	22AP	7;29	212.08
16812	14AP	4;36	23.73	16812	29AP	16;28	218.63	16887	7MY	16;32	46.82	16887	21MY	16;13	240.17
16813	13MY	17;17	52.05	16813	29MY	0;32	246.64	16888	6JN	4;34	74.94	16888	20JN	2;15	268.19
16814	12JN	7;02	80.23	16814	27JN	7;26	274.54	16889	5JL	14;37	102.99	16889	19JL	14;12	296.39
16815	11JL	21;54	108.50	16815	26JL	14;23	302.60	16890	3AU	23;22	131.22	16890	18AU	4;28	325.01
16816	10AU	13;40	137.07	16816	24AU	22;20	331.07	16891	2SE	7;46	159.83	16891	16SE	21;07	354.20
16817	9SE	5;44	166.10	16817	23SE	8;36	0.09	16892	1OC	16;45	188.95	16892	16OC	15;32	23.96
16818	8OC	21;13	195.61	16818	22OC	21;09	29.71	16893	31OC	3;01	218.60	16893	15NO	10;17	54.15
16819	7NO	11;21	225.51	16819	21NO	12;15	59.80	16894	29NO	14;48	248.62	16894	15DE	3;37	84.47
16820	6DE	23;56	255.60	16820	21DE	5;36	90.12	16895	29DE	4;06	278.76				

360 / 366

NUMBER	DATE	TIME	LONG.	NUMBER	DATE	TIME	LONG.	NUMBER	DATE	TIME	LONG.	NUMBER	DATE	TIME	LONG.
16821	5JA	11;09	285.63	16821	20JA	0;24	120.37					16895	13JA	18;21	114.59
16822	3FE	21;22	315.34	16822	18FE	19;11	150.24	16896	27JA	18;42	308.74	16896	12FE	6;18	144.26
16823	4MR	6;53	344.59	16823	19MR	12;17	179.53	16897	26FE	10;23	338.35	16897	13MR	15;59	173.37
16824	2AP	16;03	13.32	16824	18AP	2;35	208.23	16898	28MR	2;43	7.48	16898	12AP	0;09	201.92
16825	2MY	1;23	41.63	16825	17MY	13;59	236.46	16899	26AP	18;58	36.14	16899	11MY	7;33	230.05
16826	31MY	11;38	69.71	16826	15JN	23;12	264.45	16900	26MY	10;12	64.46	16900	9JN	14;59	257.98
16827	29JN	23;41	97.82	16827	15JL	7;16	292.48	16901	24JN	23;46	92.62	16901	8JL	23;21	285.98
16828	29JL	14;04	126.21	16828	13AU	15;03	320.79	16902	24JL	11;38	120.86	16902	7AU	9;37	314.30
16829	28AU	6;38	155.08	16829	11SE	23;49	349.57	16903	22AU	22;20	149.40	16903	5SE	22;45	343.15
16830	27SE	0;24	184.49	16830	11OC	9;35	18.90	16904	21SE	8;36	178.40	16904	5OC	15;08	12.65
16831	26OC	17;58	214.38	16831	9NO	20;57	48.72	16905	20OC	18;55	207.90	16905	4NO	10;08	42.71
16832	25NO	10;12	244.56	16832	9DE	10;19	78.85	16906	19NO	5;33	237.78	16906	4DE	5;59	73.10
16833	25DE	0;33	274.76					16907	18DE	16;32	267.83				

361 / 367

NUMBER	DATE	TIME	LONG.	NUMBER	DATE	TIME	LONG.	NUMBER	DATE	TIME	LONG.	NUMBER	DATE	TIME	LONG.
				16833	8JA	1;51	109.05					16907	3JA	0;35	103.45
16834	23JA	12;55	304.69	16834	6FE	19;04	139.01	16908	17JA	3;58	297.79	16908	1FE	16;39	133.44
16835	21FE	23;21	334.14	16835	8MR	12;46	168.53	16909	15FE	16;12	327.43	16909	3MR	5;50	162.87
16836	23MR	8;08	3.03	16836	7AP	5;32	197.50	16910	17MR	5;31	356.62	16910	1AP	16;22	191.69
16837	21AP	15;50	31.42	16837	6MY	20;27	225.98	16911	15AP	19;57	25.37	16911	1MY	0;48	219.98
16838	20MY	23;23	59.47	16838	5JN	9;22	254.15	16912	15MY	11;03	53.78	16912	30MY	7;53	247.95
16839	19JN	7;57	87.43	16839	4JL	20;36	282.24	16913	14JN	2;12	82.01	16913	28JN	14;34	275.84
16840	18JL	18;36	115.58	16840	3AU	6;42	310.51	16914	13JL	16;54	110.28	16914	27JL	22;01	303.94
16841	17AU	8;00	144.15	16841	1SE	16;10	339.16	16915	12AU	6;59	138.81	16915	26AU	7;22	332.47
16842	16SE	0;09	173.30	16842	1OC	1;33	8.29	16916	10SE	20;25	167.75	16916	24SE	19;35	1.60
16843	15OC	18;17	203.04	16843	30OC	11;21	37.91	16917	10OC	9;08	197.17	16917	24OC	11;01	31.36
16844	14NO	13;13	233.23	16844	28NO	22;07	67.88	16918	8NO	21;02	226.99	16918	23NO	5;07	61.59
16845	14DE	7;32	263.59	16845	28DE	10;14	97.97	16919	8DE	8;06	257.02	16919	23DE	0;32	92.00

368

NEW MOONS

NUMBER	DATE	TIME	LONG.
16920	6JA	18;33	287.01
16921	5FE	4;47	316.71
16922	5MR	15;14	345.98
16923	4AP	2;18	14.76
16924	3MY	14;15	43.16
16925	2JN	3;17	71.35
16926	1JL	17;33	99.55
16927	31JL	9;02	128.00
16928	30AU	1;18	156.88
16929	28SE	17;30	186.26
16930	28OC	8;39	216.07
16931	26NO	22;11	246.15
16932	26DE	10;05	276.24

FULL MOONS

NUMBER	DATE	TIME	LONG.
16920	21JA	19;38	122.25
16921	20FE	13;00	152.03
16922	21MR	3;40	181.20
16923	19AP	15;21	209.77
16924	19MY	0;25	237.89
16925	17JN	7;47	265.81
16926	16JL	14;39	293.79
16927	14AU	22;11	322.10
16928	13SE	7;20	350.93
16929	12OC	18;44	20.34
16930	11NO	8;36	50.29
16931	11DE	0;50	80.55

369

NEW MOONS

NUMBER	DATE	TIME	LONG.
16933	24JA	20;41	306.08
16934	23FE	6;22	335.48
16935	24MR	15;26	4.37
16936	23AP	0;22	32.78
16937	22MY	9;52	60.90
16938	20JN	20;47	88.95
16939	20JL	9;56	117.21
16940	19AU	1;35	145.90
16941	17SE	19;08	175.14
16942	17OC	13;13	204.90
16943	16NO	6;27	235.04
16944	15DE	21;54	265.28

FULL MOONS

NUMBER	DATE	TIME	LONG.
16932	9JA	18;54	110.84
16933	8FE	13;43	140.85
16934	10MR	7;40	170.35
16935	8AP	23;20	199.25
16936	8MY	12;07	227.63
16937	6JN	22;22	255.68
16938	6JL	7;00	283.68
16939	4AU	15;04	311.88
16940	2SE	23;28	340.50
16941	2OC	8;45	9.65
16942	31OC	19;23	39.32
16943	30NO	7;42	69.37
16944	29DE	22;02	99.55

370

NEW MOONS

NUMBER	DATE	TIME	LONG.
16945	14JA	11;17	295.34
16946	12FE	22;34	324.97
16947	14MR	7;59	354.05
16948	12AP	15;57	22.58
16949	11MY	23;18	50.71
16950	10JN	7;06	78.65
16951	9JL	16;31	106.70
16952	8AU	4;31	135.09
16953	6SE	19;25	164.03
16954	6OC	12;51	193.59
16955	5NO	7;43	223.66
16956	5DE	2;39	254.01

FULL MOONS

NUMBER	DATE	TIME	LONG.
16945	28JA	14;17	129.59
16946	27FE	7;39	159.26
16947	29MR	0;48	188.41
16948	27AP	16;35	217.04
16949	27MY	6;29	245.30
16950	25JN	18;37	273.40
16951	25JL	5;25	301.61
16952	23AU	15;23	330.13
16953	22SE	0;59	359.11
16954	21OC	10;42	28.59
16955	19NO	21;00	58.46
16956	19DE	8;22	88.53

371

NEW MOONS

NUMBER	DATE	TIME	LONG.
16957	3JA	20;12	284.31
16958	2FE	11;16	314.25
16959	3MR	23;22	343.62
16960	2AP	8;45	12.38
16961	1MY	16;15	40.63
16962	30MY	22;58	68.58
16963	29JN	6;05	96.49
16964	28JL	14;37	124.63
16965	27AU	1;22	153.23
16966	25SE	14;51	182.43
16967	25OC	7;13	212.24
16968	24NO	2;01	242.51
16969	23DE	21;56	272.94

FULL MOONS

NUMBER	DATE	TIME	LONG.
16957	17JA	21;04	118.54
16958	16FE	11;01	148.24
16959	18MR	1;50	177.48
16960	16AP	17;01	206.25
16961	16MY	8;14	234.65
16962	14JN	23;09	262.87
16963	14JL	13;24	291.13
16964	13AU	2;40	319.63
16965	11SE	14;48	348.54
16966	11OC	1;59	17.91
16967	9NO	12;42	47.69
16968	8DE	23;24	77.70

372

NEW MOONS

NUMBER	DATE	TIME	LONG.
16970	22JA	16;55	303.17
16971	21FE	9;14	332.89
16972	21MR	22;10	1.98
16973	20AP	8;09	30.47
16974	19MY	16;09	58.55
16975	17JN	23;14	86.45
16976	17JL	6;22	114.45
16977	15AU	14;22	142.79
16978	14SE	0;03	171.65
16979	13OC	12;11	201.10
16980	12NO	3;18	231.11
16981	11DE	21;16	261.45

FULL MOONS

NUMBER	DATE	TIME	LONG.
16969	7JA	10;18	107.71
16970	5FE	21;26	137.44
16971	6MR	8;50	166.73
16972	4AP	20;46	195.54
16973	4MY	9;41	223.97
16974	2JN	23;53	252.19
16975	2JL	15;11	280.44
16976	1AU	6;55	308.92
16977	30AU	22;16	337.77
16978	29SE	12;42	7.09
16979	29OC	2;08	36.84
16980	27NO	14;36	66.88
16981	27DE	2;08	96.95

373

NEW MOONS

NUMBER	DATE	TIME	LONG.
16982	10JA	16;46	291.79
16983	9FE	11;46	321.80
16984	11MR	4;36	351.23
16985	9AP	18;37	20.06
16986	9MY	6;01	48.37
16987	7JN	15;23	76.39
16988	6JL	23;28	104.36
16989	5AU	7;03	132.55
16990	3SE	15;03	161.17
16991	3OC	0;20	190.33
16992	1NO	11;42	220.04
16993	1DE	1;36	250.16
16994	30DE	17;48	280.42

FULL MOONS

NUMBER	DATE	TIME	LONG.
16982	25JA	12;39	126.78
16983	23FE	22;18	156.17
16984	25MR	7;33	185.05
16985	23AP	17;12	213.49
16986	23MY	4;05	241.65
16987	21JN	16;48	269.78
16988	21JL	7;24	298.10
16989	19AU	23;28	326.82
16990	18SE	16;21	356.04
16991	18OC	9;16	25.78
16992	17NO	1;23	55.87
16993	16DE	15;52	86.07

374

NEW MOONS

NUMBER	DATE	TIME	LONG.
16995	29JA	11;21	310.51
16996	28FE	5;03	340.17
16997	29MR	21;52	9.29
16998	28AP	13;04	37.89
16999	28MY	2;14	66.11
17000	26JN	13;21	94.18
17001	25JL	22;50	122.34
17002	24AU	7;30	150.82
17003	22SE	16;18	179.78
17004	22OC	2;02	209.26
17005	20NO	13;06	239.18
17006	20DE	1;36	269.30

FULL MOONS

NUMBER	DATE	TIME	LONG.
16994	15JA	4;14	116.08
16995	13FE	14;31	145.66
16996	14MR	23;15	174.70
16997	13AP	7;15	203.23
16998	12MY	15;24	231.38
16999	11JN	0;34	259.38
17000	10JL	11;26	287.48
17001	9AU	0;32	315.93
17002	7SE	16;08	344.91
17003	7OC	9;59	14.50
17004	6NO	5;00	44.58
17005	5DE	23;24	74.92

375

NEW MOONS

NUMBER	DATE	TIME	LONG.
17007	18JA	15;24	299.34
17008	17FE	6;18	329.07
17009	18MR	22;04	358.34
17010	17AP	14;15	27.13
17011	17MY	5;59	55.55
17012	15JN	20;29	83.75
17013	15JL	9;21	111.96
17014	13AU	20;49	140.40
17015	12SE	7;29	169.25
17016	11OC	17;59	198.60
17017	10NO	4;36	228.39
17018	9DE	15;26	258.41

FULL MOONS

NUMBER	DATE	TIME	LONG.
17006	4JA	15;35	105.16
17007	3FE	4;51	135.01
17008	4MR	15;28	164.31
17009	3AP	0;09	193.03
17010	2MY	7;43	221.28
17011	31MY	14;56	249.24
17012	29JN	22;40	277.18
17013	29JL	7;51	305.36
17014	27AU	19;31	334.00
17015	26SE	10;22	3.27
17016	26OC	4;21	33.16
17017	25NO	0;08	63.48
17018	24DE	19;38	93.89

376

NEW MOONS

NUMBER	DATE	TIME	LONG.
17019	8JA	2;31	288.42
17020	6FE	14;06	318.16
17021	7MR	2;33	347.48
17022	5AP	16;09	16.35
17023	5MY	6;44	44.84
17024	3JN	21;50	73.10
17025	3JL	12;51	101.34
17026	2AU	3;28	129.77
17027	31AU	17;31	158.59
17028	30SE	6;55	187.88
17029	29OC	19;29	217.60
17030	28NO	7;06	247.61
17031	27DE	17;49	277.64

FULL MOONS

NUMBER	DATE	TIME	LONG.
17019	23JA	13;02	124.04
17020	22FE	3;35	153.68
17021	22MR	15;17	182.71
17022	21AP	0;37	211.16
17023	20MY	8;13	239.22
17024	18JN	14;58	267.11
17025	17JL	21;55	295.10
17026	16AU	6;15	323.46
17027	14SE	17;02	352.38
17028	14OC	6;57	21.92
17029	12NO	23;53	52.01
17030	12DE	18;49	82.39

377

NEW MOONS

NUMBER	DATE	TIME	LONG.
17032	26JA	3;59	307.45
17033	24FE	14;03	336.85
17034	26MR	0;30	5.77
17035	24AP	11;42	34.26
17036	23MY	23;58	62.47
17037	22JN	13;30	90.63
17038	22JL	4;27	118.98
17039	20AU	20;36	147.71
17040	19SE	13;14	176.94
17041	19OC	5;19	206.65
17042	17NO	19;55	236.69
17043	17DE	8;43	266.81

FULL MOONS

NUMBER	DATE	TIME	LONG.
17031	11JA	14;12	112.72
17032	10FE	8;28	142.69
17033	12MR	0;28	172.08
17034	10AP	13;33	200.85
17035	9MY	23;49	229.11
17036	8JN	7;56	257.08
17037	7JL	14;59	285.01
17038	5AU	22;10	313.19
17039	4SE	6;33	341.83
17040	3OC	16;54	11.05
17041	2NO	5;34	40.82
17042	1DE	20;38	70.99
17043	31DE	13;46	101.29

378

NEW MOONS

NUMBER	DATE	TIME	LONG.
17044	15JA	19;53	296.77
17045	14FE	5;51	326.33
17046	15MR	14;59	355.38
17047	13AP	23;41	23.92
17048	13MY	8;36	52.09
17049	11JN	18;33	80.12
17050	11JL	6;27	108.27
17051	9AU	20;54	136.79
17052	8SE	13;47	165.83
17053	8OC	8;02	195.45
17054	7NO	2;05	225.51
17055	6DE	18;39	255.77

FULL MOONS

NUMBER	DATE	TIME	LONG.
17044	30JA	8;11	131.41
17045	1MR	2;34	161.09
17046	30MR	19;22	190.19
17047	29AP	9;33	218.74
17048	28MY	21;01	246.89
17049	27JN	6;26	274.89
17050	26JL	14;50	303.01
17051	24AU	23;10	331.49
17052	23SE	8;09	0.46
17053	22OC	18;12	29.97
17054	21NO	5;39	59.91
17055	20DE	18;53	90.06

379

NEW MOONS

NUMBER	DATE	TIME	LONG.
17056	5JA	9;09	285.93
17057	3FE	21;25	315.73
17058	5MR	7;36	345.00
17059	3AP	16;03	13.70
17060	2MY	23;26	41.93
17061	1JN	6;44	69.90
17062	30JN	15;08	97.87
17063	30JL	1;45	126.11
17064	28AU	15;16	154.84
17065	27SE	7;39	184.19
17066	27OC	2;07	214.11
17067	25NO	21;20	244.41
17068	25DE	15;48	274.77

FULL MOONS

NUMBER	DATE	TIME	LONG.
17056	19JA	10;02	120.15
17057	18FE	2;42	149.94
17058	19MR	19;51	179.25
17059	18AP	12;15	208.05
17060	18MY	3;03	236.41
17061	16JN	16;07	264.56
17062	16JL	3;42	292.72
17063	14AU	14;17	321.14
17064	13SE	0;16	349.98
17065	12OC	10;05	19.31
17066	10NO	20;09	49.08
17067	10DE	6;57	79.10

Left half

NUMBER	DATE	TIME	LONG.	NUMBER	DATE	TIME	LONG.
						380	
				17068	8JA	18;51	109.14
17069	24JA	8;13	304.88	17069	7FE	7;58	138.94
17070	22FE	21;45	334.46	17070	7MR	22;06	168.32
17071	23MR	8;19	3.43	17071	6AP	12;52	197.22
17072	21AP	16;31	31.83	17072	6MY	3;54	225.71
17073	20MY	23;24	59.85	17073	4JN	18;56	253.96
17074	19JN	6;09	87.74	17074	4JL	9;40	282.20
17075	18JL	13;52	115.77	17075	2AU	23;42	310.62
17076	16AU	23;29	144.19	17076	1SE	12;41	339.40
17077	15SE	11;38	173.18	17077	1OC	0;35	8.64
17078	15OC	2;39	202.78	17078	30OC	11;42	38.32
17079	13NO	20;24	232.91	17079	28NO	22;29	68.29
17080	13DE	16;00	263.32	17080	28DE	9;15	98.32
						381	
17081	12JA	11;40	293.66	17081	26JA	20;07	128.16
17082	11FE	5;24	323.59	17082	25FE	7;07	157.59
17083	12MR	19;57	352.90	17083	26MR	18;27	186.53
17084	11AP	7;14	21.59	17084	25AP	6;33	215.04
17085	10MY	16;03	49.79	17085	24MY	19;53	243.30
17086	8JN	23;28	77.72	17086	23JN	10;37	271.50
17087	8JL	6;31	105.66	17087	23JL	2;20	299.88
17088	6AU	14;06	133.87	17088	21AU	18;13	328.62
17089	4SE	22;58	162.54	17089	20SE	9;27	357.80
17090	4OC	9;53	191.79	17090	19OC	23;40	27.45
17091	2NO	23;31	221.61	17091	18NO	12;49	57.43
17092	2DE	16;07	251.85	17092	18DE	0;54	87.53
						382	
17093	1JA	10;58	282.22	17093	16JA	11;53	117.47
17094	31JA	6;19	312.36	17094	14FE	21;47	147.02
17095	2MR	0;13	342.01	17095	16MR	6;56	176.06
17096	31MR	15;34	11.04	17096	14AP	16;01	204.61
17097	30AP	4;10	39.51	17097	14MY	1;57	232.82
17098	29MY	14;30	67.61	17098	12JN	13;34	260.91
17099	27JN	23;14	95.59	17099	12JL	3;12	289.13
17100	27JL	7;06	123.69	17100	10AU	18;41	317.70
17101	25AU	14;57	152.16	17101	9SE	11;28	346.75
17102	23SE	23;38	181.13	17102	9OC	4;44	16.34
17103	23OC	10;00	210.67	17103	7NO	21;37	46.36
17104	21NO	22;39	240.66	17104	7DE	13;10	76.58
17105	21DE	13;40	270.88				
						383	
17106	20JA	6;29	301.04	17105	6JA	2;40	106.70
17107	19FE	0;00	330.85	17106	4FE	13;53	136.45
17108	20MR	17;09	0.15	17107	5MR	23;09	165.67
17109	19AP	9;05	28.91	17108	4AP	7;13	194.35
17110	18MY	23;17	57.25	17109	3MY	15;00	222.59
17111	17JN	11;30	85.36	17110	1JN	23;25	250.59
17112	16JL	21;54	113.48	17111	1JL	9;16	278.62
17113	15AU	7;05	141.85	17112	30JL	21;11	306.92
17114	13SE	15;56	170.65	17113	29AU	11;34	335.70
17115	13OC	1;18	199.98	17114	28SE	4;34	5.09
17116	11NO	11;47	229.77	17115	27OC	23;23	35.03
17117	10DE	23;34	259.84	17116	26NO	18;30	65.33
				17117	26DE	12;02	95.66
						384	
17118	9JA	12;36	289.92	17118	25JA	2;46	125.69
17119	8FE	2;42	319.75	17119	23FE	14;33	155.19
17120	8MR	17;47	349.15	17120	23MR	23;59	184.10
17121	7AP	9;33	18.08	17121	22AP	7;52	212.48
17122	7MY	1;28	46.60	17122	21MY	15;00	240.51
17123	5JN	16;40	74.86	17123	19JN	22;22	268.42
17124	5JL	6;30	103.06	17124	19JL	6;42	296.48
17125	3AU	18;50	131.42	17125	17AU	17;04	324.94
17126	2SE	6;06	160.15	17126	16SE	6;21	353.98
17127	1OC	16;54	189.35	17127	15OC	22;56	23.66
17128	31OC	3;38	219.02	17128	14NO	18;09	53.86
17129	29NO	14;27	249.00	17129	14DE	14;06	84.29
17130	29DE	1;21	279.03				
						385	
17131	27JA	12;27	308.87	17130	13JA	8;43	114.58
17132	26FE	0;09	338.32	17131	12FE	0;40	144.42
17133	27MR	12;51	7.31	17132	13MR	13;40	173.66
17134	26AP	2;43	35.88	17133	12AP	0;02	202.30
17135	25MY	17;28	64.18	17134	11MY	8;20	230.47
17136	24JN	8;35	92.41	17135	9JN	15;23	258.39
17137	23JL	23;34	120.77	17136	8JL	22;08	286.32
17138	22AU	14;10	149.46	17137	7AU	5;42	314.53
17139	21SE	4;14	178.61	17138	5SE	15;14	343.24
17140	20OC	17;31	208.23	17139	5OC	3;38	12.56
17141	19NO	5;48	238.18	17140	3NO	19;11	42.47
17142	18DE	17;00	268.24	17141	3DE	13;16	72.78

Right half

NUMBER	DATE	TIME	LONG.	NUMBER	DATE	TIME	LONG.
						386	
				17142	2JA	8;33	103.16
17143	17JA	3;19	298.15	17143	1FE	3;26	133.28
17144	15FE	13;11	327.69	17144	2MR	20;34	162.87
17145	16MR	23;08	356.75	17145	1AP	11;03	191.86
17146	15AP	9;39	25.39	17146	30AP	22;39	220.28
17147	14MY	21;07	53.61	17147	30MY	7;45	248.33
17148	13JN	9;51	81.74	17148	28JN	15;15	276.26
17149	13JL	0;04	109.90	17149	27JL	22;21	304.34
17150	11AU	15;48	138.58	17150	26AU	6;10	332.81
17151	10SE	8;34	167.65	17151	24SE	15;35	1.83
17152	10OC	1;21	197.24	17152	24OC	3;10	31.42
17153	8NO	17;02	227.21	17153	22NO	17;04	61.47
17154	8DE	6;53	257.36	17154	22DE	9;07	91.74
						387	
17155	6JA	18;51	287.41	17155	21JA	2;51	121.93
17156	5FE	5;16	317.12	17156	19FE	21;12	151.76
17157	6MR	14;36	346.34	17157	21MR	14;47	181.06
17158	4AP	23;15	15.03	17158	20AP	6;16	209.79
17159	4MY	7;46	43.29	17159	19MY	19;03	238.07
17160	2JN	16;54	71.32	17160	18JN	5;29	266.10
17161	2JL	3;08	99.38	17161	17JL	14;25	294.17
17162	31JL	16;46	127.74	17162	15AU	22;53	322.52
17163	30AU	8;38	156.59	17163	14SE	7;40	351.33
17164	29SE	2;35	186.03	17164	13OC	17;18	20.67
17165	28OC	21;09	215.98	17165	12NO	4;06	50.49
17166	27NO	14;46	246.22	17166	11DE	16;22	80.58
17167	27DE	6;26	276.47				
						388	
17168	25JA	19;48	306.43	17167	10JA	6;24	110.70
17169	24FE	6;53	335.89	17168	8FE	22;08	140.59
17170	24MR	16;00	4.78	17169	9MR	14;55	170.05
17171	22AP	23;40	33.14	17170	8AP	7;36	199.00
17172	22MY	6;45	61.16	17171	7MY	23;09	227.50
17173	20JN	14;22	89.08	17172	6JN	13;06	255.70
17174	19JL	23;44	117.19	17173	6JL	1;31	283.85
17175	18AU	11;47	145.73	17174	4AU	12;47	312.18
17176	17SE	2;53	174.86	17175	2SE	23;17	340.88
17177	16OC	20;35	204.60	17176	2OC	9;22	10.07
17178	15NO	15;44	234.80	17177	31OC	19;26	39.71
17179	15DE	10;51	265.20	17178	30NO	5;52	69.68
				17179	29DE	17;07	99.73
						389	
17180	14JA	4;27	295.44	17180	28JA	5;26	129.61
17181	12FE	19;25	325.23	17181	26FE	18;49	159.12
17182	14MR	7;20	354.41	17182	28MR	9;00	188.15
17183	12AP	16;31	22.99	17183	26AP	23;41	216.75
17184	11MY	23;50	51.11	17184	26MY	14;37	245.05
17185	10JN	6;28	79.01	17185	25JN	5;35	273.27
17186	9JL	13;35	106.96	17186	24JL	20;35	301.63
17187	7AU	22;12	135.23	17187	23AU	10;01	330.30
17188	6SE	9;06	164.00	17188	21SE	22;46	359.41
17189	5OC	22;46	193.39	17189	21OC	10;28	28.97
17190	4NO	15;16	223.35	17190	19NO	21;31	58.88
17191	4DE	10;07	253.69	17191	19DE	8;18	88.92
						390	
17192	3JA	5;58	284.10	17192	17JA	19;01	118.84
17193	2FE	0;49	314.20	17193	16FE	5;44	148.41
17194	3MR	16;56	343.74	17194	17MR	16;35	177.49
17195	2AP	5;43	12.64	17195	16AP	3;57	206.11
17196	1MY	15;36	40.99	17196	15MY	16;23	234.41
17197	30MY	23;34	68.99	17197	14JN	6;18	262.59
17198	29JN	6;44	96.90	17198	13JL	21;39	290.89
17199	28JL	14;03	125.00	17199	12AU	13;45	319.50
17200	26AU	22;18	153.50	17200	11SE	5;40	348.55
17201	25SE	8;13	182.55	17201	10OC	20;43	18.08
17202	24OC	20;30	212.18	17202	9NO	10;39	47.99
17203	23NO	11;39	242.29	17203	8DE	23;24	78.09
17204	23DE	5;27	272.63				
						391	
17205	22JA	0;38	302.87	17204	7JA	10;56	108.12
17206	20FE	19;17	332.71	17205	5FE	21;14	137.82
17207	22MR	11;50	1.95	17206	7MR	6;30	167.02
17208	21AP	1;43	30.60	17207	5AP	15;16	195.70
17209	20MY	13;07	58.82	17208	5MY	0;26	223.98
17210	18JN	22;39	86.82	17209	3JN	10;57	252.07
17211	18JL	6;59	114.86	17210	2JL	23;28	280.20
17212	16AU	14;55	143.20	17211	1AU	14;07	308.63
17213	14SE	23;16	172.00	17212	31AU	6;30	337.52
17214	14OC	8;50	201.35	17213	29SE	23;52	6.94
17215	12NO	20;20	231.20	17214	29OC	17;19	36.86
17216	12DE	10;09	261.36	17215	28NO	9;50	67.06
				17216	28DE	0;31	97.27

392 / 398

NEW MOONS (392)

NUMBER	DATE	TIME	LONG.
17217	11JA	2;01	291.56
17218	9FE	19;05	321.49
17219	10MR	12;16	350.95
17220	9AP	4;42	19.88
17221	8MY	19;45	48.35
17222	7JN	9;01	76.52
17223	6JL	20;26	104.63
17224	5AU	6;22	132.91
17225	3SE	15;30	161.58
17226	3OC	0;44	190.74
17227	1NO	10;46	220.39
17228	30NO	21;56	250.39
17229	30DE	10;15	280.49

FULL MOONS (392)

NUMBER	DATE	TIME	LONG.
17217	26JA	12;50	127.18
17218	24FE	22;53	156.58
17219	25MR	7;17	185.43
17220	23AP	14;54	213.78
17221	22MY	22;44	241.83
17222	21JN	7;41	269.80
17223	20JL	18;28	297.97
17224	19AU	7;39	326.57
17225	17SE	23;30	355.74
17226	17OC	17;42	25.51
17227	16NO	13;04	55.74
17228	16DE	7;43	86.11

NEW MOONS (398)

NUMBER	DATE	TIME	LONG.
17291	5JA	0;01	285.93
17292	3FE	16;22	315.91
17293	5MR	5;44	345.31
17294	3AP	16;06	14.09
17295	3MY	0;08	42.35
17296	1JN	6;55	70.30
17297	30JN	13;39	98.20
17298	29JL	21;27	126.32
17299	28AU	7;13	154.90
17300	26SE	19;33	184.07
17301	26OC	10;43	213.85
17302	25NO	4;32	244.08
17303	25DE	0;04	274.50

FULL MOONS (398)

NUMBER	DATE	TIME	LONG.
17291	19JA	3;24	120.26
17292	17FE	16;01	149.89
17293	19MR	5;31	179.06
17294	17AP	19;42	207.77
17295	17MY	10;21	236.14
17296	16JN	1;18	264.36
17297	15JL	16;16	292.65
17298	14AU	6;48	321.22
17299	12SE	20;25	350.21
17300	12OC	8;54	19.65
17301	10NO	20;25	49.48
17302	10DE	7;23	79.51

393 / 399

NEW MOONS (393)

NUMBER	DATE	TIME	LONG.
17230	28JA	23;37	310.40
17231	27FE	13;56	339.94
17232	29MR	5;06	9.00
17233	27AP	20;49	37.63
17234	27MY	12;24	65.95
17235	26JN	3;03	94.16
17236	25JL	16;20	122.46
17237	24AU	4;21	151.08
17238	22SE	15;36	180.14
17239	22OC	2;33	209.68
17240	20NO	13;27	239.58
17241	20DE	0;19	269.63

FULL MOONS (393)

NUMBER	DATE	TIME	LONG.
17229	14JA	23;57	116.29
17230	13FE	13;07	145.99
17231	14MR	23;31	175.10
17232	13AP	7;57	203.64
17233	12MY	15;17	231.76
17234	10JN	22;22	259.68
17235	10JL	6;04	287.65
17236	8AU	15;19	315.95
17237	7SE	3;07	344.77
17238	6OC	18;09	14.23
17239	5NO	12;09	44.27
17240	5DE	8;13	74.66

NEW MOONS (399)

NUMBER	DATE	TIME	LONG.
17304	23JA	19;35	304.74
17305	22FE	13;07	334.49
17306	24MR	3;29	3.62
17307	22AP	14;39	32.14
17308	21MY	23;26	60.24
17309	20JN	6;55	88.16
17310	19JL	14;09	116.17
17311	17AU	21;58	144.52
17312	16SE	7;05	173.38
17313	15OC	18;12	202.81
17314	14NO	7;55	232.77
17315	14DE	0;25	263.05

FULL MOONS (399)

NUMBER	DATE	TIME	LONG.
17303	8JA	18;04	109.50
17304	7FE	4;36	139.19
17305	8MR	15;06	168.42
17306	7AP	1;52	197.16
17307	6MY	13;26	225.52
17308	5JN	2;24	253.69
17309	4JL	17;03	281.92
17310	3AU	9;00	310.41
17311	2SE	1;23	339.33
17312	1OC	17;14	8.73
17313	31OC	8;01	38.55
17314	29NO	21;31	68.64
17315	29DE	9;43	98.73

394 / 400

NEW MOONS (394)

NUMBER	DATE	TIME	LONG.
17242	18JA	11;10	299.55
17243	16FE	22;17	329.12
17244	18MR	10;08	358.24
17245	16AP	23;08	26.91
17246	16MY	13;16	55.27
17247	15JN	4;11	83.49
17248	14JL	19;22	111.78
17249	13AU	10;25	140.37
17250	12SE	1;04	169.38
17251	11OC	15;05	198.87
17252	10NO	4;07	228.76
17253	9DE	15;57	258.83

FULL MOONS (394)

NUMBER	DATE	TIME	LONG.
17241	4JA	3;44	105.05
17242	2FE	21;03	135.07
17243	4MR	11;26	164.53
17244	2AP	22;58	193.37
17245	2MY	8;09	221.68
17246	31MY	15;42	249.66
17247	29JN	22;30	277.56
17248	29JL	5;35	305.65
17249	27AU	14;06	334.17
17250	26SE	1;05	3.28
17251	25OC	15;08	32.99
17252	24NO	8;06	63.19
17253	24DE	2;55	93.57

NEW MOONS (400)

NUMBER	DATE	TIME	LONG.
17316	12JA	18;58	293.34
17317	11FE	13;56	323.33
17318	12MR	7;30	352.78
17319	10AP	22;38	21.63
17320	10MY	11;12	49.98
17321	8JN	21;33	78.04
17322	8JL	6;37	106.06
17323	6AU	14;50	134.29
17324	4SE	23;04	162.93
17325	4OC	8;05	192.10
17326	2NO	18;38	221.80
17327	2DE	7;17	251.87
17328	31DE	22;04	282.06

FULL MOONS (400)

NUMBER	DATE	TIME	LONG.
17316	27JA	20;34	128.56
17317	26FE	6;08	157.93
17318	26MR	14;50	186.77
17319	24AP	23;27	215.15
17320	24MY	8;59	243.25
17321	22JN	20;19	271.32
17322	22JL	9;54	299.61
17323	21AU	1;36	328.33
17324	19SE	18;48	357.59
17325	19OC	12;35	27.37
17326	18NO	5;55	57.53
17327	17DE	21;44	87.79

395 / 401

NEW MOONS (395)

NUMBER	DATE	TIME	LONG.
17254	8JA	2;39	288.82
17255	6FE	12;32	318.49
17256	7MR	22;09	347.69
17257	6AP	8;04	16.41
17258	5MY	18;46	44.74
17259	4JN	6;39	72.88
17260	3JL	20;02	101.05
17261	2AU	11;07	129.50
17262	1SE	3;39	158.41
17263	30SE	20;51	187.85
17264	30OC	13;30	217.74
17265	29NO	4;31	247.89
17266	28DE	17;27	278.01

FULL MOONS (395)

NUMBER	DATE	TIME	LONG.
17254	22JA	22;04	123.80
17255	21FE	16;04	153.59
17256	23MR	7;50	182.79
17257	21AP	20;49	211.40
17258	21MY	7;05	239.56
17259	19JN	15;19	267.51
17260	18JL	22;35	295.52
17261	17AU	6;04	323.84
17262	15SE	14;45	352.67
17263	15OC	1;19	22.08
17264	13NO	14;06	51.98
17265	13DE	5;03	82.20

NEW MOONS (401)

NUMBER	DATE	TIME	LONG.
17329	30JA	14;25	312.09
17330	1MR	7;24	341.70
17331	31MR	0;04	10.80
17332	29AP	15;45	39.40
17333	29MY	5;57	67.66
17334	27JN	18;24	95.78
17335	27JL	5;13	124.01
17336	25AU	14;54	152.54
17337	24SE	0;13	181.55
17338	23OC	9;58	211.06
17339	21NO	20;37	240.96
17340	21DE	8;20	271.05

FULL MOONS (401)

NUMBER	DATE	TIME	LONG.
17328	16JA	11;17	117.84
17329	14FE	22;19	147.43
17330	16MR	7;17	176.46
17331	14AP	15;00	204.95
17332	13MY	22;28	233.06
17333	12JN	6;38	261.02
17334	11JL	16;22	289.08
17335	10AU	4;19	317.50
17336	8SE	18;55	346.46
17337	8OC	12;11	16.04
17338	7NO	7;20	46.15
17339	7DE	2;43	76.53

396 / 402

NEW MOONS (396)

NUMBER	DATE	TIME	LONG.
17267	27JA	4;32	307.86
17268	25FE	14;13	337.25
17269	25MR	22;57	6.09
17270	24AP	7;15	34.47
17271	23MY	15;48	62.54
17272	22JN	1;30	90.54
17273	21JL	13;19	118.76
17274	20AU	3;54	147.42
17275	18SE	21;06	176.66
17276	18OC	15;47	206.47
17277	17NO	10;15	236.67
17278	17DE	3;06	266.97

FULL MOONS (396)

NUMBER	DATE	TIME	LONG.
17266	11JA	21;52	112.42
17267	10FE	15;50	142.38
17268	11MR	9;46	171.86
17269	10AP	2;17	200.78
17270	9MY	16;24	229.20
17271	8JN	4;00	257.31
17272	7JL	13;42	285.35
17273	5AU	22;29	313.60
17274	4SE	7;14	342.25
17275	3OC	16;36	11.43
17276	2NO	2;54	41.10
17277	1DE	14;24	71.12
17278	31DE	3;25	101.24

NEW MOONS (402)

NUMBER	DATE	TIME	LONG.
17341	19JA	21;01	301.03
17342	18FE	10;36	330.69
17343	20MR	1;03	359.88
17344	18AP	16;17	28.62
17345	18MY	7;53	57.02
17346	16JN	23;06	85.25
17347	16JL	13;16	113.52
17348	15AU	2;08	142.04
17349	13SE	13;59	170.96
17350	13OC	1;18	200.37
17351	11NO	12;24	230.18
17352	10DE	23;21	260.22

FULL MOONS (402)

NUMBER	DATE	TIME	LONG.
17340	5JA	20;22	106.82
17341	4FE	11;03	136.72
17342	5MR	22;39	166.04
17343	4AP	7;50	194.76
17344	3MY	15;31	222.99
17345	1JN	22;33	250.95
17346	1JL	5;48	278.87
17347	30JL	14;11	307.03
17348	29AU	0;40	335.64
17349	27SE	14;07	4.87
17350	27OC	6;53	34.73
17351	26NO	2;13	65.03
17352	25DE	22;12	95.46

397 / 403

NEW MOONS (397)

NUMBER	DATE	TIME	LONG.
17279	15JA	17;39	297.07
17280	14FE	5;47	326.71
17281	15MR	15;42	355.79
17282	13AP	23;52	24.31
17283	13MY	6;59	52.41
17284	11JN	14;06	80.33
17285	10JL	22;55	108.34
17286	9AU	9;03	136.69
17287	7SE	22;42	165.60
17288	7OC	15;19	195.14
17289	6NO	10;03	225.22
17290	6DE	5;28	255.59

FULL MOONS (397)

NUMBER	DATE	TIME	LONG.
17279	29JA	18;07	131.21
17280	28FE	10;12	160.80
17281	30MR	2;48	189.91
17282	28AP	18;51	218.54
17283	28MY	9;35	246.82
17284	26JN	22;51	274.97
17285	26JL	10;51	303.24
17286	24AU	21;58	331.82
17287	23SE	8;29	0.87
17288	22OC	18;43	30.39
17289	21NO	5;01	60.27
17290	20DE	15;47	90.32

NEW MOONS (403)

NUMBER	DATE	TIME	LONG.
17353	9JA	10;07	290.20
17354	7FE	20;51	319.89
17355	9MR	8;00	349.14
17356	7AP	20;05	17.93
17357	7MY	9;25	46.35
17358	5JN	23;52	74.58
17359	5JL	15;00	102.83
17360	4AU	6;19	131.30
17361	2SE	21;28	160.18
17362	2OC	12;09	189.54
17363	1NO	1;59	219.34
17364	30NO	14;36	249.39
17365	30DE	1;51	279.45

FULL MOONS (403)

NUMBER	DATE	TIME	LONG.
17353	24JA	16;44	125.65
17354	23FE	8;31	155.33
17355	24MR	21;21	184.38
17356	23AP	7;34	212.85
17357	22MY	15;49	240.92
17358	20JN	22;53	268.83
17359	20JL	5;46	296.83
17360	18AU	13;32	325.18
17361	16SE	23;16	354.08
17362	16OC	11;50	23.58
17363	15NO	3;27	53.63
17364	14DE	21;27	83.97

404

NUMBER	DATE	TIME	LONG.	NUMBER	DATE	TIME	LONG.
				17365	13JA	16;31	114.28
17366	28JA	11;59	309.24	17366	12FE	11;06	144.24
17367	26FE	21;28	338.59	17367	13MR	3;57	173.65
17368	27MR	6;55	7.45	17368	11AP	18;16	202.45
17369	25AP	16;55	35.87	17369	11MY	5;50	230.75
17370	25MY	3;57	64.03	17370	9JN	15;02	258.76
17371	23JN	16;27	92.15	17371	8JL	22;45	286.73
17372	23JL	6;41	120.48	17372	7AU	6;08	314.93
17373	21AU	22;42	149.22	17373	5SE	14;16	343.58
17374	20SE	15;58	178.49	17374	4OC	23;59	12.80
17375	20OC	9;20	208.27	17375	3NO	11;43	42.55
17376	19NO	1;29	238.39	17376	3DE	1;35	72.67
17377	18DE	15;35	268.57				

405

NUMBER	DATE	TIME	LONG.	NUMBER	DATE	TIME	LONG.
				17377	1JA	17;23	102.91
17378	17JA	3;32	298.55	17378	31JA	10;40	132.96
17379	15FE	13;44	328.10	17379	2MR	4;33	162.60
17380	16MR	22;43	357.12	17380	31MR	21;45	191.71
17381	15AP	6;58	25.62	17381	30AP	13;03	220.28
17382	14MY	15;06	53.76	17382	30MY	1;54	248.48
17383	12JN	23;59	81.74	17383	28JN	12;35	276.54
17384	12JL	10;34	109.84	17384	27JL	21;54	304.71
17385	10AU	23;45	138.32	17385	26AU	6;47	333.22
17386	9SE	15;51	167.35	17386	24SE	16;00	2.23
17387	9OC	10;11	196.99	17387	24OC	1;57	31.76
17388	8NO	5;10	227.10	17388	22NO	12;53	61.68
17389	7DE	23;07	257.42	17389	22DE	1;02	91.79

406

NUMBER	DATE	TIME	LONG.	NUMBER	DATE	TIME	LONG.
17390	6JA	14;54	287.64	17390	20JA	14;42	121.81
17391	5FE	4;11	317.47	17391	19FE	5;25	151.51
17392	6MR	15;04	346.74	17392	20MR	22;03	180.77
17393	4AP	23;54	15.44	17393	19AP	14;16	209.54
17394	4MY	7;18	43.66	17394	19MY	5;38	237.92
17395	2JN	14;11	71.60	17395	17JN	19;41	266.10
17396	1JL	21;42	99.53	17396	17JL	8;28	294.32
17397	31JL	7;04	127.73	17397	15AU	20;14	322.80
17398	29AU	19;13	156.43	17398	14SE	7;18	351.70
17399	28SE	10;31	185.75	17399	13OC	17;52	21.09
17400	28OC	4;27	215.67	17400	12NO	4;14	50.88
17401	26NO	23;48	245.98	17401	11DE	14;46	80.90
17402	26DE	19;01	276.38				

407

NUMBER	DATE	TIME	LONG.	NUMBER	DATE	TIME	LONG.
				17402	10JA	1;49	110.90
17403	25JA	12;34	306.52	17403	8FE	13;42	140.62
17404	24FE	3;25	336.14	17404	10MR	2;29	169.93
17405	25MR	15;09	5.13	17405	8AP	16;03	198.76
17406	24AP	0;09	33.55	17406	8MY	6;15	227.21
17407	23MY	7;22	61.57	17407	6JN	20;59	255.45
17408	21JN	13;58	89.46	17408	6JL	12;03	283.70
17409	20JL	21;09	117.47	17409	5AU	3;04	312.17
17410	19AU	5;55	145.88	17410	3SE	17;29	341.03
17411	17SE	17;00	174.84	17411	3OC	6;51	10.35
17412	17OC	6;50	204.41	17412	1NO	19;03	40.09
17413	15NO	23;25	234.50	17413	1DE	6;23	70.09
17414	15DE	18;14	264.88	17414	30DE	17;10	100.12

408

NUMBER	DATE	TIME	LONG.	NUMBER	DATE	TIME	LONG.
17415	14JA	13;55	295.21	17415	29JA	3;39	129.93
17416	13FE	8;33	325.16	17416	27FE	13;55	159.30
17417	14MR	0;28	354.52	17417	28MR	0;14	188.18
17418	12AP	13;06	23.24	17418	26AP	11;03	216.63
17419	11MY	22;56	51.46	17419	25MY	23;03	244.82
17420	10JN	6;58	79.42	17420	24JN	12;46	272.99
17421	9JL	14;18	107.38	17421	24JL	4;12	301.37
17422	7AU	21;51	135.60	17422	22AU	20;01	330.14
17423	6SE	6;22	164.28	17423	21SE	13;12	359.40
17424	5OC	16;32	193.52	17424	21OC	4;51	29.12
17425	4NO	4;57	223.30	17425	19NO	19;13	59.17
17426	3DE	20;02	253.49	17426	19DE	8;11	89.31

409

NUMBER	DATE	TIME	LONG.	NUMBER	DATE	TIME	LONG.
17427	2JA	13;35	283.79	17427	17JA	19;40	119.26
17428	1FE	8;23	313.90	17428	16FE	5;42	148.79
17429	3MR	2;38	343.54	17429	17MR	14;34	177.79
17430	1AP	18;54	12.59	17430	15AP	22;53	206.29
17431	1MY	8;40	41.10	17431	15MY	7;38	234.44
17432	30MY	20;09	69.24	17432	13JN	17;50	262.48
17433	29JN	5;54	97.26	17433	13JL	6;13	290.66
17434	28JL	14;34	125.41	17434	11AU	20;58	319.21
17435	26AU	22;54	153.91	17435	10SE	13;41	348.28
17436	25SE	7;37	182.91	17436	10OC	7;32	17.91
17437	24OC	17;27	212.44	17437	9NO	1;27	47.99
17438	23NO	5;03	242.40	17438	8DE	18;19	78.27
17439	22DE	18;42	272.56				

410

NUMBER	DATE	TIME	LONG.	NUMBER	DATE	TIME	LONG.
				17439	7JA	9;06	108.44
17440	21JA	10;10	302.66	17440	5FE	21;19	138.21
17441	20FE	2;41	332.41	17441	7MR	7;07	167.43
17442	21MR	19;20	1.66	17442	5AP	15;11	196.09
17443	20AP	11;24	30.41	17443	4MY	22;29	224.29
17444	20MY	2;20	58.77	17444	3JN	6;04	252.26
17445	18JN	15;46	86.93	17445	2JL	14;51	280.25
17446	18JL	3;34	115.11	17446	1AU	1;38	308.51
17447	16AU	13;59	143.55	17447	30AU	14;57	337.27
17448	14SE	23;38	172.41	17448	29SE	7;02	6.63
17449	14OC	9;17	201.77	17449	29OC	1;32	36.58
17450	12NO	19;39	231.56	17450	27NO	21;11	66.92
17451	12DE	6;46	261.61	17451	27DE	15;59	97.30

411

NUMBER	DATE	TIME	LONG.	NUMBER	DATE	TIME	LONG.
17452	10JA	18;51	291.64	17452	26JA	8;13	127.37
17453	9FE	7;45	321.41	17453	24FE	21;15	156.90
17454	10MR	21;27	350.73	17454	26MR	7;26	185.82
17455	9AP	12;01	19.60	17455	24AP	15;39	214.20
17456	9MY	3;19	48.08	17456	23MY	22;49	242.21
17457	7JN	18;46	76.34	17457	22JN	5;50	270.12
17458	7JL	9;38	104.59	17458	21JL	13;33	298.16
17459	5AU	23;23	133.02	17459	19AU	22;55	326.60
17460	4SE	12;00	161.82	17460	18SE	10;53	355.61
17461	3OC	23;49	191.09	17461	18OC	2;05	25.25
17462	2NO	11;13	220.80	17462	16NO	20;22	55.42
17463	1DE	22;21	250.80	17463	16DE	16;19	85.85
17464	31DE	9;10	280.83				

412

NUMBER	DATE	TIME	LONG.	NUMBER	DATE	TIME	LONG.
				17464	15JA	11;45	116.16
17465	29JA	19;44	310.63	17465	14FE	4;55	146.04
17466	28FE	6;21	340.01	17466	14MR	19;07	175.31
17467	28MR	17;37	8.92	17467	13AP	6;30	203.97
17468	27AP	6;02	37.42	17468	12MY	15;36	232.16
17469	26MY	19;47	65.68	17469	10JN	23;10	260.10
17470	25JN	10;35	93.89	17470	10JL	6;05	288.04
17471	25JL	1;58	122.27	17471	8AU	13;22	316.26
17472	23AU	17;29	151.02	17472	6SE	22;06	344.95
17473	22SE	8;45	180.24	17473	6OC	9;17	14.25
17474	21OC	23;21	209.93	17474	4NO	23;27	44.12
17475	20NO	12;48	239.95	17475	4DE	16;22	74.38
17476	20DE	0;48	270.04				

413

NUMBER	DATE	TIME	LONG.	NUMBER	DATE	TIME	LONG.
				17476	3JA	10;58	104.73
17477	18JA	11;24	299.95	17477	2FE	5;48	134.83
17478	16FE	20;58	329.46	17478	3MR	23;29	164.43
17479	18MR	6;08	358.46	17479	2AP	15;02	193.44
17480	16AP	15;32	26.99	17480	2MY	3;56	221.90
17481	16MY	1;47	55.20	17481	31MY	14;17	249.99
17482	14JN	13;21	83.29	17482	29JN	22;42	277.96
17483	14JL	2;40	111.50	17483	29JL	6;16	306.07
17484	12AU	17;54	140.09	17484	27AU	14;04	334.56
17485	11SE	10;51	169.18	17485	25SE	23;04	3.58
17486	11OC	4;37	198.83	17486	25OC	9;52	33.16
17487	9NO	21;46	228.88	17487	23NO	22;42	63.18
17488	9DE	13;07	259.10	17488	23DE	13;28	93.39

414

NUMBER	DATE	TIME	LONG.	NUMBER	DATE	TIME	LONG.
17489	8JA	2;08	289.18	17489	22JA	5;53	123.51
17490	6FE	13;05	318.90	17490	20FE	23;21	153.29
17491	7MR	22;27	348.09	17491	22MR	16;49	182.57
17492	6AP	6;48	16.75	17492	21AP	9;03	211.31
17493	5MY	14;44	44.97	17493	20MY	23;09	239.63
17494	3JN	23;00	72.96	17494	19JN	10;57	267.72
17495	3JL	8;31	100.98	17495	18JL	21;00	295.84
17496	1AU	20;19	129.30	17496	17AU	6;13	324.24
17497	31AU	11;03	158.12	17497	15SE	15;26	353.09
17498	30SE	4;34	187.56	17498	15OC	1;10	22.46
17499	29OC	23;40	217.55	17499	13NO	11;40	52.27
17500	28NO	18;29	247.85	17500	12DE	23;10	82.34
17501	28DE	11;30	278.15				

415

NUMBER	DATE	TIME	LONG.	NUMBER	DATE	TIME	LONG.
				17501	11JA	11;54	112.39
17502	27JA	2;02	308.15	17502	10FE	2;05	142.20
17503	25FE	14;00	337.63	17503	11MR	17;33	171.59
17504	26MR	23;41	6.51	17504	10AP	9;36	200.50
17505	25AP	7;35	34.87	17505	10MY	1;20	228.99
17506	24MY	14;29	62.87	17506	8JN	16;04	257.22
17507	22JN	21;29	90.77	17507	8JL	5;36	285.41
17508	22JL	5;47	118.84	17508	6AU	18;05	313.81
17509	20AU	16;30	147.34	17509	5SE	5;45	342.58
17510	19SE	6;19	176.44	17510	4OC	16;48	11.83
17511	18OC	23;09	206.16	17511	3NO	3;26	41.52
17512	17NO	18;04	236.37	17512	2DE	13;54	71.49
17513	17DE	13;35	266.78				

416

NEW MOONS

NUMBER	DATE	TIME	LONG.
17514	16JA	8;07	297.05
17515	15FE	0;21	326.88
17516	15MR	13;33	356.09
17517	13AP	23;45	24.69
17518	13MY	7;41	52.83
17519	11JN	14;25	80.73
17520	10JL	21;13	108.68
17521	9AU	5;09	136.92
17522	7SE	15;06	165.68
17523	7OC	3;37	195.04
17524	5NO	18;54	224.97
17525	5DE	12;42	255.27

FULL MOONS

NUMBER	DATE	TIME	LONG.
17513	1JA	0;36	101.51
17514	30JA	11;52	131.33
17515	28FE	23;56	160.76
17516	29MR	12;49	189.73
17517	28AP	2;27	218.27
17518	27MY	16;46	246.54
17519	26JN	7;41	274.76
17520	25JL	22;57	303.15
17521	24AU	14;01	331.89
17522	23SE	4;15	1.08
17523	22OC	17;18	30.72
17524	21NO	5;12	60.67
17525	20DE	16;16	90.73

422

NEW MOONS

NUMBER	DATE	TIME	LONG.
17588	9JA	10;39	290.62
17589	7FE	20;32	320.27
17590	9MR	5;38	349.43
17591	7AP	14;36	18.09
17592	7MY	0;07	46.36
17593	5JN	10;48	74.45
17594	4JL	23;07	102.58
17595	3AU	13;25	131.01
17596	2SE	5;45	159.93
17597	1OC	23;30	189.41
17598	31OC	17;24	219.37
17599	30NO	9;58	249.59
17600	30DE	0;14	279.77

FULL MOONS

NUMBER	DATE	TIME	LONG.
17588	24JA	0;22	125.36
17589	22FE	18;35	155.14
17590	24MR	11;09	184.35
17591	23AP	1;20	212.99
17592	22MY	12;56	241.19
17593	20JN	22;18	269.19
17594	20JL	6;18	297.24
17595	18AU	14;01	325.59
17596	16SE	22;30	354.43
17597	16OC	8;30	23.83
17598	14NO	20;21	53.72
17599	14DE	10;07	83.88

417

NEW MOONS

NUMBER	DATE	TIME	LONG.
17526	4JA	8;05	285.65
17527	3FE	3;21	315.76
17528	4MR	20;39	345.33
17529	3AP	10;51	14.27
17530	2MY	21;57	42.65
17531	1JN	6;46	70.68
17532	30JN	14;24	98.61
17533	29JL	21;51	126.73
17534	28AU	5;57	155.24
17535	26SE	15;22	184.29
17536	26OC	2;40	213.90
17537	24NO	16;24	243.96
17538	24DE	8;42	274.24

FULL MOONS

NUMBER	DATE	TIME	LONG.
17526	19JA	2;48	120.63
17527	17FE	12;58	150.15
17528	18MR	22;58	179.18
17529	17AP	9;11	207.73
17530	16MY	20;17	235.96
17531	15JN	8;57	264.09
17532	14JL	23;34	292.37
17533	13AU	15;47	321.00
17534	12SE	8;40	350.11
17535	12OC	1;08	19.71
17536	10NO	16;25	49.70
17537	10DE	6;14	79.85

423

NEW MOONS

NUMBER	DATE	TIME	LONG.
17601	28JA	12;07	309.64
17602	26FE	22;05	339.01
17603	28MR	6;42	7.83
17604	26AP	14;35	36.17
17605	25MY	22;26	64.20
17606	24JN	7;06	92.17
17607	23JL	17;38	120.34
17608	22AU	6;53	148.97
17609	20SE	23;13	178.19
17610	20OC	17;55	208.02
17611	19NO	13;16	238.26
17612	19DE	7;26	268.62

FULL MOONS

NUMBER	DATE	TIME	LONG.
17600	13JA	1;35	114.04
17601	11FE	18;22	143.94
17602	13MR	11;44	173.37
17603	12AP	4;33	202.29
17604	11MY	19;43	230.74
17605	10JN	8;42	258.89
17606	9JL	19;42	287.00
17607	8AU	5;27	315.29
17608	6SE	14;49	344.00
17609	6OC	0;27	13.21
17610	4NO	10;42	42.90
17611	3DE	21;42	72.90

418

NEW MOONS

NUMBER	DATE	TIME	LONG.
17539	23JA	2;53	304.42
17540	21FE	21;24	334.23
17541	23MR	14;36	3.48
17542	22AP	5;33	32.17
17543	21MY	18;07	60.42
17544	20JN	4;45	88.46
17545	19JL	14;03	116.56
17546	17AU	22;40	144.94
17547	16SE	7;18	173.78
17548	15OC	16;38	203.14
17549	14NO	3;22	232.97
17550	13DE	15;57	263.08

FULL MOONS

NUMBER	DATE	TIME	LONG.
17538	8JA	18;28	109.90
17539	7FE	5;07	139.60
17540	8MR	14;20	168.77
17541	6AP	22;37	197.42
17542	6MY	6;49	225.64
17543	4JN	16;01	253.67
17544	4JL	3;09	281.75
17545	2AU	16;45	310.15
17546	1SE	8;41	339.04
17547	1OC	2;17	8.49
17548	30OC	20;32	38.46
17549	29NO	14;15	68.72
17550	29DE	6;16	98.98

424

NEW MOONS

NUMBER	DATE	TIME	LONG.
17613	17JA	23;16	298.76
17614	16FE	12;26	328.43
17615	16MR	23;06	357.52
17616	15AP	7;41	26.04
17617	14MY	14;52	54.13
17618	12JN	21;38	82.03
17619	12JL	5;07	110.01
17620	10AU	14;32	138.34
17621	9SE	2;50	167.21
17622	8OC	18;19	196.72
17623	7NO	12;26	226.78
17624	7DE	7;54	257.17

FULL MOONS

NUMBER	DATE	TIME	LONG.
17612	2JA	9;40	102.97
17613	31JA	22;54	132.86
17614	1MR	13;28	162.37
17615	31MR	5;03	191.42
17616	29AP	20;50	220.02
17617	29MY	12;03	248.32
17618	28JN	2;16	276.51
17619	27JL	15;28	304.84
17620	26AU	3;48	333.49
17621	24SE	15;25	2.60
17622	24OC	2;28	32.17
17623	22NO	13;06	62.08
17624	21DE	23;38	92.12

419

NEW MOONS

NUMBER	DATE	TIME	LONG.
17551	12JA	6;24	293.20
17552	10FE	22;15	323.07
17553	12MR	14;39	352.48
17554	11AP	6;51	21.38
17555	10MY	22;18	49.85
17556	9JN	12;33	78.06
17557	9JL	1;20	106.23
17558	7AU	12;37	134.59
17559	5SE	22;50	163.31
17560	5OC	8;38	192.52
17561	3NO	18;42	222.19
17562	3DE	5;29	252.18

FULL MOONS

NUMBER	DATE	TIME	LONG.
17551	27JA	19;47	128.92
17552	26FE	6;37	158.34
17553	27MR	15;17	187.18
17554	25AP	22;42	215.50
17555	25MY	5;54	243.51
17556	23JN	13;54	271.45
17557	22JL	23;35	299.59
17558	21AU	11;37	328.15
17559	20SE	2;25	357.30
17560	19OC	19;56	27.06
17561	18NO	15;22	57.30
17562	18DE	10;55	87.72

425

NEW MOONS

NUMBER	DATE	TIME	LONG.
17625	6JA	3;06	287.53
17626	4FE	20;34	317.54
17627	6MR	11;14	346.97
17628	4AP	22;49	15.78
17629	4MY	7;42	44.06
17630	2JN	14;52	72.01
17631	1JL	21;31	99.91
17632	31JL	4;49	128.03
17633	29AU	13;47	156.60
17634	28SE	1;04	185.75
17635	27OC	15;02	215.49
17636	26NO	7;38	245.68
17637	26DE	2;19	276.06

FULL MOONS

NUMBER	DATE	TIME	LONG.
17625	20JA	10;27	122.02
17626	18FE	21;51	151.57
17627	20MR	10;02	180.66
17628	18AP	23;01	209.31
17629	18MY	12;47	237.64
17630	17JN	3;21	265.84
17631	16JL	18;35	294.15
17632	15AU	10;03	322.78
17633	14SE	1;05	351.84
17634	13OC	15;03	21.36
17635	12NO	3;42	51.25
17636	11DE	15;14	81.31

420

NEW MOONS

NUMBER	DATE	TIME	LONG.
17563	1JA	17;04	282.24
17564	31JA	5;22	312.10
17565	29FE	18;22	341.55
17566	30MR	8;11	10.54
17567	28AP	22;54	39.12
17568	28MY	14;15	67.42
17569	27JN	5;33	95.66
17570	26JL	20;05	124.03
17571	25AU	9;31	152.71
17572	23SE	21;59	181.85
17573	23OC	9;49	211.44
17574	21NO	21;14	241.38
17575	21DE	8;14	271.44

FULL MOONS

NUMBER	DATE	TIME	LONG.
17563	17JA	4;36	117.94
17564	15FE	19;11	147.69
17565	16MR	6;36	176.82
17566	14AP	15;35	205.36
17567	13MY	23;05	233.47
17568	12JN	6;02	261.38
17569	11JL	13;18	289.35
17570	9AU	21;48	317.63
17571	8SE	8;26	346.42
17572	7OC	22;03	15.84
17573	6NO	14;57	45.85
17574	6DE	10;19	76.22

426

NEW MOONS

NUMBER	DATE	TIME	LONG.
17638	24JA	21;46	306.28
17639	23FE	16;07	336.06
17640	25MR	7;49	5.22
17641	23AP	20;21	33.78
17642	23MY	6;12	61.91
17643	21JN	14;21	89.86
17644	20JL	21;54	117.89
17645	19AU	5;45	146.26
17646	17SE	14;35	175.13
17647	17OC	0;59	204.55
17648	15NO	13;29	234.47
17649	15DE	4;27	264.69

FULL MOONS

NUMBER	DATE	TIME	LONG.
17637	10JA	1;59	111.29
17638	8FE	12;10	140.95
17639	9MR	22;00	170.13
17640	8AP	7;46	198.81
17641	7MY	18;05	227.11
17642	6JN	5;43	255.23
17643	5JL	19;18	283.42
17644	4AU	10;52	311.91
17645	3SE	3;47	340.86
17646	2OC	20;52	10.33
17647	1NO	13;04	40.23
17648	1DE	3;49	70.37
17649	30DE	16;55	100.50

421

NEW MOONS

NUMBER	DATE	TIME	LONG.
17576	19JA	18;48	301.33
17577	18FE	5;08	330.85
17578	19MR	15;43	359.89
17579	18AP	3;13	28.49
17580	17MY	16;05	56.78
17581	16JN	6;17	84.97
17582	15JL	21;29	113.28
17583	14AU	13;11	141.90
17584	13SE	4;54	170.98
17585	12OC	20;12	200.54
17586	11NO	10;31	230.49
17587	10DE	23;24	260.61

FULL MOONS

NUMBER	DATE	TIME	LONG.
17575	5JA	6;14	106.61
17576	4FE	0;36	136.67
17577	5MR	16;12	166.16
17578	4AP	4;52	195.03
17579	3MY	14;59	223.36
17580	1JN	23;14	251.36
17581	1JL	6;24	279.28
17582	30JL	13;28	307.38
17583	28AU	21;29	335.90
17584	27SE	7;27	4.99
17585	26OC	20;10	34.67
17586	25NO	11;47	64.81
17587	25DE	5;37	95.15

427

NEW MOONS

NUMBER	DATE	TIME	LONG.
17650	13JA	21;40	294.92
17651	12FE	16;00	324.86
17652	14MR	9;49	354.30
17653	13AP	1;49	23.17
17654	12MY	15;31	51.55
17655	11JN	3;07	79.65
17656	10JL	13;10	107.72
17657	8AU	22;14	136.00
17658	7SE	7;00	164.69
17659	6OC	16;05	193.89
17660	5NO	2;10	223.58
17661	4DE	13;47	253.61

FULL MOONS

NUMBER	DATE	TIME	LONG.
17650	29JA	4;18	130.34
17651	27FE	14;03	159.70
17652	28MR	22;31	188.50
17653	27AP	6;26	216.83
17654	26MY	14;49	244.88
17655	25JN	0;47	272.90
17656	24JL	13;05	301.15
17657	23AU	3;59	329.85
17658	21SE	21;01	359.12
17659	21OC	15;19	28.95
17660	20NO	9;39	59.15
17661	20DE	2;45	89.47

NUMBER	DATE	TIME	LONG.	NUMBER	DATE	TIME	LONG.	NUMBER	DATE	TIME	LONG.	NUMBER	DATE	TIME	LONG.
			428								434				
17662	3JA	3;13	283.74	17662	18JA	17;36	119.56					17736	11JA	9;21	112.68
17663	1FE	18;13	313.70	17663	17FE	5;40	149.18	17737	26JA	16;07	308.12	17737	9FE	20;14	142.34
17664	2MR	10;08	343.25	17664	17MR	15;12	178.21	17738	25FE	8;11	337.77	17738	11MR	7;44	171.57
17665	1AP	2;15	12.31	17665	15AP	22;59	206.68	17739	26MR	21;13	6.80	17739	9AP	20;01	200.34
17666	30AP	17;59	40.90	17666	15MY	6;01	234.77	17740	25AP	7;17	35.24	17740	9MY	9;09	228.73
17667	30MY	8;51	69.18	17667	13JN	13;25	262.69	17741	24MY	15;09	63.28	17741	7JN	23;11	256.93
17668	28JN	22;30	97.35	17668	12JL	22;08	290.72	17742	22JN	21;56	91.17	17742	7JL	14;07	285.18
17669	28JL	10;44	125.64	17669	11AU	8;57	319.11	17743	22JL	4;51	119.19	17743	6AU	5;43	313.69
17670	26AU	21;42	154.25	17670	9SE	22;25	348.04	17744	20AU	12;59	147.59	17744	4SE	21;20	342.62
17671	25SE	7;52	183.31	17671	9OC	14;44	17.60	17745	18SE	23;09	176.53	17745	4OC	12;13	12.03
17672	24OC	17;56	212.85	17672	8NO	9;29	47.70	17746	18OC	11;50	206.08	17746	3NO	1;47	41.84
17673	23NO	4;26	242.76	17673	8DE	5;19	78.11	17747	17NO	3;10	236.13	17747	2DE	14;00	71.88
17674	22DE	15;36	272.82					17748	16DE	20;52	266.46				
			429								435				
				17674	7JA	0;11	108.45					17748	1JA	1;07	101.93
17675	21JA	3;23	302.76	17675	5FE	16;21	138.39	17749	15JA	16;01	296.76	17749	30JA	11;27	131.71
17676	19FE	15;46	332.34	17676	7MR	5;13	167.74	17750	14FE	10;59	326.71	17750	28FE	21;13	161.04
17677	21MR	4;51	1.46	17677	5AP	15;12	196.47	17751	16MR	4;01	356.09	17751	30MR	6;43	189.87
17678	19AP	18;50	30.14	17678	4MY	23;15	224.71	17752	14AP	18;03	24.85	17752	28AP	16;26	218.25
17679	19MY	9;45	58.50	17679	3JN	6;19	252.66	17753	14MY	5;08	53.12	17753	28MY	3;07	246.38
17680	18JN	1;08	86.73	17680	2JL	13;20	280.57	17754	12JN	14;04	81.10	17754	26JN	15;33	274.50
17681	17JL	16;16	115.05	17681	31JL	21;09	308.72	17755	11JL	21;54	109.09	17755	26JL	6;11	302.86
17682	16AU	6;32	143.64	17682	30AU	6;41	337.32	17756	10AU	5;39	137.33	17756	24AU	22;42	331.65
17683	14SE	19;45	172.64	17683	28SE	18;49	6.52	17757	8SE	14;05	166.03	17757	23SE	16;06	0.97
17684	14OC	8;08	202.11	17684	28OC	10;10	36.33	17758	7OC	23;46	195.27	17758	23OC	9;08	30.76
17685	12NO	19;57	231.97	17685	27NO	4;30	66.60	17759	6NO	11;14	225.03	17759	22NO	0;53	60.88
17686	12DE	7;14	262.02	17686	27DE	0;22	97.02	17760	6DE	0;55	255.16	17760	21DE	14;55	91.06
			430								436				
17687	10JA	17;58	292.00	17687	25JA	19;38	127.23	17761	4JA	16;56	285.40	17761	20JA	3;08	121.03
17688	9FE	4;11	321.65	17688	24FE	12;36	156.93	17762	3FE	10;40	315.45	17762	18FE	13;34	150.56
17689	10MR	14;19	350.83	17689	26MR	2;38	186.01	17763	4MR	4;42	345.06	17763	18MR	22;26	179.54
17690	9AP	1;00	19.54	17690	24AP	13;54	214.51	17764	2AP	21;33	14.12	17764	17AP	6;19	208.00
17691	8MY	12;53	47.89	17691	23MY	22;59	242.61	17765	2MY	12;20	42.65	17765	16MY	14;10	236.11
17692	7JN	2;17	76.07	17692	22JN	6;38	270.54	17766	1JN	0;58	70.83	17766	14JN	23;05	264.09
17693	6JL	17;01	104.31	17693	21JL	13;43	298.56	17767	30JN	11;51	98.90	17767	14JL	10;05	292.22
17694	5AU	8;40	132.82	17694	19AU	21;15	326.92	17768	29JL	21;32	127.10	17768	12AU	23;44	320.74
17695	4SE	0;41	161.75	17695	18SE	6;16	355.80	17769	28AU	6;36	155.65	17769	11SE	15;55	349.81
17696	3OC	16;33	191.18	17696	17OC	17;38	25.28	17770	26SE	15;39	184.69	17770	11OC	9;55	19.47
17697	2NO	7;42	221.04	17697	16NO	7;51	55.28	17771	26OC	1;18	214.23	17771	10NO	4;35	49.59
17698	1DE	21;31	251.15	17698	16DE	0;38	85.57	17772	24NO	12;09	244.17	17772	9DE	22;36	79.92
17699	31DE	9;37	281.24					17773	24DE	0;36	274.28				
			431								437				
				17699	14JA	18;57	115.85					17773	8JA	14;44	110.14
17700	29JA	20;03	311.03	17700	13FE	13;24	145.78	17774	22JA	14;40	304.30	17774	7FE	4;09	139.94
17701	28FE	5;17	340.35	17701	15MR	6;44	175.19	17775	21FE	5;56	333.98	17775	8MR	14;47	169.18
17702	29MR	14;00	9.16	17702	13AP	22;04	204.02	17776	22MR	21;45	3.19	17776	6AP	23;11	197.83
17703	27AP	22;57	37.52	17703	13MY	10;56	232.36	17777	21AP	13;31	31.91	17777	6MY	6;19	226.01
17704	27MY	8;47	65.63	17704	11JN	21;25	260.41	17778	21MY	4;46	60.27	17778	4JN	13;19	253.95
17705	25JN	20;07	93.70	17705	11JL	6;07	288.43	17779	19JN	19;08	88.47	17779	3JL	21;13	281.91
17706	25JL	9;24	122.00	17706	9AU	14;01	316.67	17780	19JL	8;17	116.71	17780	2AU	6;55	310.14
17707	24AU	0;51	150.73	17707	7SE	22;12	345.35	17781	17AU	20;07	145.22	17781	31AU	19;05	338.87
17708	22SE	18;13	180.03	17708	7OC	7;31	14.56	17782	16SE	6;52	174.15	17782	30SE	10;04	8.21
17709	22OC	12;29	209.87	17709	5NO	18;31	44.30	17783	15OC	17;09	203.55	17783	30OC	3;49	38.14
17710	21NO	6;06	240.05	17710	5DE	7;20	74.39	17784	14NO	3;32	233.37	17784	28NO	23;26	68.48
17711	20DE	21;41	270.30					17785	13DE	14;22	263.40	17785	28DE	19;05	98.89
			432								438				
				17711	3JA	21;51	104.56	17786	12JA	1;45	293.40	17786	27JA	12;43	129.01
17712	19JA	10;44	300.31	17712	2FE	13;48	134.55	17787	10FE	13;36	323.10	17787	26FE	3;09	158.59
17713	17FE	21;29	329.87	17713	3MR	6;42	164.13	17788	12MR	2;00	352.35	17788	27MR	14;24	187.52
17714	18MR	6;33	358.86	17714	1AP	23;42	193.20	17789	10AP	15;13	21.14	17789	25AP	23;12	215.91
17715	16AP	14;34	27.34	17715	1MY	15;42	221.80	17790	10MY	5;28	49.57	17790	25MY	6;36	243.93
17716	15MY	22;11	55.44	17716	31MY	5;49	250.04	17791	8JN	20;36	77.82	17791	23JN	13;32	271.83
17717	14JN	6;13	83.39	17717	29JN	17;53	278.15	17792	8JL	12;02	106.09	17792	22JL	20;54	299.87
17718	13JL	15;39	111.45	17718	29JL	4;21	306.38	17793	7AU	3;00	134.58	17793	21AU	5;32	328.29
17719	12AU	3;28	139.89	17719	27AU	14;04	334.95	17794	5SE	17;02	163.46	17794	19SE	16;22	357.27
17720	10SE	18;23	168.90	17720	25SE	23;45	4.00	17795	5OC	6;06	192.80	17795	19OC	6;08	26.87
17721	10OC	12;12	198.53	17721	25OC	9;50	33.55	17796	3NO	18;25	222.57	17796	17NO	23;06	57.00
17722	9NO	7;38	228.67	17722	23NO	20;31	63.48	17797	3DE	6;06	252.60	17797	17DE	18;25	87.41
17723	9DE	2;42	259.04	17723	23DE	7;55	93.54								
			433								439				
17724	7JA	19;50	289.31	17724	21JA	20;18	123.50	17798	1JA	17;06	282.63	17798	16JA	14;10	117.73
17725	6FE	10;18	319.17	17725	20FE	9;56	153.12	17799	31JA	3;25	312.40	17799	15FE	8;19	147.63
17726	7MR	22;05	348.46	17726	22MR	0;46	182.30	17800	1MR	13;17	341.73	17800	16MR	23;43	176.92
17727	6AP	7;31	17.16	17727	20AP	16;18	211.03	17801	30MR	23;20	10.57	17801	15AP	12;14	205.61
17728	5MY	15;12	45.38	17728	20MY	7;45	239.41	17802	29AP	10;17	39.00	17802	14MY	22;19	233.83
17729	3JN	21;59	73.31	17729	18JN	22;32	267.62	17803	28MY	22;43	67.19	17803	13JN	6;38	261.79
17730	3JL	4;56	101.23	17730	18JL	12;23	295.89	17804	27JN	12;45	95.38	17804	12JL	13;58	289.76
17731	1AU	13;16	129.40	17731	17AU	1;23	324.43	17805	27JL	4;04	123.77	17805	10AU	21;18	318.00
17732	31AU	0;07	158.06	17732	15SE	13;39	353.41	17806	25AU	20;10	152.56	17806	9SE	5;35	346.69
17733	29SE	14;06	187.35	17733	15OC	1;14	22.85	17807	24SE	12;28	181.84	17807	8OC	15;47	15.97
17734	29OC	7;06	217.24	17734	13NO	12;13	52.69	17808	24OC	4;21	211.60	17808	7NO	4;37	45.80
17735	28NO	2;08	247.55	17735	12DE	22;49	82.71	17809	22NO	19;07	241.68	17809	6DE	20;10	76.01
17736	27DE	21;40	277.96					17810	22DE	8;11	271.82				

--------NEW MOONS--------				--------FULL MOONS--------			
NUMBER	DATE	TIME	LONG.	NUMBER	DATE	TIME	LONG.

440

NUMBER	DATE	TIME	LONG.	NUMBER	DATE	TIME	LONG.
				17810	5JA	13;45	106.31
17811	20JA	19;22	301.74	17811	4FE	8;05	136.37
17812	19FE	4;59	331.23	17812	5MR	1;55	165.96
17813	19MR	13;40	0.19	17813	3AP	18;11	194.98
17814	17AP	22;11	28.67	17814	3MY	8;16	223.48
17815	17MY	7;18	56.82	17815	1JN	19;58	251.62
17816	15JN	17;41	84.86	17816	1JL	5;34	279.63
17817	15JL	5;53	113.04	17817	30JL	13;55	307.79
17818	13AU	20;18	141.60	17818	28AU	22;02	336.31
17819	12SE	12;57	170.71	17819	27SE	6;52	5.35
17820	12OC	7;11	200.39	17820	26OC	17;07	34.93
17821	11NO	1;34	230.51	17821	25NO	5;04	64.92
17822	10DE	18;26	260.79	17822	24DE	18;39	95.08

441

NUMBER	DATE	TIME	LONG.	NUMBER	DATE	TIME	LONG.
17823	9JA	8;48	290.93	17823	23JA	9;43	125.13
17824	7FE	20;35	320.66	17824	22FE	1;57	154.84
17825	9MR	6;17	349.84	17825	23MR	18;45	184.07
17826	7AP	14;35	18.48	17826	22AP	11;13	212.81
17827	6MY	22;09	46.67	17827	22MY	2;18	241.16
17828	5JN	5;46	74.63	17828	20JN	15;28	269.30
17829	4JL	14;18	102.62	17829	20JL	2;51	297.49
17830	3AU	0;49	130.89	17830	18AU	13;05	325.94
17831	1SE	14;12	159.68	17831	16SE	22;57	354.84
17832	1OC	6;47	189.10	17832	16OC	9;01	24.25
17833	31OC	1;47	219.10	17833	14NO	19;31	54.07
17834	29NO	21;24	249.44	17834	14DE	6;32	84.12
17835	29DE	15;42	279.80				

442

NUMBER	DATE	TIME	LONG.	NUMBER	DATE	TIME	LONG.
				17835	12JA	18;15	114.12
17836	28JA	7;31	309.83	17836	11FE	6;59	143.85
17837	26FE	20;32	339.33	17837	12MR	20;57	173.16
17838	28MR	6;59	8.23	17838	11AP	11;55	202.00
17839	26AP	15;22	36.59	17839	11MY	3;19	230.47
17840	25MY	22;24	64.58	17840	9JN	18;26	258.71
17841	24JN	5;06	92.47	17841	9JL	8;53	286.95
17842	23JL	12;37	120.53	17842	7AU	22;32	315.40
17843	21AU	22;09	149.00	17843	6SE	11;28	344.25
17844	20SE	10;37	178.06	17844	5OC	23;40	13.56
17845	20OC	2;16	207.75	17845	4NO	11;09	43.31
17846	18NO	20;30	237.94	17846	3DE	22;00	73.30
17847	18DE	15;59	268.35				

443

NUMBER	DATE	TIME	LONG.	NUMBER	DATE	TIME	LONG.
				17847	2JA	8;28	103.31
17848	17JA	11;06	298.64	17848	31JA	18;59	133.08
17849	16FE	4;24	328.49	17849	2MR	5;54	162.45
17850	17MR	18;54	357.74	17850	31MR	17;29	191.34
17851	16AP	6;20	26.37	17851	30AP	5;54	219.81
17852	15MY	15;09	54.53	17852	29MY	19;18	248.04
17853	13JN	22;20	82.45	17853	28JN	9;45	276.24
17854	13JL	5;06	110.40	17854	28JL	1;11	304.65
17855	11AU	12;37	138.65	17855	26AU	17;08	333.44
17856	9SE	21;48	167.39	17856	25SE	8;47	2.71
17857	9OC	9;17	196.73	17857	24OC	23;20	32.43
17858	7NO	23;21	226.62	17858	23NO	12;24	62.44
17859	7DE	15;54	256.88	17859	23DE	0;05	92.53

444

NUMBER	DATE	TIME	LONG.	NUMBER	DATE	TIME	LONG.
17860	6JA	10;22	287.21	17860	21JA	10;43	122.42
17861	5FE	5;29	317.29	17861	19FE	20;35	151.91
17862	5MR	23;30	346.88	17862	20MR	5;57	180.89
17863	4AP	15;00	15.85	17863	18AP	15;14	209.39
17864	4MY	3;29	44.27	17864	18MY	1;06	237.56
17865	2JN	13;24	72.34	17865	16JN	12;26	265.63
17866	1JL	21;45	100.31	17866	16JL	1;55	293.87
17867	31JL	5;35	128.45	17867	14AU	17;41	322.50
17868	29AU	13;46	156.99	17868	13SE	11;01	351.65
17869	27SE	22;56	186.05	17869	13OC	4;39	21.32
17870	27OC	9;32	215.65	17870	11NO	21;21	51.37
17871	25NO	22;04	245.67	17871	11DE	12;25	81.58
17872	25DE	12;51	275.88				

445

NUMBER	DATE	TIME	LONG.	NUMBER	DATE	TIME	LONG.
				17872	10JA	1;35	111.67
17873	24JA	5;39	306.00	17873	8FE	12;49	141.37
17874	22FE	23;28	335.76	17874	9MR	22;15	170.53
17875	24MR	16;51	5.00	17875	8AP	6;21	199.14
17876	23AP	8;35	33.70	17876	7MY	13;55	227.33
17877	22MY	22;16	61.98	17877	5JN	22;01	255.31
17878	21JN	10;04	90.07	17878	5JL	7;48	283.35
17879	20JL	20;28	118.22	17879	3AU	20;06	311.70
17880	19AU	6;00	146.66	17880	2SE	11;09	340.57
17881	17SE	15;13	175.54	17881	2OC	4;32	10.04
17882	17OC	0;40	204.93	17882	31OC	23;12	40.03
17883	15NO	10;57	234.76	17883	30NO	17;52	70.34
17884	14DE	22;33	264.83	17884	30DE	11;09	100.65

446

NUMBER	DATE	TIME	LONG.	NUMBER	DATE	TIME	LONG.
17885	13JA	11;40	294.89	17885	29JA	1;58	130.63
17886	12FE	2;09	324.68	17886	27FE	13;53	160.08
17887	13MR	17;28	354.02	17887	28MR	23;10	188.91
17888	12AP	9;03	22.88	17888	27AP	6;41	217.23
17889	12MY	0;28	51.34	17889	26MY	13;31	245.22
17890	10JN	15;19	79.57	17890	24JN	20;48	273.13
17891	10JL	5;16	107.79	17891	24JL	5;30	301.24
17892	8AU	18;00	136.22	17892	22AU	16;25	329.77
17893	7SE	5;31	165.02	17893	21SE	6;03	358.89
17894	6OC	16;13	194.29	17894	20OC	22;34	28.63
17895	5NO	2;40	223.99	17895	19NO	17;31	58.86
17896	4DE	13;20	253.98	17896	19DE	13;27	89.29

447

NUMBER	DATE	TIME	LONG.	NUMBER	DATE	TIME	LONG.
17897	3JA	0;23	284.01	17897	18JA	8;17	119.56
17898	1FE	11;49	313.82	17898	17FE	0;20	149.35
17899	2MR	23;39	343.21	17899	18MR	13;01	178.51
17900	1AP	12;07	12.12	17900	16AP	22;51	207.07
17901	1MY	1;34	40.63	17901	16MY	6;46	235.18
17902	30MY	16;09	68.90	17902	14JN	13;49	263.09
17903	29JN	7;31	97.14	17903	13JL	20;54	291.06
17904	28JL	22;58	125.56	17904	12AU	4;53	319.33
17905	27AU	13;47	154.32	17905	10SE	14;36	348.11
17906	26SE	3;38	183.53	17906	10OC	2;54	17.50
17907	25OC	16;34	213.18	17907	8NO	18;21	47.45
17908	24NO	4;44	243.17	17908	8DE	12;39	77.79
17909	23DE	16;07	273.24				

448

NUMBER	DATE	TIME	LONG.	NUMBER	DATE	TIME	LONG.
				17909	7JA	8;22	108.17
17910	22JA	2;41	303.12	17910	6FE	3;24	138.24
17911	20FE	12;32	332.60	17911	6MR	20;07	167.76
17912	20MR	22;09	1.58	17912	5AP	9;58	196.65
17913	19AP	8;18	30.10	17913	4MY	21;10	225.01
17914	18MY	19;43	58.33	17914	3JN	6;18	253.04
17915	17JN	8;49	86.47	17915	2JL	14;06	280.99
17916	16JL	23;33	114.77	17916	31JL	21;27	309.12
17917	15AU	15;29	143.42	17917	30AU	5;17	337.65
17918	14SE	8;01	172.54	17918	28SE	14;33	6.73
17919	14OC	0;29	202.18	17919	28OC	2;06	36.38
17920	12NO	16;08	232.20	17920	26NO	16;20	66.47
17921	12DE	6;14	262.37	17921	26DE	8;54	96.76

449

NUMBER	DATE	TIME	LONG.	NUMBER	DATE	TIME	LONG.
17922	10JA	18;21	292.40	17922	25JA	2;50	126.92
17923	9FE	4;36	322.05	17923	23FE	20;50	156.67
17924	10MR	13;29	351.18	17924	25MR	13;49	185.88
17925	8AP	21;46	19.79	17925	24AP	4;58	214.54
17926	8MY	6;18	48.01	17926	23MY	17;51	242.79
17927	6JN	15;49	76.04	17927	22JN	4;33	270.84
17928	6JL	2;58	104.14	17928	21JL	13;35	298.94
17929	4AU	16;16	132.54	17929	19AU	21;53	327.33
17930	3SE	7;57	161.45	17930	18SE	6;27	356.20
17931	3OC	1;43	190.95	17931	17OC	16;06	25.61
17932	1NO	20;28	220.96	17932	16NO	3;15	55.48
17933	1DE	14;27	251.25	17933	15DE	15;59	85.60
17934	31DE	6;13	281.49				

450

NUMBER	DATE	TIME	LONG.	NUMBER	DATE	TIME	LONG.
				17934	14JA	6;10	115.70
17935	29JA	19;13	311.39	17935	12FE	21;36	145.52
17936	28FE	5;46	340.76	17936	14MR	13;55	174.89
17937	29MR	14;32	9.57	17937	13AP	6;27	203.78
17938	27AP	22;14	37.88	17938	12MY	22;14	232.24
17939	27MY	5;36	65.89	17939	11JN	12;27	260.44
17940	25JN	13;30	93.82	17940	11JL	0;50	288.61
17941	24JL	22;53	121.96	17941	9AU	11;47	316.97
17942	23AU	10;47	150.55	17942	7SE	22;01	345.73
17943	22SE	1;53	179.74	17943	7OC	8;10	14.98
17944	21OC	19;58	209.56	17944	5NO	18;36	44.69
17945	20NO	15;40	239.83	17945	5DE	5;23	74.69
17946	20DE	10;55	270.23				

451

NUMBER	DATE	TIME	LONG.	NUMBER	DATE	TIME	LONG.
				17946	3JA	16;38	104.73
17947	19JA	4;04	300.42	17947	2FE	4;37	134.55
17948	17FE	18;25	330.13	17948	3MR	17;40	163.98
17949	19MR	6;00	359.23	17949	2AP	7;52	192.95
17950	17AP	15;14	27.75	17950	1MY	22;54	221.51
17951	16MY	22;46	55.85	17951	31MY	14;08	249.80
17952	15JN	5;28	83.74	17952	30JN	5;00	278.03
17953	14JL	12;27	111.71	17953	29JL	15;14	306.40
17954	12AU	20;54	140.01	17954	28AU	8;48	335.12
17955	11SE	7;53	168.85	17955	26SE	21;41	4.31
17956	10OC	22;03	198.33	17956	26OC	9;47	33.94
17957	9NO	15;10	228.37	17957	24NO	21;04	63.89
17958	9DE	10;14	258.74	17958	24DE	7;42	93.93

	NEW MOONS				FULL MOONS				NEW MOONS				FULL MOONS		
NUMBER	DATE	TIME	LONG.	NUMBER	DATE	TIME	LONG.	NUMBER	DATE	TIME	LONG.	NUMBER	DATE	TIME	LONG.
	452								**458**						
17959	8JA	5;41	289.10	17959	22JA	18;02	123.79	18033	1JA	16;55	283.01	18033	15JA	21;48	117.43
17960	6FE	23;58	319.13	17960	21FE	4;29	153.29	18034	31JA	3;59	312.82	18034	14FE	15;41	147.32
17961	7MR	15;50	348.59	17961	21MR	15;25	182.31	18035	1MR	13;18	342.12	18035	16MR	9;04	176.71
17962	6AP	4;43	17.44	17962	20AP	3;08	210.89	18036	30MR	21;36	10.89	18036	15AP	1;04	205.55
17963	5MY	14;43	45.74	17963	19MY	15;48	239.16	18037	29AP	5;42	39.20	18037	14MY	15;06	233.93
17964	3JN	22;35	73.72	17964	18JN	5;36	267.33	18038	28MY	14;28	67.25	18038	13JN	2;56	262.03
17965	3JL	5;28	101.63	17965	17JL	20;37	295.64	18039	27JN	0;38	95.28	18039	12JL	12;52	290.11
17966	1AU	12;34	129.76	17966	16AU	12;36	324.30	18040	26JL	12;47	123.55	18040	10AU	21;37	318.40
17967	30AU	20;57	158.32	17967	15SE	4;48	353.43	18041	25AU	3;21	152.26	18041	9SE	6;09	347.11
17968	29SE	7;21	187.46	17968	14OC	20;17	23.04	18042	23SE	20;20	181.56	18042	8OC	15;22	16.34
17969	28OC	20;11	217.17	17969	13NO	10;21	53.00	18043	23OC	14;59	211.43	18043	7NO	1;51	46.07
17970	27NO	11;31	247.32	17970	12DE	22;48	83.10	18044	22NO	9;46	241.68	18044	6DE	13;48	76.13
17971	27DE	5;02	277.64					18045	22DE	2;53	271.99				
	453								**459**						
				17971	11JA	9;54	113.09					18045	5JA	3;09	106.25
17972	25JA	23;50	307.83	17972	9FE	19;59	142.72	18046	20JA	17;17	302.05	18046	3FE	17;44	136.16
17973	24FE	18;27	337.59	17973	11MR	5;21	171.86	18047	19FE	4;55	331.62	18047	5MR	9;22	165.67
17974	26MR	11;12	6.78	17974	9AP	14;23	200.49	18048	20MR	14;21	0.61	18048	4AP	1;38	194.70
17975	25AP	1;07	35.38	17975	8MY	23;38	228.74	18049	18AP	22;21	29.06	18049	3MY	17;46	223.28
17976	24MY	12;15	63.55	17976	7JN	9;58	256.80	18050	18MY	5;40	57.14	18050	2JN	8;49	251.56
17977	22JN	21;20	91.54	17977	6JL	22;14	284.94	18051	16JN	13;06	85.06	18051	1JL	22;14	279.73
17978	22JL	5;27	119.61	17978	5AU	12;56	313.41	18052	15JL	21;35	113.10	18052	31JL	10;03	308.02
17979	20AU	13;33	148.00	17979	4SE	5;47	342.38	18053	14AU	8;09	141.50	18053	29AU	20;50	336.65
17980	18SE	22;20	176.89	17980	3OC	23;41	11.89	18054	12SE	21;41	170.47	18054	28SE	7;13	5.76
17981	18OC	8;18	206.32	17981	2NO	17;14	41.87	18055	12OC	14;29	200.08	18055	27OC	17;41	35.34
17982	16NO	19;52	236.21	17982	2DE	9;22	72.08	18056	11NO	9;44	230.23	18056	26NO	4;23	65.28
17983	16DE	9;27	266.37	17983	31DE	23;34	102.25	18057	11DE	5;32	260.63	18057	25DE	15;21	95.33
	454								**460**						
17984	15JA	1;07	296.52	17984	30JA	11;42	132.11	18058	9JA	23;54	290.94	18058	24JA	2;45	125.23
17985	13FE	18;20	326.41	17985	28FE	21;53	161.46	18059	8FE	15;38	320.84	18059	22FE	14;58	154.77
17986	15MR	11;51	355.82	17986	30MR	6;24	190.24	18060	9MR	4;29	350.16	18060	23MR	4;18	183.87
17987	14AP	4;20	24.69	17987	28AP	13;57	218.54	18061	7AP	14;44	18.87	18061	21AP	18;42	212.53
17988	13MY	19;00	53.10	17988	27MY	21;29	246.55	18062	6MY	22;57	47.09	18062	21MY	9;44	240.89
17989	12JN	7;46	81.24	17989	26JN	6;12	274.52	18063	5JN	5;55	75.02	18063	20JN	0;49	269.11
17990	11JL	18;58	109.36	17990	25JL	17;09	302.73	18064	4JL	12;37	102.94	18064	19JL	15;32	297.42
17991	10AU	5;06	137.70	17991	24AU	6;53	331.40	18065	2AU	20;14	131.10	18065	18AU	5;42	326.03
17992	8SE	14;39	166.44	17992	22SE	23;20	0.66	18066	1SE	5;56	159.73	18066	16SE	19;14	355.08
17993	8OC	0;08	195.68	17993	22OC	17;41	30.50	18067	30SE	18;33	188.98	18067	16OC	8;01	24.59
17994	6NO	10;04	225.38	17994	21NO	12;41	60.75	18068	30OC	10;21	218.84	18068	14NO	19;54	54.48
17995	5DE	20;58	255.39	17995	21DE	6;55	91.11	18069	29NO	4;38	249.12	18069	14DE	6;53	84.52
								18070	29DE	0;02	279.52				
	455								**461**						
17996	4JA	9;13	285.46	17996	19JA	23;05	121.25					18070	12JA	17;15	114.47
17997	2FE	22;51	315.34	17997	18FE	12;23	150.90	18071	27JA	18;59	309.69	18071	11FE	3;25	144.09
17998	4MR	13;31	344.82	17998	19MR	22;47	179.94	18072	26FE	12;04	339.37	18072	12MR	13;49	173.25
17999	3AP	4;44	13.82	17999	18AP	6;57	208.42	18073	28MR	2;23	8.43	18073	11AP	0;50	201.94
18000	2MY	20;04	42.39	18000	17MY	13;53	236.48	18074	26AP	13;43	36.91	18074	10MY	12;45	230.28
18001	1JN	11;11	70.67	18001	15JN	20;45	264.38	18075	25MY	22;32	64.98	18075	9JN	1;49	258.44
18002	1JL	1;44	98.88	18002	15JL	4;38	292.39	18076	24JN	5;49	92.89	18076	8JL	16;13	286.67
18003	30JL	15;18	127.24	18003	13AU	14;24	320.75	18077	23JL	12;46	120.92	18077	7AU	7;54	315.20
18004	29AU	3;42	155.92	18004	12SE	2;42	349.66	18078	21AU	20;31	149.32	18078	6SE	0;21	344.18
18005	27SE	15;02	185.05	18005	11OC	17;53	19.19	18079	20SE	5;58	178.26	18079	5OC	16;37	13.66
18006	27OC	1;46	214.64	18006	10NO	11;49	49.27	18080	19OC	17;39	207.78	18080	4NO	7;43	43.55
18007	25NO	12;24	244.57	18007	10DE	7;32	79.67	18081	18NO	7;46	237.79	18081	3DE	21;08	73.66
18008	24DE	23;14	274.61					18082	18DE	0;10	268.07				
	456								**462**						
				18008	9JA	3;10	110.04					18082	2JA	8;53	103.72
18009	23JA	10;21	304.51	18009	7FE	20;41	140.02	18083	16JA	18;19	298.32	18083	31JA	19;21	133.49
18010	21FE	21;44	334.03	18010	8MR	10;58	169.41	18084	15FE	13;02	328.24	18084	2MR	4;52	162.79
18011	22MR	9;32	3.08	18011	6AP	22;03	198.17	18085	17MR	6;43	357.63	18085	31MR	13;47	191.57
18012	20AP	22;10	31.68	18012	6MY	6;44	226.41	18086	15AP	22;02	26.42	18086	29AP	22;37	219.91
18013	20MY	11;58	59.99	18013	4JN	14;04	254.37	18087	15MY	10;29	54.73	18087	29MY	8;07	247.98
18014	19JN	2;57	88.21	18014	3JL	21;04	282.29	18088	13JN	20;33	82.76	18088	27JN	19;12	276.05
18015	18JL	18;34	116.55	18015	2AU	4;35	310.43	18089	13JL	5;10	110.79	18089	27JL	8;39	304.37
18016	17AU	10;00	145.20	18016	31AU	13;26	339.03	18090	11AU	13;21	139.06	18090	26AU	0;38	333.16
18017	16SE	0;39	174.28	18017	30SE	0;27	8.20	18091	9SE	21;55	167.79	18091	24SE	18;24	2.51
18018	15OC	14;19	203.82	18018	29OC	14;21	37.96	18092	9OC	7;24	197.04	18092	24OC	12;33	32.37
18019	14NO	3;05	233.74	18019	28NO	7;20	68.19	18093	7NO	18;12	226.79	18093	23NO	5;42	62.55
18020	13DE	14;57	263.82	18020	28DE	2;30	98.58	18094	7DE	6;43	256.88	18094	22DE	21;00	92.79
	457								**463**						
18021	12JA	1;54	293.79	18021	26JA	21;59	128.78	18095	5JA	21;13	287.04	18095	21JA	10;10	122.78
18022	10FE	11;55	323.41	18022	25FE	15;52	158.51	18096	4FE	13;32	317.02	18096	19FE	21;12	152.32
18023	11MR	21;20	352.55	18023	27MR	7;03	187.61	18097	6MR	6;48	346.58	18097	21MR	6;20	181.29
18024	10AP	6;51	21.19	18024	25AP	19;28	216.15	18098	4AP	23;43	15.62	18098	19AP	14;06	209.73
18025	9MY	17;19	49.47	18025	25MY	5;34	244.27	18099	4MY	15;14	44.17	18099	18MY	21;22	237.79
18026	8JN	5;23	77.60	18026	23JN	14;01	272.23	18100	3JN	4;56	72.39	18100	17JN	5;15	265.74
18027	7JL	19;17	105.81	18027	22JL	21;36	300.28	18101	2JL	17;00	100.51	18101	16JL	14;55	293.82
18028	6AU	10;46	134.32	18028	21AU	5;13	328.67	18102	1AU	3;50	128.77	18102	15AU	3;15	322.31
18029	5SE	3;17	163.29	18029	19SE	13;49	357.56	18103	30AU	13;52	157.38	18103	13SE	18;30	351.36
18030	4OC	20;09	192.78	18030	19OC	0;15	27.02	18104	28SE	23;33	186.47	18104	13OC	12;11	21.02
18031	3NO	12;35	222.71	18031	17NO	13;09	56.97	18105	28OC	9;22	216.03	18105	12NO	7;12	51.16
18032	3DE	3;43	252.89	18032	17DE	4;34	87.21	18106	26NO	19;48	245.96	18106	12DE	2;06	81.54
								18107	26DE	7;17	276.03				

464

NUMBER	DATE	TIME	LONG.	NUMBER	DATE	TIME	LONG.
				18107	10JA	19;28	111.80
18108	24JA	20;03	305.98	18108	9FE	10;13	141.65
18109	23FE	9;58	335.59	18109	9MR	21;56	170.90
18110	24MR	0;40	4.73	18110	8AP	6;59	199.56
18111	22AP	15;44	33.41	18111	7MY	14;18	227.73
18112	22MY	6;52	61.76	18112	5JN	21;00	255.65
18113	20JN	21;47	89.97	18113	5JL	4;14	283.59
18114	20JL	12;04	118.27	18114	3AU	13;00	311.80
18115	19AU	1;20	146.86	18115	2SE	0;03	340.50
18116	17SE	13;28	175.86	18116	1OC	13;52	9.82
18117	17OC	0;41	205.32	18117	31OC	6;33	39.72
18118	15NO	11;28	235.17	18118	30NO	1;36	70.04
18119	14DE	22;14	265.21	18119	29DE	21;31	100.46

465

NUMBER	DATE	TIME	LONG.	NUMBER	DATE	TIME	LONG.
18120	13JA	9;07	295.17	18120	28JA	16;16	130.62
18121	11FE	20;09	324.82	18121	27FE	8;09	160.23
18122	13MR	7;26	354.00	18122	28MR	20;40	189.20
18123	11AP	19;18	22.72	18123	27AP	6;22	217.60
18124	11MY	8;14	51.08	18124	26MY	14;14	245.63
18125	9JN	22;32	79.29	18125	24JN	21;19	273.54
18126	9JL	13;57	107.57	18126	24JL	4;32	301.59
18127	8AU	5;46	136.11	18127	22AU	12;44	330.01
18128	6SE	21;09	165.07	18128	20SE	22;40	358.98
18129	6OC	11;37	194.48	18129	20OC	11;09	28.54
18130	5NO	1;04	224.31	18130	19NO	2;38	58.62
18131	4DE	13;32	254.38	18131	18DE	20;49	88.98

466

NUMBER	DATE	TIME	LONG.	NUMBER	DATE	TIME	LONG.
18132	3JA	0;58	284.43	18132	17JA	16;17	119.27
18133	1FE	11;18	314.19	18133	16FE	11;00	149.18
18134	2MR	20;46	343.48	18134	18MR	3;28	178.50
18135	1AP	5;53	12.25	18135	16AP	17;10	207.22
18136	30AP	15;31	40.61	18136	16MY	4;21	235.47
18137	30MY	2;31	68.75	18137	14JN	13;35	263.47
18138	28JN	15;25	96.89	18138	13JL	21;37	291.48
18139	28JL	6;11	125.27	18139	12AU	5;17	319.74
18140	26AU	22;26	154.08	18140	10SE	13;26	348.45
18141	25SE	15;29	183.41	18141	9OC	22;59	17.72
18142	25OC	8;30	213.24	18142	8NO	10;40	47.52
18143	24NO	0;36	243.38	18143	8DE	0;50	77.68
18144	23DE	14;56	273.58				

467

NUMBER	DATE	TIME	LONG.	NUMBER	DATE	TIME	LONG.
				18144	6JA	17;08	107.91
18145	22JA	3;01	303.52	18145	5FE	10;36	137.93
18146	20FE	13;01	333.01	18146	7MR	4;08	167.49
18147	21MR	21;33	1.94	18147	5AP	20;44	196.50
18148	20AP	5;27	30.37	18148	5MY	11;43	225.02
18149	19MY	13;37	58.47	18149	4JN	0;42	253.21
18150	17JN	22;52	86.46	18150	3JL	11;40	281.28
18151	17JL	9;55	114.61	18151	1AU	21;06	309.49
18152	15AU	23;17	143.15	18152	31AU	5;51	338.06
18153	14SE	15;13	172.24	18153	29SE	14;50	7.13
18154	14OC	9;22	201.93	18154	29OC	0;47	36.71
18155	13NO	4;31	232.10	18155	27NO	12;02	66.68
18156	12DE	22;48	262.44	18156	27DE	0;37	96.80

468

NUMBER	DATE	TIME	LONG.	NUMBER	DATE	TIME	LONG.
18157	11JA	14;40	292.64	18157	25JA	14;24	126.78
18158	10FE	3;35	322.40	18158	24FE	5;16	156.42
18159	10MR	13;54	351.59	18159	24MR	20;59	185.58
18160	8AP	22;24	20.21	18160	23AP	13;05	214.29
18161	8MY	5;50	48.38	18161	23MY	4;42	242.65
18162	6JN	13;02	76.32	18162	21JN	19;03	270.85
18163	5JL	20;50	104.28	18163	21JL	7;49	299.10
18164	4AU	6;15	132.53	18164	19AU	19;18	327.62
18165	2SE	18;16	161.28	18165	18SE	6;05	356.58
18166	2OC	9;33	190.67	18166	17OC	16;43	26.03
18167	1NO	3;52	220.65	18167	16NO	3;25	55.87
18168	30NO	23;45	251.01	18168	15DE	14;16	85.91
18169	30DE	19;04	281.40				

469

NUMBER	DATE	TIME	LONG.	NUMBER	DATE	TIME	LONG.
				18169	14JA	1;18	115.88
18170	29JA	12;10	311.48	18170	12FE	12;50	145.54
18171	28FE	2;22	341.01	18171	14MR	1;17	174.76
18172	29MR	13;46	9.92	18172	12AP	14;52	203.54
18173	27AP	22;50	38.29	18173	12MY	5;26	231.96
18174	27MY	6;17	66.30	18174	10JN	20;29	260.20
18175	25JN	12;59	94.19	18175	10JL	11;30	288.46
18176	24JL	20;03	122.24	18176	9AU	2;10	316.97
18177	23AU	4;40	150.69	18177	7SE	16;20	345.88
18178	21SE	15;50	179.72	18178	7OC	5;49	15.27
18179	21OC	6;08	209.37	18179	5NO	18;24	45.08
18180	19NO	23;20	239.53	18180	5DE	5;56	75.11
18181	19DE	18;20	269.92				

470

NUMBER	DATE	TIME	LONG.	NUMBER	DATE	TIME	LONG.
				18181	3JA	16;33	105.12
18182	18JA	13;37	300.20	18182	2FE	2;37	134.86
18183	17FE	7;39	330.07	18183	3MR	12;37	164.16
18184	18MR	23;19	359.34	18184	1AP	23;00	192.98
18185	17AP	12;04	28.01	18185	1MY	10;10	221.39
18186	16MY	22;02	56.21	18186	30MY	22;26	249.57
18187	15JN	5;59	84.15	18187	29JN	12;05	277.74
18188	14JL	13;03	112.12	18188	29JL	3;14	306.15
18189	12AU	20;24	140.38	18189	27AU	19;36	334.97
18190	11SE	5;04	169.13	18190	26SE	12;23	4.31
18191	10OC	15;42	198.45	18191	26OC	4;28	34.11
18192	9NO	4;38	228.31	18192	24NO	18;57	64.19
18193	8DE	19;54	258.52	18193	24DE	7;35	94.31

471

NUMBER	DATE	TIME	LONG.	NUMBER	DATE	TIME	LONG.
18194	7JA	13;08	288.79	18194	22JA	18;36	124.21
18195	6FE	7;32	318.83	18195	21FE	4;24	153.67
18196	8MR	1;44	348.40	18196	22MR	13;22	182.61
18197	6AP	18;13	17.40	18197	20AP	21;57	211.07
18198	6MY	8;03	45.86	18198	20MY	6;48	239.19
18199	4JN	19;17	73.97	18199	18JN	16;51	267.21
18200	4JL	4;37	101.99	18200	18JL	5;01	295.41
18201	2AU	13;05	130.17	18201	16AU	19;49	324.01
18202	31AU	21;34	158.74	18202	15SE	13;00	353.17
18203	30SE	6;44	187.82	18203	15OC	7;24	22.89
18204	29OC	16;57	217.42	18204	14NO	1;25	53.01
18205	28NO	4;35	247.41	18205	13DE	17;51	83.28
18206	27DE	17;58	277.56				

472

NUMBER	DATE	TIME	LONG.	NUMBER	DATE	TIME	LONG.
				18206	12JA	8;08	113.40
18207	26JA	9;13	307.60	18207	10FE	20;09	143.12
18208	25FE	1;52	337.30	18208	11MR	6;03	172.28
18209	25MR	18;51	6.50	18209	9AP	14;16	200.88
18210	24AP	11;00	35.20	18210	8MY	21;30	229.04
18211	24MY	1;36	63.52	18211	7JN	4;49	256.98
18212	22JN	14;32	91.65	18212	6JL	13;24	284.97
18213	22JL	2;08	119.86	18213	5AU	0;20	313.29
18214	20AU	12;46	148.36	18214	3SE	14;13	342.13
18215	18SE	22;50	177.30	18215	3OC	6;54	11.59
18216	18OC	8;43	206.73	18216	2NO	1;34	41.59
18217	16NO	18;54	236.56	18217	1DE	20;49	71.94
18218	16DE	5;48	266.60	18218	31DE	15;11	102.28

473

NUMBER	DATE	TIME	LONG.	NUMBER	DATE	TIME	LONG.
18219	14JA	17;47	296.60	18219	30JA	7;19	132.31
18220	13FE	6;54	326.32	18220	28FE	20;28	161.79
18221	14MR	20;57	355.60	18221	30MR	6;40	190.64
18222	13AP	11;35	24.40	18222	28AP	14;37	218.95
18223	13MY	2;32	52.83	18223	27MY	21;25	246.93
18224	11JN	17;34	81.06	18224	26JN	4;13	274.83
18225	11JL	8;20	109.32	18225	25JL	12;09	302.91
18226	9AU	22;24	137.82	18226	23AU	22;02	331.43
18227	8SE	11;24	166.70	18227	22SE	10;31	0.52
18228	7OC	23;19	196.03	18228	22OC	1;52	30.23
18229	6NO	10;28	225.78	18229	20NO	19;54	60.43
18230	5DE	21;17	255.79	18230	20DE	15;37	90.85

474

NUMBER	DATE	TIME	LONG.	NUMBER	DATE	TIME	LONG.
18231	4JA	8;04	285.80	18231	19JA	11;09	121.14
18232	2FE	18;52	315.56	18232	18FE	4;30	150.96
18233	4MR	5;44	344.89	18233	19MR	18;37	180.16
18234	2AP	16;57	13.73	18234	18AP	5;33	208.75
18235	2MY	5;02	42.18	18235	17MY	14;10	236.88
18236	31MY	18;28	70.39	18236	15JN	21;32	264.80
18237	30JN	9;21	98.62	18237	15JL	4;40	292.78
18238	30JL	1;12	127.06	18238	13AU	12;23	321.06
18239	28AU	17;08	155.88	18239	11SE	21;28	349.83
18240	27SE	8;24	185.17	18240	11OC	8;41	19.19
18241	26OC	22;39	214.90	18241	9NO	22;41	49.10
18242	25NO	11;47	244.93	18242	9DE	15;35	79.38
18243	24DE	23;48	275.03				

475

NUMBER	DATE	TIME	LONG.	NUMBER	DATE	TIME	LONG.
				18243	8JA	10;31	109.72
18244	23JA	10;37	304.91	18244	7FE	5;41	139.78
18245	21FE	20;18	334.36	18245	8MR	23;14	169.32
18246	23MR	5;16	3.29	18246	7AP	14;13	198.24
18247	21AP	14;17	31.75	18247	7MY	2;35	226.63
18248	21MY	0;18	59.91	18248	5JN	12;45	254.70
18249	19JN	12;04	88.01	18249	4JL	21;25	282.69
18250	19JL	1;55	116.27	18250	3AU	5;18	310.85
18251	17AU	17;36	144.92	18251	1SE	13;16	339.41
18252	16SE	10;34	174.09	18252	30SE	22;11	8.49
18253	16OC	3;58	203.78	18253	30OC	8;50	38.12
18254	14NO	20;53	233.87	18254	28NO	21;45	68.17
18255	14DE	12;20	264.10	18255	28DE	12;57	98.40

476

NEW MOONS				FULL MOONS			
NUMBER	DATE	TIME	LONG.	NUMBER	DATE	TIME	LONG.
18256	13JA	1;35	294.17	18256	27JA	5;46	128.49
18257	11FE	12;29	323.83	18257	25FE	23;08	158.21
18258	11MR	21;30	352.94	18258	26MR	16;04	187.39
18259	10AP	5;25	21.52	18259	25AP	7;49	216.07
18260	9MY	13;10	49.69	18260	24MY	21;50	244.35
18261	7JN	21;39	77.68	18261	23JN	9;53	272.45
18262	7JL	7;39	105.73	18262	22JL	20;12	300.61
18263	5AU	19;49	134.10	18263	21AU	5;24	329.07
18264	4SE	10;33	162.99	18264	19SE	14;24	357.97
18265	4OC	3;51	192.49	18265	18OC	23;58	27.40
18266	2NO	22;54	222.53	18266	17NO	10;39	57.26
18267	2DE	18;00	252.87	18267	16DE	22;33	87.34

477

NEW MOONS				FULL MOONS			
18268	1JA	11;17	283.16	18268	15JA	11;35	117.38
18269	31JA	1;39	313.11	18269	14FE	1;39	147.13
18270	1MR	13;06	342.50	18270	15MR	16;40	176.43
18271	30MR	22;17	11.30	18271	14AP	8;24	205.27
18272	29AP	6;01	39.60	18272	14MY	0;13	233.72
18273	28MY	13;09	67.59	18273	12JN	15;18	261.96
18274	26JN	20;30	95.51	18274	12JL	5;01	290.18
18275	26JL	4;59	123.62	18275	10AU	17;20	318.61
18276	24AU	15;39	152.17	18276	9SE	4;41	347.44
18277	23SE	5;21	181.33	18277	8OC	15;36	16.74
18278	22OC	22;21	211.12	18278	7NO	2;26	46.49
18279	21NO	17;46	241.39	18279	6DE	13;17	76.50
18280	21DE	13;40	271.82				

478

NEW MOONS				FULL MOONS			
				18280	5JA	0;07	106.51
18281	20JA	7;59	302.04	18281	3FE	11;10	136.28
18282	18FE	23;35	331.78	18282	4MR	22;50	165.63
18283	20MR	12;16	0.91	18283	3AP	11;32	194.52
18284	18AP	22;21	29.45	18284	3MY	1;24	223.02
18285	18MY	6;28	57.56	18285	1JN	16;08	251.29
18286	16JN	13;25	85.46	18286	1JL	7;14	279.52
18287	15JL	20;12	113.43	18287	30JL	22;16	307.94
18288	14AU	3;59	141.72	18288	29AU	12;59	336.72
18289	12SE	13;52	170.54	18289	28SE	3;09	5.98
18290	12OC	2;40	199.97	18290	27OC	16;28	35.68
18291	10NO	18;32	229.97	18291	26NO	4;41	65.68
18292	10DE	12;47	260.31	18292	25DE	15;46	95.74

479

NEW MOONS				FULL MOONS			
18293	9JA	8;01	290.66	18293	24JA	1;57	125.58
18294	8FE	2;43	320.69	18294	22FE	11;44	155.03
18295	9MR	19;33	350.18	18295	23MR	21;37	183.99
18296	8AP	9;42	19.05	18296	22AP	8;05	212.50
18297	7MY	20;59	47.40	18297	21MY	19;33	240.71
18298	6JN	5;52	75.41	18298	20JN	8;21	268.84
18299	5JL	13;18	103.35	18299	19JL	22;46	297.14
18300	3AU	20;30	131.49	18300	18AU	14;44	325.81
18301	2SE	4;33	160.06	18301	17SE	7;42	354.99
18302	1OC	14;17	189.19	18302	17OC	0;34	24.67
18303	31OC	2;08	218.88	18303	15NO	16;10	54.71
18304	29NO	16;14	248.98	18304	15DE	5;51	84.87
18305	29DE	8;26	279.25				

480

NEW MOONS				FULL MOONS			
				18305	13JA	17;37	114.88
18306	28JA	2;11	309.38	18306	12FE	3;52	144.50
18307	26FE	20;27	339.12	18307	12MR	13;02	173.61
18308	27MR	13;46	8.30	18308	10AP	21;31	202.20
18309	26AP	4;55	36.94	18309	10MY	5;57	230.39
18310	25MY	17;25	65.16	18310	8JN	15;08	258.40
18311	24JN	3;41	93.19	18311	8JL	2;04	286.50
18312	23JL	12;38	121.30	18312	6AU	15;32	314.93
18313	21AU	21;14	149.74	18313	5SE	7;45	343.89
18314	20SE	6;12	178.66	18314	5OC	1;56	13.44
18315	19OC	16;00	208.10	18315	3NO	20;33	43.47
18316	18NO	2;56	237.98	18316	3DE	14;03	73.75
18317	17DE	15;21	268.09				

481

NEW MOONS				FULL MOONS			
				18317	2JA	5;31	103.97
18318	16JA	5;30	298.17	18318	31JA	18;38	133.85
18319	14FE	21;17	327.97	18319	2MR	5;27	163.20
18320	16MR	13;58	357.33	18320	31MR	14;18	191.98
18321	15AP	6;27	26.18	18321	29AP	21;46	220.26
18322	14MY	21;46	54.61	18322	29MY	4;47	248.24
18323	13JN	11;34	82.79	18323	27JN	12;31	276.17
18324	12JL	23;58	110.97	18324	26JL	22;09	304.34
18325	11AU	11;17	139.37	18325	25AU	10;35	332.98
18326	9SE	21;51	168.17	18326	24SE	2;02	2.22
18327	9OC	8;00	197.46	18327	23OC	19;59	32.06
18328	7NO	18;08	227.18	18328	22NO	15;15	62.33
18329	7DE	4;40	257.18	18329	22DE	10;18	92.72

482

NEW MOONS				FULL MOONS			
18330	5JA	15;59	287.21	18330	21JA	3;41	122.90
18331	4FE	4;20	317.02	18331	19FE	18;18	152.59
18332	5MR	17;40	346.43	18332	21MR	5;50	181.65
18333	4AP	7;44	15.36	18333	19AP	14;42	210.14
18334	3MY	22;19	43.88	18334	18MY	21;51	238.20
18335	2JN	13;14	72.15	18335	17JN	4;29	266.09
18336	2JL	4;15	100.39	18336	16JL	11;45	294.08
18337	31JL	18;56	128.80	18337	14AU	20;38	322.43
18338	30AU	8;47	157.56	18338	13SE	7;51	351.31
18339	28SE	21;31	186.77	18339	12OC	21;50	20.81
18340	28OC	9;15	216.42	18340	11NO	14;38	50.85
18341	26NO	20;19	246.38	18341	11DE	9;42	81.23
18342	26DE	7;07	276.42				

483

NEW MOONS				FULL MOONS			
				18342	10JA	5;32	111.60
18343	24JA	17;46	306.28	18343	9FE	0;05	141.61
18344	23FE	4;22	335.75	18344	10MR	15;47	171.04
18345	24MR	15;05	4.73	18345	9AP	4;10	199.83
18346	23AP	2;24	33.26	18346	8MY	13;47	228.10
18347	22MY	14;53	61.51	18347	6JN	21;39	256.07
18348	21JN	4;57	89.69	18348	6JL	4;51	284.00
18349	20JL	20;28	118.04	18349	4AU	12;17	312.00
18350	19AU	12;41	146.73	18350	2SE	20;43	340.76
18351	18SE	4;39	175.89	18351	2OC	6;54	9.92
18352	17OC	19;43	205.51	18352	31OC	19;30	39.64
18353	16NO	9;39	235.48	18353	30NO	10;58	69.81
18354	15DE	22;20	265.60	18354	30DE	4;58	100.15

484

NEW MOONS				FULL MOONS			
18355	14JA	9;44	295.58	18355	29JA	0;05	130.33
18356	12FE	19;49	325.19	18356	27FE	18;27	160.05
18357	13MR	4;52	354.28	18357	28MR	10;38	189.18
18358	11AP	13;31	22.87	18358	27AP	0;13	217.74
18359	10MY	22;42	51.09	18359	26MY	11;26	245.91
18360	9JN	9;21	79.16	18360	24JN	20;52	273.90
18361	8JL	22;05	107.33	18361	24JL	5;12	302.00
18362	7AU	12;58	135.82	18362	22AU	13;12	330.42
18363	6SE	5;33	164.81	18363	20SE	21;43	359.33
18364	5OC	23;05	194.35	18364	20OC	7;32	28.78
18365	4NO	16;38	224.35	18365	18NO	19;19	58.70
18366	4DE	9;06	254.58	18366	18DE	9;21	88.88

485

NEW MOONS				FULL MOONS			
18367	2JA	23;34	284.76	18367	17JA	1;17	119.03
18368	1FE	11;34	314.59	18368	15FE	18;15	148.87
18369	2MR	21;19	343.89	18369	17MR	11;15	178.23
18370	1AP	5;30	12.63	18370	16AP	3;30	207.06
18371	30AP	13;03	40.90	18371	15MY	18;23	235.46
18372	29MY	20;55	68.91	18372	14JN	7;30	263.61
18373	28JN	5;59	96.90	18373	13JL	18;48	291.75
18374	27JL	16;59	125.13	18374	12AU	4;42	320.10
18375	26AU	6;28	153.82	18375	10SE	13;56	348.86
18376	24SE	22;39	183.10	18376	9OC	23;20	18.13
18377	24OC	17;10	212.98	18377	8NO	9;33	47.86
18378	23NO	12;38	243.27	18378	7DE	20;51	77.90
18379	23DE	7;07	273.64				

486

NEW MOONS				FULL MOONS			
				18379	6JA	9;13	107.97
18380	21JA	23;01	303.74	18380	4FE	22;33	137.81
18381	20FE	11;48	333.34	18381	6MR	12;48	167.25
18382	21MR	21;54	2.34	18382	5AP	3;56	196.21
18383	20AP	6;08	30.79	18383	4MY	19;36	224.76
18384	19MY	13;23	58.85	18384	3JN	11;06	253.05
18385	17JN	20;27	86.76	18385	3JL	1;39	281.27
18386	17JL	4;15	114.77	18386	1AU	14;52	309.63
18387	15AU	13;45	143.15	18387	31AU	2;55	338.33
18388	14SE	1;55	172.08	18388	29SE	14;16	7.50
18389	13OC	17;24	201.66	18389	29OC	1;21	37.13
18390	12NO	11;53	231.78	18390	27NO	12;18	67.08
18391	12DE	7;51	262.20	18391	26DE	23;07	97.12

487

NEW MOONS				FULL MOONS			
18392	11JA	3;09	292.54	18392	25JA	9;53	126.98
18393	9FE	20;07	322.48	18393	23FE	20;56	156.46
18394	11MR	10;09	351.82	18394	25MR	8;46	185.47
18395	9AP	21;23	20.55	18395	23AP	21;47	214.07
18396	9MY	6;21	48.79	18396	23MY	11;56	242.38
18397	7JN	13;46	76.74	18397	22JN	2;51	270.59
18398	6JL	20;32	104.66	18398	21JL	18;04	298.93
18399	5AU	3;46	132.81	18399	20AU	9;13	327.60
18400	3SE	12;34	161.44	18400	18SE	23;59	356.71
18401	2OC	23;56	190.65	18401	18OC	14;04	26.31
18402	1NO	14;22	220.47	18402	17NO	3;04	56.25
18403	1DE	7;33	250.71	18403	16DE	14;47	86.33
18404	31DE	2;24	281.09				

488

NUMBER	DATE	TIME	LONG.	NUMBER	DATE	TIME	LONG.
				18404	15JA	1;20	116.27
18405	29JA	21;25	311.25	18405	13FE	11;05	145.86
18406	28FE	15;10	340.94	18406	13MR	20;37	174.96
18407	29MR	6;38	10.02	18407	12AP	6;29	203.59
18408	27AP	19;17	38.54	18408	11MY	17;10	231.86
18409	27MY	5;17	66.65	18409	10JN	5;06	259.97
18410	25JN	13;24	94.59	18410	9JL	18;38	288.18
18411	24JL	20;42	122.65	18411	8AU	9;57	316.70
18412	23AU	4;21	151.07	18412	7SE	2;45	345.71
18413	21SE	13;19	180.00	18413	6OC	20;06	15.26
18414	21OC	0;10	209.51	18414	5NO	12;44	45.23
18415	19NO	13;11	239.49	18415	5DE	3;34	75.40
18416	19DE	4;17	269.71				

494

NUMBER	DATE	TIME	LONG.	NUMBER	DATE	TIME	LONG.
				18478	7JA	21;18	109.55
18479	23JA	10;09	305.28	18479	6FE	13;37	139.50
18480	21FE	20;51	334.77	18480	8MR	6;26	169.01
18481	23MR	5;34	3.69	18481	6AP	22;55	198.01
18482	21AP	13;09	32.09	18482	6MY	14;26	226.53
18483	20MY	20;35	60.15	18483	5JN	4;30	254.76
18484	19JN	4;51	88.11	18484	4JL	16;51	282.89
18485	18JL	14;47	116.21	18485	3AU	3;36	311.17
18486	17AU	3;00	144.72	18486	1SE	13;18	339.80
18487	15SE	17;56	173.79	18487	30SE	22;45	8.91
18488	15OC	11;32	203.48	18488	30OC	8;41	38.51
18489	14NO	6;54	233.66	18489	28NO	19;29	68.47
18490	14DE	2;14	264.06	18490	28DE	7;16	98.55

489

NUMBER	DATE	TIME	LONG.	NUMBER	DATE	TIME	LONG.
				18416	3JA	16;18	105.50
18417	17JA	21;10	299.90	18417	2FE	3;12	135.27
18418	16FE	15;06	329.77	18418	3MR	12;41	164.55
18419	18MR	8;52	359.14	18419	1AP	21;16	193.29
18420	17AP	1;05	27.96	18420	1MY	5;27	221.59
18421	16MY	14;53	56.31	18421	30MY	13;58	249.62
18422	15JN	2;16	84.39	18422	28JN	23;49	277.64
18423	14JL	11;56	112.47	18423	28JL	11;55	305.92
18424	12AU	20;48	140.79	18424	27AU	2;52	334.68
18425	11SE	5;43	169.54	18425	25SE	20;24	4.04
18426	10OC	15;14	198.82	18426	25OC	15;14	33.94
18427	9NO	1;41	228.58	18427	24NO	9;38	64.19
18428	8DE	13;19	258.63	18428	24DE	2;18	94.48

495

NUMBER	DATE	TIME	LONG.	NUMBER	DATE	TIME	LONG.
18491	12JA	19;36	294.31	18491	26JA	19;56	128.47
18492	11FE	9;53	324.11	18492	25FE	9;26	158.03
18493	12MR	21;08	353.31	18493	26MR	23;50	187.12
18494	11AP	6;05	21.93	18494	25AP	15;03	215.78
18495	10MY	13;37	50.10	18495	25MY	6;37	244.13
18496	8JN	20;37	78.02	18496	23JN	21;46	272.36
18497	8JL	3;57	105.97	18497	23JL	11;51	300.67
18498	6AU	12;30	134.20	18498	22AU	0;43	329.26
18499	4SE	23;18	162.92	18499	20SE	12;39	358.29
18500	4OC	13;10	192.27	18500	20OC	0;05	27.79
18501	3NO	6;20	222.22	18501	18NO	11;15	57.67
18502	3DE	1;51	252.57	18502	17DE	22;10	87.72

490

NUMBER	DATE	TIME	LONG.	NUMBER	DATE	TIME	LONG.
18429	7JA	2;26	288.73	18429	22JA	16;36	124.51
18430	5FE	17;13	318.62	18430	21FE	4;27	154.06
18431	7MR	9;16	348.12	18431	22MR	14;06	183.03
18432	6AP	1;43	17.12	18432	20AP	22;01	211.46
18433	5MY	17;33	45.67	18433	20MY	5;01	239.50
18434	4JN	8;07	73.92	18434	18JN	12;10	267.41
18435	3JL	21;19	102.08	18435	17JL	20;42	295.46
18436	2AU	9;21	130.41	18436	16AU	7;41	323.91
18437	31AU	20;32	159.08	18437	14SE	21;43	352.93
18438	30SE	7;07	188.23	18438	14OC	14;38	22.58
18439	29OC	17;24	217.83	18439	13NO	9;32	52.73
18440	28NO	3;46	247.77	18440	13DE	4;58	83.13
18441	27DE	14;37	277.81				

496

NUMBER	DATE	TIME	LONG.	NUMBER	DATE	TIME	LONG.
18503	1JA	21;44	282.98	18503	16JA	8;51	117.66
18504	31JA	15;57	313.09	18504	14FE	19;29	147.26
18505	1MR	7;23	342.65	18505	15MR	6;35	176.41
18506	30MR	19;53	11.59	18506	13AP	18;41	205.10
18507	29AP	5;51	39.98	18507	13MY	8;03	233.47
18508	28MY	13;55	68.00	18508	11JN	22;31	261.68
18509	26JN	20;55	95.91	18509	11JL	13;41	289.96
18510	26JL	3;52	123.96	18510	10AU	5;06	318.51
18511	24AU	11;52	152.41	18511	8SE	20;23	347.48
18512	22SE	21;57	181.42	18512	8OC	11;10	16.95
18513	22OC	10;55	211.03	18513	7NO	1;00	46.82
18514	21NO	2;49	241.14	18514	6DE	13;30	76.90
18515	20DE	20;56	271.50				

491

NUMBER	DATE	TIME	LONG.	NUMBER	DATE	TIME	LONG.
				18441	11JA	23;22	113.42
18442	26JA	2;16	307.70	18442	10FE	15;24	143.31
18443	24FE	14;51	337.23	18443	12MR	4;23	172.60
18444	26MR	4;16	6.30	18444	10AP	14;24	201.27
18445	24AP	18;21	34.92	18445	9MY	22;12	229.45
18446	24MY	8;57	63.24	18446	8JN	4;55	257.37
18447	22JN	23;57	91.46	18447	7JL	11;45	285.29
18448	22JL	15;00	119.80	18448	5AU	19;46	313.49
18449	21AU	5;36	148.45	18449	4SE	5;50	342.17
18450	19SE	19;13	177.54	18450	3OC	18;29	11.46
18451	19OC	7;42	207.07	18451	2NO	9;58	41.33
18452	17NO	19;14	236.97	18452	2DE	4;02	71.61
18453	17DE	6;11	267.01	18453	31DE	23;39	102.02

497

NUMBER	DATE	TIME	LONG.	NUMBER	DATE	TIME	LONG.
				18515	5JA	0;36	106.92
18516	19JA	15;55	301.76	18516	3FE	10;34	136.64
18517	18FE	10;18	331.62	18517	4MR	19;56	165.89
18518	20MR	2;52	0.91	18518	3AP	5;19	194.65
18519	18AP	16;53	29.62	18519	2MY	15;17	223.00
18520	18MY	4;09	57.85	18520	1JN	2;22	251.12
18521	16JN	13;10	85.84	18521	30JN	14;58	279.26
18522	15JL	20;51	113.84	18522	30JL	5;25	307.65
18523	14AU	4;21	142.12	18523	28AU	21;43	336.49
18524	12SE	12;43	170.88	18524	27SE	15;12	5.87
18525	11OC	22;43	200.20	18525	27OC	8;38	35.74
18526	10NO	10;42	230.03	18526	26NO	0;40	65.90
18527	10DE	0;44	260.19	18527	25DE	14;32	96.07

492

NUMBER	DATE	TIME	LONG.	NUMBER	DATE	TIME	LONG.
18454	15JA	16;50	296.95	18454	30JA	19;00	132.18
18455	14FE	3;16	326.56	18455	29FE	12;09	161.83
18456	14MR	13;37	355.68	18456	30MR	2;05	190.84
18457	13AP	0;17	24.33	18457	28AP	12;56	219.21
18458	12MY	11;52	52.63	18458	27MY	21;33	247.32
18459	11JN	0;58	80.79	18459	26JN	5;01	275.25
18460	10JL	15;48	109.05	18460	25JL	12;20	303.31
18461	9AU	7;56	137.62	18461	23AU	20;19	331.75
18462	8SE	0;23	166.63	18462	22SE	5;40	0.71
18463	7OC	16;16	196.13	18463	21OC	17;04	30.25
18464	6NO	7;03	226.03	18464	20NO	7;06	60.28
18465	5DE	20;31	256.15	18465	19DE	23;51	90.57

498

NUMBER	DATE	TIME	LONG.	NUMBER	DATE	TIME	LONG.
18528	8JA	16;38	290.40	18528	24JA	2;16	125.99
18529	7FE	9;55	320.38	18529	22FE	12;16	155.44
18530	9MR	3;42	349.92	18530	23MR	21;04	184.35
18531	7AP	20;40	18.91	18531	22AP	5;10	212.77
18532	7MY	11;40	47.41	18532	21MY	13;16	240.85
18533	6JN	0;16	75.57	18533	19JN	22;12	268.82
18534	5JL	10;50	103.64	18534	19JL	9;01	296.98
18535	3AU	20;11	131.87	18535	17AU	22;34	325.54
18536	2SE	5;13	160.48	18536	16SE	15;02	354.69
18537	1OC	14;36	189.60	18537	16OC	9;37	24.44
18538	31OC	0;41	219.21	18538	15NO	4;38	54.61
18539	29NO	11;44	249.18	18539	14DE	22;25	84.94
18540	28DE	23;59	279.28				

493

NUMBER	DATE	TIME	LONG.	NUMBER	DATE	TIME	LONG.
18466	4JA	8;35	286.22	18466	18JA	18;27	120.82
18467	2FE	19;13	315.97	18467	17FE	13;13	150.72
18468	4MR	4;34	345.23	18468	19MR	6;27	180.05
18469	2AP	13;05	13.96	18469	17AP	21;14	208.80
18470	1MY	21;40	42.27	18470	17MY	9;35	237.08
18471	31MY	7;18	70.34	18471	15JN	19;54	265.12
18472	29JN	18;50	98.43	18472	15JL	4;50	293.17
18473	29JL	8;40	126.78	18473	13AU	13;06	321.48
18474	28AU	0;36	155.60	18474	11SE	21;27	350.22
18475	26SE	17;59	184.96	18475	11OC	6;41	19.50
18476	26OC	11;54	214.84	18476	9NO	17;30	49.27
18477	25NO	5;15	245.05	18477	9DE	6;23	79.38
18478	24DE	20;54	275.30				

499

NUMBER	DATE	TIME	LONG.	NUMBER	DATE	TIME	LONG.
				18540	13JA	13;58	115.11
18541	27JA	13;43	309.25	18541	12FE	2;58	144.85
18542	26FE	4;55	338.86	18542	13MR	13;34	174.02
18543	27MR	21;00	8.01	18543	11AP	22;08	202.61
18544	26AP	13;03	36.69	18544	11MY	5;21	230.75
18545	26MY	4;14	65.02	18545	9JN	12;12	258.67
18546	24JN	18;11	93.20	18546	8JL	19;52	286.64
18547	24JL	6;58	121.47	18547	7AU	5;32	314.91
18548	22AU	18;49	150.03	18548	5SE	18;04	343.72
18549	21SE	5;57	179.04	18549	5OC	9;44	13.15
18550	20OC	16;34	208.51	18550	4NO	3;54	43.16
18551	19NO	2;59	238.37	18551	3DE	23;20	73.51
18552	18DE	13;32	268.40				

500

NEW MOON				FULL MOON			
18553	17JA	0;38	298.35	18552	2JA	18;27	103.89
18554	15FE	12;31	328.00	18553	1FE	11;46	133.95
18555	16MR	1;14	357.20	18554	2MR	2;14	163.46
18556	14AP	14;43	25.94	18555	31MR	13;35	192.34
18557	14MY	4;51	54.32	18556	29AP	22;18	220.67
18558	12JN	19;36	82.55	18557	29MY	5;22	248.65
18559	12JL	10;46	110.83	18558	27JN	12;00	276.54
18560	11AU	1;53	139.38	18559	26JL	19;22	304.62
18561	9SE	16;21	168.34	18560	25AU	4;24	333.11
18562	9OC	5;42	197.75	18561	23SE	15;48	2.18
18563	7NO	17;53	227.56	18562	23OC	5;56	31.86
18564	7DE	5;11	257.60	18563	21NO	22;48	62.02
				18564	21DE	17;47	92.41

501

NEW MOON				FULL MOON			
18565	5JA	15;57	287.60	18565	20JA	13;26	122.69
18566	4FE	2;20	317.33	18566	19FE	7;45	152.54
18567	5MR	12;28	346.60	18567	20MR	23;15	181.77
18568	3AP	22;39	15.38	18568	19AP	11;31	210.40
18569	3MY	9;25	43.76	18569	18MY	21;06	238.56
18570	1JN	21;31	71.92	18570	17JN	5;03	266.50
18571	1JL	11;26	100.10	18571	16JL	12;26	294.50
18572	31JL	3;05	128.55	18572	14AU	20;08	322.79
18573	29AU	19;43	157.42	18573	13SE	4;51	351.57
18574	28SE	12;17	186.78	18574	12OC	15;16	20.92
18575	28OC	3;56	216.59	18575	11NO	3;58	50.79
18576	26NO	18;15	246.68	18576	10DE	19;21	81.01
18577	26DE	7;07	276.81				

502

NEW MOON				FULL MOON			
18578	24JA	18;25	306.69	18577	9JA	13;03	111.29
18579	23FE	4;13	336.13	18578	8FE	7;45	141.32
18580	24MR	12;52	5.02	18579	10MR	1;43	170.85
18581	22AP	21;04	33.43	18580	8AP	17;39	199.79
18582	22MY	5;52	61.54	18581	8MY	7;09	228.22
18583	20JN	16;14	89.57	18582	6JN	18;28	256.33
18584	20JL	4;52	117.80	18583	6JL	4;09	284.36
18585	18AU	19;53	146.44	18584	4AU	12;51	312.57
18586	17SE	12;49	175.62	18585	2SE	21;15	341.17
18587	17OC	6;50	205.35	18586	2OC	6;08	10.27
18588	16NO	0;49	235.50	18587	31OC	16;11	39.90
18589	15DE	17;35	265.79	18588	30NO	4;02	69.91
				18589	29DE	17;52	100.07

503

NEW MOON				FULL MOON			
18590	14JA	8;07	295.91	18590	28JA	9;21	130.10
18591	12FE	20;00	325.59	18591	27FE	1;45	159.75
18592	14MR	5;28	354.70	18592	28MR	18;14	188.90
18593	12AP	13;20	23.25	18593	27AP	10;08	217.57
18594	11MY	20;35	51.39	18594	27MY	0;57	245.88
18595	10JN	4;13	79.34	18595	25JN	14;16	274.03
18596	9JL	13;10	107.36	18596	25JL	2;00	302.26
18597	8AU	0;11	135.70	18597	23AU	12;24	330.77
18598	6SE	13;49	164.56	18598	21SE	22;08	359.74
18599	6OC	6;15	194.04	18599	21OC	7;56	29.19
18600	5NO	1;03	224.08	18600	19NO	18;23	59.05
18601	4DE	20;47	254.45	18601	19DE	5;40	89.11

504

NEW MOON				FULL MOON			
18602	3JA	15;23	284.80	18602	17JA	17;46	119.10
18603	2FE	7;14	314.79	18603	16FE	6;35	148.78
18604	2MR	19;52	344.22	18604	16MR	20;13	178.00
18605	1AP	5;45	13.02	18605	15AP	10;46	206.78
18606	30AP	13;47	41.32	18606	15MY	2;03	235.20
18607	29MY	20;53	69.30	18607	13JN	17;28	263.44
18608	28JN	3;56	97.20	18608	13JL	8;17	291.72
18609	27JL	11;47	125.31	18609	11AU	22;00	320.22
18610	25AU	21;25	153.84	18610	10SE	10;39	349.12
18611	24SE	9;45	182.96	18611	9OC	22;34	18.49
18612	24OC	1;23	212.71	18612	8NO	10;03	48.28
18613	22NO	19;58	242.95	18613	7DE	21;12	78.30
18614	22DE	15;55	273.38				

505

NEW MOON				FULL MOON			
				18614	6JA	7;56	108.31
18615	21JA	11;07	303.63	18615	4FE	18;22	138.03
18616	20FE	3;55	333.40	18616	6MR	4;55	167.31
18617	21MR	17;46	2.56	18617	4AP	16;10	196.12
18618	20AP	4;52	31.12	18618	4MY	4;37	224.55
18619	19MY	13;47	59.25	18619	2JN	18;24	252.78
18620	17JN	21;14	87.17	18620	2JL	9;16	281.00
18621	17JL	4;09	115.16	18621	1AU	0;44	309.45
18622	15AU	11;36	143.45	18622	30AU	16;22	338.29
18623	13SE	20;38	172.25	18623	29SE	7;46	7.62
18624	13OC	8;12	201.66	18624	28OC	22;25	37.39
18625	11NO	22;42	231.62	18625	27NO	11;48	67.45
18626	11DE	15;48	261.91	18626	26DE	23;38	97.54

506

NEW MOON				FULL MOON			
18627	10JA	10;24	292.22	18627	25JA	10;02	127.38
18628	9FE	5;05	322.24	18628	23FE	19;27	156.79
18629	10MR	22;31	351.73	18629	25MR	4;31	185.69
18630	9AP	13;46	20.63	18630	23AP	13;53	214.14
18631	9MY	2;23	49.02	18631	23MY	0;09	242.49
18632	7JN	12;29	77.07	18632	21JN	11;48	270.38
18633	6JL	20;49	105.06	18633	21JL	1;18	298.65
18634	5AU	4;25	133.23	18634	19AU	16;49	327.32
18635	3SE	12;25	161.82	18635	18SE	10;03	356.53
18636	2OC	21;42	190.95	18636	18OC	3;57	26.27
18637	1NO	8;46	220.62	18637	16NO	21;03	56.39
18638	30NO	21;46	250.69	18638	16DE	12;11	86.61
18639	30DE	12;39	280.89				

507

NEW MOON				FULL MOON			
				18639	15JA	0;58	116.65
18640	29JA	5;06	310.96	18640	13FE	11;40	146.27
18641	27FE	22;31	340.64	18641	14MR	20;51	175.36
18642	29MR	15;50	9.81	18642	13AP	5;03	203.91
18643	28AP	7;48	38.46	18643	12MY	12;53	232.07
18644	27MY	21;38	66.73	18644	10JN	21;09	260.04
18645	26JN	9;15	94.82	18645	10JL	6;51	288.09
18646	25JL	19;17	122.98	18646	8AU	18;58	316.49
18647	24AU	4;36	151.47	18647	7SE	10;06	345.42
18648	22SE	13;59	180.42	18648	7OC	3;57	14.97
18649	21OC	23;52	209.89	18649	5NO	23;10	45.04
18650	20NO	10;29	239.76	18650	5DE	17;53	75.38
18651	19DE	22;03	269.84				

508

NEW MOON				FULL MOON			
				18651	4JA	10;41	105.65
18652	18JA	10;51	299.85	18652	3FE	0;56	135.57
18653	17FE	1;05	329.58	18653	3MR	12;37	164.93
18654	17MR	16;31	358.86	18654	1AP	22;01	193.71
18655	16AP	8;27	27.68	18655	1MY	5;41	221.98
18656	16MY	0;00	56.10	18656	30MY	12;30	249.95
18657	14JN	14;37	84.31	18657	28JN	19;34	277.86
18658	14JL	4;07	112.54	18658	28JL	4;06	305.99
18659	12AU	16;39	141.01	18659	26AU	15;11	334.59
18660	11SE	4;25	169.88	18660	25SE	5;23	3.80
18661	10OC	15;31	199.23	18661	24OC	22;31	33.63
18662	9NO	2;10	228.99	18662	23NO	17;35	63.90
18663	8DE	12;40	258.99	18663	23DE	13;05	94.31

509

NEW MOON				FULL MOON			
18664	6JA	23;23	288.99	18664	22JA	7;26	124.52
18665	5FE	10;39	318.74	18665	20FE	23;20	154.24
18666	6MR	22;40	348.07	18666	22MR	12;09	183.34
18667	5AP	11;29	16.93	18667	20AP	22;01	211.84
18668	5MY	1;02	45.40	18668	20MY	5;43	239.92
18669	3JN	15;21	73.64	18669	18JN	12;25	267.81
18670	3JL	6;22	101.88	18670	17JL	19;20	295.79
18671	1AU	21;45	130.33	18671	16AU	3;31	324.13
18672	31AU	12;54	159.16	18672	14SE	13;47	352.99
18673	30SE	3;09	188.45	18673	14OC	2;36	22.46
18674	29OC	16;10	218.17	18674	12NO	18;10	52.47
18675	28NO	4;02	248.17	18675	12DE	12;11	82.80
18676	27DE	15;03	278.22				

510

NEW MOON				FULL MOON			
				18676	11JA	7;37	113.15
18677	26JA	1;30	308.06	18677	10FE	2;43	143.17
18678	24FE	11;33	337.48	18678	11MR	19;37	172.63
18679	25MR	21;24	6.41	18679	10AP	9;24	201.45
18680	24AP	7;32	34.87	18680	9MY	20;12	229.76
18681	23MY	18;40	63.06	18681	8JN	4;53	257.75
18682	22JN	7;31	91.19	18682	7JL	12;31	285.71
18683	21JL	22;21	119.52	18683	5AU	20;05	313.89
18684	20AU	14;47	148.24	18684	4SE	4;22	342.50
18685	19SE	7;47	177.46	18685	3OC	13;59	11.66
18686	19OC	0;15	207.15	18686	2NO	1;33	41.36
18687	17NO	15;31	237.20	18687	1DE	15;34	71.47
18688	17DE	5;14	267.36	18688	31DE	8;05	101.75

511

NEW MOON				FULL MOON			
18689	15JA	17;18	297.36	18689	30JA	2;17	131.87
18690	14FE	3;44	326.97	18690	28FE	20;36	161.58
18691	15MR	12;42	356.03	18691	30MR	13;29	190.72
18692	13AP	20;49	24.58	18692	29AP	4;07	219.31
18693	13MY	4;59	52.74	18693	28MY	16;30	247.51
18694	11JN	14;18	80.75	18694	27JN	3;02	275.55
18695	11JL	1;41	108.88	18695	26JL	12;20	303.70
18696	9AU	15;33	137.35	18696	24AU	21;00	332.16
18697	8SE	7;45	166.34	18697	23SE	5;45	1.10
18698	8OC	1;33	195.90	18698	22OC	15;17	30.57
18699	6NO	19;55	225.95	18699	21NO	2;15	60.46
18700	6DE	13;37	256.24	18700	20DE	15;01	90.59

```
--------NEW MOONS---------        ---------FULL MOONS--------        --------NEW MOONS---------        --------FULL MOONS--------
NUMBER  DATE   TIME   LONG.     NUMBER  DATE   TIME   LONG.       NUMBER  DATE   TIME   LONG.     NUMBER  DATE   TIME   LONG.
```

				512								518			
18701	5JA	5;25	286.47	18701	19JA	5;34	120.67					18775	13JA	2;31	115.02
18702	3FE	18;36	316.33	18702	17FE	21;20	150.45	18776	27JA	9;12	309.45	18776	11FE	19;42	144.94
18703	4MR	5;06	345.64	18703	18MR	13;35	179.75	18777	25FE	20;35	338.91	18777	13MR	10;00	174.26
18704	2AP	13;30	14.37	18704	17AP	5;38	208.56	18778	27MR	8;42	7.90	18778	11AP	21;12	202.96
18705	1MY	20;47	42.62	18705	16MY	20;58	236.97	18779	25AP	21;36	36.46	18779	11MY	5;49	231.16
18706	31MY	4;00	70.59	18706	15JN	11;07	265.16	18780	25MY	11;21	64.74	18780	9JN	12;51	259.09
18707	29JN	12;07	98.55	18707	14JL	23;49	293.36	18781	24JN	1;58	92.94	18781	8JL	19;34	287.01
18708	28JL	22;01	126.74	18708	13AU	11;04	321.78	18782	23JL	17;20	121.31	18782	7AU	3;05	315.20
18709	27AU	10;21	155.40	18709	11SE	21;19	350.61	18783	22AU	8;56	150.02	18783	5SE	12;20	343.87
18710	26SE	1;29	184.67	18710	11OC	7;14	19.91	18784	21SE	0;02	179.18	18784	4OC	23;56	13.13
18711	25OC	19;21	214.53	18711	9NO	17;28	49.66	18785	20OC	13;58	208.80	18785	3NO	14;11	42.97
18712	24NO	14;58	244.83	18712	9DE	4;22	79.69	18786	19NO	2;35	238.74	18786	3DE	7;02	73.21
18713	24DE	10;27	275.24					18787	18DE	14;03	268.81				

				513								519			
				18713	7JA	15;58	109.72					18787	2JA	1;50	103.57
18714	23JA	3;48	305.40	18714	6FE	4;12	139.50	18788	17JA	0;43	298.75	18788	31JA	21;12	133.73
18715	21FE	17;57	335.04	18715	7MR	17;07	168.86	18789	15FE	10;47	328.32	18789	2MR	15;15	163.40
18716	23MR	5;01	4.05	18716	6AP	6;53	197.75	18790	16MR	20;27	357.39	18790	1AP	6;33	192.44
18717	21AP	13;46	32.50	18717	5MY	21;37	226.25	18791	15AP	6;07	25.98	18791	30AP	18;43	220.91
18718	20MY	21;10	60.56	18718	4JN	12;58	254.53	18792	14MY	16;25	54.22	18792	30MY	4;21	248.99
18719	19JN	4;06	88.46	18719	4JL	4;15	282.78	18793	13JN	4;10	82.32	18793	28JN	12;28	276.95
18720	18JL	11;28	116.47	18720	2AU	18;45	311.20	18794	12JL	17;59	110.55	18794	27JL	20;05	305.03
18721	16AU	20;10	144.83	18721	1SE	8;12	339.98	18795	11AU	9;49	139.12	18795	26AU	4;06	333.49
18722	15SE	7;07	173.74	18722	30SE	20;44	9.21	18796	10SE	2;54	168.17	18796	24SE	13;08	2.46
18723	14OC	21;09	203.27	18723	30OC	8;39	38.90	18797	9OC	20;02	197.74	18797	23OC	23;45	31.99
18724	13NO	14;26	233.36	18724	28NO	20;06	68.88	18798	8NO	12;13	227.71	18798	22NO	12;31	61.97
18725	13DE	9;57	263.76	18725	28DE	7;03	98.93	18799	8DE	2;53	257.89	18799	22DE	3;44	92.21

				514								520			
18726	12JA	5;43	294.11	18726	26JA	17;29	128.76	18800	6JA	15;50	287.99	18800	20JA	21;03	122.39
18727	10FE	23;45	324.08	18727	25FE	3;41	158.18	18801	5FE	3;00	317.75	18801	19FE	15;17	152.25
18728	12MR	15;00	353.45	18728	26MR	14;13	187.12	18802	5MR	12;29	346.99	18802	20MR	8;49	181.57
18729	11AP	3;21	22.21	18729	25AP	1;44	215.64	18803	3AP	20;44	15.69	18803	19AP	0;30	210.34
18730	10MY	13;15	50.47	18730	24MY	14;39	243.89	18804	3MY	4;34	43.95	18804	18MY	13;58	238.66
18731	8JN	21;20	78.44	18731	23JN	4;56	272.08	18805	1JN	13;01	71.97	18805	17JN	1;27	266.74
18732	8JL	4;29	106.38	18732	22JL	20;14	300.43	18806	30JN	23;11	100.00	18806	16JL	11;28	294.85
18733	6AU	11;38	134.55	18733	21AU	12;02	329.14	18807	30JL	11;46	128.32	18807	14AU	20;35	323.20
18734	4SE	19;53	163.17	18734	20SE	3;54	358.32	18808	29AU	2;57	157.12	18808	13SE	5;26	351.99
18735	4OC	6;12	192.37	18735	19OC	19;17	27.98	18809	27SE	20;14	186.50	18809	12OC	14;40	21.29
18736	2NO	19;17	222.14	18736	18NO	9;35	57.99	18810	27OC	14;41	216.42	18810	11NO	0;56	51.06
18737	2DE	11;10	252.34	18737	17DE	22;18	88.12	18811	26NO	9;03	246.68	18811	10DE	12;46	81.12
								18812	26DE	2;02	276.98				

				515								521			
18738	1JA	5;04	282.67	18738	16JA	9;21	118.07					18812	9JA	2;19	111.23
18739	30JA	23;42	312.80	18739	14FE	19;03	147.63	18813	24JA	16;34	307.01	18813	7FE	17;19	141.11
18740	1MR	17;43	342.48	18740	16MR	4;01	176.69	18814	23FE	4;17	336.52	18814	9MR	9;07	170.56
18741	31MR	10;01	11.58	18741	14AP	12;55	205.26	18815	24MR	13;30	5.44	18815	8AP	1;04	199.51
18742	29AP	23;54	40.13	18742	13MY	22;27	233.47	18816	22AP	21;05	33.82	18816	7MY	16;40	228.03
18743	29MY	11;15	68.28	18743	12JN	9;11	261.54	18817	22MY	4;05	61.85	18817	6JN	7;28	256.28
18744	27JN	20;28	96.28	18744	11JL	21;39	289.70	18818	20JN	11;34	89.77	18818	5JL	21;03	284.46
18745	27JL	4;27	124.38	18745	10AU	12;13	318.21	18819	19JL	20;28	117.85	18819	4AU	9;15	312.81
18746	25AU	12;18	152.82	18746	9SE	4;51	347.24	18820	18AU	7;33	146.33	18820	2SE	20;12	341.51
18747	23SE	21;02	181.77	18747	8OC	22;50	16.82	18821	16SE	21;21	175.37	18821	2OC	6;27	10.68
18748	23OC	7;17	211.27	18748	7NO	16;46	46.86	18822	16OC	14;01	205.04	18822	31OC	16;38	40.31
18749	21NO	19;21	241.22	18749	7DE	9;10	77.11	18823	15NO	9;03	235.22	18823	30NO	3;16	70.26
18750	21DE	9;15	271.39					18824	15DE	4;56	265.64	18824	29DE	14;28	100.32

				516								522			
				18750	5JA	23;10	107.25	18825	13JA	23;33	295.93	18825	28JA	2;13	130.19
18751	20JA	0;46	301.51	18751	4FE	10;48	137.04	18826	12FE	15;19	325.78	18826	26FE	14;30	159.68
18752	18FE	17;32	331.31	18752	4MR	20;32	166.31	18827	14MR	3;46	355.01	18827	28MR	3;31	188.69
18753	19MR	10;47	0.65	18753	3AP	4;59	195.02	18828	12AP	13;28	23.64	18828	26AP	17;30	217.29
18754	18AP	3;24	29.46	18754	2MY	12;45	223.29	18829	11MY	21;21	51.81	18829	26MY	8;27	245.61
18755	17MY	18;19	57.85	18755	31MY	20;34	251.28	18830	10JN	4;23	79.73	18830	24JN	23;51	273.84
18756	16JN	7;05	85.98	18756	30JN	5;20	279.26	18831	9JL	11;27	107.68	18831	24JL	14;59	302.20
18757	15JL	18;00	114.12	18757	29JL	16;07	307.50	18832	7AU	19;26	135.89	18832	23AU	5;14	330.87
18758	14AU	3;48	142.49	18758	28AU	5;46	336.23	18833	6SE	5;14	164.60	18833	21SE	18;30	359.98
18759	12SE	13;19	171.29	18759	26SE	22;30	5.57	18834	5OC	17;44	193.91	18834	21OC	6;58	29.54
18760	11OC	23;07	200.61	18760	26OC	17;25	35.49	18835	4NO	9;29	223.81	18835	19NO	18;50	59.46
18761	10NO	9;28	230.37	18761	25NO	12;46	65.79	18836	4DE	4;05	254.13	18836	19DE	6;05	89.52
18762	9DE	20;33	260.40	18762	25DE	6;44	96.13								

				517								523			
18763	8JA	8;34	290.45	18763	23JA	22;18	126.21	18837	2JA	23;57	284.54	18837	17JA	16;41	119.45
18764	6FE	21;50	320.26	18764	22FE	11;10	155.78	18838	1FE	18;57	314.67	18838	16FE	2;45	149.01
18765	8MR	12;25	349.68	18765	23MR	21;32	184.76	18839	3MR	11;33	344.26	18839	17MR	12;47	178.09
18766	7AP	3;55	18.63	18766	22AP	5;52	213.18	18840	2AP	1;13	13.23	18840	15AP	23;27	206.71
18767	6MY	19;34	47.16	18767	21MY	12;54	241.22	18841	1MY	12;14	41.64	18841	15MY	11;25	235.00
18768	5JN	10;39	75.42	18768	19JN	19;39	269.11	18842	30MY	21;09	69.69	18842	14JN	0;54	263.17
18769	5JL	0;49	103.63	18769	19JL	3;18	297.13	18843	29JN	4;43	97.62	18843	13JL	15;44	291.45
18770	3AU	14;02	132.01	18770	17AU	13;03	325.55	18844	28JL	11;51	125.70	18844	12AU	7;30	320.02
18771	2SE	2;27	160.76	18771	16SE	1;44	354.53	18845	26AU	19;33	154.15	18845	10SE	23;39	349.06
18772	1OC	14;09	189.97	18772	15OC	17;35	24.15	18846	25SE	4;51	183.15	18846	10OC	15;39	18.59
18773	31OC	1;13	219.63	18773	14NO	11;55	54.30	18847	24OC	16;35	212.73	18847	9NO	6;50	48.53
18774	29NO	11;52	249.58	18774	14DE	7;26	84.70	18848	23NO	7;07	242.79	18848	8DE	20;32	78.67
18775	28DE	22;23	279.60					18849	23DE	0;03	273.10				

524

NEW MOONS

NUMBER	DATE	TIME	LONG.
18850	21JA	18;19	303.32
18851	20FE	12;36	333.16
18852	21MR	5;42	2.45
18853	19AP	20;45	31.18
18854	19MY	9;22	59.46
18855	17JN	19;39	87.50
18856	17JL	4;16	115.55
18857	15AU	12;15	143.87
18858	13SE	20;37	172.64
18859	13OC	6;13	201.97
18860	11NO	17;26	231.78
18861	11DE	6;24	261.90

FULL MOONS

NUMBER	DATE	TIME	LONG.
18849	7JA	8;25	108.72
18850	5FE	18;37	138.43
18851	6MR	3;41	167.65
18852	4AP	12;19	196.35
18853	3MY	21;14	224.64
18854	2JN	7;07	252.72
18855	1JL	18;34	280.81
18856	31JL	8;04	309.17
18857	29AU	23;50	338.01
18858	28SE	17;29	7.41
18859	28OC	11;54	37.34
18860	27NO	5;25	67.57
18861	26DE	20;46	97.80

530

NEW MOONS

NUMBER	DATE	TIME	LONG.
18924	15JA	13;51	297.61
18925	14FE	2;55	327.32
18926	15MR	13;11	356.44
18927	13AP	21;20	24.99
18928	13MY	4;22	53.11
18929	11JN	11;24	81.03
18930	10JL	19;28	109.02
18931	9AU	5;25	137.33
18932	7SE	17;52	166.16
18933	7OC	9;13	195.61
18934	6NO	3;17	225.64
18935	5DE	23;04	256.02

FULL MOONS

NUMBER	DATE	TIME	LONG.
18924	29JA	13;45	131.74
18925	28FE	4;56	161.32
18926	29MR	20;36	190.42
18927	28AP	12;14	219.06
18928	28MY	3;25	247.38
18929	26JN	17;44	275.57
18930	26JL	6;51	303.87
18931	24AU	18;39	332.46
18932	23SE	5;27	1.48
18933	22OC	15;49	30.98
18934	21NO	2;19	60.86
18935	20DE	13;14	90.90

525

NEW MOONS

NUMBER	DATE	TIME	LONG.
18862	9JA	20;58	292.05
18863	8FE	12;56	321.95
18864	10MR	5;47	351.43
18865	8AP	22;38	20.41
18866	8MY	14;25	48.93
18867	7JN	4;18	77.13
18868	6JL	16;14	105.26
18869	5AU	2;42	133.55
18870	3SE	12;31	162.21
18871	2OC	22;22	191.37
18872	1NO	8;35	221.01
18873	30NO	19;20	250.98
18874	30DE	6;46	281.04

FULL MOONS

NUMBER	DATE	TIME	LONG.
18862	25JA	9;31	127.75
18863	23FE	20;01	157.20
18864	25MR	4;53	186.09
18865	23AP	12;45	214.48
18866	22MY	20;18	242.53
18867	21JN	4;23	270.47
18868	20JL	14;00	298.58
18869	19AU	2;10	327.11
18870	17SE	17;30	356.24
18871	17OC	11;39	25.98
18872	16NO	7;11	56.19
18873	16DE	2;07	86.57

531

NEW MOONS

NUMBER	DATE	TIME	LONG.
18936	4JA	18;35	286.40
18937	3FE	11;52	316.44
18938	5MR	1;52	345.90
18939	3AP	12;46	14.73
18940	2MY	21;21	43.03
18941	1JN	4;40	71.01
18942	30JN	11;37	98.91
18943	29JL	19;06	127.01
18944	28AU	3;58	155.53
18945	26SE	15;07	184.63
18946	26OC	5;17	214.33
18947	24NO	22;36	244.53
18948	24DE	18;02	274.94

FULL MOONS

NUMBER	DATE	TIME	LONG.
18936	19JA	0;35	120.85
18937	17FE	12;21	150.46
18938	19MR	0;39	179.61
18939	17AP	13;50	208.31
18940	17MY	4;07	236.68
18941	15JN	19;19	264.92
18942	15JL	10;46	293.23
18943	14AU	1;45	321.79
18944	12SE	15;48	350.77
18945	12OC	4;56	20.21
18946	10NO	17;18	50.05
18947	10DE	4;58	80.10

526

NEW MOONS

NUMBER	DATE	TIME	LONG.
18875	28JA	19;11	310.93
18876	27FE	8;50	340.46
18877	28MR	23;39	9.54
18878	27AP	15;04	38.18
18879	27MY	6;24	66.51
18880	25JN	21;06	94.72
18881	25JL	10;58	123.04
18882	24AU	0;03	151.67
18883	22SE	12;24	180.75
18884	22OC	0;02	210.29
18885	20NO	11;00	240.18
18886	19DE	21;33	270.21

FULL MOONS

NUMBER	DATE	TIME	LONG.
18874	14JA	18;59	116.78
18875	13FE	9;09	146.55
18876	14MR	20;38	175.73
18877	13AP	5;48	204.33
18878	12MY	13;17	232.47
18879	10JN	19;59	260.38
18880	10JL	3;02	288.33
18881	8AU	11;38	316.58
18882	6SE	22;51	345.35
18883	6OC	13;14	14.75
18884	5NO	6;31	44.73
18885	5DE	1;40	75.08

532

NEW MOONS

NUMBER	DATE	TIME	LONG.
18949	23JA	13;37	305.19
18950	22FE	7;24	334.99
18951	22MR	22;27	4.17
18952	21AP	10;41	32.77
18953	20MY	20;33	60.92
18954	19JN	4;45	88.87
18955	18JL	12;05	116.88
18956	16AU	19;31	145.19
18957	15SE	4;02	174.00
18958	14OC	14;34	203.38
18959	13NO	3;45	233.30
18960	12DE	19;31	263.54

FULL MOONS

NUMBER	DATE	TIME	LONG.
18948	8JA	15;52	110.10
18949	7FE	2;01	139.80
18950	7MR	11;45	169.02
18951	5AP	21;45	197.76
18952	5MY	8;44	226.12
18953	3JN	21;16	254.29
18954	3JL	11;25	282.50
18955	2AU	2;52	310.95
18956	31AU	19;07	339.83
18957	30SE	11;34	9.22
18958	30OC	3;31	39.07
18959	28NO	18;13	69.19
18960	28DE	7;05	99.32

527

NEW MOONS

NUMBER	DATE	TIME	LONG.
18887	18JA	8;05	300.13
18888	16FE	18;55	329.71
18889	18MR	6;23	358.84
18890	16AP	18;36	27.51
18891	16MY	7;41	55.84
18892	14JN	21;45	84.03
18893	14JL	12;50	112.32
18894	13AU	4;35	140.91
18895	11SE	20;20	169.94
18896	11OC	11;12	199.44
18897	10NO	0;43	229.32
18898	9DE	12;51	259.39

FULL MOONS

NUMBER	DATE	TIME	LONG.
18886	3JA	21;09	105.47
18887	2FE	15;22	135.55
18888	4MR	7;06	165.10
18889	2AP	19;45	194.01
18890	2MY	5;30	222.36
18891	31MY	13;11	250.36
18892	29JN	19;57	278.26
18893	29JL	3;01	306.34
18894	27AU	11;24	334.83
18895	25SE	21;53	3.88
18896	25OC	10;52	33.53
18897	24NO	2;27	63.64
18898	23DE	20;20	93.99

533

NEW MOONS

NUMBER	DATE	TIME	LONG.
18961	11JA	13;08	293.80
18962	10FE	7;21	323.78
18963	12MR	0;57	353.26
18964	10AP	16;59	22.18
18965	10MY	6;49	50.60
18966	8JN	18;17	78.70
18967	8JL	3;46	106.74
18968	6AU	12;07	134.96
18969	4SE	20;22	163.58
18970	4OC	5;27	192.73
18971	2NO	15;57	222.39
18972	2DE	4;04	252.43
18973	31DE	17;44	282.58

FULL MOONS

NUMBER	DATE	TIME	LONG.
18961	26JA	18;01	129.17
18962	25FE	3;26	158.56
18963	26MR	11;59	187.41
18964	24AP	20;27	215.81
18965	24MY	5;35	243.91
18966	22JN	16;04	271.95
18967	22JL	4;28	300.19
18968	20AU	19;10	328.84
18969	19SE	12;08	358.06
18970	19OC	6;36	27.84
18971	18NO	0;59	58.02
18972	17DE	17;39	88.31

528

NEW MOONS

NUMBER	DATE	TIME	LONG.
18899	7JA	23;52	289.40
18900	6FE	10;05	319.10
18901	6MR	19;43	348.34.
18902	5AP	5;04	17.06
18903	4MY	14;43	45.37
18904	3JN	1;28	73.47
18905	2JL	14;07	101.62
18906	1AU	5;01	130.04
18907	30AU	21;47	158.93
18908	29SE	15;19	188.35
18909	29OC	8;20	218.23
18910	28NO	0;01	248.39
18911	27DE	13;55	278.56

FULL MOONS

NUMBER	DATE	TIME	LONG.
18899	22JA	15;30	124.24
18900	21FE	10;16	154.09
18901	22MR	2;55	183.35
18902	20AP	16;34	212.01
18903	20MY	3;22	240.21
18904	18JN	12;11	268.19
18905	17JL	20;03	296.21
18906	16AU	3;56	324.53
18907	14SE	12;33	353.33
18908	13OC	22;27	22.68
18909	12NO	10;08	52.52
18910	12DE	0;04	82.68

534

NEW MOONS

NUMBER	DATE	TIME	LONG.
18974	30JA	8;48	312.57
18975	1MR	1;00	342.18
18976	30MR	17;44	11.31
18977	29AP	10;01	39.96
18978	29MY	0;54	68.26
18979	27JN	13;53	96.40
18980	27JL	1;13	124.63
18981	25AU	11;31	153.17
18982	23SE	21;32	182.18
18983	23OC	7;44	211.68
18984	21NO	18;19	241.57
18985	21DE	5;21	271.62

FULL MOONS

NUMBER	DATE	TIME	LONG.
18973	16JA	7;43	118.39
18974	14FE	19;12	148.03
18975	16MR	4;40	177.10
18976	14AP	12;48	205.64
18977	13MY	20;16	233.77
18978	12JN	3;52	261.71
18979	11JL	12;33	289.73
18980	9AU	23;20	318.08
18981	8SE	13;08	346.98
18982	8OC	6;07	16.52
18983	7NO	1;20	46.60
18984	6DE	20;55	76.98

529

NEW MOONS

NUMBER	DATE	TIME	LONG.
18912	26JA	1;56	308.46
18913	24FE	12;06	337.89
18914	25MR	20;43	6.77
18915	24AP	4;27	35.14
18916	23MY	12;17	63.19
18917	21JN	21;22	91.18
18918	21JL	8;39	119.36
18919	19AU	22;36	147.97
18920	18SE	15;04	177.16
18921	18OC	9;16	206.91
18922	17NO	4;01	237.10
18923	16DE	21;58	267.44

FULL MOONS

NUMBER	DATE	TIME	LONG.
18911	10JA	16;16	112.89
18912	9FE	10;00	142.86
18913	11MR	3;49	172.37
18914	9AP	20;21	201.32
18915	9MY	10;52	229.77
18916	7JN	23;21	257.92
18917	7JL	10;11	286.01
18918	5AU	19;53	314.27
18919	4SE	5;01	342.92
18920	3OC	14;11	12.06
18921	2NO	0;00	41.69
18922	1DE	11;02	71.67
18923	30DE	23;38	101.78

535

NEW MOONS

NUMBER	DATE	TIME	LONG.
18986	19JA	17;05	301.58
18987	18FE	5;50	331.22
18988	19MR	19;48	0.42
18989	18AP	10;43	29.18
18990	18MY	2;00	57.59
18991	16JN	17;03	85.81
18992	16JL	7;28	114.08
18993	14AU	21;12	142.61
18994	13SE	10;13	171.56
18995	12OC	22;29	200.97
18996	11NO	9;57	230.78
18997	10DE	20;45	260.80

FULL MOONS

NUMBER	DATE	TIME	LONG.
18985	5JA	14;59	107.30
18986	4FE	6;30	137.25
18987	5MR	19;12	166.64
18988	4AP	5;22	195.43
18989	3MY	13;30	223.70
18990	1JN	20;25	251.66
18991	1JL	3;08	279.56
18992	30JL	10;51	307.68
18993	28AU	20;43	336.25
18994	27SE	9;35	5.42
18995	27OC	1;35	35.22
18996	25NO	20;00	65.47
18997	25DE	15;30	95.88

536

New Moons

NUMBER	DATE	TIME	LONG.
18998	9JA	7;11	290.78
18999	7FE	17;39	320.48
19000	8MR	4;31	349.74
19001	6AP	16;03	18.53
19002	6MY	4;25	46.94
19003	4JN	17;50	75.14
19004	4JL	8;24	103.36
19005	3AU	0;01	131.83
19006	1SE	16;07	160.72
19007	1OC	7;50	190.10
19008	30OC	22;20	219.90
19009	29NO	11;18	249.95
19010	28DE	22;53	280.02

Full Moons

NUMBER	DATE	TIME	LONG.
18998	24JA	10;28	126.10
18999	23FE	3;28	155.85
19000	23MR	17;36	184.98
19001	22AP	4;41	213.52
19002	21MY	13;14	241.62
19003	19JN	20;20	269.53
19004	19JL	3;11	297.52
19005	17AU	10;55	325.85
19006	15SE	20;25	354.70
19007	15OC	8;12	24.15
19008	13NO	22;31	54.12
19009	13DE	15;16	84.40

537

New Moons

NUMBER	DATE	TIME	LONG.
19011	27JA	9;24	309.84
19012	25FE	19;07	339.23
19013	27MR	4;20	8.11
19014	25AP	13;30	36.52
19015	24MY	23;23	64.65
19016	23JN	10;53	92.73
19017	23JL	0;38	121.03
19018	21AU	16;41	149.75
19019	20SE	10;13	179.00
19020	20OC	3;54	208.76
19021	18NO	20;33	238.88
19022	18DE	11;30	269.10

Full Moons

NUMBER	DATE	TIME	LONG.
19010	12JA	9;50	114.70
19011	11FE	4;51	144.70
19012	12MR	22;34	174.18
19013	11AP	13;41	203.04
19014	11MY	1;49	231.39
19015	9JN	11;33	259.42
19016	8JL	19;53	287.41
19017	7AU	3;50	315.62
19018	5SE	12;11	344.26
19019	4OC	21;32	13.42
19020	3NO	8;21	43.11
19021	2DE	21;07	73.18

538

New Moons

NUMBER	DATE	TIME	LONG.
19023	17JA	0;28	299.13
19024	15FE	11;27	328.73
19025	16MR	20;38	357.79
19026	15AP	4;31	26.30
19027	14MY	12;00	54.43
19028	12JN	20;12	82.39
19029	12JL	6;13	110.46
19030	10AU	18;50	138.90
19031	9SE	10;12	167.88
19032	9OC	3;49	197.45
19033	7NO	22;39	227.53
19034	7DE	17;18	257.87

Full Moons

NUMBER	DATE	TIME	LONG.
19022	1JA	12;05	103.38
19023	31JA	4;57	133.44
19024	1MR	22;40	163.10
19025	31MR	15;46	192.23
19026	30AP	7;14	220.84
19027	29MY	20;42	249.08
19028	28JN	8;26	277.17
19029	27JL	18;50	305.37
19030	26AU	4;25	333.89
19031	24SE	13;43	2.88
19032	23OC	23;18	32.36
19033	22NO	9;45	62.25
19034	21DE	21;30	92.33

539

New Moons

NUMBER	DATE	TIME	LONG.
19035	6JA	10;25	288.15
19036	5FE	0;54	318.05
19037	6MR	12;26	347.38
19038	4AP	21;24	16.11
19039	4MY	4;44	44.34
19040	2JN	11;33	72.29
19041	1JL	18;57	100.22
19042	31JL	3;53	128.39
19043	29AU	15;05	157.03
19044	28SE	5;03	186.26
19045	27OC	21;55	216.10
19046	26NO	17;06	246.39
19047	26DE	13;02	276.82

Full Moons

NUMBER	DATE	TIME	LONG.
19035	20JA	10;42	122.35
19036	19FE	1;09	152.05
19037	20MR	16;20	181.29
19038	19AP	7;47	210.06
19039	18MY	23;07	238.46
19040	17JN	13;57	266.67
19041	17JL	3;52	294.93
19042	15AU	16;35	323.43
19043	14SE	4;07	352.32
19044	13OC	14;52	21.69
19045	12NO	1;25	51.47
19046	11DE	12;10	81.49

540

New Moons

NUMBER	DATE	TIME	LONG.
19048	25JA	7;37	307.02
19049	23FE	23;14	336.70
19050	24MR	11;31	5.74
19051	22AP	21;04	34.21
19052	22MY	4;51	62.27
19053	20JN	11;53	90.17
19054	19JL	19;03	118.18
19055	18AU	3;12	146.54
19056	16SE	13;12	175.43
19057	16OC	1;53	204.92
19058	14NO	17;42	234.96
19059	14DE	12;14	265.32

Full Moons

NUMBER	DATE	TIME	LONG.
19047	9JA	23;13	111.49
19048	8FE	10;34	141.21
19049	8MR	22;18	170.50
19050	7AP	10;42	199.32
19051	7MY	0;10	227.76
19052	5JN	14;49	256.00
19053	5JL	6;16	284.26
19054	3AU	21;45	312.74
19055	2SE	12;35	341.59
19056	2OC	2;28	10.90
19057	31OC	15;28	40.65
19058	30NO	3;38	70.67
19059	29DE	14;57	100.73

541

New Moons

NUMBER	DATE	TIME	LONG.
19060	13JA	7;54	295.66
19061	12FE	2;39	325.64
19062	13MR	19;00	355.04
19063	12AP	8;31	23.83
19064	11MY	19;29	52.12
19065	10JN	4;28	80.12
19066	9JL	12;13	108.09
19067	7AU	19;37	136.29
19068	6SE	3;38	164.92
19069	5OC	13;12	194.11
19070	4NO	1;05	223.85
19071	3DE	15;35	253.99

Full Moons

NUMBER	DATE	TIME	LONG.
19060	28JA	1;21	130.54
19061	26FE	11;01	159.92
19062	27MR	20;32	188.80
19063	26AP	6;41	217.24
19064	25MY	18;11	245.43
19065	24JN	7;26	273.58
19066	23JL	22;19	301.93
19067	22AU	14;24	330.66
19068	21SE	7;05	359.89
19069	20OC	23;40	29.62
19070	19NO	15;19	59.70
19071	19DE	5;15	89.88

542

New Moons

NUMBER	DATE	TIME	LONG.
19072	2JA	8;16	284.27
19073	1FE	2;08	314.35
19074	2MR	19;57	344.01
19075	1AP	12;42	13.11
19076	1MY	3;37	41.68
19077	30MY	16;17	69.89
19078	29JN	2;48	97.93
19079	28JL	11;47	126.08
19080	26AU	20;10	154.57
19081	25SE	4;57	183.54
19082	24OC	14;50	213.05
19083	23NO	2;11	242.98
19084	22DE	15;01	273.10

Full Moons

NUMBER	DATE	TIME	LONG.
19072	17JA	17;07	119.86
19073	16FE	3;06	149.41
19074	17MR	11;48	178.44
19075	15AP	20;00	206.96
19076	15MY	4;32	235.12
19077	13JN	14;07	263.13
19078	13JL	1;26	291.26
19079	11AU	14;59	319.74
19080	10SE	7;00	348.76
19081	10OC	1;05	18.37
19082	8NO	19;57	48.46
19083	8DE	13;48	78.77

543

New Moons

NUMBER	DATE	TIME	LONG.
19085	21JA	5;13	303.15
19086	19FE	20;38	332.89
19087	21MR	12;54	2.16
19088	20AP	5;19	30.95
19089	19MY	20;56	59.35
19090	18JN	10;57	87.54
19091	17JL	23;14	115.73
19092	16AU	10;12	144.17
19093	14SE	20;33	173.03
19094	14OC	6;52	202.39
19095	12NO	17;23	232.17
19096	12DE	4;12	262.20

Full Moons

NUMBER	DATE	TIME	LONG.
19084	7JA	5;16	108.97
19085	5FE	17;57	138.79
19086	7MR	4;14	168.06
19087	5AP	12;48	196.76
19088	4MY	20;22	225.00
19089	3JN	3;42	252.97
19090	2JL	11;39	280.92
19091	31JL	21;16	309.12
19092	30AU	9;32	337.81
19093	29SE	1;04	7.12
19094	28OC	19;29	37.04
19095	27NO	15;15	67.36
19096	27DE	10;20	97.75

544

New Moons

NUMBER	DATE	TIME	LONG.
19097	10JA	15;27	292.20
19098	9FE	3;25	321.95
19099	9MR	16;28	351.28
19100	8AP	6;39	20.15
19101	7MY	21;37	48.64
19102	6JN	12;46	76.90
19103	6JL	3;36	105.14
19104	4AU	17;53	133.58
19105	3SE	7;34	162.40
19106	2OC	20;31	191.68
19107	1NO	8;37	221.40
19108	30NO	19;51	251.39
19109	30DE	6;25	281.41

Full Moons

NUMBER	DATE	TIME	LONG.
19097	26JA	3;10	127.87
19098	24FE	17;12	157.47
19099	25MR	4;29	186.46
19100	23AP	13;28	214.89
19101	22MY	20;49	242.93
19102	21JN	3;28	270.82
19103	20JL	10;34	298.83
19104	18AU	19;18	327.22
19105	17SE	6;41	356.18
19106	16OC	21;14	25.76
19107	15NO	14;37	55.88
19108	15DE	9;46	86.27

545

New Moons

NUMBER	DATE	TIME	LONG.
19110	28JA	16;41	311.21
19111	27FE	3;05	340.61
19112	28MR	13;58	9.53
19113	27AP	1;38	38.03
19114	26MY	14;18	66.26
19115	25JN	4;11	94.44
19116	24JL	19;23	122.80
19117	23AU	11;33	151.55
19118	22SE	3;53	180.78
19119	21OC	19;22	210.48
19120	20NO	9;19	240.50
19121	19DE	21;39	270.60

Full Moons

NUMBER	DATE	TIME	LONG.
19109	14JA	5;07	116.59
19110	12FE	23;09	146.53
19111	14MR	14;42	175.88
19112	13AP	3;13	204.61
19113	12MY	12;54	232.84
19114	10JN	20;37	260.80
19115	10JL	3;30	288.74
19116	8AU	10;47	316.93
19117	6SE	19;26	345.60
19118	6OC	6;08	14.84
19119	4NO	19;15	44.65
19120	4DE	10;47	74.84

546

New Moons

NUMBER	DATE	TIME	LONG.
19122	18JA	8;37	300.54
19123	16FE	18;33	330.08
19124	18MR	3;46	359.12
19125	16AP	12;39	27.66
19126	15MY	21;52	55.84
19127	14JN	8;18	83.89
19128	13JL	20;49	112.07
19129	12AU	11;50	140.62
19130	11SE	4;57	169.69
19131	10OC	22;59	199.32
19132	9NO	16;30	229.36
19133	9DE	8;31	259.60

Full Moons

NUMBER	DATE	TIME	LONG.
19121	3JA	4;26	105.15
19122	1FE	23;15	135.27
19123	3MR	17;39	164.93
19124	2AP	10;03	194.00
19125	1MY	23;36	222.51
19126	31MY	10;28	250.64
19127	29JN	19;29	278.63
19128	29JL	3;40	306.75
19129	27AU	11;55	335.24
19130	25SE	20;53	4.23
19131	25OC	7;02	33.76
19132	23NO	18;48	63.71
19133	23DE	8;34	93.88

547

New Moons

NUMBER	DATE	TIME	LONG.
19134	7JA	22;32	289.73
19135	6FE	10;27	319.51
19136	7MR	20;21	348.75
19137	6AP	4;37	17.43
19138	5MY	12;01	45.65
19139	3JN	19;35	73.63
19140	3JL	4;29	101.62
19141	1AU	15;44	129.90
19142	31AU	5;49	158.67
19143	29SE	22;33	188.05
19144	29OC	17;06	217.98
19145	28NO	12;10	248.28
19146	28DE	6;17	278.63

Full Moons

NUMBER	DATE	TIME	LONG.
19134	22JA	0;22	123.99
19135	20FE	17;34	153.78
19136	22MR	10;53	183.08
19137	21AP	3;05	211.86
19138	20MY	17;31	240.20
19139	19JN	6;10	268.33
19140	18JL	17;21	296.49
19141	17AU	3;31	324.90
19142	15SE	13;09	353.75
19143	14OC	22;43	23.08
19144	13NO	8;47	52.85
19145	12DE	19;51	82.89

NUMBER	DATE	TIME	LONG.	NUMBER	DATE	TIME	LONG.	NUMBER	DATE	TIME	LONG.	NUMBER	DATE	TIME	LONG.
			548								554				
				19146	11JA	8;12	112.94								
19147	26JA	22;10	308.70	19147	9FE	21;50	142.74	19221	19JA	15;55	301.92	19220	4JA	23;31	107.03
19148	25FE	11;06	338.24	19148	10MR	12;24	172.12	19222	18FE	2;01	331.45	19221	3FE	18;17	137.13
19149	25MR	21;08	7.17	19149	9AP	3;29	201.02	19223	19MR	12;21	0.51	19222	5MR	11;04	166.69
19150	24AP	5;03	35.55	19150	8MY	18;44	229.52	19224	17AP	23;19	29.11	19223	4AP	1;02	195.64
19151	23MY	11;54	63.57	19151	7JN	9;49	257.77	19225	17MY	11;12	57.38	19224	3MY	12;02	224.03
19152	21JN	18;50	91.46	19152	7JL	0;22	286.00	19226	16JN	0;20	85.54	19225	1JN	20;37	252.06
19153	21JL	2;54	119.52	19153	5AU	13;56	314.42	19227	15JL	14;53	113.81	19226	1JL	3;50	279.98
19154	19AU	12;56	147.97	19154	4SE	2;20	343.20	19228	14AU	6;48	142.42	19227	30JL	10;54	308.07
19155	18SE	1;34	176.99	19155	3OC	13;41	12.43	19229	12SE	23;26	171.50	19228	28AU	18;53	336.56
19156	17OC	17;05	206.63	19156	2NO	0;29	42.10	19230	12OC	15;46	201.08	19229	27SE	4;38	5.61
19157	16NO	11;19	236.79	19157	1DE	11;12	72.07	19231	11NO	6;47	231.04	19230	26OC	16;36	35.23
19158	16DE	7;09	267.20	19158	30DE	22;04	102.10	19232	10DE	20;03	261.16	19231	25NO	6;56	65.30
												19232	24DE	23;31	95.59
			549								555				
19159	15JA	2;38	297.52	19159	29JA	9;07	131.93	19233	9JA	7;39	291.19	19233	23JA	17;43	125.79
19160	13FE	19;47	327.41	19160	27FE	20;23	161.36	19234	7FE	17;58	320.88	19234	22FE	12;19	155.61
19161	15MR	9;37	356.68	19161	29MR	8;05	190.30	19235	9MR	3;19	350.08	19235	24MR	5;43	184.89
19162	13AP	20;21	25.33	19162	27AP	20;42	218.82	19236	7AP	12;05	18.76	19236	22AP	20;40	213.58
19163	13MY	4;51	53.51	19163	27MY	10;36	247.09	19237	6MY	20;50	47.02	19237	22MY	8;49	241.83
19164	11JN	12;08	81.44	19164	26JN	1;41	275.32	19238	5JN	6;22	75.07	19238	20JN	18;44	269.85
19165	10JL	19;11	109.39	19165	25JL	17;22	303.71	19239	4JL	17;39	103.16	19239	20JL	3;21	297.91
19166	9AU	2;50	137.61	19166	24AU	8;51	332.45	19240	3AU	7;25	131.55	19240	18AU	11;40	326.27
19167	7SE	11;55	166.30	19167	22SE	23;31	1.63	19241	1SE	23;43	160.44	19241	16SE	20;24	355.09
19168	6OC	23;15	195.59	19168	22OC	13;14	31.26	19242	1OC	17;42	189.90	19242	16OC	6;04	24.45
19169	5NO	13;32	225.45	19169	21NO	2;01	61.24	19243	31OC	11;53	219.84	19243	14NO	17;02	54.28
19170	5DE	6;49	255.72	19170	20DE	13;50	91.32	19244	30NO	4;56	250.07	19244	14DE	5;44	84.39
								19245	29DE	20;04	280.29				
			550								556				
19171	4JA	2;04	286.09	19171	19JA	0;37	121.24					19245	12JA	20;23	114.53
19172	2FE	21;22	316.22	19172	17FE	10;27	150.77	19246	28JA	9;01	310.22	19246	11FE	12;45	144.42
19173	4MR	14;53	345.83	19173	18MR	19;43	179.80	19247	26FE	19;47	339.65	19247	12MR	5;53	173.88
19174	3AP	5;43	14.83	19174	17AP	5;11	208.35	19248	27MR	4;39	8.51	19248	10AP	22;34	202.82
19175	2MY	17;52	43.27	19175	16MY	15;42	236.58	19249	25AP	12;12	36.86	19249	10MY	13;50	231.29
19176	1JN	3;48	71.36	19176	15JN	3;55	264.70	19250	24MY	19;25	64.88	19250	9JN	3;23	259.48
19177	30JN	12;09	99.32	19177	14JL	17;59	292.94	19251	23JN	3;25	92.82	19251	8JL	15;25	287.62
19178	29JL	19;45	127.43	19178	13AU	9;38	321.53	19252	22JL	13;21	120.96	19252	7AU	2;16	315.95
19179	28AU	3;30	155.90	19179	12SE	2;20	350.60	19253	21AU	2;02	149.54	19253	5SE	12;22	344.65
19180	26SE	12;20	184.90	19180	11OC	19;21	20.20	19254	19SE	17;37	178.70	19254	4OC	22;08	13.84
19181	25OC	23;05	214.46	19181	10NO	11;49	50.21	19255	19OC	11;33	208.47	19255	3NO	8;02	43.49
19182	24NO	12;18	244.48	19182	10DE	2;51	80.40	19256	18NO	6;41	238.68	19256	2DE	18;36	73.46
19183	24DE	3;53	274.73					19257	18DE	1;33	269.06				
			551								557				
				19183	8JA	15;48	110.49					19257	1JA	6;12	103.52
19184	22JA	21;07	304.89	19184	7FE	2;35	140.21	19258	16JA	18;42	299.27	19258	30JA	19;00	133.41
19185	21FE	14;51	334.69	19185	8MR	11;41	169.41	19259	15FE	9;06	329.02	19259	1MR	8;52	162.92
19186	23MR	8;02	3.97	19186	6AP	19;50	198.07	19260	16MR	20;25	358.16	19260	30MR	23;27	191.96
19187	21AP	23;49	32.72	19187	6MY	3;54	226.32	19261	15AP	5;10	26.72	19261	29AP	14;24	220.55
19188	21MY	13;37	61.03	19188	4JN	12;43	254.34	19262	14MY	12;19	54.83	19262	29MY	5;30	248.86
19189	20JN	1;17	89.12	19189	3JL	23;01	282.39	19263	12JN	19;02	82.73	19263	27JN	20;26	277.08
19190	19JL	11;07	117.23	19190	2AU	11;24	310.72	19264	12JL	2;25	110.70	19264	27JL	10;44	305.43
19191	17AU	19;53	145.60	19191	1SE	2;16	339.54	19265	10AU	11;25	138.99	19265	26AU	0;00	334.10
19192	16SE	4;34	174.41	19192	30SE	19;35	8.95	19266	8SE	22;46	167.80	19266	24SE	12;08	3.20
19193	15OC	14;01	203.75	19193	30OC	14;29	38.91	19267	8OC	12;55	197.22	19267	23OC	23;24	32.76
19194	14NO	0;42	233.56	19194	29NO	9;14	69.20	19268	7NO	5;56	227.22	19268	22NO	10;15	62.66
19195	13DE	12;47	263.64	19195	29DE	2;06	99.50	19269	7DE	1;11	257.58	19269	21DE	21;03	92.70
			552								558				
19196	12JA	2;10	293.73	19196	27JA	16;09	129.48	19270	5JA	21;05	287.97	19270	20JA	7;54	122.62
19197	10FE	16;44	323.56	19197	26FE	3;28	158.95	19271	4FE	15;32	318.04	19271	18FE	18;49	152.18
19198	11MR	8;20	352.97	19198	26MR	12;40	187.83	19272	6MR	6;59	347.54	19272	20MR	5;59	181.25
19199	10AP	0;32	21.90	19199	24AP	20;31	216.20	19273	4AP	19;07	16.40	19273	18AP	17;48	209.88
19200	9MY	16;32	50.42	19200	24MY	3;45	244.23	19274	4MY	4;33	44.72	19274	18MY	6;47	238.19
19201	8JN	7;25	78.66	19201	22JN	11;13	272.15	19275	2JN	12;18	72.71	19275	16JN	21;12	266.40
19202	7JL	20;42	106.84	19202	21JL	19;51	300.23	19276	1JL	19;24	100.63	19276	16JL	12;45	294.71
19203	6AU	8;29	135.20	19203	20AU	6;44	328.72	19277	31JL	2;44	128.73	19277	15AU	4;38	323.33
19204	4SE	19;21	163.92	19204	18SE	20;41	357.81	19278	29AU	11;07	157.25	19278	13SE	20;03	352.38
19205	4OC	5;52	193.13	19205	18OC	13;53	27.53	19279	27SE	21;20	186.33	19279	13OC	10;33	21.90
19206	2NO	16;27	222.80	19206	17NO	9;20	57.75	19280	27OC	10;09	216.00	19280	12NO	0;02	51.80
19207	2DE	3;12	252.78	19207	17DE	5;04	88.16	19281	26NO	1;59	246.14	19281	11DE	12;27	81.89
19208	31DE	14;08	282.81					19282	25DE	20;22	276.50				
			553								559				
				19208	15JA	23;09	118.42					19282	9JA	23;45	111.90
19209	30JA	1;31	312.65	19209	14FE	14;33	148.22	19283	24JA	15;45	306.74	19283	8FE	9;54	141.57
19210	28FE	13;44	342.10	19210	16MR	3;05	177.43	19284	23FE	10;10	336.55	19284	9MR	19;10	170.76
19211	30MR	3;03	11.10	19211	14AP	13;03	206.04	19285	25MR	2;16	5.75	19285	8AP	4;11	199.44
19212	28AP	17;25	39.68	19212	13MY	21;03	234.19	19286	23AP	15;40	34.38	19286	7MY	13;50	227.73
19213	28MY	8;24	67.99	19213	12JN	3;55	262.10	19287	23MY	2;38	62.57	19287	6JN	0;58	255.84
19214	26JN	23;27	96.22	19214	11JL	10;40	290.04	19288	21JN	11;46	90.55	19288	5JL	14;02	284.00
19215	26JL	14;11	124.58	19215	9AU	18;31	318.27	19289	20JL	19;47	118.60	19289	4AU	5;00	312.45
19216	25AU	4;27	153.27	19216	8SE	4;33	347.02	19290	19AU	3;31	146.94	19290	2SE	21;26	341.36
19217	23SE	18;05	182.43	19217	7OC	17;35	16.38	19291	17SE	11;51	175.76	19291	2OC	14;39	10.80
19218	23OC	6;54	212.03	19218	6NO	9;42	46.33	19292	16OC	21;40	205.14	19292	1NO	7;46	40.71
19219	21NO	18;44	241.97	19219	6DE	4;08	76.65	19293	15NO	9;40	235.01	19293	30NO	23;49	70.90
19220	21DE	5;39	272.02					19294	15DE	0;04	265.20	19294	30DE	13;56	101.07

560

NUMBER	DATE	TIME	LONG.	NUMBER	DATE	TIME	LONG.
19295	13JA	16;26	295.40	19295	29JA	1;44	130.95
19296	12FE	9;49	325.33	19296	27FE	11;28	160.33
19297	13MR	3;09	354.78	19297	27MR	19;48	189.15
19298	11AP	19;33	23.70	19298	26AP	3;37	217.51
19299	11MY	10;21	52.14	19299	25MY	11;48	245.56
19300	9JN	23;08	80.30	19300	23JN	21;10	273.56
19301	9JL	9;58	108.39	19301	23JL	8;25	301.76
19302	7AU	19;22	136.67	19302	21AU	22;04	330.38
19303	6SE	4;13	165.33	19303	20SE	14;21	359.59
19304	5OC	13;24	194.51	19304	20OC	8;49	29.39
19305	3NO	23;33	224.18	19305	19NO	4;03	59.62
19306	3DE	10;58	254.19	19306	18DE	22;10	89.97

561

NUMBER	DATE	TIME	LONG.	NUMBER	DATE	TIME	LONG.
19307	1JA	23;37	284.29	19307	17JA	13;42	120.10
19308	31JA	13;23	314.21	19308	16FE	2;15	149.76
19309	2MR	4;12	343.75	19309	17MR	12;18	178.84
19310	31MR	19;53	12.81	19310	15AP	20;36	207.37
19311	30AP	11;53	41.44	19311	15MY	3;56	235.48
19312	30MY	3;22	69.76	19312	13JN	11;06	263.40
19313	28JN	17;35	97.95	19313	12JL	19;01	291.39
19314	28JL	6;17	126.25	19314	11AU	4;40	319.72
19315	26AU	17;48	154.86	19315	9SE	17;05	348.58
19316	25SE	4;43	183.92	19316	9OC	8;48	18.08
19317	24OC	15;28	213.46	19317	8NO	3;26	48.15
19318	23NO	2;14	243.37	19318	7DE	23;21	78.55
19319	22DE	13;03	273.41				

562

NUMBER	DATE	TIME	LONG.	NUMBER	DATE	TIME	LONG.
				19319	6JA	18;28	108.91
19320	21JA	0;03	303.33	19320	5FE	11;13	138.90
19321	19FE	11;33	332.90	19321	7MR	1;05	168.32
19322	20MR	23;59	2.02	19322	5AP	12;12	197.12
19323	19AP	13;34	30.71	19323	4MY	21;02	225.41
19324	19MY	4;06	59.07	19324	3JN	4;19	253.38
19325	17JN	19;07	87.30	19325	2JL	11;00	281.28
19326	17JL	10;09	115.60	19326	31JL	18;13	309.39
19327	16AU	0;54	144.18	19327	30AU	3;07	337.94
19328	14SE	15;11	173.20	19328	28SE	14;41	7.08
19329	14OC	4;44	202.69	19329	28OC	5;22	36.84
19330	12NO	17;18	232.56	19330	26NO	22;48	67.05
19331	12DE	4;44	262.61	19331	26DE	17;51	97.44

563

NUMBER	DATE	TIME	LONG.	NUMBER	DATE	TIME	LONG.
19332	10JA	15;14	292.58	19332	25JA	12;59	127.66
19333	9FE	1;12	322.24	19333	24FE	6;47	157.43
19334	10MR	11;08	351.44	19334	25MR	22;07	186.59
19335	8AP	21;28	20.17	19335	24AP	10;31	215.16
19336	8MY	8;36	48.51	19336	23MY	20;13	243.30
19337	6JN	20;55	76.66	19337	22JN	4;02	271.23
19338	6JL	10;41	104.86	19338	21JL	11;08	299.25
19339	5AU	2;03	133.33	19339	19AU	18;40	327.59
19340	3SE	18;39	162.26	19340	18SE	3;36	356.44
19341	3OC	11;34	191.70	19341	17OC	14;32	25.87
19342	2NO	3;37	221.58	19342	16NO	3;43	55.81
19343	1DE	17;58	251.70	19343	15DE	19;09	86.04
19344	31DE	6;25	281.80				

564

NUMBER	DATE	TIME	LONG.	NUMBER	DATE	TIME	LONG.
				19344	14JA	12;29	116.28
19345	29JA	17;16	311.63	19345	13FE	6;52	146.24
19346	28FE	2;54	340.99	19346	14MR	0;51	175.70
19347	28MR	11;42	9.83	19347	12AP	17;00	204.59
19348	26AP	20;10	38.20	19348	12MY	6;31	232.98
19349	26MY	5;00	66.28	19349	10JN	17;31	261.06
19350	24JN	15;11	94.31	19350	10JL	2;48	289.09
19351	24JL	3;37	122.56	19351	8AU	11;21	317.34
19352	22AU	18;47	151.26	19352	6SE	20;00	346.01
19353	21SE	12;16	180.53	19353	6OC	5;20	15.20
19354	21OC	6;46	210.34	19354	4NO	15;43	44.89
19355	20NO	0;44	240.52	19355	4DE	3;30	74.92
19356	19DE	17;01	270.80				

565

NUMBER	DATE	TIME	LONG.	NUMBER	DATE	TIME	LONG.
				19356	2JA	17;02	105.06
19357	18JA	7;04	300.86	19357	1FE	8;23	135.03
19358	16FE	18;50	330.48	19358	3MR	1;00	164.63
19359	18MR	4;27	359.53	19359	1AP	17;48	193.73
19360	16AP	12;25	28.03	19360	1MY	9;42	222.35
19361	15MY	19;32	56.13	19361	31MY	0;06	250.61
19362	14JN	2;54	84.06	19362	29JN	12;58	278.76
19363	13JL	11;42	112.09	19363	29JL	0;35	307.02
19364	11AU	22;58	140.49	19364	27AU	11;16	335.60
19365	10SE	13;12	169.44	19365	25SE	21;24	4.64
19366	10OC	6;12	199.00	19366	25OC	7;22	34.16
19367	9NO	1;02	229.10	19367	23NO	17;38	64.05
19368	8DE	20;19	259.47	19368	23DE	4;39	94.10

566

NUMBER	DATE	TIME	LONG.	NUMBER	DATE	TIME	LONG.
19369	7JA	14;32	289.78	19369	21JA	16;41	124.06
19370	6FE	6;21	319.72	19370	20FE	5;48	153.68
19371	7MR	19;07	349.09	19371	21MR	19;45	182.85
19372	6AP	4;57	17.83	19372	20AP	10;16	211.57
19373	5MY	12;41	46.07	19373	20MY	1;10	239.94
19374	3JN	19;25	74.01	19374	18JN	16;13	268.17
19375	3JL	2;19	101.92	19375	18JL	7;02	296.46
19376	1AU	10;27	130.07	19376	16AU	21;07	325.03
19377	30AU	20;38	158.68	19377	15SE	10;07	354.06
19378	29SE	9;25	187.89	19378	14OC	22;03	23.44
19379	29OC	1;06	217.70	19379	13NO	9;14	53.26
19380	27NO	19;25	247.96	19380	12DE	20;06	83.29
19381	27DE	15;13	278.38				

567

NUMBER	DATE	TIME	LONG.	NUMBER	DATE	TIME	LONG.
				19381	11JA	6;51	113.27
19382	26JA	10;34	308.60	19382	9FE	17;33	142.95
19383	25FE	3;32	338.32	19383	11MR	4;18	172.18
19384	26MR	17;13	7.40	19384	9AP	15;25	200.92
19385	25AP	3;49	35.89	19385	9MY	3;31	229.30
19386	24MY	12;16	63.97	19386	7JN	17;04	257.49
19387	22JN	19;37	91.88	19387	7JL	8;06	285.74
19388	22JL	2;49	119.90	19388	6AU	0;04	314.25
19389	20AU	10;41	148.27	19389	4SE	16;04	343.17
19390	18SE	20;01	177.15	19390	4OC	7;22	12.56
19391	18OC	7;32	206.62	19391	2NO	21;38	42.37
19392	16NO	21;53	236.61	19392	2DE	10;45	72.44
19393	16DE	15;03	266.91	19393	31DE	22;39	102.52

568

NUMBER	DATE	TIME	LONG.	NUMBER	DATE	TIME	LONG.
19394	15JA	10;02	297.22	19394	30JA	9;17	132.33
19395	14FE	4;59	327.19	19395	28FE	18;45	161.68
19396	14MR	22;11	356.60	19396	29MR	3;34	190.50
19397	13AP	12;50	25.42	19397	27AP	12;33	218.89
19398	13MY	0;57	53.74	19398	26MY	22;39	247.01
19399	11JN	10;59	81.78	19399	25JN	10;36	275.11
19400	10JL	19;35	109.80	19400	25JL	0;39	303.43
19401	9AU	3;31	138.03	19401	23AU	16;33	332.17
19402	7SE	11;37	166.68	19402	22SE	9;42	1.44
19403	6OC	20;45	195.87	19403	22OC	3;15	31.23
19404	5NO	7;41	225.59	19404	20NO	20;11	61.38
19405	4DE	20;53	255.69	19405	20DE	11;28	91.61

569

NUMBER	DATE	TIME	LONG.	NUMBER	DATE	TIME	LONG.
19406	3JA	12;13	285.90	19406	19JA	0;26	121.62
19407	2FE	4;59	315.93	19407	17FE	11;02	151.19
19408	3MR	22;12	345.54	19408	18MR	19;48	180.19
19409	2AP	14;58	14.62	19409	17AP	3;35	208.67
19410	2MY	6;31	43.21	19410	16MY	11;19	236.79
19411	31MY	20;21	71.45	19411	14JN	19;53	264.76
19412	30JN	8;16	99.56	19412	14JL	6;03	292.85
19413	29JL	18;30	127.77	19413	12AU	18;28	321.31
19414	28AU	3;45	156.30	19414	11SE	9;33	350.30
19415	26SE	12;53	185.32	19415	11OC	3;12	19.91
19416	25OC	22;40	214.83	19416	9NO	22;27	50.03
19417	24NO	9;31	244.76	19417	9DE	17;30	80.40
19418	23DE	21;30	274.85				

570

NUMBER	DATE	TIME	LONG.	NUMBER	DATE	TIME	LONG.
				19418	8JA	10;29	110.66
19419	22JA	10;32	304.84	19419	7FE	0;28	140.51
19420	21FE	0;33	334.49	19420	8MR	11;36	169.79
19421	22MR	15;32	3.69	19421	6AP	20;32	198.49
19422	21AP	7;13	32.44	19422	6MY	4;09	226.71
19423	20MY	22;57	60.84	19423	4JN	11;13	254.67
19424	19JN	13;55	89.06	19424	3JL	18;37	282.60
19425	19JL	3;33	117.32	19425	2AU	3;17	310.78
19426	17AU	15;51	145.82	19426	31AU	14;16	339.43
19427	16SE	3;17	174.75	19427	30SE	4;24	8.71
19428	15OC	14;19	204.15	19428	29OC	21;48	38.60
19429	14NO	1;15	233.97	19429	28NO	17;24	68.92
19430	13DE	12;05	264.00	19430	28DE	13;11	99.34

571

NUMBER	DATE	TIME	LONG.	NUMBER	DATE	TIME	LONG.
19431	11JA	22;52	293.98	19431	27JA	7;12	129.49
19432	10FE	9;51	323.66	19432	25FE	22;27	159.13
19433	11MR	21;29	352.91	19433	27MR	10;48	188.14
19434	10AP	10;12	21.71	19434	25AP	20;38	216.59
19435	10MY	0;03	50.15	19435	25MY	4;33	244.64
19436	8JN	14;47	78.39	19436	23JN	11;25	272.54
19437	8JL	5;53	106.64	19437	22JL	18;17	300.55
19438	6AU	21;00	135.12	19438	21AU	2;18	328.93
19439	5SE	11;49	164.01	19439	19SE	12;33	357.86
19440	5OC	2;05	193.37	19440	19OC	1;44	27.41
19441	3NO	15;25	223.15	19441	17NO	17;55	57.48
19442	3DE	3;33	253.19	19442	17DE	12;17	87.84

	NEW MOONS				FULL MOONS		
NUMBER	DATE	TIME	LONG.	NUMBER	DATE	TIME	LONG.

572

NUMBER	DATE	TIME	LONG.	NUMBER	DATE	TIME	LONG.
19443	1JA	14;30	283.22	19443	16JA	7;27	118.15
19444	31JA	0;34	313.00	19444	15FE	1;57	148.09
19445	29FE	10;15	342.35	19445	15MR	18;29	177.46
19446	29MR	20;04	11.20	19446	14AP	8;18	206.23
19447	28AP	6;30	39.63	19447	13MY	19;17	234.50
19448	27MY	17;59	67.81	19448	12JN	3;57	262.49
19449	26JN	6;53	95.95	19449	11JL	11;22	290.46
19450	25JL	21;29	124.30	19450	9AU	18;41	318.67
19451	24AU	13;43	153.06	19451	8SE	2;59	347.34
19452	23SE	6;53	182.35	19452	7OC	13;00	16.58
19453	22OC	23;48	212.13	19453	6NO	1;06	46.36
19454	21NO	15;17	242.22	19454	5DE	15;24	76.50
19455	21DE	4;46	272.37				

578

NUMBER	DATE	TIME	LONG.	NUMBER	DATE	TIME	LONG.
				19517	9JA	22;33	112.24
19518	24JA	0;31	306.49	19518	8FE	10;14	141.98
19519	22FE	17;22	336.24	19519	9MR	19;41	171.17
19520	24MR	10;12	5.49	19520	8AP	3;41	199.81
19521	23AP	2;16	34.23	19521	7MY	11;10	228.01
19522	22MY	16;59	62.57	19522	5JN	19;05	255.99
19523	21JN	5;57	90.71	19523	5JL	4;18	284.00
19524	20JL	17;10	118.88	19524	3AU	15;31	312.30
19525	19AU	3;03	147.31	19525	2SE	5;19	341.09
19526	17SE	12;22	176.17	19526	1OC	21;52	10.49
19527	16OC	21;57	205.54	19527	31OC	16;39	40.47
19528	15NO	8;21	235.35	19528	30NO	12;12	70.80
19529	14DE	19;46	265.41	19529	30DE	6;29	101.15

573

NUMBER	DATE	TIME	LONG.	NUMBER	DATE	TIME	LONG.
				19455	4JA	7;43	106.75
19456	19JA	16;21	302.32	19456	3FE	1;30	136.81
19457	18FE	2;25	331.85	19457	4MR	19;39	166.45
19458	19MR	11;25	0.86	19458	3AP	12;42	195.53
19459	17AP	19;46	29.35	19459	3MY	3;31	224.08
19460	17MY	4;07	57.49	19460	1JN	15;45	252.25
19461	15JN	13;22	85.49	19461	1JL	1;54	280.29
19462	15JL	0;32	113.62	19462	30JL	10;53	308.46
19463	13AU	14;21	142.14	19463	28AU	19;36	336.98
19464	12SE	6;55	171.21	19464	27SE	4;45	6.00
19465	12OC	1;19	200.87	19465	26OC	14;42	35.54
19466	10NO	19;57	230.97	19466	25NO	1;47	65.48
19467	10DE	13;19	261.27	19467	24DE	14;20	95.59

579

NUMBER	DATE	TIME	LONG.	NUMBER	DATE	TIME	LONG.
19530	13JA	8;09	295.44	19530	28JA	22;00	131.18
19531	11FE	21;26	325.20	19531	27FE	10;25	160.67
19532	13MR	11;38	354.54	19532	28MR	20;14	189.56
19533	12AP	2;44	23.41	19533	27AP	4;18	217.92
19534	11MY	18;21	51.89	19534	26MY	11;27	245.93
19535	10JN	9;46	80.15	19535	24JN	18;32	273.84
19536	10JL	0;14	108.39	19536	24JL	2;28	301.91
19537	8AU	13;25	136.82	19537	22AU	12;14	330.37
19538	7SE	1;31	165.61	19538	21SE	0;47	359.42
19539	6OC	12;59	194.88	19539	20OC	16;42	29.10
19540	5NO	0;09	224.59	19540	19NO	11;28	59.31
19541	4DE	11;07	254.59	19541	19DE	7;27	89.73

574

NUMBER	DATE	TIME	LONG.	NUMBER	DATE	TIME	LONG.
19468	9JA	4;34	291.45	19468	23JA	4;36	125.62
19469	7FE	17;26	321.25	19469	21FE	20;24	155.34
19470	9MR	3;58	350.50	19470	23MR	12;58	184.59
19471	7AP	12;32	19.17	19471	22AP	5;14	213.35
19472	6MY	19;49	47.37	19472	21MY	20;21	241.72
19473	5JN	2;49	75.32	19473	20JN	10;02	269.89
19474	4JL	10;42	103.27	19474	19JL	22;25	298.10
19475	2AU	20;38	131.51	19475	18AU	9;47	326.58
19476	1SE	9;25	160.25	19476	16SE	20;05	355.48
19477	1OC	1;13	189.60	19477	16OC	6;39	24.87
19478	30OC	19;24	219.54	19478	14NO	16;51	54.66
19479	29NO	14;45	249.86	19479	14DE	3;28	84.68
19480	29DE	9;45	280.24				

580

NUMBER	DATE	TIME	LONG.	NUMBER	DATE	TIME	LONG.
19542	2JA	21;53	284.61	19542	18JA	2;30	120.02
19543	1FE	8;33	314.40	19543	16FE	19;07	149.86
19544	1MR	19;33	343.78	19544	17MR	8;49	179.09
19545	31MR	7;23	12.69	19545	15AP	19;46	207.72
19546	29AP	20;24	41.21	19546	15MY	4;31	235.89
19547	29MY	10;34	69.48	19547	13JN	11;48	263.81
19548	28JN	1;30	97.70	19548	12JL	18;35	291.76
19549	27JL	16;47	126.10	19549	11AU	1;59	320.00
19550	26AU	8;02	154.85	19550	9SE	11;05	348.72
19551	24SE	22;56	184.07	19551	8OC	22;51	18.05
19552	24OC	13;04	213.75	19552	7NO	13;37	47.96
19553	23NO	2;01	243.75	19553	7DE	7;01	78.25
19554	22DE	13;35	273.83				

575

NUMBER	DATE	TIME	LONG.	NUMBER	DATE	TIME	LONG.
				19480	12JA	14;51	114.68
19481	28JA	2;52	310.35	19481	11FE	3;12	144.41
19482	26FE	17;08	339.93	19482	12MR	16;28	173.72
19483	28MR	4;16	8.88	19483	11AP	6;26	202.56
19484	26AP	12;50	37.27	19484	10MY	20;57	231.01
19485	25MY	19;51	65.28	19485	9JN	11;52	259.25
19486	24JN	2;31	93.17	19486	9JL	2;56	287.51
19487	23JL	9;58	121.21	19487	7AU	17;40	315.99
19488	21AU	19;06	149.65	19488	6SE	7;33	344.85
19489	20SE	6;37	178.64	19489	5OC	20;17	14.15
19490	19OC	20;56	208.25	19490	4NO	8;01	43.88
19491	18NO	14;02	238.37	19491	3DE	19;07	73.88
19492	18DE	9;17	268.76				

581

NUMBER	DATE	TIME	LONG.	NUMBER	DATE	TIME	LONG.
				19554	6JA	1;52	108.59
19555	20JA	23;58	303.71	19555	4FE	20;44	138.68
19556	19FE	9;37	333.21	19556	6MR	14;14	168.26
19557	20MR	19;03	2.21	19557	5AP	5;22	197.23
19558	19AP	4;52	30.74	19558	4MY	17;42	225.66
19559	18MY	15;33	58.96	19559	3JN	3;28	253.73
19560	17JN	3;33	87.07	19560	2JL	11;28	281.69
19561	16JL	17;16	115.31	19561	31JL	18;49	309.80
19562	15AU	8;50	143.92	19562	30AU	2;41	338.31
19563	14SE	1;53	173.04	19563	28SE	11;55	7.36
19564	13OC	19;23	202.69	19564	27OC	23;03	36.96
19565	12NO	11;57	232.72	19565	26NO	12;16	67.00
19566	12DE	2;36	262.91	19566	26DE	3;30	97.23

576

NUMBER	DATE	TIME	LONG.	NUMBER	DATE	TIME	LONG.
				19492	2JA	5;54	103.90
19493	17JA	5;03	299.09	19493	31JA	16;29	133.69
19494	15FE	23;18	329.00	19494	1MR	2;57	163.06
19495	16MR	14;34	358.31	19495	30MR	13;33	191.94
19496	15AP	2;33	27.00	19496	29AP	0;49	220.40
19497	14MY	11;56	55.20	19497	28MY	13;23	248.61
19498	12JN	19;44	83.15	19498	27JN	3;38	276.80
19499	12JL	2;58	111.11	19499	26JL	19;19	305.20
19500	10AU	10;32	139.34	19500	25AU	11;38	333.99
19501	8SE	19;10	168.04	19501	24SE	3;24	3.24
19502	8OC	5;37	197.31	19502	23OC	18;44	32.95
19503	6NO	18;33	227.12	19503	22NO	8;39	62.98
19504	6DE	10;20	257.34	19504	21DE	21;16	93.10

582

NUMBER	DATE	TIME	LONG.	NUMBER	DATE	TIME	LONG.
19567	10JA	15;07	292.97	19567	24JA	20;27	127.36
19568	9FE	1;48	322.66	19568	23FE	14;20	157.14
19569	10MR	11;07	351.83	19569	25MR	7;54	186.40
19570	8AP	19;32	20.48	19570	23AP	23;49	215.12
19571	8MY	3;36	48.70	19571	23MY	13;20	243.41
19572	6JN	12;09	76.70	19572	22JN	0;32	271.48
19573	5JL	22;09	104.75	19573	21JL	15;09	299.59
19574	4AU	10;34	133.10	19574	19AU	19;08	328.00
19575	3SE	1;54	161.96	19575	18SE	4;13	356.86
19576	2OC	19;44	191.43	19576	17OC	13;54	26.24
19577	1NO	14;41	221.43	19577	16NO	0;28	56.06
19578	1DE	9;00	251.71	19578	15DE	12;13	86.13
19579	31DE	1;27	281.98				

577

NUMBER	DATE	TIME	LONG.	NUMBER	DATE	TIME	LONG.
19505	5JA	4;28	287.66	19505	20JA	8;29	123.03
19506	3FE	23;29	317.76	19506	18FE	18;21	152.54
19507	5MR	17;32	347.38	19507	20MR	3;12	181.53
19508	4AP	9;23	16.39	19508	18AP	11;45	210.03
19509	3MY	22;41	44.87	19509	17MY	20;58	238.19
19510	2JN	9;44	73.00	19510	16JN	7;47	266.25
19511	1JL	19;05	101.00	19511	15JL	20;44	294.46
19512	31JL	3;25	129.15	19512	14AU	11;50	323.04
19513	29AU	11;31	157.66	19513	13SE	4;38	352.14
19514	27SE	20;12	186.67	19514	12OC	22;21	21.78
19515	27OC	6;16	216.22	19515	11NO	15;57	51.85
19516	25NO	18;20	246.21	19516	11DE	8;20	82.11
19517	25DE	8;33	276.39				

583

NUMBER	DATE	TIME	LONG.	NUMBER	DATE	TIME	LONG.
				19579	14JA	1;27	116.20
19580	29JA	15;30	311.94	19580	12FE	16;17	146.02
19581	28FE	3;04	341.39	19581	14MR	8;17	175.41
19582	29MR	12;26	10.25	19582	13AP	0;35	204.32
19583	27AP	20;08	38.58	19583	12MY	16;13	232.79
19584	27MY	3;01	66.58	19584	11JN	6;38	261.01
19585	25JN	10;15	94.50	19585	10JL	19;48	289.20
19586	24JL	19;01	122.60	19586	9AU	7;52	317.59
19587	23AU	6;21	151.14	19587	7SE	19;07	346.36
19588	21SE	20;46	180.28	19588	7OC	5;46	15.61
19589	21OC	13;59	210.03	19589	5NO	16;06	45.30
19590	20NO	9;03	240.25	19590	5DE	2;31	75.27
19591	20DE	4;28	270.66				

584

NUMBER	DATE	TIME	LONG.	NUMBER	DATE	TIME	LONG.
				19591	3JA	13;25	105.29
19592	18JA	22;41	300.90	19592	2FE	1;06	135.12
19593	17FE	14;23	330.69	19593	2MR	13;39	164.55
19594	18MR	2;59	359.86	19594	1AP	2;58	193.52
19595	16AP	12;39	28.43	19595	30AP	16;58	222.06
19596	15MY	20;14	56.54	19596	30MY	7;34	250.34
19597	14JN	2;55	84.44	19597	28JN	22;37	278.57
19598	13JL	9;52	112.40	19598	28JL	13;45	306.97
19599	11AU	18;07	140.68	19599	27AU	4;24	335.71
19600	10SE	4;29	169.47	19600	25SE	18;02	4.89
19601	9OC	17;27	198.86	19601	25OC	6;30	34.52
19602	8NO	9;14	228.82	19602	23NO	18;02	64.46
19603	8DE	3;32	259.15	19603	23DE	4;58	94.51

585

NUMBER	DATE	TIME	LONG.	NUMBER	DATE	TIME	LONG.
19604	6JA	23;13	289.53	19604	21JA	15;34	124.40
19605	5FE	18;22	319.61	19605	20FE	1;53	153.91
19606	7MR	11;07	349.15	19606	21MR	12;06	182.93
19607	6AP	0;38	18.04	19607	19AP	22;40	211.49
19608	5MY	11;10	46.39	19608	19MY	10;17	239.74
19609	3JN	19;38	74.40	19609	17JN	23;33	267.89
19610	3JL	3;06	102.34	19610	17JL	14;35	296.20
19611	1AU	10;31	130.46	19611	16AU	6;52	324.84
19612	30AU	18;41	158.99	19612	14SE	23;25	353.96
19613	29SE	4;16	188.07	19613	14OC	15;20	23.55
19614	28OC	15;57	217.70	19614	13NO	6;05	53.52
19615	27NO	6;18	247.79	19615	12DE	19;29	83.66
19616	26DE	23;17	278.09				

586

NUMBER	DATE	TIME	LONG.	NUMBER	DATE	TIME	LONG.
				19616	11JA	7;25	113.69
19617	25JA	17;54	308.29	19617	9FE	17;50	143.36
19618	24FE	12;26	338.08	19618	11MR	2;56	172.51
19619	26MR	5;19	7.30	19619	9AP	11;19	201.14
19620	24AP	19;49	35.95	19620	8MY	19;52	229.38
19621	24MY	7;56	64.18	19621	7JN	5;37	257.43
19622	22JN	18;09	92.21	19622	6JL	17;22	285.54
19623	22JL	3;04	120.30	19623	5AU	7;27	313.97
19624	20AU	11;23	148.69	19624	3SE	23;37	342.88
19625	18SE	19;52	177.53	19625	3OC	17;12	12.36
19626	18OC	5;18	206.91	19626	2NO	11;15	42.32
19627	16NO	16;23	236.76	19627	2DE	4;34	72.57
19628	16DE	5;30	266.90	19628	31DE	20;02	102.80

587

NUMBER	DATE	TIME	LONG.	NUMBER	DATE	TIME	LONG.
19629	14JA	20;30	297.03	19629	30JA	8;57	132.70
19630	13FE	12;45	326.90	19630	28FE	19;20	162.09
19631	15MR	5;25	356.31	19631	30MR	3;48	190.90
19632	13AP	21;44	25.20	19632	28AP	11;15	219.22
19633	13MY	13;06	53.66	19633	27MY	18;42	247.24
19634	12JN	3;02	81.85	19634	26JN	3;05	275.20
19635	11JL	15;16	110.01	19635	25JL	13;13	303.36
19636	10AU	1;58	138.35	19636	24AU	1;43	331.96
19637	8SE	11;44	167.08	19637	22SE	17;00	1.14
19638	7OC	21;19	196.29	19638	22OC	10;56	30.94
19639	6NO	7;25	225.97	19639	21NO	6;30	61.19
19640	5DE	18;22	255.97	19640	21DE	1;44	91.59

588

NUMBER	DATE	TIME	LONG.	NUMBER	DATE	TIME	LONG.
19641	4JA	6;11	286.03	19641	19JA	18;46	121.78
19642	2FE	18;49	315.89	19642	18FE	8;39	151.48
19643	3MR	8;15	345.35	19643	18MR	19;34	180.57
19644	1AP	22;37	14.34	19644	17AP	4;17	209.09
19645	1MY	13;48	42.93	19645	16MY	11;43	237.19
19646	31MY	5;20	71.24	19646	14JN	18;41	265.10
19647	29JN	20;25	99.47	19647	14JL	2;05	293.08
19648	29JL	10;27	127.83	19648	12AU	10;51	321.39
19649	27AU	23;19	156.51	19649	10SE	21;59	350.22
19650	26SE	11;20	185.64	19650	10OC	12;17	19.68
19651	25OC	22;52	215.24	19651	9NO	5;50	49.72
19652	24NO	10;05	245.17	19652	9DE	1;29	80.11
19653	23DE	20;58	275.21				

589

NUMBER	DATE	TIME	LONG.	NUMBER	DATE	TIME	LONG.
				19653	7JA	21;13	110.49
19654	22JA	7;32	305.10	19654	6FE	15;06	140.51
19655	20FE	18;05	334.61	19655	8MR	6;11	169.96
19656	22MR	5;08	3.65	19656	6AP	18;22	198.79
19657	20AP	17;15	32.27	19657	6MY	4;06	227.10
19658	20MY	6;39	60.58	19658	4JN	12;00	255.08
19659	18JN	21;10	88.78	19659	3JL	18;57	283.00
19660	18JL	12;24	117.10	19660	2AU	2;00	311.12
19661	17AU	3;54	145.73	19661	31AU	10;14	339.66
19662	15SE	19;19	174.81	19662	29SE	20;41	8.78
19663	15OC	10;11	204.37	19663	29OC	10;01	38.49
19664	14NO	0;00	234.31	19664	28NO	2;12	68.67
19665	13DE	12;23	264.41	19665	27DE	20;24	99.02

590

NUMBER	DATE	TIME	LONG.	NUMBER	DATE	TIME	LONG.
19666	11JA	23;18	294.39	19666	26JA	15;17	129.22
19667	10FE	9;06	324.02	19667	25FE	9;27	158.98
19668	11MR	18;22	353.17	19668	27MR	1;44	188.16
19669	10AP	3;41	21.84	19669	25AP	15;26	216.77
19670	9MY	13;38	50.12	19670	25MY	2;26	244.95
19671	8JN	0;45	78.22	19671	23JN	11;16	272.92
19672	7JL	13;30	106.38	19672	22JL	18;57	300.97
19673	6AU	4;12	134.84	19673	21AU	2;36	329.33
19674	4SE	20;47	163.78	19674	19SE	11;13	358.20
19675	4OC	14;29	193.27	19675	18OC	21;29	27.62
19676	3NO	7;56	223.23	19676	17NO	9;42	57.53
19677	2DE	23;48	253.42	19677	16DE	23;53	87.71

591

NUMBER	DATE	TIME	LONG.	NUMBER	DATE	TIME	LONG.
19678	1JA	13;27	283.57	19678	15JA	15;51	117.88
19679	31JA	0;57	313.40	19679	14FE	9;09	147.78
19680	1MR	10;45	342.76	19680	16MR	2;48	177.21
19681	30MR	19;22	11.56	19681	14AP	19;32	206.11
19682	29AP	3;21	39.89	19682	14MY	10;15	234.53
19683	28MY	11;24	67.93	19683	12JN	22;37	262.66
19684	26JN	20;26	95.92	19684	12JL	9;05	290.75
19685	26JL	7;31	124.13	19685	10AU	18;29	319.05
19686	24AU	21;26	152.79	19686	9SE	3;40	347.76
19687	23SE	14;17	182.06	19687	8OC	13;13	16.98
19688	23OC	9;04	211.90	19688	6NO	23;26	46.68
19689	22NO	4;05	242.13	19689	6DE	10;35	76.69
19690	21DE	21;41	272.46				

592

NUMBER	DATE	TIME	LONG.	NUMBER	DATE	TIME	LONG.
				19690	4JA	22;55	106.77
19691	20JA	12;59	302.57	19691	3FE	12;44	136.67
19692	19FE	1;42	332.21	19692	4MR	3;56	166.19
19693	19MR	12;01	1.27	19693	2AP	19;55	195.24
19694	17AP	20;19	29.76	19694	2MY	11;47	223.84
19695	17MY	3;23	57.85	19695	1JN	2;48	252.12
19696	15JN	10;13	85.75	19696	30JN	16;41	280.31
19697	14JL	18;03	113.75	19697	30JL	5;30	308.63
19698	13AU	4;03	142.11	19698	28AU	17;25	337.28
19699	11SE	16;59	171.03	19699	27SE	4;36	6.39
19700	11OC	8;58	200.57	19700	26OC	15;16	35.96
19701	10NO	3;22	230.66	19701	25NO	1;43	65.86
19702	9DE	22;52	261.05	19702	24DE	12;19	95.90

593

NUMBER	DATE	TIME	LONG.	NUMBER	DATE	TIME	LONG.
19703	8JA	17;52	291.39	19703	22JA	23;26	125.80
19704	7FE	10;54	321.36	19704	21FE	11;18	155.35
19705	9MR	1;00	350.76	19705	22MR	23;57	184.45
19706	7AP	11;58	19.53	19706	21AP	13;20	213.10
19707	6MY	20;24	47.78	19707	21MY	3;26	241.43
19708	5JN	3;21	75.72	19708	19JN	18;14	269.65
19709	4JL	10;03	103.63	19709	19JL	9;30	297.97
19710	2AU	17;37	131.77	19710	18AU	0;43	326.60
19711	1SE	2;56	160.37	19711	16SE	15;13	355.66
19712	30SE	14;39	189.55	19712	16OC	4;33	25.17
19713	30OC	5;05	219.33	19713	14NO	16;42	55.05
19714	28NO	22;13	249.55	19714	14DE	3;59	85.10
19715	28DE	17;21	279.94				

594

NUMBER	DATE	TIME	LONG.	NUMBER	DATE	TIME	LONG.
				19715	12JA	14;42	115.07
19716	27JA	12;54	310.15	19716	11FE	0;59	144.71
19717	26FE	6;54	339.89	19717	12MR	10;58	173.88
19718	27MR	21;58	9.01	19718	10AP	21;01	202.56
19719	26AP	9;52	37.54	19719	10MY	7;47	230.87
19720	25MY	19;14	65.64	19720	8JN	20;00	259.01
19721	24JN	3;09	93.59	19721	8JL	10;07	287.23
19722	23JL	10;35	121.63	19722	7AU	1;59	315.75
19723	21AU	18;26	150.01	19723	5SE	18;46	344.71
19724	20SE	3;22	178.89	19724	5OC	11;22	14.17
19725	19OC	14;01	208.35	19725	4NO	3;01	44.06
19726	18NO	3;02	238.29	19726	3DE	17;17	74.19
19727	17DE	18;41	268.53				

595

NUMBER	DATE	TIME	LONG.	NUMBER	DATE	TIME	LONG.
				19727	2JA	6;01	104.30
19728	16JA	12;29	298.78	19728	31JA	17;07	134.11
19729	15FE	7;04	328.72	19729	2MR	2;40	163.44
19730	17MR	0;44	358.14	19730	31MR	11;07	192.23
19731	15AP	16;20	26.97	19731	29AP	19;14	220.56
19732	15MY	5;35	55.33	19732	29MY	4;05	248.62
19733	13JN	16;46	83.41	19733	27JN	14;39	276.67
19734	13JL	2;24	111.47	19734	27JL	3;33	304.96
19735	11AU	11;07	139.75	19735	25AU	18;59	333.69
19736	9SE	19;38	168.45	19736	24SE	11;59	2.98
19737	9OC	4;40	197.66	19737	24OC	6;10	32.81
19738	7NO	14;58	227.36	19738	23NO	0;12	63.02
19739	7DE	3;02	257.42	19739	22DE	16;50	93.31

——————NEW MOONS—————— / ——————FULL MOONS—————— / ——————NEW MOONS—————— / ——————FULL MOONS——————

596 / 602

NEW MOONS NUMBER	DATE	TIME	LONG.	FULL MOONS NUMBER	DATE	TIME	LONG.	NEW MOONS NUMBER	DATE	TIME	LONG.	FULL MOONS NUMBER	DATE	TIME	LONG.
19740	5JA	17;00	287.57	19740	21JA	7;05	123.36								
19741	4FE	8;30	317.53	19741	19FE	18;36	152.94	19815	29JA	6;42	311.96	19814	13JA	22;08	116.45
19742	5MR	0;46	347.08	19742	20MR	3;47	181.94	19816	27FE	22;16	341.58	19815	12FE	9;24	146.12
19743	3AP	17;05	16.12	19743	18AP	11;28	210.40	19817	29MR	10;41	10.56	19816	13MR	21;22	175.35
19744	3MY	8;51	44.71	19744	17MY	18;40	238.48	19818	27AP	20;13	38.97	19817	12AP	10;06	204.12
19745	1JN	23;33	72.98	19745	16JN	2;23	266.42	19819	27MY	3;44	67.00	19818	11MY	23;36	232.52
19746	1JL	12;46	101.14	19746	15JL	11;30	294.48	19820	25JN	10;25	94.89	19819	10JN	13;56	260.74
19747	31JL	0;25	129.42	19747	13AU	22;47	322.90	19821	24JL	17;29	122.93	19820	10JL	5;04	289.00
19748	29AU	10;49	158.02	19748	12SE	12;44	351.87	19822	23AU	1;55	151.35	19821	8AU	20;35	317.52
19749	27SE	20;39	187.08	19749	12OC	5;32	21.46	19823	21SE	12;29	180.32	19822	7SE	11;48	346.45
19750	27OC	6;37	216.63	19750	11NO	0;36	51.59	19824	21OC	1;37	209.90	19823	7OC	2;04	15.85
19751	25NO	17;13	246.55	19751	10DE	20;22	81.99	19825	19NO	17;28	239.98	19824	5NO	15;02	45.64
19752	25DE	4;34	276.61					19826	19DE	11;41	270.33	19825	5DE	2;51	75.67

597 / 603

NEW MOONS NUMBER	DATE	TIME	LONG.	FULL MOONS NUMBER	DATE	TIME	LONG.	NEW MOONS NUMBER	DATE	TIME	LONG.	FULL MOONS NUMBER	DATE	TIME	LONG.
				19752	9JA	14;43	112.30					19826	3JA	13;49	105.70
19753	23JA	16;38	306.55	19753	8FE	6;10	142.20	19827	18JA	7;08	300.64	19827	2FE	0;11	135.47
19754	22FE	5;22	336.13	19754	9MR	18;25	171.51	19828	17FE	2;00	330.56	19828	3MR	10;05	164.79
19755	23MR	18;57	5.26	19755	8AP	4;02	200.21	19829	18MR	18;31	359.90	19829	1AP	19;47	193.62
19756	22AP	9;29	33.94	19756	7MY	11;54	228.43	19830	17AP	7;54	28.63	19830	1MY	5;51	222.01
19757	22MY	0;46	62.31	19757	5JN	18;37	256.37	19831	16MY	18;25	56.86	19831	30MY	17;03	250.16
19758	20JN	16;10	90.55	19758	5JL	2;02	284.30	19832	15JN	2;58	84.83	19832	29JN	6;06	278.30
19759	20JL	6;56	118.86	19759	3AU	10;02	312.47	19833	14JL	10;38	112.82	19833	28JL	21;11	306.69
19760	18AU	20;38	147.44	19760	1SE	19;57	341.10	19834	12AU	18;20	141.08	19834	27AU	13;49	335.50
19761	17SE	9;20	176.44	19761	1OC	8;40	10.33	19835	11SE	2;48	169.79	19835	26SE	6;54	4.82
19762	16OC	21;21	205.90	19762	31OC	0;43	40.19	19836	10OC	12;39	199.05	19836	25OC	23;23	34.61
19763	15NO	8;54	235.76	19763	29NO	19;34	70.48	19837	9NO	0;28	228.84	19837	24NO	14;36	64.71
19764	14DE	20;01	265.81	19764	29DE	15;30	100.90	19838	8DE	14;46	258.99	19838	24DE	4;13	94.87

598 / 604

NEW MOONS NUMBER	DATE	TIME	LONG.	FULL MOONS NUMBER	DATE	TIME	LONG.	NEW MOONS NUMBER	DATE	TIME	LONG.	FULL MOONS NUMBER	DATE	TIME	LONG.
19765	13JA	6;39	295.77	19765	28JA	10;25	131.08	19839	7JA	7;28	289.25	19839	22JA	16;06	124.81
19766	11FE	16;58	325.41	19766	27FE	2;51	160.75	19840	6FE	1;39	319.30	19840	21FE	2;16	154.32
19767	13MR	3;26	354.59	19767	28MR	16;23	189.79	19841	6MR	19;44	348.90	19841	21MR	11;00	183.27
19768	11AP	14;41	23.31	19768	27AP	3;13	218.26	19842	5AP	12;18	17.93	19842	19AP	18;58	211.73
19769	11MY	3;11	51.68	19769	26MY	11;55	246.34	19843	5MY	2;39	46.44	19843	19MY	3;08	239.84
19770	9JN	17;01	79.88	19770	24JN	19;17	274.26	19844	3JN	14;51	74.60	19844	17JN	12;37	267.84
19771	9JL	7;57	108.13	19771	24JL	2;14	302.28	19845	3JL	1;20	102.66	19845	17JL	0;14	296.01
19772	7AU	23;31	136.64	19772	22AU	9;51	330.66	19846	1AU	10;37	130.86	19846	15AU	14;24	324.57
19773	6SE	15;17	165.58	19773	20SE	19;13	359.58	19847	30AU	19;21	159.41	19847	14SE	6;51	353.67
19774	6OC	6;48	195.01	19774	20OC	7;09	29.09	19848	29SE	4;14	188.46	19848	14OC	0;51	23.34
19775	4NO	21;29	224.87	19775	18NO	21;58	59.12	19849	28OC	13;58	218.01	19849	12NO	19;20	53.46
19776	4DE	10;46	254.96	19776	18DE	15;14	89.44	19850	27NO	1;08	247.96	19850	12DE	12;57	83.77
								19851	26DE	14;05	278.09				

599 / 605

NEW MOONS NUMBER	DATE	TIME	LONG.	FULL MOONS NUMBER	DATE	TIME	LONG.	NEW MOONS NUMBER	DATE	TIME	LONG.	FULL MOONS NUMBER	DATE	TIME	LONG.
19777	2JA	22;24	285.02	19777	17JA	9;49	119.71					19851	11JA	4;32	113.96
19778	1FE	8;37	314.79	19778	16FE	4;19	149.63	19852	25JA	4;41	308.12	19852	9FE	17;21	143.72
19779	2MR	17;54	344.10	19779	17MR	21;30	179.01	19853	23FE	20;23	337.81	19853	11MR	3;30	172.92
19780	1AP	2;52	12.90	19780	16AP	12;28	207.81	19854	25MR	12;28	7.00	19854	9AP	11;40	201.55
19781	30AP	12;12	41.27	19781	16MY	0;47	236.13	19855	24AP	4;23	35.72	19855	8MY	18;51	229.72
19782	29MY	22;29	69.39	19782	14JN	10;40	264.16	19856	23MY	19;36	64.08	19856	7JN	2;05	257.67
19783	28JN	10;15	97.48	19783	13JL	18;55	292.16	19857	22JN	9;41	92.26	19857	6JL	10;22	285.65
19784	27JL	23;57	125.81	19784	12AU	2;36	320.41	19858	21JL	22;18	120.50	19858	4AU	20;30	313.92
19785	26AU	15;46	154.58	19785	10SE	10;49	349.10	19859	20AU	9;32	149.00	19859	3SE	9;07	342.68
19786	25SE	9;17	183.90	19786	9OC	20;22	18.34	19860	18SE	19;49	177.92	19860	3OC	0;37	12.06
19787	25OC	3;18	213.73	19787	8NO	7;40	48.10	19861	18OC	5;52	207.33	19861	1NO	18;49	42.02
19788	23NO	20;19	243.90	19788	7DE	20;51	78.20	19862	16NO	16;14	237.15	19862	1DE	14;35	72.37
19789	23DE	11;13	274.12					19863	16DE	3;13	267.19	19863	31DE	9;56	102.76

600 / 606

NEW MOONS NUMBER	DATE	TIME	LONG.	FULL MOONS NUMBER	DATE	TIME	LONG.	NEW MOONS NUMBER	DATE	TIME	LONG.	FULL MOONS NUMBER	DATE	TIME	LONG.
				19789	6JA	11;49	108.39	19864	14JA	14;49	297.18	19864	30JA	2;55	132.84
19790	21JA	23;44	304.09	19790	5FE	4;18	138.38	19865	13FE	2;59	326.88	19865	28FE	16;40	162.37
19791	20FE	10;13	333.62	19791	5MR	21;39	167.97	19866	14MR	15;49	356.14	19866	30MR	3;24	191.27
19792	20MR	19;13	2.60	19792	4AP	14;47	197.03	19867	13AP	5;34	24.93	19867	28AP	11;56	219.63
19793	19AP	3;15	31.07	19793	4MY	6;30	225.60	19868	12MY	20;19	53.38	19868	27MY	19;14	247.64
19794	18MY	11;00	59.17	19794	2JN	20;04	253.82	19869	11JN	11;41	81.63	19869	26JN	2;11	275.55
19795	16JN	19;19	87.13	19795	2JL	7;33	281.92	19870	11JL	2;56	109.91	19870	25JL	9;39	303.61
19796	16JL	5;12	115.22	19796	31JL	17;34	310.14	19871	9AU	17;25	138.40	19871	23AU	18;34	332.06
19797	14AU	17;40	143.70	19797	30AU	3;01	338.71	19872	8SE	6;54	167.27	19872	22SE	5;52	1.07
19798	13SE	9;12	172.74	19798	28SE	12;33	7.77	19873	7OC	19;31	196.60	19873	21OC	20;19	30.72
19799	13OC	3;22	202.41	19799	27OC	22;35	37.33	19874	6NO	7;30	226.37	19874	20NO	13;56	60.88
19800	11NO	22;40	232.55	19800	26NO	9;18	67.26	19875	5DE	18;58	256.39	19875	20DE	9;34	91.29
19801	11DE	17;16	262.91	19801	25DE	20;56	97.34								

601 / 607

NEW MOONS NUMBER	DATE	TIME	LONG.	FULL MOONS NUMBER	DATE	TIME	LONG.	NEW MOONS NUMBER	DATE	TIME	LONG.	FULL MOONS NUMBER	DATE	TIME	LONG.
19802	10JA	9;49	293.14	19802	24JA	9;48	127.30	19876	4JA	5;49	286.41	19876	19JA	5;10	121.59
19803	8FE	23;48	322.96	19803	23FE	0;04	156.94	19877	2FE	16;06	316.17	19877	17FE	22;51	151.46
19804	10MR	11;10	352.22	19804	24MR	15;27	186.12	19878	4MR	2;11	345.49	19878	19MR	13;44	180.72
19805	8AP	20;17	20.90	19805	23AP	7;15	214.84	19879	2AP	12;41	14.33	19879	18AP	1;48	209.38
19806	8MY	3;45	49.09	19806	22MY	22;38	243.21	19880	2MY	0;15	42.77	19880	17MY	11;28	237.57
19807	6JN	10;29	77.02	19807	21JN	13;09	271.42	19881	31MY	13;14	70.99	19881	15JN	19;26	265.52
19808	5JL	17;39	104.95	19808	21JL	2;39	299.68	19882	30JN	3;36	99.19	19882	15JL	2;32	293.49
19809	4AU	2;27	133.16	19809	19AU	15;15	328.23	19883	29JL	18;59	127.60	19883	13AU	9;49	321.73
19810	2SE	13;55	161.86	19810	18SE	3;05	357.20	19884	28AU	10;56	156.40	19884	11SE	18;19	350.46
19811	2OC	4;30	191.19	19811	17OC	14;14	26.64	19885	27SE	2;56	185.68	19885	11OC	4;59	19.77
19812	31OC	21;55	221.11	19812	16NO	0;54	56.47	19886	26OC	18;24	215.44	19886	9NO	18;25	49.63
19813	30NO	17;07	251.43	19813	15DE	11;25	86.49	19887	25NO	8;38	245.50	19887	9DE	10;32	79.86
19814	30DE	12;34	281.83					19888	24DE	21;11	275.62				

608

NEW MOONS				FULL MOONS			
NUMBER	DATE	TIME	LONG.	NUMBER	DATE	TIME	LONG.
				19888	8JA	4:29	110.17
19889	23JA	8:01	305.51	19889	6FE	23:00	140.23
19890	21FE	17:32	334.98	19890	7MR	16:47	169.80
19891	22MR	2:22	3.93	19891	6AP	8:49	198.79
19892	20AP	11:13	32.41	19892	5MY	22:26	227.26
19893	19MY	20:45	60.57	19893	4JN	9:32	255.37
19894	18JN	7:34	88.63	19894	3JL	18:36	283.37
19895	17JL	20:13	116.84	19895	2AU	2:36	311.53
19896	16AU	11:03	145.43	19896	31AU	10:37	340.06
19897	15SE	4:00	174.57	19897	29SE	19:35	9.12
19898	14OC	22:12	204.26	19898	29OC	6:06	38.72
19899	13NO	16:08	234.37	19899	27NO	18:22	68.73
19900	13DE	8:20	264.62	19900	27DE	8:21	98.90

609

NEW MOONS				FULL MOONS			
19901	11JA	22:03	294.72	19901	25JA	23:55	128.96
19902	10FE	9:25	324.42	19902	24FE	16:40	158.67
19903	11MR	18:57	353.59	19903	26MR	9:49	187.90
19904	10AP	3:14	22.20	19904	25AP	2:13	216.63
19905	9MY	10:53	50.39	19905	24MY	16:53	244.95
19906	7JN	18:40	78.36	19906	23JN	5:27	273.08
19907	7JL	3:34	106.37	19907	22JL	16:18	301.25
19908	5AU	14:39	134.68	19908	21AU	2:11	329.71
19909	4SE	4:42	163.51	19909	19SE	11:51	358.61
19910	3OC	21:49	192.97	19910	18OC	21:47	28.03
19911	2NO	16:57	222.98	19911	17NO	8:15	57.86
19912	2DE	12:15	253.32	19912	16DE	19:23	87.91

610

NEW MOONS				FULL MOONS			
19913	1JA	6:00	283.64	19913	15JA	7:27	117.92
19914	30JA	21:16	313.64	19914	13FE	20:46	147.65
19915	1MR	9:50	343.11	19915	15MR	11:20	176.97
19916	30MR	19:55	11.97	19916	14AP	2:45	205.82
19917	29AP	4:00	40.31	19917	13MY	18:15	234.28
19918	28MY	10:54	68.30	19918	12JN	9:14	262.52
19919	26JN	17:40	96.19	19919	11JL	23:21	290.75
19920	26JL	1:31	124.27	19920	10AU	12:39	319.20
19921	24AU	11:37	152.78	19921	9SE	1:09	348.05
19922	23SE	0:42	181.88	19922	8OC	12:54	17.36
19923	22OC	16:53	211.61	19923	6NO	23:59	47.09
19924	21NO	11:25	241.82	19924	6DE	10:36	77.08
19925	21DE	6:57	272.23				

611

NEW MOONS				FULL MOONS			
				19925	4JA	21:08	107.08
19926	20JA	1:53	302.49	19926	3FE	7:55	136.86
19927	18FE	18:47	332.31	19927	4MR	19:16	166.22
19928	20MR	8:42	1.52	19928	3AP	7:18	195.11
19929	18AP	19:32	30.12	19929	2MY	20:09	223.60
19930	18MY	3:53	58.25	19930	1JN	9:54	251.84
19931	16JN	10:51	86.16	19931	1JL	0:37	280.06
19932	15JL	17:38	114.12	19932	30JL	16:08	308.48
19933	14AU	1:23	142.39	19933	29AU	7:52	337.28
19934	12SE	10:55	171.17	19934	27SE	22:59	6.55
19935	11OC	22:49	200.54	19935	27OC	12:54	36.25
19936	10NO	13:21	230.46	19936	26NO	1:26	66.24
19937	10DE	6:26	260.74	19937	25DE	12:51	96.31

612

NEW MOONS				FULL MOONS			
19938	9JA	1:21	291.08	19938	23JA	23:25	126.19
19939	7FE	20:36	321.16	19939	22FE	9:21	155.66
19940	8MR	14:19	350.71	19940	22MR	18:52	184.63
19941	7AP	5:12	19.64	19941	21AP	4:25	213.13
19942	6MY	17:03	48.03	19942	20MY	14:44	241.31
19943	5JN	2:29	76.08	19943	19JN	2:38	269.42
19944	4JL	10:34	104.04	19944	18JL	16:42	297.69
19945	2AU	18:18	132.19	19945	17AU	8:48	326.35
19946	1SE	2:28	160.74	19946	16SE	2:02	355.51
19947	30SE	11:42	189.82	19947	15OC	19:13	25.17
19948	29OC	22:33	219.44	19948	14NO	11:22	55.21
19949	28NO	11:35	249.49	19949	14DE	1:56	85.40
19950	28DE	3:01	279.72				

613

NEW MOONS				FULL MOONS			
				19950	12JA	14:43	115.46
19951	26JA	20:25	309.85	19951	11FE	1:38	145.13
19952	25FE	14:31	339.61	19952	12MR	10:52	174.27
19953	27MR	7:46	8.82	19953	10AP	18:55	202.87
19954	25AP	23:09	37.50	19954	10MY	2:40	231.05
19955	25MY	12:24	65.76	19955	8JN	11:13	259.05
19956	23JN	23:47	93.84	19956	7JL	21:36	287.12
19957	23JL	9:46	121.98	19957	6AU	10:30	315.51
19958	21AU	18:56	150.42	19958	5SE	1:58	344.41
19959	20SE	3:52	179.31	19959	4OC	19:29	13.90
19960	19OC	13:16	208.71	19960	3NO	14:05	43.91
19961	17NO	23:44	238.54	19961	3DE	8:28	74.21
19962	17DE	11:45	268.63				

614

NEW MOONS				FULL MOONS			
				19962	2JA	1:16	104.49
19963	16JA	1:24	298.70	19963	31JA	15:29	134.43
19964	14FE	16:22	328.50	19964	2MR	2:50	163.84
19965	16MR	8:02	357.84	19965	31MR	11:45	192.64
19966	14AP	23:51	26.70	19966	29AP	19:09	220.94
19967	14MY	15:21	55.15	19967	29MY	2:08	248.93
19968	13JN	6:04	83.38	19968	27JN	9:43	276.86
19969	12JL	19:36	111.59	19969	26JL	18:50	305.00
19970	11AU	7:45	140.01	19970	25AU	6:11	333.57
19971	9SE	18:43	168.80	19971	23SE	20:19	2.73
19972	9OC	5:02	198.06	19972	23OC	13:21	32.50
19973	7NO	15:22	227.77	19973	22NO	8:38	62.75
19974	7DE	2:06	257.77	19974	22DE	4:31	93.17

615

NEW MOONS				FULL MOONS			
19975	5JA	13:19	287.80	19975	20JA	22:52	123.41
19976	4FE	1:00	317.60	19976	19FE	14:12	153.15
19977	5MR	13:12	346.99	19977	21MR	2:16	182.27
19978	4AP	2:09	15.91	19978	19AP	11:42	210.80
19979	3MY	16:09	44.43	19979	18MY	19:26	238.90
19980	2JN	7:08	72.71	19980	17JN	2:26	266.81
19981	1JL	22:35	100.96	19981	16JL	9:35	294.79
19982	31JL	13:42	129.38	19982	14AU	17:44	323.09
19983	30AU	3:58	158.13	19983	13SE	3:49	351.89
19984	28SE	17:17	187.33	19984	12OC	16:43	21.32
19985	28OC	5:49	216.99	19985	11NO	8:52	51.31
19986	26NO	17:43	246.96	19986	11DE	3:41	81.67
19987	26DE	4:54	277.02				

616

NEW MOONS				FULL MOONS			
				19987	9JA	23:30	112.05
19988	24JA	15:21	306.89	19988	8FE	18:12	142.09
19989	23FE	1:16	336.35	19989	9MR	10:25	171.57
19990	23MR	11:13	5.33	19990	7AP	23:47	200.42
19991	21AP	21:54	33.86	19991	7MY	10:32	228.76
19992	21MY	9:55	62.11	19992	5JN	19:17	256.77
19993	19JN	23:31	90.28	19993	5JL	2:47	284.72
19994	19JL	14:28	118.59	19994	3AU	9:58	312.85
19995	18AU	6:22	147.25	19995	1SE	17:52	341.40
19996	16SE	22:40	176.39	19996	1OC	3:28	10.51
19997	16OC	14:47	206.02	19997	30OC	15:34	40.19
19998	15NO	5:57	236.03	19998	29NO	6:23	70.31
19999	14DE	19:30	266.18	19999	28DE	23:27	100.61

617

NEW MOONS				FULL MOONS			
20000	13JA	7:10	296.18	20000	27JA	17:39	130.77
20001	11FE	17:09	325.80	20001	26FE	11:45	160.51
20002	13MR	2:03	354.92	20002	28MR	4:36	189.70
20003	11AP	10:35	23.53	20003	26AP	19:24	218.34
20004	10MY	19:30	51.76	20004	26MY	7:46	246.56
20005	9JN	5:26	79.80	20005	24JN	17:51	274.59
20006	8JL	17:02	107.92	20006	24JL	2:25	302.68
20007	7AU	6:47	136.36	20007	22AU	10:30	331.08
20008	5SE	22:52	165.30	20008	20SE	19:05	359.96
20009	5OC	16:49	194.82	20009	20OC	4:55	29.39
20010	4NO	11:20	224.83	20010	18NO	16:22	59.28
20011	4DE	4:43	255.10	20011	18DE	5:27	89.41

618

NEW MOONS				FULL MOONS			
20012	2JA	19:47	285.30	20012	16JA	20:04	119.52
20013	1FE	8:15	315.16	20013	15FE	12:02	149.34
20014	2MR	18:30	344.51	20014	17MR	4:49	178.72
20015	1AP	3:11	13.30	20015	15AP	21:32	207.60
20016	30AP	10:54	41.60	20016	15MY	13:05	236.05
20017	29MY	18:23	69.61	20017	14JN	2:46	264.23
20018	28JN	2:32	97.56	20018	13JL	14:34	292.38
20019	27JL	12:23	125.73	20019	12AU	1:04	320.74
20020	26AU	0:55	154.36	20020	10SE	11:01	349.50
20021	24SE	16:40	183.60	20021	9OC	21:00	18.76
20022	24OC	11:08	213.44	20022	8NO	7:20	48.48
20023	23NO	6:44	243.71	20023	7DE	18:08	78.48
20024	23DE	1:30	274.09				

619

NEW MOONS				FULL MOONS			
				20024	6JA	5:36	108.52
20025	21JA	18:05	304.25	20025	4FE	18:03	138.34
20026	20FE	7:57	333.92	20026	6MR	7:43	167.78
20027	21MR	19:07	2.98	20027	4AP	22:29	196.76
20028	20AP	4:01	31.48	20028	4MY	13:49	225.32
20029	19MY	11:19	59.56	20029	3JN	5:01	253.61
20030	17JN	17:58	87.46	20030	2JL	19:40	281.83
20031	17JL	1:08	115.44	20031	1AU	9:34	310.20
20032	15AU	10:02	143.78	20032	30AU	22:45	338.93
20033	13SE	21:39	172.66	20033	29SE	11:10	8.11
20034	13OC	12:24	202.17	20034	28OC	22:49	37.73
20035	12NO	5:57	232.24	20035	27NO	9:46	67.67
20036	12DE	1:12	262.61	20036	26DE	20:17	97.70

NEW MOONS / FULL MOONS / NEW MOONS / FULL MOONS

NEW MOONS (left)

NUMBER	DATE	TIME	LONG.
620			
20037	10JA	20;36	292.97
20038	9FE	14;35	322.97
20039	10MR	5;58	352.40
20040	8AP	18;14	21.20
20041	8MY	3;40	49.47
20042	6JN	11;11	77.44
20043	5JL	17;58	105.35
20044	4AU	1;12	133.50
20045	2SE	9;52	162.09
20046	1OC	20;39	191.25
20047	31OC	9;55	220.99
20048	30NO	1;45	251.17
20049	29DE	19;47	281.51
621			
20050	28JA	14;56	311.69
20051	27FE	9;29	341.44
20052	29MR	1;45	10.59
20053	27AP	15;02	39.15
20054	27MY	1;35	67.30
20055	25JN	10;18	95.27
20056	24JL	18;14	123.34
20057	23AU	2;15	151.75
20058	21SE	11;03	180.65
20059	20OC	21;09	210.10
20060	19NO	9;05	240.02
20061	18DE	23;14	270.20
622			
20062	17JA	15;35	300.37
20063	16FE	9;16	330.26
20064	18MR	2;52	359.65
20065	16AP	19;07	28.50
20066	16MY	9;23	56.89
20067	14JN	21;43	85.01
20068	14JL	8;31	113.13
20069	12AU	18;14	141.46
20070	11SE	3;27	170.21
20071	10OC	12;44	199.45
20072	8NO	22;43	229.16
20073	8DE	9;55	259.18
623			
20074	6JA	22;39	289.27
20075	5FE	12;47	319.16
20076	7MR	3;53	348.64
20077	5AP	19;24	17.64
20078	5MY	10;55	46.20
20079	4JN	2;03	74.48
20080	3JL	16;20	102.69
20081	2AU	5;24	131.03
20082	31AU	17;11	159.71
20083	30SE	4;02	188.84
20084	29OC	14;30	218.43
20085	28NO	1;06	248.36
20086	27DE	12;04	278.40
624			
20087	25JA	23;23	308.29
20088	24FE	11;02	337.81
20089	24MR	23;16	6.86
20090	23AP	12;26	35.47
20091	23MY	2;46	63.80
20092	21JN	18;02	92.03
20093	21JL	9;31	120.38
20094	20AU	0;30	149.02
20095	18SE	14;36	178.10
20096	18OC	3;47	207.63
20097	16NO	16;12	237.54
20098	16DE	3;50	267.61
625			
20099	14JA	14;36	297.57
20100	13FE	0;34	327.17
20101	14MR	10;11	356.29
20102	12AP	20;08	24.94
20103	12MY	7;11	53.24
20104	10JN	19;49	81.39
20105	10JL	10;06	109.62
20106	9AU	1;42	138.15
20107	7SE	18;06	167.13
20108	7OC	10;41	196.63
20109	6NO	2;42	226.56
20110	5DE	17;18	256.71

FULL MOONS (left)

NUMBER	DATE	TIME	LONG.
620			
20037	25JA	6;47	127.56
20038	23FE	17;35	157.06
20039	24MR	4;59	186.08
20040	22AP	17;08	214.67
20041	22MY	6;12	242.95
20042	20JN	20;20	271.14
20043	20JL	11;35	299.47
20044	19AU	3;30	328.14
20045	17SE	19;20	357.27
20046	17OC	10;12	26.86
20047	15NO	23;38	56.81
20048	15DE	11;40	86.89
621			
20049	13JA	22;36	116.86
20050	12FE	8;42	146.48
20051	13MR	18;10	175.61
20052	12AP	3;23	204.24
20053	11MY	12;59	232.48
20054	9JN	23;50	260.57
20055	9JL	12;24	288.74
20056	8AU	3;55	317.24
20057	6SE	20;55	346.23
20058	6OC	14;32	15.76
20059	5NO	7;33	45.72
20060	4DE	23;08	75.91
622			
20061	3JA	12;53	106.05
20062	2FE	0;41	135.88
20063	3MR	10;35	165.20
20064	1AP	18;57	193.97
20065	1MY	2;33	222.26
20066	30MY	10;24	250.28
20067	28JN	19;40	278.28
20068	28JL	7;14	306.52
20069	26AU	21;30	335.23
20070	25SE	14;15	4.52
20071	25OC	8;38	34.38
20072	24NO	3;28	64.62
20073	23DE	21;19	94.96
623			
20074	22JA	12;56	125.07
20075	21FE	1;37	154.68
20076	22MR	11;32	183.68
20077	20AP	19;27	212.13
20078	20MY	2;24	240.20
20079	18JN	9;30	268.11
20080	17JL	17;44	296.14
20081	16AU	3;55	324.53
20082	14SE	16;42	353.47
20083	14OC	8;24	23.04
20084	13NO	2;47	53.15
20085	12DE	22;41	83.56
624			
20086	11JA	18;03	113.90
20087	10FE	10;56	143.85
20088	11MR	0;31	173.19
20089	9AP	11;05	201.91
20090	8MY	19;29	230.14
20091	7JN	2;43	258.08
20092	6JL	9;43	286.01
20093	4AU	17;19	314.18
20094	3SE	2;25	342.80
20095	2OC	13;55	12.00
20096	1NO	4;29	41.81
20097	30NO	22;07	72.06
20098	30DE	17;37	102.46
625			
20099	29JA	12;59	132.65
20100	28FE	6;26	162.33
20101	29MR	21;07	191.40
20102	28AP	9;05	219.91
20103	27MY	18;46	248.01
20104	26JN	2;51	275.96
20105	25JL	10;11	304.02
20106	23AU	17;45	332.41
20107	22SE	2;32	1.32
20108	21OC	13;24	30.82
20109	20NO	2;54	60.80
20110	19DE	18;52	91.06

NEW MOONS (right)

NUMBER	DATE	TIME	LONG.
626			
20111	4JA	5;56	286.80
20112	2FE	16;38	316.58
20113	4MR	1;50	345.86
20114	2AP	10;15	14.62
20115	1MY	18;40	42.94
20116	31MY	3;51	71.00
20117	29JN	14;27	99.05
20118	29JL	3;03	127.35
20119	27AU	18;04	156.10
20120	26SE	11;22	185.43
20121	26OC	6;02	215.31
20122	25NO	0;24	245.54
20123	24DE	16;50	275.82
627			
20124	23JA	6;34	305.84
20125	21FE	17;46	335.38
20126	23MR	3;01	4.34
20127	21AP	10;59	32.79
20128	20MY	18;22	60.86
20129	19JN	1;58	88.79
20130	18JL	10;48	116.85
20131	16AU	21;55	145.29
20132	15SE	12;08	174.30
20133	15OC	5;30	203.95
20134	14NO	0;54	234.11
20135	13DE	20;25	264.51
628			
20136	12JA	14;14	294.78
20137	11FE	5;25	324.65
20138	11MR	17;49	353.93
20139	10AP	3;41	22.61
20140	9MY	11;36	50.81
20141	7JN	18;24	78.74
20142	7JL	1;10	106.66
20143	5AU	9;06	134.84
20144	3SE	19;20	163.52
20145	3OC	8;36	192.81
20146	2NO	0;55	222.70
20147	1DE	19;31	253.00
20148	31DE	15;00	283.39
629			
20149	30JA	9;47	313.55
20150	1MR	2;29	343.19
20151	30MR	16;14	12.21
20152	29AP	2;58	40.65
20153	28MY	11;18	68.70
20154	26JN	18;20	96.61
20155	26JL	1;18	124.65
20156	24AU	9;16	153.07
20157	22SE	19;03	182.04
20158	22OC	7;08	211.59
20159	20NO	21;42	241.63
20160	20DE	14;39	271.93
630			
20161	19JA	9;16	302.19
20162	18FE	4;10	332.10
20163	19MR	21;35	1.45
20164	18AP	12;18	30.21
20165	18MY	0;07	58.49
20166	16JN	9;42	86.50
20167	15JL	18;02	114.53
20168	14AU	2;06	142.81
20169	12SE	10;37	171.55
20170	11OC	20;09	200.82
20171	10NO	7;11	230.59
20172	9DE	20;10	260.69
631			
20173	8JA	11;19	290.88
20174	7FE	4;14	320.86
20175	8MR	21;48	350.42
20176	7AP	14;38	19.44
20177	7MY	5;50	47.97
20178	5JN	19;08	76.17
20179	5JL	6;47	104.28
20180	3AU	17;12	132.54
20181	2SE	2;50	161.15
20182	1OC	12;14	190.24
20183	30OC	21;57	219.81
20184	29NO	8;33	249.75
20185	28DE	20;26	279.83

FULL MOONS (right)

NUMBER	DATE	TIME	LONG.
626			
20111	18JA	12;29	121.28
20112	17FE	6;33	151.17
20113	18MR	23;57	180.54
20114	17AP	15;45	209.36
20115	17MY	5;19	237.71
20116	15JN	16;35	265.79
20117	15JL	1;57	293.85
20118	13AU	10;20	322.14
20119	11SE	18;45	350.86
20120	11OC	4;05	20.12
20121	9NO	14;48	49.87
20122	9DE	3;03	79.94
627			
20123	7JA	16;47	110.06
20124	6FE	7;52	139.98
20125	8MR	0;02	169.50
20126	6AP	16;40	198.53
20127	6MY	8;47	227.10
20128	4JN	23;27	255.36
20129	4JL	12;18	283.51
20130	2AU	23;35	311.80
20131	1SE	9;59	340.43
20132	30SE	20;09	9.54
20133	30OC	6;28	39.13
20134	28NO	17;06	69.06
20135	28DE	4;10	99.11
628			
20136	26JA	15;54	129.01
20137	25FE	4;40	158.57
20138	25MR	18;37	187.67
20139	24AP	9;28	216.34
20140	24MY	0;40	244.70
20141	22JN	15;38	272.92
20142	22JL	6;04	301.23
20143	20AU	19;53	329.84
20144	19SE	9;00	358.89
20145	18OC	21;18	28.40
20146	17NO	8;45	58.27
20147	16DE	19;30	88.30
629			
20148	15JA	5;53	118.24
20149	13FE	16;18	147.86
20150	15MR	3;07	177.02
20151	13AP	14;34	205.71
20152	13MY	2;55	234.06
20153	11JN	16;22	262.24
20154	11JL	7;04	290.49
20155	9AU	22;53	319.04
20156	8SE	15;09	348.03
20157	8OC	6;54	17.50
20158	6NO	21;20	47.37
20159	6DE	10;11	77.46
630			
20160	4JA	21;39	107.50
20161	3FE	8;02	137.25
20162	4MR	17;36	166.54
20163	3AP	2;39	195.31
20164	2MY	11;44	223.65
20165	31MY	21;40	251.74
20166	30JN	9;21	279.84
20167	29JL	23;24	308.20
20168	28AU	15;45	337.01
20169	27SE	9;27	6.37
20170	27OC	3;10	36.23
20171	25NO	19;45	66.39
20172	25DE	10;33	96.60
631			
20173	23JA	23;19	126.57
20174	22FE	10;02	156.08
20175	23MR	18;56	185.02
20176	22AP	2;38	213.45
20177	21MY	10;04	241.52
20178	19JN	18;23	269.48
20179	19JL	4;40	297.60
20180	17AU	17;36	326.12
20181	16SE	9;17	355.21
20182	16OC	3;09	24.89
20183	14NO	22;06	55.04
20184	14DE	16;44	85.40

632

NEW NUMBER	DATE	TIME	LONG.	FULL NUMBER	DATE	TIME	LONG.
				20185	13JA	9;38	115.63
20186	27JA	9;43	309.79	20186	11FE	23;46	145.44
20187	26FE	0;06	339.40	20187	12MR	10;54	174.66
20188	26MR	15;10	8.54	20188	10AP	19;35	203.28
20189	25AP	6;30	37.22	20189	10MY	2;46	231.45
20190	24MY	21;46	65.57	20190	8JN	9;35	259.37
20191	23JN	12;34	93.78	20191	7JL	17;07	287.33
20192	23JL	2;28	122.08	20192	6AU	2;17	315.57
20193	21AU	15;10	150.65	20193	4SE	13;46	344.30
20194	20SE	2;43	179.65	20194	4OC	4;05	13.65
20195	19OC	13;32	209.11	20195	2NO	21;17	43.59
20196	18NO	0;11	238.95	20196	2DE	16;42	73.93
20197	17DE	10;59	268.99				

633

NEW NUMBER	DATE	TIME	LONG.	FULL NUMBER	DATE	TIME	LONG.
				20197	1JA	12;36	104.34
20198	15JA	22;01	298.95	20198	31JA	6;52	134.46
20199	14FE	9;17	328.59	20199	1MR	22;03	164.03
20200	15MR	20;54	357.77	20200	31MR	9;57	192.96
20201	14AP	9;15	26.50	20201	29AP	19;15	221.34
20202	13MY	22;46	54.88	20202	29MY	2;55	249.35
20203	12JN	13;30	83.11	20203	27JN	9;57	277.26
20204	12JL	5;02	111.40	20204	26JL	17;13	305.32
20205	10AU	20;33	139.94	20205	25AU	1;33	333.77
20206	9SE	11;24	168.89	20206	23SE	11;51	2.76
20207	9OC	1;20	198.30	20207	23OC	0;54	32.37
20208	7NO	14;23	228.12	20208	21NO	17;05	62.48
20209	7DE	2;32	258.18	20209	21DE	11;49	92.86

634

NEW NUMBER	DATE	TIME	LONG.	FULL NUMBER	DATE	TIME	LONG.
20210	5JA	13;44	288.21	20210	20JA	7;23	123.15
20211	3FE	23;57	317.95	20211	19FE	1;49	153.02
20212	5MR	9;27	347.22	20212	20MR	17;48	182.31
20213	3AP	18;53	16.00	20213	19AP	7;01	211.00
20214	3MY	5;03	44.37	20214	18MY	17;46	239.22
20215	1JN	16;40	72.53	20215	17JN	2;37	267.21
20216	1JL	6;03	100.69	20216	16JL	10;20	295.21
20217	30JL	21;05	129.10	20217	14AU	17;49	323.48
20218	29AU	13;20	157.92	20218	13SE	2;01	352.21
20219	28SE	6;12	187.26	20219	12OC	11;53	21.51
20220	27OC	22;52	217.09	20220	11NO	0;06	51.33
20221	26NO	14;29	247.22	20221	10DE	14;51	81.52
20222	26DE	4;14	277.38				

635

NEW NUMBER	DATE	TIME	LONG.	FULL NUMBER	DATE	TIME	LONG.
				20222	9JA	7;37	111.76
20223	24JA	15;50	307.30	20223	8FE	1;23	141.78
20224	23FE	1;34	336.75	20224	9MR	19;01	171.32
20225	24MR	10;06	5.67	20225	8AP	11;33	200.32
20226	22AP	18;13	34.10	20226	8MY	2;13	228.82
20227	22MY	2;45	62.21	20227	6JN	14;41	256.98
20228	20JN	12;26	90.22	20228	6JL	1;03	285.04
20229	19JL	23;56	118.40	20229	4AU	10;00	313.24
20230	18AU	13;46	146.97	20230	2SE	18;30	341.82
20231	17SE	6;07	176.10	20231	2OC	3;28	10.90
20232	17OC	0;29	205.81	20232	31OC	13;36	40.50
20233	15NO	19;26	235.97	20233	30NO	1;07	70.48
20234	15DE	13;07	266.30	20234	29DE	14;02	100.61

636

NEW NUMBER	DATE	TIME	LONG.	FULL NUMBER	DATE	TIME	LONG.
20235	14JA	4;17	296.45	20235	28JA	4;15	130.60
20236	12FE	16;38	326.17	20236	26FE	19;38	160.24
20237	13MR	2;39	355.33	20237	27MR	11;51	189.41
20238	11AP	11;02	23.93	20238	26AP	4;09	218.11
20239	10MY	18;29	52.10	20239	25MY	19;35	246.46
20240	9JN	1;46	80.05	20240	24JN	9;26	274.64
20241	8JL	9;49	108.02	20241	23JL	21;38	302.88
20242	6AU	19;41	136.30	20242	22AU	8;39	331.39
20243	5SE	8;21	165.09	20243	20SE	19;08	0.36
20244	5OC	0;18	194.52	20244	20OC	5;34	29.81
20245	3NO	19;01	224.54	20245	18NO	16;10	59.66
20246	3DE	14;50	254.90	20246	18DE	3;00	89.70

637

NEW NUMBER	DATE	TIME	LONG.	FULL NUMBER	DATE	TIME	LONG.
20247	2JA	9;42	285.26	20247	16JA	14;13	119.66
20248	1FE	2;14	315.30	20248	15FE	2;12	149.32
20249	2MR	15;56	344.79	20249	16MR	15;15	178.55
20250	1AP	2;55	13.68	20250	15AP	5;24	207.34
20251	30AP	11;38	42.02	20251	14MY	20;18	235.77
20252	29MY	18;50	70.02	20252	13JN	11;23	264.01
20253	28JN	1;28	97.91	20253	13JL	2;13	292.27
20254	27JL	8;43	125.97	20254	11AU	16;34	320.78
20255	25AU	17;45	154.46	20255	10SE	6;21	349.70
20256	24SE	5;32	183.53	20256	9OC	19;22	19.08
20257	23OC	20;27	213.22	20257	8NO	7;28	48.87
20258	22NO	14;05	243.40	20258	7DE	18;38	78.89
20259	22DE	9;17	273.80				

638

NEW NUMBER	DATE	TIME	LONG.	FULL NUMBER	DATE	TIME	LONG.
				20259	6JA	5;07	108.89
20260	21JA	4;32	304.06	20260	4FE	15;20	138.62
20261	19FE	22;18	333.90	20261	6MR	1;40	167.92
20262	21MR	13;30	3.14	20262	4AP	12;29	196.74
20263	20AP	1;39	31.78	20263	4MY	0;06	225.16
20264	19MY	11;03	59.94	20264	2JN	12;46	253.35
20265	17JN	18;37	87.87	20265	2JL	2;47	281.55
20266	17JL	1;34	115.85	20266	31JL	18;11	309.98
20267	15AU	9;02	144.12	20267	30AU	10;33	338.82
20268	13SE	17;58	172.90	20268	29SE	2;59	8.16
20269	13OC	4;57	202.25	20269	28OC	18;26	37.94
20270	11NO	18;19	232.14	20270	27NO	8;17	68.00
20271	11DE	10;04	262.36	20271	26DE	20;28	98.10

639

NEW NUMBER	DATE	TIME	LONG.	FULL NUMBER	DATE	TIME	LONG.
20272	10JA	3;51	292.65	20272	25JA	7;18	127.98
20273	8FE	22;37	322.69	20273	23FE	17;05	157.42
20274	10MR	16;47	352.24	20274	25MR	2;08	186.35
20275	9AP	8;49	21.21	20275	23AP	10;54	214.80
20276	8MY	22;02	49.64	20276	22MY	20;05	242.93
20277	7JN	8;41	77.72	20277	21JN	6;39	270.98
20278	6JL	17;38	105.73	20278	20JL	19;26	299.21
20279	5AU	1;53	133.91	20279	19AU	10;47	327.85
20280	3SE	10;18	162.49	20280	18SE	4;09	357.03
20281	2OC	19;26	191.59	20281	17OC	22;17	26.76
20282	1NO	5;47	221.21	20282	16NO	15;46	56.87
20283	30NO	17;44	251.22	20283	16DE	7;39	87.11
20284	30DE	7;42	281.38				

640

NEW NUMBER	DATE	TIME	LONG.	FULL NUMBER	DATE	TIME	LONG.
				20284	14JA	21;29	117.20
20285	28JA	23;37	311.44	20285	13FE	9;08	146.89
20286	27FE	16;45	341.14	20286	13MR	18;45	176.02
20287	28MR	9;51	10.33	20287	12AP	2;48	204.60
20288	27AP	1;48	39.01	20288	11MY	10;04	232.75
20289	26MY	16;01	67.31	20289	9JN	17;41	260.71
20290	25JN	4;34	95.43	20290	9JL	2;48	288.73
20291	24JL	15;45	123.63	20291	7AU	14;22	317.08
20292	23AU	1;58	152.13	20292	6SE	4;47	345.96
20293	21SE	11;39	181.07	20293	5OC	21;48	15.45
20294	20OC	21;20	210.50	20294	4NO	16;32	45.47
20295	19NO	7;32	240.34	20295	4DE	11;38	75.81
20296	18DE	18;43	270.40				

641

NEW NUMBER	DATE	TIME	LONG.	FULL NUMBER	DATE	TIME	LONG.
				20296	3JA	5;37	106.14
20297	17JA	7;09	300.40	20297	1FE	21;12	136.12
20298	15FE	20;47	330.13	20298	3MR	9;44	165.56
20299	17MR	11;15	359.40	20299	1AP	19;26	194.38
20300	16AP	2;13	28.21	20300	1MY	3;07	222.67
20301	15MY	17;23	56.64	20301	30MY	9;54	250.64
20302	14JN	8;27	84.87	20302	28JN	16;56	278.55
20303	13JL	23;00	113.13	20303	28JL	1;12	306.67
20304	12AU	12;35	141.62	20304	26AU	11;31	335.21
20305	11SE	0;58	170.49	20305	25SE	0;27	4.34
20306	10OC	12;21	199.82	20306	24OC	16;19	34.08
20307	8NO	23;13	229.57	20307	23NO	10;50	64.31
20308	8DE	10;00	259.57	20308	23DE	6;46	94.74

642

NEW NUMBER	DATE	TIME	LONG.	FULL NUMBER	DATE	TIME	LONG.
20309	6JA	20;52	289.58	20309	22JA	2;03	125.00
20310	5FE	7;50	319.34	20310	20FE	18;48	154.78
20311	6MR	18;59	348.66	20311	22MR	8;13	183.94
20312	5AP	6;37	17.51	20312	20AP	18;38	212.49
20313	4MY	19;14	45.96	20313	20MY	2;57	240.60
20314	3JN	9;13	74.20	20314	18JN	10;12	268.52
20315	3JL	0;25	102.44	20315	17JL	17;18	296.51
20316	1AU	16;11	130.89	20316	16AU	1;06	324.81
20317	31AU	7;42	159.71	20317	14SE	10;26	353.61
20318	29SE	22;25	188.99	20318	13OC	22;06	23.00
20319	29OC	12;10	218.72	20319	12NO	12;46	52.95
20320	28NO	0;57	248.74	20320	12DE	6;20	83.25
20321	27DE	12;41	278.82				

643

NEW NUMBER	DATE	TIME	LONG.	FULL NUMBER	DATE	TIME	LONG.
				20321	11JA	1;37	113.60
20322	25JA	23;18	308.68	20322	9FE	20;41	143.64
20323	24FE	8;55	338.11	20323	11MR	13;50	173.14
20324	25MR	18;03	7.03	20324	10AP	4;21	202.02
20325	24AP	3;30	35.50	20325	9MY	16;14	230.39
20326	23MY	14;06	63.68	20326	8JN	2;00	258.44
20327	22JN	2;27	91.80	20327	7JL	10;17	286.42
20328	21JL	16;42	120.09	20328	5AU	17;55	314.59
20329	20AU	8;32	148.76	20329	4SE	1;48	343.16
20330	19SE	1;25	177.94	20330	3OC	10;53	12.27
20331	18OC	18;34	207.64	20331	1NO	21;57	41.92
20332	17NO	11;03	237.71	20332	1DE	11;27	72.00
20333	17DE	1;56	267.92	20333	31DE	3;11	102.24

644

NEW MOONS				FULL MOONS			
20334	15JA	14;37	297.96	20334	29JA	20;23	132.34
20335	14FE	1;08	327.58	20335	28FE	13;58	162.04
20336	14MR	10;00	356.68	20336	29MR	6;57	191.22
20337	12AP	18;02	25.24	20337	27AP	22;32	219.87
20338	12MY	2;05	53.42	20338	27MY	12;07	248.14
20339	10JN	10;58	81.43	20339	25JN	23;36	276.22
20340	9JL	21;24	109.50	20340	25JL	9;21	304.37
20341	8AU	10;02	137.91	20341	23AU	18;10	332.82
20342	7SE	1;15	166.83	20342	22SE	3;01	1.74
20343	6OC	18;54	196.36	20343	21OC	12;41	31.18
20344	5NO	13;59	226.41	20344	19NO	23;34	61.05
20345	5DE	8;41	256.73	20345	19DE	11;45	91.15

645

NEW MOONS				FULL MOONS			
20346	4JA	1;16	287.00	20346	18JA	1;10	121.19
20347	2FE	14;58	316.90	20347	16FE	15;43	150.94
20348	4MR	1;59	346.26	20348	18MR	7;16	180.25
20349	2AP	10;57	15.03	20349	16AP	23;24	209.09
20350	1MY	18;39	43.32	20350	16MY	15;16	237.54
20351	31MY	1;49	71.31	20351	15JN	5;59	265.76
20352	29JN	9;19	99.24	20352	14JL	19;09	293.97
20353	28JL	18;09	127.38	20353	13AU	6;56	322.39
20354	27AU	5;21	155.97	20354	11SE	17;53	351.22
20355	25SE	19;45	185.17	20355	11OC	4;33	20.53
20356	25OC	13;20	215.00	20356	9NO	15;14	50.28
20357	24NO	8;57	245.28	20357	9DE	1;59	80.28
20358	24DE	4;33	275.69				

646

NEW MOONS				FULL MOONS			
				20358	7JA	12;54	110.29
20359	22JA	22;21	305.88	20359	6FE	0;15	140.05
20360	21FE	13;25	335.58	20360	7MR	12;27	169.41
20361	23MR	1;38	4.68	20361	6AP	1;46	198.31
20362	21AP	11;20	33.19	20362	5MY	16;07	226.82
20363	20MY	19;08	61.28	20363	4JN	7;02	255.10
20364	19JN	1;54	89.18	20364	3JL	22;04	283.33
20365	18JL	8;44	117.16	20365	2AU	12;52	311.75
20366	16AU	16;49	145.47	20366	1SE	3;14	340.54
20367	15SE	3;14	174.33	20367	30SE	16;58	9.80
20368	14OC	16;39	203.80	20368	30OC	5;48	39.49
20369	13NO	9;04	233.84	20369	28NO	17;34	69.48
20370	13DE	3;38	264.19	20370	28DE	4;22	99.51

647

NEW MOONS				FULL MOONS			
20371	11JA	22;58	294.53	20371	26JA	14;34	129.35
20372	10FE	17;32	324.54	20372	25FE	0;34	158.79
20373	12MR	10;01	354.00	20373	26MR	10;51	187.74
20374	10AP	23;37	22.83	20374	24AP	21;46	216.26
20375	10MY	10;17	51.14	20375	24MY	9;39	244.49
20376	8JN	18;40	79.14	20376	22JN	22;51	272.64
20377	8JL	1;51	107.08	20377	22JL	13;36	300.96
20378	6AU	9;03	135.23	20378	21AU	5;45	329.65
20379	4SE	17;18	163.82	20379	19SE	22;33	358.84
20380	4OC	3;20	192.98	20380	19OC	14;55	28.52
20381	2NO	15;34	222.70	20381	18NO	5;50	58.54
20382	2DE	6;07	252.82	20382	17DE	18;56	88.67
20383	31DE	22;51	283.10				

648

NEW MOONS				FULL MOONS			
				20383	16JA	6;24	118.65
20384	30JA	17;06	313.24	20384	14FE	16;33	148.26
20385	29FE	11;34	342.96·	20385	15MR	1;44	177.35
20386	30MR	4;39	12.13	20386	13AP	10;21	205.93
20387	28AP	19;14	40.73	20387	12MY	19;01	234.13
20388	28MY	7;07	68.92	20388	11JN	4;37	262.16
20389	26JN	16;54	96.94	20389	10JL	16;08	290.28
20390	26JL	1;33	125.05	20390	9AU	4;38	318.75
20391	24AU	10;00	153.49	20391	7SE	22;51	347.75
20392	22SE	18;54	182.42	20392	7OC	17;01	17.31
20393	22OC	4;44	211.88	20393	6NO	11;13	47.34
20394	20NO	15;53	241.77	20394	6DE	4;10	77.59
20395	20DE	4;45	271.90				

649

NEW MOONS				FULL MOONS			
				20395	4JA	19;07	107.78
20396	18JA	19;33	302.00	20396	3FE	7;48	137.63
20397	17FE	11;56	331.81	20397	4MR	18;17	166.96
20398	19MR	4;56	1.16	20398	3AP	2;53	195.71
20399	17AP	21;22	30.00	20399	2MY	10;16	223.98
20400	17MY	12;25	58.41	20400	31MY	17;27	251.96
20401	16JN	1;50	86.58	20401	30JN	1;36	279.92
20402	15JL	13;49	114.75	20402	29JL	11;50	308.12
20403	14AU	0;43	143.14	20403	28AU	0;52	336.79
20404	12SE	10;52	171.95	20404	26SE	16;47	6.07
20405	11OC	20;43	201.23	20405	26OC	10;56	35.94
20406	10NO	6;44	230.96	20406	25NO	6;10	66.21
20407	9DE	17;24	260.97	20407	25DE	0;59	96.59

650

NEW MOONS				FULL MOONS			
20408	8JA	5;06	291.01	20408	23JA	17;53	126.73
20409	6FE	17;56	320.82	20409	22FE	7;54	156.38
20410	8MR	7;44	350.23	20410	23MR	18;50	185.41
20411	6AP	22;11	19.17	20411	22AP	3;18	213.86
20412	6MY	13;03	47.69	20412	21MY	10;19	241.91
20413	5JN	4;08	75.96	20413	19JN	17;04	269.81
20414	4JL	19;05	104.20	20414	19JL	0;36	297.82
20415	3AU	9;25	132.61	20415	17AU	9;52	326.19
20416	1SE	22;41	161.37	20416	15SE	21;31	355.11
20417	1OC	10;50	190.57	20417	15OC	12;00	24.64
20418	30OC	22;08	220.21	20418	14NO	5;20	54.72
20419	29NO	9;02	250.16	20419	14DE	0;47	85.11
20420	28DE	19;51	280.20				

651

NEW MOONS				FULL MOONS			
				20420	12JA	20;37	115.47
20421	27JA	6;38	310.05	20421	11FE	14;44	145.45
20422	25FE	17;26	339.51	20422	13MR	5;45	174.83
20423	27MR	4;29	8.49	20423	11AP	17;30	203.58
20424	25AP	16;16	37.03	20424	11MY	2;42	231.83
20425	25MY	5;20	65.30	20425	9JN	10;22	259.79
20426	23JN	19;54	93.51	20426	8JL	17;29	287.73
20427	23JL	11;33	121.87	20427	7AU	0;57	315.90
20428	22AU	3;31	150.57	20428	5SE	9;32	344.52
20429	20SE	18;58	179.72	20429	4OC	20;02	13.71
20430	20OC	9;30	209.33	20430	3NO	9;13	43.47
20431	18NO	23;00	239.29	20431	3DE	1;23	73.67
20432	18DE	11;22	269.40				

652

NEW MOONS				FULL MOONS			
				20432	1JA	19;55	104.02
20433	16JA	22;30	299.36	20433	31JA	15;10	134.19
20434	15FE	8;26	328.95	20434	1MR	9;17	163.89
20435	15MR	17;31	358.02	20435	31MR	1;01	192.98
20436	14AP	2;27	26.61	20436	29AP	14;08	221.52
20437	13MY	12;09	54.84	20437	29MY	0;55	249.66
20438	11JN	23;25	82.93	20438	27JN	9;57	277.65
20439	11JL	12;40	111.13	20439	26JL	17;57	305.74
20440	10AU	3;50	139.66	20440	25AU	1;46	334.17
20441	8SE	20;28	168.67	20441	23SE	10;18	3.09
20442	8OC	13;51	198.21	20442	22OC	20;24	32.57
20443	7NO	7;03	228.20	20443	21NO	8;42	62.52
20444	6DE	23;01	258.42	20444	20DE	23;18	92.72

653

NEW MOONS				FULL MOONS			
20445	5JA	12;54	288.56	20445	19JA	15;42	122.88
20446	4FE	0;23	318.36	20446	18FE	8;58	152.71
20447	5MR	9;51	347.63	20447	20MR	2;08	182.06
20448	3AP	18;01	16.35	20448	18AP	18;20	210.87
20449	3MY	1;46	44.63	20449	18MY	8;56	239.26
20450	1JN	10;00	72.65	20450	16JN	21;33	267.39
20451	30JN	19;29	100.66	20451	16JL	8;16	295.51
20452	30JL	6;56	128.92	20452	14AU	17;39	323.86
20453	28AU	20;54	157.64	20453	13SE	2;36	352.63
20454	27SE	13;32	186.96	20454	12OC	11;59	21.90
20455	27OC	8;17	216.86	20455	10NO	22;21	51.65
20456	26NO	3;35	247.15	20456	10DE	9;54	81.70
20457	25DE	21;29	277.49				

654

NEW MOONS				FULL MOONS			
				20457	8JA	22;34	111.77
20458	24JA	12;40	307.55	20458	7FE	12;19	141.62
20459	23FE	0;53	337.11	20459	9MR	3;06	171.06
20460	24MR	10;39	6.08	20460	7AP	18;44	200.03
20461	22AP	18;46	34.51	20461	7MY	10;40	228.58
20462	22MY	2;00	62.57	20462	6JN	2;01	256.86
20463	20JN	9;11	90.48	20463	5JL	16;07	285.07
20464	19JL	17;12	118.52	20464	4AU	4;46	313.42
20465	18AU	3;08	146.92	20465	2SE	16;20	342.12
20466	16SE	15;57	175.90	20466	2OC	3;22	11.29
20467	16OC	8;06	205.52	20467	31OC	14;13	40.92
20468	15NO	3;00	235.67	20468	30NO	1;02	70.87
20469	14DE	22;56	266.08	20469	29DE	11;50	100.90

655

NEW MOONS				FULL MOONS			
20470	13JA	17;48	296.40	20470	27JA	22;45	130.76
20471	12FE	10;14	326.29	20471	26FE	10;13	160.24
20472	13MR	23;46	355.60	20472	27MR	22;40	189.26
20473	12AP	10;35	24.30	20473	26AP	12;15	217.86
20474	11MY	19;10	52.52	20474	26MY	2;45	246.18
20475	10JN	2;19	80.45	20475	24JN	17;45	274.41
20476	9JL	9;01	108.38	20476	24JL	8;49	302.75
20477	7AU	16;24	136.56	20477	22AU	23;41	331.42
20478	6SE	1;38	165.21	20478	21SE	14;04	0.54
20479	5OC	13;35	194.47	20479	21OC	3;40	30.12
20480	4NO	4;37	224.32	20480	19NO	16;11	60.05
20481	3DE	22;15	254.58	20481	19DE	3;30	90.11

656

NUMBER	DATE	TIME	LONG.	NUMBER	DATE	TIME	LONG.
20482	2JA	17;19	284.96	20482	17JA	13;53	120.04
20483	1FE	12;20	315.11	20483	15FE	23;46	149.61
20484	2MR	5;52	344.77	20484	16MR	9;37	178.71
20485	31MR	20;52	13.82	20485	14AP	19;54	207.34
20486	30AP	8;55	42.30	20486	14MY	7;01	235.62
20487	29MY	18;21	70.38	20487	12JN	19;22	263.76
20488	28JN	2;04	98.32	20488	12JL	9;18	291.99
20489	27JL	9;13	126.38	20489	11AU	0;55	320.54
20490	25AU	16;59	154.82	20490	9SE	17;45	349.57
20491	24SE	2;11	183.78	20491	9OC	10;46	19.11
20492	23OC	13;23	213.31	20492	8NO	2;45	49.07
20493	22NO	2;48	243.31	20493	7DE	16;57	79.21
20494	21DE	18;24	273.56				

657

NUMBER	DATE	TIME	LONG.	NUMBER	DATE	TIME	LONG.
				20494	6JA	5;13	109.28
20495	20JA	11;50	303.75	20495	4FE	15;54	139.03
20496	19FE	6;09	333.62	20496	6MR	1;22	168.30
20497	20MR	23;55	2.98	20497	4AP	10;00	197.03
20498	19AP	15;44	31.76	20498	3MY	18;20	225.32
20499	19MY	4;55	60.08	20499	2JN	3;11	253.36
20500	17JN	15;44	88.15	20500	1JL	13;32	281.41
20501	17JL	1;00	116.21	20501	31JL	2;17	309.72
20502	15AU	9;38	144.54	20502	29AU	17;48	338.53
20503	13SE	18;27	173.31	20503	28SE	11;33	7.91
20504	13OC	3;57	202.60	20504	28OC	6;09	37.81
20505	11NO	14;29	232.36	20505	27NO	0;03	68.04
20506	11DE	2;26	262.43	20506	26DE	16;09	98.31

658

NUMBER	DATE	TIME	LONG.	NUMBER	DATE	TIME	LONG.
20507	9JA	16;07	292.54	20507	25JA	5;59	128.31
20508	8FE	7;32	322.45	20508	23FE	17;27	157.83
20509	10MR	0;05	351.95	20509	25MR	2;48	186.77
20510	8AP	16;41	20.94	20510	23AP	10;32	215.17
20511	8MY	8;22	49.48	20511	22MY	17;33	243.22
20512	6JN	22;36	77.71	20512	21JN	0;59	271.14
20513	6JL	11;25	105.87	20513	20JL	10;02	299.22
20514	4AU	23;02	134.19	20514	18AU	21;38	327.71
20515	3SE	9;47	162.86	20515	17SE	12;14	356.77
20516	2OC	19;59	192.01	20516	17OC	5;31	26.44
20517	1NO	6;02	221.61	20517	16NO	0;30	56.61
20518	30NO	16;23	251.55	20518	15DE	19;48	87.00
20519	30DE	3;29	281.60				

659

NUMBER	DATE	TIME	LONG.	NUMBER	DATE	TIME	LONG.
				20519	14JA	13;51	117.27
20520	28JA	15;35	311.49	20520	13FE	5;20	147.12
20521	27FE	4;39	341.03	20521	14MR	17;42	176.37
20522	28MR	18;30	10.09	20522	13AP	3;11	205.00
20523	27AP	8;55	38.72	20523	12MY	10;42	233.17
20524	26MY	23;47	67.05	20524	10JN	17;24	261.08
20525	25JN	14;52	95.27	20525	10JL	0;26	289.02
20526	25JL	5;44	123.62	20526	8AU	8;47	317.25
20527	23AU	19;51	152.27	20527	6SE	19;15	345.96
20528	22SE	8;51	181.35	20528	6OC	8;22	15.28
20529	21OC	20;47	210.87	20529	5NO	0;23	45.18
20530	20NO	8;00	240.75	20530	4DE	18;57	75.49
20531	19DE	18;53	270.79				

660

NUMBER	DATE	TIME	LONG.	NUMBER	DATE	TIME	LONG.
				20531	3JA	14;48	105.90
20532	18JA	5;36	300.73	20532	2FE	9;56	136.04
20533	16FE	16;12	330.32	20533	3MR	2;29	165.65
20534	17MR	2;48	359.44	20534	1AP	15;44	194.61
20535	15AP	13;52	28.10	20535	1MY	2;03	223.01
20536	15MY	1;59	56.41	20536	30MY	10;21	251.05
20537	13JN	15;40	84.59	20537	28JN	17;41	278.97
20538	13JL	6;52	112.88	20538	28JL	0;58	307.04
20539	11AU	22;57	141.46	20539	26AU	9;01	335.50
20540	10SE	15;01	170.47	20540	24SE	18;35	4.49
20541	10OC	6;21	199.96	20541	24OC	6;26	34.06
20542	8NO	20;37	229.85	20542	22NO	21;07	64.12
20543	8DE	9;42	259.95	20543	22DE	14;32	94.44

661

NUMBER	DATE	TIME	LONG.	NUMBER	DATE	TIME	LONG.
20544	6JA	21;29	290.00	20544	21JA	9;31	124.69
20545	5FE	7;54	319.73	20545	20FE	4;13	154.57
20546	6MR	17;09	348.97	20546	21MR	21;05	183.87
20547	5AP	1;49	17.70	20547	20AP	11;25	212.58
20548	4MY	10;48	46.01	20548	19MY	23;18	240.85
20549	2JN	21;01	74.10	20549	18JN	9;13	268.87
20550	2JL	9;09	102.22	20550	17JL	17;46	296.92
20551	31JL	23;49	130.61	20551	16AU	1;45	325.22
20552	30AU	15;31	159.44	20552	14SE	9;59	353.98
20553	29SE	8;52	188.82	20553	13OC	19;22	23.27
20554	29OC	2;32	218.70	20554	12NO	6;35	53.08
20555	27NO	19;27	248.90	20555	11DE	20;01	83.21
20556	27DE	10;33	279.12				

662

NUMBER	DATE	TIME	LONG.	NUMBER	DATE	TIME	LONG.
				20556	10JA	11;28	113.39
20557	25JA	23;13	309.06	20557	9FE	4;10	143.34
20558	24FE	9;30	338.52	20558	10MR	21;14	172.85
20559	25MR	18;03	7.42	20559	9AP	13;48	201.83
20560	24AP	1;44	35.81	20560	9MY	5;11	230.34
20561	23MY	9;27	63.88	20561	7JN	18;51	258.55
20562	21JN	18;07	91.85	20562	7JL	6;38	286.66
20563	21JL	4;28	119.99	20563	5AU	16;49	314.93
20564	19AU	17;10	148.53	20564	4SE	2;07	343.56
20565	18SE	8;36	177.64	20565	3OC	11;24	12.68
20566	18OC	2;35	207.36	20566	1NO	21;23	42.29
20567	16NO	22;01	237.55	20567	1DE	8;24	72.26
20568	16DE	16;57	267.93	20568	30DE	20;26	102.34

663

NUMBER	DATE	TIME	LONG.	NUMBER	DATE	TIME	LONG.
20569	15JA	9;37	298.14	20569	29JA	9;27	132.27
20570	13FE	23;14	327.89	20570	27FE	23;26	161.83
20571	15MR	10;02	357.06	20571	29MR	14;22	190.93
20572	13AP	18;46	25.66	20572	28AP	6;01	219.60
20573	13MY	2;14	53.82	20573	27MY	21;40	247.95
20574	11JN	9;16	81.74	20574	26JN	12;31	276.17
20575	10JL	16;44	109.71	20575	26JL	2;04	304.47
20576	9AU	1;37	137.96	20576	24AU	14;23	333.05
20577	7SE	12;57	166.72	20577	23SE	1;55	2.08
20578	7OC	3;31	196.11	20578	22OC	13;04	31.58
20579	5NO	21;17	226.10	20579	21NO	0;03	61.46
20580	5DE	17;01	256.46	20580	20DE	10;52	91.50

664

NUMBER	DATE	TIME	LONG.	NUMBER	DATE	TIME	LONG.
20581	4JA	12;39	286.85	20581	18JA	21;35	121.43
20582	3FE	6;21	316.93	20582	17FE	8;30	151.03
20583	3MR	21;15	346.45	20583	17MR	20;07	180.18
20584	2AP	9;18	15.35	20584	16AP	8;50	208.89
20585	1MY	18;52	43.72	20585	15MY	22;42	237.27
20586	31MY	2;36	71.73	20586	14JN	13;25	265.49
20587	29JN	9;25	99.63	20587	14JL	4;33	293.77
20588	28JL	16;24	127.69	20588	12AU	19;45	322.33
20589	27AU	0;40	156.17	20589	11SE	10;42	351.31
20590	25SE	11;16	185.21	20590	11OC	1;03	20.77
20591	25OC	0;51	214.86	20591	9NO	14;22	50.63
20592	23NO	17;18	245.00	20592	9DE	2;24	80.69
20593	23DE	11;45	275.37				

665

NUMBER	DATE	TIME	LONG.	NUMBER	DATE	TIME	LONG.
				20593	7JA	13;12	110.69
20594	22JA	6;51	305.62	20594	5FE	23;08	140.40
20595	21FE	1;08	335.46	20595	7MR	8;44	169.64
20596	22MR	17;23	4.73	20596	5AP	18;29	198.40
20597	21AP	6;51	33.40	20597	5MY	4;53	226.76
20598	20MY	17;31	61.60	20598	3JN	16;23	254.90
20599	19JN	2;01	89.57	20599	3JL	5;25	283.06
20600	18JL	9;25	117.57	20600	1AU	20;15	311.47
20601	16AU	16;54	145.87	20601	31AU	12;25	340.34
20602	15SE	1;27	174.64	20602	30SE	6;06	9.73
20603	14OC	11;45	203.99	20603	29OC	23;02	39.60
20604	13NO	0;06	233.85	20604	28NO	14;22	69.73
20605	12DE	14;34	264.02	20605	28DE	3;40	99.87

666

NUMBER	DATE	TIME	LONG.	NUMBER	DATE	TIME	LONG.
20606	11JA	7;00	294.25	20606	26JA	15;03	129.76
20607	10FE	0;47	324.23	20607	25FE	0;56	159.19
20608	11MR	18;48	353.76	20608	26MR	9;45	188.09
20609	10AP	11;34	22.73	20609	24AP	17;57	216.50
20610	10MY	2;03	51.20	20610	24MY	2;16	244.58
20611	8JN	14;03	79.34	20611	22JN	11;37	272.58
20612	8JL	0;06	107.39	20612	21JL	23;02	300.76
20613	6AU	9;08	135.62	20613	20AU	13;13	329.37
20614	4SE	18;01	164.24	20614	19SE	6;08	358.56
20615	4OC	3;19	193.37	20615	19OC	0;43	28.31
20616	2NO	13;25	223.00	20616	17NO	19;20	58.48
20617	2DE	0;39	252.98	20617	17DE	12;34	88.79
20618	31DE	13;19	283.09				

667

NUMBER	DATE	TIME	LONG.	NUMBER	DATE	TIME	LONG.
				20618	16JA	3;35	118.92
20619	30JA	3;41	313.07	20619	14FE	16;10	148.63
20620	28FE	19;29	342.69	20620	16MR	2;25	177.76
20621	30MR	11;56	11.84	20621	14AP	10;43	206.33
20622	29AP	3;59	40.51	20622	13MY	17;51	234.47
20623	28MY	18;55	68.82	20623	12JN	0;50	262.39
20624	27JN	8;31	96.99	20624	11JL	8;54	290.38
20625	26JL	20;54	125.25	20625	9AU	19;09	318.69
20626	25AU	8;19	153.81	20626	8SE	8;19	347.54
20627	23SE	19;01	182.82	20627	8OC	0;26	17.01
20628	23OC	5;18	212.30	20628	6NO	18;51	47.04
20629	21NO	15;34	242.15	20629	6DE	14;16	77.39
20630	21DE	2;15	272.18				

668 / 674

NUMBER	DATE	TIME	LONG.	NUMBER	DATE	TIME	LONG.	NUMBER	DATE	TIME	LONG.	NUMBER	DATE	TIME	LONG.
				20630	5JA	9;09	107.75	20705	13JA	1;18	296.09	20705	27JA	22;35	131.14
20631	19JA	13;41	302.14	20631	4FE	2;00	137.78	20706	11FE	20;00	326.09	20706	26FE	8;05	160.54
20632	18FE	2;02	331.79	20632	4MR	15;52	167.25	20707	13MR	13;14	355.56	20707	27MR	17;28	189.44
20633	18MR	15;13	0.99	20633	3AP	2;38	196.09	20708	12AP	4;03	24.43	20708	26AP	3;14	217.89
20634	17AP	5;05	29.73	20634	2MY	10;55	224.39	20709	11MY	16;04	52.78	20709	25MY	13;55	246.06
20635	16MY	19;33	58.13	20635	31MY	17;50	252.36	20710	10JN	1;36	80.81	20710	24JN	2;01	274.17
20636	15JN	10;30	86.36	20636	30JN	0;34	280.26	20711	9JL	9;31	108.79	20711	23JL	15;55	302.46
20637	15JL	1;39	114.65	20637	29JL	8;11	308.36	20712	7AU	16;58	136.97	20712	22AU	7;46	331.16
20638	13AU	16;26	143.20	20638	27AU	17;36	336.89	20713	6SE	1;03	165.58	20713	21SE	1;05	0.39
20639	12SE	6;20	172.15	20639	26SE	5;26	5.99	20714	5OC	10;34	194.73	20714	20OC	18;40	30.13
20640	11OC	19;04	201.56	20640	25OC	20;04	35.70	20715	3NO	21;56	224.43	20715	19NO	11;08	60.23
20641	10NO	6;48	231.36	20641	24NO	13;28	65.89	20716	3DE	11;21	254.51	20716	19DE	1;36	90.42
20642	9DE	17;55	261.38	20642	24DE	8;52	96.29								

669 / 675

NUMBER	DATE	TIME	LONG.	NUMBER	DATE	TIME	LONG.	NUMBER	DATE	TIME	LONG.	NUMBER	DATE	TIME	LONG.
20643	8JA	4;40	291.38	20643	23JA	4;32	126.56	20717	2JA	2;43	284.73	20717	17JA	13;54	120.43
20644	6FE	15;10	321.10	20644	21FE	22;26	156.38	20718	31JA	19;42	314.80	20718	16FE	0;23	150.03
20645	8MR	1;29	350.36	20645	23MR	13;16	185.57	20719	2MR	13;32	344.48	20719	17MR	9;30	179.09
20646	6AP	11;58	19.14	20646	22AP	0;54	214.15	20720	1AP	6;53	13.64	20720	15AP	17;45	207.64
20647	5MY	23;13	47.52	20647	21MY	10;04	242.29	20721	30AP	22;30	42.27	20721	15MY	1;44	235.80
20648	4JN	11;54	75.71	20648	19JN	17;54	270.23	20722	30MY	11;44	70.51	20722	13JN	10;18	263.79
20649	4JL	2;20	103.92	20649	19JL	1;05	298.23	20723	28JN	22;47	98.58	20723	12JL	20;30	291.86
20650	2AU	18;11	132.39	20650	17AU	8;47	326.54	20724	28JL	8;24	126.74	20724	11AU	9;16	320.30
20651	1SE	10;36	161.26	20651	15SE	17;38	355.34	20725	26AU	17;30	155.23	20725	10SE	1;00	349.27
20652	1OC	2;40	190.62	20652	15OC	4;22	24.72	20726	25SE	2;44	184.19	20726	9OC	19;07	18.85
20653	30OC	17;46	220.42	20653	13NO	17;37	54.62	20727	24OC	12;34	213.67	20727	8NO	14;07	48.92
20654	29NO	7;39	250.49	20654	13DE	9;42	84.86	20728	22NO	23;16	243.55	20728	8DE	8;20	79.24
20655	28DE	20;09	280.60					20729	22DE	11;08	273.64				

670 / 676

NUMBER	DATE	TIME	LONG.	NUMBER	DATE	TIME	LONG.	NUMBER	DATE	TIME	LONG.	NUMBER	DATE	TIME	LONG.
				20655	12JA	3;57	115.16	20730	21JA	0;28	303.66	20729	7JA	0;35	109.48
20656	27JA	7;11	310.46	20656	10FE	22;49	145.18	20731	19FE	15;20	333.40	20730	5FE	14;21	139.36
20657	25FE	16;49	339.88	20657	12MR	16;34	174.68	20732	20MR	7;17	2.69	20731	6MR	1;38	168.70
20658	27MR	1;28	8.75	20658	11AP	8;04	203.59	20733	18AP	23;24	31.50	20732	4AP	10;43	197.45
20659	25AP	9;57	37.17	20659	10MY	21;07	231.99	20734	18MY	14;50	59.91	20733	3MY	18;11	225.70
20660	24MY	19;14	65.28	20660	9JN	8;00	260.08	20735	17JN	5;09	88.11	20734	2JN	1;01	253.66
20661	23JN	6;13	93.35	20661	8JL	17;17	288.11	20736	16JL	18;17	116.33	20735	1JL	8;20	281.59
20662	22JL	19;24	121.61	20662	7AU	1;38	316.32	20737	15AU	6;25	144.80	20736	30JL	17;23	309.76
20663	21AU	10;44	150.28	20663	5SE	9;50	344.92	20738	13SE	17;43	173.67	20737	29AU	5;05	338.40
20664	20SE	3;45	179.48	20664	4OC	18;42	14.04	20739	13OC	4;25	203.01	20738	27SE	19;52	7.65
20665	19OC	21;38	209.22	20665	3NO	5;02	43.68	20740	11NO	14;48	232.77	20739	27OC	13;22	37.50
20666	18NO	15;17	239.36	20666	2DE	17;22	73.72	20741	11DE	1;16	262.77	20740	26NO	8;33	67.78
20667	18DE	7;33	269.63									20741	26DE	3;57	98.18

671 / 677

NUMBER	DATE	TIME	LONG.	NUMBER	DATE	TIME	LONG.	NUMBER	DATE	TIME	LONG.	NUMBER	DATE	TIME	LONG.
				20667	1JA	7;45	103.90	20742	9JA	12;13	292.77	20742	24JA	21;57	128.36
20668	16JA	21;29	299.71	20668	30JA	23;42	133.93	20743	7FE	23;54	322.52	20743	23FE	13;19	158.04
20669	15FE	8;50	329.35	20669	1MR	16;26	163.58	20744	9MR	12;24	351.86	20744	25MR	1;30	187.10
20670	16MR	18;00	358.43	20670	31MR	9;05	192.73	20745	8AP	1;38	20.72	20745	23AP	10;50	215.57
20671	15AP	1;50	26.97	20671	30AP	1;00	221.38	20746	7MY	15;33	49.20	20746	22MY	18;14	243.63
20672	14MY	9;16	55.11	20672	29MY	15;34	249.67	20747	6JN	6;10	77.45	20747	21JN	0;54	271.52
20673	12JN	17;15	83.08	20673	28JN	4;24	277.81	20748	5JL	21;18	105.69	20748	20JL	8;00	299.53
20674	12JL	2;37	111.12	20674	27JL	15;30	306.03	20749	4AU	12;32	134.15	20749	18AU	16;30	327.89
20675	10AU	14;05	139.49	20675	26AU	1;24	334.54	20750	3SE	3;14	162.98	20750	17SE	3;10	356.78
20676	9SE	4;12	168.39	20676	24SE	10;50	3.50	20751	2OC	16;51	192.27	20751	16OC	16;27	26.29
20677	8OC	21;07	197.90	20677	23OC	20;36	32.97	20752	1NO	5;18	221.97	20752	15NO	8;32	56.33
20678	7NO	16;11	227.97	20678	22NO	7;10	62.84	20753	30NO	16;50	251.96	20753	15DE	3;04	86.68
20679	7DE	11;46	258.33	20679	21DE	18;41	92.91	20754	30DE	3;45	282.00				

672 / 678

NUMBER	DATE	TIME	LONG.	NUMBER	DATE	TIME	LONG.	NUMBER	DATE	TIME	LONG.	NUMBER	DATE	TIME	LONG.
20680	6JA	5;47	288.65	20680	20JA	7;03	122.90					20754	13JA	22;46	117.03
20681	4FE	20;55	318.60	20681	18FE	20;17	152.58	20755	28JA	14;16	311.83	20755	12FE	17;40	147.02
20682	5MR	8;58	347.98.	20682	19MR	10;26	181.81	20756	27FE	0;26	341.24	20756	14MR	10;00	176.44
20683	3AP	18;31	16.76	20683	18AP	1;31	210.59	20757	28MR	10;31	10.16	20757	12AP	23;07	205.23
20684	3MY	2;25	45.04	20684	17MY	17;06	239.02	20758	26AP	21;02	38.63	20758	12MY	9;22	233.50
20685	1JN	9;30	73.01	20685	16JN	8;26	267.26	20759	26MY	8;43	66.84	20759	10JN	17;43	261.48
20686	30JN	16;37	100.93	20686	15JL	22;49	295.52	20760	24JN	22;09	95.00	20760	10JL	1;12	289.44
20687	30JL	0;42	129.05	20687	14AU	11;58	324.02	20761	24JL	13;24	123.35	20761	8AU	8;44	317.63
20688	28AU	10;45	157.62	20688	13SE	0;08	352.91	20762	23AU	5;51	152.09	20762	6SE	17;04	346.26
20689	26SE	23;43	186.78	20689	12OC	11;42	22.28	20763	21SE	22;28	181.31	20763	6OC	2;53	15.44
20690	26OC	16;02	216.57	20690	10NO	22;58	52.07	20764	21OC	14;23	210.99	20764	4NO	14;53	45.17
20691	25NO	11;04	246.84	20691	10DE	9;56	82.09	20765	20NO	5;08	241.02	20765	4DE	5;32	75.31
20692	25DE	7;01	277.26					20766	19DE	18;27	271.17				

673 / 679

NUMBER	DATE	TIME	LONG.	NUMBER	DATE	TIME	LONG.	NUMBER	DATE	TIME	LONG.	NUMBER	DATE	TIME	LONG.
				20692	8JA	20;37	112.08					20766	2JA	22;43	105.61
20693	24JA	1;48	307.49	20693	7FE	7;12	141.79	20767	18JA	6;13	301.15	20767	1FE	17;18	135.74
20694	22FE	18;04	337.22	20694	8MR	18;08	171.08	20768	16FE	16;23	330.72	20768	3MR	11;36	165.42
20695	24MR	7;26	6.34	20695	7AP	5;58	199.90	20769	18MR	1;16	359.77	20769	2AP	4;08	194.53
20696	22AP	18;06	34.87	20696	6MY	19;00	228.34	20770	16AP	9;30	28.31	20770	1MY	18;21	223.09
20697	22MY	2;38	62.98	20697	5JN	9;11	256.58	20771	15MY	18;03	56.48	20771	31MY	6;17	251.27
20698	20JN	9;49	90.90	20698	5JL	0;10	284.82	20772	14JN	3;57	84.52	20772	29JN	16;24	279.31
20699	19JL	16;38	118.88	20699	3AU	15;31	313.28	20773	13JL	15;56	112.67	20773	29JL	1;18	307.45
20700	18AU	0;13	147.20	20700	2SE	6;54	342.12	20774	12AU	6;16	141.17	20774	27AU	9;40	335.92
20701	16SE	9;39	176.03	20701	1OC	21;55	11.45	20775	10SE	22;40	170.19	20775	25SE	18;18	4.87
20702	15OC	21;48	205.48	20702	31OC	12;04	41.22	20776	10OC	16;28	199.77	20776	25OC	3;58	34.35
20703	14NO	12;54	235.46	20703	30NO	0;56	71.26	20777	9NO	10;36	229.82	20777	23NO	15;18	64.26
20704	14DE	6;27	265.78	20704	29DE	12;21	101.32	20778	9DE	3;53	260.10	20778	23DE	4;36	94.41

680

NEW MOONS				FULL MOONS			
NUMBER	DATE	TIME	LONG.	NUMBER	DATE	TIME	LONG.
20779	7JA	19;07	290.29	20779	21JA	19;40	124.50
20780	6FE	7;42	320.11	20780	20FE	11;50	154.28
20781	6MR	17;45	349.39	20781	21MR	4;21	183.58
20782	5AP	1;59	18.09	20782	19AP	20;31	212.37
20783	4MY	9;21	46.34	20783	19MY	11;45	240.77
20784	2JN	16;49	74.32	20784	18JN	1;33	268.95
20785	2JL	1;20	102.30	20785	17JL	13;41	297.14
20786	31JL	11;40	130.52	20786	16AU	0;21	325.55
20787	30AU	0;28	159.22	20787	14SE	10;10	354.38
20788	28SE	16;07	188.52	20788	13OC	19;54	23.69
20789	28OC	10;23	218.41	20789	12NO	6;10	53.45
20790	27NO	6;05	248.72	20790	11DE	17;14	83.48
20791	27DE	1;12	279.11				

681

NEW MOONS				FULL MOONS			
NUMBER	DATE	TIME	LONG.	NUMBER	DATE	TIME	LONG.
				20791	10JA	5;04	113.51
20792	25JA	17;52	309.23	20792	8FE	17;38	143.29
20793	24FE	7;21	338.82	20793	10MR	7;01	172.65
20794	25MR	17;57	7.80	20794	8AP	21;21	201.55
20795	24AP	2;28	36.23	20795	8MY	12;32	230.06
20796	23MY	9;47	64.28	20796	7JN	4;02	258.35
20797	21JN	16;44	92.19	20797	6JL	19;03	286.59
20798	21JL	0;14	120.21	20798	5AU	9;02	315.01
20799	19AU	9;14	148.60	20799	3SE	21;57	343.78
20800	17SE	20;44	177.54	20800	3OC	10;03	13.01
20801	17OC	11;27	207.11	20801	1NO	21;41	42.69
20802	16NO	5;21	237.24	20802	1DE	8;55	72.67
20803	16DE	1;06	267.65	20803	30DE	19;44	102.71

682

NEW MOONS				FULL MOONS			
NUMBER	DATE	TIME	LONG.	NUMBER	DATE	TIME	LONG.
20804	14JA	20;39	297.98	20804	29JA	6;11	132.52
20805	13FE	14;12	327.91	20805	27FE	16;38	161.94
20806	15MR	4;55	357.24	20806	29MR	3;40	190.89
20807	13AP	16;49	25.97	20807	27AP	15;49	219.42
20808	13MY	2;18	54.21	20808	27MY	5;15	248.09
20809	11JN	10;03	82.16	20809	25JN	19;49	275.89
20810	10JL	16;59	110.10	20810	25JL	11;07	304.25
20811	9AU	0;09	138.29	20811	24AU	2;44	332.97
20812	7SE	8;39	166.93	20812	22SE	18;17	2.16
20813	6OC	19;28	196.16	20813	22OC	9;14	31.81
20814	5NO	9;10	225.97	20814	20NO	23;00	61.81
20815	5DE	1;35	256.19	20815	20DE	11;13	91.91

683

NEW MOONS				FULL MOONS			
NUMBER	DATE	TIME	LONG.	NUMBER	DATE	TIME	LONG.
20816	3JA	19;50	286.53	20816	18JA	21;57	121.84
20817	2FE	14;37	316.66	20817	17FE	7;36	151.38
20818	4MR	8;34	346.32	20818	18MR	16;46	180.43
20819	3AP	0;34	15.39	20819	17AP	2;01	209.00
20820	2MY	13;57	43.91	20820	16MY	11;58	237.23
20821	1JN	0;41	72.04	20821	14JN	23;09	265.31
20822	30JN	9;22	100.01	20822	14JL	12;03	293.50
20823	29JL	17;03	128.11	20823	13AU	3;01	322.04
20824	28AU	0;52	156.57	20824	11SE	19;54	351.09
20825	26SE	9;45	185.54	20825	11OC	13;47	20.69
20826	25OC	20;17	215.06	20826	10NO	7;13	50.72
20827	24NO	8;42	245.03	20827	9DE	22;55	80.94
20828	23DE	23;01	275.22				

684

NEW MOONS				FULL MOONS			
NUMBER	DATE	TIME	LONG.	NUMBER	DATE	TIME	LONG.
				20828	8JA	12;19	111.05
20829	22JA	15;04	305.35	20829	6FE	23;35	140.81
20830	21FE	8;21	335.16	20830	7MR	9;12	170.05
20831	22MR	1;52	4.48	20831	5AP	17;38	198.76
20832	20AP	18;21	33.28	20832	5MY	1;29	227.01
20833	20MY	8;46	61.64	20833	3JN	9;30	255.01
20834	18JN	20;56	89.75	20834	2JL	18;41	283.02
20835	18JL	7;20	117.88	20835	1AU	6;03	311.29
20836	16AU	16;49	146.25	20836	30AU	20;22	340.06
20837	15SE	2;09	175.06	20837	29SE	13;35	9.44
20838	14OC	11;51	204.39	20838	29OC	8;33	39.38
20839	12NO	22;11	234.16	20839	28NO	3;31	69.66
20840	12DE	9;25	264.20	20840	27DE	20;55	99.98

685

NEW MOONS				FULL MOONS			
NUMBER	DATE	TIME	LONG.	NUMBER	DATE	TIME	LONG.
20841	10JA	21;51	294.25	20841	26JA	11;57	130.02
20842	9FE	11;44	324.08	20842	25FE	0;23	159.55
20843	11MR	2;55	353.50	20843	26MR	10;23	188.50
20844	9AP	18;47	22.45	20844	24AP	18;27	216.90
20845	9MY	10;29	50.97	20845	24MY	1;22	244.93
20846	8JN	1;22	79.22	20846	22JN	8;15	272.83
20847	7JL	15;12	107.43	20847	21JL	16;17	300.88
20848	6AU	4;03	135.81	20848	20AU	2;36	329.33
20849	4SE	16;02	164.55	20849	18SE	15;56	358.36
20850	4OC	3;16	193.76	20850	18OC	8;15	28.02
20851	2NO	13;58	223.41	20851	17NO	2;51	58.18
20852	2DE	0;27	253.36	20852	16DE	22;23	88.58
20853	31DE	11;04	283.38				

686

NEW MOONS				FULL MOONS			
NUMBER	DATE	TIME	LONG.	NUMBER	DATE	TIME	LONG.
				20853	15JA	17;15	118.88
20854	29JA	22;12	313.23	20854	14FE	9;59	148.76
20855	28FE	10;02	342.69	20855	15MR	23;41	178.04
20856	29MR	22;36	11.68	20856	14AP	10;17	206.70
20857	28AP	11;54	40.25	20857	13MY	18;27	234.88
20858	28MY	2;00	68.54	20858	12JN	1;20	262.80
20859	26JN	16;52	96.76	20859	11JL	8;07	290.74
20860	26JL	8;16	125.13	20860	9AU	15;53	318.95
20861	24AU	23;34	153.85	20861	8SE	1;29	347.65
20862	23SE	14;05	183.01	20862	7OC	13;30	16.95
20863	23OC	3;24	212.61	20863	6NO	4;15	46.81
20864	21NO	15;32	242.54	20864	5DE	21;39	77.08
20865	21DE	2;47	272.60				

687

NEW MOONS				FULL MOONS			
NUMBER	DATE	TIME	LONG.	NUMBER	DATE	TIME	LONG.
				20865	4JA	16;53	107.45
20866	19JA	13;25	302.52	20866	3FE	12;19	137.60
20867	17FE	23;34	332.08	20867	5MR	5;58	167.23
20868	19MR	9;25	1.14	20868	3AP	20;37	196.23
20869	17AP	19;22	29.73	20869	3MY	8;10	224.67
20870	17MY	6;08	57.98	20870	1JN	17;22	252.73
20871	15JN	18;30	86.11	20871	1JL	1;14	280.68
20872	15JL	8;51	114.37	20872	30JL	8;46	308.77
20873	14AU	0;55	142.96	20873	28AU	16;45	337.24
20874	12SE	17;49	172.03	20874	27SE	1;54	6.24
20875	12OC	10;29	201.59	20875	26OC	12;49	35.79
20876	11NO	2;06	231.55	20876	25NO	2;07	65.80
20877	10DE	16;19	261.70	20877	24DE	18;01	96.06

688

NEW MOONS				FULL MOONS			
NUMBER	DATE	TIME	LONG.	NUMBER	DATE	TIME	LONG.
20878	9JA	4;54	291.78	20878	23JA	11;54	126.25
20879	7FE	15;46	321.51	20879	22FE	6;20	156.10
20880	8MR	1;04	350.73	20880	22MR	23;41	185.40
20881	6AP	9;19	19.42	20881	21AP	14;58	214.14
20882	5MY	17;23	47.68	20882	21MY	4;00	242.44
20883	4JN	2;19	75.71	20883	19JN	15;04	270.51
20884	3JL	13;06	103.78	20884	19JL	0;40	298.60
20885	2AU	2;15	132.13	20885	17AU	9;24	326.49
20886	31AU	17;47	160.97	20886	15SE	18;01	355.75
20887	30SE	11;11	190.37	20887	15OC	3;14	25.06
20888	30OC	5;31	220.29	20888	13NO	13;46	54.85
20889	28NO	23;34	250.54	20889	13DE	2;03	84.93
20890	28DE	16;03	280.82				

689

NEW MOONS				FULL MOONS			
NUMBER	DATE	TIME	LONG.	NUMBER	DATE	TIME	LONG.
				20890	11JA	16;08	115.05
20891	27JA	5;58	310.80	20891	10FE	7;36	144.93
20892	25FE	17;08	340.28	20892	11MR	23;44	174.38
20893	27MR	2;02	9.17	20893	10AP	15;33	203.33
20894	25AP	9;34	37.54	20894	10MY	7;33	231.84
20895	24MY	16;44	65.57	20895	8JN	22;08	260.08
20896	23JN	0;33	93.51	20896	8JL	11;16	288.26
20897	22JL	9;51	121.62	20897	6AU	22;51	316.59
20898	20AU	21;24	150.13	20898	5SE	9;16	345.29
20899	19SE	11;41	179.21	20899	4OC	19;12	14.45
20900	19OC	4;51	208.91	20900	3NO	5;19	44.09
20901	18NO	0;11	239.11	20901	2DE	16;02	74.05
20902	17DE	19;56	269.52				

690

NEW MOONS				FULL MOONS			
NUMBER	DATE	TIME	LONG.	NUMBER	DATE	TIME	LONG.
				20902	1JA	3;26	104.11
20903	16JA	14;00	299.78	20903	30JA	15;27	133.98
20904	15FE	5;03	329.58	20904	1MR	4;07	163.46
20905	16MR	16;55	358.78	20905	30MR	17;40	192.49
20906	15AP	2;16	27.38	20906	29AP	8;12	221.09
20907	14MY	10;00	55.53	20907	28MY	23;29	249.42
20908	12JN	17;00	83.45	20908	27JN	14;51	277.66
20909	12JL	0;08	111.41	20909	27JL	5;35	306.02
20910	10AU	8;18	139.65	20910	25AU	19;17	334.68
20911	8SE	18;31	168.38	20911	24SE	8;03	3.78
20912	8OC	7;39	197.73	20912	23OC	20;09	33.34
20913	7NO	0;06	227.68	20913	22NO	7;45	63.26
20914	6DE	19;10	258.02	20914	21DE	18;49	93.31

691

NEW MOONS				FULL MOONS			
NUMBER	DATE	TIME	LONG.	NUMBER	DATE	TIME	LONG.
20915	5JA	15;02	288.42	20915	20JA	5;20	123.22
20916	4FE	9;40	318.52	20916	18FE	15;32	152.77
20917	6MR	1;44	348.07	20917	20MR	1;56	181.85
20918	4AP	14;56	17.00	20918	18AP	13;11	210.48
20919	4MY	1;31	45.39	20919	18MY	1;44	238.79
20920	2JN	10;02	73.42	20920	16JN	15;38	266.98
20921	1JL	17;19	101.35	20921	16JL	6;38	295.27
20922	31JL	0;20	129.43	20922	14AU	22;19	323.86
20923	29AU	8;09	157.90	20923	13SE	14;15	352.90
20924	27SE	17;50	186.93	20924	13OC	5;52	22.43
20925	27OC	6;08	216.55	20925	11NO	20;33	52.36
20926	25NO	21;16	246.64	20926	11DE	9;42	82.47
20927	25DE	14;39	276.96				

NUMBER	DATE	TIME	LONG.	NUMBER	DATE	TIME	LONG.	NUMBER	DATE	TIME	LONG.	NUMBER	DATE	TIME	LONG.
--------NEW MOONS---------				--------FULL MOONS--------				--------NEW MOONS---------				--------FULL MOONS--------			

692 / **698**

NUMBER	DATE	TIME	LONG.	NUMBER	DATE	TIME	LONG.	NUMBER	DATE	TIME	LONG.	NUMBER	DATE	TIME	LONG.
				20927	9JA	21;09	112.49					21001	2JA	13;09	105.59
20928	24JA	9;11	307.18	20928	8FE	7;10	142.18	21002	18JA	3;35	301.43	21002	1FE	3;47	135.56
20929	23FE	3;31	337.00	20929	8MR	16;18	171.39	21003	16FE	16;02	331.10	21003	2MR	19;22	165.14
20930	23MR	20;27	6.28	20930	7AP	1;12	200.09	21004	18MR	1;51	0.18	21004	1AP	11;19	194.24
20931	22AP	11;06	34.98	20931	6MY	10;30	228.39	21005	16AP	9;48	28.71	21005	1MY	3;06	222.87
20932	21MY	23;08	63.23	20932	4JN	20;50	256.48	21006	15MY	16;54	56.82	21006	30MY	18;14	251.18
20933	20JN	8;50	91.24	20933	4JL	8;43	284.59	21007	14JN	0;11	84.75	21007	29JN	8;14	279.37
20934	19JL	17;01	119.29	20934	2AU	22;40	312.99	21008	13JL	8;37	112.77	21008	28JL	20;48	307.65
20935	18AU	0;49	147.61	20935	1SE	14;47	341.85	21009	11AU	18;59	141.11	21009	27AU	8;00	336.24
20936	16SE	9;15	176.41	20936	1OC	8;33	11.28	21010	10SE	7;56	169.98	21010	25SE	18;21	5.26
20937	15OC	19;03	205.75	20937	31OC	2;40	41.21	21011	9OC	23;48	199.47	21011	25OC	4;31	34.76
20938	14NO	6;35	235.59	20938	29NO	19;33	71.42	21012	8NO	18;19	229.52	21012	23NO	15;01	64.64
20939	13DE	19;55	265.72	20939	29DE	10;13	101.62	21013	8DE	14;12	259.91	21013	23DE	2;05	94.69

693 / **699**

NUMBER	DATE	TIME	LONG.	NUMBER	DATE	TIME	LONG.	NUMBER	DATE	TIME	LONG.	NUMBER	DATE	TIME	LONG.
20940	12JA	10;58	295.88	20940	27JA	22;28	131.53	21014	7JA	9;22	290.27	21014	21JA	13;39	124.64
20941	11FE	3;28	325.79	20941	26FE	8;44	160.95	21015	6FE	1;58	320.26	21015	20FE	1;43	154.25
20942	12MR	20;45	355.27	20942	27MR	17;31	189.83	21016	7MR	15;18	349.68	21016	21MR	14;29	183.40
20943	11AP	13;42	24.24	20943	26AP	1;25	218.20	21017	6AP	1;43	18.47	21017	20AP	4;14	212.11
20944	11MY	5;10	52.73	20944	25MY	9;06	246.25	21018	5MY	10;04	46.75	21018	19MY	19;00	240.49
20945	9JN	18;29	80.92	20945	23JN	17;28	274.22	21019	3JN	17;17	74.73	21019	18JN	10;23	268.74
20946	9JL	5;50	109.03	20946	23JL	3;35	302.36	21020	3JL	0;15	102.64	21020	18JL	1;38	297.05
20947	7AU	15;53	137.31	20947	21AU	16;25	330.93	21021	1AU	7;50	130.76	21021	16AU	16;06	325.61
20948	6SE	1;27	165.98	20948	20SE	8;22	0.09	21022	30AU	17;00	159.30	21022	15SE	5;37	354.58
20949	5OC	11;09	195.15	20949	20OC	2;49	29.86	21023	29SE	4;40	188.43	21023	14OC	18;19	24.01
20950	3NO	21;18	224.79	20950	18NO	22;10	60.07	21024	28OC	19;32	218.18	21024	13NO	6;22	53.85
20951	3DE	8;05	254.76	20951	18DE	16;37	90.43	21025	27NO	13;29	248.41	21025	12DE	17;48	83.89
								21026	27DE	9;10	278.82				

694 / **700**

NUMBER	DATE	TIME	LONG.	NUMBER	DATE	TIME	LONG.	NUMBER	DATE	TIME	LONG.	NUMBER	DATE	TIME	LONG.
20952	1JA	19;48	284.83	20952	17JA	8;55	120.61					21026	11JA	4;32	113.88
20953	31JA	8;44	314.73	20953	15FE	22;36	150.34	21027	26JA	4;33	309.06	21027	9FE	14;41	143.56
20954	1MR	23;01	344.27	20954	17MR	9;40	179.49	21028	24FE	21;53	338.82	21028	10MR	0;39	172.78
20955	31MR	14;20	13.36	20955	15AP	18;30	208.06	21029	25MR	12;25	7.96	21029	8AP	11;08	201.52
20956	30AP	6;00	42.00	20956	15MY	1;47	236.19	21030	24AP	0;11	36.53	21030	7MY	22;44	229.90
20957	29MY	21;15	70.32	20957	13JN	8;28	264.10	21031	23MY	9;39	64.66	21031	6JN	11;48	258.09
20958	28JN	11;41	98.52	20958	12JL	15;46	292.06	21032	21JN	17;30	92.60	21032	6JL	2;16	286.31
20959	28JL	1;12	126.84	20959	11AU	0;52	320.34	21033	21JL	0;36	120.61	21033	4AU	17;46	314.78
20960	26AU	13;53	155.47	20960	9SE	12;42	349.16	21034	19AU	8;01	148.94	21034	3SE	9;52	343.68
20961	25SE	1;47	184.54	20961	9OC	3;40	18.60	21035	17SE	16;47	177.77	21035	3OC	2;01	13.07
20962	24OC	12;58	214.08	20962	7NO	21;20	48.61	21036	17OC	3;49	207.18	21036	1NO	17;31	42.91
20963	22NO	23;38	243.96	20963	7DE	16;38	78.96	21037	15NO	17;35	237.13	21037	1DE	7;40	73.01
20964	22DE	10;09	273.99					21038	15DE	9;54	267.39	21038	30DE	20;01	103.11

695 / **701**

NUMBER	DATE	TIME	LONG.	NUMBER	DATE	TIME	LONG.	NUMBER	DATE	TIME	LONG.	NUMBER	DATE	TIME	LONG.
				20964	6JA	12;01	109.34	21039	14JA	3;51	297.66	21039	29JA	6;37	132.93
20965	20JA	20;53	303.90	20965	5FE	5;55	139.39	21040	12FE	22;15	327.63	21040	27FE	15;57	162.30
20966	19FE	8;07	333.49	20966	6MR	21;07	168.90	21041	14MR	15;49	357.10	21041	29MR	0;41	191.15
20967	20MR	20;02	2.62	20967	5AP	9;09	197.77	21042	13AP	7;35	25.98	21042	27AP	9;29	219.55
20968	19AP	8;40	31.29	20968	4MY	18;21	226.09	21043	12MY	20;55	54.38	21043	26MY	19;02	247.66
20969	18MY	22;08	59.63	20969	3JN	1;43	254.08	21044	11JN	7;47	82.46	21044	25JN	5;57	275.73
20970	17JN	12;32	87.84	20970	2JL	8;26	281.98	21045	10JL	16;44	110.48	21045	24JL	18;48	303.99
20971	17JL	3;47	116.14	20971	31JL	15;40	310.07	21046	9AU	0;46	138.70	21046	23AU	9;57	332.68
20972	15AU	19;26	144.74	20972	30AU	0;22	338.59	21047	7SE	8;58	167.33	21047	22SE	3;12	1.92
20973	14SE	10;44	173.77	20973	28SE	11;13	7.68	21048	6OC	18;11	196.50	21048	21OC	21;35	31.72
20974	14OC	0;58	203.26	20974	28OC	0;39	37.36	21049	5NO	4;56	226.19	21049	20NO	15;29	61.88
20975	12NO	13;54	233.12	20975	26NO	16;46	67.50	21050	4DE	17;22	256.24	21050	20DE	7;27	92.14
20976	12DE	1;40	263.18	20976	26DE	11;11	97.86								

696 / **702**

NUMBER	DATE	TIME	LONG.	NUMBER	DATE	TIME	LONG.	NUMBER	DATE	TIME	LONG.	NUMBER	DATE	TIME	LONG.
20977	10JA	12;34	293.18	20977	25JA	6;37	128.11	21051	3JA	7;27	286.40	21051	18JA	20;54	122.18
20978	8FE	22;49	322.87	20978	24FE	1;14	157.93	21052	1FE	23;02	316.40	21052	17FE	8;00	151.79
20979	9MR	8;34	352.09	20979	24MR	17;21	187.16	21053	3MR	15;46	346.01	21053	18MR	17;19	180.84
20980	7AP	18;08	20.81	20980	23AP	6;20	215.78	21054	2AP	8;48	15.14	21054	17AP	1;26	209.37
20981	7MY	4;09	49.13	20981	22MY	16;36	243.95	21055	2MY	0;59	43.78	21055	16MY	8;59	237.49
20982	5JN	15;27	77.25	20982	21JN	1;04	271.92	21056	31MY	15;24	72.05	21056	14JN	16;46	265.44
20983	5JL	4;42	105.42	20983	20JL	8;47	299.95	21057	30JN	3;49	100.18	21057	14JL	1;50	293.48
20984	3AU	20;03	133.88	20984	18AU	16;36	328.28	21058	29JL	14;37	128.40	21058	12AU	13;13	321.88
20985	2SE	12;53	162.79	20985	17SE	1;15	357.10	21059	28AU	0;35	156.94	21059	11SE	3;42	350.82
20986	2OC	6;03	192.21	20986	16OC	11;19	26.46	21060	26SE	10;24	185.96	21060	10OC	21;11	20.39
20987	31OC	22;32	222.08	20987	14NO	23;25	56.33	21061	25OC	20;29	215.46	21061	9NO	16;28	50.49
20988	30NO	13;41	252.22	20988	14DE	13;59	86.52	21062	24NO	7;01	245.35	21062	9DE	11;42	80.85
20989	30DE	3;10	282.36					21063	23DE	18;12	275.41				

697 / **703**

NUMBER	DATE	TIME	LONG.	NUMBER	DATE	TIME	LONG.	NUMBER	DATE	TIME	LONG.	NUMBER	DATE	TIME	LONG.
				20989	13JA	6;50	116.74					21063	8JA	5;13	111.14
20990	28JA	14;51	312.24	20990	12FE	0;58	146.72	21064	22JA	6;19	305.37	21064	6FE	20;12	141.05
20991	27FE	0;45	341.64	20991	13MR	18;48	176.21	21065	20FE	19;39	335.02	21065	8MR	8;28	170.41
20992	28MR	9;15	10.49	20992	12AP	11;03	205.13	21066	22MR	10;13	4.23	21066	6AP	18;14	199.17
20993	26AP	17;05	38.86	20993	12MY	1;09	233.56	21067	21AP	1;32	32.99	21067	6MY	2;05	227.42
20994	26MY	1;17	66.93	20994	10JN	13;12	261.69	21068	20MY	16;55	61.40	21068	4JN	8;52	255.38
20995	24JN	10;56	94.94	20995	9JL	23;37	289.77	21069	19JN	7;48	89.62	21069	3JL	15;42	283.29
20996	23JL	22;50	123.16	20996	8AU	8;55	318.03	21070	18JL	21;55	117.89	21070	1AU	23;47	311.43
20997	22AU	13;16	151.81	20997	6SE	17;43	346.68	21071	17AU	11;16	146.42	21071	31AU	10;14	340.04
20998	21SE	5;59	181.02	20998	6OC	2;44	15.83	21072	15SE	23;51	175.36	21072	29SE	23;43	9.25
20999	21OC	0;11	210.79	20999	4NO	12;40	45.47	21073	15OC	11;39	204.77	21073	29OC	16;12	39.08
21000	19NO	18;44	240.97	21000	4DE	0;03	75.47	21074	13NO	22;44	234.57	21074	28NO	10;55	69.35
21001	19DE	12;17	271.30					21075	13DE	9;20	264.58	21075	28DE	6;27	99.75

--------NEW MOONS--------				--------FULL MOONS-------			
NUMBER	DATE	TIME	LONG.	NUMBER	DATE	TIME	LONG.
			704				
21076	11JA	19;51	294.56	21076	27JA	1;13	129.96
21077	10FE	6;37	324.25	21077	25FE	17;48	159.67
21078	10MR	17;54	353.52	21078	26MR	7;20	188.76
21079	9AP	5;53	22.31	21079	24AP	17;48	217.26
21080	8MY	18;40	50.73	21080	24MY	1;55	245.34
21081	7JN	8;26	78.94	21081	22JN	8;50	273.24
21082	6JL	23;17	107.18	21082	21JL	15;44	301.25
21083	5AU	14;58	135.67	21083	19AU	23;42	329.60
21084	4SE	6;49	164.56	21084	18SE	9;32	358.49
21085	30C	21;58	193.93	21085	17OC	21;44	27.97
21086	2NO	11;49	223.71	21086	16NO	12;33	57.96
21087	2DE	0;17	253.75	21087	16DE	5;51	88.27
21088	31DE	11;37	283.80				
			705				
				21088	15JA	0;51	118.58
21089	29JA	22;05	313.61	21089	13FE	19;57	148.57
21090	28FE	7;52	342.99	21090	15MR	13;19	178.01
21091	29MR	17;14	11.85	21091	14AP	3;48	206.83
21092	28AP	2;41	40.27	21092	13MY	15;19	235.14
21093	27MY	13;02	68.41	21093	12JN	0;37	263.16
21094	26JN	1;08	96.52	21094	11JL	8;42	291.15
21095	25JL	15;28	124.85	21095	9AU	16;31	319.37
21096	24AU	7;48	153.60	21096	8SE	0;51	348.02
21097	23SE	1;12	182.86	21097	7OC	10;17	17.20
21098	22OC	18;24	212.62	21098	5NO	21;23	46.91
21099	21NO	10;30	242.72	21099	5DE	10;39	77.00
21100	21DE	0;58	272.91				
			706				
				21100	4JA	2;18	107.23
21101	19JA	13;33	302.92	21101	2FE	19;45	137.29
21102	18FE	0;13	332.49	21102	4MR	13;41	166.95
21103	19MR	9;11	1.52	21103	3AP	6;38	196.05
21104	17AP	17;03	30.03	21104	2MY	21;44	224.64
21105	17MY	0;46	58.15	21105	1JN	10;48	252.86
21106	15JN	9;26	86.14	21106	30JN	22;06	280.94
21107	14JL	20;04	114.24	21107	30JL	8;05	309.13
21108	13AU	9;15	142.72	21108	28AU	17;17	337.65
21109	12SE	1;01	171.73	21109	27SE	2;20	6.65
21110	11OC	18;46	201.32	21110	26OC	11;53	36.14
21111	10NO	13;30	231.41	21111	24NO	22;33	66.04
21112	10DE	7;52	261.74	21112	24DE	10;44	96.14
			707				
21113	9JA	0;28	291.98	21113	23JA	0;27	126.16
21114	7FE	14;20	321.84	21114	21FE	15;22	155.87
21115	9MR	1;18	351.13	21115	23MR	6;54	185.11
21116	7AP	9;56	19.83	21116	21AP	22;36	213.87
21117	6MY	17;12	48.06	21117	21MY	14;00	242.27
21118	5JN	0;11	76.01	21118	20JN	4;40	270.48
21119	4JL	7;53	103.96	21119	19JL	18;09	298.73
21120	2AU	17;13	132.16	21120	18AU	6;15	327.22
21121	1SE	4;52	160.83	21121	16SE	17;14	356.10
21122	30SE	19;21	190.10	21122	16OC	3;39	25.47
21123	30OC	12;44	219.97	21123	14NO	14;06	55.25
21124	29NO	8;14	250.28	21124	14DE	0;55	85.27
21125	29DE	4;04	280.70				
			708				
				21125	12JA	12;09	115.27
21126	27JA	22;06	310.86	21126	10FE	23;45	145.00
21127	26FE	13;00	340.50	21127	11MR	11;51	174.28
21128	27MR	0;42	9.50	21128	10AP	0;46	203.10
21129	25AP	9;53	37.94	21129	9MY	14;48	231.56
21130	24MY	17;30	65.99	21130	8JN	5;50	259.82
21131	23JN	0;29	93.90	21131	7JL	21;18	288.09
21132	22JL	7;43	121.92	21132	6AU	12;25	316.56
21133	20AU	16;03	150.30	21133	5SE	2;42	345.41
21134	19SE	2;27	179.22	21134	4OC	16;05	14.71
21135	18OC	15;45	208.75	21135	3NO	4;42	44.45
21136	17NO	8;16	238.83	21136	2DE	16;35	74.47
21137	17DE	3;17	269.21				
			709				
				21137	1JA	3;41	104.51
21138	15JA	22;59	299.54	21138	30JA	13;58	134.31
21139	14FE	17;23	329.49	21139	28FE	23;45	163.67
21140	16MR	9;14	358.85	21140	30MR	9;37	192.55
21141	14AP	22;17	27.60	21141	28AP	20;19	221.01
21142	14MY	8;49	55.87	21142	28MY	8;26	249.21
21143	12JN	17;24	83.86	21143	26JN	22;08	277.39
21144	12JL	0;51	111.82	21144	26JL	13;12	305.75
21145	10AU	8;07	140.03	21145	25AU	5;15	334.50
21146	8SE	16;14	168.68	21146	23SE	21;44	3.74
21147	8OC	2;09	197.89	21147	23OC	13;56	33.47
21148	6NO	14;35	227.67	21148	22NO	5;04	63.54
21149	6DE	5;40	257.84	21149	21DE	18;27	93.68

--------NEW MOONS--------				--------FULL MOONS-------			
NUMBER	DATE	TIME	LONG.	NUMBER	DATE	TIME	LONG.
			710				
21150	4JA	22;49	288.12	21150	20JA	5;52	123.63
21151	3FE	16;57	318.21	21151	18FE	15;38	153.16
21152	5MR	10;51	347.85	21152	20MR	0;23	182.17
21153	4AP	3;29	16.93	21153	18AP	8;50	210.69
21154	3MY	18;01	45.48	21154	17MY	17;44	238.86
21155	2JN	6;07	73.65	21155	16JN	3;46	266.90
21156	1JL	16;03	101.68	21156	15JL	15;31	295.05
21157	31JL	0;35	129.83	21157	14AU	5;32	323.57
21158	29AU	8;46	158.32	21158	12SE	21;57	352.62
21159	27SE	17;35	187.31	21159	12OC	16;10	22.25
21160	27OC	3;40	216.83	21160	11NO	10;45	52.34
21161	25NO	15;18	246.78	21161	11DE	3;59	82.62
21162	25DE	4;29	276.92				
			711				
				21162	9JA	18;46	112.78
21163	23JA	19;09	306.98	21163	8FE	6;56	142.56
21164	22FE	11;06	336.71	21164	9MR	16;57	171.81
21165	24MR	3;50	5.99	21165	8AP	1;26	200.49
21166	22AP	20;23	34.77	21166	7MY	9;01	228.72
21167	22MY	11;43	63.16	21167	5JN	16;28	256.69
21168	21JN	1;12	91.33	21168	5JL	0;42	284.66
21169	20JL	12;55	119.51	21169	3AU	10;48	312.90
21170	18AU	23;27	147.94	21170	1SE	23;44	341.63
21171	17SE	9;32	176.81	21171	1OC	15;54	10.98
21172	16OC	19;40	206.17	21172	31OC	10;38	40.93
21173	15NO	6;05	235.96	21173	30NO	6;16	71.25
21174	14DE	16;56	265.98	21174	30DE	0;51	101.61
			712				
21175	13JA	4;25	295.99	21175	28JA	17;09	131.69
21176	11FE	16;53	325.74	21176	27FE	6;41	161.26
21177	12MR	6;34	355.08	21177	27MR	17;33	190.21
21178	10AP	21;17	23.96	21178	26AP	2;11	218.62
21179	10MY	12;31	52.46	21179	25MY	9;19	246.65
21180	9JN	3;38	80.72	21180	23JN	15;57	274.54
21181	8JL	18;14	108.95	21181	22JL	23;17	302.57
21182	7AU	8;12	137.39	21182	21AU	8;29	331.00
21183	5SE	21;28	166.21	21183	19SE	20;29	359.99
21184	5OC	9;57	195.49	21184	19OC	11;37	29.61
21185	3NO	21;35	225.20	21185	18NO	5;24	59.75
21186	3DE	8;30	255.18	21186	18DE	0;43	90.15
			713				
21187	1JA	19;00	285.19	21187	16JA	20;01	120.45
21188	31JA	5;27	314.98	21188	15FE	13;45	150.36
21189	1MR	16;13	344.38	21189	17MR	4;46	179.68
21190	31MR	3;33	13.31	21190	15AP	16;39	208.37
21191	29AP	15;38	41.81	21191	15MY	1;48	236.58
21192	29MY	4;42	70.05	21192	13JN	9;10	264.51
21193	27JN	18;57	98.25	21193	12JL	16;00	292.46
21194	27JL	10;21	126.63	21194	10AU	23;26	320.68
21195	26AU	2;27	155.39	21195	9SE	8;22	349.37
21196	24SE	18;21	184.63	21196	8OC	19;26	18.65
21197	24OC	9;11	214.31	21197	7NO	8;59	48.47
21198	22NO	22;33	244.31	21198	7DE	1;04	78.69
21199	22DE	10;29	274.39				
			714				
				21199	5JA	19;15	109.02
21200	20JA	21;18	304.31	21200	4FE	14;21	139.13
21201	19FE	7;15	333.84	21201	6MR	8;38	168.77
21202	20MR	16;34	2.86	21202	5AP	0;31	197.81
21203	19AP	1;39	31.40	21203	4MY	13;26	226.28
21204	18MY	11;13	59.59	21204	2JN	23;45	254.39
21205	16JN	22;12	87.66	21205	2JL	8;25	282.37
21206	16JL	11;21	115.87	21206	31JL	16;25	310.49
21207	14AU	2;51	144.46	21207	30AU	0;36	338.99
21208	13SE	20;03	173.56	21208	28SE	9;34	8.00
21209	13OC	13;46	203.18	21209	27OC	19;53	37.55
21210	12NO	6;45	233.21	21210	26NO	8;02	67.52
21211	11DE	22;14	263.42	21211	25DE	22;25	97.71
			715				
21212	10JA	11;49	293.53	21212	24JA	14;53	127.84
21213	8FE	23;22	323.28	21213	23FE	8;30	157.63
21214	10MR	8;59	352.49	21214	25MR	1;51	186.92
21215	8AP	17;08	21.15	21215	23AP	17;49	215.67
21216	8MY	0;37	49.37	21216	23MY	7;52	243.99
21217	6JN	8;32	77.36	21217	21JN	20;06	272.11
21218	5JL	17;59	105.38	21218	21JL	6;51	300.26
21219	4AU	5;51	133.70	21219	19AU	16;36	328.67
21220	2SE	20;27	162.51	21220	18SE	1;53	357.51
21221	2OC	13;27	191.91	21221	17OC	11;17	26.85
21222	1NO	8;02	221.86	21222	15NO	21;26	56.64
21223	1DE	2;55	252.15	21223	15DE	8;49	86.69
21224	30DE	20;38	282.48				

716 / 722

NUMBER	DATE	TIME	LONG.	NUMBER	DATE	TIME	LONG.	NUMBER	DATE	TIME	LONG.	NUMBER	DATE	TIME	LONG.
				21224	13JA	21;40	116.75					21298	7JA	14;28	110.90
21225	29JA	11;56	312.51	21225	12FE	11;47	146.56	21299	22JA	4;33	305.68	21299	6FE	9;04	140.98
21226	28FE	0;15	342.01	21226	13MR	2;46	175.94	21300	20FE	14;55	335.21	21300	8MR	1;26	170.51
21227	28MR	9;49	10.91	21227	11AP	18;10	204.83	21301	22MR	1;40	4.27	21301	6AP	14;49	199.42
21228	26AP	17;31	39.27	21228	11MY	9;36	233.33	21302	20AP	13;04	32.88	21302	6MY	1;12	227.77
21229	26MY	0;25	67.28	21229	10JN	0;41	261.58	21303	20MY	1;23	61.17	21303	4JN	9;20	255.78
21230	24JN	7;35	95.19	21230	9JL	14;56	289.81	21304	18JN	14;54	89.34	21304	3JL	16;20	283.70
21231	23JL	16;01	123.27	21231	8AU	3;58	318.22	21305	18JL	5;46	117.63	21305	1AU	23;26	311.80
21232	22AU	2;28	151.76	21232	6SE	15;45	346.99	21306	16AU	21;47	146.26	21306	31AU	7;39	340.32
21233	20SE	15;34	180.81	21233	6OC	2;38	16.21	21307	15SE	14;11	175.35	21307	29SE	17;43	9.40
21234	20OC	7;38	210.48	21234	4NO	13;12	45.89	21308	15OC	5;57	204.92	21308	29OC	6;05	39.04
21235	19NO	2;19	240.67	21235	3DE	23;54	75.86	21309	13NO	20;19	234.86	21309	27NO	20;54	69.14
21236	18DE	22;18	271.09					21310	13DE	9;03	264.96	21310	27DE	14;02	99.45

717 / 723

NUMBER	DATE	TIME	LONG.	NUMBER	DATE	TIME	LONG.	NUMBER	DATE	TIME	LONG.	NUMBER	DATE	TIME	LONG.
				21236	2JA	10;53	105.89	21311	11JA	20;24	294.97	21311	26JA	8;42	129.65
21237	17JA	17;27	301.39	21237	31JA	22;08	135.71	21312	10FE	6;39	324.64	21312	25FE	3;26	159.47
21238	16FE	9;56	331.23	21238	2MR	9;41	165.13	21313	11MR	16;03	353.83	21313	26MR	20;30	188.71
21239	17MR	23;06	0.46	21239	31MR	21;51	194.08	21314	10AP	0;56	22.50	21314	25AP	10;51	217.37
21240	16AP	9;21	29.08	21240	30AP	11;02	222.62	21315	9MY	9;57	50.76	21315	24MY	22;23	245.59
21241	15MY	17;35	57.24	21241	30MY	1;26	250.90	21316	7JN	19;56	78.83	21316	23JN	7;51	273.59
21242	14JN	0;46	85.16	21242	28JN	16;45	279.14	21317	7JL	7;51	106.95	21317	22JL	16;12	301.65
21243	13JL	7;48	113.12	21243	28JL	8;16	307.54	21318	5AU	22;13	135.38	21318	21AU	0;23	330.02
21244	11AU	15;33	141.36	21244	26AU	23;16	336.27	21319	4SE	14;50	164.30	21319	19SE	9;05	358.86
21245	10SE	0;54	170.08	21245	25SE	13;24	5.45	21320	4OC	8;42	193.77	21320	18OC	18;49	28.23
21246	9OC	12;45	199.40	21246	25OC	2;40	35.07	21321	3NO	2;26	223.71	21321	17NO	6;02	58.08
21247	8NO	3;44	229.30	21247	23NO	15;06	65.04	21322	2DE	18;56	253.91	21322	16DE	19;13	88.21
21248	7DE	21;39	259.60	21248	23DE	2;40	95.11								

718 / 724

NUMBER	DATE	TIME	LONG.	NUMBER	DATE	TIME	LONG.	NUMBER	DATE	TIME	LONG.	NUMBER	DATE	TIME	LONG.
21249	6JA	17;11	289.98	21249	21JA	13;16	125.01	21323	1JA	9;34	284.10	21323	15JA	10;32	118.36
21250	5FE	12;19	320.08	21250	19FE	23;04	154.53	21324	30JA	22;07	314.00	21324	14FE	3;29	148.27
21251	7MR	5;24	349.66	21251	21MR	8;34	183.54	21325	29FE	8;33	343.40	21325	14MR	20;52	177.72
21252	5AP	19;45	18.62	21252	19AP	18;30	212.10	21326	29MR	17;11	12.24	21326	13AP	13;27	206.64
21253	5MY	7;26	47.04	21253	19MY	5;37	240.35	21327	28AP	0;43	40.58	21327	13MY	4;24	235.09
21254	3JN	16;56	75.10	21254	17JN	18;23	268.49	21328	27MY	8;07	68.60	21328	11JN	17;33	263.27
21255	3JL	0;56	103.06	21255	17JL	8;48	296.76	21329	25JN	16;35	96.57	21329	11JL	5;10	291.40
21256	1AU	8;18	131.16	21256	16AU	0;33	325.37	21330	25JL	3;09	124.74	21330	9AU	15;35	319.72
21257	30AU	16;01	159.65	21257	14SE	17;07	354.45	21331	23AU	16;25	153.36	21331	8SE	1;17	348.42
21258	29SE	1;04	188.67	21258	14OC	9;51	24.05	21332	22SE	8;24	182.56	21332	7OC	10;46	17.61
21259	28OC	12;17	218.27	21259	13NO	1;53	54.05	21333	22OC	2;30	212.34	21333	5NO	20;37	47.27
21260	27NO	2;04	248.32	21260	12DE	16;21	84.22	21334	20NO	21;34	242.56	21334	5DE	7;22	77.25
21261	26DE	18;12	278.58					21335	20DE	16;09	272.93				

719 / 725

NUMBER	DATE	TIME	LONG.	NUMBER	DATE	TIME	LONG.	NUMBER	DATE	TIME	LONG.	NUMBER	DATE	TIME	LONG.
				21261	11JA	4;45	114.28					21335	3JA	19;23	107.32
21262	25JA	11;47	308.74	21262	9FE	15;11	143.97	21336	19JA	8;48	303.11	21336	2FE	8;41	137.22
21263	24FE	5;43	338.54	21263	11MR	0;11	173.15	21337	17FE	22;34	332.81	21337	3MR	23;01	166.73
21264	25MR	22;55	7.80	21264	9AP	8;30	201.80	21338	19MR	9;19	1.92	21338	2AP	13;57	195.76
21265	24AP	14;28	36.52	21265	8MY	16;53	230.05	21339	17AP	17;43	30.44	21339	2MY	5;11	224.36
21266	24MY	3;48	64.82	21266	7JN	2;06	258.09	21340	17MY	0;47	58.54	21340	31MY	20;24	252.67
21267	22JN	14;51	92.89	21267	6JL	12;50	286.16	21341	15JN	7;38	86.45	21341	30JN	11;12	280.89
21268	22JL	0;08	120.98	21268	5AU	1;41	314.52	21342	14JL	15;19	114.44	21342	30JL	1;05	309.24
21269	20AU	8;34	149.35	21269	3SE	17;01	343.38	21343	13AU	0;42	142.76	21343	28AU	13;46	337.90
21270	18SE	17;10	178.17	21270	3OC	10;40	12.82	21344	11SE	12;30	171.60	21344	27SE	1;20	6.99
21271	18OC	2;44	207.53	21271	2NO	5;30	42.79	21345	11OC	3;10	201.06	21345	26OC	12;13	36.55
21272	16NO	13;39	237.35	21272	1DE	23;47	73.07	21346	9NO	20;43	231.09	21346	24NO	22;57	66.45
21273	16DE	2;02	267.45	21273	31DE	15;57	103.33	21347	9DE	16;19	261.47	21347	24DE	9;47	96.49

720 / 726

NUMBER	DATE	TIME	LONG.	NUMBER	DATE	TIME	LONG.	NUMBER	DATE	TIME	LONG.	NUMBER	DATE	TIME	LONG.
21274	14JA	15;49	297.54	21274	30JA	5;22	133.27	21348	8JA	12;08	291.85	21348	22JA	20;47	126.40
21275	13FE	6;54	327.38	21275	28FE	16;17	162.70	21349	7FE	6;03	321.88	21349	21FE	7;56	155.95
21276	13MR	23;02	356.80	21276	29MR	1;19	191.57	21350	8MR	20;48	351.34	21350	22MR	19;27	185.03
21277	12AP	15;34	25.73	21277	27AP	9;08	219.92	21351	7AP	8;20	20.16	21351	21AP	7;47	213.66
21278	12MY	7;31	54.23	21278	26MY	16;26	247.95	21352	6MY	17;24	48.45	21352	20MY	21;21	241.99
21279	10JN	21;59	82.46	21279	25JN	0;04	275.88	21353	5JN	0;58	76.43	21353	19JN	12;12	270.21
21280	10JL	10;42	110.63	21280	24JL	9;05	303.99	21354	4JL	8;01	104.35	21354	19JL	3;48	298.54
21281	8AU	21;58	138.98	21281	22AU	20;33	332.52	21355	2AU	15;23	132.47	21355	17AU	19;21	327.16
21282	7SE	8;28	167.70	21282	21SE	11;12	1.65	21356	31AU	23;56	161.02	21356	16SE	10;14	356.21
21283	6OC	18;46	196.92	21283	21OC	4;56	31.41	21357	30SE	10;32	190.12	21357	16OC	0;14	25.72
21284	5NO	5;12	226.59	21284	20NO	0;29	61.64	21358	29OC	23;58	219.83	21358	14NO	13;18	55.61
21285	4DE	15;53	256.57	21285	19DE	19;52	92.04	21359	28NO	16;30	250.01	21359	14DE	1;25	85.69
								21360	28DE	11;23	280.38				

721 / 727

NUMBER	DATE	TIME	LONG.	NUMBER	DATE	TIME	LONG.	NUMBER	DATE	TIME	LONG.	NUMBER	DATE	TIME	LONG.
21286	3JA	2;57	286.60	21286	18JA	13;25	122.26	21361	27JA	6;49	310.61	21360	12JA	12;29	115.68
21287	1FE	14;41	316.44	21287	17FE	4;18	152.02	21362	26FE	0;56	340.38	21361	10FE	22;30	145.33
21288	3MR	3;28	345.89	21288	18MR	16;22	181.19	21363	27MR	16;34	9.56	21362	12MR	7;51	174.51
21289	1AP	17;24	14.90	21289	17AP	1;57	209.78	21364	26AP	5;29	38.15	21363	10AP	17;12	203.19
21290	1MY	8;11	43.49	21290	16MY	9;39	237.91	21365	25MY	16;01	66.32	21364	10MY	3;24	231.49
21291	30MY	23;18	71.80	21291	14JN	16;22	265.82	21366	24JN	0;45	94.29	21365	8JN	15;08	259.62
21292	29JN	14;14	100.03	21292	13JL	23;13	293.77	21367	23JL	8;27	122.34	21366	8JL	4;41	287.81
21293	29JL	4;42	128.39	21293	12AU	7;24	322.03	21368	21AU	16;01	150.69	21367	6AU	19;53	316.29
21294	27AU	18;36	157.09	21294	10SE	18;00	350.81	21369	20SE	0;26	179.52	21368	5SE	12;19	345.21
21295	26SE	7;48	186.24	21295	10OC	7;40	20.22	21370	19OC	10;36	208.93	21369	5OC	5;20	14.66
21296	25OC	20;07	215.84	21296	9NO	0;17	50.20	21371	17NO	23;08	238.83	21370	3NO	22;06	44.57
21297	24NO	7;32	245.77	21297	8DE	19;01	80.53	21372	17DE	14;06	269.04	21371	3DE	13;38	74.74
21298	23DE	18;13	275.80												

728

NEW MOON				FULL MOON			
				21372	2JA	3;10	104.88
21373	16JA	6;55	299.25	21373	31JA	14;29	134.72
21374	15FE	0;35	329.18	21374	29FE	23;58	164.07
21375	15MR	18;03	358.62	21375	30MR	8;21	192.88
21376	14AP	10;21	27.51	21376	28AP	16;24	221.24
21377	14MY	0;48	55.94	21377	28MY	0;57	249.30
21378	12JN	13;03	84.07	21378	26JN	10;45	277.32
21379	11JL	23;17	112.15	21379	25JL	22;26	305.55
21380	10AU	8;13	140.42	21380	24AU	12;35	334.21
21381	8SE	16;51	169.09	21381	23SE	5;17	3.45
21382	8OC	2;02	198.28	21382	22OC	23;55	33.27
21383	6NO	12;23	227.97	21383	21NO	18;55	63.50
21384	6DE	0;03	257.99	21384	21DE	12;24	93.82

729

NEW MOON				FULL MOON			
21385	4JA	13;01	288.10	21385	20JA	3;13	123.91
21386	3FE	3;14	318.02	21386	18FE	15;15	153.54
21387	4MR	18;36	347.57	21387	20MR	1;01	182.59
21388	3AP	10;46	16.64	21388	18AP	9;13	211.09
21389	3MY	2;57	45.26	21389	17MY	16;33	239.20
21390	1JN	18;12	73.57	21390	15JN	23;50	267.13
21391	1JL	7;54	101.75	21391	15JL	8;00	295.14
21392	30JL	20;03	130.03	21392	13AU	18;09	323.49
21393	29AU	7;07	158.64	21393	12SE	7;13	352.40
21394	27SE	17;44	187.71	21394	11OC	23;36	21.95
21395	27OC	4;17	217.25	21395	10NO	18;34	52.04
21396	25NO	14;57	247.16	21396	10DE	14;22	82.43
21397	25DE	1;46	277.20				

730

NEW MOON				FULL MOON			
				21397	9JA	9;01	112.76
21398	23JA	12;58	307.11	21398	8FE	1;15	142.72
21399	22FE	0;56	336.68	21399	9MR	14;37	172.10
21400	23MR	14;00	5.81	21400	8AP	1;18	200.87
21401	22AP	4;08	34.51	21401	7MY	9;46	229.14
21402	21MY	18;58	62.88	21402	5JN	16;49	257.09
21403	20JN	10;00	91.11	21403	4JL	23;28	285.00
21404	20JL	0;50	119.41	21404	3AU	6;54	313.13
21405	18AU	15;17	148.00	21405	1SE	16;16	341.71
21406	17SE	5;10	177.02	21406	1OC	4;26	10.90
21407	16OC	18;14	206.50	21407	30OC	19;42	40.69
21408	15NO	6;18	236.36	21408	29NO	13;32	70.93
21409	14DE	17;23	266.40	21409	29DE	8;46	101.32

731

NEW MOON				FULL MOON			
21410	13JA	3;48	296.35	21410	28JA	3;53	131.52
21411	11FE	13;56	326.01	21411	26FE	21;24	161.26
21412	13MR	0;12	355.21	21412	28MR	12;14	190.39
21413	11AP	10;58	23.93	21413	27AP	0;01	218.92
21414	10MY	22;32	52.29	21414	26MY	9;09	247.03
21415	9JN	11;15	80.45	21415	24JN	16;37	274.96
21416	9JL	1;24	108.67	21416	23JL	23;38	302.98
21417	7AU	17;02	137.17	21417	22AU	7;19	331.34
21418	6SE	9;36	166.11	21418	20SE	16;31	0.22
21419	6OC	2;06	195.55	21419	20OC	3;48	29.68
21420	4NO	17;30	225.42	21420	18NO	17;24	59.64
21421	4DE	7;13	255.51	21421	18DE	9;22	89.89

732

NEW MOON				FULL MOON			
21422	2JA	19;16	285.59	21422	17JA	3;15	120.14
21423	1FE	5;57	315.40	21423	15FE	21;57	150.10
21424	1MR	15;35	344.74	21424	16MR	15;52	179.54
21425	31MR	0;27	13.57	21425	15AP	7;32	208.39
21426	29AP	9;06	41.94	21426	14MY	20;25	236.75
21427	28MY	18;17	70.02	21427	13JN	6;52	264.81
21428	27JN	5;01	98.08	21428	12JL	15;47	292.84
21429	26JL	18;06	126.37	21429	11AU	0;08	321.09
21430	25AU	9;47	155.10	21430	9SE	8;42	349.77
21431	24SE	3;23	184.40	21431	8OC	18;01	18.98
21432	23OC	21;36	214.21	21432	7NO	4;32	48.68
21433	22NO	15;02	244.38	21433	6DE	16;41	78.73
21434	22DE	6;46	274.63				

733

NEW MOON				FULL MOON			
				21434	5JA	6;50	108.88
21435	20JA	20;23	304.66	21435	3FE	22;49	138.87
21436	19FE	7;46	334.25	21436	5MR	15;53	168.47
21437	20MR	17;06	3.27	21437	4AP	8;46	197.56
21438	19AP	0;55	31.75	21438	4MY	0;27	226.15
21439	18MY	8;06	59.85	21439	2JN	14;30	254.41
21440	16JN	15;47	87.79	21440	2JL	2;58	282.54
21441	16JL	1;06	115.85	21441	31JL	14;09	310.79
21442	14AU	13;02	144.29	21442	30AU	0;24	339.37
21443	13SE	3;47	173.28	21443	28SE	10;10	8.42
21444	12OC	21;05	202.87	21444	27OC	19;57	37.94
21445	11NO	15;59	232.98	21445	26NO	6;16	67.84
21446	11DE	11;07	263.34	21446	25DE	17;35	97.90

734

NEW MOON				FULL MOON			
21447	10JA	4;56	293.63	21447	24JA	6;06	127.86
21448	8FE	20;10	323.53	21448	22FE	19;41	157.49
21449	10MR	8;18	352.85	21449	24MR	10;03	186.66
21450	8AP	17;39	21.56	21450	23AP	0;54	215.37
21451	8MY	1;09	49.78	21451	22MY	16;01	243.75
21452	6JN	7;54	77.72	21452	21JN	7;06	271.98
21453	5JL	15;02	105.65	21453	20JL	21;40	300.28
21454	3AU	23;31	133.83	21454	19AU	11;13	328.84
21455	2SE	10;07	162.47	21455	17SE	23;36	357.81
21456	1OC	23;23	191.72	21456	17OC	11;02	27.23
21457	31OC	15;36	221.56	21457	15NO	21;58	57.05
21458	30NO	10;24	251.84	21458	15DE	8;48	87.08
21459	30DE	6;22	282.26				

735

NEW MOON				FULL MOON			
				21459	13JA	19;39	117.05
21460	29JA	1;24	312.46	21460	12FE	6;31	146.73
21461	27FE	17;44	342.13	21461	13MR	17;32	175.95
21462	29MR	6;44	11.17	21462	12AP	5;06	204.70
21463	27AP	16;51	39.63	21463	11MY	17;46	233.08
21464	27MY	1;02	67.69	21464	10JN	7;51	261.30
21465	25JN	8;15	95.61	21465	9JL	23;10	289.57
21466	24JL	15;26	123.64	21466	8AU	15;01	318.09
21467	22AU	23;24	152.02	21467	7SE	6;34	347.00
21468	21SE	8;58	180.93	21468	6OC	21;19	16.38
21469	20OC	21;00	210.43	21469	5NO	11;07	46.19
21470	19NO	12;02	240.46	21470	4DE	23;52	76.25
21471	19DE	5;50	270.79				

736

NEW MOON				FULL MOON			
				21471	3JA	11;30	106.31
21472	18JA	1;07	301.09	21472	1FE	21;55	136.09
21473	16FE	19;56	331.04	21473	2MR	7;21	165.42
21474	17MR	12;44	0.42	21474	31MR	16;21	194.24
21475	16AP	2;55	29.20	21475	30AP	1;47	222.63
21476	15MY	14;35	57.50	21476	29MY	12;29	250.78
21477	14JN	0;11	85.53	21477	28JN	1;00	278.90
21478	13JL	8;25	113.53	21478	27JL	15;26	307.25
21479	11AU	16;05	141.77	21479	26AU	7;28	336.02
21480	10SE	0;08	170.44	21480	25SE	0;31	5.30
21481	9OC	9;28	199.65	21481	24OC	17;48	35.09
21482	7NO	20;51	229.40	21482	23NO	10;17	65.23
21483	7DE	10;37	259.52	21483	23DE	1;00	95.43

737

NEW MOON				FULL MOON			
21484	6JA	2;28	289.74	21484	21JA	13;24	125.41
21485	4FE	19;37	319.77	21485	19FE	23;37	154.94
21486	6MR	13;03	349.37	21486	21MR	8;17	183.92
21487	5AP	5;50	18.44	21487	19AP	16;13	212.40
21488	4MY	21;12	47.01	21488	19MY	0;15	240.52
21489	3JN	10;36	75.23	21489	17JN	9;12	268.51
21490	2JL	21;55	103.32	21490	16JL	19;49	296.63
21491	1AU	7;35	131.52	21491	15AU	8;43	325.12
21492	30AU	16;29	160.06	21492	14SE	0;16	354.15
21493	29SE	1;30	189.08	21493	13OC	18;16	23.79
21494	28OC	11;23	218.62	21494	12NO	13;30	53.92
21495	26NO	22;27	248.55	21495	12DE	8;05	84.26
21496	26DE	10;42	278.65				

738

NEW MOON				FULL MOON			
				21496	11JA	0;22	114.48
21497	25JA	0;07	308.64	21497	9FE	13;43	144.30
21498	23FE	14;39	338.30	21498	11MR	0;26	173.55
21499	25MR	6;11	7.51	21499	9AP	9;12	202.22
21500	23AP	22;14	36.26	21500	8MY	16;45	230.44
21501	23MY	13;58	64.65	21501	6JN	23;52	258.39
21502	22JN	4;33	92.86	21502	6JL	7;26	286.34
21503	21JL	17;37	121.11	21503	4AU	16;28	314.54
21504	20AU	5;25	149.61	21504	3SE	4;02	343.24
21505	18SE	16;28	178.53	21505	2OC	18;52	12.56
21506	18OC	3;15	207.94	21506	1NO	12;49	42.48
21507	16NO	14;00	237.76	21507	1DE	8;32	72.81
21508	16DE	0;46	267.79	21508	31DE	4;00	103.21

739

NEW MOON				FULL MOON			
21509	14JA	11;38	297.76	21509	29JA	21;30	133.33
21510	12FE	22;57	327.44	21510	28FE	12;14	162.92
21511	14MR	11;09	356.70	21511	30MR	0;08	191.90
21512	13AP	0;28	25.51	21512	28AP	9;33	220.33
21513	12MY	14;46	53.95	21513	27MY	17;10	248.37
21514	11JN	5;40	82.20	21514	25JN	23;53	276.26
21515	10JL	20;42	110.46	21515	25JL	6;49	304.29
21516	9AU	11;34	138.95	21516	23AU	15;10	332.69
21517	8SE	2;03	167.83	21517	22SE	1;57	1.66
21518	7OC	15;51	197.19	21518	21OC	15;46	31.25
21519	6NO	4;41	226.96	21519	20NO	8;27	61.35
21520	5DE	16;22	256.98	21520	20DE	3;08	91.72

--------NEW MOONS--------				--------FULL MOONS-------			
NUMBER	DATE	TIME	LONG.	NUMBER	DATE	TIME	LONG.

740

NEW NUMBER	DATE	TIME	LONG.	FULL NUMBER	DATE	TIME	LONG.
21521	4JA	3:04	286.99	21521	18JA	22:24	122.02
21522	2FE	13:10	316.76	21522	17FE	16:45	151.93
21523	2MR	23:06	346.10	21523	18MR	8:55	181.28
21524	1AP	9:19	14.96	21524	16AP	22:09	210.01
21525	30AP	20:11	43.40	21525	16MY	8:30	238.25
21526	30MY	8:05	71.59	21526	14JN	16:42	266.22
21527	28JN	21:24	99.75	21527	13JL	23:53	294.18
21528	28JL	12:21	128.12	21528	12AU	7:14	322.41
21529	27AU	4:44	156.91	21529	10SE	15:45	351.11
21530	25SE	21:42	186.21	21530	10OC	2:03	20.38
21531	25OC	14:03	215.97	21531	8NO	14:33	50.17
21532	24NO	4:52	246.04	21532	8DE	5:19	80.34
21533	23DE	17:48	276.17				

741

NEW NUMBER	DATE	TIME	LONG.	FULL NUMBER	DATE	TIME	LONG.
				21533	6JA	22:11	110.61
21534	22JA	5:06	306.10	21534	5FE	16:27	140.67
21535	20FE	15:05	335.61	21535	7MR	10:45	170.29
21536	22MR	0:06	4.60	21536	6AP	3:31	199.35
21537	20AP	8:34	33.09	21537	5MY	17:45	227.86
21538	19MY	17:10	61.23	21538	4JN	5:23	256.01
21539	18JN	2:53	89.25	21539	3JL	15:04	284.04
21540	17JL	14:39	117.41	21540	1AU	23:47	312.21
21541	16AU	5:06	145.97	21541	31AU	8:21	340.74
21542	14SE	22:02	175.08	21542	29SE	17:26	9.77
21543	14OC	16:21	204.74	21543	29OC	3:26	39.32
21544	13NO	10:33	234.84	21544	27NO	14:45	69.27
21545	13DE	3:22	265.11	21545	27DE	3:47	99.40

742

NEW NUMBER	DATE	TIME	LONG.	FULL NUMBER	DATE	TIME	LONG.
21546	11JA	18:07	295.26	21546	25JA	18:42	129.45
21547	10FE	6:33	325.03	21547	24FE	11:04	159.18
21548	11MR	16:44	354.25	21548	26MR	3:55	188.43
21549	10AP	1:04	22.89	21549	24AP	20:07	217.17
21550	9MY	8:18	51.09	21550	24MY	10:57	245.52
21551	7JN	15:29	79.04	21551	23JN	0:17	273.68
21552	6JL	23:49	107.02	21552	22JL	12:15	301.88
21553	5AU	10:21	135.29	21553	20AU	23:11	330.36
21554	3SE	23:44	164.07	21554	19SE	9:23	359.26
21555	3OC	15:58	193.46	21555	18OC	19:18	28.65
21556	2NO	10:21	223.42	21556	17NO	5:25	58.44
21557	2DE	5:40	253.74	21557	16DE	16:12	88.47

743

NEW NUMBER	DATE	TIME	LONG.	FULL NUMBER	DATE	TIME	LONG.
21558	1JA	0:23	284.10	21558	15JA	3:59	118.48
21559	30JA	17:01	314.18	21559	13FE	16:49	148.21
21560	1MR	6:38	343.71	21560	15MR	6:32	177.52
21561	30MR	17:12	12.63	21561	13AP	20:53	206.36
21562	29AP	1:23	40.99	21562	13MY	11:41	234.82
21563	28MY	8:18	69.00	21563	12JN	2:47	263.07
21564	26JN	15:06	96.90	21564	11JL	17:46	291.33
21565	25JL	22:50	124.96	21565	10AU	8:06	319.80
21566	24AU	8:21	153.42	21566	8SE	21:23	348.65
21567	22SE	20:19	182.45	21567	8OC	9:32	17.95
21568	22OC	11:08	212.09	21568	6NO	20:52	47.67
21569	21NO	4:47	242.24	21569	6DE	7:49	77.66
21570	21DE	0:24	272.65				

744

NEW NUMBER	DATE	TIME	LONG.	FULL NUMBER	DATE	TIME	LONG.
				21570	4JA	18:38	107.68
21571	19JA	20:07	302.96	21571	3FE	5:20	137.47
21572	18FE	13:52	332.84	21572	3MR	16:01	166.83
21573	19MR	4:26	2.10	21573	2AP	2:57	195.71
21574	17AP	15:50	30.75	21574	1MY	14:44	224.17
21575	17MY	0:49	58.93	21575	31MY	3:54	252.40
21576	15JN	8:25	86.87	21576	29JN	18:36	280.62
21577	14JL	15:35	114.84	21577	29JL	10:23	309.04
21578	12AU	23:10	143.09	21578	28AU	2:24	337.83
21579	11SE	7:58	171.80	21579	26SE	17:54	7.08
21580	10OC	18:46	201.10	21580	26OC	8:28	36.78
21581	9NO	8:18	230.96	21581	24NO	21:58	66.80
21582	9DE	0:47	261.20	21582	24DE	10:14	96.90

745

NEW NUMBER	DATE	TIME	LONG.	FULL NUMBER	DATE	TIME	LONG.
21583	7JA	19:26	291.53	21583	22JA	21:12	126.81
21584	6FE	14:32	321.63	21584	21FE	6:55	156.29
21585	8MR	8:19	351.21	21585	22MR	15:50	185.27
21586	6AP	23:43	20.20	21586	21AP	0:42	213.77
21587	6MY	12:34	48.65	21587	20MY	10:28	241.95
21588	4JN	23:10	76.75	21588	18JN	21:53	270.03
21589	4JL	8:06	104.75	21589	18JL	11:20	298.27
21590	2AU	16:06	132.89	21590	17AU	2:42	326.88
21591	1SE	0:02	161.41	21591	15SE	19:32	355.99
21592	30SE	8:47	190.44	21592	15OC	13:05	25.64
21593	29OC	19:10	220.02	21593	14NO	6:20	55.71
21594	28NO	7:45	250.03	21594	13DE	22:12	85.94
21595	27DE	22:31	280.23				

746

NEW NUMBER	DATE	TIME	LONG.	FULL NUMBER	DATE	TIME	LONG.
				21595	12JA	11:48	116.04
21596	26JA	14:56	310.33	21596	10FE	22:59	145.74
21597	25FE	8:05	340.08	21597	12MR	8:11	174.91
21598	27MR	1:04	9.32	21598	10AP	16:12	203.53
21599	25AP	17:06	38.04	21599	9MY	23:54	231.74
21600	25MY	7:30	66.37	21600	8JN	8:10	259.73
21601	23JN	19:57	94.49	21601	7JL	17:48	287.77
21602	23JL	6:33	122.65	21602	6AU	5:29	316.10
21603	21AU	15:57	151.07	21603	4SE	19:47	344.92
21604	20SE	1:02	179.94	21604	4OC	12:47	14.36
21605	19OC	10:36	209.32	21605	3NO	7:47	44.35
21606	17NO	21:10	239.14	21606	3DE	3:06	74.68
21607	17DE	8:49	269.21				

747

NEW NUMBER	DATE	TIME	LONG.	FULL NUMBER	DATE	TIME	LONG.
				21607	1JA	20:45	105.00
21608	15JA	21:30	299.25	21608	31JA	11:34	134.99
21609	14FE	11:13	329.01	21609	1MR	23:26	164.44
21610	16MR	1:58	358.35	21610	31MR	8:58	193.30
21611	14AP	17:34	27.22	21611	29AP	16:55	221.64
21612	14MY	9:25	55.71	21612	29MY	0:03	249.65
21613	13JN	0:39	83.97	21613	27JN	7:14	277.57
21614	12JL	14:38	112.19	21614	26JL	15:25	305.66
21615	11AU	3:15	140.61	21615	25AU	1:39	334.16
21616	9SE	14:54	169.40	21616	23SE	14:53	3.25
21617	9OC	2:03	198.67	21617	23OC	7:27	32.97
21618	7NO	13:00	228.38	21618	22NO	2:36	63.20
21619	6DE	23:50	258.37	21619	21DE	22:29	93.62

748

NEW NUMBER	DATE	TIME	LONG.	FULL NUMBER	DATE	TIME	LONG.
21620	5JA	10:34	288.39	21620	20JA	17:05	123.87
21621	3FE	21:26	318.17	21621	19FE	9:11	153.67
21622	4MR	8:52	347.55	21622	19MR	22:23	182.87
21623	2AP	21:19	16.48	21623	18AP	8:55	211.47
21624	2MY	10:53	45.01	21624	17MY	17:17	239.62
21625	1JN	1:23	73.29	21625	16JN	0:19	267.53
21626	30JN	16:23	101.52	21626	15JL	7:03	295.49
21627	30JL	7:31	129.92	21627	13AU	14:38	323.74
21628	28AU	22:29	158.68	21628	12SE	0:11	352.50
21629	27SE	12:59	187.90	21629	11OC	12:32	21.87
21630	27OC	2:36	217.57	21630	10NO	3:54	51.81
21631	25NO	15:03	247.55	21631	9DE	21:43	82.12
21632	25DE	2:15	277.61				

749

NEW NUMBER	DATE	TIME	LONG.	FULL NUMBER	DATE	TIME	LONG.
				21632	8JA	16:46	112.47
21633	23JA	12:31	307.48	21633	7FE	11:38	142.54
21634	21FE	22:18	336.96	21634	9MR	4:53	172.09
21635	23MR	8:04	5.96	21635	7AP	19:32	201.03
21636	21AP	18:18	34.50	21636	7MY	7:16	229.42
21637	21MY	5:24	62.73	21637	5JN	16:27	257.47
21638	19JN	17:50	90.86	21638	5JL	0:05	285.41
21639	19JL	7:57	119.13	21639	3AU	7:21	313.54
21640	17AU	23:49	147.76	21640	1SE	15:19	342.06
21641	16SE	16:53	176.90	21641	1OC	0:48	11.14
21642	16OC	9:59	206.55	21642	30OC	12:16	40.77
21643	15NO	1:53	236.56	21643	29NO	1:54	70.82
21644	14DE	15:54	266.72	21644	28DE	17:39	101.07

750

NEW NUMBER	DATE	TIME	LONG.	FULL NUMBER	DATE	TIME	LONG.
21645	13JA	3:59	296.75	21645	27JA	11:09	131.21
21646	11FE	14:29	326.42	21646	26FE	5:24	160.99
21647	12MR	23:46	355.58	21647	27MR	22:55	190.23
21648	11AP	8:15	24.21	21648	26AP	14:24	218.92
21649	10MY	16:29	52.43	21649	26MY	3:18	247.18
21650	9JN	1:22	80.45	21650	24JN	13:57	275.24
21651	8JL	11:54	108.52	21651	23JL	23:12	303.35
21652	7AU	0:59	136.91	21652	22AU	7:58	331.75
21653	5SE	16:52	165.82	21653	20SE	16:56	0.63
21654	5OC	10:53	195.31	21654	20OC	2:35	30.02
21655	4NO	5:33	225.30	21655	18NO	13:16	59.85
21656	3DE	23:20	255.57	21656	18DE	1:22	89.94

751

NEW NUMBER	DATE	TIME	LONG.	FULL NUMBER	DATE	TIME	LONG.
21657	2JA	15:15	285.81	21657	16JA	15:11	120.02
21658	1FE	4:50	315.74	21658	15FE	6:39	149.85
21659	2MR	16:01	345.15	21659	16MR	23:07	179.24
21660	1AP	1:04	13.98	21660	15AP	15:31	208.14
21661	30AP	8:36	42.30	21661	15MY	6:59	236.60
21662	29MY	15:33	70.30	21662	13JN	21:05	264.81
21663	27JN	23:05	98.23	21663	13JL	9:52	292.99
21664	27JL	8:24	126.37	21664	11AU	21:31	321.38
21665	25AU	20:22	154.95	21665	10SE	8:19	350.15
21666	24SE	11:19	184.12	21666	9OC	18:34	19.39
21667	24OC	4:51	213.90	21667	8NO	4:41	49.08
21668	23NO	0:00	244.13	21668	7DE	15:08	79.05
21669	22DE	19:17	274.53				

Left half (years 752–757)

NEW MOONS NUMBER	DATE	TIME	LONG.	FULL MOONS NUMBER	DATE	TIME	LONG.
752							
				21669	6JA	2;19	109.08
21670	21JA	13;07	304.75	21670	4FE	14;26	138.91
21671	20FE	4;15	334.49	21671	5MR	3;27	168.35
21672	20MR	16;12	3.62	21672	3AP	17;12	197.32
21673	19AP	1;22	32.16	21673	3MY	7;33	225.87
21674	18MY	8;42	60.26	21674	1JN	22;24	254.15
21675	16JN	15;24	88.16	21675	1JL	13;32	282.39
21676	15JL	22;33	116.14	21676	31JL	4;27	310.79
21677	14AU	7;09	144.44	21677	29AU	18;35	339.53
21678	12SE	17;55	173.26	21678	28SE	7;35	8.70
21679	12OC	7;21	202.69	21679	27OC	19;32	38.31
21680	10NO	23;42	232.68	21680	26NO	6;47	68.25
21681	10DE	18;30	263.03	21681	25DE	17;40	98.29
753							
21682	9JA	14;22	293.41	21682	24JA	4;20	128.17
21683	8FE	9;14	323.47	21683	22FE	14;47	157.67
21684	10MR	1;22	352.96	21684	24MR	1;16	186.69
21685	8AP	14;13	21.81	21685	22AP	12;16	215.25
21686	8MY	0;15	50.13	21686	22MY	0;28	243.52
21687	6JN	8;26	78.13	21687	20JN	14;18	271.70
21688	5JL	15;46	106.07	21688	20JL	5;40	300.03
21689	3AU	23;09	134.20	21689	18AU	21;52	328.69
21690	2SE	7;22	162.75	21690	17SE	13;59	357.80
21691	1OC	17;11	191.85	21691	17OC	5;20	27.39
21692	31OC	5;22	221.52	21692	15NO	19;37	57.35
21693	29NO	20;23	251.64	21693	15DE	8;39	87.46
21694	29DE	14;01	281.96				
754							
				21694	13JA	20;16	117.47
21695	28JA	8;57	312.16	21695	12FE	6;27	147.11
21696	27FE	3;23	341.93	21696	13MR	15;30	176.25
21697	28MR	19;54	11.12	21697	12AP	0;03	204.88
21698	27AP	9;57	39.74	21698	11MY	9;02	233.12
21699	26MY	21;38	67.94	21699	9JN	19;22	261.19
21700	25JN	7;25	95.96	21700	9JL	7;43	289.34
21701	24JL	15;57	124.05	21701	7AU	22;12	317.80
21702	22AU	23;59	152.43	21702	6SE	14;32	346.73
21703	21SE	8;23	181.29	21703	6OC	8;04	16.22
21704	20OC	18;00	210.69	21704	5NO	1;51	46.19
21705	19NO	5;31	240.57	21705	4DE	18;44	76.42
21706	18DE	19;10	270.72				
755							
				21706	3JA	9;36	106.62
21707	17JA	10;40	300.87	21707	1FE	21;57	136.48
21708	16FE	3;19	330.74	21708	3MR	7;56	165.84
21709	17MR	20;13	0.14	21709	1AP	16;16	194.63
21710	16AP	12;37	29.02	21710	30AP	23;51	222.94
21711	16MY	3;50	57.46	21711	30MY	7;35	250.97
21712	14JN	17;20	85.64	21712	28JN	16;21	278.95
21713	14JL	4;59	113.78	21713	28JL	2;55	307.14
21714	12AU	15;07	142.12	21714	26AU	15;54	335.77
21715	11SE	0;30	170.84	21715	25SE	7;42	5.00
21716	10OC	9;58	200.06	21716	25OC	2;01	34.82
21717	8NO	20;08	229.76	21717	23NO	21;35	65.08
21718	8DE	7;16	259.77	21718	23DE	16;22	95.45
756							
21719	6JA	19;21	289.83	21719	22JA	8;41	125.60
21720	5FE	8;19	319.69	21720	20FE	21;56	155.26
21721	5MR	22;16	349.15	21721	21MR	8;26	184.32
21722	4AP	13;11	18.16	21722	19AP	16;57	212.82
21723	4MY	4;47	46.75	21723	19MY	0;18	240.91
21724	2JN	20;21	75.06	21724	17JN	7;19	268.83
21725	2JL	11;06	103.28	21725	16JL	14;52	296.82
21726	1AU	0;36	131.63	21726	14AU	23;59	325.15
21727	30AU	12;57	160.31	21727	13SE	11;42	354.02
21728	29SE	0;35	189.44	21728	13OC	2;42	23.53
21729	28OC	11;50	219.03	21729	11NO	20;48	53.60
21730	26NO	22;51	248.96	21730	11DE	16;37	84.00
21731	26DE	9;38	279.00				
757							
				21731	10JA	12;04	114.35
21732	24JA	20;15	308.87	21732	9FE	5;27	144.34
21733	23FE	7;07	338.38	21733	10MR	20;01	173.75
21734	24MR	18;43	7.43	21734	9AP	7;45	202.55
21735	23AP	7;27	36.06	21735	8MY	17;03	230.83
21736	22MY	21;19	64.38	21736	7JN	0;38	258.81
21737	21JN	12;02	92.60	21737	6JL	7;25	286.72
21738	21JL	3;14	120.92	21738	4AU	14;31	314.85
21739	19AU	18;32	149.56	21739	2SE	23;04	343.42
21740	18SE	9;37	178.64	21740	2OC	10;03	12.58
21741	18OC	0;01	208.20	21741	1NO	0;00	42.33
21742	16NO	13;18	238.12	21742	30NO	16;41	72.53
21743	16DE	1;12	268.20	21743	30DE	11;13	102.89

Right half (years 758–763)

NEW MOONS NUMBER	DATE	TIME	LONG.	FULL MOONS NUMBER	DATE	TIME	LONG.
758							
21744	14JA	11;52	298.16	21744	29JA	6;13	133.08
21745	12FE	21;40	327.78	21745	28FE	0;17	162.82
21746	14MR	7;10	356.92	21746	29MR	16;12	191.97
21747	12AP	16;52	25.59	21747	28AP	5;20	220.55
21748	12MY	3;14	53.88	21748	27MY	15;43	248.69
21749	10JN	14;47	82.00	21749	26JN	0;04	276.66
21750	10JL	3;58	110.18	21750	25JL	7;30	304.70
21751	8AU	19;03	138.67	21751	23AU	15;09	333.08
21752	7SE	11;50	167.64	21752	21SE	23;58	1.97
21753	7OC	5;21	197.14	21753	21OC	10;31	31.42
21754	5NO	22;15	227.08	21754	19NO	23;06	61.34
21755	5DE	13;26	257.25	21755	19DE	13;44	91.54
759							
21756	4JA	2;31	287.36	21756	18JA	6;16	121.73
21757	2FE	13;42	317.17	21757	17FE	0;03	151.63
21758	3MR	23;24	346.51	21758	18MR	17;54	181.05
21759	2AP	8;02	15.30	21759	17AP	10;23	209.92
21760	1MY	16;07	43.62	21760	17MY	0;33	238.32
21761	31MY	0;23	71.67	21761	15JN	12;19	266.43
21762	29JN	9;52	99.68	21762	14JL	22;19	294.51
21763	28JL	21;34	127.92	21763	13AU	7;26	322.81
21764	27AU	12;09	156.63	21764	11SE	16;27	351.53
21765	26SE	5;23	185.93	21765	11OC	1;54	20.76
21766	26OC	0;08	215.78	21766	9NO	12;09	50.47
21767	24NO	18;43	246.00	21767	8DE	23;30	80.49
21768	24DE	11;46	276.31				
760							
				21768	7JA	12;18	110.58
21769	23JA	2;34	306.38	21769	6FE	2;45	140.49
21770	21FE	14;51	335.99	21770	6MR	18;32	170.02
21771	22MR	0;48	5.01	21771	5AP	10;50	199.06
21772	20AP	8;51	33.49	21772	5MY	2;40	227.65
21773	19MY	15;50	61.56	21773	3JN	17;27	255.93
21774	17JN	22;51	89.47	21774	3JL	6;59	284.10
21775	17JL	7;07	117.50	21775	1AU	19;24	312.42
21776	15AU	17;42	145.90	21776	31AU	6;52	341.07
21777	14SE	7;15	174.86	21777	29SE	17;37	10.17
21778	13OC	23;41	204.44	21778	29OC	3;57	39.74
21779	12NO	18;18	234.54	21779	27NO	14;17	69.65
21780	12DE	13;47	264.93	21780	27DE	1;02	99.68
761							
21781	11JA	8;32	295.25	21781	25JA	12;31	129.58
21782	10FE	1;05	325.19	21782	24FE	0;50	159.14
21783	11MR	14;33	354.54	21783	25MR	13;56	188.24
21784	10AP	0;56	23.27	21784	24AP	3;42	216.90
21785	9MY	8;58	51.50	21785	23MY	18;08	245.24
21786	7JN	15;49	79.44	21786	22JN	9;09	273.47
21787	6JL	22;37	107.36	21787	22JL	0;22	301.80
21788	5AU	6;27	135.52	21788	20AU	15;13	330.43
21789	3SE	16;08	164.15	21789	19SE	5;07	359.48
21790	3OC	4;17	193.37	21790	18OC	17;50	28.98
21791	1NO	19;14	223.17	21791	17NO	5;35	58.84
21792	1DE	12;55	253.42	21792	16DE	16;42	88.88
21793	31DE	8;26	283.82				
762							
				21793	15JA	3;24	118.84
21794	30JA	3;58	314.02	21794	13FE	13;47	148.48
21795	28FE	21;30	343.73	21795	14MR	23;58	177.64
21796	30MR	11;54	12.80	21796	13AP	10;21	206.32
21797	28AP	23;11	41.29	21797	12MY	21;37	234.64
21798	28MY	8;10	69.38	21798	11JN	10;26	262.81
21799	26JN	15;52	97.32	21799	11JL	1;03	291.05
21800	25JL	23;14	125.36	21800	9AU	17;04	319.59
21801	24AU	7;04	153.76	21801	8SE	9;35	348.56
21802	22SE	16;08	182.67	21802	8OC	1;41	18.02
21803	22OC	3;09	212.15	21803	6NO	16;48	47.89
21804	20NO	16;44	242.12	21804	6DE	6;38	78.01
21805	20DE	9;05	272.39				
763							
				21805	4JA	19;01	108.09
21806	19JA	3;24	302.65	21806	3FE	5;50	137.88
21807	17FE	22;06	332.58	21807	4MR	15;13	167.19
21808	19MR	15;31	1.96	21808	2AP	23;42	195.96
21809	18AP	6;43	30.77	21809	2MY	8;08	224.30
21810	17MY	19;32	59.11	21810	31MY	17;30	252.38
21811	16JN	6;16	87.17	21811	30JN	4;41	280.46
21812	15JL	15;29	115.22	21812	29JL	18;06	308.77
21813	13AU	23;51	143.50	21813	28AU	9;40	337.54
21814	12SE	8;10	172.20	21814	27SE	2;54	6.85
21815	11OC	17;15	201.43	21815	26OC	20;57	36.69
21816	10NO	3;50	231.16	21816	25NO	14;38	66.88
21817	9DE	16;24	261.23	21817	25DE	6;44	97.14

Left half (periods 764–769):

NEW MOONS NUMBER	DATE	TIME	LONG.	FULL MOONS NUMBER	DATE	TIME	LONG.
764							
21818	8JA	6:55	291.40	21818	23JA	20:22	127.16
21819	6FE	22:51	321.36	21819	22FE	7:22	156.70
21820	7MR	15:27	350.91	21820	22MR	16:17	185.68
21821	6AP	7:57	19.95	21821	20AP	23:57	214.12
21822	5MY	23:42	48.52	21822	20MY	7:22	242.21
21823	4JN	14:07	76.78	21823	18JN	15:25	270.16
21824	4JL	2:50	104.92	21824	18JL	0:57	298.25
21825	2AU	13:53	133.19	21825	16AU	12:41	326.70
21826	31AU	23:47	161.79	21826	15SE	3:09	355.71
21827	30SE	9:20	190.86	21827	14OC	20:26	25.34
21828	29OC	19:17	220.41	21828	13NO	15:45	55.48
21829	28NO	6:00	250.34	21829	13DE	11:18	85.87
21830	27DE	17:34	280.41				
765							
				21830	12JA	5:02	116.14
21831	26JA	5:55	310.35	21831	10FE	19:47	146.00
21832	24FE	19:05	339.93	21832	12MR	7:28	175.27
21833	26MR	9:13	9.06	21833	10AP	16:46	203.94
21834	25AP	0:16	37.75	21834	10MY	0:31	232.15
21835	24MY	15:49	66.13	21835	8JN	7:32	260.10
21836	23JN	7:05	94.37	21836	7JL	14:42	288.03
21837	22JL	21:24	122.67	21837	5AU	22:57	316.22
21838	21AU	10:33	151.24	21838	4SE	9:19	344.88
21839	19SE	22:47	180.24	21839	3OC	22:42	14.16
21840	19OC	10:28	209.70	21840	2NO	15:26	44.06
21841	17NO	21:46	239.56	21841	2DE	10:40	74.37
21842	17DE	8:43	269.59				
766							
				21842	1JA	6:32	104.79
21843	15JA	19:18	299.54	21843	31JA	1:02	134.93
21844	14FE	5:48	329.17	21844	1MR	16:58	164.56
21845	15MR	16:41	358.36	21845	31MR	6:00	193.57
21846	14AP	4:32	27.09	21846	29AP	16:24	222.01
21847	13MY	17:36	55.47	21847	29MY	0:43	250.07
21848	12JN	7:48	83.69	21848	27JN	7:49	277.98
21849	11JL	22:50	111.95	21849	26JL	14:42	306.01
21850	10AU	14:17	140.47	21850	24AU	22:29	334.42
21851	9SE	5:48	169.42	21851	23SE	8:16	3.36
21852	8OC	20:55	198.85	21852	22OC	20:47	32.92
21853	7NO	11:04	228.69	21853	21NO	12:12	62.98
21854	6DE	23:49	258.76	21854	21DE	5:53	93.30
767							
21855	5JA	11:04	288.80	21855	20JA	0:42	123.57
21856	3FE	21:08	318.55	21856	18FE	19:13	153.48
21857	5MR	6:32	347.85	21857	20MR	12:12	182.84
21858	3AP	15:50	16.65	21858	19AP	2:41	211.61
21859	3MY	1:34	45.02	21859	18MY	14:24	239.89
21860	1JN	12:16	73.15	21860	16JN	23:43	267.90
21861	1JL	0:29	101.27	21861	16JL	7:35	295.90
21862	30JL	14:37	129.63	21862	14AU	15:09	324.15
21863	29AU	6:46	158.43	21863	12SE	23:27	352.86
21864	28SE	0:18	187.76	21864	12OC	9:14	22.13
21865	27OC	17:57	217.60	21865	10NO	20:51	51.91
21866	26NO	10:19	247.74	21866	10DE	10:26	82.03
21867	26DE	0:33	277.93				
768							
				21867	9JA	1:55	112.23
21868	24JA	12:38	307.87	21868	7FE	18:57	142.23
21869	22FE	22:54	337.38	21869	8MR	12:41	171.81
21870	23MR	7:51	6.34	21870	7AP	5:49	200.86
21871	21AP	15:56	34.80	21871	6MY	21:08	229.41
21872	20MY	23:50	62.90	21872	5JN	10:07	257.60
21873	19JN	8:28	90.87	21873	4JL	21:02	285.68
21874	18JL	18:54	118.99	21874	3AU	6:41	313.90
21875	17AU	8:02	147.51	21875	1SE	15:53	342.48
21876	16SE	0:08	176.60	21876	1OC	1:17	11.55
21877	15OC	18:31	206.29	21877	30OC	11:15	41.12
21878	14NO	13:34	236.43	21878	28NO	22:04	71.05
21879	14DE	7:39	266.76	21879	28DE	10:01	101.14
769							
21880	12JA	23:40	296.96	21880	26JA	23:27	131.11
21881	11FE	13:10	326.75	21881	25FE	14:21	160.76
21882	13MR	0:08	355.98	21882	27MR	6:13	189.94
21883	11AP	8:56	24.63	21883	25AP	22:10	218.66
21884	10MY	16:13	52.81	21884	25MY	13:26	247.02
21885	8JN	22:59	80.74	21885	24JN	3:39	275.22
21886	8JL	6:27	108.69	21886	23JL	16:48	303.48
21887	6AU	15:47	136.93	21887	22AU	4:59	332.02
21888	5SE	3:52	165.67	21888	20SE	16:20	0.99
21889	4OC	19:00	195.04	21889	20OC	3:05	30.43
21890	3NO	12:46	224.99	21890	18NO	13:30	60.25
21891	3DE	8:04	255.31	21891	18DE	0:01	90.27

Right half (periods 770–775):

NEW MOONS NUMBER	DATE	TIME	LONG.	FULL MOONS NUMBER	DATE	TIME	LONG.
770							
21892	2JA	3:24	285.70	21892	16JA	11:00	120.23
21893	31JA	21:11	315.81	21893	14FE	22:41	149.91
21894	2MR	12:10	345.38	21894	16MR	11:06	179.14
21895	31MR	23:57	14.32	21895	15AP	0:15	207.91
21896	30AP	8:58	42.70	21896	14MY	14:07	236.32
21897	29MY	16:13	70.71	21897	13JN	4:46	264.55
21898	27JN	22:54	98.61	21898	12JL	20:00	292.83
21899	27JL	6:10	126.67	21899	11AU	11:20	321.35
21900	25AU	14:55	155.12	21900	10SE	2:04	350.35
21901	24SE	1:52	184.12	21901	9OC	15:41	19.66
21902	23OC	15:28	213.73	21902	8NO	4:07	49.44
21903	22NO	7:52	243.84	21903	7DE	15:37	79.47
21904	22DE	2:36	274.21				
771							
				21904	6JA	2:30	109.48
21905	20JA	22:17	304.51	21905	4FE	12:55	139.24
21906	19FE	16:54	334.41	21906	5MR	22:57	168.55
21907	21MR	8:49	3.71	21907	4AP	8:54	197.37
21908	19AP	21:32	32.39	21908	3MY	19:22	225.77
21909	19MY	7:32	60.60	21909	2JN	7:09	253.94
21910	17JN	15:48	88.57	21910	1JL	20:47	282.11
21911	16JL	23:19	116.56	21911	31JL	12:15	310.53
21912	15AU	6:58	144.82	21912	30AU	4:50	339.36
21913	13SE	15:29	173.55	21913	28SE	21:32	8.67
21914	13OC	1:33	202.84	21914	28OC	13:28	38.45
21915	11NO	13:50	232.65	21915	27NO	4:11	68.53
21916	11DE	4:48	262.84	21916	26DE	17:23	98.67
772							
21917	9JA	22:08	293.11	21917	25JA	4:57	128.59
21918	8FE	16:40	323.17	21918	23FE	14:53	158.07
21919	9MR	10:41	352.74	21919	23MR	23:32	187.01
21920	8AP	2:54	21.74	21920	22AP	7:39	215.46
21921	7MY	16:50	50.22	21921	21MY	16:14	243.58
21922	6JN	4:37	78.37	21922	20JN	2:18	271.61
21923	5JL	14:39	106.41	21923	19JL	14:31	299.81
21924	3AU	23:32	134.61	21924	18AU	5:07	328.40
21925	2SE	7:59	163.17	21925	16SE	21:46	357.52
21926	1OC	16:46	192.22	21926	16OC	15:45	27.21
21927	31OC	2:39	221.79	21927	15NO	9:59	57.33
21928	29NO	14:13	251.77	21928	15DE	3:10	87.62
21929	29DE	3:42	281.92				
773							
				21929	13JA	18:08	117.77
21930	27JA	18:48	311.95	21930	12FE	6:22	147.50
21931	26FE	10:53	341.63	21931	13MR	16:06	176.67
21932	28MR	3:15	10.83	21932	12AP	0:08	205.27
21933	26AP	19:16	39.54	21933	11MY	7:25	233.44
21934	26MY	10:23	67.88	21934	9JN	14:56	261.40
21935	25JN	0:04	96.06	21935	8JL	23:35	289.40
21936	24JL	12:06	124.28	21936	7AU	10:09	317.70
21937	22AU	22:45	152.77	21937	5SE	23:16	346.50
21938	21SE	8:38	181.70	21938	5OC	15:17	15.91
21939	20OC	18:31	211.11	21939	4NO	9:53	45.91
21940	19NO	4:57	240.94	21940	4DE	5:41	76.26
21941	18DE	16:06	270.98				
774							
				21941	3JA	0:36	106.63
21942	17JA	3:56	300.98	21942	1FE	16:54	136.66
21943	15FE	16:26	330.67	21943	3MR	5:59	166.15
21944	17MR	5:45	359.93	21944	1AP	16:17	195.02
21945	15AP	20:05	28.74	21945	1MY	0:36	223.36
21946	15MY	11:16	57.19	21946	30MY	7:49	251.37
21947	14JN	2:43	85.45	21947	28JN	14:47	279.27
21948	13JL	17:41	113.72	21948	27JL	22:24	307.35
21949	12AU	7:39	142.21	21949	26AU	7:39	335.83
21950	10SE	20:36	171.07	21950	24SE	19:31	4.88
21951	10OC	8:48	200.41	21951	24OC	10:41	34.57
21952	8NO	20:30	230.17	21952	23NO	4:53	64.76
21953	8DE	7:44	260.18	21953	23DE	0:42	95.18
775							
21954	6JA	18:27	290.19	21954	21JA	20:01	125.46
21955	5FE	4:47	319.93	21955	20FE	13:14	155.28
21956	6MR	15:10	349.25	21956	22MR	3:37	184.51
21957	5AP	2:11	18.10	21957	20AP	15:13	213.13
21958	4MY	14:21	46.56	21958	20MY	0:28	241.31
21959	3JN	3:50	74.79	21959	18JN	8:05	269.25
21960	2JL	18:27	103.01	21960	17JL	15:01	297.22
21961	1AU	9:51	131.43	21961	15AU	22:20	325.47
21962	31AU	1:37	160.23	21962	14SE	7:07	354.23
21963	29SE	17:16	189.52	21963	13OC	18:18	23.57
21964	29OC	8:18	219.27	21964	12NO	8:20	53.47
21965	27NO	21:59	249.32	21965	12DE	0:58	83.72
21966	27DE	10:02	279.41				

776

NEW MOONS				FULL MOONS			
NUMBER	DATE	TIME	LONG.	NUMBER	DATE	TIME	LONG.
				21966	10JA	19;15	114.04
21967	25JA	20;34	309.28	21967	9FE	13;54	144.08
21968	24FE	6;04	338.73	21968	10MR	7;38	173.63
21969	24MR	15;07	7.67	21969	8AP	23;20	202.60
21970	23AP	0;20	36.16	21970	8MY	12;24	231.04
21971	22MY	10;17	64.33	21971	6JN	22;53	259.12
21972	20JN	21;32	92.41	21972	6JL	7;27	287.11
21973	20JL	10;38	120.65	21973	4AU	15;11	315.27
21974	19AU	1;54	149.27	21974	2SE	23;11	343.82
21975	17SE	19;04	178.43	21975	2OC	8;20	12.90
21976	17OC	13;07	208.13	21976	31OC	19;06	42.52
21977	16NO	6;30	238.23	21977	30NO	7;42	72.54
21978	15DE	22;00	268.45	21978	29DE	22;09	102.73

777

NEW MOONS				FULL MOONS			
21979	14JA	11;09	298.52	21979	28JA	14;15	132.80
21980	12FE	22;11	328.19	21980	27FE	7;31	162.52
21981	14MR	7;35	357.33	21981	29MR	0;53	191.74
21982	12AP	15;52	25.93	21982	27AP	17;06	220.44
21983	11MY	23;36	54.12	21983	27MY	7;15	248.75
21984	10JN	7;37	82.10	21984	25JN	19;14	276.85
21985	9JL	16;57	110.13	21985	25JL	5;36	305.01
21986	8AU	4;38	138.48	21986	23AU	15;10	333.47
21987	6SE	19;22	167.35	21987	22SE	0;39	2.39
21988	6OC	12;54	196.85	21988	21OC	10;29	31.81
21989	5NO	8;01	226.87	21989	19NO	20;56	61.64
21990	5DE	2;55	257.19	21990	19DE	8;15	91.70

778

NEW MOONS				FULL MOONS			
21991	3JA	20;08	287.48	21991	17JA	20;46	121.72
21992	2FE	10;53	317.45	21992	16FE	10;42	151.46
21993	3MR	23;01	346.88	21993	18MR	1;52	180.78
21994	2AP	8;43	15.71	21994	16AP	17;37	209.64
21995	1MY	16;32	44.03	21995	16MY	9;09	238.09
21996	30MY	23;21	72.01	21996	14JN	23;55	266.32
21997	29JN	6;16	99.92	21997	14JL	13;43	294.55
21998	28JL	14;32	128.03	21998	13AU	2;37	323.00
21999	27AU	1;13	156.57	21999	11SE	14;40	351.84
22000	25SE	14;55	185.72	22000	11OC	1;57	21.15
22001	25OC	7;34	215.47	22001	9NO	12;41	50.88
22002	24NO	2;20	245.70	22002	8DE	23;10	80.86
22003	23DE	21;53	276.11				

779

NEW MOONS				FULL MOONS			
				22003	7JA	9;49	110.86
22004	22JA	16;35	306.35	22004	5FE	20;57	140.64
22005	21FE	9;00	336.13	22005	7MR	8;44	170.00
22006	22MR	22;18	5.30	22006	5AP	21;13	198.90
22007	21AP	8;32	33.86	22007	5MY	10;27	227.39
22008	20MY	16;29	61.97	22008	4JN	0;34	255.64
22009	18JN	23;18	89.88	22009	3JL	15;32	283.88
22010	18JL	6;12	117.85	22010	2AU	7;03	312.31
22011	16AU	14;12	146.15	22011	31AU	22;26	341.11
22012	15SE	0;05	174.95	22012	30SE	12;58	10.37
22013	14OC	12;24	204.36	22013	30OC	2;15	40.06
22014	13NO	3;27	234.31	22014	28NO	14;21	70.04
22015	12DE	21;06	264.61	22015	28DE	1;33	100.09

780

NEW MOONS				FULL MOONS			
22016	11JA	16;25	294.96	22016	26JA	12;06	129.96
22017	10FE	11;41	325.02	22017	24FE	22;07	159.42
22018	11MR	4;58	354.54	22018	25MR	7;48	188.38
22019	9AP	19;12	23.43	22019	23AP	17;40	216.88
22020	9MY	6;26	51.79	22020	23MY	4;29	245.08
22021	7JN	15;28	79.81	22021	21JN	17;01	273.21
22022	6JL	23;20	107.78	22022	21JL	7;36	301.52
22023	5AU	6;57	135.93	22023	19AU	23;54	330.19
22024	3SE	15;06	164.50	22024	18SE	16;55	359.36
22025	3OC	0;27	193.60	22025	18OC	9;36	29.02
22026	1NO	11;38	223.25	22026	17NO	1;12	59.05
22027	1DE	1;14	253.31	22027	16DE	15;19	89.22
22028	30DE	17;21	283.57				

781

NEW MOONS				FULL MOONS			
				22028	15JA	3;44	119.25
22029	29JA	11;17	313.71	22029	13FE	14;21	148.89
22030	28FE	5;31	343.46	22030	14MR	23;24	178.01
22031	29MR	22;34	12.65	22031	13AP	7;28	206.59
22032	28AP	13;34	41.29	22032	12MY	15;30	234.79
22033	28MY	2;24	69.54	22033	11JN	0;34	262.80
22034	26JN	13;21	97.60	22034	10JL	11;34	290.90
22035	25JL	22;55	125.74	22035	9AU	1;00	319.33
22036	24AU	7;42	154.17	22036	7SE	16;48	348.26
22037	22SE	16;25	183.07	22037	7OC	10;25	17.77
22038	22OC	1;50	212.48	22038	6NO	4;54	47.78
22039	20NO	12;35	242.34	22039	5DE	22;57	78.07
22040	20DE	1;04	272.44				

782

NEW MOONS				FULL MOONS			
				22040	4JA	15;13	108.32
22041	18JA	15;14	302.52	22041	3FE	4;48	138.23
22042	17FE	6;39	332.32	22042	4MR	15;37	167.59
22043	18MR	22;39	1.66	22043	3AP	0;14	196.37
22044	17AP	14;41	30.52	22044	2MY	7;37	224.66
22045	17MY	6;13	58.96	22045	31MY	14;48	252.66
22046	15JN	20;42	87.18	22046	29JN	22;43	280.61
22047	15JL	9;45	115.38	22047	29JL	8;14	308.77
22048	13AU	21;17	143.79	22048	27AU	20;03	337.37
22049	12SE	7;43	172.58	22049	26SE	10;42	6.57
22050	11OC	17;46	201.84	22050	26OC	4;14	36.37
22051	10NO	4;02	231.56	22051	24NO	23;47	66.64
22052	9DE	14;52	261.56	22052	24DE	19;28	97.05

783

NEW MOONS				FULL MOONS			
22053	8JA	2;17	291.59	22053	23JA	13;13	127.25
22054	6FE	14;14	321.39	22054	22FE	3;50	156.95
22055	8MR	2;50	350.78	22055	23MR	15;21	186.02
22056	6AP	16;20	19.70	22056	22AP	0;28	214.53
22057	6MY	6;53	48.24	22057	21MY	8;04	242.62
22058	4JN	22;11	76.53	22058	19JN	15;02	270.54
22059	4JL	13;32	104.78	22059	18JL	22;14	298.52
22060	3AU	4;13	133.19	22060	17AU	6;36	326.84
22061	1SE	17;57	161.94	22061	15SE	17;08	355.69
22062	1OC	6;47	191.14	22062	15OC	6;42	25.15
22063	30OC	18;59	220.79	22063	13NO	23;32	55.19
22064	29NO	6;36	250.76	22064	13DE	18;46	85.56
22065	28DE	17;35	280.80				

784

NEW MOONS				FULL MOONS			
				22065	12JA	14;32	115.92
22066	27JA	3;58	310.65	22066	11FE	8;51	145.93
22067	25FE	14;03	340.11	22067	12MR	0;34	175.37
22068	26MR	0;24	9.09	22068	10AP	13;27	204.20
22069	24AP	11;40	37.64	22069	9MY	23;46	232.51
22070	24MY	0;16	65.90	22070	8JN	8;07	260.51
22071	22JN	14;15	94.09	22071	7JL	15;21	288.45
22072	22JL	5;21	122.42	22072	5AU	22;26	316.59
22073	20AU	21;10	151.09	22073	4SE	6;29	345.16
22074	19SE	13;14	180.23	22074	3OC	16;30	14.30
22075	19OC	4;58	209.86	22075	2NO	5;10	44.01
22076	17NO	19;36	239.86	22076	1DE	20;34	74.17
22077	17DE	8;36	269.98	22077	31DE	14;03	104.48

785

NEW MOONS				FULL MOONS			
22078	15JA	19;50	299.96	22078	30JA	8;30	134.63
22079	14FE	5;40	329.56	22079	1MR	2;39	164.36
22080	15MR	14;40	358.66	22080	30MR	19;20	193.52
22081	13AP	23;29	27.27	22081	29AP	9;42	222.13
22082	13MY	8;47	55.51	22082	28MY	21;27	250.32
22083	11JN	19;09	83.57	22083	27JN	6;58	278.33
22084	11JL	7;12	111.72	22084	26JL	15;08	306.42
22085	9AU	21;24	140.18	22085	24AU	23;03	334.83
22086	8SE	13;51	169.16	22086	23SE	7;43	3.73
22087	8OC	7;52	198.70	22087	22OC	17;47	33.18
22088	7NO	2;02	228.70	22088	21NO	5;31	63.09
22089	6DE	18;46	258.94	22089	20DE	18;58	93.23

786

NEW MOONS				FULL MOONS			
22090	5JA	9;09	289.11	22090	19JA	10;06	123.35
22091	3FE	21;09	318.94	22091	18FE	2;37	153.18
22092	5MR	7;11	348.26	22092	19MR	19;49	182.56
22093	3AP	15;48	17.03	22093	18AP	12;34	211.43
22094	2MY	23;33	45.33	22094	18MY	3;46	239.85
22095	1JN	7;11	73.34	22095	16JN	16;52	268.01
22096	30JN	15;38	101.31	22096	16JL	4;08	296.15
22097	30JL	2;01	129.51	22097	14AU	14;14	324.50
22098	28AU	15;14	158.18	22098	12SE	23;56	353.27
22099	27SE	7;35	187.46	22099	12OC	9;46	22.54
22100	27OC	2;17	217.33	22100	10NO	20;01	52.26
22101	25NO	21;39	247.60	22101	10DE	6;53	82.27
22102	25DE	15;56	277.95				

787

NEW MOONS				FULL MOONS			
				22102	8JA	18;39	112.31
22103	24JA	7;58	308.07	22103	7FE	7;38	142.15
22104	22FE	21;21	337.70	22104	8MR	21;56	171.59
22105	24MR	8;07	6.74	22105	7AP	13;11	200.58
22106	22AP	16;40	35.21	22106	7MY	4;42	229.14
22107	21MY	23;46	63.28	22107	5JN	19;50	257.42
22108	20JN	6;27	91.18	22108	5JL	10;13	285.64
22109	19JL	13;54	119.18	22109	3AU	23;47	314.01
22110	17AU	23;18	147.55	22110	2SE	12;32	342.73
22111	16SE	11;32	176.47	22111	2OC	0;29	11.91
22112	16OC	2;51	206.03	22112	31OC	11;41	41.53
22113	14NO	20;46	236.11	22113	29NO	22;21	71.46
22114	14DE	16;09	266.50	22114	29DE	8;52	101.48

NEW MOONS (left)

NUMBER	DATE	TIME	LONG.
788			
22115	13JA	11;27	296.84
22116	12FE	5;06	326.81
22117	12MR	19;55	356.20
22118	11AP	7;33	24.96
22119	10MY	16;28	53.20
22120	8JN	23;41	81.16
22121	8JL	6;27	109.08
22122	6AU	13;52	137.24
22123	4SE	22;50	165.86
22124	4OC	10;00	195.06
22125	2NO	23;43	224.83
22126	2DE	16;07	255.02
789			
22127	1JA	10;40	285.38
22128	31JA	6;04	315.56
22129	2MR	0;24	345.28
22130	31MR	16;06	14.39
22131	30AP	4;43	42.92
22132	29MY	14;46	71.04
22133	27JN	23;10	99.01
22134	27JL	6;56	127.09
22135	25AU	14;54	155.50
22136	23SE	23;44	184.42
22137	23OC	10;02	213.89
22138	21NO	22;24	243.83
22139	21DE	13;12	274.04
790			
22140	20JA	6;12	304.23
22141	19FE	0;14	334.11
22142	20MR	17;49	3.49
22143	19AP	9;45	32.31
22144	18MY	23;38	60.67
22145	17JN	11;33	88.79
22146	16JL	21;54	116.89
22147	15AU	7;13	145.22
22148	13SE	16;06	173.97
22149	13OC	1;16	203.22
22150	11NO	11;23	232.94
22151	10DE	22;58	262.98
791			
22152	9JA	12;12	293.08
22153	8FE	2;50	322.98
22154	9MR	18;19	352.46
22155	8AP	10;08	21.45
22156	8MY	1;49	50.01
22157	6JN	16;52	78.29
22158	6JL	6;47	106.49
22159	4AU	19;17	134.83
22160	3SE	6;28	163.49
22161	2OC	16;53	192.62
22162	1NO	3;11	222.21
22163	30NO	13;49	252.14
22164	30DE	0;54	282.19
792			
22165	28JA	12;26	312.08
22166	27FE	0;25	341.59
22167	27MR	13;08	10.64
22168	26AP	2;53	39.27
22169	25MY	17;41	67.61
22170	24JN	9;05	95.85
22171	24JL	0;19	124.20
22172	22AU	14;48	152.84
22173	21SE	4;22	181.91
22174	20OC	17;08	211.44
22175	19NO	5;14	241.34
22176	18DE	16;37	271.40
793			
22177	17JA	3;13	301.34
22178	15FE	13;13	330.93
22179	16MR	23;06	0.05
22180	15AP	9;33	28.70
22181	14MY	21;13	57.02
22182	13JN	10;22	85.19
22183	13JL	0;56	113.44
22184	11AU	16;34	141.99
22185	10SE	8;50	170.98
22186	10OC	1;07	200.48
22187	8NO	16;39	230.40
22188	8DE	6;40	260.53

FULL MOONS (left)

NUMBER	DATE	TIME	LONG.
788			
22115	27JA	19;36	131.34
22116	26FE	6;48	160.83
22117	26MR	18;38	189.86
22118	25AP	7;12	218.45
22119	24MY	20;40	246.74
22120	23JN	11;09	274.95
22121	23JL	2;32	303.30
22122	21AU	18;19	331.98
22123	20SE	9;40	1.11
22124	19OC	23;53	30.69
22125	18NO	12;45	60.61
22126	18DE	0;28	90.68
789			
22127	16JA	11;17	120.64
22128	14FE	21;24	150.24
22129	16MR	6;59	179.36
22130	14AP	16;25	207.98
22131	14MY	2;26	236.24
22132	12JN	13;51	264.35
22133	12JL	3;21	292.55
22134	10AU	18;58	321.08
22135	9SE	11;54	350.09
22136	9OC	5;12	19.62
22137	7NO	21;41	49.56
22138	7DE	12;45	79.74
790			
22139	6JA	2;05	109.85
22140	4FE	13;32	139.65
22141	5MR	23;10	168.95
22142	4AP	7;27	197.70
22143	3MY	15;11	225.99
22144	1JN	23;26	254.02
22145	1JL	9;17	282.04
22146	30JL	21;10	310.32
22147	29AU	12;11	339.07
22148	28SE	5;09	8.39
22149	27OC	23;33	38.26
22150	26NO	18;10	68.50
22151	26DE	11;34	98.81
791			
22152	25JA	2;34	128.88
22153	23FE	14;39	158.45
22154	25MR	0;09	187.42
22155	23AP	7;53	215.85
22156	22MY	14;56	243.91
22157	20JN	22;17	271.84
22158	20JL	6;54	299.89
22159	18AU	17;31	328.32
22160	17SE	6;47	357.30
22161	16OC	23;01	26.90
22162	15NO	17;51	57.04
22163	15DE	13;48	87.45
792			
22164	14JA	8;45	117.76
22165	13FE	0;57	147.66
22166	13MR	13;53	176.96
22167	12AP	0;00	205.65
22168	11MY	8;10	233.86
22169	9JN	15;19	261.81
22170	8JL	22;39	289.74
22171	7AU	6;03	317.92
22172	5SE	15;28	346.57
22173	5OC	3;15	15.82
22174	3NO	18;48	45.66
22175	3DE	13;01	75.94
793			
22176	2JA	8;44	106.34
22177	1FE	3;52	136.51
22178	2MR	20;51	166.16
22179	1AP	11;03	195.19
22180	30AP	22;33	223.67
22181	30MY	7;48	251.75
22182	28JN	15;33	279.69
22183	27JL	22;41	307.75
22184	26AU	6;15	336.16
22185	24SE	15;19	5.10
22186	24OC	2;41	34.62
22187	22NO	16;46	64.64
22188	22DE	9;15	94.92

NEW MOONS (right)

NUMBER	DATE	TIME	LONG.
794			
22189	6JA	18;48	290.59
22190	5FE	5;11	320.34
22191	6MR	14;21	349.61
22192	4AP	22;58	18.36
22193	4MY	7;44	46.68
22194	2JN	17;18	74.76
22195	2JL	4;21	102.83
22196	31JL	17;26	131.16
22197	30AU	8;53	159.94
22198	29SE	2;27	189.30
22199	28OC	21;00	219.19
22200	27NO	14;49	249.40
22201	27DE	6;32	279.65
795			
22202	25JA	19;41	309.63
22203	24FE	6;33	339.14
22204	25MR	15;39	8.08
22205	23AP	23;36	36.51
22206	23MY	7;03	64.59
22207	21JN	14;52	92.53
22208	21JL	0;07	120.61
22209	19AU	11;51	149.09
22210	18SE	2;45	178.15
22211	17OC	20;35	207.83
22212	16NO	16;00	238.00
22213	16DE	11;07	268.38
796			
22214	15JA	4;24	298.62
22215	13FE	19;05	328.45
22216	14MR	7;01	357.69
22217	12AP	16;30	26.35
22218	12MY	0;08	54.53
22219	10JN	6;50	82.45
22220	9JL	13;44	110.39
22221	7AU	22;05	138.60
22222	6SE	8;54	167.31
22223	5OC	22;46	196.65
22224	4NO	15;33	226.57
22225	4DE	10;25	256.88
797			
22226	3JA	5;56	287.27
22227	2FE	0;31	317.40
22228	3MR	16;45	347.01
22229	2AP	5;53	15.99
22230	1MY	16;01	44.39
22231	30MY	23;56	72.42
22232	29JN	6;49	100.33
22233	28JL	13;51	128.39
22234	26AU	22;04	156.83
22235	25SE	8;11	185.83
22236	24OC	20;40	215.41
22237	23NO	11;45	245.47
22238	23DE	5;16	275.80
798			
22239	22JA	0;19	306.06
22240	20FE	19;15	335.95
22241	22MR	12;15	5.28
22242	21AP	2;20	34.00
22243	20MY	13;34	62.25
22244	18JN	22;44	90.24
22245	18JL	6;50	118.27
22246	16AU	14;46	146.56
22247	14SE	23;16	175.31
22248	14OC	8;54	204.60
22249	12NO	20;14	234.39
22250	12DE	9;46	264.52
799			
22251	11JA	1;35	294.72
22252	9FE	19;03	324.72
22253	11MR	12;47	354.26
22254	10AP	5;27	23.26
22255	9MY	20;18	51.77
22256	8JN	9;12	79.95
22257	7JL	20;26	108.05
22258	6AU	6;25	136.30
22259	4SE	15;39	164.91
22260	4OC	0;48	194.01
22261	2NO	10;31	223.59
22262	1DE	21;22	253.54
22263	31DE	9;43	283.64

FULL MOONS (right)

NUMBER	DATE	TIME	LONG.
794			
22189	21JA	3;12	125.14
22190	19FE	21;27	155.02
22191	21MR	14;49	184.37
22192	20AP	6;18	213.16
22193	19MY	19;21	241.49
22194	18JN	6;00	269.55
22195	17JL	14;51	297.60
22196	15AU	22;57	325.89
22197	14SE	7;21	354.63
22198	13OC	16;49	23.90
22199	12NO	3;47	53.67
22200	11DE	16;21	83.75
795			
22201	10JA	6;31	113.89
22202	8FE	22;09	143.81
22203	10MR	14;50	173.33
22204	9AP	7;44	202.36
22205	8MY	23;42	230.92
22206	7JN	13;54	259.15
22207	7JL	2;09	287.29
22208	5AU	12;57	315.57
22209	3SE	23;02	344.20
22210	3OC	8;59	13.32
22211	1NO	19;11	42.91
22212	1DE	5;47	72.85
22213	30DE	17;00	102.90
796			
22214	29JA	5;09	132.81
22215	27FE	18;32	162.37
22216	28MR	9;04	191.48
22217	27AP	0;17	220.16
22218	26MY	15;33	248.51
22219	25JN	6;21	276.73
22220	24JL	20;30	305.04
22221	23AU	9;56	333.65
22222	21SE	22;35	2.69
22223	21OC	10;24	32.20
22224	19NO	21;28	62.06
22225	19DE	8;03	92.08
797			
22226	17JA	18;33	122.01
22227	16FE	5;17	151.63
22228	17MR	16;31	180.79
22229	16AP	4;24	209.49
22230	15MY	17;09	237.84
22231	14JN	6;59	266.04
22232	13JL	21;59	294.31
22233	12AU	13;50	322.88
22234	11SE	5;47	351.87
22235	10OC	20;56	21.33
22236	9NO	10;44	51.19
22237	8DE	23;07	81.25
798			
22238	7JA	10;22	111.28
22239	5FE	20;43	141.02
22240	7MR	6;20	170.29
22241	5AP	15;32	199.06
22242	5MY	0;55	227.40
22243	3JN	11;21	255.50
22244	2JL	23;39	283.63
22245	1AU	14;16	312.03
22246	31AU	6;51	340.87
22247	30SE	0;22	10.24
22248	29OC	17;36	40.09
22249	28NO	9;39	70.23
22250	27DE	23;58	100.42
799			
22251	26JA	12;21	130.36
22252	24FE	22;45	159.84
22253	26MR	7;27	188.76
22254	24AP	15;09	217.17
22255	23MY	22;51	245.25
22256	22JN	7;40	273.23
22257	21JL	18;33	301.38
22258	20AU	8;03	329.94
22259	19SE	0;06	359.06
22260	18OC	18;05	28.76
22261	17NO	12;57	58.92
22262	17DE	7;16	89.27

800 / 806

NUMBER	DATE	TIME	LONG.	NUMBER	DATE	TIME	LONG.	NUMBER	DATE	TIME	LONG.	NUMBER	DATE	TIME	LONG.
				22263	15JA	23;37	119.46					22337	8JA	17;41	112.66
22264	29JA	23;29	313.61	22264	14FE	13;07	149.23	22338	23JA	19;24	307.93	22338	7FE	4;06	142.39
22265	28FE	14;19	343.22	22265	14MR	23;43	178.41	22339	22FE	12;51	337.74	22339	8MR	14;49	171.69
22266	29MR	5;44	12.35	22266	13AP	8;05	207.01	22340	24MR	3;30	6.94	22340	7AP	2;04	200.51
22267	27AP	21;19	41.03	22267	12MY	15;13	235.16	22341	22AP	15;00	35.53	22341	6MY	14;06	228.94
22268	27MY	12;39	69.38	22268	10JN	22;13	263.10	22342	21MY	23;52	63.67	22342	5JN	3;11	257.15
22269	26JN	3;16	97.59	22269	10JL	6;04	291.07	22343	20JN	7;09	91.60	22343	4JL	17;34	285.36
22270	25JL	16;42	125.88	22270	8AU	15;37	319.34	22344	19JL	14;03	119.58	22344	3AU	9;10	313.81
22271	24AU	4;46	154.45	22271	7SE	3;35	348.11	22345	17AU	21;41	147.87	22345	2SE	1;26	342.67
22272	22SE	15;47	183.43	22272	6OC	18;25	17.50	22346	16SE	6;54	176.67	22346	1OC	17;23	12.00
22273	22OC	2;18	212.89	22273	5NO	12;09	47.47	22347	15OC	18;16	206.06	22347	31OC	8;11	41.77
22274	20NO	12;51	242.74	22274	5DE	7;51	77.82	22348	14NO	8;04	235.97	22348	29NO	21;26	71.81
22275	19DE	23;44	272.78					22349	14DE	0;23	266.22	22349	29DE	9;16	101.89

801 / 807

NUMBER	DATE	TIME	LONG.	NUMBER	DATE	TIME	LONG.	NUMBER	DATE	TIME	LONG.	NUMBER	DATE	TIME	LONG.
				22275	4JA	3;35	108.22					22350	27JA	19;58	131.75
22276	18JA	10;56	302.73	22276	2FE	21;16	138.30	22350	12JA	18;41	296.52	22351	26FE	5;46	161.18
22277	16FE	22;27	332.37	22277	4MR	11;45	167.82	22351	11FE	13;44	326.56	22352	27MR	14;55	190.10
22278	18MR	10;27	1.55	22278	2AP	23;05	196.71	22352	13MR	7;43	356.08	22353	25AP	23;52	218.54
22279	16AP	23;22	30.29	22279	2MY	8;02	225.07	22353	11AP	23;13	25.01	22354	25MY	9;27	246.68
22280	16MY	13;26	58.68	22280	31MY	15;33	253.08	22354	11MY	11;47	53.41	22355	23JN	20;36	274.76
22281	15JN	4;32	86.93	22281	29JN	22;32	280.99	22355	9JN	21;55	81.47	22356	23JL	10;01	303.02
22282	14JL	20;01	115.22	22282	29JL	5;51	309.06	22356	9JL	6;33	109.47	22357	22AU	1;49	331.70
22283	13AU	11;09	143.77	22283	27AU	14;24	337.53	22357	7AU	14;38	137.66	22358	20SE	19;14	0.90
22284	12SE	1;28	172.71	22284	26SE	1;07	6.56	22358	5SE	22;57	166.25	22359	20OC	13;00	30.62
22285	11OC	14;55	202.12	22285	25OC	14;50	36.20	22359	5OC	8;06	195.37	22360	19NO	5;58	60.72
22286	10NO	3;34	231.93	22286	24NO	7;43	66.35	22360	3NO	18;37	225.00	22361	18DE	21;19	90.94
22287	9DE	15;26	261.98	22287	24DE	2;52	96.74	22361	3DE	7;01	255.03				

802 / 808

NUMBER	DATE	TIME	LONG.	NUMBER	DATE	TIME	LONG.	NUMBER	DATE	TIME	LONG.	NUMBER	DATE	TIME	LONG.
22288	8JA	2;25	291.99	22288	22JA	22;25	127.02	22362	1JA	21;36	285.22	22362	17JA	10;43	121.00
22289	6FE	12;33	321.71	22289	21FE	16;29	156.86	22363	31JA	14;09	315.29	22363	15FE	22;00	150.66
22290	7MR	22;11	350.97	22290	23MR	8;00	186.11	22364	1MR	7;40	344.98	22364	16MR	7;20	179.76
22291	6AP	7;59	19.75	22291	21AP	20;45	214.77	22365	31MR	0;47	14.16	22365	14AP	15;16	208.32
22292	5MY	18;44	48.14	22292	21MY	7;03	242.97	22366	29AP	16;28	42.82	22366	13MY	22;40	236.48
22293	4JN	6;56	76.31	22293	19JN	15;30	270.94	22367	29MY	6;20	71.10	22367	12JN	6;39	264.45
22294	3JL	20;45	104.50	22294	18JL	22;55	298.94	22368	27JN	18;28	99.21	22368	11JL	16;21	292.50
22295	2AU	11;58	132.93	22295	17AU	6;17	327.22	22369	27JL	5;12	127.40	22369	10AU	4;31	320.89
22296	1SE	4;11	161.77	22296	15SE	14;37	355.97	22370	25AU	14;59	155.90	22370	8SE	19;25	349.80
22297	30SE	20;49	191.12	22297	15OC	0;52	25.30	22371	24SE	0;20	184.84	22371	8OC	12;42	19.32
22298	30OC	13;07	220.94	22298	13NO	13;39	55.16	22372	23OC	9;52	214.28	22372	7NO	7;27	49.35
22299	29NO	4;10	251.05	22299	13DE	4;57	85.36	22373	21NO	20;11	244.13	22373	7DE	2;22	79.69
22300	28DE	17;21	281.18					22374	21DE	7;43	274.19				

803 / 809

NUMBER	DATE	TIME	LONG.	NUMBER	DATE	TIME	LONG.	NUMBER	DATE	TIME	LONG.	NUMBER	DATE	TIME	LONG.
				22300	11JA	22;09	115.62					22374	5JA	19;55	109.98
22301	27JA	4;30	311.07	22301	10FE	16;12	145.63	22375	19JA	20;39	304.21	22375	4FE	10;53	139.93
22302	25FE	14;04	340.50	22302	12MR	9;55	175.16	22376	18FE	10;45	333.94	22376	5MR	22;48	169.32
22303	26MR	22;40	9.41	22303	11AP	2;18	204.14	22377	20MR	1;37	3.21	22377	4AP	8;03	198.11
22304	25AP	7;04	37.84	22304	10MY	16;34	232.61	22378	18AP	16;54	32.02	22378	3MY	15;34	226.39
22305	24MY	15;58	65.96	22305	9JN	4;26	260.75	22379	18MY	8;16	60.45	22379	1JN	22;25	254.36
22306	23JN	2;04	93.99	22306	8JL	14;13	288.79	22380	16JN	23;18	88.69	22380	1JL	5;42	282.29
22307	22JL	14;02	122.19	22307	6AU	22;45	316.99	22381	16JL	13;32	116.94	22381	30JL	14;19	310.43
22308	21AU	4;21	150.80	22308	5SE	7;05	345.57	22382	15AU	2;32	145.42	22382	29AU	1;03	339.00
22309	19SE	21;06	179.96	22309	4OC	16;07	14.67	22383	13SE	14;19	174.28	22383	27SE	14;29	8.17
22310	19OC	15;34	209.69	22310	3NO	2;26	44.29	22384	13OC	1;15	203.61	22384	27OC	6;55	37.95
22311	18NO	10;11	239.85	22311	2DE	14;13	74.29	22385	11NO	11;55	233.36	22385	26NO	1;53	68.20
22312	18DE	3;13	270.14					22386	10DE	22;42	263.36	22386	25DE	21;54	98.63

804 / 810

NUMBER	DATE	TIME	LONG.	NUMBER	DATE	TIME	LONG.	NUMBER	DATE	TIME	LONG.	NUMBER	DATE	TIME	LONG.
				22312	1JA	3;30	104.42	22387	9JA	9;41	293.37	22387	24JA	16;47	128.86
22313	16JA	17;41	300.26	22313	30JA	18;12	134.42	22388	7FE	20;50	323.12	22388	23FE	8;51	158.60
22314	15FE	5;34	329.94	22314	29FE	10;09	164.07	22389	9MR	8;17	352.44	22389	24MR	21;37	187.71
22315	15MR	15;20	359.08	22315	30MR	2;48	193.24	22390	7AP	20;24	21.29	22390	23AP	7;35	216.23
22316	13AP	23;38	27.66	22316	28AP	19;12	221.94	22391	7MY	9;36	49.75	22391	22MY	15;39	244.33
22317	13MY	7;06	55.83	22317	28MY	10;19	250.27	22392	6JN	0;05	78.01	22392	20JN	22;48	272.25
22318	11JN	14;31	83.77	22318	26JN	23;37	278.43	22393	5JL	15;29	106.27	22393	20JL	5;54	300.24
22319	10JL	22;52	111.77	22319	26JL	11;16	306.65	22394	4AU	7;02	134.72	22394	18AU	13;49	328.56
22320	9AU	9;15	140.08	22320	24AU	21;52	335.17	22395	2SE	22;04	163.54	22395	16SE	23;26	357.39
22321	7SE	22;36	168.92	22321	23SE	8;05	4.14	22396	2OC	12;15	192.82	22396	16OC	11;39	26.81
22322	7OC	15;11	198.39	22322	22OC	18;21	33.60	22397	1NO	1;34	222.53	22397	15NO	3;01	56.80
22323	6NO	10;10	228.43	22323	21NO	4;50	63.45	22398	30NO	14;00	252.54	22398	14DE	21;11	87.13
22324	6DE	5;46	258.78	22324	20DE	15;42	93.48	22399	30DE	1;28	282.60				

805 / 811

NUMBER	DATE	TIME	LONG.	NUMBER	DATE	TIME	LONG.	NUMBER	DATE	TIME	LONG.	NUMBER	DATE	TIME	LONG.
22325	5JA	0;09	289.12	22325	19JA	3;13	123.45					22399	13JA	16;42	117.48
22326	3FE	16;09	319.12	22326	17FE	15;42	153.12	22400	28JA	11;54	312.44	22400	12FE	11;34	147.50
22327	5MR	5;23	348.58	22327	19MR	5;23	182.36	22401	26FE	21;32	341.86	22401	14MR	4;18	176.96
22328	3AP	15;56	17.43	22328	17AP	20;03	211.15	22402	28MR	6;55	10.78	22402	12AP	18;19	205.81
22329	3MY	0;19	45.75	22329	17MY	11;11	239.58	22403	26AP	16;50	39.25	22403	12MY	5;46	234.15
22330	1JN	7;17	73.73	22330	16JN	2;13	267.82	22404	26MY	4;03	67.45	22404	10JN	15;06	262.18
22331	30JN	13;56	101.63	22331	15JL	16;50	296.09	22405	24JN	16;56	95.60	22405	9JL	23;01	290.16
22332	29JL	21;27	129.72	22332	14AU	6;52	324.59	22406	24JL	7;30	123.91	22406	8AU	6;26	318.33
22333	28AU	6;59	158.24	22333	12SE	20;13	353.51	22407	22AU	23;26	152.61	22407	6SE	14;19	346.91
22334	26SE	19;23	187.35	22334	12OC	8;44	22.89	22408	21SE	16;11	181.80	22408	5OC	23;39	16.04
22335	26OC	10;51	217.07	22335	10NO	20;22	52.67	22409	21OC	9;02	211.49	22409	4NO	11;11	45.73
22336	25NO	4;51	247.28	22336	10DE	7;14	82.68	22410	20NO	1;04	241.56	22410	4DE	1;16	75.83
22337	25DE	0;13	277.68					22411	19DE	15;21	271.74				

812

NEW MOONS NUMBER	DATE	TIME	LONG.	FULL MOONS NUMBER	DATE	TIME	LONG.
				22411	2JA	17;30	106.09
22412	18JA	3;30	301.74	22412	1FE	11;03	136.19
22413	16FE	13;41	331.34	22413	2MR	4;50	165.89
22414	16MR	22;30	0.41	22414	31MR	21;50	195.04
22415	15AP	6;42	28.98	22415	30AP	13;08	223.68
22416	14MY	15;05	57.16	22416	30MY	2;14	251.92
22417	13JN	0;21	85.18	22417	28JN	13;06	279.98
22418	12JL	11;15	113.28	22418	27JL	22;19	308.12
22419	11AU	0;21	141.72	22419	26AU	6;50	336.57
22420	9SE	16;02	170.68	22420	24SE	15;37	5.50
22421	9OC	10;00	200.24	22421	24OC	1;25	34.96
22422	8NO	4;59	230.29	22422	22NO	12;31	64.85
22423	7DE	23;09	260.60	22423	22DE	1;00	94.96

813

NEW MOONS NUMBER	DATE	TIME	LONG.	FULL MOONS NUMBER	DATE	TIME	LONG.
22424	6JA	15;01	290.82	22424	20JA	14;49	125.01
22425	5FE	4;07	320.68	22425	19FE	5;55	154.76
22426	6MR	14;46	350.01	22426	20MR	22;01	184.08
22427	4AP	23;35	18.77	22427	19AP	14;26	212.91
22428	4MY	7;14	47.05	22428	19MY	6;12	241.35
22429	2JN	14;28	75.03	22429	17JN	20;30	269.56
22430	1JL	22;10	102.97	22430	17JL	9;06	297.76
22431	31JL	7;24	131.14	22431	15AU	20;23	326.18
22432	29AU	19;14	159.77	22432	14SE	6;59	355.00
22433	28SE	10;19	189.02	22433	13OC	17;26	24.32
22434	28OC	4;23	218.88	22434	12NO	3;57	54.07
22435	27NO	0;02	249.17	22435	11DE	14;39	84.07
22436	26DE	19;17	279.56				

814

NEW MOONS NUMBER	DATE	TIME	LONG.	FULL MOONS NUMBER	DATE	TIME	LONG.
				22436	10JA	1;42	114.07
22437	25JA	12;34	309.72	22437	8FE	13;27	143.84
22438	24FE	3;07	339.38	22438	10MR	2;14	173.20
22439	25MR	14;53	8.44	22439	8AP	16;09	202.11
22440	24AP	0;11	36.93	22440	8MY	6;53	230.63
22441	23MY	7;41	65.00	22441	6JN	21;55	258.91
22442	21JN	14;20	92.90	22442	6JL	12;50	287.15
22443	20JL	21;17	120.89	22443	5AU	3;21	315.57
22444	19AU	5;45	149.24	22444	3SE	17;21	344.36
22445	17SE	16;44	178.13	22445	3OC	6;37	13.61
22446	17OC	6;46	207.65	22446	1NO	18;56	43.30
22447	15NO	23;40	237.70	22447	1DE	6;18	73.27
22448	15DE	18;31	268.07	22448	30DE	16;55	103.29

815

NEW MOONS NUMBER	DATE	TIME	LONG.	FULL MOONS NUMBER	DATE	TIME	LONG.
22449	14JA	13;55	298.40	22449	29JA	3;11	133.11
22450	13FE	8;18	328.39	22450	27FE	13;30	162.55
22451	15MR	0;20	357.81	22451	29MR	0;11	191.51
22452	13AP	13;19	26.61	22452	27AP	11;31	220.03
22453	12MY	23;23	54.89	22453	26MY	23;50	248.27
22454	11JN	7;20	82.86	22454	25JN	13;26	276.44
22455	10JL	14;21	110.80	22455	25JL	4;31	304.79
22456	8AU	21;36	138.97	22456	23AU	20;04	333.50
22457	7SE	6;05	167.59	22457	22SE	13;16	2.70
22458	6OC	16;26	196.77	22458	22OC	5;01	32.36
22459	5NO	5;03	226.52	22459	20NO	19;17	62.36
22460	4DE	20;07	256.66	22460	20DE	7;54	92.47

816

NEW MOONS NUMBER	DATE	TIME	LONG.	FULL MOONS NUMBER	DATE	TIME	LONG.
22461	3JA	13;25	286.96	22461	18JA	19;07	122.43
22462	2FE	8;06	317.10	22462	17FE	5;12	152.01
22463	3MR	2;39	346.82	22463	17MR	14;26	181.09
22464	1AP	19;22	15.95	22464	15AP	23;11	209.67
22465	1MY	9;20	44.52	22465	15MY	8;08	237.87
22466	30MY	20;38	72.68	22466	13JN	18;13	265.92
22467	29JN	5;59	100.68	22467	13JL	6;23	294.08
22468	28JL	14;23	128.80	22468	11AU	21;04	322.59
22469	26AU	22;42	157.25	22469	10SE	13;58	351.61
22470	25SE	7;33	186.19	22470	10OC	7;58	21.18
22471	24OC	17;28	215.66	22471	9NO	1;42	51.20
22472	23NO	4;54	245.58	22472	8DE	18;06	81.43
22473	22DE	18;18	275.72				

817

NEW MOONS NUMBER	DATE	TIME	LONG.	FULL MOONS NUMBER	DATE	TIME	LONG.
				22473	7JA	8;34	111.59
22474	21JA	9;45	305.84	22474	5FE	20;51	141.41
22475	20FE	2;41	335.66	22475	7MR	7;00	170.71
22476	21MR	19;53	5.00	22476	5AP	15;23	199.44
22477	20AP	12;12	33.82	22477	4MY	22;45	227.70
22478	20MY	2;56	62.21	22478	3JN	6;11	255.69
22479	18JN	15;58	90.36	22479	2JL	14;49	283.67
22480	18JL	3;33	118.52	22480	1AU	1;40	311.91
22481	16AU	13;59	146.92	22481	30AU	15;16	340.62
22482	14SE	23;44	175.72	22482	29SE	7;34	9.93
22483	14OC	9;18	205.01	22483	29OC	1;53	39.82
22484	12NO	19;17	234.74	22484	27NO	21;03	70.09
22485	12DE	6;12	264.76	22485	27DE	15;32	100.45

818

NEW MOONS NUMBER	DATE	TIME	LONG.	FULL MOONS NUMBER	DATE	TIME	LONG.
22486	10JA	18;19	294.81	22486	26JA	7;55	130.56
22487	9FE	7;38	324.63	22487	24FE	21;17	160.16
22488	10MR	21;52	354.04	22488	26MR	7;40	189.15
22489	9AP	12;41	22.97	22489	24AP	15;48	217.58
22490	9MY	3;51	51.50	22490	23MY	22;46	245.63
22491	7JN	19;03	79.78	22491	22JN	5;40	273.54
22492	7JL	9;50	108.01	22492	21JL	13;31	301.57
22493	5AU	23;43	136.42	22493	19AU	23;10	329.97
22494	4SE	12;22	165.17	22494	18SE	11;16	358.92
22495	3OC	23;57	194.36	22495	18OC	2;17	28.49
22496	2NO	10;55	224.00	22496	16NO	20;10	58.60
22497	1DE	21;43	253.95	22497	16DE	15;56	89.01
22498	31DE	8;35	283.98				

819

NEW MOONS NUMBER	DATE	TIME	LONG.	FULL MOONS NUMBER	DATE	TIME	LONG.
				22498	15JA	11;37	119.35
22499	29JA	19;31	313.83	22499	14FE	5;10	149.29
22500	28FE	6;33	343.28	22500	15MR	19;29	178.62
22501	29MR	17;58	12.26	22501	14AP	6;40	207.34
22502	28AP	6;18	40.82	22502	13MY	15;31	235.56
22503	27MY	19;57	69.10	22503	11JN	23;01	263.52
22504	26JN	10;54	97.33	22504	11JL	6;05	291.46
22505	26JL	2;34	125.70	22505	9AU	13;34	319.65
22506	24AU	18;10	154.40	22506	7SE	22;20	348.29
22507	23SE	9;06	183.55	22507	7OC	9;16	17.51
22508	22OC	23;08	213.15	22508	5NO	23;05	47.31
22509	21NO	12;14	243.11	22509	5DE	15;57	77.54
22510	21DE	0;17	273.19				

820

NEW MOONS NUMBER	DATE	TIME	LONG.	FULL MOONS NUMBER	DATE	TIME	LONG.
				22510	4JA	10;55	107.90
22511	19JA	11;11	303.14	22511	3FE	6;12	138.06
22512	17FE	21;00	332.70	22512	3MR	23;58	167.72
22513	18MR	6;12	1.77	22513	2AP	15;16	196.78
22514	16AP	15;30	30.36	22514	2MY	3;55	225.29
22515	16MY	1;46	58.60	22515	31MY	14;15	253.41
22516	14JN	13;37	86.72	22516	29JN	22;52	281.39
22517	14JL	3;20	114.95	22517	29JL	6;33	309.48
22518	12AU	18;43	143.49	22518	27AU	14;14	337.91
22519	11SE	11;20	172.52	22519	25SE	22;53	6.86
22520	11OC	4;31	202.07	22520	25OC	9;22	36.36
22521	9NO	21;21	232.06	22521	23NO	22;12	66.34
22522	9DE	12;46	262.25	22522	23DE	13;21	96.56

821

NEW MOONS NUMBER	DATE	TIME	LONG.	FULL MOONS NUMBER	DATE	TIME	LONG.
22523	8JA	2;02	292.36	22523	22JA	6;11	126.72
22524	6FE	13;04	322.12	22524	20FE	23;45	156.56
22525	7MR	22;20	351.37	22525	22MR	17;01	185.89
22526	6AP	6;34	20.08	22526	21AP	9;08	214.69
22527	5MY	14;34	48.36	22527	20MY	23;21	243.05
22528	3JN	23;09	76.39	22528	19JN	11;23	271.17
22529	3JL	9;04	104.43	22529	18JL	21;30	299.27
22530	1AU	20;59	132.71	22530	17AU	6;27	327.62
22531	31AU	11;26	161.48	22531	15SE	15;13	356.39
22532	30SE	4;31	190.83	22532	15OC	0;38	25.68
22533	29OC	23;24	220.75	22533	13NO	11;10	55.45
22534	28NO	18;23	251.02	22534	12DE	22;58	85.50
22535	28DE	11;37	281.33				

822

NEW MOONS NUMBER	DATE	TIME	LONG.	FULL MOONS NUMBER	DATE	TIME	LONG.
				22535	11JA	11;59	115.58
22536	27JA	2;06	311.36	22536	10FE	2;12	145.43
22537	25FE	13;50	340.88	22537	11MR	17;33	174.88
22538	26MR	23;11	9.82	22538	10AP	9;39	203.85
22539	25AP	7;22	38.24	22539	10MY	1;43	232.40
22540	24MY	14;36	66.29	22540	8JN	16;48	260.67
22541	22JN	21;53	94.21	22541	8JL	6;22	288.86
22542	22JL	6;12	122.27	22542	6AU	18;28	317.21
22543	20AU	16;39	150.71	22543	5SE	5;37	345.90
22544	19SE	6;09	179.73	22544	4OC	16;22	15.07
22545	18OC	22;56	209.39	22545	3NO	3;01	44.71
22546	17NO	18;08	239.56	22546	2DE	13;42	74.66
22547	17DE	13;54	269.97				

823

NEW MOONS NUMBER	DATE	TIME	LONG.	FULL MOONS NUMBER	DATE	TIME	LONG.
				22547	1JA	0;31	104.69
22548	16JA	8;18	300.25	22548	30JA	11;42	134.54
22549	15FE	0;12	330.11	22549	28FE	23;40	164.02
22550	16MR	13;15	359.38	22550	30MR	12;43	193.05
22551	14AP	23;38	28.05	22551	29AP	2;49	221.67
22552	14MY	7;52	56.24	22552	28MY	17;36	249.99
22553	12JN	14;47	84.17	22553	27JN	8;36	278.22
22554	11JL	21;28	112.10	22554	26JL	23;29	306.57
22555	10AU	5;06	140.30	22555	25AU	14;02	335.24
22556	8SE	14;49	168.99	22556	24SE	4;00	4.36
22557	8OC	3;23	198.29	22557	23OC	17;06	33.94
22558	6NO	18;58	228.18	22558	22NO	5;06	63.85
22559	6DE	13;00	258.46	22559	21DE	16;07	93.89

824

NEW MOONS NUMBER	DATE	TIME	LONG.	FULL MOONS NUMBER	DATE	TIME	LONG.
22560	5JA	8;15	288.84	22560	20JA	2;26	123.81
22561	4FE	3;13	318.97	22561	18FE	12;30	153.37
22562	4MR	20;27	348.60	22562	18MR	22;42	182.47
22563	3AP	10;55	17.61	22563	17AP	9;24	211.11
22564	2MY	22;20	46.06	22564	16MY	20;57	239.40
22565	1JN	7;13	74.12	22565	15JN	9;44	267.55
22566	30JN	14;37	102.05	22566	15JL	0;03	295.80
22567	29JL	21;44	130.12	22567	13AU	15;55	324.38
22568	28AU	5;38	158.57	22568	12SE	8;40	353.43
22569	26SE	15;07	187.56	22569	12OC	1;14	22.97
22570	26OC	2;39	217.12	22570	10NO	16;33	52.90
22571	24NO	16;31	247.14	22571	10DE	6;08	83.02
22572	24DE	8;39	277.41				

825

NEW MOONS NUMBER	DATE	TIME	LONG.	FULL MOONS NUMBER	DATE	TIME	LONG.
				22572	8JA	18;02	113.07
22573	23JA	2;37	307.61	22573	7FE	4;34	142.80
22574	21FE	21;15	337.48	22574	8MR	14;01	172.04
22575	23MR	14;53	6.81	22575	6AP	22;43	200.77
22576	22AP	6;11	35.57	22576	6MY	7;16	229.06
22577	21MY	18;45	63.85	22577	4JN	16;29	257.11
22578	20JN	5;03	91.90	22578	4JL	3;25	285.19
22579	19JL	13;58	119.96	22579	2AU	16;49	313.54
22580	17AU	22;25	148.29	22580	1SE	8;49	342.38
22581	16SE	7;08	177.07	22581	1OC	2;39	11.78
22582	15OC	16;37	206.38	22582	30OC	20;55	41.69
22583	14NO	3;18	236.16	22583	29NO	14;17	71.90
22584	13DE	15;39	266.24	22584	29DE	5;52	102.14

826

NEW MOONS NUMBER	DATE	TIME	LONG.	FULL MOONS NUMBER	DATE	TIME	LONG.
22585	12JA	5;57	296.37	22585	27JA	19;14	132.10
22586	10FE	22;01	326.29	22586	26FE	6;20	161.59
22587	12MR	14;57	355.79	22587	27MR	15;23	190.50
22588	11AP	7;37	24.77	22588	25AP	22;59	218.89
22589	10MY	23;04	53.29	22589	25MY	6;06	246.94
22590	9JN	12;58	81.50	22590	23JN	13;54	274.88
22591	9JL	1;23	109.65	22591	22JL	23;31	302.99
22592	7AU	12;34	137.97	22592	21AU	11;44	331.52
22593	5SE	22;52	166.64	22593	20SE	2;50	0.62
22594	5OC	8;42	195.78	22594	19OC	20;24	30.32
22595	3NO	18;34	225.39	22595	18NO	15;27	60.49
22596	3DE	5;02	255.34	22596	18DE	10;34	90.88

827

NEW MOONS NUMBER	DATE	TIME	LONG.	FULL MOONS NUMBER	DATE	TIME	LONG.
22597	1JA	16;27	285.39	22597	17JA	4;11	121.12
22598	31JA	5;00	315.29	22598	15FE	19;04	150.92
22599	1MR	18;32	344.83	22599	17MR	6;48	180.13
22600	31MR	8;48	13.90	22600	15AP	15;50	208.73
22601	29AP	23;34	42.53	22601	14MY	23;09	236.88
22602	29MY	14;39	70.86	22602	13JN	5;54	264.80
22603	28JN	5;45	99.09	22603	12JL	13;10	292.76
22604	27JL	20;19	127.44	22604	10AU	21;52	321.01
22605	26AU	9;53	156.08	22605	9SE	8;45	349.76
22606	24SE	22;16	185.15	22606	8OC	22;21	19.11
22607	24OC	9;44	214.66	22607	7NO	14;56	49.05
22608	22NO	20;43	244.54	22608	7DE	9;58	79.38
22609	22DE	7;35	274.58				

828

NEW MOONS NUMBER	DATE	TIME	LONG.	FULL MOONS NUMBER	DATE	TIME	LONG.
				22609	6JA	5;56	109.78
22610	20JA	18;23	304.51	22610	5FE	0;42	139.90
22611	19FE	5;09	334.10	22611	5MR	16;36	169.46
22612	19MR	16;02	3.21	22612	4AP	5;12	198.38
22613	18AP	3;34	31.87	22613	3MY	15;03	226.75
22614	17MY	16;17	60.20	22614	1JN	23;05	254.78
22615	16JN	6;30	88.41	22615	1JL	6;18	282.70
22616	15JL	21;56	116.71	22616	30JL	13;34	310.78
22617	14AU	13;51	145.30	22617	28AU	21;43	339.26
22618	13SE	5;27	174.31	22618	27SE	7;34	8.27
22619	12OC	20;15	203.79	22619	26OC	19;56	37.88
22620	11NO	10;04	233.67	22620	25NO	11;20	67.97
22621	10DE	22;47	263.76	22621	25DE	5;21	98.32

829

NEW MOONS NUMBER	DATE	TIME	LONG.	FULL MOONS NUMBER	DATE	TIME	LONG.
22622	9JA	10;16	293.78	22622	24JA	0;34	128.57
22623	7FE	20;29	323.49	22623	22FE	19;06	158.42
22624	9MR	5;43	352.72	22624	24MR	11;34	187.69
22625	7AP	14;37	21.44	22625	23AP	1;27	216.37
22626	7MY	0;04	49.76	22626	22MY	12;54	244.61
22627	5JN	10;53	77.87	22627	20JN	22;22	272.62
22628	4JL	23;34	106.02	22628	20JL	6;33	300.66
22629	3AU	14;11	134.43	22629	18AU	14;16	328.97
22630	2SE	6;25	163.29	22630	16SE	22;30	357.74
22631	1OC	23;41	192.69	22631	16OC	8;06	27.06
22632	31OC	17;04	222.57	22632	14NO	19;47	56.89
22633	30NO	9;31	252.74	22633	14DE	9;47	87.04
22634	30DE	0;00	282.93				

830

NEW MOONS NUMBER	DATE	TIME	LONG.	FULL MOONS NUMBER	DATE	TIME	LONG.
				22634	13JA	1;43	117.24
22635	28JA	12;06	312.84	22635	11FE	18;48	147.19
22636	26FE	22;03	342.27	22636	13MR	12;05	176.68
22637	28MR	6;31	11.15	22637	12AP	4;41	205.65
22638	26AP	14;21	39.54	22638	11MY	19;51	234.15
22639	25MY	22;24	67.62	22639	10JN	9;02	262.33
22640	24JN	7;27	95.61	22640	9JL	20;12	290.43
22641	23JL	18;15	123.77	22641	8AU	5;50	318.69
22642	22AU	7;26	152.35	22642	6SE	14;49	347.32
22643	20SE	23;22	181.49	22643	6OC	0;01	16.45
22644	20OC	17;41	211.24	22644	4NO	10;07	46.08
22645	19NO	13;02	241.44	22645	3DE	21;19	76.06
22646	19DE	7;28	271.79				

831

NEW MOONS NUMBER	DATE	TIME	LONG.	FULL MOONS NUMBER	DATE	TIME	LONG.
				22646	2JA	9;38	106.15
22647	17JA	23;25	301.96	22647	31JA	23;03	136.08
22648	16FE	12;25	331.68	22648	2MR	13;33	165.65
22649	17MR	22;50	0.82	22649	1AP	5;03	194.75
22650	16AP	7;24	29.39	22650	30AP	21;02	223.42
22651	15MY	14;49	57.54	22651	30MY	12;38	251.77
22652	13JN	21;54	85.47	22652	29JN	3;05	279.97
22653	13JL	5;33	113.44	22653	28JL	16;05	308.26
22654	11AU	14;49	141.73	22654	27AU	3;54	336.84
22655	10SE	2;47	170.53	22655	25SE	15;04	5.87
22656	9OC	18;03	199.96	22656	25OC	1;59	35.38
22657	8NO	12;19	229.98	22657	23NO	12;47	65.25
22658	8DE	8;07	260.35	22658	22DE	23;31	95.29

832

NEW MOONS NUMBER	DATE	TIME	LONG.	FULL MOONS NUMBER	DATE	TIME	LONG.
22659	7JA	3;24	290.72	22659	21JA	10;20	125.21
22660	5FE	20;36	320.76	22660	19FE	21;38	154.81
22661	6MR	11;00	350.24	22661	20MR	9;49	183.97
22662	4AP	22;35	19.12	22662	18AP	23;08	212.69
22663	4MY	7;44	47.45	22663	18MY	13;25	241.07
22664	2JN	15;10	75.45	22664	17JN	4;17	269.30
22665	1JL	21;51	103.35	22665	16JL	19;20	297.59
22666	31JL	4;56	131.43	22666	15AU	10;18	326.16
22667	29AU	13;34	159.93	22667	14SE	0;54	355.14
22668	28SE	0;44	189.01	22668	13OC	14;46	24.60
22669	27OC	14;54	218.71	22669	12NO	3;33	54.44
22670	26NO	7;51	248.87	22670	11DE	15;09	84.48
22671	26DE	2;36	279.25				

833

NEW MOONS NUMBER	DATE	TIME	LONG.	FULL MOONS NUMBER	DATE	TIME	LONG.
				22671	10JA	1;44	114.47
22672	24JA	21;47	309.49	22672	8FE	11;44	144.16
22673	23FE	15;55	339.31	22673	9MR	21;36	173.40
22674	25MR	7;44	8.54	22674	8AP	7;44	202.16
22675	23AP	20;37	37.17	22675	7MY	18;34	230.53
22676	23MY	6;40	65.35	22676	6JN	6;30	258.68
22677	21JN	14;44	93.30	22677	5JL	19;57	286.86
22678	20JL	21;56	121.30	22678	4AU	11;09	315.31
22679	19AU	5;28	149.61	22679	3SE	3;46	344.19
22680	17SE	14;13	178.42	22680	2OC	20;52	13.60
22681	17OC	0;49	207.79	22681	1NO	13;12	43.44
22682	15NO	13;33	237.67	22682	1DE	3;52	73.55
22683	15DE	4;30	267.86	22683	30DE	16;39	103.67

834

NEW MOONS NUMBER	DATE	TIME	LONG.	FULL MOONS NUMBER	DATE	TIME	LONG.
22684	13JA	21;30	298.10	22684	29JA	3;46	133.53
22685	12FE	15;46	328.09	22685	27FE	13;35	162.94
22686	14MR	9;53	357.61	22686	28MR	22;24	191.83
22687	13AP	2;20	26.55	22687	27AP	6;44	220.23
22688	12MY	16;13	54.99	22688	26MY	15;19	248.32
22689	11JN	3;37	83.10	22689	25JN	1;09	276.34
22690	10JL	13;15	111.14	22690	24JL	13;12	304.56
22691	8AU	22;01	139.38	22691	23AU	4;01	333.21
22692	7SE	6;44	168.01	22692	21SE	21;15	2.43
22693	6OC	15;58	197.15	22693	21OC	15;42	32.19
22694	5NO	2;08	226.78	22694	20NO	9;52	62.35
22695	4DE	13;36	256.78	22695	20DE	2;33	92.63

835

NEW MOONS NUMBER	DATE	TIME	LONG.	FULL MOONS NUMBER	DATE	TIME	LONG.
22696	3JA	2;48	286.90	22696	18JA	17;05	122.73
22697	1FE	17;50	316.90	22697	17FE	5;15	152.41
22698	3MR	10;11	346.53	22698	18MR	15;08	181.51
22699	2AP	2;51	15.67	22699	16AP	23;13	210.06
22700	1MY	18;49	44.32	22700	16MY	6;18	238.19
22701	31MY	9;29	72.62	22701	14JN	13;31	266.12
22702	29JN	22;43	100.78	22702	13JL	22;03	294.13
22703	29JL	10;42	129.04	22703	12AU	8;55	322.49
22704	27AU	21;39	157.60	22704	10SE	22;39	351.37
22705	26SE	7;55	186.60	22705	10OC	15;11	20.88
22706	25OC	17;55	216.08	22706	9NO	9;47	50.92
22707	24NO	4;07	245.93	22707	9DE	5;10	81.27
22708	23DE	15;01	275.97				

836

NEW MOONS NUMBER	DATE	TIME	LONG.	FULL MOONS NUMBER	DATE	TIME	LONG.
				22708	7JA	23;45	111.61
22709	22JA	2;51	305.94	22709	6FE	16;05	141.60
22710	20FE	15;40	335.59	22710	7MR	5;18	171.03
22711	21MR	5;17	4.79	22711	5AP	15;29	199.83
22712	19AP	19;32	33.54	22712	4MY	23;26	228.11
22713	19MY	10;18	61.94	22713	3JN	6;17	256.08
22714	18JN	1;25	90.17	22714	2JL	13;09	283.99
22715	17JL	16;27	118.47	22715	31JL	21;04	312.11
22716	16AU	6;49	147.02	22716	30AU	6;51	340.67
22717	14SE	20;05	175.96	22717	28SE	19;08	9.81
22718	14OC	8;14	205.36	22718	28OC	10;18	39.55
22719	12NO	19;37	235.15	22719	27NO	4;16	69.77
22720	12DE	6;36	265.17	22720	27DE	0;00	100.18

842

NEW MOONS NUMBER	DATE	TIME	LONG.	FULL MOONS NUMBER	DATE	TIME	LONG.
				22782	1JA	0;58	105.10
22783	15JA	16;12	299.96	22783	30JA	11;06	134.90
22784	14FE	10;53	329.95	22784	28FE	20;47	164.29
22785	16MR	3;52	359.38	22785	30MR	6;29	193.19
22786	14AP	18;10	28.22	22786	28AP	16;40	221.65
22787	14MY	5;33	56.54	22787	28MY	3;47	249.83
22788	12JN	14;32	84.55	22788	26JN	16;19	277.96
22789	11JL	22;07	112.52	22789	26JL	6;39	306.28
22790	10AU	5;29	140.71	22790	24AU	22;47	335.01
22791	8SE	13;42	169.34	22791	23SE	16;03	4.26
22792	7OC	23;27	198.52	22792	23OC	9;11	34.00
22793	6NO	11;10	228.24	22793	22NO	0;59	64.07
22794	6DE	1;00	258.34	22794	21DE	14;49	94.23

837

NEW MOONS NUMBER	DATE	TIME	LONG.	FULL MOONS NUMBER	DATE	TIME	LONG.
22721	10JA	17;23	295.16	22721	25JA	19;33	130.43
22722	9FE	4;00	324.87	22722	24FE	12;55	160.21
22723	10MR	14;32	354.13	22723	26MR	3;03	189.35
22724	9AP	1;23	22.91	22724	24AP	14;07	217.90
22725	8MY	13;10	51.30	22725	23MY	22;55	246.02
22726	7JN	2;28	79.51	22726	22JN	6;28	273.96
22727	6JL	17;19	107.75	22727	21JL	13;41	301.96
22728	5AU	9;13	136.23	22728	19AU	21;25	330.29
22729	4SE	1;19	165.11	22729	18SE	6;26	359.11
22730	3OC	16;51	194.46	22730	17OC	17;33	28.52
22731	2NO	7;27	224.24	22731	16NO	7;27	58.45
22732	1DE	20;55	254.31	22732	16DE	0;13	88.73
22733	31DE	9;05	284.40				

843

NEW MOONS NUMBER	DATE	TIME	LONG.	FULL MOONS NUMBER	DATE	TIME	LONG.
22795	4JA	16;54	288.58	22795	20JA	2;44	124.21
22796	3FE	10;27	318.66	22796	18FE	13;02	153.79
22797	5MR	4;36	348.33	22797	19MR	22;08	182.84
22798	3AP	21;52	17.47	22798	18AP	6;27	211.38
22799	3MY	13;01	46.07	22799	17MY	14;36	239.53
22800	2JN	1;38	74.28	22800	15JN	23;33	267.53
22801	1JL	12;09	102.34	22801	15JL	10;19	295.64
22802	30JL	21;26	130.49	22802	13AU	23;45	324.12
22803	29AU	6;18	158.99	22803	12SE	15;59	353.13
22804	27SE	15;26	187.97	22804	12OC	10;13	22.73
22805	27OC	1;14	217.45	22805	11NO	4;56	52.80
22806	25NO	12;03	247.35	22806	10DE	22;37	83.09
22807	25DE	0;18	277.44				

838

NEW MOONS NUMBER	DATE	TIME	LONG.	FULL MOONS NUMBER	DATE	TIME	LONG.
				22733	14JA	18;55	119.03
22734	29JA	19;51	314.23	22734	13FE	13;50	149.04
22735	28FE	5;21	343.62	22735	15MR	7;17	178.51
22736	29MR	14;06	12.49	22736	13AP	22;22	207.39
22737	27AP	22;55	40.91	22737	13MY	10;58	235.77
22738	27MY	8;46	69.04	22738	11JN	21;24	263.84
22739	25JN	20;21	97.14	22739	11JL	6;16	291.85
22740	25JL	10;00	125.43	22740	9AU	14;17	320.06
22741	24AU	1;36	154.12	22741	7SE	22;19	348.68
22742	22SE	18;38	183.34	22742	7OC	7;18	17.81
22743	22OC	12;21	213.09	22743	5NO	17;58	47.48
22744	21NO	5;38	243.22	22744	5DE	6;49	77.54
22745	20DE	21;20	273.46				

844

NEW MOONS NUMBER	DATE	TIME	LONG.	FULL MOONS NUMBER	DATE	TIME	LONG.
				22807	9JA	14;20	113.30
22808	23JA	14;14	307.48	22808	8FE	3;39	143.15
22809	22FE	5;44	337.23	22809	8MR	14;32	172.45
22810	22MR	22;06	6.52	22810	6AP	23;18	201.18
22811	21AP	14;18	35.32	22811	6MY	6;38	229.42
22812	21MY	5;34	63.72	22812	4JN	13;32	257.38
22813	19JN	19;34	91.91	22813	3JL	21;13	285.33
22814	19JL	8;20	120.13	22814	2AU	6;48	313.53
22815	17AU	20;01	148.59	22815	31AU	19;08	342.21
22816	16SE	6;52	177.45	22816	30SE	10;25	11.50
22817	15OC	17;10	206.79	22817	30OC	4;14	41.38
22818	14NO	3;21	236.55	22818	28NO	23;30	71.66
22819	13DE	13;54	266.56	22819	28DE	18;44	102.05

839

NEW MOONS NUMBER	DATE	TIME	LONG.	FULL MOONS NUMBER	DATE	TIME	LONG.
				22745	3JA	21;44	107.74
22746	19JA	10;39	303.50	22746	2FE	14;08	137.78
22747	17FE	21;31	333.11	22747	4MR	7;09	167.42
22748	19MR	6;29	2.17	22748	2AP	23;58	196.55
22749	17AP	14;21	30.70	22749	2MY	15;49	225.19
22750	16MY	22;01	58.84	22750	1JN	6;02	253.47
22751	15JN	6;21	86.82	22751	30JN	18;19	281.59
22752	14JL	16;08	114.89	22752	30JL	4;50	309.80
22753	13AU	4;05	143.29	22753	28AU	14;10	338.30
22754	11SE	18;43	172.23	22754	26SE	23;29	7.27
22755	11OC	12;05	201.78	22755	26OC	9;16	36.75
22756	10NO	7;19	231.86	22756	24NO	19;59	66.64
22757	10DE	2;36	262.21	22757	24DE	7;42	96.71

845

NEW MOONS NUMBER	DATE	TIME	LONG.	FULL MOONS NUMBER	DATE	TIME	LONG.
22820	12JA	1;08	296.56	22820	27JA	12;20	132.20
22821	10FE	13;16	326.31	22821	26FE	3;05	161.85
22822	12MR	2;13	355.65	22822	27MR	14;38	190.86
22823	10AP	15;51	24.52	22823	25AP	23;30	219.30
22824	10MY	6;09	53.00	22824	25MY	6;41	247.35
22825	8JN	21;01	81.26	22825	23JN	13;24	275.25
22826	8JL	12;13	109.52	22826	22JL	20;43	303.27
22827	7AU	3;12	137.98	22827	21AU	5;33	331.46
22828	5SE	17;20	166.80	22828	19SE	16;36	0.58
22829	5OC	6;20	196.08	22829	19OC	6;23	30.12
22830	3NO	18;17	225.77	22830	17NO	23;02	60.19
22831	3DE	5;33	255.75	22831	17DE	18;04	90.57

840

NEW MOONS NUMBER	DATE	TIME	LONG.	FULL MOONS NUMBER	DATE	TIME	LONG.
22758	8JA	19;58	292.50	22758	22JA	20;24	126.70
22759	7FE	10;24	322.40	22759	21FE	10;05	156.38
22760	7MR	21;57	351.74,	22760	22MR	0;49	185.62
22761	6AP	7;14	20.50	22761	20AP	16;23	214.41
22762	5MY	15;01	48.77	22762	20MY	8;09	242.84
22763	3JN	22;05	76.74	22763	18JN	23;16	271.07
22764	3JL	5;18	104.66	22764	18JL	13;09	299.33
22765	1AU	13;39	132.81	22765	17AU	1;46	327.82
22766	31AU	0;12	161.41	22766	15SE	13;29	356.71
22767	29SE	13;52	190.62	22767	15OC	0;45	26.08
22768	29OC	6;51	220.45	22768	13NO	11;46	55.86
22769	28NO	2;11	250.73	22769	12DE	22;36	85.88
22770	27DE	21;59	281.14				

846

NEW MOONS NUMBER	DATE	TIME	LONG.	FULL MOONS NUMBER	DATE	TIME	LONG.
22832	1JA	16;26	285.78	22832	16JA	13;54	120.91
22833	31JA	3;00	315.60	22833	15FE	8;28	150.87
22834	1MR	13;20	345.00	22834	17MR	0;10	180.25
22835	30MR	23;41	13.92	22835	15AP	12;37	208.99
22836	29AP	10;39	42.40	22836	14MY	22;24	237.24
22837	28MY	22;56	70.62	22837	13JN	6;29	265.21
22838	27JN	12;56	98.81	22838	12JL	13;51	293.18
22839	27JL	4;28	127.19	22839	10AU	21;20	321.38
22840	25AU	20;47	155.94	22840	9SE	5;45	350.02
22841	24SE	12;58	185.15	22841	8OC	15;50	19.23
22842	24OC	4;21	214.83	22842	7NO	4;20	48.99
22843	22NO	18;37	244.85	22843	6DE	19;41	79.17
22844	22DE	7;34	274.97				

841

NEW MOONS NUMBER	DATE	TIME	LONG.	FULL MOONS NUMBER	DATE	TIME	LONG.
				22770	11JA	9;16	115.86
22771	26JA	16;19	311.34	22771	9FE	20;05	145.57
22772	25FE	8;05	341.03	22772	11MR	7;30	174.85
22773	26MR	20;58	10.11	22773	9AP	19;56	203.69
22774	25AP	7;12	38.62	22774	9MY	9;31	232.14
22775	24MY	15;21	66.71	22775	8JN	0;00	260.39
22776	22JN	22;17	94.62	22776	7JL	15;01	288.64
22777	22JL	5;05	122.61	22777	6AU	6;14	317.10
22778	20AU	12;53	150.95	22778	4SE	21;19	345.95
22779	18SE	22;48	179.82	22779	4OC	11;55	15.28
22780	18OC	11;32	209.30	22780	3NO	1;33	45.04
22781	17NO	3;12	239.32	22781	2DE	13;53	75.06
22782	16DE	21;09	269.65				

847

NEW MOONS NUMBER	DATE	TIME	LONG.	FULL MOONS NUMBER	DATE	TIME	LONG.
				22844	5JA	13;29	109.48
22845	20JA	19;00	304.93	22845	4FE	8;20	139.60
22846	19FE	4;58	334.48	22846	6MR	2;29	169.27
22847	20MR	13;48	3.51	22847	4AP	18;40	198.34
22848	18AP	22;14	32.04	22848	4MY	8;26	226.88
22849	18MY	7;15	60.23	22849	2JN	19;56	255.04
22850	16JN	17;45	88.29	22850	2JL	5;37	283.06
22851	16JL	6;18	116.47	22851	31JL	14;07	311.20
22852	14AU	21;01	145.01	22852	29AU	22;14	339.67
22853	13SE	13;34	174.05	22853	28SE	6;49	8.63
22854	13OC	7;18	203.64	22854	27OC	16;41	38.13
22855	12NO	1;11	233.69	22855	26NO	4;27	68.07
22856	11DE	17;59	263.95	22856	25DE	18;18	98.24

NEW MOONS / FULL MOONS

--------NEW MOONS--------				--------FULL MOONS--------			
NUMBER	DATE	TIME	LONG.	NUMBER	DATE	TIME	LONG.

848

NUMBER	DATE	TIME	LONG.	NUMBER	DATE	TIME	LONG.
22857	10JA	8;36	294.10	22857	24JA	9;51	128.34
22858	8FE	20;36	323.89	22858	23FE	2;24	158.11
22859	9MR	6;18	353.13	22859	23MR	19;10	187.41
22860	7AP	14;26	21.82	22860	22AP	11;24	216.19
22861	6MY	21;56	50.06	22861	22MY	2;27	244.58
22862	5JN	5;43	78.06	22862	20JN	15;48	272.74
22863	4JL	14;36	106.05	22863	20JL	3;20	300.92
22864	3AU	1;23	134.31	22864	18AU	13;27	329.32
22865	1SE	14;41	163.04	22865	16SE	22;55	358.15
22866	1OC	6;51	192.38	22866	16OC	8;33	27.47
22867	31OC	1;29	222.30	22867	14NO	18;53	57.24
22868	29NO	21;09	252.61	22868	14DE	6;08	87.27
22869	29DE	15;45	282.97				

849

NUMBER	DATE	TIME	LONG.	NUMBER	DATE	TIME	LONG.
				22869	12JA	18;13	117.31
22870	28JA	7;42	313.05	22870	11FE	7;10	147.09
22871	26FE	20;34	342.60	22871	12MR	21;04	176.46
22872	28MR	6;46	11.55	22872	11AP	11;58	205.36
22873	26AP	15;06	39.96	22873	11MY	3;32	233.88
22874	25MY	22;21	68.00	22874	9JN	19;01	262.16
22875	24JN	5;21	95.91	22875	9JL	9;40	290.40
22876	23JL	13;01	123.95	22876	7AU	23;08	318.81
22877	21AU	22;23	152.37	22877	6SE	11;32	347.58
22878	20SE	10;30	181.35	22878	5OC	23;16	16.81
22879	20OC	1;56	210.97	22879	4NO	10;37	46.49
22880	18NO	20;21	241.12	22880	3DE	21;39	76.47
22881	18DE	16;12	271.54				

850

NUMBER	DATE	TIME	LONG.	NUMBER	DATE	TIME	LONG.
				22881	2JA	8;21	106.49
22882	17JA	11;26	301.84	22882	31JA	18;54	136.30
22883	16FE	4;29	331.74	22883	2MR	5;42	165.71
22884	17MR	18;43	1.03	22884	31MR	17;18	194.66
22885	16AP	6;08	29.73	22885	30AP	6;02	223.21
22886	15MY	15;12	57.94	22886	29MY	19;55	251.49
22887	13JN	22;38	85.89	22887	28JN	10;40	279.71
22888	13JL	5;25	113.83	22888	28JL	1;55	308.08
22889	11AU	12;40	142.03	22889	26AU	17;21	336.80
22890	9SE	21;31	170.70	22890	25SE	8;33	5.99
22891	9OC	8;53	199.97	22891	24OC	23;01	35.64
22892	7NO	23;10	229.82	22892	23NO	12;14	65.62
22893	7DE	16;05	260.06	22893	22DE	23;59	95.70

851

NUMBER	DATE	TIME	LONG.	NUMBER	DATE	TIME	LONG.
22894	6JA	10;39	290.40	22894	21JA	10;29	125.61
22895	5FE	5;32	320.52	22895	19FE	20;11	155.14
22896	6MR	23;22	350.16	22896	21MR	5;35	184.18
22897	5AP	14;59	19.20	22897	19AP	15;13	212.76
22898	5MY	3;46	47.68	22898	19MY	1;34	240.99
22899	3JN	13;53	75.78	22899	17JN	13;11	269.09
22900	2JL	22;07	103.75	22900	17JL	2;33	297.31
22901	1AU	5;36	131.85	22901	15AU	17;55	325.88
22902	30AU	13;26	160.32	22902	14SE	10;57	354.96
22903	28SE	22;31	189.31	22903	14OC	4;36	24.56
22904	28OC	9;19	218.86	22904	12NO	21;27	54.57
22905	26NO	22;06	248.85	22905	12DE	12;28	84.76
22906	26DE	12;54	279.06				

852

NUMBER	DATE	TIME	LONG.	NUMBER	DATE	TIME	LONG.
				22906	11JA	1;20	114.84
22907	25JA	5;31	309.19	22907	9FE	12;19	144.57
22908	23FE	23;17	339.01	22908	9MR	21;49	173.80
22909	24MR	16;57	8.32	22909	8AP	6;16	202.49
22910	23AP	9;08	37.10	22910	7MY	14;14	230.74
22911	22MY	23;00	65.43	22911	5JN	22;30	258.75
22912	21JN	10;35	93.52	22912	5JL	8;09	286.78
22913	20JL	20;33	121.64	22913	3AU	20;10	315.09
22914	19AU	5;44	150.01	22914	2SE	11;08	343.91
22915	17SE	14;54	178.83	22915	2OC	4;41	13.32
22916	17OC	0;30	208.17	22916	31OC	23;33	43.26
22917	15NO	10;53	237.95	22917	30NO	18;05	73.53
22918	14DE	22;21	267.99	22918	30DE	10;57	103.82

853

NUMBER	DATE	TIME	LONG.	NUMBER	DATE	TIME	LONG.
22919	13JA	11;16	298.06	22919	29JA	1;29	133.83
22920	12FE	1;48	327.90	22920	27FE	13;29	163.33
22921	13MR	17;32	357.32	22921	28MR	23;07	192.24
22922	12AP	9;41	26.27	22922	27AP	6;56	220.62
22923	12MY	1;20	54.78	22923	26MY	13;49	248.65
22924	10JN	15;59	83.03	22924	24JN	20;53	276.56
22925	10JL	5;28	111.22	22925	24JL	5;23	304.64
22926	8AU	17;55	139.60	22926	22AU	16;19	333.13
22927	7SE	5;26	168.34	22927	21SE	6;13	2.20
22928	6OC	16;13	197.55	22928	20OC	22;58	31.88
22929	5NO	2;36	227.20	22929	19NO	17;46	62.06
22930	4DE	13;00	257.15	22930	19DE	13;17	92.46

854

NUMBER	DATE	TIME	LONG.	NUMBER	DATE	TIME	LONG.
22931	2JA	23;48	287.17	22931	18JA	7;53	122.74
22932	1FE	11;19	317.01	22932	17FE	0;06	152.58
22933	2MR	23;35	346.48	22933	18MR	13;09	181.82
22934	1AP	12;35	15.48	22934	16AP	23;10	210.45
22935	1MY	2;17	44.05	22935	16MY	6;59	238.60
22936	30MY	16;43	72.35	22936	14JN	13;47	266.52
22937	29JN	7;48	100.58	22937	13JL	20;41	294.47
22938	28JL	23;07	128.96	22938	12AU	4;44	322.71
22939	27AU	14;01	157.68	22939	10SE	14;42	351.44
22940	26SE	3;55	186.82	22940	10OC	3;10	20.76
22941	25OC	16;37	216.41	22941	8NO	18;27	50.66
22942	24NO	4;22	246.34	22942	8DE	12;24	80.95
22943	23DE	15;28	276.38				

855

NUMBER	DATE	TIME	LONG.	NUMBER	DATE	TIME	LONG.
				22943	7JA	8;00	111.34
22944	22JA	2;07	306.29	22944	6FE	3;21	141.46
22945	20FE	12;22	335.84	22945	7MR	20;29	171.05
22946	21MR	22;24	4.90	22946	6AP	10;28	200.01
22947	20AP	8;42	33.49	22947	5MY	21;26	228.42
22948	19MY	20;01	61.75	22948	4JN	6;16	256.47
22949	18JN	8;59	89.91	22949	3JL	13;56	284.41
22950	17JL	23;48	118.19	22950	1AU	21;23	312.51
22951	16AU	15;59	146.81	22951	31AU	5;43	341.00
22952	15SE	8;36	175.88	22952	29SE	14;40	10.01
22953	15OC	0;44	205.43	22953	29OC	1;58	39.59
22954	13NO	15;51	235.38	22954	27NO	15;53	69.64
22955	13DE	5;37	265.52	22955	27DE	8;28	99.92

856

NUMBER	DATE	TIME	LONG.	NUMBER	DATE	TIME	LONG.
22956	11JA	17;51	295.57	22956	26JA	2;49	130.12
22957	10FE	4;25	325.28	22957	24FE	21;19	159.95
22958	10MR	13;35	354.48	22958	25MR	14;25	189.23
22959	8AP	21;54	23.15	22959	24AP	5;19	217.94
22960	8MY	6;18	51.42	22960	23MY	17;55	246.21
22961	6JN	15;47	79.47	22961	22JN	4;32	274.27
22962	6JL	3;10	107.57	22962	21JL	13;42	302.35
22963	4AU	16;49	135.96	22963	19AU	22;05	330.71
22964	3SE	8;38	164.81	22964	18SE	6;31	359.51
22965	3OC	2;06	194.23	22965	17OC	15;49	28.84
22966	1NO	20;17	224.16	22966	16NO	2;40	58.64
22967	1DE	13;58	254.40	22967	15DE	15;27	88.75
22968	31DE	5;52	284.65				

857

NUMBER	DATE	TIME	LONG.	NUMBER	DATE	TIME	LONG.
				22968	14JA	6;03	118.88
22969	29JA	19;10	314.60	22969	12FE	21;57	148.77
22970	28FE	5;50	344.03	22970	14MR	14;25	178.21
22971	29MR	14;30	12.90	22971	13AP	6;46	207.15
22972	27AP	22;03	41.26	22972	12MY	22;23	235.65
22973	27MY	5;27	69.30	22973	11JN	12;40	263.88
22974	25JN	13;36	97.26	22974	11JL	1;15	292.04
22975	24JL	23;19	125.39	22975	9AU	12;14	320.37
22976	23AU	11;20	153.93	22976	7SE	22;10	349.06
22977	22SE	2;09	183.05	22977	7OC	7;52	18.23
22978	21OC	19;48	212.79	22978	5NO	17;59	47.87
22979	20NO	15;20	243.00	22979	5DE	4;49	77.85
22980	20DE	10;48	273.40				

858

NUMBER	DATE	TIME	LONG.	NUMBER	DATE	TIME	LONG.
				22980	3JA	16;25	107.90
22981	19JA	4;13	303.62	22981	2FE	4;44	137.77
22982	17FE	18;34	333.38	22982	3MR	17;51	167.26
22983	19MR	5;55	2.53	22983	2AP	7;58	196.29
22984	17AP	14;59	31.11	22984	1MY	23;01	224.91
22985	16MY	22;35	59.25	22985	31MY	14;32	253.24
22986	15JN	5;34	87.18	22986	30JN	5;43	281.48
22987	14JL	12;47	115.14	22987	29JL	19;58	309.83
22988	12AU	21;13	143.40	22988	28AU	9;09	338.49
22989	11SE	7;55	172.17	22989	26SE	21;28	7.58
22990	10OC	21;45	201.57	22990	26OC	9;14	37.14
22991	9NO	14;52	231.55	22991	24NO	20;35	67.05
22992	9DE	10;16	261.91	22992	24DE	7;28	97.09

859

NUMBER	DATE	TIME	LONG.	NUMBER	DATE	TIME	LONG.
22993	8JA	6;01	292.30	22993	22JA	17;57	126.99
22994	7FE	0;13	322.36	22994	21FE	4;22	156.53
22995	8MR	15;48	351.88	22995	22MR	15;13	185.62
22996	7AP	4;31	20.78	22996	21AP	3;04	214.26
22997	6MY	14;38	49.14	22997	20MY	16;10	242.59
22998	4JN	22;47	77.15	22998	19JN	6;25	270.79
22999	4JL	5;48	105.07	22999	18JL	21;30	299.09
23000	2AU	12;46	133.16	23000	17AU	13;05	327.69
23001	31AU	20;48	161.66	23001	16SE	4;44	356.74
23002	30SE	6;56	190.72	23002	15OC	19;56	26.27
23003	29OC	19;49	220.37	23003	14NO	10;04	56.18
23004	28NO	11;30	250.50	23004	13DE	22;41	86.27
23005	28DE	5;18	280.83				

NEW MOONS / FULL MOONS / NEW MOONS / FULL MOONS

860

NUMBER	DATE	TIME	LONG.	NUMBER	DATE	TIME	LONG.
				23005	12JA	9;45	116.27
23006	27JA	0;03	311.04	23006	10FE	19;40	145.94
23007	25FE	18;24	340.85	23007	11MR	4;57	175.14
23008	26MR	11;06	10.10	23008	9AP	14;10	203.84
23009	25AP	1;16	38.77	23009	8MY	23;52	232.15
23010	24MY	12;41	66.99	23010	7JN	10;37	260.25
23011	22JN	21;49	94.98	23011	6JL	22;58	288.39
23012	22JL	5;39	123.02	23012	5AU	13;21	316.81
23013	20AU	13;21	151.36	23013	4SE	5;48	345.71
23014	18SE	21;54	180.17	23014	3OC	23;34	15.16
23015	18OC	7;56	209.54	23015	2NO	17;15	45.08
23016	16NO	19;46	239.40	23016	2DE	9;28	75.26
23017	16DE	9;31	269.55	23017	31DE	23;28	105.42

861

NUMBER	DATE	TIME	LONG.	NUMBER	DATE	TIME	LONG.
23018	15JA	1;06	299.72	23018	30JA	11;19	135.30
23019	13FE	18;09	329.64	23019	28FE	21;23	164.71
23020	15MR	11;48	359.12	23020	30MR	6;08	193.56
23021	14AP	4;42	28.06	23021	28AP	14;04	221.93
23022	13MY	19;43	56.53	23022	27MY	21;55	249.99
23023	12JN	8;27	84.70	23023	26JN	6;39	277.96
23024	11JL	19;17	112.79	23024	25JL	17;20	306.14
23025	10AU	4;58	141.08	23025	24AU	6;51	334.75
23026	8SE	14;18	169.75	23026	22SE	23;20	3.96
23027	7OC	23;51	198.93	23027	22OC	17;56	33.75
23028	6NO	9;57	228.58	23028	21NO	13;00	63.95
23029	5DE	20;51	258.56	23029	21DE	6;56	94.29

862

NUMBER	DATE	TIME	LONG.	NUMBER	DATE	TIME	LONG.
23030	4JA	8;55	288.63	23030	19JA	22;43	124.43
23031	2FE	22;26	318.54	23031	18FE	11;55	154.13
23032	4MR	13;21	348.09	23032	19MR	22;35	183.25
23033	3AP	5;06	17.18	23033	18AP	7;06	211.79
23034	2MY	20;53	45.81	23034	17MY	14;12	239.91
23035	1JN	12;01	74.13	23035	15JN	20;58	267.82
23036	1JL	2;10	102.33	23036	15JL	4;35	295.80
23037	30JL	15;19	130.64	23037	13AU	14;14	324.12
23038	29AU	3;34	159.27	23038	12SE	2;41	352.98
23039	27SE	14;58	188.34	23039	11OC	18;10	22.45
23040	27OC	1;44	217.87	23040	10NO	12;11	52.49
23041	25NO	12;12	247.75	23041	10DE	7;35	82.85
23042	24DE	22;45	277.77				

863

NUMBER	DATE	TIME	LONG.	NUMBER	DATE	TIME	LONG.
				23042	9JA	2;51	113.21
23043	23JA	9;46	307.69	23043	7FE	20;21	143.23
23044	21FE	21;25	337.27	23044	9MR	10;56	172.70
23045	23MR	9;46	6.40	23045	7AP	22;19	201.53
23046	21AP	22;49	35.08	23046	7MY	7;03	229.82
23047	21MY	12;40	63.43	23047	5JN	14;11	257.80
23048	20JN	3;23	91.66	23048	4JL	20;55	285.70
23049	19JL	18;44	119.97	23049	3AU	4;21	313.82
23050	18AU	10;09	148.58	23050	1SE	13;22	342.37
23051	17SE	0;55	177.60	23051	1OC	0;37	11.48
23052	16OC	14;31	207.08	23052	30OC	14;32	41.19
23053	15NO	2;55	236.93	23053	29NO	7;14	71.36
23054	14DE	14;24	266.97	23054	29DE	2;08	101.74

864

NUMBER	DATE	TIME	LONG.	NUMBER	DATE	TIME	LONG.
23055	13JA	1;14	296.95	23055	27JA	21;45	131.98
23056	11FE	11;32	326.63	23056	26FE	16;04	161.78
23057	11MR	21;24	355.84·	23057	27MR	7;33	190.96
23058	10AP	7;14	24.56	23058	25AP	19;54	219.55
23059	9MY	17;41	52.89	23059	25MY	5;41	247.70
23060	8JN	5;36	81.03	23060	23JN	13;52	275.65
23061	7JL	19;27	109.24	23061	22JL	21;26	303.69
23062	6AU	11;07	137.72	23062	21AU	5;13	332.03
23063	5SE	3;51	166.64	23063	19SE	13;55	0.86
23064	4OC	20;36	196.06	23064	19OC	0;14	30.26
23065	3NO	12;33	225.92	23065	17NO	12;50	60.15
23066	3DE	3;13	256.05	23066	17DE	4;04	90.36

865

NUMBER	DATE	TIME	LONG.	NUMBER	DATE	TIME	LONG.
23067	1JA	16;18	286.17	23067	15JA	21;33	120.61
23068	31JA	3;38	316.02	23068	14FE	15;58	150.58
23069	1MR	13;19	345.39	23069	16MR	9;42	180.04
23070	30MR	21;45	14.23	23070	15AP	1;37	208.94
23071	29AP	5;47	42.59	23071	14MY	15;18	237.34
23072	28MY	14;25	70.67	23072	13JN	2;56	265.46
23073	27JN	0;40	98.71	23073	12JL	12;54	293.53
23074	26JL	13;08	126.96	23074	10AU	21;47	321.79
23075	25AU	3;59	155.64	23075	9SE	6;18	350.44
23076	23SE	20;53	184.88	23076	8OC	15;15	19.60
23077	23OC	15;04	214.66	23077	7NO	1;22	49.26
23078	22NO	9;22	244.85	23078	6DE	13;10	79.28
23079	22DE	2;26	275.14				

866

NUMBER	DATE	TIME	LONG.	NUMBER	DATE	TIME	LONG.
				23079	5JA	2;48	109.42
23080	20JA	17;06	305.24	23080	3FE	17;54	139.39
23081	19FE	4;58	334.87	23081	5MR	9;52	168.97
23082	20MR	14;24	3.92	23082	4AP	2;06	198.06
23083	18AP	22;14	32.43	23083	3MY	18;00	226.69
23084	18MY	5;28	60.54	23084	2JN	8;59	254.99
23085	16JN	13;02	88.49	23085	1JL	22;33	283.17
23086	15JL	21;51	116.52	23086	31JL	10;31	311.44
23087	14AU	8;39	144.90	23087	29AU	21;09	340.01
23088	12SE	22;06	173.80	23088	28SE	7;08	9.04
23089	12OC	14;30	203.33	23089	27OC	17;10	38.54
23090	11NO	9;24	233.41	23090	26NO	3;44	68.43
23091	11DE	5;16	263.80	23091	25DE	14;56	98.49

867

NUMBER	DATE	TIME	LONG.	NUMBER	DATE	TIME	LONG.
23092	9JA	23;57	294.13	23092	24JA	2;44	128.43
23093	8FE	15;51	324.08	23093	22FE	15;10	158.04
23094	10MR	4;34	353.45	23094	24MR	4;28	187.20
23095	8AP	14;34	22.21	23095	22AP	18;47	215.92
23096	7MY	22;43	50.48	23096	22MY	9;58	244.31
23097	6JN	5;51	78.45	23097	21JN	1;24	272.56
23098	5JL	12;50	106.37	23098	20JL	16;19	300.86
23099	3AU	20;35	134.50	23099	19AU	6;16	329.42
23100	2SE	6;06	163.08	23100	17SE	19;16	358.39
23101	1OC	18;23	192.25	23101	17OC	7;34	27.82
23102	31OC	9;58	222.04	23102	15NO	19;20	57.65
23103	30NO	4;27	252.29	23103	15DE	6;32	87.68
23104	30DE	0;15	282.71				

868

NUMBER	DATE	TIME	LONG.	NUMBER	DATE	TIME	LONG.
				23104	13JA	17;08	117.66
23105	28JA	19;21	312.92	23105	12FE	3;21	147.32
23106	27FE	12;12	342.64	23106	12MR	13;39	176.54
23107	28MR	2;15	11.75	23107	11AP	0;40	205.29
23108	26AP	13;34	40.29	23108	10MY	12;53	233.69
23109	25MY	22;36	68.40	23109	9JN	2;25	261.89
23110	24JN	6;06	96.33	23110	8JL	17;06	290.13
23111	23JL	13;03	124.34	23111	7AU	8;36	318.62
23112	21AU	20;32	152.68	23112	6SE	0;32	347.52
23113	20SE	5;38	181.54	23113	5OC	16;20	16.91
23114	19OC	17;11	211.00	23114	4NO	7;21	46.75
23115	18NO	7;32	240.97	23115	3DE	20;56	76.83
23116	18DE	0;20	271.25				

869

NUMBER	DATE	TIME	LONG.	NUMBER	DATE	TIME	LONG.
				23116	2JA	8;47	106.90
23117	16JA	18;37	301.53	23117	31JA	19;09	136.70
23118	15FE	13;09	331.49	23118	2MR	4;30	166.05
23119	17MR	6;39	0.93	23119	31MR	13;27	194.89
23120	15AP	22;03	29.79	23120	29AP	22;37	223.30
23121	15MY	10;49	58.15	23121	29MY	8;34	251.43
23122	13JN	21;03	86.21	23122	27JN	19;56	279.51
23123	13JL	5;32	114.22	23123	27JL	9;15	307.80
23124	11AU	13;20	142.44	23124	26AU	0;49	336.52
23125	9SE	21;32	171.09	23125	24SE	18;17	5.79
23126	9OC	6;56	200.28	23126	24OC	12;27	35.59
23127	7NO	17;56	229.99	23127	23NO	5;46	65.74
23128	7DE	6;43	260.05	23128	22DE	21;02	95.96

870

NUMBER	DATE	TIME	LONG.	NUMBER	DATE	TIME	LONG.
23129	5JA	21;16	290.23	23129	21JA	9;56	125.97
23130	4FE	13;25	320.24	23130	19FE	20;45	155.55
23131	6MR	6;39	349.86	23131	21MR	5;56	184.59
23132	4AP	23;52	18.97	23132	19AP	14;02	213.10
23133	4MY	15;49	47.59	23133	18MY	21;40	241.22
23134	3JN	5;42	75.85	23134	17JN	5;43	269.18
23135	2JL	17;32	103.95	23135	16JL	15;14	297.25
23136	1AU	3;54	132.17	23136	15AU	3;17	325.68
23137	30AU	13;34	160.71	23137	13SE	18;25	354.67
23138	28SE	23;11	189.73	23138	13OC	12;16	24.27
23139	28OC	9;09	219.25	23139	12NO	7;30	54.38
23140	26NO	19;41	249.14	23140	12DE	2;18	84.72
23141	26DE	7;05	279.20				

871

NUMBER	DATE	TIME	LONG.	NUMBER	DATE	TIME	LONG.
				23141	10JA	19;18	114.98
23142	24JA	19;40	309.17	23142	9FE	9;46	144.86
23143	23FE	9;39	338.83	23143	10MR	21;35	174.18
23144	25MR	0;46	8.05	23144	9AP	6;59	202.91
23145	23AP	16;23	36.81	23145	8MY	14;34	231.14
23146	23MY	7;46	65.21	23146	6JN	21;18	259.09
23147	21JN	22;27	93.43	23147	6JL	4;19	287.02
23148	21JL	12;16	121.69	23148	4AU	12;50	315.19
23149	20AU	1;13	150.22	23149	2SE	23;53	343.84
23150	18SE	13;19	179.16	23150	2OC	13;57	13.09
23151	18OC	0;38	208.56	23151	1NO	6;53	42.95
23152	16NO	11;22	238.36	23152	1DE	1;50	73.23
23153	15DE	21;53	268.37	23153	30DE	21;22	103.63

872

NUMBER	DATE	TIME	LONG.	NUMBER	DATE	TIME	LONG.
23154	14JA	8;33	298.33	23154	29JA	15;54	133.81
23155	12FE	19;40	328.03	23155	28FE	7;57	163.49
23156	13MR	7;23	357.30	23156	28MR	20;51	192.54
23157	11AP	19;47	26.10	23157	27AP	6;44	221.00
23158	11MY	8;59	54.52	23158	26MY	14;29	249.06
23159	9JN	23;08	82.75	23159	24JN	21;17	276.96
23160	9JL	14;13	111.00	23160	24JL	4;18	304.98
23161	8AU	5;52	139.50	23161	22AU	12;32	333.36
23162	6SE	21;20	168.40	23162	20SE	22;42	2.28
23163	6OC	11;51	197.76	23163	20OC	11;20	31.79
23164	5NO	1;06	227.52	23164	19NO	2;41	61.81
23165	4DE	13;09	257.55	23165	18DE	20;33	92.14

873

NUMBER	DATE	TIME	LONG.	NUMBER	DATE	TIME	LONG.
23166	3JA	0;19	287.58	23166	17JA	15;56	122.46
23167	1FE	10;45	317.38	23167	16FE	11;00	152.43
23168	2MR	20;37	346.74	23168	18MR	3;53	181.83
23169	1AP	6;09	15.60	23169	16AP	17;42	210.61
23170	30AP	15;57	44.02	23170	16MY	4;39	238.90
23171	30MY	2;50	72.18	23171	14JN	13;34	266.90
23172	28JN	15;34	100.32	23172	13JL	21;26	294.89
23173	28JL	6;24	128.68	23173	12AU	5;09	323.11
23174	26AU	22;53	157.45	23174	10SE	13;28	351.78
23175	25SE	16;01	186.72	23175	9OC	23;02	20.98
23176	25OC	8;43	216.47	23176	8NO	10;29	50.72
23177	24NO	0;18	246.56	23177	8DE	0;22	80.83
23178	23DE	14;19	276.73				

874

NUMBER	DATE	TIME	LONG.	NUMBER	DATE	TIME	LONG.
				23178	6JA	16;42	111.08
23179	22JA	2;31	306.70	23179	5FE	10;37	141.15
23180	20FE	12;53	336.25	23180	7MR	4;39	170.79
23181	21MR	21;41	5.26	23181	5AP	21;24	199.87
23182	20AP	5;36	33.75	23182	5MY	12;08	228.43
23183	19MY	13;37	61.89	23183	4JN	0;47	256.63
23184	17JN	22;49	89.89	23184	3JL	11;39	284.71
23185	17JL	10;04	118.03	23185	1AU	21;11	312.89
23186	15AU	23;46	146.54	23186	31AU	6;01	341.42
23187	14SE	15;50	175.58	23187	29SE	14;51	10.41
23188	14OC	9;42	205.20	23188	29OC	0;27	39.92
23189	13NO	4;19	235.29	23189	27NO	11;25	69.84
23190	12DE	22;19	265.60	23190	27DE	0;05	99.95

875

NUMBER	DATE	TIME	LONG.	NUMBER	DATE	TIME	LONG.
23191	11JA	14;20	295.81	23191	25JA	14;19	129.99
23192	10FE	3;34	325.63	23192	24FE	5;39	159.69
23193	11MR	14;01	354.88	23193	25MR	21;32	188.93
23194	9AP	22;24	23.56	23194	24AP	13;27	217.69
23195	9MY	5;40	51.78	23195	24MY	4;53	246.08
23196	7JN	12;51	79.74	23196	22JN	19;16	274.29
23197	6JL	20;54	107.71	23197	22JL	8;13	302.52
23198	5AU	6;37	135.93	23198	20AU	19;43	331.00
23199	3SE	18;44	164.64	23199	19SE	6;12	359.89
23200	3OC	9;46	193.95	23200	18OC	16;21	29.25
23201	2NO	3;39	223.86	23201	17NO	2;46	59.04
23202	1DE	23;23	254.18	23202	16DE	13;41	89.07
23203	31DE	18;58	284.58				

876

NUMBER	DATE	TIME	LONG.	NUMBER	DATE	TIME	LONG.
				23203	15JA	1;05	119.06
23204	30JA	12;21	314.70	23204	13FE	12;58	148.78
23205	29FE	2;34	344.29	23205	14MR	1;30	178.07
23206	29MR	13;45	13.25	23206	12AP	15;00	206.90
23207	27AP	22;37	41.67	23207	12MY	5;34	235.37
23208	27MY	6;06	69.71	23208	10JN	20;53	263.64
23209	25JN	13;03	97.62	23209	10JL	12;12	291.91
23210	24JL	20;21	125.66	23210	9AU	2;53	320.38
23211	23AU	4;56	154.06	23211	7SE	16;38	349.22
23212	21SE	15;48	183.01	23212	7OC	5;33	18.53
23213	21OC	5;47	212.59	23213	5NO	17;49	48.26
23214	19NO	22;59	242.70	23214	5DE	5;26	78.27
23215	19DE	18;21	273.10				

877

NUMBER	DATE	TIME	LONG.	NUMBER	DATE	TIME	LONG.
				23215	3JA	16;19	108.29
23216	18JA	13;58	303.41	23216	2FE	2;34	138.07
23217	17FE	7;58	333.33	23217	3MR	12;31	167.43
23218	18MR	23;20	2.65	23218	1AP	22;50	196.31
23219	17AP	11;55	31.38	23219	1MY	10;08	224.78
23220	16MY	21;59	59.62	23220	30MY	22;48	253.01
23221	15JN	6;11	87.59	23221	29JN	12;52	281.20
23222	14JL	13;22	115.55	23222	29JL	4;04	309.58
23223	12AU	20;34	143.77	23223	27AU	20;03	338.34
23224	11SE	4;52	172.44	23224	26SE	12;16	7.59
23225	10OC	15;13	201.69	23225	26OC	4;04	37.32
23226	9NO	4;14	231.50	23226	24NO	18;39	67.36
23227	8DE	19;52	261.69	23227	24DE	7;27	97.48

878

NUMBER	DATE	TIME	LONG.	NUMBER	DATE	TIME	LONG.
23228	7JA	13;25	291.99	23228	22JA	18;29	127.40
23229	6FE	7;48	322.07	23229	21FE	4;07	156.91
23230	8MR	1;45	351.69	23230	22MR	13;00	185.91
23231	6AP	18;11	20.75	23231	20AP	21;45	214.43
23232	6MY	8;14	49.27	23232	20MY	7;02	242.61
23233	4JN	19;44	77.42	23233	18JN	17;28	270.67
23234	4JL	5;05	105.43	23234	18JL	5;42	298.85
23235	2AU	13;15	133.57	23235	16AU	20;12	327.40
23236	31AU	21;20	162.07	23236	15SE	12;58	356.48
23237	30SE	6;14	191.08	23237	15OC	7;13	26.13
23238	29OC	16;32	220.63	23238	14NO	1;23	56.21
23239	28NO	4;26	250.59	23239	13DE	17;56	86.46
23240	27DE	18;01	280.74				

879

NUMBER	DATE	TIME	LONG.	NUMBER	DATE	TIME	LONG.
				23240	12JA	8;03	116.59
23241	26JA	9;12	310.81	23241	10FE	19;48	146.34
23242	25FE	1;44	340.55	23242	12MR	5;36	175.55
23243	26MR	18;51	9.83	23243	10AP	14;01	204.22
23244	25AP	11;24	38.60	23244	9MY	21;38	232.45
23245	25MY	2;20	66.96	23245	8JN	5;14	260.42
23246	23JN	15;15	95.11	23246	7JL	13;49	288.41
23247	23JL	2;26	123.28	23247	6AU	0;29	316.68
23248	21AU	12;36	151.71	23248	4SE	14;06	345.46
23249	19SE	22;25	180.59	23249	4OC	6;50	14.85
23250	19OC	8;23	209.95	23250	3NO	1;46	44.82
23251	17NO	18;44	239.75	23251	2DE	21;07	75.13
23252	17DE	5;40	269.77				

880

NUMBER	DATE	TIME	LONG.	NUMBER	DATE	TIME	LONG.
				23252	1JA	15;13	105.47
23253	15JA	17;29	299.79	23253	31JA	6;59	135.51
23254	14FE	6;31	329.54	23254	29FE	20;03	165.04
23255	14MR	20;49	358.89	23255	30MR	6;30	193.97
23256	13AP	11;59	27.78	23256	28AP	14;47	222.35
23257	13MY	3;23	56.27	23257	27MY	21;44	250.37
23258	11JN	18;25	84.53	23258	26JN	4;25	278.26
23259	11JL	8;47	112.76	23259	25JL	12;04	306.32
23260	9AU	22;23	141.20	23260	23AU	21;48	334.78
23261	8SE	11;13	170.02	23261	22SE	10;25	3.81
23262	7OC	23;12	199.29	23262	22OC	2;04	33.47
23263	6NO	10;24	228.99	23263	20NO	20;13	63.63
23264	5DE	21;04	258.96	23264	20DE	15;40	94.03

881

NUMBER	DATE	TIME	LONG.	NUMBER	DATE	TIME	LONG.
23265	4JA	7;35	288.96	23265	19JA	10;51	124.32
23266	2FE	18;17	318.76	23266	18FE	4;13	154.20
23267	4MR	5;27	348.16	23267	19MR	18;38	183.47
23268	2AP	17;12	17.09	23268	18AP	5;53	212.13
23269	2MY	5;43	45.60	23269	17MY	14;31	240.31
23270	31MY	19;11	73.85	23270	15JN	21;39	268.24
23271	30JN	9;46	102.06	23271	15JL	4;29	296.19
23272	30JL	1;20	130.46	23272	13AU	12;06	324.42
23273	28AU	17;14	159.24	23273	11SE	21;21	353.15
23274	27SE	8;37	188.46	23274	11OC	8;48	22.45
23275	26OC	22;47	218.13	23275	9NO	22;49	52.31
23276	25NO	11;36	248.11	23276	9DE	15;28	82.55
23277	24DE	23;15	278.18				

882

NUMBER	DATE	TIME	LONG.	NUMBER	DATE	TIME	LONG.
				23277	8JA	10;10	112.89
23278	23JA	9;58	308.09	23278	7FE	5;29	143.00
23279	21FE	19;57	337.60	23279	8MR	23;29	172.61
23280	23MR	5;22	6.61	23280	7AP	14;47	201.61
23281	21AP	14;41	35.14	23281	7MY	3;04	230.05
23282	21MY	0;42	63.34	23282	5JN	12;54	258.13
23283	19JN	12;17	91.44	23283	4JL	21;16	286.11
23284	19JL	2;02	119.69	23284	3AU	5;06	314.24
23285	17AU	17;45	148.31	23285	1SE	13;13	342.75
23286	16SE	11;04	177.42	23286	30SE	22;13	11.77
23287	16OC	4;22	207.04	23287	30OC	8;45	41.34
23288	14NO	20;50	237.06	23288	28NO	21;24	71.34
23289	14DE	11;49	267.25	23289	28DE	12;27	101.55

883

NUMBER	DATE	TIME	LONG.	NUMBER	DATE	TIME	LONG.
23290	13JA	0;59	297.33	23290	27JA	5;31	131.69
23291	11FE	12;10	327.05	23291	25FE	23;27	161.48
23292	12MR	21;32	356.24	23292	27MR	16;45	190.75
23293	11AP	5;36	24.88	23293	26AP	8;25	219.47
23294	10MY	13;16	53.10	23294	25MY	22;05	247.78
23295	8JN	21;36	81.10	23295	24JN	9;53	275.88
23296	8JL	7;39	109.15	23296	23JL	20;12	304.02
23297	6AU	20;06	137.50	23297	22AU	5;32	332.43
23298	5SE	11;07	166.35	23298	20SE	14;30	1.28
23299	5OC	4;21	195.78	23299	19OC	23;48	30.63
23300	3NO	22;56	225.74	23300	18NO	10;08	60.43
23301	3DE	17;35	256.03	23301	17DE	21;54	90.49

NEW MOONS (left)

NUMBER	DATE	TIME	LONG.
			884
23302	2JA	10;50	286.33
23303	1FE	1;30	316.32
23304	1MR	13;12	345.78
23305	30MR	22;23	14.64
23306	29AP	5;57	42.99
23307	28MY	12;57	71.00
23308	26JN	20;24	98.93
23309	26JL	5;11	127.03
23310	24AU	16;04	155.55
23311	23SE	5;41	184.64
23312	22OC	22;18	214.35
23313	21NO	17;25	244.56
23314	21DE	13;24	274.98
			885
23315	20JA	8;04	305.25
23316	18FE	23;52	335.04
23317	20MR	12;24	4.23
23318	18AP	22;14	32.82
23319	18MY	6;14	60.96
23320	16JN	13;20	88.89
23321	15JL	20;23	116.85
23322	14AU	4;16	145.11
23323	12SE	13;59	173.86
23324	12OC	2;25	203.22
23325	10NO	18;06	233.16
23326	10DE	12;35	263.48
			886
23327	9JA	8;15	293.86
23328	8FE	3;07	323.94
23329	9MR	19;46	353.48
23330	8AP	9;38	22.40
23331	7MY	20;52	50.79
23332	6JN	5;56	78.84
23333	5JL	13;35	106.79
23334	3AU	20;45	134.90
23335	2SE	4;31	163.40
23336	1OC	13;53	192.45
23337	31OC	1;36	222.08
23338	29NO	15;58	252.16
23339	29DE	8;35	282.44
			887
23340	28JA	2;31	312.60
23341	26FE	20;37	342.39
23342	28MR	13;45	11.63
23343	27AP	4;59	40.33
23344	26MY	17;45	68.59
23345	25JN	4;12	96.64
23346	24JL	12;59	124.73
23347	22AU	21;10	153.10
23348	21SE	5;46	181.94
23349	20OC	15;28	211.32
23350	19NO	2;38	241.16
23351	18DE	15;20	271.26
			888
23352	17JA	5;34	301.37
23353	15FE	21;13	331.21
23354	16MR	13;52	0.63
23355	15AP	6;38	29.56
23356	14MY	22;23	58.05
23357	13JN	12;21	86.25
23358	13JL	0;30	114.41
23359	11AU	11;19	142.75
23360	9SE	21;30	171.48
23361	9OC	7;35	200.70
23362	7NO	17;53	230.38
23363	7DE	4;32	260.35
			889
23364	5JA	15;47	290.39
23365	4FE	3;59	320.23
23366	5MR	17;22	349.70
23367	4AP	7;52	18.71
23368	3MY	23;00	47.31
23369	2JN	14;09	75.62
23370	2JL	4;55	103.84
23371	31JL	19;07	132.20
23372	30AU	8;37	160.90
23373	28SE	21;20	190.05
23374	28OC	9;09	219.64
23375	26NO	20;12	249.56
23376	26DE	6;45	279.58

FULL MOONS (left)

NUMBER	DATE	TIME	LONG.
			884
23302	16JA	11;14	120.56
23303	15FE	1;50	150.38
23304	15MR	17;12	179.76
23305	14AP	8;54	208.65
23306	14MY	0;30	237.14
23307	12JN	15;28	265.39
23308	12JL	5;20	293.61
23309	10AU	17;46	322.01
23310	9SE	4;58	350.77
23311	8OC	15;27	20.00
23312	7NO	1;52	49.67
23313	6DE	12;36	79.65
			885
23314	4JA	23;43	109.68
23315	3FE	11;10	139.50
23316	4MR	23;04	168.91
23317	3AP	11;44	197.87
23318	3MY	1;31	226.42
23319	1JN	16;22	254.72
23320	1JL	7;47	282.97
23321	30JL	23;01	311.37
23322	29AU	13;30	340.09
23323	28SE	3;08	9.26
23324	27OC	15;59	38.89
23325	26NO	4;06	68.84
23326	25DE	15;24	98.90
			886
23327	24JA	1;51	128.78
23328	22FE	11;42	158.28
23329	23MR	21;30	187.30
23330	22AP	7;57	215.87
23331	21MY	19;41	244.13
23332	20JN	8;56	272.30
23333	19JL	23;37	300.59
23334	18AU	15;24	329.21
23335	17SE	7;50	358.31
23336	17OC	0;15	27.90
23337	15NO	15;46	57.89
23338	15DE	5;38	88.04
			887
23339	13JA	17;32	118.06
23340	12FE	3;42	147.73
23341	13MR	12;42	176.89
23342	11AP	21;13	205.54
23343	11MY	5;57	233.80
23344	9JN	15;34	261.85
23345	9JL	2;45	289.95
23346	7AU	16;05	318.34
23347	6SE	7;53	347.23
23348	6OC	1;45	16.70
23349	4NO	20;25	46.68
23350	4DE	14;07	76.93
			888
23351	3JA	5;34	107.15
23352	1FE	18;26	137.06
23353	2MR	5;02	166.46
23354	31MR	13;56	195.30
23355	29AP	21;43	223.65
23356	29MY	5;05	251.67
23357	27JN	12;58	279.62
23358	26JL	22;26	307.76
23359	25AU	10;32	336.33
23360	24SE	1;52	5.51
23361	23OC	20;01	35.29
23362	22NO	15;31	65.53
23363	22DE	10;31	95.91
			889
23364	21JA	3;32	126.10
23365	19FE	17;54	155.82
23366	21MR	5;32	184.96
23367	19AP	14;43	213.51
23368	18MY	22;09	241.63
23369	17JN	4;47	269.53
23370	16JL	11;48	297.50
23371	14AU	20;25	325.79
23372	13SE	7;37	354.62
23373	12OC	21;51	24.06
23374	11NO	14;56	54.07
23375	11DE	9;55	84.42

NEW MOONS (right)

NUMBER	DATE	TIME	LONG.
			890
23377	24JA	17;13	309.46
23378	23FE	3;55	338.99
23379	24MR	15;04	8.05
23380	23AP	2;54	36.66
23381	22MY	15;38	64.96
23382	21JN	5;33	93.15
23383	20JL	20;43	121.46
23384	19AU	12;44	150.10
23385	18SF	4;47	179.20
23386	17OC	19;55	208.76
23387	16NO	9;38	238.68
23388	15DE	21;57	268.76
			891
23389	14JA	9;06	298.75
23390	12FF	19;18	328.41
23391	14MR	4;45	357.58
23392	12AP	13;48	26.24
23393	11MY	23;08	54.51
23394	10JN	9;40	82.60
23395	9JL	22;12	110.75
23396	8AU	13;07	139.22
23397	7SE	5;56	168.16
23398	6OC	23;34	197.64
23399	5NO	16;49	227.57
23400	5DE	8;46	257.75
			892
23401	3JA	22;57	287.92
23402	2FE	11;06	317.79
23403	2MR	21;12	347.16
23404	1AP	5;40	15.97
23405	30AP	13;13	44.30
23406	29MY	20;56	72.33
23407	28JN	5;54	100.32
23408	27JL	17;04	128.54
23409	26AU	6;53	157.19
23410	24SE	23;13	186.42
23411	24OC	17;26	216.22
23412	23NO	12;24	246.44
23413	23DE	6;38	276.79
			893
23414	21JA	22;42	306.93
23415	20FE	11;49	336.60
23416	21MR	22;03	5.66
23417	20AP	6;11	34.16
23418	19MY	13;14	62.26
23419	17JN	20;16	90.18
23420	17JL	4;16	118.19
23421	15AU	14;04	146.54
23422	14SE	2;20	175.42
23423	13OC	17;32	204.91
23424	12NO	11;37	234.97
23425	12DE	7;28	265.36
			894
23426	11JA	3;03	295.73
23427	9FE	20;22	325.72
23428	11MR	10;25	355.12
23429	9AP	21;25	23.91
23430	9MY	6;10	52.18
23431	7JN	13;35	80.16
23432	6JL	20;35	108.08
23433	5AU	4;00	136.22
23434	3SE	12;47	164.78
23435	2OC	23;51	193.93
23436	1NO	13;58	223.67
23437	1DE	7;10	253.88
23438	31DE	2;25	284.27
			895
23439	29JA	21;48	314.48
23440	28FE	15;32	344.22
23441	30MR	6;43	13.36
23442	28AP	19;10	41.92
23443	28MY	5;15	70.07
23444	26JN	13;35	98.03
23445	25JL	20;59	126.07
23446	24AU	4;28	154.43
23447	22SE	13;03	183.29
23448	21OC	23;38	212.73
23449	20NO	12;44	242.66
23450	20DE	4;14	272.89

FULL MOONS (right)

NUMBER	DATE	TIME	LONG.
			890
23376	10JA	5;24	114.78
23377	8FE	23;46	144.83
23378	10MR	15;39	174.32
23379	9AP	4;23	203.19
23380	8MY	14;10	231.51
23381	6JN	21;55	259.51
23382	6JL	4;48	287.42
23383	4AU	12;00	315.54
23384	2SE	20;28	344.09
23385	2OC	6;52	13.19
23386	31OC	19;38	42.87
23387	30NO	10;59	72.99
23388	30DE	4;42	103.32
			891
23389	28JA	23;46	133.53
23390	27FE	18;29	163.32
23391	29MR	11;07	192.53
23392	28AP	0;49	221.15
23393	27MY	11;47	249.34
23394	25JN	20;51	277.33
23395	25JL	4;59	305.44
23396	23AU	13;02	333.77
23397	21SE	21;42	2.63
23398	21OC	7;32	32.02
23399	19NO	19;06	61.88
23400	19DE	8;52	92.04
			892
23401	18JA	0;52	122.21
23402	16FE	18;18	152.12
23403	17MR	11;50	181.56
23404	16AP	4;13	210.46
23405	15MY	18;50	238.89
23406	14JN	7;36	267.05
23407	13JL	18;46	295.17
23408	12AU	4;44	323.49
23409	10SE	14;02	352.19
23410	9OC	23;18	21.39
23411	8NO	9;11	51.05
23412	7DE	20;13	81.05
			893
23413	6JA	8;41	111.13
23414	4FE	22;28	141.03
23415	6MR	13;13	170.55
23416	5AP	4;31	199.58
23417	4MY	20;01	228.18
23418	3JN	11;18	256.49
23419	3JL	1;52	284.70
23420	1AU	15;15	313.04
23421	31AU	3;18	341.69
23422	29SE	14;20	10.78
23423	29OC	0;57	40.33
23424	27NO	11;37	70.24
23425	26DE	22;32	100.28
			894
23426	25JA	9;41	130.18
23427	23FE	21;06	159.73
23428	25MR	9;02	188.81
23429	23AP	21;56	217.45
23430	23MY	12;04	245.80
23431	22JN	3;14	274.04
23432	21JL	18;45	302.37
23433	20AU	9;53	330.99
23434	19SE	0;15	0.03
23435	18OC	13;46	29.54
23436	17NO	2;28	59.42
23437	16DE	14;17	89.48
			895
23438	15JA	1;07	119.46
23439	13FE	11;04	149.09
23440	14MR	20;34	178.26
23441	13AP	6;21	206.94
23442	12MY	17;08	235.26
23443	11JN	5;26	263.42
23444	10JL	19;23	291.63
23445	9AU	10;45	320.12
23446	8SE	3;09	349.06
23447	7OC	19;57	18.52
23448	6NO	12;18	48.42
23449	6DE	3;15	78.57

--------NEW MOONS--------				--------FULL MOONS--------				--------NEW MOONS--------				--------FULL MOONS--------			
NUMBER	DATE	TIME	LONG.	NUMBER	DATE	TIME	LONG.	NUMBER	DATE	TIME	LONG.	NUMBER	DATE	TIME	LONG.

896 / 902

NEW				FULL				NEW				FULL			
				23450	4JA	16;11	108.68	23525	12JA	19;10	297.48	23525	26JA	19;36	131.67
23451	18JA	21;28	303.11	23451	3FE	3;06	138.49	23526	11FE	9;46	327.34	23526	25FE	9;39	161.30
23452	17FE	15;24	333.03	23452	3MR	12;27	167.82	23527	12MR	21;17	356.62	23527	27MR	0;25	190.47
23453	18MR	8;57	2.45	23453	1AP	20;55	196.62	23528	11AP	6;14	25.30	23528	25AP	15;36	219.19
23454	17AP	1;05	31.33	23454	1MY	5;16	224.97	23529	10MY	13;34	53.50	23529	25MY	6;55	247.57
23455	16MY	15;06	59.73	23455	30MY	14;11	253.05	23530	8JN	20;25	81.45	23530	23JN	21;56	275.80
23456	15JN	2;44	87.84	23456	29JN	0;24	281.09	23531	8JL	3;49	109.39	23531	23JL	12;08	304.09
23457	14JL	12;23	115.90	23457	28JL	12;34	309.35	23532	6AU	12;39	137.59	23532	22AU	1;06	332.64
23458	12AU	20;56	144.17	23458	27AU	3;12	338.05	23533	4SE	23;40	166.27	23533	20SE	12;53	1.60
23459	11SE	5;25	172.85	23459	25SE	20;18	7.32	23534	4OC	13;27	195.55	23534	19OC	23;54	31.03
23460	10OC	14;41	202.06	23460	25OC	15;00	37.16	23535	3NO	6;14	225.43	23535	18NO	10;39	60.84
23461	9NO	1;13	231.76	23461	24NO	9;35	67.37	23536	3DE	1;28	255.74	23536	17DE	21;29	90.87
23462	8DE	13;09	261.80	23462	24DE	2;23	97.66								

897 / 903

NEW				FULL				NEW				FULL			
23463	7JA	2;30	291.92	23463	22JA	16;33	127.71	23537	1JA	21;28	286.15	23537	16JA	8;26	120.84
23464	5FE	17;14	321.85	23464	21FE	4;08	157.30	23538	31JA	16;04	316.31	23538	14FE	19;30	150.50
23465	7MR	9;10	351.40	23465	22MR	13;41	186.33	23539	2MR	7;42	345.94	23539	16MR	6;51	179.72
23466	6AP	1;45	20.47	23466	20AP	21;48	214.82	23540	31MR	20;04	14.94	23540	14AP	18;55	208.48
23467	5MY	17;58	49.09	23467	20MY	5;09	242.92	23541	30AP	5;46	43.37	23541	14MY	8;11	236.88
23468	4JN	8;52	77.37	23468	18JN	12;34	270.86	23542	29MY	13;43	71.42	23542	12JN	22;45	265.12
23469	3JL	22;01	105.54	23469	17JL	21;04	298.89	23543	27JN	20;50	99.34	23543	12JL	14;13	293.40
23470	2AU	9;38	133.81	23470	16AU	7;46	327.28	23544	27JL	4;00	127.38	23544	11AU	5;48	321.92
23471	31AU	20;19	162.42	23471	14SE	21;32	356.24	23545	25AU	12;05	155.78	23545	9SE	20;52	350.83
23472	30SE	6;40	191.49	23472	14OC	14;31	25.82	23546	23SE	22;00	184.71	23546	9OC	11;06	20.21
23473	29OC	17;02	221.04	23473	13NO	9;42	55.94	23547	23OC	10;36	214.25	23547	8NO	0;28	50.01
23474	28NO	3;35	250.95	23474	13DE	5;16	86.32	23548	22NO	2;21	244.31	23548	7DE	12;53	80.05
23475	27DE	14;28	280.98					23549	21DE	20;44	274.67				

898 / 904

NEW				FULL				NEW				FULL			
				23475	11JA	23;25	116.62					23549	6JA	0;14	110.09
23476	26JA	1;59	310.90	23476	10FE	15;07	146.53	23550	20JA	16;10	304.97	23550	4FE	10;29	139.86
23477	24FE	14;30	340.48	23477	12MR	4;01	175.88	23551	19FE	10;45	334.89	23551	4MR	19;56	169.17
23478	26MR	4;10	9.62	23478	10AP	14;16	204.62	23552	20MR	3;08	4.24	23552	3AP	5;14	197.99
23479	24AP	18;46	38.32	23479	9MY	22;23	232.86	23553	18AP	16;52	32.99	23553	2MY	15;10	226.39
23480	24MY	9;49	66.70	23480	8JN	5;14	260.81	23554	18MY	4;04	61.26	23554	1JN	2;29	254.56
23481	23JN	0;48	94.92	23481	7JL	11;55	288.73	23555	16JN	13;14	89.27	23555	30JN	15;30	282.71
23482	22JL	15;26	123.23	23482	5AU	19;39	316.88	23556	15JL	21;06	117.27	23556	30JL	6;14	311.08
23483	21AU	5;33	151.82	23483	4SE	5;32	345.50	23557	14AU	4;34	145.51	23557	28AU	22;20	339.86
23484	19SE	18;59	180.83	23484	3OC	18;19	14.72	23558	12SE	12;38	174.20	23558	27SE	15;17	9.16
23485	19OC	7;32	210.31	23485	2NO	10;07	44.55	23559	11OC	22;16	203.44	23559	27OC	8;15	38.95
23486	17NO	19;08	240.16	23486	2DE	4;19	74.81	23560	10NO	10;08	233.21	23560	26NO	0;14	69.07
23487	17DE	5;57	270.18	23487	31DE	23;43	105.20	23561	10DE	0;27	263.36	23561	25DE	14;20	99.24

899 / 905

NEW				FULL				NEW				FULL			
23488	15JA	16;21	300.13	23488	30JA	18;45	135.39	23562	8JA	16;47	293.59	23562	24JA	2;12	129.19
23489	14FE	2;43	329.77	23489	1MR	11;54	165.09	23563	7FE	10;17	323.62	23563	22FE	12;08	158.69
23490	15MR	13;22	358.97	23490	31MR	2;09	194.18	23564	9MR	3;55	353.22	23564	23MR	20;46	187.66
23491	14AP	0;33	27.71	23491	29AP	13;18	222.68	23565	7AP	20;42	22.27	23565	22AP	4;53	216.13
23492	13MY	12;32	56.07	23492	28MY	21;55	250.76	23566	7MY	11;46	50.82	23566	21MY	13;15	244.26
23493	12JN	1;41	84.25	23493	27JN	5;08	278.68	23567	6JN	0;38	79.02	23567	19JN	22;36	272.27
23494	11JL	16;13	112.49	23494	26JL	12;07	306.71	23568	5JL	11;20	107.09	23568	19JL	9;40	300.42
23495	10AU	8;01	141.01	23495	24AU	19;59	335.09	23569	3AU	20;30	135.28	23569	17AU	23;04	328.94
23496	9SE	0;26	169.96	23496	23SE	5;28	4.00	23570	2SE	5;07	163.82	23570	16SE	15;07	358.01
23497	8OC	16;26	199.40	23497	22OC	17;06	33.49	23571	1OC	14;06	192.86	23571	16OC	9;23	27.67
23498	7NO	7;09	229.24	23498	21NO	7;11	63.47	23572	31OC	0;07	222.41	23572	15NO	4;28	57.80
23499	6DE	20;19	259.32	23499	20DE	23;43	93.75	23573	29NO	11;23	252.35	23573	14DE	22;28	88.12
								23574	28DE	23;57	282.46				

900 / 906

NEW				FULL				NEW				FULL			
23500	5JA	8;02	289.38	23500	19JA	18;08	124.01					23574	13JA	14;02	118.31
23501	3FE	18;36	319.16	23501	18FE	13;04	153.96	23575	27JA	13;48	312.46	23575	12FE	2;49	148.08
23502	4MR	4;14	348.50	23502	19MR	6;44	183.37	23576	26FE	4;53	342.13	23576	13MR	13;11	177.30
23503	2AP	13;12	17.31	23503	17AP	21;51	212.19	23577	27MR	20;57	11.34	23577	11AP	21;48	205.95
23504	1MY	22;04	45.68	23504	17MY	10;06	240.52	23578	26AP	13;16	40.09	23578	11MY	5;19	234.16
23505	31MY	7;41	73.78	23505	15JN	20;04	268.56	23579	26MY	4;52	68.47	23579	9JN	12;30	262.11
23506	29JN	19;01	101.86	23506	15JL	4;41	296.58	23580	24JN	18;59	96.66	23580	8JL	20;17	290.08
23507	29JL	8;44	130.19	23507	13AU	12;51	324.85	23581	24JL	7;30	124.90	23581	7AU	5;45	318.31
23508	28AU	0;49	158.96	23508	11SE	21;20	353.54	23582	22AU	18;49	153.39	23582	5SE	17;58	347.05
23509	26SE	18;26	188.27	23509	11OC	6;40	22.76	23583	21SE	5;32	182.32	23583	5OC	9;30	16.41
23510	26OC	12;15	218.08	23510	9NO	17;23	52.47	23584	20OC	16;06	211.73	23584	4NO	3;53	46.37
23511	25NO	5;10	248.23	23511	9DE	6;00	82.54	23585	19NO	2;41	241.55	23585	3DE	23;35	76.70
23512	24DE	20;24	278.45					23586	18DE	13;24	271.57				

901 / 907

NEW				FULL				NEW				FULL			
				23512	7JA	20;47	112.72					23586	2JA	18;40	107.08
23513	23JA	9;34	308.46	23513	6FE	13;24	142.72	23587	17JA	0;26	301.54	23587	1FE	11;39	137.16
23514	21FE	20;34	338.02	23514	8MR	6;47	172.32	23588	15FE	12;11	331.23	23588	3MR	1;53	166.72
23515	23MR	5;38	7.01	23515	6AP	23;39	201.39	23589	17MR	0;59	0.49	23589	1AP	13;20	195.67
23516	21AP	13;21	35.48	23516	6MY	15;05	229.96	23590	15AP	14;52	29.31	23590	30AP	22;21	224.06
23517	20MY	20;42	63.57	23517	5JN	4;47	258.20	23591	15MY	5;32	57.76	23591	30MY	5;40	252.09
23518	19JN	4;48	91.53	23518	4JL	16;50	286.32	23592	13JN	20;31	86.01	23592	28JN	12;17	279.98
23519	18JL	14;44	119.63	23519	3AU	3;34	314.57	23593	13JL	11;26	114.28	23593	27JL	19;23	308.02
23520	17AU	3;13	148.10	23520	1SE	13;23	343.15	23594	12AU	2;02	142.77	23594	26AU	4;08	336.46
23521	15SE	18;26	177.13	23521	30SE	22;48	12.19	23595	10SE	16;08	171.65	23595	24SE	15;30	5.47
23522	15OC	11;59	206.75	23522	30OC	8;28	41.72	23596	10OC	5;27	201.00	23596	24OC	5;54	35.09
23523	14NO	6;54	236.86	23523	28NO	18;56	71.63	23597	8NO	17;45	230.77	23597	22NO	23;03	65.22
23524	14DE	1;49	267.22	23524	28DE	6;38	101.70	23598	8DE	5;02	260.77	23598	22DE	18;00	95.60

Left half:

NUMBER	DATE	TIME	LONG.	NUMBER	DATE	TIME	LONG.
						908		
23599	6JA	15;36	290.77	23599	21JA	13;20	125.89
23600	5FE	1;48	320.53	23600	20FE	7;29	155.78
23601	5MR	12;03	349.87	23601	20MR	23;10	185.09
23602	3AP	22;39	18.73	23602	19AP	11;46	213.78
23603	3MY	9;56	47.17	23603	18MY	21;31	241.99
23604	1JN	22;16	75.37	23604	17JN	5;20	269.94
23605	1JL	12;00	103.56	23605	16JL	12;22	297.91
23606	31JL	3;17	131.96	23606	14AU	19;48	326.16
23607	29AU	19;43	160.77	23607	13SE	4;32	354.88
23608	28SE	12;21	190.07	23608	12OC	15;10	24.17
23609	28OC	4;05	219.82	23609	11NO	4;03	54.00
23610	26NO	18;14	249.86	23610	10DE	19;20	84.19
23611	26DE	6;43	279.97				
						909		
23612	24JA	17;48	309.87	23611	9JA	12;47	114.47
23613	23FE	3;43	339.37	23612	8FE	7;29	144.54
23614	24MR	12;46	8.34	23613	10MR	1;48	174.14
23615	22AP	21;22	36.83	23614	8AP	18;10	203.17
23616	22MY	6;18	64.97	23615	8MY	7;40	231.65
23617	20JN	16;31	93.02	23616	6JN	18;50	259.77
23618	20JL	4;57	121.22	23617	6JL	4;08	287.79
23619	18AU	19;58	149.81	23618	4AU	12;35	315.96
23620	17SE	13;08	178.94	23619	2SE	21;02	344.50
23621	17OC	7;15	208.62	23620	2OC	6;03	13.55
23622	16NO	0;58	238.70	23621	31OC	16;08	43.11
23623	15DE	17;15	268.95	23622	30NO	3;46	73.08
				23623	29DE	17;22	103.23
						910		
23624	14JA	7;32	299.07	23624	28JA	8;57	133.29
23625	12FE	19;33	328.81	23625	27FE	1;50	163.02
23626	14MR	5;24	357.99	23626	28MR	18;51	192.26
23627	12AP	13;32	26.62	23627	27AP	10;54	220.98
23628	11MY	20;46	54.81	23628	27MY	1;27	249.32
23629	10JN	4;14	82.77	23629	25JN	14;23	277.47
23630	9JL	13;04	110.78	23630	25JL	1;56	305.67
23631	8AU	0;13	139.09	23631	23AU	12;24	334.14
23632	6SE	14;09	167.91	23632	21SE	22;11	3.04
23633	6OC	6;45	197.33	23633	21OC	7;52	32.43
23634	5NO	1;17	227.30	23634	19NO	17;59	62.23
23635	4DE	20;32	257.62	23635	19DE	5;01	92.26
						911		
23636	3JA	14;54	287.97	23636	17JA	17;14	122.28
23637	2FE	6;58	318.00	23637	16FE	6;32	152.02
23638	3MR	19;56	347.49	23638	17MR	20;40	181.33
23639	2AP	5;57	16.37	23639	16AP	11;23	210.17
23640	1MY	13;52	44.72	23640	16MY	2;30	238.63
23641	30MY	20;45	72.72	23641	14JN	17;41	266.88
23642	29JN	3;43	100.63	23642	14JL	8;29	295.15
23643	28JL	11;45	128.71	23643	12AU	22;20	323.62
23644	26AU	21;39	157.20	23644	11SE	10;59	352.46
23645	25SE	10;05	186.27	23645	10OC	22;36	21.75
23646	25OC	1;28	215.94	23646	9NO	9;37	51.47
23647	23NO	19;40	246.13	23647	8DE	20;29	81.45
23648	23DE	15;33	276.54				
						912		
23649	22JA	11;03	306.83	23648	7JA	7;21	111.47
23650	21FE	4;12	336.67	23649	5FE	18;12	141.24
23651	21MR	18;06	5.89	23650	6MR	5;06	170.60
23652	20AP	4;57	34.50	23651	4AP	16;27	199.48
23653	19MY	13;37	62.66	23652	4MY	4;48	227.96
23654	17JN	21;03	90.60	23653	2JN	18;33	256.21
23655	17JL	4;09	118.57	23654	2JL	9;37	284.45
23656	15AU	11;47	146.48	23655	1AU	1;22	312.88
23657	13SE	20;47	175.58	23656	30AU	17;00	341.66
23658	13OC	8;02	204.91	23657	29SE	7;59	10.91
23659	11NO	22;15	234.80	23658	28OC	22;04	40.60
23660	11DE	15;24	265.07	23659	27NO	11;09	70.61
				23660	26DE	23;07	100.69
						913		
23661	10JA	10;26	295.41	23661	25JA	9;50	130.58
23662	9FE	5;31	325.49	23662	23FE	19;27	160.05
23663	10MR	22;57	355.04	23663	25MR	4;30	189.01
23664	9AP	13;55	23.99	23664	23AP	13;46	217.51
23665	9MY	2;18	52.42	23665	23MY	0;07	245.71
23666	7JN	12;28	80.50	23666	21JN	12;07	273.82
23667	6JL	20;59	108.49	23667	21JL	1;59	302.09
23668	5AU	4;40	136.62	23668	19AU	17;34	330.72
23669	3SE	12;29	165.16	23669	18SE	10;24	359.85
23670	2OC	21;23	194.22	23670	18OC	3;44	29.51
23671	1NO	8;11	223.82	23671	16NO	20;35	59.56
23672	30NO	21;17	253.85	23672	16DE	11;51	89.77
23673	30DE	12;35	284.07				

Right half:

NUMBER	DATE	TIME	LONG.	NUMBER	DATE	TIME	LONG.
		914						
23674	29JA	5;25	314.18	23673	15JA	0;52	119.84
23675	27FE	22;52	343.92	23674	13FE	11;37	149.51
23676	29MR	15;58	13.15	23675	14MR	20;39	178.65
23677	28AP	7;52	41.86	23676	13AP	4;45	207.26
23678	27MY	21;52	70.16	23677	12MY	12;43	235.47
23679	26JN	9;42	98.26	23678	10JN	21;20	263.48
23680	25JL	19;43	126.41	23679	10JL	7;23	291.54
23681	24AU	4;43	154.83	23680	8AU	19;33	319.90
23682	22SE	13;38	183.71	23681	7SE	10;22	348.76
23683	21OC	23;16	213.10	23682	7OC	3;48	18.24
23684	20NO	9;59	242.94	23683	5NO	22;54	48.24
23685	19DE	21;52	273.01	23684	5DE	17;49	78.56
						915		
23686	18JA	10;55	303.05	23685	4JA	10;48	108.84
23687	17FE	1;08	332.83	23686	3FE	0;56	138.78
23688	18MR	16;28	2.17	23687	4MR	12;21	168.20
23689	17AP	8;31	31.05	23688	2AP	21;38	197.04
23690	17MY	0;27	59.53	23689	2MY	5;29	225.37
23691	15JN	15;22	87.78	23690	31MY	12;38	253.38
23692	15JL	4;49	115.99	23691	29JN	19;56	281.30
23693	13AU	16;55	144.40	23692	29JL	4;26	309.41
23694	12SE	4;09	173.19	23693	27AU	15;13	337.95
23695	11OC	15;01	202.46	23694	26SE	5;09	7.09
23696	10NO	1;45	232.18	23695	25OC	22;20	36.85
23697	9DE	12;27	262.16	23696	24NO	17;42	67.09
				23697	24DE	13;23	97.50
						916		
23698	7JA	23;14	292.17	23698	23JA	7;31	127.72
23699	6FE	10;23	321.95	23699	21FE	23;06	157.48
23700	6MR	22;21	351.34	23700	22MR	11;50	186.64
23701	5AP	11;24	20.28	23701	20AP	21;54	215.21
23702	5MY	1;28	48.82	23702	20MY	5;55	243.34
23703	3JN	16;13	77.10	23703	18JN	12;44	271.25
23704	3JL	7;12	105.34	23704	17JL	19;29	299.22
23705	1AU	22;09	133.74	23705	16AU	3;21	327.50
23706	31AU	12;49	162.50	23706	14SE	13;25	356.29
23707	30SE	2;52	191.72	23707	14OC	2;22	25.70
23708	29OC	15;58	221.39	23708	12NO	18;16	55.68
23709	28NO	3;54	251.35	23709	12DE	12;27	85.99
23710	27DE	14;49	281.39				
						917		
23711	26JA	1;03	311.25	23710	11JA	7;42	116.34
23712	24FE	11;02	340.72	23711	10FE	2;30	146.39
23713	25MR	21;10	9.72	23712	11MR	19;26	175.91
23714	24AP	7;48	38.27	23713	10AP	9;31	204.82
23715	23MY	19;20	66.51	23714	9MY	20;36	233.18
23716	22JN	8;13	94.65	23715	8JN	5;16	261.20
23717	21JL	22;44	122.95	23716	7JL	12;37	289.14
23718	20AU	14;50	151.61	23717	5AU	19;51	317.28
23719	19SE	7;46	180.76	23718	4SE	3;59	345.83
23720	19OC	0;22	210.40	23719	3OC	13;44	14.92
23721	17NO	15;36	240.40	23720	2NO	1;33	44.58
23722	17DE	5;02	270.53	23721	1DE	15;37	74.66
				23722	31DE	7;57	104.92
						918		
23723	15JA	16;46	300.54	23723	30JA	1;59	135.08
23724	14FE	3;08	330.18	23724	28FE	20;29	164.85
23725	15MR	12;24	359.32	23725	30MR	13;50	194.07
23726	13AP	20;57	27.95	23726	29AP	4;46	222.72
23727	13MY	5;24	56.17	23727	28MY	17;03	250.96
23728	11JN	14;41	84.20	23728	27JN	3;13	278.99
23729	11JL	1;50	112.31	23729	26JL	12;09	307.10
23730	9AU	15;35	140.74	23730	24AU	20;43	335.51
23731	8SE	7;54	169.68	23731	23SE	5;34	4.40
23732	8OC	1;56	199.19	23732	22OC	15;13	33.80
23733	6NO	20;15	229.18	23733	21NO	2;05	63.65
23734	6DE	13;31	259.42	23734	20DE	14;37	93.75
						919		
23735	5JA	4;55	289.64	23735	19JA	5;04	123.85
23736	3FE	18;03	319.53	23736	17FE	21;10	153.69
23737	5MR	4;51	348.91	23737	19MR	13;59	183.08
23738	3AP	13;36	17.72	23738	18AP	6;24	211.96
23739	2MY	21;01	46.02	23739	17MY	21;39	240.41
23740	1JN	4;06	74.02	23740	16JN	11;25	268.60
23741	30JN	12;03	101.97	23741	15JL	23;48	296.78
23742	29JL	21;56	130.14	23742	14AU	11;00	325.16
23743	28AU	10;30	158.76	23743	12SE	21;20	353.93
23744	27SE	1;55	187.98	23744	12OC	7;14	23.17
23745	26OC	19;44	217.78	23745	10NO	17;13	52.86
23746	25NO	14;56	248.02	23746	10DE	3;47	82.84
23747	25DE	10;01	278.40				

NEW MOONS / FULL MOONS (left)

NEW MOONS NUMBER	DATE	TIME	LONG.	FULL MOONS NUMBER	DATE	TIME	LONG.
920							
				23747	8JA	15;19	112.88
23748	24JA	3;24	308.59	23748	7FE	3;53	142.71
23749	22FE	17;53	338.29	23749	7MR	17;21	172.15
23750	23MR	5;13	7.38	23750	6AP	7;30	201.12
23751	21AP	13;57	35.89	23751	5MY	22;12	229.67
23752	20MY	21;08	63.98	23752	4JN	13;17	257.97
23753	19JN	3;54	91.89	23753	4JL	4;25	286.21
23754	18JL	11;18	119.88	23754	2AU	19;00	314.61
23755	16AU	20;14	148.21	23755	1SE	8;33	343.34
23756	15SE	7;25	177.06	23756	30SE	20;56	12.50
23757	14OC	21;22	206.53	23757	30OC	8;26	42.11
23758	13NO	14;17	236.55	23758	28NO	19;29	72.04
23759	13DE	9;33	266.92	23759	28DE	6;21	102.08
921							
23760	12JA	5;29	297.29	23760	26JA	17;05	131.95
23761	10FE	23;55	327.32	23761	25FE	3;43	161.44
23762	12MR	15;23	356.76	23762	26MR	14;30	190.46
23763	11AP	3;36	25.58	23763	25AP	2;00	219.03
23764	10MY	13;12	53.87	23764	24MY	14;48	247.31
23765	8JN	21;08	81.86	23765	23JN	5;09	275.52
23766	8JL	4;21	109.80	23766	22JL	20;43	303.86
23767	6AU	11;43	137.95	23767	21AU	12;42	332.53
23768	4SE	20;03	166.51	23768	20SE	4;21	1.64
23769	4OC	6;11	195.64	23769	19OC	19;12	31.22
23770	2NO	18;55	225.34	23770	18NO	9;02	61.17
23771	2DE	10;39	255.50	23771	17DE	21;41	91.27
922							
23772	1JA	4;51	285.84	23772	16JA	9;00	121.25
23773	30JA	23;59	316.03	23773	14FE	19;00	150.87
23774	1MR	18;13	345.77	23774	16MR	4;03	179.99
23775	31MR	10;21	14.93	23775	14AP	12;52	208.62
23776	29AP	23;56	43.52	23776	13MY	22;20	236.88
23777	29MY	11;11	71.71	23777	12JN	9;17	264.97
23778	27JN	20;31	99.71	23778	11JL	22;09	293.15
23779	27JL	4;40	127.79	23779	10AU	12;58	321.63
23780	25AU	12;28	156.18	23780	9SE	5;25	350.59
23781	23SE	20;53	185.06	23781	8OC	22;52	20.09
23782	23OC	6;47	214.48	23782	7NO	16;22	50.05
23783	21NO	18;45	244.38	23783	7DE	8;43	80.27
23784	21DE	8;57	274.56				
923							
				23784	5JA	22;58	110.43
23785	20JA	0;56	304.72	23785	4FE	10;46	140.26
23786	18FE	17;56	334.58	23786	5MR	20;26	169.58
23787	20MR	11;04	3.97	23787	4AP	4;44	198.36
23788	19AP	3;30	32.84	23788	3MY	12;29	226.67
23789	18MY	18;27	61.27	23789	1JN	20;32	254.71
23790	17JN	7;27	89.43	23790	1JL	5;42	282.71
23791	16JL	18;29	117.56	23791	30JL	16;42	310.93
23792	15AU	4;06	145.88	23792	29AU	6;12	339.60
23793	13SE	13;10	174.61	23793	27SE	22;30	8.86
23794	12OC	22;35	203.84	23794	27OC	17;08	38.71
23795	11NO	8;51	233.55	23795	26NO	12;34	68.97
23796	10DE	20;11	263.57	23796	26DE	6;47	99.32
924							
23797	9JA	8;32	293.64	23797	24JA	22;24	129.42
23798	7FE	21;56	323.50	23798	23FE	11;03	159.03
23799	8MR	12;26	352.97	23799	23MR	21;12	188.06
23800	7AP	3;54	21.98	23800	22AP	5;33	216.55
23801	6MY	19;49	50.57	23801	21MY	12;52	244.64
23802	5JN	11;17	78.87	23802	19JN	19;55	272.55
23803	5JL	1;36	107.09	23803	19JL	3;40	300.57
23804	3AU	14;33	135.43	23804	17AU	13;13	328.93
23805	2SE	2;26	164.10	23805	16SE	1;35	357.84
23806	1OC	13;42	193.23	23806	15OC	17;17	27.39
23807	31OC	0;42	222.83	23807	14NO	11;51	57.49
23808	29NO	11;32	252.75	23808	14DE	7;41	87.89
23809	28DE	22;14	282.78				
925							
				23809	13JA	2;46	118.22
23810	27JA	9;01	312.65	23810	11FE	19;38	148.17
23811	25FE	20;17	342.16	23811	13MR	9;42	177.54
23812	27MR	8;28	11.22	23812	11AP	20;59	206.31
23813	25AP	21;46	39.85	23813	11MY	5;52	234.57
23814	25MY	12;02	68.19	23814	9JN	13;09	262.53
23815	24JN	2;53	96.41	23815	8JL	19;49	290.45
23816	23JL	18;00	124.74	23816	7AU	3;04	318.59
23817	22AU	9;04	153.39	23817	5SE	12;00	347.20
23818	20SE	23;46	182.47	23818	4OC	23;33	16.39
23819	20OC	13;41	212.03	23819	3NO	14;05	46.18
23820	19NO	2;25	241.93	23820	3DE	7;15	76.40
23821	18DE	13;54	271.99				

NEW MOONS / FULL MOONS (right)

NEW MOONS NUMBER	DATE	TIME	LONG.	FULL MOONS NUMBER	DATE	TIME	LONG.
926							
				23821	2JA	2;03	106.77
23822	17JA	0;22	301.93	23822	31JA	21;09	136.94
23823	15FE	10;17	331.54	23823	2MR	15;02	166.66
23824	16MR	20;03	0.68	23824	1AP	6;31	195.78
23825	15AP	6;08	29.35	23825	30AP	19;01	224.32
23826	14MY	16;55	57.65	23826	30MY	4;48	252.44
23827	13JN	4;54	85.78	23827	28JN	12;44	280.39
23828	12JL	18;32	113.99	23828	27JL	20;00	308.44
23829	11AU	9;59	142.51	23829	26AU	3;43	336.83
23830	10SE	2;51	171.50	23830	24SE	12;45	5.74
23831	9OC	20;03	201.00	23831	23OC	23;36	35.22
23832	8NO	12;20	230.93	23832	22NO	12;33	65.17
23833	8DE	2;51	261.07	23833	22DE	3;42	95.38
927							
23834	6JA	15;27	291.16	23834	20JA	20;49	125.58
23835	5FE	2;24	320.95	23835	19FE	15;03	155.49
23836	6MR	12;01	350.26	23836	21MR	8;57	184.89
23837	4AP	20;40	19.04	23837	20AP	1;04	213.74
23838	4MY	4;52	47.36	23838	19MY	14;39	242.10
23839	2JN	13;27	75.41	23839	18JN	1;51	270.19
23840	1JL	23;27	103.44	23840	17JL	11;26	298.26
23841	31JL	11;49	131.73	23841	15AU	20;17	326.57
23842	30AU	2;59	160.48	23842	14SE	5;09	355.30
23843	28SE	20;30	189.80	23843	13OC	14;32	24.54
23844	28OC	15;04	219.66	23844	12NO	0;50	54.26
23845	27NO	9;11	249.87	23845	11DE	12;29	84.29
23846	27DE	1;42	280.15				
928							
				23846	10JA	1;50	114.40
23847	25JA	16;00	310.19	23847	8FE	16;56	144.32
23848	24FE	3;52	339.76	23848	9MR	9;14	173.85
23849	24MR	13;27	8.76	23849	8AP	1;43	202.89
23850	22AP	21;18	37.21	23850	7MY	17;29	231.46
23851	22MY	4;17	65.28	23851	6JN	7;59	259.73
23852	20JN	11;34	93.21	23852	5JL	21;10	287.90
23853	19JL	20;19	121.26	23853	4AU	9;09	316.21
23854	18AU	7;31	149.70	23854	2SE	20;09	344.85
23855	16SE	21;37	178.70	23855	2OC	6;27	13.96
23856	16OC	14;27	208.31	23856	31OC	16;31	43.52
23857	15NO	9;14	238.43	23857	30NO	2;50	73.43
23858	15DE	4;40	268.81	23858	29DE	13;49	103.47
929							
23859	13JA	23;06	299.11	23859	28JA	1;42	133.38
23860	12FE	15;05	329.01	23860	26FE	14;28	162.94
23861	14MR	3;53	358.32	23861	28MR	4;00	192.04
23862	12AP	13;43	27.02	23862	26AP	18;09	220.70
23863	11MY	21;27	55.22	23863	26MY	8;55	249.05
23864	10JN	4;15	83.16	23864	25JN	0;04	277.28
23865	9JL	11;13	111.09	23865	24JL	15;09	305.62
23866	7AU	19;20	139.28	23866	23AU	5;32	334.25
23867	6SE	5;24	167.94	23867	21SE	18;47	3.29
23868	5OC	18;00	197.19	23868	21OC	6;57	32.78
23869	4NO	9;31	227.03	23869	19NO	18;22	62.64
23870	4DE	3;46	257.30	23870	19DE	5;22	92.67
930							
23871	2JA	23;35	287.71	23871	17JA	16;06	122.62
23872	1FE	18;56	317.89	23872	16FE	2;36	152.25
23873	3MR	11;53	347.55	23873	17MR	13;00	181.40
23874	2AP	1;36	16.58	23874	15AP	23;46	210.09
23875	1MY	12;21	45.04	23875	15MY	11;37	238.42
23876	30MY	21;00	73.11	23876	14JN	1;02	266.61
23877	29JN	4;31	101.05	23877	13JL	16;03	294.88
23878	28JL	11;48	129.10	23878	12AU	8;05	323.43
23879	26AU	19;40	157.51	23879	11SE	0;15	352.41
23880	25SE	4;56	186.44	23880	10OC	15;50	21.86
23881	24OC	16;22	215.95	23881	9NO	6;27	51.72
23882	23NO	6;37	245.96	23882	8DE	19;52	81.82
23883	22DE	23;38	276.26				
931							
				23883	7JA	7;54	111.88
23884	21JA	18;22	306.52	23884	5FE	18;27	141.65
23885	20FE	13;04	336.43	23885	7MR	3;43	170.93
23886	22MR	6;11	5.79	23886	5AP	12;20	199.70
23887	20AP	20;58	34.57	23887	4MY	21;08	228.04
23888	20MY	9;20	62.88	23888	3JN	7;05	256.14
23889	18JN	19;38	90.93	23889	2JL	18;51	284.25
23890	18JL	4;25	118.97	23890	1AU	8;42	312.60
23891	16AU	12;27	147.25	23891	31AU	0;31	341.38
23892	14SE	20;38	175.96	23892	29SE	17;47	10.71
23893	14OC	5;51	205.21	23893	29OC	11;39	40.56
23894	12NO	16;49	234.96	23894	28NO	4;56	70.74
23895	12DE	5;53	265.06	23895	27DE	20;26	100.97

Left columns — New Moons / Full Moons

932

NUMBER	DATE	TIME	LONG.	NUMBER	DATE	TIME	LONG.
23896	10JA	20;55	295.23	23896	26JA	9;26	130.95
23897	9FE	13;16	325.20	23897	24FE	20;00	160.46
23898	10MR	6;10	354.74	23898	25MR	4;43	189.41
23899	8AP	22;50	23.77	23899	23AP	12;28	217.85
23900	8MY	14;30	52.33	23900	22MY	20;07	245.94
23901	7JN	4;33	80.58	23901	21JN	4;32	273.91
23902	6JL	16;40	108.71	23902	20JL	14;29	302.02
23903	5AU	3;07	136.96	23903	19AU	2;42	330.51
23904	3SE	12;36	165.56	23904	17SE	17;42	359.55
23905	2OC	21;58	194.63	23905	17OC	11;26	29.22
23906	1NO	7;56	224.20	23906	16NO	6;52	59.37
23907	30NO	18;48	254.14	23907	16DE	2;03	89.75
23908	30DE	6;35	284.21				

933

NUMBER	DATE	TIME	LONG.	NUMBER	DATE	TIME	LONG.
				23908	14JA	19;07	119.98
23909	28JA	19;16	314.15	23909	13FE	9;11	149.80
23910	27FE	8;56	343.73	23910	14MR	20;24	179.03
23911	28MR	23;38	12.87	23911	13AP	5;27	207.68
23912	27AP	15;10	41.58	23912	12MY	13;05	235.88
23913	27MY	6;51	69.95	23913	10JN	20;06	263.82
23914	25JN	21;51	98.18	23914	10JL	3;22	291.77
23915	25JL	11;39	126.47	23915	8AU	11;55	319.98
23916	24AU	0;17	155.04	23916	6SE	22;49	348.68
23917	22SE	12;06	184.03	23917	6OC	12;55	18.01
23918	21OC	23;28	213.50	23918	5NO	6;16	47.94
23919	20NO	10;33	243.35	23919	5DE	1;46	78.27
23920	19DE	21;20	273.38				

934

NUMBER	DATE	TIME	LONG.	NUMBER	DATE	TIME	LONG.
				23920	3JA	21;27	108.66
23921	18JA	7;56	303.32	23921	2FE	15;30	138.78
23922	16FE	18;42	332.95	23922	4MR	6;55	168.37
23923	18MR	6;06	2.13	23923	2AP	19;29	197.34
23924	16AP	18;32	30.88	23924	2MY	5;26	225.76
23925	16MY	8;07	59.27	23925	31MY	13;23	253.80
23926	14JN	22;36	87.50	23926	29JN	20;15	281.70
23927	14JL	13;40	115.77	23927	29JL	3;08	309.75
23928	13AU	4;58	144.31	23928	27AU	11;11	338.18
23929	11SE	20;12	173.26	23929	25SE	21;27	7.16
23930	11OC	10;52	202.68	23930	25OC	10;34	36.75
23931	10NO	0;28	232.51	23931	24NO	2;31	66.84
23932	9DE	12;42	262.56	23932	23DE	20;35	97.18

935

NUMBER	DATE	TIME	LONG.	NUMBER	DATE	TIME	LONG.
23933	7JA	23;38	292.58	23933	22JA	15;36	127.44
23934	6FE	9;39	322.31	23934	21FE	10;06	157.33
23935	7MR	19;14	351.60	23935	23MR	2;47	186.66
23936	6AP	4;52	20.40	23936	21AP	16;44	215.39
23937	5MY	15;00	48.78	23937	21MY	3;48	243.64
23938	4JN	2;08	76.93	23938	19JN	12;35	271.63
23939	3JL	14;48	105.07	23939	18JL	20;09	299.63
23940	2AU	5;22	133.46	23940	17AU	3;40	327.90
23941	31AU	21;47	162.28	23941	15SE	12;07	356.63
23942	30SE	15;15	191.63	23942	14OC	22;08	25.92
23943	30OC	8;24	221.46	23943	13NO	10;05	55.72
23944	29NO	0;04	251.58	23944	13DE	0;06	85.86
23945	28DE	13;43	281.73				

936

NUMBER	DATE	TIME	LONG.	NUMBER	DATE	TIME	LONG.
				23945	11JA	16;08	116.08
23946	27JA	1;25	311.65	23946	10FE	9;44	146.09
23947	25FE	11;33	341.13	23947	11MR	3;46	175.66
23948	25MR	20;27	10.08	23948	9AP	20;45	204.69
23949	24AP	4;36	38.53	23949	9MY	11;34	233.21
23950	23MY	12;42	66.63	23950	7JN	23;56	261.38
23951	21JN	21;44	94.63	23951	7JL	10;22	289.44
23952	21JL	8;46	122.78	23952	5AU	19;40	317.66
23953	19AU	22;34	151.34	23953	4SE	4;40	346.24
23954	18SE	15;09	180.47	23954	3OC	13;57	15.33
23955	18OC	9;35	210.17	23955	1NO	23;53	44.90
23956	17NO	4;18	240.31	23956	1DE	10;51	74.85
23957	16DE	21;52	270.62	23957	30DE	23;14	104.95

937

NUMBER	DATE	TIME	LONG.	NUMBER	DATE	TIME	LONG.
23958	15JA	13;22	300.78	23958	29JA	13;17	134.93
23959	14FE	2;24	330.54	23959	28FE	4;48	164.59
23960	15MR	12;58	359.74	23960	29MR	21;02	193.77
23961	13AP	21;28	28.36	23961	28AP	13;02	222.48
23962	13MY	4;37	56.53	23962	28MY	4;08	250.83
23963	11JN	11;31	84.46	23963	26JN	18;03	279.02
23964	10JL	19;21	112.44	23964	26JL	6;48	307.27
23965	9AU	5;15	140.71	23965	24AU	18;32	335.81
23966	7SE	17;56	169.50	23966	23SE	5;25	4.78
23967	7OC	9;34	198.90	23967	22OC	15;46	34.22
23968	6NO	3;37	228.87	23968	21NO	2;02	64.04
23969	5DE	23;01	259.20	23969	20DE	12;39	94.06

Right columns — New Moons / Full Moons

938

NUMBER	DATE	TIME	LONG.	NUMBER	DATE	TIME	LONG.
23970	4JA	18;11	289.57	23970	18JA	23;57	124.02
23971	3FE	11;30	319.65	23971	17FE	12;03	153.70
23972	5MR	1;51	349.18	23972	19MR	0;55	182.93
23973	3AP	13;00	18.08	23973	17AP	14;28	211.71
23974	2MY	21;34	46.43	23974	17MY	4;44	240.12
23975	1JN	4;39	74.43	23975	15JN	19;39	268.37
23976	30JN	11;24	102.34	23976	15JL	10;55	296.65
23977	29JL	18;53	130.41	23977	14AU	1;57	325.18
23978	28AU	3;58	158.89	23978	12SE	16;06	354.10
23979	26SE	15;20	187.93	23979	12OC	5;05	23.47
23980	26OC	5;26	217.57	23980	10NO	17;03	53.24
23981	24NO	22;25	247.71	23981	10DE	4;20	83.26
23982	24DE	17;38	278.10				

939

NUMBER	DATE	TIME	LONG.	NUMBER	DATE	TIME	LONG.
				23982	8JA	15;11	113.26
23983	23JA	13;24	308.39	23983	7FE	1;39	143.01
23984	22FE	7;37	338.26	23984	8MR	11;49	172.31
23985	23MR	22;53	7.51	23985	6AP	22;04	201.13
23986	22AP	10;59	36.16	23986	6MY	9;01	229.54
23987	21MY	20;33	64.34	23987	4JN	21;25	257.73
23988	20JN	4;33	92.30	23988	4JL	11;36	285.93
23989	19JL	11;56	120.29	23989	3AU	3;19	314.37
23990	17AU	19;32	148.57	23990	1SE	19;44	343.21
23991	16SE	4;08	177.32	23991	1OC	11;58	12.52
23992	15OC	14;30	206.63	23992	31OC	3;23	42.29
23993	14NO	3;21	236.48	23993	29NO	17;38	72.36
23994	13DE	19;00	266.69	23994	29DE	6;27	102.47

940

NUMBER	DATE	TIME	LONG.	NUMBER	DATE	TIME	LONG.
23995	12JA	12;55	296.98	23995	27JA	17;41	132.37
23996	11FE	7;40	327.03	23996	26FE	3;24	161.82
23997	12MR	1;31	356.58	23997	26MR	12;03	190.74
23998	10AP	17;23	25.55	23998	24AP	20;25	219.19
23999	10MY	6;54	54.01	23999	24MY	5;29	247.33
24000	8JN	18;14	82.13	24000	22JN	16;08	275.39
24001	8JL	3;49	110.17	24001	22JL	4;54	303.62
24002	6AU	12;18	138.36	24002	20AU	19;51	332.24
24003	4SE	20;30	166.92	24003	19SE	12;39	1.39
24004	4OC	5;16	195.99	24004	19OC	6;35	31.08
24005	2NO	15;24	225.59	24005	18NO	0;32	61.20
24006	2DE	3;25	255.58	24006	17DE	17;12	91.47
24007	31DE	17;25	285.75				

941

NUMBER	DATE	TIME	LONG.	NUMBER	DATE	TIME	LONG.
				24007	16JA	7;32	121.58
24008	30JA	9;00	315.79	24008	14FE	19;12	151.27
24009	1MR	1;27	345.47	24009	16MR	4;37	180.41
24010	30MR	18;04	14.66	24010	14AP	12;34	209.00
24011	29AP	10;09	43.36	24011	13MY	20;01	237.17
24012	29MY	1;03	71.69	24012	12JN	3;50	265.14
24013	27JN	14;15	99.85	24013	11JL	12;52	293.16
24014	27JL	1;40	128.06	24014	9AU	23;52	321.49
24015	25AU	11;47	156.54	24015	8SE	13;31	350.33
24016	23SE	21;21	185.47	24016	8OC	6;04	19.78
24017	23OC	7;09	214.89	24017	7NO	1;00	49.80
24018	21NO	17;40	244.73	24018	6DE	20;42	80.15
24019	21DE	4;59	274.78				

942

NUMBER	DATE	TIME	LONG.	NUMBER	DATE	TIME	LONG.
				24019	5JA	15;04	110.49
24020	19JA	17;04	304.78	24020	4FE	6;38	140.48
24021	18FE	5;58	334.47	24021	5MR	19;08	169.92
24022	19MR	19;51	3.74	24022	4AP	5;04	198.76
24023	18AP	10;44	32.55	24023	3MY	13;13	227.09
24024	18MY	2;16	61.01	24024	1JN	20;22	255.09
24025	16JN	17;41	89.27	24025	1JL	3;22	283.00
24026	16JL	8;15	117.53	24026	30JL	11;10	311.10
24027	14AU	21;41	146.01	24027	28AU	20;50	339.61
24028	13SE	10;09	174.88	24028	27SE	9;21	8.70
24029	12OC	21;59	204.21	24029	27OC	1;13	38.43
24030	11NO	9;24	233.97	24030	25NO	19;54	68.65
24031	10DE	20;25	263.97	24031	25DE	15;45	99.07

943

NUMBER	DATE	TIME	LONG.	NUMBER	DATE	TIME	LONG.
24032	9JA	7;02	293.96	24032	24JA	10;45	129.32
24033	7FE	17;30	323.70	24033	23FE	3;28	159.11
24034	9MR	4;15	353.02	24034	24MR	17;21	188.30
24035	7AP	15;51	21.88	24035	23AP	4;29	216.89
24036	7MY	4;36	50.35	24036	22MY	13;19	245.04
24037	5JN	18;30	78.60	24037	20JN	20;37	272.97
24038	5JL	9;17	106.83	24038	20JL	3;25	300.94
24039	4AU	0;39	135.26	24039	18AU	10;51	329.22
24040	2SE	16;13	164.07	24040	16SE	20;01	358.00
24041	2OC	7;32	193.37	24041	16OC	7;46	27.38
24042	31OC	22;00	223.10	24042	14NO	22;22	57.32
24043	30NO	11;07	253.13	24043	14DE	15;28	87.59
24044	29DE	22;43	283.19				

	NEW MOONS				FULL MOONS		
NUMBER	DATE	TIME	LONG.	NUMBER	DATE	TIME	LONG.

944

NUMBER	DATE	TIME	LONG.	NUMBER	DATE	TIME	LONG.
				24044	13JA	10;03	117.91
24045	28JA	9;04	313.04	24045	12FE	4;49	147.94
24046	26FE	18;38	342.48	24046	12MR	22;25	177.47
24047	27MR	3;57	11.42	24047	11AP	13;42	206.40
24048	25AP	13;32	39.91	24048	11MY	2;09	234.81
24049	24MY	23;53	68.09	24049	9JN	12;01	262.87
24050	23JN	11;35	96.19	24050	8JL	20;09	290.85
24051	23JL	1;10	124.46	24051	7AU	3;42	319.01
24052	21AU	16;49	153.12	24052	5SE	11;45	347.58
24053	20SE	10;07	182.30	24053	4OC	21;05	16.68
24054	20OC	3;53	212.01	24054	3NO	8;08	46.32
24055	18NO	20;39	242.08	24055	2DE	21;07	76.36
24056	18DE	11;27	272.27				

945

NUMBER	DATE	TIME	LONG.	NUMBER	DATE	TIME	LONG.
				24056	1JA	12;03	106.57
24057	17JA	0;07	302.31	24057	31JA	4;45	136.65
24058	15FE	10;53	331.95	24058	1MR	22;28	166.37
24059	16MR	20;12	1.08	24059	31MR	15;57	195.58
24060	15AP	4;28	29.67	24060	30AP	7;50	224.26
24061	14MY	12;19	57.85	24061	29MY	21;25	252.53
24062	12JN	20;37	85.84	24062	28JN	8;50	280.62
24063	12JL	6;27	113.90	24063	27JL	18;47	308.78
24064	10AU	18;49	142.29	24064	26AU	4;05	337.24
24065	9SE	10;10	171.21	24065	24SE	13;23	6.16
24066	9OC	4;00	200.73	24066	23OC	23;07	35.59
24067	7NO	22;59	230.76	24067	22NO	9;36	65.44
24068	7DE	17;26	261.06	24068	21DE	21;12	95.50

946

NUMBER	DATE	TIME	LONG.	NUMBER	DATE	TIME	LONG.
24069	6JA	10;06	291.32	24069	20JA	10;14	125.53
24070	5FE	0;22	321.25	24070	19FE	0;49	155.29
24071	6MR	12;03	350.65	24071	20MR	16;30	184.61
24072	4AP	21;23	19.45	24072	19AP	8;28	213.46
24073	4MY	4;58	47.75	24073	18MY	23;57	241.91
24074	2JN	11;46	75.73	24074	17JN	14;30	270.13
24075	1JL	18;56	103.65	24075	17JL	3;58	298.35
24076	31JL	3;41	131.79	24076	15AU	16;27	326.80
24077	29AU	14;58	160.38	24077	14SE	4;01	355.64
24078	28SE	5;14	189.56	24078	13OC	14;50	24.94
24079	27OC	22;17	219.35	24079	12NO	1;16	54.67
24080	26NO	17;15	249.59	24080	11DE	11;42	84.65
24081	26DE	12;47	279.99				

947

NUMBER	DATE	TIME	LONG.	NUMBER	DATE	TIME	LONG.
				24081	9JA	22;34	114.65
24082	25JA	7;11	310.21	24082	8FE	10;05	144.43
24083	23FE	23;03	339.95	24083	9MR	22;18	173.79
24084	25MR	11;41	9.07	24084	8AP	11;12	202.69
24085	23AP	21;21	37.60	24085	8MY	0;51	231.19
24086	23MY	4;59	65.70	24086	6JN	15;19	259.45
24087	21JN	11;45	93.60	24087	6JL	6;29	287.70
24088	20JL	18;47	121.59	24088	4AU	21;53	316.14
24089	19AU	3;03	149.91	24089	3SE	12;49	344.94
24090	17SE	13;18	178.75	24090	3OC	2;43	14.19
24091	17OC	2;05	208.18	24091	1NO	15;25	43.87
24092	15NO	17;41	238.16	24092	1DE	3;09	73.84
24093	15DE	11;54	268.49	24093	30DE	14;14	103.88

948

NUMBER	DATE	TIME	LONG.	NUMBER	DATE	TIME	LONG.
24094	14JA	7;33	298.84	24094	29JA	0;47	133.73
24095	13FE	2;40	328.88	24095	27FE	10;53	163.18
24096	13MR	19;24	358.36	24096	27MR	20;46	192.14
24097	12AP	8;58	27.21	24097	26AP	7;01	220.64
24098	11MY	19;39	55.54	24098	25MY	18;24	248.86
24099	10JN	4;20	83.55	24099	24JN	7;33	277.02
24100	9JL	12;01	111.51	24100	23JL	22;35	305.35
24101	7AU	19;32	139.68	24101	22AU	14;56	334.05
24102	6SE	3;42	168.26	24102	21SE	7;37	3.22
24103	5OC	13;13	197.38	24103	20OC	23;48	32.87
24104	4NO	0;49	227.05	24104	19NO	14;54	62.88
24105	3DE	15;03	257.15	24105	19DE	4;35	93.03

949

NUMBER	DATE	TIME	LONG.	NUMBER	DATE	TIME	LONG.
24106	2JA	7;51	287.43	24106	17JA	16;37	123.04
24107	1FE	2;12	317.58	24107	16FE	2;57	152.65
24108	2MR	20;28	347.31	24108	17MR	11;52	181.75
24109	1AP	13;15	16.47	24109	15AP	20;03	210.33
24110	1MY	3;52	45.09	24110	15MY	4;27	238.53
24111	30MY	16;16	73.31	24111	13JN	14;04	266.56
24112	29JN	2;46	101.36	24112	13JL	1;40	294.70
24113	28JL	11;54	129.49	24113	11AU	15;34	323.15
24114	26AU	20;20	157.93	24114	10SE	7;38	352.12
24115	25SE	4;54	186.83	24115	10OC	1;20	21.64
24116	24OC	14;25	216.26	24116	8NO	19;39	51.66
24117	23NO	1;31	246.14	24117	8DE	13;18	81.93
24118	22DE	14;30	276.26				

950

NUMBER	DATE	TIME	LONG.	NUMBER	DATE	TIME	LONG.
				24118	7JA	4;58	112.15
24119	21JA	5;10	306.36	24119	5FE	17;55	142.01
24120	19FE	21;00	336.15	24120	7MR	4;15	171.34
24121	21MR	13;20	5.49	24121	5AP	12;41	200.10
24122	20AP	5;34	34.34	24122	4MY	20;07	228.39
24123	19MY	21;03	62.78	24123	3JN	3;31	256.39
24124	18JN	11;12	90.98	24124	2JL	11;46	284.35
24125	17JL	23;40	119.17	24125	31JL	21;42	312.54
24126	16AU	10;36	147.57	24126	30AU	10;00	341.18
24127	14SE	20;35	176.35	24127	29SE	1;12	10.42
24128	14OC	6;25	205.62	24128	28OC	19;13	40.26
24129	12NO	16;42	235.35	24129	27NO	14;55	70.54
24130	12DE	3;39	265.35	24130	27DE	10;16	100.93

951

NUMBER	DATE	TIME	LONG.	NUMBER	DATE	TIME	LONG.
24131	10JA	15;15	295.39	24131	26JA	3;20	131.09
24132	9FE	3;31	325.18	24132	24FE	17;17	160.74
24133	10MR	16;36	354.57	24133	26MR	4;19	189.78
24134	9AP	6;41	23.51	24134	24AP	13;09	218.27
24135	8MY	21;44	52.05	24135	23MY	20;38	246.35
24136	7JN	13;13	80.35	24136	22JN	3;34	274.26
24137	7JL	4;21	108.60	24137	21JL	10;52	302.26
24138	5AU	18;34	137.01	24138	19AU	19;32	330.61
24139	4SE	7;46	165.75	24139	18SE	6;36	359.49
24140	3OC	20;10	194.95	24140	17OC	20;51	29.00
24141	2NO	8;02	224.60	24141	16NO	14;20	59.07
24142	1DE	19;23	254.56	24142	16DE	9;51	89.45
24143	31DE	6;12	284.59				

952

NUMBER	DATE	TIME	LONG.	NUMBER	DATE	TIME	LONG.
				24143	15JA	5;27	119.80
24144	29JA	16;34	314.42	24144	13FE	23;20	149.77
24145	28FE	2;53	343.87	24145	14MR	14;35	179.18
24146	28MR	13;43	12.86	24146	13AP	2;59	207.97
24147	27AP	1;36	41.42	24147	12MY	12;51	236.25
24148	26MY	14;43	69.70	24148	10JN	20;49	264.24
24149	25JN	5;01	97.90	24149	10JL	3;48	292.17
24150	24JL	20;11	126.24	24150	8AU	10;52	320.33
24151	23AU	11;54	154.93	24151	6SE	19;10	348.92
24152	22SE	3;42	184.08	24152	6OC	5;39	18.10
24153	21OC	18;59	213.70	24153	4NO	18;54	47.85
24154	20NO	9;03	243.68	24154	4DE	10;49	78.02
24155	19DE	21;30	273.78				

953

NUMBER	DATE	TIME	LONG.	NUMBER	DATE	TIME	LONG.
				24155	3JA	4;42	108.35
24156	18JA	8;24	303.73	24156	1FE	23;23	138.49
24157	16FE	18;09	333.31	24157	3MR	17;33	168.20
24158	18MR	3;19	2.41	24158	2AP	9;57	197.34
24159	16AP	12;28	31.02	24159	1MY	23;48	225.92
24160	15MY	22;09	59.26	24160	31MY	10;55	254.09
24161	14JN	8;57	87.34	24161	29JN	19;53	282.08
24162	13JL	21;28	115.52	24162	29JL	3;44	310.16
24163	12AU	12;08	144.02	24163	27AU	11;35	338.58
24164	11SE	4;53	173.02	24164	25SE	20;23	7.50
24165	10OC	22;51	202.57	24165	25OC	6;40	36.98
24166	9NO	16;32	232.57	24166	23NO	18;42	66.90
24167	9DE	8;34	262.78	24167	23DE	8;35	97.06

954

NUMBER	DATE	TIME	LONG.	NUMBER	DATE	TIME	LONG.
24168	7JA	22;21	292.91	24168	22JA	0;16	127.19
24169	6FE	9;58	322.72	24169	20FE	17;21	157.03
24170	7MR	19;49	352.02	24170	22MR	10;52	186.40
24171	6AP	4;23	20.77	24171	21AP	3;31	215.25
24172	5MY	12;10	49.06	24172	20MY	18;15	243.65
24173	3JN	19;59	77.07	24173	19JN	6;47	271.79
24174	3JL	4;50	105.06	24174	18JL	17;32	299.92
24175	1AU	15;49	133.30	24175	17AU	3;17	328.27
24176	31AU	5;43	162.02	24176	15SE	12;45	357.05
24177	29SE	22;34	191.33	24177	14OC	22;26	26.32
24178	29OC	17;22	221.22	24178	13NO	8;38	56.05
24179	28NO	12;26	251.48	24179	12DE	19;38	86.06
24180	28DE	6;12	281.81				

955

NUMBER	DATE	TIME	LONG.	NUMBER	DATE	TIME	LONG.
				24180	11JA	7;48	116.11
24181	26JA	21;43	311.89	24181	9FE	21;23	145.96
24182	25FE	10;37	341.49	24182	11MR	12;18	175.41
24183	26MR	20;58	10.49	24183	10AP	3;57	204.40
24184	25AP	5;12	38.94	24184	9MY	19;34	232.96
24185	24MY	12;10	67.00	24185	8JN	10;34	261.23
24186	22JN	18;56	94.90	24186	8JL	0;41	289.44
24187	22JL	2;46	122.93	24187	6AU	13;52	317.81
24188	20AU	12;43	151.33	24188	5SE	2;10	346.53
24189	19SE	1;33	180.30	24189	4OC	13;37	15.70
24190	18OC	17;22	209.89	24190	3NO	0;24	45.32
24191	17NO	11;36	240.00	24191	2DE	10;53	75.24
24192	17DE	7;06	270.38	24192	31DE	21;29	105.26

956

NEW MOONS

NUMBER	DATE	TIME	LONG.
24193	16JA	2;15	300.71
24194	14FE	19;28	330.64
24195	15MR	9;39	359.99
24196	13AP	20;38	28.71
24197	13MY	5;06	56.93
24198	11JN	12;08	84.88
24199	10JL	18;56	112.81
24200	9AU	2;34	140.99
24201	7SE	11;51	169.64
24202	6OC	23;24	198.86
24203	5NO	13;38	228.67
24204	5DE	6;37	258.89

FULL MOONS

NUMBER	DATE	TIME	LONG.
24193	30JA	8;30	135.12
24194	28FE	20;07	164.62
24195	29MR	8;23	193.65
24196	27AP	21;22	222.24
24197	27MY	11;13	250.55
24198	26JN	2;01	278.77
24199	25JL	17;30	307.13
24200	24AU	9;00	335.82
24201	22SE	23;46	4.94
24202	22OC	13;21	34.51
24203	21NO	1;43	64.42
24204	20DE	13;11	94.47

957

NEW MOONS

NUMBER	DATE	TIME	LONG.
24205	4JA	1;40	289.26
24206	2FE	21;11	319.44
24207	4MR	15;09	349.13
24208	3AP	6;13	18.19
24209	2MY	18;13	46.69
24210	1JN	3;49	74.79
24211	30JN	11;57	102.75
24212	29JL	19;34	130.83
24213	28AU	3;29	159.26
24214	26SE	12;23	188.20
24215	25OC	22;57	217.69
24216	24NO	11;51	247.66
24217	24DE	3;21	277.89

FULL MOONS

NUMBER	DATE	TIME	LONG.
24205	18JA	23;56	124.42
24206	17FE	10;06	154.01
24207	18MR	19;48	183.11
24208	17AP	5;31	211.74
24209	16MY	16;00	240.00
24210	15JN	4;03	268.13
24211	14JL	18;08	296.37
24212	13AU	10;01	324.93
24213	12SE	2;53	353.95
24214	11OC	19;42	23.47
24215	10NO	11;39	53.41
24216	10DE	2;15	83.56

958

NEW MOONS

NUMBER	DATE	TIME	LONG.
24218	22JA	20;55	308.09
24219	21FE	15;12	337.97
24220	23MR	8;40	7.32
24221	22AP	0;17	36.12
24222	21MY	13;45	64.46
24223	20JN	1;14	92.56
24224	19JL	11;08	120.65
24225	17AU	20;01	148.98
24226	16SE	4;38	177.73
24227	15OC	13;46	206.99
24228	14NO	0;07	236.74
24229	13DE	12;08	266.79

FULL MOONS

NUMBER	DATE	TIME	LONG.
24217	8JA	15;11	113.66
24218	7FE	2;16	143.43
24219	8MR	11;41	172.70
24220	6AP	19;56	201.43
24221	6MY	3;53	229.72
24222	4JN	12;37	257.77
24223	3JL	23;03	285.82
24224	2AU	11;46	314.13
24225	1SE	2;54	342.91
24226	30SE	20;03	12.25
24227	30OC	14;25	42.14
24228	29NO	8;46	72.37
24229	29DE	1;40	102.66

959

NEW MOONS

NUMBER	DATE	TIME	LONG.
24230	12JA	1;52	296.91
24231	10FE	16;57	326.81
24232	12MR	8;49	356.29
24233	11AP	0;55	25.28
24234	10MY	16;42	53.83
24235	9JN	7;35	82.10
24236	8JL	21;03	110.29
24237	7AU	8;56	138.61
24238	5SE	19;34	167.27
24239	5OC	5;38	196.40
24240	3NO	15;50	226.00
24241	3DE	2;31	255.94

FULL MOONS

NUMBER	DATE	TIME	LONG.
24230	27JA	16;00	132.69
24231	26FE	3;31	162.21
24232	27MR	12;39	191.16
24233	25AP	20;19	219.58
24234	25MY	3;30	247.64
24235	23JN	11;09	275.58
24236	22JL	20;07	303.66
24237	21AU	7;11	332.11
24238	19SE	20;59	1.13
24239	19OC	13;47	30.77
24240	18NO	8;58	60.93
24241	18DE	4;50	91.34

960

NEW MOONS

NUMBER	DATE	TIME	LONG.
24242	1JA	13;45	285.98
24243	31JA	1;31	315.87
24244	29FE	13;53	345.38
24245	30MR	3;08	14.44
24246	28AP	17;28	43.08
24247	28MY	8;40	71.43
24248	27JN	0;04	99.67
24249	26JL	14;57	128.02
24250	25AU	4;54	156.66
24251	23SE	17;59	185.72
24252	23OC	6;22	215.25
24253	21NO	18;09	245.14
24254	21DE	5;17	275.19

FULL MOONS

NUMBER	DATE	TIME	LONG.
24242	16JA	23;15	121.62
24243	15FE	14;45	151.48
24244	16MR	3;04	180.73
24245	14AP	12;49	209.39
24246	13MY	20;47	237.59
24247	12JN	3;51	265.53
24248	11JL	10;53	293.47
24249	9AU	18;47	321.68
24250	8SE	4;36	350.35
24251	7OC	17;17	19.64
24252	6NO	9;17	49.53
24253	6DE	4;00	79.83

961

NEW MOONS

NUMBER	DATE	TIME	LONG.
24255	19JA	15;47	305.11
24256	18FE	1;53	334.69
24257	19MR	12;07	3.81
24258	17AP	23;08	32.48
24259	17MY	11;23	60.81
24260	16JN	0;59	89.00
24261	15JL	15;45	117.27
24262	14AU	7;24	145.82
24263	12SE	23;29	174.83
24264	12OC	15;24	204.33
24265	11NO	6;25	234.23
24266	10DE	19;51	264.34

FULL MOONS

NUMBER	DATE	TIME	LONG.
24254	4JA	23;47	110.23
24255	3FE	18;36	140.36
24256	5MR	11;07	169.98
24257	4AP	0;51	198.98
24258	3MY	11;52	227.42
24259	1JN	20;42	255.49
24260	1JL	4;06	283.42
24261	30JL	11;05	311.48
24262	28AU	18;46	339.91
24263	27SE	4;11	8.88
24264	26OC	16;06	38.44
24265	25NO	6;45	68.49
24266	24DE	23;42	98.78

962

NEW MOONS

NUMBER	DATE	TIME	LONG.
24267	9JA	7;30	294.38
24268	7FE	17;40	324.10
24269	9MR	2;53	353.35
24270	7AP	11;45	22.10
24271	6MY	20;52	50.43
24272	5JN	6;51	78.52
24273	4JL	18;20	106.62
24274	3AU	7;54	134.97
24275	1SE	23;48	163.79
24276	1OC	17;32	193.18
24277	31OC	11;49	223.07
24278	30NO	5;00	253.26
24279	29DE	20;02	283.47

FULL MOONS

NUMBER	DATE	TIME	LONG.
24267	23JA	17;58	129.00
24268	22FE	12;21	158.87
24269	24MR	5;37	188.21
24270	22AP	20;43	216.97
24271	22MY	9;11	245.26
24272	20JN	19;12	273.30
24273	20JL	3;37	301.34
24274	18AU	11;30	329.64
24275	16SE	19;55	358.39
24276	16OC	5;34	27.68
24277	14NO	16;47	57.47
24278	14DE	5;42	87.57

963

NEW MOONS

NUMBER	DATE	TIME	LONG.
24280	28JA	8;41	313.42
24281	26FE	19;15	342.90
24282	28MR	4;14	11.82
24283	26AP	12;10	40.24
24284	25MY	19;43	68.31
24285	24JN	3;49	96.27
24286	23JL	13;33	124.38
24287	22AU	1;58	152.91
24288	20SE	17;31	182.01
24289	20OC	11;40	211.72
24290	19NO	6;59	241.89
24291	19DE	1;40	272.25

FULL MOONS

NUMBER	DATE	TIME	LONG.
24279	12JA	20;21	117.72
24280	11FE	12;34	147.65
24281	13MR	5;45	177.18
24282	11AP	22;47	206.19
24283	11MY	14;28	234.73
24284	10JN	4;07	262.94
24285	9JL	15;50	291.07
24286	8AU	2;12	319.34
24287	6SE	11;59	347.98
24288	5OC	21;44	17.10
24289	4NO	7;49	46.70
24290	3DE	18;26	76.64

964

NEW MOONS

NUMBER	DATE	TIME	LONG.
24292	17JA	18;25	302.46
24293	16FE	8;36	332.25
24294	16MR	20;06	1.46
24295	15AP	5;11	30.09
24296	14MY	12;34	58.25
24297	12JN	19;14	86.17
24298	12JL	2;22	114.12
24299	10AU	11;11	142.37
24300	8SE	22;35	171.12
24301	8OC	13;02	200.49
24302	7NO	6;15	230.44
24303	7DE	1;19	260.77

FULL MOONS

NUMBER	DATE	TIME	LONG.
24291	2JA	5;54	106.70
24292	31JA	18;33	136.61
24293	1MR	8;34	166.18
24294	30MR	23;38	195.30
24295	29AP	15;07	223.98
24296	29MY	6;22	252.32
24297	27JN	20;59	280.54
24298	27JL	10;49	308.85
24299	25AU	23;50	337.46
24300	24SE	11;59	6.50
24301	23OC	23;19	35.99
24302	22NO	10;04	65.85
24303	21DE	20;35	95.86

965

NEW MOONS

NUMBER	DATE	TIME	LONG.
24304	5JA	20;51	291.15
24305	4FE	15;09	321.25
24306	6MR	6;51	350.82
24307	4AP	19;19	19.76
24308	4MY	4;52	48.13
24309	2JN	12;27	76.15
24310	1JL	19;15	104.05
24311	31JL	2;26	132.13
24312	29AU	10;54	160.60
24313	27SE	21;22	189.62
24314	27OC	10;17	219.24
24315	26NO	1;56	249.33
24316	25DE	20;02	279.67

FULL MOONS

NUMBER	DATE	TIME	LONG.
24304	20JA	7;15	125.79
24305	18FE	18;21	155.41
24306	20MR	6;00	184.57
24307	18AP	18;19	213.28
24308	18MY	7;29	241.64
24309	16JN	21;42	269.85
24310	16JL	12;56	298.14
24311	15AU	4;43	326.72
24312	13SE	20;14	355.71
24313	13OC	10;45	25.16
24314	11NO	23;57	55.00
24315	11DE	11;57	85.05

966

NEW MOONS

NUMBER	DATE	TIME	LONG.
24317	24JA	15;26	309.94
24318	23FE	10;14	339.81
24319	25MR	2;44	9.10
24320	23AP	16;10	37.78
24321	23MY	2;51	66.00
24322	21JN	11;39	93.99
24323	20JL	19;33	122.01
24324	19AU	3;22	150.31
24325	17SE	11;51	179.08
24326	16OC	21;38	208.39
24327	15NO	9;22	238.20
24328	14DE	23;32	268.36

FULL MOONS

NUMBER	DATE	TIME	LONG.
24316	9JA	23;02	115.06
24317	8FE	9;22	144.79
24318	9MR	19;04	174.05
24319	8AP	4;27	202.81
24320	7MY	14;12	231.15
24321	6JN	1;11	259.28
24322	5JL	14;08	287.44
24323	4AU	5;14	315.86
24324	2SE	21;54	344.73
24325	2OC	15;08	14.10
24326	1NO	7;52	43.94
24327	30NO	23;23	74.07
24328	30DE	13;16	104.23

967

NEW MOONS

NUMBER	DATE	TIME	LONG.
24329	13JA	16;02	298.58
24330	12FE	9;55	328.58
24331	14MR	3;43	358.11
24332	12AP	20;10	27.08
24333	12MY	10;39	55.57
24334	10JN	23;09	83.73
24335	10JL	9;56	111.82
24336	8AU	19;27	140.06
24337	7SE	4;20	168.68
24338	6OC	13;18	197.78
24339	4NO	23;06	227.37
24340	4DE	10;17	257.34

FULL MOONS

NUMBER	DATE	TIME	LONG.
24329	29JA	1;15	134.14
24330	27FE	11;21	163.59
24331	28MR	19;54	192.49
24332	27AP	3;41	220.90
24333	26MY	11;44	248.98
24334	24JN	21;05	276.99
24335	24JL	8;35	305.18
24336	22AU	22;34	333.77
24337	21SE	14;55	2.92
24338	21OC	9;00	32.64
24339	20NO	3;44	62.80
24340	19DE	21;39	93.13

968

NEW NUMBER	DATE	TIME	LONG.	FULL NUMBER	DATE	TIME	LONG.
24341	2JA	23;05	287.45	24341	18JA	13;24	123.29
24342	1FE	13;21	317.43	24342	17FE	2;15	153.01
24343	2MR	4;36	347.04	24343	17MR	12;22	182.16
24344	31MR	20;22	16.17	24344	15AP	20;31	210.73
24345	30AP	12;11	44.85	24345	15MY	3;42	238.88
24346	30MY	3;31	73.19	24346	13JN	10;55	266.83
24347	28JN	17;50	101.40	24347	12JL	19;05	294.82
24348	28JL	6;42	129.68	24348	11AU	5;02	323.12
24349	26AU	18;09	158.23	24349	9SE	17;29	351.93
24350	25SE	4;42	187.22	24350	9OC	8;53	21.35
24351	24OC	14;59	216.68	24351	8NO	3;07	51.35
24352	23NO	1;31	246.53	24352	7DE	23;00	81.72
24353	22DE	12;29	276.57				

974

NEW NUMBER	DATE	TIME	LONG.	FULL NUMBER	DATE	TIME	LONG.
				24415	11JA	6;16	116.44
24416	26JA	10;13	311.80	24416	9FE	16;57	146.16
24417	25FE	3;16	341.57	24417	11MR	4;03	175.46
24418	26MR	17;17	10.73	24418	9AP	15;44	204.29
24419	25AP	4;09	39.29	24419	9MY	4;11	232.73
24420	24MY	12;33	67.40	24420	7JN	17;41	260.95
24421	22JN	19;37	95.32	24421	7JL	8;25	289.18
24422	22JL	2;32	123.31	24422	6AU	0;09	317.65
24423	20AU	10;22	151.63	24423	4SE	16;10	346.51
24424	18SE	19;53	180.46	24424	4OC	7;34	15.84
24425	18OC	7;37	209.87	24425	2NO	21;42	45.60
24426	16NO	21;56	239.81	24426	2DE	10;27	75.62
24427	16DE	14;50	270.08	24427	31DE	22;00	105.68

969

NEW NUMBER	DATE	TIME	LONG.	FULL NUMBER	DATE	TIME	LONG.
24354	20JA	23;52	306.52	24353	6JA	18;25	112.09
24355	19FE	11;40	336.16	24354	5FE	11;26	142.13
24356	21MR	0;08	5.34	24355	7MR	1;14	171.61
24357	19AP	13;38	34.09	24356	5AP	12;05	200.47
24358	19MY	4;15	62.50	24357	4MY	20;45	228.80
24359	17JN	19;34	90.75	24358	3JN	4;08	256.80
24360	17JL	10;53	119.05	24359	2JL	11;04	284.71
24361	16AU	1;33	147.59	24360	31JL	18;28	312.80
24362	14SE	15;21	176.53	24361	30AU	3;17	341.30
24363	14OC	4;21	205.93	24362	28SE	14;32	10.37
24364	12NO	16;40	235.74	24363	28OC	4;56	40.05
24365	12DE	4;14	265.77	24364	26NO	22;28	70.23
				24365	26DE	17;55	100.63

975

NEW NUMBER	DATE	TIME	LONG.	FULL NUMBER	DATE	TIME	LONG.
24428	15JA	9;39	300.40	24428	30JA	8;37	135.52
24429	14FE	4;50	330.42	24429	28FE	18;26	164.94
24430	15MR	22;30	359.92	24430	30MR	3;41	193.84
24431	14AP	13;24	28.81	24431	28AP	12;54	222.29
24432	14MY	1;21	57.17	24432	27MY	22;57	250.45
24433	12JN	11;02	85.22	24433	26JN	10;44	278.55
24434	11JL	19;22	113.21	24434	26JL	0;45	306.85
24435	10AU	3;17	141.41	24435	24AU	16;52	335.55
24436	8SE	11;32	170.01	24436	23SE	10;12	4.77
24437	7OC	20;45	199.15	24437	23OC	3;33	34.49
24438	6NO	7;31	228.80	24438	21NO	19;59	64.57
24439	5DE	20;25	258.85	24439	21DE	10;52	94.77

970

NEW NUMBER	DATE	TIME	LONG.	FULL NUMBER	DATE	TIME	LONG.
24366	10JA	15;01	295.77	24366	25JA	13;21	130.89
24367	9FE	1;06	325.47	24367	24FE	7;01	160.70
24368	10MR	10;58	354.73	24368	25MR	22;03	189.91
24369	8AP	21;15	23.52	24369	24AP	10;20	218.54
24370	8MY	8;35	51.92	24370	23MY	20;11	246.72
24371	6JN	21;19	80.11	24371	22JN	4;14	274.68
24372	6JL	11;29	108.32	24372	21JL	11;23	302.67
24373	5AU	2;49	136.76	24373	19AU	18;43	330.96
24374	3SE	18;57	165.62	24374	18SE	3;17	359.74
24375	3OC	11;21	194.97	24375	17OC	13;59	29.10
24376	2NO	3;11	224.78	24376	16NO	3;19	58.99
24377	1DE	17;40	254.88	24377	15DE	19;09	89.22
24378	31DE	6;16	284.98				

976

NEW NUMBER	DATE	TIME	LONG.	FULL NUMBER	DATE	TIME	LONG.
24440	4JA	11;41	289.06	24440	19JA	23;49	124.80
24441	3FE	4;49	319.14	24441	18FE	10;45	154.42
24442	3MR	22;36	348.83	24442	18MR	19;50	183.50
24443	2AP	15;38	17.99	24443	17AP	3;43	212.05
24444	2MY	7;02	46.63	24444	16MY	11;19	240.21
24445	31MY	20;31	74.88	24445	14JN	19;46	268.19
24446	30JN	8;13	102.99	24446	14JL	6;03	296.28
24447	29JL	18;30	131.17	24447	12AU	18;47	324.70
24448	28AU	3;51	159.66	24448	11SE	10;06	353.65
24449	26SE	12;54	188.61	24449	11OC	3;36	23.19
24450	25OC	22;22	218.06	24450	9NO	22;21	53.24
24451	24NO	8;53	247.92	24451	9DE	17;01	83.56
24452	23DE	20;50	278.00				

971

NEW NUMBER	DATE	TIME	LONG.	FULL NUMBER	DATE	TIME	LONG.
				24378	14JA	12;45	119.48
24379	29JA	17;04	314.84	24379	13FE	7;02	149.49
24380	28FE	2;32	344.25	24380	15MR	0;49	179.00
24381	29MR	11;17	13.15	24381	13AP	16;58	207.95
24382	27AP	19;59	41.58	24382	13MY	6;44	236.40
24383	27MY	5;16	69.71	24383	11JN	17;58	264.51
24384	25JN	15;48	97.76	24384	11JL	3;12	292.54
24385	25JL	4;14	126.00	24385	9AU	11;24	320.74
24386	23AU	19;02	154.63	24386	7SE	19;38	349.33
24387	22SE	12;09	183.83	24387	7OC	4;46	18.45
24388	22OC	6;36	213.58	24388	5NO	15;17	48.09
24389	21NO	0;44	243.72	24389	5DE	3;22	78.10
24390	20DE	17;04	273.98				

977

NEW NUMBER	DATE	TIME	LONG.	FULL NUMBER	DATE	TIME	LONG.
24453	22JA	10;14	308.03	24452	8JA	10;03	113.83
24454	21FE	0;47	337.76	24453	7FE	0;21	143.74
24455	22MR	16;03	7.03	24454	8MR	11;41	173.09
24456	21AP	7;39	35.84	24455	6AP	20;34	201.84
24457	20MY	23;10	64.27	24456	6MY	3;59	230.11
24458	19JN	14;05	92.50	24457	4JN	10;58	258.09
24459	19JL	3;52	120.75	24458	3JL	18;31	286.03
24460	17AU	16;15	149.22	24459	2AU	3;30	314.19
24461	16SE	3;28	178.07	24460	31AU	14;40	342.80
24462	15OC	14;03	207.40	24461	30SE	4;38	12.01
24463	14NO	0;35	237.15	24462	29OC	21;38	41.82
24464	13DE	11;23	267.15	24463	28NO	17;00	72.09
				24464	28DE	12;57	102.51

972

NEW NUMBER	DATE	TIME	LONG.	FULL NUMBER	DATE	TIME	LONG.
				24390	3JA	17;03	108.24
24391	19JA	6;54	304.06	24391	2FE	8;18	138.25
24392	17FE	18;23	333.71	24392	3MR	0;49	167.90
24393	18MR	3;58	2.83	24393	1AP	17;50	197.08
24394	16AP	12;12	31.40	24394	1MY	10;10	225.76
24395	15MY	19;42	59.55	24395	31MY	0;51	254.07
24396	14JN	3;17	87.51	24396	29JN	13;36	282.21
24397	13JL	11;53	115.53	24397	29JL	0;45	310.43
24398	11AU	23;00	143.87	24398	27AU	10;59	338.95
24399	10SE	13;02	172.76	24399	25SE	20;57	7.92
24400	10OC	6;09	202.27	24400	25OC	7;02	37.38
24401	9NO	1;15	232.32	24401	23NO	17;27	67.24
24402	8DE	20;34	262.66	24402	23DE	4;26	97.28

978

NEW NUMBER	DATE	TIME	LONG.	FULL NUMBER	DATE	TIME	LONG.
24465	11JA	22;30	297.16	24465	27JA	7;20	132.71
24466	10FE	9;52	326.90	24466	25FE	22;42	162.40
24467	11MR	21;41	356.22	24467	27MR	10;51	191.47
24468	10AP	10;19	25.08	24468	25AP	20;25	219.97
24469	10MY	0;08	53.56	24469	25MY	4;18	248.06
24470	8JN	15;03	81.83	24470	23JN	11;21	275.97
24471	8JL	6;29	110.09	24471	22JL	18;27	303.98
24472	6AU	21;43	138.55	24472	21AU	2;31	332.31
24473	5SE	12;14	167.37	24473	19SE	12;32	1.17
24474	5OC	1;56	196.64	24474	19OC	1;23	30.64
24475	3NO	14;50	226.35	24475	17NO	17;27	60.67
24476	3DE	2;57	256.35	24476	17DE	12;08	91.02

973

NEW NUMBER	DATE	TIME	LONG.	FULL NUMBER	DATE	TIME	LONG.
24403	7JA	14;28	292.97	24403	21JA	16;18	127.25
24404	6FE	5;57	322.94	24404	20FE	5;23	156.92
24405	7MR	18;41	352.36	24405	21MR	19;40	186.17
24406	6AP	4;49	21.18	24406	20AP	10;45	214.97
24407	5MY	12;51	49.47	24407	20MY	2;01	243.39
24408	3JN	19;40	77.45	24408	18JN	16;58	271.63
24409	3JL	2;24	105.36	24409	18JL	7;21	299.90
24410	1AU	10;16	133.47	24410	16AU	21;01	328.41
24411	30AU	20;21	162.02	24411	15SE	9;55	357.32
24412	29SE	9;21	191.18	24412	14OC	21;55	26.69
24413	29OC	1;20	220.94	24413	13NO	9;07	56.47
24414	27NO	19;40	251.16	24414	12DE	19;46	86.46
24415	27DE	15;10	281.56				

979

NEW NUMBER	DATE	TIME	LONG.	FULL NUMBER	DATE	TIME	LONG.
24477	1JA	14;09	286.39	24477	16JA	7;44	121.36
24478	31JA	0;26	316.21	24478	15FE	2;19	151.35
24479	1MR	10;09	345.62	24479	16MR	18;37	180.78
24480	30MR	19;52	14.53	24480	15AP	8;10	209.59
24481	29AP	6;21	43.02	24481	14MY	19;09	237.91
24482	28MY	18;09	71.24	24482	13JN	4;02	265.93
24483	27JN	7;30	99.40	24483	12JL	11;37	293.89
24484	26JL	22;19	127.74	24484	10AU	18;51	322.07
24485	25AU	14;16	156.45	24485	9SE	2;49	350.67
24486	24SE	6;53	185.65	24486	8OC	12;29	19.83
24487	23OC	23;24	215.35	24487	7NO	0;33	49.55
24488	22NO	14;53	245.40	24488	6DE	15;11	79.68
24489	22DE	4;34	275.55				

980

NEW NUMBER	DATE	TIME	LONG.	FULL NUMBER	DATE	TIME	LONG.
				24489	5JA	7;54	109.95
24490	20JA	16;13	305.52	24490	4FE	1;47	140.05
24491	19FE	2;10	335.10	24491	4MR	19;44	169.73
24492	19MR	11;00	4.15	24492	3AP	12;39	198.87
24493	17AP	19;26	32.71	24493	3MY	3;36	227.48
24494	17MY	4;09	60.91	24494	1JN	16;07	255.69
24495	15JN	13;50	88.94	24495	1JL	2;22	283.74
24496	15JL	1;10	117.07	24496	30JL	11;07	311.87
24497	13AU	14;47	145.54	24497	28AU	19;25	340.33
24498	12SE	6;56	174.54	24498	27SE	4;13	9.28
24499	12OC	1;06	204.12	24499	26OC	14;09	38.75
24500	10NO	19;50	234.18	24500	25NO	1;30	68.66
24501	10DE	13;23	264.45	24501	24DE	14;18	98.77

981

NEW NUMBER	DATE	TIME	LONG.	FULL NUMBER	DATE	TIME	LONG.
24502	9JA	4;33	294.64	24502	23JA	4;36	128.83
24503	7FE	17;08	324.47	24503	21FE	20;16	158.60
24504	9MR	3;29	353.77	24504	23MR	12;53	187.91
24505	7AP	12;10	22.51	24505	22AP	5;29	216.75
24506	6MY	19;48	50.77	24506	21MY	21;01	245.17
24507	5JN	3;07	78.76	24507	20JN	10;47	273.36
24508	4JL	11;04	106.72	24508	19JL	22;50	301.54
24509	2AU	20;47	134.92	24509	18AU	9;41	329.95
24510	1SE	9;17	163.59	24510	16SE	19;59	358.78
24511	1OC	1;02	192.88	24511	16OC	6;12	28.10
24512	30OC	19;28	222.77	24512	14NO	16;35	57.86
24513	29NO	15;02	253.06	24513	14DE	3;16	87.86
24514	29DE	9;52	283.43				

982

NEW NUMBER	DATE	TIME	LONG.	FULL NUMBER	DATE	TIME	LONG.
				24514	12JA	14;33	117.86
24515	28JA	2;37	313.55	24515	11FE	2;47	147.64
24516	26FE	16;41	343.18	24516	12MR	16;11	177.01
24517	28MR	3;59	12.20	24517	11AP	6;39	205.93
24518	26AP	12;53	40.66	24518	10MY	21;40	234.45
24519	25MY	20;07	68.72	24519	9JN	12;45	262.72
24520	24JN	2;43	96.62	24520	9JL	3;30	290.96
24521	23JL	9;53	124.63	24521	7AU	17;45	319.39
24522	21AU	18;48	153.01	24522	6SE	7;20	348.18
24523	20SE	6;22	181.94	24523	5OC	20;05	17.42
24524	19OC	20;59	211.49	24524	4NO	7;54	47.10
24525	18NO	14;19	241.58	24525	3DE	18;54	77.06
24526	18DE	9;24	271.95				

983

NEW NUMBER	DATE	TIME	LONG.	FULL NUMBER	DATE	TIME	LONG.
				24526	2JA	5;26	107.07
24527	17JA	4;50	302.28	24527	31JA	15;52	136.89
24528	15FE	22;58	332.24	24528	2MR	2;30	166.32
24529	17MR	14;28	1.62	24529	31MR	13;35	195.28
24530	16AP	2;48	30.38	24530	30AP	1;21	223.81
24531	15MY	12;17	58.63	24531	29MY	14;05	252.06
24532	13JN	19;53	86.59	24532	28JN	4;07	280.26
24533	13JL	2;48	114.53	24533	27JL	19;28	308.62
24534	11AU	10;11	142.72	24534	26AU	11;40	337.35
24535	9SE	18;54	171.36	24535	25SE	3;47	6.55
24536	9OC	5;34	200.57	24536	24OC	18;54	36.20
24537	7NO	18;38	230.34	24537	23NO	8;32	66.18
24538	7DE	10;14	260.52	24538	22DE	20;45	96.26

984

NEW NUMBER	DATE	TIME	LONG.	FULL NUMBER	DATE	TIME	LONG.
24539	6JA	4;07	290.84	24539	21JA	7;47	126.21
24540	4FE	23;11	320.98	24540	19FE	17;50	155.78
24541	5MR	17;39	350.67	24541	20MR	3;07	184.84
24542	4AP	9;54	19.76	24542	18AP	12;02	213.41
24543	3MY	23;14	48.30	24543	17MY	21;20	241.62
24544	2JN	9;58	76.44	24544	16JN	7;59	269.69
24545	1JL	18;57	104.43	24545	15JL	20;48	297.89
24546	31JL	3;08	132.55	24546	14AU	12;00	326.43
24547	29AU	11;19	161.01	24547	13SE	5;02	355.48
24548	27SE	20;09	189.96	24548	12OC	22;47	25.06
24549	27OC	6;11	219.45	24549	11NO	16;01	55.06
24550	25NO	17;59	249.39	24550	11DE	7;54	85.27
24551	25DE	8;00	279.55				

985

NEW NUMBER	DATE	TIME	LONG.	FULL NUMBER	DATE	TIME	LONG.
				24551	9JA	21;54	115.41
24552	24JA	0;07	309.68	24552	8FE	9;47	145.19
24553	22FE	17;30	339.51	24553	9MR	19;36	174.46
24554	24MR	10;48	8.84	24554	8AP	3;49	203.17
24555	23AP	2;56	37.64	24555	7MY	11;15	231.42
24556	22MY	17;20	66.01	24556	5JN	19;00	259.42
24557	21JN	5;59	94.15	24557	5JL	4;11	287.43
24558	20JL	17;06	122.30	24558	3AU	15;38	315.71
24559	19AU	3;05	150.69	24559	2SE	5;45	344.46
24560	17SE	12;26	179.49	24560	1OC	22;22	13.80
24561	16OC	21;49	208.79	24561	31OC	16;48	43.70
24562	15NO	7;52	238.53	24562	30NO	11;52	73.97
24563	14DE	19;04	268.56	24563	30DE	5;59	104.31

986

NEW NUMBER	DATE	TIME	LONG.	FULL NUMBER	DATE	TIME	LONG.
24564	13JA	7;38	298.62	24564	28JA	21;44	134.39
24565	11FE	21;25	328.44	24565	27FE	10;27	163.94
24566	13MR	12;05	357.85	24566	28MR	20;20	192.90
24567	12AP	3;16	26.79	24567	27AP	4;15	221.31
24568	11MY	18;41	55.32	24568	26MY	11;13	249.35
24569	10JN	9;56	83.60	24569	24JN.	18;20	277.27
24570	10JL	0;28	111.83	24570	24JL	2;29	305.32
24571	8AU	13;47	140.23	24571	22AU	12;32	333.75
24572	7SE	1;50	168.97	24572	21SE	1;07	2.74
24573	6OC	12;55	198.15	24573	20OC	16;43	32.35
24574	4NO	23;37	227.79	24574	19NO	11;07	62.49
24575	4DE	10;23	257.74	24575	19DE	7;05	92.90

987

NEW NUMBER	DATE	TIME	LONG.	FULL NUMBER	DATE	TIME	LONG.
24576	2JA	21;19	287.77	24576	18JA	2;29	123.22
24577	1FE	8;24	317.61	24577	16FE	19;23	153.12
24578	2MR	19;42	347.06	24578	18MR	9;01	182.41
24579	1AP	7;34	16.04	24579	16AP	19;42	211.09
24580	30AP	20;30	44.61	24580	16MY	4;15	239.29
24581	30MY	10;42	72.91	24581	14JN	11;36	267.24
24582	29JN	1;56	101.15	24582	13JL	18;37	295.19
24583	28JL	17;29	129.53	24583	12AU	2;11	323.39
24584	27AU	8;38	158.24	24584	10SE	11;12	352.06
24585	25SE	23;03	187.38	24585	9OC	22;37	21.31
24586	25OC	12;38	216.97	24586	8NO	13;08	51.15
24587	24NO	1;21	246.92	24587	8DE	6;39	81.41
24588	23DE	13;05	276.99				

988

NEW NUMBER	DATE	TIME	LONG.	FULL NUMBER	DATE	TIME	LONG.
				24588	7JA	1;57	111.79
24589	21JA	23;46	306.91	24589	5FE	21;08	141.92
24590	20FE	9;32	336.46	24590	6MR	14;32	171.56
24591	20MR	18;56	5.52	24591	5AP	5;22	200.58
24592	19AP	4;41	34.11	24592	4MY	17;33	229.06
24593	18MY	15;32	62.38	24593	3JN	3;26	257.16
24594	17JN	3;57	90.52	24594	2JL	11;39	285.13
24595	16JL	18;01	118.77	24595	31JL	19;03	313.22
24596	15AU	9;34	147.33	24596	30AU	2;40	341.66
24597	14SE	2;09	176.38	24597	28SE	11;32	10.63
24598	13OC	19;06	205.93	24598	27OC	22;27	40.17
24599	12NO	11;29	235.91	24599	26NO	11;50	70.17
24600	12DE	2;18	266.08	24600	26DE	3;30	100.41

989

NEW NUMBER	DATE	TIME	LONG.	FULL NUMBER	DATE	TIME	LONG.
24601	10JA	14;59	296.16	24601	24JA	20;44	130.58
24602	9FE	1;38	325.89	24602	23FE	14;34	160.41
24603	10MR	10;47	355.11	24603	25MR	7;55	189.73
24604	8AP	19;09	23.82	24604	23AP	23;50	218.51
24605	8MY	3;26	52.10	24605	23MY	13;35	246.84
24606	6JN	12;23	80.14	24606	22JN	1;00	274.94
24607	5JL	22;43	108.20	24607	21JL	10;33	303.03
24608	4AU	11;08	136.52	24608	19AU	19;09	331.37
24609	3SE	2;06	165.32	24609	18SE	3;48	0.16
24610	2OC	19;34	194.71	24610	17OC	13;17	29.46
24611	1NO	14;27	224.64	24611	16NO	0;01	59.24
24612	1DE	8;59	254.90	24612	15DE	12;04	89.31
24613	31DE	1;31	285.17				

990

NEW NUMBER	DATE	TIME	LONG.	FULL NUMBER	DATE	TIME	LONG.
				24613	14JA	1;28	119.40
24614	29JA	15;22	315.16	24614	12FE	16;14	149.26
24615	28FE	2;40	344.65	24615	14MR	8;09	178.71
24616	29MR	11;59	13.56	24616	13AP	0;39	207.68
24617	27AP	19;55	41.97	24617	12MY	16;42	236.23
24618	27MY	3;10	70.02	24618	11JN	7;24	264.48
24619	25JN	10;36	97.95	24619	10JL	20;25	292.66
24620	24JL	19;17	126.03	24620	9AU	8;01	320.99
24621	23AU	6;20	154.51	24621	7SE	18;48	349.69
24622	21SE	20;32	183.57	24622	7OC	5;16	18.86
24623	21OC	13;53	213.27	24623	5NO	15;43	48.50
24624	20NO	9;14	243.46	24624	5DE	2;18	78.45
24625	20DE	4;43	273.85				

991

NEW NUMBER	DATE	TIME	LONG.	FULL NUMBER	DATE	TIME	LONG.
				24625	3JA	13;12	108.47
24626	18JA	22;38	304.10	24626	2FE	0;44	138.33
24627	17FE	14;02	333.92	24627	3MR	13;16	167.82
24628	19MR	2;36	3.16	24628	2AP	2;55	196.86
24629	17AP	12;32	31.80	24629	1MY	17;28	225.48
24630	16MY	20;25	59.97	24630	31MY	8;26	253.81
24631	15JN	3;10	87.89	24631	29JN	23;23	282.04
24632	14JL	9;55	115.83	24632	29JL	14;04	310.39
24633	12AU	17;53	144.06	24633	28AU	4;16	339.06
24634	11SE	4;08	172.78	24634	26SE	17;46	8.18
24635	10OC	17;18	202.12	24635	26OC	6;19	37.75
24636	9NO	9;24	232.04	24636	24NO	17;53	67.65
24637	9DE	3;47	262.34	24637	24DE	4;39	97.68

992 / 998

NEW MOONS				FULL MOONS				NEW MOONS				FULL MOONS			
24638	7JA	23;11	292.72	24638	22JA	15;00	127.58	24712	1JA	13;15	286.74	24712	15JA	16;03	121.09
24639	6FE	18;03	322.83	24639	21FE	1;18	157.14	24713	31JA	0;50	316.62	24713	14FE	9;28	151.04
24640	7MR	10;54	352.43	24640	21MR	11;53	186.24	24714	1MR	10;32	346.02	24714	16MR	2;57	180.52
24641	6AP	0;45	21.40	24641	19AP	22;59	214.88	24715	30MR	19;00	14.89	24715	14AP	19;32	209.47
24642	5MY	11;32	49.81	24642	19MY	10;58	243.18	24716	29AP	3;03	43.28	24716	14MY	10;22	237.95
24643	3JN	19;56	77.85	24643	18JN	0;11	271.35	24717	28MY	11;25	71.36	24717	12JN	23;00	266.11
24644	3JL	3;06	105.77	24644	17JL	14;54	299.63	24718	26JN	20;51	99.37	24718	12JL	9;33	294.20
24645	1AU	10;13	133.86	24645	16AU	6;55	328.23	24719	26JL	8;07	127.56	24719	10AU	18;42	322.45
24646	30AU	18;18	162.33	24646	14SE	23;28	357.28	24720	24AU	21;49	156.17	24720	9SE	3;26	351.09
24647	29SE	4;04	191.35	24647	14OC	15;29	26.82	24721	23SE	14;14	185.36	24721	8OC	12;38	20.23
24648	28OC	15;59	220.94	24648	13NO	6;08	56.73	24722	23OC	8;48	215.13	24722	6NO	22;50	49.87
24649	27NO	6;19	250.99	24649	12DE	19;10	86.83	24723	22NO	3;56	245.33	24723	6DE	10;15	79.86
24650	26DE	23;03	281.27					24724	21DE	21;45	275.65				

993 / 999

NEW MOONS				FULL MOONS				NEW MOONS				FULL MOONS			
				24650	11JA	6;47	116.86					24724	4JA	22;53	109.96
24651	25JA	17;33	311.49	24651	9FE	17;12	146.57	24725	20JA	12;59	305.78	24725	3FE	12;45	139.90
24652	24FE	12;21	341.34	24652	11MR	2;39	175.80	24726	19FE	1;27	335.45	24726	5MR	3;50	169.47
24653	26MR	5;41	10.64	24653	9AP	11;27	204.51	24727	20MR	11;34	4.57	24727	3AP	19;52	198.58
24654	24AP	20;25	39.36	24654	8MY	20;14	232.80	24728	18AP	19;58	33.12	24728	3MY	12;04	227.25
24655	24MY	8;23	67.62	24655	7JN	5;55	260.87	24729	18MY	3;21	61.26	24729	2JN	3;29	255.58
24656	22JN	18;13	95.65	24656	6JL	17;28	288.98	24730	16JN	10;30	89.20	24730	1JL	17;27	283.77
24657	22JL	2;50	123.71	24657	5AU	7;30	317.37	24731	15JL	18;24	117.19	24731	31JL	5;54	312.05
24658	20AU	11;06	152.05	24658	3SE	23;52	346.24	24732	14AU	4;09	145.50	24732	29AU	17;17	340.63
24659	18SE	19;44	180.84	24659	3OC	17;39	15.66	24733	12SE	16;47	174.35	24733	28SE	4;07	9.66
24660	18OC	5;14	210.16	24660	2NO	11;31	45.56	24734	12OC	8;43	203.83	24734	27OC	14;47	39.17
24661	16NO	16;10	239.96	24661	2DE	4;21	75.75	24735	11NO	3;23	233.87	24735	26NO	1;25	69.05
24662	16DE	5;00	270.06	24662	31DE	19;26	105.96	24736	10DE	23;07	264.24	24736	25DE	12;07	99.07

994 / 1000

NEW MOONS				FULL MOONS				NEW MOONS				FULL MOONS			
24663	14JA	19;58	300.21	24663	30JA	8;22	135.90	24737	9JA	18;01	294.59	24737	23JA	23;09	128.99
24664	13FE	12;37	330.14	24664	28FE	19;05	165.35	24738	8FE	10;42	324.59	24738	22FE	10;54	158.60
24665	15MR	5;51	359.63	24665	30MR	3;52	194.24	24739	9MR	0;36	354.04	24739	22MR	23;42	187.76
24666	13AP	22;28	28.59	24666	28AP	11;24	222.62	24740	7AP	11;43	22.87	24740	21AP	13;33	216.49
24667	13MY	13;40	57.09	24667	27MY	18;43	250.66	24741	6MY	20;28	51.19	24741	21MY	4;10	244.89
24668	12JN	3;13	85.30	24668	26JN	2;57	278.63	24742	5JN	3;37	79.17	24742	19JN	19;06	273.12
24669	11JL	15;13	113.44	24669	25JL	13;09	306.77	24743	4JL	10;14	107.07	24743	19JL	10;03	301.42
24670	10AU	1;56	141.74	24670	24AU	1;57	335.34	24744	2AU	17;30	135.17	24744	18AU	0;45	329.98
24671	8SE	11;47	170.42	24671	22SE	17;29	4.47	24745	1SE	2;35	163.71	24745	16SE	14;57	358.97
24672	7OC	21;17	199.56	24672	22OC	11;17	34.20	24746	30SE	14;19	192.83	24746	16OC	4;17	28.42
24673	6NO	7;05	229.17	24673	21NO	6;22	64.38	24747	30OC	5;04	222.56	24747	14NO	16;33	58.25
24674	5DE	17;43	259.13	24674	21DE	1;15	94.75	24748	28NO	22;27	252.75	24748	14DE	3;46	88.27
								24749	28DE	17;28	283.13				

995 / 1001

NEW MOONS				FULL MOONS				NEW MOONS				FULL MOONS			
24675	4JA	5;31	289.19	24675	19JA	18;21	124.96					24749	12JA	14;14	118.24
24676	2FE	18;32	319.10	24676	18FE	8;34	154.72	24750	27JA	12;43	313.36	24750	11FE	0;23	147.93
24677	4MR	8;31	348.64	24677	19MR	19;42	183.89	24751	26FE	6;37	343.15	24751	12MR	10;33	177.17
24678	2AP	23;10	17.71	24678	18AP	4;21	212.47	24752	27MR	21;56	12.34	24752	10AP	21;05	205.93
24679	2MY	14;16	46.34	24679	17MY	11;34	240.60	24753	26AP	10;09	40.94	24753	10MY	8;19	234.30
24680	1JN	5;34	74.68	24680	15JN	18;26	268.53	24754	25MY	19;37	69.09	24754	8JN	20;42	262.47
24681	30JN	20;34	102.91	24681	15JL	1;57	296.50	24755	24JN	3;19	97.03	24755	8JL	10;36	290.68
24682	30JL	10;44	131.25	24682	13AU	11;00	324.78	24756	23JL	10;24	125.04	24756	7AU	2;07	319.15
24683	28AU	23;41	159.89	24683	11SE	22;19	353.56	24757	21AU	18;02	153.36	24757	5SE	18;45	348.05
24684	27SE	11;29	188.94	24684	11OC	12;28	22.95	24758	20SE	3;01	182.19	24758	5OC	11;27	17.45
24685	26OC	22;33	218.46	24685	10NO	5;37	52.92	24759	19OC	13;55	211.59	24759	4NO	3;08	47.29
24686	25NO	9;24	248.34	24686	10DE	1;04	83.28	24760	18NO	3;04	241.49	24760	3DE	17;10	77.37
24687	24DE	20;16	278.37					24761	17DE	18;34	271.71				

996 / 1002

NEW MOONS				FULL MOONS				NEW MOONS				FULL MOONS			
				24687	8JA	21;01	113.67					24761	2JA	5;31	107.46
24688	23JA	7;10	308.29	24688	7FE	15;16	143.75	24762	16JA	12;10	301.97	24762	31JA	16;26	137.30
24689	21FE	18;07	337.87	24689	8MR	6;29	173.26	24763	15FE	6;49	331.96	24763	2MR	2;11	166.70
24690	22MR	5;22	6.98	24690	6AP	18;28	202.15	24764	17MR	0;54	1.45	24764	31MR	11;03	195.57
24691	20AP	17;25	35.65	24691	6MY	3;56	230.50	24765	15AP	16;54	30.37	24765	29AP	19;32	223.97
24692	20MY	6;45	64.00	24692	4JN	11;45	258.51	24766	15MY	6;11	58.77	24766	29MY	4;27	252.07
24693	18JN	21;25	92.23	24693	3JL	18;52	286.43	24767	13JN	17;02	86.86	24767	27JN	14;51	280.12
24694	18JL	12;57	120.54	24694	2AU	2;07	314.52	24768	13JL	2;17	114.90	24768	27JL	3;34	308.37
24695	17AU	4;35	149.14	24695	31AU	10;24	343.02	24769	11AU	10;49	143.13	24769	25AU	18;55	337.06
24696	15SE	19;41	178.14	24696	29SE	20;37	12.07	24770	9SE	19;23	171.77	24770	24SE	12;20	6.30
24697	15OC	9;59	207.62	24697	29OC	9;36	41.71	24771	9OC	4;34	200.92	24771	24OC	6;33	36.07
24698	13NO	23;23	237.49	24698	28NO	1;42	71.84	24772	7NO	14;49	230.57	24772	23NO	0;14	66.22
24699	13DE	11;45	267.56	24699	27DE	20;15	102.20	24773	7DE	2;40	260.59	24773	22DE	16;23	96.47

997 / 1003

NEW MOONS				FULL MOONS				NEW MOONS				FULL MOONS			
24700	11JA	22;57	297.57	24700	26JA	15;35	132.44	24774	5JA	16;27	290.73	24774	21JA	6;27	126.54
24701	10FE	9;00	327.26	24701	25FE	9;53	162.26	24775	4FE	8;08	320.74	24775	19FE	18;11	156.18
24702	11MR	18;18	356.47	24702	27MR	1;55	191.50	24776	6MR	0;56	350.37	24776	21MR	3;44	185.26
24703	10AP	3;31	25.19	24703	25AP	15;21	220.15	24777	4AP	17;45	19.50	24777	19AP	11;37	213.78
24704	9MY	13;30	53.52	24704	25MY	2;20	248.37	24778	4MY	9;34	48.14	24778	18MY	18;46	241.91
24705	8JN	0;54	81.66	24705	23JN	11;21	276.36	24779	2JN	23;56	76.42	24779	17JN	2;18	269.86
24706	7JL	14;05	109.83	24706	22JL	19;10	304.40	24780	2JL	12;48	104.57	24780	16JL	11;21	297.90
24707	6AU	4;59	138.27	24707	21AU	2;43	332.71	24781	1AU	0;20	132.82	24781	14AU	22;49	326.29
24708	4SE	21;17	167.14	24708	19SE	11;00	1.50	24782	30AU	10;49	161.38	24782	13SE	13;05	355.21
24709	4OC	14;25	196.55	24709	18OC	20;56	30.85	24783	28SE	20;40	190.38	24783	13OC	5;58	24.75
24710	3NO	7;29	226.43	24710	17NO	9;07	60.71	24784	28OC	6;26	219.86	24784	12NO	0;43	54.81
24711	2DE	23;23	256.59	24711	16DE	23;39	90.88	24785	26NO	16;41	249.72	24785	11DE	20;01	85.16
								24786	26DE	3;51	279.77				

1004

NEW — NUMBER	DATE	TIME	LONG.	FULL — NUMBER	DATE	TIME	LONG.
				24786	10JA	14;14	115.47
24787	24JA	16;07	309.74	24787	9FE	5;57	145.42
24788	23FE	5;23	339.39	24788	9MR	18;30	174.80
24789	23MR	19;26	8.60	24789	8AP	4;11	203.57
24790	22AP	10;03	37.35	24790	7MY	11;53	231.83
24791	22MY	1;08	65.75	24791	5JN	18;43	259.80
24792	20JN	16;20	93.99	24792	5JL	1;47	287.72
24793	20JL	7;08	122.29	24793	3AU	10;00	315.87
24794	18AU	20;58	150.83	24794	1SE	20;10	344.46
24795	17SE	9;36	179.77	24795	1OC	8;56	13.63
24796	16OC	21;15	209.16	24796	31OC	0;41	43.41
24797	15NO	8;21	238.95	24797	29NO	19;11	73.66
24798	14DE	19;15	268.96	24798	29DE	15;08	104.08

1005

NEW — NUMBER	DATE	TIME	LONG.	FULL — NUMBER	DATE	TIME	LONG.
24799	13JA	6;05	298.94	24799	28JA	10;26	134.30
24800	11FE	16;50	328.64	24800	27FE	3;10	164.03
24801	13MR	3;37	357.90	24801	28MR	16;38	193.14
24802	11AP	14;54	26.68	24802	27AP	3;12	221.65
24803	11MY	3;18	55.09	24803	26MY	11;41	249.76
24804	9JN	17;10	83.32	24804	24JN	19;05	277.69
24805	9JL	8;20	111.58	24805	24JL	2;14	305.70
24806	8AU	0;10	140.07	24806	22AU	10;00	334.04
24807	6SE	15;51	168.95	24807	20SE	19;15	2.89
24808	6OC	6;53	198.29	24808	20OC	6;52	32.33
24809	4NO	21;00	228.07	24809	18NO	21;27	62.30
24810	4DE	10;05	258.12	24810	18DE	14;52	92.60

1006

NEW — NUMBER	DATE	TIME	LONG.	FULL — NUMBER	DATE	TIME	LONG.
24811	2JA	21;54	288.19	24811	17JA	9;55	122.91
24812	1FE	8;25	318.00	24812	16FE	4;46	152.90
24813	2MR	17;51	347.37	24813	17MR	21;52	182.34
24814	1AP	2;47	16.24	24814	16AP	12;31	211.19
24815	30AP	12;02	44.66	24815	16MY	0;40	239.54
24816	29MY	22;27	72.81	24816	14JN	10;39	267.59
24817	28JN	10;37	100.93	24817	13JL	19;05	295.60
24818	28JL	0;40	129.25	24818	12AU	2;48	323.81
24819	26AU	16;27	157.97	24819	10SE	10;46	352.43
24820	25SE	9;29	187.21	24820	9OC	19;55	21.59
24821	25OC	2;59	216.96	24821	8NO	7;01	51.29
24822	23NO	19;50	247.08	24822	7DE	20;23	81.37
24823	23DE	10;55	277.29				

1007

NEW — NUMBER	DATE	TIME	LONG.	FULL — NUMBER	DATE	TIME	LONG.
				24823	6JA	11;49	111.58
24824	21JA	23;38	307.30	24824	5FE	4;37	141.62
24825	20FE	10;06	336.87	24825	6MR	21;56	171.26
24826	21MR	18;55	5.91	24826	5AP	14;52	200.38
24827	20AP	2;54	34.43	24827	5MY	6;34	229.01
24828	19MY	10;50	62.58	24828	3JN	20;20	257.27
24829	17JN	19;32	90.57	24829	3JL	8;00	285.37
24830	17JL	5;44	118.66	24830	1AU	17;56	313.56
24831	15AU	18;10	147.10	24831	31AU	2;59	342.07
24832	14SE	9;20	176.08	24832	29SE	12;05	11.05
24833	14OC	3;08	205.66	24833	28OC	21;55	40.54
24834	12NO	22;24	235.75	24834	27NO	8;48	70.43
24835	12DE	17;14	266.09	24835	26DE	20;46	100.52

1008

NEW — NUMBER	DATE	TIME	LONG.	FULL — NUMBER	DATE	TIME	LONG.
24836	11JA	9;55	296.33	24836	25JA	9;50	130.51
24837	9FE	23;42	326.20	24837	24FE	0;02	160.20
24838	10MR	10;49	355.50	24838	24MR	15;22	189.44
24839	8AP	19;52	24.24	24839	23AP	7;21	218.23
24840	8MY	3;34	52.49	24840	22MY	23;09	246.66
24841	6JN	10;38	80.46	24841	21JN	13;55	274.88
24842	5JL	17;59	108.40	24842	21JL	3;16	303.13
24843	4AU	2;41	136.57	24843	19AU	15;23	331.61
24844	2SE	13;49	165.21	24844	18SE	2;43	0.50
24845	2OC	4;12	194.46	24845	17OC	13;41	29.87
24846	31OC	21;45	224.33	24846	16NO	0;29	59.66
24847	30NO	17;16	254.62	24847	15DE	11;11	89.66
24848	30DE	12;50	285.02				

1009

NEW — NUMBER	DATE	TIME	LONG.	FULL — NUMBER	DATE	TIME	LONG.
				24848	13JA	21;55	119.64
24849	29JA	6;42	315.18	24849	12FE	9;03	149.35
24850	27FE	21;57	344.84	24850	13MR	21;01	178.64
24851	29MR	10;21	13.89	24851	12AP	10;04	207.48
24852	27AP	20;08	42.36	24852	12MY	0;06	235.95
24853	27MY	3;56	70.43	24853	10JN	14;48	264.21
24854	25JN	10;40	98.34	24854	10JL	5;49	292.46
24855	24JL	17;31	126.34	24855	8AU	20;51	320.93
24856	23AU	1;39	154.71	24856	7SE	11;38	349.79
24857	21SE	12;04	183.62	24857	7OC	1;45	19.11
24858	21OC	1;24	213.14	24858	5NO	14;49	48.85
24859	19NO	17;35	243.19	24859	5DE	2;41	78.86
24860	19DE	11;54	273.53				

1010

NEW — NUMBER	DATE	TIME	LONG.	FULL — NUMBER	DATE	TIME	LONG.
				24860	3JA	13;29	108.88
24861	18JA	7;07	303.84	24861	1FE	23;38	138.67
24862	17FE	1;45	333.80	24862	3MR	9;33	168.05
24863	18MR	18;21	3.21	24863	1AP	19;36	196.95
24864	17AP	8;04	32.00	24864	1MY	6;10	225.42
24865	16MY	18;48	60.29	24865	30MY	17;43	253.61
24866	15JN	3;17	88.28	24866	29JN	6;42	281.76
24867	14JL	10;37	116.25	24867	28JL	21;28	310.12
24868	12AU	18;00	144.45	24868	27AU	13;49	338.87
24869	11SE	2;22	173.10	24869	26SE	6;54	8.12
24870	10OC	12;23	202.31	24870	25OC	23;30	37.85
24871	9NO	0;27	232.05	24871	24NO	14;37	67.90
24872	8DE	14;45	262.18	24872	24DE	3;54	98.04

1011

NEW — NUMBER	DATE	TIME	LONG.	FULL — NUMBER	DATE	TIME	LONG.
24873	7JA	7;14	292.43	24873	22JA	15;29	127.99
24874	6FE	1;20	322.52	24874	21FE	1;40	157.55
24875	7MR	19;41	352.19	24875	22MR	10;44	186.58
24876	6AP	12;43	21.30	24876	20AP	19;07	215.11
24877	6MY	3;18	49.87	24877	20MY	3;31	243.27
24878	4JN	15;20	78.05	24878	18JN	12;54	271.29
24879	4JL	1;23	106.09	24879	18JL	0;18	299.44
24880	2AU	10;21	134.25	24880	16AU	14;23	327.95
24881	31AU	19;02	162.75	24881	15SE	7;02	357.00
24882	30SE	4;02	191.74	24882	15OC	1;14	26.61
24883	29OC	13;51	221.24	24883	13NO	19;34	56.68
24884	28NO	0;53	251.15	24884	13DE	12;44	86.95
24885	27DE	13;36	281.26				

1012

NEW — NUMBER	DATE	TIME	LONG.	FULL — NUMBER	DATE	TIME	LONG.
				24885	12JA	3;57	117.13
24886	26JA	4;11	311.31	24886	10FE	16;48	146.94
24887	24FE	20;16	341.07	24887	11MR	3;17	176.21
24888	25MR	12;57	10.35	24888	9AP	11;46	204.91
24889	24AP	5;09	39.14	24889	8MY	19;01	233.14
24890	23MY	20;12	67.53	24890	7JN	2;06	261.11
24891	22JN	9;53	95.71	24891	6JL	10;12	289.08
24892	21JL	22;14	123.92	24892	4AU	20;23	317.31
24893	20AU	9;27	152.37	24893	3SE	9;17	346.03
24894	18SE	19;49	181.24	24894	3OC	1;02	15.36
24895	18OC	5;47	210.58	24895	1NO	19;06	45.26
24896	16NO	15;52	240.34	24896	1DE	14;26	75.55
24897	16DE	2;33	270.35	24897	31DE	9;27	105.93

1013

NEW — NUMBER	DATE	TIME	LONG.	FULL — NUMBER	DATE	TIME	LONG.
24898	14JA	14;09	300.36	24898	30JA	2;33	136.05
24899	13FE	2;43	330.11	24899	28FE	16;38	165.64
24900	14MR	16;06	359.45	24900	30MR	3;34	194.62
24901	13AP	6;10	28.32	24901	28AP	12;02	223.03
24902	12MY	20;09	56.81	24902	27MY	19;06	251.07
24903	11JN	11;55	85.08	24903	26JN	1;54	278.97
24904	11JL	3;05	113.34	24904	25JL	9;28	307.01
24905	9AU	17;41	141.80	24905	23AU	18;38	335.43
24906	8SE	7;13	170.62	24906	22SE	6;07	4.39
24907	7OC	19;36	199.88	24907	21OC	20;26	33.97
24908	6NO	7;09	229.57	24908	20NO	13;41	64.07
24909	5DE	18;14	259.55	24909	20DE	9;08	94.46

1014

NEW — NUMBER	DATE	TIME	LONG.	FULL — NUMBER	DATE	TIME	LONG.
24910	4JA	5;06	289.57	24910	19JA	4;59	124.79
24911	2FE	15;45	319.38	24911	17FE	23;04	154.72
24912	4MR	2;15	348.77	24912	19MR	14;06	184.05
24913	2AP	12;56	17.68	24913	18AP	1;57	212.76
24914	2MY	0;26	46.18	24914	17MY	11;20	240.98
24915	31MY	13;20	74.42	24915	15JN	19;11	268.95
24916	30JN	3;50	102.63	24916	15JL	2;25	296.91
24917	29JL	19;31	131.03	24917	13AU	9;53	325.12
24918	28AU	11;34	159.78	24918	11SE	18;25	353.79
24919	27SE	3;16	188.99	24919	11OC	4;51	23.03
24920	26OC	18;10	218.67	24920	9NO	17;57	52.82
24921	25NO	7;59	248.67	24921	9DE	10;01	83.03
24922	24DE	20;33	278.77				

1015

NEW — NUMBER	DATE	TIME	LONG.	FULL — NUMBER	DATE	TIME	LONG.
				24922	8JA	4;20	113.36
24923	23JA	7;41	308.71	24923	6FE	23;20	143.47
24924	21FE	17;27	338.23	24924	8MR	17;16	173.11
24925	23MR	2;20	7.25	24925	7AP	9;04	202.15
24926	21AP	11;05	35.79	24926	6MY	22;24	230.66
24927	20MY	20;37	63.99	24927	5JN	9;26	258.80
24928	19JN	7;42	92.07	24928	4JL	18;40	286.81
24929	18JL	20;45	120.28	24929	3AU	2;47	314.94
24930	17AU	11;47	148.85	24930	1SE	10;41	343.42
24931	16SE	4;27	177.91	24931	30SE	19;19	12.40
24932	15OC	22;06	207.51	24932	30OC	5;29	41.93
24933	14NO	15;40	237.56	24933	28NO	17;44	71.89
24934	14DE	7;54	267.79	24934	28DE	8;06	102.07

1016 / 1022

NUMBER	DATE	TIME	LONG.	NUMBER	DATE	TIME	LONG.	NUMBER	DATE	TIME	LONG.	NUMBER	DATE	TIME	LONG.
24935	12JA	21;53	297.91	24935	27JA	0;07	132.18	25009	5JA	12;36	290.96	25009	20JA	22;24	126.59
24936	11FE	9;21	327.66	24936	25FE	17;01	161.96	25010	4FE	0;31	320.81	25010	19FE	14;01	156.40
24937	11MR	18;46	356.88	24937	26MR	10;01	191.24	25011	5MR	13;14	350.28	25011	21MR	2;24	185.59
24938	10AP	2;54	25.55	24938	25AP	2;16	220.02	25012	4AP	2;39	19.28	25012	19AP	11;53	214.18
24939	9MY	10;35	53.79	24939	24MY	17;02	248.39	25013	3MY	16;45	47.85	25013	18MY	19;27	242.32
24940	7JN	18;40	81.80	24940	23JN	5;51	276.53	25014	2JN	7;31	76.16	25014	17JN	2;13	270.24
24941	7JL	3;57	109.81	24941	22JL	16;45	304.69	25015	1JL	22;44	104.40	25015	16JL	9;18	298.21
24942	5AU	15;11	138.10	24942	21AU	2;22	333.09	25016	31JL	13;52	132.79	25016	14AU	17;38	326.47
24943	4SE	5;01	166.87	24943	19SE	11;34	1.92	25017	30AU	4;15	161.50	25017	13SE	3;58	355.23
24944	3OC	21;42	196.25	24944	18OC	21;09	31.25	25018	28SE	17;30	190.64	25018	12OC	16;54	24.59
24945	2NO	16;37	226.19	24945	17NO	7;36	61.03	25019	28OC	5;41	220.22	25019	11NO	8;46	54.52
24946	2DE	12;05	256.50	24946	16DE	19;02	91.08	25020	26NO	17;07	250.13	25020	11DE	3;17	84.84
								25021	26DE	4;08	280.17				

1017 / 1023

NUMBER	DATE	TIME	LONG.	NUMBER	DATE	TIME	LONG.	NUMBER	DATE	TIME	LONG.	NUMBER	DATE	TIME	LONG.
24947	1JA	6;04	286.83	24947	15JA	7;25	121.11					25021	9JA	23;09	115.23
24948	30JA	21;19	316.86	24948	13FE	20;48	150.90	25022	24JA	14;48	310.07	25022	8FE	18;15	145.32
24949	1MR	9;38	346.37	24949	15MR	11;17	180.27	25023	23FE	1;09	339.61	25023	10MR	10;47	174.88
24950	30MR	19;31	15.30	24950	14AP	2;44	209.19	25024	24MR	11;25	8.66	25024	9AP	0;06	203.79
24951	29AP	3;41	43.69	24951	13MY	18;34	237.71	25025	22AP	22;09	37.26	25025	8MY	10;34	232.17
24952	28MY	10;53	71.73	24952	12JN	9;54	265.98	25026	22MY	10;03	65.54	25026	6JN	19;04	260.20
24953	26JN	17;56	99.64	24953	12JL	0;07	294.21	25027	20JN	23;38	93.72	25027	6JL	2;34	288.14
24954	26JL	1;49	127.70	24954	10AU	13;02	322.61	25028	20JL	14;49	122.03	25028	4AU	9;56	316.25
24955	24AU	11;40	156.15	24955	9SE	0;58	351.38	25029	19AU	6;58	150.65	25029	2SE	17;57	344.76
24956	23SE	0;26	185.18	24956	8OC	12;22	20.61	25030	17SE	23;11	179.73	25030	2OC	3;27	13.80
24957	22OC	16;34	214.84	24957	6NO	23;27	50.29	25031	17OC	14;49	209.28	25031	31OC	15;14	43.41
24958	21NO	11;23	245.02	24958	6DE	10;17	80.26	25032	16NO	5;27	239.21	25032	30NO	5;50	73.48
24959	21DE	7;12	275.43					25033	15DE	18;49	269.33	25033	29DE	23;04	103.79

1018 / 1024

NUMBER	DATE	TIME	LONG.	NUMBER	DATE	TIME	LONG.	NUMBER	DATE	TIME	LONG.	NUMBER	DATE	TIME	LONG.
				24959	4JA	20;56	110.27	25034	14JA	6;40	299.36	25034	28JA	17;47	133.99
24960	20JA	2;04	305.71	24960	3FE	7;39	140.07	25035	12FE	16;59	329.04	25035	27FE	12;14	163.80
24961	18FE	18;38	335.56	24961	4MR	18;54	169.49	25036	13MR	2;03	358.22	25036	28MR	5;02	193.05
24962	20MR	8;21	4.82	24962	3AP	7;05	198.45	25037	11AP	10;31	26.89	25037	26AP	19;31	221.73
24963	18AP	19;19	33.49	24963	2MY	20;23	227.01	25038	10MY	19;20	55.16	25038	26MY	7;41	249.99
24964	18MY	3;58	61.68	24964	1JN	10;37	255.30	25039	9JN	5;24	83.24	25039	24JN	17;50	278.02
24965	16JN	11;06	89.61	24965	1JL	1;29	283.53	25040	8JL	17;21	111.37	25040	24JL	2;34	306.10
24966	15JL	17;48	117.56	24966	30JL	16;40	311.91	25041	7AU	7;26	139.78	25041	22AU	10;38	334.46
24967	14AU	1;14	145.77	24967	29AU	7;52	340.64	25042	5SE	23;29	168.67	25042	20SE	18;59	3.27
24968	12SE	10;30	174.48	24968	27SE	22;41	9.83	25043	5OC	16;58	198.11	25043	20OC	4;26	32.62
24969	11OC	22;26	203.79	24969	27OC	12;36	39.47	25044	4NO	10;58	228.04	25044	18NO	15;40	62.45
24970	10NO	13;17	233.67	24970	26NO	1;15	69.43	25045	4DE	4;13	258.26	25045	18DE	4;58	92.58
24971	10DE	6;38	263.93	24971	25DE	12;37	99.49								

1019 / 1025

NUMBER	DATE	TIME	LONG.	NUMBER	DATE	TIME	LONG.	NUMBER	DATE	TIME	LONG.	NUMBER	DATE	TIME	LONG.
24972	9JA	1;29	294.28	24972	23JA	22;58	129.38	25046	2JA	19;29	288.48	25046	16JA	20;04	122.72
24973	7FE	20;28	324.38	24973	22FE	8;47	158.90	25047	1FE	8;10	318.38	25047	15FE	12;22	152.61
24974	9MR	14;06	353.99	24974	23MR	18;29	187.94	25048	2MR	18;25	347.79	25048	17MR	5;09	182.05
24975	8AP	5;13	23.00	24975	22AP	4;29	216.52	25049	1AP	2;55	16.63	25049	15AP	21;40	210.98
24976	7MY	17;22	51.45	24976	21MY	15;16	244.76	25050	30AP	10;34	44.99	25050	15MY	13;10	239.47
24977	6JN	2;54	79.53	24977	20JN	3;19	272.88	25051	29MY	18;13	73.03	25051	14JN	3;02	267.68
24978	5JL	10;44	107.48	24978	19JL	17;09	301.13	25052	28JN	2;43	101.01	25052	13JL	15;01	295.82
24979	3AU	18;05	135.59	24979	18AU	8;52	329.73	25053	27JL	12;51	129.17	25053	12AU	1;24	324.14
24980	2SE	2;01	164.08	24980	17SE	1;58	358.82	25054	26AU	1;22	157.75	25054	10SE	10;57	352.83
24981	1OC	11;18	193.10	24981	16OC	19;15	28.43	25055	24SE	16;45	186.90	25055	9OC	20;29	22.01
24982	30OC	22;24	222.67	24982	15NO	11;27	58.42	25056	24OC	10;50	216.67	25056	8NO	6;38	51.66
24983	29NO	11;34	252.68	24983	15DE	1;48	88.58	25057	23NO	6;26	246.90	25057	7DE	17;37	81.65
24984	29DE	2;54	282.90					25058	23DE	1;29	277.28				

1020 / 1026

NUMBER	DATE	TIME	LONG.	NUMBER	DATE	TIME	LONG.	NUMBER	DATE	TIME	LONG.	NUMBER	DATE	TIME	LONG.
				24984	13JA	14;13	118.63					25058	6JA	5;26	111.70
24985	27JA	20;07	313.05	24985	12FE	0;59	148.34	25059	21JA	18;12	307.46	25059	4FE	18;06	141.57
24986	26FE	14;18	342.87	24986	12MR	10;25	177.55	25060	20FE	7;54	337.17	25060	6MR	7;44	171.07
24987	27MR	7;58	12.16	24987	10AP	18;53	206.23	25061	21MR	18;49	6.29	25061	4AP	22;26	200.11
24988	25AP	23;45	40.91	24988	10MY	2;58	234.48	25062	20AP	3;38	34.85	25062	4MY	13;56	228.73
24989	25MY	13;02	69.21	24989	8JN	11;35	262.50	25063	19MY	11;08	62.98	25063	3JN	5;32	257.07
24990	24JN	0;04	97.29	24990	7JL	21;46	290.56	25064	17JN	18;06	90.90	25064	2JL	20;26	285.30
24991	23JL	9;38	125.39	24991	6AU	10;28	318.91	25065	17JL	1;26	118.88	25065	1AU	10;11	313.64
24992	21AU	18;34	153.78	24992	5SE	2;00	347.76	25066	15AU	10;12	147.17	25066	30AU	22;50	342.29
24993	20SE	3;34	182.61	24993	4OC	19;46	17.19	25067	13SE	21;29	175.98	25067	29SE	10;45	11.38
24994	19OC	13;06	211.95	24994	3NO	14;26	47.14	25068	13OC	12;02	205.42	25068	28OC	22;13	40.94
24995	17NO	23;33	241.74	24995	3DE	8;29	77.40	25069	12NO	5;44	235.44	25069	27NO	9;18	70.85
24996	17DE	11;21	271.80					25070	12DE	1;20	265.81	25070	26DE	20;03	100.88

1021 / 1027

NUMBER	DATE	TIME	LONG.	NUMBER	DATE	TIME	LONG.	NUMBER	DATE	TIME	LONG.	NUMBER	DATE	TIME	LONG.
				24996	2JA	0;50	107.66	25071	10JA	20;53	296.17	25071	25JA	6;34	130.77
24997	16JA	0;51	301.88	24997	31JA	14;53	137.63	25072	9FE	14;38	326.21	25072	23FE	17;17	160.31
24998	14FE	16;01	331.73	24998	2MR	2;27	167.10	25073	11MR	5;43	355.68	25073	25MR	4;40	189.39
24999	16MR	8;14	1.16	24999	31MR	11;43	195.98	25074	9AP	17;57	24.55	25074	23AP	17;07	218.05
25000	15AP	0;33	30.10	25000	29AP	19;20	224.35	25075	9MY	3;37	52.88	25075	23MY	6;42	246.39
25001	14MY	16;06	58.60	25001	29MY	2;15	252.37	25076	7JN	11;23	80.88	25076	21JN	21;12	274.61
25002	13JN	6;29	86.83	25002	27JN	9;37	280.30	25077	6JL	18;12	108.80	25077	21JL	12;19	302.92
25003	12JL	19;38	115.02	25003	26JL	18;38	308.40	25078	5AU	1;12	136.90	25078	20AU	3;45	331.53
25004	11AU	7;37	143.39	25004	25AU	6;10	336.93	25079	3SE	9;32	165.42	25079	18SE	19;06	0.58
25005	9SE	18;39	172.13	25005	23SE	20;36	6.04	25080	2OC	20;10	194.52	25080	18OC	9;50	30.10
25006	9OC	5;00	201.33	25006	23OC	13;44	35.76	25081	1NO	9;38	224.21	25081	16NO	23;24	60.00
25007	7NO	15;08	230.98	25007	22NO	8;42	65.95	25082	1DE	1;50	254.36	25082	16DE	11;29	90.07
25008	7DE	1;32	260.93	25008	22DE	4;09	96.34	25083	30DE	20;00	284.70				

1028 / 1034

NEW #	DATE	TIME	LONG.	FULL #	DATE	TIME	LONG.	NEW #	DATE	TIME	LONG.	FULL #	DATE	TIME	LONG.
				25083	14JA	22;17	120.05					25157	7JA	16;32	113.25
25084	29JA	14;57	314.91	25084	13FE	8;10	149.70	25158	23JA	6;25	309.04	25158	6FE	8;06	143.22
25085	28FE	9;16	344.70	25085	13MR	17;40	178.89	25159	21FE	17;44	338.64	25159	8MR	0;26	172.80
25086	29MR	1;38	13.92	25086	12AP	3;12	207.60	25160	23MR	2;52	7.66	25160	6AP	16;56	201.89
25087	27AP	15;14	42.55	25087	11MY	13;18	235.91	25161	21AP	10;41	36.16	25161	6MY	8;53	230.51
25088	27MY	2;00	70.74	25088	10JN	0;29	264.03	25162	20MY	18;04	64.27	25162	4JN	23;38	258.80
25089	25JN	10;37	98.72	25089	9JL	13;18	292.19	25163	19JN	1;57	92.23	25163	4JL	12;41	286.96
25090	24JL	18;12	126.76	25090	8AU	4;09	320.65	25164	18JL	11;08	120.29	25164	3AU	0;00	315.22
25091	23AU	1;52	155.11	25091	6SE	20;51	349.57	25165	16AU	22;24	148.69	25165	1SE	10;07	343.79
25092	21SE	10;34	183.94	25092	6OC	14;28	19.03	25166	15SE	12;23	177.64	25166	30SE	19;49	12.82
25093	20OC	20;50	213.34	25093	5NO	7;37	48.94	25167	15OC	5;20	207.21	25167	30OC	5;47	42.33
25094	19NO	9;00	243.21	25094	4DE	23;08	79.10	25168	14NO	0;32	237.31	25168	28NO	16;26	72.23
25095	18DE	23;12	273.38					25169	13DE	20;14	267.69	25169	28DE	3;48	102.28

1029 / 1035

NEW #	DATE	TIME	LONG.	FULL #	DATE	TIME	LONG.	NEW #	DATE	TIME	LONG.	FULL #	DATE	TIME	LONG.
				25095	3JA	12;34	109.23								
25096	17JA	15;22	303.56	25096	2FE	0;05	139.08	25170	12JA	14;19	297.99	25170	26JA	15;52	132.23
25097	16FE	8;59	333.49	25097	3MR	10;00	168.46	25171	11FE	5;31	327.89	25171	25FE	4;44	161.84
25098	18MR	2;52	2.96	25098	1AP	18;43	197.30	25172	12MR	17;40	357.22	25172	26MR	18;36	191.00
25099	16AP	19;34	31.89	25099	1MY	2;43	225.67	25173	11AP	3;20	25.96	25173	25AP	9;29	219.73
25100	16MY	10;04	60.33	25100	30MY	10;46	253.72	25174	10MY	11;18	54.21	25174	25MY	0;59	248.14
25101	14JN	22;13	88.47	25101	28JN	19;56	281.72	25175	8JN	18;22	82.17	25175	23JN	16;19	276.38
25102	14JL	8;34	116.56	25102	28JL	7;15	309.93	25176	8JL	1;24	110.10	25176	23JL	6;49	304.68
25103	12AU	17;56	144.84	25103	26AU	21;26	338.59	25177	6AU	9;21	138.26	25177	21AU	20;14	333.23
25104	11SE	3;04	173.52	25104	25SE	14;21	7.83	25178	4SE	19;20	166.87	25178	20SE	8;47	2.19
25105	10OC	12;29	202.71	25105	25OC	8;58	37.63	25179	4OC	8;16	196.08	25179	19OC	20;43	31.62
25106	8NO	22;33	232.37	25106	24NO	3;41	67.83	25180	3NO	0;33	225.91	25180	18NO	8;11	61.45
25107	8DE	9;39	262.36	25107	23DE	21;06	98.14	25181	2DE	19;28	256.19	25181	17DE	19;09	91.47

1030 / 1036

NEW #	DATE	TIME	LONG.	FULL #	DATE	TIME	LONG.	NEW #	DATE	TIME	LONG.	FULL #	DATE	TIME	LONG.
25108	6JA	22;10	292.44	25108	22JA	12;22	128.25	25182	1JA	15;16	286.60	25182	16JA	5;41	121.43
25109	5FE	12;18	322.37	25109	21FE	1;06	157.91	25183	31JA	10;00	316.78	25183	14FE	16;04	151.09
25110	7MR	3;48	351.92	25110	22MR	11;21	187.00	25184	1MR	2;23	346.46	25184	15MR	2;47	180.31
25111	5AP	19;54	21.01	25111	20AP	19;34	215.52	25185	30MR	15;57	15.54	25185	13AP	14;23	209.08
25112	5MY	11;44	49.63	25112	20MY	2;35	243.62	25186	29AP	2;47	44.04	25186	13MY	3;09	237.48
25113	4JN	2;41	77.94	25113	18JN	9;30	271.55	25187	28MY	11;24	72.13	25187	11JN	17;04	265.70
25114	3JL	16;32	106.13	25114	17JL	17;32	299.56	25188	26JN	18;35	100.06	25188	11JL	7;55	293.95
25115	2AU	5;18	134.44	25115	16AU	3;45	327.91	25189	26JL	1;26	128.07	25189	9AU	23;23	322.45
25116	31AU	17;04	163.06	25116	14SE	16;48	356.80	25190	24AU	9;04	156.44	25190	8SE	15;06	351.36
25117	30SE	3;59	192.13	25117	14OC	8;45	26.32	25191	22SE	18;34	185.33	25191	8OC	6;32	20.76
25118	29OC	14;23	221.65	25118	13NO	3;02	56.37	25192	22OC	6;41	214.82	25192	6NO	21;00	50.58
25119	28NO	0;42	251.53	25119	12DE	22;31	86.74	25193	20NO	21;36	244.82	25193	6DE	10;00	80.64
25120	27DE	11;23	281.56					25194	20DE	14;51	275.12				

1031 / 1037

NEW #	DATE	TIME	LONG.	FULL #	DATE	TIME	LONG.	NEW #	DATE	TIME	LONG.	FULL #	DATE	TIME	LONG.
				25120	11JA	17;35	117.08					25194	4JA	21;26	110.68
25121	25JA	22;44	311.48	25121	10FE	10;36	147.07	25195	19JA	9;26	305.40	25195	3FE	7;37	140.46
25122	24FE	10;48	341.06	25122	12MR	0;32	176.49	25196	18FE	4;04	335.35	25196	4MR	17;04	169.80
25123	25MR	23;35	10.20	25123	10AP	11;18	205.28	25197	19MR	21;24	4.76	25197	3AP	2;17	198.64
25124	24AP	13;03	38.88	25124	9MY	19;37	233.55	25198	18AP	12;21	33.59	25198	2MY	11;49	227.05
25125	24MY	3;18	67.24	25125	8JN	2;36	261.52	25199	18MY	0;29	61.92	25199	31MY	22;11	255.19
25126	22JN	18;17	95.48	25126	7JL	9;25	289.43	25200	16JN	10;07	89.96	25200	30JN	10;00	283.30
25127	22JL	9;39	123.80	25127	5AU	17;05	317.57	25201	15JL	18;12	117.96	25201	29JL	23;49	311.62
25128	21AU	0;43	152.41	25128	4SE	2;26	346.15	25202	14AU	1;51	146.19	25202	28AU	15;46	340.38
25129	19SE	14;52	181.42	25129	3OC	14;05	15.29	25203	12SE	10;07	174.86	25203	27SE	9;20	9.67
25130	19OC	3;51	210.89	25130	2NO	4;32	45.04	25204	11OC	19;42	204.07	25204	27OC	3;09	39.47
25131	17NO	15;49	240.73	25131	1DE	21;50	75.24	25205	10NO	6;58	233.80	25205	25NO	19;49	69.59
25132	17DE	3;06	270.76	25132	31DE	17;12	105.63	25206	9DE	20;08	263.88	25206	25DE	10;25	99.78

1032 / 1038

NEW #	DATE	TIME	LONG.	FULL #	DATE	TIME	LONG.	NEW #	DATE	TIME	LONG.	FULL #	DATE	TIME	LONG.
25133	15JA	13;54	300.74	25133	30JA	12;50	135.86	25207	8JA	11;12	294.07	25207	23JA	22;51	129.76
25134	14FE	0;14	330.40	25134	29FE	6;42	165.62	25208	7FE	3;58	324.09	25208	22FE	9;25	159.32
25135	14MR	10;16	359.60	25135	29MR	21;32	194.76	25209	8MR	21;38	353.71	25209	23MR	18;31	188.33
25136	12AP	20;24	28.32	25136	28AP	9;17	223.31	25210	7AP	14;53	22.81	25210	22AP	2;37	216.83
25137	12MY	7;23	56.66	25137	27MY	18;39	251.44	25211	7MY	6;28	51.40	25211	21MY	10;22	244.95
25138	10JN	19;55	84.83	25138	26JN	2;36	279.39	25212	5JN	19;48	79.63	25212	19JN	18;43	272.93
25139	10JL	10;18	113.06	25139	25JL	10;02	307.43	25213	5JL	7;05	107.73	25213	19JL	4;47	301.03
25140	9AU	2;10	141.57	25140	23AU	17;46	335.78	25214	3AU	17;02	135.93	25214	17AU	17;31	329.50
25141	7SE	18;41	170.50	25141	22SE	2;34	4.64	25215	2SE	2;26	164.48	25215	16SE	9;15	358.53
25142	7OC	10;59	199.91	25142	21OC	13;13	34.06	25216	1OC	11;53	193.51	25216	16OC	3;22	28.16
25143	6NO	2;25	229.76	25143	20NO	2;24	63.98	25217	30OC	21;44	223.03	25217	14NO	22;24	58.26
25144	5DE	16;38	259.87	25144	19DE	18;20	94.22	25218	29NO	8;20	252.93	25218	14DE	16;45	88.59
								25219	28DE	20;03	283.00				

1033 / 1039

NEW #	DATE	TIME	LONG.	FULL #	DATE	TIME	LONG.	NEW #	DATE	TIME	LONG.	FULL #	DATE	TIME	LONG.
25145	4JA	5;19	289.97	25145	18JA	12;21	124.48					25219	13JA	9;13	118.81
25146	2FE	16;19	319.79	25146	17FE	6;56	154.44	25220	27JA	9;11	312.98	25220	11FE	23;12	148.66
25147	4MR	1;47	349.14	25147	19MR	0;30	183.88	25221	25FE	23;48	342.66	25221	13MR	10;34	177.95
25148	2AP	10;15	17.96	25148	17AP	16;03	212.75	25222	27MR	15;24	11.88	25222	11AP	19;36	206.65
25149	1MY	18;33	46.33	25149	17MY	5;20	241.13	25223	26AP	7;14	40.64	25223	11MY	2;58	234.86
25150	31MY	3;42	74.43	25150	15JN	16;30	269.23	25224	25MY	22;33	69.02	25224	9JN	9;42	262.81
25151	29JN	14;32	102.49	25151	15JL	1;00	297.28	25225	24JN	13;00	97.24	25225	8JL	17;00	290.75
25152	29JL	3;32	130.78	25152	13AU	10;29	325.54	25226	24JL	2;29	125.50	25226	7AU	1;02	318.96
25153	27AU	18;44	159.49	25153	11SE	18;47	354.20	25227	22AU	15;00	154.02	25227	5SE	13;40	347.65
25154	26SE	11;46	188.75	25154	11OC	3;45	23.37	25228	21SE	2;36	182.96	25228	5OC	4;17	16.94
25155	26OC	5;54	218.54	25155	9NO	14;08	53.05	25229	20OC	13;27	212.36	25229	3NO	21;37	46.83
25156	24NO	23;53	248.72	25156	9DE	2;24	83.10	25230	18NO	23;55	242.15	25230	3DE	16;45	77.12
25157	24DE	16;24	278.99					25231	18DE	10;25	272.15				

1040

NEW NUMBER	DATE	TIME	LONG.	FULL NUMBER	DATE	TIME	LONG.
				25231	2JA	12;16	107.52
25232	16JA	21;19	302.12	25232	1FE	6;27	137.67
25233	15FE	8;49	331.82	25233	1MR	21;55	167.30
25234	15MR	20;57	1.09	25234	31MR	10;08	196.31
25235	14AP	9;47	29.89	25235	29AP	19;29	224.74
25236	13MY	23;23	58.32	25236	29MY	2;57	252.78
25237	12JN	13;54	86.56	25237	27JN	9;43	280.69
25238	12JL	5;11	114.83	25238	26JL	16;54	308.72
25239	10AU	20;41	143.34	25239	25AU	1;24	337.13
25240	9SE	11;39	172.24	25240	23SE	11;56	6.08
25241	9OC	1;31	201.58	25241	23OC	1;02	35.62
25242	7NO	14;12	231.33	25242	21NO	16;57	65.67
25243	7DE	1;56	261.34	25243	21DE	11;24	96.03

1041

NEW NUMBER	DATE	TIME	LONG.	FULL NUMBER	DATE	TIME	LONG.
25244	5JA	12;58	291.37	25244	20JA	7;04	126.34
25245	3FE	23;25	321.15	25245	19FE	1;55	156.28
25246	5MR	9;22	350.50	25246	20MR	18;14	185.65
25247	3AP	19;06	19.36	25247	19AP	7;24	214.39
25248	3MY	5;19	47.78	25248	18MY	17;50	242.65
25249	1JN	16;48	75.96	25249	17JN	2;25	270.64
25250	1JL	6;09	104.13	25250	16JL	10;05	298.63
25251	30JL	21;23	132.52	25251	14AU	17;43	326.86
25252	29AU	13;53	161.31	25252	13SE	2;03	355.54
25253	28SE	6;40	190.58	25253	12OC	11;48	24.77
25254	27OC	22;52	220.32	25254	10NO	23;43	54.53
25255	26NO	13;57	250.39	25255	10DE	14;15	84.68
25256	26DE	3;32	280.54				

1042

NEW NUMBER	DATE	TIME	LONG.	FULL NUMBER	DATE	TIME	LONG.
				25256	9JA	7;14	114.94
25257	24JA	15;21	310.49	25257	8FE	1;32	145.02
25258	23FE	1;26	340.01	25258	9MR	19;34	174.64
25259	24MR	10;07	9.00	25259	8AP	12;02	203.70
25260	22AP	18;11	37.48	25260	8MY	2;24	232.23
25261	22MY	2;36	65.63	25261	6JN	14;37	260.41
25262	20JN	12;22	93.66	25262	6JL	1;01	288.47
25263	20JL	0;11	121.83	25263	4AU	10;06	316.65
25264	18AU	14;21	150.37	25264	2SE	18;36	345.17
25265	17SE	6;41	179.44	25265	2OC	3;19	14.18
25266	17OC	0;36	209.07	25266	31OC	13;03	43.71
25267	15NO	19;02	239.17	25267	30NO	0;24	73.65
25268	15DE	12;36	269.46	25268	29DE	13;32	103.77

1043

NEW NUMBER	DATE	TIME	LONG.	FULL NUMBER	DATE	TIME	LONG.
25269	14JA	4;00	299.64	25269	28JA	4;15	133.82
25270	12FE	16;35	329.41	25270	26FE	20;01	163.52
25271	14MR	2;36	358.63	25271	28MR	12;14	192.76
25272	12AP	10;49	27.29	25272	27AP	4;20	221.51
25273	11MY	18;10	55.50	25273	26MY	19;42	249.90
25274	10JN	1;35	83.47	25274	25JN	9;43	278.09
25275	9JL	9;58	111.46	25275	24JL	22;04	306.31
25276	7AU	20;06	139.72	25276	23AU	8;58	334.78
25277	6SE	8;43	168.46	25277	21SE	19;01	3.67
25278	6OC	0;19	197.81	25278	21OC	5;00	33.04
25279	4NO	18;41	227.75	25279	19NO	15;25	62.83
25280	4DE	14;30	258.07	25280	19DE	2;27	92.86

1044

NEW NUMBER	DATE	TIME	LONG.	FULL NUMBER	DATE	TIME	LONG.
25281	3JA	9;41	288.45	25281	17JA	14;03	122.86
25282	2FE	2;23	318.53	25282	16FE	2;16	152.57
25283	2MR	15;57	348.08	25283	16MR	15;18	181.87
25284	1AP	2;40	17.01	25284	15AP	5;24	210.71
25285	30AP	11;17	45.41	25285	14MY	20;27	239.19
25286	29MY	18;39	73.44	25286	13JN	11;54	267.47
25287	28JN	1;34	101.35	25287	13JL	2;58	295.73
25288	27JL	8;58	129.40	25288	11AU	17;09	324.20
25289	25AU	17;52	157.83	25289	10SE	6;24	353.04
25290	24SE	5;19	186.82	25290	9OC	18;54	22.34
25291	23OC	20;01	216.44	25291	8NO	6;49	52.07
25292	22NO	13;49	246.59	25292	7DE	18;10	82.06
25293	22DE	9;25	276.99				

1045

NEW NUMBER	DATE	TIME	LONG.	FULL NUMBER	DATE	TIME	LONG.
				25293	6JA	4;53	112.07
25294	21JA	4;50	307.29	25294	4FE	15;08	141.84
25295	19FE	22;24	337.17	25295	6MR	1;23	171.20
25296	21MR	13;18	6.46	25296	4AP	12;12	200.08
25297	20AP	1;24	35.15	25297	4MY	0;06	228.57
25298	19MY	11;01	63.36	25298	2JN	13;16	256.81
25299	17JN	18;49	91.32	25299	2JL	3;37	285.02
25300	17JL	1;47	119.28	25300	31JL	18;54	313.41
25301	15AU	8;59	147.51	25301	30AU	10;46	342.19
25302	13SE	17;34	176.21	25302	29SE	2;43	11.44
25303	13OC	4;25	205.49	25303	28OC	18;02	41.16
25304	11NO	17;59	235.33	25304	27NO	8;01	71.19
25305	11DE	10;08	265.55	25305	26DE	20;17	101.28

1046

NEW NUMBER	DATE	TIME	LONG.	FULL NUMBER	DATE	TIME	LONG.
25306	10JA	4;04	295.86	25306	25JA	7;00	131.18
25307	8FE	22;40	325.93	25307	23FE	16;36	160.67
25308	10MR	16;38	355.54	25308	25MR	1;40	189.66
25309	9AP	8;45	24.56	25309	23AP	10;44	218.18
25310	8MY	22;16	53.06	25310	22MY	20;24	246.37
25311	7JN	9;07	81.18	25311	21JN	7;17	274.44
25312	6JL	17;57	109.18	25312	20JL	20;00	302.65
25313	5AU	1;50	137.32	25313	19AU	10;58	331.24
25314	3SE	9;52	165.83	25314	18SE	4;02	0.35
25315	2OC	18;53	194.86	25315	17OC	22;10	30.01
25316	1NO	5;24	224.42	25316	16NO	15;49	60.08
25317	30NO	17;38	254.41	25317	16DE	7;39	90.30
25318	30DE	7;39	284.57				

1047

NEW NUMBER	DATE	TIME	LONG.	FULL NUMBER	DATE	TIME	LONG.
				25318	14JA	21;11	120.39
25319	28JA	23;25	314.65	25319	13FE	8;34	150.11
25320	27FE	16;30	344.40	25320	14MR	18;13	179.30
25321	29MR	9;54	13.67	25321	13AP	2;34	207.96
25322	28AP	2;17	42.42	25322	12MY	10;14	236.17
25323	27MY	16;45	70.76	25323	10JN	18;02	264.16
25324	26JN	5;05	98.89	25324	10JL	3;02	292.17
25325	25JL	15;48	127.05	25325	8AU	14;20	320.48
25326	24AU	1;37	155.49	25326	7SE	4;39	349.30
25327	22SE	11;13	184.36	25327	6OC	21;51	18.73
25328	21OC	21;01	213.74	25328	5NO	16;49	48.71
25329	20NO	7;20	243.54	25329	5DE	11;50	79.01
25330	19DE	18;25	273.57				

1048

NEW NUMBER	DATE	TIME	LONG.	FULL NUMBER	DATE	TIME	LONG.
				25330	4JA	5;25	109.32
25331	18JA	6;40	303.59	25331	2FE	20;41	139.33
25332	16FE	20;19	333.36	25332	3MR	9;16	168.82
25333	17MR	11;12	2.71	25333	1AP	19;17	197.72
25334	16AP	2;45	31.60	25334	1MY	3;16	226.08
25335	15MY	18;13	60.09	25335	30MY	10;06	254.08
25336	14JN	9;07	88.34	25336	28JN	16;55	281.99
25337	13JL	23;13	116.57	25337	28JL	0;58	310.07
25338	12AU	12;27	145.01	25338	26AU	11;16	338.57
25339	11SE	0;47	173.82	25339	25SE	0;28	7.65
25340	10OC	12;15	203.09	25340	24OC	16;36	37.34
25341	8NO	23;04	232.78	25341	23NO	11;03	67.52
25342	8DE	9;34	262.75	25342	23DE	6;36	97.92

1049

NEW NUMBER	DATE	TIME	LONG.	FULL NUMBER	DATE	TIME	LONG.
25343	6JA	20;11	292.75	25343	22JA	1;37	128.19
25344	5FE	7;13	322.54	25344	20FE	18;31	158.03
25345	6MR	18;46	351.94	25345	22MR	8;17	187.26
25346	5AP	6;57	20.87	25346	20AP	18;54	215.88
25347	4MY	19;52	49.39	25347	20MY	3;07	244.03
25348	3JN	9;45	77.65	25348	18JN	10;05	271.96
25349	3JL	0;40	105.89	25349	17JL	16;59	299.92
25350	1AU	16;17	134.30	25350	16AU	0;49	328.18
25351	31AU	7;52	163.08	25351	14SE	10;22	356.93
25352	29SE	22;38	192.30	25352	13OC	22;13	26.27
25353	29OC	12;11	221.95	25353	12NO	12;46	56.16
25354	28NO	0;32	251.92	25354	12DE	6;01	86.43
25355	27DE	11;57	281.97				

1050

NEW NUMBER	DATE	TIME	LONG.	FULL NUMBER	DATE	TIME	LONG.
				25355	11JA	1;12	116.78
25356	25JA	22;37	311.86	25356	9FE	20;34	146.87
25357	24FE	8;37	341.36	25357	11MR	14;10	176.45
25358	25MR	18;10	10.36	25358	10AP	4;49	205.41
25359	24AP	3;47	38.90	25359	9MY	16;30	233.82
25360	23MY	14;18	67.11	25360	8JN	1;55	261.88
25361	22JN	2;32	95.24	25361	7JL	10;01	289.85
25362	21JL	16;51	123.52	25362	5AU	17;43	317.99
25363	20AU	8;57	152.16	25363	4SE	1;46	346.51
25364	19SE	1;57	181.28	25364	3OC	10;52	15.55
25365	18OC	18;48	210.90	25365	1NO	21;42	45.14
25366	17NO	10;45	240.91	25366	1DE	10;54	75.17
25367	17DE	1;16	271.08	25367	31DE	2;39	105.41

1051

NEW NUMBER	DATE	TIME	LONG.	FULL NUMBER	DATE	TIME	LONG.
25368	15JA	14;01	301.13	25368	29JA	20;16	135.56
25369	14FE	0;51	330.82	25369	28FE	14;23	165.34
25370	15MR	10;00	359.98	25370	30MR	7;33	194.58
25371	13AP	18;04	28.61	25371	28AP	22;54	223.28
25372	13MY	1;59	56.83	25372	28MY	12;10	251.57
25373	11JN	10;49	84.86	25373	26JN	23;32	279.66
25374	10JL	21;27	112.94	25374	26JL	9;22	307.79
25375	9AU	10;27	141.32	25375	24AU	18;17	336.20
25376	8SE	1;51	170.20	25376	23SE	3;00	5.05
25377	7OC	19;15	199.65	25377	22OC	12;18	34.41
25378	6NO	13;48	229.64	25378	20NO	22;53	64.22
25379	6DE	8;09	259.90	25379	20DE	11;05	94.31

```
---------NEW MOONS---------        ---------FULL MOONS--------        ---------NEW MOONS---------        ---------FULL MOONS--------
NUMBER   DATE   TIME   LONG.       NUMBER   DATE   TIME   LONG.       NUMBER   DATE   TIME   LONG.       NUMBER   DATE   TIME   LONG.
```

1052 / 1058

NUMBER	DATE	TIME	LONG.		NUMBER	DATE	TIME	LONG.		NUMBER	DATE	TIME	LONG.		NUMBER	DATE	TIME	LONG.
25380	5JA	0;51	290.17		25380	19JA	0;55	124.39							25454	12JA	20;18	118.66
25381	3FE	14;51	320.12		25381	17FE	15;57	154.21		25455	27JA	5;57	313.24		25455	11FE	14;21	148.68
25382	4MR	1;59	349.54		25382	18MR	7;43	183.58		25456	25FE	17;00	342.76		25456	13MR	5;40	178.13
25383	2AP	10;51	18.38		25383	16AP	23;42	212.48		25457	27MR	4;34	11.83		25457	11AP	17;43	206.96
25384	1MY	18;22	46.71		25384	16MY	15;23	240.96		25458	25AP	16;49	40.45		25458	11MY	2;58	235.25
25385	31MY	1;32	74.73		25385	15JN	6;10	269.21		25459	25MY	5;58	68.75		25459	9JN	10;24	263.23
25386	29JN	9;16	102.67		25386	14JL	19;31	297.41		25460	23JN	20;17	96.96		25460	8JL	17;15	291.15
25387	28JL	18;25	130.80		25387	13AU	7;20	325.80		25461	23JL	11;41	125.29		25461	7AU	0;35	319.29
25388	27AU	5;46	159.35		25388	11SE	18;00	354.56		25462	22AU	3;36	153.95		25462	5SE	9;18	347.86
25389	25SE	19;56	188.48		25389	11OC	4;11	23.78		25463	20SE	19;10	183.04		25463	4OC	20;03	16.99
25390	25OC	13;07	218.23		25390	9NO	14;31	53.46		25464	20OC	9;39	212.59		25464	3NO	9;17	46.70
25391	24NO	8;33	248.46		25391	9DE	1;18	83.44		25465	18NO	22;47	242.49		25465	3DE	1;13	76.85
25392	24DE	4;23	278.87							25466	18DE	10;45	272.56					

1053 / 1059

NUMBER	DATE	TIME	LONG.		NUMBER	DATE	TIME	LONG.		NUMBER	DATE	TIME	LONG.		NUMBER	DATE	TIME	LONG.
					25392	7JA	12;33	113.47							25466	1JA	19;30	107.20
25393	22JA	22;29	309.10		25393	6FE	0;15	143.29		25467	16JA	21;45	302.53		25467	31JA	14;53	137.40
25394	21FE	13;34	338.85		25394	7MR	12;33	172.70		25468	15FE	7;56	332.17		25468	2MR	9;25	167.17
25395	23MR	1;32	8.00		25395	6AP	1;48	201.67		25469	16MR	17;27	1.33		25469	1AP	1;30	196.35
25396	21AP	11;01	36.56		25396	5MY	16;09	230.23		25470	15AP	2;42	29.99		25470	30AP	14;34	224.93
25397	20MY	18;50	64.69		25397	4JN	7;21	258.54		25471	14MY	12;26	58.27		25471	30MY	1;01	253.10
25398	19JN	1;51	92.62		25398	3JL	22;44	286.79		25472	12JN	23;33	86.38		25472	28JN	9;44	281.08
25399	18JL	8;55	120.59		25399	2AU	13;35	315.19		25473	12JL	12;44	114.57		25473	27JL	17;40	309.14
25400	16AU	17;01	148.87		25400	1SE	3;33	343.91		25474	11AU	4;05	143.06		25474	26AU	1;37	337.53
25401	15SE	3;09	177.65		25401	30SE	16;42	13.08		25475	9SE	20;57	172.02		25475	24SE	10;16	6.40
25402	14OC	16;15	207.05		25402	30OC	5;10	42.70		25476	9OC	14;16	201.50		25476	23OC	20;16	35.81
25403	13NO	8;39	237.03		25403	28NO	16;58	72.64		25477	8NO	7;01	231.42		25477	22NO	8;17	65.70
25404	13DE	3;34	267.37		25404	28DE	4;01	102.69		25478	7DE	22;29	261.59		25478	21DE	22;42	95.88

1054 / 1060

NUMBER	DATE	TIME	LONG.		NUMBER	DATE	TIME	LONG.		NUMBER	DATE	TIME	LONG.		NUMBER	DATE	TIME	LONG.
25405	11JA	23;15	297.74		25405	26JA	14;23	132.55		25479	6JA	12;12	291.73		25479	20JA	15;20	126.07
25406	10FE	17;48	327.79		25406	25FE	0;22	162.04		25480	4FE	23;56	321.57		25480	19FE	9;10	155.98
25407	12MR	9;59	357.29		25407	26MR	10;33	191.06		25481	5MR	9;45	350.91		25481	20MR	2;43	185.40
25408	10AP	23;23	26.19		25408	24AP	21;35	219.64		25482	3AP	18;04	19.70		25482	18AP	18;53	214.28
25409	10MY	10;08	54.55		25409	24MY	9;53	247.92		25483	3MY	1;45	48.03		25483	18MY	9;09	242.69
25410	8JN	18;46	82.58		25410	22JN	23;33	276.11		25484	1JN	9;50	76.08		25484	16JN	21;30	270.83
25411	8JL	2;06	110.52		25411	22JL	14;26	304.41		25485	30JN	19;23	104.09		25485	16JL	8;12	298.94
25412	6AU	9;09	138.64		25412	21AU	6;13	333.05		25486	30JL	7;08	132.34		25486	14AU	17;43	327.25
25413	4SE	17;03	167.16		25413	19SE	22;28	2.16		25487	28AU	21;25	161.03		25487	13SE	2;40	355.96
25414	4OC	2;47	196.24		25414	19OC	14;30	31.76		25488	27SE	14;02	190.28		25488	12OC	11;46	25.16
25415	2NO	15;03	225.90		25415	18NO	5;28	61.73		25489	27OC	8;21	220.11		25489	10NO	21;46	54.84
25416	2DE	5;58	256.01		25416	17DE	18;44	91.85		25490	26NO	3;10	250.33		25490	10DE	9;09	84.85
25417	31DE	23;02	286.30							25491	25DE	20;58	280.65					

1055 / 1061

NUMBER	DATE	TIME	LONG.		NUMBER	DATE	TIME	LONG.		NUMBER	DATE	TIME	LONG.		NUMBER	DATE	TIME	LONG.
					25417	16JA	6;11	121.85							25491	8JA	22;05	114.95
25418	30JA	17;17	316.47		25418	14FE	16;10	151.49		25492	24JA	12;24	310.76		25492	7FE	12;21	144.86
25419	1MR	11;31	346.24		25419	16MR	1;14	180.64		25493	23FE	0;52	340.38		25493	9MR	3;31	174.37
25420	31MR	4;32	15.46		25420	14AP	10;00	209.29		25494	24MR	10;39	9.41		25494	7AP	19;10	203.40
25421	29AP	19;20	44.13		25421	13MY	19;05	237.55		25495	22AP	18;35	37.89		25495	7MY	10;53	232.00
25422	29MY	7;30	72.37		25422	12JN	5;07	265.62		25496	22MY	1;42	65.98		25496	6JN	2;09	260.30
25423	27JN	17;19	100.39		25423	11JL	16;45	293.74		25497	20JN	8;58	93.91		25497	5JL	16;23	288.51
25424	27JL	1;42	128.47		25424	10AU	6;36	322.17		25498	19JL	17;18	121.95		25498	4AU	5;10	316.84
25425	25AU	9;43	156.85		25425	8SE	22;50	351.09		25499	18AU	3;29	150.32		25499	2SE	16;37	345.48
25426	23SE	18;21	185.71		25426	8OC	16;50	20.58		25500	16SE	16;15	179.24		25500	2OC	3;13	14.57
25427	23OC	4;13	215.11		25427	7NO	11;10	50.55		25501	16OC	8;03	208.78		25501	31OC	13;37	44.12
25428	21NO	15;38	244.97		25428	7DE	4;13	80.78		25502	15NO	2;37	238.86		25502	30NO	0;16	74.03
25429	21DE	4;42	275.08							25503	14DE	22;36	269.26		25503	29DE	11;17	104.07

1056 / 1062

NUMBER	DATE	TIME	LONG.		NUMBER	DATE	TIME	LONG.		NUMBER	DATE	TIME	LONG.		NUMBER	DATE	TIME	LONG.
					25429	5JA	18;59	110.97		25504	13JA	17;48	299.60		25504	27JA	22;36	133.97
25430	19JA	19;27	305.20		25430	4FE	7;22	140.84		25505	12FE	10;26	329.55		25505	26FE	10;19	163.51
25431	18FE	11;42	335.06		25431	4MR	17;43	170.22		25506	13MR	23;50	358.91		25506	27MR	22;45	192.60
25432	19MR	4;49	4.48		25432	3AP	3;21	199.04		25507	12AP	10;22	27.66		25507	26AP	12;16	221.26
25433	17AP	21;40	33.39		25433	2MY	10;16	227.38		25508	11MY	18;51	55.92		25508	26MY	2;54	249.62
25434	17MY	13;05	61.86		25434	31MY	17;45	255.40		25509	10JN	2;08	83.89		25509	24JN	18;15	277.86
25435	16JN	2;31	90.04		25435	30JN	1;55	283.37		25510	9JL	9;05	111.81		25510	24JL	9;33	306.20
25436	15JL	14;07	118.19		25436	29JL	11;55	311.54		25511	7AU	16;37	139.96		25511	23AU	0;14	334.82
25437	14AU	0;31	146.53		25437	28AU	0;43	340.15		25512	6SE	1;41	168.56		25512	21SE	14;05	3.85
25438	12SE	10;25	175.26		25438	26SE	16;40	9.37		25513	5OC	13;18	197.74		25513	21OC	3;10	33.35
25439	11OC	20;18	204.49		25439	26OC	11;06	39.18		25514	4NO	4;08	227.53		25514	19NO	15;30	63.23
25440	10NO	6;28	234.17		25440	25NO	6;27	69.42		25515	3DE	21;58	257.76		25515	19DE	3;01	93.28
25441	9DE	17;10	264.15		25441	25DE	1;00	99.78										

1057 / 1063

NUMBER	DATE	TIME	LONG.		NUMBER	DATE	TIME	LONG.		NUMBER	DATE	TIME	LONG.		NUMBER	DATE	TIME	LONG.
25442	8JA	4;42	294.18		25442	23JA	17;30	129.93		25516	2JA	17;28	288.16		25516	17JA	13;39	123.23
25443	6FE	17;25	324.03		25443	22FE	7;23	159.62		25517	1FE	12;41	318.35		25517	15FE	23;36	152.86
25444	8MR	7;27	353.51		25444	23MR	18;32	188.72		25518	3MR	6;01	348.06		25518	17MR	9;22	182.01
25445	6AP	22;27	22.53		25445	22AP	3;21	217.25		25519	1AP	20;43	17.16		25519	15AP	19;38	210.70
25446	6MY	13;48	51.13		25446	21MY	10;32	245.35		25520	1MY	8;43	45.69		25520	15MY	7;01	239.04
25447	5JN	4;57	79.43		25447	19JN	17;10	273.25		25521	30MY	18;20	73.81		25521	13JN	19;50	267.22
25448	4JL	19;32	107.65		25448	19JL	0;27	301.24		25522	29JN	2;15	101.77		25522	13JL	10;06	295.45
25449	3AU	9;24	136.02		25449	17AU	9;34	329.57		25523	28JL	9;25	129.81		25523	12AU	1;35	323.96
25450	1SE	22;29	164.71		25450	15SE	21;21	358.43		25524	26AU	16;53	158.18		25524	10SE	17;55	352.91
25451	1OC	10;40	193.85		25451	15OC	12;08	27.91		25525	25SE	1;44	187.07		25525	10OC	10;27	22.37
25452	30OC	22;00	223.44		25452	14NO	5;37	57.95		25526	24OC	12;47	216.53		25526	9NO	2;19	52.27
25453	29NO	8;45	253.34		25453	14DE	0;50	88.30		25527	23NO	2;26	246.26		25527	8DE	16;40	82.39
25454	28DE	19;16	283.36							25528	22DE	18;27	276.75					

NUMBER	DATE	TIME	LONG.	NUMBER	DATE	TIME	LONG.	NUMBER	DATE	TIME	LONG.	NUMBER	DATE	TIME	LONG.
			1064								**1070**				
				25528	7JA	5;03	112.47	25603	15JA	9;13	301.32	25603	29JA	9;13	135.48
25529	21JA	12;04	306.97	25529	5FE	15;37	142.25	25604	13FE	23;09	331.14	25604	27FE	23;42	165.12
25530	20FE	6;15	336.89	25530	6MR	0;54	171.57	25605	15MR	10;05	0.37	25605	29MR	14;52	194.29
25531	20MR	23;50	6.29	25531	4AP	9;33	200.36	25606	13AP	18;42	29.03	25606	28AP	6;22	223.01
25532	19AP	15;43	35.14	25532	3MY	18;11	228.72	25607	13MY	1;59	57.22	25607	27MY	21;49	251.39
25533	19MY	5;11	63.52	25533	2JN	3;29	256.81	25608	11JN	8;59	85.17	25608	26JN	12;41	279.61
25534	17JN	16;11	91.60	25534	1JL	14;08	284.87	25609	10JL	16;39	113.13	25609	26JL	2;24	307.90
25535	17JL	1;18	119.65	25535	31JL	2;48	313.16	25610	9AU	1;50	141.37	25610	24AU	14;45	336.44
25536	15AU	9;33	147.92	25536	29AU	17;56	341.89	25611	7SE	13;18	170.08	25611	23SE	1;59	5.39
25537	13SE	17;58	176.62	25537	28SE	11;23	11.20	25612	7OC	3;39	199.39	25612	22OC	12;39	34.82
25538	13OC	3;21	205.84	25538	28OC	5;59	41.04	25613	5NO	21;01	229.31	25613	20NO	23;18	64.64
25539	11NO	14;04	235.56	25539	27NO	0;04	71.24	25614	5DE	16;36	259.63	25614	20DE	10;10	94.66
25540	11DE	2;18	265.61	25540	26DE	16;09	101.50								
			1065								**1071**				
25541	9JA	16;04	295.74	25541	25JA	5;43	131.51	25615	4JA	12;29	290.04	25615	18JA	21;14	124.62
25542	8FE	7;22	325.68	25542	23FE	16;56	161.07	25616	3FE	6;31	320.16	25616	17FE	8;31	154.29
25543	9MR	23;53	355.24	25543	25MR	2;17	190.08	25617	4MR	21;27	349.74	25617	18MR	20;15	183.50
25544	8AP	16;46	24.31	25544	23AP	10;20	218.55	25618	3AP	9;15	18.70	25618	17AP	8;53	212.27
25545	8MY	8;53	52.91	25545	22MY	17;43	246.65	25619	2MY	18;35	47.11	25619	16MY	22;45	240.69
25546	6JN	23;21	81.18	25546	21JN	1;19	274.59	25620	1JN	2;20	75.15	25620	15JN	13;43	268.94
25547	6JL	11;56	109.32	25547	20JL	10;14	302.65	25621	30JN	9;21	103.06	25621	15JL	5;11	297.23
25548	4AU	23;04	137.59	25548	18AU	21;34	331.09	25622	29JL	16;32	131.12	25622	13AU	20;26	325.75
25549	3SE	9;24	166.20	25549	17SE	12;02	0.08	25623	28AU	0;48	159.54	25623	12SE	10;59	354.66
25550	2OC	19;29	195.28	25550	17OC	5;30	29.70	25624	26SE	11;08	188.51	25624	12OC	0;45	24.03
25551	1NO	5;41	224.83	25551	16NO	0;45	59.83	25625	26OC	0;23	218.09	25625	10NO	13;42	53.82
25552	30NO	16;09	254.74	25552	15DE	20;00	90.20	25626	24NO	16;50	248.19	25626	10DE	1;47	83.86
25553	30DE	3;11	284.77					25627	24DE	11;40	278.55				
			1066								**1072**				
				25553	14JA	13;40	120.47					25627	8JA	12;51	113.88
25554	28JA	15;06	314.69	25554	13FE	4;51	150.34	25628	23JA	7;10	308.85	25628	6FE	22;59	143.62
25555	27FE	4;13	344.28	25555	14MR	17;16	179.65	25629	22FE	1;27	338.74	25629	7MR	8;33	172.93
25556	28MR	18;29	13.43	25556	13AP	3;05	208.37	25630	22MR	17;24	8.05	25630	5AP	18;13	201.75
25557	27AP	9;29	42.14	25557	12MY	10;53	236.59	25631	21AP	6;40	36.77	25631	5MY	4;43	230.16
25558	27MY	0;38	70.51	25558	10JN	17;36	264.53	25632	20MY	17;24	65.02	25632	3JN	16;36	258.35
25559	25JN	15;32	98.74	25559	10JL	0;24	292.46	25633	19JN	2;07	93.01	25633	3JL	6;04	286.52
25560	25JL	5;56	127.05	25560	8AU	8;30	320.64	25634	18JL	9;38	121.01	25634	1AU	21;02	314.92
25561	23AU	19;41	155.64	25561	6SE	18;56	349.29	25635	16AU	16;58	149.25	25635	31AU	13;11	343.72
25562	22SE	8;37	184.65	25562	6OC	8;19	18.56	25636	15SE	1;09	177.96	25636	30SE	5;58	13.02
25563	21OC	20;38	214.11	25563	5NO	0;36	48.42	25637	14OC	11;09	207.23	25637	29OC	22;34	42.81
25564	20NO	7;49	243.95	25564	4DE	19;08	78.69	25638	12NO	23;32	237.04	25638	28NO	13;59	72.91
25565	19DE	18;27	273.96					25639	12DE	14;24	267.20	25639	28DE	3;27	103.05
			1067								**1073**				
				25565	3JA	14;39	109.08					25627	8JA	12;51	113.88
25566	18JA	4;56	303.91	25566	2FE	9;32	139.25	25640	11JA	7;11	297.45	25640	26JA	14;51	132.97
25567	16FE	15;35	333.55	25567	4MR	2;15	168.92	25641	10FE	1;01	327.48	25641	25FE	0;35	162.44
25568	18MR	2;37	2.75	25568	2AP	15;51	197.96	25642	11MR	18;49	357.06	25642	26MR	9;17	191.40
25569	16AP	14;13	31.48	25569	2MY	2;22	226.43	25643	10AP	11;31	26.09	25643	24AP	17;38	219.87
25570	16MY	2;38	59.86	25570	31MY	10;33	254.49	25644	10MY	2;11	54.62	25644	24MY	2;20	248.01
25571	14JN	16;13	88.06	25571	29JN	17;34	282.41	25645	8JN	14;26	82.79	25645	22JN	12;05	276.04
25572	14JL	7;06	116.32	25572	29JL	0;37	310.45	25646	8JL	0;32	110.84	25646	21JL	23;37	304.21
25573	12AU	23;00	144.86	25573	27AU	8;40	338.85	25647	6AU	9;16	139.03	25647	20AU	13;32	332.77
25574	11SE	15;08	173.81	25574	25SE	18;27	7.79	25648	4SE	17;41	167.58	25648	19SE	6;03	1.87
25575	11OC	6;32	203.24	25575	25OC	6;29	37.30	25649	4OC	2;42	196.63	25649	19OC	0;29	31.56
25576	9NO	20;36	233.07	25576	23NO	21;05	67.32	25650	2NO	12;51	226.20	25650	17NO	19;15	61.69
25577	9DE	9;16	263.12	25577	23DE	14;13	97.61	25651	2DE	0;21	256.16	25651	17DE	12;37	91.98
								25652	31DE	13;16	286.28				
			1068								**1074**				
25578	7JA	20;45	293.17	25578	22JA	9;07	127.89					25652	16JA	3;29	122.12
25579	6FE	7;14	322.94	25579	21FE	4;09	157.82	25653	30JA	3;36	316.28	25653	14FE	15;46	151.86
25580	6MR	16;52	352.25	25580	21MR	21;27	187.21	25654	28FE	19;18	345.96	25654	16MR	1;52	181.05
25581	5AP	1;57	21.06	25581	20AP	11;57	216.53	25655	30MR	11;51	15.17	25655	14AP	10;21	209.69
25582	4MY	11;06	49.43	25582	19MY	23;36	244.28	25656	29AP	4;18	43.92	25656	13MY	17;51	237.89
25583	2JN	21;13	77.54	25583	18JN	9;09	272.31	25657	28MY	19;37	72.28	25657	12JN	1;07	265.84
25584	2JL	9;12	105.66	25584	17JL	17;29	300.33	25658	27JN	9;12	100.46	25658	11JL	9;11	293.83
25585	31JL	23;31	134.02	25585	16AU	1;30	328.60	25659	26JL	21;11	128.68	25659	9AU	19;11	322.10
25586	30AU	15;52	162.82	25586	14SE	9;53	357.30	25660	25AU	8;06	157.18	25660	8SE	8;06	350.88
25587	29SE	9;21	192.14	25587	13OC	19;17	26.54	25661	23SE	18;31	186.11	25661	8OC	0;15	20.28·
25588	29OC	2;45	221.95	25588	12NO	6;18	56.28	25662	23OC	4;50	215.52	25662	6NO	18;57	50.26
25589	27NO	19;08	252.08	25589	11DE	19;28	86.37	25663	21NO	15;17	245.34	25663	6DE	14;32	80.59
25590	27DE	9;53	282.28					25664	21DE	2;00	275.36				
			1069								**1075**				
				25590	10JA	10;56	116.57					25664	5JA	9;12	110.95
25591	25JA	22;37	312.25	25591	9FE	4;05	146.58	25665	19JA	13;18	305.33	25665	4FE	1;40	140.99
25592	24FE	9;15	341.78	25592	10MR	21;42	176.16	25666	18FE	1;34	335.02	25666	5MR	15;24	170.51
25593	25MR	18;05	10.75	25593	9AP	14;28	205.21	25667	19MR	14;58	4.30	25667	4AP	2;22	199.43
25594	24AP	1;47	39.20	25594	9MY	5;37	233.77	25668	18AP	5;22	33.12	25668	3MY	10;59	227.80
25595	23MY	9;22	67.30	25595	7JN	18;56	261.99	25669	17MY	20;19	61.58	25669	1JN	18;03	255.81
25596	21JN	17;57	95.29	25596	7JL	6;33	290.10	25670	16JN	11;19	89.83	25670	1JL	0;47	283.70
25597	21JL	4;28	123.41	25597	5AU	16;48	318.34	25671	16JL	2;05	118.09	25671	30JL	8;00	311.77
25598	19AU	17;30	151.92	25598	4SE	2;11	346.92	25672	14AU	16;24	146.59	25672	28AU	17;14	340.24
25599	18SE	9;07	180.98	25599	3OC	11;20	15.96	25673	13SE	6;04	175.47	25673	27SE	5;11	9.29
25600	18OC	2;52	210.63	25600	1NO	20;58	45.50	25674	12OC	18;51	204.82	25674	26OC	20;08	38.95
25601	16NO	21;47	240.75	25601	1DE	7;40	75.42	25675	11NO	6;38	234.57	25675	25NO	13;42	69.10
25602	16DE	16;25	271.09	25602	30DE	19;46	105.51	25676	10DE	17;36	264.56	25676	25DE	8;54	99.48

Left half (periods 1076–1081)

NUMBER	DATE	TIME	LONG.	NUMBER	DATE	TIME	LONG.
1076							
25677	9JA	4;06	294.55	25677	24JA	4;14	129.76
25678	7FE	14;29	324.30	25678	22FE	22;06	159.63
25679	8MR	1;04	353.64	25679	23MR	13;14	188.89
25680	6AP	12;04	22.50	25680	22AP	1;10	217.55
25681	5MY	23;46	50.95	25681	21MY	10;22	245.73
25682	4JN	12;33	79.17	25682	19JN	17;51	273.67
25683	4JL	2;43	107.37	25683	19JL	0;49	301.65
25684	2AU	18;16	135.80	25684	17AU	8;23	329.91
25685	1SE	10;38	164.62	25685	15SE	17;21	358.66
25686	1OC	2;48	193.92	25686	15OC	4;19	27.98
25687	30OC	17;52	223.65	25687	13NO	17;39	57.83
25688	29NO	7;25	253.68	25688	13DE	9;31	88.05
25689	28DE	19;32	283.77				
1077							
				25689	12JA	3;33	118.34
25690	27JA	6;27	313.65	25690	10FE	22;34	148.41
25691	25FE	16;20	343.13	25691	12MR	16;45	177.99
25692	27MR	1;25	12.09	25692	11AP	8;37	206.98
25693	25AP	10;13	40.57	25693	10MY	21;36	235.43
25694	24MY	19;31	68.73	25694	9JN	8;08	263.53
25695	23JN	6;20	96.79	25695	8JL	17;04	291.53
25696	22JL	19;32	125.03	25696	7AU	1;19	319.71
25697	21AU	10;55	153.66	25697	5SE	9;37	348.26
25698	20SE	4;10	182.82	25698	4OC	18;37	17.32
25699	19OC	22;00	212.49	25699	3NO	4;51	46.91
25700	18NO	15;14	242.56	25700	2DE	16;54	76.89
25701	18DE	7;00	272.79				
1078							
				25701	1JA	7;08	107.06
25702	16JA	20;49	302.88	25702	30JA	23;22	137.14
25703	15FE	8;25	332.58	25703	1MR	16;39	166.87
25704	16MR	17;56	1.74	25704	31MR	9;44	196.09
25705	15AP	1;55	30.35	25705	30AP	1;36	224.80
25706	14MY	9;16	58.53	25706	29MY	15;49	253.12
25707	12JN	17;06	86.51	25707	28JN	4;22	281.25
25708	12JL	2;29	114.55	25708	27JL	15;27	309.45
25709	10AU	14;13	142.89	25709	26AU	1;26	337.91
25710	9SE	4;39	171.75	25710	24SE	10;51	6.81
25711	8OC	21;33	201.20	25711	23OC	20;21	36.21
25712	7NO	16;13	231.19	25712	22NO	6;33	66.02
25713	7DE	11;19	261.51	25713	21DE	17;55	96.07
1079							
25714	6JA	5;17	291.83	25714	20JA	6;34	126.09
25715	4FE	20;42	321.82	25715	18FE	20;20	155.83
25716	6MR	9;01	351.27	25716	20MR	10;53	185.14
25717	4AP	18;34	20.11	25717	19AP	1;59	213.98
25718	4MY	2;16	48.44	25718	18MY	17;21	242.45
25719	2JN	9;12	76.44	25719	17JN	8;35	270.71
25720	1JL	16;23	104.36	25720	16JL	23;04	298.96
25721	31JL	0;44	132.47	25721	15AU	12;21	327.42
25722	29AU	11;01	160.99	25722	14SE	0;22	356.26
25723	27SE	23;57	190.09	25723	13OC	11;31	25.54
25724	27OC	15;56	219.81	25724	11NO	22;19	55.26
25725	26NO	10;39	250.02	25725	11DE	9;08	85.25
25726	26DE	6;41	280.44				
1080							
				25726	9JA	20;04	115.25
25727	25JA	1;50	310.70	25727	8FE	7;03	145.03
25728	23FE	18;19	340.49	25728	8MR	18;15	174.38
25729	24MR	7;33	9.67	25729	7AP	6;05	203.26
25730	22AP	17;57	38.25	25730	6MY	19;03	231.76
25731	22MY	2;20	66.40	25731	5JN	9;20	260.03
25732	20JN	9;37	94.33	25732	5JL	0;38	288.28
25733	19JL	16;40	122.31	25733	3AU	16;13	316.71
25734	18AU	0;22	150.59	25734	2SE	7;25	345.50
25735	16SE	9;39	179.36	25735	1OC	21;53	14.74
25736	15OC	21;27	208.72	25736	31OC	11;31	44.43
25737	14NO	12;22	238.65	25737	30NO	0;14	74.42
25738	14DE	6;09	268.95	25738	29DE	11;52	104.49
1081							
25739	13JA	1;27	299.30	25739	27JA	22;22	134.35
25740	11FE	20;23	329.35	25740	26FE	7;58	163.80
25741	13MR	13;28	358.87	25741	27MR	17;15	192.76
25742	12AP	3;59	27.79	25742	26AP	2;59	221.27
25743	11MY	15;54	56.19	25743	25MY	13;55	249.49
25744	10JN	1;35	84.25	25744	24JN	2;27	277.63
25745	9JL	9;42	112.23	25745	23JL	16;41	305.92
25746	7AU	17;07	140.38	25746	22AU	8;24	334.57
25747	6SE	0;55	168.92	25747	21SE	1;12	3.71
25748	5OC	10;03	198.00	25748	20OC	18;18	33.37
25749	3NO	21;17	227.63	25749	19NO	10;40	63.42
25750	3DE	10;56	257.69	25750	19DE	1;18	93.60

Right half (periods 1082–1087)

NUMBER	DATE	TIME	LONG.	NUMBER	DATE	TIME	LONG.
1082							
25751	2JA	2;45	287.93	25751	17JA	13;44	123.63
25752	31JA	19;58	318.04	25752	16FE	0;08	153.27
25753	2MR	13;41	347.77	25753	17MR	9;05	182.39
25754	1AP	6;52	16.98	25754	15AP	17;20	211.00
25755	30AP	22;32	45.67	25755	15MY	1;35	239.21
25756	30MY	12;01	73.95	25756	13JN	10;35	267.24
25757	28JN	23;14	102.03	25757	12JL	21;03	295.32
25758	28JL	8;42	130.17	25758	11AU	9;44	323.71
25759	26AU	17;22	158.59	25759	10SE	1;04	352.62
25760	25SE	2;12	187.48	25760	9OC	18;52	22.12
25761	24OC	11;55	216.89	25761	8NO	13;55	52.14
25762	22NO	22;49	246.74	25762	8DE	8;20	82.43
25763	22DE	10;58	276.82				
1083							
				25763	7JA	0;36	112.67
25764	21JA	0;25	306.87	25764	5FE	14;08	142.58
25765	19FE	15;12	336.65	25765	7MR	1;09	171.97
25766	21MR	7;07	6.00	25766	5AP	10;14	200.78
25767	19AP	23;31	34.88	25767	4MY	18;00	229.10
25768	19MY	15;23	63.36	25768	3JN	1;10	257.10
25769	18JN	5;54	91.58	25769	2JL	8;39	285.04
25770	17JL	18;49	119.78	25770	31JL	17;32	313.18
25771	16AU	6;26	148.19	25771	30AU	4;57	341.75
25772	14SE	17;18	176.98	25772	28SE	19;35	10.94
25773	14OC	3;53	206.26	25773	28OC	13;17	40.74
25774	12NO	14;25	235.97	25774	27NO	8;46	70.99
25775	12DE	1;01	265.95	25775	27DE	4;09	101.38
1084							
25776	10JA	11;55	295.95	25776	25JA	21;49	131.57
25777	8FE	23;27	325.74	25777	24FE	12;53	161.29
25778	9MR	12;00	355.14	25778	25MR	1;07	190.42
25779	8AP	1;38	24.08	25779	23AP	10;45	218.96
25780	7MY	16;07	52.63	25780	22MY	18;25	247.07
25781	6JN	7;01	80.92	25781	21JN	1;05	274.97
25782	5JL	21;58	109.16	25782	20JL	7;57	302.95
25783	4AU	12;43	137.57	25783	18AU	16;10	331.26
25784	3SE	3;01	166.33	25784	17SE	2;47	0.09
25785	2OC	16;34	195.55	25785	16OC	16;19	29.54
25786	1NO	5;07	225.20	25786	15NO	8;43	59.54
25787	30NO	16;37	255.15	25787	15DE	3;14	89.88
25788	30DE	3;19	285.17				
1085							
				25788	13JA	22;38	120.22
25789	28JA	13;37	315.02	25789	12FE	17;19	150.25
25790	26FE	23;51	344.49	25790	14MR	9;49	179.74
25791	28MR	10;21	13.49	25791	12AP	23;17	208.60
25792	26AP	21;24	42.04	25792	12MY	9;43	236.93
25793	26MY	9;22	70.29	25793	10JN	17;56	264.93
25794	24JN	22;41	98.46	25794	10JL	1;05	292.87
25795	24JL	13;37	126.78	25795	8AU	8;21	321.02
25796	23AU	5;51	155.47	25796	6SE	16;40	349.59
25797	21SE	22;31	184.62	25797	6OC	2;42	18.72
25798	21OC	14;32	214.25	25798	4NO	14;52	48.40
25799	20NO	5;05	244.23	25799	4DE	5;28	78.50
25800	19DE	18;01	274.34				
1086							
25801	18JA	5;29	304.32	25800	2JA	22;24	108.79
25802	16FE	15;45	333.95	25801	1FE	16;56	138.95
25803	18MR	1;00	3.07	25802	3MR	11;34	168.70
25804	16AP	9;39	31.69	25803	2AP	4;34	197.89
25805	15MY	18;22	59.91	25804	1MY	18;56	226.52
25806	14JN	4;09	87.97	25805	31MY	6;38	254.72
25807	13JL	15;57	116.10	25806	29JN	16;21	282.75
25808	12AU	6;18	144.57	25807	29JL	0;59	310.85
25809	10SE	22;57	173.54	25808	27AU	9;22	339.27
25810	10OC	16;53	203.06	25809	25SE	18;09	8.17
25811	9NO	10;47	233.05	25810	25OC	3;50	37.59
25812	9DE	3;32	263.27	25811	23NO	14;58	67.45
				25812	23DE	4;01	97.57
1087							
25813	7JA	18;27	293.46	25813	21JA	19;08	127.69
25814	6FE	7;08	323.32	25814	20FE	11;47	157.53
25815	7MR	17;32	352.67	25815	22MR	4;51	186.92
25816	6AP	2;02	21.45	25816	20AP	21;45	215.78
25817	5MY	9;25	49.75	25817	20MY	12;13	244.22
25818	3JN	16;44	77.75	25818	19JN	1;39	272.40
25819	3JL	1;08	105.73	25819	18JL	13;35	300.57
25820	1AU	11;36	133.93	25820	17AU	0;18	328.94
25821	31AU	0;44	162.59	25821	15SE	10;11	357.71
25822	29SE	16;34	191.84	25822	14OC	19;47	26.95
25823	29OC	10;38	221.66	25823	13NO	5;43	56.64
25824	28NO	5;50	251.91	25824	12DE	16;29	86.64
25825	28DE	0;40	282.28				

Left half

	NEW MOONS				FULL MOONS		
NUMBER	DATE	TIME	LONG.	NUMBER	DATE	TIME	LONG.
1088							
				25825	11JA	4;24	116.68
25826	26JA	17;30	312.43	25826	9FE	17;25	146.52
25827	25FE	7;18	342.09	25827	10MR	7;19	175.96
25828	25MR	18;03	11.14	25828	8AP	21;53	204.93
25829	24AP	2;26	39.62	25829	8MY	12;56	233.49
25830	23MY	9;32	67.70	25830	7JN	4;12	261.79
25831	21JN	16;26	95.61	25831	6JL	19;13	290.03
25832	21JL	0;06	123.63	25832	5AU	9;21	318.43
25833	19AU	9;23	151.99	25833	3SE	22;16	347.15
25834	17SE	21;00	180.87	25834	3OC	10;05	16.30
25835	17OC	11;31	210.38	25835	1NO	21;13	45.91
25836	16NO	5;02	240.43	25836	1DE	8;08	75.84
25837	16DE	0;40	270.82	25837	30DE	19;01	105.87
1089							
25838	14JA	20;30	301.18	25838	29JA	5;51	135.73
25839	13FE	14;24	331.16	25839	27FE	16;40	165.22
25840	15MR	5;11	0.56	25840	29MR	3;50	194.23
25841	13AP	16;49	29.34	25841	27AP	15;54	222.82
25842	13MY	2;03	57.61	25842	27MY	5;19	251.12
25843	11JN	9;47	85.59	25843	25JN	20;06	279.34
25844	10JL	16;53	113.53	25844	25JL	11;42	307.70
25845	9AU	0;15	141.69	25845	24AU	3;23	336.37
25846	7SE	8;44	170.28	25846	22SE	18;32	5.48
25847	6OC	19;16	199.44	25847	22OC	8;54	35.05
25848	5NO	8;39	229.18	25848	20NO	22;19	64.99
25849	5DE	1;05	259.37	25849	20DE	10;36	95.07
1090							
25850	3JA	19;45	289.72	25850	18JA	21;37	125.03
25851	2FE	14;57	319.90	25851	17FE	7;29	154.63
25852	4MR	8;57	349.62	25852	18MR	16;37	183.74
25853	3AP	0;40	18.74	25853	17AP	1;47	212.37
25854	2MY	13;48	47.30	25854	16MY	11;49	240.64
25855	1JN	0;34	75.47	25855	14JN	23;20	268.76
25856	30JN	9;27	103.46	25856	14JL	12;40	296.96
25857	29JL	17;14	131.53	25857	13AU	3;45	325.47
25858	28AU	0;53	159.93	25858	11SE	20;17	354.45
25859	26SE	9;24	188.83	25859	11OC	13;36	23.96
25860	25OC	19;38	218.28	25860	10NO	6;44	53.92
25861	24NO	8;07	248.21	25861	9DE	22;31	84.11
25862	23DE	22;50	278.40				
1091							
				25862	8JA	12;08	114.24
25863	22JA	15;16	308.57	25863	6FE	23;26	144.03
25864	21FE	8;37	338.43	25864	8MR	8;53	173.33
25865	23MR	1;57	7.81	25865	6AP	17;12	202.09
25866	21AP	18;20	36.66	25866	6MY	1;11	230.41
25867	21MY	8;56	65.07	25867	4JN	9;33	258.45
25868	19JN	21;20	93.21	25868	3JL	19;06	286.47
25869	19JL	7;45	121.32	25869	2AU	6;35	314.72
25870	17AU	16;54	149.64	25870	31AU	20;38	343.43
25871	16SE	1;46	178.38	25871	30SE	13;26	12.73
25872	15OC	11;11	207.62	25872	30OC	8;15	42.60
25873	13NO	21;35	237.35	25873	29NO	3;24	72.85
25874	13DE	9;07	267.37	25874	28DE	20;59	103.17
1092							
25875	11JA	21;47	297.45	25875	27JA	11;54	133.23
25876	10FE	11;40	327.31	25876	26FE	0;02	162.81
25877	11MR	2;46	356.79	25877	26MR	9;54	191.81
25878	9AP	18;45	25.81	25878	24AP	18;07	220.28
25879	9MY	10;50	54.40	25879	24MY	1;22	248.36
25880	8JN	2;05	82.69	25880	22JN	8;30	276.28
25881	7JL	15;54	110.89	25881	21JL	16;32	304.32
25882	6AU	4;19	139.22	25882	20AU	2;35	332.71
25883	4SE	15;46	167.89	25883	18SE	15;38	1.67
25884	4OC	2;43	197.03	25884	18OC	8;00	31.26
25885	2NO	13;28	226.62	25885	17NO	2;54	61.39
25886	2DE	0;08	256.54	25886	16DE	22;38	91.78
25887	31DE	10;49	286.57				
1093							
				25887	15JA	17;19	122.08
25888	29JA	21;50	316.43	25888	14FE	9;41	152.00
25889	28FE	9;35	345.95	25889	15MR	23;16	181.33
25890	29MR	22;23	15.02	25890	14AP	10;03	210.07
25891	28AP	12;12	43.66	25891	13MY	18;32	238.30
25892	28MY	2;46	72.00	25892	12JN	1;33	266.25
25893	26JN	17;42	100.23	25893	11JL	8;11	294.17
25894	26JL	8;42	128.57	25894	9AU	15;39	322.34
25895	24AU	23;30	157.22	25895	8SE	1;04	350.98
25896	23SE	13;47	186.30	25896	7OC	13;11	20.21
25897	23OC	3;08	215.85	25897	6NO	4;16	50.04
25898	21NO	15;20	245.74	25898	5DE	21;51	80.28
25899	21DE	2;28	275.77				

Right half

	NEW MOONS				FULL MOONS		
NUMBER	DATE	TIME	LONG.	NUMBER	DATE	TIME	LONG.
1094							
				25899	4JA	16;56	110.65
25900	19JA	12;52	305.70	25900	3FE	12;03	140.81
25901	17FE	22;56	335.30	25901	5MR	5;41	170.50
25902	19MR	9;01	4.45	25902	3AP	20;38	199.58
25903	17AP	19;28	33.11	25903	3MY	8;29	228.08
25904	17MY	6;41	61.42	25904	1JN	17;41	256.18
25905	15JN	19;08	89.58	25905	1JL	1;17	284.12
25906	15JL	9;13	117.81	25906	30JL	8;28	312.18
25907	14AU	0;58	146.36	25907	28AU	16;18	340.59
25908	12SE	17;48	175.36	25908	27SE	1;33	9.53
25909	12OC	10;34	204.86	25909	26OC	12;42	39.03
25910	11NO	2;10	234.77	25910	25NO	2;06	69.00
25911	10DE	16;05	264.88	25911	24DE	17;49	99.24
1095							
25912	9JA	4;17	294.95	25912	23JA	11;31	129.45
25913	7FE	15;03	324.72	25913	22FE	6;07	159.35
25914	9MR	0;36	354.01	25914	23MR	23;55	188.74
25915	7AP	9;17	22.78	25915	22AP	15;34	217.55
25916	6MY	17;39	51.10	25916	22MY	4;32	245.88
25917	5JN	2;36	79.16	25917	20JN	15;13	273.95
25918	4JL	13;11	107.22	25918	20JL	0;26	302.02
25919	3AU	2;14	135.54	25919	18AU	9;03	330.33
25920	1SE	17;54	164.33	25920	16SE	17;45	359.06
25921	1OC	11;33	193.68	25921	16OC	3;05	28.32
25922	31OC	5;51	223.54	25922	14NO	13;32	58.05
25923	29NO	23;29	253.74	25923	14DE	1;34	88.10
25924	29DE	15;30	283.99				
1096							
				25924	12JA	15;32	118.23
25925	28JA	5;19	313.99	25925	11FE	7;17	148.16
25926	26FE	16;45	343.53	25926	12MR	0;00	177.69
25927	27MR	2;00	12.50	25927	10AP	16;35	206.72
25928	25AP	9;40	40.94	25928	10MY	8;12	235.28
25929	24MY	16;45	69.00	25929	8JN	22;25	263.53
25930	23JN	0;22	96.95	25930	8JL	11;13	291.69
25931	22JL	9;40	125.03	25931	6AU	22;44	319.99
25932	20AU	21;27	153.51	25932	5SE	9;14	348.64
25933	19SE	12;03	182.54	25933	4OC	19;09	17.74
25934	19OC	5;14	212.18	25934	3NO	5;01	47.31
25935	18NO	0;10	242.32	25935	2DE	15;24	77.22
25936	17DE	19;29	272.69				
1097							
				25936	1JA	2;40	107.27
25937	16JA	13;31	302.97	25937	30JA	14;59	137.18
25938	15FE	4;52	332.83	25938	1MR	4;12	166.75
25939	16MR	17;00	2.09	25939	30MR	18;08	195.85
25940	15AP	2;21	30.75	25940	29AP	8;42	224.51
25941	14MY	9;52	58.94	25941	28MY	23;46	252.87
25942	12JN	16;41	86.88	25942	27JN	14;59	281.11
25943	11JL	23;51	114.83	25943	27JL	5;48	309.45
25944	10AU	8;17	143.04	25944	25AU	19;37	338.07
25945	8SE	18;43	171.73	25945	24SE	8;14	7.10
25946	8OC	7;50	201.02	25946	23OC	19;55	36.58
25947	6NO	23;57	230.90	25947	22NO	7;05	66.43
25948	6DE	18;43	261.19	25948	21DE	18;01	96.46
1098							
25949	5JA	14;43	291.60	25949	20JA	4;48	126.41
25950	4FE	9;44	321.75	25950	18FE	15;25	156.02
25951	6MR	2;03	351.37	25951	20MR	2;05	185.17
25952	4AP	15;07	20.36	25952	18AP	13;20	213.86
25953	4MY	1;24	48.79	25953	18MY	1;47	242.22
25954	2JN	9;45	76.85	25954	16JN	15;47	270.43
25955	1JL	17;06	104.78	25955	16JL	7;04	298.71
25956	31JL	0;19	132.84	25956	14AU	22;59	327.27
25957	29AU	8;15	161.27	25957	13SE	14;43	356.25
25958	27SE	17;46	190.23	25958	13OC	5;48	25.70
25959	27OC	5;44	219.77	25959	11NO	19;58	55.55
25960	25NO	20;41	249.82	25960	11DE	8;59	85.63
25961	25DE	14;20	280.14				
1099							
				25961	9JA	20;40	115.67
25962	24JA	9;21	310.40	25962	8FE	6;58	145.41
25963	23FE	3;57	340.29	25963	9MR	16;12	174.68
25964	24MR	20;44	9.62	25964	8AP	1;01	203.44
25965	23AP	11;05	38.36	25965	7MY	10;17	231.79
25966	22MY	23;00	66.65	25966	5JN	20;48	259.92
25967	21JN	8;49	94.68	25967	5JL	9;07	288.05
25968	20JL	17;10	122.72	25968	3AU	23;22	316.42
25969	19AU	0;56	151.00	25969	2SE	15;22	345.24
25970	17SE	9;04	179.72	25970	2OC	8;38	14.58
25971	16OC	18;30	208.99	25971	1NO	2;15	44.43
25972	15NO	5;53	238.77	25972	30NO	19;04	74.60
25973	14DE	19;29	268.89	25973	30DE	9;55	104.80

1100

New #	Date	Time	Long.	Full #	Date	Time	Long.
25974	13JA	11:00	299.08	25974	28JA	22:20	134.74
25975	12FE	3:46	329.05	25975	27FE	8:31	164.22
25976	12MR	20:58	358.59	25976	27MR	17:09	193.15
25977	11AP	13:44	27.60	25977	26AP	1:01	221.58
25978	11MY	5:13	56.15	25978	25MY	8:57	249.68
25979	9JN	18:47	84.37	25979	23JN	17:42	277.67
25980	9JL	6:17	112.48	25980	23JL	4:05	305.80
25981	7AU	16:10	140.73	25981	21AU	16:49	334.32
25982	6SE	1:18	169.32	25982	20SE	8:23	3.41
25983	5OC	10:34	198.41	25983	20OC	2:31	33.10
25984	3NO	20:35	227.99	25984	18NO	21:55	63.27
25985	3DE	7:36	257.94	25985	18DE	16:37	93.62

1101

New #	Date	Time	Long.	Full #	Date	Time	Long.
25986	1JA	19:38	288.02	25986	17JA	8:58	123.82
25987	31JA	8:42	317.96	25987	15FE	22:25	153.59
25988	1MR	22:55	347.55	25988	17MR	9:14	182.79
25989	31MR	14:13	16.70	25989	15AP	18:04	211.42
25990	30AP	6:08	45.40	25990	15MY	1:36	239.60
25991	29MY	21:49	73.78	25991	13JN	8:37	267.55
25992	28JN	12:27	101.99	25992	12JL	16:02	295.51
25993	28JL	1:44	130.28	25993	11AU	0:58	323.75
25994	26AU	13:52	158.84	25994	9SE	12:30	352.49
25995	25SE	1:18	187.83	25995	9OC	3:19	21.86
25996	24OC	12:22	217.30	25996	7NO	21:12	51.82
25997	22NO	23:13	247.15	25997	7DE	16:49	82.16
25998	22DE	9:53	277.17				

1102

New #	Date	Time	Long.	Full #	Date	Time	Long.
				25998	6JA	12:14	112.54
25999	20JA	20:35	307.10	25999	5FE	5:50	142.62
26000	19FE	7:42	336.73	26000	6MR	20:45	172.17
26001	20MR	19:40	5.92	26001	5AP	8:49	201.11
26002	19AP	8:42	34.67	26002	4MY	18:18	229.50
26003	18MY	22:42	63.08	26003	3JN	1:54	257.52
26004	17JN	13:24	91.32	26004	2JL	8:36	285.43
26005	17JL	4:27	119.60	26005	31JL	15:34	313.49
26006	15AU	19:35	148.14	26006	29AU	23:59	341.94
26007	14SE	10:28	177.09	26007	28SE	10:46	10.96
26008	14OC	0:39	206.51	26008	28OC	0:28	40.59
26009	12NO	13:40	236.33	26009	26NO	16:54	70.71
26010	12DE	1:26	266.36	26010	26DE	11:20	101.06

1103

New #	Date	Time	Long.	Full #	Date	Time	Long.
26011	10JA	12:08	296.36	26011	25JA	6:31	131.32
26012	8FE	22:11	326.08	26012	24FE	0:56	161.19
26013	10MR	8:01	355.36	26013	25MR	17:13	190.48
26014	8AP	17:59	24.16	26014	24AP	6:33	219.18
26015	8MY	4:31	52.55	26015	23MY	16:58	247.40
26016	6JN	16:05	80.71	26016	22JN	1:17	275.37
26017	6JL	5:13	108.88	26017	21JL	8:38	303.37
26018	4AU	20:14	137.29	26018	19AU	16:10	331.65
26019	3SE	12:50	166.14	26019	18SE	0:47	0.40
26020	3OC	6:03	195.50	26020	17OC	11:04	29.72
26021	1NO	22:38	225.32	26021	15NO	23:22	59.54
26022	1DE	13:37	255.41	26022	15DE	13:53	89.70
26023	31DE	2:44	285.54				

1104

New #	Date	Time	Long.	Full #	Date	Time	Long.
				26023	14JA	6:32	119.93
26024	29JA	14:09	315.43	26024	13FE	0:38	149.95
26025	28FE	0:09	344.89	26025	13MR	18:50	179.51
26026	28MR	9:01	13.82	26026	12AP	11:32	208.52
26027	26AP	17:15	42.26	26027	12MY	1:47	237.00
26028	26MY	1:36	70.37	26028	10JN	13:35	265.15
26029	24JN	11:07	98.39	26029	9JL	23:33	293.20
26030	23JL	22:49	126.58	26030	8AU	8:34	321.42
26031	22AU	13:15	155.18	26031	6SE	17:22	350.01
26032	21SE	6:12	184.34	26032	6OC	2:31	19.10
26033	21OC	0:34	214.06	26033	4NO	12:29	48.69
26034	19NO	18:53	244.19	26034	3DE	23:41	78.65
26035	19DE	11:56	274.47				

1105

New #	Date	Time	Long.	Full #	Date	Time	Long.
				26035	2JA	12:34	108.76
26036	18JA	2:56	304.61	26036	1FE	3:16	138.77
26037	16FE	15:30	334.33	26037	2MR	19:21	168.42
26038	18MR	1:40	3.49	26038	1AP	11:52	197.61
26039	16AP	9:53	32.09	26039	1MY	3:51	226.30
26040	15MY	16:59	60.25	26040	30MY	18:45	254.64
26041	14JN	0:06	88.19	26041	29JN	8:20	282.82
26042	13JL	8:23	116.19	26042	28JL	20:40	311.07
26043	11AU	18:52	144.50	26043	27AU	7:54	339.60
26044	10SE	8:07	173.33	26044	25SE	18:19	8.57
26045	10OC	0:11	202.76	26045	25OC	4:20	38.00
26046	8NO	18:30	232.75	26046	23NO	14:31	67.83
26047	8DE	13:56	263.09	26047	23DE	1:19	97.85

1106

New #	Date	Time	Long.	Full #	Date	Time	Long.
26048	7JA	8:51	293.44	26048	21JA	12:59	127.82
26049	6FE	1:39	323.48	26049	20FE	1:31	157.50
26050	7MR	15:19	352.97	26050	21MR	14:49	186.74
26051	6AP	1:52	21.83	26051	20AP	4:47	215.51
26052	5MY	10:05	50.16	26052	19MY	19:25	243.94
26053	3JN	17:03	78.15	26053	18JN	10:34	272.19
26054	2JL	23:55	106.07	26054	18JL	1:46	300.48
26055	1AU	7:39	134.16	26055	16AU	16:22	329.01
26056	30AU	17:04	162.67	26056	15SE	5:54	357.92
26057	29SE	4:52	191.74	26057	14OC	18:18	27.28
26058	28OC	19:32	221.42	26058	13NO	5:52	57.05
26059	27NO	13:07	251.59	26059	12DE	17:00	87.05
26060	27DE	8:44	282.00				

1107

New #	Date	Time	Long.	Full #	Date	Time	Long.
				26060	11JA	3:50	117.05
26061	26JA	4:26	312.27	26061	9FE	14:22	146.78
26062	24FE	22:09	342.10	26062	11MR	0:43	176.08
26063	26MR	12:44	11.31	26063	9AP	11:19	204.89
26064	25AP	0:15	39.92	26064	8MY	22:50	233.32
26065	24MY	9:26	68.08	26065	7JN	11:52	261.53
26066	22JN	17:12	96.03	26066	7JL	2:31	289.76
26067	22JL	0:28	124.03	26067	5AU	18:19	318.21
26068	20AU	8:04	152.32	26068	4SE	10:28	347.06
26069	18SE	16:48	181.09	26069	4OC	2:13	16.37
26070	18OC	3:33	210.44	26070	2NO	17:09	46.13
26071	16NO	17:02	240.32	26071	2DE	6:57	76.18
26072	16DE	9:23	270.56	26072	31DE	19:23	106.27

1108

New #	Date	Time	Long.	Full #	Date	Time	Long.
26073	15JA	3:47	300.86	26073	30JA	6:18	136.14
26074	13FE	22:37	330.90	26074	28FE	15:51	165.57
26075	14MR	16:16	0.42	26075	29MR	0:34	194.49
26076	13AP	7:45	29.36	26076	27AP	9:16	222.94
26077	12MY	20:49	57.79	26077	26MY	18:52	251.09
26078	11JN	7:41	85.90	26078	25JN	6:07	279.18
26079	10JL	16:48	113.92	26079	24JL	19:22	307.44
26080	9AU	0:55	142.11	26080	23AU	10:37	336.08
26081	7SE	8:56	170.68	26081	22SE	3:32	5.25
26082	6OC	17:46	199.77	26082	21OC	21:22	34.96
26083	5NO	4:14	229.38	26083	20NO	14:58	65.07
26084	4DE	16:44	259.40	26084	20DE	7:03	95.31

1109

New #	Date	Time	Long.	Full #	Date	Time	Long.
26085	3JA	7:15	289.58	26085	18JA	20:44	125.39
26086	1FE	23:16	319.63	26086	17FE	7:53	155.04
26087	3MR	16:04	349.31	26087	18MR	17:03	184.15
26088	2AP	8:55	18.49	26088	17AP	1:02	212.73
26089	2MY	1:01	47.18	26089	16MY	8:40	240.90
26090	31MY	15:35	75.50	26090	14JN	16:48	268.89
26091	30JN	4:13	103.64	26091	14JL	2:13	296.93
26092	29JL	15:01	131.83	26092	12AU	13:42	325.29
26093	28AU	0:39	160.31	26093	11SE	3:53	354.17
26094	26SE	9:59	189.24	26094	10OC	20:58	23.66
26095	25OC	19:46	218.68	26095	9NO	16:08	53.70
26096	24NO	6:22	248.53	26096	9DE	11:34	84.04
26097	23DE	17:52	278.58				

1110

New #	Date	Time	Long.	Full #	Date	Time	Long.
				26097	8JA	5:18	114.34
26098	22JA	6:15	308.58	26098	6FE	20:11	144.29
26099	20FE	19:38	338.28	26099	8MR	8:09	173.69
26100	22MR	10:06	7.55	26100	6AP	17:47	202.51
26101	21AP	1:32	36.38	26101	6MY	1:46	230.82
26102	20MY	17:17	64.84	26102	4JN	8:52	258.81
26103	19JN	8:31	93.09	26103	3JL	15:56	286.74
26104	18JL	22:37	121.34	26104	1AU	23:59	314.85
26105	17AU	11:32	149.82	26105	31AU	10:09	343.40
26106	15SE	23:33	178.68	26106	29SE	23:22	12.54
26107	15OC	11:03	208.01	26107	29OC	15:54	42.31
26108	13NO	22:11	237.76	26108	28NO	10:56	72.55
26109	13DE	9:00	267.76	26109	28DE	6:42	102.96

1111

New #	Date	Time	Long.	Full #	Date	Time	Long.
26110	11JA	19:36	297.75	26110	27JA	1:19	133.18
26111	10FE	6:16	327.48	26111	25FE	17:34	162.93
26112	11MR	17:29	356.80	26112	27MR	6:57	192.08
26113	10AP	5:41	25.67	26113	25AP	17:36	220.65
26114	9MY	18:58	54.15	26114	25MY	2:01	248.77
26115	8JN	9:12	82.41	26115	23JN	9:03	276.69
26116	8JL	0:06	110.65	26116	22JL	15:47	304.68
26117	6AU	15:23	139.09	26117	20AU	23:26	332.97
26118	5SE	6:42	167.91	26118	19SE	9:02	1.79
26119	4OC	21:36	197.20	26119	18OC	21:21	31.21
26120	3NO	11:31	226.93	26120	17NO	12:30	61.17
26121	3DE	0:05	256.93	26121	17DE	6:02	91.47

```
--------NEW MOONS--------        --------FULL MOONS--------        --------NEW MOONS--------        --------FULL MOONS--------
NUMBER  DATE   TIME   LONG.      NUMBER  DATE   TIME   LONG.      NUMBER  DATE   TIME   LONG.      NUMBER  DATE   TIME   LONG.
```

1112 / 1118

NEW MOONS				FULL MOONS				NEW MOONS				FULL MOONS			
26122	1JA	11;18	286.98	26122	16JA	0;54	121.78					26196	9JA	18;29	115.97
26123	30JA	21;33	316.81	26123	14FE	19;45	151.81	26197	23JA	19;12	310.19	26197	8FE	6;50	145.80
26124	29FE	7;15	346.24	26124	15MR	13;06	181.31	26198	22FE	11;26	339.99	26198	9MR	16;47	175.10
26125	29MR	16;52	15.18	26125	14AP	3;52	210.20	26199	24MR	4;05	9.33	26199	8AP	1;05	203.84
26126	28AP	2;48	43.67	26126	13MY	15;40	238.58	26200	22AP	20;28	38.17	26200	7MY	8;38	232.11
26127	27MY	13;34	71.87	26127	12JN	0;57	266.61	26201	22MY	11;49	66.59	26201	5JN	16;18	260.12
26128	26JN	1;45	99.99	26128	11JL	8;44	294.59	26202	21JN	1;30	94.78	26202	5JL	0;54	288.11
26129	25JL	15;48	128.28	26129	9AU	16;12	322.75	26203	20JL	13;21	122.95	26203	3AU	11;13	316.33
26130	24AU	7;48	156.97	26130	8SE	0;20	351.34	26204	18AU	23;42	151.34	26204	2SE	0;04	345.00
26131	23SE	1;07	186.17	26131	7OC	9;52	20.47	26205	17SE	9;20	180.13	26205	1OC	15;51	14.28
26132	22OC	18;27	215.87	26132	5NO	21;13	50.13	26206	16OC	19;02	209.41	26206	31OC	10;17	44.15
26133	21NO	10;33	245.93	26133	5DE	10;36	80.19	26207	15NO	5;21	239.14	26207	30NO	5;59	74.43
26134	21DE	0;44	276.09					26208	14DE	16;25	269.15	26208	30DE	0;52	104.80

1113 / 1119

NEW MOONS				FULL MOONS				NEW MOONS				FULL MOONS			
				26134	4JA	2;06	110.41	26209	13JA	4;15	299.19	26209	28JA	17;14	134.92
26135	19JA	12;58	306.10	26135	2FE	19;24	140.51	26210	11FE	16;53	328.98	26210	27FE	6;33	164.53
26136	17FE	23;32	335.72	26136	4MR	13;30	170.23	26211	13MR	6;30	358.38	26211	28MR	17;10	193.54
26137	19MR	8;46	4.82	26137	3AP	6;56	199.41	26212	11AP	21;10	27.33	26212	27AP	1;47	222.00
26138	17AP	17;03	33.40	26138	2MY	22;22	228.07	26213	11MY	12;41	55.88	26213	26MY	9;09	250.07
26139	17MY	1;03	61.59	26139	1JN	11;22	256.32	26214	10JN	4;12	84.18	26214	24JN	16;04	277.99
26140	15JN	9;42	89.59	26140	30JN	22;16	284.39	26215	9JL	19;00	112.42	26215	23JL	23;33	306.01
26141	14JL	20;07	117.68	26141	30JL	7;51	312.54	26216	8AU	8;42	140.82	26216	22AU	8;32	334.38
26142	13AU	9;10	146.11	26142	28AU	16;53	341.01	26217	6SE	21;25	169.56	26217	20SE	20;13	3.30
26143	12SE	1;03	175.07	26143	27SE	2;01	9.94	26218	6OC	9;26	198.75	26218	20OC	11;12	32.85
26144	11OC	19;04	204.61	26144	26OC	11;41	39.38	26219	4NO	20;58	228.40	26219	19NO	5;13	62.96
26145	10NO	13;48	234.64	26145	24NO	22;17	69.23	26220	4DE	8;03	258.35	26220	19DE	0;54	93.35
26146	10DE	7;46	264.92	26146	24DE	10;14	99.31								

1114 / 1120

NEW MOONS				FULL MOONS				NEW MOONS				FULL MOONS			
26147	8JA	23;56	295.16	26147	22JA	23;52	129.35	26221	2JA	18;44	288.37	26221	17JA	20;15	123.67
26148	7FE	13;43	325.05	26148	21FE	15;04	159.12	26222	1FE	5;11	318.20	26222	16FE	13;42	153.61
26149	9MR	0;57	354.41	26149	23MR	7;12	188.45	26223	1MR	15;50	347.65	26223	17MR	4;27	182.98
26150	7AP	9;56	23.19	26150	21AP	23;19	217.29	26224	31MR	3;13	16.64	26224	15AP	16;21	211.74
26151	6MY	17;20	51.47	26151	21MY	14;42	245.72	26225	29AP	15;40	45.21	26225	15MY	1;46	240.00
26152	5JN	0;12	79.45	26152	20JN	4;59	273.94	26226	29MY	5;15	73.51	26226	13JN	9;22	267.96
26153	4JL	7;42	107.39	26153	19JL	18;06	302.15	26227	27JN	19;47	101.72	26227	12JL	16;09	295.90
26154	2AU	16;58	135.56	26154	18AU	6;07	330.60	26228	27JL	11;00	130.08	26228	10AU	23;18	324.07
26155	1SE	4;51	164.19	26155	16SE	17;10	359.43	26229	26AU	2;34	158.77	26229	9SE	7;55	352.69
26156	30SE	19;38	193.41	26156	16OC	3;34	28.73	26230	24SE	18;03	187.92	26230	8OC	18;55	21.90
26157	30OC	13;03	223.23	26157	14NO	13;46	58.45	26231	24OC	8;49	217.54	26231	7NO	8;44	51.69
26158	29NO	8;11	253.48	26158	14DE	0;15	88.44	26232	22NO	22;18	247.50	26232	7DE	1;10	81.89
26159	29DE	3;38	283.87					26233	22DE	10;15	277.57				

1115 / 1121

NEW MOONS				FULL MOONS				NEW MOONS				FULL MOONS			
				26159	12JA	11;23	118.44	26234	20JA	20;53	307.51	26233	5JA	19;25	112.22
26160	27JA	21;39	314.06	26160	10FE	23;18	148.22	26235	19FE	6;39	337.07	26234	4FE	14;17	142.36
26161	26FE	12;52	343.76	26161	12MR	11;57	177.59	26236	20MR	16;03	6.16	26235	6MR	8;24	172.05
26162	28MR	0;50	12.84	26162	11AP	1;16	206.49	26237	19AP	1;34	34.77	26236	5AP	0;27	201.16
26163	26AP	10;01	41.34	26163	10MY	15;20	235.00	26238	18MY	11;35	63.03	26237	4MY	13;41	229.70
26164	25MY	17;24	69.42	26164	9JN	6;08	263.27	26239	16JN	22;50	91.13	26238	3JN	0;09	257.84
26165	24JN	0;11	97.33	26165	8JL	21;25	291.53	26240	16JL	11;50	119.33	26239	2JL	8;38	285.82
26166	23JL	7;24	125.33	26166	7AU	12;36	319.98	26241	15AU	2;58	147.86	26240	31JL	16;15	313.90
26167	21AU	15;58	153.68	26167	6SE	2;59	348.77	26242	13SE	19;57	176.89	26241	30AU	0;07	342.34
26168	20SE	2;35	182.54	26168	5OC	16;14	18.01	26243	13OC	13;43	206.45	26242	28SE	9;03	11.28
26169	19OC	15;51	212.01	26169	4NO	4;25	47.67	26244	12NO	6;49	236.43	26243	27OC	19;34	40.78
26170	18NO	8;04	242.03	26170	3DE	15;53	77.64	26245	11DE	22;10	266.61	26244	26NO	7;56	70.72
26171	18DE	2;50	272.38									26245	25DE	22;18	100.90

1116 / 1122

NEW MOONS				FULL MOONS				NEW MOONS				FULL MOONS			
				26171	2JA	2;52	107.67	26246	10JA	11;24	296.72	26246	24JA	14;35	131.04
26172	16JA	22;41	302.73	26172	31JA	13;27	137.51	26247	8FE	22;42	326.49	26247	23FE	8;12	160.89
26173	15FE	17;30	332.74	26173	29FE	23;39	166.95	26248	10MR	8;25	355.77	26248	25MR	1;56	190.25
26174	16MR	9;36	2.18	26174	30MR	9;47	195.90	26249	8AP	16;56	24.51	26249	23AP	18;20	219.08
26175	14AP	22;32	30.99	26175	28AP	20;30	224.41	26250	8MY	0;47	52.79	26250	23MY	8;32	247.45
26176	14MY	8;44	59.29	26176	28MY	8;30	252.65	26251	6JN	8;50	80.81	26251	21JN	20;30	275.57
26177	12JN	17;07	87.29	26177	26JN	22;15	280.83	26252	5JL	18;09	108.83	26252	21JL	6;48	303.68
26178	12JL	0;37	115.25	26178	26JL	13;36	309.19	26253	4AU	5;47	137.10	26253	19AU	16;14	332.04
26179	10AU	8;04	143.43	26179	25AU	5;51	337.90	26254	2SE	20;21	165.86	26254	18SE	1;29	0.82
26180	8SE	16;16	172.02	26180	23SE	22;09	7.07	26255	2OC	13;36	195.21	26255	17OC	11;01	30.11
26181	8OC	2;01	201.17	26181	23OC	13;50	36.72	26256	1NO	8;21	225.11	26256	15NO	21;13	59.84
26182	6NO	14;08	230.87	26182	22NO	4;28	66.72	26257	1DE	3;03	255.36	26257	15DE	8;26	89.86
26183	6DE	5;04	261.01	26183	21DE	17;44	96.84	26258	30DE	20;18	285.66				

1117 / 1123

NEW MOONS				FULL MOONS				NEW MOONS				FULL MOONS			
26184	4JA	22;29	291.31	26184	20JA	5;23	126.82					26258	13JA	21;05	119.93
26185	3FE	17;09	321.45	26185	18FE	15;28	156.41	26259	29JA	11;20	315.71	26259	12FE	11;18	149.78
26186	5MR	11;21	351.16	26186	20MR	0;19	185.49	26260	27FE	23;45	345.27	26260	14MR	2;47	179.24
26187	4AP	3;50	20.29	26187	18AP	8;41	214.06	26261	29MR	9;40	14.24	26261	12AP	18;44	208.23
26188	3MY	18;03	48.88	26188	17MY	17;31	242.27	26262	27AP	17;38	42.67	26262	12MY	10;23	236.78
26189	2JN	6;00	77.09	26189	16JN	3;43	270.34	26263	27MY	0;31	70.72	26263	11JN	1;13	265.04
26190	1JL	16;01	105.12	26190	15JL	15;52	298.50	26264	25JN	7;29	98.63	26264	10JL	15;02	293.25
26191	31JL	0;42	133.25	26191	14AU	6;11	326.99	26265	24JL	15;44	126.69	26265	9AU	3;49	321.62
26192	29AU	8;50	161.69	26192	12SE	22;29	355.98	26266	23AU	2;17	155.13	26266	7SE	15;36	350.33
26193	27SE	17;21	190.60	26193	12OC	16;12	25.52	26267	21SE	15;41	184.13	26267	7OC	2;33	19.50
26194	27OC	3;03	220.05	26194	11NO	10;19	55.54	26268	21OC	7;57	213.75	26268	5NO	13;00	49.11
26195	25NO	14;34	249.95	26195	11DE	3;29	85.79	26269	20NO	2;28	243.88	26269	4DE	23;23	79.03
26196	25DE	4;02	280.09					26270	19DE	22;02	274.27				

----------NEW MOONS--------- ---------FULL MOONS-------- ----------NEW MOONS--------- ---------FULL MOONS--------

1124

NEW NUMBER	DATE	TIME	LONG.	FULL NUMBER	DATE	TIME	LONG.
				26270	3JA	10;08	109.05
26271	18JA	16;58	304.58	26271	1FE	21;29	138.91
26272	17FE	9;39	334.47	26272	2MR	9;31	168.41
26273	17MR	23;09	3.78	26273	31MR	22;12	197.44
26274	16AP	9;32	32.46	26274	30AP	11;36	226.04
26275	15MY	17;37	60.66	26275	30MY	1;52	254.36
26276	14JN	0;33	88.60	26276	28JN	16;56	282.59
26277	13JL	7;27	116.54	26277	28JL	8;23	310.96
26278	11AU	15;18	144.75	26278	26AU	23;30	339.65
26279	10SE	0;55	173.42	26279	25SE	13;39	8.76
26280	9OC	12;53	202.68	26280	25OC	2;36	38.32
26281	8NO	3;41	232.52	26281	23NO	14;35	68.22
26282	7DE	21;16	262.77	26282	23DE	1;51	98.27

1125

NEW NUMBER	DATE	TIME	LONG.	FULL NUMBER	DATE	TIME	LONG.
26283	6JA	16;45	293.15	26283	21JA	12;35	128.20
26284	5FE	12;14	323.31	26284	19FE	22;47	157.77
26285	7MR	5;43	352.96	26285	21MR	8;39	186.87
26286	5AP	20;08	21.99	26286	19AP	18;42	215.49
26287	5MY	7;33	50.45	26287	19MY	5;44	243.78
26288	3JN	16;45	78.53	26288	17JN	18;26	271.94
26289	3JL	0;39	106.49	26289	17JL	9;00	300.20
26290	1AU	8;08	134.57	26290	16AU	1;03	328.78
26291	30AU	16;00	163.02	26291	14SE	17;40	357.81
26292	29SE	1;01	191.97	26292	14OC	10;01	27.33
26293	28OC	11;58	221.49	26293	13NO	1;28	57.25
26294	27NO	1;29	251.49	26294	12DE	15;37	87.38
26295	26DE	17;40	281.75				

1126

NEW NUMBER	DATE	TIME	LONG.	FULL NUMBER	DATE	TIME	LONG.
				26295	11JA	4;08	117.45
26296	25JA	11;44	311.96	26296	9FE	14;53	147.20
26297	24FE	6;08	341.83	26297	11MR	0;07	176.44
26298	25MR	23;25	11.16	26298	9AP	8;25	205.16
26299	24AP	14;41	39.92	26299	8MY	16;41	233.46
26300	24MY	3;44	68.24	26300	7JN	1;56	261.52
26301	22JN	14;45	96.33	26301	6JL	12;57	289.61
26302	22JL	0;11	124.41	26302	5AU	2;11	317.95
26303	20AU	8;41	152.74	26303	3SE	17;38	346.76
26304	18SE	17;05	181.50	26304	3OC	10;56	16.13
26305	18OC	2;16	210.77	26305	2NO	5;14	46.02
26306	16NO	12;55	240.54	26306	1DE	23;14	76.24
26307	16DE	1;23	270.61	26307	31DE	15;33	106.50

1127

NEW NUMBER	DATE	TIME	LONG.	FULL NUMBER	DATE	TIME	LONG.
26308	14JA	15;37	300.74	26308	30JA	5;14	136.49
26309	13FE	7;09	330.64	26309	28FE	16;12	165.98
26310	14MR	23;24	0.12	26310	30MR	1;05	194.90
26311	13AP	15;45	29.11	26311	28AP	8;45	223.31
26312	13MY	7;35	57.65	26312	27MY	16;07	251.37
26313	11JN	22;11	85.91	26313	26JN	0;04	279.32
26314	11JL	11;05	114.08	26314	25JL	9;25	307.43
26315	9AU	22;21	142.39	26315	23AU	20;57	335.92
26316	8SE	8;29	171.05	26316	22SE	11;19	4.97
26317	7OC	18;18	200.18	26317	22OC	4;40	34.65
26318	6NO	4;27	229.79	26318	21NO	0;06	64.83
26319	5DE	15;13	259.73	26319	20DE	19;44	95.22

1128

NEW NUMBER	DATE	TIME	LONG.	FULL NUMBER	DATE	TIME	LONG.
26320	4JA	2;37	289.78	26320	19JA	13;31	125.47
26321	2FE	14;39	319.66	26321	18FE	4;19	155.28
26322	3MR	3;28	349.18	26322	18MR	16;07	184.50
26323	1AP	17;19	18.25	26323	17AP	1;18	213.14
26324	1MY	8;13	46.90	26324	16MY	9;21	241.32
26325	30MY	23;40	75.25	26325	14JN	16;21	269.26
26326	29JN	14;56	103.50	26326	13JL	23;25	297.21
26327	29JL	5;23	131.84	26327	12AU	7;34	325.43
26328	27AU	18;50	160.47	26328	10SE	17;52	354.15
26329	26SE	7;27	189.53	26329	10OC	7;15	23.48
26330	25OC	19;28	219.06	26330	8NO	23;55	53.40
26331	24NO	6;58	248.95	26331	8DE	19;01	83.73
26332	23DE	17;52	278.98				

1129

NEW NUMBER	DATE	TIME	LONG.	FULL NUMBER	DATE	TIME	LONG.
				26332	7JA	14;44	114.11
26333	22JA	4;18	308.89	26333	6FE	9;13	144.22
26334	20FE	14;36	338.46	26334	8MR	1;16	173.80
26335	22MR	1;17	7.58	26335	6AP	14;30	202.76
26336	20AP	12;53	36.26	26336	6MY	1;02	231.18
26337	20MY	1;41	64.61	26337	4JN	9;26	259.22
26338	18JN	15;38	92.81	26338	3JL	16;33	287.15
26339	18JL	6;34	121.09	26339	1AU	23;27	315.22
26340	16AU	22;10	149.67	26340	31AU	7;20	343.67
26341	15SE	14;02	178.68	26341	29SE	17;10	12.67
26342	15OC	5;33	208.17	26342	29OC	5;38	42.27
26343	13NO	19;59	238.06	26343	27NO	20;50	72.34
26344	13DE	8;50	268.15	26344	27DE	14;13	102.65

1130

NEW NUMBER	DATE	TIME	LONG.	FULL NUMBER	DATE	TIME	LONG.
26345	11JA	20;06	298.16	26345	26JA	8;47	132.88
26346	10FE	6;08	327.86	26346	25FE	3;16	162.73
26347	11MR	15;27	357.11	26347	26MR	20;20	192.04
26348	10AP	0;36	25.85	26348	25AP	10;57	220.77
26349	9MY	10;04	54.18	26349	24MY	22;46	249.03
26350	7JN	20;28	82.29	26350	23JN	8;12	277.05
26351	7JL	8;26	110.41	26351	22JL	16;14	305.08
26352	5AU	22;30	138.80	26352	21AU	0;01	333.39
26353	4SE	14;47	167.65	26353	19SE	8;30	2.16
26354	4OC	8;34	197.05	26354	18OC	18;20	31.48
26355	3NO	2;26	226.94	26355	17NO	5;49	61.28
26356	2DE	18;58	257.11	26356	16DE	19;08	91.40

1131

NEW NUMBER	DATE	TIME	LONG.	FULL NUMBER	DATE	TIME	LONG.
26357	1JA	9;21	287.29	26357	15JA	10;20	121.56
26358	30JA	21;33	317.20	26358	14FE	3;10	151.51
26359	1MR	7;54	346.66	26359	15MR	20;45	181.03
26360	30MR	16;47	15.57	26360	14AP	13;47	210.02
26361	29AP	0;43	43.98	26361	14MY	5;04	238.54
26362	28MY	8;24	72.05	26362	12JN	18;09	266.73
26363	26JN	16;50	100.02	26363	12JL	5;20	294.84
26364	26JL	3;10	128.17	26364	10AU	15;19	323.11
26365	24AU	16;16	156.73	26365	9SE	0;50	351.75
26366	23SE	8;23	185.87	26366	8OC	10;24	20.88
26367	23OC	2;44	215.61	26367	6NO	20;22	50.49
26368	21NO	21;50	245.78	26368	6DE	7;04	80.44
26369	21DE	16;03	276.12				

1132

NEW NUMBER	DATE	TIME	LONG.	FULL NUMBER	DATE	TIME	LONG.
				26369	4JA	18;53	110.50
26370	20JA	8;18	306.29	26370	3FE	8;07	140.43
26371	18FE	21;59	336.05	26371	3MR	22;45	170.01
26372	19MR	9;01	5.22	26372	2AP	14;16	199.13
26373	17AP	17;44	33.82	26373	2MY	5;56	227.80
26374	17MY	0;56	61.97	26374	31MY	21;08	256.14
26375	15JN	7;38	89.90	26375	30JN	11;32	284.35
26376	14JL	15;05	117.87	26376	30JL	1;01	312.66
26377	13AU	0;25	146.14	26377	28AU	13;34	341.26
26378	11SE	12;25	174.94	26378	27SE	1;12	10.30
26379	11OC	3;23	204.35	26379	26OC	12;05	39.79
26380	9NO	20;59	234.32	26380	24NO	22;35	69.64
26381	9DE	16;15	264.66	26381	24DE	9;07	99.65

1133

NEW NUMBER	DATE	TIME	LONG.	FULL NUMBER	DATE	TIME	LONG.
26382	8JA	11;43	295.03	26382	22JA	20;02	129.58
26383	7FE	5;39	325.11	26383	21FE	7;30	159.19
26384	8MR	20;43	354.63	26384	22MR	19;34	188.36
26385	7AP	8;31	23.52	26385	21AP	8;19	217.07
26386	6MY	17;34	51.91	26386	20MY	21;55	245.44
26387	5JN	0;54	79.87	26387	19JN	12;30	273.67
26388	4JL	7;41	107.78	26388	19JL	3;54	301.98
26389	2AU	15;02	135.87	26389	17AU	19;29	330.56
26390	31AU	23;47	164.37	26390	16SE	10;29	359.55
26391	30SE	10;36	193.43	26391	16OC	0;20	28.99
26392	30OC	0;01	223.07	26392	14NO	13;00	58.81
26393	28NO	16;16	253.20	26393	14DE	0;42	88.85
26394	28DE	10;55	283.56				

1134

NEW NUMBER	DATE	TIME	LONG.	FULL NUMBER	DATE	TIME	LONG.
				26394	12JA	11;41	118.85
26395	27JA	6;33	313.82	26395	10FE	22;00	148.56
26396	26FE	1;06	343.67	26396	12MR	7;47	177.81
26397	27MR	16;59	12.91	26397	10AP	17;24	206.56
26398	26AP	5;47	41.56	26398	10MY	3;36	234.91
26399	25MY	15;59	69.75	26399	8JN	15;12	263.07
26400	24JN	0;28	97.73	26400	8JL	4;46	291.26
26401	23JL	8;10	125.75	26401	6AU	20;13	319.71
26402	21AU	15;54	154.06	26402	5SE	12;52	348.59
26403	20SE	0;25	182.85	26403	5OC	5;43	17.96
26404	19OC	10;25	212.18	26404	3NO	21;57	47.79
26405	17NO	22;38	242.02	26405	3DE	13;00	77.91
26406	17DE	13;29	272.20				

1135

NEW NUMBER	DATE	TIME	LONG.	FULL NUMBER	DATE	TIME	LONG.
				26406	2JA	2;26	108.04
26407	16JA	6;36	302.45	26407	31JA	14;02	137.93
26408	15FE	0;49	332.44	26408	1MR	23;50	167.35
26409	16MR	18;35	1.95	26409	31MR	8;19	196.23
26410	15AP	10;46	30.90	26410	29AP	16;16	224.63
26411	15MY	0;53	59.36	26411	29MY	0;44	252.73
26412	13JN	12;57	87.51	26412	27JN	10;40	280.76
26413	12JL	23;14	115.58	26413	26JL	22;44	308.98
26414	11AU	8;18	143.82	26414	25AU	13;09	337.61
26415	9SE	16;52	172.44	26415	24SE	5;45	6.79
26416	9OC	1;45	201.55	26416	23OC	23;54	36.52
26417	7NO	11;43	231.17	26417	22NO	18;26	66.68
26418	6DE	23;17	261.15	26418	22DE	11;53	96.99

NEW MOONS / FULL MOONS

1136

NEW NUMBER	DATE	TIME	LONG.	FULL NUMBER	DATE	TIME	LONG.
26419	5JA	12;34	291.28	26419	21JA	2;58	127.11
26420	4FE	3;18	321.26	26420	19FE	15;12	156.79
26421	4MR	18;58	350.87	26421	20MR	0;54	185.90
26422	3AP	11;04	20.00	26422	18AP	8;55	214.46
26423	3MY	3;04	48.67	26423	17MY	16;11	242.61
26424	1JN	18;19	77.01	26424	15JN	23;39	270.56
26425	1JL	8;12	105.20	26425	15JL	8;09	298.58
26426	30JL	20;28	133.47	26426	13AU	18;32	326.91
26427	29AU	7;20	162.02	26427	12SE	7;29	355.75
26428	27SE	17;29	191.00	26428	11OC	23;29	25.22
26429	27OC	3;37	220.47	26429	10NO	18;11	55.25
26430	25NO	14;10	250.33	26430	10DE	14;05	85.61
26431	25DE	1;15	280.37				

1137

NEW NUMBER	DATE	TIME	LONG.	FULL NUMBER	DATE	TIME	LONG.
				26431	9JA	9;03	115.96
26432	23JA	12;49	310.32	26432	8FE	1;22	145.96
26433	22FE	0;58	339.95	26433	9MR	14;32	175.39
26434	23MR	13;59	9.14	26434	8AP	0;58	204.22
26435	22AP	4;05	37.90	26435	7MY	9;24	232.53
26436	21MY	19;08	66.32	26436	5JN	16;39	260.53
26437	20JN	10;33	94.58	26437	4JL	23;34	288.45
26438	20JL	1;35	122.87	26438	3AU	7;05	316.55
26439	18AU	15;46	151.41	26439	1SE	16;15	345.08
26440	17SE	5;04	180.34	26440	1OC	4;06	14.18
26441	16OC	17;40	209.74	26441	30OC	19;14	43.91
26442	15NO	5;38	239.55	26442	29NO	13;19	74.12
26443	14DE	16;55	269.57	26443	29DE	8;57	104.52

1138

NEW NUMBER	DATE	TIME	LONG.	FULL NUMBER	DATE	TIME	LONG.
26444	13JA	3;32	299.55	26444	28JA	4;10	134.76
26445	11FE	13;40	329.24	26445	26FE	21;25	164.54
26446	12MR	23;51	358.50	26446	28MR	11;59	193.71
26447	11AP	10;39	27.29	26447	26AP	23;46	222.31
26448	10MY	22;35	55.70	26448	26MY	9;08	250.46
26449	9JN	11;48	83.92	26449	24JN	16;48	278.41
26450	9JL	2;14	112.14	26450	23JL	23;46	306.41
26451	7AU	17;38	140.60	26451	22AU	7;08	334.71
26452	6SE	9;40	169.47	26452	20SE	16;01	3.52
26453	6OC	1;45	198.83	26453	20OC	3;13	32.91
26454	4NO	17;05	228.63	26454	18NO	17;07	62.83
26455	4DE	6;57	258.70	26455	18DE	9;26	93.08

1139

NEW NUMBER	DATE	TIME	LONG.	FULL NUMBER	DATE	TIME	LONG.
26456	2JA	19;02	288.78	26456	17JA	3;25	123.35
26457	1FE	5;33	318.61	26457	15FE	21;56	153.35
26458	2MR	15;01	348.00	26458	17MR	15;41	182.85
26459	31MR	23;58	16.89	26459	16AP	7;30	211.77
26460	30AP	8;58	45.33	26460	15MY	20;41	240.18
26461	29MY	18;38	73.47	26461	14JN	7;17	268.27
26462	28JN	5;37	101.55	26462	13JL	16;00	296.28
26463	27JL	18;33	129.81	26463	11AU	23;57	324.49
26464	26AU	9;51	158.48	26464	10SE	8;10	353.10
26465	25SE	3;13	187.70	26465	9OC	17;26	22.23
26466	24OC	21;30	217.46	26466	8NO	4;10	51.89
26467	23NO	15;04	247.59	26467	7DE	16;34	81.92
26468	23DE	6;42	277.82				

1140

NEW NUMBER	DATE	TIME	LONG.	FULL NUMBER	DATE	TIME	LONG.
				26468	6JA	6;42	112.08
26469	21JA	19;59	307.86	26469	4FE	22;33	142.10
26470	20FE	7;08	337.49	26470	5MR	15;38	171.75
26471	20MR	16;33	6.57	26471	4AP	8;52	200.92
26472	19AP	0;43	35.13	26472	4MY	1;00	229.58
26473	18MY	8;16	63.28	26473	2JN	15;12	257.87
26474	16JN	16;05	91.25	26474	2JL	3;23	286.00
26475	16JL	1;16	119.29	26475	31JL	14;04	314.20
26476	14AU	12;55	147.68	26476	29AU	23;59	342.72
26477	13SE	3;38	176.61	26477	28SE	9;42	11.70
26478	12OC	21;10	206.15	26478	27OC	19;38	41.17
26479	11NO	16;16	236.21	26479	26NO	6;01	71.03
26480	11DE	11;13	266.54	26480	25DE	17;12	101.07

1141

NEW NUMBER	DATE	TIME	LONG.	FULL NUMBER	DATE	TIME	LONG.
26481	10JA	4;37	296.82	26481	24JA	5;32	131.05
26482	8FE	19;35	326.75	26482	22FE	19;14	160.74
26483	10MR	7;51	356.13	26483	24MR	10;05	189.99
26484	8AP	17;33	24.92	26484	23AP	1;30	218.79
26485	8MY	1;17	53.20	26485	22MY	16;50	247.21
26486	6JN	8;01	81.17	26486	21JN	7;39	275.45
26487	5JL	14;55	109.09	26487	20JL	21;45	303.71
26488	3AU	23;12	137.23	26488	19AU	11;02	332.22
26489	2SE	9;51	165.82	26489	17SE	23;25	1.13
26490	1OC	23;25	195.01	26490	17OC	10;54	30.49
26491	31OC	15;52	224.81	26491	15NO	21;43	60.26
26492	30NO	10;31	255.05	26492	15DE	8;15	90.25
26493	30DE	6;06	285.45				

1142

NEW NUMBER	DATE	TIME	LONG.	FULL NUMBER	DATE	TIME	LONG.
				26493	13JA	18;53	120.23
26494	29JA	0;57	315.66	26494	12FE	5;53	149.95
26495	27FE	17;30	345.40	26495	13MR	17;23	179.25
26496	29MR	6;50	14.51	26496	12AP	5;28	208.08
26497	27AP	17;05	43.03	26497	11MY	18;22	236.52
26498	27MY	1;06	71.13	26498	10JN	8;18	264.76
26499	25JN	8;02	99.04	26499	9JL	23;20	293.01
26500	24JL	15;02	127.05	26500	8AU	15;05	321.49
26501	22AU	23;05	155.39	26501	7SE	6;44	350.36
26502	21SE	8;55	184.25	26502	6OC	21;30	19.68
26503	20OC	21;04	213.69	26503	5NO	11;01	49.41
26504	19NO	11;55	243.66	26504	4DE	23;20	79.42
26505	19DE	5;27	273.96				

1143

NEW NUMBER	DATE	TIME	LONG.	FULL NUMBER	DATE	TIME	LONG.
				26505	3JA	10;41	109.47
26506	18JA	0;42	304.28	26506	1FE	21;14	139.29
26507	16FE	19;53	334.29	26507	3MR	7;04	168.70
26508	18MR	13;06	3.75	26508	1AP	16;28	197.60
26509	17AP	3;21	32.60	26509	1MY	2;01	226.04
26510	16MY	14;44	60.93	26510	30MY	12;37	254.22
26511	15JN	0;01	88.97	26511	29JN	1;02	282.35
26512	14JL	8;06	116.96	26512	28JL	15;36	310.68
26513	12AU	15;52	145.16	26513	27AU	7;54	339.41
26514	11SE	0;04	173.78	26514	26SE	1;01	8.63
26515	10OC	9;22	202.93	26515	25OC	17;56	38.35
26516	8NO	20;29	232.61	26516	24NO	9;51	68.41
26517	8DE	9;59	262.69	26517	24DE	0;16	98.59

1144

NEW NUMBER	DATE	TIME	LONG.	FULL NUMBER	DATE	TIME	LONG.
26518	7JA	1;56	292.92	26518	22JA	12;48	128.60
26519	5FE	19;35	323.01	26519	20FE	23;22	158.19
26520	6MR	13;31	352.69	26520	21MR	8;15	187.24
26521	5AP	6;24	21.82	26521	19AP	16;09	215.78
26522	4MY	21;29	50.43	26522	19MY	0;04	243.94
26523	3JN	10;34	78.67	26523	17JN	9;01	271.95
26524	2JL	21;49	106.76	26524	16JL	19;53	300.07
26525	1AU	7;36	134.94	26525	15AU	9;09	328.53
26526	30AU	16;33	163.43	26526	14SE	0;49	357.51
26527	29SE	1;22	192.38	26527	13OC	18;29	27.07
26528	28OC	10;52	221.84	26528	12NO	13;12	57.12
26529	26NO	21;40	251.72	26529	12DE	7;32	87.43
26530	26DE	10;03	281.82				

1145

NEW NUMBER	DATE	TIME	LONG.	FULL NUMBER	DATE	TIME	LONG.
				26530	10JA	23;59	117.67
26531	24JA	23;56	311.85	26531	9FE	13;37	147.53
26532	23FE	14;55	341.58	26532	11MR	0;23	176.85
26533	25MR	6;34	10.86	26533	9AP	9;00	205.58
26534	23AP	22;28	39.66	26534	8MY	16;24	233.83
26535	23MY	14;04	68.09	26535	6JN	23;33	261.82
26536	22JN	4;44	96.32	26536	6JL	7;23	289.77
26537	21JL	17;59	124.55	26537	4AU	16;44	317.96
26538	20AU	5;45	153.01	26538	3SE	4;23	346.61
26539	18SE	16;27	181.86	26539	2OC	18;56	15.85
26540	18OC	2;45	211.19	26540	1NO	12;29	45.71
26541	16NO	13;13	240.94	26541	1DE	8;08	75.99
26542	16DE	0;04	270.95	26542	31DE	3;53	106.40

1146

NEW NUMBER	DATE	TIME	LONG.	FULL NUMBER	DATE	TIME	LONG.
26543	14JA	11;19	300.95	26543	29JA	21;38	136.56
26544	12FE	22;56	330.69	26544	28FE	12;19	166.21
26545	14MR	11;11	0.01	26545	29MR	23;56	195.24
26546	13AP	0;26	28.88	26546	28AP	9;11	223.71
26547	12MY	14;49	57.37	26547	27MY	16;52	251.79
26548	11JN	6;02	85.66	26548	25JN	23;50	279.70
26549	10JL	21;23	113.92	26549	25JL	6;59	307.72
26550	9AU	12;13	142.38	26550	23AU	15;16	336.08
26551	8SE	2;14	171.19	26551	22SE	1;45	4.97
26552	7OC	15;28	200.46	26552	21OC	15;17	34.48
26553	6NO	4;00	230.16	26553	20NO	8;03	64.54
26554	5DE	15;46	260.15	26554	20DE	3;07	94.91

1147

NEW NUMBER	DATE	TIME	LONG.	FULL NUMBER	DATE	TIME	LONG.
26555	4JA	2;43	290.18	26555	18JA	22;41	125.24
26556	2FE	12;56	319.98	26556	17FE	16;57	155.20
26557	3MR	22;49	349.38	26557	19MR	8;48	184.59
26558	2AP	8;58	18.30	26558	17AP	21;53	213.38
26559	1MY	20;01	46.80	26559	17MY	8;22	241.67
26560	31MY	8;22	75.03	26560	15JN	16;49	269.66
26561	29JN	22;07	103.22	26561	15JL	0;05	297.63
26562	29JL	13;07	131.58	26562	13AU	7;13	325.81
26563	28AU	5;05	160.30	26563	11SE	15;22	354.44
26564	26SE	21;30	189.51	26564	11OC	1;26	23.63
26565	26OC	13;36	219.20	26565	9NO	14;03	53.38
26566	25NO	4;30	249.23	26566	9DE	5;12	83.53
26567	24DE	17;35	279.36				

1148

NUMBER	DATE	TIME	LONG.	NUMBER	DATE	TIME	LONG.
				26567	7JA	22;21	113.81
26568	23JA	4;49	309.30	26568	6FE	16;34	143.91
26569	21FE	14;36	338.85	26569	7MR	10;39	173.59
26570	21MR	23;33	7.90	26570	6AP	3;25	202.70
26571	20AP	8;14	36.46	26571	5MY	17;54	231.28
26572	19MY	17;17	64.66	26572	4JN	5;47	259.47
26573	18JN	3;23	92.71	26573	3JL	15;26	287.49
26574	17JL	15;12	120.87	26574	1AU	23;47	315.62
26575	16AU	5;21	149.38	26575	31AU	7;57	344.09
26576	14SE	21;55	178.41	26576	29SE	16;48	13.05
26577	14OC	16;10	208.01	26577	29OC	2;54	42.54
26578	13NO	10;31	238.06	26578	27NO	14;30	72.47
26579	13DE	3;23	268.31	26579	27DE	3;41	102.59

1149

NUMBER	DATE	TIME	LONG.	NUMBER	DATE	TIME	LONG.
26580	11JA	17;54	298.46	26580	25JA	18;31	132.67
26581	10FE	6;01	328.25	26581	24FE	10;48	162.44
26582	11MR	16;08	357.52	26582	26MR	3;50	191.76
26583	10AP	0;42	26.24	26583	24AP	20;29	220.58
26584	9MY	8;19	54.50	26584	24MY	11;39	248.98
26585	7JN	15;45	82.49	26585	23JN	0;53	277.15
26586	7JL	0;02	110.47	26586	22JL	12;25	305.32
26587	5AU	10;20	138.70	26587	20AU	22;52	333.73
26588	3SE	23;32	167.42	26588	19SE	8;53	2.57
26589	3OC	15;52	196.75	26589	18OC	18;53	31.89
26590	2NO	10;32	226.66	26590	17NO	5;08	61.65
26591	2DE	5;54	256.95	26591	16DE	15;54	91.65

1150

NUMBER	DATE	TIME	LONG.	NUMBER	DATE	TIME	LONG.
26592	1JA	0;18	287.30	26592	15JA	3;30	121.66
26593	30JA	16;33	317.38	26593	13FE	16;16	151.44
26594	1MR	6;06	346.97	26594	15MR	6;18	180.82
26595	30MR	16;55	15.96	26595	13AP	21;13	209.75
26596	29AP	1;27	44.39	26596	13MY	12;28	238.27
26597	28MY	8;28	72.44	26597	12JN	3;31	266.54
26598	26JN	15;06	100.34	26598	11JL	18;05	294.78
26599	25JL	22;34	128.37	26599	10AU	8;01	323.20
26600	24AU	7;59	156.79	26600	8SE	21;09	351.99
26601	22SE	20;09	185.76	26601	8OC	9;21	21.23
26602	22OC	11;17	215.35	26602	6NO	20;41	50.89
26603	21NO	5;01	245.46	26603	6DE	7;26	80.84
26604	21DE	0;19	275.84				

1151

NUMBER	DATE	TIME	LONG.	NUMBER	DATE	TIME	LONG.
				26604	4JA	17;57	110.85
26605	19JA	19;43	306.16	26605	3FE	4;36	140.67
26606	18FE	13;30	336.08	26606	4MR	15;36	170.10
26607	20MR	4;24	5.42	26607	3AP	3;05	199.06
26608	18AP	16;04	34.14	26608	2MY	15;16	227.60
26609	18MY	1;01	62.36	26609	1JN	4;27	255.86
26610	16JN	8;21	90.31	26610	30JN	18;54	284.08
26611	15JL	15;15	118.26	26611	30JL	10;27	312.46
26612	13AU	22;46	146.47	26612	29AU	2;30	341.20
26613	12SE	7;44	175.14	26613	27SE	18;06	10.39
26614	11OC	18;46	204.38	26614	27OC	8;32	40.03
26615	10NO	8;18	234.18	26615	25NO	21;38	69.99
26616	10DE	0;31	264.38	26616	25DE	9;31	100.06

1152

NUMBER	DATE	TIME	LONG.	NUMBER	DATE	TIME	LONG.
26617	8JA	18;58	294.72	26617	23JA	20;25	129.99
26618	7FE	14;17	324.85	26618	22FE	6;27	159.54
26619	8MR	8;32	354.52	26619	22MR	15;47	188.59
26620	7AP	0;12	23.57	26620	21AP	0;55	217.16
26621	6MY	12;55	52.07	26621	20MY	10;40	245.38
26622	4JN	23;10	80.19	26622	18JN	21;56	273.48
26623	4JL	7;50	108.18	26623	18JL	11;22	301.71
26624	2AU	15;48	136.30	26624	17AU	2;59	330.28
26625	31AU	23;52	164.77	26625	15SE	20;02	359.34
26626	30SE	8;41	193.74	26626	15OC	13;25	28.92
26627	29OC	18;56	223.25	26627	14NO	6;10	58.92
26628	28NO	7;13	253.20	26628	13DE	21;33	89.11
26629	27DE	21;53	283.40				

1153

NUMBER	DATE	TIME	LONG.	NUMBER	DATE	TIME	LONG.
				26629	12JA	11;06	119.21
26630	26JA	14;38	313.54	26630	10FE	22;34	148.97
26631	25FE	8;21	343.36	26631	12MR	8;05	178.21
26632	27MR	1;40	12.68	26632	10AP	16;12	206.90
26633	25AP	17;34	41.45	26633	9MY	23;48	235.15
26634	25MY	7;38	69.80	26634	8JN	7;57	263.17
26635	23JN	19;51	97.93	26635	7JL	17;42	291.20
26636	23JL	6;29	126.07	26636	6AU	5;43	319.51
26637	21AU	16;00	154.46	26637	4SE	20;17	348.30
26638	20SE	1;00	183.26	26638	4OC	13;12	17.67
26639	19OC	10;16	212.57	26639	3NO	7;43	47.59
26640	17NO	20;28	242.32	26640	3DE	2;36	77.86
26641	17DE	8;02	272.37				

1154

NUMBER	DATE	TIME	LONG.	NUMBER	DATE	TIME	LONG.
				26641	1JA	20;15	108.17
26642	15JA	21;03	302.43	26642	31JA	11;21	138.20
26643	14FE	11;18	332.27	26643	1MR	23;25	167.72
26644	16MR	2;23	1.68	26644	31MR	8;53	196.64
26645	14AP	17;56	30.61	26645	29AP	16;38	225.03
26646	14MY	9;35	59.14	26646	28MY	23;42	253.08
26647	13JN	0;47	87.42	26647	27JN	7;02	281.01
26648	12JL	14;55	115.64	26648	26JL	15;31	309.08
26649	11AU	3;38	144.03	26649	25AU	1;58	337.55
26650	9SE	15;04	172.76	26650	23SE	15;05	6.57
26651	9OC	1;45	201.94	26651	23OC	7;17	36.22
26652	7NO	12;17	231.58	26652	22NO	2;09	66.39
26653	6DE	23;01	261.54	26653	21DE	22;11	96.80

1155

NUMBER	DATE	TIME	LONG.	NUMBER	DATE	TIME	LONG.
26654	5JA	10;02	291.56	26654	20JA	17;09	127.09
26655	3FE	21;17	321.40	26655	19FE	9;22	156.94
26656	5MR	8;56	350.84	26656	20MR	22;22	186.19
26657	3AP	21;20	19.83	26657	19AP	8;37	214.84
26658	3MY	10;52	48.41	26658	18MY	16;55	243.03
26659	2JN	1;34	76.73	26659	17JN	0;08	270.97
26660	1JL	16;56	104.98	26660	16JL	7;06	298.93
26661	31JL	8;14	133.37	26661	14AU	14;46	327.14
26662	29AU	22;56	162.06	26662	13SE	0;07	355.84
26663	28SE	12;51	191.20	26663	12OC	12;08	25.13
26664	28OC	2;00	220.79	26664	11NO	3;23	55.01
26665	26NO	14;21	250.73	26665	10DE	21;28	85.30
26666	26DE	1;47	280.78				

1156

NUMBER	DATE	TIME	LONG.	NUMBER	DATE	TIME	LONG.
				26666	9JA	16;58	115.68
26667	24JA	12;16	310.68	26667	8FE	11;57	145.79
26668	22FE	22;04	340.22	26668	9MR	4;58	175.39
26669	23MR	7;45	9.28	26669	7AP	19;20	204.38
26670	21AP	18;01	37.88	26670	7MY	7;03	232.83
26671	21MY	5;27	66.16	26671	5JN	16;27	260.91
26672	19JN	18;21	94.32	26672	5JL	0;15	288.86
26673	19JL	8;44	122.59	26673	3AU	7;27	316.95
26674	18AU	0;24	151.18	26674	1SE	15;06	345.42
26675	16SE	16;55	180.23	26675	1OC	0;15	14.41
26676	16OC	9;35	209.80	26676	30OC	11;38	43.98
26677	15NO	1;26	239.76	26677	29NO	1;34	74.01
26678	14DE	15;37	269.90	26678	28DE	17;43	104.27

1157

NUMBER	DATE	TIME	LONG.	NUMBER	DATE	TIME	LONG.
26679	13JA	3;46	299.95	26679	27JA	11;21	134.44
26680	11FE	14;07	329.65	26680	26FE	5;25	164.27
26681	12MR	23;15	358.86	26681	27MR	22;48	193.57
26682	11AP	7;46	27.56	26682	26AP	14;25	222.32
26683	10MY	16;22	55.84	26683	26MY	3;36	250.63
26684	9JN	1;41	83.90	26684	24JN	14;22	278.70
26685	8JL	12;28	111.98	26685	23JL	23;25	306.78
26686	7AU	1;24	140.34	26686	22AU	7;44	335.13
26687	5SE	16;54	169.17	26687	20SE	16;21	3.92
26688	5OC	10;39	198.59	26688	20OC	1;57	33.25
26689	4NO	5;24	228.53	26689	18NO	12;52	63.05
26690	3DE	23;22	258.77	26690	18DE	1;13	93.12

1158

NUMBER	DATE	TIME	LONG.	NUMBER	DATE	TIME	LONG.
26691	2JA	15;12	289.00	26691	16JA	15;04	123.22
26692	1FE	4;28	318.95	26692	15FE	6;25	153.09
26693	2MR	15;26	348.41	26693	16MR	22;55	182.55
26694	1AP	0;34	17.31	26694	15AP	15;40	211.52
26695	30AP	8;26	45.70	26695	15MY	7;34	240.05
26696	29MY	15;43	73.74	26696	13JN	21;49	268.28
26697	27JN	23;22	101.69	26697	13JL	10;17	296.44
26698	27JL	8;29	129.79	26698	11AU	21;25	324.78
26699	25AU	20;11	158.32	26699	10SE	7;51	353.47
26700	24SE	11;05	187.43	26700	9OC	18;03	22.65
26701	24OC	4;52	217.16	26701	8NO	4;20	52.29
26702	23NO	0;15	247.35	26702	7DE	14;51	82.24
26703	22DE	19;24	277.73				

1159

NUMBER	DATE	TIME	LONG.	NUMBER	DATE	TIME	LONG.
				26703	6JA	1;55	112.26
26704	21JA	12;50	307.95	26704	4FE	13;54	142.12
26705	20FE	3;43	337.73	26705	6MR	3;02	171.62
26706	21MR	15;47	6.93	26706	4AP	17;15	200.67
26707	20AP	1;17	35.54	26707	4MY	8;09	229.30
26708	19MY	8;51	63.69	26708	2JN	23;14	257.63
26709	17JN	15;30	91.61	26709	2JL	14;06	285.86
26710	16JL	22;25	119.57	26710	1AU	4;32	314.21
26711	15AU	6;47	147.82	26711	30AU	18;22	342.89
26712	13SE	17;35	176.59	26712	29SE	7;21	12.00
26713	13OC	7;19	205.97	26713	28OC	19;22	41.55
26714	11NO	23;54	235.91	26714	27NO	6;31	71.44
26715	11DE	18;36	266.23	26715	26DE	17;07	101.46

```
--------NEW MOONS---------        --------FULL MOONS--------        --------NEW MOONS---------        --------FULL MOONS--------
NUMBER  DATE   TIME   LONG.       NUMBER  DATE   TIME   LONG.       NUMBER  DATE   TIME   LONG.       NUMBER  DATE   TIME   LONG.

                      1160                                                                1166
26716   10JA  14;07  296.60      26716   25JA   3;35  131.36       26790    4JA   2;18  290.55       26790   18JA   6;27  124.95
26717    9FE   8;49  326.69      26717   23FE  14;11  160.92       26791    2FE  13;26  320.40       26791   17FE   0;12  154.89
26718   10MR   1;10  356.25      26718   24MR   1;08  190.02       26792    3MR  22;57  349.77       26792   18MR  17;51  184.37
26719    8AP  14;22   25.18      26719   22AP  12;39  218.66       26793    2AP   7;31   18.63       26793   17AP  10;19  213.30
26720    8MY   0;31   53.56      26720   22MY   1;04  246.97       26794    1MY  15;47   47.02       26794   17MY   0;44  241.75
26721    6JN   8;31   81.58      26721   20JN  14;44  275.16       26795   31MY   0;29   75.11       26795   15JN  12;44  269.89
26722    5JL  15;32  109.50      26722   20JL   5;48  303.46       26796   29JN  10;20  103.14       26796   14JL  22;40  297.96
26723    3AU  22;43  137.60      26723   18AU  21;53  332.08       26797   28JL  22;05  131.36       26797   13AU   7;25  326.21
26724    2SE   7;00  166.10      26724   17SE  14;06    1.14       26798   27AU  12;20  160.01       26798   11SE  15;59  354.85
26725   10C   17;03  195.14      26725   17OC   5;29   30.66       26799   26SE   5;13  189.23       26799   11OC   1;13   24.01
26726   31OC   5;22  224.76      26726   15NO  19;30   60.56       26800   25OC  23;53  219.02       26800    9NO  11;35   53.67
26727   29NO  20;15  254.84      26727   15DE   8;05   90.63       26801   24NO  18;40  249.21       26801    8DE  23;13   83.67
26728   29DE  13;37  285.14                                        26802   24DE  11;48  279.50

                      1161                                                                1167
26729   28JA   8;34  315.36      26728   13JA  19;28  120.64       26803   23JA   2;22  309.59       26802    7JA  12;12  113.77
26730   27FE   3;24  345.20      26729   12FE   5;48  150.33       26804   21FE  14;22  339.23       26803    6FE   2;35  143.72
26731   28MR  20;20   14.47      26730   13MR  15;15  179.55       26805   23MR   0;13    8.32       26804    7MR  18;18  173.30
26732   27AP  10;26   43.16      26731   12AP   0;11  208.25       26806   21AP   8;30   36.86       26805    6AP  10;47  202.42
26733   26MY  21;50   71.39      26732   11MY   9;16  236.55       26807   20MY  15;51   64.99       26806    6MY   3;04  231.08
26734   25JN   7;15   99.40      26733    9JN  19;30  264.64       26808   18JN  23;06   92.93       26807    4JN  18;10  259.39
26735   24JL  15;36  127.46      26734    9JL   7;43  292.78       26809   18JL   7;19  120.94       26808    4JL   7;37  287.57
26736   22AU  23;43  155.81      26735    7AU  22;18  321.21       26810   16AU  17;38  149.29       26809    2AU  19;33  315.84
26737   21SE   8;15  184.61      26736    6SE  14;55  350.10       26811   15SE   6;58  178.18       26810    1SE   6;32  344.42
26738   20OC  17;51  213.95      26737    6OC   8;31   19.53       26812   14OC  23;31  207.70       26811   30SE  17;04   13.45
26739   19NO   5;06  243.77      26738    5NO   1;57   49.42       26813   13NO  18;26  237.77       26812   30OC   3;29   42.96
26740   18DE  18;31  273.89      26739    4DE  18;17   79.60       26814   13DE  14;01  268.13       26813   28NO  13;58   72.84
                                                                                                     26814   28DE   0;42  102.86

                      1162                                                                1168
26741   17JA  10;09  304.06      26740    3JA   8;53  109.78       26815   12JA   8;29  298.45       26815   26JA  12;02  132.79
26742   16FE   3;18  333.99      26741    1FE  21;22  139.69       26816   11FE   0;40  328.42       26816   25FE   0;19  162.39
26743   17MR  20;44    3.47      26742    3MR   7;42  169.11       26817   11MR  14;04  357.83       26817   25MR  13;43  191.57
26744   16AP  13;14   32.42      26743    1AP  16;16  197.98       26818   10AP   0;42   26.63       26818   24AP   4;04  220.31
26745   16MY   4;10   60.90      26744   30AP  23;49  226.34       26819    9MY   9;03   54.92       26819   23MY  18;55  248.70
26746   14JN  17;20   89.09      26745   30MY   7;24  254.40       26820    7JN  15;58   82.89       26820   22JN   9;54  276.94
26747   14JL   4;53  117.22      26746   28JN  16;08  282.38       26821    6JL  22;36  110.80       26821   22JL   0;41  305.24
26748   12AU  15;06  145.52      26747   28JL   2;55  310.56       26822    5AU   6;09  138.92       26822   20AU  15;06  333.81
26749   11SE   0;31  174.19      26748   26AU  16;16  339.16       26823    3SE  15;43  167.49       26823   19SE   4;50    2.79
26750   10OC   9;47  203.34      26749   25SE   8;11    8.33       26824   3OC    4;03  196.65       26824   18OC  17;37   32.23
26751    8NO  19;35  232.96      26750   25OC   2;11   38.08       26825    1NO  19;20  226.42       26825   17NO   5;22   62.05
26752    8DE   6;28  262.93      26751   23NO  21;14   68.27       26826    1DE  13;07  256.63       26826   16DE  16;17   92.06
                                 26752   23DE  15;49   98.62       26827   31DE   8;22  287.01

                      1163                                                                1169
26753    6JA  18;41  293.00      26753   22JA   8;20  128.80                                          26827   15JA   2;44  122.02
26754    5FE   8;09  322.92      26754   20FE  21;52  158.52       26828   30JA   3;36  317.23       26828   13FE  13;05  151.70
26755    6MR  22;34  352.46      26755   22MR   8;26  187.64       26829   28FE  21;11  346.99       26829   14MR  23;35  180.94
26756    5AP  13;37   21.53      26756   20AP  16;48  216.20       26830   30MR  11;55   16.15       26830   13AP  10;30  209.70
26757    5MY   5;03   50.17      26757   19MY  23;58  244.33       26831   28AP  23;28   44.70       26831   12MY  21;10  238.08
26758    3JN  20;28   78.50      26758   18JN   6;59  272.26       26832   28MY   8;24   72.82       26832   11JN  11;00  266.27
26759    3JL  11;17  106.73      26759   17JL  14;47  300.25       26833   26JN  15;48  100.76       26833   11JL   1;20  294.50
26760    2AU   5;07  135.06      26760   16AU   0;11  328.56       26834   25JL  22;57  128.77       26834    9AU  17;07  323.00
26761   31AU  13;15  163.69      26761   14SE  11;58  357.37       26835   24AU   6;37  157.12       26835    8SE   9;38  351.91
26762   30SE   0;31  192.74      26762   14OC   2;42   26.80       26836   22SE  15;51  185.97       26836    8OC   1;50   21.31
26763   29OC  11;17  222.25      26763   12NO  20;26   56.81       26837   22OC   3;05  215.40       26837    6NO  16;50   51.12
26764   27NO  22;02  252.13      26764   12DE  16;12   87.18       26838   20NO  16;41  245.33       26838    6DE   6;18   81.19
26765   27DE   8;56  282.16                                        26839   20DE   8;47  275.57

                      1164                                                                1170
26766   25JA  19;56  312.08      26765   11JA  11;57  117.55                                          26839    4JA  18;18  111.26
26767   24FE   7;07  341.65      26766   10FE   5;38  147.59       26840   19JA   2;58  305.84       26840    3FE   5;04  141.08
26768   24MR  18;47   10.77      26767   10MR  20;09  177.06       26841   17FE  21;54  335.83       26841    4MR  14;46  170.46
26769   23AP   7;26   39.45      26768    9AP   7;36  205.91       26842   19MR  15;48    5.30       26842    2AP  23;41  199.31
26770   22MY  21;22   67.81      26769    8MY  16;43  234.24       26843   18AP   7;15   34.18       26843    2MY   8;22  227.71
26771   21JN  12;23   96.06      26770    7JN   0;20  262.24       26844   17MY  19;56   62.55       26844   31MY  17;42  255.82
26772   21JL   3;53  124.38      26771    6JL   7;21  290.16       26845   16JN  10;17   90.62       26845   30JN   4;43  283.90
26773   19AU  19;09  152.97      26772    4AU  14;38  318.27       26846   15JL  15;12  118.65       26846   29JL  18;05  312.20
26774   18SE   9;46  181.98      26773    2SE  23;07  346.79       26847   13AU  23;30  146.89       26847   28AU   9;53  340.92
26775   17OC  23;36  211.45      26774    2OC   9;47   15.87       26848   12SE   7;57  175.54       26848   27SE   3;20   10.17
26776   16NO  12;35  241.31      26775   31OC  23;27   45.55       26849   11OC  17;06  204.70       26849   26OC  21;14   39.95
26777   16DE   0;36  271.36      26776   30NO  16;15   75.71       26850   10NO   3;33  234.37       26850   25NO  14;26   70.08
                                 26777   30DE  11;12  106.09       26851    9DE  15;50  264.41       26851   25DE   6;05  100.31

                      1165                                                                1171
26778   14JA  11;31  301.35      26778   29JA   6;32  136.32       26852    8JA   6;16  294.57       26852   23JA  19;40  130.35
26779   12FE  21;28  331.02      26779   28FE   0;32  166.11       26853    6FE  22;34  324.59       26853   22FE   6;59  159.95
26780   14MR   6;55    0.22      26780   29MR  16;09  195.31       26854    8MR  15;45  354.22       26854   23MR  16;12  189.00
26781   12AP  16;32   28.95      26781   28AP   5;07  223.94       26855    7AP   8;36   23.33       26855   21AP  23;59  217.51
26782   12MY   3;05   57.29      26782   27MY  15;36  252.13       26856    7MY   0;14   51.95       26856   21MY   7;16  245.63
26783   10JN  15;03   85.45      26783   26JN   0;10  280.11       26857    5JN  14;17   80.23       26857   19JN  15;11  273.60
26784   10JL   4;39  113.65      26784   25JL   7;40  308.14       26858    5JL   2;44  108.36       26858   19JL   0;48  301.67
26785    8AU  19;46  142.10      26785   23AU  15;06  336.46       26859    3AU  13;47  136.60       26859   17AU  12;51  330.10
26786    7SE  12;07  171.00      26786   21SE  23;32    5.27       26860    1SE  23;47  165.15       26860   16SE   3;35  359.06
26787    7OC   5;06  200.42      26787   21OC   9;51   34.65       26861    1OC   9;16  194.16       26861   15OC  20;47   28.62
26788    5NO  21;45  230.29      26788   19NO  22;33   64.53       26862   30OC  18;54  223.64       26862   14NO  15;38   58.70
26789    5DE  13;03  260.43      26789   19DE  13;36   94.73       26863   29NO   5;16  253.51       26863   14DE  10;47   89.04
                                                                   26864   28DE  16;46  283.57
```

1172

NUMBER	DATE	TIME	LONG.	NUMBER	DATE	TIME	LONG.
				26864	13JA	4;33	119.33
26865	27JA	5;29	313.55	26865	11FE	19;36	149.24
26866	25FE	19;11	343.21	26866	12MR	7;30	178.57
26867	26MR	9;39	12.41	26867	10AP	16;44	207.31
26868	25AP	0;40	41.16	26868	10MY	0;16	235.56
26869	24MY	16;01	69.57	26869	8JN	7;11	263.52
26870	23JN	7;13	97.82	26870	7JL	14;27	291.46
26871	22JL	21;40	126.11	26871	5AU	22;59	319.63
26872	21AU	10;54	154.64	26872	4SE	9;33	348.25
26873	19SE	22;55	183.57	26873	3OC	22;51	17.47
26874	19OC	10;08	212.95	26874	2NO	15;12	47.28
26875	17NO	21;01	242.74	26875	2DE	10;12	77.55
26876	17DE	7;54	272.75				

1173

NUMBER	DATE	TIME	LONG.	NUMBER	DATE	TIME	LONG.
				26876	1JA	6;15	107.97
26877	15JA	18;47	302.73	26877	31JA	1;08	138.17
26878	14FE	5;40	332.42	26878	1MR	17;12	167.85
26879	15MR	16;46	1.67	26879	31MR	6;02	196.91
26880	14AP	4;35	30.46	26880	29AP	16;09	225.40
26881	13MY	17;36	58.89	26881	29MY	0;23	253.49
26882	12JN	7;59	87.14	26882	27JN	7;37	281.42
26883	11JL	23;20	115.41	26883	26JL	14;43	309.44
26884	10AU	14;58	143.91	26884	24AU	22;35	337.80
26885	9SE	6;12	172.78	26885	23SE	8;09	6.68
26886	8OC	20;44	202.13	26886	22OC	20;20	36.15
26887	7NO	10;25	231.90	26887	21NO	11;38	66.16
26888	6DE	23;06	261.93	26888	21DE	5;38	96.49

1174

NUMBER	DATE	TIME	LONG.	NUMBER	DATE	TIME	LONG.
26889	5JA	10;36	291.98	26889	20JA	0;54	126.80
26890	3FE	20;54	321.77	26890	18FE	19;35	156.76
26891	5MR	6;20	351.13	26891	20MR	12;20	186.17
26892	3AP	15;32	19.99	26892	19AP	2;33	214.98
26893	3MY	1;17	48.42	26893	18MY	14;13	243.31
26894	1JN	12;18	76.60	26894	16JN	23;43	271.34
26895	1JL	0;58	104.74	26895	16JL	7;44	299.34
26896	30JL	15;21	133.08	26896	14AU	15;13	327.55
26897	29AU	7;17	161.82	26897	12SE	23;11	356.19
26898	28SE	0;17	191.07	26898	12OC	8;37	25.38
26899	27OC	17;31	220.83	26899	10NO	20;10	55.10
26900	26NO	9;50	250.93	26900	10DE	10;04	85.21
26901	26DE	0;16	281.11				

1175

NUMBER	DATE	TIME	LONG.	NUMBER	DATE	TIME	LONG.
				26901	9JA	1;58	115.43
26902	24JA	12;26	311.08	26902	7FE	19;10	145.48
26903	22FE	22;34	340.63	26903	9MR	12;46	175.11
26904	24MR	7;21	9.65	26904	8AP	5;45	204.22
26905	22AP	15;29	38.17	26905	7MY	21;11	232.82
26906	21MY	23;43	66.32	26906	6JN	10;26	261.06
26907	20JN	8;46	94.33	26907	5JL	21;27	289.14
26908	19JL	19;25	122.45	26908	4AU	6;52	317.32
26909	18AU	8;23	150.92	26909	2SE	15;37	345.83
26910	17SE	0;06	179.94	26910	2OC	0;39	14.82
26911	16OC	18;13	209.55	26911	31OC	10;34	44.33
26912	15NO	13;23	239.64	26912	29NO	21;37	74.24
26913	15DE	7;40	269.96	26913	29DE	9;51	104.33

1176

NUMBER	DATE	TIME	LONG.	NUMBER	DATE	TIME	LONG.
26914	13JA	23;38	300.17	26914	27JA	23;21	134.33
26915	12FE	12;50	329.99	26915	26FE	14;09	164.02
26916	12MR	23;35	359.27	26916	27MR	6;03	193.27
26917	11AP	8;27	27.98	26917	25AP	22;21	222.07
26918	10MY	16;03	56.22	26918	25MY	14;03	250.48
26919	8JN	23;09	84.19	26919	24JN	4;23	278.69
26920	8JL	6;42	112.14	26920	23JL	17;13	306.92
26921	6AU	15;50	140.34	26921	22AU	4;51	335.40
26922	5SE	3;37	169.02	26922	20SE	15;50	4.29
26923	4OC	18;42	198.32	26923	20OC	2;31	33.67
26924	3NO	12;43	228.22	26924	18NO	13;06	63.45
26925	3DE	8;17	258.52	26925	17DE	23;43	93.45

1177

NUMBER	DATE	TIME	LONG.	NUMBER	DATE	TIME	LONG.
26926	2JA	3;32	288.90	26926	16JA	10;36	123.43
26927	31JA	20;56	319.03	26927	14FE	22;09	153.14
26928	2MR	11;41	348.65	26928	16MR	10;43	182.44
26929	31MR	23;35	17.66	26929	15AP	0;19	211.29
26930	30AP	8;55	46.11	26930	14MY	14;45	239.77
26931	29MY	16;23	74.16	26931	13JN	5;36	268.03
26932	27JN	23;00	102.06	26932	12JL	20;34	296.29
26933	27JL	5;59	130.08	26933	11AU	11;24	324.76
26934	25AU	14;30	158.48	26934	10SE	1;48	353.62
26935	24SE	1;28	187.42	26935	9OC	15;23	22.93
26936	23OC	15;22	216.98	26936	8NO	3;54	52.66
26937	22NO	8;02	247.06	26937	7DE	15;20	82.65
26938	22DE	2;41	277.41				

1178

NUMBER	DATE	TIME	LONG.	NUMBER	DATE	TIME	LONG.
				26938	6JA	1;58	112.66
26939	20JA	22;03	307.72	26939	4FE	12;12	142.44
26940	19FE	16;32	337.66	26940	5MR	22;22	171.82
26941	21MR	8;41	7.03	26941	4AP	8;47	200.72
26942	19AP	21;44	35.79	26942	3MY	19;46	229.19
26943	19MY	7;51	64.04	26943	2JN	7;45	257.40
26944	17JN	15;54	92.02	26944	1JL	21;13	285.57
26945	16JL	23;05	119.98	26945	31JL	12;22	313.95
26946	15AU	6;30	148.20	26946	30AU	4;49	342.73
26947	13SE	15;03	176.87	26947	28SE	21;36	11.98
26948	13OC	1;21	206.11	26948	28OC	13;34	41.70
26949	11NO	13;47	235.88	26949	27NO	4;02	71.73
26950	11DE	4;37	266.03	26950	26DE	16;50	101.84

1179

NUMBER	DATE	TIME	LONG.	NUMBER	DATE	TIME	LONG.
26951	9JA	21;44	296.30	26951	25JA	4;10	131.78
26952	8FE	16;18	326.39	26952	23FE	14;15	161.31
26953	10MR	10;44	356.05	26953	24MR	23;19	190.33
26954	9AP	3;23	25.13	26954	23AP	7;48	218.85
26955	8MY	17;23	53.66	26955	22MY	16;29	247.02
26956	7JN	4;51	81.82	26956	21JN	2;24	275.06
26957	6JL	14;29	109.85	26957	20JL	14;29	303.24
26958	4AU	23;10	138.01	26958	19AU	5;09	331.79
26959	3SE	7;40	166.52	26959	17SE	22;04	0.87
26960	2OC	16;34	195.51	26960	17OC	16;08	30.49
26961	1NO	2;26	225.03	26961	16NO	10;02	60.55
26962	30NO	13;47	254.95	26962	16DE	2;43	90.80
26963	30DE	3;03	285.08				

1180

NUMBER	DATE	TIME	LONG.	NUMBER	DATE	TIME	LONG.
				26963	14JA	17;26	120.95
26964	28JA	18;18	315.16	26964	13FE	5;49	150.73
26965	27FE	10;54	344.91	26965	13MR	15;55	179.97
26966	28MR	3;48	14.19	26966	12AP	0;09	208.64
26967	26AP	19;56	42.96	26967	11MY	7;24	236.86
26968	26MY	10;45	71.33	26968	9JN	14;45	264.84
26969	25JN	0;04	99.50	26969	8JL	23;20	292.84
26970	24JL	11;59	127.70	26970	7AU	10;05	321.11
26971	22AU	22;41	156.15	26971	5SE	23;33	349.87
26972	21SE	8;36	185.02	26972	5OC	15;42	19.22
26973	20OC	18;18	214.36	26973	4NO	10;00	49.15
26974	19NO	4;21	244.13	26974	4DE	5;20	79.44
26975	18DE	15;17	274.14				

1181

NUMBER	DATE	TIME	LONG.	NUMBER	DATE	TIME	LONG.
				26975	3JA	0;03	109.80
26976	17JA	3;16	304.16	26976	1FE	16;34	139.88
26977	15FE	16;16	333.92	26977	3MR	5;58	169.43
26978	17MR	6;05	3.26	26978	1AP	16;20	198.37
26979	15AP	20;33	32.14	26979	1MY	0;29	226.76
26980	15MY	11;34	60.63	26980	30MY	7;30	254.79
26981	14JN	2;51	88.90	26981	28JN	14;26	282.71
26982	13JL	17;51	117.17	26982	27JL	22;16	310.77
26983	12AU	7;58	145.62	26983	26AU	7;47	339.21
26984	10SE	20;52	174.43	26984	24SE	19;43	8.21
26985	10OC	8;42	203.68	26985	24OC	10;37	37.82
26986	8NO	19;55	233.37	26986	23NO	4;29	67.95
26987	8DE	6;53	263.34	26987	23DE	0;16	98.36

1182

NUMBER	DATE	TIME	LONG.	NUMBER	DATE	TIME	LONG.
26988	6JA	17;45	293.36	26988	21JA	19;57	128.67
26989	5FE	4;29	323.16	26989	20FE	13;28	158.56
26990	6MR	15;11	352.54	26990	22MR	3;49	187.84
26991	5AP	2;17	21.46	26991	20AP	15;07	216.52
26992	4MY	14;22	49.97	26992	20MY	0;09	244.72
26993	3JN	3;53	78.23	26993	18JN	7;47	272.68
26994	2JL	18;47	106.46	26994	17JL	14;55	300.65
26995	1AU	10;28	134.87	26995	15AU	22;23	328.87
26996	31AU	2;12	163.63	26996	14SE	7;06	357.56
26997	29SE	17;25	192.84	26997	13OC	17;59	26.83
26998	29OC	7;49	222.50	26998	12NO	7;45	56.67
26999	27NO	21;14	252.49	26999	12DE	0;29	86.90
27000	27DE	9;25	282.58				

1183

NUMBER	DATE	TIME	LONG.	NUMBER	DATE	TIME	LONG.
				27000	10JA	19;14	117.24
27001	25JA	20;15	312.48	27001	9FE	14;16	147.34
27002	24FE	5;54	341.99	27002	11MR	7;57	176.95
27003	25MR	14;54	11.00	27003	9AP	23;21	205.96
27004	24AP	0;02	39.54	27004	9MY	12;13	234.44
27005	23MY	10;07	67.75	27005	7JN	22;47	262.56
27006	21JN	21;46	95.86	27006	7JL	7;32	290.56
27007	21JL	11;16	124.10	27007	5AU	15;19	318.69
27008	20AU	2;34	152.69	27008	3SE	23;05	347.18
27009	18SE	19;19	181.77	27009	3OC	7;50	16.18
27010	18OC	12;49	211.39	27010	1NO	18;23	45.73
27011	17NO	5;59	241.42	27011	1DE	7;08	75.72
27012	16DE	21;37	271.63	27012	30DE	22;00	105.92

1184

NUMBER	DATE	TIME	LONG.	NUMBER	DATE	TIME	LONG.
27013	15JA	10;57	301.72	27013	29JA	14;27	136.03
27014	13FE	21;58	331.43	27014	28FE	7;43	165.80
27015	14MR	7;11	0.62	27015	29MR	0;54	195.08
27016	12AP	15;22	29.29	27016	27AP	17;05	223.84
27017	11MY	23;17	57.53	27017	27MY	7;27	252.19
27018	10JN	7;41	85.55	27018	25JN	19;39	280.31
27019	9JL	17;22	113.59	27019	25JL	5;57	308.46
27020	8AU	5;05	141.90	27020	23AU	15;08	336.85
27021	6SE	19;30	170.71	27021	22SE	0;09	5.69
27022	6OC	12;40	200.13	27022	21OC	9;45	35.04
27023	5NO	7;44	230.09	27023	19NO	20;20	64.83
27024	5DE	2;51	260.39	27024	19DE	7;57	94.88

1185

NUMBER	DATE	TIME	LONG.	NUMBER	DATE	TIME	LONG.
27025	3JA	20;10	290.68	27025	17JA	20;40	124.92
27026	2FE	10;44	320.67	27026	16FE	10;34	154.72
27027	3MR	22;34	350.15	27027	18MR	1;40	184.10
27028	2AP	8;11	19.04	27028	16AP	17;36	213.02
27029	1MY	16;12	47.42	27029	16MY	9;34	241.54
27030	30MY	23;21	75.45	27030	15JN	0;38	269.80
27031	29JN	6;30	103.37	27031	14JL	14;21	298.02
27032	28JL	14;41	131.46	27032	13AU	2;45	326.41
27033	27AU	1;05	159.94	27033	11SE	14;17	355.17
27034	25SE	14;34	189.02	27034	11OC	1;21	24.41
27035	25OC	7;20	218.72	27035	9NO	12;10	54.09
27036	24NO	2;26	248.91	27036	8DE	22;50	84.05
27037	23DE	22;07	279.31				

1186

NUMBER	DATE	TIME	LONG.	NUMBER	DATE	TIME	LONG.
				27037	7JA	9;29	114.05
27038	22JA	16;34	309.57	27038	5FE	20;29	143.86
27039	21FE	8;38	339.39	27039	7MR	8;15	173.28
27040	22MR	21;52	8.61	27040	5AP	21;02	202.25
27041	21AP	8;20	37.24	27041	5MY	10;50	230.82
27042	20MY	16;34	65.41	27042	4JN	1;21	259.12
27043	18JN	23;28	93.33	27043	3JL	16;17	287.35
27044	18JL	6;10	121.29	27044	2AU	7;21	315.74
27045	16AU	13;51	149.53	27045	31AU	22;17	344.48
27046	14SE	23;36	178.27	27046	30SE	12;38	13.66
27047	14OC	12;06	207.62	27047	30OC	1;59	43.29
27048	13NO	3;29	237.53	27048	28NO	14;07	73.24
27049	12DE	21;16	267.81	27049	28DE	1;09	103.27

1187

NUMBER	DATE	TIME	LONG.	NUMBER	DATE	TIME	LONG.
27050	11JA	16;22	298.16	27050	26JA	11;27	133.15
27051	10FE	11;22	328.25	27051	24FE	21;26	162.66
27052	12MR	4;42	357.83	27052	26MR	7;27	191.70
27053	10AP	19;16	26.80	27053	24AP	17;50	220.28
27054	10MY	6;45	55.22	27054	24MY	5;02	248.54
27055	8JN	15;43	83.27	27055	22JN	17;34	276.68
27056	7JL	23;16	111.22	27056	22JL	7;52	304.96
27057	6AU	6;33	139.33	27057	20AU	23;55	333.58
27058	4SE	14;35	167.84	27058	19SE	16;53	2.69
27059	4OC	0;06	196.88	27059	19OC	9;41	32.29
27060	2NO	11;31	226.48	27060	18NO	1;12	62.26
27061	2DE	1;08	256.51	27061	17DE	14;58	92.40
27062	31DE	17;04	286.76				

1188

NUMBER	DATE	TIME	LONG.	NUMBER	DATE	TIME	LONG.
				27062	16JA	3;01	122.43
27063	30JA	10;52	316.92	27063	14FE	13;37	152.11
27064	29FE	5;22	346.73	27064	14MR	22;58	181.30
27065	29MR	22;53	16.00	27065	13AP	7;28	209.97
27066	28AP	14;10	44.72	27066	12MY	15;45	238.22
27067	28MY	2;51	72.99	27067	11JN	0;45	266.25
27068	26JN	13;23	101.05	27068	10JL	11;34	294.34
27069	25JL	22;36	129.15	27069	9AU	0;56	322.73
27070	24AU	7;19	157.54	27070	7SE	16;56	351.62
27071	22SE	16;09	186.38	27071	7OC	10;47	21.08
27072	22OC	1;38	215.73	27072	6NO	5;10	51.02
27073	20NO	12;15	245.54	27073	5DE	22;44	81.26
27074	20DE	0;29	275.61				

1189

NUMBER	DATE	TIME	LONG.	NUMBER	DATE	TIME	LONG.
				27074	4JA	14;35	111.49
27075	18JA	14;37	305.71	27075	3FE	4;09	141.43
27076	17FE	6;23	335.57	27076	4MR	15;16	170.87
27077	18MR	23;00	5.00	27077	3AP	0;12	199.72
27078	17AP	15;23	33.92	27078	2MY	7;41	228.07
27079	17MY	6;47	62.41	27079	31MY	14;42	256.09
27080	15JN	20;54	90.63	27080	29JN	22;28	284.04
27081	15JL	9;39	118.82	27081	29JL	8;01	312.18
27082	13AU	21;09	147.18	27082	27AU	20;08	340.75
27083	12SE	7;41	175.92	27083	26SE	11;04	9.89
27084	11OC	17;38	205.12	27084	26OC	4;31	39.64
27085	10NO	3;37	234.77	27085	24NO	23;38	69.84
27086	9DE	14;07	264.73	27086	24DE	18;58	100.23

1190

NUMBER	DATE	TIME	LONG.	NUMBER	DATE	TIME	LONG.
27087	8JA	1;29	294.76	27087	23JA	12;45	130.45
27088	6FE	13;49	324.61	27088	22FE	3;42	160.21
27089	8MR	2;57	354.08	27089	23MR	15;26	189.36
27090	6AP	16;48	23.08	27090	22AP	0;28	217.92
27091	6MY	7;19	51.67	27091	21MY	7;50	246.05
27092	4JN	22;23	79.98	27092	19JN	14;39	273.97
27093	4JL	13;39	108.23	27093	18JL	21;57	301.95
27094	3AU	4;28	136.62	27094	17AU	6;35	330.24
27095	1SE	18;16	165.32	27095	15SE	17;18	359.03
27096	1OC	6;53	194.45	27096	15OC	6;46	28.43
27097	30OC	18;36	224.02	27097	13NO	23;15	58.40
27098	29NO	5;49	253.93	27098	13DE	18;17	88.74
27099	28DE	16;46	283.96				

1191

NUMBER	DATE	TIME	LONG.	NUMBER	DATE	TIME	LONG.
				27099	12JA	14;15	119.12
27100	27JA	3;28	313.85	27100	11FE	8;58	149.18
27101	25FE	13;57	343.38	27101	13MR	0;52	178.69
27102	27MR	0;30	12.43	27102	11AP	13;33	207.57
27103	25AP	11;44	41.03	27103	10MY	23;34	235.92
27104	25MY	0;17	69.33	27104	9JN	7;47	263.94
27105	23JN	14;24	97.54	27105	8JL	15;07	291.88
27106	23JL	5;49	125.87	27106	6AU	22;25	320.00
27107	21AU	21;48	154.50	27107	5SE	6;31	348.52
27108	20SE	13;37	183.58	27108	4OC	16;19	17.59
27109	20OC	4;45	213.12	27109	3NO	4;39	47.23
27110	18NO	18;55	243.04	27110	2DE	19;57	77.34
27111	18DE	7;52	273.14				

1192

NUMBER	DATE	TIME	LONG.	NUMBER	DATE	TIME	LONG.
				27111	1JA	13;47	107.66
27112	16JA	19;22	303.15	27112	31JA	8;44	137.87
27113	15FE	5;27	332.80	27113	1MR	3;04	167.66
27114	15MR	14;30	1.97	27114	30MR	19;32	196.88
27115	13AP	23;14	30.63	27115	29AP	9;36	225.53
27116	13MY	8;31	58.92	27116	28MY	21;17	253.76
27117	11JN	19;09	87.02	27117	27JN	6;57	281.78
27118	11JL	7;38	115.17	27118	26JL	15;15	309.85
27119	9AU	22;05	143.62	27119	24AU	23;05	338.21
27120	8SE	14;20	172.53	27120	23SE	7;24	7.04
27121	8OC	7;48	201.98	27121	22OC	17;06	36.41
27122	7NO	1;33	231.91	27122	21NO	4;47	66.27
27123	6DE	18;16	262.12	27123	20DE	18;35	96.41

1193

NUMBER	DATE	TIME	LONG.	NUMBER	DATE	TIME	LONG.
27124	5JA	8;53	292.30	27124	19JA	10;10	126.57
27125	3FE	20;59	322.17	27125	18FE	2;52	156.45
27126	5MR	6;54	351.54	27126	19MR	19;57	185.89
27127	3AP	15;20	20.37	27127	18AP	12;34	214.81
27128	2MY	23;07	48.72	27128	18MY	3;51	243.28
27129	1JN	7;02	76.77	27129	16JN	17;12	271.47
27130	30JN	15;53	104.77	27130	16JL	4;33	299.60
27131	30JL	2;28	132.95	27131	14AU	14;24	327.91
27132	28AU	15;31	161.57	27132	12SE	23;37	356.60
27133	27SE	7;28	190.77	27133	12OC	9;04	25.79
27134	27OC	1;56	220.57	27134	10NO	19;17	55.46
27135	25NO	21;27	250.80	27135	10DE	6;25	85.45
27136	25DE	15;57	281.15				

1194

NUMBER	DATE	TIME	LONG.	NUMBER	DATE	TIME	LONG.
				27136	8JA	18;29	115.51
27137	24JA	7;58	311.29	27137	7FE	7;33	145.38
27138	22FE	21;04	340.96	27138	8MR	21;46	174.88
27139	24MR	7;36	10.05	27139	7AP	13;04	203.93
27140	22AP	16;13	38.59	27140	7MY	4;55	232.56
27141	21MY	23;37	66.71	27141	5JN	20;27	260.89
27142	20JN	6;35	94.63	27142	5JL	10;58	289.11
27143	19JL	14;06	122.62	27143	4AU	0;11	317.44
27144	17AU	23;18	150.94	27144	2SE	12;23	346.09
27145	16SE	11;14	179.80	27145	1OC	23;55	15.18
27146	16OC	2;29	209.28	27146	31OC	11;04	44.74
27147	14NO	20;41	239.33	27147	29NO	21;56	74.64
27148	14DE	16;21	269.70	27148	29DE	8;34	104.66

1195

NUMBER	DATE	TIME	LONG.	NUMBER	DATE	TIME	LONG.
27149	13JA	11;36	300.05	27149	27JA	19;13	134.55
27150	12FE	4;54	330.05	27150	26FE	6;19	164.09
27151	13MR	19;29	359.49	27151	27MR	18;16	193.18
27152	12AP	7;13	28.32	27152	26AP	7;18	221.85
27153	11MY	16;26	56.62	27153	25MY	21;17	250.20
27154	9JN	23;51	84.61	27154	24JN	11;58	278.43
27155	9JL	6;32	112.52	27155	24JL	3;05	306.75
27156	7AU	13;39	140.64	27156	22AU	18;21	335.37
27157	5SE	22;21	169.20	27157	21SE	9;21	4.42
27158	5OC	9;31	198.33	27158	20OC	23;33	33.94
27159	3NO	23;33	228.06	27159	19NO	12;31	63.82
27160	3DE	16;14	258.23	27160	19DE	0;09	93.87

```
                        1196                                                         1202
27161   2JA   10;46  288.58   27161  17JA   10;45  123.83                            27235  10JA    6;22  117.09
27162   1FE    5;53  318.79   27162  15FE   20;42  153.47   27236  25JA  19;31  312.84  27236   8FE   22;22  147.07
27163   2MR    0;05  348.56   27163  16MR    6;26  182.65   27237  24FE   6;21  342.40  27237  10MR   15;05  176.65
27164  31MR   16;01   17.74   27164  14AP   16;20  211.36   27238  25MR  15;17   11.40  27238   9AP    7;48  205.72
27165  30AP    4;58   46.34   27165  14MY    2;49  239.68   27239  23AP  23;08   39.89  27239   8MY   23;44  234.33
27166  29MY   15;06   74.50   27166  12JN   14;27  267.81   27240  23MY   6;44   68.01  27240   7JN   14;07  262.61
27167  27JN   23;17  102.46   27167  12JL    3;46  296.01   27241  21JN  14;55   95.98  27241   7JL    2;34  290.75
27168  27JL    6;40  130.50   27168  10AU   19;02  324.49   27242  21JL   0;29  124.06  27242   5AU   13;17  319.00
27169  25AU   14;23  158.86   27169   9SE   11;53  353.44   27243  19AU  12;15  152.50  27243   3SE   22;57  347.56
27170  23SE   23;14  187.72   27170   9OC    5;13   22.90   27244  18SE   2;49  181.49  27244   3OC    8;26   16.60
27171  23OC    9;46  217.14   27171   7NO   21;45   52.80   27245  17OC  20;17  211.09  27245   1NO   18;24   46.12
27172  21NO   22;18  247.04   27172   7DE   12;36   82.93   27246  16NO  15;40  241.21  27246   1DE    5;08   76.03
27173  21DE   13;01  277.22                                 27247  16DE  11;02  271.57  27247  30DE   16;41  106.09

                        1197                                                         1203
                              27173   6JA    1;32  113.03   27248  15JA   4;28  301.84  27248  29JA    5;04  136.03
27174  20JA    5;49  307.42   27174   4FE   12;47  142.86   27249  13FE  18;58  331.70  27249  27FE   18;26  165.64
27175  18FE   23;55  337.36   27175   5MR   22;35  172.22   27250  15MR   6;38    0.99  27250  29MR    8;54  194.82
27176  20MR   17;55    6.81   27176   4AP    7;15  201.04   27251  13AP  16;01   29.70  27251  28AP    0;18  223.56
27177  19AP   10;17   35.71   27177   3MY   15;21  229.40   27252  12MY  23;49   57.94  27252  27MY   15;59  251.96
27178  19MY    0;13   64.12   27178   1JN   23;41  257.47   27253  11JN   6;50   85.90  27253  26JN    7;05  280.20
27179  17JN   11;49   92.24   27179   1JL    9;22  285.49   27254  10JL  13;56  113.84  27254  25JL   21;07  308.49
27180  16JL   21;44  120.32   27180  30JL   21;22  313.74   27255   8AU  22;11  142.02  27255  24AU   10;03  337.04
27181  15AU    6;48  148.60   27181  29AU   12;10  342.44   27256   7SE   8;42  170.66  27256  22SE   22;10    6.00
27182  13SE   15;44  177.29   27182  28SE    5;24   11.71   27257   6OC  22;21  199.92  27257  22OC    9;45   35.43
27183  13OC    1;01  206.49   27183  27OC   23;53   41.52   27258   5NO  15;16  229.79  27258  20NO   20;55   65.25
27184  11NO   11;07  236.16   27184  26NO   18;12   71.71   27259   5DE  10;29  260.08  27259  20DE    7;42   95.26
27185  10DE   22;30  266.16   27185  26DE   11;07  101.99

                        1198                                                         1204
27186   9JA   11;33  296.26   27186  25JA    1;53  132.07   27260   4JA   6;10  290.48  27260  18JA   18;13  125.21
27187   8FE    2;21  326.20   27187  23FE   14;09  161.70   27261   3FE   0;32  320.64  27261  17FE    4;51  154.87
27188   9MR   18;22  355.76   27188  24MR   23;59  190.75   27262   3MR  16;26  350.28  27262  17MR   16;03  184.09
27189   8AP   10;43   24.84   27189  23AP    7;57  219.25   27263   2AP   5;30   19.32  27263  16AP    4;15  212.87
27190   8MY    2;31   53.45   27190  22MY   14;56  247.35   27264   1MY  15;51   47.79  27264  15MY   17;32  241.28
27191   6JN   17;16   81.75   27191  20JN   22;06  275.28   27265  31MY   0;02   75.87  27265  14JN    7;46  269.52
27192   6JL    6;48  109.93   27192  20JL    6;36  303.32   27266  29JN   6;58  103.78  27266  13JL   22;43  297.78
27193   4AU   19;07  138.23   27193  18AU   17;24  331.71   27267  28JL  13;47  131.81  27267  12AU   14;07  326.29
27194   3SE    6;22  166.85   27194  17SE    7;00    0.64   27268  26AU  21;10  160.20  27268  11SE    5;35  355.21
27195   2OC   16;49  195.92   27195  16OC   23;22   30.19   27269  25SE   7;38  189.12  27269  10OC   20;33   24.60
27196   1NO    2;55  225.44   27196  15NO   17;55   60.26   27270  24OC  20;18  218.65  27270   9NO   10;26   54.41
27197  30NO   13;11  255.32   27197  15DE   13;26   90.63   27271  23NO  11;45  248.68  27271   8DE   22;52   84.44
27198  30DE    0;05  285.35                                 27272  23DE   5;25  279.00

                        1199                                                         1205
                              27198  14JA    8;13  120.95                            27272   7JA    9;58  114.47
27199  28JA   11;47  315.27   27199  13FE    0;40  150.90   27273  22JA   0;17  309.27  27273   5FE   20;05  144.23
27200  27FE    0;17  344.86   27200  14MR   13;55  180.27   27274  20FE  18;59  339.21  27274   7MR    5;41  173.56
27201  28MR   13;29   14.00   27201  13AP    0;06  209.03   27275  22MR  12;03    8.60  27275   5AP   15;12  202.40
27202  27AP    3;23   42.69   27202  12MY    8;05  237.28   27276  21AP   2;27   37.39  27276   5MY    1;05  230.81
27203  26MY   18;01   71.06   27203  10JN   15;00  265.24   27277  20MY  13;55   65.69  27277   3JN   11;53  258.97
27204  25JN    9;13   99.30   27204   9JL   21;56  293.17   27278  18JN  23;00   93.70  27278   3JL    0;11  287.10
27205  25JL    0;28  127.63   27205   8AU    5;51  321.33   27279  18JL   6;45  121.70  27279   1AU   14;29  315.45
27206  23AU   15;05  156.24   27206   6SE   15;32  349.93   27280  16AU  14;20  149.94  27280  31AU    6;47  344.24
27207  22SE    4;35  185.24   27207   6OC    3;39   19.11   27281  14SE  22;42  178.63  27281  30SE    0;18   13.54
27208  21OC   16;59  214.70   27208   4NO   18;41   48.89   27282  14OC   8;29  207.86  27282  29OC   17;39   43.33
27209  20NO    4;36  244.53   27209   4DE   12;35   79.13   27283  12NO  20;03  237.61  27283  28NO    9;38   73.44
27210  19DE   15;45  274.56                                 27284  12DE   9;38  267.71  27284  27DE   23;37  103.60

                        1200                                                         1206
                              27210   3JA    8;18  109.53   27285  11JA   1;18  297.92  27285  26JA   11;40  133.55
27211  18JA    2;31  304.52   27211   2FE    3;49  139.74   27286   9FE  18;40  327.95  27286  24FE   22;02  163.08
27212  16FE   12;56  334.17   27212   2MR   21;08  169.46   27287  11MR  12;40  357.56  27287  26MR    7;03  192.08
27213  16MR   23;09    3.37   27213   1AP   11;19  198.55   27288  10AP   5;49   26.65  27288  24AP   15;10  220.56
27214  15AP    9;41   32.09   27214  30AP   22;31  227.07   27289   9MY  20;56   55.21  27289  23MY   23;05  248.69
27215  14MY   21;15   60.44   27215  30MY    7;31  255.18   27290   8JN   9;41   83.42  27290  22JN    7;50  276.68
27216  13JN   10;25   88.64   27216  28JN   15;14  283.13   27291   7JL  20;29  111.49  27291  21JL   18;31  304.81
27217  13JL    1;13  116.89   27217  27JL   22;32  311.17   27292   6AU   6;04  139.70  27292  20AU    7;56  333.33
27218  11AU   17;09  145.41   27218  26AU    6;16  339.54   27293   4SE  15;13  168.26  27293  19SE    0;10    2.40
27219  10SE    9;22  174.35   27219  24SE   15;14    8.41   27294   4OC   0;29  197.29  27294  18OC   18;24   32.04
27220  10OC    1;11  203.76   27220  24OC    2;18   37.86   27295   2NO  10;16  226.82  27295  17NO   13;10   62.15
27221   8NO   16;08  233.60   27221  22NO   16;08   67.82   27296   1DE  21;01  256.73  27296  17DE    7;03   92.45
27222   8DE    5;54  263.69   27222  22DE    8;45   98.09   27297  31DE   9;07  286.81

                        1201                                                         1207
27223   6JA   18;11  293.77   27223  21JA    3;12  128.36                            27297  15JA   23;00  122.65
27224   5FE    4;53  323.56   27224  19FE   21;51  158.30   27298  29JA  22;52  316.81  27298  14FE   12;29  152.46
27225   6MR   14;12  352.89   27225  21MR   15;11  187.72   27299  28FE  14;06  346.49  27299  15MR   23;24  181.71
27226   4AP   22;46   21.70   27226  20AP    6;23  216.55   27300  30MR   6;06   15.71  27300  14AP    8;04  210.38
27227   4MY    7;27   50.08   27227  19MY   19;13  244.91   27301  28AP  22;03   44.46  27301  13MY   15;18  238.59
27228   2JN   17;08   78.19   27228  18JN    5;54  272.99   27302  28MY  13;16   72.84  27302  11JN   22;08  266.54
27229   2JL    4;33  106.29   27229  17JL   14;55  301.04   27303  27JN   3;28  101.04  27303  11JL    5;48  294.50
27230  31JL   18;00  134.60   27230  15AU   23;03  329.29   27304  26JL  16;35  129.30  27304   9AU   15;21  322.74
27231  30AU    9;30  163.34   27231  14SE    7;11  357.96   27305  25AU   4;36  157.82  27305   8SE    3;35  351.47
27232  29SE    2;39  192.62   27232  13OC   16;16   27.15   27306  23SE  15;41  186.75  27306   7OC   18;42   20.80
27233  28OC   20;39  222.42   27233  12NO    3;00   56.86   27307  23OC   2;07  216.15  27307   6NO   12;23   50.71
27234  27NO   14;16  252.59   27234  11DE   15;45   86.92   27308  21NO  12;24  245.94  27308   6DE    7;41   81.01
27235  27DE    6;09  282.83                                 27309  20DE  22;58  275.94
```

--------NEW MOONS--------				--------FULL MOONS-------			
NUMBER	DATE	TIME	LONG.	NUMBER	DATE	TIME	LONG.

1208

NUMBER	DATE	TIME	LONG.	NUMBER	DATE	TIME	LONG.
				27309	5JA	3;05	111.40
27310	19JA	10;09	305.91	27310	3FE	20;51	141.52
27311	17FE	22;03	335.61	27311	4MR	11;40	171.11
27312	18MR	10;36	4.88	27312	2AP	23;13	200.07
27313	16AP	23;51	33.69	27313	2MY	8;05	228.48
27314	16MY	13;54	62.13	27314	31MY	15;21	256.51
27315	15JN	4;45	90.38	27315	29JN	22;09	284.42
27316	14JL	20;07	118.67	27316	29JL	5;31	312.47
27317	13AU	11;20	147.18	27317	27AU	14;19	340.90
27318	12SE	1;44	176.07	27318	26SE	1;13	9.88
27319	11OC	14;58	205.40	27319	25OC	14;50	39.46
27320	10NO	3;10	235.14	27320	24NO	7;23	69.55
27321	9DE	14;38	265.14	27321	24DE	2;22	99.92

1209

NUMBER	DATE	TIME	LONG.	NUMBER	DATE	TIME	LONG.
27322	8JA	1;35	295.16	27322	22JA	22;10	130.23
27323	6FE	12;03	324.93	27323	21FE	16;40	160.14
27324	7MR	22;06	354.27	27324	23MR	8;21	189.46
27325	6AP	8;08	23.12	27325	21AP	20;54	218.17
27326	5MY	18;50	51.55	27326	21MY	6;53	246.40
27327	4JN	6;57	79.76	27327	19JN	15;10	274.38
27328	3JL	20;52	107.95	27328	18JL	22;40	302.37
27329	2AU	12;24	136.36	27329	17AU	6;12	330.61
27330	1SE	4;46	165.16	27330	15SE	14;35	359.31
27331	30SE	21;08	194.44	27331	15OC	0;38	28.56
27332	30OC	12;51	224.17	27332	13NO	13;05	58.36
27333	29NO	3;28	254.23	27333	13DE	4;19	88.53
27334	28DE	16;37	284.35				

1210

NUMBER	DATE	TIME	LONG.	NUMBER	DATE	TIME	LONG.
				27334	11JA	21;54	118.82
27335	27JA	4;04	314.27	27335	10FE	16;27	148.89
27336	25FE	13;53	343.77	27336	12MR	10;24	178.49
27337	26MR	22;32	12.74	27337	11AP	2;35	207.52
27338	25AP	6;50	41.22	27338	10MY	16;32	236.03
27339	24MY	15;42	69.38	27339	9JN	4;18	264.19
27340	23JN	2;03	97.44	27340	8JL	14;12	292.23
27341	22JL	14;25	125.64	27341	6AU	22;51	320.41
27342	21AU	4;58	154.21	27342	5SE	7;04	348.93
27343	19SE	21;31	183.31	27343	4OC	15;45	17.95
27344	19OC	15;28	212.96	27344	3NO	1;43	47.50
27345	18NO	9;40	243.05	27345	2DE	13;28	77.46
27346	18DE	2;43	273.32				

1211

NUMBER	DATE	TIME	LONG.	NUMBER	DATE	TIME	LONG.
				27346	1JA	3;06	107.60
27347	16JA	17;26	303.46	27347	30JA	18;17	137.65
27348	15FE	5;25	333.19	27348	1MR	10;27	167.36
27349	16MR	15;05	2.38	27349	31MR	2;59	196.59
27350	14AP	23;13	31.02	27350	29AP	19;14	225.34
27351	14MY	6;41	59.23	27351	29MY	10;25	253.71
27352	12JN	14;22	87.21	27352	27JN	23;57	281.89
27353	11JL	23;05	115.22	27353	27JL	11;40	310.10
27354	10AU	9;39	143.51	27354	25AU	22;01	338.56
27355	8SE	22;50	172.28	27355	24SE	7;44	7.44
27356	8OC	15;01	201.67	27356	23OC	17;36	36.82
27357	7NO	9;46	231.64	27357	22NO	4;04	66.63
27358	7DE	5;32	261.97	27358	21DE	15;13	96.66

1212

NUMBER	DATE	TIME	LONG.	NUMBER	DATE	TIME	LONG.
27359	6JA	0;11	292.32	27359	20JA	3;03	126.66
27360	4FE	16;12	322.36	27360	18FE	15;39	156.38
27361	5MR	5;09	351.86	27361	19MR	5;15	185.68
27362	3AP	15;29	20.76	27362	17AP	19;57	214.53
27363	2MY	23;53	49.14	27363	17MY	11;24	243.02
27364	1JN	7;08	77.17	27364	16JN	2;50	271.29
27365	30JN	14;03	105.08	27365	15JL	17;33	299.55
27366	29JL	21;37	133.15	27366	14AU	7;15	328.01
27367	28AU	6;55	161.61	27367	12SE	20;01	356.85
27368	26SE	19;00	190.64	27368	12OC	8;07	26.14
27369	26OC	10;25	220.31	27369	10NO	19;43	55.87
27370	25NO	4;43	250.48	27370	10DE	6;47	85.86
27371	25DE	0;25	280.88				

1213

NUMBER	DATE	TIME	LONG.	NUMBER	DATE	TIME	LONG.
				27371	8JA	17;23	115.85
27372	23JA	19;35	311.16	27372	7FE	3;45	145.61
27373	22FE	12;43	341.00	27373	8MR	14;21	174.97
27374	24MR	3;08	10.26	27374	7AP	1;44	203.86
27375	22AP	14;43	38.91	27375	6MY	14;12	232.36
27376	21MY	23;52	67.10	27376	5JN	3;48	260.62
27377	20JN	7;19	95.05	27377	4JL	18;23	288.84
27378	19JL	14;07	123.01	27378	3AU	9;42	317.25
27379	17AU	21;26	151.26	27379	2SE	1;25	346.03
27380	16SE	6;22	179.99	27380	1OC	17;01	15.29
27381	15OC	17;43	209.31	27381	31OC	7;48	45.00
27382	14NO	7;51	239.18	27382	29NO	21;10	75.00
27383	14DE	0;29	269.42	27383	29DE	8;58	105.07

1214

NUMBER	DATE	TIME	LONG.	NUMBER	DATE	TIME	LONG.
27384	12JA	18;47	299.73	27384	27JA	19;28	134.95
27385	11FE	13;35	329.80	27385	26FE	5;06	164.43
27386	13MR	7;28	359.38	27386	27MR	14;23	193.41
27387	11AP	23;11	28.38	27387	25AP	23;47	221.94
27388	11MY	12;04	56.84	27388	25MY	9;50	250.13
27389	9JN	22;17	84.93	27389	23JN	21;10	278.23
27390	9JL	6;40	112.92	27390	23JL	10;24	306.47
27391	7AU	14;20	141.06	27391	22AU	1;50	335.09
27392	5SE	22;23	169.59	27392	20SE	19;06	4.22
27393	5OC	7;33	198.64	27393	20OC	12;59	33.89
27394	3NO	18;18	228.23	27394	19NO	6;00	63.94
27395	3DE	6;52	258.23	27395	18DE	21;10	94.13

1215

NUMBER	DATE	TIME	LONG.	NUMBER	DATE	TIME	LONG.
27396	1JA	21;24	288.41	27396	17JA	10;11	124.19
27397	31JA	13;48	318.50	27397	15FE	21;16	153.88
27398	2MR	7;24	348.26	27398	17MR	6;46	183.06
27399	1AP	0;56	17.51	27399	15AP	15;05	211.69
27400	30AP	17;02	46.25	27400	14MY	22;50	239.91
27401	30MY	6;57	74.56	27401	13JN	6;54	267.90
27402	28JN	18;45	102.67	27402	12JL	16;24	295.94
27403	28JL	5;02	130.82	27403	11AU	4;23	324.29
27404	26AU	14;32	159.26	27404	9SE	19;20	353.15
27405	24SE	23;55	188.14	27405	9OC	12;52	22.61
27406	24OC	9;34	217.52	27406	8NO	7;45	52.60
27407	22NO	19;53	247.33	27407	8DE	2;24	82.89
27408	22DE	7;14	277.37				

1216

NUMBER	DATE	TIME	LONG.	NUMBER	DATE	TIME	LONG.
				27408	6JA	19;29	113.17
27409	20JA	20;00	307.40	27409	5FE	10;14	143.14
27410	19FE	10;17	337.18	27410	5MR	22;20	172.60
27411	20MR	1;42	6.54	27411	4AP	7;56	201.46
27412	18AP	17;31	35.43	27412	3MY	15;38	229.80
27413	18MY	9;00	63.91	27413	1JN	22;26	257.81
27414	16JN	23;44	92.15	27414	1JL	5;29	285.73
27415	16JL	13;32	120.38	27415	30JL	13;59	313.84
27416	15AU	2;21	148.81	27416	29AU	0;52	342.37
27417	13SE	14;10	177.62	27417	27SE	14;37	11.49
27418	13OC	1;08	206.88	27418	27OC	7;12	41.21
27419	11NO	11;37	236.57	27419	26NO	1;56	71.41
27420	10DE	22;03	266.53	27420	25DE	21;32	101.81

1217

NUMBER	DATE	TIME	LONG.	NUMBER	DATE	TIME	LONG.
27421	9JA	8;51	296.54	27421	24JA	16;18	132.06
27422	7FE	20;13	326.33	27422	23FE	8;37	161.86
27423	9MR	8;10	355.73	27423	24MR	21;42	191.04
27424	7AP	20;46	24.66	27424	23AP	7;43	219.63
27425	7MY	10;08	53.19	27425	22MY	15;36	247.76
27426	6JN	0;26	81.47	27426	20JN	22;29	275.68
27427	5JL	15;36	109.72	27427	20JL	5;30	303.66
27428	4AU	7;09	138.14	27428	18AU	13;34	331.94
27429	2SE	22;18	166.92	27429	16SE	23;26	0.72
27430	2OC	12;26	196.13	27430	16OC	11;43	30.09
27431	1NO	1;23	225.77	27431	15NO	2;51	60.01
27432	30NO	13;21	255.72	27432	14DE	20;44	90.31
27433	30DE	0;36	285.77				

1218

NUMBER	DATE	TIME	LONG.	NUMBER	DATE	TIME	LONG.
				27433	13JA	16;17	120.67
27434	28JA	11;13	315.64	27434	12FE	11;34	150.75
27435	26FE	21;16	345.13	27435	14MR	4;38	180.28
27436	28MR	6;59	14.12	27436	12AP	18;39	209.20
27437	26AP	16;59	42.66	27437	12MY	5;47	237.58
27438	26MY	4;05	70.89	27438	10JN	14;49	265.62
27439	24JN	16;58	99.05	27439	9JL	22;42	293.59
27440	24JL	7;45	127.36	27440	8AU	6;15	321.73
27441	22AU	23;57	156.01	27441	6SE	14;15	350.27
27442	21SE	16;40	185.14	27442	5OC	23;30	19.33
27443	21OC	9;04	214.76	27443	4NO	10;45	48.95
27444	20NO	0;32	244.75	27444	4DE	0;36	79.01
27445	19DE	14;35	274.90				

1219

NUMBER	DATE	TIME	LONG.	NUMBER	DATE	TIME	LONG.
				27445	2JA	17;01	109.27
27446	18JA	2;54	304.93	27446	1FE	11;05	139.43
27447	16FE	13;24	334.59	27447	3MR	5;17	169.19
27448	17MR	22;23	3.73	27448	1AP	22;16	198.41
27449	16AP	6;32	32.35	27449	1MY	13;16	227.09
27450	15MY	14;49	60.58	27450	31MY	2;07	255.35
27451	14JN	0;10	88.62	27451	29JN	13;00	283.43
27452	13JL	11;24	116.73	27452	28JL	22;21	311.55
27453	12AU	0;52	145.15	27453	27AU	6;53	339.96
27454	10SE	16;35	174.05	27454	25SE	15;25	8.81
27455	10OC	10;08	203.53	27455	25OC	0;49	38.19
27456	9NO	4;36	233.51	27456	23NO	11;43	68.03
27457	8DE	22;35	263.77	27457	23DE	0;23	98.13

1220

NEW NUMBER	DATE	TIME	LONG.	FULL NUMBER	DATE	TIME	LONG.
27458	7JA	14;39	294.01	27458	21JA	14;41	128.22
27459	6FE	3;58	323.92	27459	20FE	6;10	158.04
27460	6MR	14;37	353.30	27460	20MR	22;18	187.42
27461	4AP	23;15	22.11	27461	19AP	14;34	216.31
27462	4MY	6;48	50.44	27462	19MY	6;16	244.79
27463	2JN	14;09	78.46	27463	17JN	20;43	273.02
27464	1JL	22;11	106.42	27464	17JL	9;29	301.21
27465	31JL	7;43	134.58	27465	15AU	20;41	329.59
27466	29AU	19;33	163.16	27466	14SE	6;52	358.34
27467	28SE	10;19	192.33	27467	13OC	16;50	27.57
27468	28OC	4;02	222.12	27468	12NO	3;07	57.26
27469	26NO	23;41	252.36	27469	11DE	13;59	87.24
27470	26DE	19;13	282.76				

1221

NEW NUMBER	DATE	TIME	LONG.	FULL NUMBER	DATE	TIME	LONG.
				27470	10JA	1;23	117.27
27471	25JA	12;39	312.95	27471	8FE	13;23	147.08
27472	24FE	3;04	342.65	27472	10MR	2;10	176.50
27473	25MR	14;32	11.76	27473	8AP	16;02	205.47
27474	23AP	23;43	40.30	27474	8MY	6;55	234.05
27475	23MY	7;22	68.42	27475	6JN	22;21	262.37
27476	21JN	14;19	96.34	27476	6JL	13;33	290.62
27477	20JL	21;27	124.33	27477	5AU	3;57	319.01
27478	19AU	5;48	152.63	27478	3SE	17;26	347.73
27479	17SE	16;28	181.45	27479	3OC	6;09	16.89
27480	17OC	6;17	210.90	27480	1NO	18;15	46.51
27481	15NO	23;19	240.91	27481	1DE	5;43	76.45
27482	15DE	18;34	271.27	27482	30DE	16;34	106.47

1222

NEW NUMBER	DATE	TIME	LONG.	FULL NUMBER	DATE	TIME	LONG.
27483	14JA	14;10	301.62	27483	29JA	2;53	136.33
27484	13FE	8;22	331.64	27484	27FE	13;06	165.81
27485	15MR	0;05	1.11	27485	28MR	23;45	194.83
27486	13AP	12;59	29.97	27486	27AP	11;22	223.43
27487	12MY	23;15	58.30	27487	27MY	0;11	251.72
27488	11JN	7;27	86.31	27488	25JN	14;12	279.92
27489	10JL	14;29	114.25	27489	25JL	5;14	308.25
27490	8AU	21;30	142.38	27490	23AU	20;59	336.90
27491	7SE	5;38	170.93	27491	22SE	13;00	6.01
27492	6OC	15;49	200.04	27492	22OC	4;35	35.61
27493	5NO	4;38	229.73	27493	20NO	18;57	65.56
27494	4DE	20;04	259.87	27494	20DE	7;39	95.65

1223

NEW NUMBER	DATE	TIME	LONG.	FULL NUMBER	DATE	TIME	LONG.
27495	3JA	13;34	290.17	27495	18JA	18;43	125.63
27496	2FE	8;06	320.34	27496	17FE	4;36	155.25
27497	4MR	2;26	350.10	27497	18MR	13;49	184.39
27498	2AP	19;13	19.29	27498	16AP	22;52	213.04
27499	2MY	9;30	47.93	27499	16MY	8;18	241.30
27500	31MY	21;00	76.14	27500	14JN	18;44	269.39
27501	30JN	6;16	104.14	27501	14JL	6;53	297.54
27502	29JL	14;18	132.22	27502	12AU	21;15	326.00
27503	27AU	22;13	160.61	27503	11SE	13;51	354.96
27504	26SE	6;56	189.49	27504	11OC	7;50	24.46
27505	25OC	16;59	218.90	27505	10NO	1;43	54.43
27506	24NO	4;40	248.78	27506	9DE	18;04	84.63
27507	23DE	18;09	278.91				

1224

NEW NUMBER	DATE	TIME	LONG.	FULL NUMBER	DATE	TIME	LONG.
				27507	8JA	8;14	114.78
27508	22JA	9;28	309.04	27508	6FE	20;12	144.63
27509	21FE	2;20	338.91	27509	7MR	6;20	173.98
27510	21MR	19;49	8.33	27510	5AP	15;01	202.78
27511	20AP	12;36	37.22	27511	4MY	22;47	231.11
27512	20MY	3;36	65.67	27512	3JN	6;25	259.14
27513	18JN	16;29	93.83	27513	2JL	14;58	287.12
27514	18JL	3;35	121.96	27514	1AU	1;35	315.33
27515	16AU	13;37	150.30	27515	30AU	15;05	343.99
27516	14SE	23;14	179.04	27516	29SE	7;33	13.24
27517	14OC	8;55	208.27	27517	29OC	2;08	43.08
27518	12NO	19;00	237.96	27518	27NO	21;14	73.31
27519	12DE	5;48	267.94	27519	27DE	15;20	103.64

1225

NEW NUMBER	DATE	TIME	LONG.	FULL NUMBER	DATE	TIME	LONG.
27520	10JA	17;43	297.99	27520	26JA	7;20	133.76
27521	9FE	7;02	327.85	27521	24FE	20;42	163.42
27522	10MR	21;40	357.34	27522	26MR	7;24	192.48
27523	9AP	13;05	26.36	27523	24AP	15;50	220.98
27524	9MY	4;36	54.95	27524	23MY	22;51	249.07
27525	7JN	19;41	83.25	27525	22JN	5;35	276.98
27526	7JL	10;03	111.47	27526	21JL	13;12	304.99
27527	5AU	23;34	139.83	27527	19AU	22;50	333.35
27528	4SE	12;10	168.52	27528	18SE	11;12	2.25
27529	3OC	23;41	197.65	27529	18OC	2;30	31.77
27530	2NO	10;42	227.23	27530	16NO	20;22	61.83
27531	1DE	21;14	257.13	27531	16DE	15;45	92.20
27532	31DE	7;49	287.15				

1226

NEW NUMBER	DATE	TIME	LONG.	FULL NUMBER	DATE	TIME	LONG.
				27532	15JA	11;09	122.54
27533	29JA	18;45	317.03	27533	14FE	4;47	152.53
27534	28FE	6;10	346.55	27534	15MR	19;27	181.94
27535	29MR	18;08	15.61	27535	14AP	6;51	210.72
27536	28AP	6;49	44.24	27536	13MY	15;36	238.99
27537	27MY	20;26	72.56	27537	11JN	22;50	266.96
27538	26JN	11;07	100.78	27538	11JL	5;41	294.89
27539	26JL	2;39	129.13	27539	9AU	13;12	323.04
27540	24AU	18;19	157.79	27540	7SE	22;11	351.64
27541	23SE	9;19	186.88	27541	7OC	9;18	20.80
27542	22OC	23;09	216.41	27542	5NO	23;02	50.54
27543	21NO	11;47	246.30	27543	5DE	15;36	80.72
27544	20DE	23;28	276.36				

1227

NEW NUMBER	DATE	TIME	LONG.	FULL NUMBER	DATE	TIME	LONG.
				27544	4JA	10;26	111.09
27545	19JA	10;22	306.32	27545	3FE	5;59	141.28
27546	17FE	20;32	335.94	27546	5MR	0;12	171.02
27547	19MR	6;09	5.09	27547	3AP	15;41	200.15
27548	17AP	15;39	33.75	27548	3MY	4;08	228.71
27549	17MY	1;52	62.04	27549	1JN	14;07	256.85
27550	15JN	13;37	90.17	27550	30JN	22;32	284.82
27551	15JL	3;25	118.39	27551	30JL	6;16	312.89
27552	13AU	19;05	146.91	27552	28AU	14;07	341.29
27553	12SE	11;52	175.89	27553	26SE	22;48	10.17
27554	12OC	4;48	205.37	27554	26OC	9;04	39.61
27555	10NO	21;03	235.28	27555	24NO	21;36	69.52
27556	10DE	12;02	265.42	27556	24DE	12;42	99.73

1228

NEW NUMBER	DATE	TIME	LONG.	FULL NUMBER	DATE	TIME	LONG.
27557	9JA	1;19	295.54	27557	23JA	5;56	129.93
27558	7FE	12;39	325.35	27558	22FE	0;02	159.84
27559	7MR	22;11	354.66	27559	22MR	17;33	189.24
27560	6AP	6;28	23.44	27560	21AP	9;28	218.09
27561	5MY	14;21	51.76	27561	20MY	23;21	246.48
27562	3JN	22;53	79.83	27562	19JN	11;15	274.61
27563	3JL	9;00	107.87	27563	18JL	21;28	302.71
27564	1AU	21;18	136.15	27564	17AU	6;31	331.02
27565	31AU	12;00	164.87	27565	15SE	15;09	359.73
27566	30SE	4;53	194.16	27566	15OC	0;13	28.94
27567	29OC	23;14	223.99	27567	13NO	10;24	58.64
27568	28NO	17;50	254.21	27568	12DE	22;11	88.67
27569	28DE	11;08	284.51				

1229

NEW NUMBER	DATE	TIME	LONG.	FULL NUMBER	DATE	TIME	LONG.
				27569	11JA	11;35	118.77
27570	27JA	1;53	314.57	27570	10FE	2;18	148.69
27571	25FE	13;44	344.15	27571	11MR	17;53	178.20
27572	26MR	23;09	13.15	27572	10AP	9;53	207.23
27573	25AP	6;59	41.62	27573	10MY	1;47	235.83
27574	24MY	14;12	69.71	27574	8JN	16;56	264.12
27575	22JN	21;42	97.65	27575	8JL	6;42	292.32
27576	22JL	6;22	125.71	27576	6AU	18;51	320.64
27577	20AU	16;59	154.12	27577	5SE	5;44	349.27
27578	19SE	6;19	183.07	27578	4OC	15;58	18.36
27579	18OC	22;43	212.65	27579	3NO	2;14	47.92
27580	17NO	17;42	242.76	27580	2DE	12;54	77.83
27581	17DE	13;39	273.16				

1230

NEW NUMBER	DATE	TIME	LONG.	FULL NUMBER	DATE	TIME	LONG.
				27581	1JA	0;01	107.87
27582	16JA	8;21	303.46	27582	30JA	11;32	137.76
27583	15FE	0;17	333.37	27583	28FE	23;38	167.30
27584	16MR	13;05	2.69	27584	30MR	12;38	196.40
27585	14AP	23;13	31.41	27585	29AP	2;45	225.07
27586	14MY	7;28	59.65	27586	28MY	17;49	253.44
27587	12JN	14;37	87.61	27587	27JN	9;12	281.69
27588	11JL	21;33	115.55	27588	27JL	0;12	310.03
27589	10AU	5;13	143.72	27589	25AU	14;23	338.64
27590	8SE	14;41	172.34	27590	24SE	3;46	7.67
27591	8OC	2;56	201.56	27591	23OC	16;26	37.17
27592	6NO	18;29	231.39	27592	22NO	4;26	67.04
27593	6DE	12;50	261.66	27593	21DE	15;39	97.07

1231

NEW NUMBER	DATE	TIME	LONG.	FULL NUMBER	DATE	TIME	LONG.
27594	5JA	8;27	292.05	27594	20JA	2;08	127.01
27595	4FE	3;26	322.21	27595	18FE	12;10	156.62
27596	5MR	20;22	351.89	27596	19MR	22;17	185.78
27597	4AP	10;36	20.95	27597	18AP	9;05	214.48
27598	3MY	22;05	49.46	27598	17MY	21;03	242.83
27599	2JN	7;13	77.56	27599	16JN	10;19	271.02
27600	1JL	14;46	105.50	27600	16JL	0;51	299.28
27601	30JL	21;46	133.54	27601	14AU	16;24	327.80
27602	29AU	5;19	161.93	27602	13SE	8;37	356.77
27603	27SE	14;30	190.85	27603	13OC	0;49	26.23
27604	27OC	2;04	220.35	27604	11NO	16;08	56.11
27605	25NO	16;15	250.35	27605	11DE	5;51	86.21
27606	25DE	8;44	280.61				

```
--------NEW MOONS--------    --------FULL MOONS-------        --------NEW MOONS--------    --------FULL MOONS-------
NUMBER  DATE   TIME   LONG.   NUMBER  DATE   TIME   LONG.     NUMBER  DATE   TIME   LONG.   NUMBER  DATE   TIME   LONG.

                        1232                                                          1238
                                27606   9JA  17;44  116.26                                    27680   2JA   9;00  109.32
27607  24JA   2;44  310.84      27607   8FE   4;04  146.02    27681  17JA  23;04  305.16      27681  31JA  22;55  139.31
27608  22FE  21;09  340.75      27608   8MR  13;23  175.32    27682  16FE  12;18  334.93      27682   2MR  13;50  168.94
27609  23MR  14;41   10.14      27609   6AP  22;14  204.11    27683  17MR  22;44    4.13      27683   1AP   5;24  198.12
27610  22AP   6;12   38.96      27610   6MY   7;11  232.47    27684  16AP   7;07   32.76      27684  30AP  21;12  226.83
27611  21MY  19;04   67.30      27611   4JN  16;51  260.57    27685  15MY  14;24   60.95      27685  30MY  12;43  255.21
27612  20JN   5;25   95.36      27612   4JL   3;57  288.66    27686  13JN  21;34   88.91      27686  29JN   3;18  283.43
27613  19JL  14;04  123.40      27613   2AU  17;09  316.98    27687  13JL   5;31  116.89      27687  28JL  16;27  311.70
27614  17AU  22;05  151.68      27614   1SE   8;48  345.75    27688  11AU  15;05  145.14      27688  27AU   4;10  340.24
27615  16SE   6;30  180.39      27615   1OC   2;27   15.08    27689  10SE   3;02  173.89      27689  25SE  14;55    9.18
27616  15OC  16;00  209.63      27616  30OC  20;51   44.93    27690   9OC  17;59  203.25      27690  25OC   1;20   38.61
27617  14NO   2;57  239.37      27617  29NO  14;18   75.11    27691   8NO  11;55  233.20      27691  23NO  11;55   68.43
27618  13DE  15;29  269.43      27618  29DE   5;42  105.33    27692   8DE   7;44  263.54      27692  22DE  22;50   98.45

                        1233                                                          1239
27619  12JA   5;45  299.57      27619  27JA  18;44  135.31    27693   7JA   3;20  293.92      27693  21JA  10;02  128.41
27620  10FE  21;41  329.52      27620  26FE   5;39  164.84    27694   5FE  20;44  324.00      27694  19FE  21;35  158.07
27621  12MR  14;43  359.09      27621  27MR  14;51  193.82    27695   7MR  11;00  353.54      27695  21MR   9;47  187.29
27622  11AP   7;48   28.15      27622  25AP  22;50  222.29    27696   5AP  22;18   22.46      27696  19AP  23;03  216.07
27623  10MY  23;40   56.73      27623  25MY   6;16  250.38    27697   5MY   7;19   50.85      27697  19MY  13;28  244.51
27624   9JN  13;37   84.98      27624  23JN  14;08  278.34    27698   3JN  14;52   78.88      27698  18JN   4;42  272.77
27625   9JL   1;41  113.11      27625  22JL  23;32  306.42    27699   2JL  21;49  106.80      27699  17JL  20;02  301.06
27626   7AU  12;22  141.37      27626  21AU  11;32  334.90    27700   1AU   5;02  134.86      27700  16AU  10;52  329.58
27627   5SE  22;22  169.98      27627  20SE   2;40    3.94    27701  30AU  13;33  163.31      27701  15SE   0;56  358.49
27628   5OC   8;13  199.06      27628  19OC  20;30   33.59    27702  29SE   0;24  192.31      27702  14OC  14;15   27.85
27629   3NO  18;14  228.62      27629  18NO  15;43   63.72    27703  28OC  14;22  221.93      27703  13NO   2;49   57.64
27630   3DE   4;42  258.53      27630  18DE  10;36   94.07    27704  27NO   7;28  252.07      27704  12DE  14;33   87.66
                                                             27705  27DE   2;39  282.45

                        1234                                                          1240
27631   1JA  15;58  288.57      27631  17JA   3;46  124.31                                    27705  11JA   1;23  117.66
27632  31JA   4;23  318.50      27632  15FE  18;27  154.16    27706  25JA  22;04  312.72      27706   9FE  11;27  147.39
27633   1MR  18;07  348.10      27633  17MR   6;22  183.43    27707  24FE  16;02  342.59      27707   9MR  21;13  176.69
27634  31MR   8;54   17.25      27634  15AP  15;44  212.11    27708  25MR   7;33   11.87      27708   8AP   7;21  205.51
27635  30AP   0;12   45.96      27635  14MY  23;15  240.31    27709  23AP  20;19   40.55      27709   7MY  18;25  233.94
27636  29MY  15;25   74.33      27636  13JN   5;55  268.25    27710  23MY   6;33   68.78      27710   6JN   6;51  262.14
27637  28JN   6;12  102.56      27637  12JL  12;56  296.19    27711  21JN  14;50   96.75      27711   5JL  20;41  290.34
27638  27JL  20;19  130.86      27638  10AU  21;29  324.40    27712  20JL  22;03  124.74      27712   4AU  11;40  318.75
27639  26AU   9;39  159.45      27639   9SE   8;29  353.10    27713  19AU   5;19  153.00      27713   3SE   3;59  347.57
27640  24SE  22;04  188.46      27640   8OC  22;25   22.40    27714  17SE  13;43  181.73      27714   2OC  20;34   16.89
27641  24OC   9;33  217.92      27641   7NO  15;09   52.29    27715  17OC   0;08  211.03      27715   1NO  12;43   46.67
27642  22NO  20;23  247.74      27642   7DE   9;59   82.58    27716  15NO  13;04  240.87      27716   1DE   3;31   76.75
27643  22DE   6;56  277.75                                   27717  15DE   4;26  271.06      27717  30DE  16;23  106.86

                        1235                                                          1241
                                27643   6JA   5;35  112.97    27718  13JA  21;39  301.32      27718  29JA   3;24  136.74
27644  20JA  17;34  307.69      27644   5FE   0;15  143.12    27719  12FE  15;48  331.34      27719  27FE  13;01  166.20
27645  19FE   4;32  337.33      27645   6MR  16;24  172.74    27720  14MR   9;44    0.91      27720  28MR  21;49  195.14
27646  20MR  15;57    6.53      27646   5AP   5;20  201.75    27721  13AP   2;14   29.92      27721  27AP   6;26  223.62
27647  19AP   3;57   35.27      27647   4MY  15;14  230.17    27722  12MY  16;25   58.42      27722  26MY  15;28  251.76
27648  18MY  16;50   63.65      27648   2JN  23;03  258.22    27723  11JN   4;01   86.57      27723  25JN   1;38  279.81
27649  17JN   6;51   91.87      27649   2JL   5;59  286.14    27724  10JL  13;32  114.60      27724  24JL  13;40  308.01
27650  16JL  22;02  120.15      27650  31JL  13;07  314.19    27725   8AU  21;54  142.78      27725  23AU   4;08  336.61
27651  15AU  13;55  148.70      27651  29AU  21;24  342.62    27726   7SE   6;13  171.34      27726  21SE  21;04    5.75
27652  14SE   5;38  177.66      27652  28SE   7;29   11.58    27727   6OC  15;17  200.41      27727  21OC  15;31   35.45
27653  13OC  20;23  207.08      27653  27OC  19;56   41.13    27728   5NO   1;36  230.00      27728  20NO   9;51   65.57
27654  12NO   9;51  236.89      27654  26NO  11;07   71.17    27729   4DE  13;21  259.97      27729  20DE   2;31   95.83
27655  11DE  22;07  266.93      27655  26DE   4;53  101.49

                        1236                                                          1242
27656  10JA   9;24  296.95      27656  25JA   0;11  131.77    27730   3JA   2;39  290.10      27730  18JA  16;46  125.94
27657   8FE  19;49  326.71      27657  23FE  19;09  161.69    27731   1FE  17;34  320.12      27731  17FE   4;38  155.64
27658   9MR   5;29  356.01      27658  24MR  11;58  191.04    27732   3MR   9;52  349.80      27732  18MR  14;30  184.81
27659   7AP  14;43   24.81      27659  23AP   1;50  219.78    27733   2AP   2;49   19.02      27733  16AP  22;52  213.43
27660   7MY   0;14   53.18      27660  22MY  12;57  248.05    27734   1MY  19;15   47.75      27734  16MY   6;20  241.61
27661   5JN  10;56   81.32      27661  20JN  22;06  276.06    27735  31MY  10;11   76.09      27735  14JN  13;45  269.58
27662   4JL  23;34  109.47      27662  20JL   6;11  304.08    27736  29JN  23;15  104.25      27736  13JL  22;11  297.58
27663   3AU  14;23  137.86      27663  18AU  14;03  332.35    27737  29JL  10;44  132.47      27737  12AU   8;47  325.89
27664   2SE   6;53  166.68      27664  16SE  22;23    1.07    27738  27AU  21;15  160.96      27738  10SE  22;24  354.71
27665   2OC   0;07  196.01      27665  16OC   7;54   30.32    27739  26SE   7;22  189.90      27739  10OC  15;07   24.16
27666  31OC  17;04  225.81      27666  14NO  19;18   60.09    27740  25OC  17;28  219.32      27740   9NO   9;59   54.16
27667  30NO   8;58  255.93      27667  14DE   9;05   90.21    27741  24NO   3;48  249.13      27741   9DE   5;21   84.48
27668  29DE  23;14  286.10                                   27742  23DE  14;37  279.15

                        1237                                                          1243
                                27668  13JA   1;13  120.42    27742   7JA  23;34  114.81
27669  28JA  11;32  316.05      27669  11FE  18;51  150.44    27743  22JA   2;16  309.13      27743   6FE  15;32  144.82
27670  26FE  21;48  345.54      27670  13MR  12;34  180.01    27744  20FE  15;06  338.83      27744   8MR   4;46  174.30
27671  28MR   6;27   14.49      27671  12AP   5;11  209.04    27745  22MR   5;07    8.11      27745   6AP  15;15  203.18
27672  26AP  14;13   42.93      27672  11MY  20;02  237.57    27746  20AP  19;57   36.94      27746   5MY  23;29  231.53
27673  25MY  22;08   71.04      27673  10JN   8;57  265.78    27747  20MY  11;05   65.40      27747   4JN   6;22  259.53
27674  24JN   7;14   99.05      27674   9JL  20;06  293.87    27748  19JN   2;04   93.65      27748   3JL  13;03  287.43
27675  23JL  18;21  127.21      27675   8AU   5;51  322.11    27749  18JL  16;39  121.92      27749   1AU  20;43  315.52
27676  22AU   7;53  155.75      27676   6SE  14;49  350.68    27750  17AU   6;39  150.41      27750  31AU   6;28  344.03
27677  20SE  23;51  184.84      27677   5OC  23;46   19.74    27751  15SE  19;49  179.29      27751  29SE  18;59   13.11
27678  20OC  17;46  214.51      27678   4NO   9;28   49.29    27752  15OC   8;03  208.63      27752  29OC  10;27   42.81
27679  19NO  12;37  244.64      27679   3DE  20;28   79.23    27753  13NO  19;22  238.37      27753  28NO   4;25   72.99
27680  19DE   6;54  274.97                                   27754  13DE   6;06  268.35      27754  27DE  23;49  103.38
```

1244 / 1250

NEW				FULL				NEW				FULL			
27755	11JA	16;37	298.33	27755	26JA	19;06	133.64	27829	4JA	16;59	291.78	27829	20JA	2;27	127.42
27756	10FE	3;14	328.08	27756	25FE	12;35	163.47	27830	3FE	10;36	321.90	27830	18FE	12;35	157.03
27757	10MR	14;10	357.42	27757	26MR	3;04	192.69	27831	5MR	4;33	351.63	27831	19MR	21;32	186.14
27758	9AP	1;34	26.28	27758	24AP	14;20	221.31	27832	3AP	21;44	20.82	27832	18AP	5;58	214.74
27759	8MY	13;42	54.74	27759	23MY	23;02	249.46	27833	3MY	13;04	49.48	27833	17MY	14;31	242.96
27760	7JN	2;56	82.97	27760	22JN	6;17	277.40	27834	2JN	1;58	77.74	27834	15JN	23;53	271.00
27761	6JL	17;32	111.20	27761	21JL	13;16	305.38	27835	1JL	12;32	105.80	27835	15JL	10;49	299.11
27762	5AU	9;16	139.64	27762	19AU	20;59	333.67	27836	30JL	21;31	133.92	27836	14AU	0;03	327.53
27763	4SE	1;25	168.47	27763	18SE	6;12	2.44	27837	29AU	5;57	162.35	27837	12SE	15;54	356.48
27764	3OC	17;02	197.77	27764	17OC	17;31	31.79	27838	27SE	14;45	191.25	27838	12OC	9;58	26.00
27765	2NO	7;26	227.49	27765	16NO	7;21	61.67	27839	27OC	0;34	220.68	27839	11NO	4;48	56.03
27766	1DE	20;28	257.49	27766	15DE	23;50	91.91	27840	25NO	11;35	250.54	27840	10DE	22;38	86.29
27767	31DE	8;16	287.56					27841	25DE	0;07	280.64				

1245 / 1251

NEW				FULL				NEW				FULL			
				27767	14JA	18;26	122.23					27841	9JA	14;12	116.50
27768	29JA	19;03	317.43	27768	13FE	13;39	152.29	27842	23JA	14;03	310.70	27842	8FE	3;11	146.37
27769	28FE	4;54	346.89	27769	15MR	7;34	181.84	27843	22FE	5;27	340.48	27843	9MR	13;53	175.73
27770	29MR	14;04	15.84	27770	13AP	22;50	210.79	27844	23MR	21;54	9.84	27844	7AP	22;47	204.52
27771	27AP	23;06	44.32	27771	13MY	11;14	239.20	27845	22AP	14;31	38.72	27845	7MY	6;29	232.83
27772	27MY	8;53	72.49	27772	11JN	21;18	267.28	27846	22MY	6;12	67.18	27846	5JN	13;41	260.83
27773	25JN	20;20	100.59	27773	11JL	5;55	295.28	27847	20JN	20;15	95.39	27847	4JL	21;24	288.78
27774	25JL	10;03	128.86	27774	9AU	13;57	323.46	27848	20JL	8;38	123.57	27848	3AU	6;47	316.95
27775	24AU	1;54	157.52	27775	7SE	22;08	352.03	27849	18AU	19;48	151.97	27849	1SE	18;52	345.57
27776	22SE	19;07	186.69	27776	7OC	7;09	21.10	27850	17SE	6;19	180.77	27850	1OC	10;11	14.80
27777	22OC	12;35	216.36	27777	5NO	17;37	50.70	27851	16OC	16;38	210.05	27851	31OC	4;17	44.63
27778	21NO	5;19	246.42	27778	5DE	6;11	80.72	27852	15NO	2;58	239.76	27852	29NO	23;44	74.88
27779	20DE	20;37	276.63					27853	14DE	13;33	269.74	27853	29DE	18;46	105.26

1246 / 1252

NEW				FULL				NEW				FULL			
				27779	3JA	21;05	110.91	27854	13JA	0;39	299.75	27854	28JA	11;57	135.42
27780	19JA	9;57	306.69	27780	2FE	13;54	141.00	27855	11FE	12;40	329.54	27855	27FE	2;31	165.10
27781	17FE	21;08	336.36	27781	4MR	7;29	170.73	27856	12MR	1;49	358.94	27856	27MR	14;15	194.19
27782	19MR	6;22	5.49	27782	3AP	0;33	199.93	27857	10AP	15;59	27.90	27857	25AP	23;26	222.70
27783	17AP	14;17	34.08	27783	2MY	16;12	228.62	27858	10MY	6;48	56.45	27858	25MY	6;48	250.79
27784	16MY	21;49	62.26	27784	1JN	6;05	256.92	27859	8JN	21;48	84.74	27859	23JN	13;24	278.70
27785	15JN	6;05	90.26	27785	30JN	18;11	285.04	27860	8JL	12;40	112.98	27860	22JL	20;27	306.69
27786	14JL	16;02	118.33	27786	30JL	4;46	313.22	27861	7AU	3;10	141.39	27861	21AU	5;06	335.03
27787	13AU	4;20	146.71	27787	28AU	14;17	341.68	27862	5SE	17;04	170.16	27862	19SE	16;16	3.90
27788	11SE	19;12	175.60	27788	26SE	23;22	10.59	27863	5OC	6;05	199.37	27863	19OC	6;22	33.39
27789	11OC	12;23	205.08	27789	26OC	8;47	39.99	27864	3NO	18;05	229.01	27864	17NO	23;13	63.42
27790	10NO	7;07	235.08	27790	24NO	19;11	69.81	27865	3DE	5;12	258.94	27865	17DE	18;04	93.77
27791	10DE	2;02	265.39	27791	24DE	6;55	99.88								

1247 / 1253

NEW				FULL				NEW				FULL			
27792	8JA	19;29	295.68	27792	22JA	20;00	129.91	27866	1JA	15;47	288.96	27866	16JA	13;34	124.11
27793	7FE	10;13	325.63	27793	21FE	10;13	159.66	27867	31JA	2;13	318.80	27867	15FE	8;03	154.11
27794	8MR	21;54	355.04	27794	23MR	1;12	188.96	27868	1MR	12;45	348.27	27868	17MR	0;02	183.56
27795	7AP	7;04	23.85	27795	21AP	16;40	217.81	27869	30MR	23;37	17.26	27869	15AP	12;48	212.38
27796	6MY	14;39	52.17	27796	21MY	8;15	246.27	27870	29AP	11;04	45.82	27870	14MY	22;38	240.68
27797	4JN	21;41	80.17	27797	19JN	23;24	274.53	27871	28MY	23;29	74.09	27871	13JN	6;28	268.66
27798	4JL	5;05	108.10	27798	19JL	13;27	302.78	27872	27JN	13;17	102.28	27872	12JL	13;30	296.61
27799	2AU	13;45	136.24	27799	18AU	2;07	331.23	27873	27JL	4;33	130.62	27873	10AU	20;51	324.77
27800	1SE	0;28	164.79	27800	16SE	13;33	0.05	27874	25AU	20;48	159.32	27874	9SE	5;22	353.36
27801	30SE	13;58	193.93	27801	16OC	0;19	29.34	27875	24SE	13;06	188.48	27875	8OC	15;41	22.52
27802	30OC	6;33	223.68	27802	14NO	10;57	59.06	27876	24OC	4;27	218.09	27876	7NO	4;17	52.23
27803	29NO	1;43	253.92	27803	13DE	21;46	89.04	27877	22NO	18;23	248.05	27877	6DE	19;26	82.36
27804	28DE	21;45	284.33					27878	22DE	6;53	278.14				

1248 / 1254

NEW				FULL				NEW				FULL			
				27804	12JA	8;46	119.05					27878	5JA	13;00	112.66
27805	27JA	16;25	314.56	27805	10FE	19;56	148.81	27879	20JA	18;09	308.11	27879	4FE	7;58	142.83
27806	26FE	8;13	344.31	27806	11MR	7;30	178.16	27880	19FE	4;19	337.72	27880	6MR	2;35	172.56
27807	26MR	20;51	13.45	27807	9AP	9;53	207.05	27881	20MR	13;35	6.82	27881	4AP	19;08	201.72
27808	25AP	6;49	42.00	27808	9MY	9;28	235.56	27882	18AP	22;22	35.43	27882	4MY	8;52	230.31
27809	24MY	14;58	70.13	27809	8JN	0;13	263.85	27883	18MY	7;26	63.67	27883	2JN	20;02	258.49
27810	22JN	22;06	98.06	27810	7JL	15;36	292.11	27884	16JN	17;47	91.74	27884	2JL	5;22	286.50
27811	22JL	5;07	126.05	27811	6AU	6;55	320.55	27885	16JL	6;15	119.91	27885	31JL	13;44	314.61
27812	20AU	12;57	154.34	27812	4SE	21;39	349.33	27886	14AU	21;08	148.42	27886	29AU	21;57	343.03
27813	18SE	22;37	183.15	27813	4OC	11;38	18.57	27887	13SE	13;58	177.41	27887	28SE	6;38	11.94
27814	18OC	11;02	212.55	27814	3NO	0;51	48.25	27888	13OC	7;42	206.94	27888	27OC	16;26	41.38
27815	17NO	2;40	242.52	27815	2DE	13;11	78.23	27889	12NO	1;09	236.92	27889	26NO	3;56	71.26
27816	16DE	20;58	272.84					27890	11DE	17;25	267.12	27890	25DE	17;36	101.41

1249 / 1255

NEW				FULL				NEW				FULL			
				27816	1JA	0;30	108.28	27891	10JA	7;50	297.28	27891	24JA	9;22	131.54
27817	15JA	16;25	303.18	27817	30JA	10;49	138.12	27892	8FE	20;03	327.11	27892	23FE	2;29	161.39
27818	14FE	11;10	333.21	27818	28FE	20;29	167.56	27893	10MR	6;05	356.43	27893	24MR	19;42	190.76
27819	16MR	3;51	2.70	27819	30MR	6;05	196.52	27894	8AP	14;24	25.18	27894	23AP	11;58	219.61
27820	14AP	17;55	31.59	27820	28AP	16;22	225.04	27895	7MY	21;49	53.47	27895	23MY	2;41	248.02
27821	14MY	5;20	59.96	27821	28MY	3;52	253.28	27896	6JN	5;27	81.49	27896	21JN	15;44	276.19
27822	12JN	14;32	88.00	27822	26JN	16;52	281.43	27897	5JL	14;22	109.49	27897	21JL	3;13	304.35
27823	11JL	22;15	115.97	27823	26JL	7;24	309.75	27898	4AU	1;25	137.73	27898	19AU	13;25	332.72
27824	10AU	5;29	144.12	27824	24AU	23;14	338.42	27899	2SE	15;03	166.42	27899	17SE	22;52	1.48
27825	8SE	13;21	172.68	27825	23SE	15;57	7.58	27900	2OC	7;16	195.70	27900	17OC	8;14	30.73
27826	7OC	22;48	201.78	27826	23OC	8;44	37.24	27901	1NO	1;32	225.55	27901	15NO	18;13	60.44
27827	6NO	10;31	231.45	27827	22NO	0;33	67.27	27902	30NO	20;42	255.80	27902	15DE	5;16	90.44
27828	6DE	0;42	261.53	27828	21DE	14;32	97.42	27903	30DE	15;11	286.15				

1256 / 1262

NUMBER	DATE	TIME	LONG.	NUMBER	DATE	TIME	LONG.	NUMBER	DATE	TIME	LONG.	NUMBER	DATE	TIME	LONG.
				27903	13JA	17;35	120.49					27977	7JA	7;50	114.54
27904	29JA	7;23	316.26	27904	12FE	7;03	150.34	27978	22JA	1;21	309.48	27978	6FE	2;57	144.69
27905	27FE	20;30	345.88	27905	12MR	21;23	179.78	27979	20FE	11;38	339.08	27979	7MR	20;13	174.34
27906	28MR	6;43	14.89	27906	11AP	12;21	208.75	27980	21MR	22;03	8.21	27980	6AP	10;32	203.38
27907	26AP	14;51	43.35	27907	11MY	3;44	237.31	27981	20AP	8;54	36.89	27981	5MY	21;42	231.84
27908	25MY	21;56	71.42	27908	9JN	19;07	265.62	27982	19MY	20;32	65.21	27982	4JN	6;24	259.92
27909	24JN	5;00	99.35	27909	9JL	9;53	293.86	27983	18JN	9;28	93.38	27983	3JL	13;45	287.85
27910	23JL	12;56	127.38	27910	7AU	23;29	322.24	27984	18JL	0;00	121.64	27984	1AU	20;55	315.92
27911	21AU	22;34	155.77	27911	6SE	11;46	350.95	27985	16AU	15;59	150.21	27985	31AU	4;53	344.35
27912	20SE	10;41	184.69	27912	5OC	23;04	20.10	27986	15SE	8;39	179.22	27986	29SE	14;22	13.31
27913	20OC	1;48	214.23	27913	4NO	9;56	49.70	27987	15OC	0;52	208.72	27987	29OC	1;52	42.84
27914	18NO	19;54	244.32	27914	3DE	20;46	79.63	27988	13NO	15;48	238.61	27988	27NO	15;44	72.84
27915	18DE	15;48	274.72					27989	13DE	5;09	268.70	27989	27DE	8;05	103.10

1257 / 1263

NUMBER	DATE	TIME	LONG.	NUMBER	DATE	TIME	LONG.	NUMBER	DATE	TIME	LONG.	NUMBER	DATE	TIME	LONG.
				27915	2JA	7;40	109.66	27990	11JA	17;02	298.74	27990	26JA	2;22	133.33
27916	17JA	11;23	305.06	27916	31JA	18;36	139.52	27991	10FE	3;39	328.49	27991	24FE	21;11	163.22
27917	16FE	4;40	335.00	27917	2MR	5;41	169.00	27992	11MR	13;09	357.77	27992	26MR	14;45	192.58
27918	17MR	18;46	4.36	27918	31MR	17;18	198.01	27993	9AP	21;53	26.52	27993	25AP	5;51	221.36
27919	16AP	5;54	33.10	27919	30AP	5;58	226.61	27994	9MY	6;29	54.84	27994	24MY	18;13	249.66
27920	15MY	14;48	61.35	27920	29MY	19;58	254.93	27995	7JN	15;53	82.92	27995	23JN	4;26	277.71
27921	13JN	22;20	89.32	27921	28JN	11;04	283.17	27996	7JL	3;07	111.02	27996	22JL	13;20	305.77
27922	13JL	5;21	117.27	27922	28JL	2;36	311.54	27997	5AU	16;48	139.37	27997	20AU	21;43	334.09
27923	11AU	12;44	145.44	27923	26AU	17;54	340.21	27998	4SE	8;53	168.19	27998	19SE	6;17	2.83
27924	9SE	21;27	174.05	27924	25SE	8;34	9.31	27999	4OC	2;31	197.56	27999	18OC	15;37	32.10
27925	9OC	8;30	203.25	27925	24OC	22;28	38.88	28000	2NO	20;59	227.42	28000	17NO	2;16	61.85
27926	7NO	22;34	233.03	27926	23NO	11;28	68.80	28001	2DE	13;38	257.60	28001	16DE	14;47	91.92
27927	7DE	15;40	263.25	27927	22DE	23;23	98.87								

1258 / 1264

NUMBER	DATE	TIME	LONG.	NUMBER	DATE	TIME	LONG.	NUMBER	DATE	TIME	LONG.	NUMBER	DATE	TIME	LONG.
27928	6JA	10;42	293.61	27928	21JA	10;08	128.81	28002	1JA	5;09	287.82	28002	15JA	5;25	122.07
27929	5FE	5;52	323.77	27929	19FE	19;55	158.39	28003	30JA	18;29	317.80	28003	13FE	21;45	152.01
27930	6MR	23;33	353.46	27930	21MR	5;14	187.50	28004	29FE	5;29	347.30	28004	14MR	14;47	181.54
27931	5AP	14;51	22.55	27931	19AP	14;50	216.13	28005	29MR	14;25	16.24	28005	13AP	7;24	210.55
27932	5MY	3;31	51.09	27932	19MY	1;26	244.41	28006	27AP	22;00	44.66	28006	12MY	22;50	239.10
27933	3JN	13;47	79.22	27933	17JN	13;30	272.56	28007	27MY	5;15	72.73	28007	11JN	12;44	267.33
27934	2JL	22;13	107.20	27934	17JL	3;14	300.78	28008	25JN	13;18	100.70	28008	11JL	1;07	295.48
27935	1AU	5;41	135.28	27935	15AU	18;33	329.31	28009	24JL	23;09	128.81	28009	9AU	12;08	323.78
27936	30AU	13;15	163.69	27936	14SE	11;07	358.31	28010	23AU	11;30	157.33	28010	7SE	22;09	352.42
27937	28SE	21;57	192.60	27937	14OC	4;15	27.83	28011	22SE	2;34	186.40	28011	7OC	7;42	21.52
27938	28OC	8;35	222.09	27938	12NO	20;57	57.78	28012	21OC	20;03	216.06	28012	5NO	17;28	51.09
27939	26NO	21;35	252.04	27939	12DE	12;06	87.95	28013	20NO	15;06	246.21	28013	5DE	4;00	81.02
27940	26DE	12;48	282.25					28014	20DE	10;14	276.58				

1259 / 1265

NUMBER	DATE	TIME	LONG.	NUMBER	DATE	TIME	LONG.	NUMBER	DATE	TIME	LONG.	NUMBER	DATE	TIME	LONG.
				27940	11JA	1;05	118.04					28014	3JA	15;37	111.07
27941	25JA	5;41	312.43	27941	9FE	11;59	147.81	28015	19JA	3;46	306.82	28015	2FE	4;21	140.99
27942	23FE	23;22	342.29	27942	10MR	21;17	177.08	28016	17FE	18;25	336.64	28016	3MR	18;00	170.56
27943	25MR	16;51	11.66	27943	9AP	5;43	205.83	28017	19MR	5;55	5.86	28017	2AP	8;22	199.66
27944	24AP	9;05	40.49	27944	8MY	22;38	234.15	28018	17AP	14;52	34.49	28018	1MY	23;20	228.33
27945	23MY	23;13	68.87	27945	6JN	22;38	262.21	28019	16MY	22;15	62.67	28019	31MY	14;40	256.69
27946	22JN	10;59	96.99	27946	6JL	8;36	290.25	28020	15JN	5;09	90.61	28020	30JN	5;51	284.94
27947	21JL	20;49	125.08	27947	4AU	20;35	318.53	28021	14JL	12;32	118.57	28021	29JL	20;15	313.27
27948	20AU	5;36	153.40	27948	3SE	11;12	347.28	28022	12AU	21;16	146.81	28022	28AU	9;28	341.88
27949	18SE	14;19	182.15	27949	3OC	4;26	16.61	28023	11SE	8;07	175.53	28023	26SE	21;29	10.90
27950	17OC	23;46	211.41	27950	1NO	23;19	46.50	28024	10OC	21;47	204.86	28024	26OC	8;45	40.38
27951	16NO	10;18	241.15	27951	1DE	18;03	76.74	28025	9NO	14;32	234.77	28025	24NO	19;44	70.23
27952	15DE	22;04	271.18	27952	31DE	10;56	107.02	28026	9DE	9;47	265.09	28026	24DE	6;38	100.26

1260 / 1266

NUMBER	DATE	TIME	LONG.	NUMBER	DATE	TIME	LONG.	NUMBER	DATE	TIME	LONG.	NUMBER	DATE	TIME	LONG.
27953	14JA	11;07	301.26	27953	30JA	1;12	137.04	28027	8JA	5;47	295.49	28027	22JA	17;28	130.19
27954	13FE	1;33	331.14	27954	28FE	12;55	166.59	28028	7FE	0;21	325.61	28028	21FE	4;14	159.80
27955	13MR	17;16	0.63	27955	28MR	22;31	195.56	28029	8MR	15;59	355.19	28029	22MR	15;15	188.95
27956	12AP	9;41	29.64	27956	27AP	6;34	224.01	28030	7AP	4;27	24.14	28030	21AP	3;03	217.65
27957	12MY	1;48	58.22	27957	26MY	13;50	252.09	28031	6MY	14;19	52.54	28031	20MY	16;08	246.02
27958	10JN	16;42	86.50	27958	24JN	21;06	280.02	28032	4JN	22;24	80.58	28032	19JN	6;37	274.25
27959	10JL	6;00	114.69	27959	24JL	5;28	308.08	28033	4JL	5;36	108.51	28033	18JL	22;03	302.55
27960	8AU	17;56	143.01	27960	22AU	16;07	336.51	28034	2AU	12;46	136.59	28034	17AU	13;43	331.12
27961	7SE	4;59	171.68	27961	21SE	5;53	5.51	28035	31AU	20;49	165.03	28035	16SE	5;01	0.09
27962	6OC	15;38	200.82	27962	20OC	22;50	35.14	28036	30SE	6;41	194.02	28036	15OC	19;37	29.53
27963	5NO	2;08	230.42	27963	19NO	17;56	65.29	28037	29OC	19;16	223.60	28037	14NO	9;20	59.38
27964	4DE	12;38	260.34	27964	19DE	13;28	95.67	28038	28NO	10;56	253.68	28038	13DE	21;57	89.44
								28039	28DE	5;07	284.02				

1261 / 1267

NUMBER	DATE	TIME	LONG.	NUMBER	DATE	TIME	LONG.	NUMBER	DATE	TIME	LONG.	NUMBER	DATE	TIME	LONG.
27965	2JA	23;24	290.35	27965	18JA	7;43	125.95					28039	12JA	9;18	119.46
27966	1FE	10;45	320.22	27966	16FE	23;36	155.82	28040	27JA	0;18	314.28	28040	10FE	19;24	149.18
27967	2MR	23;03	349.74	27967	18MR	12;40	185.12	28041	25FE	18;44	344.14	28041	12MR	4;40	178.43
27968	1AP	12;27	18.82	27968	16AP	22;58	213.82	28042	27MR	11;09	13.44	28042	10AP	13;48	207.20
27969	1MY	2;43	47.47	27969	16MY	7;04	242.03	28043	26AP	1;04	42.16	28043	9MY	23;34	235.56
27970	30MY	17;31	75.82	27970	14JN	13;52	269.97	28044	25MY	12;29	70.42	28044	8JN	10;41	263.70
27971	29JN	8;28	104.05	27971	13JL	20;34	297.91	28045	23JN	21;49	98.43	28045	7JL	23;29	291.86
27972	28JL	23;19	132.40	27972	12AU	4;20	326.10	28046	23JL	5;46	126.46	28046	6AU	14;03	320.26
27973	27AU	13;49	161.05	27973	10SE	14;14	354.77	28047	21AU	13;19	154.75	28047	5SE	6;12	349.09
27974	26SE	3;36	190.13	27974	10OC	2;56	24.04	28048	19SE	21;30	183.49	28048	4OC	23;25	18.45
27975	25OC	16;23	219.66	27975	8NO	18;32	53.90	28049	19OC	7;13	212.78	28049	3NO	16;45	48.30
27976	24NO	4;06	249.54	27976	8DE	12;32	84.16	28050	17NO	19;04	242.59	28050	3DE	9;00	78.45
27977	23DE	14;57	279.56					28051	17DE	9;11	272.73				

---------NEW MOONS---------				---------FULL MOONS---------			
NUMBER	DATE	TIME	LONG.	NUMBER	DATE	TIME	LONG.

1268

NUMBER	DATE	TIME	LONG.	NUMBER	DATE	TIME	LONG.
				28051	1JA	23;12	108.62
28052	16JA	1;10	302.93	28052	31JA	11;04	138.52
28053	14FE	18;20	332.90	28053	29FE	20;58	167.97
28054	15MR	11;48	2.43	28054	30MR	5;34	196.89
28055	14AP	4;37	31.44	28055	28AP	13;37	225.32
28056	13MY	19;48	59.96	28056	27MY	21;49	253.42
28057	12JN	8;48	88.16	28057	26JN	6;57	281.43
28058	11JL	19;40	116.25	28058	25JL	17;47	309.59
28059	10AU	5;02	144.49	28059	24AU	7;05	338.15
28060	8SE	13;54	173.09	28060	22SE	23;11	7.28
28061	7OC	23;07	202.19	28061	22OC	17;37	37.00
28062	6NO	9;14	231.79	28062	21NO	12;51	67.16
28063	5DE	20;25	261.74	28063	21DE	6;57	97.49

1269

NUMBER	DATE	TIME	LONG.	NUMBER	DATE	TIME	LONG.
28064	4JA	8;44	291.83	28064	19JA	22;36	127.65
28065	2FE	22;15	321.77	28065	18FE	11;30	157.38
28066	4MR	13;05	351.38	28066	19MR	21;59	186.55
28067	3AP	4;57	20.53	28067	18AP	6;37	215.16
28068	2MY	21;08	49.24	28068	17MY	14;04	243.33
28069	1JN	12;40	77.60	28069	15JN	21;06	271.28
28070	1JL	2;51	105.80	28070	15JL	4;45	299.25
28071	30JL	15;37	134.08	28071	13AU	14;09	327.52
28072	29AU	3;18	162.64	28072	12SE	2;21	356.31
28073	27SE	14;23	191.63	28073	11OC	17;52	25.73
28074	27OC	1;09	221.10	28074	10NO	12;11	55.72
28075	25NO	11;47	250.94	28075	10DE	7;48	86.06
28076	24DE	22;23	280.96				

1270

NUMBER	DATE	TIME	LONG.	NUMBER	DATE	TIME	LONG.
				28076	9JA	2;53	116.42
28077	23JA	9;17	310.89	28077	7FE	20;01	146.46
28078	21FE	20;50	340.52	28078	9MR	10;26	175.98
28079	23MR	9;23	9.72	28079	7AP	21;59	204.88
28080	21AP	22;58	38.48	28080	7MY	7;01	233.24
28081	21MY	13;20	66.90	28081	5JN	14;18	261.25
28082	20JN	4;10	95.14	28082	4JL	20;55	289.15
28083	19JL	19;11	123.43	28083	3AU	4;03	317.23
28084	18AU	10;06	151.97	28084	1SE	12;52	345.72
28085	17SE	0;36	180.93	28085	1OC	0;13	14.78
28086	16OC	14;13	210.34	28086	30OC	14;27	44.44
28087	15NO	2;41	240.14	28087	29NO	7;22	74.58
28088	14DE	14;02	270.16	28088	29DE	2;08	104.94

1271

NUMBER	DATE	TIME	LONG.	NUMBER	DATE	TIME	LONG.
28089	13JA	0;35	300.14	28089	27JA	21;27	135.20
28090	11FE	10;45	329.85	28090	26FE	15;42	165.04
28091	12MR	20;51	359.13	28091	28MR	7;28	194.30
28092	11AP	7;10	27.93	28092	26AP	20;08	222.96
28093	10MY	18;06	56.33	28093	26MY	5;57	251.14
28094	9JN	6;08	84.50	28094	24JN	13;52	279.11
28095	8JL	19;46	112.70	28095	23JL	21;05	307.11
28096	7AU	11;10	141.14	28096	22AU	4;41	335.40
28097	6SE	3;49	170.01	28097	20SE	13;28	4.18
28098	5OC	20;41	199.37	28098	20OC	0;02	33.52
28099	4NO	12;37	229.17	28099	18NO	12;43	63.37
28100	4DE	2;58	259.24	28100	18DE	3;48	93.55

1272

NUMBER	DATE	TIME	LONG.	NUMBER	DATE	TIME	LONG.
28101	2JA	15;38	289.34	28101	16JA	21;05	123.80
28102	1FE	2;49	319.22	28102	15FE	15;38	153.82
28103	1MR	12;42	348.66	28103	16MR	9;50	183.37
28104	30MR	21;34	17.57	28104	15AP	2;08	212.34
28105	29AP	5;55	46.00	28105	14MY	15;48	240.79
28106	28MY	14;36	74.12	28106	13JN	3;03	268.92
28107	27JN	0;40	102.16	28107	12JL	12;38	296.96
28108	26JL	13;03	130.39	28108	10AU	21;49	325.18
28109	25AU	4;03	159.03	28109	9SE	5;58	353.78
28110	23SE	21;13	188.22	28110	8OC	15;01	22.88
28111	23OC	15;24	217.94	28111	7NO	1;04	52.48
28112	22NO	9;18	248.06	28112	6DE	12;37	82.46
28113	22DE	1;52	278.32				

1273

NUMBER	DATE	TIME	LONG.	NUMBER	DATE	TIME	LONG.
				28113	5JA	2;05	112.59
28114	20JA	16;22	308.42	28114	3FE	17;26	142.61
28115	19FE	4;27	338.11	28115	5MR	9;59	172.27
28116	20MR	14;14	7.24	28116	4AP	2;41	201.44
28117	18AP	22;14	35.82	28117	3MY	18;36	230.13
28118	18MY	5;22	63.97	28118	2JN	9;26	258.45
28119	16JN	12;46	91.93	28119	1JL	22;30	286.62
28120	15JL	21;33	119.95	28120	31JL	10;22	314.86
28121	14AU	8;37	148.30	28121	29AU	21;05	343.39
28122	12SE	22;24	177.16	28122	28SE	7;02	12.35
28123	12OC	14;51	206.63	28123	27OC	16;49	41.79
28124	11NO	9;24	236.64	28124	26NO	3;01	71.62
28125	11DE	4;49	266.98	28125	25DE	14;04	101.65

1274

NUMBER	DATE	TIME	LONG.	NUMBER	DATE	TIME	LONG.
28126	9JA	23;24	297.32	28126	24JA	2;07	131.63
28127	8FE	15;34	327.31	28127	22FE	15;05	161.31
28128	10MR	4;33	356.75	28128	24MR	4;48	190.54
28129	8AP	14;34	25.58	28129	22AP	19;12	219.33
28130	7MY	22;30	53.89	28130	22MY	10;12	247.76
28131	6JN	5;27	81.88	28131	21JN	1;30	276.02
28132	5JL	12;27	109.80	28132	20JL	16;30	304.31
28133	3AU	20;27	137.92	28133	19AU	6;35	332.83
28134	2SE	6;13	166.46	28134	17SE	19;28	1.73
28135	1OC	18;30	195.56	28135	17OC	7;20	31.08
28136	31OC	9;47	225.28	28136	15NO	18;36	60.85
28137	30NO	3;58	255.48	28137	15DE	5;37	90.85
28138	29DE	23;51	285.90				

1275

NUMBER	DATE	TIME	LONG.	NUMBER	DATE	TIME	LONG.
				28138	13JA	16;27	120.84
28139	28JA	19;20	316.15	28139	12FE	3;05	150.56
28140	27FE	12;27	345.93	28140	13MR	13;40	179.85
28141	29MR	2;23	15.10	28141	12AP	0;42	208.67
28142	27AP	13;23	43.68	28142	11MY	12;50	237.11
28143	26MY	22;13	71.82	28143	10JN	2;28	265.34
28144	25JN	5;47	99.77	28144	9JL	17;28	293.59
28145	24JL	12;57	127.77	28145	8AU	9;14	322.06
28146	22AU	20;33	156.08	28146	7SE	1;02	350.90
28147	21SE	5;30	184.87	28147	6OC	16;18	20.21
28148	20OC	16;44	214.25	28148	5NO	6;46	49.96
28149	19NO	6;53	244.17	28149	4DE	20;09	80.00
28150	18DE	23;54	274.44				

1276

NUMBER	DATE	TIME	LONG.	NUMBER	DATE	TIME	LONG.
				28150	3JA	8;11	110.07
28151	17JA	18;41	304.74	28151	1FE	18;49	139.92
28152	16FE	13;31	334.76	28152	2MR	4;16	169.33
28153	17MR	6;53	4.26	28153	31MR	13;08	198.23
28154	15AP	21;59	33.17	28154	29AP	22;15	226.69
28155	15MY	10;36	61.57	28155	29MY	8;25	254.86
28156	13JN	20;58	89.66	28156	27JN	20;13	282.97
28157	13JL	5;37	117.67	28157	27JL	9;54	311.26
28158	11AU	13;23	145.86	28158	26AU	1;24	339.93
28159	9SE	21;18	174.44	28159	24SE	18;23	9.12
28160	9OC	6;19	203.55	28160	24OC	12;04	38.84
28161	7NO	17;09	233.19	28161	23NO	5;14	68.93
28162	7DE	6;09	263.23	28162	22DE	20;40	99.15

1277

NUMBER	DATE	TIME	LONG.	NUMBER	DATE	TIME	LONG.
28163	5JA	21;09	293.43	28163	21JA	9;43	129.19
28164	4FE	13;37	323.48	28164	19FE	20;26	158.81
28165	6MR	6;47	353.16	28165	21MR	5;27	187.89
28166	4AP	23;49	22.33	28166	19AP	13;30	216.46
28167	4MY	15;48	51.01	28167	18MY	21;22	244.64
28168	3JN	5;57	79.30	28168	17JN	5;49	272.64
28169	2JL	17;57	107.42	28169	16JL	15;38	300.71
28170	1AU	4;09	135.60	28170	15AU	3;38	329.10
28171	30AU	13;23	164.08	28171	13SE	18;25	358.02
28172	28SE	22;34	193.02	28172	13OC	11;58	27.54
28173	28OC	8;22	222.47	28173	12NO	7;13	57.59
28174	26NO	19;05	252.33	28174	12DE	2;15	87.92
28175	26DE	6;47	282.39				

1278

NUMBER	DATE	TIME	LONG.	NUMBER	DATE	TIME	LONG.
				28175	10JA	19;18	118.19
28176	24JA	19;31	312.39	28176	9FE	9;32	148.10
28177	23FE	9;26	342.10	28177	10MR	21;03	177.46
28178	25MR	0;32	11.38	28178	9AP	6;25	206.25
28179	23AP	16;25	40.21	28179	8MY	14;16	234.55
28180	23MY	8;15	68.67	28180	6JN	21;19	262.54
28181	21JN	23;11	96.91	28181	6JL	4;29	290.47
28182	21JL	12;48	125.15	28182	4AU	12;53	318.61
28183	20AU	1;13	153.62	28183	2SE	23;38	347.20
28184	18SE	12;50	182.48	28184	2OC	13;33	16.38
28185	18OC	0;00	211.81	28185	1NO	6;41	46.19
28186	16NO	10;51	241.56	28186	1DE	1;58	76.45
28187	15DE	21;30	271.55	28187	30DE	21;33	106.84

1279

NUMBER	DATE	TIME	LONG.	NUMBER	DATE	TIME	LONG.
28188	14JA	8;08	301.53	28188	29JA	15;46	137.04
28189	12FE	19;07	331.26	28189	28FE	7;31	166.75
28190	14MR	6;53	0.59	28190	29MR	20;24	195.87
28191	12AP	19;40	29.47	28191	28AP	6;33	224.40
28192	12MY	9;25	57.96	28192	27MY	14;34	252.51
28193	10JN	23;55	86.23	28193	25JN	21;22	280.43
28194	10JL	14;53	114.48	28194	25JL	4;09	308.41
28195	9AU	6;03	142.93	28195	23AU	12;05	336.74
28196	7SE	21;05	171.75	28196	21SE	22;11	5.59
28197	7OC	11;30	201.04	28197	21OC	11;03	35.04
28198	6NO	0;49	230.75	28198	20NO	2;43	65.04
28199	5DE	12;52	260.74	28199	19DE	20;40	95.35

1280

NUMBER	DATE	TIME	LONG.	NUMBER	DATE	TIME	LONG.
28200	3JA	23;48	290.77	28200	18JA	15;48	125.67
28201	2FE	10;01	320.58	28201	17FE	10;38	155.67
28202	2MR	19;55	350.01	28202	18MR	3;40	185.14
28203	1AP	5;50	18.94	28203	16AP	17;49	214.00
28204	30AP	16;09	47.44	28204	16MY	4;57	242.34
28205	30MY	3;21	75.65	28205	14JN	13;44	270.36
28206	28JN	16;01	103.79	28206	13JL	21;14	298.32
28207	28JL	6;33	132.12	28207	12AU	4;39	326.50
28208	26AU	22;50	160.83	28208	10SE	12;55	355.11
28209	25SE	16;00	190.04	28209	9OC	22;40	24.26
28210	25OC	8;49	219.74	28210	8NO	10;20	53.95
28211	24NO	0;13	249.77	28211	8DE	0;11	84.03
28212	23DE	13;50	279.91				

1281

NUMBER	DATE	TIME	LONG.	NUMBER	DATE	TIME	LONG.
				28212	6JA	16;19	114.27
28213	22JA	1;44	309.89	28213	5FE	10;12	144.38
28214	20FE	12;08	339.49	28214	7MR	4;34	174.09
28215	21MR	21;17	8.57	28215	5AP	21;47	203.25
28216	20AP	5;36	37.14	28216	5MY	12;43	231.87
28217	19MY	13;49	65.33	28217	4JN	1;08	260.09
28218	17JN	22;55	93.35	28218	3JL	11;33	288.15
28219	17JL	9;59	121.47	28219	1AU	20;47	316.30
28220	15AU	23;41	149.94	28220	31AU	5;35	344.77
28221	14SE	16;01	178.93	28221	29SE	14;33	13.71
28222	14OC	10;03	208.49	28222	29OC	0;12	43.16
28223	13NO	4;28	238.52	28223	27NO	10;59	73.03
28224	12DE	21;58	268.79	28224	26DE	23;24	103.12

1282

NUMBER	DATE	TIME	LONG.	NUMBER	DATE	TIME	LONG.
28225	11JA	13;38	298.99	28225	25JA	13;40	133.18
28226	10FE	2;55	328.85	28226	24FE	5;28	162.96
28227	11MR	13;42	358.18	28227	25MR	21;56	192.28
28228	9AP	22;21	26.93	28228	24AP	14;07	221.11
28229	9MY	5;39	55.20	28229	24MY	5;22	249.54
28230	7JN	12;40	83.18	28230	22JN	19;22	277.74
28231	6JL	20;34	111.14	28231	22JL	8;05	305.95
28232	5AU	6;24	139.35	28232	20AU	19;35	334.39
28233	3SE	18;51	168.01	28233	19SE	6;07	3.22
28234	3OC	10;06	197.27	28234	18OC	16;09	32.52
28235	2NO	3;51	227.11	28235	17NO	2;13	62.24
28236	1DE	23;08	257.37	28236	16DE	12;50	92.23
28237	31DE	18;25	287.76				

1283

NUMBER	DATE	TIME	LONG.	NUMBER	DATE	TIME	LONG.
				28237	15JA	0;17	122.24
28238	30JA	11;56	317.91	28238	13FE	12;36	152.02
28239	1MR	2;28	347.57	28239	15MR	1;40	181.39
28240	30MR	13;48	16.60	28240	13AP	15;26	210.30
28241	28AP	22;33	45.07	28241	13MY	5;55	238.81
28242	28MY	5;47	73.14	28242	11JN	21;01	267.10
28243	26JN	12;37	101.06	28243	11JL	12;19	295.36
28244	25JL	20;03	129.08	28244	10AU	3;08	323.81
28245	24AU	4;54	157.45	28245	8SE	16;55	352.59
28246	22SE	15;56	186.35	28246	8OC	5;32	21.82
28247	22OC	5;45	215.86	28247	6NO	17;18	51.48
28248	20NO	22;36	245.91	28248	6DE	4;33	81.43
28249	20DE	17;51	276.28				

1284

NUMBER	DATE	TIME	LONG.	NUMBER	DATE	TIME	LONG.
				28249	4JA	15;29	111.46
28250	19JA	13;45	306.62	28250	3FE	2;05	141.29
28251	18FE	8;08	336.60	28251	3MR	12;20	170.72
28252	18MR	23;36	5.99	28252	1AP	22;53	199.67
28253	17AP	11;55	34.76	28253	1MY	10;08	228.19
28254	16MY	21;42	63.04	28254	30MY	22;46	256.45
28255	15JN	5;49	91.03	28255	29JN	13;03	284.66
28256	14JL	13;08	118.99	28256	29JL	4;35	313.03
28257	12AU	20;31	147.17	28257	27AU	20;39	341.75
28258	11SE	4;49	175.79	28258	26SE	12;31	10.92
28259	10OC	14;54	204.97	28259	26OC	3;43	40.56
28260	9NO	3;37	234.71	28260	24NO	17;53	70.55
28261	8DE	19;15	264.87	28261	24DE	6;44	100.65

1285

NUMBER	DATE	TIME	LONG.	NUMBER	DATE	TIME	LONG.
28262	7JA	13;13	295.19	28262	22JA	18;02	130.61
28263	6FE	8;04	325.32	28263	21FE	3;53	160.17
28264	8MR	2;08	355.01	28264	22MR	12;45	189.24
28265	6AP	18;18	24.11	28265	20AP	21;25	217.81
28266	6MY	8;05	52.68	28266	20MY	6;45	246.03
28267	4JN	19;34	80.86	28267	18JN	17;31	274.12
28268	4JL	5;05	108.88	28268	18JL	6;10	302.31
28269	2AU	13;21	137.00	28269	16AU	20;51	330.83
28270	31AU	21;15	165.44	28270	15SE	12;31	359.84
28271	30SE	5;46	194.37	28271	15OC	7;01	29.40
28272	29OC	15;45	223.85	28272	14NO	0;51	59.41
28273	28NO	3;42	253.77	28273	13DE	17;28	89.64
28274	27DE	17;41	283.93				

1286

NUMBER	DATE	TIME	LONG.	NUMBER	DATE	TIME	LONG.
				28274	12JA	7;48	119.79
28275	26JA	9;18	314.04	28275	10FE	19;34	149.58
28276	25FE	1;57	343.84	28276	12MR	5;13	178.85
28277	26MR	18;54	13.17	28277	10AP	13;29	207.57
28278	25AP	11;22	42.00	28278	9MY	21;12	235.85
28279	25MY	2;27	70.41	28279	8JN	5;07	263.87
28280	23JN	15;36	98.58	28280	7JL	14;05	291.87
28281	23JL	2;48	126.73	28281	6AU	0;53	320.12
28282	21AU	12;38	155.11	28282	4SE	14;17	348.83
28283	19SE	21;59	183.90	28283	4OC	6;38	18.15
28284	19OC	7;37	213.19	28284	3NO	1;24	48.05
28285	17NO	17;59	242.94	28285	2DE	20;57	78.33
28286	17DE	5;12	272.96				

1287

NUMBER	DATE	TIME	LONG.	NUMBER	DATE	TIME	LONG.
				28286	1JA	15;14	108.67
28287	15JA	17;18	302.99	28287	31JA	6;55	138.74
28288	14FE	6;22	332.79	28288	1MR	19;40	168.31
28289	15MR	20;36	2.20	28289	31MR	5;56	197.29
28290	14AP	11;52	31.15	28290	29AP	14;20	225.74
28291	14MY	3;39	59.71	28291	28MY	21;36	253.80
28292	12JN	19;04	88.00	28292	27JN	4;32	281.72
28293	12JL	9;28	116.24	28293	26JL	12;11	309.76
28294	10AU	22;40	144.62	28294	24AU	21;40	338.16
28295	9SE	10;55	173.36	28295	23SE	10;01	7.13
28296	8OC	22;33	202.56	28296	23OC	1;42	36.72
28297	7NO	9;47	232.20	28297	21NO	20;10	66.85
28298	6DE	20;38	262.15	28298	21DE	15;52	97.24

1288

NUMBER	DATE	TIME	LONG.	NUMBER	DATE	TIME	LONG.
28299	5JA	7;13	292.16	28299	20JA	10;56	127.54
28300	3FE	17;49	321.98	28300	19FE	3;55	157.45
28301	4MR	4;54	351.43	28301	19MR	18;10	186.78
28302	2AP	16;51	20.43	28302	18AP	5;34	215.50
28303	2MY	5;52	49.01	28303	17MY	14;31	243.74
28304	31MY	19;50	77.32	28304	15JN	21;47	271.69
28305	30JN	10;33	105.54	28305	15JL	4;28	299.63
28306	30JL	1;45	133.91	28306	13AU	11;46	327.82
28307	28AU	17;09	162.61	28307	11SE	20;47	356.48
28308	27SE	8;15	191.77	28308	11OC	8;19	25.72
28309	26OC	22;27	221.38	28309	9NO	22;41	55.54
28310	25NO	11;20	251.32	28310	9DE	15;34	85.76
28311	24DE	22;51	281.37				

1289

NUMBER	DATE	TIME	LONG.	NUMBER	DATE	TIME	LONG.
				28311	8JA	10;11	116.10
28312	23JA	9;20	311.28	28312	7FE	5;13	146.24
28313	21FE	19;12	340.84	28313	8MR	23;11	175.90
28314	23MR	4;50	9.92	28314	7AP	14;45	204.97
28315	21AP	14;38	38.53	28315	7MY	3;19	233.48
28316	21MY	1;06	66.80	28316	5JN	13;11	261.59
28317	19JN	12;49	94.92	28317	4JL	21;16	289.56
28318	19JL	2;20	123.14	28318	3AU	4;43	317.65
28319	17AU	17;54	151.71	28319	1SE	12;37	346.10
28320	16SE	10;59	180.76	28320	30SE	21;43	15.07
28321	16OC	4;24	210.32	28321	30OC	8;29	44.58
28322	14NO	20;52	240.29	28322	28NO	21;15	74.54
28323	14DE	11;33	270.44	28323	28DE	12;10	104.74

1290

NUMBER	DATE	TIME	LONG.	NUMBER	DATE	TIME	LONG.
28324	13JA	0;19	300.52	28324	27JA	5;05	134.90
28325	11FE	11;22	330.27	28325	25FE	23;10	164.75
28326	12MR	20;57	359.52	28326	27MR	16;56	194.10
28327	11AP	5;26	28.25	28327	26AP	8;58	222.90
28328	10MY	13;24	56.53	28328	25MY	22;37	251.24
28329	8JN	21;46	84.56	28329	24JN	10;02	279.34
28330	8JL	7;38	112.60	28330	23JL	19;56	307.45
28331	6AU	19;58	140.92	28331	22AU	5;04	335.81
28332	5SE	11;07	169.72	28332	20SE	14;06	4.60
28333	5OC	4;37	199.09	28333	19OC	23;32	33.89
28334	3NO	23;14	229.00	28334	18NO	9;47	63.64
28335	3DE	17;30	259.23	28335	17DE	21;19	93.67

1291

NUMBER	DATE	TIME	LONG.	NUMBER	DATE	TIME	LONG.
28336	2JA	10;17	289.51	28336	16JA	10;32	123.75
28337	1FE	0;47	319.52	28337	15FE	1;24	153.62
28338	2MR	12;43	349.05	28338	16MR	17;21	183.08
28339	31MR	22;14	17.99	28339	15AP	9;31	212.06
28340	30AP	5;58	46.39	28340	15MY	1;09	240.60
28341	29MY	12;52	74.44	28341	13JN	15;56	268.86
28342	27JN	20;07	102.37	28342	13JL	5;16	297.06
28343	27JL	4;51	130.45	28343	11AU	17;35	325.42
28344	25AU	15;58	158.93	28344	10SE	4;51	354.13
28345	24SE	5;55	187.98	28345	9OC	15;19	23.29
28346	23OC	22;36	217.63	28346	8NO	1;29	52.90
28347	22NO	17;22	247.78	28347	7DE	11;52	82.82
28348	22DE	12;56	278.16				

1292

NUMBER	DATE	TIME	LONG.	NUMBER	DATE	TIME	LONG.
				28348	5JA	22;50	112.85
28349	21JA	7;33	308.44	28349	4FE	10;33	142.71
28350	19FE	23;37	338.30	28350	4MR	23;00	172.21
28351	20MR	12;26	7.56	28351	3AP	12;06	201.24
28352	18AP	22;16	36.21	28352	3MY	1;57	229.85
28353	18MY	6;03	64.39	28353	1JN	16;37	258.18
28354	16JN	12;56	92.33	28354	1JL	7;53	286.42
28355	15JL	19;58	120.28	28355	30JL	23;10	314.80
28356	14AU	4;05	148.51	28356	29AU	13;47	343.48
28357	12SE	14;01	177.21	28357	28SE	3;17	12.59
28358	12OC	2;28	206.51	28358	27OC	15;42	42.13
28359	10NO	17;52	236.38	28359	26NO	3;21	72.02
28360	10DE	12;04	266.66	28360	25DE	14;29	102.06

1293

NUMBER	DATE	TIME	LONG.	NUMBER	DATE	TIME	LONG.
28361	9JA	7;52	297.05	28361	24JA	1;11	131.98
28362	8FE	3;09	327.18	28362	22FE	11;27	161.55
28363	9MR	20;03	356.79	28363	23MR	21;31	190.64
28364	8AP	9;49	25.78	28364	22AP	8;01	219.26
28365	7MY	20;44	54.21	28365	21MY	19;39	247.57
28366	6JN	5;35	82.27	28366	20JN	8;58	275.75
28367	5JL	13;15	110.22	28367	19JL	23;56	304.04
28368	3AU	20;37	138.32	28368	18AU	15;59	332.63
28369	2SE	4;28	166.78	28369	17SE	8;18	1.67
28370	1OC	13;41	195.76	28370	17OC	0;10	31.18
28371	31OC	1;06	225.31	28371	15NO	15;09	61.09
28372	29NO	15;17	255.34	28372	15DE	4;50	91.21
28373	29DE	8;08	285.62				

1294

NUMBER	DATE	TIME	LONG.	NUMBER	DATE	TIME	LONG.
				28373	13JA	16;57	121.25
28374	28JA	2;35	315.84	28374	12FE	3;24	150.97
28375	26FE	21;01	345.69	28375	13MR	12;30	180.19
28376	28MR	14;04	14.99	28376	11AP	20;56	208.91
28377	27AP	4;58	43.73	28377	11MY	5;37	237.21
28378	26MY	17;35	72.03	28378	9JN	15;24	265.29
28379	25JN	4;06	100.09	28379	9JL	2;59	293.41
28380	24JL	13;02	128.17	28380	7AU	16;40	321.78
28381	22AU	21;11	156.49	28381	6SE	8;25	350.62
28382	21SE	5;29	185.26	28382	6OC	1;49	20.00
28383	20OC	14;48	214.56	28383	4NO	19;59	49.90
28384	19NO	1;48	244.34	28384	4DE	13;33	80.11
28385	18DE	14;45	274.44				

1295

NUMBER	DATE	TIME	LONG.	NUMBER	DATE	TIME	LONG.
				28385	3JA	5;13	110.34
28386	17JA	5;28	304.58	28386	1FE	18;14	140.28
28387	15FE	21;26	334.48	28387	3MR	4;46	169.74
28388	17MR	14;03	3.96	28388	1AP	13;29	198.64
28389	16AP	6;39	32.94	28389	30AP	21;12	227.04
28390	15MY	22;25	61.48	28390	30MY	4;47	255.11
28391	14JN	12;36	89.72	28391	28JN	13;02	283.08
28392	14JL	0;54	117.87	28392	27JL	22;47	311.20
28393	12AU	11;33	146.17	28393	26AU	10;50	339.73
28394	10SE	21;17	174.83	28394	25SE	1;48	8.83
28395	10OC	6;55	203.97	28395	24OC	19;39	38.54
28396	8NO	17;03	233.58	28396	23NO	15;13	68.73
28397	8DE	3;54	263.53	28397	23DE	10;28	99.11

1296

NUMBER	DATE	TIME	LONG.	NUMBER	DATE	TIME	LONG.
28398	6JA	15;29	293.58	28398	22JA	3;34	129.32
28399	5FE	3;51	323.47	28399	20FE	17;43	159.08
28400	5MR	17;12	352.99	28400	21MR	5;03	188.27
28401	4AP	7;41	22.07	28401	19AP	14;12	216.88
28402	3MY	23;03	50.72	28402	18MY	21;50	245.05
28403	2JN	14;39	79.08	28403	17JN	4;47	272.99
28404	2JL	5;39	107.32	28404	16JL	11;57	300.95
28405	31JL	19;38	135.65	28405	14AU	20;25	329.20
28406	30AU	8;35	164.28	28406	13SE	7;17	357.95
28407	28SE	20;48	193.34	28407	12OC	21;23	27.33
28408	28OC	8;28	222.87	28408	11NO	14;40	57.29
28409	26NO	19;39	252.75	28409	11DE	10;01	87.62
28410	26DE	6;22	282.76				

1297

NUMBER	DATE	TIME	LONG.	NUMBER	DATE	TIME	LONG.
				28410	10JA	5;36	118.00
28411	24JA	16;49	312.67	28411	8FE	23;41	148.07
28412	23FE	3;24	342.24	28412	10MR	15;16	177.61
28413	24MR	14;36	11.36	28413	9AP	3;59	206.54
28414	23AP	2;47	40.05	28414	8MY	14;02	234.93
28415	22MY	16;04	68.41	28415	6JN	22;00	262.96
28416	21JN	6;19	96.63	28416	6JL	4;53	290.88
28417	20JL	21;21	124.93	28417	4AU	11;49	318.96
28418	19AU	12;53	153.51	28418	2SE	19;57	347.44
28419	18SE	4;29	182.53	28419	2OC	6;16	16.48
28420	17OC	19;31	212.02	28420	31OC	19;17	46.10
28421	16NO	9;20	241.89	28421	30NO	10;59	76.20
28422	15DE	21;39	271.95	28422	30DE	4;48	106.53

1298

NUMBER	DATE	TIME	LONG.	NUMBER	DATE	TIME	LONG.
28423	14JA	8;35	301.94	28423	28JA	23;39	136.75
28424	12FE	18;35	331.63	28424	27FE	18;11	166.59
28425	14MR	4;05	0.86	28425	29MR	10;57	195.87
28426	12AP	13;31	29.61	28426	28AP	0;58	224.56
28427	11MY	23;21	57.95	28427	27MY	12;07	252.80
28428	10JN	10;10	86.07	28428	25JN	21;01	280.79
28429	9JL	22;39	114.22	28429	25JL	4;46	308.83
28430	8AU	13;14	142.64	28430	23AU	12;30	337.14
28431	7SE	5;49	171.52	28431	21SE	21;05	5.93
28432	6OC	23;30	200.93	28432	21OC	7;06	35.27
28433	5NO	16;51	230.81	28433	19NO	18;54	65.10
28434	5DE	8;41	260.95	28434	19DE	8;40	95.23

1299

NUMBER	DATE	TIME	LONG.	NUMBER	DATE	TIME	LONG.
28435	3JA	22;29	291.10	28435	18JA	0;29	125.41
28436	2FE	10;20	320.99	28436	16FE	17;54	155.37
28437	3MR	20;29	350.42	28437	18MR	11;47	184.88
28438	2AP	5;18	19.31	28438	17AP	4;39	213.86
28439	1MY	13;14	47.71	28439	16MY	19;27	242.35
28440	30MY	21;07	75.79	28440	15JN	7;59	270.52
28441	29JN	5;59	103.78	28441	14JL	18;41	298.61
28442	28JL	16;57	131.96	28442	13AU	4;19	326.88
28443	27AU	6;44	160.57	28443	11SE	13;34	355.53
28444	25SE	23;19	189.75	28444	10OC	22;57	24.66
28445	25OC	17;44	219.50	28445	9NO	8;52	54.28
28446	24NO	12;32	249.67	28446	8DE	19;45	84.24
28447	24DE	6;18	279.98				

1300

NUMBER	DATE	TIME	LONG.	NUMBER	DATE	TIME	LONG.
				28447	7JA	8;00	114.31
28448	22JA	22;02	310.13	28448	5FE	21;51	144.25
28449	21FE	11;12	339.84	28449	6MR	13;04	173.84
28450	21MR	21;46	8.98	28450	5AP	4;58	202.96
28451	20AP	6;10	37.55	28451	4MY	20;44	231.62
28452	19MY	13;14	65.69	28452	3JN	11;49	259.95
28453	17JN	20;05	93.62	28453	3JL	1;58	288.16
28454	17JL	3;55	121.62	28454	1AU	15;05	316.46
28455	15AU	13;47	149.93	28455	31AU	3;07	345.06
28456	14SE	2;21	178.76	28456	29SE	14;13	14.09
28457	13OC	17;48	208.21	28457	29OC	0;42	43.58
28458	12NO	11;45	238.21	28458	27NO	11;02	73.43
28459	12DE	7;12	268.55	28459	26DE	21;41	103.44

1301

NUMBER	DATE	TIME	LONG.	NUMBER	DATE	TIME	LONG.
28460	11JA	2;31	298.91	28460	25JA	8;54	133.37
28461	9FE	19;59	328.95	28461	23FE	20;45	162.99
28462	11MR	10;22	358.43	28462	25MR	9;13	192.15
28463	9AP	21;31	27.28	28463	23AP	22;24	220.87
28464	9MY	6;07	55.60	28464	23MY	12;27	249.25
28465	7JN	13;17	83.60	28465	22JN	3;23	277.50
28466	6JL	20;08	111.52	28466	21JL	18;50	305.81
28467	5AU	3;40	139.63	28467	20AU	10;06	334.40
28468	3SE	12;41	168.15	28468	19SE	0;29	3.38
28469	2OC	23;54	197.24	28469	18OC	13;42	32.81
28470	1NO	13;52	226.92	28470	17NO	1;55	62.62
28471	1DE	6;45	257.07	28471	16DE	13;23	92.65
28472	31DE	1;55	287.45				

1302

NUMBER	DATE	TIME	LONG.	NUMBER	DATE	TIME	LONG.
				28472	15JA	0;17	122.63
28473	29JA	21;37	317.70	28473	13FE	10;36	152.33
28474	28FE	15;46	347.52	28474	14MR	20;30	181.57
28475	30MR	7;02	16.72	28475	13AP	6;26	210.32
28476	28AP	19;14	45.33	28476	12MY	17;09	238.69
28477	28MY	5;00	73.50	28477	11JN	5;24	266.86
28478	26JN	13;12	101.47	28478	10JL	19;31	295.08
28479	25JL	20;43	129.50	28479	9AU	11;13	323.56
28480	24AU	4;22	157.82	28480	8SE	3;42	352.45
28481	22SE	12;57	186.62	28481	7OC	20;09	21.82
28482	21OC	23;16	215.98	28482	6NO	11;54	51.64
28483	20NO	12;04	245.85	28483	6DE	2;28	81.74
28484	20DE	3;36	276.06				

1303

NUMBER	DATE	TIME	LONG.	NUMBER	DATE	TIME	LONG.
				28484	4JA	15;28	111.85
28485	18JA	21;17	306.32	28485	3FE	2;41	141.71
28486	17FE	15;43	336.30	28486	4MR	12;15	171.10
28487	19MR	9;23	5.80	28487	2AP	20;43	199.97
28488	18AP	1;16	34.72	28488	2MY	4;57	228.37
28489	17MY	15;00	63.16	28489	31MY	13;53	256.49
28490	16JN	2;35	91.29	28490	30JN	0;24	284.54
28491	15JL	12;22	119.35	28491	29JL	12;58	312.80
28492	13AU	21;00	147.58	28492	28AU	3;47	341.46
28493	12SE	5;18	176.20	28493	26SE	20;36	10.66
28494	11OC	14;11	205.33	28494	26OC	14;46	40.41
28495	10NO	0;24	234.96	28495	25NO	9;01	70.56
28496	9DE	12;23	264.97	28496	25DE	1;55	100.84

1304

NUMBER	DATE	TIME	LONG.	NUMBER	DATE	TIME	LONG.
28497	8JA	2;09	295.11	28497	23JA	16;19	130.93
28498	6FE	17;20	325.10	28498	22FE	3;58	160.57
28499	7MR	9;25	354.71	28499	22MR	13;21	189.65
28500	6AP	1;52	23.84	28500	20AP	21;18	218.19
28501	5MY	17;59	52.50	28501	20MY	4;43	246.34
28502	4JN	9;01	80.83	28502	18JN	12;26	274.31
28503	3JL	22;23	109.00	28503	17JL	21;17	302.34
28504	2AU	10;00	137.25	28504	16AU	8;06	330.70
28505	31AU	20;20	165.80	28505	14SE	21;39	359.59
28506	30SE	6;10	194.78	28506	14OC	14;14	29.09
28507	29OC	16;12	224.26	28507	13NO	9;17	59.15
28508	28NO	2;48	254.13	28508	13DE	5;04	89.51
28509	27DE	14;00	284.17				

1305

NUMBER	DATE	TIME	LONG.	NUMBER	DATE	TIME	LONG.
				28509	11JA	23;28	119.83
28510	26JA	1;48	314.12	28510	10FE	15;05	149.78
28511	24FE	14;22	343.75	28511	12MR	3;42	179.17
28512	26MR	3;59	12.95	28512	10AP	13;45	207.97
28513	24AP	18;41	41.72	28513	9MY	21;57	236.27
28514	24MY	10;05	70.15	28514	8JN	5;06	264.26
28515	23JN	1;27	98.40	28515	7JL	12;01	292.18
28516	22JL	16;07	126.70	28516	5AU	19;44	320.30
28517	21AU	5;48	155.23	28517	4SE	5;20	348.86
28518	19SE	18;39	184.16	28518	3OC	17;51	18.01
28519	19OC	6;51	213.55	28519	2NO	9;41	47.78
28520	17NO	18;28	243.35	28520	2DE	4;14	78.01
28521	17DE	5;30	273.36	28521	31DE	23;55	108.41

1306

NUMBER	DATE	TIME	LONG.	NUMBER	DATE	TIME	LONG.
28522	15JA	16;00	303.33	28522	30JA	18;51	138.63
28523	14FE	2;17	333.01	28523	1MR	11;41	168.37
28524	15MR	12;51	2.27	28524	31MR	1;45	197.51
28525	14AP	0;13	31.07	28525	29AP	13;02	226.07
28526	13MY	12;42	59.50	28526	28MY	21;55	254.20
28527	12JN	2;20	87.73	28527	27JN	5;15	282.14
28528	11JL	16;59	115.97	28528	26JL	12;05	310.14
28529	10AU	8;26	144.44	28529	24AU	19;36	338.47
28530	9SE	0;19	173.32	28530	23SE	4;51	7.30
28531	8OC	16;01	202.68	28531	22OC	16;34	36.73
28532	7NO	6;47	232.47	28532	21NO	7;00	66.69
28533	6DE	20;02	262.51	28533	20DE	23;48	96.95

1307

NUMBER	DATE	TIME	LONG.	NUMBER	DATE	TIME	LONG.
28534	5JA	7;39	292.57	28534	19JA	18;09	127.23
28535	3FE	18;00	322.38	28535	18FE	12;51	157.21
28536	5MR	3;31	351.76	28536	20MR	6;30	186.69
28537	3AP	12;42	20.65	28537	18AP	21;52	215.58
28538	2MY	22;02	49.09	28538	18MY	10;24	243.96
28539	1JN	8;05	77.24	28539	16JN	20;22	272.02
28540	30JN	19;32	105.34	28540	16JL	4;41	300.03
28541	30JL	9;01	133.63	28541	14AU	12;27	328.24
28542	29AU	0;46	162.34	28542	12SE	20;41	356.86
28543	27SE	18;17	191.58	28543	12OC	6;06	26.02
28544	27OC	12;15	221.34	28544	10NO	17;04	55.70
28545	26NO	5;10	251.45	28545	10DE	5;49	85.74
28546	25DE	20;08	281.65				

1308

NUMBER	DATE	TIME	LONG.	NUMBER	DATE	TIME	LONG.
				28546	8JA	20;30	115.92
28547	24JA	8;56	311.66	28547	7FE	12;59	145.95
28548	22FE	19;48	341.26	28548	8MR	6;33	175.61
28549	23MR	5;05	10.32	28549	6AP	23;53	204.76
28550	21AP	13;12	38.87	28550	6MY	15;41	233.40
28551	20MY	20;50	67.01	28551	5JN	5;21	261.67
28552	19JN	4;57	94.99	28552	4JL	17;00	289.78
28553	18JL	14;41	123.07	28553	3AU	3;16	317.98
28554	17AU	3;01	151.50	28554	1SE	12;52	346.50
28555	15SE	18;21	180.47	28555	30SE	22;22	15.49
28556	15OC	12;11	210.04	28556	30OC	8;08	44.96
28557	14NO	7;10	240.10	28557	28NO	18;34	74.82
28558	14DE	1;44	270.42	28558	28DE	6;02	104.88

1309

NUMBER	DATE	TIME	LONG.	NUMBER	DATE	TIME	LONG.
28559	12JA	18;37	300.67	28559	26JA	18;55	134.86
28560	11FE	9;05	330.56	28560	25FE	9;14	164.56
28561	12MR	20;51	359.91	28561	27MR	0;35	193.82
28562	11AP	6;07	28.67	28562	25AP	16;15	222.61
28563	10MY	13;37	56.93	28563	25MY	7;36	251.03
28564	8JN	20;20	84.89	28564	23JN	22;16	279.26
28565	8JL	3;30	112.83	28565	23JL	12;04	307.52
28566	6AU	12;16	141.00	28566	22AU	0;53	336.03
28567	4SE	23;29	169.63	28567	20SE	12;43	4.93
28568	4OC	13;36	198.86	28568	19OC	23;43	34.29
28569	3NO	6;28	228.68	28569	18NO	10;14	64.05
28570	3DE	1;24	258.94	28570	17DE	20;44	94.04

1310

NUMBER	DATE	TIME	LONG.	NUMBER	DATE	TIME	LONG.
28571	1JA	21;01	289.34	28571	16JA	7;34	124.02
28572	31JA	15;35	319.53	28572	14FE	18;54	153.74
28573	2MR	7;31	349.22	28573	16MR	6;47	183.04
28574	31MR	20;10	18.29	28574	14AP	19;18	211.87
28575	30AP	5;51	46.78	28575	14MY	8;39	240.33
28576	29MY	13;33	74.86	28576	12JN	23;01	268.58
28577	27JN	20;26	102.77	28577	12JL	14;18	296.85
28578	27JL	3;33	130.79	28578	11AU	5;56	325.34
28579	25AU	11;50	159.16	28579	9SE	21;06	354.20
28580	23SE	21;59	188.04	28580	9OC	11;13	23.51
28581	23OC	10;35	217.52	28581	8NO	0;09	53.23
28582	22NO	2;04	247.52	28582	7DE	12;07	83.23
28583	21DE	20;12	277.85				

1311

NUMBER	DATE	TIME	LONG.	NUMBER	DATE	TIME	LONG.
				28583	5JA	23;19	113.26
28584	20JA	15;48	308.17	28584	4FE	9;49	143.07
28585	19FE	10;49	338.16	28585	5MR	19;42	172.46
28586	21MR	3;30	7.58	28586	4AP	5;17	201.35
28587	19AP	17;07	36.39	28587	3MY	15;15	229.81
28588	19MY	3;59	64.69	28588	2JN	2;27	258.00
28589	17JN	12;53	92.71	28589	1JL	15;31	286.16
28590	16JL	20;45	120.70	28590	31JL	6;30	314.52
28591	15AU	4;22	148.91	28591	29AU	22;52	343.27
28592	13SE	12;32	177.54	28592	28SE	15;42	12.50
28593	12OC	22;01	206.72	28593	28OC	8;09	42.21
28594	11NO	9;35	236.42	28594	26NO	23;35	72.26
28595	10DE	23;44	266.53	28595	26DE	13;32	102.41

1312

NUMBER	DATE	TIME	LONG.	NUMBER	DATE	TIME	LONG.
28596	9JA	16;20	296.79	28596	25JA	1;37	132.39
28597	8FE	10;23	326.87	28597	23FE	11;51	161.95
28598	9MR	4;23	356.54	28598	23MR	20;36	190.99
28599	7AP	21;04	25.65	28599	22AP	4;38	219.52
28600	7MY	11;49	54.24	28600	21MY	12;55	247.69
28601	6JN	0;29	82.46	28601	19JN	22;25	275.72
28602	5JL	11;14	110.53	28602	19JL	9;51	303.87
28603	3AU	20;32	138.70	28603	17AU	23;35	332.36
28604	2SE	5;05	167.19	28604	16SE	15;35	1.37
28605	1OC	13;47	196.15	28605	16OC	9;23	30.95
28606	30OC	23;24	225.63	28606	15NO	3;59	61.01
28607	29NO	10;32	255.53	28607	14DE	21;54	91.31
28608	28DE	23;22	285.64				

1313

NUMBER	DATE	TIME	LONG.	NUMBER	DATE	TIME	LONG.
				28608	13JA	13;42	121.51
28609	27JA	13;42	315.69	28609	12FE	2;39	151.33
28610	26FE	5;08	345.42	28610	13MR	18;07	180.60
28611	27MR	21;11	14.69	28611	11AP	21;23	209.31
28612	26AP	13;20	43.49	28612	11MY	4;49	237.56
28613	26MY	4;55	71.91	28613	9JN	12;11	265.55
28614	24JN	19;14	100.13	28614	8JL	20;18	293.53
28615	24JL	7;53	128.35	28615	7AU	6;03	321.74
28616	22AU	19;02	156.80	28616	5SE	18;12	350.43
28617	21SE	5;17	185.65	28617	5OC	9;22	19.71
28618	20OC	15;23	214.98	28618	4NO	3;28	49.60
28619	19NO	1;49	244.74	28619	3DE	23;15	79.90
28620	18DE	12;44	274.75				

1314

NUMBER	DATE	TIME	LONG.	NUMBER	DATE	TIME	LONG.
				28620	2JA	18;38	110.28
28621	17JA	0;08	304.74	28621	1FE	11;44	140.40
28622	15FE	12;04	334.48	28622	3MR	1;45	170.01
28623	17MR	0;50	3.81	28623	1AP	12;54	199.00
28624	15AP	14;43	32.69	28624	30AP	21;51	227.45
28625	15MY	5;37	61.20	28625	30MY	5;22	255.52
28626	13JN	21;00	89.49	28626	28JN	12;16	283.43
28627	13JL	12;09	117.75	28627	27JL	19;29	311.46
28628	12AU	2;33	146.20	28628	26AU	4;05	339.85
28629	10SE	16;04	175.01	28629	24SE	15;07	8.78
28630	10OC	4;52	204.27	28630	24OC	5;22	38.34
28631	8NO	17;01	233.97	28631	22NO	22;45	68.43
28632	8DE	4;28	263.95	28632	22DE	18;06	98.81

1315

NUMBER	DATE	TIME	LONG.	NUMBER	DATE	TIME	LONG.
28633	6JA	15;13	293.96	28633	21JA	13;33	129.12
28634	5FE	1;25	323.75	28634	20FE	7;27	159.05
28635	6MR	11;33	353.15	28635	21MR	22;51	188.40
28636	4AP	22;12	22.07	28636	20AP	11;25	217.16
28637	4MY	9;50	50.58	28637	19MY	21;24	245.42
28638	2JN	22;41	78.84	28638	18JN	5;25	273.40
28639	2JL	12;46	107.04	28639	17JL	12;26	301.36
28640	1AU	3;54	135.41	28640	15AU	19;35	329.56
28641	30AU	19;50	164.15	28641	14SE	3;58	358.21
28642	29SE	12;01	193.37	28642	13OC	14;30	27.43
28643	29OC	3;38	223.06	28643	12NO	3;39	57.22
28644	27NO	17;54	253.06	28644	11DE	19;19	87.39
28645	27DE	6;25	283.16				

1316 / 1322

NEW NUMBER	DATE	TIME	LONG.	FULL NUMBER	DATE	TIME	LONG.	NEW NUMBER	DATE	TIME	LONG.	FULL NUMBER	DATE	TIME	LONG.
				28645	10JA	12;54	117.68					28719	4JA	10;20	112.03
28646	25JA	17;19	313.08	28646	9FE	7;24	147.78	28720	18JA	10;34	306.26	28720	3FE	0;44	142.02
28647	24FE	3;02	342.61	28647	10MR	1;33	177.44	28721	17FE	1;16	336.09	28721	4MR	12;13	171.49
28648	24MR	12;08	11.65	28648	8AP	18;03	206.53	28722	18MR	16;46	5.51	28722	2AP	21;20	200.38
28649	22AP	21;05	40.21	28649	8MY	7;59	235.08	28723	17AP	8;41	34.44	28723	2MY	5;01	228.76
28650	22MY	6;30	68.42	28650	6JN	19;12	263.24	28724	17MY	0;29	62.97	28724	31MY	12;12	256.81
28651	20JN	17;01	96.49	28651	6JL	4;19	291.24	28725	15JN	15;31	91.24	28725	29JN	19;46	284.75
28652	20JL	5;21	124.68	28652	4AU	12;22	319.37	28726	15JL	5;10	119.45	28726	29JL	4;35	312.85
28653	18AU	20;02	153.22	28653	2SE	20;27	347.85	28727	13AU	17;15	147.82	28727	27AU	15;29	341.35
28654	17SE	12;58	182.28	28654	2OC	5;23	16.83	28728	12SE	4;08	176.55	28728	26SE	5;11	10.41
28655	17OC	7;08	211.89	28655	31OC	15;40	46.34	28729	11OC	14;29	205.73	28729	25OC	22;00	40.10
28656	16NO	0;59	241.93	28656	30NO	3;32	76.28	28730	10NO	0;53	235.38	28730	24NO	17;16	70.29
28657	15DE	17;10	272.15	28657	29DE	17;09	106.42	28731	9DE	11;39	265.33	28731	24DE	13;11	100.69

1317 / 1323

NEW NUMBER	DATE	TIME	LONG.	FULL NUMBER	DATE	TIME	LONG.	NEW NUMBER	DATE	TIME	LONG.	FULL NUMBER	DATE	TIME	LONG.
28658	14JA	7;05	302.27	28658	28JA	8;35	136.51	28732	7JA	22;45	295.36	28732	23JA	7;36	130.95
28659	12FE	18;49	332.03	28659	27FE	1;28	166.29	28733	6FE	10;13	325.19	28733	21FE	23;07	160.76
28660	14MR	4;43	1.28	28660	28MR	18;50	195.60	28734	7MR	22;16	354.64	28734	23MR	11;34	189.97
28661	12AP	13;11	29.98	28661	27AP	11;23	224.41	28735	6AP	11;15	23.63	28735	21AP	21;26	218.59
28662	11MY	20;48	58.23	28662	27MY	2;06	252.79	28736	6MY	1;24	52.23	28736	21MY	5;30	246.76
28663	10JN	4;25	86.23	28663	25JN	14;47	280.94	28737	4JN	16;29	80.56	28737	19JN	12;35	274.70
28664	9JL	13;07	114.23	28664	25JL	1;51	309.10	28738	4JL	7;51	108.81	28738	18JL	19;33	302.66
28665	8AU	0;03	142.50	28665	23AU	11;56	337.51	28739	2AU	22;49	137.20	28739	17AU	3;23	330.90
28666	6SE	13;57	171.26	28666	21SE	21;40	6.35	28740	1SE	13;02	165.89	28740	15SE	13;10	359.63
28667	6OC	6;47	200.63	28667	21OC	7;28	35.68	28741	1OC	2;29	195.02	28741	15OC	1;50	28.97
28668	5NO	1;32	230.56	28668	19NO	17;38	65.44	28742	30OC	15;14	224.61	28742	13NO	17;47	58.89
28669	4DE	20;39	260.84	28669	19DE	4;32	95.45	28743	29NO	3;13	254.54	28743	13DE	12;21	89.19
								28744	28DE	14;21	284.57				

1318 / 1324

NEW NUMBER	DATE	TIME	LONG.	FULL NUMBER	DATE	TIME	LONG.	NEW NUMBER	DATE	TIME	LONG.	FULL NUMBER	DATE	TIME	LONG.
28670	3JA	14;35	291.16	28670	17JA	16;33	125.47					28744	12JA	7;55	119.57
28671	2FE	6;19	321.21	28671	16FE	5;56	155.26	28745	27JA	0;42	314.46	28745	11FE	2;40	149.65
28672	3MR	19;22	350.76	28672	17MR	20;33	184.65	28746	25FE	10;37	343.98	28746	11MR	19;16	179.22
28673	2AP	5;43	19.72	28673	16AP	11;51	213.57	28747	25MR	20;41	13.04	28747	10AP	9;10	208.17
28674	1MY	13;52	48.13	28674	16MY	3;14	242.09	28748	24AP	7;30	41.66	28748	9MY	20;22	236.59
28675	30MY	20;46	76.16	28675	14JN	18;14	270.36	28749	23MY	19;30	69.95	28749	8JN	5;17	264.65
28676	29JN	3;31	104.07	28676	14JL	8;35	298.60	28750	22JN	8;50	98.13	28750	7JL	12;44	292.60
28677	28JL	11;21	132.13	28677	12AU	22;09	327.02	28751	21JL	23;28	126.42	28751	5AU	19;46	320.70
28678	26AU	21;18	160.58	28678	11SE	10;46	355.80	28752	20AU	15;12	155.03	28752	4SE	3;33	349.18
28679	25SE	10;02	189.59	28679	10OC	22;25	25.03	28753	19SE	7;36	184.09	28753	3OC	13;03	18.20
28680	25OC	1;40	219.22	28680	9NO	9;20	54.69	28754	18OC	23;54	213.66	28754	2NO	0;56	47.80
28681	23NO	19;46	249.35	28681	8DE	19;53	84.64	28755	17NO	15;11	243.61	28755	1DE	15;24	77.86
28682	23DE	15;16	279.73					28756	17DE	4;44	273.72	28756	31DE	8;01	108.13

1319 / 1325

NEW NUMBER	DATE	TIME	LONG.	FULL NUMBER	DATE	TIME	LONG.	NEW NUMBER	DATE	TIME	LONG.	FULL NUMBER	DATE	TIME	LONG.
				28682	7JA	6;30	114.64	28757	15JA	16;24	303.74	28757	30JA	2;02	138.31
28683	22JA	10;33	310.03	28683	5FE	17;25	144.46	28758	14FE	2;33	333.42	28758	28FE	20;20	168.13
28684	21FE	3;52	339.93	28684	7MR	4;47	173.89	28759	15MR	11;43	2.61	28759	30MR	13;38	197.41
28685	22MR	18;06	9.22	28685	5AP	16;40	202.85	28760	13AP	20;28	31.31	28760	29AP	4;50	226.13
28686	21AP	5;06	37.90	28686	5MY	5;17	231.39	28761	13MY	5;21	59.60	28761	28MY	17;23	254.41
28687	20MY	13;37	66.10	28687	3JN	18;56	259.67	28762	11JN	15;04	87.67	28762	27JN	3;32	282.46
28688	18JN	20;45	94.04	28688	3JL	9;45	287.91	28763	11JL	2;19	115.77	28763	26JL	12;08	310.53
28689	18JL	3;41	122.00	28689	2AU	1;25	316.31	28764	9AU	15;45	144.16	28764	24AU	20;16	338.89
28690	16AU	11;23	150.22	28690	31AU	17;10	345.05	28765	8SE	7;47	173.04	28765	23SE	4;53	7.70
28691	14SE	20;37	178.92	28691	30SE	8;10	14.23	28766	8OC	1;44	202.48	28766	22OC	14;36	37.04
28692	14OC	8;01	208.19	28692	29OC	21;58	43.86	28767	6NO	20;11	232.42	28767	21NO	1;43	66.86
28693	12NO	22;05	238.03	28693	28NO	10;35	73.80	28768	6DE	13;31	262.63	28768	20DE	14;25	96.95
28694	12DE	14;57	268.26	28694	27DE	22;13	103.86								

1320 / 1326

NEW NUMBER	DATE	TIME	LONG.	FULL NUMBER	DATE	TIME	LONG.	NEW NUMBER	DATE	TIME	LONG.	FULL NUMBER	DATE	TIME	LONG.
28695	11JA	9;56	298.61	28695	26JA	9;01	133.77	28769	5JA	4;40	292.83	28769	19JA	4;48	127.06
28696	10FE	5;22	328.73	28696	24FE	19;01	163.31	28770	3FE	17;27	322.75	28770	17FE	20;47	156.94
28697	10MR	23;14	358.36	28697	25MR	4;27	192.35	28771	5MR	4;07	352.18	28771	19MR	13;47	186.40
28698	9AP	14;18	27.38	28698	23AP	13;53	220.92	28772	3AP	13;05	21.05	28772	18AP	6;40	215.36
28699	9MY	2;25	55.85	28699	23MY	0;08	249.15	28773	2MY	20;53	49.43	28773	17MY	22;17	243.87
28700	7JN	12;14	83.94	28700	21JN	12;04	277.28	28774	1JN	4;15	77.48	28774	16JN	12;01	272.00
28701	6JL	20;36	111.92	28701	21JL	2;05	305.54	28775	30JN	12;11	105.43	28775	15JL	23;58	300.23
28702	5AU	4;22	140.05	28702	19AU	17;58	334.14	28776	29JL	21;50	133.57	28776	14AU	10;41	328.56
28703	3SE	12;20	168.53	28703	18SE	10;54	3.21	28777	28AU	10;14	162.14	28777	12SE	20;47	357.26
28704	2OC	21;13	197.52	28704	18OC	3;54	32.79	28778	27SE	1;46	191.29	28778	12OC	6;44	26.44
28705	1NO	7;45	227.05	28705	16NO	20;10	62.77	28779	26OC	19;53	221.05	28779	10NO	16;50	56.08
28706	30NO	20;35	257.03	28706	16DE	11;04	92.94	28780	25NO	15;10	251.25	28780	10DE	3;23	86.03
28707	30DE	11;57	287.25					28781	25DE	9;56	281.61				

1321 / 1327

NEW NUMBER	DATE	TIME	LONG.	FULL NUMBER	DATE	TIME	LONG.	NEW NUMBER	DATE	TIME	LONG.	FULL NUMBER	DATE	TIME	LONG.
				28707	15JA	0;09	123.02					28781	8JA	14;43	116.06
28708	29JA	5;15	317.41	28708	13FE	11;13	152.75	28782	24JA	2;53	311.79	28782	7FE	3;12	145.93
28709	27FE	23;13	347.22	28709	14MR	20;29	181.96	28783	22FE	17;15	341.54	28783	8MR	16;57	175.44
28710	29MR	16;28	16.51	28710	13AP	4;34	210.63	28784	24MR	4;49	10.70	28784	7AP	7;42	204.49
28711	28AP	8;06	45.27	28711	12MY	12;25	238.89	28785	22AP	13;53	39.28	28785	6MY	22;52	233.12
28712	27MY	21;48	73.60	28712	10JN	21;02	266.92	28786	21MY	21;11	67.42	28786	5JN	14;00	261.45
28713	26JN	9;33	101.71	28713	10JL	7;21	294.99	28787	20JN	3;49	95.34	28787	5JL	4;45	289.68
28714	25JL	19;41	129.84	28714	8AU	19;54	323.33	28788	19JL	10;58	123.31	28788	3AU	18;54	318.03
28715	24AU	4;44	158.22	28715	7SE	10;53	352.15	28789	17AU	19;48	151.60	28789	2SE	8;17	346.70
28716	22SE	13;28	187.03	28716	7OC	4;02	21.55	28790	16SE	7;09	180.40	28790	1OC	20;43	15.81
28717	21OC	22;42	216.35	28717	5NO	22;36	51.48	28791	15OC	21;27	209.82	28791	31OC	8;12	45.36
28718	20NO	9;07	246.12	28718	5DE	17;14	81.74	28792	14NO	14;28	239.79	28792	29NO	19;02	75.24
28719	19DE	21;05	276.18					28793	14DE	9;28	270.12	28793	29DE	5;35	105.25

NUMBER	DATE	TIME	LONG.	NUMBER	DATE	TIME	LONG.	NUMBER	DATE	TIME	LONG.	NUMBER	DATE	TIME	LONG.
			1328								1334				
28794	13JA	5;03	300.49	28794	27JA	16;14	135.14	28868	6JA	15;09	294.35	28868	20JA	20;56	128.81
28795	11FE	23;29	330.55	28795	26FE	3;09	164.70	28869	5FE	1;56	324.17	28869	19FE	15;01	158.76
28796	12MR	15;15	0.07	28796	26MR	14;28	193.80	28870	6MR	11;22	353.53	28870	21MR	8;46	188.22
28797	11AP	3;44	28.96	28797	25AP	2;24	222.45	28871	4AP	20;03	22.37	28871	20AP	1;00	217.13
28798	10MY	13;20	57.30	28798	24MY	15;17	250.77	28872	4MY	4;36	50.76	28872	19MY	14;53	245.55
28799	8JN	21;00	85.31	28799	23JN	5;25	278.98	28873	2JN	13;38	78.87	28873	18JN	2;14	273.66
28800	8JL	3;56	113.23	28800	22JL	20;46	307.30	28874	1JL	23;54	106.92	28874	17JL	11;37	301.71
28801	6AU	11;14	141.35	28801	21AU	12;47	335.93	28875	31JL	12;11	135.17	28875	15AU	20;02	329.97
28802	4SE	19;44	169.87	28802	20SE	4;32	4.99	28876	30AU	3;00	163.86	28876	14SE	4;30	358.62
28803	4OC	6;05	198.94	28803	19OC	19;16	34.50	28877	28SE	20;15	193.11	28877	13OC	13;48	27.80
28804	2NO	18;51	228.59	28804	18NO	8;41	64.38	28878	28OC	14;54	222.91	28878	12NO	0;19	57.47
28805	2DE	10;20	258.70	28805	17DE	20;54	94.44	28879	27NO	9;11	253.09	28879	11DE	12;12	87.48
								28880	27DE	1;37	283.35				
			1329								1335				
28806	1JA	4;19	289.02	28806	16JA	8;06	124.43					28880	10JA	1;37	117.60
28807	30JA	23;38	319.25	28807	14FE	18;22	154.11	28881	25JA	15;35	313.40	28881	8FE	16;36	147.56
28808	1MR	18;20	349.07	28808	16MR	3;50	183.31	28882	24FE	3;11	343.01	28882	10MR	8;55	177.15
28809	31MR	10;46	18.30	28809	14AP	12;56	212.00	28883	25MR	12;49	12.07	28883	9AP	1;45	206.26
28810	30AP	0;15	46.94	28810	13MY	22;26	240.31	28884	23AP	20;58	40.59	28884	8MY	17;59	234.91
28811	29MY	11;08	75.15	28811	12JN	9;15	268.42	28885	23MY	4;19	68.72	28885	7JN	8;41	263.21
28812	27JN	20;11	103.15	28812	11JL	22;08	296.59	28886	21JN	11;44	96.67	28886	6JL	21;35	291.37
28813	27JL	4;17	131.21	28813	10AU	13;12	325.06	28887	20JL	20;20	124.71	28887	5AU	9;03	319.63
28814	25AU	12;14	159.56	28814	9SE	5;53	353.97	28888	19AU	7;17	153.09	28888	3SE	19;39	348.21
28815	23SE	20;43	188.38	28815	8OC	23;14	23.40	28889	17SE	21;19	182.03	28889	3OC	5;53	17.24
28816	23OC	6;28	217.74	28816	7NO	16;13	53.29	28890	17OC	14;24	211.59	28890	1NO	16;04	46.76
28817	21NO	18;09	247.58	28817	7DE	8;03	83.45	28891	16NO	9;27	241.67	28891	1DE	2;27	76.63
28818	21DE	8;12	277.73					28892	16DE	4;46	272.02	28892	30DE	13;19	106.65
			1330								1336				
				28818	5JA	22;10	113.60	28893	14JA	22;48	302.31	28893	29JA	1;02	136.58
28819	20JA	0;30	307.92	28819	4FE	10;13	143.48	28894	13FE	14;29	332.24	28894	27FE	13;54	166.20
28820	18FE	18;04	337.85	28820	5MR	20;12	172.87	28895	14MR	3;22	1.61	28895	28MR	3;54	195.38
28821	20MR	11;34	7.32	28821	4AP	4;36	201.71	28896	12AP	13;30	30.38	28896	26AP	18;39	224.12
28822	19AP	3;56	36.25	28822	3MY	12;15	230.08	28897	11MY	21;29	58.65	28897	26MY	9;41	252.52
28823	18MY	18;33	64.71	28823	1JN	20;12	258.14	28898	10JN	4;16	86.61	28898	25JN	0;38	280.76
28824	17JN	7;19	92.88	28824	1JL	5;28	286.15	28899	9JL	11;00	114.53	28899	24JL	15;15	309.06
28825	16JL	18;22	121.00	28825	30JL	16;49	314.36	28900	7AU	18;54	142.68	28900	23AU	5;18	337.63
28826	15AU	4;05	149.29	28826	29AU	6;39	343.00	28901	6SE	4;59	171.29	28901	21SE	18;31	6.61
28827	13SE	13;06	177.96	28827	27SE	22;55	12.20	28902	5OC	17;52	200.49	28902	21OC	6;44	36.04
28828	12OC	22;12	207.12	28828	27OC	17;05	41.97	28903	4NO	9;40	230.28	28903	19NO	18;02	65.85
28829	11NO	8;06	236.75	28829	26NO	12;04	72.16	28904	4DE	3;50	260.52	28904	19DE	4;45	95.85
28830	10DE	19;17	266.74	28830	26DE	6;14	102.50								
			1331								1337				
28831	9JA	7;56	296.82	28831	24JA	22;05	132.63	28905	2JA	23;18	290.90	28905	17JA	15;15	125.81
28832	7FE	21;51	326.74	28832	23FE	10;56	162.31	28906	1FE	18;27	321.11	28906	16FE	1;51	155.48
28833	9MR	12;42	356.28	28833	24MR	21;02	191.40	28907	3MR	11;36	350.83	28907	17MR	12;41	184.71
28834	8AP	4;11	25.36	28834	23AP	5;11	219.93	28908	2AP	1;40	19.94	28908	16AP	0;00	213.48
28835	7MY	19;55	53.99	28835	22MY	12;23	248.05	28909	1MY	12;33	48.46	28909	15MY	12;07	241.87
28836	6JN	11;22	82.33	28836	20JN	19;35	275.99	28910	30MY	21;02	76.56	28910	14JN	1;26	270.08
28837	6JL	1;51	110.55	28837	20JL	3;39	304.01	28911	29JN	4;13	104.49	28911	13JL	16;11	298.33
28838	4AU	14;55	138.87	28838	18AU	13;27	332.34	28912	28JL	11;18	132.52	28912	12AU	8;06	326.85
28839	3SE	2;36	167.48	28839	17SE	1;44	1.19	28913	26AU	19;13	160.88	28913	11SE	0;21	355.77
28840	2OC	13;24	196.53	28840	16OC	17;06	30.67	28914	25SE	4;42	189.76	28914	10OC	15;59	25.16
28841	31OC	23;57	226.05	28841	15NO	11;23	60.71	28915	24OC	16;17	219.22	28915	9NO	6;19	54.96
28842	30NO	10;38	255.93	28842	15DE	7;19	91.08	28916	23NO	6;25	249.18	28916	8DE	19;17	85.00
28843	29DE	21;34	285.96					28917	22DE	23;11	279.44				
			1332								1338				
				28843	14JA	2;44	121.43	28918	21JA	17;53	309.72	28917	7JA	7;01	115.05
28844	28JA	8;43	315.87	28844	12FE	19;45	151.43	28919	20FE	12;58	339.70	28918	5FE	17;39	144.86
28845	26FE	20;11	345.44	28845	13MR	9;37	180.86	28920	22MR	6;32	9.14	28919	7MR	3;19	174.22
28846	27MR	8;22	14.56	28846	11AP	20;36	209.67	28921	20AP	21;24	37.98	28920	5AP	12;18	203.06
28847	25AP	21;39	43.25	28847	11MY	5;24	237.98	28922	20MY	9;30	66.32	28921	4MY	21;15	231.46
28848	25MY	12;07	71.63	28848	9JN	12;51	265.97	28923	18JN	19;25	94.38	28922	3JN	7;06	259.59
28849	24JN	3;22	99.89	28849	8JL	19;47	293.90	28924	18JL	4;01	122.40	28923	2JL	18;46	287.70
28850	23JL	18;42	128.21	28850	7AU	3;07	322.02	28925	16AU	12;06	150.65	28924	1AU	8;45	316.03
28851	22AU	9;33	156.80	28851	5SE	11;53	350.56	28926	14SE	20;25	179.30	28925	31AU	0;52	344.78
28852	20SE	23;40	185.81	28852	4OC	23;07	19.68	28927	14OC	5;37	208.49	28926	29SE	18;14	14.05
28853	20OC	13;03	215.27	28853	3NO	13;29	49.40	28928	12NO	16;20	238.18	28927	29OC	11;46	43.82
28854	19NO	1;39	245.12	28854	3DE	6;54	79.60	28929	12DE	5;10	268.23	28928	28NO	4;29	73.94
28855	18DE	13;19	275.17									28929	27DE	19;39	104.14
			1333								1339				
				28855	2JA	2;09	109.98	28930	10JA	20;16	298.42	28930	26JA	8;45	134.15
28856	17JA	0;00	305.13	28856	31JA	21;24	140.19	28931	9FE	13;07	328.44	28931	24FE	19;38	163.72
28857	15FE	9;55	334.78	28857	2MR	15;04	169.95	28932	11MR	6;34	358.07	28932	26MR	4;35	192.74
28858	16MR	19;36	3.98	28858	1AP	6;15	199.12	28933	9AP	23;23	27.17	28933	24AP	12;19	221.24
28859	15AP	5;42	32.71	28859	30AP	18;43	227.71	28934	9MY	14;48	55.77	28934	23MY	19;50	249.37
28860	14MY	16;49	61.07	28860	30MY	4;42	255.88	28935	8JN	4;31	84.02	28935	22JN	4;13	277.35
28861	13JN	5;19	89.25	28861	28JN	12;49	283.85	28936	7JL	16;31	112.15	28936	21JL	14;23	305.46
28862	12JL	19;16	117.47	28862	27JL	20;02	311.87	28937	6AU	3;03	140.38	28937	20AU	2;58	333.92
28863	11AU	10;34	145.95	28863	26AU	3;27	340.21	28938	4SE	12;34	168.93	28938	18SE	18;09	2.92
28864	10SE	2;55	174.86	28864	24SE	12;07	9.05	28939	3OC	21;45	197.93	28939	18OC	11;37	32.50
28865	9OC	19;40	204.28	28865	23OC	22;52	38.46	28940	2NO	7;20	227.43	28940	17NO	6;33	62.59
28866	8NO	11;51	234.15	28866	22NO	12;06	68.37	28941	1DE	17;55	257.31	28941	17DE	1;28	92.93
28867	8DE	2;30	264.26	28867	22DE	3;39	98.59	28942	31DE	5;47	287.38				

1340 / **1346**

NUMBER	DATE	TIME	LONG.	NUMBER	DATE	TIME	LONG.	NUMBER	DATE	TIME	LONG.	NUMBER	DATE	TIME	LONG.
				28942	15JA	18;40	123.18					29016	8JA	14;25	116.44
28943	29JA	18;56	317.37	28943	14FE	9;01	153.05	29017	23JA	12;59	311.60	29017	7FE	0;48	146.22
28944	28FE	9;05	347.03	28944	14MR	20;19	182.34	29018	22FE	7;13	341.52	29018	8MR	11;16	175.60
28945	28MR	23;58	16.23	28945	13AP	5;12	211.05	29019	23MR	22;48	10.85	29019	6AP	22;03	204.49
28946	27AP	15;23	44.99	28946	12MY	12;39	239.29	29020	22AP	11;10	39.56	29020	6MY	9;26	232.97
28947	27MY	6;55	73.40	28947	10JN	19;39	267.25	29021	21MY	20;42	67.79	29021	4JN	21;54	261.20
28948	25JN	22;00	101.64	28948	10JL	3;10	295.21	29022	20JN	4;25	95.75	29022	4JL	11;51	289.40
28949	25JL	11;59	129.93	28949	8AU	12;01	323.40	29023	19JL	11;30	123.72	29023	3AU	3;20	317.80
28950	24AU	0;35	158.45	28950	6SE	23;01	352.06	29024	17AU	19;00	151.96	29024	1SE	19;45	346.59
28951	22SE	12;03	187.37	28951	6OC	12;54	21.31	29025	16SE	3;45	180.65	29025	1OC	12;06	15.84
28952	21OC	22;54	216.75	28952	5NO	5;53	51.16	29026	15OC	14;20	209.91	29026	31OC	3;25	45.55
28953	20NO	9;39	246.54	28953	5DE	1;18	81.46	29027	14NO	3;13	239.71	29027	29NO	17;16	75.56
28954	19DE	20;30	276.55					29028	13DE	18;39	269.89	29028	29DE	5;40	105.65

1341 / **1347**

NUMBER	DATE	TIME	LONG.	NUMBER	DATE	TIME	LONG.	NUMBER	DATE	TIME	LONG.	NUMBER	DATE	TIME	LONG.
				28954	3JA	21;16	111.86					29029	27JA	16;48	135.56
28955	18JA	7;28	306.52	28955	2FE	15;37	142.02	29029	12JA	12;24	300.18	29030	26FE	2;47	165.08
28956	16FE	18;33	336.21	28956	4MR	7;00	171.67	29030	11FE	7;21	330.27	29031	27MR	11;51	194.08
28957	18MR	6;02	5.46	28957	2AP	19;16	200.69	29031	13MR	1;41	359.90	29032	25AP	20;30	222.60
28958	16AP	18;25	34.26	28958	2MY	5;00	229.15	29032	11AP	17;52	28.95	29033	25MY	5;34	250.78
28959	16MY	8;04	62.70	28959	31MY	12;59	257.23	29033	11MY	7;16	57.45	29034	23JN	16;05	278.84
28960	14JN	22;52	90.96	28960	29JN	20;04	285.15	29034	9JN	18;13	85.59	29035	23JL	4;50	307.06
28961	14JL	14;17	119.25	28961	29JL	3;09	313.18	29035	9JL	3;28	113.60	29036	21AU	20;01	335.64
28962	13AU	5;37	147.74	28962	27AU	11;10	341.56	29036	7AU	11;53	141.76	29037	20SE	13;04	4.74
28963	11SE	20;23	176.62	28963	25SE	21;08	10.47	29037	5SE	20;12	170.28	29038	20OC	6;54	34.37
28964	11OC	10;27	205.96	28964	25OC	9;58	39.99	29038	5OC	5;03	199.29	29039	19NO	0;22	64.42
28965	9NO	23;43	235.72	28965	24NO	1;59	70.03	29039	3NO	15;02	228.82	29040	18DE	16;32	94.65
28966	9DE	12;00	265.74	28966	23DE	20;28	100.38	29040	3DE	2;47	258.77				

1342 / **1348**

NUMBER	DATE	TIME	LONG.	NUMBER	DATE	TIME	LONG.	NUMBER	DATE	TIME	LONG.	NUMBER	DATE	TIME	LONG.
28967	7JA	23;11	295.77	28967	22JA	15;50	130.68	29041	1JA	16;41	288.92	29041	17JA	6;45	124.77
28968	6FE	9;20	325.54	28968	21FE	10;19	160.62	29042	31JA	8;34	319.01	29042	15FE	18;41	154.51
28969	7MR	18;51	354.89	28969	23MR	2;41	189.99	29043	1MR	1;36	348.77	29043	16MR	4;24	183.72
28970	6AP	4;24	23.75	28970	21AP	16;26	218.78	29044	30MR	18;37	18.03	29044	14AP	12;29	212.37
28971	5MY	14;42	52.19	28971	21MY	3;36	247.07	29045	29AP	10;39	46.79	29045	13MY	19;48	240.59
28972	4JN	2;16	80.39	28972	19JN	12;36	275.09	29046	29MY	1;11	75.14	29046	12JN	3;30	268.58
28973	3JL	15;23	108.55	28973	18JL	20;14	303.08	29047	27JN	14;07	103.30	29047	11JL	12;36	296.60
28974	2AU	6;04	136.91	28974	17AU	3;33	331.30	29048	27JL	1;32	131.49	29048	9AU	23;56	324.91
28975	31AU	22;07	165.67	28975	15SE	11;38	359.96	29049	25AU	11;44	159.93	29049	8SE	13;53	353.71
28976	30SE	15;02	194.94	28976	14OC	21;23	29.17	29050	23SE	21;14	188.80	29050	8OC	6;24	23.10
28977	30OC	7;54	224.70	28977	13NO	9;26	58.93	29051	23OC	6;44	218.14	29051	7NO	0;55	53.04
28978	28NO	23;38	254.78	28978	12DE	23;50	89.06	29052	21NO	16;52	247.92	29052	6DE	20;11	83.34
28979	28DE	13;25	284.93					29053	21DE	4;04	277.95				

1343 / **1349**

NUMBER	DATE	TIME	LONG.	NUMBER	DATE	TIME	LONG.	NUMBER	DATE	TIME	LONG.	NUMBER	DATE	TIME	LONG.
				28979	11JA	16;13	119.29					29053	5JA	14;30	113.67
28980	27JA	1;04	314.87	28980	10FE	9;49	149.34	29054	19JA	16;28	307.97	29054	4FE	6;22	143.71
28981	25FE	11;00	344.39	28981	12MR	3;39	178.97	29055	18FE	5;54	337.74	29055	5MR	19;04	173.22
28982	26MR	19;48	13.40	28982	10AP	20;37	208.06	29056	19MR	20;09	7.08	29056	4AP	4;57	202.12
28983	25AP	4;08	41.91	28983	10MY	11;40	236.63	29057	18AP	11;03	35.96	29057	3MY	12;52	230.49
28984	24MY	12;39	70.07	28984	9JN	0;18	264.84	29058	18MY	2;24	64.45	29058	1JN	19;54	258.52
28985	22JN	22;04	98.10	28985	8JL	10;41	292.91	29059	16JN	17;46	92.73	29059	1JL	3;00	286.44
28986	22JL	9;12	126.24	28986	6AU	19;38	321.08	29060	16JL	8;29	120.99	29060	30JL	11;06	314.52
28987	20AU	22;44	154.75	28987	5SE	4;11	349.60	29061	14AU	22;01	149.44	29061	28AU	20;59	343.00
28988	19SE	14;59	183.80	28988	4OC	13;12	18.60	29062	13SE	10;17	178.24	29062	27SE	9;26	12.03
28989	19OC	9;19	213.44	28989	2NO	23;13	48.12	29063	12OC	21;38	207.49	29063	27OC	0;58	41.68
28990	18NO	4;13	243.54	28990	2DE	10;27	78.05	29064	11NO	8;36	237.17	29064	25NO	19;24	71.85
28991	17DE	21;52	273.82	28991	31DE	23;01	108.14	29065	10DE	19;29	267.14	29065	25DE	15;23	102.26

1344 / **1350**

NUMBER	DATE	TIME	LONG.	NUMBER	DATE	TIME	LONG.	NUMBER	DATE	TIME	LONG.	NUMBER	DATE	TIME	LONG.
28992	16JA	13;09	303.99	28992	30JA	13;01	138.16	29066	9JA	6;22	297.15	29066	24JA	10;45	132.55
28993	15FE	1;50	333.78	28993	29FE	4;27	167.86	29067	7FE	17;13	326.94	29067	23FE	3;38	162.39
28994	15MR	12;17	3.03	28994	29MR	20;52	197.12	29068	9MR	4;12	356.32	29068	24MR	17;20	191.63
28995	13AP	20;58	31.72	28995	28AP	13;20	225.90	29069	7AP	15;47	25.24	29069	23AP	4;10	220.28
28996	13MY	4;30	59.95	28996	28MY	4;48	254.30	29070	7MY	4;30	53.76	29070	22MY	12;53	248.46
28997	11JN	11;39	87.92	28997	26JN	18;40	282.50	29071	5JN	18;35	82.05	29071	20JN	20;19	276.41
28998	10JL	19;28	115.89	28998	26JL	6;59	310.72	29072	5JL	9;44	110.30	29072	20JL	3;20	304.39
28999	9AU	5;07	144.12	28999	24AU	18;11	339.19	29073	4AU	1;19	138.71	29073	18AU	10;51	332.63
29000	7SE	17;36	172.85	29000	23SE	4;49	8.09	29074	2SE	16;39	167.47	29074	16SE	19;50	1.34
29001	7OC	9;20	202.19	29001	22OC	15;13	37.46	29075	2OC	7;23	196.67	29075	16OC	7;15	30.65
29002	6NO	3;42	232.11	29002	21NO	1;37	67.24	29076	31OC	21;21	226.33	29076	14NO	21;44	60.52
29003	5DE	23;13	262.42	29003	20DE	12;13	97.25	29077	30NO	10;20	256.31	29077	14DE	15;06	90.78
								29078	29DE	22;08	286.37				

1345 / **1351**

NUMBER	DATE	TIME	LONG.	NUMBER	DATE	TIME	LONG.	NUMBER	DATE	TIME	LONG.	NUMBER	DATE	TIME	LONG.
29004	4JA	18;06	292.78	29004	18JA	23;21	127.22					29078	13JA	10;09	121.13
29005	3FE	11;02	322.87	29005	17FE	11;24	156.94	29079	28JA	8;43	316.26	29079	12FE	5;08	151.21
29006	5MR	1;16	352.45	29006	19MR	0;34	186.25	29080	26FE	18;19	345.75	29080	13MR	22;30	180.79
29007	3AP	12;39	21.42	29007	17AP	14;41	215.10	29081	28MR	3;32	14.75	29081	12AP	13;30	209.77
29008	2MY	21;32	49.84	29008	17MY	5;25	243.58	29082	26AP	13;07	43.30	29082	12MY	1;53	238.22
29009	1JN	4;43	77.88	29009	15JN	20;22	271.85	29083	25MY	23;47	71.53	29083	10JN	11;56	266.32
29010	30JN	11;18	105.79	29010	15JL	11;15	300.11	29084	24JN	11;58	99.66	29084	9JL	20;14	294.30
29011	29JL	18;31	133.83	29011	14AU	1;50	328.59	29085	24JL	1;51	127.93	29085	8AU	3;43	322.43
29012	28AU	3;28	162.26	29012	12SE	15;48	357.45	29086	22AU	17;21	156.54	29086	6SE	11;27	350.93
29013	26SE	15;00	191.24	29013	12OC	4;49	26.75	29087	21SE	10;08	185.64	29087	5OC	20;25	19.96
29014	26OC	5;26	220.84	29014	10NO	16;47	56.47	29088	21OC	3;26	215.26	29088	4NO	7;22	49.54
29015	24NO	22;34	250.94	29015	10DE	3;52	86.45	29089	19NO	20;08	245.28	29089	3DE	20;37	79.55
29016	24DE	17;32	281.30					29090	19DE	11;06	275.47				

	NEW MOONS				FULL MOONS		
NUMBER	DATE	TIME	LONG.	NUMBER	DATE	TIME	LONG.

1352

NEW MOONS				FULL MOONS			
				29090	2JA	11;59	109.77
29091	17JA	23;50	305.52	29091	1FE	4;53	139.89
29092	16FE	10;28	335.20	29092	1MR	22;29	169.66
29093	16MR	19;35	4.37	29093	31MR	15;49	198.92
29094	15AP	3;53	33.03	29094	30AP	7;49	227.67
29095	14MY	12;02	61.27	29095	29MY	21;41	255.99
29096	12JN	20;46	89.30	29096	28JN	9;14	284.09
29097	12JL	6;52	117.36	29097	27JL	18;57	312.22
29098	10AU	19;08	145.72	29098	26AU	3;48	340.62
29099	9SE	10;07	174.57	29099	24SE	12;42	9.46
29100	9OC	3;43	204.01	29100	23OC	22;20	38.83
29101	7NO	22;46	233.99	29101	22NO	9;03	68.63
29102	7DE	17;24	264.27	29102	21DE	20;54	98.69

1353

NEW MOONS				FULL MOONS			
29103	6JA	10;02	294.53	29103	20JA	10;01	128.74
29104	4FE	23;59	324.48	29104	19FE	0;30	158.54
29105	6MR	11;24	353.92	29105	20MR	16;13	187.93
29106	4AP	20;47	22.79	29106	19AP	8;32	216.85
29107	4MY	4;40	51.15	29107	19MY	0;29	245.36
29108	2JN	11;47	79.18	29108	17JN	15;12	273.61
29109	1JL	19;05	107.11	29109	17JL	4;24	301.82
29110	31JL	3;40	135.22	29110	15AU	16;20	330.21
29111	29AU	14;41	163.75	29111	14SE	3;28	358.97
29112	28SE	4;53	192.87	29112	13OC	14;13	28.21
29113	27OC	22;11	222.60	29113	12NO	0;46	57.88
29114	26NO	17;26	252.81	29114	11DE	11;18	87.84
29115	26DE	12;54	283.20				

1354

NEW MOONS				FULL MOONS			
				29115	9JA	22;04	117.84
29116	25JA	6;55	313.43	29116	8FE	9;26	147.65
29117	23FE	22;30	343.21	29117	9MR	21;45	177.08
29118	25MR	11;12	12.39	29118	8AP	11;07	206.06
29119	23AP	21;11	40.99	29119	8MY	1;21	234.63
29120	23MY	5;02	69.14	29120	6JN	16;05	262.94
29121	21JN	11;45	97.06	29121	6JL	7;02	291.18
29122	20JL	18;32	125.02	29122	4AU	21;58	319.57
29123	19AU	2;34	153.29	29123	3SE	12;33	348.31
29124	17SE	12;49	182.07	29124	3OC	2;24	17.49
29125	17OC	1;53	211.46	29125	1NO	15;09	47.11
29126	15NO	17;46	241.40	29126	1DE	2;48	77.04
29127	15DE	11;57	271.70	29127	30DE	13;36	107.06

1355

NEW MOONS				FULL MOONS			
29128	14JA	7;17	302.05	29128	28JA	23;57	136.93
29129	13FE	2;14	332.12	29129	27FE	10;09	166.44
29130	14MR	19;10	1.67	29130	28MR	20;29	195.47
29131	13AP	9;04	30.59	29131	27AP	7;15	224.06
29132	12MY	19;53	58.97	29132	26MY	18;54	252.32
29133	11JN	4;23	87.01	29133	25JN	7;56	280.49
29134	10JL	11;42	114.95	29134	24JL	22;41	308.79
29135	8AU	19;00	143.08	29135	23AU	14;54	337.44
29136	7SE	3;11	171.61	29136	22SE	7;41	6.56
29137	6OC	12;55	200.68	29137	21OC	23;59	36.15
29138	5NO	0;41	230.30	29138	20NO	14;45	66.10
29139	4DE	14;49	260.36	29139	20DE	3;59	96.21

1356

NEW MOONS				FULL MOONS			
29140	3JA	7;23	290.62	29140	18JA	15;45	126.22
29141	2FE	1;45	320.80	29141	17FE	2;11	155.88
29142	2MR	20;25	350.60	29142	17MR	11;29	185.05
29143	1AP	13;39	19.84	29143	15AP	20;03	213.71
29144	1MY	4;22	48.52	29144	15MY	4;34	241.96
29145	30MY	16;29	76.77	29145	13JN	14;05	270.02
29146	29JN	2;34	104.81	29146	13JL	1;33	298.14
29147	28JL	11;28	132.91	29147	11AU	15;25	326.57
29148	26AU	19;57	161.30	29148	10SE	7;55	355.49
29149	25SE	4;38	190.15	29149	10OC	1;43	24.96
29150	24OC	14;08	219.52	29150	8NO	19;44	54.90
29151	23NO	1;00	249.34	29151	8DE	12;50	85.12
29152	22DE	13;45	279.44				

1357

NEW MOONS				FULL MOONS			
				29152	7JA	4;11	115.33
29153	21JA	4;32	309.55	29153	5FE	17;15	145.23
29154	19FE	20;53	339.42	29154	7MR	3;55	174.63
29155	21MR	13;46	8.84	29155	5AP	12;34	203.46
29156	20AP	6;10	37.76	29156	4MY	19;59	231.80
29157	19MY	21;24	66.23	29157	3JN	3;14	259.83
29158	18JN	11;11	94.44	29158	2JL	11;25	287.80
29159	17JL	23;30	122.61	29159	31JL	21;32	315.97
29160	16AU	10;29	150.97	29160	30AU	10;12	344.57
29161	14SE	20;31	179.70	29161	29SE	1;35	13.75
29162	14OC	6;09	208.90	29162	28OC	19;21	43.52
29163	12NO	16;03	238.56	29163	27NO	14;34	73.74
29164	12DE	2;44	268.52	29164	27DE	9;40	104.11

1358

NEW MOONS				FULL MOONS			
29165	10JA	14;27	298.57	29165	26JA	2;55	134.30
29166	9FE	3;12	328.42	29166	24FE	17;10	164.01
29167	10MR	16;46	357.89	29167	26MR	4;17	193.12
29168	9AP	7;04	26.89	29168	24AP	12;57	221.66
29169	8MY	21;59	55.49	29169	23MY	20;12	249.77
29170	7JN	13;18	83.81	29170	22JN	3;07	277.70
29171	7JL	4;29	112.06	29171	21JL	10;37	305.70
29172	5AU	18;51	140.44	29172	19AU	19;34	334.01
29173	4SE	8;02	169.14	29173	18SE	6;44	2.84
29174	3OC	20;04	198.26	29174	17OC	20;46	32.28
29175	2NO	7;24	227.82	29175	16NO	13;53	62.28
29176	1DE	18;27	257.73	29176	16DE	9;22	92.64
29177	31DE	5;22	287.76				

1359

NEW MOONS				FULL MOONS			
				29177	15JA	5;17	123.01
29178	29JA	16;07	317.64	29178	13FE	23;30	153.04
29179	28FE	2;46	347.16	29179	15MR	14;43	182.50
29180	29MR	13;41	16.20	29180	14AP	2;50	211.34
29181	28AP	1;30	44.82	29181	13MY	12;27	239.67
29182	27MY	14;40	73.14	29182	11JN	20;25	267.67
29183	26JN	5;16	101.37	29183	11JL	3;33	295.62
29184	25JL	20;47	129.71	29184	9AU	10;50	323.75
29185	24AU	12;30	158.35	29185	7SE	19;05	352.29
29186	23SE	3;52	187.42	29186	7OC	5;16	21.39
29187	22OC	18;31	216.96	29187	5NO	18;14	51.07
29188	21NO	8;15	246.87	29188	5DE	10;15	81.21
29189	20DE	20;47	276.95				

1360

NEW MOONS				FULL MOONS			
				29189	4JA	4;34	111.55
29190	19JA	7;57	306.93	29190	2FE	23;39	141.75
29191	17FE	17;51	336.57	29191	3MR	17;49	171.51
29192	18MR	2;58	5.72	29192	2AP	9;55	200.70
29193	16AP	12;02	34.39	29193	1MY	23;33	229.32
29194	15MY	21;51	62.68	29194	31MY	10;44	257.53
29195	14JN	9;03	90.80	29195	29JN	19;54	285.53
29196	13JL	22;01	118.99	29196	29JL	3;48	313.60
29197	12AU	12;47	147.46	29197	27AU	11;26	341.96
29198	11SE	5;10	176.39	29198	25SE	19;50	10.81
29199	10OC	22;35	205.86	29199	25OC	5;52	40.21
29200	9NO	16;00	235.79	29200	23NO	18;00	70.09
29201	9DE	8;08	265.97	29201	23DE	8;18	100.25

1361

NEW MOONS				FULL MOONS			
29202	7JA	22;04	296.11	29202	22JA	0;21	130.42
29203	6FE	9;39	325.95	29203	20FE	17;28	160.31
29204	7MR	19;19	355.30	29204	22MR	10;49	189.73
29205	6AP	3;45	24.11	29205	21AP	3;26	218.64
29206	5MY	11;43	52.46	29206	20MY	18;23	247.10
29207	3JN	19;55	80.52	29207	19JN	7;09	275.26
29208	3JL	5;08	108.53	29208	18JL	17;51	303.37
29209	1AU	16;12	136.75	29209	17AU	3;13	331.68
29210	31AU	5;50	165.40	29210	15SE	12;13	0.38
29211	29SE	22;20	194.64	29211	14OC	21;38	29.58
29212	29OC	17;03	224.46	29212	13NO	7;55	59.26
29213	28NO	12;19	254.69	29213	12DE	19;13	89.25
29214	28DE	6;12	285.01				

1362

NEW MOONS				FULL MOONS			
				29214	11JA	7;34	119.32
29215	26JA	21;31	315.11	29215	9FE	21;08	149.20
29216	25FE	10;06	344.74	29216	11MR	11;59	178.71
29217	26MR	20;19	13.81	29217	10AP	3;49	207.77
29218	25AP	4;44	42.32	29218	9MY	19;53	236.39
29219	24MY	12;02	70.43	29219	8JN	11;15	264.71
29220	22JN	19;03	98.36	29220	8JL	1;18	292.92
29221	22JL	2;50	126.38	29221	6AU	14;02	321.24
29222	20AU	12;32	154.73	29222	5SE	1;47	349.89
29223	19SE	1;09	183.63	29223	4OC	12;58	18.99
29224	18OC	17;05	213.15	29224	2NO	23;49	48.54
29225	17NO	11;39	243.23	29225	2DE	10;27	78.44
29226	17DE	7;17	273.60	29226	31DE	21;03	108.45

1363

NEW MOONS				FULL MOONS			
29227	16JA	2;12	303.92	29227	30JA	7;55	138.33
29228	14FE	19;02	333.88	29228	28FE	19;29	167.88
29229	16MR	9;07	3.29	29229	30MR	8;02	196.98
29230	14AP	20;20	32.08	29230	28AP	21;35	225.66
29231	14MY	5;05	60.36	29231	28MY	11;55	254.02
29232	12JN	12;12	88.33	29232	27JN	2;44	282.25
29233	11JL	18;55	116.25	29233	26JL	17;49	310.58
29234	10AU	2;10	144.39	29234	25AU	8;51	339.20
29235	8SE	11;17	172.98	29235	23SE	23;25	8.26
29236	7OC	23;00	202.15	29236	23OC	13;02	37.77
29237	6NO	13;34	231.91	29237	22NO	1;26	67.63
29238	6DE	6;43	262.11	29238	21DE	12;42	97.66

1364

NUMBER	DATE	TIME	LONG.	NUMBER	DATE	TIME	LONG.
29239	5JA	1;35	292.47	29239	19JA	23;11	127.61
29240	3FE	20;49	322.66	29240	18FE	9;17	157.24
29241	4MR	14;48	352.41	29241	18MR	19;17	186.42
29242	3AP	6;12	21.55	29242	17AP	5;31	215.12
29243	2MY	18;28	50.11	29243	16MY	16;25	243.45
29244	1JN	4;00	78.24	29244	15JN	4;32	271.61
29245	30JN	11;50	106.20	29245	14JL	18;22	299.83
29246	29JL	19;06	134.24	29246	13AU	10;00	328.34
29247	28AU	2;53	162.62	29247	12SE	2;51	357.31
29248	26SE	11;55	191.51	29248	11OC	19;47	26.77
29249	25OC	22;44	220.95	29249	10NO	11;39	56.65
29250	24NO	11;41	250.87	29250	10DE	1;52	86.75
29251	24DE	2;59	281.08				

1365

NUMBER	DATE	TIME	LONG.	NUMBER	DATE	TIME	LONG.
				29251	8JA	14;24	116.84
29252	22JA	20;25	311.29	29252	7FE	1;25	146.64
29253	21FE	14;56	341.23	29253	8MR	11;06	175.98
29254	23MR	8;53	10.67	29254	6AP	19;46	204.79
29255	22AP	0;48	39.53	29255	6MY	4;00	233.14
29256	21MY	14;10	67.91	29256	4JN	12;42	261.22
29257	20JN	1;14	96.01	29257	3JL	22;58	289.27
29258	19JL	10;47	124.08	29258	2AU	11;39	317.56
29259	17AU	19;34	152.37	29259	1SE	2;59	346.30
29260	16SE	4;17	181.07	29260	30SE	20;24	15.59
29261	15OC	13;29	210.27	29261	30OC	14;41	45.40
29262	13NO	23;42	239.95	29262	29NO	8;34	75.58
29263	13DE	11;28	269.97	29263	29DE	1;00	105.84

1366

NUMBER	DATE	TIME	LONG.	NUMBER	DATE	TIME	LONG.
29264	12JA	1;07	300.09	29264	27JA	15;15	135.89
29265	10FE	16;33	330.04	29265	26FE	3;02	165.48
29266	12MR	9;01	359.61	29266	27MR	12;29	194.50
29267	11AP	1;31	28.68	29267	25AP	20;15	222.98
29268	10MY	17;14	57.28	29268	25MY	3;18	251.07
29269	9JN	7;45	85.56	29269	23JN	10;48	279.02
29270	8JL	20;55	113.73	29270	22JL	19;49	307.09
29271	7AU	8;46	142.03	29271	21AU	7;10	335.51
29272	5SE	19;29	170.64	29272	19SE	21;17	4.49
29273	5OC	5;29	199.70	29273	19OC	14;03	34.06
29274	3NO	15;21	229.23	29274	18NO	8;50	64.15
29275	3DE	1;41	259.12	29275	18DE	4;19	94.52

1367

NUMBER	DATE	TIME	LONG.	NUMBER	DATE	TIME	LONG.
29276	1JA	12;51	289.15	29276	16JA	22;43	124.82
29277	31JA	0;56	319.08	29277	15FE	14;31	154.73
29278	1MR	13;50	348.67	29278	17MR	3;03	184.05
29279	31MR	3;29	17.81	29279	15AP	12;44	212.77
29280	29AP	17;49	46.50	29280	14MY	20;29	241.01
29281	29MY	8;50	74.88	29281	13JN	3;23	268.97
29282	28JN	0;09	103.13	29282	12JL	10;29	296.91
29283	27JL	15;09	131.47	29283	10AU	18;39	325.09
29284	26AU	5;12	160.06	29284	9SE	4;41	353.72
29285	24SE	18;05	189.06	29285	8OC	17;18	22.94
29286	24OC	5;59	218.50	29286	7NO	8;59	52.76
29287	22NO	17;19	248.33	29287	7DE	3;28	83.02
29288	22DE	4;21	278.35				

1368

NUMBER	DATE	TIME	LONG.	NUMBER	DATE	TIME	LONG.
				29288	5JA	23;25	113.42
29289	20JA	15;07	308.31	29289	4FE	18;39	143.61
29290	19FE	1;37	337.95	29290	5MR	11;21	173.29
29291	19MR	12;05	7.14	29291	4AP	0;53	202.34
29292	17AP	23;06	35.86	29292	3MY	11;36	230.83
29293	17MY	11;18	64.24	29293	1JN	20;17	258.92
29294	16JN	1;03	92.46	29294	1JL	3;47	286.86
29295	15JL	16;10	120.73	29295	30JL	10;58	314.91
29296	14AU	8;01	149.26	29296	28AU	18;43	343.30
29297	12SE	23;53	178.20	29297	27SE	3;56	12.20
29298	12OC	15;14	207.61	29298	26OC	15;32	41.68
29299	11NO	5;43	237.44	29299	25NO	6;04	71.68
29300	10DE	19;02	267.51	29300	24DE	23;19	101.97

1369

NUMBER	DATE	TIME	LONG.	NUMBER	DATE	TIME	LONG.
29301	9JA	6;56	297.56	29301	23JA	18;05	132.24
29302	7FE	17;20	327.34	29302	22FE	12;42	162.16
29303	9MR	2;35	356.65	29303	24MR	5;47	191.55
29304	7AP	11;21	25.45	29304	22AP	20;35	220.36
29305	6MY	20;28	53.83	29305	22MY	8;57	248.69
29306	5JN	6;43	81.97	29306	20JN	19;07	276.75
29307	4JL	18;40	110.09	29307	20JL	3;40	304.79
29308	3AU	8;32	138.43	29308	18AU	11;29	333.04
29309	2SE	0;17	167.19	29309	16SE	19;33	1.72
29310	1OC	17;30	196.49	29310	16OC	4;50	30.94
29311	31OC	11;20	226.30	29311	14NO	15;58	60.67
29312	30NO	4;28	256.45	29312	14DE	5;11	90.75
29313	29DE	19;41	286.66				

1370

NUMBER	DATE	TIME	LONG.	NUMBER	DATE	TIME	LONG.
				29313	12JA	20;17	120.93
29314	28JA	8;25	316.64	29314	11FE	12;44	150.92
29315	26FE	18;52	346.16	29315	13MR	5;49	180.49
29316	28MR	3;40	15.15	29316	11AP	22;42	209.56
29317	26AP	11;36	43.62	29317	11MY	14;29	238.16
29318	25MY	19;26	71.75	29318	10JN	4;24	266.41
29319	24JN	3;56	99.73	29319	9JL	16;13	294.53
29320	23JL	13;56	127.84	29320	8AU	2;21	322.77
29321	22AU	2;13	156.32	29321	6SE	11;39	351.33
29322	20SE	17;24	185.34	29322	5OC	21;00	20.38
29323	20OC	11;19	214.98	29323	4NO	6;59	49.91
29324	19NO	6;44	245.11	29324	3DE	17;50	79.83
29325	19DE	1;38	275.46				

1371

NUMBER	DATE	TIME	LONG.	NUMBER	DATE	TIME	LONG.
				29325	2JA	5;36	109.89
29326	17JA	18;22	305.68	29326	31JA	18;21	139.84
29327	16FE	8;16	335.50	29327	2MR	8;17	169.46
29328	17MR	19;29	4.76	29328	31MR	23;23	198.65
29329	16AP	4;37	33.45	29329	30AP	15;13	227.39
29330	15MY	12;17	61.67	29330	30MY	6;55	255.79
29331	13JN	19;15	89.63	29331	28JN	21;42	284.02
29332	13JL	2;29	117.58	29332	28JL	11;15	312.30
29333	11AU	11;06	145.78	29333	26AU	23;42	340.84
29334	9SE	22;14	174.47	29334	25SE	11;24	9.80
29335	9OC	12;36	203.78	29335	24OC	22;39	39.23
29336	8NO	6;05	233.68	29336	23NO	9;32	69.05
29337	8DE	1;29	263.98	29337	22DE	20;10	99.05

1372

NUMBER	DATE	TIME	LONG.	NUMBER	DATE	TIME	LONG.
29338	6JA	20;58	294.37	29338	21JA	6;46	129.00
29339	5FE	14;55	324.49	29339	19FE	17;44	158.65
29340	6MR	6;21	354.10	29340	20MR	5;29	187.88
29341	4AP	18;53	23.10	29341	18AP	18;15	216.67
29342	4MY	4;44	51.55	29342	18MY	7;59	245.09
29343	2JN	12;31	79.60	29343	16JN	22;28	273.34
29344	1JL	19;16	107.51	29344	16JL	13;29	301.61
29345	31JL	2;09	135.55	29345	15AU	4;47	330.13
29346	29AU	10;21	163.96	29346	13SE	19;56	359.05
29347	27SE	20;48	192.93	29347	13OC	10;23	28.44
29348	27OC	10;01	222.49	29348	11NO	23;39	58.23
29349	26NO	1;59	252.55	29349	11DE	11;35	88.25
29350	25DE	20;04	282.88				

1373

NUMBER	DATE	TIME	LONG.	NUMBER	DATE	TIME	LONG.
				29350	9JA	22;25	118.25
29351	24JA	15;12	313.16	29351	8FE	8;33	148.00
29352	23FE	9;52	343.07	29352	9MR	18;21	177.33
29353	25MR	2;33	12.43	29353	8AP	4;11	206.17
29354	23AP	16;19	41.19	29354	7MY	14;26	234.58
29355	23MY	3;07	69.45	29355	6JN	1;40	262.75
29356	21JN	11;43	97.45	29356	5JL	14;30	290.91
29357	20JL	19;14	125.44	29357	4AU	5;17	319.29
29358	19AU	2;47	153.69	29358	2SE	21;49	348.10
29359	17SE	11;17	182.40	29359	2OC	15;08	17.42
29360	16OC	21;17	211.66	29360	1NO	7;56	47.20
29361	15NO	9;10	241.43	29361	30NO	23;13	77.28
29362	14DE	23;16	271.55	29362	30DE	12;41	107.41

1374

NUMBER	DATE	TIME	LONG.	NUMBER	DATE	TIME	LONG.
29363	13JA	15;34	301.78	29363	29JA	0;24	137.34
29364	12FE	9;30	331.81	29364	27FE	10;36	166.85
29365	14MR	3;42	1.42	29365	28MR	19;33	195.82
29366	12AP	20;37	30.48	29366	27AP	3;42	224.30
29367	12MY	11;12	59.02	29367	26MY	11;52	252.43
29368	10JN	23;24	87.20	29368	24JN	21;05	280.45
29369	10JL	9;44	115.26	29369	24JL	8;26	308.61
29370	8AU	18;59	143.47	29370	22AU	22;30	337.17
29371	7SE	3;53	172.03	29371	21SE	15;07	6.27
29372	6OC	12;59	201.07	29372	21OC	9;21	35.93
29373	4NO	22;46	230.61	29373	20NO	3;47	66.03
29374	4DE	9;44	260.53	29374	19DE	21;12	96.32

1375

NUMBER	DATE	TIME	LONG.	NUMBER	DATE	TIME	LONG.
29375	2JA	22;20	290.63	29375	18JA	12;39	126.48
29376	1FE	12;43	320.64	29376	17FE	1;37	156.25
29377	3MR	4;30	350.33	29377	18MR	12;04	185.47
29378	1AP	20;50	19.55	29378	16AP	20;27	214.12
29379	1MY	12;49	48.29	29379	16MY	3;35	242.31
29380	31MY	3;55	76.66	29380	14JN	10;37	270.27
29381	29JN	17;50	104.85	29381	13JL	18;42	298.26
29382	29JL	6;31	133.10	29382	12AU	4;49	326.53
29383	27AU	18;01	161.62	29383	10SE	17;36	355.29
29384	26SE	4;35	190.54	29384	10OC	9;11	24.66
29385	25OC	14;50	219.94	29385	9NO	3;12	54.60
29386	24NO	0;50	249.73	29386	8DE	22;38	84.91
29387	23DE	11;34	279.74				

1376 / 1382

NUMBER	DATE	TIME	LONG.	NUMBER	DATE	TIME	LONG.	NUMBER	DATE	TIME	LONG.	NUMBER	DATE	TIME	LONG.
				29387	7JA	17;50	115.28								
29388	21JA	23;04	309.72	29388	6FE	11;03	145.36	29462	15JA	9;34	303.61	29462	30JA	7;53	138.72
29389	20FE	11;21	339.41	29389	7MR	1;10	174.91	29463	14FE	4;31	333.67	29463	28FE	17;38	168.19
29390	21MR	0;20	8.68	29390	5AP	12;05	203.83	29464	15MR	22;13	3.23	29464	30MR	3;11	197.17
29391	19AP	14;02	37.49	29391	4MY	20;35	232.21	29465	14AP	13;25	32.19	29465	28AP	12;55	225.70
29392	19MY	4;31	65.94	29392	3JN	3;43	260.24	29466	14MY	1;38	60.61	29466	27MY	23;22	253.91
29393	17JN	19;40	94.21	29393	2JL	10;36	288.15	29467	12JN	11;15	88.68	29467	26JN	11;11	282.02
29394	17JL	11;00	122.50	29394	31JL	18;10	316.22	29468	11JL	19;15	116.66	29468	26JL	0;57	310.30
29395	16AU	1;49	151.01	29395	30AU	3;15	344.68	29469	10AU	2;48	144.81	29469	24AU	16;48	338.95
29396	14SE	15;34	179.89	29396	28SE	14;35	13.69	29470	8SE	10;54	173.36	29470	23SE	10;06	8.10
29397	14OC	4;13	209.21	29397	28OC	4;47	43.31	29471	7OC	20;14	202.43	29471	23OC	3;36	37.76
29398	12NO	16;00	238.95	29398	26NO	21;59	73.43	29472	6NO	7;14	232.04	29472	21NO	19;58	67.79
29399	12DE	3;17	268.94	29399	26DE	17;25	103.82	29473	5DE	20;12	262.06	29473	21DE	10;29	97.96

1377 / 1383

NUMBER	DATE	TIME	LONG.	NUMBER	DATE	TIME	LONG.	NUMBER	DATE	TIME	LONG.	NUMBER	DATE	TIME	LONG.
29400	10JA	14;11	298.95	29400	25JA	13;13	134.11	29474	4JA	11;19	292.26	29474	19JA	23;04	127.99
29401	9FE	0;40	328.71	29401	24FE	7;14	163.99	29475	3FE	4;21	322.36	29475	18FE	9;55	157.66
29402	10MR	10;52	358.04	29402	25MR	22;16	193.27	29476	4MR	22;22	352.12	29476	19MR	19;17	186.81
29403	8AP	21;14	26.89	29403	24AP	10;14	221.94	29477	3AP	15;54	21.36	29477	18AP	3;34	215.43
29404	8MY	8;30	55.34	29404	23MY	19;48	250.14	29478	3MY	7;37	50.06	29478	17MY	11;26	243.64
29405	6JN	21;16	83.56	29405	22JN	3;49	278.12	29479	1JN	20;58	78.35	29479	15JN	19;51	271.65
29406	6JL	11;42	111.78	29406	21JL	11;09	306.11	29480	1JL	8;15	106.44	29480	15JL	5;56	299.72
29407	5AU	3;21	140.21	29407	19AU	18;38	334.36	29481	30JL	18;08	134.59	29481	13AU	18;36	328.11
29408	3SE	19;31	169.02	29408	18SE	3;08	3.08	29482	29AU	3;21	163.03	29482	12SE	10;07	357.01
29409	3OC	11;27	198.29	29409	17OC	13;33	32.37	29483	27SE	12;30	191.92	29483	12OC	3;53	26.50
29410	2NO	2;42	228.02	29410	16NO	2;37	62.20	29484	26OC	22;03	221.31	29484	10NO	22;35	56.49
29411	1DE	16;51	258.06	29411	15DE	18;34	92.40	29485	25NO	8;27	251.13	29485	10DE	16;48	86.76
29412	31DE	5;33	288.16					29486	24DE	20;09	281.18				

1378 / 1384

NUMBER	DATE	TIME	LONG.	NUMBER	DATE	TIME	LONG.	NUMBER	DATE	TIME	LONG.	NUMBER	DATE	TIME	LONG.
				29412	14JA	12;38	122.70					29486	9JA	9;24	117.02
29413	29JA	16;38	318.05	29413	13FE	7;21	152.76	29487	23JA	9;30	311.23	29487	7FE	23;38	146.96
29414	28FE	2;16	347.52	29414	15MR	1;09	182.33	29488	22FE	0;24	341.02	29488	8MR	11;14	176.37
29415	29MR	10;58	16.48	29415	13AP	17;00	211.33	29489	22MR	16;17	10.38	29489	6AP	20;26	205.21
29416	27AP	19;34	44.97	29416	13MY	6;32	239.82	29490	21AP	8;17	39.26	29490	6MY	3;56	233.53
29417	27MY	4;58	73.14	29417	11JN	17;48	267.96	29491	20MY	23;44	67.73	29491	4JN	10;47	261.53
29418	25JN	15;52	101.22	29418	11JL	3;12	295.99	29492	19JN	14;17	95.97	29492	3JL	18;09	289.47
29419	25JL	4;44	129.46	29419	9AU	11;26	324.16	29493	19JL	3;44	124.19	29493	2AU	3;08	317.61
29420	23AU	19;38	158.06	29420	7SE	19;26	352.69	29494	17AU	16;04	152.62	29494	31AU	14;35	346.18
29421	22SE	12;22	187.18	29421	7OC	4;11	21.73	29495	16SE	3;20	181.42	29495	30SE	4;52	15.34
29422	22OC	6;17	216.84	29422	5NO	14;26	51.30	29496	15OC	13;55	210.68	29496	29OC	21;51	45.09
29423	21NO	0;10	246.92	29423	5DE	2;38	81.28	29497	14NO	0;04	240.36	29497	28NO	16;50	75.31
29424	20DE	16;37	277.17					29498	13DE	10;32	270.33	29498	28DE	12;26	105.70

1379 / 1385

NUMBER	DATE	TIME	LONG.	NUMBER	DATE	TIME	LONG.	NUMBER	DATE	TIME	LONG.	NUMBER	DATE	TIME	LONG.
				29424	3JA	16;45	111.44					29499	27JA	6;50	135.92
29425	19JA	6;39	307.27	29425	2FE	8;24	141.50	29499	11JA	21;35	300.33	29500	25FE	22;30	165.68
29426	17FE	18;06	336.97	29426	4MR	0;59	171.21	29500	10FE	9;18	330.13	29501	27MR	10;53	194.82
29427	19MR	3;30	6.13	29427	2AP	17;50	200.43	29501	11MR	21;39	359.53	29502	25AP	20;23	223.37
29428	17AP	11;37	34.76	29428	2MY	10;07	229.17	29502	10AP	10;41	28.46	29503	25MY	4;00	251.49
29429	16MY	19;15	62.97	29429	1JN	1;01	257.53	29503	10MY	0;31	57.00	29504	23JN	10;52	279.41
29430	15JN	3;12	90.96	29430	30JN	13;58	285.69	29504	8JN	15;13	85.29	29505	22JL	18;01	307.40
29431	14JL	12;16	118.99	29431	30JL	1;03	313.88	29505	8JL	6;33	113.55	29506	21AU	2;20	335.71
29432	12AU	23;19	147.30	29432	28AU	10;54	342.33	29506	6AU	21;53	141.98	29507	19SE	12;33	4.52
29433	11SE	13;05	176.12	29433	26SE	20;22	11.23	29507	5SE	12;30	170.75	29508	19OC	1;20	33.92
29434	11OC	5;51	205.55	29434	26OC	6;11	40.62	29508	5OC	1;59	199.95	29509	17NO	17;06	63.88
29435	10NO	0;53	235.54	29435	24NO	16;42	70.43	29509	3NO	14;25	229.58	29510	17DE	11;35	94.20
29436	9DE	20;26	265.87	29436	24DE	3;59	100.46	29510	3DE	2;05	259.53				

1380 / 1386

NUMBER	DATE	TIME	LONG.	NUMBER	DATE	TIME	LONG.	NUMBER	DATE	TIME	LONG.	NUMBER	DATE	TIME	LONG.
29437	8JA	14;29	296.19	29437	22JA	16;05	130.46	29511	1JA	13;12	289.56	29511	16JA	7;24	124.56
29438	7FE	5;47	326.18	29438	21FE	5;10	160.19	29512	30JA	23;48	319.42	29512	15FE	2;25	154.61
29439	7MR	18;13	355.64	29439	21MR	19;24	189.49	29513	1MR	9;54	348.90	29513	16MR	18;54	184.11
29440	6AP	4;12	24.52	29440	20AP	10;40	218.36	29514	30MR	19;52	17.88	29514	15AP	8;17	212.98
29441	5MY	12;24	52.88	29441	20MY	2;22	246.85	29515	29AP	6;19	46.43	29515	14MY	18;55	241.33
29442	3JN	19;33	80.90	29442	18JN	17;40	275.12	29516	28MY	18;03	74.68	29516	13JN	3;38	269.37
29443	3JL	2;30	108.81	29443	18JL	7;58	303.37	29517	27JN	7;32	102.86	29517	12JL	11;16	297.33
29444	1AU	10;18	136.90	29444	16AU	21;10	331.82	29518	26JL	22;41	131.20	29518	10AU	18;41	325.48
29445	30AU	20;06	165.40	29445	15SE	9;29	0.66	29519	25AU	14;51	159.87	29519	9SE	2;43	354.03
29446	29SE	8;52	194.48	29446	14OC	21;13	29.95	29520	24SE	7;15	189.00	29520	8OC	12;11	23.12
29447	29OC	0;58	224.18	29447	13NO	8;29	59.67	29521	23OC	23;11	218.61	29521	6NO	23;56	52.77
29448	27NO	19;40	254.38	29448	12DE	19;19	89.65	29522	22NO	14;09	248.59	29522	6DE	14;27	82.86
29449	27DE	15;21	284.78					29523	22DE	3;45	278.72				

1381 / 1387

NUMBER	DATE	TIME	LONG.	NUMBER	DATE	TIME	LONG.	NUMBER	DATE	TIME	LONG.	NUMBER	DATE	TIME	LONG.
				29449	11JA	5;50	119.64					29523	5JA	7;30	113.14
29450	26JA	10;12	315.03	29450	9FE	16;24	149.39	29524	20JA	15;39	308.72	29524	4FE	1;55	143.30
29451	25FE	2;54	344.84	29451	11MR	3;28	178.75	29525	19FE	1;51	338.35	29525	5MR	20;08	173.05
29452	26MR	16;48	14.05	29452	9AP	15;25	207.65	29526	20MR	10;45	7.47	29526	4AP	12;53	202.24
29453	25AP	3;52	42.68	29453	9MY	4;25	236.16	29527	18AP	19;05	36.09	29527	4MY	3;31	230.89
29454	24MY	12;33	70.84	29454	7JN	18;23	264.43	29528	18MY	3;46	64.33	29528	2JN	15;55	259.14
29455	22JN	19;41	98.78	29455	7JL	9;08	292.67	29529	16JN	13;41	92.39	29529	2JL	2;17	287.19
29456	22JL	2;25	126.75	29456	6AU	0;27	321.09	29530	16JL	1;27	120.53	29530	31JL	11;09	315.31
29457	20AU	9;55	155.01	29457	4SE	15;59	349.88	29531	14AU	15;13	148.98	29531	29AU	19;21	343.71
29458	18SE	19;15	183.78	29458	4OC	7;10	19.14	29532	13SE	7;22	177.92	29532	28SE	3;48	12.58
29459	18OC	7;09	213.13	29459	2NO	21;21	48.83	29533	13OC	1;01	207.41	29533	27OC	13;22	41.98
29460	16NO	21;21	243.04	29460	2DE	10;08	78.82	29534	11NO	19;19	237.39	29534	26NO	0;38	71.84
29461	16DE	14;55	273.30	29461	31DE	21;32	108.87	29535	11DE	12;50	267.64	29535	25DE	13;46	101.96

1388   **1394**

NUMBER	DATE	TIME	LONG.	NUMBER	DATE	TIME	LONG.	NUMBER	DATE	TIME	LONG.	NUMBER	DATE	TIME	LONG.
29536	10JA	4;13	297.84	29536	24JA	4;32	132.06	29610	2JA	20;23	290.94	29610	18JA	1;55	126.42
29537	8FE	16;54	327.71	29537	22FE	20;27	161.88	29611	1FE	7;37	320.82	29611	16FE	19;03	156.37
29538	9MR	3;08	357.06	29538	23MR	12;59	191.26	29612	2MR	19;25	350.34	29612	18MR	9;00	185.73
29539	7AP	11;38	25.85	29539	22AP	5;27	220.14	29613	1AP	7;48	19.41	29613	16AP	19;46	214.48
29540	6MY	19;15	54.17	29540	21MY	21;04	248.61	29614	30AP	20;55	48.04	29614	16MY	4;07	242.72
29541	5JN	2;49	82.20	29541	20JN	11;05	276.82	29615	30MY	11;00	76.37	29615	14JN	11;12	270.68
29542	4JL	11;09	110.18	29542	19JL	23;13	305.00	29616	29JN	2;01	104.62	29616	13JL	18;07	298.62
29543	2AU	21;06	138.36	29543	18AU	9;49	333.37	29617	28JL	17;34	132.98	29617	12AU	1;49	326.80
29544	1SE	9;29	166.98	29544	16SE	19;38	2.12	29618	27AU	8;52	161.64	29618	10SE	11;06	355.42
29545	1OC	0;51	196.19	29545	16OC	5;25	31.36	29619	25SE	23;14	190.71	29619	9OC	22;37	24.62
29546	30OC	19;04	226.01	29546	14NO	15;43	61.05	29620	25OC	12;27	220.23	29620	8NO	12;56	54.39
29547	29NO	14;45	256.26	29547	14DE	2;39	91.04	29621	24NO	0;40	250.11	29621	8DE	6;09	84.61
29548	29DE	9;51	286.64					29622	23DE	12;08	280.16				

1389   **1395**

NUMBER	DATE	TIME	LONG.	NUMBER	DATE	TIME	LONG.	NUMBER	DATE	TIME	LONG.	NUMBER	DATE	TIME	LONG.
				29548	12JA	14;15	121.07					29622	7JA	1;27	114.98
29549	28JA	2;37	316.79	29549	11FE	2;36	150.88	29623	21JA	22;56	310.10	29623	5FE	21;01	145.17
29550	26FE	16;23	346.45	29550	12MR	15;57	180.31	29624	20FE	9;07	339.71	29624	7MR	14;48	174.87
29551	28MR	3;26	15.53	29551	11AP	6;26	209.29	29625	21MR	18;51	8.85	29625	6AP	5;38	203.96
29552	26AP	12;20	44.04	29552	10MY	21;47	237.88	29626	20AP	4;42	37.51	29626	5MY	17;31	232.48
29553	25MY	19;50	72.15	29553	9JN	13;18	266.20	29627	19MY	15;28	65.81	29627	4JN	3;06	260.60
29554	24JN	2;43	100.07	29554	9JL	4;13	294.44	29628	18JN	3;53	93.97	29628	3JL	11;14	288.57
29555	23JL	9;58	128.07	29555	7AU	18;09	322.83	29629	17JL	18;12	122.22	29629	1AU	18;47	316.64
29556	21AU	18;41	156.40	29556	6SE	7;10	351.54	29630	16AU	10;03	150.77	29630	31AU	2;33	345.04
29557	20SE	5;57	185.27	29557	5OC	19;27	20.71	29631	15SE	2;39	179.75	29631	29SE	11;20	13.95
29558	19OC	20;29	214.75	29558	4NO	7;11	50.32	29632	14OC	19;11	209.23	29632	28OC	21;57	43.41
29559	18NO	14;06	244.80	29559	3DE	18;21	80.25	29633	13NO	10;57	239.13	29633	27NO	11;05	73.36
29560	18DE	9;33	275.17					29634	13DE	1;28	269.26	29634	27DE	2;53	103.59

1390   **1396**

NUMBER	DATE	TIME	LONG.	NUMBER	DATE	TIME	LONG.	NUMBER	DATE	TIME	LONG.	NUMBER	DATE	TIME	LONG.
				29560	2JA	5;01	110.26	29635	11JA	14;17	299.34	29635	25JA	20;38	133.81
29561	17JA	4;59	305.51	29561	31JA	15;23	140.11	29636	10FE	1;14	329.12	29636	24FE	14;54	163.70
29562	15FE	22;47	335.49	29562	2MR	1;55	169.59	29637	10MR	10;33	358.41	29637	25MR	8;19	193.09
29563	17MR	14;02	4.93	29563	31MR	13;06	198.62	29638	8AP	18;51	27.18	29638	23AP	23;56	221.91
29564	16AP	2;25	33.75	29564	30AP	1;18	227.22	29639	8MY	3;03	55.51	29639	23MY	13;25	250.28
29565	15MY	12;10	62.05	29565	29MY	14;35	255.53	29640	6JN	12;05	83.59	29640	22JN	0;50	278.39
29566	13JN	19;57	90.05	29566	28JN	4;53	283.74	29641	5JL	22;45	111.66	29641	21JL	10;32	306.47
29567	13JL	2;48	117.98	29567	27JL	20;01	312.08	29642	4AU	11;34	139.97	29642	19AU	19;09	334.78
29568	11AU	9;52	146.12	29568	26AU	11;42	340.75	29643	3SE	2;38	168.72	29643	18SE	3;33	3.49
29569	9SE	18;17	174.70	29569	25SE	3;25	9.87	29644	2OC	19;44	198.03	29644	17OC	12;39	32.72
29570	9OC	4;57	203.85	29570	24OC	18;29	39.45	29645	1NO	14;06	227.88	29645	15NO	23;07	62.44
29571	7NO	18;18	233.58	29571	23NO	8;13	69.39	29646	1DE	8;23	258.09	29646	15DE	11;19	92.49
29572	7DE	10;15	263.73	29572	22DE	20;23	99.46	29647	31DE	1;04	288.36				

1391   **1397**

NUMBER	DATE	TIME	LONG.	NUMBER	DATE	TIME	LONG.	NUMBER	DATE	TIME	LONG.	NUMBER	DATE	TIME	LONG.
29573	6JA	4;10	294.05	29573	21JA	7;11	129.41					29647	14JA	1;10	122.61
29574	4FE	22;59	324.21	29574	19FE	17;03	159.01	29648	29JA	15;08	318.38	29648	12FE	16;21	152.52
29575	6MR	17;20	353.95	29575	21MR	2;26	188.15	29649	28FE	2;26	347.92	29649	14MR	8;21	182.04
29576	5AP	9;46	23.12	29576	19AP	11;47	216.79	29650	29MR	11;33	16.90	29650	13AP	0;42	211.07
29577	4MY	23;26	51.72	29577	18MY	21;34	245.07	29651	27AP	19;22	45.35	29651	12MY	16;42	239.66
29578	3JN	10;16	79.90	29578	17JN	8;28	273.17	29652	27MY	2;43	73.44	29652	11JN	7;34	267.94
29579	2JL	19;02	107.89	29579	16JL	21;08	301.35	29653	25JN	10;29	101.40	29653	10JL	20;47	296.12
29580	1AU	2;48	135.97	29580	15AU	12;01	329.85	29654	24JL	19;30	129.48	29654	9AU	8;19	324.43
29581	30AU	10;41	164.37	29581	14SE	4;54	358.83	29655	23AU	6;36	157.92	29655	7SE	18;40	353.05
29582	28SE	19;30	193.26	29582	13OC	22;44	28.35	29656	21SE	20;31	186.91	29656	7OC	4;38	22.14
29583	28OC	5;46	222.70	29583	12NO	16;03	58.30	29657	21OC	13;31	216.53	29657	5NO	14;49	51.71
29584	26NO	17;45	252.60	29584	12DE	7;43	88.47	29658	20NO	8;49	246.67	29658	5DE	1;31	81.63
29585	26DE	7;43	282.75					29659	20DE	4;35	277.05				

1392   **1398**

NUMBER	DATE	TIME	LONG.	NUMBER	DATE	TIME	LONG.	NUMBER	DATE	TIME	LONG.	NUMBER	DATE	TIME	LONG.
				29585	10JA	21;19	118.60					29659	3JA	12;45	111.66
29586	24JA	23;40	312.89	29586	9FE	8;57	148.41	29660	18JA	22;41	307.33	29660	2FE	0;31	141.56
29587	23FE	17;07	342.77	29587	9MR	18;53	177.74	29661	17FE	13;55	337.19	29661	3MR	13;04	171.10
29588	24MR	10;50	12.18	29588	8AP	3;29	206.53	29662	19MR	2;11	6.47	29662	2AP	2;41	200.21
29589	23AP	3;25	41.06	29589	7MY	11;17	234.84	29663	17AP	11;58	35.16	29663	1MY	17;24	228.89
29590	22MY	17;56	69.47	29590	5JN	19;08	262.88	29664	16MY	20;00	63.39	29664	31MY	8;46	257.27
29591	21JN	6;15	97.62	29591	5JL	4;09	290.89	29665	15JN	3;02	91.34	29665	30JN	0;04	285.52
29592	20JL	16;54	125.74	29592	3AU	15;26	319.13	29666	14JL	9;59	119.29	29666	29JL	14;40	313.85
29593	19AU	2;35	154.08	29593	2SE	5;36	347.83	29667	12AU	17;52	147.48	29667	28AU	4;23	342.46
29594	17SE	11;56	182.82	29594	1OC	22;30	17.12	29668	11SE	3;49	176.13	29668	26SE	17;18	11.49
29595	16OC	21;26	212.06	29595	31OC	17;06	46.97	29669	10OC	16;45	205.40	29669	26OC	5;35	40.98
29596	15NO	7;29	241.75	29596	30NO	11;54	77.19	29670	9NO	8;59	235.27	29670	24NO	17;13	70.84
29597	14DE	18;30	271.74	29597	30DE	5;31	107.50	29671	9DE	3;45	265.55	29671	24DE	4;10	100.87

1393   **1399**

NUMBER	DATE	TIME	LONG.	NUMBER	DATE	TIME	LONG.	NUMBER	DATE	TIME	LONG.	NUMBER	DATE	TIME	LONG.
29598	13JA	6;53	301.80	29598	28JA	21;01	137.59	29672	7JA	23;23	295.94	29672	22JA	14;34	130.79
29599	11FE	20;49	331.68	29599	27FE	9;52	167.21	29673	6FE	18;05	326.08	29673	21FE	0;47	160.39
29600	13MR	12;00	1.17	29600	28MR	20;05	196.24	29674	8MR	10;35	355.72	29674	22MR	11;19	189.55
29601	12AP	3;46	30.19	29601	27AP	4;12	224.71	29675	7AP	0;19	24.75	29675	20AP	22;41	218.26
29602	11MY	19;22	58.77	29602	26MY	11;08	252.79	29676	6MY	11;17	53.22	29676	20MY	11;11	246.63
29603	10JN	10;21	87.07	29603	24JN	18;02	280.71	29677	4JN	19;57	81.30	29677	19JN	0;51	274.84
29604	10JL	0;28	115.28	29604	24JL	2;04	308.75	29678	4JL	3;10	109.23	29678	18JL	15;35	303.11
29605	8AU	13;35	143.64	29605	22AU	12;15	337.14	29679	2AU	10;04	137.28	29679	17AU	7;11	331.65
29606	7SE	1;38	172.33	29606	21SE	1;10	6.09	29680	31AU	17;48	165.70	29680	15SE	23;14	0.63
29607	6OC	12;45	201.45	29607	20OC	16;57	35.64	29681	30SE	3;22	194.64	29681	15OC	15;02	30.09
29608	4NO	23;16	231.03	29608	19NO	11;09	65.72	29682	29OC	15;27	224.18	29682	14NO	5;45	59.95
29609	4DE	9;40	260.93	29609	19DE	6;42	96.09	29683	28NO	6;09	254.20	29683	13DE	18;51	90.02
								29684	27DE	23;07	284.48				

1400

NEW NUMBER	DATE	TIME	LONG.	FULL NUMBER	DATE	TIME	LONG.
				29684	12JA	6;19	120.05
29685	26JA	17;29	314.72	29685	10FE	16;29	149.79
29686	25FE	12;04	344.61	29686	11MR	1;53	179.08
29687	26MR	5;28	13.98	29687	9AP	10;58	207.86
29688	24AP	20;29	42.77	29688	8MY	20;14	236.23
29689	24MY	8;41	71.08	29689	7JN	6;18	264.34
29690	22JN	18;27	99.12	29690	6JL	17;54	292.45
29691	22JL	2;42	127.15	29691	5AU	7;40	320.81
29692	20AU	10;35	155.43	29692	3SE	23;45	349.61
29693	18SE	19;02	184.16	29693	3OC	17;30	18.96
29694	18OC	4;40	213.42	29694	2NO	11;31	48.81
29695	16NO	15;50	243.18	29695	2DE	4;19	78.97
29696	16DE	4;46	273.26	29696	31DE	19;04	109.16

1401

NEW NUMBER	DATE	TIME	LONG.	FULL NUMBER	DATE	TIME	LONG.
29697	14JA	19;36	303.41	29697	30JA	7;38	139.11
29698	13FE	12;10	333.38	29698	28FE	18;17	168.61
29699	15MR	5;39	2.94	29699	30MR	3;20	197.57
29700	13AP	22;46	31.99	29700	28AP	11;17	226.02
29701	13MY	14;17	60.55	29701	27MY	18;50	254.12
29702	12JN	3;42	88.77	29702	26JN	3;01	282.09
29703	11JL	15;15	116.89	29703	25JL	13;00	310.21
29704	10AU	1;32	145.15	29704	24AU	1;42	338.72
29705	8SE	11;14	173.76	29705	22SE	17;25	7.81
29706	7OC	20;50	202.85	29706	22OC	11;30	37.48
29707	6NO	6;43	232.41	29707	21NO	6;34	67.62
29708	5DE	17;14	262.32	29708	21DE	1;03	97.95

1402

NEW NUMBER	DATE	TIME	LONG.	FULL NUMBER	DATE	TIME	LONG.
29709	4JA	4;50	292.38	29709	19JA	17;44	128.16
29710	2FE	17;48	322.31	29710	18FE	7;53	157.97
29711	4MR	8;09	351.92	29711	19MR	19;18	187.20
29712	2AP	23;25	21.08	29712	18AP	4;15	215.85
29713	2MY	14;56	49.79	29713	17MY	11;33	244.04
29714	1JN	6;10	78.15	29714	15JN	18;14	271.98
29715	30JN	20;47	106.38	29715	15JL	1;33	299.94
29716	30JL	10;36	134.68	29716	13AU	10;35	328.18
29717	28AU	23;27	163.27	29717	11SE	22;09	356.92
29718	27SE	11;18	192.26	29718	11OC	12;36	26.26
29719	26OC	22;18	221.71	29719	10NO	5;47	56.17
29720	25NO	8;51	251.54	29720	10DE	0;53	86.48
29721	24DE	19;24	281.54				

1403

NEW NUMBER	DATE	TIME	LONG.	FULL NUMBER	DATE	TIME	LONG.
				29721	8JA	20;30	116.86
29722	23JA	6;15	311.48	29722	7FE	14;49	146.98
29723	21FE	17;34	341.12	29723	9MR	6;21	176.56
29724	23MR	5;22	10.32	29724	7AP	18;34	205.52
29725	21AP	17;48	39.06	29725	7MY	3;56	233.92
29726	21MY	7;08	67.46	29726	5JN	11;29	261.95
29727	19JN	21;36	95.69	29727	4JL	18;22	289.87
29728	19JL	13;00	123.99	29728	3AU	1;38	317.94
29729	18AU	4;43	152.55	29729	1SE	10;08	346.39
29730	16SE	19;55	181.51	29730	30SE	20;34	15.38
29731	16OC	10;01	210.91	29731	30OC	9;30	44.96
29732	14NO	22;56	240.71	29732	29NO	1;19	75.04
29733	14DE	10;53	270.74	29733	28DE	19;41	105.38

1404

NEW NUMBER	DATE	TIME	LONG.	FULL NUMBER	DATE	TIME	LONG.
29734	12JA	22;01	300.75	29734	27JA	15;16	135.66
29735	11FE	8;23	330.48	29735	26FE	10;01	165.55
29736	11MR	18;05	359.77	29736	27MR	2;17	194.86
29737	10AP	3;32	28.56	29737	25AP	15;31	223.56
29738	9MY	13;29	56.95	29738	25MY	2;08	251.80
29739	8JN	0;49	85.11	29739	23JN	10;57	279.81
29740	7JL	14;05	113.29	29740	22JL	18;48	307.83
29741	6AU	5;18	141.71	29741	21AU	2;30	336.11
29742	4SE	21;49	170.54	29742	19SE	10;50	4.84
29743	4OC	14;44	199.88	29743	18OC	20;34	34.12
29744	3NO	7;14	229.67	29744	17NO	8;26	63.91
29745	2DE	22;39	259.77	29745	16DE	22;54	94.06

1405

NEW NUMBER	DATE	TIME	LONG.	FULL NUMBER	DATE	TIME	LONG.
29746	1JA	12;26	289.92	29746	15JA	15;39	124.29
29747	31JA	0;18	319.84	29747	14FE	9;38	154.31
29748	1MR	10;15	349.30	29748	16MR	3;24	183.86
29749	30MR	18;47	18.23	29749	14AP	19;50	212.87
29750	29AP	2;43	46.67	29750	14MY	10;20	241.38
29751	28MY	11;02	74.79	29751	12JN	22;49	269.56
29752	26JN	20;40	102.82	29752	12JL	9;27	297.65
29753	26JL	8;20	131.01	29753	10AU	18;42	325.87
29754	24AU	22;19	159.59	29754	9SE	3;19	354.45
29755	23SE	14;36	188.71	29755	8OC	12;10	23.52
29756	23OC	8;40	218.40	29756	6NO	22;01	53.08
29757	22NO	3;23	248.53	29757	6DE	9;22	83.04
29758	21DE	21;12	278.84				

1406

NEW NUMBER	DATE	TIME	LONG.	FULL NUMBER	DATE	TIME	LONG.
				29758	4JA	22;20	113.15
29759	20JA	12;41	308.99	29759	3FE	12;42	143.14
29760	19FE	1;16	338.72	29760	5MR	4;04	172.78
29761	20MR	11;16	7.89	29761	3AP	20;02	201.95
29762	18AP	19;28	36.50	29762	3MY	12;05	230.66
29763	18MY	2;49	64.68	29763	2JN	3;33	259.03
29764	16JN	10;12	92.64	29764	1JL	17;44	287.24
29765	15JL	18;26	120.64	29765	31JL	6;16	315.50
29766	14AU	4;24	148.93	29766	29AU	17;23	344.03
29767	12SE	16;54	177.72	29767	28SE	3;43	12.97
29768	12OC	8;29	207.12	29768	27OC	13;57	42.40
29769	11NO	2;55	237.10	29769	26NO	0;31	72.23
29770	10DE	22;49	267.44	29770	25DE	11;29	102.25

1407

NEW NUMBER	DATE	TIME	LONG.	FULL NUMBER	DATE	TIME	LONG.
29771	9JA	18;01	297.80	29771	23JA	22;51	132.21
29772	8FE	10;44	327.84	29772	22FE	10;44	161.87
29773	10MR	0;22	357.34	29773	23MR	23;29	191.09
29774	8AP	11;13	26.22	29774	22AP	13;23	219.88
29775	7MY	19;57	54.59	29775	22MY	4;17	248.33
29776	6JN	3;20	82.61	29776	20JN	19;39	276.60
29777	5JL	10;12	110.53	29777	20JL	10;45	304.89
29778	3AU	17;32	138.61	29778	19AU	1;09	333.41
29779	2SE	2;23	167.09	29779	17SE	14;44	2.31
29780	1OC	13;50	196.13	29780	17OC	3;37	31.68
29781	31OC	4;31	225.79	29781	15NO	15;48	61.45
29782	29NO	22;12	255.96	29782	15DE	3;12	91.46
29783	29DE	17;36	286.34				

1408

NEW NUMBER	DATE	TIME	LONG.	FULL NUMBER	DATE	TIME	LONG.
				29783	13JA	13;49	121.44
29784	28JA	12;53	316.60	29784	11FE	23;55	151.16
29785	27FE	6;30	346.43	29785	12MR	10;00	180.46
29786	27MR	21;33	15.68	29786	10AP	20;37	209.29
29787	26AP	9;49	44.33	29787	10MY	8;16	237.73
29788	25MY	19;32	72.53	29788	8JN	21;10	265.95
29789	24JN	3;23	100.49	29789	8JL	11;20	294.16
29790	23JL	10;23	128.48	29790	7AU	2;37	322.59
29791	21AU	17;41	156.75	29791	5SE	18;44	351.43
29792	20SE	2;21	185.51	29792	5OC	11;03	20.75
29793	19OC	13;14	214.84	29793	4NO	2;41	50.52
29794	18NO	2;41	244.71	29794	3DE	16;50	80.57
29795	17DE	18;33	274.92				

1409

NEW NUMBER	DATE	TIME	LONG.	FULL NUMBER	DATE	TIME	LONG.
				29795	2JA	5;09	110.66
29796	16JA	12;12	305.19	29796	31JA	15;51	140.52
29797	15FE	6;39	335.21	29797	2MR	1;26	169.96
29798	17MR	0;38	4.76	29798	31MR	10;24	198.90
29799	15AP	16;49	33.75	29799	29AP	19;17	227.37
29800	15MY	6;25	62.21	29800	29MY	4;41	255.52
29801	13JN	17;22	90.33	29801	27JN	15;17	283.59
29802	13JL	2;21	118.35	29802	27JL	3;52	311.83
29803	11AU	10;27	146.54	29803	25AU	18;53	340.46
29804	9SE	18;42	175.11	29804	24SE	12;07	9.63
29805	9OC	3;52	204.20	29805	24OC	6;27	39.34
29806	7NO	14;21	233.80	29806	23NO	0;14	69.44
29807	7DE	2;23	263.79	29807	22DE	16;13	99.67

1410

NEW NUMBER	DATE	TIME	LONG.	FULL NUMBER	DATE	TIME	LONG.
29808	5JA	16;10	293.93	29808	21JA	5;53	129.74
29809	4FE	7;42	323.96	29809	19FE	17;23	159.42
29810	6MR	0;35	353.65	29810	21MR	3;03	188.56
29811	4AP	17;49	22.86	29811	19AP	11;19	217.17
29812	4MY	10;06	51.58	29812	18MY	18;48	245.35
29813	3JN	0;34	79.90	29813	17JN	2;25	273.32
29814	2JL	13;06	108.04	29814	16JL	11;18	301.35
29815	1AU	0;07	136.25	29815	14AU	22;34	329.69
29816	30AU	10;17	164.74	29816	13SE	12;52	358.56
29817	28SE	20;07	193.68	29817	13OC	6;02	28.04
29818	28OC	6;00	223.11	29818	12NO	0;57	58.06
29819	26NO	16;16	252.93	29819	11DE	20;02	88.37
29820	26DE	3;16	282.95				

1411

NEW NUMBER	DATE	TIME	LONG.	FULL NUMBER	DATE	TIME	LONG.
				29820	10JA	13;48	118.67
29821	24JA	15;22	312.94	29821	9FE	5;16	148.65
29822	23FE	4;48	342.65	29822	10MR	17;57	178.09
29823	24MR	19;23	11.94	29823	9AP	3;58	206.94
29824	23AP	10;35	40.77	29824	8MY	11;52	235.26
29825	23MY	1;50	69.22	29825	6JN	18;39	263.25
29826	21JN	16;47	97.47	29826	6JL	1;28	291.17
29827	21JL	7;08	125.74	29827	4AU	9;32	319.28
29828	19AU	20;44	154.23	29828	2SE	19;49	347.82
29829	18SE	9;23	183.11	29829	2OC	8;54	16.95
29830	17OC	21;02	212.43	29830	1NO	0;51	46.68
29831	16NO	7;57	242.16	29831	30NO	19;11	76.88
29832	15DE	18;32	272.14	29832	30DE	14;46	107.27

NUMBER	DATE	TIME	LONG.	NUMBER	DATE	TIME	LONG.	NUMBER	DATE	TIME	LONG.	NUMBER	DATE	TIME	LONG.

1412 / 1418

NUMBER	DATE	TIME	LONG.	NUMBER	DATE	TIME	LONG.	NUMBER	DATE	TIME	LONG.	NUMBER	DATE	TIME	LONG.
29833	14JA	5;10	302.12	29833	29JA	9;54	137.51	29907	7JA	7;18	295.65	29907	22JA	15;01	131.20
29834	12FE	16;04	331.87	29834	28FE	2;53	167.31	29908	6FE	1;18	325.77	29908	21FE	0;59	160.79
29835	13MR	3;21	1.20	29835	28MR	16;40	196.49	29909	7MR	19;28	355.48	29909	22MR	10;01	189.89
29836	11AP	15;09	30.07	29836	27AP	3;18	225.06	29910	6AP	12;32	24.66	29910	20AP	18;39	218.49
29837	11MY	3;44	58.54	29837	26MY	11;35	253.20	29911	6MY	3;25	53.29	29911	20MY	3;31	246.71
29838	9JN	17;27	86.79	29838	24JN	18;41	281.13	29912	4JN	15;40	81.52	29912	18JN	13;16	274.77
29839	9JL	8;25	115.03	29839	24JL	1;42	309.12	29913	4JL	1;38	109.56	29913	18JL	0;42	302.90
29840	8AU	0;13	143.49	29840	22AU	9;35	337.43	29914	2AU	10;12	137.68	29914	16AU	14;30	331.37
29841	6SE	16;02	172.33	29841	20SE	19;05	6.23	29915	31AU	18;27	166.11	29915	15SE	6;51	0.35
29842	6OC	7;01	201.61	29842	20OC	6;47	35.61	29916	30SE	3;17	195.03	29916	15OC	1;02	29.90
29843	4NO	20;48	231.31	29843	18NO	21;11	65.52	29917	29OC	13;13	224.48	29917	13NO	19;32	59.91
29844	4DE	9;22	261.31	29844	18DE	14;20	95.79	29918	28NO	0;31	254.35	29918	13DE	12;41	90.16
								29919	27DE	13;20	284.46				

1413 / 1419

NUMBER	DATE	TIME	LONG.	NUMBER	DATE	TIME	LONG.	NUMBER	DATE	TIME	LONG.	NUMBER	DATE	TIME	LONG.
29845	2JA	20;57	291.36	29845	17JA	9;26	126.12					29919	12JA	3;36	120.33
29846	1FE	7;37	321.21	29846	16FE	4;42	156.16	29920	26JA	3;49	314.53	29920	10FE	16;06	150.17
29847	2MR	17;27	350.65	29847	17MR	22;12	185.68	29921	24FE	19;51	344.33	29921	12MR	2;32	179.49
29848	1AP	2;43	19.59	29848	16AP	12;51	214.59	29922	26MR	12;48	13.69	29922	10AP	11;17	208.27
29849	30AP	12;04	48.07	29849	16MY	0;41	242.97	29923	25AP	5;30	42.56	29923	9MY	18;54	236.56
29850	29MY	22;24	76.26	29850	14JN	10;20	271.04	29924	24MY	20;52	71.00	29924	8JN	2;12	264.57
29851	28JN	10;31	104.39	29851	13JL	18;39	299.03	29925	23JN	10;24	99.19	29925	7JL	10;14	292.54
29852	28JL	0;47	132.70	29852	12AU	2;29	327.22	29926	22JL	22;16	127.36	29926	5AU	20;16	320.74
29853	26AU	16;53	161.38	29853	10SE	10;34	355.79	29927	21AU	9;02	155.76	29927	4SE	8;58	349.40
29854	25SE	9;57	190.56	29854	9OC	19;39	24.88	29928	19SE	19;13	184.56	29928	4OC	0;54	18.67
29855	25OC	3;01	220.23	29855	8NO	6;28	54.51	29929	19OC	5;16	213.84	29929	2NO	19;16	48.52
29856	23NO	19;17	250.28	29856	7DE	19;36	84.55	29930	17NO	15;27	243.55	29930	2DE	14;36	78.78
29857	23DE	10;04	280.46					29931	17DE	2;04	273.54				

1414 / 1420

NUMBER	DATE	TIME	LONG.	NUMBER	DATE	TIME	LONG.	NUMBER	DATE	TIME	LONG.	NUMBER	DATE	TIME	LONG.
				29857	6JA	11;11	114.77					29931	1JA	9;15	109.13
29858	21JA	22;56	310.49	29858	5FE	4;31	144.87	29932	15JA	13;28	303.55	29932	31JA	1;57	139.26
29859	20FE	9;43	340.13	29859	6MR	22;19	174.58	29933	14FE	2;01	333.35	29933	29FE	16;00	168.91
29860	21MR	18;43	9.23	29860	5AP	15;19	203.77	29934	14MR	15;46	2.76	29934	30MR	3;13	197.96
29861	20AP	2;38	37.81	29861	5MY	6;43	232.43	29935	13AP	6;26	31.72	29935	28AP	11;58	226.44
29862	19MY	10;28	66.00	29862	3JN	20;13	260.71	29936	12MY	21;30	60.27	29936	27MY	19;06	254.52
29863	17JN	19;12	94.02	29863	3JL	7;51	288.82	29937	11JN	12;33	88.56	29937	26JN	1;43	282.43
29864	17JL	5;43	122.11	29864	1AU	17;54	316.99	29938	11JL	3;18	116.80	29938	25JL	9;02	310.44
29865	15AU	18;32	150.53	29865	31AU	2;57	345.45	29939	9AU	17;31	145.22	29939	23AU	18;09	338.81
29866	14SE	9;48	179.45	29866	29SE	11;48	14.36	29940	8SE	6;56	173.98	29940	22SE	5;53	7.72
29867	14OC	3;15	208.96	29867	28OC	21;13	43.77	29941	7OC	19;23	203.18	29941	21OC	20;30	37.25
29868	12NO	22;00	238.97	29868	27NO	7;52	73.62	29942	6NO	6;51	232.81	29942	20NO	13;48	67.30
29869	12DE	16;38	269.27	29869	26DE	19;59	103.69	29943	5DE	17;41	262.74	29943	20DE	8;57	97.66

1415 / 1421

NUMBER	DATE	TIME	LONG.	NUMBER	DATE	TIME	LONG.	NUMBER	DATE	TIME	LONG.	NUMBER	DATE	TIME	LONG.
29870	11JA	9;29	299.53	29870	25JA	9;32	133.73	29944	4JA	4;14	292.74	29944	19JA	4;29	127.99
29871	9FE	23;30	329.44	29871	24FE	0;11	163.48	29945	2FE	14;52	322.58	29945	17FE	22;39	157.97
29872	11MR	10;37	358.81	29872	25MR	15;36	192.79	29946	4MR	1;43	352.05	29946	19MR	14;01	187.37
29873	9AP	19;29	27.59	29873	24AP	7;26	221.64	29947	2AP	12;57	21.04	29947	18AP	2;05	216.15
29874	9MY	3;01	55.89	29874	23MY	23;10	250.10	29948	2MY	0;49	49.61	29948	17MY	11;23	244.42
29875	7JN	10;11	83.90	29875	22JN	14;06	278.35	29949	31MY	13;44	77.89	29949	15JN	18;56	272.40
29876	6JL	17;50	111.85	29876	22JL	3;38	306.59	29950	30JN	4;00	106.10	29950	15JL	1;54	300.34
29877	5AU	2;50	140.01	29877	20AU	15;39	335.03	29951	29JL	19;32	134.47	29951	13AU	9;22	328.52
29878	3SE	14;01	168.60	29878	19SE	2;33	3.84	29952	28AU	11;40	163.18	29952	11SE	18;05	357.14
29879	3OC	4;07	197.78	29879	18OC	13;01	33.13	29953	27SE	3;27	192.33	29953	11OC	4;43	26.32
29880	1NO	21;20	227.57	29880	16NO	23;33	62.85	29954	26OC	18;09	221.93	29954	9NO	17;47	56.06
29881	1DE	16;50	257.82	29881	16DE	10;22	92.84	29955	25NO	7;31	251.87	29955	9DE	9;35	86.22
29882	31DE	12;41	288.23					29956	24DE	19;40	281.95				

1416 / 1422

NUMBER	DATE	TIME	LONG.	NUMBER	DATE	TIME	LONG.	NUMBER	DATE	TIME	LONG.	NUMBER	DATE	TIME	LONG.
				29882	14JA	21;28	122.84					29956	8JA	3;46	116.55
29883	30JA	6;47	318.42	29883	13FE	8;52	152.60	29957	23JA	6;45	311.90	29957	6FE	23;03	146.71
29884	28FE	21;54	348.13	29884	13MR	20;52	181.95	29958	21FE	16;51	341.48	29958	8MR	17;28	176.42
29885	29MR	9;59	17.22	29885	12AP	9;52	210.86	29959	23MR	2;08	10.58	29959	7AP	9;29	205.54
29886	27AP	19;36	45.75	29886	12MY	0;03	239.38	29960	21AP	11;07	39.18	29960	6MY	22;38	234.09
29887	27MY	3;31	73.86	29887	10JN	15;09	267.68	29961	20MY	20;37	67.43	29961	5JN	9;17	262.25
29888	25JN	10;31	101.79	29888	10JL	6;29	295.94	29962	19JN	7;35	95.53	29962	4JL	18;15	290.25
29889	24JL	17;32	129.79	29889	8AU	21;27	324.38	29963	18JL	20;43	123.73	29963	3AU	2;23	318.36
29890	23AU	1;34	158.11	29890	7SE	11;43	353.16	29964	17AU	12;02	152.27	29964	1SE	10;25	346.80
29891	21SE	11;41	186.94	29891	7OC	1;15	22.40	29965	16SE	4;55	181.28	29965	30SE	19;05	15.72
29892	21OC	0;47	216.39	29892	5NO	14;02	52.07	29966	15OC	22;22	210.82	29966	30OC	5;04	45.17
29893	19NO	17;07	246.40	29893	5DE	2;00	82.04	29967	14NO	15;22	240.78	29967	28NO	17;01	75.08
29894	19DE	11;51	276.74					29968	14DE	7;09	270.97	29968	28DE	7;20	105.25

1417 / 1423

NUMBER	DATE	TIME	LONG.	NUMBER	DATE	TIME	LONG.	NUMBER	DATE	TIME	LONG.	NUMBER	DATE	TIME	LONG.
				29894	3JA	13;01	112.07	29969	12JA	21;04	301.10	29969	26JA	23;44	135.40
29895	18JA	7;20	307.07	29895	1FE	23;14	141.90	29970	11FE	8;50	330.90	29970	25FE	17;14	165.25
29896	17FE	1;50	337.07	29896	3MR	9;03	171.33	29971	12MR	18;31	0.18	29971	27MR	10;31	194.61
29897	18MR	18;07	6.53	29897	1AP	19;00	200.29	29972	11AP	2;42	28.92	29972	26AP	2;37	223.44
29898	17AP	7;41	35.38	29898	1MY	5;53	228.82	29973	10MY	10;16	57.21	29973	25MY	17;03	251.83
29899	16MY	18;36	63.72	29899	30MY	17;56	257.07	29974	8JN	18;17	85.24	29974	24JN	5;40	279.99
29900	15JN	3;18	91.74	29900	29JN	7;21	285.25	29975	8JL	3;44	113.26	29975	23JL	16;38	308.13
29901	14JL	10;40	119.70	29901	28JL	22;08	313.58	29976	6AU	15;20	141.54	29976	22AU	2;20	336.49
29902	12AU	17;48	147.86	29902	27AU	14;03	342.27	29977	5SE	5;27	170.27	29977	20SE	11;24	5.26
29903	11SE	1;48	176.44	29903	26SE	6;37	11.44	29978	4OC	22;01	199.58	29978	19OC	20;39	34.51
29904	10OC	11;38	205.58	29904	25OC	23;01	41.10	29979	3NO	16;26	229.44	29979	18NO	6;44	64.23
29905	8NO	23;51	235.27	29905	24NO	14;13	71.11	29980	3DE	11;30	259.69	29980	17DE	18;07	94.25
29906	8DE	14;33	265.38	29906	24DE	3;34	101.23								

1424

NEW MOONS				FULL MOONS			
29981	2JA	5;31	290.02	29981	16JA	6;52	124.31
29982	31JA	21;02	320.09	29982	14FE	20;46	154.16
29983	1MR	9;29	349.66	29983	15MR	11;32	183.60
29984	30MR	19;15	18.64	29984	14AP	2;56	212.58
29985	29AP	3;13	47.08	29985	13MY	18;37	241.14
29986	28MY	10;21	75.15	29986	12JN	9;59	269.44
29987	26JN	17;35	103.09	29987	12JL	0;23	297.68
29988	26JL	1;49	131.14	29988	10AU	13;22	326.05
29989	24AU	11;51	159.56	29989	9SE	1;03	354.75
29990	23SE	0;29	188.52	29990	8OC	11;55	23.90
29991	22OC	16;16	218.10	29991	6NO	22;34	53.50
29992	21NO	10;53	248.22	29992	6DE	9;21	83.43
29993	21DE	6;54	278.62				

1425

NEW MOONS				FULL MOONS			
				29993	4JA	20;17	113.45
29994	20JA	2;05	308.93	29994	3FE	7;22	143.30
29995	18FE	18;43	338.84	29995	4MR	18;46	172.78
29996	20MR	8;11	8.15	29996	3AP	6;55	201.81
29997	18AP	18;52	36.86	29997	2MY	20;14	230.42
29998	18MY	3;28	65.09	29998	1JN	10;45	258.76
29999	16JN	10;49	93.06	29999	1JL	2;01	287.01
30000	15JL	17;44	121.01	30000	30JL	17;21	315.37
30001	14AU	1;13	149.19	30001	29AU	8;14	344.05
30002	12SE	10;15	177.84	30002	27SE	22;26	13.15
30003	11OC	21;52	207.07	30003	27OC	11;53	42.71
30004	10NO	12;40	236.89	30004	26NO	0;28	72.62
30005	10DE	6;21	267.13	30005	25DE	12;02	102.67

1426

NEW MOONS				FULL MOONS			
30006	9JA	1;37	297.50	30006	23JA	22;34	132.59
30007	7FE	20;41	327.65	30007	22FE	8;21	162.16
30008	9MR	14;02	357.30	30008	23MR	17;57	191.26
30009	8AP	4;54	26.36	30009	22AP	4;02	219.90
30010	7MY	17;04	54.86	30010	21MY	15;12	248.20
30011	6JN	2;49	82.98	30011	20JN	3;46	276.36
30012	5JL	10;48	110.94	30012	19JL	17;51	304.61
30013	3AU	18;02	139.02	30013	18AU	9;21	333.16
30014	2SE	1;36	167.44	30014	17SE	1;54	2.17
30015	1OC	10;34	196.39	30015	16OC	18;48	31.70
30016	30OC	21;39	225.90	30016	15NO	10;58	61.64
30017	29NO	11;09	255.88	30017	15DE	1;27	91.78
30018	29DE	2;52	286.11				

1427

NEW MOONS				FULL MOONS			
				30018	13JA	13;52	121.84
30019	27JA	20;11	316.29	30019	12FE	0;26	151.58
30020	26FE	14;11	346.15	30020	13MR	9;42	180.84
30021	28MR	7;46	15.50	30021	11AP	18;16	209.58
30022	26AP	23;44	44.31	30022	11MY	2;44	237.90
30023	26MY	13;18	72.67	30023	9JN	11;47	265.96
30024	25JN	0;24	100.76	30024	8JL	22;11	294.03
30025	24JL	9;42	128.84	30025	7AU	10;43	322.35
30026	22AU	18;11	157.16	30026	6SE	1;54	351.13
30027	21SE	2;50	185.92	30027	5OC	19;30	20.50
30028	20OC	12;21	215.20	30028	4NO	14;17	50.39
30029	18NO	23;02	244.95	30029	4DE	8;29	80.61
30030	18DE	11;03	275.00				

1428

NEW MOONS				FULL MOONS			
				30030	3JA	0;40	110.86
30031	17JA	0;34	305.09	30031	1FE	14;22	140.85
30032	15FE	15;37	334.98	30032	2MR	1;41	170.36
30033	16MR	7;56	4.47	30033	31MR	11;05	199.31
30034	15AP	0;39	33.49	30034	29AP	19;02	227.75
30035	14MY	16;40	62.05	30035	29MY	2;16	255.82
30036	13JN	7;09	90.31	30036	27JN	9;43	283.76
30037	12JL	19;56	118.49	30037	26JL	18;32	311.84
30038	11AU	7;24	146.81	30038	25AU	5;51	340.32
30039	9SE	18;05	175.48	30039	23SE	20;18	9.37
30040	9OC	4;24	204.61	30040	23OC	13;43	39.03
30041	7NO	14;40	234.21	30041	22NO	8;55	69.19
30042	7DE	1;06	264.13	30042	22DE	4;11	99.55

1429

NEW MOONS				FULL MOONS			
30043	5JA	12;01	294.15	30043	20JA	21;59	129.80
30044	3FE	23;47	324.02	30044	19FE	13;22	159.64
30045	5MR	12;41	353.55	30045	21MR	1;54	188.90
30046	4AP	2;38	22.64	30046	19AP	11;42	217.57
30047	3MY	17;18	51.29	30047	18MY	19;27	245.76
30048	2JN	8;15	79.64	30048	17JN	2;08	273.70
30049	1JL	23;11	107.88	30049	16JL	8;58	301.64
30050	31JL	13;51	136.23	30050	14AU	17;07	329.86
30051	30AU	3;59	164.88	30051	13SE	3;33	358.57
30052	28SE	17;14	193.96	30052	12OC	16;48	27.89
30053	28OC	5;26	223.47	30053	11NO	8;53	57.77
30054	26NO	16;42	253.34	30054	11DE	3;16	88.05
30055	26DE	3;24	283.35				

1430

NEW MOONS				FULL MOONS			
				30055	9JA	22;47	118.43
30056	24JA	13;53	313.27	30056	8FE	17;46	148.56
30057	23FE	0;25	342.86	30057	10MR	10;33	178.18
30058	24MR	11;10	11.99	30058	9AP	0;11	207.17
30059	22AP	22;24	40.67	30059	8MY	10;43	235.60
30060	22MY	10;30	69.00	30060	6JN	18;59	263.65
30061	20JN	23;56	97.19	30061	6JL	2;10	291.59
30062	20JL	14;52	125.48	30062	4AU	9;21	319.66
30063	19AU	6;58	154.06	30063	2SE	17;29	348.12
30064	17SE	23;19	183.09	30064	2OC	3;13	17.11
30065	17OC	14;55	212.57	30065	31OC	15;06	46.66
30066	16NO	5;13	242.44	30066	30NO	5;31	76.69
30067	15DE	18;06	272.51	30067	29DE	22;32	106.97

1431

NEW MOONS				FULL MOONS			
30068	14JA	5;43	302.54	30068	28JA	17;19	137.21
30069	12FE	16;12	332.27	30069	27FE	12;13	167.09
30070	14MR	1;40	1.52	30070	29MR	5;25	196.42
30071	12AP	10;29	30.27	30071	27AP	19;55	225.16
30072	11MY	19;23	58.59	30072	27MY	7;44	253.43
30073	10JN	5;20	86.69	30073	25JN	17;31	281.47
30074	9JL	17;14	114.82	30074	25JL	2;07	309.53
30075	8AU	7;30	143.21	30075	23AU	10;17	337.85
30076	6SE	23;51	172.06	30076	21SE	18;44	6.61
30077	6OC	17;22	201.43	30077	21OC	4;07	35.89
30078	5NO	10;58	231.29	30078	19NO	15;05	65.66
30079	5DE	3;39	261.46	30079	19DE	4;10	95.75

1432

NEW MOONS				FULL MOONS			
30080	3JA	18;39	291.66	30080	17JA	19;27	125.92
30081	2FE	7;30	321.60	30081	16FE	12;18	155.87
30082	2MR	18;04	351.07	30082	17MR	5;35	185.39
30083	1AP	2;45	19.98	30083	15AP	22;10	214.38
30084	30AP	10;20	48.39	30084	15MY	13;23	242.91
30085	29MY	17;50	76.47	30085	14JN	2;56	271.13
30086	28JN	2;21	104.45	30086	13JL	14;51	299.27
30087	27JL	12;46	132.60	30087	12AU	1;20	327.56
30088	26AU	1;39	161.15	30088	10SE	10;52	356.19
30089	24SE	17;08	190.26	30089	9OC	20;09	25.30
30090	24OC	10;54	219.95	30090	8NO	5;54	54.88
30091	23NO	6;00	250.11	30091	7DE	16;39	84.82
30092	23DE	0;52	280.46				

1433

NEW MOONS				FULL MOONS			
				30092	6JA	4;39	114.88
30093	21JA	17;48	310.67	30093	4FE	17;49	144.81
30094	20FE	7;45	340.44	30094	6MR	7;54	174.38
30095	21MR	18;40	9.62	30095	4AP	22;43	203.48
30096	20AP	3;17	38.23	30096	4MY	14;04	232.16
30097	19MY	10;37	66.40	30097	3JN	5;35	260.52
30098	17JN	17;37	94.34	30098	2JL	20;36	288.76
30099	17JL	1;14	122.32	30099	1AU	10;31	317.09
30100	15AU	10;18	150.59	30100	30AU	23;04	345.69
30101	13SE	21;37	179.35	30101	29SE	10;33	14.70
30102	13OC	11;53	208.71	30102	28OC	21;30	44.18
30103	12NO	5;16	238.66	30103	27NO	8;20	74.03
30104	12DE	0;53	269.00	30104	26DE	19;14	104.05

1434

NEW MOONS				FULL MOONS			
30105	10JA	20;45	299.39	30105	25JA	6;07	133.98
30106	9FE	14;45	329.47	30106	23FE	17;06	163.58
30107	11MR	5;43	359.07	30107	25MR	4;32	192.73
30108	9AP	17;38	27.91	30108	23AP	16;56	221.44
30109	9MY	3;07	56.29	30109	23MY	6;39	249.83
30110	7JN	10;58	84.32	30110	21JN	21;31	278.08
30111	6JL	18;02	112.25	30111	21JL	12;58	306.39
30112	5AU	1;11	140.33	30112	20AU	4;19	334.96
30113	3SE	9;24	168.80	30113	18SE	19;09	3.93
30114	2OC	19;43	197.82	30114	18OC	9;18	33.37
30115	1NO	8;58	227.44	30115	16NO	22;34	63.20
30116	1DE	1;19	257.56	30116	16DE	10;47	93.25
30117	30DE	19;57	287.91				

1435

NEW MOONS				FULL MOONS			
				30117	14JA	21;49	123.25
30118	29JA	15;12	318.16	30118	13FE	7;47	152.95
30119	28FE	9;25	348.00	30119	14MR	17;12	182.19
30120	30MR	1;28	17.27	30120	13AP	2;42	210.96
30121	28AP	14;55	45.95	30121	12MY	13;01	239.33
30122	28MY	1;48	74.19	30122	11JN	0;40	267.49
30123	26JN	10;38	102.18	30123	10JL	13;55	295.67
30124	25JL	18;13	130.21	30124	9AU	4;46	324.10
30125	24AU	1;38	158.50	30125	7SE	21;02	352.95
30126	22SE	9;57	187.26	30126	7OC	14;09	22.33
30127	21OC	20;01	216.58	30127	6NO	7;05	52.17
30128	20NO	8;22	246.42	30128	5DE	22;43	82.30
30129	19DE	22;59	276.58				

NEW MOONS NUMBER	DATE	TIME	LONG.	FULL MOONS NUMBER	DATE	TIME	LONG.
			1436				
				30129	4JA	12:15	112.43
30130	18JA	15:26	306.79	30130	2FE	23:39	142.31
30131	17FE	8:59	336.76	30131	3MR	9:22	171.73
30132	18MR	2:42	6.29	30132	1AP	18:01	200.63
30133	16AP	19:27	35.27	30133	1MY	2:16	229.06
30134	16MY	10:13	63.77	30134	30MY	10:45	257.17
30135	14JN	22:35	91.94	30135	28JN	20:16	285.20
30136	14JL	8:49	120.02	30136	28JL	7:36	313.39
30137	12AU	17:46	148.25	30137	26AU	21:29	341.99
30138	11SE	2:27	176.86	30138	25SE	14:06	11.15
30139	10OC	11:40	205.98	30139	25OC	8:42	40.89
30140	8NO	21:52	235.58	30140	24NO	3:37	71.05
30141	8DE	9:14	265.55	30141	23DE	21:03	101.35
			1437				
30142	6JA	21:53	295.64	30142	22JA	12:03	131.47
30143	5FE	11:58	325.60	30143	21FE	0:27	161.16
30144	7MR	3:25	355.21	30144	22MR	10:38	190.31
30145	5AP	19:48	24.37	30145	20AP	19:06	218.90
30146	5MY	12:06	53.07	30146	20MY	2:28	247.06
30147	4JN	3:22	81.42	30147	18JN	9:35	275.01
30148	3JL	17:04	109.61	30148	17JL	17:32	303.01
30149	2AU	5:20	137.87	30149	16AU	3:30	331.31
30150	31AU	16:36	166.43	30150	14SE	16:24	0.15
30151	30SE	3:20	195.43	30151	14OC	8:32	29.60
30152	29OC	13:49	224.90	30152	13NO	3:09	59.62
30153	28NO	0:15	254.74	30153	12DE	22:41	89.95
30154	27DE	10:53	284.75				
			1438				
				30154	11JA	17:24	120.29
30155	25JA	22:03	314.68	30155	10FE	10:03	150.30
30156	24FE	10:08	344.32	30156	11MR	23:57	179.78
30157	25MR	23:16	13.53	30157	10AP	10:59	208.64
30158	24AP	13:21	42.30	30158	9MY	19:34	236.98
30159	24MY	4:00	70.72	30159	8JN	2:36	264.97
30160	22JN	18:55	98.96	30160	7JL	9:13	292.88
30161	22JL	9:51	127.26	30161	5AU	16:37	320.98
30162	21AU	0:31	155.81	30162	4SE	1:53	349.50
30163	19SE	14:33	184.76	30163	3OC	13:46	18.60
30164	19OC	3:34	214.16	30164	2NO	4:33	48.30
30165	17NO	15:29	243.95	30165	1DE	21:55	78.46
30166	17DE	2:31	273.95	30166	31DE	17:00	108.84
			1439				
30167	15JA	13:02	303.92	30167	30JA	12:23	139.08
30168	13FE	23:22	333.63	30168	1MR	6:20	168.90
30169	15MR	9:45	2.90	30169	30MR	21:31	198.11
30170	13AP	20:26	31.70	30170	29AP	9:29	226.73
30171	13MY	7:47	60.11	30171	28MY	18:44	254.89
30172	11JN	20:19	88.31	30172	27JN	2:21	282.84
30173	11JL	10:27	116.52	30173	26JL	9:30	310.85
30174	10AU	2:09	144.99	30174	24AU	17:11	339.16
30175	8SE	18:43	173.87	30175	23SE	2:11	7.96
30176	8OC	11:06	203.23	30176	22OC	13:01	37.33
30177	7NO	2:22	233.01	30177	21NO	2:11	67.20
30178	6DE	16:08	263.06	30178	20DE	17:53	97.41
			1440				
30179	5JA	4:26	293.15	30179	19JA	11:48	127.68
30180	3FE	15:25	323.00	30180	18FE	6:41	157.70
30181	4MR	1:12	352.42	30181	19MR	0:44	187.22
30182	2AP	10:05	21.32	30182	17AP	16:32	216.16
30183	1MY	18:37	49.75	30183	17MY	5:37	244.58
30184	31MY	3:43	77.88	30184	15JN	16:22	272.68
30185	29JN	14:25	105.95	30185	15JL	1:35	300.72
30186	29JL	3:27	134.22	30186	13AU	10:03	328.94
30187	27AU	18:55	162.89	30187	11SE	18:28	357.55
30188	26SE	12:11	192.10	30188	11OC	3:28	26.66
30189	26OC	6:07	221.82	30189	9NO	13:40	56.28
30190	24NO	23:35	251.93	30190	9DE	1:40	86.28
30191	24DE	15:38	282.17				
			1441				
				30191	7JA	15:46	116.43
30192	23JA	5:38	312.24	30192	6FE	7:44	146.46
30193	21FE	17:15	341.89	30193	8MR	0:40	176.12
30194	23MR	2:40	10.99	30194	6AP	17:29	205.28
30195	21AP	10:31	39.55	30195	6MY	9:17	233.95
30196	20MY	17:46	67.70	30196	4JN	23:41	262.26
30197	19JN	1:33	95.68	30197	4JL	12:31	290.41
30198	18JL	10:52	123.73	30198	2AU	23:52	318.65
30199	16AU	22:29	152.11	30199	1SE	10:03	347.17
30200	15SE	12:45	181.01	30200	30SE	19:37	16.13
30201	15OC	5:35	210.51	30201	30OC	5:14	45.57
30202	14NO	0:19	240.54	30202	28NO	15:32	75.41
30203	13DE	19:39	270.87	30203	28DE	2:53	105.46

NEW MOONS NUMBER	DATE	TIME	LONG.	FULL MOONS NUMBER	DATE	TIME	LONG.
			1442				
30204	12JA	13:48	301.18	30204	26JA	15:19	135.44
30205	11FE	5:16	331.14	30205	25FE	4:43	165.12
30206	12MR	17:34	0.53	30206	26MR	18:53	194.36
30207	11AP	3:07	29.33	30207	25AP	9:44	223.15
30208	10MY	10:52	57.62	30208	25MY	1:03	251.58
30209	8JN	17:50	85.61	30209	23JN	16:24	279.85
30210	8JL	1:02	113.54	30210	23JL	7:04	308.14
30211	6AU	9:17	141.68	30211	21AU	20:33	336.65
30212	4SE	19:26	170.25	30212	20SE	8:49	5.54
30213	4OC	8:15	199.40	30213	19OC	20:14	34.89
30214	3NO	0:11	229.15	30214	18NO	7:16	64.65
30215	2DE	18:55	259.38	30215	17DE	18:13	94.65
			1443				
30216	1JA	14:57	289.80	30216	16JA	5:03	124.63
30217	31JA	10:04	320.02	30217	14FE	15:47	154.34
30218	2MR	2:32	349.76	30218	16MR	2:40	183.63
30219	31MR	15:50	18.89	30219	14AP	14:14	212.45
30220	30AP	2:23	47.44	30220	14MY	3:01	240.91
30221	29MY	10:55	75.57	30221	12JN	17:11	269.17
30222	27JN	18:17	103.51	30222	12JL	8:25	297.43
30223	27JL	1:20	131.51	30223	11AU	0:02	325.90
30224	25AU	9:00	159.83	30224	9SE	15:26	354.75
30225	23SE	18:16	188.66	30225	9OC	6:14	24.05
30226	23OC	6:04	218.07	30226	7NO	20:15	53.80
30227	21NO	20:56	248.03	30227	7DE	9:11	83.82
30228	21DE	14:32	278.32				
			1444				
				30228	5JA	20:51	113.87
30229	20JA	9:35	308.63	30229	4FE	7:14	143.69
30230	19FE	4:20	338.63	30230	4MR	16:40	173.09
30231	19MR	21:25	8.10	30231	3AP	1:48	201.99
30232	18AP	12:06	36.98	30232	2MY	11:23	230.45
30233	18MY	0:13	65.35	30233	31MY	22:07	258.64
30234	16JN	10:03	93.42	30234	30JN	10:25	286.78
30235	15JL	18:15	121.42	30235	30JL	0:28	315.09
30236	14AU	1:46	149.61	30236	28AU	16:12	343.79
30237	12SE	9:40	178.20	30237	27SE	9:13	12.99
30238	11OC	18:55	207.34	30238	27OC	2:39	42.71
30239	10NO	6:10	237.01	30239	25NO	19:18	72.80
30240	9DE	19:40	267.07	30240	25DE	10:04	102.98
			1445				
30241	8JA	11:09	297.28	30241	23JA	22:31	132.98
30242	7FE	4:03	327.34	30242	22FE	8:54	162.57
30243	8MR	21:34	357.01	30243	23MR	17:50	191.64
30244	7AP	14:44	26.17	30244	22AP	2:01	220.20
30245	7MY	6:29	54.83	30245	21MY	10:07	248.39
30246	5JN	20:05	83.10	30246	19JN	18:54	276.40
30247	5JL	7:26	111.20	30247	19JL	5:09	304.49
30248	3AU	17:06	139.37	30248	17AU	17:43	332.92
30249	2SE	2:01	167.85	30249	16SE	9:05	1.88
30250	1OC	11:06	196.80	30250	16OC	3:02	31.44
30251	30OC	20:56	226.26	30251	14NO	22:13	61.49
30252	29NO	7:47	256.13	30252	14DE	16:44	91.80
30253	28DE	19:44	286.20				
			1446				
				30253	13JA	9:04	122.03
30254	27JA	8:54	316.21	30254	11FE	22:43	151.90
30255	25FE	23:35	345.93	30255	13MR	9:51	181.23
30256	27MR	15:08	15.22	30256	11AP	18:59	210.00
30257	26AP	7:22	44.05	30257	11MY	2:41	238.28
30258	25MY	23:08	72.49	30258	9JN	9:43	266.27
30259	24JN	13:41	100.72	30259	8JL	17:04	294.21
30260	24JL	2:48	128.96	30260	7AU	1:53	322.38
30261	22AU	14:45	157.42	30261	5SE	13:17	351.01
30262	21SE	1:59	186.28	30262	5OC	3:55	20.24
30263	20OC	12:49	215.61	30263	3NO	21:33	50.08
30264	18NO	23:25	245.36	30264	3DE	16:56	80.35
30265	18DE	9:57	275.34				
			1447				
				30265	2JA	12:17	110.73
30266	16JA	20:43	305.32	30266	1FE	6:05	140.89
30267	15FE	8:06	335.05	30267	2MR	21:20	170.57
30268	16MR	20:26	4.39	30268	1AP	9:41	199.65
30269	15AP	9:46	33.27	30269	30AP	19:20	228.15
30270	14MY	23:56	61.78	30270	30MY	2:58	256.24
30271	13JN	14:38	90.05	30271	28JN	9:38	284.15
30272	13JL	5:38	118.31	30272	27JL	16:32	312.15
30273	11AU	20:39	146.76	30273	26AU	0:50	340.51
30274	10SE	11:20	175.59	30274	24SE	11:26	9.39
30275	10OC	1:12	204.87	30275	24OC	0:51	38.89
30276	8NO	13:55	234.57	30276	22NO	17:02	68.91
30277	8DE	1:30	264.54	30277	22DE	11:22	99.23

1448

NEW MOON	DATE	TIME	LONG.	FULL MOON	DATE	TIME	LONG.
30278	6JA	12;14	294.55	30278	21JA	6;43	129.55
30279	4FE	22;31	324.36	30279	20FE	1;29	159.54
30280	5MR	8;39	353.78	30280	20MR	18;03	188.97
30281	3AP	18;53	22.71	30281	19AP	7;32	217.79
30282	3MY	5;34	51.21	30282	18MY	18;02	246.09
30283	1JN	17;15	79.44	30283	17JN	2;21	274.10
30284	1JL	6;26	107.60	30284	16JL	9;40	302.06
30285	30JL	21;25	135.96	30285	14AU	17;06	330.25
30286	29AU	13;51	164.70	30286	13SE	1;30	358.88
30287	28SE	6;44	193.91	30287	12OC	11;30	28.06
30288	27OC	22;56	223.59	30288	10NO	23;31	57.77
30289	26NO	13;41	253.61	30289	10DE	13;55	87.88
30290	26DE	2;49	283.72				

1449

NEW MOON	DATE	TIME	LONG.	FULL MOON	DATE	TIME	LONG.
				30290	9JA	6;42	118.54
30291	24JA	14;25	313.68	30291	8FE	1;06	148.25
30292	23FE	0;40	343.26	30292	9MR	19;35	177.95
30293	24MR	9;46	12.32	30293	8AP	12;29	207.09
30294	22AP	18;10	40.88	30294	8MY	2;51	235.67
30295	22MY	2;39	69.08	30295	6JN	14;43	263.87
30296	20JN	12;17	97.12	30296	6JL	0;43	291.92
30297	20JL	0;01	125.28	30297	4AU	9;37	320.07
30298	18AU	14;21	153.78	30298	2SE	18;11	348.54
30299	17SE	6;58	182.81	30299	2OC	3;01	17.49
30300	17OC	0;57	212.38	30300	31OC	12;41	46.95
30301	15NO	19;00	242.40	30301	29NO	23;46	76.84
30302	15DE	12;01	272.65	30302	29DE	12;44	106.95

1450

NEW MOON	DATE	TIME	LONG.	FULL MOON	DATE	TIME	LONG.
30303	14JA	3;11	302.82	30303	28JA	3;39	137.03
30304	12FE	15;57	332.65	30304	26FE	19;58	166.81
30305	14MR	2;18	1.94	30305	28MR	12;42	196.13
30306	12AP	10;40	30.67	30306	27AP	4;53	224.94
30307	11MY	17;57	58.92	30307	26MY	19;57	253.35
30308	10JN	1;13	86.92	30308	25JN	9;38	281.55
30309	9JL	9;34	114.90	30309	24JL	21;53	309.75
30310	7AU	19;57	143.14	30310	23AU	8;51	338.18
30311	6SE	8;56	171.84	30311	21SE	18;54	7.01
30312	6OC	0;38	201.13	30312	21OC	4;37	36.30
30313	4NO	18;41	231.00	30313	19NO	14;39	66.03
30314	4DE	14;03	261.27	30314	19DE	1;28	96.03

1451

NEW MOON	DATE	TIME	LONG.	FULL MOON	DATE	TIME	LONG.
30315	3JA	9;05	291.64	30315	17JA	13;16	126.05
30316	2FE	2;01	321.76	30316	16FE	2;00	155.83
30317	3MR	15;50	351.37	30317	17MR	15;29	185.20
30318	2AP	2;34	20.36	30318	16AP	5;43	214.11
30319	1MY	10;59	48.81	30319	15MY	20;37	242.63
30320	30MY	18;09	76.87	30320	14JN	11;57	270.93
30321	29JN	1;05	104.79	30321	14JL	3;07	299.19
30322	28JL	8;43	132.83	30322	12AU	17;28	327.63
30323	26AU	17;54	161.23	30323	11SE	6;36	356.41
30324	25SE	5;22	190.16	30324	10OC	18;40	25.63
30325	24OC	19;48	219.71	30325	9NO	6;04	55.28
30326	23NO	13;18	249.80	30326	8DE	17;10	85.24
30327	23DE	8;57	280.18				

1452

NEW MOON	DATE	TIME	LONG.	FULL MOON	DATE	TIME	LONG.
				30327	7JA	4;04	115.25
30328	22JA	4;44	310.51	30328	5FE	14;42	145.07
30329	20FE	22;35	340.45	30329	6MR	1;14	174.49
30330	21MR	13;23	9.80	30330	4AP	12;06	203.44
30331	20AP	1;09	38.53	30331	3MY	23;56	231.98
30332	19MY	10;33	66.78	30332	2JN	13;13	260.26
30333	17JN	18;24	94.77	30333	2JL	3;55	288.49
30334	17JL	1;34	122.73	30334	31JL	19;30	316.88
30335	15AU	8;55	150.92	30335	30AU	11;18	345.60
30336	13SE	17;22	179.56	30336	29SE	2;43	14.77
30337	13OC	3;54	208.77	30337	28OC	17;27	44.40
30338	11NO	17;16	238.55	30338	27NO	7;10	74.38
30339	11DE	9;35	268.74	30339	26DE	19;35	104.46

1453

NEW MOON	DATE	TIME	LONG.	FULL MOON	DATE	TIME	LONG.
30340	10JA	4;01	299.07	30340	25JA	6;33	134.39
30341	8FE	22;57	329.20	30341	23FE	16;15	163.94
30342	10MR	16;50	358.86	30342	25MR	1;13	192.99
30343	9AP	8;39	27.93	30343	23AP	10;15	221.56
30344	8MY	21;59	56.47	30344	22MY	20;07	249.80
30345	7JN	8;57	84.63	30345	21JN	7;26	277.91
30346	6JL	17;57	112.63	30346	20JL	20;33	306.13
30347	5AU	1;50	140.75	30347	19AU	11;32	334.67
30348	3SE	9;35	169.20	30348	18SE	4;11	3.71
30349	2OC	18;13	198.15	30349	17OC	21;48	33.28
30350	1NO	4;32	227.65	30350	16NO	15;15	63.29
30351	30NO	16;57	257.60	30351	16DE	7;14	93.49
30352	30DE	7;24	287.77				

1454

NEW MOON	DATE	TIME	LONG.	FULL MOON	DATE	TIME	LONG.
				30352	14JA	20;53	123.60
30353	28JA	23;29	317.89	30353	13FE	8;10	153.36
30354	27FE	16;33	347.69	30354	14MR	17;37	182.60
30355	29MR	9;47	17.02	30355	13AP	1;54	211.31
30356	28AP	2;13	45.83	30356	12MY	9;48	239.59
30357	27MY	16;55	74.22	30357	10JN	18;00	267.62
30358	26JN	5;27	102.37	30358	10JL	3;20	295.64
30359	25JL	16;02	130.51	30359	8AU	14;38	323.92
30360	24AU	1;26	158.88	30360	7SE	4;38	352.68
30361	22SE	10;33	187.68	30361	6OC	21;32	22.03
30362	21OC	20;10	216.98	30362	5NO	16;30	51.95
30363	20NO	6;36	246.74	30363	5DE	11;45	82.22
30364	19DE	18;00	276.76				

1455

NEW MOON	DATE	TIME	LONG.	FULL MOON	DATE	TIME	LONG.
				30364	4JA	5;23	112.53
30365	18JA	6;24	306.80	30365	2FE	20;23	142.56
30366	16FE	20;00	336.61	30366	4MR	8;39	172.09
30367	18MR	10;52	6.03	30367	2AP	18;36	201.05
30368	17AP	2;40	34.99	30368	2MY	2;49	229.47
30369	16MY	18;37	63.54	30369	31MY	9;59	257.53
30370	15JN	9;49	91.83	30370	29JN	17;00	285.46
30371	14JL	23;45	120.05	30371	29JL	0;56	313.51
30372	13AU	12;28	148.43	30372	27AU	10;57	341.95
30373	12SE	0;17	177.17	30373	26SE	0;01	10.96
30374	11OC	11;34	206.36	30374	25OC	16;20	40.60
30375	9NO	22;28	236.00	30375	24NO	11;08	70.75
30376	9DE	9;06	265.94	30376	24DE	6;46	101.14

1456

NEW MOON	DATE	TIME	LONG.	FULL MOON	DATE	TIME	LONG.
30377	7JA	19;41	295.94	30377	23JA	1;29	131.42
30378	6FE	6;33	325.77	30378	21FE	18;01	161.29
30379	6MR	18;07	355.22	30379	22MR	7;45	190.58
30380	5AP	6;40	24.23	30380	20AP	18;37	219.27
30381	4MY	20;10	52.82	30381	20MY	3;05	247.48
30382	3JN	10;28	81.14	30382	18JN	10;05	275.42
30383	3JL	1;18	109.37	30383	17JL	16;45	303.36
30384	1AU	16;28	137.75	30384	16AU	0;18	331.58
30385	31AU	7;38	166.46	30385	14SE	9;45	0.27
30386	29SE	22;16	195.61	30386	13OC	21;50	29.55
30387	29OC	11;52	225.21	30387	12NO	12;43	59.40
30388	28NO	0;11	255.13	30388	12DE	6;04	89.64
30389	27DE	11;22	285.16				

1457

NEW MOON	DATE	TIME	LONG.	FULL MOON	DATE	TIME	LONG.
				30389	11JA	1;01	119.99
30390	25JA	21;46	315.06	30390	9FE	20;09	150.11
30391	24FE	7;46	344.61	30391	11MR	13;51	179.75
30392	25MR	17;41	13.69	30392	10AP	4;51	208.78
30393	24AP	3;50	42.30	30393	9MY	16;44	237.25
30394	23MY	14;43	70.57	30394	8JN	2;02	265.34
30395	22JN	2;56	98.72	30395	7JL	9;47	293.30
30396	21JL	16;59	126.97	30396	5AU	17;09	321.39
30397	20AU	8;53	155.57	30397	4SE	1;08	349.86
30398	19SE	1;55	184.63	30398	3OC	10;24	18.85
30399	18OC	18;53	214.19	30399	1NO	21;27	48.39
30400	17NO	10;40	244.14	30400	1DE	10;39	78.38
30401	17DE	0;46	274.27	30401	31DE	2;11	108.60

1458

NEW MOON	DATE	TIME	LONG.	FULL MOON	DATE	TIME	LONG.
30402	15JA	13;08	304.32	30402	29JA	19;44	138.77
30403	13FE	23;58	334.05	30403	28FE	14;11	168.62
30404	15MR	9;27	3.28	30404	30MR	7;51	197.95
30405	13AP	17;55	31.99	30405	28AP	23;26	226.71
30406	13MY	2;03	60.27	30406	28MY	12;30	255.03
30407	11JN	10;49	88.32	30407	26JN	23;25	283.11
30408	10JL	21;18	116.39	30408	26JL	8;56	311.22
30409	9AU	10;18	144.74	30409	24AU	17;48	339.58
30410	8SE	1;58	173.58	30410	23SE	2;37	8.37
30411	7OC	19;36	202.98	30411	22OC	11;58	37.68
30412	6NO	13;59	232.88	30412	20NO	22;22	67.44
30413	6DE	7;50	263.11	30413	20DE	10;19	97.49

1459

NEW MOON	DATE	TIME	LONG.	FULL MOON	DATE	TIME	LONG.
30414	5JA	0;06	293.35	30414	19JA	0;09	127.58
30415	3FE	14;05	323.33	30415	17FE	15;37	157.46
30416	5MR	1;32	352.82	30416	19MR	7;59	186.92
30417	3AP	10;40	21.73	30417	18AP	0;18	215.89
30418	2MY	18;14	50.12	30418	17MY	15;51	244.42
30419	1JN	1;14	78.17	30419	16JN	6;15	272.67
30420	30JN	8;50	106.12	30420	15JL	19;21	300.86
30421	29JL	18;06	134.23	30421	14AU	7;10	329.22
30422	28AU	5;46	162.75	30422	12SE	17;53	357.92
30423	26SE	20;13	191.83	30423	12OC	3;56	27.08
30424	26OC	13;18	221.51	30424	10NO	13;55	56.69
30425	25NO	8;17	251.68	30425	10DE	0;22	86.62
30426	25DE	3;48	282.06				

Left half

1460

NEW MOONS NUMBER	DATE	TIME	LONG.	FULL MOONS NUMBER	DATE	TIME	LONG.
				30426	8JA	11:37	116.65
30427	23JA	21:59	312.31	30427	6FE	23:42	146.51
30428	22FE	13:22	342.12	30428	7MR	12:33	176.01
30429	23MR	1:30	11.33	30429	6AP	2:07	205.05
30430	21AP	10:51	39.95	30430	5MY	16:25	233.66
30431	20MY	18:26	68.12	30431	4JN	7:27	262.00
30432	19JN	1:19	96.06	30432	3JL	22:48	290.26
30433	18JL	8:31	124.03	30433	2AU	13:48	318.63
30434	16AU	16:53	152.28	30434	1SE	3:50	347.31
30435	15SE	3:12	181.01	30435	30SE	16:42	16.40
30436	14OC	16:10	210.34	30436	30OC	4:39	45.94
30437	13NO	8:14	240.25	30437	28NO	16:02	75.83
30438	13DE	3:00	270.56	30438	28DE	3:04	105.86

1461

NEW MOONS NUMBER	DATE	TIME	LONG.	FULL MOONS NUMBER	DATE	TIME	LONG.
30439	11JA	22:58	300.95	30439	26JA	13:45	135.76
30440	10FE	17:54	331.05	30440	25FE	0:07	165.32
30441	12MR	10:11	0.62	30441	26MR	10:28	194.41
30442	10AP	23:20	29.56	30442	24AP	21:28	223.04
30443	10MY	9:47	57.96	30443	24MY	9:45	251.36
30444	8JN	18:18	86.02	30444	22JN	23:38	279.57
30445	8JL	1:46	113.96	30445	22JL	14:53	307.88
30446	6AU	9:01	142.06	30446	21AU	6:49	336.48
30447	4SE	16:55	170.54	30447	19SE	22:45	5.52
30448	4OC	2:25	199.55	30448	19OC	14:10	35.03
30449	2NO	14:23	229.13	30449	18NO	4:41	64.93
30450	2DE	5:16	259.20	30450	17DE	17:55	95.03
30451	31DE	22:43	289.50				

1462

NEW MOONS NUMBER	DATE	TIME	LONG.	FULL MOONS NUMBER	DATE	TIME	LONG.
				30451	16JA	5:37	125.05
30452	30JA	17:27	319.72	30452	14FE	15:48	154.74
30453	1MR	11:50	349.55	30453	16MR	0:52	183.94
30454	31MR	4:37	18.82	30454	14AP	9:32	212.65
30455	29AP	19:08	47.53	30455	13MY	18:40	240.96
30456	29MY	7:16	75.81	30456	12JN	5:01	269.07
30457	27JN	17:15	103.85	30457	11JL	17:07	297.21
30458	27JL	1:43	131.92	30458	10AU	7:12	325.62
30459	25AU	9:35	160.25	30459	8SE	23:12	354.48
30460	23SE	17:51	189.03	30460	8OC	16:40	23.88
30461	23OC	3:23	218.35	30461	7NO	10:37	53.78
30462	21NO	14:48	248.16	30462	7DE	3:41	83.98
30463	21DE	4:13	278.28				

1463

NEW MOONS NUMBER	DATE	TIME	LONG.	FULL MOONS NUMBER	DATE	TIME	LONG.
				30463	5JA	18:39	114.17
30464	19JA	19:24	308.42	30464	4FE	7:04	144.08
30465	18FE	11:49	338.33	30465	5MR	17:14	173.50
30466	20MR	4:48	7.81	30466	4AP	1:51	202.38
30467	18AP	21:33	36.78	30467	3MY	9:41	230.77
30468	18MY	13:08	65.30	30468	1JN	17:29	258.85
30469	17JN	2:49	93.52	30469	1JL	2:03	286.83
30470	16JL	14:28	121.65	30470	30JL	12:14	314.99
30471	15AU	0:34	149.95	30471	29AU	0:52	343.55
30472	13SE	9:57	178.61	30472	27SE	16:27	12.69
30473	12OC	19:28	207.75	30473	27OC	10:42	42.44
30474	11NO	5:37	237.38	30474	26NO	6:14	72.64
30475	10DE	16:35	267.34	30475	26DE	0:59	102.99

1464

NEW MOONS NUMBER	DATE	TIME	LONG.	FULL MOONS NUMBER	DATE	TIME	LONG.
30476	9JA	4:23	297.39	30476	24JA	17:23	133.16
30477	7FE	17:09	327.28	30477	23FE	6:56	162.88
30478	8MR	7:06	356.81	30478	23MR	17:52	192.03
30479	6AP	22:12	25.89	30479	22AP	2:46	220.62
30480	6MY	13:58	54.56	30480	21MY	10:16	248.78
30481	5JN	5:33	82.91	30481	19JN	17:10	276.72
30482	4JL	20:13	111.14	30482	19JL	0:29	304.69
30483	3AU	9:42	139.46	30483	17AU	9:22	332.97
30484	1SE	22:12	168.09	30484	15SE	20:53	1.77
30485	1OC	10:00	197.15	30485	15OC	11:42	31.19
30486	30OC	21:19	226.67	30486	14NO	5:30	61.18
30487	29NO	8:13	256.54	30487	14DE	0:59	91.52
30488	28DE	18:48	286.55				

1465

NEW MOONS NUMBER	DATE	TIME	LONG.	FULL MOONS NUMBER	DATE	TIME	LONG.
				30488	12JA	20:21	121.88
30489	27JA	5:22	316.45	30489	11FE	14:02	151.93
30490	25FE	16:19	346.02	30490	13MR	5:07	181.43
30491	27MR	4:04	15.15	30491	11AP	17:09	210.32
30492	25AP	16:49	43.85	30492	11MY	2:51	238.68
30493	25MY	6:31	72.22	30493	9JN	10:26	266.69
30494	23JN	21:01	100.45	30494	8JL	17:09	294.61
30495	23JL	12:07	128.76	30495	7AU	0:11	322.70
30496	22AU	3:32	157.35	30496	5SE	8:40	351.21
30497	20SE	18:48	186.37	30497	4OC	19:29	20.29
30498	20OC	9:16	215.86	30498	3NO	9:02	49.95
30499	18NO	22:28	245.71	30499	3DE	1:14	80.08
30500	18DE	10:18	275.75				

Right half

1466

NEW MOONS NUMBER	DATE	TIME	LONG.	FULL MOONS NUMBER	DATE	TIME	LONG.
				30500	1JA	19:28	110.41
30501	16JA	21:01	305.73	30501	31JA	14:34	140.63
30502	15FE	7:03	335.40	30502	2MR	9:02	170.45
30503	16MR	16:45	4.63	30503	1AP	1:23	199.70
30504	15AP	2:29	33.37	30504	30AP	14:45	228.35
30505	14MY	12:42	61.72	30505	30MY	1:15	256.56
30506	12JN	23:59	89.86	30506	28JN	9:41	284.54
30507	12JL	13:00	118.03	30507	27JL	17:14	312.57
30508	11AU	4:04	146.49	30508	26AU	0:57	340.90
30509	9SE	20:51	175.39	30509	24SE	9:40	9.71
30510	9OC	14:17	204.81	30510	23OC	19:54	39.07
30511	8NO	7:02	234.67	30511	22NO	8:02	68.92
30512	7DE	22:12	264.79	30512	21DE	22:20	99.07

1467

NEW MOONS NUMBER	DATE	TIME	LONG.	FULL MOONS NUMBER	DATE	TIME	LONG.
30513	6JA	11:30	294.91	30513	20JA	14:49	129.27
30514	4FE	23:02	324.78	30514	19FE	8:46	159.24
30515	6MR	9:01	354.18	30515	21MR	2:48	188.74
30516	4AP	17:45	23.06	30516	19AP	19:23	217.69
30517	4MY	1:45	51.45	30517	19MY	9:39	246.15
30518	2JN	9:54	79.53	30518	17JN	21:38	274.30
30519	1JL	19:17	107.55	30519	17JL	7:54	302.38
30520	31JL	6:55	135.77	30520	15AU	17:12	330.65
30521	29AU	21:21	164.42	30521	14SE	2:12	359.30
30522	28SE	14:16	193.63	30522	13OC	11:25	28.45
30523	28OC	8:39	223.39	30523	11NO	21:21	58.07
30524	27NO	3:06	253.55	30524	11DE	8:29	88.04
30525	26DE	20:24	283.84				

1468

NEW MOONS NUMBER	DATE	TIME	LONG.	FULL MOONS NUMBER	DATE	TIME	LONG.
				30525	9JA	21:16	118.13
30526	25JA	11:37	313.96	30526	8FE	11:45	148.08
30527	24FE	0:16	343.63	30527	9MR	3:30	177.68
30528	24MR	10:23	12.74	30528	7AP	19:40	206.80
30529	22AP	18:29	41.29	30529	7MY	11:29	235.45
30530	22MY	1:30	69.42	30530	6JN	2:26	263.77
30531	20JN	8:35	97.36	30531	5JL	16:18	291.97
30532	19JL	16:52	125.38	30532	4AU	4:58	320.27
30533	18AU	3:17	153.73	30533	2SE	16:27	348.86
30534	16SE	16:24	182.60	30534	2OC	3:02	17.89
30535	16OC	8:18	212.08	30535	31OC	13:11	47.37
30536	15NO	2:35	242.10	30536	29NO	23:28	77.22
30537	14DE	22:08	272.45	30537	29DE	10:17	107.24

1469

NEW MOONS NUMBER	DATE	TIME	LONG.	FULL MOONS NUMBER	DATE	TIME	LONG.
30538	13JA	17:13	302.80	30538	27JA	21:50	137.17
30539	12FE	10:06	332.79	30539	26FE	10:04	166.79
30540	13MR	23:47	2.23	30540	27MR	22:58	195.96
30541	12AP	10:20	31.04	30541	26AP	12:36	224.68
30542	11MY	18:35	59.34	30542	26MY	3:06	253.07
30543	10JN	1:39	87.33	30543	24JN	18:18	281.33
30544	9JL	8:35	115.25	30544	24JL	9:41	309.65
30545	7AU	16:19	143.38	30545	23AU	0:30	338.23
30546	6SE	1:38	171.94	30546	21SE	14:15	7.21
30547	5OC	13:17	201.05	30547	21OC	2:53	36.62
30548	4NO	3:51	230.77	30548	19NO	14:43	66.43
30549	3DE	21:25	260.96	30549	19DE	2:01	96.45

1470

NEW MOONS NUMBER	DATE	TIME	LONG.	FULL MOONS NUMBER	DATE	TIME	LONG.
30550	2JA	16:59	291.35	30550	17JA	12:51	126.42
30551	1FE	12:37	321.59	30551	15FE	23:11	156.11
30552	3MR	6:15	351.36	30552	17MR	9:15	185.34
30553	1AP	20:52	20.52	30553	15AP	19:34	214.09
30554	1MY	8:31	49.10	30554	15MY	6:53	242.47
30555	30MY	17:54	77.25	30555	13JN	19:47	270.67
30556	29JN	1:50	105.21	30556	13JL	10:22	298.92
30557	28JL	9:10	133.24	30557	12AU	2:09	327.41
30558	26AU	16:46	161.57	30558	10SE	18:24	356.30
30559	25SE	1:29	190.40	30559	10OC	10:25	25.68
30560	24OC	12:14	219.79	30560	9NO	1:42	55.49
30561	23NO	1:40	249.70	30561	8DE	15:48	85.57
30562	22DE	17:53	279.93				

1471

NEW MOONS NUMBER	DATE	TIME	LONG.	FULL MOONS NUMBER	DATE	TIME	LONG.
				30562	7JA	4:20	115.66
30563	21JA	12:01	310.20	30563	5FE	15:11	145.48
30564	20FE	6:34	340.17	30564	7MR	0:35	174.86
30565	22MR	0:05	9.64	30565	5AP	9:09	203.71
30566	20AP	15:40	38.54	30566	4MY	17:43	232.12
30567	20MY	4:57	66.95	30567	3JN	3:11	260.25
30568	18JN	16:02	95.06	30568	2JL	14:15	288.34
30569	18JL	1:18	123.11	30569	1AU	3:18	316.62
30570	16AU	9:31	151.34	30570	30AU	18:27	345.31
30571	14SE	17:39	179.97	30571	29SE	11:28	14.53
30572	14OC	2:37	209.11	30572	29OC	5:34	44.30
30573	12NO	13:09	238.77	30573	27NO	23:28	74.44
30574	12DE	1:35	268.80	30574	27DE	15:44	104.69

1472

NUMBER	DATE	TIME	LONG.	NUMBER	DATE	TIME	LONG.
30575	10JA	15;49	298.95	30575	26JA	5;26	134.74
30576	9FE	7;27	328.94	30576	24FE	16;34	164.34
30577	9MR	23;58	358.55	30577	25MR	1;44	193.40
30578	8AP	16;43	27.68	30578	23AP	9;42	221.93
30579	8MY	8;51	56.33	30579	22MY	17;16	250.08
30580	6JN	23;33	84.64	30580	21JN	1;15	278.06
30581	6JL	12;19	112.80	30581	20JL	10;29	306.11
30582	4AU	23;18	141.04	30582	18AU	21;48	334.51
30583	3SE	9;11	169.58	30583	17SE	11;57	3.43
30584	2OC	18;47	198.57	30584	17OC	5;07	32.98
30585	1NO	4;46	228.05	30585	16NO	0;24	63.05
30586	30NO	15;24	257.93	30586	15DE	19;54	93.41
30587	30DE	2;45	287.97				

1473

NUMBER	DATE	TIME	LONG.	NUMBER	DATE	TIME	LONG.
				30587	14JA	13;39	123.69
30588	28JA	14;51	317.92	30588	13FE	4;37	153.60
30589	27FE	3;56	347.56	30589	14MR	16;43	182.95
30590	28MR	18;10	16.77	30590	13AP	2;26	211.72
30591	27AP	9;25	45.55	30591	12MY	10;27	240.00
30592	27MY	1;02	73.98	30592	10JN	17;29	267.99
30593	25JN	16;14	102.23	30593	10JL	0;26	295.92
30594	25JL	6;28	130.52	30594	8AU	8;25	324.06
30595	23AU	19;41	159.04	30595	6SE	18;34	352.66
30596	22SE	8;05	187.97	30596	6OC	7;47	21.85
30597	21OC	19;54	217.37	30597	5NO	0;16	51.66
30598	20NO	7;11	247.16	30598	4DE	19;10	81.91
30599	19DE	17;58	277.16				

1474

NUMBER	DATE	TIME	LONG.	NUMBER	DATE	TIME	LONG.
				30599	3JA	14;49	112.31
30600	18JA	4;26	307.11	30600	2FE	9;26	142.49
30601	16FE	14;57	336.79	30601	4MR	1;49	172.20
30602	18MR	2;00	6.05	30602	2AP	15;22	201.31
30603	16AP	13;57	34.86	30603	2MY	2;07	229.83
30604	16MY	2;56	63.30	30604	31MY	10;32	257.94
30605	14JN	16;54	91.55	30605	29JN	17;34	285.87
30606	14JL	7;44	119.80	30606	29JL	0;22	313.88
30607	12AU	23;11	148.29	30607	27AU	8;06	342.23
30608	11SE	14;51	177.17	30608	25SE	17;47	11.10
30609	11OC	6;06	206.53	30609	25OC	6;01	40.56
30610	9NO	20;15	236.30	30610	23NO	20;59	70.54
30611	9DE	8;55	266.33	30611	23DE	14;15	100.83

1475

NUMBER	DATE	TIME	LONG.	NUMBER	DATE	TIME	LONG.
30612	7JA	20;10	296.36	30612	22JA	8;57	131.11
30613	6FE	6;25	326.15	30613	21FE	3;47	161.09
30614	7MR	16;03	355.53	30614	22MR	21;12	190.54
30615	6AP	1;29	24.41	30615	21AP	12;01	219.39
30616	5MY	11;09	52.85	30616	20MY	23;53	247.74
30617	3JN	21;37	81.01	30617	19JN	9;16	275.78
30618	3JL	9;35	109.14	30618	18JL	17;14	303.77
30619	1AU	23;36	137.46	30619	17AU	0;54	331.99
30620	31AU	15;45	166.20	30620	15SE	9;12	0.64
30621	30SE	9;16	195.46	30621	14OC	18;46	29.81
30622	30OC	2;47	225.22	30622	13NO	5;59	59.51
30623	28NO	19;02	255.30	30623	12DE	19;10	89.57
30624	28DE	9;23	285.47				

1476

NUMBER	DATE	TIME	LONG.	NUMBER	DATE	TIME	LONG.
				30624	11JA	10;28	119.77
30625	26JA	21;46	315.45	30625	10FE	3;35	149.81
30626	25FE	8;24	345.03	30626	10MR	21;32	179.47
30627	25MR	17;33	14.07	30627	9AP	14;48	208.60
30628	24AP	1;39	42.60	30628	9MY	6;11	237.22
30629	23MY	9;26	70.75	30629	7JN	19;18	265.46
30630	21JN	17;56	98.75	30630	7JL	6;27	293.55
30631	21JL	4;16	126.86	30631	5AU	16;21	321.75
30632	19AU	17;18	155.33	30632	4SE	1;39	350.28
30633	18SE	9;10	184.34	30633	3OC	10;54	19.27
30634	18OC	3;09	213.93	30634	1NO	20;35	48.75
30635	16NO	21;56	243.99	30635	1DE	7;07	78.62
30636	16DE	16;05	274.29	30636	30DE	18;59	108.69

1477

NUMBER	DATE	TIME	LONG.	NUMBER	DATE	TIME	LONG.
30637	15JA	8;29	304.51	30637	29JA	8;27	138.69
30638	13FE	22;26	334.37	30638	27FE	23;23	168.40
30639	15MR	9;41	3.68	30639	29MR	15;09	197.66
30640	13AP	18;34	32.41	30640	28AP	7;00	226.44
30641	13MY	1;52	60.65	30641	27MY	22;19	254.86
30642	11JN	8;41	88.62	30642	26JN	12;47	283.08
30643	10JL	16;11	116.58	30643	26JL	2;13	311.34
30644	9AU	1;27	144.78	30644	24AU	14;33	339.84
30645	7SE	13;14	173.45	30645	23SE	1;49	8.73
30646	7OC	3;51	202.72	30646	22OC	12;22	38.09
30647	5NO	21;09	232.57	30647	20NO	22;40	67.84
30648	5DE	16;19	262.84	30648	20DE	9;13	97.83

1478

NUMBER	DATE	TIME	LONG.	NUMBER	DATE	TIME	LONG.
30649	4JA	11;54	293.23	30649	18JA	20;18	127.81
30650	3FE	6;03	323.39	30650	17FE	8;00	157.53
30651	4MR	21;19	353.04	30651	18MR	20;17	186.83
30652	3AP	9;16	22.06	30652	17AP	9;14	215.67
30653	2MY	18;28	50.52	30653	16MY	23;03	244.14
30654	1JN	1;56	78.59	30654	15JN	13;49	272.41
30655	30JN	8;49	106.51	30655	15JL	5;15	300.69
30656	29JL	16;05	134.54	30656	13AU	20;37	329.18
30657	28AU	0;36	162.93	30657	12SE	11;14	358.04
30658	26SE	11;06	191.84	30658	12OC	0;42	27.33
30659	26OC	0;14	221.36	30659	10NO	13;09	57.05
30660	24NO	16;22	251.40	30660	10DE	0;50	87.03
30661	24DE	11;06	281.74				

1479

NUMBER	DATE	TIME	LONG.	NUMBER	DATE	TIME	LONG.
				30661	8JA	11;54	117.05
30662	23JA	6;53	312.07	30662	6FE	22;22	146.85
30663	22FE	1;36	342.02	30663	8MR	8;19	176.23
30664	23MR	17;40	11.40	30664	6AP	18;10	205.11
30665	22AP	6;41	40.17	30665	6MY	4;37	233.58
30666	21MY	17;04	68.45	30666	4JN	16;28	261.80
30667	20JN	1;40	96.46	30667	4JL	6;08	289.99
30668	19JL	9;17	124.45	30668	2AU	21;26	318.37
30669	17AU	16;47	152.66	30669	1SE	13;44	347.13
30670	16SE	0;58	181.31	30670	1OC	6;13	16.36
30671	15OC	10;43	210.51	30671	30OC	22;12	46.06
30672	13NO	22;49	240.25	30672	29NO	13;10	76.10
30673	13DE	13;39	270.39	30673	29DE	2;38	106.23

1480

NUMBER	DATE	TIME	LONG.	NUMBER	DATE	TIME	LONG.
30674	12JA	6;52	300.66	30674	27JA	14;18	136.18
30675	11FE	1;13	330.75	30675	26FE	0;15	165.72
30676	11MR	19;11	0.39	30676	26MR	8;57	194.73
30677	10AP	11;39	29.48	30677	24AP	17;11	223.26
30678	10MY	2;02	58.04	30678	24MY	1;55	251.44
30679	8JN	14;14	86.24	30679	22JN	11;57	279.50
30680	8JL	0;27	114.30	30680	21JL	23;55	307.67
30681	6AU	9;16	142.46	30681	20AU	14;05	336.20
30682	4SE	17;31	170.96	30682	19SE	6;22	5.24
30683	4OC	2;09	199.93	30683	19OC	0;16	34.84
30684	2NO	11;58	229.43	30684	17NO	18;41	64.90
30685	1DE	23;29	259.35	30685	17DE	12;05	95.17
30686	31DE	12;46	289.47				

1481

NUMBER	DATE	TIME	LONG.	NUMBER	DATE	TIME	LONG.
				30686	16JA	3;10	125.33
30687	30JA	3;34	319.52	30687	14FE	15;30	155.12
30688	28FE	19;27	349.26	30688	16MR	1;26	184.36
30689	30MR	11;53	18.53	30689	14AP	9;45	213.05
30690	29AP	4;14	47.33	30690	13MY	17;16	241.30
30691	28MY	19;41	75.74	30691	12JN	0;50	269.29
30692	27JN	9;31	103.94	30692	11JL	9;17	297.29
30693	26JL	21;32	132.14	30693	9AU	19;27	325.53
30694	25AU	8;07	160.58	30694	8SE	8;10	354.26
30695	23SE	18;00	189.43	30695	7OC	23;58	23.58
30696	23OC	3;58	218.77	30696	6NO	18;31	53.50
30697	21NO	14;23	248.54	30697	6DE	14;18	83.80
30698	21DE	1;24	278.55				

1482

NUMBER	DATE	TIME	LONG.	NUMBER	DATE	TIME	LONG.
				30698	5JA	9;11	114.16
30699	19JA	12;59	308.55	30699	4FE	1;36	144.24
30700	18FE	1;19	338.29	30700	5MR	15;01	173.80
30701	19MR	14;40	7.62	30701	4AP	1;45	202.77
30702	18AP	5;10	36.51	30702	3MY	10;26	231.19
30703	17MY	20;29	65.03	30703	1JN	17;47	259.25
30704	16JN	11;55	93.32	30704	1JL	0;38	287.17
30705	16JL	2;46	121.58	30705	30JL	8;00	315.21
30706	14AU	16;41	150.02	30706	28AU	16;59	343.63
30707	13SE	5;45	178.83	30707	27SE	4;40	12.60
30708	12OC	18;08	208.09	30708	26OC	19;38	42.20
30709	11NO	5;55	237.78	30709	25NO	13;32	72.32
30710	10DE	17;03	267.75	30710	25DE	9;03	102.70

1483

NUMBER	DATE	TIME	LONG.	NUMBER	DATE	TIME	LONG.
30711	9JA	3;38	297.75	30711	24JA	4;19	133.00
30712	7FE	13;55	327.53	30712	22FE	21;50	162.90
30713	9MR	0;25	356.92	30713	24MR	12;45	192.21
30714	7AP	11;35	25.85	30714	23AP	0;48	220.93
30715	6MY	23;47	54.38	30715	22MY	10;16	249.17
30716	5JN	13;05	82.65	30716	20JN	17;53	277.14
30717	5JL	3;26	110.86	30717	20JL	0;43	305.09
30718	3AU	18;41	139.25	30718	18AU	7;56	333.31
30719	2SE	10;32	168.01	30719	16SE	16;40	1.99
30720	2OC	2;23	197.23	30720	16OC	3;40	31.25
30721	31OC	17;27	226.90	30721	14NO	17;21	61.06
30722	30NO	7;05	256.89	30722	14DE	9;30	91.26
30723	29DE	19;05	286.96				

1484 / 1490

NEW MOONS				FULL MOONS				NEW MOONS				FULL MOONS			
NUMBER	DATE	TIME	LONG.	NUMBER	DATE	TIME	LONG.	NUMBER	DATE	TIME	LONG.	NUMBER	DATE	TIME	LONG.
		1484		30723	13JA	3;31	121.56			**1490**		30797	7JA	0;11	115.87
30724	28JA	5;44	316.86	30724	11FE	22;17	151.66	30798	21JA	0;10	310.09	30798	5FE	13;53	145.82
30725	26FE	15;29	346.38	30725	12MR	16;26	181.30	30799	19FE	15;19	339.93	30799	7MR	0;50	175.26
30726	27MR	0;45	15.41	30726	11AP	8;32	210.36	30800	21MR	7;15	9.34	30800	5AP	9;43	204.13
30727	25AP	10;00	43.97	30727	10MY	21;50	238.87	30801	19AP	23;30	38.28	30801	4MY	17;23	232.50
30728	24MY	19;47	72.18	30728	9JN	8;24	267.00	30802	19MY	15;23	66.80	30802	3JN	0;43	260.54
30729	23JN	6;45	100.27	30729	8JL	17;02	294.99	30803	18JN	6;07	95.05	30803	2JL	8;33	288.50
30730	22JL	19;39	128.49	30730	7AU	0;51	323.12	30804	17JL	19;11	123.25	30804	31JL	17;44	316.62
30731	21AU	10;51	157.07	30731	5SE	8;55	351.61	30805	16AU	6;38	151.61	30805	30AU	5;07	345.16
30732	20SE	4;01	186.16	30732	4OC	17;57	20.62	30806	14SE	17;02	180.33	30806	28SE	19;27	14.26
30733	19OC	21;58	215.78	30733	3NO	4;26	50.15	30807	14OC	3;08	209.53	30807	28OC	12;51	43.99
30734	18NO	15;14	245.80	30734	2DE	16;37	80.10	30808	12NO	13;28	239.18	30808	27NO	8;23	74.20
30735	18DE	6;43	275.99					30809	12DE	0;14	269.13	30809	27DE	4;03	104.59

1485 / 1491

NEW MOONS				FULL MOONS				NEW MOONS				FULL MOONS			
		1485		30735	1JA	6;46	110.26			**1491**		30810	25JA	21;50	134.81
30736	16JA	20;07	306.08	30736	30JA	22;51	140.36	30810	10JA	11;28	299.16	30811	24FE	12;41	164.57
30737	15FE	7;32	335.81	30737	1MR	16;17	170.15	30811	8FE	23;12	328.99	30812	26MR	0;37	193.74
30738	16MR	17;14	5.04	30738	31MR	9;51	199.46	30812	10MR	11;44	358.44	30813	24AP	10;09	222.34
30739	15AP	1;37	33.72	30739	30AP	2;08	228.24	30813	9AP	1;22	27.45	30814	23MY	18;00	250.50
30740	14MY	9;17	61.97	30740	29MY	16;22	256.59	30814	8MY	16;05	56.05	30815	22JN	0;57	278.43
30741	12JN	17;08	89.97	30741	28JN	4;31	284.72	30815	7JN	7;25	84.39	30816	21JL	7;57	306.40
30742	12JL	2;21	118.00	30742	27JL	15;08	312.88	30816	6JL	22;40	112.65	30817	19AU	16;02	334.67
30743	10AU	13;57	146.31	30743	26AU	0;53	341.29	30817	5AU	13;14	141.02	30818	18SE	2;20	3.43
30744	9SE	4;30	175.12	30744	24SE	10;20	10.13	30818	4SE	2;59	169.71	30819	17OC	15;43	32.81
30745	8OC	21;43	204.52	30745	23OC	19;56	39.47	30819	3OC	16;00	198.85	30820	16NO	8;19	62.77
30746	7NO	16;28	234.45	30746	22NO	6;06	69.23	30820	2NO	4;20	228.43	30821	16DE	3;15	93.09
30747	7DE	11;15	264.72	30747	21DE	17;14	99.25	30821	1DE	15;57	258.34				
								30822	31DE	2;49	288.37				

1486 / 1492

NEW MOONS				FULL MOONS				NEW MOONS				FULL MOONS			
		1486		30748	20JA	5;45	129.28			**1492**		30822	14JA	22;48	123.45
30748	6JA	4;43	295.02	30749	18FE	19;46	159.08	30823	29JA	13;07	318.24	30823	13FE	17;16	153.52
30749	4FE	19;57	325.04	30750	20MR	10;54	188.48	30824	27FE	23;15	347.76	30824	14MR	9;27	183.04
30750	6MR	8;27	354.55	30751	19AP	2;31	217.40	30825	28MR	9;45	16.82	30825	12AP	22;51	211.97
30751	4AP	18;20	23.47	30752	18MY	17;59	245.91	30826	26AP	21;08	45.44	30826	12MY	9;30	240.35
30752	4MY	2;12	51.86	30753	17JN	8;53	274.18	30827	26MY	9;39	73.75	30827	10JN	17;56	268.39
30753	2JN	9;01	79.89	30754	16JL	22;53	302.42	30828	24JN	23;22	101.95	30828	10JL	1;04	296.33
30754	1JL	15;59	107.81	30755	15AU	12;07	330.84	30829	24JL	14;13	130.26	30829	8AU	8;04	324.44
30755	31JL	0;15	135.89	30756	14SE	0;10	359.61	30830	23AU	5;59	158.88	30830	6SE	16;02	352.95
30756	29AU	10;45	164.38	30757	13OC	11;18	28.84	30831	21SE	22;12	187.96	30831	6OC	1;57	22.00
30757	28SE	0;01	193.43	30758	11NO	21;51	58.49	30832	21OC	14;04	217.52	30832	4NO	14;21	51.64
30758	27OC	16;07	223.09	30759	11DE	8;19	88.43	30833	20NO	4;42	247.45	30833	4DE	5;19	81.72
30759	26NO	10;34	253.24					30834	19DE	17;39	277.54				
30760	26DE	6;12	283.63												

1487 / 1493

NEW MOONS				FULL MOONS				NEW MOONS				FULL MOONS			
		1487		30760	9JA	19;04	118.43			**1493**		30834	2JA	22;25	112.01
30761	25JA	1;17	313.91	30761	8FE	6;18	148.25	30835	18JA	4;55	307.53	30835	1FE	16;48	142.19
30762	23FE	18;03	343.77	30762	9MR	18;02	177.68	30836	16FE	14;57	337.19	30836	3MR	11;15	171.99
30763	25MR	7;33	13.01	30763	8AP	6;20	206.64	30837	18MR	0;13	6.37	30837	2AP	4;22	201.25
30764	23AP	17;57	41.65	30764	7MY	19;25	235.20	30838	16AP	9;12	35.06	30838	1MY	19;03	229.94
30765	23MY	2;06	69.83	30765	6JN	9;33	263.49	30839	15MY	18;25	63.35	30839	31MY	6;56	258.19
30766	21JN	9;07	97.77	30766	6JL	0;41	291.74	30840	14JN	4;32	91.44	30840	29JN	16;30	286.21
30767	20JL	16;07	125.74	30767	4AU	16;19	320.15	30841	13JL	16;18	119.57	30841	29JL	0;44	314.29
30768	19AU	0;00	153.99	30768	3SE	7;38	348.90	30842	12AU	6;21	148.00	30842	27AU	8;44	342.65
30769	17SE	9;32	182.71	30769	2OC	22;00	18.07	30843	10SE	22;46	176.91	30843	25SE	17;24	11.48
30770	16OC	21;22	212.02	30770	1NO	11;13	47.68	30844	10OC	16;45	206.37	30844	25OC	3;15	40.84
30771	15NO	12;02	241.88	30771	30NO	23;25	77.61	30845	9NO	10;47	236.30	30845	23NO	14;37	70.67
30772	15DE	5;34	272.14	30772	30DE	10;51	107.66	30846	9DE	3;26	266.49	30846	23DE	3;43	100.77

1488 / 1494

NEW MOONS				FULL MOONS				NEW MOONS				FULL MOONS			
		1488		30773	28JA	21;33	137.55			**1494**		30847	21JA	18;41	130.90
30773	14JA	0;59	302.50	30774	27FE	7;34	167.08	30847	7JA	17;58	296.66	30848	20FE	11;19	160.79
30774	12FE	20;22	332.61	30775	27MR	17;09	196.11	30848	6FE	6;18	326.54	30849	22MR	4;43	190.25
30775	13MR	13;45	2.21	30776	26AP	2;56	224.68	30849	7MR	16;43	355.94	30850	20AP	21;36	219.20
30776	12AP	4;11	31.18	30777	25MY	13;47	252.93	30850	6AP	1;33	24.80	30851	20MY	12;50	247.68
30777	11MY	15;46	59.62	30778	24JN	2;22	281.09	30851	5MY	9;18	53.16	30852	19JN	2;02	275.88
30778	10JN	1;10	87.69	30779	23JL	16;53	309.37	30852	3JN	16;43	81.21	30853	18JL	13;30	304.02
30779	9JL	9;15	115.67	30780	22AU	8;55	337.99	30853	3JL	1;06	109.19	30854	16AU	23;49	332.34
30780	7AU	16;51	143.80	30781	21SE	1;38	7.08	30854	1AU	11;22	137.36	30855	15SE	9;36	1.04
30781	6SE	0;44	172.29	30782	20OC	18;14	36.66	30855	31AU	0;27	165.97	30856	14OC	19;18	30.23
30782	5OC	9;44	201.30	30783	19NO	10;01	66.63	30856	29SE	16;32	195.17	30857	13NO	5;17	59.87
30783	3NO	20;40	230.86	30784	19DE	0;26	96.77	30857	29OC	10;51	224.94	30858	12DE	15;55	89.83
30784	3DE	10;08	260.88					30858	28NO	5;58	255.14				
								30859	28DE	0;20	285.48				

1489 / 1495

NEW MOONS				FULL MOONS				NEW MOONS				FULL MOONS			
		1489		30785	17JA	13;03	126.83			**1495**		30859	11JA	3;37	119.87
30785	2JA	2;10	291.12	30786	15FE	23;44	156.52	30860	26JA	16;48	315.64	30860	9FE	16;41	149.75
30786	31JA	19;56	321.28	30787	17MR	8;48	185.71	30861	25FE	6;38	345.35	30861	11MR	7;02	179.26
30787	2MR	14;03	351.08	30788	15AP	16;57	214.37	30862	26MR	17;41	14.47	30862	9AP	22;12	208.32
30788	1AP	7;11	20.36	30789	15MY	1;08	242.63	30863	25AP	2;20	43.02	30863	9MY	13;35	236.95
30789	30AP	22;33	49.09	30790	13JN	10;16	270.68	30864	24MY	9;27	71.14	30864	8JN	4;44	265.27
30790	30MY	11;49	77.40	30791	12JL	21;08	298.78	30865	22JN	16;08	99.07	30865	7JL	19;19	293.50
30791	28JN	23;05	105.49	30792	11AU	10;10	327.16	30866	21JL	23;37	127.06	30866	6AU	9;09	321.85
30792	28JL	8;41	133.61	30793	10SE	1;32	356.01	30867	20AU	8;56	155.38	30867	4SE	22;01	350.52
30793	26AU	17;18	161.99	30794	9OC	18;54	25.43	30868	18SE	20;51	184.22	30868	4OC	9;52	19.62
30794	25SE	1;50	190.80	30795	8NO	13;27	55.37	30869	18OC	11;39	213.67	30869	2NO	20;53	49.16
30795	24OC	11;09	220.13	30796	8DE	7;44	85.62	30870	17NO	5;09	243.68	30870	2DE	7;28	79.03
30796	22NO	21;51	249.93					30871	17DE	0;22	274.02	30871	31DE	18;04	109.04
30797	22DE	10;14	280.00												

	NEW MOONS				FULL MOONS				NEW MOONS				FULL MOONS		
NUMBER	DATE	TIME	LONG.	NUMBER	DATE	TIME	LONG.	NUMBER	DATE	TIME	LONG.	NUMBER	DATE	TIME	LONG.

1496 / 1502

NUMBER	DATE	TIME	LONG.	NUMBER	DATE	TIME	LONG.	NUMBER	DATE	TIME	LONG.	NUMBER	DATE	TIME	LONG.
30872	15JA	19;57	304.38	30872	30JA	4;56	138.93	30946	9JA	3;50	298.15	30946	23JA	11;30	132.68
30873	14FE	13;59	334.41	30873	28FE	16;10	168.49	30947	7FE	14;22	327.94	30947	22FE	5;53	162.62
30874	15MR	5;05	3.88	30874	29MR	3;53	197.59	30948	8MR	23;47	357.29	30948	23MR	23;39	192.07
30875	13AP	16;54	32.73	30875	27AP	16;15	226.24	30949	7AP	8;39	26.12	30949	22AP	15;32	220.95
30876	13MY	1;58	61.04	30876	27MY	5;38	254.58	30950	6MY	17;27	54.52	30950	22MY	4;48	249.34
30877	11JN	9;24	89.04	30877	25JN	20;12	282.81	30951	5JN	2;50	82.63	30951	20JN	15;30	277.43
30878	10JL	16;20	116.97	30878	25JL	11;44	311.15	30952	4JL	13;35	110.70	30952	20JL	0;23	305.47
30879	8AU	23;45	145.10	30879	24AU	3;32	339.78	30953	3AU	2;25	138.99	30953	18AU	8;34	333.72
30880	7SE	8;28	173.65	30880	22SE	18;44	8.83	30954	1SE	17;47	167.72	30954	16SE	17;00	2.39
30881	6OC	19;10	202.75	30881	22OC	8;49	38.33	30955	1OC	11;19	197.00	30955	16OC	2;23	31.59
30882	5NO	8;27	232.43	30882	20NO	21;43	68.20	30956	31OC	5;46	226.80	30956	14NO	13;03	61.27
30883	5DE	0;36	262.57	30883	20DE	9;38	98.24	30957	29NO	23;28	256.96	30957	14DE	1;15	91.30
								30958	29DE	15;13	287.19				

1497 / 1503

NUMBER	DATE	TIME	LONG.	NUMBER	DATE	TIME	LONG.	NUMBER	DATE	TIME	LONG.	NUMBER	DATE	TIME	LONG.
30884	3JA	19;11	292.92	30884	18JA	20;40	128.22					30958	12JA	15;09	121.43
30885	2FE	14;42	323.14	30885	17FE	6;53	157.88	30959	28JA	4;39	317.21	30959	11FE	6;47	151.40
30886	4MR	9;08	352.93	30886	18MR	16;25	187.07	30960	26FE	15;54	346.79	30960	12MR	23;40	181.00
30887	3AP	0;59	22.12	30887	17AP	1;45	215.76	30961	28MR	1;20	15.82	30961	11AP	16;44	210.11
30888	2MY	13;53	50.73	30888	16MY	11;43	244.08	30962	26AP	9;23	44.33	30962	11MY	8;46	238.73
30889	1JN	0;17	78.91	30889	14JN	23;12	272.22	30963	25MY	16;45	72.45	30963	9JN	23;00	267.01
30890	30JN	9;00	106.90	30890	14JL	12;41	300.42	30964	24JN	0;24	100.41	30964	9JL	11;23	295.16
30891	29JL	16;51	134.96	30891	13AU	4;07	328.91	30965	23JL	9;30	128.48	30965	7AU	22;25	323.41
30892	28AU	0;39	163.32	30892	11SE	20;47	357.84	30966	21AU	21;08	156.91	30966	6SE	8;39	351.99
30893	26SE	9;09	192.16	30893	11OC	13;46	27.27	30967	20SE	11;50	185.89	30967	5OC	18;35	21.03
30894	25OC	19;09	221.54	30894	10NO	6;20	57.15	30968	20OC	5;19	215.48	30968	4NO	4;34	50.55
30895	24NO	7;20	251.41	30895	9DE	21;41	87.29	30969	19NO	0;23	245.57	30969	3DE	14;54	80.42
30896	23DE	22;04	281.58					30970	18DE	19;25	275.91				

1498 / 1504

NUMBER	DATE	TIME	LONG.	NUMBER	DATE	TIME	LONG.	NUMBER	DATE	TIME	LONG.	NUMBER	DATE	TIME	LONG.
				30896	8JA	11;19	117.42					30970	2JA	1;59	110.45
30897	22JA	14;57	311.79	30897	6FE	22;54	147.27	30971	17JA	12;59	306.17	30971	31JA	14;11	140.39
30898	21FE	8;50	341.72	30898	8MR	8;35	176.63	30972	16FE	4;09	336.06	30972	1MR	3;39	170.02
30899	23MR	2;22	11.17	30899	6AP	16;54	205.45	30973	16MR	16;29	5.40	30973	30MR	18;10	199.21
30900	21AP	18;32	40.07	30900	6MY	0;45	233.81	30974	15AP	2;09	34.13	30974	29AP	9;16	227.95
30901	21MY	8;50	68.51	30901	4JN	9;08	261.89	30975	14MY	9;49	62.38	30975	29MY	0;25	256.35
30902	19JN	21;08	96.67	30902	3JL	18;57	289.93	30976	12JN	16;31	90.34	30976	27JN	15;19	284.59
30903	19JL	7;39	124.77	30903	2AU	6;50	318.17	30977	11JL	23;26	118.27	30977	27JL	5;43	312.89
30904	17AU	16;52	153.06	30904	31AU	21;06	346.84	30978	10AU	7;45	146.45	30978	25AU	19;20	341.46
30905	16SE	1;34	181.73	30905	30SE	13;41	16.07	30979	8SE	18;22	175.09	30979	24SE	8;00	10.43
30906	15OC	10;35	210.90	30906	30OC	7;59	45.86	30980	8OC	7;48	204.33	30980	23OC	19;40	39.85
30907	13NO	20;39	240.55	30907	29NO	2;48	76.05	30981	7NO	0;04	234.16	30981	22NO	6;35	69.65
30908	13DE	8;12	270.55	30908	28DE	20;27	106.36	30982	6DE	18;37	264.41	30982	21DE	17;11	99.64

1499 / 1505

NUMBER	DATE	TIME	LONG.	NUMBER	DATE	TIME	LONG.	NUMBER	DATE	TIME	LONG.	NUMBER	DATE	TIME	LONG.
30909	11JA	21;17	300.65	30909	27JA	11;36	136.46	30983	5JA	14;15	294.80	30983	20JA	3;49	129.59
30910	10FE	11;39	330.57	30910	25FE	23;49	166.09	30984	4FE	9;13	324.98	30984	18FE	14;40	159.26
30911	12MR	2;57	0.12	30911	27MR	9;30	195.15	30985	6MR	1;49	354.67	30985	20MR	1;52	188.50
30912	10AP	18;50	29.19	30912	25AP	17;32	223.66	30986	4AP	15;10	23.73	30986	18AP	13;35	217.27
30913	10MY	10;49	57.83	30913	25MY	0;48	251.79	30987	4MY	1;27	52.21	30987	18MY	2;10	245.67
30914	9JN	2;10	86.15	30914	23JN	8;12	279.74	30988	2JN	9;32	80.30	30988	16JN	15;59	273.90
30915	8JL	16;13	114.36	30915	22JL	16;35	307.77	30989	1JL	16;36	108.22	30989	16JL	7;06	302.17
30916	7AU	4;39	142.67	30916	21AU	2;47	336.13	30990	30JL	23;44	136.26	30990	14AU	23;02	330.70
30917	5SE	15;45	171.28	30917	19SE	15;39	5.03	30991	29AU	7;50	164.65	30991	13SE	14;54	359.62
30918	5OC	2;11	200.32	30918	19OC	7;39	34.54	30992	27SE	17;34	193.56	30992	13OC	5;53	29.00
30919	3NO	12;33	229.84	30919	18NO	2;25	64.61	30993	27OC	5;35	223.04	30993	11NO	19;37	58.79
30920	2DE	23;12	259.73	30920	17DE	22;23	94.98	30994	25NO	20;19	253.03	30994	11DE	8;09	88.82
								30995	25DE	13;44	283.33				

1500 / 1506

NUMBER	DATE	TIME	LONG.	NUMBER	DATE	TIME	LONG.	NUMBER	DATE	TIME	LONG.	NUMBER	DATE	TIME	LONG.
30921	1JA	10;12	289.76	30921	16JA	17;20	125.31					30995	9JA	19;39	118.85
30922	30JA	21;31	319.66	30922	15FE	9;40	155.26	30996	24JA	8;54	313.61	30996	8FE	6;11	148.64
30923	29FE	9;22	349.23	30923	15MR	22;56	184.65	30997	23FE	3;58	343.57	30997	9MR	15;49	177.98
30924	29MR	22;07	18.36	30924	14AP	9;29	213.43	30998	24MR	21;05	12.98	30998	8AP	0;56	206.81
30925	28AP	12;01	47.07	30925	13MY	18;01	241.72	30999	23AP	11;21	41.78	30999	7MY	10;15	235.22
30926	28MY	2;57	75.46	30926	12JN	1;17	269.70	31000	22MY	22;54	70.09	31000	5JN	20;40	263.37
30927	26JN	18;17	103.72	30927	11JL	8;09	297.63	31001	21JN	8;25	98.13	31001	5JL	9;00	291.51
30928	26JL	9;21	132.04	30928	9AU	15;36	325.77	31002	20JL	16;43	126.15	31002	3AU	23;31	319.87
30929	24AU	23;46	160.63	30929	8SE	0;45	354.34	31003	19AU	0;36	154.40	31003	2SE	15;49	348.64
30930	23SE	13;26	189.63	30930	7OC	12;36	23.50	31004	17SE	8;50	183.07	31004	2OC	9;01	17.92
30931	23OC	2;23	219.10	30931	6NO	3;42	53.27	31005	16OC	18;08	212.27	31005	1NO	2;09	47.69
30932	21NO	14;34	248.94	30932	5DE	21;39	83.49	31006	15NO	5;13	241.99	31006	30NO	18;23	77.79
30933	21DE	1;54	278.96					31007	14DE	18;39	272.07	31007	30DE	9;03	107.98

1501 / 1507

NUMBER	DATE	TIME	LONG.	NUMBER	DATE	TIME	LONG.	NUMBER	DATE	TIME	LONG.	NUMBER	DATE	TIME	LONG.
				30933	4JA	17;05	113.87	31008	13JA	10;25	302.28	31008	28JA	21;40	137.95
30934	19JA	12;24	308.91	30934	3FE	12;11	144.07	31009	12FE	3;45	332.31	31009	27FE	8;09	167.49
30935	17FE	22;23	338.56	30935	5MR	5;29	173.80	31010	13MR	21;22	1.92	31010	28MR	16;53	196.49
30936	19MR	8;24	7.75	30936	3AP	20;12	202.93	31011	12AP	14;07	31.00	31011	27AP	0;41	224.98
30937	17AP	19;01	36.49	30937	3MY	8;09	231.49	31012	12MY	5;17	59.59	31012	26MY	8;30	253.11
30938	17MY	6;42	64.86	30938	1JN	17;36	259.63	31013	10JN	18;37	87.82	31013	24JN	17;22	281.12
30939	15JN	19;39	93.06	30939	1JL	1;19	287.58	31014	10JL	6;07	115.94	31014	24JL	4;06	309.25
30940	15JL	9;54	121.30	30940	30JL	8;20	315.61	31015	8AU	16;07	144.16	31015	22AU	17;12	337.75
30941	14AU	1;22	149.80	30941	28AU	15;49	343.97	31016	7SE	1;11	172.70	31016	21SE	8;46	6.78
30942	12SE	17;40	178.72	30942	27SE	0;47	12.83	31017	6OC	10;09	201.71	31017	21OC	2;30	36.39
30943	12OC	10;06	208.15	30943	26OC	12;00	42.27	31018	4NO	19;46	231.22	31018	19NO	21;26	66.49
30944	11NO	1;44	238.00	30944	25NO	1;45	72.22	31019	4DE	6;37	261.12	31019	19DE	16;00	96.81
30945	10DE	15;43	268.09	30945	24DE	17;47	102.45								

1508

NEW MOON				FULL MOON			
31020	2JA	18;53	291.20	31020	18JA	8;34	127.03
31021	1FE	8;28	321.19	31021	16FE	22;12	156.85
31022	1MR	23;03	350.85	31022	17MR	8;58	186.11
31023	31MR	14;24	20.06	31023	15AP	17;35	214.79
31024	30AP	6;11	48.82	31024	15MY	1;00	243.02
31025	29MY	21;50	77.23	31025	13JN	8;09	270.99
31026	28JN	12;39	105.47	31026	12JL	15;54	298.96
31027	28JL	2;05	133.74	31027	11AU	1;06	327.18
31028	26AU	14;03	162.25	31028	9SE	12;36	355.87
31029	25SE	1;00	191.16	31029	9OC	3;06	25.16
31030	24OC	11;35	220.55	31030	7NO	20;43	55.06
31031	22NO	22;13	250.34	31031	7DE	16;25	85.36
31032	22DE	9;05	280.35				

1509

NEW MOON				FULL MOON			
				31032	6JA	12;09	115.76
31033	20JA	20;08	310.31	31033	5FE	5;54	145.88
31034	19FE	7;28	339.99	31034	6MR	20;37	175.47
31035	20MR	19;26	9.25	31035	5AP	8;22	204.46
31036	19AP	8;27	38.06	31036	4MY	17;44	232.90
31037	18MY	22;41	66.52	31037	3JN	1;29	260.96
31038	17JN	13;47	94.80	31038	2JL	8;27	288.88
31039	17JL	5;07	123.08	31039	31JL	15;32	316.93
31040	15AU	20;05	151.58	31040	29AU	23;47	345.33
31041	14SE	10;25	180.45	31041	28SE	10;16	14.28
31042	14OC	0;01	209.79	31042	27OC	23;48	43.83
31043	12NO	12;51	239.54	31043	26NO	16;28	73.92
31044	12DE	0;45	269.55	31044	26DE	11;20	104.27

1510

NEW MOON				FULL MOON			
31045	10JA	11;38	299.56	31045	25JA	6;43	134.56
31046	8FE	21;43	329.32	31046	24FE	0;56	164.47
31047	10MR	7;26	358.65	31047	25MR	16;54	193.82
31048	8AP	17;25	27.51	31048	24AP	6;10	222.57
31049	8MY	4;15	55.97	31049	23MY	16;47	250.84
31050	6JN	16;21	84.18	31050	22JN	1;18	278.84
31051	6JL	5;52	112.37	31051	21JL	8;37	306.82
31052	4AU	20;48	140.75	31052	19AU	15;51	335.05
31053	3SE	12;56	169.53	31053	18SE	0;06	3.73
31054	3OC	5;41	198.81	31054	17OC	10;16	32.98
31055	1NO	22;07	228.56	31055	15NO	22;48	62.76
31056	1DE	13;13	258.62	31056	15DE	13;42	92.91
31057	31DE	2;22	288.74				

1511

NEW MOON				FULL MOON			
				31057	14JA	6;33	123.16
31058	29JA	13;36	318.65	31058	13FE	0;32	153.21
31059	27FE	23;23	348.15	31059	14MR	18;34	182.82
31060	29MR	8;16	17.15	31060	13AP	11;23	211.89
31061	27AP	16;49	45.66	31061	13MY	1;57	240.44
31062	27MY	1;38	73.83	31062	11JN	13;55	268.62
31063	25JN	11;28	101.87	31063	10JL	23;43	296.67
31064	24JL	23;07	130.04	31064	9AU	8;17	324.84
31065	23AU	13;15	158.59	31065	7SE	16;41	353.36
31066	22SE	5;57	187.68	31066	7OC	1;43	22.39
31067	22OC	0;22	217.34	31067	5NO	11;51	51.92
31068	20NO	18;51	247.42	31068	4DE	23;18	81.86
31069	20DE	11;50	277.68				

1512

NEW MOON				FULL MOON			
				31069	3JA	12;15	111.97
31070	19JA	2;28	307.82	31070	2FE	2;50	141.99
31071	17FE	14;44	337.57	31071	2MR	18;55	171.71
31072	18MR	0;54	6.79	31072	1AP	11;46	200.96
31073	16AP	9;25	35.46	31073	1MY	4;16	229.74
31074	15MY	16;53	63.68	31074	30MY	19;24	258.12
31075	14JN	0;09	91.66	31075	29JN	8;45	286.30
31076	13JL	8;19	119.65	31076	28JL	20;35	314.51
31077	11AU	18;34	147.92	31077	27AU	7;23	342.98
31078	10SE	7;46	176.69	31078	25SE	17;41	11.88
31079	10OC	0;05	206.07	31079	25OC	3;49	41.26
31080	8NO	18;41	236.01	31080	23NO	14;04	71.04
31081	8DE	14;02	266.31	31081	23DE	0;44	101.04

1513

NEW MOON				FULL MOON			
31082	7JA	8;33	296.65	31082	21JA	12;13	131.02
31083	6FE	0;59	326.71	31083	20FE	0;48	160.75
31084	7MR	14;41	356.25	31084	21MR	14;33	190.06
31085	6AP	1;32	25.19	31085	20AP	5;08	218.93
31086	5MY	10;00	53.58	31086	19MY	20;06	247.41
31087	3JN	16;58	81.61	31087	18JN	11;06	275.67
31088	2JL	23;37	109.52	31088	18JL	1;53	303.94
31089	1AU	7;07	137.58	31089	16AU	16;09	332.42
31090	30AU	16;34	166.04	31090	15SE	5;36	1.27
31091	29SE	4;38	195.07	31091	14OC	18;02	30.57
31092	28OC	19;36	224.70	31092	13NO	5;30	60.28
31093	27NO	13;09	254.82	31093	12DE	16;19	90.24
31094	27DE	8;26	285.20				

1514

NEW MOON				FULL MOON			
				31094	11JA	2;52	120.23
31095	26JA	3;54	315.49	31095	9FE	13;28	150.00
31096	24FE	21;46	345.37	31096	11MR	0;14	179.38
31097	26MR	12;42	14.66	31097	9AP	11;23	208.27
31098	25AP	0;22	43.33	31098	8MY	23;13	236.76
31099	24MY	9;24	71.53	31099	7JN	12;11	265.00
31100	22JN	16;51	99.48	31100	7JL	2;36	293.23
31101	21JL	23;53	127.46	31101	5AU	18;19	321.65
31102	20AU	7;31	155.71	31102	4SE	10;34	350.45
31103	18SE	16;28	184.43	31103	4OC	2;22	19.70
31104	18OC	3;23	213.72	31104	2NO	17;02	49.39
31105	16NO	16;46	243.55	31105	2DE	6;21	79.38
31106	16DE	8;51	273.75	31106	31DE	18;25	109.45

1515

NEW MOON				FULL MOON			
31107	15JA	3;13	304.06	31107	30JA	5;22	139.35
31108	13FE	22;24	334.15	31108	28FE	15;17	168.84
31109	15MR	16;31	3.76	31109	30MR	0;23	197.83
31110	14AP	8;08	32.76	31110	28AP	9;15	226.35
31111	13MY	20;57	61.23	31111	27MY	18;47	254.54
31112	12JN	7;25	89.35	31112	26JN	5;57	282.63
31113	11JL	16;20	117.36	31113	25JL	19;20	310.88
31114	10AU	0;30	145.52	31114	24AU	10;55	339.50
31115	8SE	8;38	174.05	31115	23SE	3;59	8.62
31116	7OC	17;28	203.07	31116	22OC	21;31	38.25
31117	6NO	3;41	232.62	31117	21NO	14;32	68.29
31118	5DE	15;56	262.59	31118	21DE	6;13	98.49

1516

NEW MOON				FULL MOON			
31119	4JA	6;29	292.77	31119	19JA	19;56	128.58
31120	2FE	22;58	322.87	31120	18FE	7;23	158.29
31121	3MR	16;20	352.62	31121	18MR	16;47	187.47
31122	2AP	9;24	21.87	31122	17AP	0;45	216.11
31123	2MY	1;17	50.61	31123	16MY	8;16	244.33
31124	31MY	15;31	78.95	31124	14JN	16;21	272.33
31125	30JN	4;01	107.09	31125	14JL	2;00	300.38
31126	29JL	14;54	135.27	31126	12AU	13;52	328.72
31127	28AU	0;35	163.71	31127	11SE	4;18	357.56
31128	26SE	9;43	192.57	31128	10OC	21;10	26.97
31129	25OC	19;07	221.93	31129	9NO	15;50	56.93
31130	24NO	5;24	251.72	31130	9DE	10;57	87.23
31131	23DE	16;57	281.76				

1517

NEW MOON				FULL MOON			
				31131	8JA	4;47	117.54
31132	22JA	5;45	311.79	31132	6FE	19;55	147.53
31133	20FE	19;37	341.56	31133	8MR	7;59	176.99
31134	22MR	10;19	10.90	31134	6AP	17;26	205.87
31135	21AP	1;40	39.78	31135	6MY	1;13	234.22
31136	20MY	17;18	68.28	31136	4JN	8;18	262.25
31137	19JN	8;36	96.56	31137	3JL	15;36	290.19
31138	18JL	22;55	124.81	31138	1AU	23;59	318.29
31139	17AU	11;50	153.25	31139	31AU	10;17	346.79
31140	15SE	23;30	182.04	31140	29SE	23;18	15.87
31141	15OC	10;27	211.29	31141	29OC	15;29	45.56
31142	13NO	21;13	240.97	31142	28NO	10;25	75.75
31143	13DE	8;04	270.94	31143	28DE	6;27	106.16

1518

NEW MOON				FULL MOON			
31144	11JA	18;58	300.94	31144	27JA	1;22	136.42
31145	10FE	5;58	330.73	31145	25FE	17;35	166.22
31146	11MR	17;18	0.11	31146	27MR	6;41	195.42
31147	10AP	5;26	29.04	31147	25AP	17;05	224.03
31148	9MY	18;48	57.57	31148	25MY	1;30	252.20
31149	8JN	9;22	85.88	31149	23JN	8;45	280.15
31150	8JL	0;40	114.13	31150	22JL	15;42	308.12
31151	6AU	16;01	142.55	31151	20AU	23;20	336.38
31152	5SE	6;56	171.31	31152	19SE	8;40	5.13
31153	4OC	21;13	200.51	31153	18OC	20;41	34.47
31154	3NO	10;44	230.16	31154	17NO	11;53	64.39
31155	2DE	23;17	260.13	31155	17DE	5;48	94.67

1519

NEW MOON				FULL MOON			
31156	1JA	10;44	290.17	31156	16JA	1;04	125.02
31157	30JA	21;06	320.04	31157	14FE	19;54	155.08
31158	1MR	6;44	349.51	31158	16MR	12;58	184.63
31159	30MR	16;16	18.51	31159	15AP	3;30	213.58
31160	29AP	2;21	47.07	31160	14MY	15;23	242.00
31161	28MY	13;34	75.32	31161	13JN	0;53	270.07
31162	27JN	2;14	103.47	31162	12JL	8;46	298.05
31163	26JL	16;27	131.76	31163	10AU	16;02	326.18
31164	25AU	8;09	160.39	31164	8SE	23;48	354.70
31165	24SE	0;56	189.51	31165	8OC	9;03	23.75
31166	23OC	17;57	219.14	31166	6NO	20;27	53.36
31167	22NO	10;04	249.14	31167	6DE	10;13	83.40
31168	22DE	0;22	279.29				

1520

NUMBER	DATE	TIME	LONG.	NUMBER	DATE	TIME	LONG.
				31168	5JA	2;04	113.63
31169	20JA	12;32	309.31	31169	3FE	19;24	143.76
31170	18FE	22;53	338.97	31170	4MR	13;20	173.52
31171	19MR	7;59	8.12	31171	3AP	6;43	202.77
31172	17AP	16;26	36.77	31172	2MY	22;24	231.49
31173	17MY	0;50	65.02	31173	1JN	11;40	259.78
31174	15JN	9;55	93.06	31174	30JN	22;34	287.86
31175	14JL	20;28	121.15	31175	30JL	7;47	315.98
31176	13AU	9;19	149.54	31176	28AU	16;22	344.38
31177	12SE	0;52	178.43	31177	27SE	1;12	13.24
31178	11OC	18;47	207.90	31178	26OC	10;55	42.62
31179	10NO	13;41	237.89	31179	24NO	21;46	72.44
31180	10DE	7;45	268.14	31180	24DE	9;54	102.51

1521

NUMBER	DATE	TIME	LONG.	NUMBER	DATE	TIME	LONG.
31181	8JA	23;41	298.37	31181	22JA	23;30	132.57
31182	7FE	13;05	328.28	31182	21FE	14;37	162.38
31183	9MR	0;09	357.69	31183	23MR	6;55	191.78
31184	7AP	9;18	26.54	31184	21AP	23;30	220.69
31185	6MY	17;04	54.89	31185	21MY	15;18	249.19
31186	5JN	0;12	82.91	31186	20JN	5;35	277.42
31187	4JL	7;42	110.86	31187	19JL	18;17	305.62
31188	2AU	16;46	138.99	31188	18AU	5;46	334.00
31189	1SE	4;27	167.57	31189	16SE	16;32	2.77
31190	30SE	19;20	196.73	31190	16OC	2;57	32.00
31191	30OC	13;04	226.50	31191	14NO	13;16	61.68
31192	29NO	8;22	256.71	31192	13DE	23;44	91.63
31193	29DE	3;33	287.09				

1522

NUMBER	DATE	TIME	LONG.	NUMBER	DATE	TIME	LONG.
				31193	12JA	10;42	121.64
31194	27JA	21;09	317.28	31194	10FE	22;31	151.45
31195	26FE	12;12	347.03	31195	12MR	11;25	180.89
31196	28MR	0;22	16.17	31196	11AP	1;20	209.87
31197	26AP	9;51	44.74	31197	10MY	15;55	238.45
31198	25MY	17;22	72.87	31198	9JN	6;49	266.76
31199	24JN	0;00	100.79	31199	8JL	21;46	295.01
31200	23JL	6;57	128.77	31200	7AU	12;30	323.41
31201	21AU	15;22	157.07	31201	6SE	2;40	352.14
31202	20SE	2;09	185.88	31202	5OC	15;57	21.32
31203	19OC	15;45	215.30	31203	4NO	4;07	50.92
31204	18NO	8;08	245.27	31204	3DE	15;22	80.84
31205	18DE	2;42	275.59				

1523

NUMBER	DATE	TIME	LONG.	NUMBER	DATE	TIME	LONG.
				31205	2JA	2;02	110.85
31206	16JA	22;14	305.94	31206	31JA	12;28	140.71
31207	15FE	17;01	335.99	31207	1MR	22;56	170.21
31208	17MR	9;26	5.50	31208	31MR	9;36	199.25
31209	15AP	22;38	34.38	31209	29AP	20;46	227.84
31210	15MY	8;50	62.73	31210	29MY	8;53	256.12
31211	13JN	16;56	90.74	31211	27JN	22;27	284.31
31212	13JL	0;06	118.69	31212	27JL	13;36	312.64
31213	11AU	7;26	146.83	31213	26AU	5;52	341.30
31214	9SE	15;47	175.38	31214	24SE	22;17	10.42
31215	9OC	1;46	204.47	31215	24OC	13;52	40.00
31216	7NO	13;56	234.12	31216	23NO	4;06	69.94
31217	7DE	4;39	264.21	31217	22DE	16;54	100.02

1524

NUMBER	DATE	TIME	LONG.	NUMBER	DATE	TIME	LONG.
31218	5JA	21;54	294.50	31218	21JA	4;24	130.01
31219	4FE	16;43	324.68	31219	19FE	14;42	159.66
31220	5MR	11;25	354.46	31220	19MR	23;58	188.81
31221	4AP	4;14	23.68	31221	18AP	8;38	217.46
31222	3MY	18;23	52.32	31222	17MY	17;30	245.72
31223	2JN	5;57	80.54	31223	16JN	3;34	273.79
31224	1JL	15;38	108.57	31224	15JL	15;43	301.95
31225	31JL	0;13	136.67	31225	14AU	6;17	330.42
31226	29AU	8;28	165.07	31226	12SE	22;52	359.36
31227	27SE	17;03	193.93	31227	12OC	16;32	28.84
31228	27OC	2;38	223.31	31228	11NO	10;10	58.78
31229	25NO	13;51	253.15	31229	11DE	2;48	88.98
31230	25DE	3;11	283.27				

1525

NUMBER	DATE	TIME	LONG.	NUMBER	DATE	TIME	LONG.
				31230	9JA	17;38	119.16
31231	23JA	18;37	313.41	31231	8FE	6;11	149.03
31232	22FE	11;27	343.28	31232	9MR	16;26	178.40
31233	24MR	4;33	12.69	31233	8AP	0;52	207.20
31234	22AP	20;54	41.59	31234	7MY	8;19	235.53
31235	22MY	11;55	70.04	31235	5JN	15;51	263.57
31236	21JN	1;21	98.24	31236	5JL	0;31	291.56
31237	20JL	13;11	126.40	31237	3AU	11;11	319.77
31238	18AU	23;37	154.75	31238	2SE	0;22	348.41
31239	17SE	9;11	183.48	31239	1OC	16;10	17.62
31240	16OC	18;34	212.69	31240	31OC	10;13	47.42
31241	15NO	4;29	242.35	31241	30NO	5;28	77.64
31242	14DE	15;25	272.33	31242	30DE	0;15	108.00

1526

NUMBER	DATE	TIME	LONG.	NUMBER	DATE	TIME	LONG.
31243	13JA	3;30	302.38	31243	28JA	16;52	138.14
31244	11FE	16;39	332.24	31244	27FE	6;24	167.81
31245	13MR	6;40	1.71	31245	28MR	16;57	196.88
31246	11AP	21;25	30.72	31246	27AP	1;20	225.39
31247	11MY	12;45	59.31	31247	26MY	8;34	253.50
31248	10JN	4;14	87.64	31248	24JN	15;36	281.44
31249	9JL	19;12	115.89	31249	23JL	23;19	309.45
31250	8AU	9;02	144.26	31250	22AU	8;36	337.79
31251	6SE	21;34	172.95	31251	20SE	20;15	6.66
31252	6OC	9;05	202.06	31252	20OC	10;55	36.13
31253	4NO	20;07	231.63	31253	19NO	4;41	66.17
31254	4DE	7;02	261.53	31254	19DE	0;28	96.54

1527

NUMBER	DATE	TIME	LONG.	NUMBER	DATE	TIME	LONG.
31255	2JA	17;56	291.55	31255	17JA	20;11	126.90
31256	1FE	4;45	321.43	31256	16FE	13;49	156.89
31257	2MR	15;37	350.94	31257	18MR	4;23	186.31
31258	1AP	3;01	19.99	31258	16AP	15;58	215.11
31259	30AP	15;27	48.62	31259	16MY	1;13	243.41
31260	30MY	5;14	76.96	31260	14JN	8;57	271.41
31261	28JN	20;09	105.20	31261	13JL	15;59	299.35
31262	28JL	11;38	133.55	31262	11AU	23;13	327.50
31263	27AU	3;02	162.19	31263	10SE	7;41	356.06
31264	25SE	17;57	191.26	31264	9OC	18;21	25.19
31265	25OC	8;09	220.80	31265	8NO	8;01	54.92
31266	23NO	21;26	250.70	31266	8DE	0;41	85.09
31267	23DE	9;33	280.76				

1528

NUMBER	DATE	TIME	LONG.	NUMBER	DATE	TIME	LONG.
				31267	6JA	19;24	115.44
31268	21JA	20;24	310.72	31268	5FE	14;31	145.63
31269	20FE	6;13	340.33	31269	6MR	8;27	175.36
31270	20MR	15;30	9.48	31270	5AP	0;11	204.52
31271	19AP	0;58	38.15	31271	4MY	13;20	233.11
31272	18MY	11;20	66.46	31272	2JN	23;59	261.29
31273	16JN	23;04	94.60	31273	2JL	8;39	289.28
31274	16JL	12;27	122.81	31274	31JL	16;12	317.34
31275	15AU	3;30	151.30	31275	29AU	23;45	345.72
31276	13SE	20;00	180.26	31276	28SE	8;19	14.59
31277	13OC	13;18	209.74	31277	27OC	18;42	44.01
31278	12NO	6;17	239.66	31278	26NO	7;19	73.92
31279	11DE	21;45	269.82	31279	25DE	22;06	104.11

1529

NUMBER	DATE	TIME	LONG.	NUMBER	DATE	TIME	LONG.
31280	10JA	11;02	299.92	31280	24JA	14;37	134.28
31281	8FE	22;11	329.73	31281	23FE	8;08	164.17
31282	10MR	7;41	359.05	31282	25MR	1;43	193.59
31283	8AP	16;12	27.85	31283	23AP	18;14	222.48
31284	8MY	0;21	56.20	31284	23MY	8;44	250.91
31285	6JN	8;51	84.27	31285	21JN	20;50	279.05
31286	5JL	18;28	112.30	31286	21JL	6;57	307.14
31287	4AU	6;03	140.55	31287	19AU	15;55	335.44
31288	2SE	20;17	169.25	31288	18SE	0;45	4.15
31289	2OC	13;17	198.53	31289	17OC	10;10	33.37
31290	1NO	8;07	228.37	31290	15NO	20;32	63.06
31291	1DE	2;59	258.58	31291	15DE	8;01	93.06
31292	30DE	20;12	288.87				

1530

NUMBER	DATE	TIME	LONG.	NUMBER	DATE	TIME	LONG.
				31292	13JA	20;45	123.14
31293	29JA	10;54	318.93	31293	12FE	10;53	153.03
31294	27FE	23;01	348.53	31294	14MR	2;23	182.55
31295	29MR	8;56	17.56	31295	12AP	18;41	211.61
31296	27AP	17;11	46.06	31296	12MY	10;50	240.23
31297	27MY	0;25	74.16	31297	11JN	1;53	268.53
31298	25JN	7;31	102.10	31298	10JL	15;28	296.73
31299	24JL	15;38	130.14	31299	9AU	3;43	325.05
31300	23AU	1;56	158.53	31300	7SE	15;03	353.69
31301	21SE	15;16	187.47	31301	7OC	1;52	22.79
31302	21OC	7;47	217.04	31302	5NO	12;25	52.34
31303	20NO	2;36	247.13	31303	4DE	22;53	82.23
31304	19DE	22;08	277.49				

1531

NUMBER	DATE	TIME	LONG.	NUMBER	DATE	TIME	LONG.
				31304	3JA	9;32	112.24
31305	18JA	16;40	307.79	31305	1FE	20;43	142.13
31306	17FE	9;02	337.72	31306	3MR	8;49	171.68
31307	18MR	22;34	7.08	31307	1AP	21;57	200.79
31308	17AP	9;15	35.84	31308	1MY	11;58	229.47
31309	16MY	17;34	64.10	31309	31MY	2;34	257.84
31310	15JN	0;28	92.06	31310	29JN	17;29	286.08
31311	14JL	7;07	119.99	31311	29JL	8;29	314.42
31312	12AU	14;44	148.15	31312	27AU	23;14	343.04
31313	11SE	0;20	176.77	31313	26SE	13;18	12.09
31314	10OC	12;35	205.98	31314	26OC	2;19	41.59
31315	9NO	3;41	235.77	31315	24NO	14;11	71.44
31316	8DE	21;16	265.99	31316	24DE	1;10	101.45

1532

NUMBER	DATE	TIME	LONG.	NUMBER	DATE	TIME	LONG.
31317	7JA	16;27	296.36	31317	22JA	11;38	131.39
31318	6FE	11;44	326.54	31318	20FE	21;54	161.02
31319	7MR	5;23	356.26	31319	21MR	8;11	190.19
31320	5AP	20;09	25.36	31320	19AP	18;47	218.90
31321	5MY	7;43	53.88	31321	19MY	6;07	247.24
31322	3JN	16;44	81.99	31322	17JN	18;45	275.41
31323	3JL	0;17	109.94	31323	17JL	9;05	303.66
31324	1AU	7;31	137.99	31324	16AU	1;00	332.20
31325	30AU	15;24	166.39	31325	14SE	17;43	1.18
31326	29SE	0;37	195.29	31326	14OC	10;07	30.63
31327	28OC	11;44	224.76	31327	13NO	1;19	60.49
31328	27NO	1;10	254.71	31328	12DE	14;59	90.58
31329	26DE	17;08	284.94				

1533

NUMBER	DATE	TIME	LONG.	NUMBER	DATE	TIME	LONG.
				31329	11JA	3;10	120.64
31330	25JA	11;11	315.17	31330	9FE	13;59	150.42
31331	24FE	5;58	345.10	31331	10MR	23;34	179.74
31332	25MR	23;43	14.52	31332	9AP	8;15	208.54
31333	24AP	15;08	43.35	31333	8MY	16;41	236.89
31334	24MY	3;55	71.70	31334	7JN	1;51	264.98
31335	22JN	14;31	99.78	31335	6JL	12;46	293.06
31336	21JL	23;42	127.84	31336	5AU	2;06	321.39
31337	20AU	8;13	156.13	31337	3SE	17;52	350.16
31338	18SE	16;45	184.84	31338	3OC	11;19	19.47
31339	18OC	1;55	214.05	31339	2NO	5;21	49.29
31340	16NO	12;20	243.76	31340	1DE	22;48	79.45
31341	16DE	0;33	273.80	31341	31DE	14;43	109.69

1534

NUMBER	DATE	TIME	LONG.	NUMBER	DATE	TIME	LONG.
31342	14JA	14;51	303.93	31342	30JA	4;27	139.70
31343	13FE	6;52	333.89	31343	28FE	15;45	169.25
31344	14MR	23;41	3.46	31344	30MR	0;51	198.24
31345	13AP	16;17	32.51	31345	28AP	8;31	226.71
31346	13MY	7;54	61.10	31346	27MY	15;44	254.81
31347	11JN	22;09	89.37	31347	25JN	23;36	282.77
31348	11JL	10;54	117.53	31348	25JL	9;09	310.87
31349	9AU	22;12	145.82	31349	23AU	21;04	339.33
31350	8SE	8;23	174.43	31350	22SE	11;39	8.34
31351	7OC	18;00	203.49	31351	22OC	4;48	37.94
31352	6NO	3;45	233.02	31352	20NO	23;46	68.05
31353	5DE	14;13	262.91	31353	20DE	19;07	98.41

1535

NUMBER	DATE	TIME	LONG.	NUMBER	DATE	TIME	LONG.
31354	4JA	1;41	292.96	31354	19JA	13;02	128.68
31355	2FE	14;09	322.89	31355	18FE	4;07	158.55
31356	4MR	3;29	352.48	31356	19MR	16;00	187.83
31357	2AP	17;34	21.62	31357	18AP	1;15	216.52
31358	2MY	8;23	50.32	31358	17MY	8;50	244.74
31359	31MY	23;42	78.71	31359	15JN	15;47	272.70
31360	30JN	15;02	106.97	31360	14JL	23;03	300.65
31361	30JL	5;39	135.29	31361	13AU	7;29	328.86
31362	28AU	19;06	163.88	31362	11SE	17;56	357.52
31363	27SE	7;22	192.87	31363	11OC	7;07	26.78
31364	26OC	18;51	222.31	31364	9NO	23;27	56.64
31365	25NO	5;58	252.14	31365	9DE	18;28	86.92
31366	24DE	16;55	282.15				

1536

NUMBER	DATE	TIME	LONG.	NUMBER	DATE	TIME	LONG.
				31366	8JA	14;29	117.33
31367	23JA	3;42	312.10	31367	7FE	9;18	147.48
31368	21FE	14;19	341.73	31368	8MR	1;21	177.11
31369	22MR	1;07	10.92	31369	6AP	14;17	206.13
31370	20AP	12;40	39.65	31370	6MY	0;34	234.58
31371	20MY	1;31	68.04	31371	4JN	8;56	262.66
31372	18JN	15;48	96.28	31372	3JL	16;14	290.60
31373	18JL	7;06	124.57	31373	1AU	23;20	318.65
31374	16AU	22;46	153.11	31374	31AU	7;10	347.06
31375	15SE	14;14	182.05	31375	29SE	16;44	15.99
31376	15OC	5;07	211.46	31376	29OC	4;55	45.51
31377	13NO	19;09	241.28	31377	27NO	20;09	75.54
31378	13DE	8;02	271.33	31378	27DE	13;58	105.86

1537

NUMBER	DATE	TIME	LONG.	NUMBER	DATE	TIME	LONG.
31379	11JA	19;31	301.36	31379	26JA	8;58	136.12
31380	10FE	5;42	331.11	31380	25FE	3;29	166.03
31381	11MR	14;59	0.40	31381	26MR	20;16	195.39
31382	10AP	0;02	29.21	31382	25AP	10;39	224.16
31383	9MY	9;38	57.60	31383	24MY	22;30	252.47
31384	7JN	20;26	85.75	31384	23JN	8;08	280.51
31385	7JL	8;52	113.90	31385	22JL	16;15	308.54
31386	5AU	23;07	142.26	31386	20AU	23;49	336.80
31387	4SE	15;05	171.05	31387	19SE	7;55	5.49
31388	4OC	8;20	200.37	31388	18OC	17;27	34.73
31389	3NO	1;54	230.18	31389	17NO	5;01	64.49
31390	2DE	18;28	260.31	31390	16DE	18;43	94.60

1538

NUMBER	DATE	TIME	LONG.	NUMBER	DATE	TIME	LONG.
31391	1JA	8;59	290.49	31391	15JA	10;18	124.79
31392	30JA	21;09	320.43	31392	14FE	3;11	154.77
31393	1MR	7;17	349.93	31393	15MR	20;37	184.35
31394	30MR	16;03	18.89	31394	14AP	13;37	213.41
31395	29AP	0;07	47.37	31395	14MY	5;07	241.98
31396	28MY	8;11	75.49	31396	12JN	18;28	270.21
31397	26JN	17;01	103.50	31397	12JL	5;38	298.31
31398	26JL	3;28	131.63	31398	10AU	15;15	326.54
31399	24AU	16;21	160.14	31399	9SE	0;16	355.11
31400	23SE	8;08	189.21	31400	8OC	9;32	24.16
31401	23OC	2;24	218.88	31401	6NO	19;33	53.71
31402	21NO	21;40	249.01	31402	6DE	6;31	83.64
31403	21DE	16;01	279.33				

1539

NUMBER	DATE	TIME	LONG.	NUMBER	DATE	TIME	LONG.
				31403	4JA	18;32	113.70
31404	20JA	8;04	309.52	31404	3FE	7;46	143.66
31405	18FE	21;23	339.30	31405	4MR	22;19	173.29
31406	20MR	8;15	8.53	31406	3AP	14;01	202.48
31407	18AP	17;08	37.19	31407	3MY	6;09	231.23
31408	18MY	0;40	65.40	31408	1JN	21;45	259.62
31409	16JN	7;38	93.36	31409	1JL	12;09	287.84
31410	15JL	15;04	121.32	31410	31JL	1;12	316.11
31411	14AU	0;09	149.56	31411	29AU	13;12	344.65
31412	12SE	11;57	178.29	31412	28SE	0;31	13.61
31413	12OC	3;01	207.64	31413	27OC	11;25	43.04
31414	10NO	20;57	237.57	31414	25NO	22;03	72.85
31415	10DE	16;25	267.89	31415	25DE	8;35	102.85

1540

NUMBER	DATE	TIME	LONG.	NUMBER	DATE	TIME	LONG.
31416	9JA	11;39	298.25	31416	23JA	19;21	132.79
31417	8FE	5;11	328.34	31417	22FE	6;45	162.44
31418	8MR	20;06	357.92	31418	22MR	19;04	191.68
31419	7AP	8;06	26.88	31419	21AP	8;23	220.47
31420	6MY	17;26	55.29	31420	20MY	22;30	248.91
31421	5JN	0;52	83.33	31421	19JN	13;11	277.16
31422	4JL	7;30	111.24	31422	19JL	4;14	305.45
31423	2AU	14;33	139.30	31423	17AU	19;22	333.97
31424	31AU	23;08	167.74	31424	16SE	10;07	2.90
31425	30SE	10;06	196.74	31425	16OC	0;00	32.27
31426	29OC	23;51	226.34	31426	14NO	12;40	62.04
31427	28NO	16;18	256.43	31427	14DE	0;10	92.05
31428	28DE	10;47	286.77				

1541

NUMBER	DATE	TIME	LONG.	NUMBER	DATE	TIME	LONG.
				31428	12JA	10;50	122.04
31429	27JA	6;08	317.04	31429	10FE	21;03	151.78
31430	26FE	0;40	346.94	31430	12MR	7;05	181.10
31431	27MR	16;52	16.26	31431	10AP	17;13	209.93
31432	26AP	5;57	44.97	31432	10MY	3;52	238.36
31433	25MY	16;07	73.21	31433	8JN	15;35	266.54
31434	24JN	0;18	101.19	31434	8JL	4;57	294.73
31435	23JL	7;39	129.18	31435	6AU	20;11	323.14
31436	21AU	15;14	157.45	31436	5SE	12;49	351.97
31437	19SE	23;52	186.18	31437	5OC	5;48	21.29
31438	19OC	10;06	215.46	31438	3NO	21;58	51.06
31439	17NO	22;22	245.25	31439	3DE	12;37	81.12
31440	17DE	13;02	275.40				

1542

NUMBER	DATE	TIME	LONG.	NUMBER	DATE	TIME	LONG.
				31440	2JA	1;37	111.23
31441	16JA	6;00	305.65	31441	31JA	13;03	141.13
31442	15FE	0;26	335.69	31442	1MR	23;06	170.61
31443	16MR	18;42	5.28	31443	31MR	7;59	199.57
31444	15AP	11;14	34.31	31444	29AP	16;14	228.04
31445	15MY	1;16	62.82	31445	29MY	0;43	256.18
31446	13JN	12;56	90.97	31446	27JN	10;30	284.22
31447	12JL	22;51	119.03	31447	26JL	22;32	312.24
31448	11AU	7;47	147.23	31448	25AU	13;11	341.02
31449	9SE	16;26	175.80	31449	24SE	6;04	10.15
31450	9OC	1;24	204.85	31450	24OC	0;11	39.82
31451	7NO	11;15	234.40	31451	22NO	18;16	69.91
31452	6DE	22;32	264.35	31452	22DE	11;12	100.18

1543

NUMBER	DATE	TIME	LONG.	NUMBER	DATE	TIME	LONG.
31453	5JA	11;42	294.46	31453	21JA	2;08	130.31
31454	4FE	2;44	324.49	31454	19FE	14;35	160.04
31455	5MR	19;01	354.18	31455	21MR	0;36	189.23
31456	4AP	11;34	23.39	31456	19AP	8;43	217.85
31457	4MY	3;34	52.12	31457	18MY	15;53	246.04
31458	2JN	18;28	80.48	31458	16JN	23;11	274.01
31459	2JL	8;08	108.66	31459	16JL	7;44	302.02
31460	31JL	20;16	136.90	31460	14AU	18;25	330.33
31461	30AU	7;13	165.41	31461	13SE	7;43	359.13
31462	28SE	17;27	194.33	31462	12OC	23;44	28.53
31463	28OC	3;06	223.72	31463	11NO	18;04	58.50
31464	26NO	13;16	253.52	31464	11DE	13;33	88.81
31465	26DE	0;13	283.54				

1544 / 1550

NUMBER	DATE	TIME	LONG.	NUMBER	DATE	TIME	LONG.	NUMBER	DATE	TIME	LONG.	NUMBER	DATE	TIME	LONG.
				31465	10JA	8;27	119.16					31539	3JA	10;00	112.66
31466	24JA	12;04	313.52	31466	9FE	1;03	149.20	31540	18JA	0;25	307.50	31540	1FE	20;19	142.50
31467	23FE	0;45	343.22	31467	9MR	14;26	178.70	31541	16FE	19;27	337.55	31541	3MR	6;13	171.96
31468	23MR	14;10	12.49	31468	8AP	0;48	207.59	31542	18MR	12;50	7.08	31542	1AP	16;02	200.94
31469	22AP	4;20	41.31	31469	7MY	8;59	235.95	31543	17AP	3;26	36.00	31543	1MY	2;06	229.47
31470	21MY	19;14	69.77	31470	5JN	16;04	263.97	31544	16MY	14;57	64.38	31544	30MY	12;59	257.69
31471	20JN	10;36	98.05	31471	4JL	23;03	291.89	31545	15JN	0;02	92.43	31545	29JN	1;20	285.83
31472	20JL	1;46	126.34	31472	3AU	6;51	319.98	31546	14JL	7;44	120.40	31546	28JL	15;38	314.13
31473	18AU	16;04	154.84	31473	1SE	16;15	348.47	31547	12AU	15;13	148.56	31547	27AU	7;48	342.81
31474	17SE	5;11	183.71	31474	1OC	4;04	17.51	31548	10SE	23;24	177.13	31548	26SE	1;01	11.98
31475	16OC	17;17	213.02	31475	30OC	18;53	47.17	31549	10OC	8;54	206.22	31549	25OC	18;00	41.63
31476	15NO	4;45	242.76	31476	29NO	12;44	77.32	31550	8NO	20;12	235.85	31550	24NO	9;41	71.64
31477	14DE	15;53	272.74	31477	29DE	8;31	107.72	31551	8DE	9;38	265.89	31551	23DE	23;38	101.78

1545 / 1551

NUMBER	DATE	TIME	LONG.	NUMBER	DATE	TIME	LONG.	NUMBER	DATE	TIME	LONG.	NUMBER	DATE	TIME	LONG.
31478	13JA	2;44	302.74	31478	28JA	4;08	138.00	31552	7JA	1;24	296.12	31552	22JA	11;51	131.79
31479	11FE	13;15	332.49	31479	26FE	21;35	167.84	31553	5FE	19;03	326.24	31553	20FE	22;29	161.43
31480	12MR	23;40	1.81	31480	28MR	11;58	197.07	31554	7MR	13;23	355.99	31554	22MR	7;44	190.56
31481	11AP	10;29	30.66	31481	26AP	23;26	225.71	31555	6AP	6;45	25.21	31555	20AP	16;01	219.17
31482	10MY	22;22	59.13	31482	26MY	8;37	253.89	31556	5MY	21;59	53.88	31556	20MY	0;04	247.39
31483	9JN	11;46	87.38	31483	24JN	16;23	281.86	31557	4JN	10;48	82.14	31557	18JN	8;55	275.41
31484	9JL	2;34	115.62	31484	23JL	23;33	309.85	31558	3JL	21;36	110.22	31558	17JL	19;39	303.52
31485	7AU	18;15	144.06	31485	22AU	7;01	338.12	31559	2AU	7;06	138.36	31559	16AU	9;00	331.95
31486	6SE	10;06	172.87	31486	20SE	15;42	6.86	31560	31AU	16;02	166.81	31560	15SE	0;58	0.88
31487	6OC	1;37	202.14	31487	20OC	2;35	36.18	31561	30SE	0;59	195.70	31561	14OC	18;49	30.39
31488	4NO	16;23	231.86	31488	18NO	16;20	66.04	31562	29OC	10;28	225.10	31562	13NO	13;17	60.38
31489	4DE	6;04	261.88	31489	18DE	8;56	96.28	31563	27NO	21;03	254.93	31563	13DE	7;05	90.64
								31564	27DE	9;12	285.00				

1546 / 1552

NUMBER	DATE	TIME	LONG.	NUMBER	DATE	TIME	LONG.	NUMBER	DATE	TIME	LONG.	NUMBER	DATE	TIME	LONG.
31490	2JA	18;21	291.97	31490	17JA	3;25	126.58					31564	11JA	23;10	120.86
31491	1FE	5;06	321.83	31491	15FE	22;12	156.63	31565	25JA	23;10	315.06	31565	10FE	12;52	150.76
31492	2MR	14;36	351.28	31492	17MR	15;48	186.18	31566	24FE	14;40	344.86	31566	10MR	23;58	180.15
31493	31MR	23;27	20.23	31493	16AP	7;19	215.15	31567	25MR	6;54	14.22	31567	9AP	8;48	208.95
31494	30AP	8;27	48.73	31494	15MY	20;24	243.61	31568	23AP	23;02	43.09	31568	8MY	16;11	237.26
31495	29MY	18;22	76.92	31495	14JN	7;07	271.73	31569	23MY	14;25	71.55	31569	6JN	23;10	265.26
31496	28JN	5;49	105.02	31496	13JL	16;00	299.74	31570	22JN	4;44	99.78	31570	6JL	6;54	293.22
31497	27JL	19;07	133.28	31497	11AU	23;52	327.91	31571	21JL	17;47	128.00	31571	4AU	16;25	321.39
31498	26AU	10;20	161.91	31498	10SE	7;46	356.46	31572	20AU	5;35	156.42	31572	3SE	4;24	350.00
31499	25SE	3;14	191.05	31499	9OC	16;39	25.52	31573	18SE	16;17	185.21	31573	2OC	19;11	19.19
31500	24OC	21;02	220.72	31500	8NO	3;15	55.11	31574	18OC	2;24	214.47	31574	1NO	12;34	48.98
31501	23NO	14;30	250.80	31501	7DE	15;54	85.11	31575	16NO	12;29	244.16	31575	1DE	7;46	79.21
31502	23DE	6;17	281.02					31576	15DE	23;03	274.12	31576	31DE	3;16	109.59

1547 / 1553

NUMBER	DATE	TIME	LONG.	NUMBER	DATE	TIME	LONG.	NUMBER	DATE	TIME	LONG.	NUMBER	DATE	TIME	LONG.
				31502	6JA	6;30	115.29	31577	14JA	10;22	304.14	31577	29JA	21;11	139.79
31503	21JA	19;39	311.08	31503	4FE	22;36	145.35	31578	12FE	22;27	333.93	31578	28FE	12;09	169.50
31504	20FE	6;39	340.74	31504	6MR	15;36	175.06	31579	14MR	11;13	3.33	31579	29MR	23;52	198.59
31505	21MR	15;52	9.88	31505	5AP	8;43	204.28	31580	13AP	0;43	32.27	31580	28AP	8;56	227.11
31506	20AP	0;01	38.50	31506	5MY	0;57	233.00	31581	12MY	15;01	60.82	31581	27MY	16;22	255.22
31507	19MY	7;50	66.71	31507	3JN	15;25	261.34	31582	11JN	6;05	89.12	31582	25JN	23;15	283.15
31508	17JN	16;04	94.71	31508	3JL	3;44	289.48	31583	10JL	21;28	117.39	31583	25JL	6;34	311.15
31509	17JL	1;32	122.76	31509	1AU	14;13	317.65	31584	9AU	12;28	145.82	31584	23AU	15;08	339.48
31510	15AU	13;07	151.11	31510	30AU	23;39	346.11	31585	8SE	2;28	174.58	31585	22SE	1;44	8.32
31511	14SE	3;30	179.97	31511	29SE	8;56	15.01	31586	7OC	15;20	203.77	31586	21OC	15;05	37.76
31512	13OC	20;47	209.44	31512	28OC	18;43	44.41	31587	6NO	3;19	233.39	31587	20NO	7;32	67.76
31513	12NO	15;58	239.45	31513	27NO	5;18	74.27	31588	5DE	14;45	263.33	31588	20DE	2;33	98.11
31514	12DE	11;10	269.76	31514	26DE	16;46	104.27								

1548 / 1554

NUMBER	DATE	TIME	LONG.	NUMBER	DATE	TIME	LONG.	NUMBER	DATE	TIME	LONG.	NUMBER	DATE	TIME	LONG.
31515	11JA	4;32	300.04	31515	25JA	5;12	134.27	31589	4JA	1;46	293.36	31589	18JA	22;28	128.46
31516	9FE	19;12	330.00	31516	23FE	18;50	164.01	31590	2FE	12;20	323.21	31590	17FE	17;05	158.48
31517	10MR	7;09	359.42	31517	24MR	9;43	193.32	31591	3MR	22;33	352.67	31591	19MR	8;57	187.93
31518	8AP	16;50	28.27	31518	23AP	1;28	222.19	31592	2AP	8;49	21.65	31592	17AP	21;44	216.77
31519	8MY	0;51	56.61	31519	22MY	17;18	250.68	31593	1MY	19;49	50.21	31593	17MY	7;56	245.09
31520	6JN	7;54	84.63	31520	21JN	8;21	278.94	31594	31MY	8;13	78.48	31594	15JN	16;19	273.11
31521	5JL	14;56	112.55	31521	20JL	22;12	307.18	31595	29JN	22;14	106.69	31595	14JL	23;45	301.07
31522	3AU	23;03	140.66	31522	19AU	10;55	335.63	31596	29JL	13;36	135.04	31596	13AU	7;03	329.23
31523	2SE	9;26	169.20	31523	17SE	22;49	4.46	31597	28AU	5;39	163.73	31597	11SE	15;09	357.80
31524	1OC	22;56	198.32	31524	17OC	10;11	33.76	31598	26SE	21;39	192.86	31598	11OC	0;57	26.92
31525	31OC	15;38	228.07	31525	15NO	21;07	63.48	31599	26OC	13;08	222.46	31599	9NO	13;16	56.60
31526	30NO	10;37	258.28	31526	15DE	7;44	93.45	31600	25NO	3;39	252.43	31600	9DE	4;30	86.73
31527	30DE	6;12	288.67					31601	24DE	16;46	282.54				

1549 / 1555

NUMBER	DATE	TIME	LONG.	NUMBER	DATE	TIME	LONG.	NUMBER	DATE	TIME	LONG.	NUMBER	DATE	TIME	LONG.
				31527	13JA	18;18	123.43					31601	7JA	22;06	117.03
31528	29JA	0;42	318.89	31528	12FE	5;09	153.18	31602	23JA	4;15	312.51	31602	6FE	16;46	147.18
31529	27FE	16;56	348.67	31529	13MR	16;43	182.54	31603	21FE	14;12	342.12	31603	8MR	10;55	176.91
31530	29MR	6;19	17.85	31530	12AP	5;15	211.45	31604	22MR	23;06	11.22	31604	7AP	3;24	206.07
31531	27AP	16;50	46.44	31531	11MY	18;43	239.97	31605	21AP	7;42	39.84	31605	6MY	17;39	234.70
31532	27MY	1;04	74.58	31532	10JN	9;00	268.25	31606	20MY	16;51	68.09	31606	5JN	5;32	262.92
31533	25JN	7;57	102.51	31533	9JL	23;53	296.50	31607	19JN	3;19	96.18	31607	4JL	15;22	290.96
31534	24JL	14;42	130.49	31534	8AU	15;10	324.93	31608	18JL	15;36	124.34	31608	2AU	23;46	319.06
31535	22AU	22;28	158.78	31535	7SE	6;26	353.73	31609	17AU	5;54	152.82	31609	1SE	7;42	347.47
31536	21SE	8;16	187.57	31536	6OC	21;07	22.98	31610	15SE	22;10	181.78	31610	30SE	16;10	16.36
31537	20OC	20;41	216.97	31537	5NO	10;41	52.66	31611	15OC	15;53	211.30	31611	30OC	1;59	45.78
31538	19NO	11;52	246.90	31538	4DE	22;55	82.63	31612	14NO	9;56	241.28	31612	28NO	13;39	75.66
31539	19DE	5;25	277.18					31613	14DE	2;53	271.51	31613	28DE	3;14	105.79

NUMBER	DATE	TIME	LONG.	NUMBER	DATE	TIME	LONG.	NUMBER	DATE	TIME	LONG.	NUMBER	DATE	TIME	LONG.
			1556								**1562**				
31614	12JA	17;34	301.67	31614	26JA	18;28	135.90	31688	5JA	9;00	294.74	31688	20JA	16;34	130.30
31615	11FE	5;39	331.50	31615	25FE	10;51	165.73	31689	3FE	20;33	324.62	31689	19FE	9;05	160.21
31616	11MR	15;33	0.82	31616	26MR	3;46	195.10	31690	5MR	8;44	354.14	31690	20MR	22;19	189.53
31617	9AP	23;59	29.59	31617	24AP	20;22	223.98	31691	3AP	21;33	23.21	31691	19AP	8;30	218.23
31618	9MY	7;43	57.91	31618	24MY	11;45	252.43	31692	3MY	11;09	51.85	31692	18MY	16;33	246.46
31619	7JN	15;31	85.95	31619	23JN	1;13	280.63	31693	2JN	1;41	80.20	31693	16JN	23;33	274.41
31620	7JL	0;11	113.94	31620	22JL	12;43	308.79	31694	1JL	16;58	108.45	31694	16JL	6;34	302.36
31621	5AU	10;35	142.15	31621	20AU	22;47	337.14	31695	31JL	8;23	136.82	31695	14AU	14;28	330.56
31622	3SE	23;33	170.81	31622	19SE	8;17	5.90	31696	29AU	23;11	165.48	31696	13SE	0;03	359.21
31623	3OC	15;33	200.07	31623	18OC	17;58	35.15	31697	28SE	12;55	194.54	31697	12OC	12;02	28.44
31624	2NO	10;08	229.91	31624	17NO	4;17	64.85	31698	28OC	1;35	224.05	31698	11NO	2;59	58.25
31625	2DE	5;44	260.17	31625	16DE	15;19	94.85	31699	26NO	13;27	253.92	31699	10DE	20;51	88.50
								31700	26DE	0;44	283.96				
			1557								**1563**				
31626	1JA	0;17	290.52	31626	15JA	3;08	124.88					31700	9JA	16;32	118.88
31627	30JA	16;21	320.62	31627	13FE	15;56	154.69	31701	24JA	11;28	313.89	31701	8FE	11;57	149.05
31628	1MR	5;34	350.25	31628	15MR	5;54	184.13	31702	22FE	21;40	343.49	31702	10MR	5;12	178.71
31629	30MR	16;12	19.29	31629	13AP	21;00	213.12	31703	24MR	7;35	12.62	31703	8AP	19;24	207.76
31630	29AP	0;52	47.79	31630	13MY	12;42	241.71	31704	22AP	17;52	41.28	31704	8MY	6;45	236.25
31631	28MY	8;13	75.88	31631	12JN	4;09	270.03	31705	22MY	5;15	69.60	31705	6JN	15;57	264.35
31632	26JN	15;05	103.81	31632	11JL	18;43	298.27	31706	20JN	18;18	97.79	31706	5JL	23;49	292.31
31633	25JL	22;31	131.82	31633	10AU	8;11	326.64	31707	20JL	9;02	126.06	31707	4AU	7;12	320.39
31634	24AU	7;40	160.18	31634	8SE	20;44	355.36	31708	19AU	0;58	154.62	31708	2SE	14;55	348.81
31635	22SE	19;37	189.09	31635	8OC	8;37	24.52	31709	17SE	17;18	183.61	31709	1OC	23;53	17.73
31636	22OC	10;51	218.62	31636	6NO	19;59	54.13	31710	17OC	9;24	213.09	31710	31OC	10;57	47.23
31637	21NO	4;56	248.70	31637	6DE	6;52	84.05	31711	16NO	0;42	242.98	31711	30NO	0;45	77.21
31638	21DE	0;28	279.07					31712	15DE	14;44	273.09	31712	29DE	17;12	107.46
			1558								**1564**				
				31638	4JA	17;25	114.05	31713	14JA	3;05	303.15	31713	28JA	11;22	137.69
31639	19JA	19;41	309.39	31639	3FE	3;56	143.89	31714	12FE	13;41	332.90	31714	27FE	5;44	167.57
31640	18FE	13;05	339.34	31640	4MR	14;52	173.37	31715	12MR	22;51	2.17	31715	27MR	22;59	196.93
31641	20MR	3;50	8.74	31641	3AP	2;37	202.41	31716	11AP	7;17	30.92	31716	26AP	14;17	225.73
31642	18AP	15;41	37.52	31642	2MY	15;21	231.02	31717	10MY	15;51	59.26	31717	26MY	3;20	254.07
31643	18MY	0;55	65.80	31643	1JN	5;02	259.34	31718	9JN	1;24	87.35	31718	24JN	14;13	282.16
31644	16JN	8;20	93.78	31644	30JN	19;34	287.57	31719	8JL	12;37	115.45	31719	23JL	23;23	310.23
31645	15JL	15;03	121.71	31645	30JL	10;46	315.92	31720	7AU	1;54	143.79	31720	22AU	7;37	338.54
31646	13AU	22;15	149.87	31646	29AU	2;20	344.60	31721	5SE	17;19	172.58	31721	20SE	15;53	7.26
31647	12SE	7;02	178.48	31647	27SE	17;41	13.72	31722	5OC	10;36	201.91	31722	20OC	1;06	36.51
31648	11OC	18;12	207.67	31648	27OC	8;10	43.29	31723	4NO	4;53	231.77	31723	18NO	11;54	66.25
31649	10NO	8;05	237.43	31649	25NO	21;16	73.21	31724	3DE	22;46	261.97	31724	18DE	0;32	96.32
31650	10DE	0;31	267.60	31650	25DE	8;58	103.26								
			1559								**1565**				
31651	8JA	18;51	297.93	31651	23JA	19;35	133.19	31725	2JA	14;47	292.21	31725	16JA	14;51	126.44
31652	7FE	13;55	328.09	31652	22FE	5;31	162.78	31726	1FE	4;10	322.19	31726	15FE	6;29	156.37
31653	9MR	8;10	357.81	31653	23MR	15;07	191.91	31727	2MR	14;59	351.69	31727	16MR	22;56	185.88
31654	8AP	0;08	26.95	31654	22AP	0;56	220.56	31728	31MR	23;55	20.64	31728	15AP	15;34	214.91
31655	7MY	13;07	55.51	31655	21MY	10;56	248.84	31729	30AP	7;45	49.09	31729	15MY	7;33	243.49
31656	5JN	23;20	83.66	31656	19JN	22;18	276.96	31730	29MY	15;17	77.18	31730	13JN	22;03	271.76
31657	5JL	7;40	111.64	31657	19JL	11;32	305.17	31731	27JN	23;19	105.15	31731	13JL	10;38	299.92
31658	3AU	15;15	139.72	31658	18AU	2;53	333.70	31732	27JL	8;42	133.25	31732	11AU	21;33	328.21
31659	1SE	23;09	168.13	31659	16SE	19;55	2.71	31733	25AU	20;19	161.73	31733	10SE	7;29	356.84
31660	1OC	8;05	197.05	31660	16OC	13;27	32.23	31734	24SE	10;53	190.77	31734	9OC	17;14	25.93
31661	30OC	18;33	226.51	31661	15NO	6;09	62.16	31735	24OC	4;26	220.42	31735	8NO	3;22	55.51
31662	29NO	6;54	256.42	31662	14DE	21;10	92.31	31736	22NO	23;55	250.58	31736	7DE	14;07	85.43
31663	28DE	21;26	286.60					31737	22DE	19;20	280.94				
			1560								**1566**				
				31663	13JA	10;17	122.41					31737	6JA	1;29	115.47
31664	27JA	14;03	316.76	31664	11FE	21;37	152.20	31738	21JA	12;47	311.18	31738	4FE	13;35	145.36
31665	26FE	8;00	346.64	31665	12MR	7;23	181.50	31739	20FE	3;22	341.00	31739	6MR	2;40	174.92
31666	27MR	1;49	16.03	31666	10AP	15;54	210.27	31740	21MR	15;09	10.24	31740	4AP	16;56	204.03
31667	25AP	18;05	44.88	31667	9MY	23;46	238.58	31741	20AP	0;58	38.91	31741	4MY	8;09	232.72
31668	25MY	8;04	73.27	31668	8JN	7;56	266.63	31742	19MY	8;26	67.12	31742	2JN	23;42	261.10
31669	23JN	19;52	101.40	31669	7JL	17;30	294.66	31743	17JN	15;23	95.07	31743	2JL	14;47	289.35
31670	23JL	6;05	129.51	31670	6AU	5;28	322.94	31744	16JL	22;24	123.03	31744	1AU	4;58	317.67
31671	21AU	15;26	157.85	31671	4SE	20;14	351.69	31745	15AU	6;35	151.24	31745	30AU	18;13	346.28
31672	20SE	0;32	186.60	31672	4OC	13;26	21.01	31746	13SE	17;06	179.94	31746	29SE	6;43	15.31
31673	19OC	9;52	215.84	31673	3NO	7;58	50.86	31747	13OC	6;45	209.25	31747	28OC	18;35	44.80
31674	17NO	19;57	245.55	31674	3DE	2;25	81.08	31748	11NO	23;36	239.15	31748	27NO	5;52	74.65
31675	17DE	7;16	275.55					31749	11DE	18;40	269.45	31749	26DE	16;36	104.66
			1561								**1567**				
				31675	1JA	19;34	111.36					31750	25JA	3;00	134.57
31676	15JA	20;12	305.63	31676	31JA	10;32	141.42	31750	10JA	14;14	299.83	31751	23FE	13;28	164.17
31677	14FE	10;45	335.51	31677	1MR	22;51	170.99	31751	9FE	8;37	329.94	31752	25MR	0;30	193.34
31678	16MR	2;27	5.01	31678	31MR	8;37	199.99	31752	11MR	0;40	359.55	31753	23AP	12;26	222.06
31679	14AP	18;28	34.02	31679	29AP	16;29	228.44	31753	9AP	13;53	28.54	31754	23MY	1;26	250.44
31680	14MY	10;07	62.60	31680	28MY	23;24	256.52	31754	9MY	0;18	56.98	31755	21JN	15;26	278.66
31681	13JN	0;58	90.89	31681	27JN	6;33	284.45	31755	7JN	8;30	85.04	31756	21JL	6;20	306.94
31682	12JL	14;47	119.10	31682	26JL	15;03	312.52	31756	6JL	15;27	112.96	31757	19AU	21;56	335.50
31683	11AU	3;25	147.45	31683	25AU	1;46	340.95	31757	4AU	22;21	141.03	31758	18SE	13;45	4.48
31684	9SE	14;54	176.13	31684	23SE	15;13	9.93	31758	3SE	6;20	169.46	31759	18OC	5;03	33.94
31685	9OC	1;31	205.25	31685	23OC	7;28	39.51	31759	2OC	16;21	198.44	31760	16NO	19;08	63.79
31686	7NO	11;44	234.82	31686	22NO	2;00	69.62	31760	1NO	4;56	228.01	31761	16DE	7;39	93.84
31687	6DE	22;06	264.72	31687	21DE	21;38	99.99	31761	30NO	20;09	258.06				
								31762	30DE	13;35	288.36				

1568

NEW NUMBER	DATE	TIME	LONG.	FULL NUMBER	DATE	TIME	LONG.
				31762	14JA	18;47	123.84
31763	29JA	8;18	318.60	31763	13FE	4;54	153.56
31764	28FE	3;00	348.48	31764	13MR	14;26	182.84
31765	28MR	20;07	17.82	31765	11AP	23;45	211.62
31766	27AP	10;34	46.57	31766	11MY	9;21	239.99
31767	26MY	22;05	74.85	31767	9JN	19;51	268.12
31768	25JN	7;17	102.87	31768	9JL	7;59	296.26
31769	24JL	15;14	130.90	31769	7AU	22;18	324.65
31770	22AU	23;01	159.19	31770	6SE	14;45	353.48
31771	21SE	7;32	187.94	31771	6OC	8;27	22.85
31772	20OC	17;19	217.22	31772	5NO	1;58	52.69
31773	19NO	4;46	246.99	31773	4DE	18;06	82.82
31774	18DE	18;09	277.09				

1569

NEW NUMBER	DATE	TIME	LONG.	FULL NUMBER	DATE	TIME	LONG.
				31774	3JA	8;15	112.98
31775	17JA	9;37	307.27	31775	1FE	20;27	142.90
31776	16FE	2;48	337.24	31776	3MR	6;51	172.38
31777	17MR	20;38	6.80	31777	1AP	15;46	201.32
31778	16AP	13;38	35.83	31778	30AP	23;41	229.76
31779	16MY	4;43	64.36	31779	30MY	7;25	257.86
31780	14JN	17;36	92.57	31780	28JN	16;02	285.85
31781	14JL	4;40	120.67	31781	28JL	2;39	314.00
31782	12AU	14;35	148.93	31782	26AU	16;03	342.56
31783	10SE	23;58	177.54	31783	25SE	8;15	11.68
31784	10OC	9;20	206.63	31784	25OC	2;27	41.38
31785	8NO	19;07	236.20	31785	23NO	21;18	71.51
31786	8DE	5;48	266.13	31786	23DE	15;22	101.82

1570

NEW NUMBER	DATE	TIME	LONG.	FULL NUMBER	DATE	TIME	LONG.
31787	6JA	17;49	296.19	31787	22JA	7;33	132.00
31788	5FE	7;24	326.14	31788	20FE	21;10	161.77
31789	6MR	22;20	355.76	31789	22MR	8;03	190.97
31790	5AP	13;59	24.92	31790	20AP	16;38	219.59
31791	5MY	5;40	53.62	31791	19MY	23;47	247.76
31792	3JN	20;52	81.98	31792	18JN	6;36	275.71
31793	3JL	11;17	110.20	31793	17JL	14;16	303.69
31794	2AU	0;43	138.50	31794	15AU	23;48	331.97
31795	31AU	13;02	167.08	31795	14SE	11;55	0.74
31796	30SE	0;19	196.07	31796	14OC	2;53	30.12
31797	29OC	10;53	225.51	31797	12NO	20;28	60.06
31798	27NO	21;16	255.33	31798	12DE	15;49	90.38
31799	27DE	7;54	285.34				

1571

NEW NUMBER	DATE	TIME	LONG.	FULL NUMBER	DATE	TIME	LONG.
				31799	11JA	11;21	120.75
31800	25JA	19;01	315.28	31800	10FE	5;12	150.84
31801	24FE	6;39	344.92	31801	11MR	20;02	180.37
31802	25MR	18;51	14.12	31802	10AP	7;35	209.29
31803	24AP	7;45	42.87	31803	9MY	16;30	237.66
31804	23MY	21;35	71.27	31804	7JN	23;51	265.68
31805	22JN	12;26	99.53	31805	7JL	6;45	293.61
31806	22JL	3;57	127.84	31806	5AU	14;10	321.69
31807	20AU	19;22	156.40	31807	3SE	22;54	350.17
31808	19SE	9;58	185.34	31808	3OC	9;42	19.19
31809	18OC	23;26	214.74	31809	1NO	23;12	48.81
31810	17NO	11;53	244.52	31810	1DE	15;41	78.92
31811	16DE	23;34	274.54	31811	31DE	10;37	109.28

1572

NEW NUMBER	DATE	TIME	LONG.	FULL NUMBER	DATE	TIME	LONG.
31812	15JA	10;34	304.54	31812	30JA	6;20	139.55
31813	13FE	20;53	334.26	31813	29FE	0;43	169.41
31814	14MR	6;41	3.54	31814	29MR	16;22	198.68
31815	12AP	16;25	32.33	31815	28AP	5;02	227.35
31816	12MY	2;54	60.72	31816	27MY	15;12	255.56
31817	10JN	14;53	88.91	31817	25JN	23;41	283.55
31818	10JL	4;44	117.11	31818	25JL	7;18	311.57
31819	8AU	20;13	145.56	31819	23AU	14;53	339.86
31820	7SE	12;38	174.41	31820	21SE	23;16	8.61
31821	7OC	5;13	203.74	31821	21OC	9;18	37.92
31822	5NO	21;15	233.53	31822	19NO	21;44	67.74
31823	5DE	12;11	263.62	31823	19DE	12;52	97.92

1573

NEW NUMBER	DATE	TIME	LONG.	FULL NUMBER	DATE	TIME	LONG.
31824	4JA	1;29	293.74	31824	18JA	6;12	128.17
31825	2FE	12;54	323.62	31825	17FE	0;27	158.18
31826	3MR	22;35	353.06	31826	18MR	18;11	187.72
31827	2AP	7;06	21.97	31827	17AP	10;23	216.70
31828	1MY	15;17	50.42	31828	17MY	0;31	245.18
31829	31MY	0;03	78.55	31829	15JN	12;31	273.35
31830	29JN	10;15	106.60	31830	14JL	22;36	301.42
31831	28JL	22;25	134.83	31831	13AU	7;22	329.64
31832	27AU	12;50	163.44	31832	11SE	15;42	358.22
31833	26SE	5;24	192.58	31833	11OC	0;32	27.30
31834	25OC	23;33	222.29	31834	9NO	10;36	56.88
31835	24NO	18;03	252.42	31835	8DE	22;20	86.86
31836	24DE	11;17	282.70				

1574

NEW NUMBER	DATE	TIME	LONG.	FULL NUMBER	DATE	TIME	LONG.
				31836	7JA	11;45	116.98
31837	23JA	2;03	312.81	31837	6FE	2;34	146.97
31838	21FE	14;02	342.50	31838	7MR	18;24	176.62
31839	22MR	23;41	11.64	31839	6AP	10;45	205.79
31840	21AP	7;49	40.23	31840	6MY	3;00	234.50
31841	20MY	15;16	68.41	31841	4JN	18;17	262.86
31842	18JN	22;51	96.39	31842	4JL	7;57	291.05
31843	18JL	7;25	124.41	31843	2AU	19;50	319.30
31844	16AU	17;49	152.72	31844	1SE	6;25	347.81
31845	15SE	6;55	181.55	31845	30SE	16;25	16.76
31846	14OC	23;08	210.99	31846	30OC	2;31	46.20
31847	13NO	18;00	241.00	31847	28NO	13;04	76.03
31848	13DE	13;49	271.34	31848	28DE	0;07	106.06

1575

NEW NUMBER	DATE	TIME	LONG.	FULL NUMBER	DATE	TIME	LONG.
31849	12JA	8;28	301.68	31849	26JA	11;41	136.01
31850	11FE	0;31	331.67	31850	25FE	0;00	165.67
31851	12MR	13;35	1.13	31851	26MR	13;22	194.91
31852	11AP	0;02	29.99	31852	25AP	3;52	223.71
31853	10MY	8;30	58.33	31853	24MY	19;10	252.16
31854	8JN	15;43	86.34	31854	23JN	10;31	280.43
31855	7JL	22;33	114.26	31855	23JL	1;18	308.72
31856	6AU	6;03	142.36	31856	21AU	15;15	337.23
31857	4SE	15;20	170.87	31857	20SE	4;23	6.13
31858	4OC	3;26	199.96	31858	19OC	16;51	35.49
31859	2NO	18;49	229.66	31859	18NO	4;38	65.26
31860	2DE	13;00	259.85	31860	17DE	15;43	95.26

1576

NEW NUMBER	DATE	TIME	LONG.	FULL NUMBER	DATE	TIME	LONG.
31861	1JA	8;31	290.24	31861	16JA	2;12	125.23
31862	31JA	3;36	320.48	31862	14FE	12;26	154.94
31863	29FE	20;50	350.28	31863	14MR	22;53	184.24
31864	30MR	11;24	19.48	31864	13AP	10;03	213.07
31865	28AP	23;07	48.11	31865	12MY	22;15	241.53
31866	28MY	8;19	76.28	31866	11JN	11;34	269.76
31867	26JN	15;47	104.23	31867	11JL	2;00	297.99
31868	25JL	22;38	132.22	31868	9AU	17;24	326.44
31869	24AU	6;03	160.51	31869	8SE	9;25	355.29
31870	22SE	15;05	189.30	31870	8OC	1;23	24.61
31871	22OC	2;26	218.66	31871	6NO	16;25	54.37
31872	20NO	16;24	248.56	31872	6DE	5;55	84.40
31873	20DE	8;46	278.79				

1577

NEW NUMBER	DATE	TIME	LONG.	FULL NUMBER	DATE	TIME	LONG.
				31873	4JA	17;45	114.46
31874	19JA	2;51	309.07	31874	3FE	4;16	144.30
31875	17FE	21;34	339.09	31875	4MR	13;52	173.72
31876	19MR	15;29	8.62	31876	2AP	23;02	202.65
31877	18AP	7;15	37.57	31877	2MY	8;12	231.13
31878	17MY	20;11	66.00	31878	31MY	17;58	259.29
31879	16JN	6;29	94.09	31879	30JN	5;04	287.38
31880	15JL	15;02	122.10	31880	29JL	18;13	315.65
31881	13AU	22;56	150.29	31881	28AU	9;44	344.32
31882	12SE	7;10	178.88	31882	27SE	3;10	13.51
31883	11OC	16;26	207.99	31883	26OC	21;13	43.23
31884	10NO	3;07	237.61	31884	25NO	14;23	73.31
31885	9DE	15;30	267.62	31885	25DE	5;42	103.51

1578

NEW NUMBER	DATE	TIME	LONG.	FULL NUMBER	DATE	TIME	LONG.
31886	8JA	5;49	297.77	31886	23JA	18;53	133.55
31887	6FE	22;01	327.82	31887	22FE	6;04	163.20
31888	8MR	15;26	357.52	31888	23MR	15;32	192.32
31889	7AP	8;47	26.72	31889	21AP	23;42	220.90
31890	7MY	0;48	55.40	31890	21MY	7;15	249.08
31891	5JN	14;46	83.71	31891	19JN	15;09	277.07
31892	5JL	2;47	111.83	31892	19JL	0;34	305.12
31893	3AU	13;22	140.03	31893	17AU	12;32	333.51
31894	1SE	23;11	168.53	31894	16SE	3;28	2.42
31895	1OC	8;44	197.47	31895	15OC	20;58	31.93
31896	30OC	18;27	226.90	31896	14NO	15;50	61.95
31897	29NO	4;43	256.72	31897	14DE	10;35	92.26
31898	28DE	15;59	286.76				

1579

NEW NUMBER	DATE	TIME	LONG.	FULL NUMBER	DATE	TIME	LONG.
				31898	13JA	3;53	122.53
31899	27JA	4;37	316.75	31899	11FE	18;49	152.47
31900	25FE	18;40	346.47	31900	13MR	6;59	181.87
31901	27MR	9;45	15.77	31901	11AP	16;31	210.69
31902	26AP	1;14	44.60	31902	11MY	0;08	238.99
31903	25MY	16;35	73.05	31903	9JN	6;54	266.98
31904	24JN	7;25	101.30	31904	8JL	13;58	294.91
31905	23JL	21;31	129.56	31905	6AU	22;28	323.05
31906	22AU	10;39	158.05	31906	5SE	9;17	351.63
31907	20SE	22;43	186.92	31907	4OC	22;54	20.80
31908	20OC	9;51	216.24	31908	3NO	15;20	50.56
31909	18NO	20;26	245.96	31909	3DE	10;01	80.77
31910	18DE	6;57	275.93				

```
--------NEW MOONS--------   --------FULL MOONS--------   --------NEW MOONS--------   --------FULL MOONS--------
NUMBER  DATE   TIME  LONG.   NUMBER  DATE   TIME  LONG.   NUMBER  DATE   TIME  LONG.   NUMBER  DATE   TIME  LONG.
```

1580 / 1586

NUMBER	DATE	TIME	LONG.	NUMBER	DATE	TIME	LONG.	NUMBER	DATE	TIME	LONG.	NUMBER	DATE	TIME	LONG.
31911	16JA	17;45	305.92	31910	2JA	5;42	111.17	31985	9JA	21;42	299.52	31985	25JA	3;30	134.99
31912	15FE	4;57	335.66	31911	1FE	0;35	141.39	31986	8FE	16;05	329.64	31986	23FE	13;22	164.56
31913	15MR	16;36	4.99	31912	1MR	16;58	171.14	31987	10MR	10;24	359.35	31987	24MR	22;31	193.65
31914	14AP	4;49	33.86	31913	31MR	6;02	200.28	31988	9AP	3;14	28.50	31988	23AP	7;23	222.24
31915	13MY	17;54	62.34	31914	29AP	16;05	228.82	31989	8MY	17;33	57.10	31989	22MY	16;34	250.47
31916	12JN	8;06	90.61	31915	29MY	0;02	256.93	31990	7JN	5;08	85.29	31990	21JN	2;45	278.54
31917	11JL	23;21	118.87	31916	27JN	7;02	284.86	31991	6JL	14;32	113.32	31991	20JL	14;43	306.71
31918	10AU	15;05	147.34	31917	26JL	14;08	312.87	31992	4AU	22;47	141.43	31992	19AU	5;06	335.21
31919	9SE	6;26	176.17	31918	24AU	22;13	341.20	31993	3SE	6;56	169.88	31993	17SE	21;51	4.22
31920	8OC	20;47	205.45	31919	23SE	8;00	10.02	31994	2OC	15;48	198.81	31994	17OC	16;01	33.79
31921	7NO	9;59	235.14	31920	22OC	20;09	39.43	31995	1NO	1;52	228.28	31995	16NO	10;02	63.80
31922	6DE	22;11	265.12	31921	21NO	11;11	69.39	31996	30NO	13;24	258.17	31996	16DE	2;31	94.01
				31922	21DE	5;00	99.68	31997	30DE	2;39	288.29				

1581 / 1587

NUMBER	DATE	TIME	LONG.	NUMBER	DATE	TIME	LONG.	NUMBER	DATE	TIME	LONG.	NUMBER	DATE	TIME	LONG.
31923	5JA	9;33	295.16	31923	20JA	0;29	130.01					31997	14JA	16;49	124.15
31924	3FE	20;08	324.99	31924	18FE	19;38	160.04	31998	28JA	17;46	318.38	31998	13FE	4;56	153.96
31925	5MR	5;58	354.42	31925	20MR	12;38	189.52	31999	27FE	10;26	348.19	31999	14MR	15;06	183.26
31926	3AP	15;24	23.35	31926	19AP	2;40	218.39	32000	29MR	3;44	17.54	32000	12AP	23;42	212.01
31927	3MY	1;10	51.84	31927	18MY	13;59	246.74	32001	27AP	20;23	46.39	32001	12MY	7;17	240.29
31928	1JN	12;06	80.05	31928	16JN	23;14	274.79	32002	27MY	11;21	74.81	32002	10JN	14;45	268.31
31929	1JL	0;53	108.20	31929	16JL	7;17	302.78	32003	26JN	0;22	102.99	32003	9JL	23;11	296.30
31930	30JL	15;36	136.54	31930	14AU	14;55	330.97	32004	25JL	11;46	131.15	32004	8AU	9;46	324.53
31931	29AU	7;48	165.24	31931	12SE	22;57	359.56	32005	23AU	22;08	159.55	32005	6SE	23;15	353.24
31932	28SE	0;38	194.43	31932	12OC	8;11	28.67	32006	22SE	8;00	188.35	32006	6OC	15;42	22.55
31933	27OC	17;18	224.10	31933	10NO	19;26	58.33	32007	21OC	17;47	217.63	32007	5NO	10;13	52.42
31934	26NO	9;05	254.13	31934	10DE	9;13	88.40	32008	20NO	3;52	247.35	32008	5DE	5;22	82.67
31935	25DE	23;23	284.29					32009	19DE	14;36	277.34				

1582 / 1588

NUMBER	DATE	TIME	LONG.	NUMBER	DATE	TIME	LONG.	NUMBER	DATE	TIME	LONG.	NUMBER	DATE	TIME	LONG.
31936	24JA	11;45	314.29	31935	9JA	1;27	118.64					32009	3JA	23;37	113.00
31937	22FE	22;10	343.90	31936	7FE	19;13	148.74	32010	18JA	2;25	307.36	32010	2FE	15;49	143.10
31938	24MR	7;00	12.98	31937	9MR	13;08	178.44	32011	16FE	15;32	337.17	32011	3MR	5;18	172.71
31939	22AP	15;02	41.55	31938	8AP	6;00	207.61	32012	17MR	5;52	6.59	32012	1AP	16;00	201.72
31940	21MY	23;12	69.75	31939	7MY	21;07	236.25	32013	15AP	20;57	35.55	32013	1MY	0;21	230.18
31941	20JN	8;27	97.79	31940	6JN	10;12	264.51	32014	15MY	12;13	64.09	32014	30MY	7;19	258.24
31942	19JL	19;31	125.91	31941	5JL	21;18	292.60	32015	14JN	3;17	92.39	32015	28JN	14;02	286.16
31943	18AU	8;49	154.36	31942	4AU	6;49	320.76	32016	13JL	17;52	120.63	32016	27JL	21;42	314.20
31944	17SE	0;28	183.32	31943	2SE	15;28	349.22	32017	12AU	7;43	149.04	32017	26AU	7;20	342.60
31945	16OC	18;07	212.85	31944	2OC	0;09	18.14	32018	10SE	20;36	177.80	32018	24SE	19;36	11.55
31946	15NO	12;50	242.87	31945	31OC	9;40	47.57	32019	10OC	8;27	206.99	32019	24OC	10;44	41.11
31947	15DE	7;04	273.16	31946	29NO	20;37	77.43	32020	8NO	19;29	236.61	32020	23NO	4;28	71.19
				31947	29DE	9;09	107.52	32021	8DE	6;06	266.53	32021	22DE	23;52	101.56

1583 / 1589

NUMBER	DATE	TIME	LONG.	NUMBER	DATE	TIME	LONG.	NUMBER	DATE	TIME	LONG.	NUMBER	DATE	TIME	LONG.
31948	13JA	23;14	303.38	31948	27JA	23;09	137.56	32022	6JA	16;43	296.54	32022	21JA	19;22	131.88
31949	12FE	12;34	333.24	31949	26FE	14;15	167.32	32023	5FE	3;34	326.37	32023	20FE	13;05	161.82
31950	13MR	23;12	2.57	31950	28MR	6;07	196.63	32024	6MR	14;45	355.84	32024	22MR	3;46	191.18
31951	12AP	7;51	31.34	31951	26AP	22;17	225.47	32025	5AP	2;22	24.83	32025	20AP	19;09	219.92
31952	11MY	15;24	59.63	31952	26MY	14;03	253.93	32026	4MY	14;42	53.41	32026	19MY	23;59	248.16
31953	9JN	22;42	87.64	31953	25JN	4;38	282.17	32027	3JN	4;07	81.70	32027	18JN	7;19	276.13
31954	9JL	6;37	115.60	31954	24JL	17;34	310.39	32028	2JL	18;49	109.93	32028	17JL	14;18	304.08
31955	7AU	15;49	143.79	31955	23AU	4;58	338.82	32029	1AU	10;30	138.32	32029	15AU	21;52	332.28
31956	6SE	3;42	172.41	31956	21SE	15;25	7.63	32030	31AU	2;22	167.03	32030	14SE	6;50	0.92
31957	5OC	18;26	201.64	31957	21OC	1;39	36.93	32031	29SE	17;34	196.18	32031	13OC	17;49	30.13
31958	4NO	12;14	231.47	31958	19NO	12;06	66.65	32032	29OC	7;37	225.77	32032	12NO	7;26	59.91
31959	4DE	7;56	261.73	31959	18DE	22;57	96.64	32033	27NO	20;30	255.69	32033	11DE	23;54	90.10
								32034	27DE	8;22	285.75				

1584 / 1590

NUMBER	DATE	TIME	LONG.	NUMBER	DATE	TIME	LONG.	NUMBER	DATE	TIME	LONG.	NUMBER	DATE	TIME	LONG.
31960	3JA	3;28	292.12	31960	17JA	10;10	126.64					32034	10JA	18;40	120.44
31961	1FE	20;55	322.27	31961	15FE	21;52	156.40	32035	25JA	19;18	315.68	32035	9FE	14;06	150.59
31962	2MR	11;24	351.94	31962	16MR	10;23	185.75	32036	24FE	5;20	345.25	32036	11MR	8;11	180.28
31963	31MR	23;00	21.00	31963	15AP	0;01	214.67	32037	25MR	14;41	14.34	32037	9AP	23;38	209.36
31964	30AP	8;17	49.50	31964	14MY	14;45	243.21	32038	23AP	23;56	42.94	32038	9MY	12;12	237.88
31965	29MY	15;58	77.60	31965	13JN	6;04	271.51	32039	23MY	9;57	71.20	32039	7JN	22;24	266.01
31966	27JN	22;52	105.52	31966	12JL	21;15	299.78	32040	21JN	21;35	99.33	32040	7JL	7;02	294.01
31967	27JL	5;56	133.53	31967	11AU	11;49	328.21	32041	21JL	11;19	127.56	32041	5AU	14;55	322.11
31968	25AU	14;15	161.88	31968	10SE	1;37	356.99	32042	20AU	2;57	156.13	32042	3SE	22;49	350.56
31969	24SE	0;55	190.75	31969	9OC	14;43	26.22	32043	18SE	19;47	185.15	32043	3OC	7;30	19.50
31970	23OC	14;44	220.24	31970	8NO	3;05	55.89	32044	18OC	12;53	214.69	32044	1NO	17;46	48.97
31971	22NO	7;41	250.28	31971	7DE	14;40	85.85	32045	17NO	5;27	244.65	32045	1DE	6;16	78.91
31972	22DE	2;44	280.63					32046	16DE	20;44	274.81	32046	30DE	21;15	109.11

1585 / 1591

NUMBER	DATE	TIME	LONG.	NUMBER	DATE	TIME	LONG.	NUMBER	DATE	TIME	LONG.	NUMBER	DATE	TIME	LONG.
				31972	6JA	1;26	115.86	32047	15JA	10;09	304.91	32047	29JA	14;13	139.27
31973	20JA	22;11	310.96	31973	4FE	11;37	145.67	32048	13FE	21;27	334.68	32048	28FE	7;59	169.11
31974	19FE	16;23	340.93	31974	5MR	21;41	175.10	32049	15MR	6;51	3.94	32049	30MR	1;17	198.45
31975	21MR	8;15	10.35	31975	4AP	8;10	204.06	32050	13AP	14;59	32.66	32050	28AP	17;12	227.26
31976	19AP	21;18	39.17	31976	3MY	19;33	232.61	32051	12MY	22;47	60.95	32051	28MY	7;18	255.64
31977	19MY	7;40	67.48	31977	2JN	6;49	260.87	32052	11JN	7;14	88.99	32052	26JN	19;27	283.78
31978	17JN	15;54	95.48	31978	1JL	21;53	289.07	32053	10JL	17;14	117.05	32053	26JL	5;51	311.91
31979	16JL	22;59	123.44	31979	31JL	12;52	317.42	32054	9AU	5;22	145.35	32054	24AU	15;03	340.26
31980	15AU	6;05	151.61	31980	30AU	4;50	346.13	32055	7SE	19;55	174.12	32055	22SE	23;49	9.03
31981	13SE	14;19	180.22	31981	28SE	21;12	15.31	32056	7OC	12;48	203.45	32056	22OC	9;01	38.30
31982	13OC	0;34	209.39	31982	28OC	13;06	44.96	32057	6NO	7;21	233.34	32057	20NO	19;19	68.03
31983	11NO	13;17	239.11	31983	27NO	3;39	74.95	32058	6DE	2;12	263.59	32058	20DE	7;03	98.06
31984	11DE	4;29	269.25	31984	26DE	16;24	105.04								

NEW MOONS (left)

NUMBER	DATE	TIME	LONG.
		1592	
32059	4JA	19;40	293.88
32060	3FE	10;27	323.91
32061	3MR	22;17	353.44
32062	2AP	7;42	22.39
32063	1MY	15;33	50.82
32064	30MY	22;46	78.89
32065	29JN	6;13	106.83
32066	28JL	14;44	134.91
32067	27AU	1;12	163.36
32068	25SE	14;27	192.36
32069	25OC	6;53	221.98
32070	24NO	1;57	252.13
32071	23DE	21;54	282.53
		1593	
32072	22JA	16;35	312.81
32073	21FE	8;32	342.66
32074	22MR	21;26	11.94
32075	21AP	7;42	40.62
32076	20MY	16;02	68.83
32077	18JN	23;12	96.79
32078	18JL	6;05	124.74
32079	16AU	13;42	152.95
32080	14SE	23;10	181.63
32081	14OC	11;26	210.90
32082	13NO	2;56	240.76
32083	12DE	21;07	271.03
		1594	
32084	11JA	16;32	301.40
32085	10FE	11;25	331.52
32086	12MR	4;25	1.14
32087	10AP	18;49	30.16
32088	10MY	6;26	58.64
32089	8JN	15;39	86.73
32090	7JL	23;14	114.68
32091	6AU	6;18	142.76
32092	4SE	13;59	171.21
32093	3OC	23;16	200.18
32094	2NO	10;49	229.72
32095	2DE	0;49	259.73
32096	31DE	17;01	289.98
		1595	
32097	30JA	10;46	320.16
32098	1MR	5;04	350.02
32099	30MR	22;38	19.35
32100	29AP	14;12	48.14
32101	29MY	3;08	76.46
32102	27JN	13;36	104.53
32103	26JL	22;26	132.60
32104	25AU	6;42	160.93
32105	23SE	15;19	189.70
32106	23OC	0;55	218.99
32107	21NO	11;46	248.76
32108	21DE	0;07	278.82
		1596	
32109	19JA	14;09	308.92
32110	18FE	5;52	338.82
32111	18MR	22;43	8.32
32112	17AP	15;37	37.33
32113	17MY	7;23	65.88
32114	15JN	21;24	94.12
32115	15JL	9;41	122.28
32116	13AU	20;43	150.60
32117	12SE	7;02	179.27
32118	11OC	17;03	208.41
32119	10NO	3;07	238.00
32120	9DE	13;32	267.93
		1597	
32121	8JA	0;42	297.95
32122	6FE	12;58	327.83
32123	8MR	2;28	357.37
32124	6AP	16;55	26.46
32125	6MY	7;54	55.12
32126	4JN	22;59	83.47
32127	4JL	13;52	111.71
32128	3AU	4;18	140.06
32129	1SE	17;58	168.71
32130	1OC	6;38	197.77
32131	30OC	18;17	227.28
32132	29NO	5;12	257.14
32133	28DE	15;49	287.15

FULL MOONS (left)

NUMBER	DATE	TIME	LONG.
		1592	
32059	18JA	20;13	128.14
32060	17FE	10;34	157.99
32061	18MR	1;48	187.43
32062	16AP	17;37	216.41
32063	16MY	9;31	244.98
32064	15JN	0;46	273.27
32065	14JL	14;41	301.49
32066	13AU	3;01	329.85
32067	11SE	14;08	358.55
32068	11OC	0;40	27.69
32069	9NO	11;10	57.30
32070	8DE	21;54	87.23
		1593	
32071	7JA	8;53	117.25
32072	5FE	20;09	147.10
32073	7MR	7;57	176.58
32074	5AP	20;43	205.61
32075	5MY	10;39	234.24
32076	4JN	1;35	262.59
32077	3JL	16;54	290.84
32078	2AU	7;58	319.21
32079	31AU	22;24	347.88
32080	30SE	12;09	16.98
32081	30OC	1;10	46.53
32082	28NO	13;20	76.44
32083	28DE	0;34	106.47
		1594	
32084	26JA	10;56	136.37
32085	24FE	20;49	165.93
32086	26MR	6;46	195.03
32087	24AP	17;23	223.67
32088	24MY	5;06	251.99
32089	22JN	18;07	280.17
32090	22JL	8;30	308.44
32091	21AU	0;08	337.01
32092	19SE	16;39	6.04
32093	19OC	9;11	35.57
32094	18NO	0;46	65.49
32095	17DE	14;35	95.60
		1595	
32096	16JA	2;30	125.63
32097	14FE	12;50	155.35
32098	15MR	22;07	184.59
32099	14AP	6;51	213.33
32100	13MY	15;35	241.65
32101	12JN	1;00	269.73
32102	11JL	11;53	297.82
32103	10AU	1;01	326.17
32104	8SE	16;44	355.00
32105	8OC	10;34	24.39
32106	7NO	5;06	54.29
32107	6DE	22;41	84.48
		1596	
32108	5JA	14;12	114.70
32109	4FE	3;23	144.65
32110	4MR	14;23	174.14
32111	2AP	23;33	203.07
32112	2MY	7;25	231.48
32113	31MY	14;41	259.55
32114	29JN	22;27	287.51
32115	29JL	7;45	315.62
32116	27AU	19;45	344.14
32117	26SE	10;52	13.23
32118	26OC	4;38	42.93
32119	24NO	23;48	73.09
32120	24DE	18;46	103.44
		1597	
32121	23JA	12;08	133.66
32122	22FE	2;59	163.46
32123	23MR	14;57	192.68
32124	22AP	0;17	221.32
32125	21MY	7;44	249.49
32126	19JN	14;22	277.43
32127	18JL	21;26	305.39
32128	17AU	6;00	333.64
32129	15SE	16;58	2.39
32130	15OC	6;45	31.73
32131	13NO	23;19	61.65
32132	13DE	18;04	91.95

NEW MOONS (right)

NUMBER	DATE	TIME	LONG.
		1598	
32134	27JA	2;26	317.05
32135	25FE	13;15	346.64
32136	27MR	0;21	15.77
32137	25AP	11;59	44.45
32138	25MY	0;35	72.80
32139	23JN	14;32	101.01
32140	23JL	5;49	129.32
32141	21AU	21;53	157.92
32142	20SE	13;47	186.94
32143	20OC	4;45	216.42
32144	18NO	18;27	246.27
32145	18DE	6;56	276.33
		1599	
32146	16JA	18;20	306.33
32147	15FE	4;42	336.04
32148	16MR	14;09	5.28
32149	14AP	23;07	34.02
32150	14MY	8;24	62.35
32151	12JN	18;57	90.48
32152	12JL	7;31	118.63
32153	10AU	22;17	147.06
32154	9SE	14;47	175.93
32155	9OC	8;06	205.31
32156	8NO	1;18	235.17
32157	7DE	17;30	265.31
		1600	
32158	6JA	8;00	295.48
32159	4FE	20;20	325.39
32160	5MR	6;31	354.83
32161	3AP	15;01	23.72
32162	2MY	22;41	52.13
32163	1JN	6;32	80.21
32164	30JN	15;33	108.22
32165	30JL	2;31	136.40
32166	28AU	15;53	164.99
32167	27SE	7;46	194.13
32168	27OC	1;47	223.84
32169	25NO	20;52	254.01
32170	25DE	15;21	284.34
		1601	
32171	24JA	7;36	314.51
32172	22FE	20;51	344.24
32173	24MR	7;16	13.38
32174	22AP	15;39	41.97
32175	21MY	22;58	70.13
32176	20JN	6;07	98.08
32177	19JL	13;59	126.08
32178	17AU	23;23	154.37
32179	16SE	11;14	183.17
32180	16OC	2;09	212.57
32181	14NO	20;08	242.56
32182	14DE	15;59	272.91
		1602	
32183	13JA	11;34	303.28
32184	12FE	4;56	333.31
32185	13MR	19;16	2.80
32186	12AP	6;41	31.68
32187	11MY	15;50	60.04
32188	9JN	23;27	88.06
32189	9JL	6;22	115.98
32190	7AU	13;33	144.08
32191	5SE	22;03	172.58
32192	5OC	8;55	201.63
32193	3NO	22;51	231.30
32194	3DE	15;51	261.45
		1603	
32195	2JA	10;48	291.81
32196	1FE	6;03	322.04
32197	3MR	0;00	351.85
32198	1AP	15;38	21.09
32199	1MY	4;35	49.74
32200	30MY	14;56	77.95
32201	28JN	23;16	105.93
32202	28JL	6;33	133.95
32203	26AU	13;56	162.26
32204	24SE	22;27	191.04
32205	24OC	8;56	220.39
32206	22NO	21;45	250.26
32207	22DE	12;51	280.44

FULL MOONS (right)

NUMBER	DATE	TIME	LONG.
		1598	
32133	12JA	13;44	122.32
32134	11FE	8;29	152.43
32135	13MR	0;41	182.01
32136	11AP	13;36	210.96
32137	10MY	23;33	239.35
32138	9JN	7;28	267.39
32139	8JL	14;32	295.32
32140	6AU	21;48	323.42
32141	5SE	6;05	351.89
32142	4OC	16;06	20.91
32143	3NO	4;25	50.49
32144	2DE	19;28	80.55
		1599	
32145	1JA	13;09	110.86
32146	31JA	8;20	141.10
32147	2MR	3;10	170.96
32148	31MR	19;54	200.25
32149	30AP	9;48	228.95
32150	29MY	21;06	257.21
32151	28JN	6;29	285.23
32152	27JL	14;47	313.29
32153	25AU	22;44	341.61
32154	24SE	7;06	10.38
32155	23OC	16;37	39.68
32156	22NO	4;00	69.48
32157	21DE	17;43	99.60
		1600	
32158	20JA	9;39	129.78
32159	19FE	2;56	159.73
32160	19MR	20;22	189.24
32161	18AP	12;52	218.22
32162	18MY	3;50	246.73
32163	16JN	17;00	274.93
32164	16JL	4;23	303.06
32165	14AU	14;19	331.34
32166	12SE	23;26	359.97
32167	12OC	8;31	29.08
32168	10NO	18;21	58.68
32169	10DE	5;23	88.63
		1601	
32170	8JA	17;46	118.70
32171	7FE	7;21	148.63
32172	8MR	21;53	178.20
32173	7AP	13;10	207.31
32174	7MY	4;54	235.99
32175	5JN	20;29	264.35
32176	5JL	11;12	292.59
32177	4AU	0;32	320.90
32178	2SE	12;28	349.49
32179	1OC	23;28	18.50
32180	31OC	10;10	47.98
32181	29NO	20;54	77.83
32182	29DE	7;47	107.85
		1602	
32183	27JA	18;47	137.77
32184	26FE	6;02	167.37
32185	27MR	17;58	196.53
32186	26AP	7;01	225.25
32187	25MY	21;18	253.66
32188	24JN	12;25	281.92
32189	24JL	3;45	310.23
32190	22AU	18;45	338.80
32191	21SE	9;08	7.77
32192	20OC	22;49	37.20
32193	19NO	11;40	67.03
32194	18DE	23;29	97.06
		1603	
32195	17JA	10;14	127.04
32196	15FE	20;09	156.72
32197	17MR	5;47	185.96
32198	15AP	15;44	214.72
32199	15MY	2;36	243.11
32200	13JN	14;46	271.29
32201	13JL	4;24	299.50
32202	11AU	19;31	327.95
32203	10SE	11;51	356.82
32204	10OC	4;47	26.20
32205	8NO	21;14	56.04
32206	8DE	12;12	86.14

Column headers for each block: NUMBER DATE TIME LONG. (New Moons) and NUMBER DATE TIME LONG. (Full Moons)

1604

New #	Date	Time	Long.	Full #	Date	Time	Long.
				32207	7JA	1;07	116.24
32208	21JA	5;47	310.66	32208	5FE	12;08	146.09
32209	19FE	23;44	340.63	32209	5MR	21;44	175.49
32210	20MR	17;37	10.14	32210	4AP	6;29	204.38
32211	19AP	10;11	39.11	32211	3MY	14;57	232.81
32212	19MY	0;25	67.58	32212	1JN	23;45	260.93
32213	17JN	12;07	95.72	32213	1JL	9;40	288.97
32214	16JL	21;47	123.78	32214	30JL	21;34	317.20
32215	15AU	6;24	152.02	32215	29AU	12;03	345.84
32216	13SE	14;57	180.64	32216	28SE	5;07	15.04
32217	13OC	0;11	209.76	32217	27OC	23;43	44.79
32218	11NO	10;30	239.39	32218	26NO	18;10	74.94
32219	10DE	22;05	269.37	32219	26DE	10;56	105.21

1605

New #	Date	Time	Long.	Full #	Date	Time	Long.
32220	9JA	11;09	299.47	32220	25JA	1;19	135.29
32221	8FE	1;50	329.44	32221	23FE	13;18	164.95
32222	9MR	17;56	359.06	32222	24MR	23;12	194.07
32223	8AP	10;42	28.22	32223	23AP	7;30	222.64
32224	8MY	3;00	56.91	32224	22MY	14;50	250.79
32225	6JN	17;54	85.24	32225	20JN	22;05	278.75
32226	6JL	7;06	113.41	32226	20JL	6;26	306.77
32227	4AU	18;54	141.67	32227	18AU	17;01	335.11
32228	3SE	5;46	170.22	32228	17SE	6;38	3.99
32229	2OC	16;09	199.22	32229	16OC	23;18	33.49
32230	1NO	2;22	228.69	32230	15NO	18;06	63.52
32231	30NO	12;40	258.53	32231	15DE	13;27	93.85
32232	29DE	23;24	288.54				

1606

New #	Date	Time	Long.	Full #	Date	Time	Long.
				32232	14JA	7;48	124.16
32233	28JA	10;56	318.48	32233	12FE	23;57	154.14
32234	26FE	23;34	348.13	32234	14MR	13;18	183.57
32235	28MR	13;18	17.35	32235	12AP	23;48	212.40
32236	27AP	3;47	46.12	32236	12MY	7;59	240.71
32237	26MY	18;41	74.54	32237	10JN	14;50	268.70
32238	25JN	9;39	102.79	32238	9JL	21;31	296.62
32239	25JL	0;28	131.09	32239	8AU	5;15	324.74
32240	23AU	14;48	159.64	32240	6SE	15;00	353.30
32241	22SE	4;17	188.59	32241	6OC	3;26	22.43
32242	21OC	16;42	217.98	32242	4NO	18;44	52.16
32243	20NO	4;09	247.76	32243	4DE	12;32	82.35
32244	19DE	14;57	277.75				

1607

New #	Date	Time	Long.	Full #	Date	Time	Long.
				32244	3JA	7;55	112.73
32245	18JA	1;29	307.71	32245	2FE	3;16	142.96
32246	16FE	12;02	337.41	32246	3MR	20;49	172.75
32247	17MR	22;43	6.68	32247	2AP	11;19	201.92
32248	16AP	9;46	35.48	32248	1MY	22;36	230.49
32249	15MY	21;35	63.90	32249	31MY	7;22	258.63
32250	14JN	10;38	92.12	32250	29JN	14;46	286.58
32251	14JL	1;15	120.36	32251	28JL	23;15	314.59
32252	12AU	17;08	148.85	32252	27AU	5;41	342.93
32253	11SE	9;29	177.73	32253	25SE	14;53	11.75
32254	11OC	1;18	207.09	32254	25OC	2;05	41.14
32255	9NO	15;55	236.85	32255	23NO	15;46	71.05
32256	9DE	5;10	266.89	32256	23DE	8;09	101.29

1608

New #	Date	Time	Long.	Full #	Date	Time	Long.
32257	7JA	17;09	296.95	32257	22JA	2;39	131.57
32258	6FE	3;57	326.78	32258	20FE	21;43	161.58
32259	6MR	13;40	356.18	32259	21MR	15;29	191.07
32260	4AP	22;35	25.07	32260	20AP	6;29	219.97
32261	4MY	7;22	53.50	32261	19MY	19;15	248.36
32262	2JN	16;58	81.65	32262	18JN	5;33	276.45
32263	2JL	4;20	109.75	32263	17JL	14;25	304.48
32264	31JL	18;00	138.05	32264	15AU	22;37	332.70
32265	30AU	9;49	166.75	32265	14SE	6;52	1.32
32266	29SE	3;04	195.97	32266	13OC	15;53	30.45
32267	28OC	20;41	225.70	32267	12NO	2;21	60.09
32268	27NO	13;42	255.80	32268	11DE	14;51	90.11
32269	27DE	5;16	286.01				

1609

New #	Date	Time	Long.	Full #	Date	Time	Long.
				32269	10JA	5;37	120.28
32270	25JA	18;44	316.05	32270	8FE	22;09	150.32
32271	24FE	5;53	345.67	32271	10MR	15;24	179.98
32272	25MR	14;59	14.74	32272	9AP	8;14	209.12
32273	23AP	22;47	43.28	32273	8MY	23;54	237.77
32274	23MY	6;15	71.44	32274	7JN	13;59	266.07
32275	21JN	14;27	99.43	32275	7JL	2;21	294.21
32276	21JL	0;18	127.51	32276	5AU	13;10	322.43
32277	19AU	12;27	155.93	32277	3SE	22;50	350.95
32278	18SE	3;10	184.87	32278	3OC	8;04	19.91
32279	17OC	20;21	214.39	32279	1NO	17;37	49.36
32280	16NO	15;15	244.44	32280	1DE	4;05	79.22
32281	16DE	10;23	274.77	32281	30DE	15;46	109.27

1610

New #	Date	Time	Long.	Full #	Date	Time	Long.
32282	15JA	3;59	305.05	32282	29JA	4;37	139.26
32283	13FE	18;44	334.96	32283	27FE	18;27	168.94
32284	15MR	6;24	4.31	32284	29MR	9;05	198.18
32285	13AP	15;34	33.07	32285	28AP	0;22	226.98
32286	12MY	23;12	61.35	32286	27MY	15;58	255.42
32287	11JN	6;15	89.34	32287	26JN	7;13	283.68
32288	10JL	13;37	117.29	32288	25JL	21;26	311.96
32289	8AU	22;10	145.45	32289	24AU	10;17	340.46
32290	7SE	8;45	174.05	32290	22SE	21;59	9.35
32291	6OC	22;09	203.24	32291	22OC	9;01	38.69
32292	5NO	14;46	233.04	32292	20NO	19;53	68.45
32293	5DE	9;58	263.29	32293	20DE	6;46	98.45

1611

New #	Date	Time	Long.	Full #	Date	Time	Long.
32294	4JA	5;58	293.70	32294	18JA	17;37	128.42
32295	3FE	0;36	323.89	32295	17FE	4;32	158.13
32296	4MR	16;24	353.59	32296	18MR	15;48	187.41
32297	3AP	5;08	22.68	32297	17AP	3;57	216.25
32298	2MY	15;16	51.19	32298	16MY	17;22	244.72
32299	31MY	23;30	79.31	32299	15JN	7;59	273.00
32300	30JN	6;41	107.24	32300	14JL	23;19	301.27
32301	29JL	13;40	135.26	32301	13AU	14;42	329.75
32302	27AU	21;28	163.60	32302	12SE	5;40	358.59
32303	26SE	7;08	192.45	32303	11OC	20;02	27.89
32304	25OC	19;34	221.90	32304	10NO	9;35	57.63
32305	24NO	11;08	251.90	32305	9DE	22;05	87.63
32306	24DE	5;15	282.21				

1612

New #	Date	Time	Long.	Full #	Date	Time	Long.
				32306	8JA	9;23	117.67
32307	23JA	0;28	312.52	32307	6FE	19;35	147.47
32308	21FE	19;05	342.50	32308	7MR	5;05	176.85
32309	22MR	11;50	11.93	32309	5AP	14;33	205.75
32310	21AP	2;03	40.78	32310	5MY	0;39	234.22
32311	20MY	13;39	69.13	32311	3JN	11;56	262.43
32312	18JN	22;56	97.17	32312	3JL	0;42	290.59
32313	18JL	6;43	125.16	32313	1AU	15;05	318.93
32314	16AU	14;03	153.36	32314	31AU	7;00	347.65
32315	14SE	22;02	181.97	32315	30SE	0;00	16.87
32316	14OC	7;35	211.13	32316	29OC	17;07	46.59
32317	12NO	19;18	240.83	32317	28NO	9;10	76.65
32318	12DE	9;17	270.92	32318	27DE	23;14	106.81

1613

New #	Date	Time	Long.	Full #	Date	Time	Long.
32319	11JA	1;15	301.14	32319	26JA	11;09	136.77
32320	9FE	18;35	331.21	32320	24FE	21;17	166.34
32321	11MR	12;26	0.87	32321	26MR	6;13	195.39
32322	10AP	5;37	30.02	32322	24AP	14;33	223.95
32323	9MY	21;01	58.65	32323	23MY	22;55	252.14
32324	8JN	10;00	86.89	32324	22JN	8;03	280.16
32325	7JL	20;42	114.97	32325	21JL	18;47	308.28
32326	6AU	5;53	143.14	32326	20AU	7;57	336.75
32327	4SE	14;34	171.62	32327	18SE	23;54	5.75
32328	3OC	23;36	200.59	32328	18OC	18;07	35.33
32329	2NO	9;30	230.06	32329	17NO	13;04	65.39
32330	1DE	20;29	259.94	32330	17DE	6;59	95.67
32331	31DE	8;44	290.02				

1614

New #	Date	Time	Long.	Full #	Date	Time	Long.
				32331	15JA	22;39	125.86
32332	29JA	22;25	320.04	32332	14FE	11;46	155.70
32333	28FE	13;36	349.76	32333	15MR	22;33	185.00
32334	30MR	5;52	19.06	32334	14AP	7;28	213.75
32335	28AP	22;18	47.89	32335	13MY	15;03	242.02
32336	28MY	13;54	76.32	32336	11JN	22;06	270.00
32337	27JN	4;00	104.54	32337	11JL	5;42	297.96
32338	26JL	16;38	132.76	32338	9AU	15;02	326.16
32339	25AU	4;09	161.22	32339	8SE	3;08	354.83
32340	23SE	15;00	190.08	32340	7OC	18;26	24.11
32341	23OC	1;30	219.41	32341	6NO	12;27	53.98
32342	21NO	11;52	249.16	32342	6DE	7;50	84.25
32343	20DE	22;22	279.14				

1615

New #	Date	Time	Long.	Full #	Date	Time	Long.
				32343	5JA	2;54	114.61
32344	19JA	9;22	309.12	32344	3FE	20;15	144.75
32345	17FE	21;13	338.86	32345	5MR	10;59	174.39
32346	19MR	10;08	8.20	32346	3AP	22;46	203.42
32347	17AP	23;59	37.09	32347	3MY	7;56	231.89
32348	17MY	14;30	65.60	32348	1JN	15;15	259.97
32349	16JN	5;22	93.88	32349	30JN	21;52	287.88
32350	15JL	20;20	122.14	32350	30JL	4;58	315.90
32351	14AU	11;09	150.60	32351	28AU	13;41	344.29
32352	13SE	1;24	179.43	32352	27SE	0;48	13.21
32353	12OC	14;40	208.70	32353	26OC	14;45	42.74
32354	11NO	2;48	238.39	32354	25NO	7;25	72.79
32355	10DE	14;00	268.34	32355	25DE	2;09	103.13

NUMBER	DATE	TIME	LONG.	NUMBER	DATE	TIME	LONG.

1616

NUMBER	DATE	TIME	LONG.	NUMBER	DATE	TIME	LONG.
32356	9JA	0;38	298.34	32356	23JA	21;40	133.44
32357	7FE	11;03	328.14	32357	22FE	16;13	163.41
32358	7MR	21;25	357.55	32358	23MR	8;14	192.80
32359	6AP	7;59	26.49	32359	21AP	21;01	221.58
32360	5MY	19;05	54.99	32360	21MY	6;54	249.85
32361	4JN	7;16	83.23	32361	19JN	14;52	277.84
32362	3JL	20;59	111.43	32362	18JL	22;04	305.80
32363	2AU	12;21	139.81	32363	17AU	5;33	334.01
32364	1SE	4;47	168.56	32364	15SE	14;06	2.66
32365	30SE	21;16	197.78	32365	15OC	0;20	31.86
32366	30OC	12;49	227.45	32366	13NO	12;47	61.60
32367	29NO	2;58	257.44	32367	13DE	3;48	91.74
32368	28DE	15;40	287.53				

1617

NUMBER	DATE	TIME	LONG.	NUMBER	DATE	TIME	LONG.
				32368	11JA	21;15	122.02
32369	27JA	3;02	317.47	32369	10FE	16;05	152.14
32370	25FE	13;09	347.03	32370	12MR	10;32	181.81
32371	26MR	22;13	16.08	32371	11AP	2;59	210.92
32372	25AP	6;44	44.63	32372	10MY	16;47	239.47
32373	24MY	15;36	72.83	32373	9JN	4;08	267.65
32374	23JN	1;50	100.90	32374	8JL	13;44	295.68
32375	22JL	14;15	129.09	32375	6AU	22;20	323.83
32376	21AU	5;06	157.64	32376	5SE	6;40	352.31
32377	19SE	21;55	186.68	32377	4OC	15;24	21.27
32378	19OC	15;43	216.26	32378	3NO	1;11	50.75
32379	18NO	9;23	246.28	32379	2DE	12;39	80.66
32380	18DE	1;56	276.51				

1618

NUMBER	DATE	TIME	LONG.	NUMBER	DATE	TIME	LONG.
				32380	1JA	2;13	110.79
32381	16JA	16;34	306.65	32381	30JA	17;46	140.88
32382	15FE	4;49	336.44	32382	1MR	10;32	170.66
32383	16MR	14;45	5.70	32383	31MR	3;27	199.97
32384	14AP	22;56	34.40	32384	29AP	19;36	228.77
32385	14MY	6;16	62.66	32385	29MY	10;27	257.17
32386	12JN	13;51	90.66	32386	27JN	23;45	285.35
32387	11JL	22;42	118.67	32387	27JL	11;30	313.55
32388	10AU	9;38	146.94	32388	25AU	21;54	341.97
32389	8SE	23;07	175.68	32389	24SE	7;30	10.79
32390	8OC	15;15	205.01	32390	23OC	17;01	40.09
32391	7NO	9;34	234.90	32391	22NO	3;06	69.83
32392	7DE	4;57	265.17	32392	21DE	14;10	99.84

1619

NUMBER	DATE	TIME	LONG.	NUMBER	DATE	TIME	LONG.
32393	5JA	23;36	295.52	32393	20JA	2;20	129.86
32394	4FE	15;52	325.60	32394	18FE	15;28	159.65
32395	6MR	4;58	355.16	32395	20MR	5;25	189.02
32396	4AP	15;11	24.12	32396	18AP	20;06	217.93
32397	3MY	23;21	52.54	32397	18MY	11;25	246.46
32398	2JN	6;30	80.60	32398	17JN	2;52	274.76
32399	1JL	13;34	108.53	32399	16JL	17;47	303.03
32400	30JL	21;26	136.59	32400	15AU	7;33	331.45
32401	29AU	6;56	165.02	32401	13SE	20;04	0.22
32402	27SE	18;56	193.99	32402	13OC	7;38	29.44
32403	27OC	10;01	223.58	32403	11NO	18;46	59.09
32404	26NO	4;08	253.69	32404	11DE	5;44	89.04
32405	26DE	0;02	284.09				

1620

NUMBER	DATE	TIME	LONG.	NUMBER	DATE	TIME	LONG.
				32405	9JA	16;36	119.05
32406	24JA	19;34	314.40	32406	8FE	3;19	148.86
32407	23FE	12;48	344.29	32407	8MR	14;06	178.27
32408	24MR	2;58	13.60	32408	7AP	1;27	207.23
32409	22AP	14;14	42.30	32409	6MY	13;56	235.78
32410	21MY	23;17	70.53	32410	5JN	3;48	264.08
32411	20JN	6;54	98.50	32411	4JL	18;48	292.33
32412	19JL	13;55	126.47	32412	3AU	10;20	320.72
32413	17AU	21;17	154.68	32413	2SE	1;47	349.45
32414	16SE	6;00	183.34	32414	1OC	16;46	18.62
32415	15OC	17;03	212.59	32415	31OC	7;02	48.24
32416	14NO	7;06	242.40	32416	29NO	20;18	78.20
32417	14DE	0;04	272.63	32417	29DE	8;17	108.27

1621

NUMBER	DATE	TIME	LONG.	NUMBER	DATE	TIME	LONG.
32418	12JA	18;50	302.96	32418	27JA	18;57	138.17
32419	11FE	13;47	333.07	32419	26FE	4;35	167.70
32420	13MR	7;27	2.71	32420	27MR	13;46	196.74
32421	11AP	22;52	31.75	32421	25AP	23;13	225.33
32422	11MY	11;44	60.26	32422	25MY	9;37	253.58
32423	9JN	22;08	88.39	32423	23JN	21;28	281.71
32424	9JL	6;38	116.39	32424	23JL	11;00	309.95
32425	7AU	14;12	144.50	32425	22AU	2;16	338.52
32426	5SE	21;53	172.96	32426	20SE	19;01	7.58
32427	5OC	6;43	201.94	32427	20OC	12;29	37.16
32428	3NO	17;25	231.46	32428	19NO	5;28	67.16
32429	3DE	6;17	261.43	32429	18DE	20;45	97.34

1622

NUMBER	DATE	TIME	LONG.	NUMBER	DATE	TIME	LONG.
32430	1JA	21;14	291.63	32430	17JA	9;46	127.41
32431	31JA	13;47	321.75	32431	15FE	20;40	157.13
32432	2MR	7;15	351.55	32432	17MR	5;58	186.36
32433	1AP	0;42	20.87	32433	15AP	14;21	215.06
32434	30AP	16;59	49.66	32434	14MY	22;26	243.33
32435	30MY	7;11	78.03	32435	13JN	6;56	271.37
32436	28JN	19;04	106.15	32436	12JL	16;40	299.42
32437	28JL	5;04	134.28	32437	11AU	4;32	327.73
32438	26AU	14;06	162.65	32438	9SE	19;09	356.53
32439	24SE	23;05	191.46	32439	9OC	12;32	25.92
32440	24OC	8;41	220.78	32440	8NO	7;32	55.85
32441	22NO	19;13	250.54	32441	8DE	2;21	86.11
32442	22DE	6;48	280.57				

1623

NUMBER	DATE	TIME	LONG.	NUMBER	DATE	TIME	LONG.
				32442	6JA	19;19	116.39
32443	20JA	19;36	310.62	32443	5FE	9;42	146.38
32444	19FE	9;48	340.44	32444	6MR	21;31	175.88
32445	21MR	1;18	9.87	32445	5AP	7;11	204.80
32446	19AP	17;32	38.83	32446	4MY	15;13	233.21
32447	19MY	9;30	67.38	32447	2JN	22;19	261.26
32448	18JN	0;23	95.65	32448	2JL	5;27	289.20
32449	17JL	13;52	123.86	32449	31JL	13;46	317.28
32450	16AU	2;07	152.23	32450	30AU	0;25	345.76
32451	14SE	13;31	180.97	32451	28SE	14;11	14.81
32452	14OC	0;25	210.17	32452	28OC	7;04	44.49
32453	12NO	11;01	239.80	32453	27NO	2;04	74.65
32454	11DE	21;30	269.74	32454	26DE	21;33	105.03

1624

NUMBER	DATE	TIME	LONG.	NUMBER	DATE	TIME	LONG.
32455	10JA	8;10	299.74	32455	25JA	15;54	135.28
32456	8FE	19;23	329.56	32456	24FE	7;57	165.12
32457	9MR	7;29	359.02	32457	24MR	21;08	194.37
32458	7AP	20;36	28.04	32458	23AP	7;28	223.03
32459	7MY	10;33	56.64	32459	22MY	15;31	251.21
32460	6JN	1;06	84.96	32460	20JN	22;19	279.15
32461	5JL	16;03	113.20	32461	20JL	5;03	307.10
32462	4AU	7;08	141.59	32462	18AU	12;54	335.34
32463	2SE	21;59	170.30	32463	16SE	22;50	4.07
32464	2OC	12;05	199.45	32464	16OC	11;26	33.38
32465	1NO	1;03	229.03	32465	15NO	2;50	63.27
32466	30NO	12;52	258.93	32466	14DE	20;39	93.53
32467	29DE	23;47	288.96				

1625

NUMBER	DATE	TIME	LONG.	NUMBER	DATE	TIME	LONG.
				32467	13JA	15;54	123.88
32468	28JA	10;12	318.84	32468	12FE	11;03	153.99
32469	26FE	20;23	348.39	32469	14MR	4;22	183.60
32470	28MR	6;35	17.46	32470	12AP	18;42	212.59
32471	26AP	17;05	46.08	32471	12MY	5;55	241.02
32472	26MY	4;26	74.36	32472	10JN	14;43	269.08
32473	24JN	17;11	102.53	32473	9JL	22;13	297.04
32474	24JL	7;44	130.81	32474	8AU	5;33	325.15
32475	22AU	23;53	159.43	32475	6SE	13;38	353.63
32476	21SE	16;44	188.50	32476	5OC	23;06	22.65
32477	21OC	9;08	218.06	32477	4NO	10;28	52.21
32478	20NO	0;16	247.99	32478	4DE	0;11	82.22
32479	19DE	13;50	278.09				

1626

NUMBER	DATE	TIME	LONG.	NUMBER	DATE	TIME	LONG.
				32479	2JA	16;23	112.47
32480	18JA	1;52	308.12	32480	1FE	10;32	142.65
32481	16FE	12;30	337.82	32481	3MR	5;12	172.49
32482	17MR	21;53	7.04	32482	1AP	22;37	201.79
32483	16AP	6;22	35.74	32483	1MY	13;41	230.53
32484	15MY	14;44	64.02	32484	31MY	2;12	258.82
32485	14JN	0;00	92.08	32485	29JN	12;40	286.88
32486	13JL	11;09	120.19	32486	28JL	21;49	314.98
32487	12AU	0;48	148.58	32487	27AU	6;24	343.35
32488	10SE	16;50	177.45	32488	25SE	15;03	12.15
32489	10OC	10;29	206.86	32489	25OC	0;22	41.46
32490	9NO	4;36	236.77	32490	23NO	11;00	71.24
32491	8DE	22;00	266.98	32491	22DE	23;28	101.31

1627

NUMBER	DATE	TIME	LONG.	NUMBER	DATE	TIME	LONG.
32492	7JA	13;46	297.20	32492	21JA	13;56	131.42
32493	6FE	3;13	327.14	32493	20FE	5;58	161.31
32494	7MR	14;11	356.59	32494	21MR	22;39	190.77
32495	5AP	23;00	25.48	32495	20AP	15;03	219.73
32496	5MY	6;28	53.85	32496	20MY	6;29	248.24
32497	3JN	13;40	81.91	32497	18JN	20;37	276.48
32498	2JL	21;41	109.87	32498	18JL	9;17	304.66
32499	1AU	7;28	138.02	32499	16AU	20;32	333.01
32500	30AU	19;41	166.57	32500	15SE	6;43	1.71
32501	29SE	10;36	195.69	32501	14OC	16;25	30.86
32502	29OC	4;03	225.40	32502	13NO	2;18	60.48
32503	27NO	23;14	255.58	32503	12DE	12;54	90.42
32504	27DE	18;34	285.95				

Left panel — 1628–1633

1628

NM #	NM Date	NM Time	NM Long	FM #	FM Date	FM Time	FM Long
				32504	11JA	0:28	120.46
32505	26JA	12:12	316.17	32505	9FE	12:56	150.33
32506	25FE	2:52	345.94	32506	10MR	2:12	179.82
32507	25MR	14:21	15.10	32507	8AP	16:14	208.86
32508	23AP	23:19	43.70	32508	8MY	7:01	237.49
32509	23MY	6:46	71.85	32509	6JN	22:21	265.84
32510	21JN	13:43	99.79	32510	6JL	13:40	294.09
32511	20JL	21:05	127.78	32511	5AU	4:15	322.47
32512	19AU	5:43	156.05	32512	3SE	17:39	351.13
32513	17SE	16:27	184.82	32513	3OC	5:56	20.22
32514	17OC	6:01	214.19	32514	1NO	17:28	49.75
32515	15NO	22:46	244.14	32515	1DE	4:39	79.63
32516	15DE	18:01	274.47	32516	30DE	15:37	109.66

1629

NM #	NM Date	NM Time	NM Long	FM #	FM Date	FM Time	FM Long
32517	14JA	13:59	304.85	32517	29JA	2:18	139.55
32518	13FE	8:28	334.92	32518	27FE	12:48	169.10
32519	15MR	0:06	4.44	32519	28MR	23:31	198.18
32520	13AP	12:41	33.35	32520	27AP	11:06	226.83
32521	12MY	22:42	61.72	32521	27MY	0:02	255.17
32522	11JN	6:56	89.76	32522	25JN	14:24	283.40
32523	10JL	14:11	117.70	32523	25JL	5:47	311.73
32524	8AU	21:21	145.81	32524	23AU	21:32	340.34
32525	7SE	5:22	174.31	32525	22SE	13:04	9.37
32526	6OC	15:15	203.35	32526	22OC	4:01	38.88
32527	5NO	3:50	232.97	32527	20NO	18:04	68.77
32528	4DE	19:25	263.07	32528	20DE	6:50	98.84

1630

NM #	NM Date	NM Time	NM Long	FM #	FM Date	FM Time	FM Long
32529	3JA	13:23	293.39	32529	18JA	18:09	128.84
32530	2FE	8:18	323.60	32530	17FE	4:07	158.50
32531	4MR	2:35	353.41	32531	18MR	13:15	187.70
32532	2AP	19:04	22.66	32532	16AP	22:14	216.41
32533	2MY	9:09	51.34	32533	16MY	7:52	244.73
32534	31MY	20:45	79.59	32534	14JN	18:46	272.86
32535	30JN	6:12	107.61	32535	14JL	7:21	301.02
32536	29JL	14:14	135.67	32536	12AU	21:47	329.46
32537	27AU	21:54	164.01	32537	11SE	14:01	358.34
32538	26SE	6:13	192.80	32538	11OC	7:30	27.77
32539	25OC	16:02	222.15	32539	10NO	1:08	57.67
32540	24NO	3:52	251.99	32540	9DE	17:36	87.84
32541	23DE	17:46	282.12				

1631

NM #	NM Date	NM Time	NM Long	FM #	FM Date	FM Time	FM Long
				32541	8JA	7:51	117.99
32542	22JA	9:25	312.28	32542	6FE	19:44	147.87
32543	21FE	2:18	342.19	32543	8MR	5:37	177.26
32544	22MR	19:38	11.66	32544	6AP	14:13	206.13
32545	21AP	12:27	40.62	32545	5MY	22:11	234.51
32546	21MY	3:43	69.12	32546	4JN	6:14	262.60
32547	19JN	16:48	97.32	32547	3JL	15:09	290.60
32548	19JL	3:49	125.44	32548	2AU	1:48	318.78
32549	17AU	13:25	153.72	32549	31AU	15:02	347.39
32550	15SE	22:33	182.39	32550	30SE	7:14	16.57
32551	15OC	7:59	211.54	32551	30OC	1:48	46.34
32552	13NO	18:11	241.18	32552	28NO	21:07	76.54
32553	13DE	5:15	271.14	32553	28DE	15:16	106.86

1632

NM #	NM Date	NM Time	NM Long	FM #	FM Date	FM Time	FM Long
32554	11JA	17:20	301.20	32554	27JA	7:00	136.99
32555	10FE	6:36	331.10	32555	25FE	20:01	166.68
32556	10MR	21:12	0.64·	32556	26MR	6:36	195.80
32557	9AP	12:53	29.73	32557	24AP	15:15	224.37
32558	9MY	4:53	58.40	32558	23MY	22:36	252.51
32559	7JN	20:20	86.74	32559	22JN	5:32	280.45
32560	7JL	10:36	114.96	32560	21JL	13:05	308.45
32561	5AU	23:37	143.28	32561	19AU	22:28	336.76
32562	4SE	11:40	171.90	32562	18SE	10:41	5.60
32563	3OC	23:04	200.96	32563	18OC	2:09	35.06
32564	2NO	10:02	230.48	32564	16NO	20:22	65.08
32565	1DE	20:40	260.34	32565	16DE	15:53	95.43
32566	31DE	7:12	290.35				

1633

NM #	NM Date	NM Time	NM Long	FM #	FM Date	FM Time	FM Long
				32566	15JA	10:59	125.76
32567	29JA	17:58	320.24	32567	14FE	4:15	155.78
32568	28FE	5:22	349.82	32568	15MR	18:49	185.24
32569	29MR	17:41	18.95	32569	14AP	6:27	214.10
32570	28AP	6:57	47.66	32570	13MY	15:29	242.42
32571	27MY	21:02	76.04	32571	11JN	22:45	270.42
32572	26JN	11:44	104.28	32572	11JL	5:23	298.34
32573	26JL	2:52	132.60	32573	9AU	12:36	326.73
32574	24AU	18:06	161.20	32574	7SE	21:29	355.00
32575	23SE	8:56	190.22	32575	7OC	8:48	24.10
32576	22OC	22:48	219.69	32576	5NO	22:53	53.80
32577	21NO	11:24	249.53	32577	5DE	15:35	83.95
32578	20DE	22:49	279.55				

Right panel — 1634–1639

1634

NM #	NM Date	NM Time	NM Long	FM #	FM Date	FM Time	FM Long
				32578	4JA	10:12	114.30
32579	19JA	9:25	309.51	32579	3FE	5:31	144.52
32580	17FE	19:33	339.18	32580	4MR	23:48	174.32
32581	19MR	5:29	8.40	32581	3AP	15:37	203.52
32582	17AP	15:32	37.14	32582	3MY	4:17	232.14
32583	17MY	2:08	65.49	32583	1JN	14:10	260.31
32584	15JN	13:56	93.65	32584	30JN	22:15	288.28
32585	15JL	3:30	121.86	32585	30JL	5:39	316.32
32586	13AU	19:00	150.34	32586	28AU	13:24	344.66
32587	12SE	11:50	179.27	32587	26SE	22:15	13.50
32588	12OC	4:53	208.69	32588	26OC	8:43	42.88
32589	10NO	20:59	238.53	32589	24NO	21:15	72.75
32590	10DE	11:32	268.63	32590	24DE	12:10	102.93

1635

NM #	NM Date	NM Time	NM Long	FM #	FM Date	FM Time	FM Long
32591	9JA	0:23	298.72	32591	23JA	5:19	133.15
32592	7FE	11:39	328.56	32592	21FE	23:43	163.11
32593	8MR	21:29	357.95	32593	23MR	17:44	192.60
32594	7AP	6:09	26.80	32594	22AP	9:56	221.52
32595	6MY	14:17	55.19	32595	21MY	23:39	249.95
32596	4JN	22:47	83.29	32596	20JN	11:08	278.08
32597	4JL	8:45	111.33	32597	19JL	20:59	306.15
32598	2AU	21:05	139.59	32598	18AU	5:58	334.42
32599	1SE	12:03	168.27	32599	16SE	14:43	3.08
32600	1OC	5:12	197.51	32600	15OC	23:48	32.23
32601	30OC	23:27	227.28	32601	14NO	9:50	61.87
32602	29NO	17:32	257.43	32602	13DE	21:20	91.86
32603	29DE	10:21	287.70				

1636

NM #	NM Date	NM Time	NM Long	FM #	FM Date	FM Time	FM Long
				32603	12JA	10:41	121.96
32604	28JA	1:02	317.78	32604	11FE	1:48	151.93
32605	26FE	13:09	347.42	32605	11MR	18:00	181.53
32606	26MR	22:50	16.49	32606	10AP	10:24	210.63
32607	25AP	6:44	45.02	32607	10MY	2:12	239.28
32608	24MY	13:48	73.15	32608	8JN	16:59	267.59
32609	22JN	21:11	101.10	32609	8JL	6:30	295.78
32610	22JL	5:56	129.15	32610	6AU	18:39	324.07
32611	20AU	16:54	157.53	32611	5SE	5:35	352.66
32612	19SE	6:31	186.44	32612	4OC	15:41	21.68
32613	18OC	22:53	215.96	32613	3NO	1:36	51.16
32614	17NO	17:28	246.00	32614	2DE	11:54	81.02
32615	17DE	13:03	276.35	32615	31DE	22:57	111.04

1637

NM #	NM Date	NM Time	NM Long	FM #	FM Date	FM Time	FM Long
32616	16JA	7:47	306.67	32616	30JA	10:50	140.98
32617	15FE	0:00	336.63	32617	28FE	23:28	170.60
32618	16MR	12:57	6.02	32618	30MR	12:48	199.77
32619	14AP	22:58	34.80	32619	29AP	2:56	228.50
32620	14MY	6:58	63.07	32620	28MY	17:52	256.90
32621	12JN	13:59	91.06	32621	27JN	9:14	285.17
32622	11JL	21:02	119.00	32622	27JL	0:24	313.49
32623	10AU	4:59	147.15	32623	25AU	14:40	342.06
32624	8SE	14:38	175.73	32624	24SE	3:47	11.03
32625	8OC	2:48	204.88	32625	23OC	15:55	40.44
32626	6NO	18:02	234.64	32626	22NO	3:27	70.24
32627	6DE	12:13	264.86	32627	21DE	14:35	100.25

1638

NM #	NM Date	NM Time	NM Long	FM #	FM Date	FM Time	FM Long
32628	5JA	8:04	295.26	32628	20JA	1:22	130.21
32629	4FE	3:27	325.47	32629	18FE	11:45	159.88
32630	5MR	20:31	355.20	32630	19MR	22:03	189.11
32631	4AP	10:30	24.32	32631	18AP	8:50	217.87
32632	3MY	21:40	52.87	32632	17MY	20:48	246.27
32633	2JN	6:40	81.00	32633	16JN	10:18	274.49
32634	1JL	14:21	108.95	32634	16JL	1:14	302.76
32635	30JL	21:32	136.99	32635	14AU	17:00	331.26
32636	29AU	5:07	165.33	32636	13SE	8:57	0.16
32637	27SE	14:05	194.18	32637	13OC	0:32	29.53
32638	27OC	1:19	223.61	32638	11NO	15:21	59.34
32639	25NO	15:27	253.55	32639	11DE	4:58	89.40
32640	25DE	8:18	283.81				

1639

NM #	NM Date	NM Time	NM Long	FM #	FM Date	FM Time	FM Long
				32640	9JA	17:04	119.46
32641	24JA	2:48	314.08	32641	8FE	3:35	149.26
32642	22FE	21:24	344.04	32642	9MR	12:53	178.61
32643	24MR	14:43	13.49	32643	7AP	21:38	207.46
32644	23AP	5:57	42.36	32644	7MY	6:37	235.88
32645	22MY	18:45	70.74	32645	5JN	16:37	264.03
32646	21JN	5:17	98.83	32646	5JL	4:12	292.14
32647	20JL	14:02	126.86	32647	3AU	17:42	320.45
32648	18AU	21:55	155.10	32648	2SE	9:10	349.16
32649	17SE	5:58	183.73	32649	2OC	2:20	18.41
32650	16OC	15:06	212.90	32650	31OC	20:18	48.19
32651	15NO	2:00	242.58	32651	30NO	13:44	78.32
32652	14DE	14:52	272.63	32652	30DE	5:18	108.53

NUMBER	DATE	TIME	LONG.	NUMBER	DATE	TIME	LONG.

1640

NUMBER	DATE	TIME	LONG.	NUMBER	DATE	TIME	LONG.
32653	13JA	5;34	302.79	32653	28JA	18;21	138.54
32654	11FE	21;41	332.79	32654	27FE	5;04	168.11
32655	12MR	14;38	2.41	32655	27MR	14;05	197.15
32656	11AP	7;37	31.53	32656	25AP	22;06	225.67
32657	10MY	23;39	60.17	32657	25MY	5;52	253.82
32658	9JN	13;53	88.46	32658	23JN	14;08	281.81
32659	9JL	2;00	116.59	32659	22JL	23;45	309.89
32660	7AU	12;24	144.82	32660	21AU	11;38	338.32
32661	5SE	21;54	173.35	32661	20SE	2;26	7.29
32662	5OC	7;20	202.36	32662	19OC	20;06	36.87
32663	3NO	17;18	231.85	32663	18NO	15;27	66.96
32664	3DE	4;00	261.73	32664	18DE	10;32	97.30

1641

NUMBER	DATE	TIME	LONG.	NUMBER	DATE	TIME	LONG.
32665	1JA	15;31	291.78	32665	17JA	3;37	127.54
32666	31JA	3;59	321.73	32666	15FE	17;57	157.41
32667	1MR	17;39	351.38	32667	17MR	5;36	186.73
32668	31MR	8;33	20.60	32668	15AP	15;02	215.48
32669	30AP	0;15	49.38	32669	14MY	22;51	243.74
32670	29MY	15;57	77.81	32670	13JN	5;48	271.72
32671	28JN	6;52	106.06	32671	12JL	12;52	299.66
32672	27JL	20;39	134.33	32672	10AU	21;13	327.83
32673	26AU	9;24	162.86	32673	9SE	7;58	356.47
32674	24SE	21;23	191.78	32674	8OC	21;54	25.71
32675	24OC	8;48	221.18	32675	7NO	14;57	55.55
32676	22NO	19;45	250.96	32676	7DE	10;05	85.82
32677	22DE	6;21	280.95				

1642

NUMBER	DATE	TIME	LONG.	NUMBER	DATE	TIME	LONG.
				32677	6JA	5;37	116.20
32678	20JA	16;53	310.90	32678	4FE	23;54	146.36
32679	19FE	3;43	340.58	32679	6MR	15;48	176.03
32680	20MR	15;17	9.85	32680	5AP	4;49	205.10
32681	19AP	3;48	38.66	32681	4MY	15;00	233.59
32682	18MY	17;15	67.12	32682	2JN	23;00	261.68
32683	17JN	7;31	95.37	32683	2JL	5;49	289.60
32684	16JL	22;29	123.64	32684	31JL	12;39	317.62
32685	15AU	13;53	152.13	32685	29AU	20;41	346.00
32686	14SE	5;16	181.02	32686	28SE	6;49	14.90
32687	13OC	19;59	210.38	32687	27OC	19;35	44.40
32688	12NO	9;30	240.13	32688	26NO	11;03	74.41
32689	11DE	21;37	270.14	32689	26DE	4;47	104.71

1643

NUMBER	DATE	TIME	LONG.	NUMBER	DATE	TIME	LONG.
32690	10JA	8;36	300.15	32690	24JA	23;49	135.00
32691	8FE	18;50	329.93	32691	23FE	18;41	164.96
32692	10MR	4;38	359.29	32692	25MR	11;45	194.38
32693	8AP	14;20	28.17	32693	24AP	1;56	223.19
32694	8MY	0;20	56.62	32694	23MY	13;07	251.51
32695	6JN	11;15	84.80	32695	21JN	22;00	279.53
32696	5JL	23;46	112.95	32696	21JL	5;42	307.52
32697	4AU	14;20	141.31	32697	19AU	13;39	335.75
32698	3SE	6;46	170.07	32698	17SE	21;42	4.41
32699	3OC	0;08	199.35	32699	17OC	7;26	33.61
32700	1NO	17;06	229.09	32700	15NO	18;58	63.33
32701	1DE	8;41	259.15	32701	15DE	8;39	93.42
32702	30DE	22;29	289.29				

1644

NUMBER	DATE	TIME	LONG.	NUMBER	DATE	TIME	LONG.
				32702	14JA	0;36	123.63
32703	29JA	10;31	319.25	32703	12FE	18;20	153.69
32704	27FE	20;56	348.80	32704	13MR	12;31	183.34
32705	28MR	5;58	17.83	32705	12AP	5;36	212.45
32706	26AP	14;04	46.34	32706	11MY	20;30	241.03
32707	25MY	22;04	74.50	32707	10JN	9;04	269.25
32708	24JN	7;03	102.51	32708	9JL	19;46	297.33
32709	23JL	18;03	130.66	32709	8AU	5;17	325.52
32710	22AU	7;44	159.17	32710	6SE	14;17	354.05
32711	21SE	0;02	188.21	32711	5OC	23;21	23.05
32712	20OC	18;04	217.82	32712	4NO	8;59	52.54
32713	19NO	12;35	247.89	32713	3DE	19;44	82.43
32714	19DE	6;19	278.17				

1645

NUMBER	DATE	TIME	LONG.	NUMBER	DATE	TIME	LONG.
				32714	2JA	8;05	112.51
32715	17JA	22;12	308.36	32715	31JA	22;10	142.53
32716	16FE	11;35	338.18	32716	2MR	13;39	172.24
32717	17MR	22;20	7.45	32717	1AP	5;46	201.50
32718	16AP	6;53	36.15	32718	30AP	21;44	230.28
32719	15MY	14;06	64.38	32719	30MY	12;59	258.68
32720	13JN	21;05	92.36	32720	29JN	3;13	286.90
32721	13JL	4;59	120.33	32721	28JL	16;14	315.15
32722	11AU	14;46	148.57	32722	27AU	3;59	343.64
32723	10SE	3;05	177.28	32723	25SE	14;43	12.53
32724	9OC	18;12	206.58	32724	25OC	0;52	41.88
32725	8NO	11;53	236.46	32725	23NO	11;04	71.64
32726	8DE	7;16	266.74	32726	22DE	21;44	101.63

1646

NUMBER	DATE	TIME	LONG.	NUMBER	DATE	TIME	LONG.
32727	7JA	2;42	297.12	32727	21JA	9;07	131.61
32728	5FE	20;19	327.24	32728	19FE	21;09	161.34
32729	7MR	10;52	356.85	32729	21MR	9;51	190.64
32730	5AP	22;10	25.83	32730	19AP	23;17	219.48
32731	5MY	6;57	54.26	32731	19MY	13;35	247.96
32732	3JN	14;17	82.32	32732	18JN	4;42	276.24
32733	2JL	21;12	110.24	32733	17JL	20;08	304.53
32734	1AU	4;38	138.29	32734	16AU	11;08	333.02
32735	30AU	13;24	166.71	32735	15SE	1;07	1.87
32736	29SE	0;19	195.66	32736	14OC	14;00	31.16
32737	28OC	14;02	225.21	32737	13NO	2;01	60.86
32738	27NO	6;52	255.28	32738	12DE	13;28	90.84
32739	27DE	2;05	285.65				

1647

NUMBER	DATE	TIME	LONG.	NUMBER	DATE	TIME	LONG.
				32739	11JA	0;26	120.85
32740	25JA	21;55	315.96	32740	9FE	10;53	150.63
32741	24FE	16;12	345.88	32741	10MR	20;57	180.00
32742	26MR	7;39	15.22	32742	9AP	7;08	208.88
32743	24AP	20;05	43.95	32743	8MY	18;10	237.36
32744	24MY	6;03	72.21	32744	7JN	6;41	265.60
32745	22JN	14;19	100.21	32745	6JL	20;51	293.82
32746	21JL	21;43	128.19	32746	5AU	12;21	322.22
32747	20AU	5;07	156.42	32747	4SE	4;29	350.99
32748	18SE	13;24	185.09	32748	3OC	20;35	20.22
32749	17OC	23;31	214.31	32749	2NO	12;07	49.92
32750	16NO	12;13	244.09	32750	2DE	2;36	79.94
32751	16DE	3;45	274.26	32751	31DE	15;35	110.05

1648

NUMBER	DATE	TIME	LONG.	NUMBER	DATE	TIME	LONG.
32752	14JA	21;28	304.54	32752	30JA	2;51	139.97
32753	13FE	16;02	334.62	32753	28FE	12;34	169.48
32754	14MR	9;56	4.25	32754	28MR	21;17	198.48
32755	13AP	2;08	33.31	32755	27AP	5;50	227.01
32756	12MY	16;07	61.85	32756	26MY	15;02	255.20
32757	11JN	3;47	90.02	32757	25JN	1;37	283.28
32758	10JL	13;27	118.07	32758	24JL	14;05	311.49
32759	8AU	21;49	146.22	32759	23AU	4;38	340.04
32760	7SE	5;51	174.72	32760	21SE	21;11	9.11
32761	6OC	14;31	203.71	32761	21OC	15;07	38.73
32762	5NO	0;36	233.23	32762	20NO	9;15	68.79
32763	4DE	12;31	263.17	32763	20DE	2;02	99.03

1649

NUMBER	DATE	TIME	LONG.	NUMBER	DATE	TIME	LONG.
32764	3JA	2;15	293.31	32764	18JA	16;25	129.16
32765	1FE	17;31	323.37	32765	17FE	4;11	158.90
32766	3MR	9;52	353.11	32766	18MR	13;50	188.12
32767	2AP	2;41	22.38	32767	16AP	22;06	216.79
32768	1MY	19;09	51.17	32768	16MY	5;45	245.04
32769	31MY	10;19	79.56	32769	14JN	13;32	273.04
32770	29JN	23;35	107.74	32770	13JL	22;18	301.05
32771	29JL	10;57	135.93	32771	12AU	8;57	329.33
32772	27AU	21;01	164.36	32772	10SE	22;18	358.09
32773	26SE	6;38	193.22	32773	10OC	14;43	27.46
32774	25OC	16;30	222.56	32774	9NO	9;36	57.41
32775	24NO	2;56	252.34	32775	9DE	5;12	87.71
32776	23DE	14;02	282.35				

1650

NUMBER	DATE	TIME	LONG.	NUMBER	DATE	TIME	LONG.
				32776	7JA	23;31	118.04
32777	22JA	1;53	312.35	32777	6FE	15;15	148.07
32778	20FE	14;41	342.10	32778	8MR	4;08	177.59
32779	22MR	4;41	11.44	32779	6AP	14;30	206.53
32780	20AP	19;46	40.34	32780	5MY	22;56	234.94
32781	20MY	11;23	68.86	32781	4JN	6;08	262.98
32782	19JN	2;44	97.15	32782	3JL	12;59	290.90
32783	18JL	17;12	125.40	32783	1AU	20;33	318.96
32784	17AU	6;40	153.84	32784	31AU	6;02	347.41
32785	15SE	19;18	182.64	32785	29SE	18;24	16.43
32786	15OC	7;16	211.91	32786	29OC	10;02	46.07
32787	13NO	18;39	241.59	32787	28NO	4;23	76.22
32788	13DE	5;31	271.55	32788	27DE	23;56	106.60

1651

NUMBER	DATE	TIME	LONG.	NUMBER	DATE	TIME	LONG.
32789	11JA	16;00	301.54	32789	26JA	18;59	136.88
32790	10FE	2;29	331.32	32790	25FE	12;06	166.74
32791	11MR	13;24	0.71	32791	27MR	2;30	196.02
32792	10AP	1;08	29.65	32792	25AP	13;59	224.71
32793	9MY	13;51	58.18	32793	24MY	22;56	252.92
32794	8JN	3;32	86.46	32794	23JN	6;12	280.87
32795	7JL	18;08	114.69	32795	22JL	12;57	308.83
32796	6AU	9;27	143.10	32796	20AU	20;21	337.06
32797	5SE	1;10	171.86	32797	19SE	5;26	5.78
32798	4OC	16;36	201.08	32798	18OC	16;57	35.07
32799	3NO	7;03	230.74	32799	17NO	7;08	64.92
32800	2DE	20;03	260.71	32800	16DE	23;47	95.14